Industrial and Municipal Sludge

Emerging Concerns and Scope for Resource Recovery

Industrial and Municipal Sludge

Emerging Concerns and Scope for Resource Recovery

Edited by

Majeti Narasimha Vara Prasad
Paulo Jorge de Campos Favas
Meththika Vithanage
S.Venkata Mohan

ELSEVIER

Butterworth-Heinemann
An imprint of Elsevier

Butterworth-Heinemann is an imprint of Elsevier
The Boulevard, Langford Lane, Kidlington, Oxford OX5 1GB, United Kingdom
50 Hampshire Street, 5th Floor, Cambridge, MA 02139, United States

Library of Congress Cataloging-in-Publication Data
A catalog record for this book is available from the Library of Congress

British Library Cataloguing-in-Publication Data
A catalogue record for this book is available from the British Library

ISBN: 978-0-12-815907-1

For information on all Butterworth-Heinemann publications
visit our website at https://www.elsevier.com/books-and-journals

Publisher: Matthew Deans
Acquisition Editor: Ken McCombs
Editorial Project Manager: Andrae Akeh
Production Project Manager: Sruthi Satheesh
Cover Designer: Mark Rogers

Typeset by SPi Global, India

Contents

Section D Valorization of sludge as a resource

Contributors

Kotamraju Amulya Bioengineering and Environmental Sciences Lab, EEFF Department, CSIR-Indian Institute of Chemical Technology (CSIR-IICT), Hyderabad, India

Azize Ayol Department of Environmental Engineering, Dokuz Eylül University, Izmir, Turkey

Jianfeng Bai WEEE Research Center of Shanghai Polytechnic University; Shanghai Collaborative Innovation Centre for WEEE Recycling, Shanghai, China

Sartaj Ahmad Bhat Department of Botanical and Environmental Sciences, Guru Nanak Dev University, Amritsar, India; River Basin Research Center, Gifu University, Gifu, Japan

Jayanta Kumar Biswas International Centre for Ecological Engineering, Department of Ecological Studies, University of Kalyani, Nadia, India

Karla Liliana Tarango Bustamante Department of Nanotechnology, University Technical of City Juarez, Juarez, Mexico

John Campbell Everse Limited, Hamilton, New Zealand

Roberto Canziani DICA—Sezione Ambientale, Politecnico di Milano, Milano, Italy

Piotr Celary Central Laboratory for Measurements and Research, Jastrzebie-Zdroj, Poland

Debkumar Chakraborty Bioengineering and Environmental Sciences Lab, EEFF Department, CSIR-Indian Institute of Chemical Technology (CSIR-IICT), Hyderabad, India

Diyun Chen Guangdong Provincial Key Laboratory of Radionuclides Pollution Control and Resources, School of Environmental Science and Engineering, Guangzhou University, Guangzhou, China

Pauline Courtois Campus ULille Sciences and Technologies, University of Lille, Lille, France

Karolina Czerwińska Department of Environmental Engineering, Czestochowa University of Technology, Czestochowa, Poland

Shikha Dahiya Bioengineering and Environmental Sciences Lab, EEFF Department, CSIR-Indian Institute of Chemical Technology (CSIR-IICT), Hyderabad, India

Bin Dong Department of Environmental Science and Engineering, Tongji University, Shanghai, China

Xiaomin Dou College of Environmental Science and Engineering, Beijing Forestry University, Beijing, PR China

Mariusz Dudziak Institute of Water and Wastewater Engineering, Silesian University of Technology, Gliwice, Poland

Krzysztof Fijalkowski Institute of Environmental Engineering, Czestochowa University of Technology, Czestochowa, Poland

Martina Grifoni Institute of Ecosystem Study, Pisa, Italy

Anna Grobelak Department of Environmental Engineering, Czestochowa University of Technology, Czestochowa, Poland

Anna Grosser Institute of Environmental Engineering, Czestochowa University of Technology, Czestochowa, Poland

Weihua Gu WEEE Research Center of Shanghai Polytechnic University; Shanghai Collaborative Innovation Centre for WEEE Recycling, Shanghai; College of Environment, Zhejiang University of Technology, Hangzhou, China

Marta Jaskulak Department of Environmental Engineering, Czestochowa University of Technology, Czestochowa, Poland

Małgorzata Kacprzak Department of Environmental Engineering, Czestochowa University of Technology, Czestochowa, Poland

Tomasz Kamizela Institute of Environmental Engineering, Czestochowa University of Technology, Częstochowa, Poland

Ranaprathap Katakojwala Bioengineering and Environmental Sciences Lab, EEFF Department, CSIR-Indian Institute of Chemical Technology (CSIR-IICT); Academy of Scientific and Innovative Research (AcSIR), Hyderabad, India

Paulina Kokot Department of Environmental Engineering, Czestochowa University of Technology, Czestochowa, Poland

Lingjun Kong Guangdong Provincial Key Laboratory of Radionuclides Pollution Control and Resources, School of Environmental Science and Engineering, Guangzhou University, Guangzhou; Department of Civil Engineering, The University of Hong Kong, Hong Kong SAR, China

Mariusz Kowalczyk Institute of Environmental Engineering, Czestochowa University of Technology, Częstochowa, Poland

Anna Kwarciak-Kozłowska Faculty of Infrastructure and Environment, Institute of Environmental Engineering, Czestochowa University of Technology, Czestochowa, Poland

Sébastien Lemiere Campus ULille Sciences and Technologies, University of Lille, Lille, France

Changzhong Liao Department of Civil Engineering, The University of Hong Kong, Hong Kong SAR; Guangdong Key Laboratory of Integrated Agro-Environmental Pollution Control and Management, Guangdong Institute of Eco-Environmental Science & Technology, Guangzhou, China

Sanchita Mandal Future Industries Institute, University of South Australia, Mawson Lakes, SA, Australia

Vladimir Matichenkov Department of Ecology and Physiology of Autotrophic Organisms, Institute Basic Biological Problems, Pushchino, Russia

Aneta Murtaś Department of Environmental Engineering, Czestochowa University of Technology, Czestochowa, Poland

A. Naresh Kumar Bioengineering and Environmental Sciences Lab, EEFF Department, CSIR-Indian Institute of Chemical Technology (CSIR-IICT); Academy of Scientific and Innovative Research (AcSIR), Hyderabad, India

Ewa Neczaj Institute of Environmental Engineering, Czestochowa University of Technology, Czestochowa, Poland

Gauravarapu Navlur Nikhil Department of Biotechnology, Dr. B.R. Ambedkar National Institute of Technology Jalandhar, Punjab, India

Yong Sik Ok Korea Biochar Research Centre, O-Jeong Eco-Resilience Institute & Division of Environmental Science and Ecological Engineering, Korea University, Seoul, Republic of Korea

Francesca Pedron Institute of Ecosystem Study, Pisa, Italy

Giannantonio Petruzzelli Institute of Ecosystem Study, Pisa, Italy

Majeti Narasimha Vara Prasad School of Life Sciences, University of Hyderabad, Hyderabad, India

Agnieszka Rorat Campus ULille Sciences and Technologies, University of Lille, Lille, France

Irene Rosellini Institute of Ecosystem Study, Pisa, Italy

Agata Rosińska Faculty of Infrastructure and Environment, Czestochowa University of Technology, Częstochowa, Poland

Binoy Sarkar Department of Animal and Plant Sciences, University of Sheffield, Sheffield, United Kingdom; Future Industries Institute, University of South Australia, Mawson Lakes, SA, Australia

Hemen Sarma Department of Botany, N N Saikia College, Titabar, India

Kaimin Shih Department of Civil Engineering, The University of Hong Kong, Hong Kong SAR, China

Marzena Smol Mineral and Energy Economy Research Institute, Polish Academy of Sciences, Cracow, Poland

Jolanta Sobik-Szołtysek Institute of Environmental Engineering, Czestochowa University of Technology, Częstochowa, Poland

Ludovico Spinosa ISO and CEN Expert, Bari, Italy

Venu Srivastav Bioengineering and Environmental Sciences Lab, EEFF Department, CSIR-Indian Institute of Chemical Technology (CSIR-IICT), Hyderabad, India

Minhua Su Guangdong Provincial Key Laboratory of Radionuclides Pollution Control and Resources, School of Environmental Science and Engineering, Guangzhou University, Guangzhou, China

Yuanyuan Tang School of Environmental Science and Engineering, Southern University of Science and Technology, Shenzhen, China

Yiu Fai Tsang Department of Science and Environmental Studies, The Education University of Hong Kong, Tai Po, Hong Kong

Franck Vandenbulcke Campus ULille Sciences and Technologies, University of Lille, Lille, France

S. Venkata Mohan Bioengineering and Environmental Sciences Lab, EEFF Department, CSIR-Indian Institute of Chemical Technology (CSIR-IICT), Hyderabad, India

Adarsh Pal Vig Department of Botanical and Environmental Sciences, Guru Nanak Dev University, Amritsar, India

Meththika Vithanage Ecosphere Resilience Research Centre, University of Sri Jayewardenepura, Nugegoda, Sri Lanka

Jingwei Wang WEEE Research Center of Shanghai Polytechnic University; Shanghai Collaborative Innovation Centre for WEEE Recycling, Shanghai, China

Sebastian Werle Institute of Thermal Technology, Silesian University of Technology, Gliwice, Poland

Ewa Wiśniowska Department of Environmental Engineering, Czestochowa University of Technology, Czestochowa, Poland

Małgorzata Worwag Faculty of Infrastructure and Environment, Institute of Environmental Engineering, Czestochowa University of Technology, Czestochowa, Poland

Pengfei Wu School of Environmental Science and Engineering, Southern University of Science and Technology, Shenzhen, China

Katarzyna Wystalska Institute of Environmental Engineering, Czestochowa University of Technology, Częstochowa, Poland

Haakrho Yi Department of Gwangyang Research Group, RIST, Gwangyang, Republic of Korea

Wenyi Yuan WEEE Research Center of Shanghai Polytechnic University; Shanghai Collaborative Innovation Centre for WEEE Recycling, Shanghai, China

Iwona Zawieja Faculty of Infrastructure and Environment, Institute of Environmental Engineering, Czestochowa University of Technology, Czestochowa, Poland

Chenglong Zhang WEEE Research Center of Shanghai Polytechnic University; Shanghai Collaborative Innovation Centre for WEEE Recycling, Shanghai, China

Biographies

Majeti Narasimha Vara Prasad, Emeritus Professor, School of Life Sciences, University of Hyderabad, Hyderabad, India

Prasad earned a masters in botany from Andhra University, Waltair and a doctorate in botany from Lucknow University. He has worked as a lecturer at North Eastern Hill University, and the University of Hyderabad. He has been a professor in the Department of Plant Sciences, for the past 20 years. He has made significant contributions to the field of plant-metal interactions and bioremediation. He has contributed 205 research articles in peer-reviewed journals and 21 books (as author, coauthor, editor, and coeditor) by international publishers while earning ~14075 Google Scholar citations with an h-index of 60. Prasad was the recipient of the Pitambar Pant National Environment Fellowship in 2007 and was awarded the Professor K.S. Bilgrami Memorial Award for 2015 by the Society for Plant Research, India. He has also been a visiting professor at various universities, including the University of Québec, Canada; Jagiellonian University, Krakow, Poland; the University of Coimbra, Portugal; Stockholm University, Sweden; the University of Oulu, Finland; the University of South Australia, Australia; Al-Farabi Kazakh National University, Kazakhstan; Ural Federal University, Russia; Ghent University, Belgium; Mahasarakham University, Thailand; the University of Santa Cruz, Brazil; and the Asian Institute of Technology, Thailand.

Paulo Jorge de Campos Favas is a professor at the University of Trás-os-Montes e Alto Douro (UTAD), Vila Real, Portugal. He holds a bachelors in biology and geology (UTAD), a masters in geosciences (University of Coimbra, Portugal), and a doctorate in geology (UTAD). His research interests range from biogeochemistry to environmental geochemistry. He specializes in pollutant behavior in soil-plant and water-plant interfaces, and works in the areas of hazardous metal(loid)s, phytoremediation of contaminated sites, and abandoned mine site reclamation. He serves as editor for the *Annals of Agricultural Science* journal and as a reviewer for several other scientific journals. Within the environmental science subjects, he is the author or coauthor of about 50 peer-reviewed papers in scientific journals, book chapters, and proceedings (Web of Science and/or Scopus indexed), and more than 140 papers in conference proceedings. He has received four awards for scientific and academic merit.

Meththika Vithanage is the leader of the Ecosphere Resilience Research Center, Faculty of Applied Science at the University of Sri Jayewardenepura, Sri Lanka; a Visiting Associate Research Professor, National Institute of Fundamental Studies, Sri Lanka; and an Adjunct Associate Research Professor at the University of Southern Queensland, Australia. Dr. Vithanage's research approach builds on enabling measurements of concentrations and reaction rates in environmental samples for key/emerging pollutants in the environment in order to assess their fate and transport to discover solutions to remediate those using different geo/bio/nano materials. Dr. Vithanage is particularly interested in elucidating the mechanistic understanding of release/remediation of pollutants with contrasting physical and chemical properties that can be used to obtain insights into environmental partitioning, chemical persistence, and ultimate accumulation in biota. To this end, her current research interests are focused on monitoring atmospheric deposition, landfill leachate, water and soil for pollutants, developing and enhancing the properties of biochar, nano/geo substances and composites as material for environmental remediation. She has received several awards, including the TWAS-NSF Award for the Young Scientist, American Geophysical Union, Presidential Awards for Publications, Awards for science popularization and for postgraduate supervision by the National Science Foundation, Sri Lanka. She is a Young Affiliate of the Third World Academy of Sciences. She was the Chairperson of the Young Scientists Forum on the National Science and Technology Commission of Sri Lanka in 2017. Currently, she is the President of the Engineering, Architecture, Surveying Section of the Sri Lanka Association for the Advancement of Science. She has contributed more than 80 Science Citation Indexed journal articles and 25 book chapters. Her citation record has now passed 2300 with an H index of 24.

S. Venkata Mohan is a Tata Innovation Fellow and Principal Scientist at CSIR-Indian Institute of Chemical Technology, Hyderabad, India. He was a Visiting Professor at Kyoto University (2005) and an Alexander von Humboldt (AvH) Fellow at the Technical University of Munich, Germany (2001–02). Dr. Mohan is the recipient of coveted Shanti Swarup Bhatnagar (SSB) Prize for 2014 in Engineering Sciences from Government of India. His main research interests are in the domain of environmental bioengineering, specifically in the areas of advanced waste remediation, aciodogenesis, microbial electrogenesis, photosynthesis, industrial symbiosis, and waste biorefinery. He has authored more than 300 research articles and 36 chapters for books as well as edited four books and earned nine patents. His publication has more than 13,300 citations with an h-index of 65 (Google Scholar). Recently, Dr. Mohan successfully demonstrated a pilot plant for biohydrogen production from waste for MNRE and a waste-fed biorefinery platform for CSIR. He carried out various industrial and consultancy projects in the area of environmental management. He also received several awards and honors, including Most outstanding Researcher in the field of Environmental Science in India-2018 by Carrer360, as "National Bioscience award-2012" by DBT, Government of India, SERB-IGCW-2017 for "Biohydrogen Technology" from DST-SERB, Environmental Engineering Design Award 2017 by the National Design and Research Forum (NDRF) of Institute of Engineers, India (2017), "Prosper.net-Scopus Young Researcher Award in Sustainable Development-2010" under Energy Category by United Nations University and Elsevier, "NASI-Scopus Young Scientist Award-2010" in Earth, Oceanographic & Environmental Sciences by NASI and Elsevier, Nawab Zain Yar Jung Bahadur Memorial Prize-1994 by The Institution of Engineers (India), etc. Dr. Mohan is an elected Fellow of the National Academy of Engineering, Biotech Research Society of India, Telangana and Andhra Pradesh Akademy of Sciences, International Forum on Industrial Bioprocesses, Institution of Engineers, International Society for Energy, Environment and Sustainability, etc. Dr Mohan is National Editor for Science Portal (Environmental Science), subject Editor for the *Journal of Energy*, Associate Editor for *Frontiers in Environmental Science* and *Frontiers in Energy Research* and is serving on the Editorial Board of several journals viz., Bioresource Technology, Advances in Energy Research, Carbon Resources Conversion, etc. Dr. Mohan is at present working on research projects funded by MNRE, DBT, DST-EU, and CSIR. He has carried out various industrial and consultancy projects in the area of environment and management.

Foreword

Majeti Narasimha Vara Prasad
School of Life Sciences, University of Hyderabad, Hyderabad, India

Municipal sludge is "*waste from water treatment plants that originates predominantly from the human metabolism and from household activities.*"

Industrial sludge is "*waste from water and/or run-off that originates from different types of industries along with industrial waste.*"

Sludge from a variety of sources is a potential threat to the quality of the environment. Management of a variety of sludge originating from various domestic and industrial sources is a complex subject in terms of legal, economic, technical, and environmental criteria. Often, sludge is deliberately disposed of into watercourses or landfills. Industrial and municipal sludge generation and management in the contemporary world is a subject of concern for resource generation and pollution prevention.

Three principal categories of sludge can be identified: (i) primary from mechanical and preliminary treatments, (ii) secondary/activated from biological treatments, and (iii) tertiary/chemical from final physical-chemical treatments. Sludge management require technical, professional infrastructural facilities are required such as sludge thickening, digestion, dewatering, hygienisation, drying and incineration as well as disposal need need special attention (Anonymous, 2012).

The practical and technical challenges of sludge handling are the following:

1. Sludge thickening.
2. Stabilization.
3. Sludge dewatering.
4. Sludge hygienisation.
5. Sludge incineration and ash disposal.
6. Disposal of sewage sludge.
7. Reject water treatment.
8. Recovery of phosphorus from sludge (Kabbe and Kraus, 2017; Leinweber et al., 2018).
9. Relevant legislations—national policies.
10. Drivers and obstacles for different sludge management.

Efficient municipal wastewater treatment produces vast amounts of sludge. Sludge management is an integral part of any modern municipal wastewater treatment plant. It is important not to lose the nutrients in the sludge, to make use of its material and energy, and to dispose of it efficiently and sustainably. During recent years, much effort has been put into efficient nutrient removal from the municipal wastewater. The aim is to reduce the eutrophication burden and other negative impacts on biodiversity (Manzetti and van der Spoel, 2015).

If phosphorus is removed from sewage by chemical precipitation, the amount of sludge increases by the amount of chemicals used for the precipitation. The nutrients removed from wastewater are contained in the sludge, which must be handled so that the nutrients are not released back into the watercourses while the material and energy content of the sludge need to be recovered for appropriate utilization. The large amounts of sewage sludge generated in wastewater treatment plants provide numerous opportunities for beneficial use; for example, in power generation, soil improvement, and even nutrient recycling. The possibilities of use are dependent on the quality and amount of sludge in question, the processes used in a particular treatment plant, and the national legislation and policies.

Many modern wastewater treatment plants have some difficulties in their sludge disposal. Problems include the lack of machinery or chemicals to dewater or thicken the sludge and the reduced possibility to use metal-polluted sludge on agricultural land, for example. The sludge-handling strategy of an individual wastewater treatment plant is customized, for example, according to its location, transportation costs, the quality of incoming waters, the used nutrient removal technology, legal restrictions concerning sludge disposal, the availability and price of conditioning agents, and the possibilities to outsource the treated sludge products. The flexibility of any sludge management system allows adaptation to the possible changes in national energy, waste, or nutrient regulations, and market-based policy instruments such as feed-in tariffs or tax regulations. Regular upgrading of the system is needed to keep pace with changing circumstances.

An accurate and unbiased assessment of the risks connected to the use of industrial and municipal sludge is needed. The question is how to deal with various chemical substances present in municipal sludge originating from industry and households and how to recycle the nutrients. The issue of the hazardous chemicals of sludge is currently under debate and regulatory issues and a marketing prospectus are gaining momentum. There are international dimensions to the sludge-handling issue, not only differing in legislation, energy strategies, and sludge handling costs, but also because the world's mineral resources of phosphorus (rock phosphate) are depleting rapidly. Scarcity, abundance, and necessity of nutrients such as phosphorus has become a paradox in the field of nutrient recycling. According to some estimates, phosphorus resources may be sufficient for the next 77 years. Therefore, handling the phosphorus paradox in agriculture and aquatic ecosystems is an emerging field of investigation. Municipal sewage sludge contains large amounts of precious phosphorus; however, the possibilities for recycling should be taken into account already when planning sludge management alternatives. Moreover, some sewage treatment methods do not allow the phosphorus to exist in an easily usable form in the sludge (Leinweber et al., 2018).

Decades ago, when sewage treatment was only taking its first steps, sewage sludge was dumped into international waters. In some countries, this is still a practice. Besides sewage sludge, wastewater treatment has such side products as solids from the screening and sand out of the mechanical treatment. Solids can be dewatered and burnt in an incineration plant, for example, while sand has to be washed and can be used in landfills as structure material.

Sludge ash

Acid digestion Thermal process

There are differences between countries with regard to the abundance and coverage of different sludge-handling methods, some of which are obviously mutually exclusive such as using a centrifuge or a belt filter press in sludge dewatering. Sludge drying with incineration is widely applied only in Germany and to a lesser extent in Poland and Sweden. In the coming years, the amount of sewage sludge generated is estimated to increase, mainly because enhanced wastewater treatment methods will increase.

There are different types of sludge with different physical and biological properties. Typical sewage sludge is made up of primary and secondary sludges. Primary sludge is taken from the primary sedimentation tank. Its amount depends on the retention time and the volume of the tank. Primary sludge is rich in organic compounds and optimal for anaerobic treatment. Sometimes treatment plants have only a small or no primary sedimentation tank to increase the substrate for the denitrifying part of the biological treatment process. Primary sludge has on average a dry solid (DS) content of about 4%. The organic fraction is on average 67%.

Secondary sludge is taken from the clarifier (secondary sedimentation tank). In the aeration basin, the microbial content is high. The retention time of the bacteria in the aeration is between 10 and 20 days, depending on temperature, the population equivalent of the wastewater treatment plant, and the nitrogen removal technique. Especially for nitrogen removal purposes, the bacteria need retention time for growing and thus the sludge must be circulated (return activated sludge). Part of the secondary sludge is not needed anymore (excess sludge, i.e., surplus sludge or waste-activated sludge). The amount depends on the sludge retention time, the use of substrates such as methanol, phosphorus precipitation, biological phosphorus remodal, and, of course, a proceeding treatment such as biological filtration. The chemical precipitation of phosphorus increases the inorganic fraction of the sludge.

Sludge that has been stabilized anaerobically is called digested sludge. Anaerobic digestion is the main stabilization method in municipal sewage sludge treatment. The main benefits are the reduction of organic matter and the production of methane gas, which can be used to achieve the required treatment temperature (often 35–40°C) and the production of electricity. After digestion, the sludge has a much lower volatile solid content and odor.

A maximum volume reduction is attempted after digestion (if applied) to reduce transport and disposal costs. Dewatering reduces the water content from more than 95% to 60%–80%. The dewatered sludge is not pumpable anymore. It can be used in agriculture (after possible treatment improving the hygienic quality of the sludge, or dried and burnt in an incineration plant). Also, disposal to a landfill is possible in certain countries. The applied disposal method mostly depends on the national regulatory framework. Sewage sludge treatment is more than only thickening, digestion, dewatering, and disposal. It has consequences for the whole treatment plant:

- With sludge-originated biogas, it is possible to increase energy production (electrical and thermal) to more than 100% of the power needed in the plant. Energy production and energy efficiency are thus very important issues. It is also possible to increase biogas production with certain pretreatment methods.
- The retention time in primary sedimentation has a direct positive effect on biogas production. On the other hand, a higher retention time decreases the BOD load in the biological treatment; this decreases the denitrification capacity and may require an additional carbon source. Other possible effects are better dewatering ability and lower disposal costs.

In digestion, nitrogen is reduced to ammonia, which is in high concentration in the reject water that is separated from the sludge in dewatering.

- Stabilizing—sludge is not inert and can have an unpleasant odor.
- Reducing the water content and sludge volume to the minimum.
- Utilizing the energy potential when economically possible.
- Reducing the amount of harmful microorganisms if people, animals, or plants are in contact with the sludge.
- Recovering phosphorus for agriculture.

Multipurpose plantations capable of growing on treated sludge-disposed soils with limited inputs need to be expanded/investigated. In this context, the castor bean emerged as a prominent environmental crop. Coupling energy plantations for remediation of cocontaminated substrates and coproduction of biomaterials is a feasible approach (Pandey et al., 2016; Prasad, 2015, 2016; Tripathi et al., 2016).

The European Union has developed the draft of a "working document on sludge", to promote the use of sewage sludge in agriculture while improving the safety and harmonizing quality standards. However, a wide variety of undesired chemicals may be found in sludge, which could have adverse effects on the environment. They also may affect soils, plant, animals, and human health, and have impacts on the environment.

Treated sludge: sludge that has undergone one of the treatment processes or a combination of these processes, so as to significantly reduce its potential to cause a nuisance (health and environmental hazards) when it is used on land.

Contaminants detected in municipal and industrial sludge

AOX	Sum of adsorbable organic halogen compounds
BaP	Benzo[a]pyrene
CB	Chlorobenzene
DBP	Dibutylphthalate
DEHP	Di-2-(ethylhexyl)phthalate
DEP	Diethylphthalate
DNBP	Di-n-butylphthalate
EDs	Endocrine disruptors
EDTA	Ethylene diaminetetraacetic acid
HCH	Hexachlorocyclohexane
LAS	linear alkylbenzene sulphonates (LAS)
NP	Nonylphenole
NPE	Nonylphenole and nonylphenole ethoxylates with 1 or 2 ethoxy groups)
PAH	Polynuclear aromatic hydrocarbons
PBB	Polybrominated biphenyls
PBDE	Polybrominated diphenyl ether (flame retardants)
PCA	Chlorinated paraffins
PCB	Chlorinated biphenyle
PCDD/F	Polychlorinated dibenzo-p-dioxins and -furans
PCP	Pentachlorphenole
POP	Persistent organic pollutants
TBT	Tributyltin
TBTO	Bis-tributyltin oxide
VOC	volatile organic chem.

In addition to the above organotins, musk ketone and musk xylenes are also detected in sludge.

When treated sludge is disposed in a landfill, such areas could be utilized for non-food, nonfodder short rotating plantations (SRPs) (Fig. 1). Willows and poplar trees have been found as suitable species for cultivation on treated sludge-disposed lands as their water and nutrient uptake is high. The idea of the SRPs is that the utilization of pretreated sludge is outsourced to farmers who utilize the nutrients and the water of the sludge in biomass production (BIOPROS, 2008).

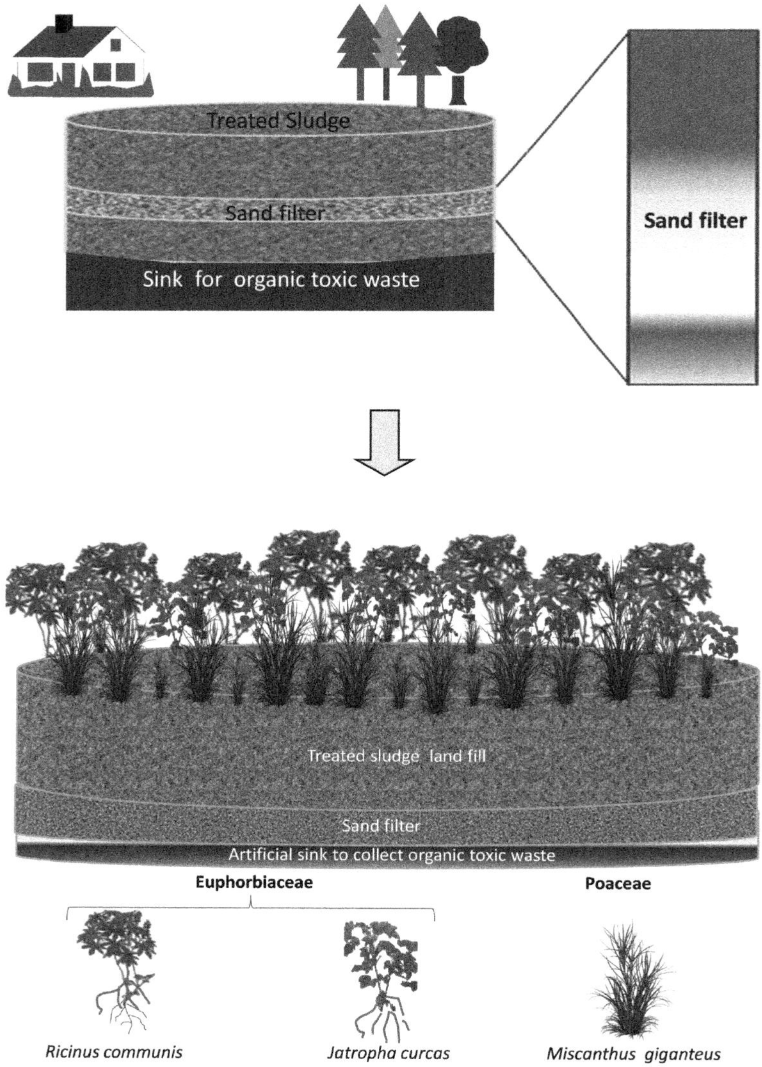

Fig. 1 Coupling energy plantations for remediation of cocontaminated substrates and coproduction of biomaterials.

References

Anonymous, 2012. Good practices in sludge management. Project on Urban Reduction of Eutrophication (PURE) c/o Union of the Baltic Cities Environment Commission, Vanha Suurtori 7, FIN-20500 Turku, Finland; www.purebalticsea.eu. 123. 978-952-5725-91-9 (Print), 978-952-5725-92-6.

BIOPROS, 2008. Short Rotation Plantations. Guidelines for efficient biomass production with the safe application of wastewater and sewage sludge. Available at: www.biopros.info. Accessed 18 September 2012.

Kabbe, C., Kraus, F., 2017. P recovery: from evolution to revolution. Fertil. Int. 479, 37–41.

Leinweber, P., Bathmann, U., Buczko, U., Douhaire, C., Eichler-Lobermann, B., Frossard, E., Ekardt, F., Jarvie, H., Kramer, I., Kabbe, C., Lennartz, B., Mellander, P.E., Nausch, G., Ohtake, H., Tranckner, J., 2018. Handling the phosphorus paradox in agriculture and natural ecosystems: scarcity, necessity, and burden of P. Ambio 47 (Suppl. 1), S3–S19. https://doi.org/10.1007/s13280-017-0968-9.

Manzetti, S., van der Spoel, D., 2015. Impact of sludge deposition on biodiversity. Ecotoxicology 24, 1799–1814.

Pandey, V.C., Bajpai, O., Singh, N., 2016. Energy crops in sustainable phytoremediation. Renew. Sust. Energy Rev. 54, 58–73.

Prasad, M.N.V., 2015. Phytoremediation and biofuels. In: Lichtfouse, E. (Ed.), Sustainable Agriculture Reviews. vol. 17. Springer International Publishing, Switzerland, pp. 159–261.

Prasad, M.N.V. (Ed.), 2016. Bioremediation and Bioeconomy. Elsevier, USA, p. 698.

Tripathi, V., Edrisi, S.A., Abhilash, P.C., 2016. Towards the coupling of phytoremediation with bioenergy production. Renew. Sustain. Energy Rev. 57, 1386–1389.

Further reading

Ohtake, H., Okano, K., Kunisada, M., Takano, H., Toda, M., 2018. Simple technology for recycling phosphate from wastewater tofarmland in rural areas. Ambio 47, 83–92. https://doi.org/10.1007/s13280-017-0976-9.

Preface

Industrial and municipal sludge generation and management in the contemporary world is a subject of concern for resource generation and pollution prevention. The notable feature of this book is its focus on sludge management and resource recovery from different sludges. This book highlights the magnitude of sludge and its management by an expert group of leading practitioners in the field of sludge management who have contributed to this book, which is organized into five sections.

Section A: 6 Chapters Sludge: Sources and characterization	1. Sludge from waste water treatment plants 2. Tannery industry sludge 3. Valorisation of sugar cane waste: Prospects of biorefinery 4. Vermistabilization and detoxification of sugar industrial sludges by earthworms 5. Earthworms converting dairy industry sludge into biomanure 6. Oil recovery from refinery sludge using bio surfactant producing bacteria
Section B: 10 Chapters Sludge management	7. General considerations on sludge disposal 8. Sanitary and environmental aspects of sewage sludge management 9. Activation, conditioning and engineering of sludge 10. Sludge legislation (comparison between different countries) 11. Rheological characterization of sludge 12. Industrial sludge for ceramic products and its benefit for metal stabilization 13. Detoxification of the organic-based sludge from water-treatment plants by active forms of silica 14. Sludge multifunctions in a phytobiome 15. Co-composting of sewage sludge and phytomass from constructed 16. Co-processing of sewagesludge in cement kiln
Section C: 8 Chapters Contaminants in Sludge & treatment strategies	17. Enhanced sludge degradation process using a microbial electrolysis cell 18. Sludge dewatering—Processes for enhanced performance 19. Traditional contaminants in sludge (Poly Aromatic Hydrocarbons (PAH) and Poly Chlorinated Biphenyls (PCBs), dioxins, perfluorinated 20. Emerging contaminants in sludge 21. Carbonization of sewage sludge as adsorbent for organic pollutants 22. Stabilization of cadmium in industrial sludge—Co-generation of crystalline products 23. Bioleaching for extracting heavy metals from electronic waste sludge 24. Treatment of emerging contaminants in urban wastewater with carbon nanotubes
Section D: 4 Chapters Valorisation of sludge as a resource	25. Gasification of sewage sludge 26. From waste to resource: Sorption properties of biological and industrial sludges 27. Valorisation of paper and pulp waste in the framework of biorefinery 28. Phosphorus extraction and sludge dissolution
Section E: 4 Chapters Energy recovery from sludge	29. Volatile fatty acid yield from sludge anaerobic fermentation through biotechnological approach 30. Biogas (methane production) and energy recovery from different sludges 31. Biogas production by continuous thermal hydrolysis 32. Microbial fuel cell application for sludge remediation and minimization

Section A: Sludge: Sources and characterization

Chapter 1 deals with sustainable procedures for sludge management that require the reduction of the amount of sludge to the minimum compatible with the final destination, and the production of good quality sludge, thus reducing the impact on the environment and favoring the application of reuse options instead of disposal ones. This chapter reports the amount of sludge produced in treatment plants and discusses the parameters to evaluate sludge management practices.

Chapter 2 summarizes sludge from tannery industries and management. Tannery sludge characters are highly dependent on the tanning technology adopted as well as the waste treatment methods. A common feature is high chromium content, which contributes approximately 40% of the pollution. This chapter highlights methods for the final management of sludge from the tanning industry, which must be preceded by a thorough analysis of the process and the environmental effects caused by it.

Chapter 3 explains the possibilities for valorization of sugar cane waste, including the prospects for a biorefinery. Currently, most sugarcane industries are using molasses waste as a resource for bioethanol production. Also, a variety of residual waste can be utilized by various biological and chemical processes for multiple product synthesis such as microcrystalline cellulose (MCC), nanocrystalline cellulose (NCC), xylitol, hydrogen, methane, hythane, phosphate fertilizers (struvite), etc.

Chapter 4 is about vermistabilization and detoxification of sugar industrial sludges by earthworms. The potential of earthworms and their ability for detoxification of toxic chemicals in sludges because of their strong metabolic system and the action of earthworm gut microorganisms and chloragocyte cells are presented.

Chapter 5 Earthworms convert milk-processing dairy industry sludge into biomanure by earthwords. Vermicompost generated from the dairy industry has several beneficial properties, including making products for application in agriculture, gardening, and remediation of polluted areas.

Chapter 6 This chapter deals with oil recovery from refinery sludge, a mixture of hazardous hydrocarbon. This waste residue is dumped into sludge pits, consisting of a leachate collection system and a polymer lining system. Several types of bacteria and fungi produce biosurfactants by using the carbon source of crude oil. The biosurfactants offer greater potential in recovery of oil from sludge, as they exhibit low cost alternatives over the chemical surfactant. Furthermore, magnetic nanomaterials assisted purification of biosurfactant for oil recovery improvement from refinery sludge are also presented.

Section B: Sludge management

Chapter 7 presents various issues on general considerations on sludge disposal *sine t*he quality of the sewage sludge depends on many factors and the final disposal is also very diverse. This chapter provides characterization of possible dehydration, stabilization, hygienisation and final disposal of municipal sewage sludge, including indication of barriers in the use of specific technologies, providing common reuse strategies. The chapter also covers potential of sewage sludge which is still largely untapped while the resource recovery from sewage sludge is promising. The existing problems are also highlighted with regard to the management of sewage sludge for legal, economic, technical, and environmental issues.

Chapter 8 covers sanitary and environmental aspects of sewage sludge management. The environmental risks of sludge disposal, treatment scenarios, and acceptable reuse of sewage sludge in a circular economy are discussed.

Chapter 9 summarizes the state of the art in sludge activation, conditioning, and engineering. It presents an overview of biological, chemical, and physical processes, especially oxidative and nonoxidative thermal conditioning methods, freeze-thaw methods, and chemical (alkaline and acid, AOPs) conditioning and stabilization are described. Information on sludge minimization technologies is also included.

Chapter 10 deals with sludge legislation (comparison between different countries) in EU and other world's biggest countries.

Chapter 11 explains the rheological characterization of sludge. As sludge is a complex mixture, the rheological parameters are of paramount importance, such as the solid content that often remains the main standard to distinguish different materials.

Chapter 12 highlights the economic importance and resource recovery from industrial sludge for ceramic products and its benefit for metal stabilization.

Chapter 13 explains the detoxification of the organic-based sludge from water-treatment plants by active forms of silica. This chapter presents evidence that the quick lime and monosilicic acid treatment of municipal sludge could be promising with regard to its utilization and use in agriculture, the greenery of large cities, covering mine tailings, and other purposes.

Chapter 14 highlights the sludge multifunctions in a phytobiome such as forest and plantation applications, including the underlying microbial aspects.

Chapter 15 is about cocomposting of sewage sludge and wetland plant material from a constructed wetland treating domestic wastewater. An interesting solution suggested in this chapter is the use of plant materials derived from wetland plants such as the hyacinth (*Eichhornia crassipes*), water lettuce (*Pistia stratiotes*), and cattail (*Typha angustata*) because of organic origin and thus will be good feedstock for cocomposting.

Chapter 16 suggests coprocessing of sewage sludge in cement. This method of sewage sludge utilization allows for reaching integrated environmental benefits through a more effective and environmentally friendly treatment combined with the protection of natural resources, which is consistent with the principles of sustainable development and the introduction of closed matter circulation.

Section C: Contaminants in sludge and treatment strategies

Chapter 17 presents the enhanced sludge degradation process using a microbial electrolysis cell (MEC). It is a technologically promising solution in terms of energy recovery from organic matter with the participation of microorganisms. The advantage of the microbial electrolysis cell is generating the biohydrogen through naturally occurring bacterial consortia without the involvement of exogenous mediators, making this technology easier to conduct and more promising in terms of electricity generation. Generated hydrogen is a clean and renewable energy source because its combustion does not cause greenhouse gas emissions.

Chapter 18 deals with sludge dewatering and processes for enhanced performance. Dehydration of sewage sludge consists in lowering the water content of sediments and increasing their dry matter content. This is the key process for posttreatment and further processing of sewage sludge. This process reduces the volume of sewage sludge. Depending on the type of water contained in the sewage sludge, it is removed by various methods. There are several types of sludge that have specific characteristics that will influence the choice in the conditioning chemical (cationic flocculent, ferric chloride, lime, etc.) and the choice in the dewatering equipment to be used (filtration, centrifuge, etc.). These choices will also depend on the final use of the sludge (incineration, agricultural spreading, etc.) as well as equipment resources.

Chapter 19 comprehensively deals with traditional contaminants in sludge, including polycyclic aromatic hydrocarbons (PAH), polychlorinated dibenzo-p-dioxins (PCDD), polychlorinated dibenzofurans (PCDF), polychlorinated biphenyls (PCB), polybrominated biphenyls (PBB), polybrominated diphenylethers (PBDE), absorbable organic halogens (AOX), per fluorinated compounds (PFCs), pesticides (e.g. organ chlorine pesticides OCP), and intermediary products of detergent biodegradation. Therefore, efficient processes are needed that would allow eliminating the contaminants and removing the harmful substances from sludge.

Chapter 20 summarized emerging contaminants in sludge such as endocrine disruptors, pesticides, and pharmaceutical residues, including illicit drugs/controlled substances, etc. As a result of progressive civilization development a large number of emerging contaminants may transfer to the sludge during wastewater treatment. Emerging contaminants in sludge have shown adverse effects on the environment.

Chapter 21 deals with carbonization of sewage sludge as an adsorbent for organic pollutants. The silicon element derived from sludge plays complexation to –COOH of the benzoic acid, leading to preferable adsorption ability. Granular sludge carbon is also developed as an adsorbent for the removal of organic pollutants, considering its efficient recovery and separation. Special hollow-like and hollow spherical sludge carbon was prepared for adsorption of the dye in batch and column experiments. Thus, carbonization of sludge into carbonaceous production could be a promising method for resource utilization of sludge as well as sludge disposal treatment.

Chapter 22 presents stabilization of cadmium in industrial sludge - cogeneration of crystalline products. Sewage sludge is a carbon-rich material and has the potential of carbonaceous adsorbents for removal of organic pollutants. Thus, carbonization of sludge into carbonaceous products could be a promising method for resource utilization of sludge, boosting the sustainable circular economy.

Chapter 23 is about bioleaching of heavy metals from electronic waste sludge. The electronic waste recycling industry generates the largest amount of sludge in the world. Application of microorganisms, namely *Acidithiobacillus ferrooxidans,* for bioleaching heavy metals from electronic waste sludge is presented. The bioleaching mechanisms and safe disposal of the electronic waste sludge are also discussed.

Chapter 24 highlights the treatment of emerging contaminants in urban wastewater with carbon nanotubes. The presence of emerging contaminants, including pharmaceuticals, personal care products (PPCPs), pesticides, and endocrine-disrupting chemicals (EDCs), in water bodies is a serious problem in many countries.

Therefore, this chapter discusses the scope of novel carbon nanotube-based materials and technologies in removing emerging contaminants from urban wastewater sludge.

Section D: Valorization of sludge as a resource

Chapter 25 is about gasification of sewage sludge. Increasing problems associated with sewage sludge disposal are observed nowadays. Thermal conversion of sewage sludge (combustion, cocombustion, gasification, and pyrolysis) appears to be the most promising alternative for its management. Among these methods, gasification is the most advantageous. Therefore, this chapter presents a multicriterial investigation of sewage sludge gasification in the fixed-bed gasifier.

Chapter 26 highlights sludge as a resource. The choice of using sludge as an adsorbent and related mechanistic are is covered in this chapters.

Chapter 27 is about valorization of paper and pulp industry produced sludge and liquor. This discharge is a potential feedstock containing a high percentage of lignin, cellulose, and hemicellulose, which can be used as a valuable raw material for the production of bioenergy, platform chemicals, and biofuels following a sustainable biorefinery concept.

Chapter 28 explains phosphorus extraction and sludge dissolution. Demand for P is expected to increase because of a growing world population and the need for food; however, the global high-grade phosphate rock reserves are dwindling. Therefore, phosphorus extraction and sludge dissolution for P recovery assumes importance.

Section E: Energy recovery from sludge

Chapter 29 narrates the biotechnological approaches for volatile fatty acid production from sludge through anaerobic fermentation. Volatile fatty acids can be obtained from natural and anthropogenic sources. Volatile fatty acids occur in large quantities in sewage sludge, sewage generated in animal farms, and in food industry effluents.

Chapters 30 and 31 deal with biogas (methane production) and energy recovery from different sludges with other organic waste in the process of anaerobic codigestion and biogas production by continuous thermal hydrolysis and thermohilic anaerobic digestion of waste-activated sludge.

Chapter 32 gives a state-of-the-art summary of the microbial fuel cell application for sludge remediation and minimization.

Overall, the book will be a useful resource for academicians, students, risk assessors, regulators, and entrepreneurs who are engaged in handling diverse issues of industrial and sewage sludge including resource recovery to prevent pollution.

Acknowledgments

The editors would like to thank all the authors in this volume for their cogent and comprehensive contributions.

The editors also would like to place on record their appreciation and thanks to Kenneth P. McCombs, Senior Acquisitions Editor (ELS-NYC), for inspiration and vital thoughts as key input for this work. We thank Andréa Akeh, Acting Senior Editorial Project Manager Oxford, for excellent coordination of this fascinating project, suggestions, and help in many ways that resulted in a record time of publication for this volume. Thanks are also due to the production project manager Sruthi Satheesh.

Last but not least we wish to thank numerous colleagues for sharing knowledge, ideas and assistance, which helped in shaping this book.

Section A

Sludge: Sources and characterization

Sludge from wastewater treatment plants

1

Roberto Canziani, Ludovico Spinosa†*
*DICA—Sezione Ambientale, Politecnico di Milano, Milano, Italy, †ISO and CEN Expert, Bari, Italy

Nomenclature

bCOD biodegradable chemical oxygen demand
BOD biochemical oxygen demand
COD chemical oxygen demand
CS tertiary/chemical sludge
DM dry matter
DS dissolved solids
FS fixed solids
LOI loss on ignition
PS primary sludge
SS suspended solids
TS total solids
VS volatile solids
VSS volatile suspended solids
WAS waste-activated sludge

1 Introduction

The management of sewage sludge in an economically, environmentally, and socially acceptable manner—a sustainable manner—is one of the critical issues facing modern society. This is due to the fast increase in sludge production as a result of (1) growing availability of household running water, (2) extended sewerage, (3) new work installations, and (4) upgrading of existing facilities. This is coupled with increased difficulties in properly locating disposal works and complying with even more stringent environmental quality requirements imposed by legislation, which force higher levels of treatment before the sludge can be sent from the plant to its final disposal or reuse.

In addition, during the last decade, there has been a worldwide movement toward a common strategy, applied to any kind of waste, that gives priority to reusing waste materials and taking advantage of their energy content (Spinosa, 2005).

Generally speaking, sludge management policy is addressed to both the development of treatment methods able to reduce the mass production of sludge and to its impact on the environment. Consequently, sludge reuse options are preferred to simple disposal as they can add value to the energy and material content of sludge.

Industrial and Municipal Sludge. https://doi.org/10.1016/B978-0-12-815907-1.00001-5

The low concentration of solids in the raw sludge (2%–3%), and the presence of pathogens, heavy metals, organic micropollutants, and potentially putrescible organic matter are the main problems to be faced and solved. Further, although sludge accounts for about 2% by volume of the processed sewage, its handling accounts for up to 50% of total operating costs.

This means that the establishment of defined procedures/routes for the final destination/outlet is needed and a good sludge quality must be guaranteed in order to (1) properly perform the utilization and disposal operations, (2) correctly fulfill the legal requirements, and (3) build stakeholder and public confidence.

Characterization is, therefore, a tool of primary importance in sewage sludge management as it allows the evaluation of the chemical, biological, and physical properties of sludge, and often the prediction of its behavior in treatment processes.

The fundamentals of procedures adopted in characterization methods are often well known and accepted everywhere, but each laboratory or textbook can follow and/or propose different equipment size, accessories, and methodologies. Consequently, results obtained in different places and at different times, even if correct, are difficult to be compared as they have been obtained under different, not standardized, conditions. In addition, a statistical analysis of errors is often not available.

Consequently, the development of standardized characterization methods and procedures becomes an additional, necessary prerequisite.

This chapter reports the sludge amount produced in wastewater treatment plants and discusses the main chemical, biological, and physical parameters to be evaluated, according to the main practices for sludge management and its final destination/outlet, that is, land application, in particular agricultural utilization; incineration and other thermal processes; and landfilling.

2 Sludge types

Sludge characteristics vary widely as they depend on influent wastewater, mainly on the type of industrial effluents discharged into the sewage systems.

Three main categories of sewage sludge can be identified: primary, waste-activated or secondary, and chemical or tertiary (Vesilind and Spinosa, 2001; Spinosa, 2005).

2.1 Primary sludge

Primary sludge (PS) originates from mechanical and preliminary treatments. Generally speaking, it is rich in readily biodegradable organic matter, has a high biogas production potential if treated in anaerobic digestion processes, and, usually, has good dewaterability.

Typical operations used for preliminary treatment include bar screens, which protect the plant from very large objects, and degritting/degreasing units, which protect the plant from sand, grit, and oily matter. These are all residues not to be handled as sludge.

Primary treatment often consists of a gravity-settling unit for the removal of the settleable solids and the collection of residual floating matter, such as grease and scum, that are produced in small amounts. The settled solids represent a major source of solids that have to be properly managed. This sludge, produced in the primary clarifier or settling tank, is known as PS; it has a strong stinking odor and contains identifiable solid matter that makes it aesthetically unpleasant. In addition, as most pathogenic organisms in wastewater attach to solid particles, PS may contain all the known human pathogens (Vesilind and Spinosa, 2001).

2.2 Secondary/waste activated sludge

After primary treatments, wastewater still has a high content of biodegradable organic matter that must be reduced to avoid polluting the receiving waterbody. The removal of biodegradable organic matter, which is expressed in terms of biochemical oxygen demand (BOD) or biodegradable chemical oxygen demand (bCOD), is the objective of secondary treatments.

The most popular secondary treatment method is the activated sludge process. In this process, wastewater is mixed with biomass, which has the form of biological aggregates named flocs. The active biomass assimilates the organic matter, thus reducing the oxygen demand. The process produces an excess growth of biomass, known as waste-activated sludge (WAS) or secondary sludge. WAS is significantly different from PS, being a complex aquatic culture of bacteria, protozoans and helminths, mixed together with organic detritus deriving from cell decay and some inert materials (silt, clay, carbonates, silicates). WAS contains 99% water, part of which is bound by chemical and physical means to the surface area provided by the floc particles. These are rich in volatile solids (VS), thus making it difficult to be dewatered.

The biodegradability of WAS is widely variable and strongly depends on the sludge age. WAS does not contain the concentration of pathogens found in PS (Vesilind and Spinosa, 2001).

2.3 Tertiary/chemical sludge

Tertiary/chemical sludge (CS) is formed during chemical nutrient removal or tertiary/advanced treatments aimed at improving the quality of the effluent. These treatment systems, such as coagulation and flocculation followed by sedimentation or, more frequently, by filtration, generally produce solids that should not be managed with other types of sludge.

The characteristics of this sludge are very different, as they depend on the treatments to which they have been subjected.

The last step in wastewater treatment is typically disinfection by chlorine or, increasingly, with ultraviolet radiation; these do not produce any sludge.

3 Production

Typical sludge quantities and concentrations of solids and nutrients are reported in Table 1. Specific production of sludge ranges from 0.2 to 5.0 L/cap/d with typical concentrations in the range 0.7%–10.0%, being 2 L/cap/d at 4% solids concentration the typical production of primary+activated sludge from municipal plants (Spinosa, 2005).

Table 1 Typical sludge quantities and characteristics

Type	Quantity (L/cap/d)	Solids concentration (%)	Nitrogen (% DM[a])	Phosphorus (% DM)	Potassium (% DM)
Raw primary	0.9–2.2	2.0–8.0	1.5–5.0	0.3–2.8	<1.0
Raw activated	1.4–7.3	0.2–1.5	3.0–10.0	1.0–7.0	0.1–0.9
Raw pr. + act.	1.8–2.8	3.0–6.0	4.0–6.0	1.0–1.2	–
Dig. pr. + act.	0.6–1.0	2.0–12.0	1.0–6.8	0.2–5.7	<4.0
Tertiary	0.2–8.0	3.0–10.0	–	–	–

[a]*DM*, dry matter.
Source: Spinosa, L., 2005. Sludge treatment and disposal. In: Lehr, J.H. (Ed.), Water Encyclopedia—Waste Water Treatment. John Wiley & Sons, Inc.; Sanin, F.D., Clarkson, W.W., Vesilind, P.A., 2011. Sludge Engineering: The Treatment and Disposal of Wastewater Sludges. DEStech Publications, Lancaster, p. 393.

Several approaches are available to estimate sludge production in a wastewater treatment plant.

Following a simple "rule of thumb" approach, in the earliest textbooks wastewater treatment plants were expected to produce about 80 g-dry-solids/capita/d, and data from 26 wastewater treatment plants in the United States showed that solids were produced at a rate between 0.2 and 0.3 kg/m^3, with an average of about 0.25 kg/m^3 (Vesilind and Spinosa, 2001).

More recent data from European countries showed an average sludge generation rate of 58.9 g/capita/d, ranging from 19.9 in Greece to 107.6 in Portugal. Differences can be explained in terms of water availability, population served by treatment facilities, and degree of treatment applied (Sanin et al., 2011).

The production of PS can be estimated by assuming that a certain fraction of the suspended solids (SS) will settle out. It follows Eq. (1):

$$W_{PS} = Q_i \times E_{SS} \times C_{SS} \times 10^{-5} \qquad (1)$$

where W_{PS} is the production of PS in kg/d of dry matter (DM), Q_i is the influent to the treatment plant in m^3/d, E_{SS} is the removal rate in percentage of SS removed, and C_{SS} is the concentration of SS in mg/L.

In general, primary clarifiers seldom accomplish more than about 70% removal of SS for a commonly adopted hydraulic retention time of about 2 h at the design flow rate and an overflow rate (flow rate, m^3/d, divided clarifier surface area, m^2) of 45 m/d (Vesilind and Spinosa, 2001; Sanin et al., 2011; Tchobanoglous et al., 2014).

Solid removal, and therefore PS production, can be enhanced by up to 90% by adding chemicals to the clarifier influent, which improves flocculation. Ferric chloride, lime, aluminum sulfate, and also organic polyelectrolytes can be used for such purpose, but the amount of added flocculants has to be added to the total sludge produced.

The addition of these flocculants changes the sludge characteristics, so preliminary careful laboratory characterization work is required prior to the designing of subsequent treatments.

In the secondary or biological treatment, most of the SS remaining after the primary treatment is removed together with the excess biological sludge produced by the biological process itself. The amount of sludge produced is obviously dependent on the type of process.

For activated sludge processes, the amount of WAS produced can be estimated using either empirical data or growth kinetics. Koch et al. (1990) suggested a simplified approach to estimate the amount of WAS, expressed by Eq. (2):

$$W_{WAS} = W_I + (a \times W_{VSS}) + (b \times BOD_{sol}) \quad (2)$$

where:

- W_{WAS} is the production of waste-activated sludge in kg-dry-solids/d.
- W_I are the inert or fixed solids estimated as the fixed suspended solids in the primary clarifier effluent in kg/d.
- W_{VSS} are the volatile suspended solids (VSS) in the primary clarifier effluent in kg/d.
- BOD_{sol} is the soluble BOD in the primary clarifier effluent in kg/d.
- a is constant ranging from 0.6 to 0.8 and b is another constant ranging from 0.3 to 0.5.

The growth kinetics approach uses the concept of yield, or the mass of solids produced per mass of BOD removed. As a rule of thumb, the yield from an activated sludge process is about 0.5 kg of the waste solids produced per kg of BOD removed. This includes SS not removed in the primary clarifier, and takes into account that some of the organisms produced in the activated sludge process are removed through death and decay. Thus, this yield is often termed "observed yield."

A different yield is the "true yield," measured using laboratory chemostats. This does not include death and decay, and only incorporates the dissolved fraction of the feed (dissolved or soluble BOD). In this case, the Water Environment Federation (1992) recommends Eq. (3):

$$W_{WAS} = Q\{[Y(S_o - S_e)/(1 + b_d\theta)] + FSS_{io} + FSS_{no}\} \quad (3)$$

where:

- W_{WAS} is the mass of total WAS produced per day in g/d.
- Q is the flow rate in m^3/d.
- Y is the true yield often expressed as mass of VSS produced per unit mass of soluble BOD removed (average: 0.4 g-VSS/g-soluble BOD).
- S_o and S_e are the influent and effluent soluble BOD, respectively, in g/m^3.
- b_d is the endogenous decay coefficient ranging from 0.04 to $0.75\,d^{-1}$ (or an average of $0.6\,d^{-1}$).
- θ is the mean cell residence time, or sludge solids retention time, or sludge age, in days (d).
- FSS_{io} is the influent fixed (or nonvolatile) suspended solids in g/m^3.
- FSS_{no} is the influent nonbiodegradable VSS in g/m^3.

Tchobanoglous et al. (2014), suggested a more complex approach that takes into account the accumulation of unbiodegradable cell debris, according to Eq. (4):

$$W_{WAS} = Q\{[Y(\mathrm{bCOD_o} - \mathrm{bCOD_e})\,(1 + f_d \times b_h \times \mathrm{SRT}]/(1 + b_h\theta) + \mathrm{FSS_{io}} + \mathrm{FSS_{no}}\} \tag{4}$$

where:

- Y is the true yield, expressed as 0.45 g-VSS/g-bCOD.
- $\mathrm{bCOD_o}$ and $\mathrm{bCOD_e}$ are the influent and effluent biodegradable COD, respectively, in g/m^3 (bCOD can be estimated as 1.6 times BOD).
- f_d is the unbiodegradable fraction of the activated sludge (0.15, dimensionless).
- b_h is the endogenous decay coefficient (0.12 d^{-1}).
- All other terms as previously defined.

In nitrification systems, an additional term that accounts for the nitrifying organisms has to be added. It accounts generally for 2%–5% of the total sludge mass, and depends mainly on the influent bCOD/TKN (biodegradable COD to total Kjeldahl nitrogen ratio): the higher this ratio, the lower the fraction of nitrifying biomass. Moreover, the longer the sludge age, the higher the removal of the organic (or volatile) matter, thus reducing the amount of produced sludge.

Low-loaded attached-growth processes, such as trickling filters and rotating biological contactors, produce less sludge than suspended growth biological processes. A rule of thumb for attached growth systems is 0.1 kg-VSS/kg-BOD removed. Submerged biofilters and high-loaded attached-growth processes produce approximately the same amount of sludge as activated sludge processes.

The estimated production of tertiary/chemical sludge is a function of the dosage of reagents, the stoichiometric excess applied and SS removed. The precipitation reactions involved include the formation of metal hydroxides, phosphates and all insoluble or precipitated salts. Tchobanoglous et al. (2014) reported an example for chemical phosphorus removal.

A more precise technique for estimating sludge production in wastewater treatment consists of first drawing the process diagram for the plant, and then evaluating the production of solids using mass balances.

Fig. 1 shows the flowsheet of a general wastewater treatment plant.

Calculation of sludge production requires appropriate values for all these parameters and coefficients, which should be carefully evaluated from field data.

Sanin et al. (2011) reported the following typical values for domestic wastewaters:

- $S_0 = 250 \times 10^{-3} \times Q$ (kg/h if Q is in m^3/h).
- $X_0 = 225 \times 10^{-3} \times Q$ (kg/h).
- $k = 0.6$.
- $h = 0.7$.
- $n = 0.3$.
- $X_f = 20 \times 10^{-3} \times Q$ (kg/h).
- $j = 0.8$ assuming no supernatant withdrawal.
- $i = 0.1$ for well-operated activated sludge.

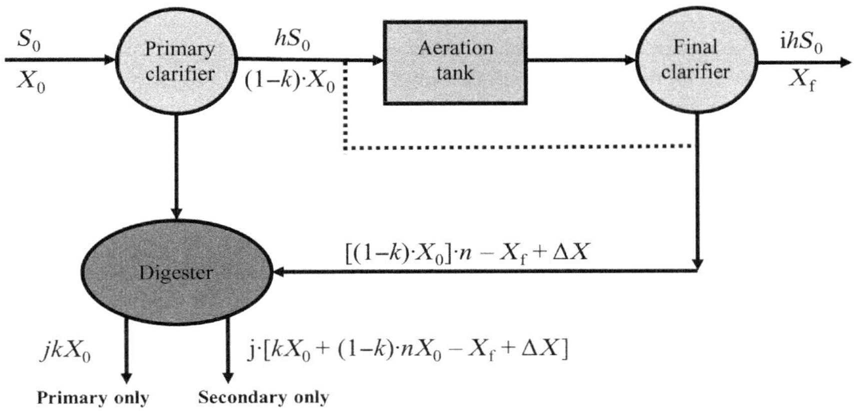

Fig. 1 Mass balance for a generalized wastewater treatment system.
Where:

- S_0 is the influent BOD, in kg/h (5-day, 20°C).
- X_0 is the influent SS, in kg/h.
- h is the fraction of BOD not removed in primary clarifier, dimensionless.
- i is the fraction of BOD not removed in the activated sludge process or trickling filter, dimensionless.
- X_f is the plant effluent SS, in kg/h.
- k is the fraction of X_0 removed in the primary clarifier, dimensionless.
- j is the fraction of solids not destroyed in digestion, dimensionless.
- ΔX are the net solids produced by biological reaction, in kg/h.
- Y (yield) $= \Delta X/\Delta S$, where $\Delta S = hS_0 - ihS_0$.

Source: Sanin, F.D., Clarkson, W.W., Vesilind, P.A., 2011. Sludge Engineering: The Treatment and Disposal of Wastewater Sludges. DEStech Publications, Lancaster, p. 393.

- $i = 0.2$ for trickling filters.
- $Y = 0.5$ for activated sludge.
- $Y = 0.2$ for trickling filters.

These values may vary greatly from plant to plant and should be considered only to achieve an initial approximate estimate.

4 Characterization parameters

The selection of the best processing sequence preliminarily requires the knowledge of sludge characteristics. This will allow the proper choice of management operations, the fulfillment of legal requirements, and building the confidence of stakeholders and the general public.

As already mentioned, characterization plays a role of primary importance in sewage sludge management, as it allows evaluating the chemical, biological, and physical properties of sludge as well as possibly predicting its behavior. However, different methodologies are adopted in different places and different times, often without a statistical analysis of errors. These factors make it difficult, if not impossible, to compare results that are not derived from standardized procedures.

Therefore, the development of standardized characterization methods and procedures becomes an additional necessary prerequisite (Spinosa, 2013).

The total solids (TS), suspended solids (SS), and volatile solids (VS) contents are the characterization parameters of interest for all sludge treatment and management operations, and often other parameters are necessarily related to them (Kopp and Dichtl, 2001).

Other chemical, biological, and physical parameters involved in sludge characterization can be categorized according to the different options available for the final sludge destination, such as material/energy recovery or disposal.

The main available practices for the management of sludge/biosolids include (1) land application, agricultural utilization in particular, (2) incineration and other thermal processes, and (3) landfilling (Spinosa, 2016).

With reference to the agricultural use of sludge, the content of organic carbon, nutrients, heavy metals, biological and biochemical properties, water content, and dry residue must be considered. Furthermore, nowadays much greater attention must be paid to organic micropollutants and other emerging contaminants.

As far as thermal processes are concerned, the general philosophy behind regulations is to fix limits for the emission of different pollutants. Therefore, their determination is of fundamental importance, together with that of liquid and solid process residues in view of their safe recycle and/or disposal.

As for landfilling, legislation generally requires wastes be previously subjected to treatment, where treatment means the physical, thermal, chemical, or biological processes that change the characteristics of the waste in order to reduce its volume or hazardous nature, facilitate its handling, or enhance recovery. Within this framework, the evaluation of biological stability and physical stability, or consistency, is of primary importance, in addition to that of chemical pollutants.

4.1 Solids content

4.1.1 Total solids

The TS content is determined by drying the sludge sample at 105°C. The TS content can be calculated in % by Eq. (5):

$$TS = (m_{dm}/m_{sl}) \times 100 \tag{5}$$

where m_{dm} is the mass of dried matter, and m_{sl} is the mass of the sludge suspension.

TS can be divided into dissolved solids (DS) and suspended solids (SS) as well as into fixed solids (FS) and volatile solids (VS).

4.1.2 Suspended solids

Suspended solids (SS) are the solids separated after filtration (or centrifugation) of the sample. They are calculated in % using Eq. (6):

$$SS = (m_{dmf}/m_{sl}) \times 100 \quad (6)$$

where m_{dmf} is the mass of dried matter separated and dried at 105°C, and m_{sl} is the mass of the sludge.

4.1.3 Volatile solids

Volatile solids (VS, also referred to as loss on ignition, LOI) generally represent the organic solid content. Consequently, they can be used as a measure of the degree of stabilization of sludge and their reduction is strongly related to biogas production.

For measurement, the dried sample is burned at 550°C and the VS content is calculated in % by Eq. (7):

$$VS = \left[1 - \left(m_{gr}/m_{dm}\right)\right] \times 100 \quad (7)$$

where m_{gr} is the residual mass after burning. The residues are inorganic minerals and sand.

4.2 Chemical parameters

4.2.1 General chemical parameters

General chemical parameters of interest for sludge management are pH, alkalinity, and volatile fatty acids.

The pH value of sewage sludge is, in general, neutral at 7.0. Slightly higher values are typical in digested sludge or sludge in the methanogenic phase of anaerobic digestion while PS and sludge in the acidogenic phase show slightly lower values.

Total alkalinity (TA), often expressed as mg-$CaCO_3$/L, is a very important parameter in controlling slightly alkaline conditions during methane fermentation, and, as such, is a good parameter for monitoring anaerobic digestion.

Volatile fatty acids (VFA) are low molecular carbonic acids generated as intermediate products during the digestion process. A concentration of both acetic acid and propionic acid is often used to characterize VFA, and high concentrations of propionic acid are an indication of some disturbance in the anaerobic digestion process.

A frequently adopted operational parameter in anaerobic digestion is the VFA/TA ratio (also known as FOS/TAC, the acronym of "Flüchtige Organische Säuren/Totales Anorganisches Carbonat" in the German technical literature), although it cannot be the only parameter for optimal process control (e.g., Pontoni et al., 2015).

4.2.2 *Chemical parameters related to land application*

The chemical parameters mainly related to the agricultural use of sludge are nutrients such as phosphorus and nitrogen and pollutants such as heavy metals and harmful or toxic organic substances, the determination of which is very important as their value can limit or even prevent spreading sludge on land.

4.2.2.1 Nutrients

Sewage sludge contains significant amounts of nitrogen and phosphorus, but only small amounts of potassium. The analysis of 6000 sewage sludge samples showed an average content of nitrogen, as N, of 38.4 g/kg-SS (range 0.1–246.0); phosphorus, as P_2O_5, of 36.5 (range 0.2–344.0); potassium, as K_2O, of 4.2 (range 0.1–95.0); magnesium, as MgO of 9.7 (range 0.1–122.0); and calcium, as CaO, of 73.7 (range 0.1–727.0) (Kopp and Dichtl, 2001).

4.2.2.2 Pollutants

Pollutants in sludge include selenium, arsenic, and heavy metals such as mercury, copper, cadmium, chromium, lead, zinc, and nickel. When sludge is used as a fertilizer or soil conditioner, the maximum allowable concentrations of all or some of these elements are defined by legal regulations as well as organic micropollutants such as AOX (adsorbable organic halogenated substances), LAS (linear alkylsulfonates), NP/NPE (nonylphenol and its ethoxylates), PAH (polycyclic aromatic hydrocarbons), PCB (polychlorinated biphenyls), and dioxins.

Other parameters of interest in land application are the organic carbon, water, and VS content.

4.2.3 *Chemical parameters related to thermal processes*

As far as incineration and other thermal processes are concerned, the general philosophy behind regulations is to fix limits at the emission for different pollutants, so their determination is of fundamental importance, together with that of liquid and solid residues of the process.

The main chemical parameters to take into account are (Spinosa, 2013, 2016):

- Chlorine compounds, both organic and inorganic, due to the tendency of the chlorine radicals to bind to active radicals such as O*, H* and OH* with the possibility of toxic compound formation. Chlorine and other halogens are also responsible for the presence in the exhaust gases of acidic compounds that are undesirable for the corrosion problems they involve.
- Sulfur, because sulfur dioxide combines with moisture to form sulfuric and sulfurous acids, thus affecting the cost of acid gas removal and gas utilization.
- Phosphorus, which is a finite resource and its recovery from ashes should be considered. Recent estimates (Cordell and White, 2011) report that in 100 years all known apatite mines will run empty and there will be no reliable phosphorous sources left. This makes it essential to recover phosphorus from sludge, as this is one of the major available phosphorous streams in society today.
- Nitrogen, due to conversion of most of it during combustion to NO_x.

- Organic micropollutants, with main regard to chlorinated hydrocarbons, phenols and polyphenols, polychlorinated biphenyls (PCB), pesticides, and polycyclic aromatic hydrocarbons (PAH).
- Trace elements, for their potential tendency to be transferred in the gaseous phase (especially for mercury). Except mercury, they can be concentrated in fly ash and collected in bag and electrofilters (arsenic, lead, cadmium, and zinc).

Chemical analysis should also include, for particular cases of contaminated sludge, the compounds that are recognized to be recalcitrant to a thermal degradation.

Further, the elemental analysis of LOI (C, H, N, S, O) is important to predict the flow rate and composition of flue gas and, therefore, to design its treatment line.

Other parameters not to be forgotten are those related to combustion conditions and to gaseous and solid discharges, for example, dry matter and organic matter content, and those related to transport, storage, and transfer such as moisture content and pH.

4.2.4 Chemical parameters related to landfilling

As for landfilling operations, regulations generally require that waste to be landfilled must not be liquid, and that it should be previously subjected to treatments, where treatment means the physical, thermal, chemical, or biological processes that change the characteristics of the waste in order to reduce its volume or hazardous nature, facilitate its handling, or enhance recovery. Therefore, all chemical parameters that may classify the sludge as hazardous or nonhazardous, according to local regulations, have to be considered in sludge landfilling, together with sludge physical consistency and moisture or DM content.

4.3 Biological parameters

The evaluation of the biological stability of sludge, or products derived from its treatment, such as compost, is of great importance considering that the level of biological stability strongly influences the handling of such materials. This is because it gives indications on the effectiveness of subsequent treatments, including risks of developments of bad odors, effective acceptability in a landfill, possible reuse for environmental purposes, etc.

The biological stability of sludge is important as its concept has been included in several regulations, particularly with reference to both landfilling and use on land. Sludge should reach a level of biological stability so that it will not affect the environment and public health.

However, a widely accepted parameter and/or procedure to evaluate the biological stability of sludge has not yet been defined, although several have been proposed. Some of them are discussed in the following paragraphs.

4.3.1 Volatile solids reduction

This parameter is a simple one to evaluate the level of biological stability. It is calculated in % by Eq. (8):

$$\rho_{VSS} = [1 - (VSS_1/VSS_0)] \times 100 \tag{8}$$

where VSS_0 represents the VSS value before digestion and VSS_1 that after digestion.

Typical values of VS reduction are 40%–45% for anaerobic digestion of sewage sludge and 25%–30% for aerobic stabilization (Kopp and Dichtl, 2001).

4.3.2 Respiration activity

The "Working Document on Sludge and Biowaste" issued by the European Union (DG ENVIRONMENT Directorate C—Industry ENV.C.2—Sustainable Production and Consumption, Brussels, September 21, 2010) states that the sludge should be sufficiently stabilized so as not to cause an unreasonable odor nuisance to the nearest dwellings.

Following that definition, parameters such as either the respiration activity after 4 days (AT_4, mg-O_2/g-DM) or the dynamic respiration index (mg-O_2/kgVS/h) are useful for evaluating the reduction of the decomposition properties to such an extent that offensive odors are minimized. The same document lists, as possible indicators, the following: volatile solid (VS) reduction of 38% or specific oxygen uptake rate of less than 1.5 mg/h/g-TS.

4.3.3 Microbiological parameters

The microbiological parameters are important for the evaluation of hygienic aspects. We must consider that pathogenic organisms are weakened or killed as a function of time and temperature of treatment, and as the consequence of microbial competition with other much more numerous nonpathogenic organisms.

4.3.4 Biochemical methane potential

Biochemical methane potential (BMP) is a procedure developed to determine the methane production of any biodegradable organic substrate, including sewage sludge under anaerobic digestion (Owen et al., 1979; Raposo et al., 2011; Labatut et al., 2011). BMP can be used as a short-term test (5–7 days) to measure the rate of methane production as an indicator of stability or as a long-term test (>30 days) to measure the overall residual methane production. Reference values for assessing sewage sludge stabilization are not available yet.

4.4 Physical parameters

The knowledge of physical properties allows the prediction of sludge behavior when handled and submitted to almost all treatment, storage, and utilization/disposal operations, including storage, pumping, transportation, land-spreading, dewatering, drying, incineration, and landfilling (Spinosa, 2013, 2016).

The main goal in physical processing is to increase the solid content by removing water, thus decreasing the volume of the sludge to be handled.

4.4.1 Density

Knowledge of density is critical to most unit operations in wastewater treatment and sludge management. A higher sludge floc density enhances the effectiveness of secondary clarifiers and thickeners, and increases cake concentration in mechanical dewatering.

Density also affects the sludge volume, and, hence, transport costs. In addition density also affects sludge fluid-dynamic behavior, as it is a factor in the Reynolds number and, consequently, it affects the head losses in a pipeline. It also affects thermal conductivity in turbulent flow.

Density can also give an indication of the content of low-density components such as grease and fats, which, in turn, affect the stability of sludge.

The evaluation of density is obtained through the use of a pycnometer, a volume-calibrated glass flask with a close-fitting ground glass stopper.

The method is based on the measurement of the weight and volume of the sludge sample under standard conditions.

4.4.2 Settling properties

These properties include evaluation of both settleability and thickenability.

4.4.2.1 Settleability

The settleability determination is used for calculating the rate of sludge settling and the sludge volume index, and for evaluating the performance of settling tanks.

The settled sludge volume and the sludge volume index are determined by 30min settling of a sludge suspension in a graduated cylinder of nominal volume of 1000mL (CEN, 2006a)

The sludge volume index (SVI) is calculated in mL/g by Eq. (9):

$$\mathrm{SVI} = V_{\mathrm{S}}/\rho_{\mathrm{T}} \tag{9}$$

where V_{S} is the sludge volume in mL/L after 30min settling, and ρ_{T} is the concentration of solids in sludge in g/L.

4.4.2.2 Thickenability

The determination of thickenability allows the amenability of sludge to further concentrate SS in gravity thickeners.

It is evaluated through a graduated cylinder (diameter 10cm, height 50cm, nominal volume 3500mL) equipped with a low speed stirrer at 1rpm (Fig. 2) to avoid the negative influence of particle bridging, wall effects, and particle size effects (White, 1976; CEN, 2006b).

The stirred sludge volume index (I_{ssv}, mL/g) is calculated from Eq. (10):

$$I_{\mathrm{SSV}} = (1000 \times H)/(50 \times C_0) \tag{10}$$

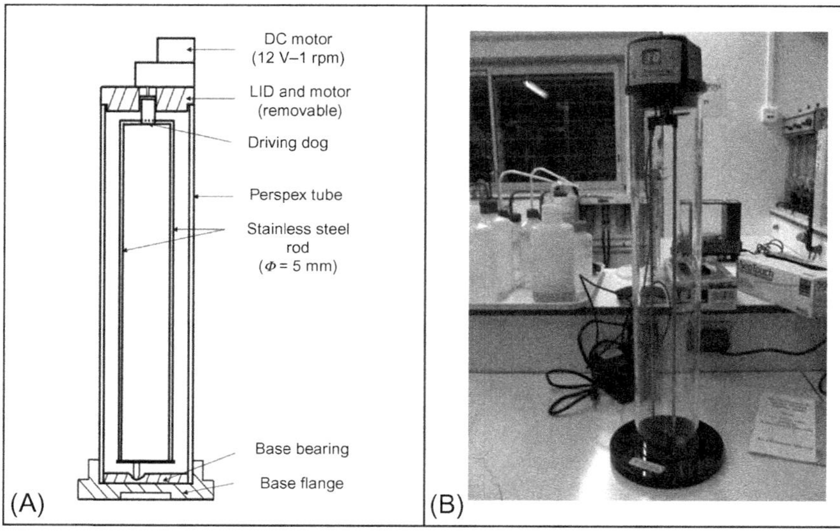

Fig. 2 Stirred cylinder: (A) Scheme. (B) View.
(A) *Source*: White, M.J.D., 1976. Design and control of secondary settlement tanks. J. Inst. Wat. Poll. Control, 461.

where H is the final sludge level in cm, C_0 the initial solids concentration in g/L, 50 the starting height of sludge suspension in cm, and 1000 a conversion factor.

I_{SSV} values below 80 mL/g are indicative of good thickenability while values higher than 120 mL/g of poor thickenability

4.4.3 Filterability properties

Within this framework, the determination of capillary suction time (CST), specific resistance to filtration, compressibility, drainability, and centrifugability are of primary importance.

4.4.3.1 Capillary suction time (CST)

The CST is a fast and simple way to evaluate sludge dewaterability by filtration. Only qualitative and comparative evaluations can be obtained and correlated to the specific resistance to filtration for a given type of sludge.

The principle of the method is that dewatering is achieved by the suction applied to the sludge by the capillary action of an absorbent filter paper. The rate at which the paper becomes wetted with filtrate is a measure correlated to sludge dewaterability: it is measured by the time necessary for the filtrate to cover the space between two probes that detect the advancement of the liquid front on the paper (CEN, 2006c).

The CST apparatus is shown in Fig. 3.

CST values of raw sewage sludge generally range from 60 to 300 s, using the 18 mm diam. reservoir. With the same reservoir, the CST of water is around 5–6 s.

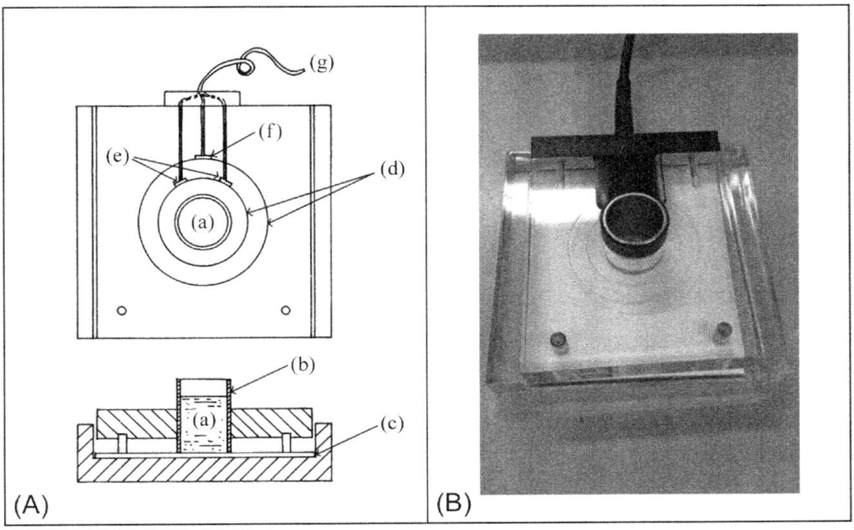

Fig. 3 CST apparatus. (A) Scheme (a: sludge; b: reservoir; c: absorbent filter paper; d: engraved circles; e: internal probes; f: external probe; g: cable). (B) View.
(A) *Source*: CNR-IRSA (Nat. Res. Council—Water Res. Inst.), 1984. Metodi analitici per i fanghi—Parametri tecnologici (Analytical Methods for Sludge—Technological Parameters). Quaderno (Report) 64. CNR, Rome. ISSN: 0390-6329 (in Italian).

4.4.3.2 Specific resistance to filtration (SRF)

The specific resistance to filtration is the typical parameter indicating the suitability of sludge to be dewatered by means of a filtration process. The value of the specific resistance to filtration has great importance in dewatering processes, as it can be useful for estimating the performance of full-scale filtering devices, mainly pressure filters, and comparing the dewaterability characteristics of sludge produced in different plants.

The principle of the method is based on the flow of a liquid through a porous medium according to Darcy's law, determined by pouring a reasonable volume of sludge into a filtering device (Buchner funnel, Fig. 4) and recording the amount of filtrate with time.

The SRF (m/kg) can be calculated with Eq. (11):

$$\mathrm{SRF} = \left(2 \times A^2 \times \Delta P \times b\right) / (\eta \times \mathrm{TS}) \tag{11}$$

where

- ΔP (N/m^2) is the pressure drop across the filter cake (applied pressure or vacuum).
- A (m^2) the filtration area.
- η the dynamic viscosity (kg/m/s).
- TS in kg/m^3.

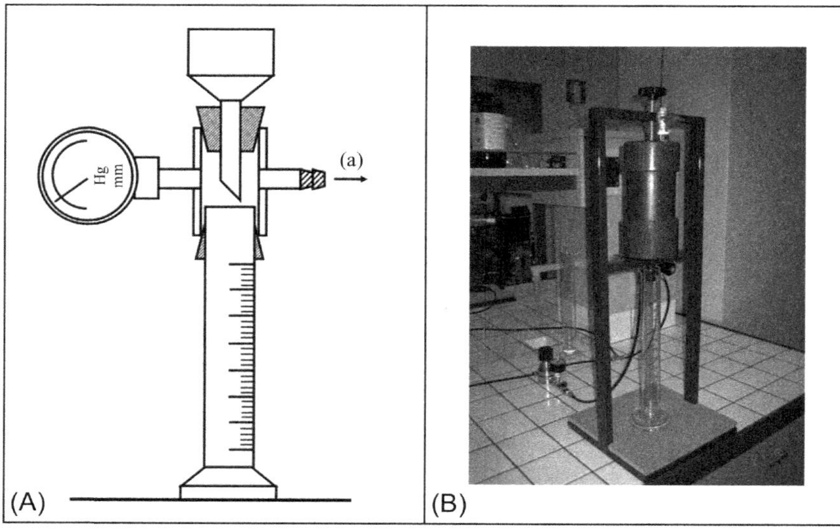

Fig. 4 (A) Buchner funnel scheme (a: vacuum). (B) Pressure filter cell view.
(A) *Source*: CNR-IRSA (Nat. Res. Council—Water Res. Inst.), 1984. Metodi analitici per i fanghi—Parametri tecnologici (Analytical Methods for Sludge—Technological Parameters). Quaderno (Report) 64. CNR, Rome. ISSN: 0390-6329 (in Italian).

The coefficient b (s/cm^6) is measured as the slope of the curve obtained by plotting, in a laboratory filtration test, the time of filtration to volume of filtrate ratio (t/V) versus V itself (CEN, 2006d).

Sludge can be considered as filterable on an industrial scale when its specific resistance to filtration is lower than 5×10^{12} m/kg.

4.4.3.3 Compressibility

Compressibility is complementary to that of specific resistance and gives information on the best range of pressure to be adopted in filtration processes.

It consists of measuring in a filter cell the specific resistance to filtration of sludge at different pressures.

The coefficient of compressibility, s, is related to the applied pressure p by Eq. (12):

$$r = r_0 \times p^s \tag{12}$$

where r_0 is the specific resistance to filtration at pressure of 100 kPa, r the specific resistance to filtration at pressure p, and p the applied pressure in Pa (CEN, 2006e).

4.4.3.4 Drainability

Drainability of flocculated sludge is an important parameter for evaluating its suitability to be thickened by means of a draining process, and for giving indications on the choice of flocculant and its dosage for sludge thickening through a filtering medium.

A given volume of flocculated sludge is poured in a filter cell, the mass of filtrate collected is recorded versus time, and the corresponding wet and dry mass of the sludge retained on the filtering medium and the undissolved solids remaining in the filtrate are measured.

The drainage index (E_g) can be then calculated following the procedure given in Ginisty et al. (2012).

4.4.4 Centrifugability properties

Centrifugation is one of the most common methods utilized for sludge dewatering. The main advantages of this technology consist of the fact that solid-liquid separation takes place in complete isolation from the outside and that the machine is relatively small, versatile, and simple to operate.

Centrifugability is defined as the aptitude of the sludge to be dewatered under the action of the centrifugal force. The mechanism of solid/liquid separation is similar to that of sedimentation, but solids are subjected to forces many times greater than gravity.

Moreover, in full-scale centrifugation, the effects induced on the sludge by the screw movement have to be taken into account. This means that the consistency of the sludge to be centrifuged must be such that it can be easily conveyed by the screw. Centrifugation performance is also affected by the aptitude of the sludge flocs to resist the mechanical stresses they are subjected to in the machine.

Centrifugability is, therefore, seen to be affected by three sludge characteristics: settleability, scrollability, and floc strength.

It is difficult to define a parameter for assessing the sludge suitability to centrifugation in a laboratory test. As a matter of fact, it is not possible to reproduce on a laboratory scale all the conditions actually occurring in an industrial centrifuge.

A reliable method to evaluate this property should be carrying out tests on a pilot scale, but this procedure is expensive, time consuming, and not always possible, so several laboratory procedures have been proposed.

The main laboratory procedures available in practice for evaluating the feasibility of industrial centrifugation and selecting the type and dosage of polymer required for conditioning are: (1) centrifugation by a lab centrifuge, followed by a penetrability test, and (2) measurement of floc strength (Spinosa, 1985).

4.4.4.1 Lab-scale centrifugation and penetrability test

This method, based on lab-scale centrifugation followed by a penetrability test, allows both the sludge settleability and scrollability features to be evaluated.

It consists of centrifuging sludge by a laboratory centrifuge under various conditions of centrifugal force and centrifugation time. Settling properties are measured by determining the SS concentration in the centrate while scrolling ones are determined by a penetration test through the apparatus consisting of a polished brass rod (diam. 3 mm, length 234 mm, weight 13.4 g) contained in a plastic support fitted with a graduated scale window (Fig. 5). The test is carried out by measuring the height of cake

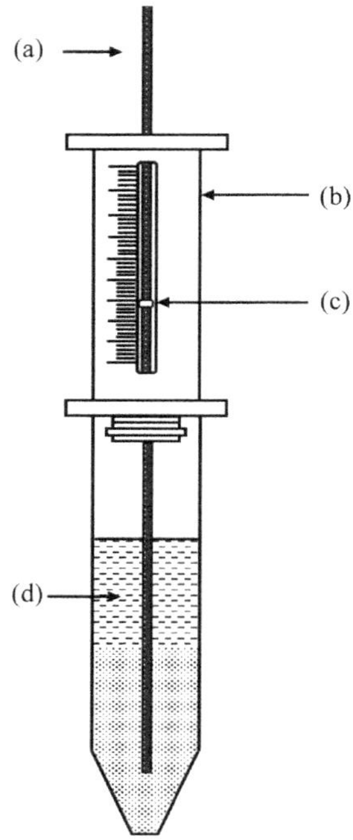

Fig. 5 Penetrometer (a: brass rod; b: plastic support; c: graduated scale window; d: sludge). *Source*: CNR-IRSA (Nat. Res. Council—Water Res. Inst.), 1984. Metodi analitici per i fanghi—Parametri tecnologici (Analytical Methods for Sludge—Technological Parameters). Quaderno (Report) 64. CNR, Rome. ISSN: 0390-6329 (in Italian).

penetrated by the rod after pouring off the centrate, and the scrollability can be evaluated from the percentage of sludge not penetrated.

This method has been applied on a larger scale, but it gave reliable results only for sludges with a high consistency, such as sludge produced by water treatment plants, which are mainly made of inorganic materials. On the contrary, they did not provide satisfactory results with activated sludge, due to its poor consistency.

Bach settling tests can be conducted also at high gravitational forces using a lab stroboscopic centrifuge, which allows following the progress of the solid-liquid interface.

Alternatively, Vesilind and Zhang (1984) have shown that sludge centrifugability can be characterized by plotting the final compacted solid concentration versus $\log z^{1.5} \times t$ (compactivity), where z is the number of gravities and t the time of centrifugation.

4.4.4.2 Floc strength evaluation

From results available in the literature, it appears that the floc strength method should be applicable in all cases.

This method allows the mechanical resistance of the flocs to be evaluated and thus their aptitude to resist the stresses they are subjected to in industrial machines. The test is carried out by subjecting the sludge to stirring at 1000 rpm for different times using a standard apparatus (Fig. 6) and measuring the CST of the stirred sludge. The floc

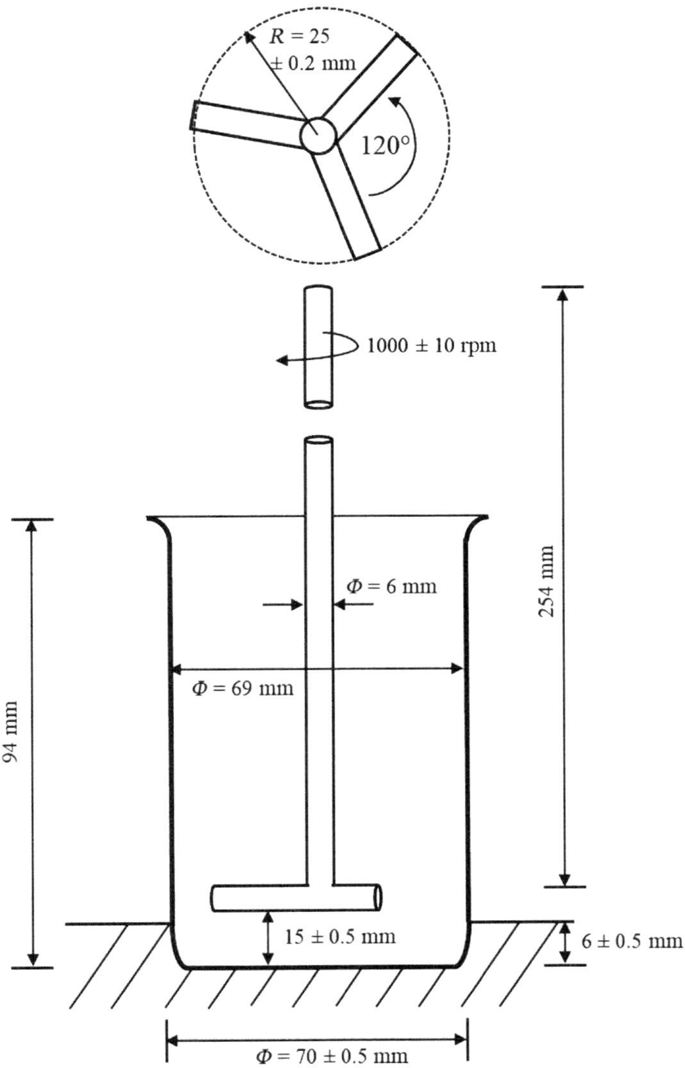

Fig. 6 Standard stirrer.
Source: CNR-IRSA (Nat. Res. Council—Water Res. Inst.), 1984. Metodi analitici per i fanghi—Parametri tecnologici (Analytical Methods for Sludge—Technological Parameters). Quaderno (Report) 64. CNR, Rome. ISSN: 0390-6329 (in Italian).

strength can be evaluated by plotting CST values versus the time of stirring. Good industrial centrifuging results can be expected when:

- The pattern of CST versus stirring time is linear, with a slight slope between 10 and 100 s stirring.
- The CST value at 10 s stirring is around 10–12 s (with 0.18 mm reservoir).

Many other completely empirical tests are utilized from manufacturers, including:

- Sludge spinning in a bottle centrifuge for 30 s and examination of the centrate clarity.
- Sludge centrifugation in a small disc-type machine.

4.4.5 Conditioning chemicals demand

The main purpose of the conditioning process is to enhance the effectiveness of the solid-liquid separation processes that will follow. Conditioning can be carried out by chemical and/or physical means.

Chemical conditioning includes coagulation (charge neutralization) and flocculation (bridging to form larger flocs) (Fig. 7).

A suspension of dispersed particles is stabilized by electrical charges on the particle surface, causing it to repel neighboring particles. This prevents charged particles from aggregating to form larger flocs, so solid-liquid separation is difficult. Coagulation is the destabilization of these suspensions by neutralizing the charges that keep them apart. Coagulants can be synthetic or natural products. Because sludge has a surface charge, generally negative, the chemical that is used to neutralize that charge should be the opposite.

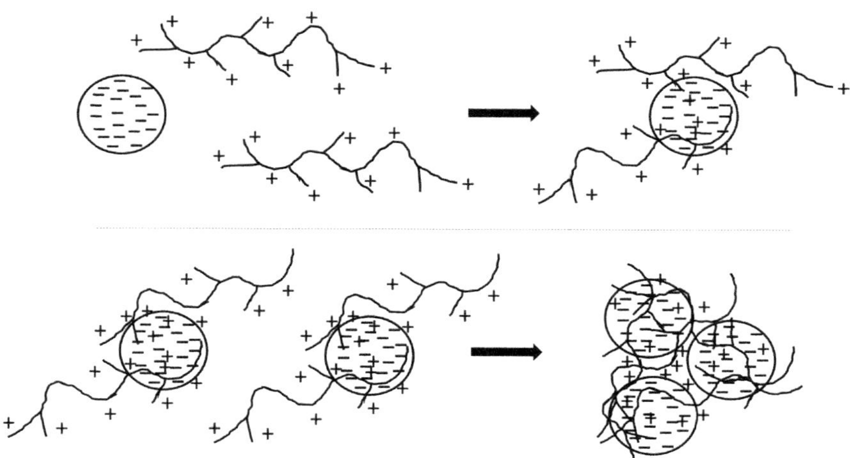

Fig. 7 Mechanism of chemical conditioning.
Source: Dentel, S.K., 2001. Conditioning. In: Spinosa, L., Vesilind, P.A. (Eds.), Sludge into Biosolids: Processing, Disposal, Utilization. IWA Publishing, London, p. 280.

While coagulation neutralizes electrostatic forces, the flocculation process bridges particles. This leads to bigger flocs that are easily separable from the liquid phase with different mechanisms.

However, the results of laboratory tests used to characterize the amenability of sludge to several treatments, such as thickening and dewaterability, are highly sensitive to the operating procedure adopted for the chemical conditioning step.

In particular, variables and/or parameters that can have significant effects on the sludge conditioning operation are either chemical or physical.

Chemical variables include (1) characteristics of the sludge such as pH, DM, and LOI, (2) characteristics of the conditioning product such as charge density, molecular weight, chemical structure, (3) concentration and dosage of the chemical product, and (4) water used for onsite product preparation when polyelectrolytes are added.

Physical variables include (1) the method of preparing the solution and its storage, (2) the intensity and duration of stirring, (3) the characteristics of the stirrer, that is, type, dimensions, and position, (4) the method of injecting the conditioning product into the sludge, and (v) (separation-) time between the end of mixing and the dewatering procedure itself.

The method of mixing the conditioning product into the sludge is particularly important because inadequate mixing, leading to poor initial dispersion or strong floc shearing, can result in very poor test performances. A large portion of the product may react instantly with particles upon first contact without dispersing throughout the mixture. It follows that adopting different procedures does not allow results obtained at different places, in different laboratories, and at different times to be compared, so the definition of a standard procedure for the chemical conditioning operation when selecting the type and dosage of a conditioning product at laboratory scale has been considered necessary by CEN (2015).

The application of this standardized procedure will allow the effectiveness of different sludge chemical conditioning agents to be evaluated. The procedure will also give information about the energy needed by the operation, mainly related to the mixing power, therefore allowing the costs of the conditioning operation to be optimized.

Physical conditioning methods include thermal conditioning, freezing, and the use of inorganic admixtures. Thermal conditioning involves heating sludge at 180–220°C for 30–90min; sludge sterilization also occurs. Conditioning by freezing seems to be justified only if accomplished by natural means. The use of inorganic substances (e.g., ash, diatomaceous earth) can produce a mixture that has improved filtering characteristics and is usually less compressible than the sludge alone (Spinosa, 2005).

4.4.6 Calorific value

The determination of calorific value is of fundamental importance in sludge treatment by thermal processes.

Sewage sludge usually contains much water and nonburnable solids, so for most purposes it may be sufficient to determine the gross calorific value only, and not the net calorific value for which additional determinations are necessary.

For calculation, a weighed portion of the solid sludge is burned in high-pressure oxygen in a bomb calorimeter under specified conditions. The effective heat capacity of the calorimeter is determined in calibration experiments by the combustion of certified benzoic acid under similar conditions, accounted for in the certificate. The corrected temperature rise is established from observations of the temperature before, during, and after the combustion reaction takes place.

The duration and frequency of the temperature observations depend on the type of calorimeter used.

Water is added to the bomb initially to give a saturated vapor phase prior to combustion, thereby allowing all the water formed from the hydrogen and moisture in the sample to be regarded as liquid water.

The gross calorific value is calculated from the corrected temperature rise and the effective heat capacity of the calorimeter, with allowances made for contributions from ignition energy, combustion of the fuse(s), and thermal effects from side reactions such as the formation of nitric acid.

Furthermore, a correction is applied to account for the difference in energy between the aqueous sulfuric acid formed in the bomb reaction and gaseous sulfur dioxide, that is, the required reaction product of sulfur in the sludge. The corresponding energy effect between aqueous and gaseous hydrochloric acid can be neglected due to the usually low chlorine content of most sludge.

The results of the calculation are reported in MJ/kg (CEN, 2006f)

5 Intrinsic characteristics

Most of the above discussed characterization parameters are somehow specific to the method of treatment and not able to give information that is fundamental for sludge handling (pumping, storing, transporting, spreading on land, injecting into a furnace, etc.).

The main parameters able to give intrinsic information for a correct sludge handling are the rheological characteristics, the particle size distribution, and the water distribution.

5.1 Rheology and physical consistency

Rheology has been widely applied in the past to study slurry flows, in particular for pumping, but some correlations exist between the rheological properties and other characteristics.

A detailed discussion of sludge rheological aspects is developed in another chapter, so here only the application of rheological concepts to the characterization of sludge will be considered. Therefore, the discussion will be addressed toward the evaluation of the physical consistency, which is a characteristic strictly linked to the rheological properties and of fundamental importance in sludge characterization.

As a matter of fact, the selection of the most suitable equipment and procedure for land application, storage, and transportation of sludge is strongly connected to its consistency; similarly, compacting sludge in a landfill or forming a pile in composting is dependent on sludge shear strength rather than on its solids concentration (Spinosa and Lotito, 1999).

It is also to be pointed out that in many analytical methods for sludge characterization (e.g., pH, DM, etc.), different procedures are indicated depending on whether the sample to be examined is liquid or solid, but no procedure is given for evaluating such a characteristic.

However, the evaluation of this property is a very complex matter that requires the preliminary definition of the physical states of the sludge and the standardization of the relevant measuring procedures capable to lead to comparable and reliable results. These procedures must also be simple to operate and applicable on the field (Leschber and Spinosa, 1998).

According to CEN (2007), sludge shows three different behaviors: liquid, paste-like, and solid. Liquid is the state in which sludge is still able to flow under the effect of gravity or pressure below a certain threshold, solid is the state in which sludge shows a shear resistance above a certain threshold, and the paste-like physical state occurs when the sludge is neither liquid nor solid.

The parameter defined as flowability allows the boundary between the liquid and the paste-like physical state to be defined while the parameter defined as solidity is that between the solid and the paste-like physical state.

5.1.1 Flowability

A simple way to evaluate sludge flowability, at both laboratory and field scale, consists of evaluating the flowability index F_i, which is related to the yield stress and is expressed in the same units, by means of a cylindrical container equipped with standardized discharge pipes at its bottom (extrusion tube apparatus, Fig. 8) (CEN, 2016).

F_i is determined by measuring the height of the sludge remaining in the cylinder above the discharge pipe fitted at its bottom when the flow through the pipe becomes intermittent. The measured height value allows the corresponding value of F_i to be calculated through the simplified flow equation in straight pipes, often referred to as the Bingham model (Eq. 13):

$$F_i = 3/8 \times \rho \times g \times h \times r/l \tag{13}$$

where F_i is the flowability index (Pa), 3/8 is the hydraulic equation ratio for laminar flow (this is generally the case of sludge having yield stress), ρ the sludge density (kg/m^3), g the gravity acceleration (m/s^2), h the height of the suspension remaining in the cylinder (m), r the pipe radius (m), and l the pipe length (m).

For this apparatus, the limit of flowability has to be possibly established at approximately 30 Pa (CEN, 2016).

5.1.2 Solidity

Solidity is the state in which a substance has no tendency to flow under moderate stress, resists shear stresses (such as compression) that tend to deform it, and retains a definite size and shape.

For determination of solidity, the undrained shear strength, c_u, is an important parameter.

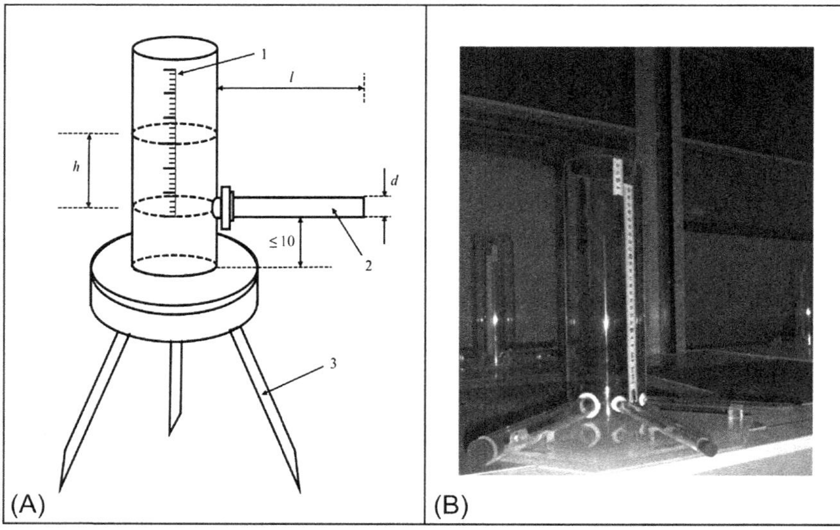

Fig. 8 Extrusion tube apparatus. (A) Scheme (1: graduated scale; 2: extrusion pipe, length 200 mm, diam. 5-10-20 mm; 3: Support). (B) View.
(A) *Source*: CEN (European Committee for Standardization), 2016. Characterization of Sludges—Physical Consistency—Part 1: Determination of Flowability—Method by Extrusion Tube Apparatus. EN 14742, Brussels.

Generally, the cohesive and frictional parameters, as defined in soil mechanics, have to be determined on the basis of evaluation of the shear strength τ as a function of the effective stress σ' normal to the plane of potential failure, using the Mohr-Coulomb criterion (Eq. 14):

$$\tau = c + \sigma' \times \tan \varphi \tag{14}$$

where c is the cohesion and φ is the angle of shearing resistance or friction angle.

It has been observed that the mechanical behavior of sludge is almost purely cohesive, as for soils in undrained conditions, so it is reasonable to consider $\tau = c$. Considering that the yield stress τ_f, that is, the stress above which the sludge is deformed, is equal to τ, it follows that $\tau_f = c_u$, where c_u is the undrained cohesion or the undrained shear strength.

According to Spinosa and Wichmann (2008), appropriate devices for evaluating solidity seem to be the laboratory vane shear apparatus and the pocket vane shear one. The vane apparatus, which is equipped with rectangular blades and is connected with the display over a torque spring, can be manually or electrically rotated. The vane rotates in a cylindrical vessel containing the sample or directly on the surface of the material to be tested. The torque is determined by the angle of rotation and the spring constant. The scale has the angle of rotation and the applied max torque when the sample shears. The torque is then converted to a shearing strength with the aid of a constant.

More recent experiences have shown that, in practice, the above methods are not fully reliable when applied to sludge and further studies are in progress.

5.2 Particle size distribution

In the literature, evidence is given that several factors such as pH, mixing, biological degradation, conditioning, etc., affect dewaterability to a wide extent. This may be explained by the effect of these factors on the size distribution of the supracolloidal particles (size in the range 1–100 μm), which are influencing the dewatering characteristics the most.

This means that the evaluation of sludge particle size should allow management operations and relevant costs to be optimized.

Three principal methods have been reported for the measurement of sludge particle size and/or shape distributions: (1) fractionation, (2) direct examination, and (3) counting. In all cases, each method could modify the sludge before or during measurements, which raises the concern that the measurement technique may alter the original distributions (Lotito et al., 1993).

The fractionation with a series of filters or screens is a simple system that can provide less accurate results. The retention of smaller particles by layers where larger particles prevail can take place, thus affecting measurements.

The measuring procedure proposed by CEC (1975) involves the use of a sieving machine equipped with a set of five filters of 1000, 500, 100, 50, and 20 μm, respectively, which gives the weight fraction of particles having a diameter higher than each filter mesh.

The primary problem when dealing with direct examination (e.g., image analysis systems) is to fix the sludge in an inert medium to allow nondisruptive examination while counting methods (e.g., laser systems) involve the use of different instruments that produce signals proportional to particle size.

Such instruments are convenient to use, but generally induce mixing with possible change in particle size distribution because the original shapes, sizes, and structure of flocs are possibly changed. A reproducibility problem is also due to the irregular shapes of sludge flocs, even though an ellipsoidal shape has been evidenced.

5.3 Water distribution

Water distribution represents the different types of water that can be distinguished according to how water is bound to the sludge particles.

The water types are: (1) free water, not associated with solid particles and that can be mostly removed by gravity, (2) interstitial and (3) vicinal water (physically bound, but some of the interstitial water can become free water when the floc is destroyed), which require mechanical forces to be removed, and (4) chemically bound water, which is removable by thermal processes only.

Therefore, water distribution gives important information on the level of energy to be spent, and on the most appropriate system to be adopted for an efficient solid/liquid separation process.

Analyses are commonly carried out by thermogravimetry, which consists of submitting a small sample of material to drying under constant conditions.

The procedure adopted by Lotito et al. (1993) consists of measuring the weight loss as a function of time of a thin layer (4 mm) of sludge sample in a controlled air stream dryer (80 m^3/h) at constant temperature (25°C) and constant humidity (dew point −20°C) until a constant weight is reached. Further drying at 105°C allows the bound water to be evaluated. Results are presented as drying rate curves showing three distinct stages: the constant rate period and the first and the second falling rate periods.

Other procedures to be possibly applied are those based on dilatometric methods (Sanin et al., 2011).

It has also been observed that bound water is an operationally defined value that differs greatly when different measurement techniques (e.g., centrifugal settling, expression, differential scanning calorimetry, drying tests) are applied.

6 Conclusions

The fast increase in sludge production resulting from the (1) growing availability of household running water, (2) installation of more extended sewerage, new facilities, and upgrading of existing ones, and (3) increasing environmental constraints and stringent rules, requires the assessment of standardized characterization procedures addressed to the development of sustainable sludge regulation, thus allowing:

- The legal requirements to be fulfilled in a correct and uniform manner.
- The sludge utilization and disposal operations properly performed.
- Stakeholder and public confidence built.

A few characterization parameters, such as TS, SS, and VS or LOI, are of interest for all treatments/operations available for sludge management, but most can be categorized according to the different options available for the final sludge disposal/utilization outlets.

Many other chemical, biological, and physical parameters are involved in sludge characterization for the optimization and routine monitoring of processes, in particular digestion, dewatering, and utilization in agriculture.

However, many of these parameters are specific to the method of treatment. Attention must be addressed to rheological properties, particle size distribution and water distribution which are able to give more fundamental information on sludge management. In particular, the physical consistency is a characteristic strictly linked to the rheological properties and has a great influence in almost all management operations, including storage, pumping, transportation, land-spreading, dewatering, drying, incineration, landfilling.

Acknowledgment

Thanks are due to Simone Visigalli of Politecnico di Milano for his invaluable help and assistance in the editing of this chapter.

References

CEC (Commission of the European Communities), 1975. Sewage Sludge Processing. COST Project 68, EUCO/SP/48/75. pp. 40–41.

CEN (European Committee for Standardization), 2006a. Characterisation of Sludges—Settling Properties—Part 1: Determination of Settleability (Determination of the Proportion of Sludge Volume and Sludge Volume Index). EN 14702-1, Brussels.

CEN (European Committee for Standardization), 2006b. Characterisation of Sludges—Settling Properties—Part 2: Determination of Thickenability. EN 14702-2, Brussels.

CEN (European Committee for Standardization), 2006c. Characterisation of Sludges—Filtration Properties—Part 1: Capillary Suction Time (CST). EN 14701-1, Brussels.

CEN (European Committee for Standardization), 2006d. Characterization of Sludges—Filtration Properties—Part 2: Determination of the Specific Resistance to Filtration. EN 14701-2, Brussels.

CEN (European Committee for Standardization), 2006e. Characterization of Sludges—Filtration Properties—Part 3: Determination of the Compressibility. EN 14701-3, Brussels.

CEN (European Committee for Standardization), 2006f. Characterization of Sludges—Determination of Calorific Value. EN 15170, Brussels.

CEN (European Committee for Standardization), 2007. Characterization of Sludges—Physical Consistency, Thixotropic Behaviour and Piling Behaviour (Technical Report 15463), Brussels.

CEN (European Committee for Standardization), 2015. Characterization of Sludges—Laboratory Chemical Conditioning Procedure. EN 14742, Brussels.

CEN (European Committee for Standardization), 2016. Characterization of Sludges—Physical Consistency—Part 1: Determination of Flowability—Method by Extrusion Tube Apparatus. EN 14742, Brussels.

Cordell, D., White, S., 2011. Peak phosphorus: clarifying the key issues of a vigorous debate about long-term phosphorus security. Sustainablity 3 (10), 2027–2049.

Ginisty, P., Olivier, J., Vaxelaire, J., Lafitte, V., 2012. Correlations between flocs characteristics and drainability properties: how to define a good flocculation? Proceedings of 11th World Filtration Congress, Graz, April.

Koch, C., Cavallos, M.A., Pascucci, J., Lindner, G., 1990. Spreadsheets for estimating sludge production. Water Environ. Technol. 2 (11), 65–69.

Kopp, J., Dichtl, N., 2001. Characterisation. In: Spinosa, L., Vesilind, P.A. (Eds.), Sludge Into Biosolids: Processing, Disposal, Utilization. IWA Publishing, London, pp. 19–39.

Labatut, R.A., Angenent, L.T., Scott, N.R., 2011. Biochemical methane potential and biodegradability of complex organic substrates. Bioresour. Technol. 102, 2255–2264.

Leschber, R., Spinosa, L., 1998. Developments in sludge characterisation in Europe. Water Sci. Technol. 38 (2), 1–7.

Lotito, V., Mininni, G., Spinosa, L., Lore', F., 1993. Developments in laboratory evaluation of sewage sludges dewaterability. Water Sci. Technol. 28 (1), 103–108.

Owen, W.F., Stuckey, D.C., Healy Jr., J.B., Young, L.Y., McCarty, P.L., 1979. Bioassay for monitoring biochemical methane potential and anaerobic toxicity. Water Res. 13, 485–492.

Pontoni, L., Panico, A., Salzano, E., Frunzo, L., Iodice, P., Pirozzi, F., 2015. Innovative parameters to control the efficiency and safety of anaerobic digestion process. Chem. Eng. Trans. 43, 2089–2094. https://doi.org/10.3303/CET1543349.

Raposo, F., Fernandez-Cegr, V., De la Rubia, M.A., Borja, R., Beline, F., Cavinato, C., Demirer, G., Fernandez, B., Fernandez-Polanco, M., Frigon, J.C., Ganesh, R., Kaparaju, P., Koubova, J., Mendez, R., Menin, G., Peene, A., Scherer, P., Torrijos, M., Uellendahl, H., Wierinckm, I., de Wildep, V., 2011. Biochemical methane potential (BMP) of solid organic substrates: evaluation of anaerobic biodegradability using data from an international interlaboratory study. J. Chem. Technol. Biotechnol. 86, 1088–1098.

Sanin, F.D., Clarkson, W.W., Vesilind, P.A., 2011. Sludge Engineering: The Treatment and Disposal of Wastewater Sludges. DEStech Publications, Lancaster, p. 393.

Spinosa, L., 1985. Technological characterization of sewage sludge. Waste Manag. Res. 3 (4), 389–398.

Spinosa, L., 2005. Sludge treatment and disposal. In: Lehr, J.H. (Ed.), Water Encyclopedia—Waste Water Treatment. John Wiley & Sons, Inc., Hoboken, NJ.

Spinosa, L., 2013. Characterization: a necessary tool in sludge management. Water Sci. Technol. 68 (4), 748–755.

Spinosa, L., 2016. Standardized characterization procedures: a necessary support to regulations. Water Sci. Technol. 71 (4), 220–228.

Spinosa, L., Lotito, V., 1999. Rheological measurements and their application in biosolids management. Water 21, 28–29.

Spinosa, L., Wichmann, K., 2008. Physical consistency: an useful tool for sludge characterization. Ingénieries Special 2008, 9–22.

Tchobanoglous, G., Stensel, H.D., Tsuchihashi, R., Burton, F., 2014. Wastewater Engineering Treatment and Resource Recovery, fifth ed. Metcalf & Eddy and AECOM, McGraw Hill Education, New York. ISBN: 978-0-07-340118-8.

Vesilind, P.A., Spinosa, L., 2001. Production and regulations. In: Spinosa, L., Vesilind, P.A. (Eds.), Sludge into Biosolids: Processing, Disposal, Utilization. IWA Publishing, London, pp. 3–18.

Vesilind, P.A., Zhang, G., 1984. Technique for estimating sludge compactability in centrifugal dewatering. Wat. Poll. Control Fed. 56 (12), 1231–1237.

WEF (Water Environment Federation), 1992. Design of municipal wastewater treatment plants. In: Manual of Practice. MOP No. 8, Alexandria, VA.

White, M.J.D., 1976. Design and control of secondary settling tanks. Wat. Pollut. Control 75 (4), 459–467.

Further reading

CNR-IRSA (Nat. Res. Council—Water Res. Inst.), 1984. Metodi analitici per i fanghi—Parametri tecnologici (Analytical Methods for Sludge—Technological Parameters). Quaderno (Report) 64, Rome, CNR. ISSN: 0390-6329 (in Italian).

Sludge from tannery industries

2

Katarzyna Wystalska, Jolanta Sobik-Szołtysek
Institute of Environmental Engineering, Czestochowa University of Technology, Częstochowa, Poland

1 Introduction

The tanning industry, where leather is manufactured, is characterized by a large amount of generated waste, contaminated mainly with chromium compounds (Kanagaraj et al., 2006; BAT, 2013; Chen et al., 2015; Juel et al., 2017). It is reported that this industry is responsible for about 40% of chromium pollution (Swarnalatha et al., 2008; Lopez-Luna et al., 2009; Jiang et al., 2016; Pantazopoulou et al., 2015; Wu et al., 2017). Among the waste, there is sewage sludge generated in the process of technological sewage treatment. The quality of tannery sludge is determined by the nature of the treated sewage.

2 Tannery sewage

Sewage in the tanning industry is generated at various stages of the technological process. Tanning (treatment and processing) of raw hides includes the beamhouse processes, proper tanning, and finishing.

The technological processes of treatment and processing of raw hides are characterized by high water absorption and significant consumption of used chemicals in the form of acids, alkali, and salts. The amount of consumed water results from the necessity of frequent rinsing of the raw material and the use of chemical reagents, most often in the form of aqueous solutions. According to literature sources (BAT, 2013; Lofrano et al., 2013; Jiang et al., 2016; Wu et al., 2017) 15–50 m^3 of water is used for processing 1 Mg of raw hides (= 1000 kg). High water consumption translates into a significant production of sewage that requires treatment before entering the environment.

The quality, quantity, and color of tannery sewage depend on the type and duration of the technological process. Rawhide preparation is done under the so-called beamhouse processes (Wu et al., 2017; Jiang et al., 2016; Stoop, 2003; Cassano et al., 2001), where they are soaked in solutions containing sodium hydroxide, potassium sulfide, or other wetting agents to loosen and purify fibers. In the next stage, the skin is subjected to a calcification treatment in limewash, which facilitates the removal of the bristles, the epidermis (Cassano et al., 2001), and subcutaneous tissue, resulting in the so-called pelt. The pelt is then decalcified using acids or acidic salts and etched (Cassano et al., 2001). The next treatment is skimming the skins, that is, removing excessive, naturally occurring fatty substances. This treatment is carried out

Industrial and Municipal Sludge. https://doi.org/10.1016/B978-0-12-815907-1.00002-7

using organic solvents (Cassano et al., 2001; Lofrano et al., 2013). In the next stage, using the mixture of acids, the so-called pickling is carried out, which aims to remove calcification residue, partially dehydrate the tissues, stop the digestion processes, and correct pH to about 3, which allows chromium salts to penetrate the skin and thus facilitates the tanning process. The pickle bath should contain a minimum of 6% sodium chloride in order to protect the collagen fibers from swelling under the influence of acids (Cassano et al., 2001; Lofrano et al., 2013; Zhang et al., 2016). After the preparatory treatments, the skins are sent to tanning. As a result of this process, the stabilization of collagen fibers occurs through the cross-links of used tannins. At present, combined tanning is commonly used involving semichrome and chrome tanning with the use of chromium salts (alkaline Cr(III) compounds), facilitating the penetration of tannins into the skin (Sreeram and Ramasami, 2003). After the tanning process, the skin is subjected to processes for reducing the water content to facilitate machining, that is, planing or skiving (BAT, 2013). The next treatments for leather processing are:

- So-called wet finishing, during which neutralization is carried out to correct the pH for subsequent processes and rinsing.
- Retannage.
- Dyeing (most often with the use of acid dyes based on water).
- Stuffing.
- Washing and hanging skins to facilitate the migration of fat from the skin surface to the inside.
- Drying and finishing, guaranteeing specific product features (BAT for tanning leather, 2013).

Implementation of subsequent technological operations in the whole process of leather tanning is related to the use of various chemical agents, the consumption of which may amount to approx. 400–500 kg per 1 mg of raw hides. The high content of chemicals in tannery sewage results from excessive dosing to ensure full penetration of the skin and reaction with collagen and the low level of their depletion. Therefore, the variability of their composition is a special feature of chrome tanning sewage (Religa et al., 2011). The degree of sewage pollution varies, and the maximum load of contaminants characterizes baths from soaking, calcification, and decalcification as well as tannery worts. The skin calcification process causes significant amounts of sulfuric and hydrochloric acid, lime, and detergents to enter the sewage. The sewage contains a significant load of organic pollutants from the processes of dissolving and leaching the organic matter of the hides and chemical compounds that are reagents at subsequent stages of tanning. There are also mineral substances, including toxic chromium (III) compounds in the amount of 2.5–4.0 g/L and sulfides in the amount of 2.0–6.5 g/L as well as altered protein and inorganic poisonous substances.

Table 1 presents average loads and concentrations of pollutants in general sewage from tannery and fur workshops.

Sewage produced in tanneries can be classified into three streams: general sewage, calcification sewage, and chromium sewage. General sewage is generally cleaned first on gratings and then undergoes a flotation process, after which it is directed to the

Table 1 Composition of tannery sewage according to literature data (mg/L)

References	Chemical oxygen demand	Biological oxygen demand	Total suspension	pH	Sulfides	Total Kjehdal nitrogen	Total dissolved substances	Cr
Przywara (2017)[a]	2700.00–5870.00	–	–	4.00–4.25	–	–	–	23.00–45.00
Munz et al. (2007)[b]	4051.00	–	976.00	7.20	27.00	220.00	–	
George et al. (2015)[f]	19,400.00	12,100.00	8430.00	3.65	–	–	–	552.00
Ranganathan and Kabadgi (2011)[c]	5659.00	2375.00	–	4.60	2.20	–	7610.00	52.04
Fahim et al. (2006)[d]	2625.50	197.50	–	3.54	1.68	–	–	5500.00
Haydar and Aziz (2009)[e]	1760.00–3320.00	390.00–1320.00	5330.00–15,912.00	7.55–9.51	14.80–424.50	104.40–140.60	4466.00–14,572.00	22.90–122.40

[a]Przywara, L., 2017. Application of Fenton's process for treatment of tannery wastewater. Ecol. Eng. 2, 117–122.
[b]Munz, G., Gori, R., Mori, G., Lubello, C., 2007. Powdered activated carbon and membrane bioreactors (MBRPAC) for tannery wastewater treatment: long term effect on biological and filtration process performances. Desalination 207, 349–360.
[c]Ranganathan, K., Kabadgi, S.D., 2011. Studies on feasibility of reverse osmosis (membrane) technology for treatment of tannery wastewater. J. Environ. Protect. 2, 37–46.
[d]Fahim, N.F., Barsoum, B.N., Eid, A.E., Khalil, M.S., 2006. Removal of chromium (III) from tannery wastewater using activated carbon from sugar industrial waste. J. Hazard. Mater. B136, 303–309.
[e]Haydar, S., Aziz, J.A., 2009. Characterization and treatability studies of tannery wastewater using chemically enhanced primary treatment (CEPT)—a case study of Saddiq leather works. J. Hazard. Mater. 163, 1076–1083.
[f]George, J.S., Ramos, A., Heather, J. Shipley, H.J., 2015. Tanning facility wastewater treatment: analysis of physical–chemical and reverse osmosis methods. J. Environ. Chem. Eng. 3, 969–976.

reaction chamber. Coagulation takes place in the chamber as a result of mixing sewage with the wort containing chromium hydroxide. Sewage from calcification is also pretreated in the float tank, then cleaned in filters and recycled. In the treatment of this sewage stream, the coagulating effect of chromium compounds from the wort is also used. Chromium sewage constitutes about 5%–6% of the volume of general sewage. The subsequent processes are cleaning in an electroflotation chamber that is equipped with insoluble electrodes and cleaning in filters. Averaged tanning sewage is often subjected to biological treatment (the activated silt method).

The search for and the use of more effective methods of tannery sewage treatment, the use of innovative chemicals, and reducing their consumption (Cassano et al., 2001; Stoop, 2003; Krishnamoorthy et al., 2012; Lofrano et al., 2013; Bacardit et al., 2015; Dixit et al., 2015; Zhang et al., 2016; Liu et al., 2016; Wu et al., 2017), however, does not solve the problem of formation and management of generated sewage sludge.

3 Characteristics of tannery sewage sludge

The amount and composition of the formed sewage sludge are variable and depend on the tanning technology, the type of processed leather, and the method of sewage treatment and sludge treatment technology in sewage treatment plants. The use of chromium compounds for tanning hides causes contamination of sewage and sludge generated during the process of their purification. Considering the content of chromium, two types of sewage sludge can be distinguished in tanneries (Fela and Wieczorek-Ciurowa, 2010):

- Sewage sludge from chromium-free sewage (from the beamhouse), which after initial mechanical cleaning can be sent directly to the municipal sewage treatment plant.
- Sewage sludge from chromium sewage (from the tanning processes), which requires purification from chromium compounds, passing to sewage sludge in the factories of sewage treatment plants.

According to literature data (Juel et al., 2017), as a result of processing 1 mg of raw hides, approximately 100–150 kg of sewage sludge is generated. Sewage sludge from the tanning industry often contains a significant amount of heavy metals (Juel et al., 2016), for example, Cr, Mn, Fe, Co, Ni, Cu, Zn, Se, Cd, Pb, and Hg. The forms of occurrence of heavy metals in sewage sludge may be different. The following types of fractions associated with heavy metals can be distinguished in sewage sludge:

- Fraction I—exchangeable metals, including metals bound to the surface of sludge due to physical adsorption and chemisorption; the transition of metals from a solid to liquid phase may occur when the equilibrium in the sorption–desorption system changes or the ionic composition of the water changes.
- Fraction II—metals in the form of carbonates or coprecipitated with carbonates; their release into the solution may occur as a result of a drop in pH.
- Fraction III—metals adsorbed or precipitated with hydroxides and iron and manganese oxides, fraction thermodynamically unstable under oxygen deficit conditions and sensitive to changes in the oxidation and reduction potential.

- Fraction IV—metals bound to organic compounds, released as a result of their mineralization.
- Fraction V—metals embedded in the crystal lattice of primary and secondary minerals (quartz, aluminosilicate) and bound to other stable compounds, fraction thermodynamically stable.

The content of heavy metals in sludge from the tanning industry according to literature data is presented in Table 2.

Dehydrated sludge from chemical pretreatment may contain 5%–20% of dry weight. The chromium content in the sludge varies and depends on factors such as sewage treatment technology, tannery production profile, and tanning method. In the absence of sewage recirculation, this amount may exceed 1000mg Cr/kg of dry weight while recirculation is much lower. If the sewage pretreatment process is effective, then in the sludge from biological treatment, the amount of chromium ranges from trace amounts to several hundred mg Cr/kg of dry weight. For example, the content of chromium in sewage sludge from the tannery treatment plant in Melaka (Malaysia) is about 500mg Cr/kg while in the sludge from the installation in Leon Guanajuato in Mexico, a 130-fold higher content of chromium was found, amounting to approximately 65,016mg Cr/kg (Lopez-Luna et al., 2009; Haroun et al., 2009).

4 Tanning sewage sludge management

Sewage sludge from the tanning industry, due to the presence of chromium, constitutes a waste that is difficult to manage. Processes commonly used for the processing and management of municipal sewage sludge, such as anaerobic digestion, composting, agricultural use, reclamation, and thermal processes, in the case of tannery sludge may prove ineffective or even impossible to use.

Sewage sludge from the tanning industry is characterized by an increased content of chromium (Swarnalatha et al., 2008; Kilic et al., 2011a; Barajas-Aceves et al., 2014; Kavouras et al., 2015), which on the one hand is a toxic component and on the other hand an element necessary for the proper functioning of higher organisms. The basic form of chromium present in this sludge, Cr(III), was previously considered by most scientists to be stable in the environment with low toxicity. Therefore, the basic method of its management was storage (Lopez-Luna et al., 2009; Altas, 2009; Chandra and Kulshreshtha, 2004). However, the thesis about the limited mobility of this form of chromium in the environment is being discussed (Thompson et al., 2002; Murray et al., 2005; Pereira et al., 2005) because, in the presence of manganese oxides, it can transform into Cr(VI) (Apte et al., 2005). According to many authors (Panda and Choudhury, 2005; Speranza et al., 2007; Mandal et al., 2011; Nakatani et al., 2012; Juel et al., 2017) Cr(III) may also exhibit harmful, mutagenic, and carcinogenic effects on mammalian cells at prolonged exposure and cause weakening of the immune system. In addition, for many plants, chromium on the third oxidation stage may be more toxic than Cr(VI), being the cause of germination inhibition (Speranza et al., 2007; Nakatani et al., 2012; Juel et al., 2017). According to Mandal et al. (2011), as a result of changes in the pH and redox potential, chromium

Table 2 The content of heavy metals in sewage sludge from the tanning industry according to literature data

References	Cr	Fe	Cd	Ni	Zn	Pb
	mg/kg d.m.					
Gondek (2005)[a]	875.00	–	0.39	24.00	195.00	20.30
Li et al. (2005)	11,320.00	–	–	–	–	–
Wang et al. (2007)[b]	35,000.00	6870.00	–	–	–	–
Fang and Zhou (2007)	14,180.00	–	–	–	126.00	87.00
Gupta and Sinha (2007)	7489.76	19,401.44	95.10	219.16	340.69	143.43
Haroun et al. (2009)	500.00	1062.00	8	–	200.00	10.00
Trebien et al. (2011)[c]	8869.00	–	–	–	–	–
Lopez-Luna et al. (2009)	65,016.00	–	–	–	–	–
Shukla et al. (2011)[d]	628.00	1667.00	125.00	76.00	592.00	427.00
Kilic et al. (2011b)	8041.00	13,200.00	18.50	34.50	1300.00	98.50
Fongsatitkul et al. (2011)	474,000.00	–	–	–	–	–
Tang et al. (2013b)	17,140.00	–	10.00	–	260.00	1610.00
Ghosh et al. (2013)[e]	14,212.00	683.00	<0.01	272.00	241.00	58.20
Montanes et al. (2014)	7903.00	–	–	–	–	–
Sobiecka and Szymański (2014)	106,800.00	–	–	–	–	–
Allue et al. (2014)[f]	6030.00	–	52.80	20.00	645.00	70.00
Kavouras et al. (2015)	86,000.00	–	–	–	–	–
Arthy and Phanikumar (2015)[g]	24,200.00	–	–	–	–	–
Alibardi and Cossu (2016)	44,900.00	12,166.00	–	26.00	1120.00	–
Juel et al. (2017)	19,299.00	–	<0.01	139.50	250.60	63.80

[a]Gondek, K., 2005. The influence of soil treatment by untreated and composted tannery sludge on yield, nutrient status, and chromium content in selected crops. Plant Soil Environ. 51 (4), 179–192.
[b]Wang, Y-S., Pan, Z-Y., Lang, J.-M., Xu, J.-M., Zheng, Y.-G., 2007. Bioleaching of chromium from tannery sludge by indigenous Acidithiobacillusthiooxidans. J. Hazard. Mater. 147, 319–324.
[c]Trebien, D.O.P., Bortolon, L., Tedesco, M.J., Bissani, C.A., Camargo, F.A.O., 2011. Environmental factors affecting chromium-manganese oxidation–reduction reactions in soil. Pedosphere 21 (1), 84–89.
[d]Shukla, O.P., Juwarkar, A.A., Singh, S.K., Khan, S., Rai, U.N., 2011. Growth responses and metal accumulation capabilities of woody plants during the phytoremediation of tannery sludge. Waste Manage. 31, 115–123.
[e]Ghosh, S., Mukherjee, S., AlHamdan, A., Reddy, K., 2013. Efficacy of fine-grained soil as landfill liner material for containment of chrome tannery sludge. Geotech. Geol. Eng. 31, 493–500.
[f]Allue, J., Moya Garces, A., Bech, J., Barcelo, J. Poschenrieder Ch., 2014. Fractionation of chromium in tannery sludge-amended soil and its availability to fenugreek plants. J. Soil Sediment. 14, 697–702.
[g]Arthy, M., Phanikumar, B.R., 2015. Immobilization of chromium in tannery sludge using iron-based nanoparticles and nanobiocomposites. Water Air Soil Poll. 226 (7).

can be rapidly actuated, leading to pollution of the ground and water environment as well as groundwater (Gupta and Sinha, 2007). In addition, the presence of chromium in the stored sewage sludge may disrupt the processes of carbon and nitrogen mineralization and affect the dehydrogenase activity, increasing soil erosion (Barajas-Aceves et al., 2014). Those premises resulted in the search for new possibilities of managing this type of waste and have become a field of research for many scientists.

The use of biological methods for processing tannery sludge was undertaken by a wide group of researchers (Haroun et al., 2007a,b, 2009; Li et al., 2005; Araujo et al., 2015; Silva et al., 2014; Vig et al., 2011; Appels et al., 2008; Priebe et al., 2016; Chen et al., 2008; Alibardi and Cossu, 2016; Zhou et al., 2005; Fang and Zhou, 2006). The results of research conducted by these authors indicate that the chromium concentrations present in the converted sludge may inhibit the used biological processes, and the use of the obtained products may be limited due to their specific properties.

Alibardi (2008) analyzed the effectiveness of biological processes to transform tannery sewage sludge. According to the author, the aerobic stabilization of mechanically dried sludge has contributed to the reduction of water and organic matter content and created the foundation for the development of oxygen treatment technology to stabilize tannery sludge. In their research, Alibardi and Cossu (2016) used a mixture of tannery sludge with a chromium content of 44,900 mg/kg. The product obtained was to be stored in the so-called sustainable landfills. Alibardi (2008) also indicated that the production of hydrogen from tannery deposits was at a level comparable to the results obtained for kitchen waste, however, it was characterized by a lower production speed. In the light of research (Appels et al., 2008; Priebe et al., 2016), the methane fermentation process seems to be an unreasonable method of processing tannery sludge, due to the potentially high content of heavy metals, mainly chromium, resulting in inhibitory effects. For the microorganisms involved in the fermentation process, the dominant Cr(III) in the sewage sludge exhibits inhibition at lower concentrations than Cr(VI), which is more toxic to all other processes. The relative toxicity of heavy metals towards microorganisms is as follows: Cr > Ni > Cu > Zn, and the presence of other heavy metals enhances the effect of chromium toxicity (Chen et al., 2008).

The quality of compost produced based on tannery sludge creates a potential risk of soil and water contamination due to the possible migration of metals to the soil and their bioaccumulation in vegetation (Li et al., 2005; Ociepa et al., 2007; Fijałkowski et al., 2011). Examples of composting of sludge from the tanning industry quoted in the literature (Li et al., 2005; Araujo et al., 2015) indicate the exceeding of the permissible local standards regarding the content of chromium and the negative impact of compost from tannery sludge on the soil environment.

In the literature, there are examples to be found of conducting composting processes for sewage sludge from the tanning industry. The concentration of chromium in the used sludge did not exceed 10,000 mg/kg. Haroun et al. (2007a,b, 2009) conducted composting of sewage sludge mixtures with a chromium content of 350–500 mg/kg, and chicken droppings and rice bran in the amount of 100 kg, 50 kg, 30 kg, and 20 kg, respectively. The material subjected to composting was placed in the form of piles that were moved every 10days. After 60days, compost having a chromium concentration of 100mg/kg of dry weight was obtained, meeting the legal requirements applicable in the country of the authors. Li et al. (2005) conducted composting of sewage tannery sludge characterized by a much higher, 11,320mg/kg chromium concentration. The compost from a mixture of sewage sludge and straw in the ratio of 3:5 obtained after a month was characterized by a significant concentration of chromium at the level of 10,360mg/kg of dry weight. Therefore, according to this author, to meet the national standards of chromium content in rice grown on soils fertilized with this compost, it is necessary to adjust the dose of compost, translating into 350mg of chromium per kilogram of soil.

Composts not meeting the legal requirements for their agricultural use have also been obtained by Araujo et al. (2015). In their research, they used sewage sludge from the tanning industry mixed with straw from sugarcane and cow dung in a volume ratio of 1:3:1, respectively. Composting was carried out for a period of 85days in the form of an aerated pile with dimensions of $2 \times 1 \times 1.5$ m. The compost pile was moved at twice a week for the first 30days and twice a month for the remaining period. The amount of chromium in the obtained compost was 1943mg/kg. The long-term use of compost in the amount of 10mg/ha resulted in negative changes in the form of biomass reduction of microorganisms and reduction of enzymatic activity in the fertilized soil.

The addition of sediments, composts, or vermicompost to the soil obtained from tannery sludge has a small influence on changes in the content of available and basic forms of Cu, Cd, and Pb in the soil. However, the proportion of soluble and basic forms of chromium is significantly increased. In soils fertilized with tannery materials, the content of basic chromium forms is increased, whereas there is no clear increase in the content of chromium extracted with an HCl solution with the concentration of 1 mol/L. Sludges from tannery sewage treatment can be used for the production of composts; however, due to the increased content of chromium compounds, they cannot be composted directly. The addition of tannery sludge to the compost mass up to 20% does not inhibit biochemical processes nor does it result in increased chromium migration to the soil and water environment. Compost obtained with the use of sewage tannery sludge can therefore be used for the reclamation of degraded areas, light soil fertilization, and forestry nursery.

In Mexico, research has been conducted on the possibility of using sludge as a potential material for changing the semidesert soil structure. It was demonstrated (Barajas-Aceves et al., 2014) that such a solution poses a serious threat due to soil erosion as well as increased chromium contamination in the vicinity of the sludge application site and in aquifers.

High chromium content in tannery sludge creates a potential for recovery of this metal. Chromium extraction studies presented in the literature were carried out using microorganisms in the bioleaching process and using specific vegetation (Zhou et al., 2005; Bareen and Khilji, 2008). The use of bacteria *Acidithiobacillus thiooxidans* (TS6 strain) allowed the researchers (Zhou et al., 2005) to extract chromium from acidified sewage sludge up to 96% within 6–8days. However, the correct course of the life cycle of microorganisms is disturbed by the presence of organic matter in the sludge. According to Fang and Zhou (2006), the concentration of soluble organic carbon on the level of 150 mg/L may cause significant inhibition (at the level of 70%) in the process of sulfur oxidation by these bacteria. The bioleaching time can be significantly shortened by using mechanisms of mutualist cooperation between the group of autotrophic bacteria responsible for this process and acidophilic heterotrophic organisms, for example, *Brettanomyces*, whose B65 strain was used in the research by Fang and Zhou (2007). Shortening the bioleaching time to three days was achieved while maintaining high efficiency at the level of 94%. However, the method of converting the recovered chromium to the commercially useful form and solving the problem of pH neutralization of the bioleaching deposits was not indicated.

Bareen and Khilji (2008) conducted research on chromium extraction using lesser bulrush (*Typhaangustifoli* L.) from aqueous solutions with variable sewage sludge content (0%–60%). The maximum reduction of chromium content in sludge (42%) was obtained after a period of 90 days for a solution containing 30% sludge. According to the authors, the time necessary for the total recovery of chromium should be about 270 days. Concentrations exceeding 60% of the sludge in solution during the week of exposure caused withering and plant chlorosis. Thus, the effectiveness of this method is strictly determined by the chromium content in the sludge.

Chromium recovery using the biohydrometallurgical method was conducted by Ma et al. (2017). The experiment showed that the eluate obtained in the bioleaching process of tannery sludge contained large amounts of Cr(III), Fe, Ca, Na, Mg, K, and small amounts of other elements. The use of the chromium-iron preparation, based on the obtained eluate from the tannery sludge, allowed for the reduction of the use of chromium in the process of leather processing. Physical and mechanical properties and air permeability of skin tanned with this product were better than the properties of skin tanned with only chromium. Compared to chrome-tanned leather, leather tanned with a chrome-iron preparation obtained from tannery sludge did not show a significant difference in thermal properties after aging in hot air. It is therefore possible to reuse the tannery sludge bioeluate and obtain a clean tannin for treating the skin. This method seems to be a possible and effective technology for removing Cr(III) from tannery sludge.

The use of chemical methods in chromium recovery processes was analyzed by Kilic et al. (2011b). Researchers used an oxidative chromium recovery process. The process consisted of three stages of rinsing and filtration. The first and second rinse was alkaline and the last one was acidic. The use of hydrogen peroxide caused the oxidation of Cr(III) to Cr(VI). During the acidic stage, a reduction to Cr(III) took place. The process yield was about 70%. According to the authors, the recovery method is interesting in the case of tannery sludge containing a high concentration of chromium.

The sewage sludge remaining after the extraction process, however, may still be an environmental problem requiring a proper solution whereas, the raw materials obtained are often characterized by low quality and the presence of numerous pollutants (Cassano et al., 2001; Basegio et al., 2009; Kilic et al., 2011b).

In addition to biological and chemical methods to neutralize tannery sludge, a method of solidification was proposed, consisting in binding impurities by adding mineral binders such as cement (Filibeli et al., 2000; Montanes et al., 2014; Fongsatitkul et al., 2011). However, the specificity of tannery sludge (high chromium content) generates the need for large amounts of cement (Filibeli et al., 2000; Montanes et al., 2014). This may in turn make it impossible to solidify with other wastes, such as fly ash, and make the materials unusable in construction due to the low compressive strength with the only advantage being the low cost of storage. Fongsatitkul et al. (2011) subjected to solidification a mixture of undefined cement, ashes from the combustion of rice husks, and sewage sludge from the chemical precipitation process, characterized by a high chromium content of 474,000 mg/kg. The obtained materials had diversified compressive strength, which was inversely proportional to the content of sewage sludge in the batch. The highest compressive strength after 28 days was obtained for a mixture containing 17% sewage sludge, 16% ash from the combustion of rice husks, and 67% cement (Fongsatitkul et al., 2011). In the case of a mixture with sewage sludge and cement content of approximately 44.5% each, compressive strength below 15 MPa was obtained and the material did not meet the national requirements for leaching of heavy metals.

Another alternative for the management of tannery sludge is thermal processes (Wolski et al., 2011) such as incineration, gasification, or pyrolysis. Thermal methods, due to the limited volatility of chromium at elevated temperatures, seem to be the most prospective methods of management of tanning sewage sludge (Jiang et al., 2010; Tang et al., 2013a, 2013b). However, according to many authors (Mao et al., 2016a; Mao et al., 2016b; Tang et al., 2008), there is a risk of oxidation of Cr(III) present in the sludge to the dangerous form of Cr(VI). The possibilities were investigated of preventing the chromium oxidation and limitation of heavy metal oxidation in the process of combustion of sewage sludge from the tanning industry by means of for example, sludge conditioning with orthophosphoric acid or the use of additives in the form of hydroxyapatites (Jiang et al., 2010; Tang et al., 2008, 2013a,b; Mao et al., 2016a,b). Celary et al. (2016) also investigated the volatilization of heavy metals (Cr, Cd, Ni, Pb, Zn, Fe) during vitrification of tannery sludge with various mineral wastes such as fly ash from brown coal combustion, mold sands, and carbonate flotation waste from Zn-Pb ore enrichment. It was also analyzed how various mineral additives affected the volatilization of metals. Fig. 1 presents the experimental procedures scheme.

Mixtures containing 35% by weight of the tannery sludge and 65% by weight of mineral sludge were subjected to vitrification. During the tests, it was found that vitrification of tannery sludge with specific mineral additives may be accompanied by a relatively low chromium emission (noted min. $22.4 \pm 1.4\%$), similar to the values quoted for the combustion process. The lowest level of chromium emission was determined during vitrification of sewage sludge mixtures with fly ash and sewage sludge with a combination of fly ash with postflotation wastes in a ratio of 3:1. However, regardless of the mineral waste used, there was a high degree of volatilization for cadmium, lead, and zinc, amounting to $\geq 73.4\%$, $\geq 79.1\%$, and $\geq 87.8\%$, respectively. It is possible to neutralize tannery sludge by vitrification while maintaining low levels of chromium emission and permissible release rates of other metals (Ni, Fe). Due to the almost complete evaporation of Cd, Pb, and Zn, the thermally processed sludge is free of heavy metals.

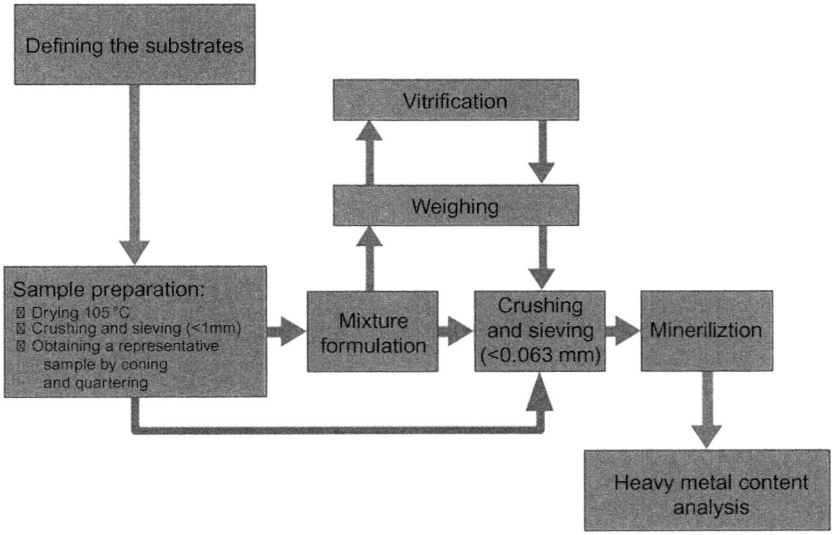

Fig. 1 Experimental procedures scheme based on Celary et al. (2016).

Vitrification seems to be an effective method of thermal treatment of wastes in the context of the immobilization of significant amounts of heavy metals (Pelino et al., 2001; Lay et al., 2009; Dunnett et al., 2012; Celary and Sobik-Szołtysek, 2014). This method is characterized by the stability and high chemical resistance of the products obtained under varying environmental conditions. Its effectiveness for immobilization of chromium has been confirmed by numerous studies (Huang et al., 2004; Basegio et al., 2009; Lazau and Vancea, 2014). Varitis et al. (2017) also studied the vitrification process of a mixture of ashes from the combustion of tannery sludge, calcium carbonate, and silica. Sobiecka and Szymański (2014) confirmed the possibility of obtaining hard products with negligible leachability during vitrification of sewage sludge from the tanning industry using fly ash from municipal waste incineration.

The vitrification of chromium-rich waste with the use of appropriate additives allows the metal to be immobilized by introducing it into the silica matrix without the risk of emissions to the atmosphere in this process (Bień et al., 2007, 2013; Basegio et al., 2009). Celary and Sobik-Szołtysek (2014) conducted research on the plasma vitrification process of sewage sludge from the tannery industry with broken glass, molding sand, and mineral waste materials (fly ash, Zn-Pb flotation waste). Mixtures of sewage sludge and waste were vitrified in a plasma furnace and a chamber furnace. The test results confirmed the possibility of obtaining compact, homogeneous, and vitrified materials (Photo 1), characterized by low leachability and a hardness characteristic of glass (about 6 on the Mohs scale).

Vitrification of a mixture containing 50% of tanning sewage sludge from the flotation purification process and 20% by volume of flotation waste from the Zn-Pb industry used as flux and 30% by volume of waste mass allowed achieving an optimal balance between the hardness and leaching of heavy metals for the resulting vitrificate. Application of conditioning of blends by burning them prior to the

Photo 1 Vitrificates obtained from tanning sludge (own material).

vitrification process resulted in a decrease of leaching of heavy metals, an increase of Cr and Zn content, and increased hardness. The obtained vitrificate is characterized by minimal leachability of heavy metals and can be safely stored. Proper selection of the composition of the mixture of waste subjected to vitrification also allows for the reduction of heavy metal emissions as well as weight reduction for tannery mixtures of organic sludge of 45%–50%. The use of only waste materials for the preparation of vitrificates is very significant for environmental reasons. It enables the associated removal of threats to the environment, resulting from various degrees of nuisance of individual wastes. An additional benefit is the indication of new directions for their management.

Many researchers (Appendino et al., 2004; Singh et al., 2007) list numerous advantages of the vitrification process, including high process efficiency, low space requirements compared to conventional incineration plants, and reduction of weight and volume of transformed waste. In addition, the inert nature of vitrificate after solidification significantly limits the migration of chemical compounds and their leachability to the environment. Therefore, it is possible to use the product obtained, for example, in the construction industry as a coarse aggregate in the case of compact, homogeneous aggregates or as fine aggregates, acting as filling materials, in the case of loose materials. Vitrificates can therefore be a substitute for gravel and other natural materials (Ferraris et al., 2009). However, the main disadvantage of the vitrification process is the high energy consumption, up to 2 MWh/Mg of recycled waste.

The decision on the choice of method for the final management of sewage sludge from the tanning industry must be preceded by a thorough analysis of the process and the environmental effects caused by it.

References

Alibardi, L., 2008. Innovative treatment of tannery sludge. https://www.researchgate.net/publication/37677896_Innovative_treatment_of_tannery_sludge.

Alibardi, L., Cossu, R., 2016. Pre-treatment of tannery sludge for sustainable landfilling. Waste Manag. 52, 202–211.

Altas, L., 2009. Inhibitory effect of heavy metals on methane-producing anaerobic granular sludge. J. Hazard. Mater. 162, 1551–1556.

Appels, L., Baeyens, J., Degreve, J., Dewil, R., 2008. Principles and potential of the anaerobic digestion of waste-activated sludge. Prog. Energ. Combust. 34, 755–781.

Appendino, P., Ferraris, M., Matekovits, I., Salvo, M., 2004. Production of glass–ceramic bodies from the bottom ashes of municipal solid waste incinerators. J. Eur. Ceram. Soc. 24, 803–810.

Apte, A., Verma, S., Tare, V., Bose, P., 2005. Oxidation of Cr (III) in tannery sludge to Cr (VI): field observations and theoretical assessment. J. Hazard. Mater. 121, 215–222.

Araujo, A.S.F., Miranda, A.R.L., Oliveira, M.L.J., Santos, V.M., Nunes, L.A.P.L., Melo, W.J., 2015. Soil microbial properties after 5 years of consecutive amendment with composted tannery sludge. Environ. Monit. Assess. 187, 41–53.

Bacardit, A., Baquero, G., Sorolla, S., Olle, L., 2015. Evaluation of a new sustainable continuous system for processing bovine leather. J. Clean. Prod. 101, 197–204.

Barajas-Aceves, M., Rios Berber, J.D., Oropeza Mota, J.L., Rodríguez Vazquez, R., 2014. Assessment of tannery waste in semi-arid soils under a simulated rainfall system. Soil Sediment Contam. 23 (8), 954–964.

Bareen, F., Khilji, S., 2008. Bioaccumulation of metals from tannery sludge by *Typha angustifolia* L. Afr. J. Biotechnol. 7 (18), 3314–3320.

Basegio, T., Leão, A., Bernardes, A., Bergmann, C., 2009. Vitrification: an alternative to minimize environmental impact caused by leather industry wastes. J. Hazard. Mater. 165, 604–611.

BAT, 2013. Best Available Techniques Reference Document for the Tanning of Hides and Skins. European IPPC Bureau, Seville.

Bień, J., Białczak, W., Wystalska, K., 2007. Solid waste vitrification using a direct current plasma arc. In: Pawłowski, L., Dudzińska, M.R. (Eds.), Environmental Engineering. Taylor & Francis Group, London, pp. 307–310.

Bień, J., Celary, P., Morzyk, B., Sobik-Szołtysek, J., Wystalska, K., 2013. Effect of additives on heavy metal immobilization during vitrification of tannery sewage sludge. Environ. Prot. Eng. 39 (2), 33–40.

Cassano, A., Molinari, R., Romano, M., Drioli, E., 2001. Treatment of aqueous effluents of the leather industry by membrane processes. A review. J. Membr. Sci. 181, 111–126.

Celary, P., Sobik-Szołtysek, J., 2014. Vitrification as an alternative to landfilling of tannery sewage sludge. Waste Manag. 34, 2520–2527.

Celary, P., Sobik-Szołtysek, J., Tajchman, A., 2016. Heavy metal volatilization during Vitrification of tannery sewage sludge with mineral waste. Eng. Protect. Environ. 19 (4), 503–515.

Chandra, P., Kulshreshtha, K., 2004. Chromium accumulation and toxicity in aquatic vascular plants. Bot. Rev. 70 (3), 313–327.

Chen, Y., Cheng, J.J., Creamer, K.S., 2008. Inhibition of anaerobic digestion process: a review. Bioresour. Technol. 99 (10), 4044–4064.

Chen, J., Dun, Y., Li, Z., Sun, D., Zhu, C., 2015. Synthesis of black pigments containing chromium from leather sludge. Ceram. Int. 4, 19455–19460.

Dixit, S., Yadav, A., Dwivedi, P.D., Das, M., 2015. Toxic hazards of leather industry and technologies to combat threat: a review. J. Clean. Prod. 87, 39–49.

Dunnett, B.F., Gribble, N.R., Short, R., Turner, E., 2012. Vitrification of high moiybdenum waste. Glass Technol. A 53 (4), 166–171.

Fang, D., Zhou, L.X., 2006. Effect of sludge dissolved organic matter on oxidation of ferrous iron and sulfur by Acidithiobacillusferroxidans and Acidithiobacillus thiooxidans. Water Air Soil Poll. 171, 81–94.

Fang, D., Zhou, L.-X., 2007. Enhanced Cr bioleaching efficiency from tannery sludge with coinoculation of Acidithiobacillus thiooxidans TS6 and Brettanomyces B65 in an air-lift reactor. Chemosphere 69, 303–310.

Fela, K., Wieczorek-Ciurowa, K., 2010. Heavy metals in ashes from thermal treatment of leather industry solid wastes. In: Bargańska, Ż. et al., (Ed.), Proceedings of 15th International Conference on Heavy Metals in the Environment. Chem. Faculty, Univ. Technol., Gdańsk, pp. 1002–1005.

Ferraris, M., Salvo, M., Ventrella, A., Buzzi, L., Veglia, M., 2009. Use of vitrified MSWI bottom ashes for concrete production. Waste Manag. 29, 1041–1047.

Fijałkowski, K., Rosikoń, K., Grobelak, A., Kacprzak, M., 2011. Migration of various chemical compounds in soil solution during inducted phytoremediation. Arch. Environ. Prot. 37 (4), 49–59.

Filibeli, A., Buyukkamaci, N., Senol, H., 2000. Solidification of tannery wastes. Resour. Conserv Recy. 29, 251–261.

Fongsatitkul, P., Elefsiniotis, P., Kitkaew, D., Rungsipanodorn, C., 2011. Use of rice husk ash as an admixture to remove chromium from a tannery waste. Water Air Soil Poll. 220, 81–88.

Gupta, A.K., Sinha, S., 2007. Assessment of single extraction methods for the prediction of bioavailability of metals to *Brassica juncea* L. Czern. (var. Vaibhav) grown on tannery waste contaminated soil. J. Hazard. Mater. 149, 144–150.

Haroun, M., Idris, A., Omar, S.R.S., 2007a. A study of heavy metals and their fate in the composting of tannery sludge. Waste Manag. 27, 1541–1550.

Haroun, M., Idris, A., Omar, S.R.S., 2007b. Characterisation and composting of tannery sludge. Malaysian J. Soil Sci. 11, 71–80.

Haroun, M., Idris, A., Omar, S.R.S., 2009. Analysis of heavy metals during composting of the tannery sludge using physicochemical and spectroscopic techniques. J. Hazard. Mater. 165, 111–119.

Huang, D., Drummond, C.H., Wang, J., Blume, R.D., 2004. Incorporation of chromium (III) and chromium (VI) oxides in a simulated basaltic. Ind. Waste Glass. J. Am. Ceram. Soc. 87 (11), 2047–2052.

Jiang, X.-G., Li, C.-Y., Fei, Z.-W., Chi, Y., Yan, J.-H., 2010. Combustion characteristics of tannery sludge and volatilization of heavy metals in combustion. J. Zhejiang Univ. Sci. A. 11 (7), 530–537.

Jiang, H., Liu, J., Han, W., 2016. The status and developments of leather solid waste treatment: a mini-review. Waste Manage. Res. 34 (5), 399–408.

Juel, M.A.I., Chowdhury, Z.U.M., Ahmed, T., 2016. Heavy metal speciation and toxicity characteristics of tannery sludge. AIP Conf. Proc. 1754. 060009. https://doi.org/10.1063/1.4958450.

Juel, M.A.I., Mizan, A., Ahmed, T., 2017. Sustainable use of tannery sludge in brick manufacturing in Bangladesh. Waste Manag. 60 (2017), 259–269.

Kanagaraj, J., Velappan, K.C., Chandra Babu, N.K., Sadulla, S., 2006. Solid wastes generation in the leather industry and its utilization for cleaner environment – a review. J. Sci. Ind. Res. India 65, 541–548.

Kavouras, P., Pantazopoulou, E., Varitis, S., Vourlias, G., Chrissa, K., Dimitrakopulos, G.P., Mitrakas, M., Zouboulis, A.I., Karaostas, T., Xenidis, A., 2015. Incineration of tannery sludge under oxic and anoxic conditions: study of chromium speciation. J. Hazard. Mater. 283, 672–679.

Kilic, E., Font, J., Puig, R., Colak, S., Celik, D., 2011a. Chromium recovery from tannery sludge with saponin and oxidative remediation. J. Hazard. Mater. 185, 456–462.

Kilic, E., Puig, R., Baquero, G., Font, J., Colak, S., Celak, D., Gürler, D., 2011b. Environmental optimization of chromium recovery from tannery sludge using alife cycle assessment approach. J. Hazard. Mater. 192, 393–401.

Krishnamoorthy, G., Sadulla, S., Sehgal, P.K., Mandal, A.B., 2012. Green chemistry approaches to leather tanning process for making chrome-free leather by unnatural amino acids. J. Hazard. Mater. 215–216, 173–182.

Lay, T., Rockwell, M., Westreich, P., 2009. Production and characteristics of glass-ceramics derived from manganese crust tailings. J. Mater. Sci. 44, 3508–3513.

Lazau, I., Vancea, C., 2014. New vitreous matrix for chromium waste immobilization. Cent. Eur. J. Chem. 12 (7), 763–768.

Li, G.-J., Zhang, C., Zhan, J., Wang, F., 2005. Chrome sludge compost – effect on crops and soils. J. Soc. Leath. Tech. Ch. 90, 10–13.

Liu, M., Ma, J., Lyu, B., Gao, D., Zhang, J., 2016. Enhancement of chromium uptake in tanning process of goat garment leather using nanocomposite. J. Clean. Prod. 133, 487–494.

Lofrano, G., Meric, S., Zengin, G.E., Orhon, D., 2013. Chemical and biological treatment technologies for leather tannery chemicals and wastewaters: a review. Sci. Total Environ. 461–462, 265–281.

Lopez-Luna, J., Gonzalez-Chavez, M.C., Esparza-Garcia, F.J., Rodriguez-Vazquez, R., 2009. Toxicity assessment of soil amended with tannery sludge, trivalent chromium and hexavalent chromium using wheat, oat and sorghum plants. J. Hazard. Mater. 163, 829–834.

Ma, H., Zhou, J., Hua, L., Cheng, F., Zhou, L., Qiao, X., 2017. Chromium recovery from tannery sludge by bioleaching and its reuse in tanning process. J. Clean. Prod. 142, 2752–2760.

Mandal, B.K., Vankayala, R., Kumar, L.U., 2011. Speciation of chromium in soil and sludge in the surrounding tannery region, Ranipet, Tamil Nadu. ISRN Toxicol. 2011. 697980. https://doi.org/10.5402/2011/697980.

Mao, L., Deng, N., Liu, L., Cui, H., Zhang, W., 2016a. Effects of Al_2O_3, Fe_2O_3, and SiO_2 on Cr (VI) formation during heating of solid waste containing Cr (III). Chem. Eng. J. 304, 216–222.

Mao, L., Deng, N., Liu, L., Cui, H., Zhang, W., 2016b. Inhibition of Cr (III) oxidation during thermal treatment of simulated tannery sludge: the role of phosphate. Chem. Eng. J. 294, 1–8.

Montanes, M.T., Sanchez-Tovar, R., Roux, M.S., 2014. The effectiveness of the stabilization/solidification process on the leachability and toxicity of the tannery sludge chromium. J. Environ. Manag. 143, 71–79.

Murray, K., Mozafarzadeh, M., Tebo, B., 2005. Cr (III) oxidation and Cr toxicity in cultures of the manganese(ii)-oxidizing *Pseudomonas putida* strain GB-1. Geomicrobiol J. 22, 151–159.

Nakatani, A.S., Nogueira, M.A., Martines, A.M., Dos Santos, C.A., Baldesin, L.F., Marschner, P., Cardoso, E.J.B.N., 2012. Effects of tannery sludge application on physiological and fatty acid profiles of the soil microbial community. Appl. Soil Ecol. 61, 92–99.

Ociepa, E., Lach, J., Stępień, W., 2007. Influence of diversified fertilization on the bioaccumulation of heavy metals and on the crop yields of plants. Ecol. Chem. Eng. S 14 (2), 223–231.

Panda, S.K., Choudhury, S., 2005. Chromium stress in plants. Toxic metals in plants. Braz. J. Plant Physiol. 17 (1), 95–102.

Pantazopoulou, E., Zebiliadou, O., Mitrakas, M., Zouboulis, A., 2015. Stabilization of tannery sludge by co-treatment with aluminum anodizing sludge and phytotoxicity of end-products. Waste Manag. 61, 327–336.

Pelino, M., Pisciella, P., Crisucci, S., Karamanov, A., 2001. Chemical durability of glasses obtained by vitrification of industrial wastes. Waste Manag. 21, 1–9.

Pereira, M., Resende, P., Azeiteiro, U., Oliveira, J., de Figueiredo, D., 2005. Differences in the effects of metals on growth of two freshwater green algae (Pseudokirchneriellasubcapitata

(Korshikov) Hindak and Gonium pectorale Müller). Bull. Environ. Contam. Toxicol. 75, 515–522.

Priebe, G.P.S., Kipper, E., Gusmao, A.L., Marcilio, N.R., Gutterres, M., 2016. Anaerobic digestion of chrome-tanned leather waste for biogas production. J. Clean. Prod. 129, 410–416.

Religa, P., Kowalik, A., Gierycz, P., 2011. Application of nanofiltration for chromium concentration in the tannery wastewater. J. Hazard. Mater. 186, 288–292.

Silva, M.D.M., Barajas-Aceves, M., Araujo, A.S.F., Araujo, F.F., Melo, J., 2014. Soil microbial biomass after three-year consecutive composted tannery sludge amendment. Pedosphere 24 (4), 469–475.

Singh, I.B., Chaturvedi, K., Morchhale, R.K., Yegneswaran, A.H., 2007. Thermal treatment of toxic metals of industrial hazardous wastes with fly ash and clay. J. Hazard. Mater. 141, 215–222.

Sobiecka, E., Szymański, Ł., 2014. Thermal plasma vitrification process as an effective technology for fly ash and chromium-rich sewage sludge utilization. J. Chem. Technol. Biotechnol. 89, 1115–1117.

Speranza, A., Ferri, P., Battistelli, M., Falcieri, E., Crinelli, R., Scoccianti, V., 2007. Both trivalent and hexavalent chromium strongly alter in vitro germination and ultrastructure of kiwi fruit pollen. Chemosphere 66, 1165–1174.

Sreeram, K.J., Ramasami, T., 2003. Sustaining tanning process through conservation, recovery and better utilization of chromium. Resour. Conserv. Recy. 38, 185–212.

Stoop, M.L.M., 2003. Water management of production systems optimized by environmentally oriented integral chain management: case study of leather manufacturing in developing countries. Technovation 23, 265–278.

Swarnalatha, S., Srinivasulu, T., Srimurali, M., Sekaran, G., 2008. Safe disposal of toxic chrome buffing dust generated from leather industries. J. Hazard. Mater. 150, 290–299.

Tang, P., Zhao, Y., Xia, F., 2008. Thermal behaviors and heavy metal vaporization of phosphatized tannery sludge in incineration process. J. Environ. Sci. 20, 1146–1152.

Tang, P., Zhao, Y.C., Xie, Z., 2013a. Effects of hydroxyapatite addition on heavy metal volatility during tannery sludge incineration. Environ. Sci. Pollut. R. 20, 4405–4413.

Tang, P., Zhao, Y.C., Xie, Z., 2013b. Immobilization of heavy metals in sludge using phosphoric acid and monobasic calcium phosphate. J. Zhejiang Univ. Sci. A. 14 (3), 177–186.

Thompson, S., Manning, F., McColl, S., 2002. Comparison of the toxicity of chromium III and chromium VI to cyanobacteria. Bull. Environ. Contam. Toxicol. 69, 286–293.

Varitis, S., Kavouras, P., Pavlidou, E., Pantazopoulou, E., Vourlias, G., Chrissafis, K., Zouboulis, A.I., Karakostas, T., Komninou, P., 2017. Vitrification of incinerated tannery sludge in silicate matrices for chromium stabilization. Waste Manag. 59, 237–246.

Vig, A.P., Singh, J., Wani, S.H., Dhaliwal, S.S., 2011. Vermicomposting of tannery sludge mixed with cattle dung into valuable manure using earthworm Eisenia fetida (Savigny). Bioresour. Technol. 102, 7941–7945.

Wolski, P., Zawieja, I., Stańczyk-Mazanek, E., 2011. Thermal utilization and agricultural use of sewage sludge in Poland. Eng. Protect. Environ. 14 (4), 411–418.

Wu, J., Zhao, L., Liu, X., Chen, W., Gu, H., 2017. Recent progress in cleaner preservation of hides and skins. J. Clean. Prod. 148, 158–173.

Zhang, C., Lin, J., Jia, X., Peng, B., 2016. A salt-free and chromium discharge minimizing tanning technology: the novel cleaner integrated chrome tanning process. J. Clean. Prod. 112, 1055–1063.

Zhou, L.X., Fang, D., Wang, S.M., Wong, J.W.C., Wang, D.Z., 2005. Bioleaching of Cr from tannery sludge: the effects of initial acid addition and recycling of acidified bioleached sludge. Environ. Technol. 26, 277–284.

Valorization of sugarcane waste: Prospects of a biorefinery

3

Ranaprathap Katakojwala[*,†], *A. Naresh Kumar*[*,†], *Debkumar Chakraborty*[*], *S. Venkata Mohan*[*]
[*]Bioengineering and Environmental Sciences Lab, EEFF Department, CSIR-Indian Institute of Chemical Technology (CSIR-IICT), Hyderabad, India, [†]Academy of Scientific and Innovative Research (AcSIR), Hyderabad, India

1 Introduction

Sugarcane is one of the most important agricultural crops and the day-to-day population increment, demands greater sugar consumption, therefore increasing the sugarcane cultivation area (Rabelo et al., 2011a,b) (Fig. 1). Sugarcane cultivation requires a tropical or temperate climate, with a minimum of 60 cm (24 in.) of annual moisture. It is one of the most efficient photosynthesizers in the plant kingdom. It is a C_4 plant, able to convert up to 1% of incident solar energy into biomass (Whitmarsh and Govindjee, 1995). Sugarcane has also been shown to utilize a wider range of wavelengths of solar radiation within the visible spectrum than most other plants. Physiological adaptations also include sucrose as a principal photosynthetic storage product which facilitates easy translocation of carbohydrates for growth of the plant. As a consequence of such adaptations, sugarcane fixes around four times as much solar energy as most temperate crops, and can consequently yield around 50 t dry matter/ha/year (Paturau, 1989).

Globally, Brazil stands in first place followed by India with a production rate of 170.7 and 167 metric tons, respectively (Fig. 2). Sugarcane fields produce mainly two types of biomass: cane trash and bagasse. Cane trash is the field residue remaining after harvesting the cane stalk and bagasse is the milling byproduct that remains after extracting the sugar from the stalk (Rabelo et al., 2011a,b). The potential energy value of these residues has traditionally been ignored by various industries. However, unavoidable rises in fossil fuel prices and depletion of resources have led to viewing the waste as a valuable renewable energy resource (Margeot et al., 2009). In this context, sugarcane molasses is being used as a resource for bioethanol production (Luo et al., 2011). Additionally, during the course of sugarcane processing, various kinds of other wastes are also being produced, such as bagasse, press mud, washwater, and vinasse. Utilization of this waste as a resource facilitates additional benefits to sugarcane industries. In this context, the present chapter discusses the composition of various sugarcane industrial wastes and their significant potential toward various bio-based product synthesis in the biorefinery framework.

Industrial and Municipal Sludge. https://doi.org/10.1016/B978-0-12-815907-1.00003-9

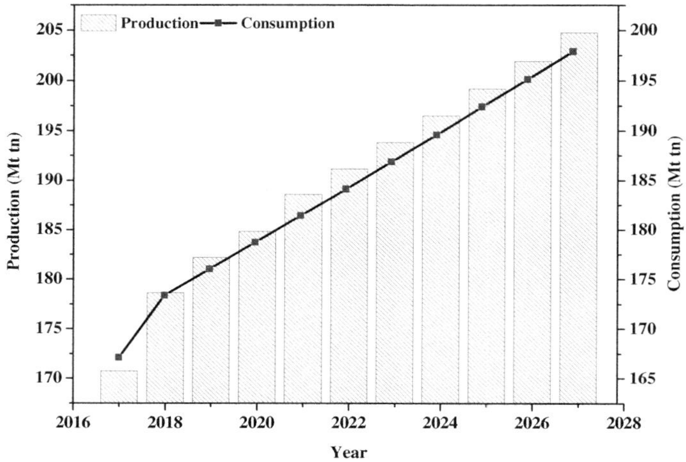

Fig 1 Sugar production and consumption profiles.
Source: OECD-FAO Agricultural Outlook 2018–2027.

2 Characterization of sugarcane waste (SCW)

The sugarcane industry produces various waste materials at the different processing stages of sugar production (Fig. 3). To utilize this waste as a resource, prior characterization needs to be understood. Therefore, in this section the detailed characterization of sugarcane industrial waste (SCW) is discussed (Table 1). Sugarcane bagasse (SCB), which is an abundant waste fibrous pulp material, results after the juice extraction from sugarcane stalks. The typical chemical composition of SCB is cellulose 40%–45%, hemicellulose 27%–29%, lignin 19%–21%, extractives 3%–4%, and other ashes 2%–4% (Karp et al., 2013). The cellulose, hemicellulose, and lignin fractions of SCB are considered natural polymers. Polysaccharides such as cellulose is composed of branched glucose (C_6) units and hemicellulose is composed of hexoses (C_6) and pentoses (C_5) units, respectively. The most significant character of SCB is its complete biodegradable and compostable nature. Additionally, the sugar content of the SCB (cellulose/hemicellulose) could be extracted and used as a pillar material for various biological and chemical processes. Lignin is a constituent of the cell walls of almost all dry land plant cell walls, the second-most abundant natural polymer in the world, which is a complex polymer of aromatic alcohols.

The high organic content of wastewater from the sugarcane industry makes it a good organic resource for biological valorization. In general in the sugarcane industry, wastewater contains (1–4.3 g/L) of COD, but the maximum COD standard for discharge of effluents is 150–250 mg/L (Hampannavar and Shivayogimath, 2010; Rais and Sheoran, 2015). Sugarcane press mud is the residue resulting after the filtration of sugarcane juice. It is soft, spongy, amorphous in nature, and dark brown in color. Press mud is characterized (on dry weight basis) by 9%–14% of wax, 10%–18% protein, 11%–17% cellulose, 15%–27% hemicelluloses, 9%–14% lignin, oil, resin, etc.

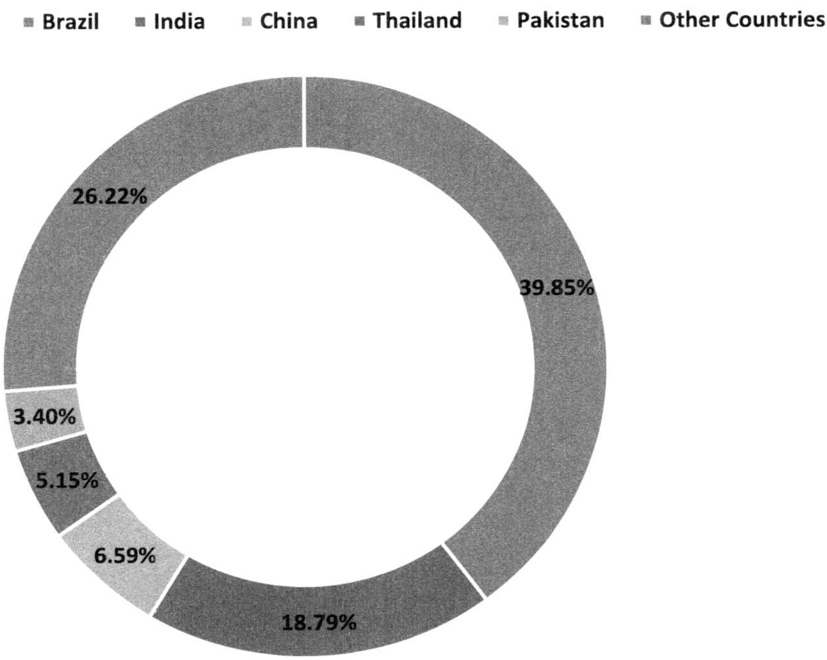

Fig 2 Global contribution in sugarcane production with respect to country.
Source: FAO report.

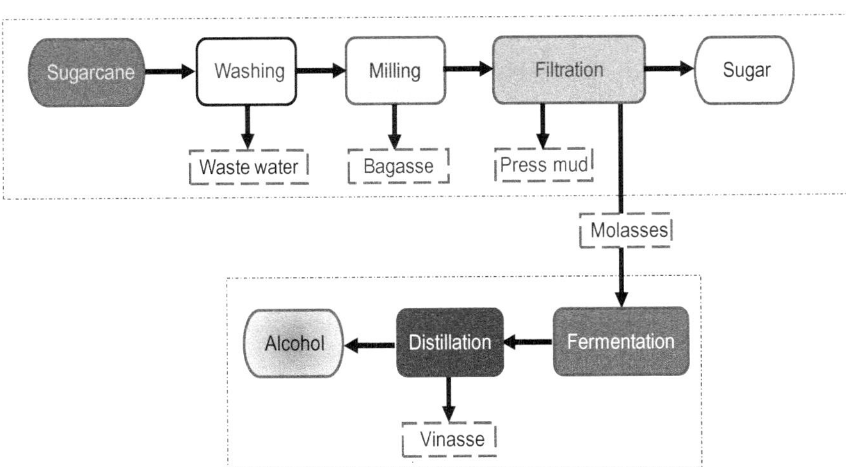

Fig 3 Various processing methods for the sugarcane industry and the corresponding waste generation.

Table 1 Physical properties and chemical composition of sugarcane bagasse, press mud, molasses, and vinasse

S. no.	Parameters	Bagasse	Press mud	Molasses	Vinasse	References
1	pH	4.5–5.5	4.95	5–5.5	4.8	España-Gamboa et al. (2011), Gupta et al. (2011), Radjaram and Saravanane (2011a,b), Saleh-e-In et al. (2012), Janke et al. (2015), Janke et al. (2016), Talha et al. (2016), Sundaranayagi et al. (2017); González et al. (2017), Ansari and Gaikar (2014), and Kumar and Thankamani (2016)
2	Total solid (%)	59	47.87	46	5.3	
3	Total VS (%)	97.6	34.8	48.1	78.6	
4	C.O.D.	NA	284.5(g/kg)	70–100 (g/L)	62.3 (g/L)	
5	B.O.D. (5 d, 27°C)	NA	222.0 (g/kg)	40–40 (g/L)	5.05 mg/L	
6	Total nitrogen (%)	1.86	1.68	<0.2	0.6	
7	Cellulose (%)	40–45	11.4	–	–	
8	Hemi cellulose (%)	27–29	27.1	–	–	
9	Lignin (%)	19–21	9.3	–	–	
10	Protein (%)	1.4	3.08	1.25	0.4	
11	Sugars (g/kg)	NA	22.5	49.9%	31.6	
12	Fiber (%)	35	15–30	–	–	
13	$N-NH_4^+$ ($g\,kg^{-1}$)	–	–	200–215	5.6×10^{-3}	
14	Phosphorous (%)	0.345	2.5–3.0	3	0.1	
15	Potassium (%)	2.1	0.30–1.80	2.89	2.06	
16	Crude wax (%)	<1	5–14	–	–	
17	Ash	2%–4%	19.2%	10.25%	21.4%	

(González et al., 2017). Press mud also contains a sizable quantity of macro- and micronutrients, besides 20%–25% of organic carbon. The X-ray fluorescent spectrometric analysis of the elemental composition carried out by Saleh-e-In et al. (2012) showed the presence of calcium (Ca), magnesium (Mg), silicon (Si), phosphorus (P), sulphur (S), aluminium (Al), and potassium (K) in press mud. The chemical composition may vary based on the cane variety, soil condition, nutrients applied in the field, the process of clarification adopted, and other environmental factors. The disposal of press mud is one of the major issues in the sugarcane industries because higher amounts of press mud are left in piles and their drainage causes the pollution of nearby water bodies. However, due to its high organic content, press mud can be used as a viable resource for secondary product synthesis using chemical and biological processes.

Molasses is the syrup remaining after the crystallization of sugar from cane or beet juice. It is separated from the sugar crystals by the repeated centrifugation process. There are several grades of molasses and the final extraction yield is known as black strap molasses, which is heavy, viscous, and dark colored. The chemical composition of black strap molasses is primarily composed of total soluble solids (83.2%), followed by total sugars (49.9%), ash (10.25%), calcium (54%), sodium (54%), and potassium (2.89%) (pH 5.6–6.0) (Abubaker et al., 2012). Molasses is widely used as animal feed and for the synthesis of various bio-based product synthesis such as ethanol, vinegar, glycerin, inositol, lactic acid, succinic acid, glutamic acid, aconitic acid, kojic acid, and single cell proteins (SCP). Additionally, vinasse is a liquid fraction discharged during the refinement and distillation of ethanol, which is 20-fold higher than the ethanol produced. Vinasse is sulfur-rich in nature with a low pH (3.8–4.0), dark in color and an odorous effluent. The typical chemical composition of vinasse is like BOD (5046 mg/L), potassium (2056 mg/L), sodium (50.2 mg/L), sulfate (710 mg/L), calcium (719 mg/L), magnesium (237 mg/L), phosphorus (190 mg/L), hardness (2493 mg/L), and trace metals (As, Ba, Cd, Cr, Cu, Hg, Mo, Ni, Pb, Se, and Zn). Vinasse has a great potential to serve as a soil conditioner, animal feed, combustible material, etc. (Robertiello, 1982; España-Gamboa et al., 2011).

3 Pretreatment of SCW

3.1 Physical methods

Common physical pretreatment methods used for the pretreatment of sugarcane waste are discussed in this section (Fig. 4), starting with mechanical grinding, which is used for reducing the crystallinity of cellulose. It mostly includes chipping, grinding, and/or milling techniques. Chipping can reduce the biomass size to 10–30 mm only while grinding and milling can reduce the particle size up to 0.2–0.4 mm, which is significant for hydrolysis (Chang et al., 1997). Ultrasonication is a relatively new technique used for the pretreatment of lignocellulosic biomass and it is a feasible pretreatment option at the lab scale. Ultrasound treatment leads to the formation of small cavitation bubbles, which rupture the cellulose and hemicellulose fractions and thereby increase

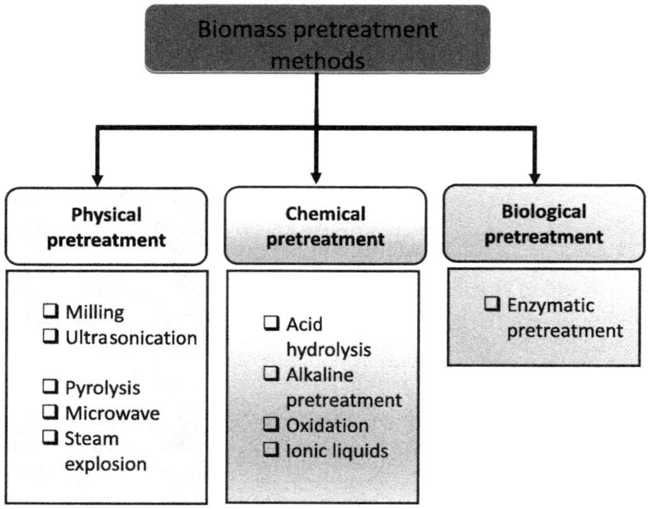

Fig 4 Various pretreatment methods of sugarcane waste.

the accessibility to cellulose-degrading enzymes for effective breakdown into simpler reducing sugars (Kumar and Sharma, 2017). Additionally, microwave irradiation is also a widely used method for lignocellulosic feedstock pretreatment because of its low energy requirement, high heating capacity in a short duration of time, minimum generation of inhibitors, and the ability to degrade the structural organization of the cellulose fraction. Moreover, addition of mild-alkali reagents is preferred for more effective breakdown. A study on microwave-based alkali pretreatment of switch grass yielded nearly 70%–90% sugars (Hu and Wen, 2008). Steam explosion is also a promising method of pretreatment. It is a thermomechanical process (200–270°C) that exposes biomass to a high pressure of steam (14–16 bar). As a result, it penetrates the biomass by diffusion for short periods of time (20 s to 20 min), followed by a sudden decompression generating shear forces that hydrolyze the glycosidic and hydrogen bonds between the glucose chains (Rojas et al., 2015).

3.2 Chemical methods

Acid-catalyzed pretreatment is the most widely used conventional pretreatment of lignocellulosic biomass. The corrosive and toxic nature of most acids requires a suitable material for building the reactor, which can sustain the required experimental conditions and corrosive nature of acids (Saha et al., 2005). Still it is the most widely employed pretreatment method on an industrial scale. In some cases, an enzymatic hydrolysis step could easily be avoided as acid itself hydrolyzes the biomass into fermentable sugars. However, extensive washing is necessary to remove acid before fermentation of sugars (Sassner et al., 2008). Alkali-catalyzed pre treatment methods can also be used at ambient temperature and pressure. The most commonly used alkali reagents are the hydroxyl derivatives of sodium, potassium, calcium, and ammonium salts. Among these hydroxyl derivatives, sodium hydroxide was found to be the most effective (Kumar and Wyman, 2009). Alkali reagents degrade the side chains of esters and glycosides, leading to structural modification of lignin, cellulose swelling, cellulose decrystallization, and hemicellulose (Cheng et al., 2010; Ibrahim et al., 2011; McIntosh and Vancov, 2010; Sills and Gossett, 2011). Additionally, studies were reported using lime-catalyzed conditions for SCW pretreatment due to its slower reactivity, which produces a lower amount of fermentation inhibitors (Rabelo et al., 2011a,b). This method is considered as a low-cost alternative for lignin solubilization, removing approximately 33% of lignin and 100% of acetyl groups. The action of lime is of low cost and its safe handling makes it attractive.

Oxidizing agents such as hydrogen peroxide, ozone, oxygen, or air were used for the treatment of the lignocellulosic biomass (Nakamura et al., 2004). A number of chemical reactions such as electrophilic substitution, side chain displacements, and oxidative cleavage of the aromatic ring ether linkages may take place during oxidative pretreatment. This process causes delignification by converting lignin to acids. A major downside of oxidative pretreatment is that it damages a significant amount of hemicellulose, making it unavailable for fermentation (Lucas et al., 2012). The most commonly employed oxidizing agent is hydrogen peroxide and it has the great advantage of not leaving residue in the biomass as it degrades into oxygen and water. It has been found that hydrolysis of hydrogen peroxide leads to the formation of hydroxyl radicals that are responsible for the degradation of lignin and the production of low molecular weight products; removal of lignin from lignocellulose exposes cellulose and hemicellulose, leading to increased enzymatic hydrolysis (Hammel et al., 2002; Rabelo et al., 2011a,b). Some studies were also reported with the use of ionic liquids for SCW pretreatment (Zavrel et al., 2009; Behera et al., 2014). The ionic liquid breaks intramolecular hydrogen bonds, whereas the cations attack the O atom of the —OH, and anions attack the hydrogen atoms of the —OH group. Imidazolium salts are the most commonly used ILs. ILs are assumed to compete with lignocellulosic components for hydrogen bonding, thereby disrupting its network (Moultrop et al., 2005).

3.3 Enzymatic method

In contrast to conventional chemical and physical pretreatment methods, biological pretreatment is considered an efficient, environmentally friendly, and cost-effective process. Nature has abundant bacterial species that can abundantly produce the cellulolytic and hemicellulolytic enzymes. The use of enzymes specially targets the β-glycosidic bonds of cellulose or hemicellulose and results in relatively effective pretreatment than other methods (Vats et al., 2013). Biological pretreatments are carried out by microorganisms such as brown, white, and soft-rot fungi, which mainly degrade lignin and hemicellulose and small amounts of cellulose (Sánchez, 2009). Though the biological pretreatment is highly intriguing, the rate of hydrolysis of lignocellulosic fractions is too slow, which influences adversely during pretreatment at an industrial scale (Sun and Cheng, 2002). In order to make biological techniques on par with other pretreatment methods, integration with mechanical treatment or chemical pretreatments is to be performed for sustainable processes.

3.4 Biological processes

3.4.1 Acidogenic fermentation

Hydrogen is a promising alternative to carbon-based fuels because it is clean, renewable, and has a high energy yield of 122 kJ/g (Lin and Chang, 2004). To explore the viability of biohydrogen production from sugarcane waste, for example, press mud (acidic fibrous waste with 5%–15% sugar), various studies were performed (Partha and Sivasubramanian, 2006). Press mud was codigested with sewage (1:10) in UASB reactors for hydrogen production and a maximum biohydrogen production of 7960 mL/day (187 mL g VS^{-1}) was observed (Radjaram and Saravanane, 2011a). In another study, press mud mixed with water for biohydrogen production was performed and a maximum gas yield of 2240 mL d^{-1} was reported on the 78th day of the UASB operation (Radjaram and Saravanane, 2011b). Pretreated press mud slurry (10% of total solids) was used as a feedstock with enriched mixed microbial consortia and 0.04 g/g; 0.27 g/g of bioethanol and VFA were respectively obtained through acidogenic anaerobic fermentation (Kuruti et al., 2015). The pretreatment of sugarcane filter cake (FC) was carried out using NaOH and an improved VFA yield was observed with the productivity of 37% (0.44 g VFA g VS^{-1}), mainly consisted of *n*-butyric and acetic acids (Janke et al., 2016).

3.4.2 Biomethanization

Another alternative for the use of SCW is biogas production, which can be a sustainable solution for organic matter removal from the sugar industry, for example, sugarcane waste effluent and press mud (77% volatile solids). Sugar molasses also has a methane potential (i.e., CH_4 per ton of raw material) of 230 m^3. The volume of biogas generated of press mud through conventional digestion is around 65 m^3 while through integrated biomethanation, it is about 80 m^3/t of press mud. Press mud, along with bagasse, generates biogas that contains 52% methane content. The rate of biogas production was increased from press mud by the supplementation of trace elements and by codigestion with kitchen waste (Sundaranayagi et al., 2017). Rouf et al. (2010) evaluated the biogas production using mixing ratio of 1:1 press mud; sugarcane straw and biogas production were increased by 58%, respective to the press mud in monodigestion. The anaerobic digestion from vinasse and press mud separately has been reported previously as reaching methane yields of 0.344 m^3 kg^{-1} COD removed and 0.250 m^3 kg^{-1} COD removed, respectively (Harada et al., 1996; Sánchez et al., 1996). The conversion efficiency of sugarcane press mud and methane yield was further increased up to 64% by the codigestion with the vinasse in stirred-tank reactors (González et al., 2017). To enhance the codigestion of degradation and improve the biomethane production potential, sugarcane bagasse and filter mud were pretreated (using heat and alkali) and a cumulative yield of biomethane (195.8 mL g VS^{-1}) was observed (Talha et al., 2016). 0.68 m^3 biogas production (CH_4—67%) from press mud was reported using a continuous-process floating drum-type anaerobic digester (1 m^3) (Sathish and Vivekanandan, 2015).

3.4.3 Solventogenesis

Pretreatment techniques generally improved the formation of sugars, avoiding the degradation of carbohydrates and the formation of byproducts inhibitory to the subsequent hydrolysis and fermentation process of sugarcane waste (Nimbalkar et al., 2017). After detoxification/neutralization, the effective utilization of press mud for bioethanol production was successfully achieved using acid-hydrolyzed press mud, resulting in a sugar release of 21 g/L that was further converted into 11.65 g/L of ethanol using *Saccharomyces cerevisiae* NRRL Y-12632 (Pawar et al., 2017). Ethanol production generates great amounts of residuals, both liquid and solid. These byproducts can be used for bioresource and energy recovery. Bioethanol production and coproduction of electrical power and biogas using dry bagasse as a substrate was also investigated in several studies reported before (Rabelo et al., 2011a,b).

3.5 Fertilizer production

Sugarcane waste, mainly press mud, can also be used as a soil conditioner and soil fertilizer and is very useful for agricultural crops and horticulture because of its richness in various micronutrients and nitrogen. Composting of organic wastes from the sugarcane industry is necessary to reduce the lignin and cellulose contents, thereby the nutrient availability can be improved. Processes used to improve the fertilizer value include composting, treatment with microorganisms, and mixing with distillery effluents (Nasir, 2006). Generally, vermicomposting of press mud is an efficient method of waste disposal and is used as manure in India by sugarcane farmers, enabling recycling of organic matter for solid waste treatment. Nowadays, toxic distillery effluent, spent wash having high BOD and COD content, is being utilized by spraying and mixing with press mud to enrich the press mud in its nutritional values and make one of the best organic manures. The organic manure made out of press mud maintains soil health, sustains sugarcane and sugar production, improves soils physical properties, retains soil moisture, and reduces erosion hazards (Baskar et al., 2003; Rath et al., 2010; Sangwan et al., 2010).

3.6 Animal feed

In animal husbandry, cane bagasse and press mud have been used as feed ingredients, notably for ruminants, because of their sugar and mineral content. Filter press mud was used as a filler in ruminant maintenance diets at a level of 10%–30%, together with poultry manure, final molasses, ground cane, urea, and minerals (containing

13% crude fiber, 8.8% crude protein, and 31.7% ash) (Rodriguez and Gonzalez, 1973). Sugarcane press mud was considered to be a potential low-energy feed ingredient (6.6% crude protein) in poultry diets in countries such as Sri Lanka, the Philippines, etc. (Rajaguru and Ravindran, 1985; Abrigo and Gerpacio, 1986). Press mud, which has a chemical composition similar to that of cattle dung, was used as a fish pond fertilizer in India (Keshavanath et al., 2005). In China, a carp and dace feed ingredient has been produced from press mud (Yu, 1990).

3.7 Other applications

Other industrial applications are reported for sugarcane waste and press mud in cement and paint manufacturing, as a foaming agent, in wax production, etc. (Van der Poel et al., 1998). Wax and its hydrolysis product, fatty alcohols ($C_{25}-C_{30}$), have applications as nutraceutical. Press mud is also a useful biosorbent for metal adsorption from wastewater. Filter press mud can also be used as a compacting and wetting agent in surface silos (Perez, 1990; Ansari and Gaikar, 2014). Press mud is also used for the production of citric acid by *Aspergillus niger* in a solid-state fermentation system that initially contained 12.5% (w/w) effective sugar and 0.1% (w/w) NH_4NO_3 (Show et al., 2015).

3.8 Struvite precipitation

Increased consumption and no alternative sources other than the rock phosphate of mineral phosphorus make phosphorus limited (Rahman et al., 2011). Alternatively, struvite can be used as a phosphate fertilizer to provide sustainable phosphorus usage. Struvite is a crystalline substance consisting of magnesium, ammonium, and phosphorus in equal molar concentrations ($MgNH_4PO_4$) (Munir et al., 2017). Struvite forms orthorhombic (straight prisms with a rectangular base) crystals in the ratio of 1:1:1 among the magnesium, ammonium, and phosphate (Eq. 1).

$$NH_4{}^{+} + Mg^{2+} + HPO_4{}^{-2} + 6H_2O \rightarrow NH_4MgPO_4 + 6H_2O \qquad (1)$$

Apart from the chemical synthesis, waste/wastewater rich in phosphorus and nitrogen could be used as a resource for struvite synthesis. Press mud/wastewater contains a good amount of nitrates and phosphates, the recovery of struvite from which offers sustainability to the current sugarcane industries. However, the recovery of struvite from press mud offers the recovery of an agriculturally important fertilizer (P) apart from waste remediation. Additionally, struvite does not burn the roots due to its slow-releasing characteristics, which is common with traditional ammonium-phosphate fertilizers (Yan et al., 2018). Conventional waste treatment technologies limit the P removal and the recovery of P in the form of struvite offers sustainability to the current wastewater treatment units and accounts for economical benefits.

4 Limitations and future prospective

Increasing sugar consumption demands greater sugarcane cultivation. Simultaneously, higher sugarcane cultivation is increasing the load on the cane-processing industries, thus leading to higher waste discharge. In this direction, the waste originating from sugar industries needs to be addressed with sustainable processes to avoid further environmental pollution. The presence of fiber, wax, and proteins makes sugarcane bagasse the most recalcitrant in nature, which limits its degradability and conversion efficiency. Hence, it cannot be easily separated into readily utilizable components and the available pretreatment techniques are the most expensive and least technologically mature steps available to convert biomass to fermentable sugars. Liquid residuals from pretreatment processes are rich in pentoses, soluble and insoluble lignin, etc. This offers a great potential for improvement in efficiency and the reduction of costs through future research and development. Currently, sugarcane molasses is widely used as a resource for bioethanol production and this leaves behind other secondary materials such as pentoses, vinasse, and effluent. The best use for the pentoses in the pretreatment liquor is ethanol production. However, the current processes using existing microorganisms lead to extremely low yields as a waste. In this context, utilization of sugarcane waste originating from different processes could be considered a renewable resource for the synthesis of various products such as MCC, NCC, xylitol, ethanol, hydrogen, methane, hythane, struvite, etc. This multiproduct approach using sugarcane waste as a resource in the biorefinery framework facilitates sustainability to the existing sugarcane industries.

References

Abrigo, C.S., Gerpacio, A.L., 1986. No title. Res. Los Banos 6 (4), 25–26.

Abubaker, H.O., Sulieman, A.M.E., Elamin, H.B., 2012. Utilization of *Schizosaccharomyces* pombe for production of ethanol from cane molasses. J. Microbiol. Res. 2, 36–40.

Ansari, K.B., Gaikar, V.G., 2014. Pressmud as an alternate resource for hydrocarbons and chemicals by thermal pyrolysis. Ind. Eng. Chem. Res. 53 (5), 1878–1889.

Baskar, M., Kayalvizhi, C., Bose, M.S.C., 2003. Eco-friendly utilisation of distillery effluent in agriculture—a review. Agri. Rev. Agri. Res. Commun. Centre India 24 (1), 16–30.

Behera, S., Arora, R., Nandhagopal, N., Kumar, S., 2014. Importance of chemical pretreatment for bioconversion of lignocellulosic biomass. Renew. Sust. Energ. Rev. 36, 91–106.

Chang, V.S., Burr, B., Holtzapple, M.T., 1997. Lime pretreatment of switchgrass. Appl. Biochem. Biotechnol. 63–65, 3–19.

Cheng, Y.S., Zheng, Y., Yu, C.W., Dooley, T.M., Jenkins, B.M., Gheynst, J.S.V., 2010. Evaluation of high solids alkaline pretreatment of rice straw. Appl. Biochem. Biotechnol. 162, 1768–1784.

España-Gamboa, E., Mijangos-Cortes, J., Barahona-Perez, L., Dominguez-Maldonado, J., Hernández-Zarate, G., Alzate-Gaviria, L., 2011. Vinasse: characterization and treatments. Waste Manage. 29, 1235–1250.

González, L.M.L., Reyes, I.P., Romero, O.R., 2017. Anaerobic co-digestion of sugarcane press mud with vinasse on methane yield. Waste Manage. 68, 139–145.

Gupta, N., Tripathi, S., Balomajumder, C., 2011. Characterization of pressmud: a sugar industry waste. Fuel 90 (1), 389–394.
Hammel, K.E., Kapich, A.N., Jensen Jr., K.A., Ryan, Z.C., 2002. Reactive oxygen species as agents of wood decay by fungi. Enzym. Microb. Technol. 30 (4), 445–453.
Hampannavar, U.S., Shivayogimath, C.B., 2010. Anaerobic treatment of sugar industry wastewater by upflow anaerobic sludge blanket reactor at ambient temperature. Int. J. Environ. Sci. 1 (4), 631.
Harada, H., Uemura, S., Chen, A.C., Jayadevan, J., 1996. Anaerobic treatment of a recalcitrant distillery wastewater by a thermophilic UASB reactor. Bioresour. Technol. 55, 215–221.
Hu, Z., Wen, Z., 2008. Enhancing enzymatic digestibility of switchgrass by microwave-assisted alkali pretreatment. Biochem. Eng. J. 38 (3), 369–378.
Ibrahim, M.M., El-Zawawy, W.K., Abdel-Fattah, Y.R., Soliman, N.A., Agblevor, F.A., 2011. Comparison of alkaline pulping with steam explosion for glucose production from rice straw. Carbohydr. Polym. 83, 720–726.
Janke, L., Leite, A., Nikolausz, M., Schmidt, T., Liebetrau, J., Nelles, M., Stinner, W., 2015. Biogas production from sugarcane waste: assessment on kinetic challenges for process designing. Int. J. Mol. Sci. 16 (9), 20685–20703.
Janke, L., Leite, A., Batista, K., Weinrich, S., Sträuber, H., Nikolausz, M., Nelles, M., Stinner, W., 2016. Optimization of hydrolysis and volatile fatty acids production from sugarcane filter cake: effects of urea supplementation and sodium hydroxide pretreatment. Bioresour. Technol. 199, 235–244.
Karp, S.G., Woiciechowski, A.L., Soccol, V.T., Soccol, C.R., 2013. Pretreatment strategies for delignification of sugarcane bagasse: a review. Braz. Arch. Biol. Technol. 56 (4), 679–689.
Keshavanath, P., Shivanna, Gangadhara, B., 2005. Evaluation of sugarcane by-product pressmud as a manure in carp culture. Bioresour. Technol. 97 (4), 628–634.
Kumar, A.K., Sharma, S., 2017. Recent updates on different methods of pretreatment of lignocellulosic feedstocks: a review. Bioresour. Bioprocess. 4 (1), 7.
Kumar, N.S., Thankamani, V., 2016. Characterization of molasses spentwash collected from United Spirits Ltd., Aleppey, India: a preliminary report. Int. J. Biotechnol. Biochem 12 (2), 103–110.
Kumar, R., Wyman, C.E., 2009. Effects of cellulase and xylanase enzymes on the deconstruction of solids from pretreatment of poplar by leading technologies. Biotechnol. Prog. 25, 302–314.
Kuruti, K., Rao, A.G., Gandu, B., Kiran, G., Mohammad, S., Sailaja, S., Swamy, Y.V., 2015. Generation of bioethanol and VFA through anaerobic acidogenic fermentation route with press mud obtained from sugar mill as a feedstock. Bioresour. Technol. 192, 646–653.
Lin, C.Y., Chang, R.C., 2004. Fermentative hydrogen production at ambient temperature. Int. J. Hydrog. Energy 29 (7), 715–720.
Lucas, M., Hanson, S.K., Wagner, G.L., Kimball, D.B., Rector, K.D., 2012. Evidence for room temperature delignification of wood using hydrogen peroxide and manganese acetate as a catalyst. Bioresour. Technol. 119, 174–180.
Luo, G., Talebnia, F., Karakashev, D., Xie, L., Zhou, Q., Angelidaki, I., 2011. Enhanced bioenergy recovery from rapeseed plant in a biorefinery concept. Bioresour. Technol. 102 (2), 1433–1439.
Margeot, A., Hahn-Hagerdal, B., Edlund, M., Slade, R., Monot, F., 2009. New improvements for lignocellulosic ethanol. Curr. Opin. Biotechnol. 20 (3), 372–380.
McIntosh, S., Vancov, T., 2010. Enhanced enzyme saccharification of *Sorghum bicolor* straw using dilute alkali pretreatment. Bioresour. Technol. 101, 6718–6727.

Moultrop, J.S., Swatloski, R.P., Moyna, G., Rogers, R.D., 2005. High resolution 13-C NMR studies of cellulose and cellulose oligomers in ionic liquid solutions. Chem. Commun, 1557–1559.

Munir, M.T., Li, B., Boiarkina, I., Baroutian, S., Yu, W., Young, B.R., 2017. Phosphate recovery from hydrothermally treated sewage sludge using struvite precipitation. Bioresour. Technol. 239, 171–179. https://doi.org/10.1016/j.biortech.2017.04.129.

Nakamura, Y., Daidai, M., Kobayashi, F., 2004. Ozonolysis mechanism of lignin model compounds and microbial treatment of organic acids produced. Water Sci. Technol. 50 (3), 167–172.

Nasir, N.M., 2006. Better Management Practices for Cotton and Sugarcane. WWF—Pakistan, Pakistan.

Nimbalkar, P.R., Khedkar, M.A., Gaikwad, S.G., Chavan, P.V., Bankar, S.B., 2017. New insight into sugarcane industry waste utilization (press mud) for cleaner biobutanol production by using *C. acetobutylicum* NRRL B-527. Appl. Biochem. Biotechnol. 183 (3), 1008–1025.

Partha, N., Sivasubramanian, V., 2006. Recovery of chemicals from pressmud—a sugar industry waste. Indian Chem. Eng. 48 (3), 160–163.

Paturau, J.M., 1989. By-Products of Sugarcane Industry, third ed. Elsevier, Amsterdam.

Pawar, S.S., Chavan, P., Bankar, S., 2017. Production of bio-ethanol from biomass (press mud). Int. J. Adv. Res. Ideas Innov. Technol. 3 (3), 1550–1555.

Perez, R., 1990. Intensive livestock production systems based on local resources in Cuba. In: Speedy, A.W. (Ed.), Developing World Agriculture. Grosvenor Press International, London, pp. 230–237.

Rabelo, H.T., Bezerra, L.A., Terra, D.F., Lima, R.M., Silva, M.A., Leite, T.K., de Oliveira, R.J., 2011a. Effects of 24 weeks of progressive resistance training on knee extensors peak torque and fat-free mass in older women. J. Strength Cond. Res. 25 (8), 2298–2303.

Rabelo, S.C., Carrere, H., Maciel Filho, R., Costa, A.C., 2011b. Production of bioethanol, methane and heat from sugarcane bagasse in a biorefinery concept. Bioresour. Technol. 102 (17), 7887–7895.

Radjaram, B., Saravanane, R., 2011a. Assessment of optimum dilution ratio for biohydrogen production by anaerobic co-digestion of press mud with sewage and water. Bioresour. Technol. 102 (3), 2773–2780.

Radjaram, B., Saravanane, R., 2011b. Start up study of UASB reactor treating press mud for biohydrogen production. Biomass Bioenergy 35 (7), 2721–2728.

Rahman, M.M., Liu, Y., Kwag, J., Ra, C., 2011. Recovery of struvite from animal wastewater and its nutrient leaching loss in soil. J. Hazarous Mater. 186, 2026–2030.

Rais, M., Sheoran, A., 2015. Treatment of sugarcane industry effluents: science & technology issues. Int. J. Eng. Res. Appl. 5(1).

Rajaguru, A.S.B., Ravindran, V., 1985. Metabolisable energy values for growing chicks of some feedstuffs from Sri Lanka. J. Sci. Food Agric. 36 (1), 1057–1064.

Rath, P., Pradhan, G., Mishra, M.K., 2010. Effect of sugar factory distillery spent wash (dsw) on the growth pattern of sugarcane (*Saccharum officinarum*) crop. J. Phytology 2 (5), 33–39.

Robertiello, A., 1982. Upgrading of agricultural and agroindustrial wastes: the treatment of distillery effluents (vinasses) in Italy. Agri. Wastes 4, 387–395.

Rodriguez, V., Gonzalez, S., 1973. The use of filter cake mud in integral diets for milk production. Cuba. J. Agric. Sci. 7 (1), 29–32.

Rojas, J., Bedoya, M., Ciro, Y., 2015. Current trends in the production of cellulose nanoparticles and nanocomposites for biomedical applications. https://doi.org/10.5772/61334.

Rouf, M.A., Bajpai, P.K., Jotshi, C.K., 2010. Optimization of biogas generation from press mud in batch reactor. Bangladesh J. Scient. Ind. Res. 45 (4), 371–376.

Saha, B.C., Iten, L.B., Cotta, M.A., Wu, Y.V., 2005. Dilute acid pretreatment, enzymatic saccharification and fermentation of wheat straw to ethanol. Process Biochem. 40 (12), 3693–3700.

Saleh-e-In, M.M., Yeasmin, S., Paul, B.K., Ahsan, M., Rahman, M.Z., Roy, S.K., 2012. Chemical studies on press mud: a sugar industries waste in Bangladesh. Sugar Tech. 14 (2), 109–118.

Sánchez, C., 2009. Lignocellulosic residues: biodegradation and bioconversion by fungi. Biotechnol. Adv. 27, 185–194.

Sánchez, E., Borja, R., Lopez, M., 1996. Determination of the kinetic constants of anaerobic digestion of sugar mill-mud-waste (SMMW). Bioresour. Technol. 56, 245–249.

Sangwan, P., Kaushik, C.P., Garg, V.K., 2010. Vermicomposting of sugar industry waste (press mud) mixed with cow dung employing an epigeic earthworm *Eisenia fetida*. Waste Manage. Res. 28 (1), 71–75.

Sassner, P., Galbe, M., Zacchi, G., 2008. Techno-economic evaluation of bioethanol production from three different lignocellulosic materials. Biomass Bioenergy 32 (5), 422–430.

Sathish, S., Vivekanandan, S., 2015. Experimental investigation on biogas production using industrial waste (press mud) to generate renewable energy. Int. J. Innovat. Res. Sci. Eng. 4 (2), 388–392.

Show, P.L., Oladele, K.O., Siew, Q.Y., Aziz Zakry, F.A., Lan, J.C.W., Ling, T.C., 2015. Overview of citric acid production from *Aspergillus niger*. Front Life Sci 8 (3), 271–283.

Sills, D.L., Gossett, J.M., 2011. Assessment of commercial hemicellulases for saccharification of alkaline pretreated perennial biomass. Bioresour. Technol. 102, 1389–1398.

Sun, Y., Cheng, J., 2002. Hydrolysis of lignocellulosic materials for ethanol production: a review. Bioresour. Technol. 83, 1–11.

Sundaranayagi, S., Sirajunnisa, A., Ramasamy, V., 2017. Production of biogas by anaerobic digestion of press mud using iron, cobalt and nickel. Indian J. Scientific Res. 14 (1), 374–377.

Talha, Z., Ding, W., Mehryar, E., Hassan, M., Bi, J., 2016. Alkaline pretreatment of sugarcane bagasse and filter mud codigested to improve biomethane production. Biomed. Res. Int.

Van der Poel, P.W., Schiweck, H., Schwartz, T., 1998. Sugar Technology. Beet and Cane Sugar Manufacture. Verlag Dr. Albert Martens KG, Berlin, p. 1005.

Vats, S., Maurya, D.P., Shaimoon, M., Negi, S., 2013. Development of a microbial consortium for the production of blend enzymes for the hydrolysis of agricultural waste into sugars. Indian J. Scientific Res. 72, 585–590.

Whitmarsh, J., Govindjee, 1995. Singhal, G.S., Renger, G., Sopory, S.K., Irrgang, K.-D., Govindjee, (Eds.), Concepts in Photobiology: Photosynthesis and Photomorphogenesis. Narosa Publishers/Kluwer Academic/Dordrecht, New Delhi, pp. 11–51.

Yan, T., Ye, Y., Ma, H., Zhang, Y., Guo, W., Du, B., Wei, Q., Wei, D., Ngo, H.H., 2018. A critical review on membrane hybrid system for nutrient recovery from wastewater. Chem. Eng. J. 348, 143–156. https://doi.org/10.1016/j.cej.2018.04.166.

Yu, B.G., 1990. Protein feed granules from bagacillo for fish farming. Sugar y Azucar 85 (11), 30–40.

Zavrel, M., Bross, D., Funke, M., Buchs, J., Spiess, A.C., 2009. High-throughput screening for ionic liquids dissolving (ligno-)cellulose. Bioresour. Technol. 100, 2580–2587.

Vermistabilization and detoxification of sugar industry sludges by earthworms

4

Sartaj Ahmad Bhat*,†, Adarsh Pal Vig*
*Department of Botanical and Environmental Sciences, Guru Nanak Dev University, Amritsar, India, †River Basin Research Center, Gifu University, Gifu, Japan

1 Introduction

Sugar production is a large agro-based factory worldwide. India is the second-largest producer after Brazil and contributes about 10%–12% of the world's sugar production. Sugar can be produced by two types of crops: sugarcane and sugar beets. Sugarcane (*Saccharum officinarum* L.) grows in the warm temperate to tropical regions whereas sugar beets (*Beta vulgaris* L.) are generally grown in temperate regions. Globally, the annual sugarcane and sugar beet production is 1600 and 271 million tons, respectively (Chandel et al., 2012). Sugarcane contributes about 75% of the world's sugar production whereas sugar beets account for approximately 25% (Bhatnagar et al., 2016). Sugarcane generates bagasse and pressmud sludge as waste products, whereas sugar beets generate sugar beet pulp and sugar beet mud as wastes (Bhat et al., 2014, 2015a). Most of the sugar mills store the sludge (pressmud and sugar beet mud) in open fields, which degrade the agricultural land and water of that area.

One eco-friendly technology method for the recycling of sugar industrial sludges is vermiremediation using earthworms to get high-quality organic manure from it. Vermicomposting involves the decomposition of organic wastes/sludges through earthworms and microorganisms, allowing the bioconversion of organic wastes/sludges into biomanure (vermicompost) (Bhat et al., 2017a; Huang and Xia, 2018). Vermicompost is one of the most nutrient-rich organic manures in the world and has been shown to have positive impacts on soil fertility and crop productivity (Bhat et al., 2018). Vermicompost contains micro- and macronutrients (nitrogen, potassium, phosphorus, calcium) in plant-available forms (Atiyeh et al., 2001). Earthworms, in association with the microbial community, modify the composition of organic waste through digestion, burrowing, mucus excretion, and casting activities, thus accelerating the process of decomposition (Brown et al., 2000; Domínguez et al., 2010). Vermicomposting is a suitable technology for bioconversion of various industrial sludges such as beverage industry sludge (Singh et al., 2010), distillery sludge (Singh et al., 2014), dyeing sludge (Bhat et al., 2013), milk processing sludge (Suthar et al., 2012; Singh et al., 2017), olive-mill waste (Vivas et al., 2009; Lim and Wu, 2016), paper mill sludge (Kaur et al., 2010), sewage sludge (Hait and Tare, 2011; Liu et al., 2012;

Industrial and Municipal Sludge. https://doi.org/10.1016/B978-0-12-815907-1.00004-0

Molina et al., 2013; Xing et al., 2015), sugar industrial sludge (Sen and Chandra, 2007; Sangwan et al., 2010; Bhat et al., 2014, 2016a, 2017a), and tannery industry sludge (Ravindran et al., 2008, 2014; Vig et al., 2011).

Vermicomposting is also beneficial for industrial sludge as it reduces the toxicity of the sludge (Bhat et al., 2017b). It is an eco-friendly and low-cost technique used for the breakdown of complex chemicals in sludges/wastes to nontoxic and usable forms (Srivastava et al., 2005; Bhat et al., 2016b). In earthworms, the intestinal microbial community and chloragocyte cells have the ability to detoxify the heavy metals in sludge (Jain et al., 2004; Srivastava et al., 2005). Various earthworm species such as *Eisenia fetida*, *Lampito mauritii*, and *Allolobophora rosea* are known to bioaccummulate toxic heavy metals in their intestines by inducing cysteine-rich metal binding metallothioneins (Sturzenbaum et al., 2012; Goswami et al., 2016). A metallothionein is a protein that regulates the bioavailability-detoxification dynamics of heavy metals in the earthworm gut (Maity et al., 2009). The metals bioaccumulate in chloragogen cells of the earthworm intestine, which floats in coelomic fluid (Vijver et al., 2003). Hence, the final vermicompost is not only a good source of available nutrients but is also capable of remediating toxic metal contamination. Plant bioassays (*Allium cepa* L., *Vicia faba* L.) are sensitive and cost effective, used for the evaluation of the toxicity of industrial sludge (Feng et al., 2007; Bhat et al., 2017b). These bioassays are generally used for the evaluation of environmental pollution. Many researchers have used plant bioassay tests for monitoring the toxicity of industrial sludges/leachates such as electronic waste leachate (Bakare et al., 2012), flyash (Jain et al., 2004), fresh and stabilized agroindustrial organic sludge (Chiochetta et al., 2014), metal and dye leachate (Chandra et al., 2005), municipal sludge (Srivastava et al., 2005), pressmud sludge (Bhat et al., 2014), sewage sludge (Mazzeo et al., 2015), sugar beet mud (Bhat et al., 2015a), and tannery waste leachate (Chandra et al., 2004).

Previous studies have focused on the bioconversion of sugar industrial sludges, but few have monitored the toxicity reduction of sugar industrial sludges during the vermiconversion process. The purpose of this chapter is to first discuss how sugar industrial sludge (pressmud and sugar beet mud) is generated and then summarize the chemical composition of these sludges. Then we review the vermicomposting and detoxification of these sludges by earthworms. Vermiconversion and toxicity reduction of sugar industrial sludges by earthworms will provide a sustainable alternative to the growing problem of industrial waste management.

2 Characterization of sugar industrial sludge

2.1 Generation of pressmud sludge and sugar beet mud

The sugar industry is a major agro-based mill worldwide (Contreras et al., 2009). During the production process, it generates sugar as well as various sludges/wastes. The byproducts of sugarcane and sugarbeets that are generated during the industrial processing include sugarcane bagasse, pressmud, molasses, sugar beet mud, and pulp. Generation of sugar industrial wastes/sludges is presented in Fig. 1.

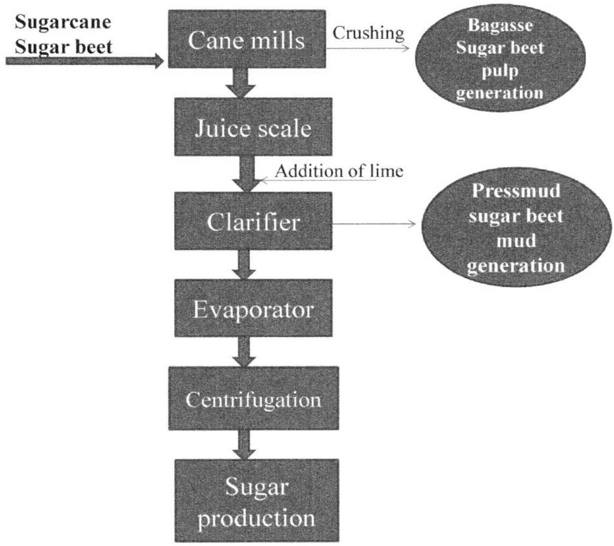

Fig. 1 Schematic representation of sugar industrial waste/sludge generation in the sugar industry.

Sugar industrial sludges are solid waste byproducts formed after the purification of sugar by the carbonation or sulfitation process. The process separates the clear juice on top and the sludge (mud) at the bottom. About three tons of pressmud are generated from 100 tons of sugarcane processed (Solomon, 2011), whereas 0.03 ton of sugar beet mud is generated per ton of sugar beet processed.

2.2 Chemical compositions of pressmud sludge and sugar beet mud

Pressmud sludge is a soft, spongy, amorphous, dark brown to black colored material, as depicted in Fig. 2. It generally contains moisture (50%–60%), fiber (20%–30%), sugar (5%–12%), coagulated colloids, wax (7%–15%), albuminoids, and organic salts (Dotaniya et al., 2016). It also contains significant amounts of iron, manganese, calcium, magnesium, silicon, and phosphorus (Yadav and Solomon, 2006; Rouf et al., 2010). Sugar beet mud is compact and grey in color, as depicted in Fig. 3. The final vermicompost of sugar beet mud is dark grey and black in color. The initial physicochemical parameters of pressmud sludge and sugar beet mud are presented in Table 1. The compositions of various element oxides in the weight percentage (Wt.%) of pressmud sludge and sugar beet mud were reported by Afshar et al. (2014). Through XRF analysis, SiO_2, Al_2O_3, Fe_2O_3, CaO, Na_2O, K_2O, and MgO were detected in pressmud sludge and sugar beet mud. The sugar industrial sludges contain up to 70% mineral matter, a large part of it being calcium oxide. These sludges have a highly variable composition and depend on the sugar beet or cane variety, temperature, soil conditions, available nutrients, the nature of precipitation, and the process of clarification.

Fig. 2 Pressmud sludge.

Fig. 3 Sugar beet mud.

3 Vermicomposting of sugar industrial sludges

Sugar industrial sludges are a good source of fertilizer. The usefulness of pressmud sludge and sugar beet mud as a fertilizer is based on their nutrient content. Vermicomposting could be a cost-effective and eco-friendly technique for converting sugar industrial sludges into high-grade manure. After vermicomposting, the manure produced by earthworms is very useful for agricultural crops because of its richness in available nutrients (nitrogen, phosphorus, potassium, calcium). Earthworms ingest,

Table 1 Initial physicochemical properties of pressmud sludge and sugar beet mud (Mean ± SE)

Physicochemical parameters	Pressmud sludge	Sugar beet mud
pH	6.93 ± 0.07	8.58 ± 0.05
EC (mS/cm)	2.50 ± 0.15	2.36 ± 0.05
TKN (%)	3.33 ± 0.82	0.34 ± 0.02
TOC (%)	19.74 ± 0.18	13.10 ± 0.81
C/N ratio	5.92 ± 0.12	38.52 ± 3.71
TAP (%)	3.02 ± 0.98	5.87 ± 0.24
TK (%)	1.65 ± 0.28	3.05 ± 0.03
TNa (%)	1.16 ± 0.16	4.03 ± 0.22
Cu (mg/kg)	10.25 ± 0.1	59.67 ± 2.02
Fe (mg/kg)	849 ± 2.0	619 ± 15.31
Zn (mg/kg)	21.25 ± 1.0	29.25 ± 1.47

EC, electrical conductivity; *TKN*, total Kjeldhal nitrogen; *TOC*, total organic carbon, *C/N ration*, carbon to nitrogen ratio; *TAP*, total available phosphorus, *TK*, total potassium, *TNa*, total sodium; *Cu*, copper; *Fe*, iron, *Zn*, zinc.
Sources: Bhat, S.A., Singh, J., Vig, A.P., 2014. Genotoxic assessment and optimization of pressmud with the help of exotic earthworm Eisenia fetida. Environ. Sci. Pollut. Res. 21, 8112–8123; Bhat, S.A., Singh, J., Vig, A.P., 2015a. Vermistabilization of sugar beet (Beta vulgaris L) waste produced from sugar factory using earthworm Eisenia fetida: Genotoxic assessment by Allium cepa test. Environ. Sci. Pollut. Res. 22, 11236–11254.

break, and digest organic sludges/wastes and convert them into finer, humified, microbially active material through the activities of both earthworms and microbes (Khwairakpam and Bhargava, 2009; Huang and Xia, 2018). The major advantage of vermicomposting lies in the conversion of sludge into stabilized forms and the detoxification of sludge through the accumulation of toxic/heavy metals in the body tissues of earthworms (Bhat et al., 2014; Goswami et al., 2014). Globally, many researchers have recycled sugar industrial sludges for large-scale vermiconversion by earthworms (Reddy and Shantaram, 2005; Parthasarathi, 2006, 2007; Sen and Chandra, 2007; Sangwan et al., 2008a,b, 2010; Khwairakpam and Bhargava, 2009; Prakash and Karmegam, 2010; Honarvar et al., 2011; Niyazi and Chaurasia, 2014; Bhat et al., 2014, 2015a, 2016a; Shah et al., 2015).

Reddy and Shantaram (2005) vermicomposted sugar mill sludge (pressmud) using the earthworm *Eisenia fetida*. The results revealed that the final vermicompost has more nitrogen content (2.0-fold increase) than traditional composting (without earthworms), which suggested that vermicomposting is an efficient technology for the biodegradation of sugar industrial sludges. The bioconversion of pressmud sludge by the earthworms *Lampito mauritii* and *Eudrilus eugeniae* has been studied by Parthasarathi (2006). The results showed that the final vermicast obtained from pressmud sludge showed a higher microbial population and increased mineralization than the initial pressmud. Sen and Chandra (2007) reported a rapid decrease in the C:N ratio during vermiconversion of pressmud sludge. This C:N ratio decrease in the final vermicompost indicates compost maturity and may be due to the microbial respiration and decomposition of organic matter. Nitrogen content was also higher in the final vermicompost. This indicates that the application of pressmud vermicast to the soil will enrich the soil and also improve the soil fertility of the agricultural land.

Vermiconversion of pressmud sludge spiked with horse dung substrate using earthworm *E. fetida* has been studied by Sangwan et al. (2008a). The study observed that the earthworm growth and cocoon production were a maximum in 10% of pressmud sludge. However, increasing the pressmud sludge ratio affected the growth and reproduction of *E. fetida*. The increase in earthworm growth and cocoon production may be due to the uptake of nutrients available in the pressmud sludge and horse dung mixture. The nutrient parameters such as nitrogen, phosphorus, and potassium were increased in the final vermicompost as compared to the initial pressmud sludge. In the final vermicompost, the increase in available nutrients may be due to the earthworm-related functions such as the addition of mucus, hormones, enzymes, nitrogenous excretory substances, and microbe-mediated transformations (Sangwan et al., 2010). The study concluded that pressmud sludge can be recycled into organic manure, if spiked at a maximum 50% with horse dung mixture. Sangwan et al. (2008b) also investigated the vermiconversion of pressmud sludge mixed with biogas plant slurry by the earthworm *E. fetida*. The best growth and reproduction of earthworms were reported in the 20% pressmud sludge + 80% biogas plant slurry ratio. The final vermicompost of pressmud sludge showed a reduction in pH, total organic carbon, carbon to nitrogen ratio, and total potassium, whereas there was an increase in the total nitrogen and total phosphorus content. The study concluded that pressmud sludge can be managed into organic manure if spiked with 30%–50% of biogas plant slurry.

Khwairakpam and Bhargava (2009) reported vermiconversion of pressmud sludge by using the earthworms *E. fetida, E. eugeniae, and Perionyx excavatus* in mono and polycultures. The final vermicompost of pressmud sludge showed a reduction in pH, total organic carbon, and the carbon to nitrogen ratio whereas there was increase in the nitrogen, phosphorus, potassium, sodium, and calcium content. The authors concluded that the maximum growth and reproduction, available nutrients, and decomposition of pressmud sludge was reported in polycultures as compared to monoculture treatments. Vermiconversion of pressmud sludge mixed with cattle dung mixture by using the earthworm *P ceylanensis* has been studied by Prakash and Karmegam (2010). In the final vermicompost of pressmud sludge, the authors observed a significant reduction in pH, total organic carbon, and carbon to nitrogen ratio, whereas there was an increase in nitrogen, phosphorus, and potassium content over compost without earthworms (traditional compost). The study further observed an increase in the populations of bacteria, fungi, and actinomycetes in the final vermicompost of pressmud sludge. The authors concluded that by using *P. ceylanensis,* the pressmud sludge can be managed into nutrient- and microbe-rich biomanure if spiked in a 1:1 ratio with cow dung mixture. Both earthworms and microorganisms use a portion of carbon as energy that brings about the degradation of organic carbon (Venkatesh and Eevera, 2008). Sangwan et al. (2010) also investigated the vermiconversion of pressmud sludge mixed with cow dung mixture using the earthworm *E. fetida*. In the final vermicompost of pressmud sludge, the authors observed a reduction in the C:N ratio and total organic carbon, whereas there was an increase in the nutrient content of nitrogen, phosphorus, and calcium. They concluded that pressmud sludge will be converted into high-quality manure if spiked up to 50% with cow dung.

Vermiconversion of sugar beet mud from sugar beets has been studied by Honarvar et al. (2011). The authors observed that the final vermicompost was superior (in terms of available nutrients and chemical characteristics) to the traditional compost produced from sugar beet mud. The final vermicompost of sugar beet mud showed an increase in the nutrient content of phosphorus, zinc, and nitrogen, whereas there was a reduction in saltiness by 50%. The authors concluded that the vermiconversion of sugar beet mud produced quality manure with enriched nutrients as compared to traditional composting. Vermiconversion of pressmud sludge mixed with flyash using the earthworm *E. fetida* has been investigated by Niyazi and Chaurasia (2014). The authors observed a reduction in pH, temperature, total organic carbon, carbon to nitrogen ratio, and fecal coliforms, whereas there was an increase in n the utrient content of nitrogen, phosphorus, and potassium in the final vermicompost. Heavy metals in the final vermicompost also saw a reduction of 35%–50%. The study showed that the flyash and pressmud sludge can be recycled into organic manure if mixed in the ratio of 40%–60%.

Shah et al. (2015) performed composting and vermicomposting of sugar industrial wastes mixed with farmyard manure using the local earthworm *Lumbricus rubellus*. The vermiconversion of sugar industrial wastes produced good-quality manure with enriched nutrient content as compared to the composting process. The authors concluded that the use of local earthworm species for vermicomposting should be recommended to the farming community. Bhat et al. (2015a) reported the vermiconversion of sugar beet mud spiked with cattle dung using the earthworm *E. fetida*. The authors observed a maximum earthworm growth in a 20% ratio of sugar beet mud. The final vermicompost obtained from sugar beet mud showed an increase in the nutrient content of nitrogen, phosphorus, and sodium, whereas the organic carbon and C:N ratio declined. The C:N ratio was less than 20, which indicates an advanced degree of stabilization and maturity of sugar beet mud vermicompost. Bhat et al. (2016a) also assessed the growth of the earthworm *Eisenia fetida* during vermiconversion of pressmud sludge spiked with cattle dung in 135 days of experiments. The best growth and reproduction of *E. fetida* was observed in 25% of pressmud sludge. Vermiconversion of pressmud sludge showed an increase in total kjeldahl nitrogen, total phosphorus, and total sodium, whereas there was a decrease in the content of total organic carbon, carbon to nitrogen ratio, total potassium, and heavy metals. The sugar industrial sludges spiked with nutrient-rich substrates (cattle dung) caused significant growth and reproduction of earthworms as well as significantly increased the available nutrients in the final vermicompost. Thus, earthworms are playing a major role in the management of sugar industrial wastes/sludges. Modifications in sugar industrial waste characteristics after bioconversion with different species of earthworms have been summarized in Table 2.

4 Detoxification of sugar industrial sludges

A plant bioassay (*Allium cepa*, *Hordeum vulgare*, *Lactuca sativa*, *Tradescantia* sp., *Vicia faba*) is a sensitive and low-cost method used to study the toxicity of industrial sludges and effluents (Grant, 1994; Souza et al., 2016). The US Environmental

Table 2 Vermicomposting of sugar industrial wastes

Sr. No	Sugar industrial waste	Amendments used	Earthworm species used	Vermicomposting time period (days)	Nutrient status	References
1	Pressmud	Cattle dung	*Eisenia fetida*	135	Nutrients such as total kjeldahl nitrogen (TKN), total phosphorus (TP), total sodium (TNa) increased from initial pressmud mixtures to final vermicompost, while total organic carbon (TOC), carbon to nitrogen ration (C:N ratio) and total potassium (TP) showed an opposite trend. Heavy metals such as copper (Cu), chromium (Cr) and iron (Fe) decreased from initial except zinc (Zn) and manganese (Mn), which increased in the final vermicompost.	Bhat et al. (2016a)
2	Sugar beet mud, sugar beet pulp	Cattle dung	*Eisenia fetida*	135	Vermicomposting in the sugar beet mud and pulp resulted in increase in the nutrient content of TKN, TP and TNa, while TOC, C:N ratio, and EC declined in all the final feed mixtures of sugar beet mud and pulp.	Bhat et al. (2015a)

3	Bagasse	Cattle dung	*Eisenia fetida*	135	The nutrient content increased in the final vermicompost of bagasse waste such as TKN (230%), TAP (88%), and TNa (134%) while as TOC (35%) and C:N ratio (73%) decreased in all the final mixtures of bagasse. The authors indicated that the vermiconversion could be introduced as an excellent technology to recycle bagasse waste into high quality biomanure.	Bhat et al. (2015b)
4	Sugarcane baggase, pressmud	Farm manure	*Lumbricus rubellus*	45	Vermicomposting of sugar industry wastes (bagasse, pressmud) mixed with farm manure produced good quality biomanure with enriched nutrients such as nitrogen (3%), phosphorus (2%), potassium (7%), calcium (3.5%), and sodium (2.5%).	Shah et al. (2015)
5	Pressmud	Cow dung	*Eisenia fetida*	120	Vermicomposting resulted in the increase of TKN, TAP, TNa, EC, and pH, while TOC C/N ratio and TK declined in all the final feed mixtures of pressmud.	Bhat et al. (2014)

Continued

Table 2 Continued

Sr. No	Sugar industrial waste	Amendments used	Earthworm species used	Vermicomposting time period (days)	Nutrient status	References
6 7	Pressmud, bagasse and sugarcane trash	Cow dung and jeevanrutham (an effective microbial suspension)	*Eisenia fetida*	60	The final vermicompost was rich in TKN, TP, K, Na, Ca, and Mg while pH, EC, TOC, and C/N ratio decreased from pre to post mixtures of vermicomposting.	Pandit and Maheshwaria (2014)
8	Pressmud and bagasse	Jeevamrutham	*Eisenia fetida*	42	Vermicomposting of sugar industrial wastes mixed with Jeevamrutham resulted in increase of nitrogen, phosphorus, potassium, calcium, magnesium, sodium, and micronutrients (Zn, Mn, Fe, Cu, Al). The authors concluded that the pretreatment with jeevamrutham at 2:1 ratio converts sugar industrial wastes into a nutrient rich biomanure.	Pandit and Maheshwaria, 2012
9	Pressmud	Leaf litters of *Polyalthia longifolia*, sugarcane trash	*Perionyx ceylanensis*	100	Vermicomposting of pressmud mixed with other substrates produced good quality biomanure with enriched nutrients such as nitrogen (35%), phosphorus	Karmegam et al. (2012)

					(22%), potassium (50%), calcium (33%), and sodium (28%). C/N ration (−109%) as compared to initial waste was decreased after vermicomposting.	
10	Pressmud	Cow dung	*Perionyx ceylanensis*	100	The nutrient content increased in the final vermicompost of pressmud such as nitrogen (36.94%), phosphorus (28.56%), potassium (20.82%), and calcium (34.92%), whereas organic carbon (−50.52%), C/N ratio (−94.41%), and C/P ratio (−84.84%) showed an opposite trend. The authors concluded that the combination of pressmud with cow dung (1:1 ratio) is suitable to produce a good-quality biofertilizer within 60 days with earthworm *P. ceylanensis*.	Prakash and Karmegam (2010)
11	Bagasse, pressmud and sugarcane trash	Inoculated with combination of *Pleurotus sajorcaju*, *Trichoderma viridae*, *Aspergillus niger* and *Pseudomonas striatum*	*Drawida willsi.*	40	The authors first predecomposed the sugar mill wastes (bagasse, pressmud and trash) with microbes for 30 days. The results indicated that both processes (microbe predecomposting and	Kumar et al. (2010)

Continued

Table 2 Continued

Sr. No	Sugar industrial waste	Amendments used	Earthworm species used	Vermicomposting time period (days)	Nutrient status	References
					vermicomposting) resulted in increase in NPK content as well as reduction in TOC and C/N ratio.	
12	Pressmud	Cow dung	*Eisenia fetida*	91	Vermicomposting of pressmud mixed with cow dung resulted in the increase of nitrogen, phosphorus, and calcium as well as a reduction in C:N ratio and total organic carbon. The authors suggested that if pressmud is mixed with cow dung (50%), it will be recycled into good-quality biomanure.	Sangwan et al. (2010)
13	Pressmud	-	*Eudrilus eugeniae*, *Lampito mauritii* and *Perionyx ceylanensis*	30	The results indicated that the vermicompost of *E. eugeniae*, *L. mauritii* and *P. ceylanensis* showed higher values for total nitrogen (40.80%), phosphorus (25.30%), and potassium (30.53%) than press mud without	Jayakumar et al. (2009)

					earthworms. The organic carbon (−44.30%), C/N ratio (−143.73), and C/P ratio (−93.18) in pressmud vermicompost were lower than in the pressmud without earthworms.	
14	Pressmud	Saw dust	*Eisenia fetida*, *Eudrilus eugeniae* and *Perionyx excavates*	45	Vermicomposting resulted in increase of nitrogen, phosphorus, potassium, calcium, and sodium, whereas there was reduction in pH, total organic carbon, and carbon to nitrogen ratio. The results further revealed that the maximum nutrient content and bioconversion of pressmud was observed in polyculture reactors as compared to monoculture reactors.	Khwairakpam and Bhargava (2009)
15	Pressmud	Horse dung	*Eisenia fetida*	98	Vermicomposting of pressmud resulted in the increase of nutrient content (nitrogen, phosphorus, potassium) as compared to initial presssmud. The authors concluded that pressmud can be converted into a good-quality	Sangwan et al. (2008a)

Continued

Table 2 Continued

Sr. No	Sugar industrial waste	Amendments used	Earthworm species used	Vermicomposting time period (days)	Nutrient status	References
					biomanure, if mixed 50% with horse dung.	
16	Pressmud	Biogas plant slurry	*Eisenia fetida*	91	Vermicomposting of pressmud with biogas plant slurry resulted in reduction of increase of total nitrogen and total phosphorus, whereas there was reduction of pH, total organic carbon, carbon to nitrogen ratio, and total potassium in the final vermicompost.	Sangwan et al. (2008b)
17	Pressmud, bagasse and cane trash	Cow dung	*Eudrilus eugeniae*	60	The authors reported that the vermicomposting increased the nitrogen content. There was a rapid decrease in lignocellulosic content and C:N ratio of final vermicompost. This decrease indicates decomposition of sugar industrial wastes and compost maturity.	Sen and Chandra (2007)

18	Pressmud	-	*Lampito mauritii* and *Eudrilus eugeniae*	30	Vermicompost produced from fresh pressmud showed higher microbiota population and increased mineralization than initial pressmud. The final vermicompost also supported soil and improved soil fertility.	Parthasarathi (2006)
19	Pressmud, bagasse and cane trash	-	*Eisenia fetida*	90	The results revealed that the vermicompost produced from sugar industrial wastes has two-fold more nitrogen content than traditional composting and suggested vermicomposting for efficient utilization of sugar industrial wastes.	Reddy and Shantaram (2005)

Protection Agency (USEPA) and the World Health Organization (WHO) have validated plant bioassay tests as efficient models to monitor the genotoxicity of industrial sludges/effluents (Souza et al., 2016). The main advantages of using plant bioassays as indicators of environmental pollution are that it is easy to handle, has low cost, direct exposure on test samples (sludge/effluent), and efficient response to chemical substances (Fiskesjo, 1985; Rank and Nielsen, 1993; Geras'kin et al., 2011). The *Allium cepa* root chromosomal aberration assay is one of the most-used bioassays in environmental monitoring and assessment because it allows the detection of different endpoint such as the root growth test, related to toxicity; the mitotic index, related to cytotoxicity; chromosomal aberrations, related to genotoxicity; and micronuclei, related to mutagenicity (Maluszynska and Juchimiuk, 2005; Leme and Marin-Morales, 2009). A schematic representation of an *A. cepa* root chromosomal aberration assay is presented in Fig. 4. The vermiremediation potential of earthworms and their ability in the detoxification of toxic chemicals in sludges are because of their strong metabolic system and the action of earthworm gut microorganisms and chloragocyte cells (Srivastava et al., 2005). Bhat et al. (2014) used *A. cepa* to access the genotoxicity reduction of pressmud sludge by the vermicomposting process. An aqueous extract of the pressmud sludge was prepared and applied in the experiments. Root growth test results showed that the initial extracts of pressmud sludge suppressed root growth, whereas after vermicomposting, root length was increased in all the extracts of pressmud sludge. The mitotic index was also increased in the final vermicomposted extracts of pressmud sludge. Roots of *A. cepa* exposed to the initial extracts of pressmud sludge showed significant frequencies of chromosomal aberrations (30.8%) while postvermicomposted pressmud sludge extracts showed lower chromosomal aberrations (20.3%). The authors concluded that vermicomposting was effective to detoxify the pressmud sludge as well as the *A. cepa* was efficient to assess the entire process. Bhat et al. (2015a) also applied the vermicomposting

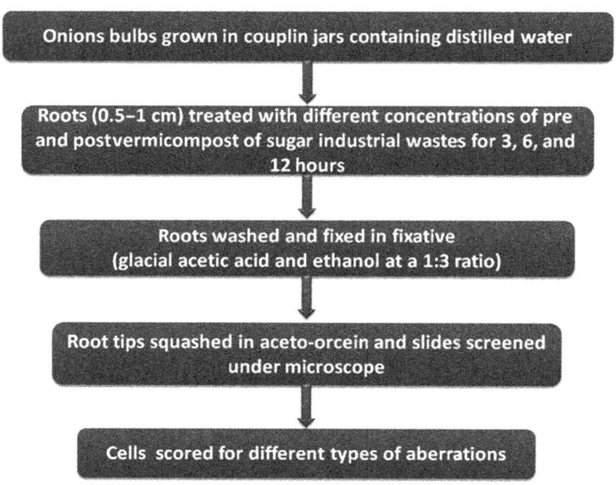

Fig. 4 Schematic representation of *Allium cepa* root chromosomal aberration assay.

process for sugar beet mud waste to know the reduction of toxicity after bioconversion. The vermiremediation efficiency was verified by an *A. cepa* bioassay. These authors observed a decrease of root length and mitotic index parameters when *A. cepa* bulbs were exposed to the extracts of sugar beet mud prior to vermicomposting, whereas the root length and mitotic index were increased after vermicomposting of sugar beet mud. Roots of *A. cepa* exposed to initial extracts of sugar beet mud showed higher chromosomal aberrations (17.7%) while post-vermicomposted extracts showed lower chromosomal aberrations (4.4%). These authors related the detoxification of sugar beet mud with possible biodegradation by earthworms. Recently, Anacleto et al. (2017) studied the toxicological effects of sugarcane filtercake sludge using *A. cepa* bioassay after three and six months of composting. The physicochemical parameters of filtercake sludge showed a decrease in Cd, Co, Cr, Ni, Pb, Zn, Cu, K, P, Mg, Na, TOC, and N after six months of degradation. The initial samples of sugarcane filtercake sludge inhibited the mitotic index and induced chromosomal aberrations, whereas the genotoxic and mutagenic effects of filtercake sludge were decreased after six months of natural attenuation. The authors revealed that the microbiota present in the filtercake sludge may have eliminated most of the toxicity. The *A. cepa* bioassay proved its usefulness to evaluate the detoxification of industrial wastes/sludges. Vermiremediation and toxicity reduction of sugar industrial sludges by earthworms will provide a sustainable alternative to the growing problem of industrial solid waste management.

5 Conclusions

Vermicomposting is a process based on earthworms and microorganisms, whose joint action provides degradation and detoxification of organic waste as well as conversion into a product to be used for agronomic purposes. This eco-friendly method is cost effective and is the best among other remediation processes. In the earthworm gut, enzymatic activities lead to toxic metal immobilization, which suggests that vermitechnology is as efficient process for the remediation of heavy metals from industrial organic wastes/sludges. It can be concluded that vermiremediation potentially converts sugar industrial sludges to nutrient-rich organic manure for agricultural applications with reduced toxicity.

Acknowledgments

The authors acknowledge University Grants Commission (UGC) New Delhi for financial support under the University with Potential for Excellence (UPE) and Basic Scientific Research (BSR) schemes.

References

Afshar, P.G., Honarvar, M., Gharachorloo, M., Eshratabadi, P., Bazyar, B., 2014. Investigation of the physico-chemical properties of press mud: a sugar industry waste. Adv. Environ. Biol. 8, 1053–1058.

Anacleto, L.R., Roberto, M.M., Marin-Morales, M.A., 2017. Toxicological effects of the waste of the sugarcane industry, used as agricultural fertilizer, on the test system *Allium cepa*. Chemosphere 173, 31–42.

Atiyeh, R.M., Arancon, N., Edwards, C.A., Metzger, J.D., 2001. The influence of earthworm processed pig manure on the growth and productivity of marigolds. Bioresour. Technol. 81, 103–108.

Bakare, A.A., Adeyemi, A.O., Adeyemi, A., Alabi, O.A., Osibanjo, O., 2012. Cytogenotoxic effects of electronic waste leachate in *Allium cepa*. Caryologia 65, 94–100.

Bhat, S.A., Singh, J., Vig, A.P., 2013. Vermiremediation of dyeing sludge from textile mill with the help of exotic earthworm *Eisenia fetida* Savigny. Environ. Sci. Pollut. Res. 20, 5975–5982.

Bhat, S.A., Singh, J., Vig, A.P., 2014. Genotoxic assessment and optimization of pressmud with the help of exotic earthworm *Eisenia fetida*. Environ. Sci. Pollut. Res. 21, 8112–8123.

Bhat, S.A., Singh, J., Vig, A.P., 2015a. Vermistabilization of sugar beet (*Beta vulgaris* L) waste produced from sugar factory using earthworm *Eisenia fetida*: Genotoxic assessment by *Allium cepa* test. Environ. Sci. Pollut. Res. 22, 11236–11254.

Bhat, S.A., Singh, J., Vig, A.P., 2015b. Potential utilization of bagasse as feed material for earthworm *Eisenia fetida* and production of vermicompost. Springerplus 4, 11.

Bhat, S.A., Singh, J., Vig, A.P., 2016a. Effect on growth of earthworm and chemical parameters during vermicomposting of pressmud sludge mixed with cattle dung mixture. Procedia Environ. Sci. 35, 425–434.

Bhat, S.A., Singh, J., Vig, A.P., 2016b. Genotoxicity reduction in bagasse waste of sugar industry by earthworm technology. Springerplus 5, 1186.

Bhat, S.A., Singh, J., Vig, A.P., 2017a. Amelioration and degradation of pressmud and bagasse wastes using vermitechnology. Bioresour. Technol. 243, 1097–1104.

Bhat, S.A., Singh, J., Singh, K., Vig, A.P., 2017b. Genotoxicity monitoring of industrial wastes using plant bioassays and management through vermitechnology: A review. Agri. Nat. Res. 51, 325–337.

Bhat, S.A., Singh, S., Singh, J., Kumar, S., Bhawana, Vig, A.P., 2018. Bioremediation and detoxification of industrial wastes by earthworms: vermicompost as powerful crop nutrient in sustainable agriculture. Bioresour. Technol. 252, 172–179.

Bhatnagar, A., Kesari, K.K., Shurpali, N., 2016. Multidisciplinary approaches to handling wastes in sugar industries. Water Air Soil Pollut. 227, 11.

Brown, G.G., Barois, I., Lavelle, P., 2000. Regulation of soil organic matter dynamics and microbial activity in the drilosphere and the role of interactions with other edaphic functional domains. Eur. J. Soil Biol. 36, 177–198.

Chandel, A.K., da Silva, S.S., Carvalho, W., Singh, O.V., 2012. Sugarcane bagasse and leaves: foreseeable biomass of biofuel and bio-products. J. Chem. Technol. Biotechnol. 87, 11–20.

Chandra, S., Chauhan, L.K.S., Pande, P.N., Gupta, S.K., 2004. Cytogenetic effects of leachates from tannery solid waste on the somatic cells of *Vicia faba*. Environ. Toxicol. 19, 129–133.

Chandra, S.C., Chauhan, L.K.S., Murthy, R.C., Saxena, A.N., Pande, P.N., Gupta, S.K., 2005. Comparative biomonitoring of leachates from hazardous solid waste of two industries using *Allium test*. Sci. Total Environ. 347, 46–52.

Chiochetta, C.G., Goetten, L.C., Almeida, S.M., Quaranta, G., Cotelle, S., Radetski, C.M., 2014. Leachates from solid wastes: chemical and eco(geno)toxicological differences between leachates obtained from fresh and stabilized industrial organic sludge. Environ. Sci. Pollut. Res. 21, 1090–1098.

Contreras, A.M., Rosa, E., Perez, M., Van Langenhove, H., Dewulf, J., 2009. Comparative life cycle assessment of four alternatives for using by-products of cane sugar production. J. Clean. Prod. 17, 772–779.

Domínguez, J., Aira, M., Gómez-Brandón, M., 2010. Vermicomposting: earthworms enhance the work of microbes. In: Insam, H., Franke-Whittle, I., Goberna, M. (Eds.), Microbes at Work: From Wastes to Resources. Springer, Berlin, Heidelberg, pp. 93–114.

Dotaniya, M.L., Datta, S.C., Biswas, D.R., Dotaniya, C.K., Meena, B.L., Rajendiran, S., Regar, K.L., Lata, M., 2016. Use of sugarcane industrial by-products for improving sugarcane productivity and soil health. Int. J. Recycling Organic Waste Agri. 5, 185–194.

Feng, S., Wang, X., Wei, G., Peng, P., Yang, Y., Cao, Z., 2007. Leachates of municipal solid waste incineration bottom ash from Macao: heavy metal concentrations and genotoxicity. Chemosphere 67, 1133–1137.

Fiskesjo, G., 1985. The *Allium* test as a standard for environmental monitoring. Hereditas 102, 99–112.

Geras'kin, S., Evseeva, T., Oudalova, A., 2011. Plants as a tool for the environmental health assessment. In: Nriagu, J.O. (Ed.), Encyclopedia of Environmental Health. Elsevier Science, San Diego, CA, pp. 571–579.

Goswami, L., Sarkar, S., Mukherjee, S., Das, S., Barman, S., Raul, P., Bhattacharyya, P., Mandal, N.C., Bhattacharya, S., Bhattacharya, S.S., 2014. Vermicomposting of tea factory coal ash: metal accumulation and metallothionein response in *Eisenia fetida* (Savigny) and *Lampito mauritii* (Kinberg). Bioresour. Technol. 166, 96–102.

Goswami, L., Pratihar, S., Dasgupta, S., Bhattacharyya, P., Mudoi, P., Bora, J., Bhattacharya, S.S., Kim, K.H., 2016. Exploring metal detoxification and accumulation potential during vermicomposting of Tea factory coal ash: sequential extraction and fluorescence probe analysis. Sci. Rep. 6, 30402. https://doi.org/10.1038/srep30402.

Grant, W.F., 1994. The present status of higher plant bioassays for the detection of environmental mutagens. Mutat. Res. 310, 175–185.

Hait, S., Tare, V., 2011. Vermistabilization of primary sewage sludge. Bioresour. Technol. 102, 2812–2820.

Honarvar, M., Samavat, S., Davoodi, M.H., Karimi, K.H., 2011. Possibility of producing compost and vermicompost from sugar beet waste in the sugar factory. Food Technol. Nutr. 8, 46–53.

Huang, K., Xia, H., 2018. Role of earthworms' mucus in vermicomposting system: biodegradation tests based on humification and microbial activity. Sci. Total Environ. 610–611, 703–708.

Jain, K., Singh, J., Chauhan, L.K.S., Murthy, R.C., Gupta, S.K., 2004. Modulation of flyash-induced genotoxicity in *Vicia faba* by vermicomposting. Ecotoxicol. Environ. Saf. 59, 89–94.

Jayakumar, M., Karthikeyan, V., Karmegam, N., 2009. Comparative Studies on physico-chemical, microbiological and enzymatic activities of vermicasts of the earthworms, *Eudrilus Eugeniae*, *Lampito Mauritii* and *Perionyx Ceylanensis* cultured in press mud. Int. J. Appl. Agric. Res. 4, 75–85.

Karmegam, N., Karthikeyan, V., Ambika, D., 2012. Vermicomposting of sugarcane trash and leaf litter in combination with pressmud using the earthworm, *Perionyx ceylanensis*. Dynamic Soil Dynamic Plant 6, 57–64.

Kaur, A., Singh, J., Vig, A.P., Dhaliwal, S.S., Rup, P.J., 2010. Cocomposting with and without *Eisenia fetida* for conversion of toxic paper mill sludge to a soil conditioner. Bioresour. Technol. 101, 8192–8198.

Khwairakpam, M., Bhargava, R., 2009. Bioconversion of filter mud using vermicomposting employing two exotic and one local earthworm species. Bioresour. Technol. 100, 5846–5852.

Kumar, R., Verma, D., Singh, B.L., Kumar, U., Shweta, 2010. Composting of sugar-cane waste by-products through treatment with microorganisms and subsequent vermicomposting. Bioresour. Technol. 101, 6707–6711.

Leme, D.M., Marin-Morales, M.A., 2009. *Allium cepa* test in environmental monitoring: a review on its application. Mutat. Res. 682, 71–81.

Lim, S.L., Wu, T.Y., 2016. Characterization of matured vermicompost derived from valorization of palm oil mill byproduct. J. Agric. Food Chem. 64, 761–769.

Liu, F., Zhu, P., Xue, J., 2012. Comparative study on physical and chemical characteristics of sludge vermicomposted by *Eisenia fetida*. Procedia Environ. Sci. 16, 418–423.

Maity, S., Bhattacharya, S., Chaudhury, S., 2009. Metallothionein response in earthworms *Lampito mauritii* (Kinberg) exposed to fly ash. Chemosphere 77, 319–324.

Maluszynska, J., Juchimiuk, J., 2005. Plant genotoxicity: a molecular cytogenetic approach in plant bioassays. Arh Hig Rada Toksikol 56, 177–184.

Mazzeo, D.E.C., Fernandes, T.C.C., Levy, C.E., Fontanetti, C.S., Marin-Morales, M.A., 2015. Monitoring the natural attenuation of a sewage sludge toxicity using the *Allium cepa* test. Ecol. Indic. 56, 60–69.

Molina, M.J., Soriano, M.D., Ingelmo, F., Llinares, J., 2013. Stabilisation of sewage sludge and vinasse bio-wastes by vermicomposting with rabbit manure using *Eisenia fetida*. Bioresour. Technol. 137, 88–97.

Niyazi, R., Chaurasia, S., 2014. Vermistabilization of fly ash amended with pressmud by employing *eisenia foetida*. Int. J. Pharma Chem. Biol. Sci. 4, 85–95.

Pandit, N.P., Maheshwaria, S.K., 2012. Optimization of vermicomposting technique for sugarcane waste management by using *Eisenia fetida*. Int. J. Biosci. 10, 143–155.

Pandit, N.P., Maheshwaria, S.K., 2014. Vermiremediation of sugarcane by-products into nutrient rich vermicompost through enhancing the bioconversion efficiency of *Eisenia fetida* by developing vermireactors. J. Bioproc. Technol. 99, 327–337.

Parthasarathi, K., 2006. Aging of Press mud vermicast of *Lampito mauritti* (Kinberg) and *Eudrilus eugeniae* (Kinberg)-Reduction in microbial population and activity. J. Environ. Biol. 27, 221–223.

Parthasarathi, K., 2007. Influence of moisture on the activity of *Perionyx excavatus* (Perrier) and microbial–nutrient dynamics of pressmud vermicompost. Iranian J. Environ. Health Sci. Engr. 4, 147–156.

Prakash, M., Karmegam, N., 2010. Vermistabilization of pressmud using *Perionyx ceylanensis* Mich. Bioresour. Technol. 101, 8464–8468.

Rank, J., Nielsen, M.H., 1993. A modified *allium* test as a tool in the screening of the genotoxicity of complex mixtures. Hereditas 118, 49–53.

Ravindran, B., Dinesh, S.L., John Kennedy, L., Sekaran, G., 2008. Vermicomposting of solid waste generated from leather industries using epigeic earthworm *Eisenia foetida*. Appl. Biochem. Biotechnol. 151, 480–488.

Ravindran, B., Contreras-Ramos, S.M., Wong, J.W.C., Selvam, A., Sekaran, G., 2014. Nutrient and enzymatic changes of hydrolysed tannery solid waste treated with epigeic earthworm *Eudrilus eugeniae* and phytotoxicity assessment on selected commercial crops. Environ. Sci. Pollut. Res. 21, 641–651.

Reddy, K.S., Shantaram, M.V., 2005. Potentiality of earthworms in composting of sugarcane by products. Asian J. Microbiol. Biotechnol. Environ. Sci. 7, 483–487.

Rouf, M.A., Bajpai, P.K., Jotshi, C.K., 2010. Optimization of biogas generation from press mud in batch reactor. Bangladesh J. Sci. Indus. Res. 45, 371–376.

Sangwan, P., Kaushik, C.P., Garg, V.K., 2008a. Feasibility of utilization of horse dung spiked filter cake in vermicomposters using exotic earthworm *Eisenia foetida*. Bioresour. Technol. 99, 2442–2448.

Sangwan, P., Kaushik, C.P., Garg, V.K., 2008b. Vermiconversion of industrial sludge for recycling the nutrients. Bioresour. Technol. 99, 8699–8704.

Sangwan, P., Kaushik, C.P., Garg, V.K., 2010. Vermicomposting of sugar industry waste (pressmud) mixed with cow dung employing an epigeic earthworm *Eisenia foetida*. Waste Manag. Res. 28, 71–75.

Sen, B., Chandra, T.S., 2007. Chemolytic and solid-state spectroscopic evaluation of organic matter transformation during vermicomposting of sugar industry wastes. Bioresour. Technol. 98, 1680–1683.

Shah, R.U., Abid, M., Qayyum, M.F., Ullah, R., 2015. Dynamics of chemical changes through production of various composts/vermicompost such as farm manure and sugar industry wastes. Int. J. Recycling Organic Waste Agri. 4, 39–51.

Singh, J., Kaur, A., Vig, A.P., Rup, P.J., 2010. Role of *Eisenia fetida* in rapid recycling of nutrients from bio sludge of beverage industry. Ecotoxicol. Environ. Saf. 73, 430–435.

Singh, J., Kaur, A., Vig, A.P., 2014. Bioremediation of distillery sludge into soil-enriching material through vermicomposting with the help of *Eisenia fetida*. Appl. Biochem. Biotechnol. 174, 1403–1419.

Singh, S., Bhat, S.A., Singh, J., Kaur, R., Vig, A.P., 2017. Earthworms converting milk processing industry sludge into Biomanure. Open Waste Manage. J. 10, 30–40.

Solomon, S., 2011. Sugarcane by-products based industries in India. Sugar Tech. 13, 408–416.

Souza, C.P., de Andrade Guedes, T., Fontanetti, C.S., 2016. Evaluation of herbicides action on plant bioindicators by genetic biomarkers: a review. Environ. Monit. Assess. 188, 694–706.

Srivastava, R., Kumar, D., Gupta, S.K., 2005. Bioremediation of municipal sludge by vermitechnology and toxicity assessment by *Allium cepa*. Bioresour. Technol. 96, 1867–1871.

Sturzenbaum, S.R., et al., 2012. Biosynthesis of luminescent quantum dots in an earthworm. Nat. Nanotechnol. 8, 57–60.

Suthar, S., Mutiyar, P.K., Singh, S., 2012. Vermicomposting of milk processing industry sludge spiked with plant wastes. Bioresour. Technol. 116, 214–219.

Venkatesh, R.M., Eevera, T., 2008. Mass reduction and recovery of nutrients through vermicomposting of fly ash. Appl. Ecol. Environ. Res. 6, 77–84.

Vig, A.P., Singh, J., Wani, S.H., Dhaliwal, S.S., 2011. Vermicomposting of tannery sludge mixed with cattle dung into valuable manure using earthworm *Eisenia fetida* (Savigny). Bioresour. Technol. 102, 7941–7945.

Vijver, M.G., Vink, J.P.M., Miermans, C.J.H., Van Gestel, C.A.M., 2003. Oral sealing using glue: a new method to distinguish between intestinal and dermal uptake of metals in earthworms. Soil Biol. Biochem. 35, 125–132.

Vivas, A., Moreno, B., Garcia-Rodriguez, S., Benitez, E., 2009. Assessing the impact of composting and vermicomposting on bacterial community size and structure, and functional diversity of an olive-mill waste. Bioresour. Technol. 100, 1319–1326.

Xing, M., Lv, B., Zhao, C., Yang, J., 2015. Towards understanding the effects of additives on the vermicomposting of sewage sludge. Environ. Sci. Pollut. Res. 22, 4644–4653.

Yadav, R.L., Solomon, S., 2006. Potential of developing sugarcane by-products based industries in India. Sugar Tech. 8, 104–111.

Earthworms converting domestic and food industry wastes into biofertilizer

5

Agnieszka Rorat, Franck Vandenbulcke
Campus ULille Sciences and Technologies, University of Lille, Lille, France

1 Introduction

The global quantity of wastes produced worldwide rises constantly. As the world's population increases and becomes more urban, in just the last century the production of waste has risen 10-fold; it was estimated that it may double again by 2025 (Hoornweg and Bhada-Tata, 2012). According to Hoornweg et al. (2013), the trend of waste production will peak at the moment when highly developed and rich societies with high living standards stabilize globally in the terms of population size. Nevertheless, the authors are rather pessimistic, showing highly increasing population growth and consumption rates, especially in developing countries (Table 1).

The Organization for Economic Cooperation and Development (OECD) defines municipal solid wastes as "waste collected and treated by or for municipalities," including waste from households (including bulky waste), commerce and trade, office buildings, institutions and small businesses, yard and garden wastes, street sweepings, the contents of litter containers, and market cleansing waste, but not the waste from construction and demolition activities. Other principal waste streams can be divided into industrial wastes, hazardous wastes, construction and demolition wastes, mining wastes, wastes from electrical and electronic equipment, packaging wastes, end-of-life vehicles and tires, and agricultural wastes. All mentioned types should be managed with respect to the following hierarchy: avoidance, reuse, recycling, recovery of energy, treatment, containment, and disposal. Nevertheless, in many cases such as food wastes, the prevention methods seem to be insufficient, as nowadays one-third of all produced food is still lost or wasted worldwide (FAO, 2011). That corresponds approximately to 1.3 billion tons per year and shall be treated as biowaste (biodegradable wastes). Next to food waste, the biodegradable fraction of municipal solid wastes, forestry residues, sewage sludge and the waste from the agriculture, food, and beverage industries can also be recognized as biowastes (Table 2).

Following the present trend of a circular economy and sustainable development, it is preferable to recover all the nutrients, carbon, and energy that could possibly be hidden in biowaste. The existing strategies of waste management are often dated and thus new or greatly improved infrastructures are needed. Modern ("smart," intelligent")

Industrial and Municipal Sludge. https://doi.org/10.1016/B978-0-12-815907-1.00005-2

Table 1 Municipal solid waste (MSW) generation and total urban population in selected countries worldwide according to the World Bank

	2012			2025			
Country	**Total urban population ($*10^6$)**	**MSW per capita (kg/day)**	**Total MSW (tons/day)**	**Total population ($*10^6$)**	**Total urban population ($*10^6$)**	**MSW per capita (kg/day)**	**Total MSW (tons/day)**
China	511	1.02	520,548	1445	822	1.7	1,397,755
Germany	60	2.11	127,816	80	61	2.05	126,633
India	321	0.34	109,589	1447	538	0.7	376,639
France	47	1.92	90,493	65	53	2.00	107,318
Poland	23	0.88	2063	36	23	1.2	27,883
United States	241	2.58	6247	354	305	2.3	701,709

Data shown as total MSW generated per day in 2012 and prognosis for 2025 (OECD, 2015).

Table 2 Types of biodegradable waste (Saveyn and Eder, 2014)

Type	Items
Municipal solid waste	Biodegradable fractions of MSW (kitchen/garden waste)
Commercial food waste (not collected as a part of the MSW)	Waste from markets, catering waste
Forestry residues	Bark, wood residues
Waste from agriculture	Animal husbandry excrements (solid and liquid manure), straw residues, sugar beet and potato haulm, residues of growing of beans, peas, flax and vegetables, spent mushroom compost
Waste from the food and beverage industry	Breweries and malt houses, wineries, fruit and vegetable production industry, potato industry including starch, residues of beet sugar production, slaughterhouse residues, meat production, whey
Sewage sludge	

cities shall thus take into account the improvements at the stage of collection (e.g., radio frequency identification, tagging, pneumatic tubes), processing (advanced material recovery facilities and mechanical-biological treatments), recovery (waste-to-energy, waste-to-fuels, waste-to-matter), and disposal of wastes (Rorat and Kacprzak, 2017). Among the possibilities in matter recovery, composting has been recognized as a strategy of turning organic waste into humus, so called "brown gold" that can be then used as a fertilizer. Lately, its modification, vermicomposting, has been greatly recognized all over the world. Vermicomposting is an inexpensive, modern biotechnology for the biodegradation of biosolids/solid wastes, in which earthworms are employed as natural bioreactors in the process of decomposition of organic matter (Wang et al., 2009). The first experiments using *Eisenia fetida* earthworm for the bioconvertion of wastes were performed by Mitchell and coworkers in 1977 (Mitchell et al., 1977). Indeed, some earthworm species are able to transfer a variety of organic wastes, such as animal waste, agroindustrial waste, domestic waste, sewage sludge, and many others (Lim et al., 2016), into a valuable product called vermicompost. Earthworms influence the substratum indirectly by fragmentation, turnover, and aeration. By acting like natural "blenders," they cooperate with microorganisms and increase the surface area for them. Furthermore, their metabolic activities, including enzyme activity (that is proteases, lipases, amylases, cellulases), nitrogen enrichment, and transport of materials enhance the decomposition of organic matter (Edwards and Lofty, 1977; Ndegwa and Thompson, 2001). Nowadays, vermicomposting contributes powerfully to the "waste-to-matter" strategy, in which organic wastes can be recycled, that is transferred into compost that could be sold in the open market if it fulfills legal requirements. Vermicomposting may constitute an interesting alternative or complementation for the classical composting process

(Ndegwa and Thompson, 2001). Combined composting-vermicomposting processes have been investigated for various organic wastes (Alidadi and Shamansouri, 2005; Yadav et al., 2012; Suleiman et al., 2017). It was shown that two to four weeks of precomposting steps may be crucial to obtain the proper parameters of the substratum and to remove ammonia, inorganic salts, or other substances that are potentially toxic for earthworms.

The popularity of this eco-biotechnology has increased greatly in the last decade, which is reflected in the important number of scientific articles published each year. The aim of this review is to gather the information coming from the research papers from recent years (published in the period 2013–2018) in order to analyze the current economic and environmental impact of this process. The major substrates and earthworm species used in the process will be listed to point out the most possible scenarios for application of this strategy. Moreover, the current state of the art in the field of the possible applications of vermicompost will be presented. Those steps will allow us to indicate the strong and weak points of the process, showing its limits according to the technology itself and to the quality of the obtained product.

2 Substrates used in the vermicomposting process

Vermicomposting is often an interesting solution in poor socioeconomic conditions, were ecofriendly technologies can be easily adapted to the local conditions. For instance, Choudhary and Suri (2018) have recently reported the utility of obnoxious weed flora of NW Himalaya as a substrate for the process, and the further use of vermicompost as a source of nutrients. Similarly, *E. eugeniae* and *P. excavatus* successfully converted the locally abundant agrowaste, a banana waste mixed with farm waste and cow dung in India (Kamalraj et al., 2017). Consequently, it can be treated as a suitable ecotechnology in agreement with sustainable agrowaste management and nutrient recovery from the farm and agricultural wastes.

Furthermore, many types of organic wastes have been tested in vermicomposting. This technology is, at the same time, a method for the management of some of them as well as a technique of biofertilizer production. The quality of the obtained product depends strongly on the composition of the primary mixture. Consequently, the use of those components that contain dangerous pollutants shall be limited and the fate of contaminants shall be carefully studied throughout the process. For instance, sewage sludge can potentially be contaminated with heavy metals, different chemicals (polyaromatic hydrocarbons (PAH), hydrocarbons, polychlorinated biphenyl (PCB), perfluorinated surfactants (PFCs), personal care products (PCPs), pharmaceuticals (PhCs), and benzotriazoles), and pathogens (*Legionella, Yersinia, Escherichia coli* O157:H7) (Fijalkowski et al., 2017). At the same time, sewage sludge was recognized as a source of organic matter and nutrients, making it potentially a good growing media for the earthworms. Many studies proposed using dewatered sewage sludge coming from municipal wastewater treatment plants, mostly as a cosubstrate in the process (Domínguez et al., 2000; Alidadi and Shamansouri, 2005;

Yadav and Garg, 2016; Rorat et al., 2015). The proper composition of the mixture can assure the dilution of the pollutants, which is crucial for earthworm fitness (Rorat et al., 2013). This also allows adjusting the physical and chemical parameters in order to obtain a quality product in a possibly short time. The variety of the cosubstrates includes the organic fraction of municipal solid wastes, green wastes (e.g., *Miscanthus*), sawdust (or wood chips) (Rorat et al., 2017; Suleiman et al., 2017; Malińska et al., 2017), and cattle/pig manure (Yadav and Garg, 2016). Similarly, some industrial sludge has been proposed as a principal ingredient, for example, sludge from the pulp and paper industry, the sugar industry (agro-based industry), the food industry, the milk processing industry, and the tannery industry (see review by Lee et al., 2018). The variety of different organic wastes applied in vermicomposting in the past 2 years is shown in Table 3. The earthworms were successfully employed to convert different organic wastes, that is, mixtures of kitchen, green, and paper wastes; the residues from the oil palm and sugarcane industries; trout aquaculture; sawdust; medicinal herbal residues; different kinds of manures; and coal fly ash.

Lately, many studies have shown the advantages of using different additives next to organic substrates (Barthod et al., 2018). For instance, biochar addition causes an increase in the growth and reproduction of earthworms as well as their enzymatic activities while enhancing the process of decomposition of organic matter and decreasing the bioavailability of some metals (Malińska et al., 2017). This specific byproduct can be defined as a solid, carbon-rich material obtained in the process of a zero or low oxygen environment from different C-based feedstocks. When applied to the soils, the byproduct sustainably sequesters carbon and thus improves soil quality in the long term (Verheijen et al., 2010). Gong et al. (2018) applied biochar from bamboo biomass to the mixture of green wastes, which positively influenced the earthworm, accelerated the decomposition of lignin, and dissolved organic carbon. The authors confirmed the economic viability of adding 6% of biochar, as it reduces the time required for compost maturity and thus makes it usable as a fertilizer. Sometimes, the adsorbent materials, for example, zeolites, can be used in order to decrease the salinity of the final compost when the substrates cause a sharp peak of salinity during the process. Zeolites are porous structured compounds with an aluminosilicate framework (AlO4 and SiO4) that binds different cations. This additive was successfully applied by Alavi et al. (2017) in vermicomposting of vinasse with the mixture of cow manure waste and bagasse. Rodrigues et al. (2017) found an application for the ornamental rock dust, as it can constitute the source of macronutrients such as nitrogen, phosphorus, potassium, calcium, magnesium, and sulfur as well as micronutrients such as iron, manganese, copper, zinc, and sodium, which are suitable for plant growth. Similarly, the positive effects of the addition of cattle's blood powder were observed while vermicomposting sawdust (Najjari and Ghasemi, 2018).

The addition of the consortia of bacterial strains (such as *Bacillus anthracis*, *B. endophyticus*, *B. funiculus*, *B. thuringiensis*, *B. cereus*, *B. toyonensis*, *Virigibacillius chiquenigi*, *Acinetobacter baumanni*, and *Lactobacillus pantheries*) for vermicomposting paper cup waste allowed obtaining a better-quality compost in a shorter time (Arumugam et al., 2018).

Table 3 The bibliographic summary of the research on the vermicomposting of different organic wastes using *E. fetida* earthworms from 2017 to 2018

Main substrate	Additives/ cosubstrates	Time	Main conclusions	Reference
Distillery sludge waste	Tea leaf residues	45 days	Increase of pH, Total N, P and K, Available N, P and K, TCa and decrease of TMg, EC, TOC and C/N ratio; successful process	Mahaly et al. (2018)
Obnoxious weed-flora of north-western (NW) Himalayas *Artemisia vulgaris, Ageratum conyzoides, Erigeron canadensis Bidens pilosa*	Farm-yard manure	3 periods of 4 months	Successful for bioconversion of obnoxious weed flora for, substantially meeting the plant nutrient demands of hill agriculture in NW Himalayas. *A. vulgaris and A. conyzoides* vermicompost contained significantly higher plant nutrient content Production of vermicompost—role in supplying plant nutrients under resource-poor farms of developing world	Choudhary and Suri (2018)
Green waste consisting of fallen leaves, grass clippings, and branch cuttings	Biochar from bamboo biomass	14 days of pre composting + 60 days	Biochar addition caused: an increase of growth and reproduction of earthworms, increased their enzymatic activities, accelerated the decomposition of lignin and dissolved organic carbon, enhanced nitrification and humification, reduced the time	Gong et al. (2018)

Table 3 Continued

Main substrate	Additives/ cosubstrates	Time	Main conclusions	Reference
			required for the compost to be toxicity-free for seed germination 6% addition of biochar is economically viable	
Sewage sludge	Cattle dung	80 days	Earthworms accelerated OM degradation, improved stabilization, affected the growth of bacteria, inhibited the growth of pathogens and promoted of growth of *Flavobacteria, Acidbacteria, Planctomycetes*	Lv et al. (2018)
Sawdust	Cattle blood powder	4 months	The addition of blood powder accelerated the stabilization and maturity of sawdust—the obtained vermicomposts were characterized by higher N, P, K, Fe, Zn and Mn contents and lower pH and C/N ratio; the obtained product stimulated the cucumber growth	Najjari and Ghasemi (2018)
Municipal waste and anaerobic digestate		45 days	Earthworms successfully transferred biosolids into dark brown compost that can be possibly used as fertilizer	Manyuchi et al. (2018)

Continued

Table 3 Continued

Main substrate	**Additives/ cosubstrates**	**Time**	**Main conclusions**	**Reference**
Medicinal herbal residues	Cow dung	49 days	Successful stabilization of medicinal herbal residues with cow dung Earthworms promoted the decomposition process compared to traditional composting and exerted a remarkable effect on the microbial composition. The abundance of the dominant phyla Proteobacteria, Bacteroidetes, Basidiomycota, and Ascomycota were affected by the earthworm inoculating density. Earthworms improved the microbial community by enhancing the growth of lignocelluloytic bacteria and fungi.	Chen et al. (2018)
Pernicious aquatic weed *Salvinia molesta*	–	n.d.	Successful management of S. molesta phytomass Earthworms helped to reduce the possible negative effects of the plant The obtained compost may serve as a fertilizer (tested on *Abelmoschus esculentus*, *Cucumis sativus*, *Vigna radiata*	Hussain et al. (2018a)

Table 3 Continued

Main substrate	Additives/ cosubstrates	Time	Main conclusions	Reference
Paper Mill Sludge	Cow/cattle dung	48 days	Covermicomposting of paper mill sludge and animal manures are beneficial from nutrient recovery point of view	Mohan (2017)
Vinasse (liquid waste comming from sugercane Agroindustry	Cow manure, chopped bagasse zeolites	60 days	Natural calcium zeolites adsorbed the potassium cations, thus they eliminated the high salinity problem The vinasse is a good source of nutrients for earthworms. The obtained product can be classified as a potential fertilizer.	Alavi et al. (2017)
Tomato stems	Cow dung	50 days	Precomposting is crucial to solve the problem of pathogens in the vermicompost. Application of earthworms enhances the process of degradation of organic matter. Vermicomposts released much less greenhouse gas and ammoniathen traditional thermophilic compost	Yang et al. (2017)
Excreta of different ruminants, viz., sheep, cow, buffalo, and goat		20 days precomposting 90 days	Ruminant excreta are good sources of nutrients for earthworms; the good parameters of final product encourage the future research on the recycling of nutrients present un manures	Sharma and Garg (2017)

Continued

Table 3 Continued

Main substrate	Additives/ cosubstrates	Time	Main conclusions	Reference
Cattle manure mixed with tannery sludge	Rock dust with treated domestic sludge	120 days	Organic wastes (manure and tannery sludge mix) and inorganic wastes(rock dust) inoculated with treated sewage wastewater can be applied as substrates in vermicomposting process. Obtained vermicomposts were feasible for R. brittoniana seedling production and for commercial substrate (humus) replacement.	Rodrigues et al. (2017)
Municipal sewage sludge and wood chips mixtures	Sewage sludge derived biochar	4 weeks of precomposting; 6 weeks	Biochar amendment: significantly increased number of cocoons and juveniles, reduced bioavailability of Cd and Zn to *E. fetida* Biochar-added vermicompost was recognized as a potential amendment for calcareous soils.	Malińska et al. (2017)
Water lettuce (*Pistia* sp.) biomass	Cow dung	28 days	Earthworms caused a decrease in pH, TOC, volatile solids, and C/N ratio and increase in EC, totN, totP, totK, totCa, totZn, totFe, and totCu. The mixture with cow dung is crucial for the worms' fitness. biomass of water lettuce can be utilized effectively for production of valuable manure through vermicomposting system.	Suthar et al. (2017)

3 Earthworm species used in the vermicomposting process

Earthworms are detritivorous dwelling organisms that participate in the circulation of organic matter in nature, increase the bioavailability of nutrients for plants, and provide the proper aeration, bioturbation, and humidity of the matrix. They play a major role in the ecosystem through participation in the water, carbon, and nutrient circulation. Simultaneously, they constitute an important element of the food chain as an aliment for many predators. Earthworms are oligochetae annelids that exhibit metameric segmentation; they are hermaphrodites that reproduce by cross- or self-fertilization. Depending on the ecological niches that they occupy, and thus their feeding and burrowing strategies, earthworms can be classified as epigeic, anecic, or endogeic (Bouché, 1977). While endogeic and anecic species inhabit the soil, the epigeic ones are considered litter dwellers and transformers that are more likely to live near the surface. Among more than 8300 oligochetae species (Reynolds and Wetzel, 2004), only a few earthworm species are being used in vermicomposting. Recently, most of the studies have highlighted the efficiency of *Eisenia fetida* (Table 3) and five other epigeic species: *Eisenia andrei*, *Dendrobaena veneta*, *Eudrilus eugeniae*, *Perionyx excavates*, and *Dichogaster annae* (Tables 4 and 5). Only one representative of epiendogeic worms, *Lumbricus rubellus*, can be found in a few studies (Azizi et al., 2013; Bakar et al., 2011; Rupani et al., 2017). Generally, the selected species are characterized with high reproduction rates, a short life cycle, low body weight, a short maturation time, and well-developed defense mechanisms that help them survive in hostile environments (Dominguez and Edwards, 2011). The particular species' characteristics are presented in Fig. 1. Lately, *D. veneta* appears rarely in studies on the bioconversion of wastes as it was recognized as more fragile and thus less effective in the process (Suleiman et al., 2017).

Rarely, researchers assess the performance of the process using a mixture of earthworm species. For instance, Hussain et al. (2018b) proposed a consortium of three—*Eisenia fetida*, *Eudrilus eugeniae*, and *Perionyx excavates*—to ameliorate the vermitechnology of transferring kitchen vegetable waste and paddy straw. Authors have shown that the synergic action of three species accelerates the process. Similarly, Rorat et al. (2016) applied a mixture of *E. fetida* and *E. andrei* to transform sewage sludge, but no significant difference was shown compared to the single species application, probably because of the close relation of those two species.

4 The influence of earthworms on the product's quality

The earthworms influence the physical and chemical properties of the substratum. Many studies have shown the biochemical changes that appear in the process and tried to explain the mechanisms involved in particular processes (Yadav and Garg, 2011). While the ideal range of pH during vermicomposting is 6.8–7.4 according to Manna et al. (2003), a significant decrease can be caused by the release of CO_2 by the

Table 4 The bibliographic summary of the research on the vermicomposting of different organic wastes using *E. andrei* earthworms from 2017 to 2018

Species	Main substrate	Additives/ cosubstrates	Time	Main conclusions	Reference
E. andrei	Sludge obtained from a trout aquaculture	shredded wheat straw	18 weeks	Positive evaluation of the process; obtained vermicomposts suitable to use in agriculture, the earthworms can serve as feed for livestock	Kouba et al. (2018)
E. andrei	Pig manure (fresh or precomposted)		70 days + 42 days of maturation	Earthworms accelerated the process of degradation of organic material—enzymatic activity	Villar et al. (2017)
E. andrei	Organic fraction of municipal solid waste, green waste, bulking agent (*Miscanthus* biomass), waste from markets, sewage sludge		30 days precomposting, 5 weeks vermicomposting	*E. andrei* successfully remediated polluted substratum by reducing significantly the content of 16 priority PAHs and selected herbicide. Heavy metals body accumulation factors could be ranked as follows (Cd > Cu > Zn > Ni > Cr > Pb)	Rorat et al. (2017)
E. eugeniae	Paper cup waste + vermicompost with bacterial consortium	Cow dung	19 weeks	Vermicomposting is suitable for management of paper cup wastes. The application of microbial consortia allowed obtaining the final product faster (12 weeks instead of 19). Moreover, the obtained compost is characterized by the increased nutrient content.	Arumugam et al. (2018)

E. eugeniae	Organic fraction of municipal solid waste (food waste, grass clippings, dry leaves and small branches, market waste, office shredded paper and newspaper)	Cow dung	7 weeks +3 weeks of precomposting	Earthworms reduced the number of total coliform bacteria and gave Salmonella-free vermicompost; Better quality of vermicomposts then their corresponding composts	Soobhany (2018)
Eudrilus sp	Coconut leaves	cow dung slurry	105 days; worms added after 16 days	Increase of microbial diversity after 75 days of the process	Gopal et al. (2017)
E. eugeniae	Oil palm empty fruit bunches	Cow dung	12 weeks	The empty fruit bunches can be successfully vermicomposted with cow dung The initial C/N ratio determines the process	Lim et al. (2015)
L. rubellus	Palm industry wastes acidic palm oil mill effluent mixed with the palm pressed fiber		45 days	The substrates used in the study constitute a good source of nutrients for earthworms; the obtained vermicompost (75% addition) enhanced the shoot length of the mung bean plant	Rupani et al. (2017)

Table 5 The bibliographic summary of the research on the vermicomposting of different organic wastes using *earthworms' consortia or comparing the species* from 2017 to 2018

Species	Main substrate	Additives/ cosubstrates	Time	Main conclusions	Reference
E. fetida *E.andrei* *D. veneta*	Organic fraction of municipal solid wastes (MSW), grass clippings and sawdust		45 days	*Eisenia* sp. shows much stronger defense mechanisms than *D. veneta*. One-month precomposting is a crucial step. Earthworms accumulate some heavy metals. BAFs were arranged as follows: Cd > Co > Cu > Zn > Ni > Pb > Cr for all three species. *E. andrei* showed the highest accumulation capabilities. The process proposed in this study is a viable solution for the cotreatment of several groups of organic wastes.	Suleiman et al. (2017)
E. eugeniae *P. excavatus*	local abundant agrowastes- banana waste mixed farm waste	Cow dung	90 days	*E. eugeniae* is slightly more efficient than *P. excavatus* All proposed substrates can be effectively used in the process	Kamalraj et al. (2017)
E. fetida *E. eugeniae* *L. rubellus*	Coal fly ash	Cow dung	90 days	*E. eugeniae* removed more metals compared to *E. fetida* and *L. rubellus* (Cr, Cu, Zn, Cd, Pb accumulation). All species	Usmani et al. (2017)
Consortium: *E. fetida,* *E. eugeniae,* *P. excavatus*	Kitchen vegetable waste paddy straw	Cow dung	120 days	The beneficial effect of using the consortium of earthworms: the increase of N, P, K availability, enzyme activity, microbial proliferation, biomass humification. Higher growth rate in consortium; highly advantageous presence of *E. fetida*	Hussain et al. (2018b)

D. annae *E. eugeniae* *P. excavatus*	Fed shredded paper, lawn clippings	Cattle manure	8 weeks	All species were capable of converting feedstock as well as partitioning energy toward growth and reproduction at different C/N ratios. Increasing C/N ratio decreased juvenile population biomass. *D. annae* has shown tremendous potential to vermicompost feedstock ranging in C/N ratio but operates best at C/N ratios >25. High C/N ratio and low feed rate favors the best product parameters, while the opposite stimulates reproductive activity and juvenile production.	Martin and Eudoxie (2018)

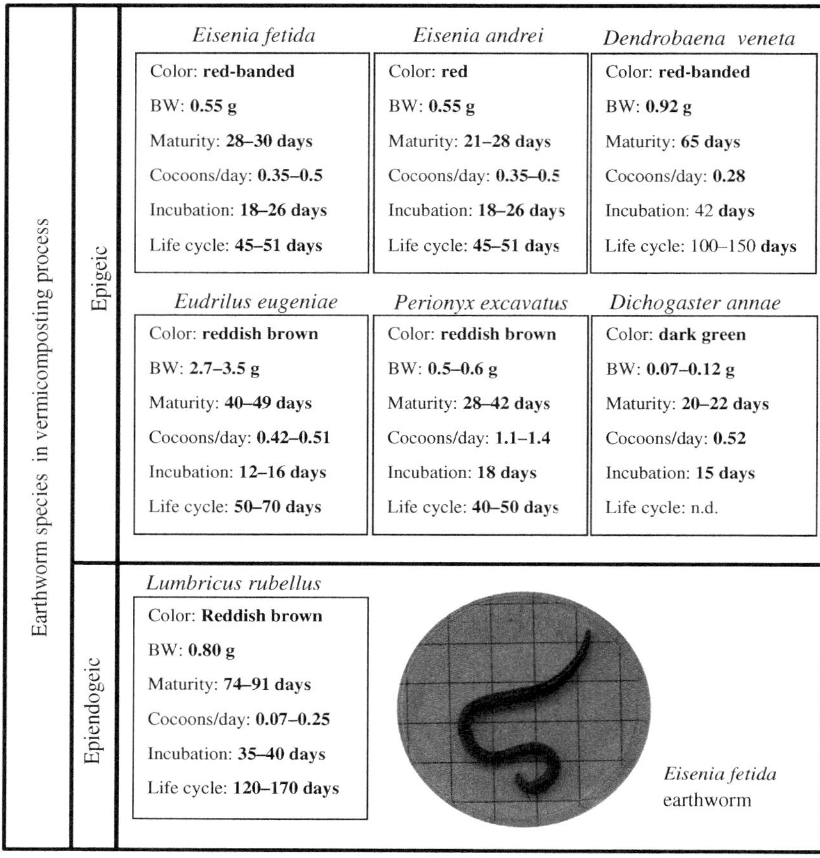

Fig. 1 Comparison of seven major earthworm species used in the process of vermicomposting of different substrates (Dominguez and Edwards, 2010; Rorat et al., 2014; Saratchandran, 2009); *BW*, mean body weight, *maturity*, the time needed to obtain maturity from the hatchling, *cocoons/day*, the average number of cocoons per one earthworm per day; *incubation*, the average time needed to hatch.
Photo: A. Rorat.

metabolic activity of earthworms and bacteria. Those metabolic processes cause consequently a decrease of total organic carbon (Prakash and Karmegam, 2010). Moreover, increasing acidity can be related to the mineralization of nitrogen and phosphorus, forming nitrites/nitrates and orthophosphates and decomposition of organic matter into organic acids (Ndegwa and Thompson, 2000; Yadav and Garg, 2009). Those organisms also use the carbon as a source of energy and nitrogen, which is necessary as a cell component. Both the C:N ratio and the volatile solids decrease as an effect of the constant decomposition of organic matter (Hait and Tare, 2011).

Earthworms can accumulate some heavy metals in their bodies in a nonlinear way, with higher body accumulation factors (BAFs) at lower concentrations (Neuhauser et al., 1994). The accumulation is metal- and species-specific and depends strongly on earthworm metabolic pathways, food selectivity, and mechanisms of detoxification (Morgan and Morgan, 1992). As an important element of the food chain, earthworms can easily incorporate dangerous contaminants into the trophic chain. Special attention should be paid to the mechanisms of detoxification and accumulation of some metallic trace elements. It has recently been claimed that earthworms enhance the biodegradation of other dangerous contaminants, polycyclic aromatic hydrocarbons, known as ubiquitous pollutants showing cancerogenicity, teratogenicity, and ecotoxicity (Lu and Lu, 2015; Rorat et al., 2017) .

Some beneficial impacts of earthworms on the bacterial profiles during the process of transformation of sewage sludge and cattle dung have recently been shown. Lv et al. (2018) have shown that the activity of earthworms promoted the growth of some bacteria (such as Acidobacteria, Actinobacteria, and Planctomycetes), which influences the physicochemical properties of substrates and accelerates the decomposition of organic matter. Moreover, authors have proved that vermicomposting can reduce the presence of human bacteria pathogens, namely *Ochrobactrum anthropi*, *Brevundimonas diminuta*, *Brevundimonas vesicularis*, *Eubacterium tenue*, and *Bacillus thuringiensis*. Similarly, Soobhany (2018) performed a preliminary study to estimate the risks related to the presence of *Escherichia coli* and *Salmonella* during land spreading of vermicompost produced from the organic fraction of municipal organic wastes. The *E. eugeniae* earthworms significantly reduced both groups of indicator pathogens.

5 Worms versus composts

Composting earthworms are highly resistant to changing environmental factors and defense mechanisms help them survive in hostile conditions. Nevertheless, some parameters of the substrates can be crucial for the vermicomposting process and thus shall be controlled as they may influence the process (Yadav and Garg, 2011; Lee et al., 2018). Briefly, the success depends on the moisture content that should be kept in the range of 60%–80% (Edwards, 1988) in order to provide the environment for chemical reactions. Acceptable pH for earthworms and microorganisms ranges from 5.5 to 8.5 with an optimum around 7 (Kaushik and Garg, 2004) while heavy metals can be absorbed from the substratum in pH 6.8–7.4. Optimal temperature for earthworm metabolic activity and reproduction ranges from 12°C to 28°C and depends strongly on the selected species of earthworms. As aerobic organisms, those invertebrates require aerobic conditions throughout the process and complete protection from light should be provided. Another parameter that affects vermicomposting is the stocking density of earthworms used in the process that depends strongly on the earthworm species and type of used substrate. It is crucial to choose an optimal worm density that allows obtaining a product of quality in a short time, but with no influence on the worms' growth and reproduction (Ndegwa et al., 2000; Sosnecka et al., 2016).

Similarly, the proper C/N ratio is critical to ensure a favorable environment for earthworm growth and reproduction. Ndegwa and Thompson (2000) estimated that the optimal value for vermicomposting of sewage sludge is around 25 while some species can tolerate a much higher ratio. For instance, (Lim et al., 2015) have performed the experiment on the bioconversion of empty fruit bunches mixed with cow dung using the *E. eugeniae* species in the mixtures characterized by the vast range of C/N ratio (from 24.03 to 148.96). The earthworms survived in all conditions. Their metabolic processes have led to the C/N reduction in relation to the drop of the organic carbon levels and simultaneous production of mucus and nitrogenous excreta. The C/N ratio influences directly the cocoon production and juvenile development as well as the adult biomass, as shown by Martin and Eudoxie (2018). Sometimes, the C/N has to be adjusted, for example, the cow dung was used for this purpose in combination with herbal residues in order to obtain the values suitable for the *E. fetida* species (Chen et al., 2018). Those favorable conditions can allow worms to maintain the aerobic conditions and thus ingest organic waste materials and produce a humus-like substance.

6 Vermicompost as a fertilizer

Many researchers have shown that vermicomposts, mostly those produced from noncontaminated food waste, can serve as a fertilizer as they clearly enhance plant growth. Indeed, vermicomposts are rich in microbial population, hold enzymes and hormones that can increase plant growth and at the same time reduce their diseases, and are particularly rich in humic substances (Bhat et al., 2018). First, vermicomposts contain phenols, hydroxyquinoline, alcohols, acetone, and ethyl ether, which break seed dormancy and thus enhance germination success up to certain concentrations. Hussain et al. (2018a) examined the ability to enhance the germination of vermicomposting produced by *E. fetida* from the *Salvinia molesta* aquatic weed. The authors proved that the process allowed degrading the phenols and other chemicals that can negatively impact the plants as well as extensive lignin content. Finally, the test performed on three common food plants—ladies finger (*Abelmoschus esculentus*), cucumber (*Cucumis sativus*), and green gram (*Vigna radiata*)—confirmed that the vermicompost significantly enhanced germination success, imparted plant-friendly physical features to the soil, increased the biomass carbon content of the soil, and promoted the early growth of selected plants.

Lately, Hussain and Abbasi (2018) reviewed the efficacy of different vermicomposts as fertilizers. A positive effect of vermicompost on seed germination, plant growth and yield as well as general soil health was highlighted. Many benefits of the application of vermicompost with cow manure on the grain yield and quality of rice (*Oryza sativa l.*) were shown by (Taheri Rahimabadi et al., 2018), but authors did not assess the sanitary risk of such a practice.

Rekha et al. (2018) compared the effectiveness of vermicompost and selected plant growth regulators (namely gibberellic acid and indole acetic acid) in promoting soil health and thus the growth of *Capsicum annum* (Linn.) Hepper. The compost obtained from cow dung with *Eudrilus eugeniae*, *Perionyx excavatus*, and *Eisenia foetida* earthworms enhanced the growth in a higher rate than the other candidates and thus confirmed its importance in organic farming as a replacement for fertilizers. Similarly, the compost obtained from sawdust and cattle's blood powder promoted the growth of cucumber (Najjari and Ghasemi, 2018). Nevertheless, the addition of any animal-derived additive can be controversial, so it should not be used for human food production. Moreover, the potential sanitary risk of introducing such products into the ecosystem shall be carefully studied.

The leachate coming from vermicomposting, the so-called "worm tea," contains important quantities of nutrients and humic acids. Thus, it can also be applied directly as fertilizers (or additives), for example, for the cultivation of sugarcane (*Saccharum* sp.) (Gutiérrez-Miceli et al., 2017) or maize (*Zea mays* L.) (Carlos et al., 2008).

7 The environmental sustainability of the process

The positive environmental impacts of vermicomposting as a method of organic waste management can be found directly, for example, in the reduction of the concentration of heavy metals or a decrease in the number of pathogenic bacteria (Hussain and Abbasi, 2018). The authors considered only the products of transformation of relatively "clean" products, showing numerous advantages of the application of vermicomposting, including the economic point of view. An important positive aspect related to vermicomposting is related to the replacement of chemical fertilizers and pesticides (Bloem et al., 2017).

The direct impact on global warming related to vermicomposting is related mainly to the reduction of emissions of CH_4, N_20, and NH_3. These are connected to earthworm movements and, consequently, the homogenization of the product as well as the ventilation of the composting piles (Yang et al., 2017; Nigussie et al., 2017). Moreover, soil application of compost and vermicompost enhances carbon sequestration and thus reduces the release of greenhouse gases (GHGs) into the environment (Lim et al., 2016).

An interesting idea on the introduction of a kitchen waste vermicomposting system at the production sites was proposed by Ori et al. (2017). The authors projected to adapt the process using *E. fetida* earthworms from the direct collection and disposal of wet kitchen waste from high-rise buildings in order to reduce the transport cost and environmental impact of those wastes. Similarly, the introduction of onsite solid waste management via vermicomposting directly at student housing was proposed and tested in Central Mexico (Aguilar-Virgen et al., 2017). This strategy allowed cutting GHG emissions related to the transport of wastes as well as possible emissions.

8 Conclusions

As proven by many authors, vermicomposting is an effective strategy for converting a variety of organic wastes into biofertilizers. Consequently, it fits well into the circular economy trend, showing many advantages compared to other options of waste management, for example, landfilling or incineration. The low-cost installation and relatively easy course of the process make vermicomposting an interesting solution for treatment of organic wastes in the place that they were produced, e.g., sewage sludge at the watewater treatment plant. Most of the studies also show a positive impact of the application of the obtained vermicompost on soils, if applied as an amendment. The results are encouraging as the substrates in the process are relatively "clean," for example, they come from the food industry or directly from households. Nevertheless, some information on the long-term impact of the application of vermicomposts shall be supplemented, mostly in relation to the presence of some emerging contaminants such as heavy metals and their nanoparticles.

Bibliography

Aguilar-Virgen, Q., Taboada-González, P., Baltierra-Trejo, E., Marquez-Benavides, L., 2017. Cutting GHG emissions at student housing in Central Mexico through solid waste management. Sustainability 9, 1415.

Alavi, N., Daneshpajou, M., Shirmardi, M., Goudarzi, G., Neisi, A., Babaei, A.A., 2017. Investigating the efficiency of co-composting and vermicomposting of vinasse with the mixture of cow manure wastes, bagasse, and natural zeolite. Waste Manage. 69, 117–126.

Alidadi, H., Shamansouri, A.P.M., 2005. Combined compost and vermicomposting process in the treatment and bioconversion of sludge. Iranian J. Environ. Health Sci. Eng. 2, 251–254.

Arumugam, K., Renganathan, S., Babalola, O.O., Muthunarayanan, V., 2018. Investigation on paper cup waste degradation by bacterial consortium and Eudrillus eugeinea through vermicomposting. Waste Manage. 74, 185–193.

Azizi, A.B., Lim, M.P., Noor, Z.M., Abdullah, N., 2013. Vermiremoval of heavy metal in sewage sludge by utilising *Lumbricus rubellus*. Ecotoxicol. Environ. Saf. 90, 13–20.

Bakar, A.A., Mahmood, N.Z., Teixeira Da Silva, J.A., Abdullah, N., Jamaludin, A.A., 2011. Vermicomposting of sewage sludge by *Lumbricus rubellus* using spent mushroom compost as feed material: effect on concentration of heavy metals. Biotechnol. Bioprocess Eng. 16, 1036–1043.

Barthod, J., Rumpel, C., Dignac, M.F., 2018. Composting with additives to improve organic amendments: a review. Agron. Sustain. Dev. 38.

Bhat, S.A., Singh, S., Singh, J., Kumar, S., Bhawana, Vig, A.P., 2018. Bioremediation and detoxification of industrial wastes by earthworms: vermicompost as powerful crop nutrient in sustainable agriculture. Bioresour. Technol. 252, 172–179.

Bloem, E., Albihn, A., Elving, J., Hermann, L., Lehmann, L., Sarvi, M., Schaaf, T., Schick, J., Turtola, E., Ylivainio, K., 2017. Contamination of organic nutrient sources with potentially toxic elements, antibiotics and pathogen microorganisms in relation to P fertilizer potential and treatment options for the production of sustainable fertilizers: a review. Sci. Total Environ. 607, 225–242.

Bouché, M., 1977. Strategies lombriciennes. In: Lohm, U., Persson, T. (Eds.), Soil Organisms as Components of Ecosystems. Swedish Natural Science Research Council, Stockholm, NFR, pp. 122–132 Ecological Bulletins 25.

Carlos, G.-G.R., Dendooven, L., Antonio, G.-M.F., 2008. Vermicomposting leachate (worm tea) as liquid fertilizer for maize (*Zea mays* L.) forage production. Asian J. Plant Sci. 7, 360–367.

Chen, Y., Chang, S.K.C., Chen, J., Zhang, Q., Yu, H., 2018. Characterization of microbial community succession during vermicomposting of medicinal herbal residues. Bioresour. Technol. 249, 542–549.

Choudhary, A.K., Suri, V.K., 2018. Low-cost vermi-composting technology and its application in bio-conversion of obnoxious weed flora of North-Western Himalayas into Vermicompost. Commun. Soil. Plant Anal. 49 (12), 1429–1441.
Dominguez, J., Edwards, C.A., 2010. Biology and ecology of earthworm species used for vermicomposting. In: Vermiculture Technology: Earthworms, Organic Waste and Environmental Management. CRC Press, USA, pp. 25–37.
Dominguez, J., Edwards, C.A., 2011. Biology and ecology of earthworm species used for vermicomposting. In: Edwards, C.A., N. Q. A, , Sherman, R.L. (Eds.), Vermiculture Technology: Earthworms, Organic Wastes, and Environmental Management. CRC Press, USA.
DomÍnguez, J., Edwards, C.A., Webster, M., 2000. Vermicomposting of sewage sludge: effect of bulking materials on the growth and reproduction of the earthworm *Eisenia andrei*. Pedobiologia 44, 24–32.
Edwards, C.A., 1988. Examines the earthworm through morphology, taxonomy, biology and look at its environmental role. In: Neuhauser, C.A.E.A.E.F. (Ed.), Earthworms in Waste and Environmental Management. SPB Academic Publ. Co, The Hague, Netherlands.
Edwards, C.A., Lofty, J.R., 1977. Examines the Earthworm Through Morphology, Taxonomy, Biology and Look at Its Environmental Role. Biology of Earthoworm. Chapman and Hall, London.
FAO, 2011. Global Food Losses and Food Waste—Extent, Causes and Prevention. FAO, Rome.
Fijalkowski, K., Rorat, A., Grobelak, A., Kacprzak, M.J., 2017. The presence of contaminations in sewage sludge—the current situation. J. Environ. Manag. 203, 1126–1136.
Gong, X., Cai, L., Li, S., Chang, S.X., Sun, X., An, Z., 2018. Bamboo biochar amendment improves the growth and reproduction of *Eisenia fetida* and the quality of green waste vermicompost. Ecotoxicol. Environ. Saf. 156, 197–204.
Gopal, M., Bhute, S.S., Gupta, A., Prabhu, S.R., Thomas, G.V., Whitman, W.B., Jangid, K., 2017. Changes in structure and function of bacterial communities during coconut leaf vermicomposting. Anton. Leeuw. Int. J. Gen. Mol. Microbiol. 110, 1339–1355.
Gutiérrez-Miceli, F.A., García-Gómez, R.C., Oliva-Llaven, M.A., Montes-Molina, J.A., Dendooven, L., 2017. Vermicomposting leachate as liquid fertilizer for the cultivation of sugarcane (Saccharum sp.). J. Plant Nutr. 40, 40–49.
Hait, S., Tare, V., 2011. Vermistabilization of primary sewage sludge. Bioresour. Technol. 102, 2812–2820.
Hoornweg, D., Bhada-Tata, P., 2012. What a Waste: A Global Review of Solid Waste Management. In: Urban Development Series Knowledge Papers, World Bank, Washington, USA.
Hoornweg, D., Bhada-Tata, P., Kennedy, C., 2013. Environment: waste production must peak this century. Nat. News 502, 615.
Hussain, N., Abbasi, S.A., 2018. Efficacy of the vermicomposts of different organic wastes as "clean" fertilizers: state-of-the-art. Sustainability 10, 1205.
Hussain, N., Abbasi, T., Abbasi, S.A., 2018a. Generation of highly potent organic fertilizer from pernicious aquatic weed Salvinia molesta. Environ. Sci. Pollut. Res. 25, 4989–5002.
Hussain, N., Das, S., Goswami, L., Das, P., Sahariah, B., Bhattacharya, S.S., 2018b. Intensification of vermitechnology for kitchen vegetable waste and paddy straw employing earthworm consortium: assessment of maturity time, microbial community structure, and economic benefit. J. Clean. Prod. 182, 414–426.
Kamalraj, R., Nandhivarman, M., Thongni, P., Pradheeps, M., Poyyamoli, G., 2017. Utilization of agrowastes for vermicomposting and its impact on growth and reproduction of selected earthworm species in Puducherry, India. Nat. Environ. Pollut. Technol. 16, 1125–1133.
Kaushik, P., Garg, V.K., 2004. Dynamics of biological and chemical parameters during vermicomposting of solid textile mill sludge mixed with cow dung and agricultural residues. Bioresour. Technol. 94, 203–209.

Kouba, A., Lunda, R., Hlaváč, D., Kuklina, I., Hamáčková, J., Randák, T., Kozák, P., KoubovÁ, A., Buřič, M., 2018. Vermicomposting of sludge from recirculating aquaculture system using *Eisenia andrei*: technological feasibility and quality assessment of end-products. J. Clean. Prod. 177, 665–673.

Lee, L.H., Wu, T.Y., Shak, K.P.Y., Lim, S.L., Ng, K.Y., Nguyen, M.N., Teoh, W.H., 2018. Sustainable approach to biotransform industrial sludge into organic fertilizer via vermicomposting: a mini-review. J. Chem. Technol. Biotechnol. 93, 925–935.

Lim, P., Wu, T., Clarke, C., Daud, N.N., 2015. A potential bioconversion of empty fruit bunches into organic fertilizer using *Eudrilus eugeniae*. Int. J. Environ. Sci. Technol. 12, 2533–2544.

Lim, S.L., Lee, L.H., Wu, T.Y., 2016. Sustainability of using composting and vermicomposting technologies for organic solid waste biotransformation: recent overview, greenhouse gases emissions and economic analysis. J. Clean. Prod. 111 (Part A), 262–278.

Lu, Y.-F., Lu, M., 2015. Remediation of Pah-contaminated soil by the combination of tall fescue, arbuscular mycorrhizal fungus and epigeic earthworms. J. Hazard. Mater. 285, 535–541.

Lv, B., Xing, M., Yang, J., 2018. Exploring the effects of earthworms on bacterial profiles during vermicomposting process of sewage sludge and cattle dung with high-throughput sequencing. Environ. Sci. Pollut. Res. 25, 12528–12537.

Mahaly, M., Senthilkumar, A.K., Arumugam, S., Kaliyaperumal, C., Karupannan, N., 2018. Vermicomposting of distillery sludge waste with tea leaf residues. Sustain. Environ. Res. 28, 223–227.

Malińska, K., Golańska, M., Caceres, R., Rorat, A., Weisser, P., Ślęzak, E., 2017. Biochar amendment for integrated composting and vermicomposting of sewage sludge—the effect of biochar on the activity of *Eisenia fetida* and the obtained vermicompost. Bioresour. Technol. 225, 206–214.

Manna, M.C., Jha, S., Ghosh, P.K., Acharya, C.L., 2003. Comparative efficacy of three epigeic earthworms under different deciduous forest litters decomposition. Bioresour. Technol. 88, 197–206.

Manyuchi, M.M., Mbohwa, C., Muzenda, E., 2018. Resource recovery from municipal waste and bio solids (digestate) through vermicomposting: a waste management initiative. In: IEEE International Conference on Industrial Engineering and Engineering Managementpp. 2054–2057.

Martin, M., Eudoxie, G., 2018. Feedstock composition influences vermicomposting performance of *Dichogaster annae* relative to *Eudrilus eugeniae* and *Perionyx excavatus*. Environ. Sci. Pollut. Res. 1–10.

Mitchell, M.J., Mulligan, R.M., Hartenstein, R., Neuhauser, E.F., 1977. Conversion of sludges into "topsoils" by earthworms. Compost. Sci. 18, 28–32.

Mohan, S.M., 2017. Vermicomposting of paper mill sludge with *Eisenia fetida* for its conversion to nutrient using different seed materials. J. Instit. Eng. (India) A 98, 545–553.

Morgan, J.E., Morgan, A.J., 1992. Heavy metal concentrations in the tissues, ingesta and faeces of ecophysiologically different earthworm species. Soil Biol. Biochem. 24, 1691–1697.

Najjari, F., Ghasemi, S., 2018. Changes in chemical properties of sawdust and blood powder mixture during vermicomposting and the effects on the growth and chemical composition of cucumber. Sci. Hortic. 232, 250–255.

Ndegwa, P.M., Thompson, S.A., 2000. Effects of C-to-N ratio on vermicomposting of biosolids. Bioresour. Technol. 75, 7–12.

Ndegwa, P.M., Thompson, S.A., 2001. Integrating composting and vermicomposting in the treatment and bioconversion of biosolids. Bioresour. Technol. 76, 107–112.

Ndegwa, P.M., Thompson, S.A., Das, K.C., 2000. Effects of stocking density and feeding rate on vermicomposting of biosolids. Bioresour. Technol. 71, 5–12.

Neuhauser, E.F., Cukic, Z.V., Malecki, M.R., Loehr, R.C., Durkin, P.R., 1994. Bioconcentration and biokinetics of heavy metals in the earthworm. Environ. Pollut. 89, 293–301.

Nigussie, A., Bruun, S., De Neergaard, A., Kuyper, T.W., 2017. Earthworms change the quantity and composition of dissolved organic carbon and reduce greenhouse gas emissions during composting. Waste Manag. 62, 43–51.

OECD, 2015. Municipal Waste.

Ori, K., Bharti, A., Kumar, S., 2017. Disposal of kitchen waste from high rise apartment. J. Instit. Eng. (India) A 98, 237–243.

Prakash, M., Karmegam, N., 2010. Vermistabilization of pressmud using Perionyx ceylanensis Mich. Bioresour. Technol. 101, 8464–8468.

Rekha, G.S., Kaleena, P.K., Elumalai, D., Srikumaran, M.P., Maheswari, V.N., 2018. Effects of vermicompost and plant growth enhancers on the exo-morphological features of *Capsicum annum* (Linn.) Hepper. Int. J. Recycl. Org. Waste Agric. 7, 83–88.

Reynolds, J.W., Wetzel, M.J., 2004. Nomenclatura Oligochaetologica. Supplementum Quartum. A Catalogue of Names, Descriptions and Type Specimens of the Oligochaeta. Illinois Natural History Survey Special Publication, Illinois, USA.

Rodrigues, A.S.D.L., Mesak, C., Silva, M.L.G., Silva, G.S., Leandro, W.M., Malafaia, G., 2017. Organic waste vermicomposting through the addition of rock dust inoculated with domestic sewage wastewater. J. Environ. Manage. 196, 651–658.

Rorat, A., Kacprzak, M., 2017. Eco-innovations in sustainable waste management strategies for smart cities. In: Brdulak, A., Brdulak, H. (Eds.), Happy City—How to Plan and Create the Best Livable Area for the People. Springer International Publishing, Cham.

Rorat, A., Kacprzak, M., Vandenbulcke, F., Płytycz, B., 2013. Soil amendment with municipal sewage sludge affects the immune system of earthworms Dendrobaena veneta. Appl. Soil Ecol. 64, 237–244.

Rorat, A., Kachamakova-Trojanowska, N., Jozkowicz, A., Kruk, J., Cocquerelle, C., Vandenbulcke, F., Santocki, M., Plytycz, B., 2014. Coelomocyte-derived fluorescence and Dna markers of composting earthworm species. J. Exp. Zool. A: Ecol. Genet. Physiol. 321, 28–40.

Rorat, A., Suleiman, H., Grobelak, A., Grosser, A., Kacprzak, M., PŁytycz, B., Vandenbulcke, F., 2015. Interactions between sewage sludge-amended soil and earthworms—comparison between *Eisenia fetida* and Eisenia andrei composting species. Environ. Sci. Pollut. Res. 1–10.

Rorat, A., Suleiman, H., Grobelak, A., Grosser, A., Kacprzak, M., Płytycz, B., Vandenbulcke, F., 2016. Interactions between sewage sludge-amended soil and earthworms—comparison between *Eisenia fetida* and Eisenia andrei composting species. Environ. Sci. Pollut. Res. 23, 3026–3035.

Rorat, A., Wloka, D., Grobelak, A., Grosser, A., Sosnecka, A., Milczarek, M., Jelonek, P., Vandenbulcke, F., Kacprzak, M., 2017. Vermiremediation of polycyclic aromatic hydrocarbons and heavy metals in sewage sludge composting process. J. Environ. Manag. 187, 347–353.

Rupani, P.F., Embrandiri, A., Ibrahim, M.H., Shahadat, M., Hansen, S.B., Mansor, N.N.A., 2017. Bioremediation of palm industry wastes using vermicomposting technology: its environmental application as green fertilizer. Biotech *3*, 7.

Saratchandran, R., 2009. Studies on the Ecobiological Aspects of Two Species of Earthworms Lampito Mauritii *Dichogaster Annae*, Ph.D. Thesis, University of Kerala, India http://hdl.handle.net/10603/130731.

Saveyn, H., Eder, P., 2014. End-of-Waste Criteria for Biodegradable Waste Subjected to Biological Treatment (Compost & Digestate): Technical Proposals. Publications Office of the European Union, Luxembourg.

Sharma, K., Garg, V.K., 2017. Vermi-modification of ruminant excreta using *Eisenia fetida*. Environ. Sci. Pollut. Res. 24, 19938–19945.

Soobhany, N., 2018. Preliminary evaluation of pathogenic bacteria loading on organic municipal solid waste compost and vermicompost. J. Environ. Manag. 206, 763–767.

Sosnecka, A., Kacprzak, M., Rorat, A., 2016. Vermicomposting as an alternative way of biodegradable waste management for small municipalities. J. Ecol. Eng. 17, 91–96.

Suleiman, H., Rorat, A., Grobelak, A., Grosser, A., Milczarek, M., Płytycz, B., Kacprzak, M., Vandenbulcke, F., 2017. Determination of the performance of vermicomposting process applied to sewage sludge by monitoring of the compost quality and immune responses in three earthworm species: *Eisenia fetida*, *Eisenia andrei* and *Dendrobaena veneta*. Bioresour. Technol. 241, 103–112.

Suthar, S., Pandey, B., Gusain, R., Gaur, R.Z., Kumar, K., 2017. Nutrient changes and biodynamics of *Eisenia fetida* during vermicomposting of water lettuce (Pistia sp.) biomass: a noxious weed of aquatic system. Environ. Sci. Pollut. Res. 24, 199–207.

Taheri Rahimabadi, E., Ansari, M.H., Razavi Nematollahi, A., 2018. Influence of cow manure and its vermicomposting on the improvement of grain yield and quality of rice (*Oryza sativa* L.) in field conditions. Appl. Ecol. Environ. Res. 16, 97–110.

Usmani, Z., Kumar, V., Mritunjay, S.K., 2017. Vermicomposting of coal fly ash using epigeic and epi-endogeic earthworm species: nutrient dynamics and metal remediation. RSC Adv. 7, 4876–4890.

Verheijen, F., Jeffery, S., Bastos, A., Van Der Velde, M., Diafas, I., 2010. Biochar Application to Soils. *A Critical Scientific Review of Effects on Soil Properties, Processes, and Functions*. (Eur, 24099)p. 162.

Villar, I., Alves, D., Mato, S., 2017. Product quality and microbial dynamics during vermicomposting and maturation of compost from pig manure. Waste Manag. 69, 498–507.

Wang, L., Hung, Y.-T., Li, K., 2009. Vermicomposting process. In: Wang, L., Pereira, N., Hung, Y.-T. (Eds.), Biological Treatment Processes. Humana Press, USA.

Yadav, A., Garg, V.K., 2009. Feasibility of nutrient recovery from industrial sludge by vermicomposting technology. J. Hazard. Mater. 168, 262–268.

Yadav, A., Garg, V.K., 2011. Industrial wastes and sludges management by vermicomposting. Rev. Environ. Sci. Biotechnol. 10, 243–276.

Yadav, A., Garg, V.K., 2016. Influence of stocking density on the vermicomposting of an effluent treatment plant sludge amended with cow dung. Environ. Sci. Pollut. Res. 23, 13317–13326.

Yadav, K.D., Tare, V., Ahammed, M.M., 2012. Integrated composting—vermicomposting process for stabilization of human faecal slurry. Ecol. Eng. 47, 24–29.

Yang, F., Li, G., Zang, B., Zhang, Z., 2017. The maturity and CH_4, N_2O, NH_3 emissions from vermicomposting with agricultural waste. Compost. Sci. Util. 25, 262–271.

Biosurfactants for oil recovery from refinery sludge: Magnetic nanoparticles assisted purification

6

Hemen Sarma, Karla Liliana Tarango Bustamante†, Majeti Narasimha Vara Prasad‡*
*Department of Botany, N N Saikia College, Titabar, India, †Department of Nanotechnology, University Technical of City Juarez, Juarez, Mexico, ‡School of Life Sciences, University of Hyderabad, Hyderabad, India

1 Refinery sludge—Emerging threat to environment: Composition and residue

The petrochemical industries produce huge quantities of oily sludge during the crude oil refining process that includes different sources such as slop oil emulsions, crude oil tank bottoms, onsite effluent treatment ponds, and oil or water separators (Hu et al., 2013). These oily sludges consist of unruly mixtures of total petroleum hydrocarbon (TPHs), drilling cutting fluid, mud, heavy metals, and water. That makes them harder to dissolve and sustain onsite for prolonged periods during the development of oilfields and refineries (Hu et al., 2013). The cleaning process of oil storage tanks in China alone generates about 1,000,000 tons of oil sludge annually—and this annual output is a significant concern for the management of oil sludge (Liu et al., 2011). The refineries in India generate about 28,220 tons of sludge per annum containing phenol (90–100 mg/kg), nickel (17–25 mg/kg), chromium (27–80 mg/kg), zinc (7–80 mg/kg), manganese (19–24 mg/kg), cadmium (0.8–2 mg/kg), copper (32–120 mg/kg), and lead (0.001–0.12 mg/kg) (Bhattacharyya and Shekdar, 2003). The oily sludge is rich in hydrocarbons with high potential energy and the average calorific value is above 3900 kcal/kg (Hou et al., 2013). This larger volume of sludge is good enough to contaminate soil, air, and ground water. Therefore, the US Environmental Protection Agency (USEPA) has listed oily sludge as a hazardous waste (USEPA, 1989). Mitigating the oily sludge that is generated in large quantities by petrochemical industries is a real challenge confronted during management (Fig. 1). The oily sludge contains toxic substances like aromatic hydrocarbons and metals that infiltrate into bio chain through life support system (air, water and soil) resulting degraded the environmental quality (Chang et al., 2000). There are several methods presently applied for oily sludge treatment via various physical and chemical techniques such as stabilization, dewatering and incineration, washing by hot water and surfactant, pyrolysis, solvent extraction, and microbial biodegradation (Jing et al., 2011a,b). All the methods have their own merits and demerits. Physical and chemical methods require expensive

Industrial and Municipal Sludge. https://doi.org/10.1016/B978-0-12-815907-1.00006-4

Fig. 1 The exploration of crude oil from an underground reserve (A), accumulation of oily sludge in a storage tank (B), physical methods of remediation of an oil spill through a landfill, (C) and contamination of refinery sites due to oil spill (D).

solvents and equipment, which increases the operating costs (Fig. 2). Moreover, this process is also complicated and adds too many pollutants during oil recovery from sludge in the form of chemicals that contaminate the environment. Bioremediation strategies are a low-cost oily sludge treatment that saves operating costs, energy, and the requirement of additional chemicals. However, sludge contains oil, so it can therefore be recycled as fuel for energy. In that case, the application of microorganisms to degrade it would be inadvisable (Shie et al., 2000). The synthetic surfactants that are needed to enhance the solubility of sludge are often harmful to the environment (Banat et al., 2004). Presently, extensive efforts have been made to identify the potential surfactant-producing microorganisms for the benefit of humans (Saeki et al., 2009; Lai et al., 2009; Sarachat et al., 2010; Ferhat et al., 2011). These potential bacteria are reportedly used in remediation processes because of their biosurfactant production capacities. The previous research on biosurfactants showed that they also help by cleaning oil storage tanks, increasing flow though pipelines, and enhancing oil recovery from oil reservoirs. The oily sludge is a heterogeneous complex mixture of water-in-oil (W/O) emulsion, typically including 30%–50% of water, 30%–50% of oil, and 10%–12% of solids by mass. It was estimated that more than 10% of crude oil containing sludge has an economic benefit from oil recovery (Ramaswamy et al., 2007; Zhanga et al., 2012). However, the oil, water, and solid content of the four samples were measured separately by using the solvent extraction

Fig. 2 Crude oil spill in a paddy field of Assam (A), high-cost physical and chemical methods were used for remediation (B–D).

method by Hou et al. (2013) in China. The results are slightly different as indicated by Ramaswamy et al. (2007), such as the oil content is 8%–15%, the moisture content is 70%–90%, and the residue content is 4%–15% in the refinery oil sludge. The variation of locations or the use of different extraction processes could be the reason.

In general, the term oil sludge refers to waste generated due to the storage of crude oil—the mixture of total petroleum hydrocarbon (TPH), water, and solids. The ratio of different fractions is subject to the origin of the crude oil or location to some extent. Sludge that contains recoverable oil of less than 40% is considered as low oil sludge, but such sludge is also harmful for the environment. Therefore, sludge needs to be treated before disposal and there are several methods being used to separate the oil, water, and solids. Decanters, tricanters, centrifuges, and hydrocyclones—some of the instruments help to recover oil from sludge but, require large amounts of solvent for these operations and usually contaminate the environment again. Three solvents are currently considered ideal based on their oil recovery performances: cyclohexane, methyl ethyl ketone, and ethyl acetate. Through these solvents, 40% of oil can be recovered after 30 min of extraction. However, the TPH content in the recovered oil was relatively low in concentration because of the presence of emulsified water (Guangji et al., 2015).

The high cost of machinery, the additional requirement of solvents, and technical operations are necessary to manage such oily sludge. But the application of microbial biotechnology to recover the oil from sludge is one of the low-cost alternative processes for this purpose. For example, two *Bacillus* strains effect the separation of oil (97%) from sludge within 48h, which has proven to be an outstanding role of bacteria-mediated removal of oil from sludge (Joseph and Joseph, 2009). Similarly, *Pseudomonas aeruginosa* F-2 was used to recover oil from refinery oily sludge, which resulted with an oil recovery of up to 91.5% (Yan et al., 2012).

2 Oil recovery from refinery sludge: Process and facilities

Many methods have been developed and examined for their efficacy in the recovery of total petroleum hydrocarbons (TPHs) in oily sludge. The process and facilities of oil recovery from sludge include recent research interest due to disposal and the reduction of environmental risk. Although many processes were developed to deal with oil recovery, little attention has been paid to how to handle its various inorganic constituents such as heavy metals. These methods were recognized as efficient for dealing with oil sludge, but at the same, no single specific method has been considered as an elixir. This is because all the methods have different advantages and limitations. The recent development of an oily sludge treatment technique from the petroleum industry is shown in Table 1.

Table 1 The different physical and chemical oil recovery methods for refinery sludge: features and benefits

Methods	Process and facilities	Reference
Solvent extraction	Easy to apply, fast and efficient and suitable for field-scale application. Solvent extraction process used for removing organic compounds (semivolatile and nonvolatile) from refinery sludge contains soil/water/crude oil matrices. In this process, required proportions of organic solvent are added in the sludge and mixed together and extraction was carried out. Overall, 67.5% of PHCs in sludge could be recovered, with most of them in the range of C_9 to C_{25}. Large amounts of organic solvents were used, high cost (US$/m^3): 100–200 and not environmentally sound, unable to treat heavy metals.	Meyer et al. (2006); Taiwo and Otolorin (2009); Al-Zahrani and Putra (2013); Zubaidy and Abouelnasr (2010); El et al. (2010)

Table 1 Continued

Methods	**Process and facilities**	**Reference**
Centrifugation	In this process, a special high-speed rotation device was used to generate a strong centrifugal force; this force separated the components with different densities (water, solids, oil, and pasty mixtures in oily sludge) within a short time. Easy to apply in field conditions. Through this process, 50%–75% of PHCs in sludge could be recovered. Cost (US$/m3): 100–200	da Silva et al. (2012); Nahmad (2012)
Chemical Surfactant	This process helps to remove the water from water/oil emulsion by separating oil and water into two phases, called emulsification. Easy to apply in field conditions. Through this process, 75%–90% of PHCs in sludge could be recovered. Cost (US$/m3): 100–200	Chen and He (2003); Jean and Lee (1999); Lin et al. (2007); Lima et al. (2011); Long et al. (2013); Calvo et al. (2009)
Pyrolysis	The decomposition of organic materials at the absence of oxygen at elevated temperatures (500–1000°C) is called pyrolysis. The pyrolysis process produces hydrocarbons with lower molecular weight in condensation (i.e. liquid) and/or noncondensable gases. It also generates a solid product called char. Easy to apply in field conditions. Through this process, 50%–75% of PHCs in sludge could be recovered. Cost (US$/m3): 100–200	Fonts et al. (2012); Wang et al. (2007); Punnaruttanakun et al. (2003); Shie et al. (2003); Shie et al. (2004);
Microwave irradiation	The microwave irradiation usually performed at a frequency close to 900 MHz or near 2450 MHz. The energy penetrated the sludge directly through molecular interaction with the electromagnetic field, resulting in a heating effect that can be used for the	Appleton et al. (2005); Tan et al. (2007); Chan and Chen (2002); Shang et al. (2006)

Continued

Table 1 Continued

Methods	Process and facilities	Reference
	demulsification of W/O emulsion. The rapidly increasing temperature of emulsions, leading to the reduction of viscosity, which could accelerate the settlement of water droplets in emulsion. The rapid temperature increase can also break heavy hydrocarbons into lighter ones, which helps remove oil from sludge. Through this process, $\geq$90% of PHCs in sludge could be recovered.	
Electro kinetic method	Three main mechanisms involved in electrokinetic separation (1) oily sludge was broken under the influence of an electrical field, leading to the movement of colloidal particles of oily sludge (2) solid phase toward the anode area as a result of electrophoresis (3) the movement of the separated liquid phase (water and oil) towards the cathode area as a result of electroosmosis.	Kuo and Lee (2010); Yang et al. (2005)
Ultrasonic irradiation	Ultrasonic irradiation is a better oil recovery process from refinery sludge and more than 60% of PHCs can be recovered at an ultrasonic power of 75 W.	Song et al. (2012); Kim and Wang (2003);Ye et al. (2008); Hu et al. (2014)
Froth flotation	Froth flotation was used to separate fine solid particles from an aqueous suspension. It involves the capture of oil droplets or small solids by air bubbles in an aqueous slurry, followed by their levitation and collection in a froth layer. Froth flotation has been successfully used to treat oily wastewater from refineries, and its utilization for oily sludge treatment.	Urbina (2003); Ramaswamy et al. (2007)

To reuse the resources and reduce the environmental risk, the research focus on the advancement of existing technologies and the improvement of oil recovery with sludge disposal is important in order to satisfy environmental regulations and resource management.

The presently available processes and facilities that can be good for oily sludge treatment include: (1) landfarming, (2) incineration, (3) solidification/stabilization, (4) solvent extraction, (5) ultrasonic treatment, (6) pyrolysis, (7) photocatalysis, (8) thermochemical cleaning, and (9) biodegradation (Xu et al., 2009; Liu et al., 2009a,b; Mater et al., 2006; Roldan-Carrillo et al., 2012; da Rocha et al., 2010; Zubaidy and Abouelnasr, 2010; Yan et al., 2012). Oil recovery was lower in the single process as compared to combined methods. For example, the oil recovery of the combined ultrasound and thermochemical cleaning treatment was 17.65% higher than only thermochemical cleaning (Jin et al., 2012). Therefore, the application of combined methods in oil recovery from refinery sludge is more productive than the single process. The requirement of a higher amount of solvent to recover oil from sludge is considered expensive and cannot be afforded by developing countries (Taiwo and Otolorin, 2009). However, the thermochemistry procedures such as boiling/heating can help to recover the solvent from treated sludge. This process is low cost and high efficiency, which also reduces the volume of sludge that can be safely disposed (Jing et al., 2011a,b). Furthermore, this treatment helps to knock down the solid sludge by altering properties such as viscosity and surface tension and helping to recover the oil (Li et al., 2011).

The process of biological treatment of refinery sludge for oil recovery might involve a low cost. Even though this low-cost biological treatment has now become popular (Liu et al., 2009a,b), but these bacteria meditate oil recovery is questionable due to the degradation of oil present in the sludge, in a number of cases.

The application of a chemical surfactant (surface modification reagents) in oil recovery from refinery sludge is extensive, and few of them have shown excellent performance under optimal conditions such as: (1) fatty alcohol ethoxylate (AEO-9); (2) peregal O; (3) triton X-100; (4) sodium dodecylbenzene sulfonate; (5) sodium metasilicate nonahydrate; (6) petroleum sulfonate; and (7) sodium silicate (Jing et al., 2011a,b; Li et al., 2011). The sludge washing effect of sodium metasilicate nonahydrate ($Na_2SiO_3 \cdot 9H_2O$) was outstanding and an estimated residual oil of 1.6% was left after treatment (Jing et al., 2011a,b).

The combined application of ultrasound and thermochemical cleaning treatment also helps in oil recovery from the sludge treatment process (Xu et al., 2009; Zhang et al., 2012). Presently, the ultrasound treatment of oily sludge significantly helps the oil recovery process, which has been established to be an ideal treatment method. This is because of the combination of both cavitation effects and mechanical vibration. The case studies on oil recovery from refinery oil sludge through a combined process (ultrasound, freeze/thaw treatment, biosurfactant, and salt amendment) show that oil recovery of up to 80.0% helped establish that biosurfactant enhanced the recovery efficacy of oil from sludge (Zhang et al., 2012).

3 Diversity of biosurfactant producing microorganism—Beneficial uses in oil recovery

Biosurfactants are surface tension-active compounds produced by microorganisms consisting of hydrophilic and hydrophobic domains (Saikia et al., 2012; Vasconcellos et al., 2011; Hoskova et al., 2013; Moya et al., 2015; Varjani, 2017). Chemical characterization of biosurfactants proved that they are structurally diverse molecules that are less harmful and biodegradable than chemical surfactants. Therefore, they are particularly applied for a wide range of environmental applications, including the enhancement of oil recovery processes (Saikia et al., 2012). Biosurfactants have drawn much attention in recent years owing to their unique functional properties such as specificity and low toxicity. They are used in many industries such as petrochemical, mining, organic chemical, metallurgy (mainly bioleaching), agrochemical, food, cosmetic, and pharmaceutical. As of now, many biosurfactants have been classified based on their chemical structure and microbial origin (Yan et al., 2012; Jadhav et al., 2013). The biosurfactants, being a compound consisting of molecules having a polar water-soluble group (acid, peptide cations, anions, mono, di or polysaccharides) attached to a water insoluble hydrocarbon chain (or fatty acids), are an interesting amphiphilic molecule (Raza et al., 2009; Hoskova et al., 2013; Yan et al., 2012; Moya et al., 2015). The most significant advantage of a biosurfactant is the aptness to reduce the surface and interfacial tension of liquids (Banat et al., 2010; Souza et al., 2012; Zhao et al., 2016) due to its diverse structures consisting of various functional properties.

In an environment, bioemulsifiers and biosurfactants are two types of compounds that are amphiphilic in nature. They are produced by a wide range of microorganisms; each exhibits structural differences. Moreover, these compounds require different screening approaches for identification and are broadly divided into two main classes (Banat et al., 2010; Yan et al., 2012; Moya et al., 2015; Uzoigwe et al., 2015). These includes (1) Biosurfactant (Perfumo et al., 2009; Satpute et al., 2010; Uzoigwe et al. 2015) and (2) bio emulsifiers (Smyth et al., 2010a,b; Markande et al., 2013; Sekhon-Randhawa, 2014). Both compounds have microbial origins but they have important differences between their physicochemical properties and physiological roles (Chibuzo Uzoigwe et al., 2015). The composition of biosurfactants is mainly (1) glycolipids (rhamnolipids, sophorolipids, trehalose lipids) consisting of different sugars linked to β-hydroxy fatty acids; and (2) lipopeptides (surfactin, iturin, fengycin) consisting of cycloheptapeptides with amino acids linked to fatty acids. The composition of bioemulsifiers includes complex mixtures of proteins, lipoproteins, lipopolysaccharides, and heteropolysaccharides (Sekhon-Randhawa, 2014; Uzoigwe et al., 2015). The bioemulsifiers possess only emulsifying activity and no surface activity because they can efficiently emulsify two immiscible liquids such as hydrocarbons or other hydrophobic substrates. However, they are less effective at surface tension reduction. In many environmental applications such as oil pollution remediation, these compounds help to solubilize the substrates (hydrocarbons) and thus make them available for biodegradation (Uzoigwe et al., 2015). It is interesting to note that, although bioemulsifiers and biosurfactants are both amphiphilic in nature, bioemulsifiers are more effective in stabilizing emulsions (oil-in-water) while biosurfactants have the capacity to reduce surface and interfacial tension (Uad et al., 2010; Varjani et al., 2014). Present research shows that some of the promising microbial genera have the distinct capacity to produce biosurfactants from the body in different substrates, including: (1) *Pseudomonas* sp., (2) *Bacillus* sp., (3) *Azotobacter* sp., (4) *Acinetobacter* sp., (5) *Stenotrophomonas* sp., (6) *Enterobacter* sp., (7) *Oceanobacillus* sp., (8) *Rhodococcus* sp., (9) *Serratia* sp., and (10) *Candida* sp. (Saikia et al., 2012; Hoskova et al., 2013; Jadhav et al., 2013; Nalini and Parthasarathi, 2013; Kavitha et al., 2014; Tiwari et al., 2016; Varjani, 2017). Many of the biosurfactant-producing microorganisms are also reported to degrade aromatic hydrocarbons. However, in recent years, studies have shown that biosurfactants not only help the bioremediation of a crude oil degraded site but also enhance oil recovery from sludge (Tabatabaee et al., 2005).

Recently, a large variety of biosurfactants and biosurfactant-producing microorganisms have been explored (see Table 2). There is substantial evidence of microorganisms utilizing organic compounds as a source of carbon and energy for their growth. In the presence of an insoluble carbon source (e.g., hydrocarbon), microorganisms facilitate their dissemination and assimilation into the cell through a surface-active substance such as biosurfactants. A few of the bacteria and fungi secrete ionic surfactants that emulsify the hydrocarbons in a cultural medium.

Table 2 Biosurfactants, producing organisms along with their beneficial uses

Microorganism	Type of biosurfactant	Application/ substrate	Study
Azobacter chroococcum	Lipopeptide	Environmental uses	Makkar et al. (2011)
Bacillus amyloliquefaciens	Lipopeptides	Rice straw and soybean flour	Zhu et al. (2013a,b)
Bacillus amyloliquefaciens XZ-173	Surfactins	Okara and sugarcane bagasse	Zhu et al. (2013a,b)
Bacillus pumilus UFPEDA448	Lipopeptide	Potato peels	Slivinsk et al. (2012)
Bacillus subtilis BSS	Lipopeptide	Bioremediation of soil contaminated by hydrocarbon	Makkar et al. (2011)
Bacillus subtilis DM-03 and DM-04		Millet	Das and Mukherjee (2007a,b)
Bacillus subtilis HOB2	Lipopeptide	Petroleum recovery, bioremediation of soil and sea environments and food industry	Makkar et al. (2011)

Continued

Table 2 Continued

Microorganism	Type of biosurfactant	Application/ substrate	Study
Bacillus subtilis SPB1	Lipopeptide	Olive leaf residue flour and olive cake flour	Ghribi et al. (2012)
Bacillus subtilis SPB1	Lipopeptide	Tuna fish flour and potato	Zouari et al. (2014)
Bacillus subtilis SPB1	Lipopeptides	Waste flour	Mnif et al. (2013)
Bacillus subtilis ZW-3	Lipopeptide	Pharmaceutical, environment protection, cosmetics and petroleum recovery	Makkar et al. (2011)
Bacillus velezensis H3	Lipopeptide	Industrial strain for lipopeptide production	Makkar et al. (2011)
Brevibacterium aureum MSA13	Lipopeptide	Petroleum recovery	Makkar et al. (2011)
Brevibacterium aureum MSA13	Glycolipid	Pretreated molasses-based medium	Kiran et al. (2010)
Burkholderia plantari DSM 9509	Rhamnolipid	Pharmaceutical and detergent industries	Makkar et al. (2011)
Calyptogena soyoae	Mannosylerythritol lipid	Bioremediation in sea environment	Makkar et al. (2011)
Candida bombicola	Sorolipídeos	Environmental uses	Makkar et al. (2011)
Micrococcus luteus BN56	Trehalose tetraester	Bioremediation of soil contaminated by petroleum	Makkar et al. (2011)
Nocardiopsis alba MSA10	Lipopeptide	Bioremediation	Makkar et al. (2011)
Nocardiopsis lucentencisMSA04	Glycolipid	Bioremediation in sea environment	Makkar et al. (2011)
Nocardiopsis lucentensis MSA04	Protein-polysaccharide-lipid	Tannery treated sludge, and wheat bran	Kiran et al. (2014)
Nocardiopsis sp. MSA13 A	Glycolipid	Treated molasses (distillery waste), tannery pretreated sludge, pretreated molasses	Kiran et al. (2014)
P. aeruginosa UFPEDA 614	Rhamnolipid	Bioremediation	Makkar et al. (2011)

Table 2 Continued

Microorganism	Type of biosurfactant	Application/ substrate	Study
Phialemonium sp.	Surfactins	Rice straw and soybean flour	Zhu et al. (2013a,b)
Pseudomonas aeruginosa BS20	Rhamnolipid	Bioremediation of soil contaminated by hydrocarbon	Makkar et al. (2011)
Pseudomonas aeruginosa S2	Rhamnolipid	Bioremediation of soil contaminated by petroleum	Makkar et al. (2011)
Pseudomonas aeruginosa UFPEDA 614	–	Corn bran	Camilos-Neto et al. (2011)
Pseudomonas alcaligenes	Rhamnolipid	Environmental uses	Makkar et al. (2011)
Pseudomonas fluorescens BDS	Lipopeptide	Bioremediation and Biomedicine	Makkar et al. (2011)
Pseudomonas koreensis	Lipopeptide	Biologic control agent	Makkar et al. (2011)
Pseudomonas libanesis M9-3	Lipopeptide	Environmental and biomedical uses	Makkar et al. (2011)
Pseudoxanthomonas sp. *PNK-04*	Rhamnolipid	Environmental uses	Makkar et al. (2011)
Pseudozyma graminicola CBS 10092	Mannosylerythritol lipid	Washing detergent	Makkar et al. (2011)
Pseudozyma hubeiensis	Glycolipid	Bioremediation in sea environment	Makkar et al. (2011)
Pseudozyma parantarctica	Mannosylmannitol lipid	Detergent or washing emulsifiers	Makkar et al. (2011)
Pseudozyma siamesis CBS 9960	Mannosylerythritol lipid	Promising yeast biosurfactant	Makkar et al. (2011)
R. wratislaviensis BN 38	Glycolipid	Bioremediation uses	Makkar et al. (2011)
Rhodococcus erythropolis 3C-9	Glycolipid and Trehalose Lipid	Oil spill cleaning operations	Makkar et al. (2011)
Rhodococcus sp. TW53	Lipopeptide	Bioremediation in sea environment	Makkar et al. (2011)
Starmerella bombicola	Lipopeptides	Oil cake (refining industry)	Jiménez Peñalver et al. (2016)
Starmerella bombicola NRRL Y-17069	Surfactins	Wheat bran, isabgol husk	Parekh and Pandit (2012)

Although a number of bacterial strains have been investigated for the production of biosurfactants, the investigation on fungi capable of producing biosurfactants is meager. Some of the reported fungi include *Candida bombicola* (Casas et al., 1997), *Candida ishiwadae* (Thanomsub et al., 2004), *Candida lipolytica* (Sarubbo et al., 2007), *Candida batistae* (Konishi et al., 2008), *Aspergillus ustus* (Alejandro et al., 2011), and *Trichosporonashii* (Chandran and Das, 2010). These fungi have been reported to produce the sophorolipid (glycolipids) type of biosurfactant on low-cost raw materials.

This organism has been produced by different types of biosurfactants for human benefit, or, in other words, to be used for rebuilding the environment. The classified biosurfactants based on their structure include: (1) glycolipids, (2) lipopeptides, (3) phospholipids, (4) polymeric biosurfactants, and (5) particulate biosurfactants (Vasconcellos et al., 2011; Jadhav et al., 2013; Gudina et al., 2016). However, the production of a biosurfactant is influenced by the growth of the microbe on a particular substrate. Therefore, to improve production of the biosurfactant, there is a need to optimize different ecological, physiological, and nutritional conditions.

The extensive research that has been conducted in the last few decades ensures the production and activity of biosurfactants, emphasizing the need for these compounds (Uad et al., 2010; Yan et al., 2012; Sajna et al., 2015; Gudina et al., 2016; Varjani and Upasani, 2016a,b,c). Presently, various industries such as pharmaceutical, food, cosmetic, detergent, petroleum, and tannery have used biosurfactants (Yan et al., 2012; Sajna et al., 2015; Varjani and Upasani, 2016a,b,c). Therefore, it could be of great significance to explore the beneficial uses of biosurfactant-producing bacteria. Some of the bacteria are shown in Table 3 that produced biosurfactants. Biosurfactants have been applied as foaming agents, emulsifiers, and demulsifiers, and they play an important role in enhanced oil recovery processes from sludge (Volkering et al., 1998).

In oil-polluted environments (oily sludge), biosurfactants enhanced the dispersion and bioavailability of hydrophobic hydrocarbon for microbial access; thus, microbial access helps in degradation through micelle solubilization (Uzoigwe et al., 2015). The critical micelle concentration (CMC) of biosurfactants is several times lower than chemical surfactants, making them more effective then chemical surfactants (Desai and Banat, 1997). Thus, biosurfactants increase the supposed solubility of these hydrocarbons at concentrations above the CMC. This leads to their availability for microbial catabolism and biodegradation (Chang et al., 2008). For this reason, the application of biosurfactants is considered a promising alternative in the petroleum industry, not only for the treatment of hydrocarbon-contaminated soil (Calvo et al., 2009) but also for the recovery of oil from sludge.

Moreover, they also help in cleansing, foaming, phase separation, and the reduction in viscosity of crude oil, allowing them to become feasible for many applications as well as oil recovery from sludge (Uad et al., 2010; Ron and Rosenberg, 2014; Gudina et al., 2016; Varjani and Upasani, 2016a,b,c). Biosurfactants decrease the interfacial surface tension (Volkering et al., 1998) of hydrocarbon, making them readily bioavailable (Banat, 1995; Karanth et al., 1999). Due to this unique property, they have found application in cleaning up the contamination in crude oil-infiltrated soil. They are one of the most preferred surfactants because of their specificity and availability as raw materials for commercial production (Olivera et al., 2003). Therefore, the petroleum industry

Table 3 Types of biosurfactants produced potential for oil recovery from refinery sludge

Biosurfactant	Microorganisms	Study
Alasan	*Acinetobacter radiodurans*	Barkay et al. (1999)
Alasan	*Acinetobacter radioresistens KA53*	Toren et al. (2002)
Di rhamnolipids	*Burkholderia pseudomallei*	Dubeau et al. (2009)
Lichenysins	*Bacillus sp.*	Yakimov et al. (1995)
Mono and di rhamnolipids	*Acinetobacter calcoaceticus*	Rooney et al. (2009); El Azhari et al. (2010)
Mono and di rhamnolipids	*Enterobacter asburiae*	Rooney et al. (2009)
Mono and di rhamnolipids	*Enterobacter hormaechei*	Rooney et al. (2009)
Mono and di rhamnolipids	*Pantoea stewartii*	Rooney et al. (2009)
Mono and di rhamnolipids	*Renibacterium salmoninarum*	Christova et al. (2004)
Rhamnolipids	*Pseudomonas aeruginosa*	Bai et al. (1997); Vandyke et al. (1993)
Sphorose lipids	*Torulopsis bombicola*	Oberbremer et al. (1990)
Surfactin, fengycin, lichenysin, iturin, pumilacidin, bacillomycin	*Bacillus subtilis*	Vater et al. (2002)
Trehalose dimycolipilipids	*Rhodococcus erythropolis*	Bruheim et al. (1997)
Unidentified rhamnolipids	*Nocardioides sp.*	Vasileva-Tonkova and Gesheva (2005)
Unidentified rhamnolipids	*Pseudomonas cepacia*	Onbasli and Aslim (2009)

paid increasing interest to the possible uses of biosurfactants in removing well clogging, heavy crude oil mobilization, oil spill management, oil sludge clean up from tank bottom storage facilities, and enhanced oil recovery from sludge (Manif et al., 2014; Sajna et al., 2015; Varjani and Upasani, 2016a,b,c; Zhao et al., 2016).

The enhanced oil recovery from sludge through thermal and chemical methods is costly and pollutes the environment because of the requirement of additional chemicals (Brown, 2010; Varjani, 2017). Oil recovery from refinery sludge is often expensive; the oil persists in the soil along with toxic aromatic compounds for prolonged periods due to incomplete decomposition of pollutants (Das and Mukherjee, 2007a,b; Farhadian et al., 2008; Manif et al., 2014; Ron and Rosenberg, 2014; Sajna et al., 2015; Tiwari et al., 2016). The application of microbial biotechnology in the oil recovery process is considered the third-most important strategy (tertiary) nowadays after the primary (mechanical) and secondary (physical) oil recovery (Banat, 1995; Pacwa-Plociniczak et al., 2011).

Oil recovery from refinery sludge using biosurfactant-producing bacteria is the current area of research. It was suggested that the optimal environmental condition to determine the efficacy of such technology includes pH, pressure, temperature, locations, porosity, nutrient availability, salinity, and the presence of native microbial flora (Banat, 1995 and Lazar et al., 2007). Although some technical issues in biosurfactant-enhanced oil recovery still exist, it seems to be one of the most ideal low-cost methods that has been inventoried recently to recover a significant proportion of the residual oil from oil sledge (Banat, 1995; Varjani and Upasani, 2016a,b,c).

4 Biosurfactant purification—Using nanomaterials for enhanced oil recovery from sludge

This process established that high-quality biosurfactant allows for more efficient oil recovery from sludge. The quality of the biosurfactant determines the oil recovery from sludge. Hence, many researchers investigate how to produce high-quality biosurfactant. However, for large-scale biosurfactant production, the low production yield and complex processing are considered the main disadvantages in the downstream process (Jain et al., 2013; Al-Bahry et al., 2013). Effort has been made to reduce the limitation in the downstream process, and research was conducted to improve the high yield production of biosurfactant at the upstream process (Mukherjee, 2007). There are different methods that have been discovered in the downstream process related to the purification of biosurfactants (Gautam and Tyagi, 2006). Presently, the filtration method is considered one of the best methods for the production of highly purified biosurfactant solutions. However, this method has some disadvantages that include membrane fouling and operational problems.

The recent advancements of nanotechnology, especially the synthesis of magnetic nanoparticles, has made it possible to utilize them for biosurfactant purification (Grass et al., 2007; Gu et al., 2006; Horak et al., 2007) for enhanced oil recovery from sludge. The magnetic nanoparticle-mediated purification process is comprised of two successive phases: the target compounds such as the biosurfactant bind to the magnetic nanoparticles, and after binding, the whole magnetic complex is separated by an external magnetic field (Grass et al., 2007; Horak et al., 2007; Gupta and Gupta, 2015). This is fully an automated separation process and has no need for expensive equipment (Horak et al., 2007).

The application of zero-valent iron nanoparticles has shown outstanding performance in the purification of rhamnolipid biosurfactant (Sahebnazar et al., 2017). In this study, zero-valent iron nanoparticles were synthesized using 8 g of $FeSO_4.7H_2O$ by dissolving them in 400 mL of methanol/deoxygenated water solution (30% v/v), with a pH of 6. 20 mL of NaBH4 (2.1 M) being added drop-wise to the mixture while stirring. The mixture was centrifuged for 20 min at 8000 rpm, then the sediment deposited at the bottom was cleaned several times with an aqueous methanol solution (50% v/v). This produced the desired magnetic nanoparticles for the purification of the biosurfactant. The purified rhamnolipid biosurfactant exhibited improved viscosity reduction of oily sludge, which consequently enhanced oil recovery from oily sludge by 27.2% more than that by unpurified biosurfactant (Sahebnazar et al., 2017). The impurities present in the biosurfactants were bound to the magnetic nanoparticles and removed easily from the biosurfactant solution under an external magnetic field. Through this process, the purity of the biosurfactant solution increased significantly, which enhanced the oil recovery process. This is because of the higher ability of the purified biosurfactant in the reduction of the interfacial tension compared to unpurified biosurfactant. Therefore, biosurfactant purification using magnetic nanomaterials for enhanced oil recovery could improve oil recovery from petroleum industry sludge for human benefit while reducing contamination (Fig. 3). El-Sheshtawy and Ahmed (2017) studied the effect of iron oxide nanoparticles on biosurfactant production as well as on the effect of biosurfactant and nanoparticle concentrations on the biodegradation process of petroleum crude oil. The *Bacillus licheniformis* and crude oil from Esh-El-Malaha, Alexandria, Egypt, were used for the investigation. The best degradation was observed in microcosm experiments containing 0.2 g of nanoparticles with biosurfactants.

In addition to the application of biosurfactants, there are other methods for oil recovery from refinery sludge. One such approach is the application of superparamagnetic iron oxide nanoparticles for water purification. Superparamagnetism occurs in small-sized materials (~ nanometers) that are ferromagnetic/ferrimagnetic in their bulk form. The magnetic hysteresis exhibited by ferromagnetic/ferrimagnetic materials is not exhibited by superparamagnetic materials, although they retaining the high susceptibility. The important aspect of superparamagnetic materials in water purification is their nonexhibition of magnetic hysteresis that prevents agglomeration

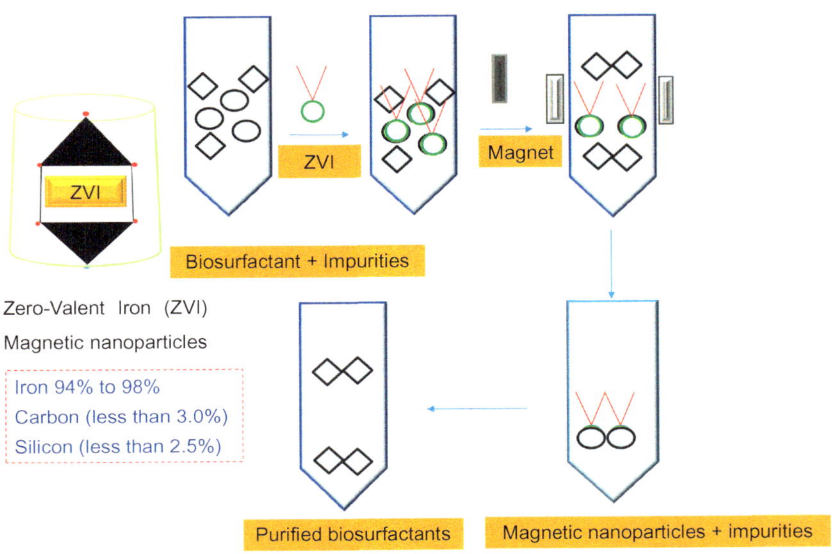

Fig. 3 Schematic view of setup used in biosurfactant purification using nanomaterials (see Sahebnazar et al., 2017).

after the removal of the magnetic field. For use in oil recovery, it is desirable that the magnetic materials exhibit hydrophobic and oleophilic properties (Calcagnile et al., 2012).

Liang et al. (2015) investigated the potential application of oleic acid-coated magnetite nanoparticles for magnetically demulsifying crude-oil-in-water nanoemulsions from oilfields. The nanoparticles were synthesized by the coprecipitation method. The crude oil was diluted with cyclohexane, followed by the addition of deionized water to produce crude oil nanoemulsion. The magnetite nanoparticles were mixed with crude oil nanoemulsions. A 5000 G magnet was used to remove the magnetic nanoparticles along with the adhered oil. The residual oil content was measured by monitoring the absorbance of it using a UV-vis spectrophotometer. The demulsification efficiency of the oleic acid-coated nanoparticles was significantly higher compared to the bare nanoparticles. The various factors such as dosage and wettability of the demulsifier and the the pH of the cyclohexane nanoemulsion and the demulsification operation process were found to affect the demulsification efficiency. A water contact angle of $\sim$90 degrees, pH in the range 4.0–7.5, and a multistep demulsification process were found to lead to higher demulsification efficiency. The oleic acid-coated magnetite nanoparticles showed good recyclability with no significant change after five cycles of operation. Pavia-Sanders et al. (2013) reported on the application of magnetic/polymer hybrid nanoparticles for crude oil recovery in aqueous environments. Iron oxide nanoparticles were synthesized by the thermal decomposition method with the use of oleic acid and oleylamine as the surfactant and cosurfactant, respectively. A diblock copolymer PAA_{20}-b-PS_{280} was synthesized followed by coassembling with the iron oxide nanoparticles, leading to the formation of amphibilic magnetic shell cross-linked knedel-like nanoparticles (MSCKs). The crude oil from the Texas-Oklahoma Enbridge pipeline was weathered and deionized water was added to it in order to mimic contaminated water. The MSCKs were added to the crude oil contaminated water. An external magnetic field was used to attract the MSCKs to allow the decanting of the contaminated water. The capacity of oil sequestration by MSCKs was quantified by gel permeation chromatography. The sorption capacity of the MSCKs and the percentage of oil recovery were investigated for different MSCK to oil ratios. An increased sorption capacity and oil recovery was observed with increasing ratios up to 1:16.8. With the further increase of the initial oil concentration, a decreased efficiency occurred as the viscosity of the oil layer resulted in magnetic immobilization of the MSCKs. Palchoudhury et al (2014). prepared polyvinylpyrrolidone-coated iron oxide nanoparticles that were used for carrying out oil recovery experiments. Crude oil from the Deepwater Horizontal spill (Aecom Environment) was mixed with ultrapure water (Milipore) to prepare the oil samples. Iron oxide nanoparticles were added to the oil samples, followed by magnetic separation. The magnetically separated oil-coated nanoparticles were sonicated in tert-butyl ether for oil recovery. Various techniques such as UV-vis spectroscopy, gas-chromatography-mass spectrometry dynamic light scattering and atomic force microscopy were used to characterize the oil sample, oil-coated nanoparticles, and the clear solution (oil sample after magnetic separation) to measure the oil recovery. The oil separation experiments were carried out at various oil to nanoparticle weight ratios and approximately 100% oil recovery under optimum conditions was observed. Nguyen et al. (2016) utilized buried red mud waste from groundwater refineries for the synthesis of Fe_3O_4 magnetic nanoparticles, followed by coating with oleic acid. The oleic acid-coated magnetic nanoparticles were encapsulated by a copolymer of methyl methacrylate and sodium 2-acry-lamido-2-methyl-propanesulfonate. The nanofluid of the copolymer/magnetic nanocomposite thus obtained was applied as an oil recovery agent in reservoirs of the White Tiger Oligocene oilfield of Vietnam. Recently, Esmaeilnezhad et al. (2018) reported on the potential application of polymer-coated magnetic-based magnetorheological fluid for oil production. The chemically synthesized Fe_3O_4 nanoparticles were coated with polyacrylamide. Magnetorheological fluids were prepared in two carrier fluids, silicone oil and water, and their possible applications for oil production were discussed. Chemical oxidation is another emerging route for the remediation of oil hydrocarbon contaminants (Yen et al., 2011).

Desalegn et al. (2018) evaluated the performance of green mango peel-nanozerovalent iron as an activator of persulfate oxidation of petroleum hydrocarbons. The oil sludge contaminated mango peel was air dried, ground to powder, boiled in deionized water followed by centrifugation. The mango peel extract thus obtained was added drop-wise into the aqueous solution of iron chloride to obtain the mango peel nanozerovalent iron. Sieved oil sludge contaminated oil was mixed with persulfate solution, followed by the addition of the green mango peel-nanozerovalent iron. The concentrations of total petroleum hydrocarbon and persulfate were monitored at various time intervals as a measure of the rate of degradation. The green mango peel-nanozerovalent iron demonstrated >90% degradation over one week of treatment.

Current opinion is that the neon-purification of biosurfactant could have the potential to provide a solution for recent challenges faced by petroleum refineries for oil recovery from sludge. This technology has some new trends in a low-cost filtration process for removing impurities from biosurfactants. Despite their enormous potential, the application of biosurfactants for oil recovery in a large scale requires detailed studies to identify their potential toxic effects on the environment. Even though they are thought to be ecofriendly and safe, under certain conditions they may exhibit undesirable toxic effects to the environment. However, cautious and controlled use of these biologically active molecules will certainly help in the enhanced recovery of crude oil from refinery sludge.

5 Conclusion

This chapter summarizes the different biosurfactant-producing bacteria, the types of different biosurfactants they produce, and their application in sludge treatment. The use of biosurfactants for oil recovery has gained popularity due to their environmentally safe use. However, due to their high production cost, the present commercial success of biosurfactants in the oil recovery process is limited. Therefore, the production of high-quality biosurfactants using low-cost economically feasible renewable substrates is a current research challenge. The multistep downstream process and optimized environmental conditions would help to produce high-quality biosurfactants for a more profitable solution. It has been established that purified biosurfactants (magnetic nanomaterials assisted) are able to decrease the viscosity of oily sludge, which is higher than unpurified biosurfactants. Therefore, effective screening methodologies and improved purification techniques play a vital role in obtaining the desired quality and quantity of biosurfactant. It was assumed that in future use, the biosurfactant-based oil recovery could overcome the limitations of traditional methods such as low efficiency and lack of flexibility. Because of this, more biotechnological research has been started in order to successfully overcome these limitations.

Acknowledgment

The research was supported by Grant BT/489/NE/TBP/2013, financed by the Ministry of Science and Technology (India). The authors thank Jessica Canales for technical support.

References

Al-Bahry, Y., Al-Wahaibi, A., Elshafie, A., Al-Bemani, S., Joshi, H., Al-Makhmari, H., Al-Sulaimani, 2013. Biosurfactant production by Bacillus subtilis B20 using date molasses and its possible application in enhanced oil recovery. Int. Biodeterior. Biodegrad. 81, 141–146.

Alejandro, C.S., Humberto, H.S., Maria, J.F., 2011. Production of glycolipids with antimicrobial activity by Ustilago maydis FBD12 in submerged culture. Afr. J. Microbiol. Res. 5, 2512–2523.

Al-Zahrani, S.M., Putra, M.D., 2013. Used lubricating oil regeneration by various sol-vent extraction techniques. J. Ind. Eng. Chem. 19, 536–539.

Appleton, T.J., Colder, R.I., Kingman, S.W., Lowndes, I.S., Read, A.G., 2005. Microwave technology for energy-efficient processing of waste. Appl. Energy 81, 85–113.

Bai, G., Brusseau, M.L., Miller, R.M., 1997. Biosurfactant-enhanced removal of residual hydrocarbon from soil. J. Contam. Hydrol. 25, 157–170.

Banat, I.M., 1995. Biosurfactants production and possible uses in microbial enhanced oil recovery and oil pollution remediation: a review. Bioresour. Technol. 51, 1–12.

Banat, I.M., Marchant, R., Rahman, T.J., 2004. *Geobacillus debilis* sp. nov., a novel obligately thermophilic bacterium isolated from a cool soil environment, and reassignment of *Bacillus pallidus* to *Geobacillus pallidus* comb. nov. Int. J. Syst.Evol. Microbiol. 54, 2197–2201.

Banat, I.M., Franzetti, A., Gandolfi, I., Bestetti, G., Martinotti, M.G., Fracchia, L., Smyth, T.J., Marchant, R., 2010. Microbial biosurfactants production, applications and future potential. Appl. Microbiol. Biotechnol. 87, 427–444.

Barkay, T., Navon-Venezia, S., Ron, E.,.Z., Rosenberg, E., 1999. Enhancement of solubilisation and biodegradation of polyaromatic hydrocarbons by the bioemulsifier alasan. Appl. Environ. Microbiol. 65, 2697–2702.

Bhattacharyya, J.K., Shekdar, A.V., 2003. Treatment and disposal of refinery sludges: Indian scenario. Waste Manag. Res. 21 (3), 249–261.

Brown, L., 2010. Microbial enhanced oil recovery (MEOR). Curr. Opin. Microbiol. 13, 316–320.

Bruheim, P., Bredholt, H., Eimhjellen, K., 1997. Bacterial degradation of emulsified crude oil and the effect of various surfactants. Can. J. Microbiol. 43, 17–22.

Calcagnile, P., Fragouli, D., Bayer, I.S., Anyfantis, G.C., Martiradonna, L., Cozzoli, P.D., Cingolani, R., Athanassiou, A., 2012. Magnetically driven floating foams for the removal of oil contaminants from water. ACS Nano 6, 5413–5419.

Calvo, C., Manzanera, M., Silva-Castro, G.A., Uad, I., Gonzalez-Lopez, J., 2009. Application of bioemulsifiers in soil oil bioremediation processes. Future prospects. Sci. Total Environ. 407, 3634–3640.

Camilos-Neto, D., Bugay, C., Santana-Filho, A.P., Justin, T., Souza, L.M., Sassaki, G.L., Mitchell, D.A., Krieger, N., 2011. Production of rhamnolipids in solid-state cultivation using a mixture of sugarcane bagasse and corn bran supplemented with glycerol and soybean oil. Appl. Microbiol. Biotechnol. 89, 1395–1403.

Casas, J.A., de Lara, S.G., Garcia-Ochoa, F., 1997. Optimization of a synthetic medium for *Candida bombicola* growth using factorial design of experiments. Enzym. Microb. Technol. 21, 221–229.

Chan, C.C., Chen, Y.C., 2002. Demulsification of W/O emulsions by microwave radiation. Sep. Sci. Technol. 37, 3407–3420.

Chandran, P., Das, N., 2010. Biosurfactant production and diesel oil degradation by yeast species *Trichosporon asahii* isolated from petroleum hydrocarbon contaminated soil. Int. J. Eng. Sci. Technol. 2, 6942–6953.

Chang, C.Y., Shie, J.L., Lin, J.P., Wu, C.H., Lee, D.J., Chang, C.F., 2000. Major products obtained from the pyrolysis of oil sludge. Energy Fuel 14, 1176–1183.

Chang, M.W., Holoman, T.P., Yi, H., 2008. Molecular characterization of surfactant-driven microbial community changes in anaerobic phenanthrene-degrading cultures under methanogenic conditions. Biotechnol. Lett. 30, 1595–1601.

Chen, G.H., He, G.H., 2003. Separation of water and oil from water-in-oil emulsion by freeze/thaw method. Sep. Purif. Technol. 31, 83–89.

Christova, N., Tuleva, B., Lalchev, Z., Jordanova, A., Jordanov, B., 2004. Rhamnolipid biosurfactants produced by *Renibacterium salmoninarum* 27BN during growth on n-hexadecane. Z. Nat. Forsch. C. J. Biosci. 59, 70–74.

da Rocha, O.R.S., Dantas, R.F., Duarte, M.M.M.B., Duarte, M.M.L., da Silva, V.L., 2010. Oil sludge treatment by photocatalysis applying black and white light. Chem. Eng. J. 157, 80–85.

da Silva, V.L., Alves, F.C., de Franc, F.P., 2012. A review of the technological solutions for the treatment of oily sludges from petroleum refineries. Waste Manage. Res. 30, 1016–1030.

Das, K., Mukherjee, A.K., 2007a. Comparison of lipopeptide biosurfactants production by *Bacillus subtilis* strains in submerged and solid-state fermentation systems using a cheap carbon source: some industrial applications of biosurfactants. Process Biochem. 42, 1191–1199.

Das, K., Mukherjee, A.K., 2007b. Crude petroleum oil biodegradation efficiency of *Bacillus subtilis* and *Pseudomonas aeruginosa* strains isolated from a petroleum oil contaminated soil from North-East India. Bioresour. Technol. 98 (7), 1339–1345.

Desai, J.D., Banat, I.M., 1997. Microbial production of surfactants and their commercial potential. Microbiol. Mol. Biol. Rev. 61, 47–64.

Desalegn, B., Maharaj, M., Chen, Z., Naidu, R., 2018. Green mango peel-nanozerovalent ironactivated persulfate oxidation of petroleum hydrocarbons in oil sludge contaminated soil. Environ. Technol. Innov. 11, 142–152.

Dubeau, D., Déziel, E., Woods, D., Lépine, F., 2009. Burkholderia thailandensis harbors two identical rhl gene clusters responsible for the biosynthesis of rhamnolipids. BMC Microbiol. 9, 263.

El Azhari, N., Devers-Lamrani, M., Chatagnier, G., Rouard, N., Martin-Laurent, F., 2010. Molecular analysis of the catechol-degrading bacterial community in a coal wasteland heavily contaminated with PAHs. J. Hazard. Mater. 117, 593–601.

El, A., Saad, Y.E.A., Kandil, A.T., Elmoher, H.O., 2010. Petroleum cuts as solvent extractor for oil recovery Naggar from petroleum sludge. J. Petrol. Technol. Altern. Fuels 1, 10–19.

El-Sheshtawy, H.S., Ahmed, W., 2017. Bioremediation of crude oil by *Bacillus licheniformis* in the presence of different concentration of nanoparticles and produced biosurfactant. Int. J. Environ. Sci. Technol. 14, 1603–1614.

Esmaeilnezhad, E., Choi, H.J., Schaffie, M., Gholizadeh, M., Ranjbar, M., 2018. Polymer coated magnetite-based magnetorheological fluid and its potential clean procedure applications to oil production. J. Clean. Prod. 171, 45–56.

Farhadian, M., Vachelard, C., Duchez, D., Larroche, C., 2008. In situ bioremediation of monoaromatic pollutants in groundwater: a review. Bioresour. Technol. 9, 5296–5308.

Ferhat, S., Mnif, S., Badis, A., Eddouaouda, K., Alouaoui, R., Boucherit, A., Mhiri, N., Moulai-Mostefa, N., Sayadi, S., 2011. Screening and preliminary characterization of biosurfactants produced by Ochrobactrum sp. 1C and Brevibacterium sp. 7 Gisolated from hydrocarbon-contaminated soils. Int. Biodeterior. Biodegrad. 65, 1182–1188.

Fonts, I., Gea, G., Azuara, M., Ábrego, J., Arauzo, J., 2012. Sewage sludge pyrolysis for liquid production: a review. Renew. Sust. Energ. Rev. 16, 2781–2805.

Gautam, K., Tyagi, V., 2006. Microbial surfactants: a review. J. Oleo Sci. 55, 155–166.

Ghribi, D., Abdelkefi-Mesrati, L., Mnif, I., Kammoun, R., Ayadi, I., Saadaoui, I., Maktouf, S., Chaabouni-Ellouze, S., 2012. Investigation of antimicrobial activity and statistical optimization of Bacillus subtilis SPB1biosurfactant production in solid-state fermentation. J. Biomed. Biotechnol. 20, 1–12.

Grass, R.N., Athanassiou, E.K., Stark, W.J., 2007. Covalently functionalized cobalt nanoparticles as a platform for magnetic separations in organic synthesis. Angew. Chem. Int. Ed. 46, 4909–4912.

Gu, H., Xu, K., Xu, C., Xu, B., 2006. Bifunctional magnetic nanoparticles for protein separation and pathogen detection. Chem. Commun. 941–949.

Guangji, H., Jianbing, L., Haobo, H., 2015. A combination of solvent extraction and freeze thaw for oil recovery from petroleum refinery waste water treatment pond sludge. J. Hazard. Mater. 283, 832–840.

Gudina, E.J., Rodrigues, A.I., de Freitas, V., Azevedo, Z., Teixeira, J.A., Rodrigues, L.R., 2016. Valorization of agro-industrial wastes towards the production of rhamnolipids. Bioresour. Technol. 212, 144–150.

Gupta, A.K., Gupta, M., 2015. Synthesis and surface engineering of iron oxide nanoparticles for biomedical applications. Biomaterials 26, 3995–4021.

Horak, D., Babič, M., Mackova, H., Beneš, M.J., 2007. Preparation and properties of magnetic nano-and microsized particles for biological and environmental separations. J. Sep. Sci. 30, 1751–1772.

Hoskova, M., Schreiberova, O., Jezdik, R., Chudoba, J., Masak, J., Sigler, K., Rezanka, T., 2013. Characterization of rhamnolipids produced by nonpathogenic Acinetobacter and Enterobacter bacteria. Bioresour. Technol. 130, 510–516.

Hou, S.X., Xie, M., Chen, Y.J., Hao, D., Wang, R.S., 2013. The treatment of refinery heavy oil sludge. Pet. Sci. Technol. 31 (5), 458–464.

Hu, G., Li, J., Zeng, G., 2013. Recent development in the treatment of oily sludge from petroleum industry: a review J. Hazard. Mater. 261, 470–490.

Hu, G., Li, J., Thring, R.W., Arocena, J., 2014. Ultrasonic oil recovery and salt removal from refinery tank bottom sludge. J. Environ. Sci. Health A Tox. Hazard. Subst. Environ. Eng. 49 (12), 1425–1435.

Jadhav, V.V., Yadav, A., Shouche, Y.S., Aphale, S., Moghe, A., Pillai, S., Arora, A., Bhadekar, R.K., 2013. Studies on biosurfactant from Oceanobacillus sp. BRI 10 isolated from antarctic sea water. Desalination. 318, 64–71.

Jain, K., Mody, N., Joshi, A., Mishra, B., 2013. Effect of unconventional carbon sources on biosurfactant production and its application in bioremediation. Intern. J. Boil. Macromol. 62, 52–58.

Jean, D.S., Lee, D.J., 1999. Expression deliquoring of oily sludge from a petroleumrefinery plant. Waste Manag. 19, 349–354.

Jime´nez-Pen˜alver, P., Gea, T., Sa´nchez, A., Font, X., 2016. Production of sophorolipids from winterization oil cake by solid-state fermentation: optimization, monitoring and effect of mixing. Biochem. Eng. J. 115, 93–100.

Jin, Y., Xiaoyuan, Z., Xiaoliang, C., Yong, C., Jianhua, Y., Kefa, C., 2012. Oil recovery from oil sludge through combined ultrasound and thermochemical cleaning treatment. Ind. Eng. Chem. Res. 51, 9213–9217.

Jing, G., Chen, T., Luan, M., 2011a. Studying oily sludge treatment by thermo chemistry. Arab. J. Chem. 9, S457–S460. https://doi.org/10.1016/j.arabjc.2011.06.007.

Jing, G., Luan, M., Chen, T., 2011b. Prospects for development of oily sludge treatment. Chem. Technol. Fuels Oils 47, 312–326.

Joseph, P., Joseph, J., 2009. Microbial enhanced separation of oil from a petroleum refinery sludge. J. Hazard. Mater. 161, 522–525.

Karanth, N.G.K., Deo, P.G., Veenanadig, N.K., 1999. Microbial production of biosurfactants and their importance. Curr. Sci. 77, 116–126.

Kavitha, V., Mandal, A.B., Gnanamani, A., 2014. Microbial biosurfactant mediated removal and/or solubilization of crude oil contamination from soil and aqueous phase: an approach with *Bacillus licheniformis* MTCC 5514. Int. Biodeterior. Biodegrad. 94, 24–30.

Kim, Y.U., Wang, M.C., 2003. Effect of ultrasound on oil removal from solids. Ultrason. Sonochem. 41, 539–542.

Kiran, G.S., Thomas, T.A., Selvin, J., Sabarathnam, B., Lipton, A.P., 2010. *Optimization and characterization of a new lipopeptide biosurfactant produced by marine Brevibacterium aureum* MSA13 in solid state culture. Bioresour. Technol. 101, 2389–2396.

Kiran, G.A., Nishanth, L.A., Priyadharshini, S., Anitha, K., Selvin, J., 2014. Effect of Fe nanoparticle on growth and glycolipid biosurfactant production under solid state culture by marine Nocardiopsis sp. MSA13A. BMC Biotechnol. 14, 48.

Konishi, M., Fukuoka, T., Morita, T., Imura, T., Kitamoto, D., 2008. Production of new types of sophorolipids by Candida batistae. J. Oleo Sci. 57, 359–369.

Kuo, C.H., Lee, C.L., 2010. Treatment of oil/water emulsions using seawater-assisted microwave irradiation. Sep. Purif. Technol. 74, 288–293.

Lai, C.C., Huang, Y.C., Wei, Y.H., Chang, J.S., 2009. Biosurfactant-enhanced removal of total petroleum hydrocarbons from contaminated soil. J. Hazard. Mater. 167, 609–614.

Lazar, I., Petrisor, I.G., Yen, T.F., 2007. Microbial Enhanced Oil Recovery (MEOR). Pet. Sci. Technol. 25 (11), 1353–1366.

Li, X.B., Liu, J.T., Xiao, Y.Q., Xiao, X., 2011. Modification technology for separation of oily sludge. J. Cent. South Univ. Technol. 18, 367.

Liang, J., Du, N., Song, S., Hou, W., 2015. Magnetic emulsification of diluted crude oil-in-water nanoemulsions using oleic acid-coated magnetite nanoparticles. Colloids Surf. A Physicochem. Eng. Asp. 466, 197–202.

Lima, T.M.S., Fonseca, A.F., Leão, B.A., Mounteer, A.H., Tótola, M.R., Borges, A.C., 2011. Oil recovery from fuel oil storage tank sludge using biosurfactants. J. Bioremed. Biodegrad. 2, 1–5.

Lin, G., He, G., Li, X., Peng, L., Dong, S., Gu, S., Xiao, G., 2007. Freeze/thaw induced demulsification of water-in-oil emulsions with loosely packed droplets. Sep. Purif. Technol. 56, 175–183.

Liu, J., Jiang, X., Zhou, L., Han, X., Cui, Z., 2009a. Pyrolysis treatment of oil sludge and model-free kinetics analysis. J. Hazard. Mater. 161, 1208–1215.

Liu, W.X., Luo, Y.M., Teng, Y., Li, Z.G., Christie, P., 2009b. Prepared bed bioremediation oily sludge in an oilfield in northern China. J. Hazard. Mater. 161, 479.

Liu, J., Jiang, X., Han, X., 2011. Devolatilization of oil sludge in a lab-scale bubbling fluidized bed. J. Hazard. Mater. 185, 1205–1213.

Long, X., Zhang, G., Shen, C., Sun, G., Wang, R., Yin, L., Meng, Q., 2013. Application of rhamnolipid as a novel biodemulsifier for destabilizing waste crude oil. Bioresour. Technol. 131, 1–5.

Makkar, R.S., Cameotra, S.S., Banat, I.M., 2011. Advances in utilization of renewable substrates for biosurfactant production. AMB Express 1, 1–19.

Manif, S., Sayadi, S., Chamkha, M., 2014. Biodegradative potential and characterization of a novel aromatic-degrading bacterium isolated from a geothermal oil field under saline and thermophilic conditions. Int. Biodeterior. Biodegrad. 86, 258–264.

Markande, A.R., Acharya, S.R., Nerurkar, A.S., 2013. Physicochemical characterization of thermostable glycoprotein bioemulsifier from *Solibacillus silvestris* AM1. Process Biochem. 48, 1800–1808.

Mater, L., Sperb, R.M., Madureira, L., Rosin, A., Correa, A., Radetski, C.M., 2006. Proposal of a sequential treatment methodology for the safe reuse of oil sludge-contaminated soil. J. Hazard. Mater. B 136, 967–971.

Meyer D.S., Brons G.B., Perry R., Wildemeersch S.L.A., Kennedy R.J., 2006. Oil tank sludge removal method, United States Patent, US 2006/0042661 A1

Mnif, I., Elleuch, M., Chaabouni, S.E., Ghribi, D., 2013. Bacillus subtilis SPB1 biosurfactant: production optimization and insecticidal activity against the carob moth Ectomyelois ceratoniae. Crop Prot. 50, 66–72.

Moya, R.I., Tsaousi, K., Rudden, M., Marchant, R., Alameda, E.J., Garcia, R.M., Banat, I.M., 2015. Rhamnolipid and surfactin production from olive oil mill waste as sole carbon source. Bioresour. Technol. 198, 231–236.

Mukherjee, A., 2007. Potential application of cyclic lipopeptide biosurfactants produced by *Bacillus subtilis* strains in laundry detergent formulations. Lett. Appl. Microbiol. 45, 330–335.

Nahmad, D.G., 2012. Method to recover crude oil from sludge or emulsion, United States Patent, US 8,197,667 B2

Nalini, S., Parthasarathi, R., 2013. Biosurfactant production by Serratia rubidaea SNAU02 isolated from hydrocarbon contaminated soil and its physicochemical characterization. Bioresour. Technol. 147, 619–622.

Nguyen, T.P., Le, U.T.P., Ngo, K.T., Pham, K.D., Dinh, L.X., 2016. Synthesis of polymer-coated magnetic nanoparticles from red mud waste for enhanced oil recovery in offshore reservoirs. J. Electron. Mater. 45, 3801–3808.

Oberbremer, A., Müller-Hurtig, R., Wagner, F., 1990. Effect of the addition of microbial surfactants on hydrocarbon degradation in a soil population in a stirred reactor. Appl. Microbiol. Biotechnol. 32, 485–489.

Olivera, N.L., Commendatore, M.G., Delgado, O., Esteves, J.L., 2003. Microbial characterization and hydrocarbon biodegradation potential of natural bilge waste microflora. J. Ind. Microbiol. Biotechnol. 30, 542–548.

Onbasli, D., Aslim, B., 2009. Biosurfactant production in sugar beet molasses by some Pseudomonas spp. J. Environ. Biol. 30, 161–163.

Pacwa-Plociniczak, M., Plaza, G.A., Piotrowska-Seget, Z., Singh, S.C., 2011. Environmental applications of biosurfactants: recent advances. Int. J. Mol. Sci. 12, 633–654.

Palchoudhury, S., Lead, J.R., Facile, A., 2014. Cost-effective method for separation of oil-water mixtures using polymer-coated iron oxide nanoparticles. Environ. Sci. Technol. 48, 14558–14563.

Parekh, V., Pandit, A.B., 2012. Solid-state fermentation (SSF) for the production of sophorolipids from Starmerella bombicola NRRL Y-17069 using glucose, wheat bran and oleic acid. Curr. Trend. Biotechnol. Pharm. 6, 418–424.

Pavia-Sanders, A., Zhang, S., Flores, J.A., Sanders, J.E., Raymond, J.E., Wooley, K.L., 2013. Robust magnetic/polymer hybrid nanoparticles designed for crude oil entrapment and recovery in aqueous environments. ACS Nano 7, 7552–7561.

Perfumo, A., Smith, T.J.P., Marchant, R., Banat, I.M., 2009. Production and roles of biosurfactant and bioemulsifiers in accessing hydrophobic substrates. In: Timmis, K.N. (Ed.), Microbiology of Hydrocarbons, Oils, Lipids and Derived Compounds. Springer-Verlag, Berlin; Heidelberg, pp. 1502–1512.

Punnaruttanakun, P., Meeyoo, V., Kalambaheti, C., Rangsunvigit, P., Rirksom-boon, T., Kitiyanan, B., 2003. Pyrolysis of API separator sludge. J. Anal. Appl. Pyrolysis 68–69, 547–560.

Ramaswamy, D., Kar, D.D., De, S., 2007. A study on recovery of oil from sludge containing oil using froth flotation. J. Environ. Manag. 85, 150–154.

Raza, Z.A., Khalid, Z.M., Banat, I.M., 2009. Characterization of rhamnolipids produced by a *Pseudomonas aeruginosa* mutant strain grown on waste oils. J. Environ. Sci. Health Part A. 44, 367–1373.

Roldan-Carrillo, T., Castorena-Cortés, G., Zapata-Peñasco, I., Reyes-Avila, J., Olguín-Lora, P., 2012. Aerobic biodegradation of sludge with high hydrocarbon content generated by a Mexican natural gas processing facility. J. Environ. Manag. 95 (Supplement), S93–S98.

Ron, E.Z., Rosenberg, E., 2014. Enhanced bioremediation of oil spills in the sea. Curr. Opin. Biotechnol. 27, 191–194.

Rooney, A.P., Price, N.P., Ray, K.J., Kuo, T.M., 2009. Isolation and characterization of rhamnolipid-producing bacterial strains from biodiesel facility. FEMS Microbiol. Lett. 295, 82–87.

Saeki, H., Sasaki, M., Komatsu, K., Miura, A., Matsuda, H., 2009. Oil spill remediation by using the remediation agent JE1058BS that contains a biosurfactant produced by Gordonia sp. strain JE-1058. Bioresour. Technol. 100, 572–577.

Sahebnazar, Z., Dariush, M., Gholamreza, K., Fatemeh, Y., 2017. Zero-valent iron nanoparticles assisted purification of Rhamnolipid for oil recovery improvement from oily sludge. J. Environ. Chem. Eng. 6, 917–922. https://doi.org/10.1016/j.jece.2017.11.043.

Saikia, R.R., Deka, S., Deka, M., Sarma, H., 2012. Optimization of environmental factors for improved production of rhamnolipid biosurfactant by *Pseudomonas aeruginosa* RS29 on glycerol. J. Basic Microbiol. 52, 446–457.

Sajna, K.V., Sukumaran, R.K., Gottumukkala, L.D., Pandey, A., 2015. Crude oil biodegradation aided by biosurfactants from Pseudozyma sp. NII 08165 or its culture broth. Bioresour. Technol. 191, 133–139.

Sarachat, T., Pornsunthorntawee, O., Chavadej, S., Rujiravanit, R., 2010. Purification and concentration of a rhamnolipid biosurfactant produced by *Pseudomonas aeruginosa* SP4 using foam fractionation. Bioresour. Technol. 101, 324–330.

Sarubbo, L.A., Farias, C.B.B., Campos-Takaki, G.M., 2007. Co-utilization of canola oil and glucose on the production of a surfactant by *Candida lipolytica*. Curr. Microbiol. 54, 68–73.
Satpute, S.K., Banpurkar, A.G., Dhakephalkar, P.K., Banat, I.M., Chopade, B., 2010. Methods of investigating biosurfactants and bioemulsifiers: a review. Crit. Rev. Biotechnol. 30, 127–144.
Sekhon-Randhawa, K.K., 2014. Biosurfactants produced genetically manipulated microorganisms: challenges and opportunities. In: Kosaric, N., Sukan, F.V. (Eds.), Biosurfactants. CRC Press, Boca Raton, FL, pp. 49–67.
Shang, H., Snape, C.E., Kingman, S.W., Robinson, J.P., 2006. Microwave treatment of oil-contaminated North Sea drill cuttings in a high-power multimode cavity. Sep. Purif. Technol. 49, 84–90.
Shie, J.L., Chang, C.Y., Lin, J.P., Wu, C.H., Lee, D.J., 2000. Resources recovery of oil sludge by pyrolysis: kinetics study. J. Chem. Technol. Biotechnol. 75, 443–450.
Shie, J.L., Lin, J.P., Chang, C.Y., Lee, D.J., Wu, C.H., 2003. Pyrolysis of oil sludge with additives of sodium and potassium compounds. Resour. Conserv. Recycl. 39, 51–64.
Shie, J.L., Lin, J.P., Chang, C.Y., Shih, S.M., Lee, D.J., Wu, C.H., 2004. Pyrolysis of oil sludge with additives of catalytic solid wastes. J. Anal. Appl. Pyrolysis 71, 695–707.
Slivinsk, C.T.I., Mallmann, E., Araujo, J.M., Mitchell, D.A., Krieger, N., 2012. Production of surfactin by Bacilluspumilus UFPEDA 448 in solid-state fermentation using a medium based on okara with sugarcane bagasse as a bulking agent. Process Biochem. 47, 1848–1855.
Smyth, T.J.P., Perfumo, A., Marchant, R., Banat, I.M., 2010a. Isolation and analysis of low molecular weight microbial glycolipids. In: Timmis, K.N. (Ed.), Hand Book of Hydrocarbon and Lipid Microbiology. Springer-Verlag, Berlin, Heidelberg, pp. 3706–3723. https://doi.org/10.1007/978-3-540-77587-4_291.
Smyth, T.J.P., Perfumo, A., Marchant, R., Banat, I.M., 2010b. Isolation and Analysis of Lipopeptide and high molecular weight biosurfactant. In: Timmis, K.N. (Ed.), Hand Book of Hydrocarbon and Lipid Microbiology. Springer-Verlag, Berlin; Heidelberg, pp. 3687–3704. https://doi.org/10.1007/978-3-540-77587-4-290.
Song, W., Li, J., Zhang, W., Hu, X., Wang, L., 2012. An experimental study on the reme-diation of phenanthrene in soil using ultrasound and soil washing. Environ. Earth Sci. 66, 1487–1496.
Souza, F.A.S.D., Salgueiro, A.A., Albuquerque, C.D.C., 2012. Production of bioemulsifier by *Yarrowiali polytica* in sea water using diesel oil as the carbon source. Braz. J. Chem. Eng. 29, 61–67. https://doi.org/10.1590/S0104-66322012000 100007.
Tabatabaee, A., Mazaheri, M.A., Noohi, A.A., Sajadian, V.A., 2005. Isolation of biosurfactant producing bacteria from oil reservoirs. Iran J. Environ. Health Sci. Eng. 2, 6–12.
Taiwo, E.A., Otolorin, J.A., 2009. Oil recovery from petroleum sludge by solvent extraction. Pet. Sci. Technol. 27, 836–844.
Tan, W., Yang, X., Tan, X., 2007. Study on demulsification of crude oil emulsions by microwave chemical method. Sep. Sci. Technol. 42, 1367–1377.
Thanomsub, B., Watcharachaipong, T., Chotelersak, K., Arunrattiyakorn, P., Nitoda, T., Kanzaki, H., 2004. Monoacylglycerols: Glycolipid biosurfactants produced by a thermotolerant yeast *Candida ishiwadae*. J. Appl. Microbiol. 96, 588–592.
Tiwari, B., Manickam, N., Kumari, S., Tiwari, A., 2016. Biodegradation and dissolution of polyaromatic hydrocarbons by Stenotrophomonas sp. Bioresour. Technol. 216, 1102–1105.
Toren, A., Ron, E.Z., Bekerman, R., Rosenberg, E., 2002. Solubilization of polyaromatic hydrocarbons by recombinant bioemulsifier AlnA. Appl. Microbiol. Biotechnol. 59, 580–584.
Uad, I., Silva-Castro, G.A., Pozo, C., Gonzalez-Lopez, J., Calvo, C., 2010. Biodegradative potential and characterization of bioemulsifiers of marine bacteria isolated from samples

of seawater, sediment and fuel extracted at 4000 m of depth Prestige wreck. Int. Biodeterior. Biodegrad. 64, 511–518.

Urbina, R.H., 2003. Recent developments and advances in formulations and appli-cations of chemical reagents. Miner. Process. Extr. Metall. Rev. 24, 139–182.

USEPA, 1989. Guidance Manual for Hazardous Waste Permits. Office of Solid Waste and Emergency Response, Washington, DC PB. 84-10057.

Uzoigwe, C., Burgess, J.G., Ennis, C.J., Rahman, P.K., 2015. Bioemulsifiers are not biosurfactants and require different screening approaches. Front. Microbiol. 6, 245.

Vandyke, M.I., Couture, P., Brauer, M., Lee, H., Trevors, T.J., 1993. Pseudomonas aeruginosa UG2 rhamnolipid biosurfactants: structural characterization and their use in removing hydrophobic compounds from soil. Can. J. Microbiol. 39, 1071–1078.

Varjani, S.J., 2017. Microbial degradation of petroleum hydrocarbons. Bioresour. Technol. 223, 277–286.

Varjani, S.J., Upasani, V.N., 2016a. Core flood study for enhanced oil recovery through ex-situ bioaugmentation with thermo- and halo-tolerant rhamnolipid produced by *Pseudomonas aeruginosa* NCIM 5514. Bioresour. Technol. 220, 175–182.

Varjani, S.J., Upasani, V.N., 2016b. Carbon spectrum utilization by an indigenous strain of *Pseudomonas aeruginosa* NCIM 5514: production, characterization and surface-active properties of biosurfactant. Bioresour. Technol. 221, 510–516.

Varjani, S.J., Upasani, V.N., 2016c. Biodegradation of petroleum hydrocarbons by oleophilic strain of *Pseudomonas aeruginosa* NCIM 5514. Bioresour. Technol. 222, 195–201.

Varjani, S.J., Rana, D.P., Bateja, S., Sharma, M.C., Upasani, V.N., 2014. Screening and identification of biosurfactant (bioemulsifier) producing bacteria from crude oil contaminated sites of Gujarat, India. Int. J. Innovative Res. Sci. Eng. Technol. 3 (2), 9205–9213.

Vasconcellos, S.P., Dellagnezze, B.M., Wieland, A., Klock, J.H., Santos Neto, E.V., Marsaioli, A.J., Oliveira, V.M., Michaelis, W., 2011. The potential for hydrocarbon biodegradation and production of extracellular polymeric substances by aerobic bacteria isolated from a Brazilian petroleum reservoir. W. J. Microbiol. Biotechnol. 7 (6), 1513–1518.

Vasileva-Tonkova, E., Gesheva, V., 2005. Glycolipids produced by antarctic Nocardioides sp during growth on n-paraffin. Process Biochem. 40, 2387–2391.

Vater, J., Kablitz, B., Wilde, C., Franke, P., Mehta, N., Cameotra, S.S., 2002. Matrix-assisted laser desorption ionization-time of flight mass spectrometry of lipopeptide biosurfactants in whole cells and culture filtrates of *Bacillus subtilis* C-1 isolated from petroleum sludge. Appl. Environ. Microbiol. 68, 6210–6219.

Volkering, F., Breure, A.M., Rulkens, W.H., 1998. Microbiological aspects of surfactant use for biological soil remediation. Biodegradation 8, 401–417.

Wang, Z., Guo, Q., Liu, X., Cao, C., 2007. Low temperature pyrolysis characteristics of oil sludge under various heating conditions. Energy Fuel 21, 957–962.

Xu, N., Wang, W., Han, P., Lu, X., 2009. Effects of ultrasound on oily sludge deoiling. J. Hazard. Mater. 171, 914–917.

Yakimov, M.M., Timmis, K.N., Wray, V., Fredrickson, H.L., 1995. Characterization of a new lipopeptide surfactant produced by thermos tolerant and halotolerant subsurface *Bacillus licheniformis* BAS50. Appl. Environ. Microbiol. 58, 3276–3282.

Yan, P., Lu, M., Yang, Q., Zhang, H., Zhang, Z., Chen, R., 2012. Oil recovery from refinery oily sludge using a rhamnolipid biosurfactant-producing Pseudomonas. Bioresour. Technol. 116, 24–28.

Yang, L., Nakhla, G., Bassi, A., 2005. Electro-kinetic dewatering of oily sludges. J. Hazard. Mater. 125, 130–140.

Ye, G., Lu, X., Han, P., Peng, F., Wang, Y., Shen, X., 2008. Application of ultrasound on crude oil pre-treatment. Chem. Eng. Process. 47, 2346–2350.

Yen, C., Chen, K., Kao, C., Liang, S., Chen, T., 2011. Application of persulfate to remediate petroleum hydrocarbon-contaminated soil: Feasibility and comparison with common oxidants. J. Hazard. Mater. 186, 2097–2102.
Zhang, J., Li, J.B., Thring, R.W., Hu, X., Song, X.Y., 2012. Oil recovery from refinery oily sludge via ultrasound and freeze/thaw. J. Hazard. Mater. 203, 195–204.
Zhanga, J., Li, J., Ronald, W., Thringa, X.H., Xinyua, S., 2012. Oil recovery from refinery oily sludge via ultrasound and freeze/thaw. J. Hazard. Mater. 203– 204, 195–203.
Zhao, F., Zhou, J.-D., Ma, F., Shi, R.-J., Han, S.-Q., Zhang, J., Zhang, Y., 2016. Simultaneous inhibition of sulfate-reducing bacteria, removal of H2S and production of rhamnolipid by recombinant *Pseudomonas stutzeri* Rhl: applications for microbial enhanced oil recovery. Bioresour. Technol. 207, 24–30. 12. R.M.
Zhu, Z., Zhang, F., Wei, Z., Ran, W., Shen, Q., 2013a. The usage of rice straw as a major substrate for the production of surfactin by *Bacillus amyloliquefaciens* XZ-173 in solid state fermentation. J. Environ. Manag. 127, 96–102.
Zhu, Z., Li, R., Yu, G., Ran, W., Shen, Q., 2013b. Enhancement of lipopeptides production in a two-temperaturestageprocess under SSF conditions and its bioprocess in the fermenter. Bioresour. Technol. 127, 209–215.
Zouari, R., Ellouze-Chaabouni, S., Ghribi-Aydi, D., 2014. Optimization of *Bacillus subtilis* SPB1 biosurfactant production under solid-state fermentation using by-products of a traditional olive mill factory. Achievement. Life Sci. 8, 162–169.
Zubaidy, E.A.H., Abouelnasr, D.M., 2010. Fuel recovery from waste oily sludge using solvent extraction. Process. Saf. Environ. 88, 318–326.

Further reading

Das, P., Mukherjee, S., Sen, R., 2008. Improved bioavailability and biodegradation of a model polyaromatic hydrocarbon by a biosurfactant producing bacterium of marine origin. Chemosphere 72, 1229–1234.
Lang, S., Wagner, F., 1993. Biological activities of biosurfactants. In: Kosaric, N. (Ed.), Biosurfactants, Surfactant Science Series. Marcel Dekker, New York, p. 251. 48.
Liu, Y., Majetich, S.A., Tilton, R.D., Sholl, D.S., Lowry, G.V., 2005. TCE dechlorination rates, pathways, and efficiency of nanoscale iron particles with different properties. Environ. Sci. Technol. 39, 1338–1345.
Nithyadevi, D., Kumar, P.S., Mangalaraj, D., Ponpandian, N., Viswanathan, C., Meena, P., 2015. Improved microbial growth inhibition activity of bio-surfactant induced Ag-TiO_2 core shell nanoparticles. Appl. Surf. Sci. 327, 504–516.
Ping, Y., Mang, L., Qin, Y., Hai-Ling, Z., Zhong-Zhi, Z., Rong, C., 2012. Oil recovery from refinery oily sludge using a rhamnolipid biosurfactant-producing Pseudomonas. Bioresour. Technol. 116, 24–28.
Satpute, S.K., Bhawsar, B.D., Dhakephalkar, P.K., Chopade, B.A., 2008. Assessment of different screening methods for selecting biosurfactant producing marine bacteria. Ind. J. Marine Sci. 37, 243–250.
Shie, J., Chang, C., Lin, J., Lee, D., Wu, C., 2012. Use of inexpensive additives in pyrolysis of oil sludge. Energy Fuel 16, 102–108.
Varjani, S.J., Rana, D.P., Bateja, S., Upasani, V.N., 2013. Isolation and screening for hydrocarbon utilizing bacteria (HUB) from petroleum samples. Int. J. Curr. Microbiol. App. Sci. 2.
Velioglu, Z., Urek, R.O., 2014. Concurrent biosurfactant and ligninolytic enzyme production by Pleurotus spp. in solid-state fermentation. Appl. Biochem. Biotechnol. 174, 1354–1364.

Section B

Sludge management

General considerations on sludge disposal, industrial and municipal sludge

7

Anna Grobelak, Karolina Czerwińska, Aneta Murtaś
Department of Environmental Engineering, Czestochowa University of Technology, Czestochowa, Poland

1 Introduction

Sewage sludge is a byproduct of wastewater treatment and is produced worldwide. In the literature, municipal sewage sludge is defined as sewage sludge from the municipal wastewater treatment process. Due to huge sewage sludge production all around the world, this organic waste must be managed properly. The two main disposal strategies for municipal sewage sludge management are reuse, including agriculture or landscaping purposes, or final disposal. There are many strategies to reuse sewage sludge but also many restrictions on the use of the given management method (Kacprzak et al., 2017). The preparation of sewage sludge for reuse, defined as sludge processing, becomes another alternative. If it is not possible to use sewage sludge in recovery processes, then it should be ultimately neutralized. The most important concept associated with the transformation of sewage sludge is the recovery, that is, the recycling or recovery of organic substances. One of the fundamental issues concerning good practices applied in the field of municipal sewage sludge is information on the amount of sewage sludge produced in sewage treatment plants worldwide. Without detailed knowledge of the amount of generated sewage sludge, it will not be possible to make the right decisions or investments aimed at the processing, recovery, or disposal of sewage sludge.

2 Sewage sludge worldwide production

Recently, the World Water Environment Federation (WEF) has adopted the "NEW" paradigm (Nutriety-Energy-Water) for wastewater treatment plants. This idea is based on the transformation of individual wastewater treatment plants into plants for the production of recovered resources. However, it is advisable to extend it to wastewater treatment systems operating in river basins, and in the case of the Baltic countries, also in the Baltic Sea basin. The premises of this paradigm are climate change and demographic changes, growing pollution of water with biogenic substances causing acceleration of the water eutrophication process, and the necessity to implement sustainable development. Currently, the highest production of sewage sludge is noted for

Industrial and Municipal Sludge. https://doi.org/10.1016/B978-0-12-815907-1.00007-6

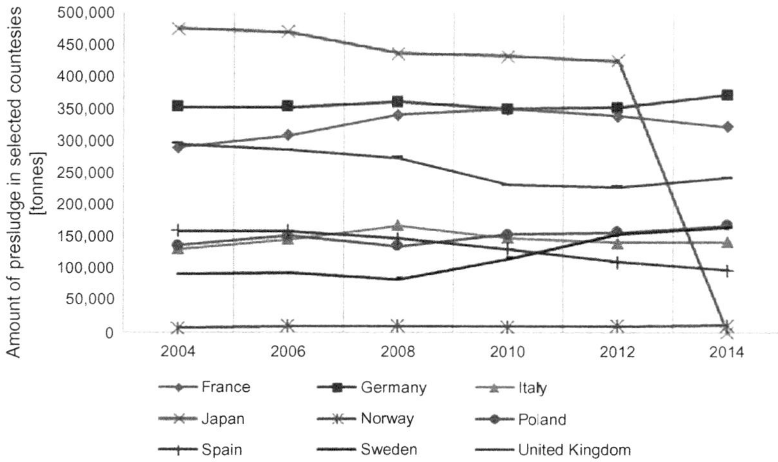

Fig. 1 The total amounts of sewage sludge production in selected countries, 2004–2014. Based on OECD.Stat.

developed countries. The worldwide production of sewage sludge based on the statistical data of OECD is shown in Fig. 1.

In Europe, North America, Australia, and New Zealand, there is more focus on how to improve the management of sewage sludge. In these places, wastewater is generally treated at the secondary, and, in many cases, tertiary levels, and the biosolid technologies and regulatory systems are complex. There are numerous details being addressed, including pharmaceuticals and personal care products (PPCPs) in treated water and biosolids, other trace chemicals and heavy metals, reactivation of pathogens in some particular biosolid treatment scenarios, energy efficiency, and greenhouse gases. From the least developed to the most developed countries, there are inevitable public concerns about how best to manage sewage sludge. Putting biosolids to their best uses in each local situation is the challenge (LeBlanc et al., 2008).

3 Characteristics of sewage sludge

The properties of sewage sludge are highly dependent on its origin and treatment. The main sources of sewage sludge are mainly municipal, industrial, or domestic wastewaters (Grosser and Neczaj, 2017). Due to various parameters of raw sewage sludge, it requires special processing to make this residue reusable. Sewage sludge mainly consists of large-sized solids such as grit or sand, organic and inorganic substances, and pathogenic microorganisms, heavy metals, and toxic or nontoxic contaminants (Bresters et al., 1997). For the treatment of industrial wastewater, depending on the type of pollution, different methods are used that can generally be divided into mechanical (gravity sedimentation, centrifugal sedimentation, pressure filtration, vacuum filtration, centrifugal filtration, reservoir filtration, centrifugal gravitational separation in hydrocyclones, and others), physicochemical (coagulation, coprecipitation, sorption, flotation, ion exchange on ionites, electrolysis, electrodialysis, reverse osmosis, ultrafiltration, and others), biological treatment on biological beds, the activated sludge method, and others (Kwarciak-Kozlowska and Krzywicka, 2014; Grosser et al., 2013a,b).

3.1 Sewage sludge physical, chemical, and biological parameters

Sewage sludge can be characterized by three main parameters:

- Physical, which determines the ability of sewage sludge to be processed and handled.
- Chemical, which involves the presence of nutrients and toxic/dangerous compounds.
- Biological, which gives information on microbial activity (mainly pathogenic) and organic matter (Grosser, 2017).

There are two main and desirable microelements in sewage sludge that are necessary in the wastewater treatment process-phosphorus and nitrogen. Due to the limited resources of phosphorus, new methods for its recovery are required. The resources of nitrogen are inexhaustible, but the demand for nitrogen in agriculture is huge (Kacprzak et al., 2014a,b). Sewage sludge contains toxic and less toxic pollutants, organic matter, pathogenic microorganisms, inorganic substances, and heavy metals (cadmium, chromium, copper, lead, mercury, nickel, zinc, platinum and platinum group metals (PGMs), silver, etc.) (Kacprzak et al., 2014a,b). The quality and quantity of sewage sludge (Fig. 2) produced on each treatment stage depends on the specific treatment as well as the content of pollutants, cleaning technologies, the method of processing the sludge in order to stabilize, the reduction of mass and volume, and the used reagents in the process of wastewater treatment (Kwarciak-Kozlowska and Krzywicka, 2014; Mateo-Sagasta et al., 2015).

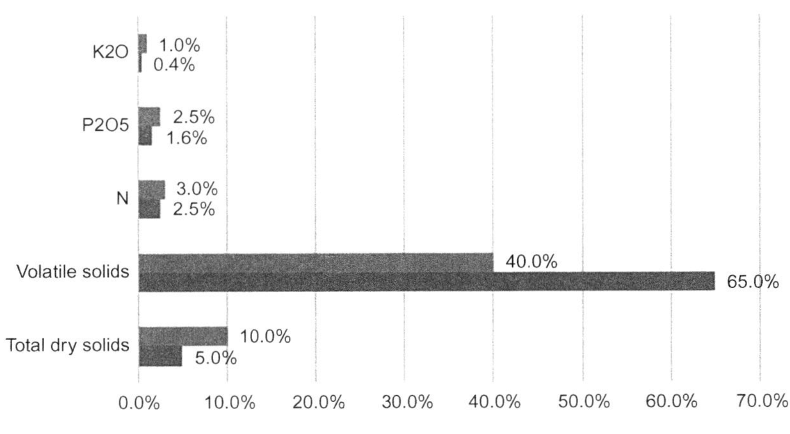

Fig. 2 Quality and quantity of sewage sludge produced on each treatment stage (Kwarciak-Kozlowska and Krzywicka, 2014; Mateo-Sagasta et al., 2015; Grosser, 2017).

Table 1 Parameters of primary sludge

Parameter	Primary sludge		Secondary sludge	
	Value	**Unit**	**Value**	**Unit**
Total solids (TS)	3.0–7.0	%	0.5–3.0	%
Volatile solids	65–90	% of TS	55–70	% of TS
Nitrogen	1.5–5.0	% of TS	2.5–6.0	% of TS
Phosphorus	0.6–3.5	% of TS	0.5–1.5	% of TS
Potassium	0.1–0.9	% of TS	0.6–1.0	% of TS
Alkalinity	500–1500	mg/L as $CaCO_3$	1000–5100	mg/L as $CaCO_3$
pH	5.0–8.0	–	6.0–8.0	–

Based on the data of the Polish Statistical Yearbook (2015).

The primary sewage sludge chemical composition and properties are highly dependent on the wastewater origin. Sewage sludge is made up of primary and secondary sludges, and the typical characteristics are included in Table 1.

Sewage sludge can be characterized by different parameters including solid concentration, which is the suspension of solid phase in liquid. A basic characteristic is the ratio of solids to the liquid and it can be expressed as grams of dry solids to grams of sludge. Many other parameters characterizing sludge can be specified: particle size, distribution of water, and flow properties. The first parameter (particle size) influences the properties of sewage sludge to the dewatering process. Distribution of water determines the adhesion of water to particles. The criteria for classification of sewage sludge result from its characteristics and properties, which are determined by the origin (from industrial or municipal waste water treatment plants), and the resulting composition, that is, the content of organic substances, the presence of pathogenic organisms, and the presence of toxic substances. The rheological properties indicate that the sludge can be defined as a non-Newtonian pseudoplastic fluid. Other parameters are dewatering properties and stability (Ahmad et al., 2016; Foladori et al., 2010).

3.2 Types of sludges

Due to the origin of the sludge, fecal sludge is specified that comes from households or septic tanks and sewage sludge that arises in wastewater treatment plants (Mateo-Sagasta et al., 2015). Due to the type of sludge formed at each stage of wastewater treatment, there are different types of sludge with different properties. Typical sewage sludge is made up of raw sludge, primary sludge, secondary sludge, mixed sludge, concentrated sludge, digested sludge, dehydrated sludge, hygienized sludge, dried sludge, and stabilized sludge. The raw sludges are crude and untreated sludges, very odorous and highly watered. Primary sludges occur in the sedimentation of easily falling organic and inorganic slurries. Secondary sludges are produced after biological treatment in flow devices with activated sludge or on biological beds, can be returned to the biological reactor and then they are recirculated sludge while excessive sludge occur because of growth of the bacterial biomass which is removing from treatment system. Mixed sludges occur from mixing primary with secondary sludges, and they need a proper treatment such as stabilization, dehydration, and hygienization. Concentrated sludges are formed after the dewatering and concentration processes. Digested sludges arise after anaerobic or aerobic stabilization. Dehydrated sludges are produced after dewatering. Hygienized sludges are formed after removing pathogenic organisms. Dried sludges are formed after drying processes. Stabilized sludges are subjected to various processes aimed at reducing reducing organic matter (Lundin et al., 2004; Grosser et al., 2013a,b).

3.3 Treatment of wastewater and sewage sludge production

One sewage sludge formation method is mechanical wastewater treatment in which large solids such as sand or gravel are separated. This step allows removing 50%–70% of the suspended solids and about 25%–40% BOD5. The next stage is biological treatment, with processes of coagulation, removal of insoluble colloidal solids, and stabilization of organic matter. This phase uses many species of microorganisms. At this stage, a secondary sludge is produced. The last step of biological treatment consists of the removal of nitrogen and phosphorus. At the beginning, ammonia is oxidized to the nitrate in the process of nitrification and then transformed to nitrogen. In the next step, the nitrate is converted to free nitrogen in a process called denitrification. The most common way of eliminating phosphorus is chemical precipitation (Smol et al., 2015; Guan et al., 2005). Generally, a sludge production consists of a complicated process of five stages. The general activities are included in Table 1. The first step (preliminary) consists of the elimination of big solids such as sand, gravel, or plastic elements. The main goal of this stage is an initial preparation of the wastewater for further treatment; it does not have a significant influence on the quality of wastewater. In the second stage, primary treatment, organic particles and suspended soilids are removed. The removing of compounds is based on gravity sedimentation, and causes the reduction of the concentration of pathogens, nutrients, elements, and toxic particles. As an effect, the primary sludge is formed. The next step is advanced primary treatment, which improves the remaining compounds from the previous stage (Metcalf and Eddy, 1991). Secondary—removing compounds which disintegrate easily. This process is carried out due to microbial activity. This stage is based on oxidizing organic matter and forming carbon dioxide and other organic products that provide energy and substrates for further operations. At this stage, secondary sludge is formed. Secondary treatment removes nutrients such as nitrogen and phosphorus. The main part of tertiary treatment consists elimination of pathogens. The filtration and coagulation allows reducing the concentrations of suspended solids and persistent compounds may be eliminated by using activated carbon (Kwarciak-Kozlowska et al., 2016; Council, 1996) (Table 2).

Table 2 Typical processes during wastewater treatment

Treatment stage	Action
Preliminary treatment	Removing grit and other large solids
Primary treatment	Sedimentation
Secondary treatment	Activated sludge or trickling filter
Tertiary treatment	Disinfection, removing nutrients, and solids

4 Sewage sludge treatment

Nowadays, there are various methods of sewage sludge treatment available. Each of them is highly focused on the hygienization and removal of toxic compounds as well as limitation of sewage sludge volume. Sewage sludge treatment should contain mechanical, biological, physical, and chemical steps in order to achieve a proper treatment process.

4.1 Stabilization as a process of reducing sewage sludge hazards

Stabilization is a process where sludge is treated to make it disposable and reusable. Three main processes are used: digestion, lime stabilization, and heat/thermal treatment. In the anaerobic digestion treatment, the organic matter is converted into a gas and stabilized residues due to aerobic or anaerobic bacterial processes. The anaerobic digestion process is mainly used in big wastewater treatment plants. The sewage sludge digestion process is one of the stabilization method. This process is carried out mainly in mesophilic conditions (30–42°C). Mesophilic digestion ensures stabilization, but without sufficient hygienization. The thermophilic fermentation (50–60°C) gives better results in lowering the level of pathogens. Conducting stabilization in thermophilic conditions also shortens the digestion time. However, the thermophilic digestion process compared to the mesophilic digestion is more sensitive to small changes in the process parameters and requires more energy to heat the sewage sludge to the required temperature. Properly maintained methane digestion should ensure obtaining biochemically stabilized sewage sludge and the production of high-caloric biogas. Increasing the degree of sewage sludge digestion and biogas production can be obtained due to the use of various methods, but codigestion is the most effective (Grosser, 2017, 2018). The mechanisms of anaerobic digestion are based on the enzyme secretion by microorganisms that allows achieving dissolved organic solids. Those particles can be used by acid formers to produce organic acids, which are very important for methanogenic bacteria for producing methane and carbon dioxide (Neczaj et al., 2013). This process reduces odors, eliminates pathogens, removes toxins, and decreases the water content, but only in specific conditions. In turn, aerobic digestion leads to the elimination of dissolved organic matters marked as BOD, due to the presence of aerobic microorganisms, creating biomass (Grosser et al., 2013a,b). The process consists of long-term diffused aeration or mechanical surface aeration and biomass storage for reducing total volatile solids. The reduction of volatile solids is usually about 40%. Another aerobic process is composting, in which decomposing of organic matters by using aerobic microorganisms occurs. Composting of sewage sludge is a process aimed at obtaining a material with the properties of an organic fertilizer. A properly conducted composting process allows stabilizing sewage sludge, decreasing the number of pathogenic organisms, and reducing the mass and volume of sewage sludge (Caceres et al., 2018). The presence of oxygen is maintained due to mixing or blowing air through the sludge. Obtained heat during composting affects the destruction of pathogens that allows achieving hygienized waste (Kacprzak et al., 2015). In lime stabilization, lime or calcium hydroxide is added to sludge in order to increase the pH to 12 or higher, eliminating the odors. Moreover, the amount of pathogens decreases, making the sewage sludge not hazardous for human health until the pH stays at the same level (Manu et al., 2017). The heat treatment of sewage sludge includes two steps. The first oscillates from 30–75°C, the second from 75–190°C; both are at a pressure of 26 bar for a short time. This process causes coagulation of the solids, breaking down the gel structure and reducing the water affinity of the sludge solids. One thermal treatment of sewage sludge is incineration and this process is based on the fully thermal oxidation of the organic matter, meaning that pathogenic microorganisms and toxic organics are destroyed (Bień et al., 2013). On the other hand, during incineration many toxic compounds such as dioxins or furans are produced. Thermal treatment causes about 90% volume reduction of sludge (Werthera and Ogadab, 1999), and the reduction concerns mainly water content. Table 3 contains the popular methods of water removal associated with flocks in sewage sludge. The more water trapped in sewage sludge, the more special treatment is needed (Table 4).

Table 3 Water content and methods of removal for different types of water associated with flocs

Type of water associated with flocs	Percentage content	Method of removing
Free water	70%–75%	Thickening
Flow water	20%–25%	Mechanical dewatering
Capillary and bound water	1%	Mechanical dewatering after chemical conditioning

Table 4 Water content in sewage sludge after treatment

Process	Content of water after treatment
Thickening	90%–95%
Mechanical dewatering	75%–87%
Conditioning	65%–70%
Drying	5%–10%

4.2 Other sewage sludge treatment processes

The aim of the thickening process is the limitation of sludge volume and it can be specified as gravity thickening and flotation (Andreoli et al., 2007). In the gravity thickening, the difference between weights of specific liquids and sludge particles is very crucial. This process can be used periodically or continuously. With gravity thickening of presludge, about 4%–6% of dry matter can be obtained (Bień et al., 2013; Zsirai, 2010). Flotation thickening is usually used for activated sludge. This process is based on the delivery of pressurized air and mixing with sewage sludge. With lowering of pressure, the sludge particles combine with air bubbles and migrate upward in the flotation thickeners (Arora et al., 1995). Examples of usage of thickening process are centrifuges, gravity drainage belts, and perforated rotating drums (Council, 1996). Dewatering as well as thickening increase the concentration of the sludge. This process causes that received product has solids properties. Mechanical dewatering helps to obtain 20%–45% solids by weight (Manu et al., 2017). Examples of the dewatering processes are sludge drying beds, filter and belt presses, and centrifuges (Andreoli et al., 2007). Conditioning is based on chemical and physical techniques that change the properties of sludge, making the dewatering process easier. In chemical processes, the organic polymers or inorganic compounds are applied. The physical methods are based on thermal treatment or freeze-thaw treatment. During the drying process, the water is evaporating using thermal energy. It can be achieved with the usage of flash dryers or rotary kilns. In the drying process, thermal drying can be differentiated with intermediate or direct drying. In some countries, solar energy can also be used for this method of sewage sludge drying. Sometimes the energy produced by microorganisms in the biological process is also utilized. Other types of sludge treatment can be solidification/immobilization, metal stripping, destruction of toxic organic matter, and combustion (Thomsen et al., 2017; Mateo-Sagasta et al., 2015).

Due to the huge sewage sludge production worldwide, it should be properly managed. The main purpose of municipal sewage sludge management is to reuse it, including reintroduction to the environment or final disposal. There are many ways to reuse sludge but also many restrictions in the use of the given management methods (Kacprzak et al., 2017; Fytili and Zabaniotou, 2008).

5 General sewage sludge disposal strategies

The limited resources of water and fertilizers in many parts of the world are a reason for sewage sludge reuse. In many countries, wastewater is the end waste and is used directly to soils (Raschid-Sally, 2013). Generally, in low-income countries where wastewater collection and treatment has limited attention, wastewater and sewage sludge are likely to be used directly and usually informally. In high-income countries, the tendency is the opposite and wastewaters and sewage sludges are generally treated, and their application and usage are controlled and scheduled (Scott et al., 2010). While any use of untreated sludge occurs in a very informal way, little data is available.

Many landowners consider sewage sludge, even untreated, to be a valuable nutrient resource, but they mainly prefer the septic sewage sludge application after drying or another processing type such as fertilizer (Kvarnström et al., 2012). The distribution of sewage sludge from onsite sanitation plants to farmers as a safe product may be an interesting model (Keraita et al., 2014). this is currently realized in Norway and Canada. In Europe, the landfilling of sewage sludge is prohibited and this is implemented in most EU member countries. Thus, the processing of sewage sludge in order to achieve valuable products is becoming more popular. The formal use of biosolids in agriculture is strictly regulated in developed countries. However, in many developed countries, there is an increasing disapproval for the agricultural use of biosolids due to potential toxicity (Bondarczuk et al., 2016).

Sewage sludge is used in agriculture (37%), incineration (11%), and landfilled (40%) while the remaining 12% is used in forestry or land reclamation (Fytili and Zabaniotou, 2008). In Canada, of the biosolids produced each year (660,000 tons of dried material), about half are applied to land while the rest is incinerated or sent to a landfill. The application of sludge in agriculture varies among EU member countries. In some EU15 countries (Belgium, Denmark, France, Ireland, Spain, and the UK), also more than half of the produced sewage sludge is being land spread. However, in three of the EU27 countries, no sludge is applied in the agricultural soils (Romania, Brussels Region, and the Flemish Region of Belgium), and in other countries (Finland, Netherland, Romania, Slovakia, Greece, and Slovenia) only 5% of the produced sewage sludge is used for agricultural purposes (Lamastra et al., 2018). The method of sewage sludge management can also be specified as organic recycling with agriculture use, reclamation, composting, and energy and material recycling with incineration, pyrolysis, and gasification. The sludge must meet specific criteria of content of heavy metals or pathogens, or otherwise it can be disqualified for use in agriculture (Fytili and Zabaniotou, 2008). *Salmonella* can be treated as an indicator for sludge disinfection and be removed by composting or pasteurization. Usually this bacterium lives about 20 weeks (Kacprzak et al., 2015). Other pathogens in the sludge are viruses, parasites, nematodes, or the ascaris species. Parasites can survive longer than bacteria.

One of the benefits of using sewage sludge in agriculture is the recovery of phosphorus, which occurs in the form of ferric phosphate and is less available for plants (Lundin et al., 2004). Sludge fertilizers can be also used in parks or green areas. Due to the content of microorganisms in the sludge, it can be used for land reclamation on damaged areas (Zsirai, 2010). Methods of using liquid sludge on the surface of the soil are fixed, such as portable sprinklers, overland flow and flooding, ridge and furrows, tank trucks, subsurface injection, or surface discharge (Fytili and Zabaniotou, 2008).

The most important criterion for the safe usage of sewage sludge is the level of heavy metals. Sewage sludge contains different levels of those metals which are determining criterion for land application (Fijalkowski et al., 2018). The second parameter is pathogen reduction, which indicates the presence or absence of pathogenic microorganisms such as *Salmonella*, viruses, or parasites. There are class A and class B levels. Class A says that the sewage sludge is almost free from pathogens and class

B says pathogens are reduced but not eliminated. Depending on the place and period of sewage sludge application, time restriction of public access, for parks and playgrounds should last about 1 year and for farm lands or remote lands public access should last 1 month. Sludge can be used in the form of pellets as fertilizer in forestry and for industrial lands (Kacprzak et al., 2014a,b). One of the criteria for the land application is soil conditions for biosolid use as well as the soil's depth, texture, erosion, topography, groundwater level, drainage and hydromorphism, slope, and structure (Grobelak et al., 2018). Biosolids used in a land application can be presented in five classes: Class I, which is very high potential, class II, high potential; class III, moderate potential; class IV, which can be a risk if the cultivation practices are improper; and class V, which is harmful to the environment (Andreoli et al., 2007).

6 Thermal methods in sewage sludge reuse

Incineration is becoming the best way for sewage sludge utilization because of the significant reduction of volume (about 10%) due to the thermal decomposition of organic matter (Fytili and Zabaniotou, 2008). After the incineration process, ash remains, which still can contain heavy metals. In the thermal process, multiple hearth and fluidized beds furnaces are used. The difference between those methods is mainly the different humidity of used sludges. Multiple hearth furnaces incinerate dewatered but still wet sludge and fluidized bed furnaces also burn semidried sludge. During this process, dioxins, furans, NO_x, N_2O, SO_2, HCl, and HF are all produced. They all effect human health and may be the dominant barriers for use of this technique (Fytili and Zabaniotou, 2008). An alternative to the classic thermal methods of sludge processing is the vitrification process. This is very effective in the disposal of hazardous and industrial waste and sewage sludge due to the stability and high chemical resistance of the obtained products. Due to the proper selection of materials in the vitrification process, glass and ceramic materials with chemical and physical properties comparable or better than commercial products can be obtained. The use of sewage sludge as raw material makes it possible to reuse it and market it as a fully fledged product without having to store any process residues. The vitrification of sewage sludge containing a significant content of chromium or other hazardous metals allows it to be immobilized by incorporation into a silica matrix without emission to the atmosphere (Celary et al., 2016; Celary and Sobik-Szołtysek, 2014).

Despite the development of sewage sludge management methods, the most emphasis is put on energy recovery and improvement of the heating value. The main sewage sludge treatment for energy recovery may be based on the pyrolysis and gasification processes. The first one is carried out by thermal treatment of dry sludge pellets with the temperatures oscillated between 300–500°C (mostly about 450°C). This treatment leads to the degradation of chemical molecules of fuel into oil, char, noncondensed gas (NCG), and reaction water (RW) (Bridle and Pritchard, 2004).

The gasification process can be used for sewage sludge with the TS between 70%–95%. This process has a huge potential for reuse of waste as an energy source and it can be valuable mainly for the cheap raw materials such as sewage sludge. The energy accumulated in the produced gas is highly dependent on agents such as either the feed fuel or reaction type, but also many others (Fytili and Zabaniotou, 2008). In comparison to the incineration, gasification is energetically self-sustaining (does not need any other fuels as supplements) and the emission of toxic gases such as sulfur dioxides and nitrogen oxides is reduced. Moreover, it has been noticed that the gasification process can potentially significantly decrease the emission of CDDs and PCDF (Marrero et al., 2003). In the chemical point of view, the gasification consists of a cascade of sequential processes in cooperation with thermal subprocesses (Marrero et al., 2003). The highest disadvantage of sewage sludge gasification is the problem with the concentration of heavy metals in the final product (Dogru et al., 2002). Using this method, it is possible to produce either heat or electricity (Groß et al., 2008).

7 Potential of sewage sludge land application

The amount of sewage sludge constantly increases, making it necessary to search for possible ways to reuse this raw material. Dried sewage sludge, depending on the stabilization methods and its composition, consists of a valuable substrate for agriculture as well as the energy yield. A high content of organic compounds (50%–70%) with macro- and micronutrients, especially nitrogen and phosphorus, makes sewage sludge practically an excellent fertilizer, a cheap and rich soil enhancer (Kacprzak et al., 2017; Linderholm et al., 2012). The usability of sewage sludge as a fertilizer is compatible with ecology and leads to the reuse of potentially dangerous wastes. Moreover, it contributes to the reduction of wastes (Cieslik et al., 2015). At present, sewage sludge is used in supporting plant growth in agriculture, and in promoting plant growth during assisted phytoremediation processes (Grobelak et al., 2013). The addition of sewage sludge to the soil contaminated with various heavy metals caused a significant increase of biomass yield and stimulation of shoot growth (for pine) as well as root growth (for spruce) (Grobelak et al., 2017). It was also found that sewage sludge fertilizer increases the sustainability of a eucalyptus plantation and improves wood production (Abreu-Junior et al., 2017). A soil amendment with sewage sludge has a positive influence on biological properties, indicated by increased microorganisms (Kulling et al., 2001). According to the symbiosis between plants and microorganisms, that effect can also easily and indirectly promote plant growth (Grobelak et al., 2015). The effects of sewage sludge application into soil on plant growth and the bioaccumulation of heavy metals in plant biomass is shown in Table 5.

Sewage sludge can be a rich source of biogenic elements and carbon, thus the land application is justified. The use of sludge cannot reduce the quality of soils, air and waters (Smol et al., 2015). Hence there are numerous limitations to the agricultural use of sewage sludge, the period of application, and the shape and location of the land. The usefulness of fertilizer for municipal sewage sludge is primarily determined by the content of organic matter, nitrogen, phosphorus, potassium, and the presence of impurities, which is most often a limiting factor in its use (Smol et al., 2016). The potassium content in sewage sludge ranges from 0.23% to 0.68% d.w. Comparing the fertilizer values of sewage sludge with organic fertilizer such as manure, it is

Table 5 Influence of sewage sludge application into a soil on development, biomass, and accumulation of heavy metals in selected plants

Plant	Effect on development and biomass	Effect on heavy metal accumulation in plant biomass	Reference
Festuca arundinacea Schreb.	Improvement of yield (30%)	Higher accumulation of Cd, Cr, Cu, Mn, Zn, Pb	Boswell (1975)
Willow	Improvement of yield, chlorophyll content, increased willow leaf length	–	Urbaniak et al. (2017)
Eucalyptus	Improvement of yield	–	Abreu-Junior et al. (2017)
Pinus sylvestris L.	Increase of biomass yield and stimulation of shoots growth	–	Grobelak et al. (2017)
Ryegrass (*Lolium perenne* L.)	Growth stimulation	–	Bai et al. (2014)
Sunflower (*Helianthus annus* L.)	Increase of dry weight biomass	Lower accumulation of Zn, Cu, Pb, and Ni for acidic, and higher accumulation of Zn and Cu for alkaline soil	Morera et al. (2002)
Willow	–	Higher accumulation of Cd, Cu, Ni, and Zn	–
Zea mays	Significant increase of shoot dry weigh	–	Kidd et al. (2007)
Alyssum serpyllifolium	Significant increase of shoot dry weigh	Reduced accumulation of Cu	Kidd et al. (2007)
Cistus ladanifer	Decrease of biomass	–	Kidd et al. (2007)
Oryza sativa	Increase of biomass and grain yield	Increased Cd accumulation with higher sludge doses	Latare et al. (2014)
Sudan grass (*Sorghum* bicolor × *Sorghum sudanense* var. Rocket)	Increase of plant production	Decrease of bioaccumulation Lower metal concentration in above-ground plant material	Alvarenga et al. (2016)
French bean (*Phaseolus vulgaris* L.)	Increase of plant biomass	Higher metal accumulation	Kumar and Chopra (2014)

observed that sewage sludge in terms of nitrogen and phosphorus content shows similar parameters. Sewage sludge applied in the soil improves the physicochemical and biological properties of soils, which in turn promotes better plant development as well as in the aspect of trace elements such as boron, manganese, copper, molybdenum, and zinc (Wołejko et al., 2018) Sewage sludge also increases the content of humus in the soil. The increase in the organic matter content reduces the bulk density, increases the aggregate stability, increases the water retention capacity in soils, and promotes increased water absorption. Organic substances also affect the circulation of elements, biomass, and soil diversity, and reduce the susceptibility to soil erosion. The use of sewage sludge in the soil increases the ability to infiltrate as well as air recirculation, contributing to the improvement of soil parameters. The processing of sewage sludge and its certification allow for the requalification of this waste into a product that can be placed on the mark et. Although large wastewater treatment plants allowed made the utilization of sewage sludge for thermal methods (drying, incineration), composting is the most frequently chosen method (Suleiman et al., 2017). Among organic fertilizers produced from sewage sludge, mainly compost predominates. However, there are very large social concerns about composites based on "biosolids," so-called biocomponents and/or sewage sludge. Many composting plants could not cope with long-term certification procedures, or the quality of the produced compost did not allow for applying for fertilizer status at all.

Nevertheless, municipal sewage sludge may comprise undesirable and even dangerous substances from an agricultural point of view (either organic and inorganic contaminants) that have not been removed in the sewage sludge processing (Sobik-Szołtysek et al., 2017). It was found that hormones in municipal wastes are one of the most dangerous substances that disrupts the growth and development of plants and animals (Thomsen et al., 2017). The potential toxicity of sewage sludge can be caused by xenobiotics, antibiotics, or heavy metals (Wołejko et al., 2017). Pharmaceuticals absorbed on the sewage sludge particles also make this a huge risk for the ecosystem due to the eventual application into a soil (Bondarczuk et al., 2016). Heavy metals in the sewage sludge may pose a huge risk for ecosystems and human health (Rorat et al., 2017; Placek et al., 2016). Thus, reuse of sewage sludge as a fertilizer needs to be strongly monitored from the ecological and human health points of view.

8 Alternative strategies

One of the alternatives for sewage sludge management and utilization is the production of hydrogen from sewage sludge by applying the downdraft gasification technique. This is an innovative method due to the fact that, at present, practically all the hydrogen produced comes from natural gases. The methods of hydrogens production are highly similar to the production of fossil fuels (Vierrath and Greil, 2001). Sewage sludge is converted by high temperature to the gas. This product consists mainly of H_2, CO, and methane. Produced methane is then converted to hydrogen and carbon oxide. In turn, the carbon oxide is next converted to achieve a higher amount of hydrogen. The electrical efficiency in this method can be more than

30% (Fytili and Zabaniotou, 2008). The hydrogen yield is about 10%–11% H in the produced gas and it could be utilized in, for example, fuel cells (Midilli et al., 2002).

Moreover, it has been shown that sewage sludge has the potential for biodiesel production by *Lipomyces starkeyi* (Angerbauer et al., 2008).

The possibility of sewage sludge disposal expands even on the industrial field. It has been found that sewage sludge, waste water sludge, and oxygen furnace sludge can be effectively used as conventional substitutes for substrates in cement production, replacing limestone, sand, and other raw materials (Ahmad et al., 2016). Nevertheless, this usability is limited by retardation of the setting time and has an influence on the mechanical characteristics of the final product (Sobik-Szołtysek and Jabłońska, 2010).

It has been reported also that sewage sludge has a huge potential in manufacturing lightweight aggregates, brick and ceramic products (Ahmad et al., 2016). Moreover, Basibuyuk and Kalat (2004) showed the usability of waterworks sludge in the treatment of vegetable oil refinery water. Therefore, the reuse of sewage sludge as a coagulant in wastewater treatment plants has gained interest. It has been investigated that alum sludge effectively increases the removal of SS and COD in primary sewage due to the high concentration of insoluble aluminium hydroxides in alum sludge and their role as a coagulant (Guan et al., 2005). Moreover, Jangkorn et al. (2011) showed that the application of alum sludge only unquestionably influences the removal of turbidity as well as COD, and anionic surfactants in the wastewater treatment of industrial and chemical wastes. It has been found that this method may also increase TSS; however this can be easily offset by the addition of fresh alum. Sewage sludge has the potential for removal of contaminants and heavy metal in wastewater treatment. It has been reported to remove of Pb, As, Cu, hydrogen sulfide etc. as well as phosphorus (Yang et al., 2014). The efficiency of the absorption processes is highly regulated by the pH value of the solution (Razali et al., 2007).

9 Conclusion

The amount of sludge as a byproduct of wastewater treatment still has an increasing tendency. The total volume of produced sewage sludge needs to be controlled. Due to the high sewage sludge potential, it can be useful in many branches of industry such as energy production, fertilizer or organic soil additives, and the cement industry. Sewage sludge may consist of valuable raw material for nutrient recovery and coagulation processes. The utility of sewage sludge in a few next years is going to increase mainly as energy source supporting politics of CO_2 emission reduction. It is well known that sewage sludge, like other wastes, must be treated as a source of potential raw materials. Sewage sludge management should be developed toward the efficient use and circulation of elements in the environment, the processing of sewage sludge, limiting storage, and landfilling. Wastewater treatment plants must select the appropriate technologies concerning the local situation, market, and legal issues to produce "product," not "waste." The use of sewage sludges as components for the production of mineral-organic fertilizers or soil substitutes is one of the methods of their management.

As with every technology, this has many pluses and minuses. However, it is undoubtedly a valuable alternative for sewage sludge if the aspects of the circular economy are considered.

Acknowledgments

The authors declare have no conflict of interest. Research funded from internal grant BS PB 401/304/11 and from the Polish Ministry of Science and Higher Education MNISW/2017/89 DiR/NN2.

References

Abreu-Junior, C.H., Firme, L.P., Maldonado, C.A.B., de Moraes Neto, S.P., Alves, M.C., Muraoka, T., Capra, G.F., 2017. Fertilization using sewage sludge in unfertile tropical soils increased wood production in Eucalyptus plantations. J. Environ. Manag. 203, 51–58.

Ahmad, T., Ahmad, K., Alam, M., 2016. Sustainable management of water treatment sludge through 3'R' concept. J. Clean. Prod. 124, 1–13.

Alvarenga, P., Farto, M., Mourinha, C., Plama, P., 2016. Beneficial use of dewatered and composted sewage sludge as soil amendments: behaviour of metals in soils and their uptake by plants. Waste Biomass Valori. 7, 1189–1201.

Andreoli, V.C., Von Sperling, M., Fernandes, F., Ronteltap, M., 2007. Sludge Treatment and Disposal. IWA Publishing, United Kingdom.

Angerbauer, C., Siebenhofer, M., Mittelbach, M., Guebitz, G.M., 2008. Conversion of sewage sludge into lipids by *Lipomyces starkeyi* for biodiesel production. Bioresour. Technol. 99 (8), 3051–3056.

Arora, H., DeWolfe, J.R., Lee, R.G., Grubb, T.P., 1995. Evaluation of dissolved air flotation process for water clarification and sludge thickening. Water Sci. Technol. 31 (3–4), 137–147.

Bai, Y.C., Gu, C.H., Tao, T.Y., Zhu, X.W., Xu, Y.R., Shan, Y.H., Feng, K., 2014. Responses of Ryegrass (*Lolium perenne* L.) grown in mudflats to sewage sludge amendment. J. Integr. Agric. 13 (2), 426–433.

Basibuyuk, M., Kalat, D.G., 2004. The use of waterworks sludge for the treatment of vegetable oil refinery industry wastewater. J. Environ. Technol. 25 (3), 373–380.

Bień, J., Celary, P., Morzyk, B., Sobik-Szołtysek, J., Wystalska, K., 2013. Effect of additives on heavy metal immobilization during vitrification of tannery sewage sludge. Environ. Prot. Eng. 39 (2), 33–40.

Bondarczuk, K., Markowicz, A., Piotrowska-Seget, Z., 2016. The urgent need for risk assessment on the antibiotic resistance spread via sewage sludge land application. Environ. Int. 87, 49–55.

Boswell, F.C., 1975. Municipal sewage sludge and selected element application to soil: effect on soil and fescue. J. Environ. Qual. 4, 267–272.

Bresters, A.R., Coulomb, I., Deak, B., Matter, B., Saabye, A., Spinosa, L., Utvik, A.O., 1997. Sludge Treatment and Disposal. Management Approaches and Experiences. IBWA, pp. 13–42.

Bridle, T.R., Pritchard, D., 2004. Energy and nutrient recovery from sewage sludge via pyrolysis. Water Sci. Technol. 50 (9), 169–175.

Caceres, R., Malińska, K., Marfa, O., 2018. Nitrification within composting: a review. Waste Manag. 72, 119–137.

Celary, P., Sobik-Szołtysek, J., 2014. Vitrification as an alternative to landfilling of tannery sewage sludge. Waste Manag. 34, 2520–2527.

Celary, P., Sobik-Szołtysek, J., Tajchman, A., 2016. Heavy metal volatilization during vitrification of tannery sewage sludge with mineral waste. Energy Prot. Environ. 19 (4), 503–515.

Cieslik, B.M., Namiesnik, J., Konieczka, P., 2015. Review of sewage sludge management: standards, regulations and analytical methods. J. Clean. Prod. 90, 1–15.

Council, N.R., 1996. Use of Reclaimed Water and Sludge in Food Crop Production. The National Academies Press, Washington, DC, pp. 48–55.

Dogru, M., Midilli, A., Howarth, C.R., 2002. Gasification of sewage sludge using a throated dowdraft gasifier and uncertainty analysis. Fuel Process. Technol. 75, 55–82.

Fijalkowski, K., Rosikon, K., Grobelak, A., Hutchison, D., Kacprzak, M.J., 2018. Modification of properties of energy crops under polish condition as an effect of sewage sludge application onto degraded soil. J. Environ. Manag. 217, 509–519. https://doi.org/10.1016/j.jenvman.2018.03.132.

Foladori, P., Andreottola, G., Ziglio, G., 2010. Sludge Reduction Technologies in Wastewater Treatment Plants. 9IWA Publishing, London, UK.

Fytili, D., Zabaniotou, A., 2008. Utilization of sewage sludge in EU application of old and new methods-A review. Renew. Sust. Energ. Rev. 12 (1), 116–140.

Grobelak, A., Kacprzak, M., Grosser, A., Napora, A., 2013. Chemophytostabilisation of soil contaminated with Cadmium, Lead and Zinc. Annuala Set Environ. Prot 15 (2), 1982–2002.

Grobelak, A., Napora, A., Kacprzak, M., 2015. Using plant growth-promoting rhizobacteria (PGPR) to improve plant growth. Ecol. Eng. 84, 22–28.

Grobelak, A., Placek, A., Grosser, A., Singh, B.R., Almås, Å.R., Napora, A., Kacprzak, M., 2017. Effects of single sewage sludge application on soil phytoremediation. J. Clean. Prod. 155, 189–197.

Grobelak, A., Rorat, A., Kokot, P., Singh, B.R., Kacprzak, M., 2018. Mine waste rehabilitation case studies from poland. In: Prasad, M.N.V., de Campos, P.J., Subodh, F., Maiti, K. (Eds.), Bio-Geotechnologies for Mine Site Rehabilitation. first ed. Elsevier, pp. 515–529 (Chapter 28).

Groß, B., Eder, C., Grziwa, P., Horst, J., Kimmerle, K., 2008. Energy recovery from sewage sludge by means of fluidised bed gasification. Waste Manag. 28 (10), 1819–1826.

Grosser, A., 2017. The influence of decreased hydraulic retention time on the performance and stability of codigestion of sewage sludge with grease trap sludge and organic fraction of municipal waste. J. Eviron. Manage. 203 (3), 1143–1157.

Grosser, A., 2018. Determination of methane potential of mixtures composed of sewage sludge, organic fraction of municipal waste and grease trap sludge using biochemical methane potential assays. A comparison of BMP tests and semi-continuous trial results. Energy 143, 488–499.

Grosser, A., Neczaj, E., 2017. Pretreatment methods as a means of boosting methane production from sewage sludge and its mixtures with grease trap sludge. In: E3S Web of Conferences 22p. 00058.

Grosser, A., Worwag, M., Neczaj, E., Grobelak, A., 2013a. Semi-continuous anaerobic codigestion of mixed sewage sludge and waste fats of vegetable origin. Rocz. Ochr. Sr. 15, 2108–2125.

Grosser, A., Worwag, M., Neczaj, E., Kamizela, T., 2013b. Codigestion of organic fraction of municipal solid waste with different organic wastes: a review. In: Environmental Engineering IV—Proceedings of the Conference on Environmental Engineering IV.

Guan, X.-H., Chen, G.-H., Shang, C., 2005. Re-use of water treatment works sludge to enhance particulate pollutant removal from sewage. Water Res. 39, 3433–3440.

Jangkorn, S., Kuhakaew, S., Theantanoo, S., Klinla-or, H., Sriwiriyarat, T., 2011. Evaluation of reusing alum sludge for the coagulation of industrial wastewater containing mixed anionic surfactants. J. Environ. Sci. 23, 587–594.

Kacprzak, M., Grobelak, A., Grosser, A., Napora, A., 2014a. The potential of biosolid application for the phytostabilisation of metals. Desalin. Water Treat. 52, 19–21 3955–3964.

Kacprzak, M., Grobelak, A., Grosser, A., Prasad, M.N.V., 2014b. Efficacy of biosolids in assisted phytostabilization of metalliferous acidic sandy soils with five grass species. Int. J. Phytoremediation 16 (6), 593–608.

Kacprzak, M., Fijałkowski, K., Grobelak, A., Rosikoń, K., Rorat, A., 2015. Escherichia coli and salmonella spp. early diagnosis and seasonal monitoring in the sewage treatment process by EMA-qPCR method. Pol. J. Microbiol. 64 (2), 143–148.

Kacprzak, M., Neczaj, E., Fijałkowski, K., Grobelak, A., Grosser, A., Worwag, M., Rorat, A., Brattebo, H., Almas, A., Singh, B.R., 2017. Sewage sludge disposal strategies for sustainable development. Environ. Res. 156, 39–46.

Keraita, B., Drechsel, P., Klutse, A., Cofie, O., 2014. On-farm Treatment options for Wastewater, Greywater and Fecal Sludge with Special Reference to West Africa. Resource Recovery and Reuse Series, vol. 1. International Water Management Institute (IWMI).CGIAR Research Program on Water, Land and Ecosystems (WLE), Colombo, Sri Lanka, p. 32.

Kidd, P.S., Domínguez-Rodríguez, M.J., Díez, J., Monterroso, C., 2007. Bioavailability and plant accumulation of heavy metals and phosphorus in agricultural soils amended by long-term application of sewage sludge. Chemosphere 66 (8), 1458–1467.

Kulling, D., Stadelmann, F., Herter, U., 2001. Sewage sludge—fertilizer or waste? In: UKWIR Conference, Brussels.

Kumar, V., Chopra, A.K., 2014. Accumulation and translocation of metals in soil and different parts of French Bean (*Phaseolus vulgaris* L.) amended with sewage sludge, bull. Environ. Contam. Toxicol. 92, 103–108.

Kvarnström, E., Verhagen, J., Nilsson, M., Srikantaiah, V., Ramachandran, S., Singh, K., 2012. The business of the honey-suckers in Bengaluru (India): the potentials and limitations of commercial fecal sludge recycling—an explorative study. (Occasional Paper 48) IRC International Water and Sanitation Centre, The Hague.

Kwarciak-Kozlowska, A., Krzywicka, A., 2014. Advanced oxidation processes with coke plant wastewater treatment. Water Sci. Technol. 69 (9), 1875–1878.

Kwarciak-Kozlowska, A., Krzywicka, A., Galwa-Widera, M., 2016. The use of ozonation process in coke wastewater treatment. Rocz. Ochro. Śr. 18 (2), 61–73.

Lamastra, L., Suciu, N.A., Trevisan, M., 2018. Sewage sludge for sustainable agriculture: contaminants' contents and potential use as fertilizer. Chem. Biol. Technol. Agric. 5 (1), 10.

Latare, A.M., Kumar, O., Singh, S.K., Gupta, A., 2014. Direct and residual effect of sewage sludge on yield, heavy metals content and soil fertility under rice-wheat system. Ecol. Eng. 69, 17–24.

LeBlanc, R.J., Matthews, P., Richard, P.R., 2008. Global Atlas of Excreta. In: Wastewater, Sludge and Biosolids Management: Moving Forward the Sustainable and Welcome Uses of a Global Resource. United Nations Human Settlements Programme (UN-HABITAT).

Linderholm, K., Tillman, A.M., Mattsson, J.E., 2012. Life cycle assessment of phosphorus alternatives for Swedish agriculture. Resour. Conserv. Recycl. 66, 27–39.

Lundin, M., Olofsson, M., Pettersson, G.J., Zetterlund, H., 2004. Environmental and economic assessment of sewage sludge handling options. Resour. Conserv. Recycl. 41, 255–278.

Manu, M.K., Kumar, R., Garg, A., 2017. Performance assessment of improved composting system for food waste with varying aeration and use of microbial inoculum. Bioresour. Technol. 234, 167–177.

Marrero, T., McAuley, B., Sutterlin, W., Morris, S., Manahan, S., 2003. Fate of heavy metals and radioactive metals in gasification of sewage sludge. Waste Manag. 24, 193–198.

Mateo-Sagasta, J., Raschid-Sally, L., Thebo, A., 2015. Global wastewater and sludge production, treatment and use. In: Drechsel, P., Qadir, M., Wichelns, D. (Eds.), Wastewater. Springer, Dordrecht, pp. 63–67.

Metcalf, K., Eddy, A., 1991. Wastewater Engineering: Treatment, Disposal and Reuse. McGraw-Hill, New York.

Midilli, A., Dogru, M., Akay, G., Howarth, C.R., 2002. Hydrogen production from sewage sludge via a fixed bed gasifier product. Int. J. Hydrog. Energy 27, 1035–1041.

Morera, M.T., Echeverria, J., Garrido, J., 2002. Bioavailability of heavy metals in soils amended with sewage sludge. Can. J. Soil Sci. 82, 433–438.

Neczaj, E., Grosser, A., Worwa¸g, M., 2013. Boosting production of methane from sewage sludge by addition of grease trap sludge. Environ. Prot. Eng.. 39.

Placek, A., Grobelak, A., Kacprzak, M., 2016. Improving the phytoremediation of heavy metals contaminated soil by use of sewage sludge. Int. J. Phytoremediation. 18 (6), 605–618.

Polish Statistical Yearbook, 2015. Main Statistical Office. Warsaw, 2015p. 445. www.stat.gov.pl. Accessed 17 May 2018.

Raschid-Sally, L., 2013. City waste for agriculture: emerging priorities which influence agenda setting. Aqua. Procedia 1, 88–99.

Razali, M., Zhao, Y.Q., Bruen, M., 2007. Effectiveness of a drinking water treatment sludge in removing different phosphorus species from aqueous solution. Sep. Purif. Technol. 55, 300–306.

Rorat, A., Wloka, D., Grobelak, A., Grosser, A., Sosnecka, A., Milczarek, M., Jelonek, P., Vandenbulcke, F., Kacprzak, M., 2017. Vermiremediation of polycyclic aromatic hydrocarbons and heavy metals in sewage sludge composting process. J. Environ. Manag. 187, 347–353.

Scott, C.A., Drechsel, P., Raschid-Sally, L., Bahri, A., Mara, D., Redwood, M., Jiménez, B., 2010. Wastewater Irrigation and Health: Challenges and Outlook for Mitigating Risks in Low-Income Countries. Earthscan, pp. 381–394.

Smol, M., Kulczycka, J., Henclik, A., Gorazda, K., Wzorek, Z., 2015. The possible use of sewage sludge ash (SSA) in the construction industry as a way towards a circular economy. J. Clean. Prod. 95, 45–54.

Smol, M., Kulczycka, J., Kowalski, Z., 2016. Sewage sludge ash (SSA) from large and small incineration plants as a potential source of phosphorus–Polish case study. J. Environ. Manag. 184, 617–628.

Sobik-Szołtysek, J., Jabłońska, B.B., 2010. Possibilities of joint management of sewage sludge and dolomite post-flotation waste. Ecol. Chem. Eng. S. 17 (2), 149–159.

Sobik-Szołtysek, J., Wystalska, K., Grobelak, A., 2017. Effect of addition of sewage sludge and coal sludge on bioavailability of selected metals in the waste from the zinc and lead industry. Environ. Res. 156, 588–596.

Suleiman, H., Rorat, A., Grobelak, A., Grosser, A., Milczarek, M., Płytycz, B., Kacprzak, M., Vandenbulcke, F., 2017. Determination of the performance of vermicomposting process applied to sewage sludge by monitoring of the compost quality and immune responses in three earthworm species: Eisenia fetida, Eisenia andrei and Dendrobaena veneta. Bioresour. Technol. 241, p103–p112.

Thomsen, T.P., Sárossy, Z., Ahrenfeldt, J., Henriksen, U.B., Frandsen, F.J., Müller-Stöver, D.S., 2017. Changes imposed by pyrolysis, thermal gasification and incineration on composition and phosphorus fertilizer quality of municipal sewage sludge. J. Environ. Manag. 198, 308–318.

Urbaniak, M., Wyrwicka, A., Tołoczko, W., Serwecińska, L., Zieliński, M., 2017. The effect of sewage sludge application on soil properties and willow (Salix sp.) cultivation. Sci. Total Environ. 586, 66–75.

Vierrath, H., Greil, C., 2001. Energy and electricity from biomass, forestry and agricultural waste. In: Kyritsis, S., Beenackers, A.A.C.M., Helm, P., Grassi, P., Chiaramonti, D. (Eds.), Proceedings of the First World Biomass Conference, Proceedings of the First World Conference and Exhibition on Biomass for Energy and Industry. James & James.

Werthera, J., Ogadab, T., 1999. Sewage sludge combustion. Prog. Energ. Combust. 29, 55–116.

Wołejko, E., Łozowicka, B., Kaczyński, P., Konecki, G.M., 2017. The influence of chemical protection on the content of heavy metals in wheat (*Triticum aestivum* L.) growing on the soil enriched with granular sludge. Environ. Monit. Assess. 189 (8), 424.

Wołejko, E., Wydro, U., Jabłońska-Trypuć, A., Butarewicz, A., Łoboda, T., 2018. The effect of sewage sludge fertilization on the concentration of pahs in urban soils. Environ. Pollut. 232, 347–357.

Yang, L., Wei, J., Zhang, Y., Wang, J., Wang, D., 2014. Reuse of acid coagulant-recovered drinking waterworks sludge residual to remove phosphorus from wastewater. Appl. Surf. Sci. 305, 337–346.

Zsirai, I., 2010. Sewage sludge as renewable energy. In: European Conference on Sludge Management. Budapest, Hungarypp. 166–169.

Sanitary and environmental aspects of sewage sludge management

8

Agnieszka Rorat, Pauline Courtois, Franck Vandenbulcke, Sébastien Lemiere
Campus ULille Sciences and Technologies, University of Lille, Lille, France

1 Introduction

Globally, 80% of wastewater is discharged untreated into the world's waterways (Opec, 2018). Failure to treat effluents constitutes a serious threat to the environment, the climate, and human health, but this can also be considered a waste of resources. According to the goals of sustainable development and a circular economy, the wastewater shall be considered primarily as a source of water, then energy, organics, metals, and other resources. Consequently, sewage sludge is recognized as a source of renewable energy and material recovery (Christodoulou and Stamatelatou, 2016). Thus, no longer considered a "waste," it became a byproduct to be posttreated in order to be recycled into nature as energy, matter, or both. Sewage sludge is a reservoir of organic matter and nutrients, so it constitutes a potential substrate for a variety of possible reuse scenarios. Therefore, all potentially applied strategies shall fulfill the requirements of ecoinnovation. Thus, they shall lead to an important reduction of the negative environmental impacts by decreasing the consumption of natural resources or the release of harmful substances (Rorat and Kacprzak, 2017). Consequently, contemporary trends in organic waste treatment follow the sustainable development strategy in terms of environmental, economic, and social impacts. The composition of sewage sludge is highly variable and may depend on many various factors such as the seasons, the technology applied in wastewater treatment plants (WWTPs), the specificity of the source area of the influent, etc. On average, dewatered sewage sludge contains 50%–70% organic matter and 30%–50% mineral components (including 1%–4% of inorganic carbon), 3.4%–4.0% nitrogen (N), 0.5%–2.5% phosphorus (P), and significant amounts of other nutrients, including micronutrients that could be recovered. For instance, extractable resources such as phosphorus (P) are predicted to become scarce or exhausted in the next 50–100 years, thus P recovery from wastewater is becoming an increasingly viable alternative (Connor et al., 2017). Nevertheless, a relatively low content of lignin and cellulose makes the organic matter easy to decompose. Hence, it degrades fast and can cause a sharp peak in the nitrate and pollutant concentration in the soil if applied without pretreatment. The biggest challenge, though, is seen in the contaminants, (1) organic (such as polycyclic aromatic hydrocarbons (PAH), polychlorinated biphenyls (PCB),

Industrial and Municipal Sludge. https://doi.org/10.1016/B978-0-12-815907-1.00008-8

adsorbable organohalogens (AOX), pesticides, surfactants, hormones, pharmaceuticals) and (2) inorganic (metals and their nanoparticles), and (3) pathogenic species of living organisms for example, bacteria, viruses, protozoa, and parasitic helminths (see review by Fijalkowski et al., 2017).

In this context, this chapter focuses on the sanitary and environmental dangers of the presence of the above-mentioned contaminants in sludge. The environmental risks of sludge spreading on soils will be presented as well as their possible treatment scenarios to propose an acceptable reuse of sewage sludge in a circular economy.

2 The global production of sewage sludge and the main directions of its management

At the European scale, the 91/271/ECC urban wastewater treatment directive adopted in May 1991 imposed the collection and treatment of wastewater in agglomerations with a population equivalent (PE) of more than 2000. Consequently, the substantial and constant increase of wastewater sludge and its disposal are becoming a growing challenge for municipalities in Europe. The annual sludge production in EU-27 will grow from 11.5 million tons of dry solids (DS) in 2010 to 13 million tons DS in 2020 (EC, 2008).

Table 1 shows the more recent data considering the production and disposal of sewage sludge for selected countries, according to OECD. While legislation more or less compels European countries to improve their sewage sludge management policies, many poor and developing countries are still studying possible wastewater treatment practices. For instance, the Federated States of Micronesia dumped almost 30% of produced sludge into the Pacific Ocean without any pretreatment in 2012 (Rouse, 2013).

The global population exceeded 7.5 billion this year and is expected to surpass 9 billion by 2050. Urban populations may rise nearly twice as fast as they are projected to nearly double from 3.4 billion in 2012 to 6.4 billion by 2050, especially in developing countries, where the number of people living in slums may rise even faster, from 1 billion to 1.4 billion in just a decade (Matiasi, 2012).

2.1 Sewage sludge as a substrate—Characteristics

The composition of sewage sludge is highly changeable during the process and also varies a lot between wastewater treatment facilities. Typically, raw (untreated) sewage sludge contains 2.0%–8.0% total dry solids (TS), 60%–80% of TS of volatile solids (VS), grease and fats, proteins, nitrogen, phosphorus, potassium, cellulose, iron, silica, alkalinity (mg/L as CaCO3), and organic acids (mg/L as Hac) (Metcalf and Eddy, 1991).

The potential danger of using raw sewage sludge (not stabilized, only mechanically treated) for the sewage-to-matter final disposal strategies is huge due to the presence of pathogenic organisms and other contaminants. Therefore, some stabilization processes shall be applied at the WWTPs. The choice of applied technology depends strongly on the characteristics of raw sludge. Some parameters are crucial for the processes of stabilization, for example, pH, organic acid content, and alkalinity limit the

Table 1 Sewage sludge production and disposal in selected countries in 2012

Country	Produced sewage sludge	Total disposal	Agricultural use	Compost and other applications	Landfill	Dumping at sea	Incineration
Austria	266	266	40	74	14	0	139
Belgium	157	107	19	n.d.	n.d.	0	89
Czech Republic	263	263	72	154	13	n.d.	8
Denmark	141	115	74	n.d.	1	0	34
Estonia	16	16	14	n.d.	2	0	..
Finland	141	141	7	93	10	0	32
France	987	932	684	n.d.	40	0	207
Germany	1849	1844	542	294	0	0	1009
Greece	119	119	14	0	40	0	39
Ireland	72	72	68	4	0	0	0
Israel	118350	n.d.	0	69311	3928	45111	0
Luxembourg	8	5	4	n.d.	0	0	1
Netherlands	346	325	0	0	0	0	321
Poland	533	533	115	33	47	0	57
Portugal	339	113	102	n.d.	11	0	0
Slovenia	26	26	0	2	1	0	13
Spain	2757	2577	1922	n.d.	384	0	100
Sweden	207	196	48	67	7	0	1
United Kingdom	1137	1078	844	n.d.	5	0	229

n.d., no data.
Data extracted on 13 May 2018, 16:08 UTC (GMT) from OECD.Stat.

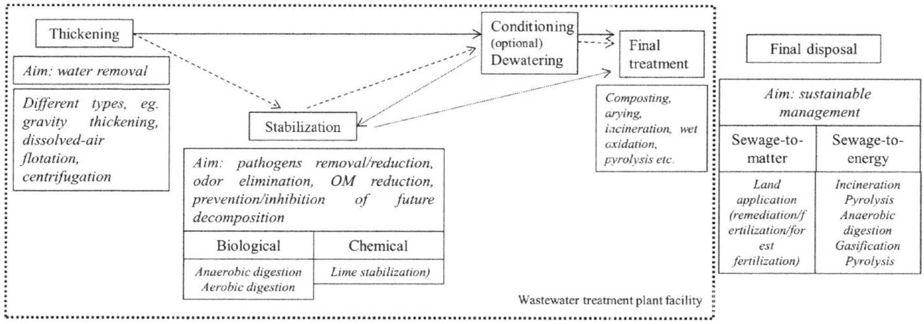

Fig. 1 Typical sewage sludge treatment process including processes at wastewater treatment plants (thickening/stabilization/dewatering) and the most common final disposal strategies applied worldwide.

anaerobic digestion process (Metcalf et al., 2013) while pH, ammonia, and other parameters are important for the composting and vermicomposting processes (Suleiman et al., 2017).

Fig. 1 presents the most typical pathways of sewage sludge treatment, including the processes at the WWTPs (thickening/stabilization/dewatering) and final disposal strategies. After thickening, sludge stabilization is usually performed. It is crucial for the further applications and aims primarily to reduce the potential risks by lowering the number of pathogens in organic matters. Two types of stabilization of liquid sewage sludge shall be distinguished:

1) Chemical stabilization by the addition of lime to alter the value of the pH to >11; eliminates the microbiological risk.
2) Biological stabilization, meaning anaerobic or aerobic digestion. Aerobic digestion is a process of treating the secondary sludge from the biological wastewater treatment process as activated sludge or trickling filters; anaerobic digestion can be conducted in low (<10%) as well as medium (15%–20%) and high 22%–40% solid anaerobic digestion systems; this technology is described in detail in the next section.

After stabilization, the sludge shall be dewatered. Usually, this process is carried out using filter presses or centrifuges. As proper dewatering is crucial for the further disposal of sewage sludge, often an additional step of conditioning is required. The conditioners, synthetic organic polymers, or metal ions (typically iron salts) are used in order to coagulate the colloids in sludge and thus fasten the dewatering process (Novak, 2006). Only an efficient stage of water removal allows applying efficiently the selected techniques of final treatment and disposal of sewage sludge.

2.2 Sewage sludge final treatment and disposal

The main directions for sustainable sewage sludge management are:

a) Matter recovery (sewage-to-matter): use in agriculture (directly as a fertilizer) and remediation of devastated or degraded lands.
b) Energy recovery (sewage-to-energy) by incineration and alternate thermal methods as pyrolysis, quasipyrolysis and gasification or coincineration (in cement plants).

Wastewater contains a chemical energy that shall be converted to the usable form and thus fulfill a part of the worldwide need for renewable energy sources (Puyol et al., 2016). Different technologies can be used to transfer the excess sewage sludge into energy. This strategy has lately been of great interest. Primarily, it allows using the potential of sewage sludge without the environmental risks often discussed for land applications that could introduce contaminants into the soil. Moreover, it responds to a global call for renewable energy sources.

c) Others, such as landfilling or dumping at sea, which are forbidden by most countries but still practiced in some parts of the world, mostly in developing countries.

2.2.1 Anaerobic digestion

At first recognized mainly as a process of the stabilization of sewage sludge with the main aim of pathogen bacteria removal, now anaerobic digestion (AD) is often considered as a technically mature and cost-effective process that converts sludge into biogas (Cao and Pawłowski, 2012). The produced biogas can be used for the self-purposes of WWTPs that are characterized by a high demand for electricity, up to 0.78 kWh per m^3 of treated wastewater (Cano et al., 2015). Ideally, the process of anaerobic digestion shall fulfill this high demand. In order to increase efficiency and thus biogas yields, it was proposed to introduce the other ingredients in the so-called codigestion. For instance, Grosser (2018) used grease trap sludge and an organic fraction of municipal waste as cosubstrates in the process, which enhanced the efficiency of sewage sludge anaerobic digestion. Similarly, other organic wastes have been tested with success, for example, food waste, cheese whey, and olive mill wastewater (Maragkaki et al., 2018). Thus, codigestion of sewage sludge can be understood as a method of management of different organic wastes. Nevertheless, it cannot be fully seen as a final disposal of sludge, as it generates another byproduct, digested sludge (digestate), that still contains a high quantity of nutrients and contaminants and must be treated. It was proven that the utilization of digestates may replace or reduce the use of mineral fertilizer in agronomic plant production, as it is rich in plant-available nutrients (ammonium, phosphate, and potassium) (Sogn et al., 2018). Yet, in-land use as a biofertilizer is possible only if the product can be qualified according to applicable norms, usually regulated by soil protection legislation, fertilizer, or waste legislation. Otherwise, other options shall be considered. Lately, Peng et al. (2018) proposed using the digestate in landfill bioreactors in order to remove the nitrogen of old landfill leachate. Digestate was also successfully applied with other organic wastes as the organic fraction of municipal wastes, sawdust, and green wastes in the process of vermicomposting (Rorat et al., 2017). As the chemical composition of digestate corresponds to the composition of the used substrates, the long-term effects of its introduction to the soil shall be studied in order to appreciate the impact on soil functions (soil biodiversity and microbial cycles). The existing studies focus mainly on the fertilizer properties of the digestates produced from different substrates. According to Nkoa (2014), the most common risks associated with the application of digestate in land are related to:

a) Risks of atmospheric pollution (ammonia emission and fallout, nitrous oxide emission).
b) Risks of nutrient pollution (excess nitrogen and phosphorus).
c) Risk of soil contamination (chemical/biological contamination).

2.2.2 Composting and vermicomposting

Composting as a method of biological decomposition of biowastes in the presence of oxygen contributes powerfully to the recycling and conservation of several macro- and micronutrients from sewage sludge in the soil. Its alternative, vermicomposting, is a modern, inexpensive, and eco-friendly biotechnology in which earthworms are employed as natural bioreactors in order to decompose the organic matter (Suleiman et al., 2017). Their metabolic activity and cooperation with microorganisms lead to a 40%–60% reduction of volume, an increase of bioavailability of nutrients to plants, a reduction in the C/N ratio, and a decrease of the availability of some dangerous contaminants such as metals (Rorat et al., 2015). Although composting can be considered highly beneficial and a low cost sewage-to-matter strategy that allows recycling organic nutrients into the ecosystem, it still causes some important problems from an environmental point of view. Due to the rapid degradation of nitrogenous organic matter, important nitrogen losses and greenhouse gas (GHG) emissions can be noted (Sánchez-Monedero et al., 2010). Those effects can be partially reduced by the introduction of different bulking agents, for example, agricultural wastes and alkaline amendments such as lime, zeolite, and bentonite. Recently, biochar has also been considered an efficient agent causing a reduction of greenhouse gases, ammonia, and extractable ammonia emissions (Malińska et al., 2014; Awasthi et al., 2016). Although these effects can be partially eliminated, researchers are concerned about the input of potentially toxic metal elements and therefore their possible accumulation for several in the soil horizon (Fang et al., 2017). The same type of risk is related to the presence of other chemical compounds and pharmaceuticals as well as some pathogens that can survive the process.

2.2.3 Thermal processes

All thermal processes are considered sludge-to-energy systems that lead to the complete oxidation of the volatile matter with production of a residue (ash). Generally, the most famous technologies are incineration, gasification, pyrolysis, and plasma gasification. Combustion and/or incineration are considered the most attractive disposal methods for sewage sludge in Europe (EC, 2008), as they replace potentially dangerous landfilling and agricultural strategies. This also allows largely reducing the volume of sewage sludge to destroy the microbiological danger, minimize the odors, and simultaneously recover the renewable energy. Three main variants of this process are used: incineration in dedicated plants, coincineration with municipal solid wastes, and incineration in cement kilns. The environmental cost related to those systems is mostly related to the high energy consumption and production of harmful gaseous emissions (i.e., dioxins and furans) (Garrido-Baserba et al., 2015). Moreover, the ashes coming from the process can be considered a concentrated pollutant that accumulates chemical contaminants. The interesting alternative for ash disposal is a cement replacement. After incineration, sewage sludge is still rich in silica, alumina, calcium oxide, and iron oxide, so it can be used in the production of building materials. Moreover, in this form, metals are stabilized and solidified, so the potential risk is reduced (Samolada and Zabaniotou, 2014).

Pyrolysis is being recognized as a relatively expensive but highly effective technology. Basically, the process converts the organic matter into bioenergy (oil/gas) with a byproduct in the form of so-called biochar. Thus, pyrolysis allows yielding a major bio-oil fraction potentially useful as a fuel or as a source of chemical products. Generally, pyrolysis and similar processes of combustion of sewage sludge are regarded as endothermic. Nevertheless, Atienza-Martínez et al. (2018) have shown lately that the necessity of pretreatment of this substrate (dehydration) moves it more to the exothermic processes, although it is still the most cost-consuming throughout the scenario. Thus, improvement of the steps allowing water removal is crucial for the future potential of the process. Independently, numerous studies have shown many advantages of using biochar for environmental management. For instance, it can be applied for soil improvement, to improve the efficiency of the resources, for remediation and/or protection of lands, and future greenhouse gas mitigation (Joseph and Lehmann, 2015). In general, biochar can thus be defined as a solid, carbon-rich material obtained in the process of zero or low oxygen pyrolysis from different C-based feedstocks, which, applied to the soils, sustainably sequesters carbon and thus improves soil quality in the long term (Verheijen et al., 2010).

As far as the process of combustion eliminates the microbiological risk related to the land application of sewage sludge, still some questions considering chemical pollutants are being posed. These need to be evaluated on a case-by-case basis, not only with concern to the biochar product itself but also to soil type and environmental conditions (Verheijen et al., 2010). Nevertheless, lately it has been recognized that the addition of activated carbon or biochar to sewage sludge immobilizes the bioavailable fractions of polycyclic aromatic hydrocarbons and metal elements (Kończak and Oleszczuk, 2018). Moreover, Frišták and Soja (2015) recognized that the addition of biochar produced from wood chips and garden residues into the sewage sludge and its application as soil amendments have increased the content of available forms of phosphorus. The positive effects of the addition of sewage sludge-derived biochar were also observed during the process of vermicomposting of sewage sludge, where it significantly reduced the bioavailability of Cd and Zn for *Eisenia fetida* earthworms (Malińska et al., 2017).

3 Sewage sludge as sources and drive pathways for contaminants

Considering all valuable resources present in sewage sludge (organic matter, plant culture available nutrients), many countries recognized this byproduct as a potential substrate for fertilization in agriculture or remediation of polluted areas. Nevertheless, sewage sludge applications on agricultural land might contribute to the dispersal of a broad range of unwanted constituents on soils possibly used for food production. Such undesirable contaminants (potentially toxic elements (PTE) such as metals, trace organic compounds (TrOC), and pathogenic organisms) may pose sanitary and environmental risks (Andreoli et al., 2017). Toxic pollutants in sewage sludge could even, in some cases, increase preexisting environmental problems and lead to secondary environmental contaminants and poisonings if not properly managed.

As a reflection of our chemical-based consumer society, our wastewater and sewage sludge mirror compounds we use, produce, release, and discharge. Sludge composition, its agronomic interests, and contaminations may so differ greatly. Such parameters depend on the wastewater origins (agricultural, industrial areas or urban), the local household and consumer habits, the sewer collection (separation or not for wastewater and runoff), the regional environmental regulations, the season and obviously of the size and the process used by the considered WWTP.

3.1 Chemical contaminants in sludge

The environmental risk of sludge contaminants and their concentrations in soil after land application is dependent on their initial concentrations (in both soils and sludge) and the application rate (cumulative effects), management practices, and losses. Therefore, volatile and easily degradable contaminants may still pose environmental risks in the case of high initial concentrations and repeated applications (Harrison et al., 2006).

There are two environmental and public health issues involved with the use or disposal of WWTP biosolids: potentially toxic elements (PTEs) and organic contaminants (OCs), chiefly persistent chemicals. Concerns relate to potential trophic transfers (via cultivated plants) and possible contamination of groundwater. PTE is a common general term that includes metal elements, formerly known as "heavy metals" or "metal trace elements." They are naturally present in soils and the current concerns come more from anthropogenic soil contamination by these elements through the use of fertilizers (including sludge, slurries, and manures), pesticides, and poor waste management. They classically comprise the following metals and metalloids: arsenic (As), cadmium (Cd), chromium (Cr), copper (Cu), lead (Pb), mercury (Hg), molybdenum (Mo), nickel (Ni), and zinc (Zn). Potential environmental risks associated with these PTEs in the context of sludge land applications have been extensively studied and environmental guidelines and regulations defined. For several OCs, such a regulatory framework exists as for polycyclic aromatic hydrocarbons and several persistent organic pollutants (POPs, such as PCBs, dioxins, and related compounds (PCDD/F)). The list and limit values for concentrations of metal elements and OCs that should restrict the use of sewage sludge in agriculture have been updated recently in Europe (EC, 2000). They have suggested the regulation of linear alkylbenzene sulfonates (LAS; extensively used as surfactants), Di(2-ethylhexyl) phthalate (DEHP), nonylphenol and nonylphenol ethoxylates (NP(E)), halogenated organic compounds (i.e., adsorbable organic halides (AOX)), PAH, PCBs, and polychlorinated dibenzo-p-dioxins and dibenzo-furans (PCDD/F). Studies on sludge contaminants started in the 1970s. Most of the chemicals we use in our everyday lives and those with industrial applications are likely to end their lifecycle in biosolids, not to mention byproducts of human activities. We could only exclude highly volatile products and those that are rapidly degraded. In addition to the OCs in urban wastewaters, surface run-off of atmospheric-deposited environmental contaminants onto artificialized areas (concreted and paved, consequently impervious) contribute to the accumulation in sewage sludge of lipophilic compounds that tend to adsorb to solid particles. This situation is like that of sediments in aquatic ecosystems. It is the nonpolar and some persistent compounds that could represent an environmental risk with sludge recycling. One of the most concrete examples may be that of surfactants such as LAS (little worrying, even considering present uses but above all realized risk assessments, (Schowanek et al., 2007)) and more worrisome fluorosurfactants such as perfluorooctanesulfonic acid (PFOS), perfluorooctanoic acid (PFOA), and perfluorononanoic acid (PFNA) (Smith, 2009).

As an introduction to their review of emerging OC in sludge, Clarke and Smith (2011) considered that of the 100,000 chemicals registered in the EU, all are likely to be found in WWTP sludge. In 1996, Wilson et al. (1996) proposed a list of 300 products to prioritize in sludge research work and surveys (priority organic pollutants on environmental agency and country lists and compounds identified in sludge worldwide). This number is close to the one (332) from (Drescher-Kaden et al., 1992) on toxic or suspected toxic OC residues in German sludge, based on a review of the literature on more than 900 articles. Harrison et al. (2006) reported 516 OCs for which concentration data were available in the peer-reviewed literature and official government reports. More recently, Eriksson et al. (2008) concluded after a literature review that 541 OCs could potentially be present in sewage sludge due to their use in construction materials, pharmaceuticals, and personal care products. However, some OC concentrations are not sufficiently characterized in such matrices as sludge and biosolids because their analyses are confined on regulated or identified contaminants on priority lists that represent only a small fraction of the present OCs (Harrison et al., 2006). These authors highlight the lack of knowledge for OCs as nitrosamines with high environmental risks while the knowledge is more comprehensive for some families such as pesticides, PAHs, and PCBs. Thus, a global vision of the contaminants present in the sludge is possible for some as metals, PCB, PAH ... Besides, the concentrations for these compounds in sludge tend to decrease. As an illustration of sludge contaminations around the world, we propose Tables 2–3. Such an overall worldwide inventory could also be made for PCBs and PCDD/Fs, but it would not be possible for most other TrOCs.

The first lesson of Table 2 on metal concentrations in sludge is that there is a real scientific consensus for these elements. Zn is always the predominant metal in terms of concentrations. PTE is one of the few families of contaminants for which sequential or selective extraction approaches could be used to approach the key issue in contaminant risk assessment, which is bioavailability. The origins of PTEs in wastewater are also well described and understood and even some technical solutions for metal removal from sludge have been developed and proposed (see (Babel and Del Mundo Dacera, 2006) for an example of review).

The fact that PAHs (Table 3) are priority environmental pollutants relies mainly on their possible harmful effects on biota as well as carcinogenicity in humans. They are too lipophilic with low biodegradability and they accumulate in sludge, sediments, and soils. The main sources of sludge PAH are industrial wastes as well as domestic sewage, atmospheric rainfall, precipitation of airborne pollutants, and road surface and tire abrasion products (PAHs) (Bomboi and Hernandez, 1991). We will emphasize the study of (Stefaniuk et al., 2018), which proposes to analyze the freely

Table 2 Total concentrations (averages of selected metals present in sewage sludge of different countries worldwide)

	Total element concentrations (mg/kg of dry sludge solids; except specific units for several values)											
	Ag	Al	As	Cd	Cr	Cu	Fe	Hg	Ni	Pb	Zn	Comments and references
Brazil	n.r.	n.r.	14.69	10.75	143.72	255.39	n.r.	2.35	41.99	80.37	688.83	Averaged values in wastewater sludge (UN-HABITAT, 2008)
Canada	n.r.	n.r.	n.r.	2.3–10	66–2021	180–2300	n.r.	n,r.	37–179	26–465	354–640	(Tyagi et al., 1988; Benmoussa et al., 1997; Filali-Meknassi et al., 2000) In (Pathak et al., 2009)
	n.r. n.r. n.r.	n.r. n.r. n.r	1 4.6 n.r.	1 2.3 0.5	50 50.7 n.r.	460 888 137	n.r. n.r. n.r.	1 3.1 0.3	16 26.4 9	51 56 27	593 588 223	Averaged values in Biosolids, Ottawa; British Columbia and Greater Moncton Sewerage Commission (UN-HABITAT, 2008)
China	n.r.	n.r.	16.7–26	5.9–13	45.8–78.4	131.2–394.5	n.r.	17–24	49.3–95.5	57.5–109.3	783.4–3096	Min and max six plants (Dai et al., 2007)
	n.r.	n.r.	n.r.	2.1–19.4	22.2–453.2	210.6–1191.3	n.r.	n.r.	25.1–106.6	41.2–452.2	1406.2–3699.2	Min and max 12 plants in Zhejiang Province (Hua et al., 2008)
	n.r.	n.r.	n.r.	1.65	1983.8	3323.9	n.r	n.r.	422.0	69.7	2424.2	Xiamen WWTP, Fujian Province (Wang et al., 2016)
	n.r.	n.r.	n.r.	11.72, 9.63–15.13	112.5, 50–125	383.47, 62.75–796.63	n.r..	n.r.	692.94, 98.63–2180.13	113.19, 86.25–136.75	609.44, 290.38–831	Mean, min and max four plants (Yang et al., 2017)
	n.r	n.r.	n.r	2.12	514.24	266.06	n.r	n.r	84.64	n.r.	1345.51	Raw sludge from the Shihezi WWTP, Xinjiang (Li et al., 2018)
Colombia	n.r.	n.r.	18.6	76	72.5	163.4	n.r.	8	42.9	87.5	1014.2	Averaged values in biosolids from El Salitre WWTP, Bogota (UN-HABITAT, 2008)
Finland	n.r.	n.r.	n.r.	0.6	18.1	244	n.r.	0.37	30.3	8.8	332	Averaged values in 2005 wastewater sludge (UN-HABITAT, 2008)
France	n.r.	n.r.	n.r.	4.5	64	286	n.r.	2.1	35	107	761	Median of 237 mainly domestic WWTP, (ADEME, 1995)
Germany	n.r.	n.r.	n.r.	1.5	50	275	n.r.	n.r.	23.3	67.7	834	(CEC, 1999). In (Pathak et al., 2009)
	n.r.	n.r.	n.r.	1.52	60.5	380.2	n.r.	0.92	32.2	61.7	955.7	Averaged values in 2003 wastewater sludge—DWA survey (UN-HABITAT, 2008)

Hong-Kong	n.r.	n.r.	n.r.	n.r.	663	112–255	n.r.	n.r.	44.5–622	52.5–57	1009–2823	(Xiang et al., 2000; Wong and Selvam, 2006) In (Pathak et al., 2009)
India	n.r.	n.r.	n.r.	41–54	102–8810	280–543	n.r.	n.r.	192–293	91–129	870–1510	(Singh et al., 2004; Pathak et al., 2008) In (Pathak et al., 2009)
Iran	n.r.	60–259	n.r.	6.1–15.3	2782–8071	57.5–163	n.r.	n.r.	17.9–59.3	n.r.	260–2077	Mean, min and max four plants (Feizi et al., 2018)
Ireland	n.r.	n.r.	n.d.	12	35	520	n.r.	n.d.	18	252	886	Averaged values from 16 plants (Healy et al., 2016)
Italy	n.r.	n.r.	7.3	2.7	62.1	601	n.r.	n.r.	29.2	16.3	961	Sludge from the municipal WWTP of Brindisi (Nissim et al., 2018)
	n.r.	n.r.	n.r	2.1	n.r.	370	n.r.	n.r.	19	72	1500	(Lazzari et al., 2000) In (Pathak et al., 2009)
	n.r.	n.r.	n.r	0.3–0.9	18–65	90–206	n.r.	0.2–0.9	11–15	80–126	283–895	Range three plants (Gianico et al., 2013)
	n.r.	n.r.	n.r	1.6	22.3	261	n.r.	0.2	15.6	76.2	577	Averaged values in 2006 Sardinia biosolids used in agriculture (UN-HABITAT, 2008)
Japan	n.r.	n.r.	8.2	2.2	19.5	n.r.	n.r.	1.1	32.3	5.2	n.r.	Averaged values in dried wastewater sludge, Suzu (UN-HABITAT, 2008)
Poland	n.r.	n.r.	n.r.	n.d.	58	194	n.r.	1.04	22	23.5	1459	Mean values in sludge from the treatment plant of Sokółka (Wołejko et al., 2018)
	n.r. n.r. n.r.	n.r. n.r. n.r.	n.r. n.r. n.r.	1.63 n.d. 1.36	16.1 17.2 37.2	191.4 158.0 55.8	n.r. n.r. n.r.	0.457 0.297 0.234	15.6 17.6 n.d.	32.4 50.4 n.d.	1248.5 962.9 344.8	Sludge from a small, a medium and a large WWTP in Central Poland (Urbaniak et al., 2017)
Netherlands	n.r.	n.r.	**7.3**; *n.d.*	**3.8;** *n.d.*	**66**; *20–40*	**420**; *10–100*	n.r.	**1.9**; *n.d.*	**36**; *7–22*	**210**; *2–12*	**1100**; *60–400*	Range values from Dutch **municipal** and *agro-industrial* WWTP sludge (Veeken and Hamelers, 1999)
Russia	n.r.	n.r.	n.r.	n.r.	305–310	200–300	n.r.	11.35	75–77	34.7	0.07–0.08 (%)	(Nikovski and Kalinichenko, 2014) In (Fijalkowski et al., 2017)
	n.r.	n.r.	0–24	0–300	18.2–1280	0.9–1200	n.r.	0–11.35	1.4–306	0.8–1070	3.0–3820	Range values in 2017 sludge from Moscow Area (UN-HABITAT, 2008)
Slovenia	n.r.	n.r.	2	1	90	200	n.r.	2	35	150	600	Averaged values in wastewater sludge (UN-HABITAT, 2008)

Continued

Table 2 Continued

	Total element concentrations (mg/kg of dry sludge solids; except specific units for several values)											Comments and references
	Ag	Al	As	Cd	Cr	Cu	Fe	Hg	Ni	Pb	Zn	
Spain	n.r.	n.r.	n.r.	2.37–18.3	54.4–3809	204–337	n.r.	n.r.	23.2–36.5	167–223	871–1626	(Alvarez et al., 2002) In (Pathak et al., 2009)
Sweden	1.98	n.r.	n.r.	2.10	n.r.	323	n.r.	1.45	17.3	45	720	Mean, min and max 11 plants (Östman et al., 2017)
	0.72–3.26			0.59–37		110–640		0.19–10	7.3–36	10.9–560	396–1500	
South Africa	n.r.	n.r.	n.r.	0.82–3.10	35.07–134.08	263.68–626.00	n.r.	n.r.	31.34–51.43	21.28–171.87	1031.75–1732.00	Min and max 5 plants, Limpopo province (Shamuyarira and Gumbo, 2014)
Turkey	n.r.	n.r.	n.r.	1.24	34.2	70.2	n.r.	n.r.	62.1	34.2	300	Averaged values in Izmir et al., 2008) Guneybati WWTP sludge (UN-HABITAT, 2008)
UE countries	0.1–14.7	0.1–60 (%)	5.1–56.1	0.3–5.1	10.8–1542.2	27.3–578.1	0.2–14.9 (%)	0.1–1.1	8.6–310	4.0–429.8	0–0.1 (%)	Range values adapted from (Gawlik, 2012) In (Fijalkowski et al., 2017)
	n.r.	n.r.	n.r.	0.4–3.8	16–275	39–641	n.r.	0.3–3	9–90	13–221	142–2000	Typical ranges from (ICON, 2001) In (Gianico et al., 2013)
UK	n.r.	n.r.	n.r.	3.5	159.5	562	n.r.	n.r.	58.5	221.5	778	(CEC, 1999; EU, 1999) In (Pathak et al., 2009)
USA	n.r.	n.r.	n.r.	25	178	616	n.r.	n.r.	71	170	1285	(Bastian, 1997) In (Pathak et al., 2009)
	n.r	n.r.	2.6	2	n.r	670	n.r.	1.3	16	39	743	Averaged values in class
	n.r	n.r.	6.06	10.2	84	1060	n.r.	1.91	50.8	38.5	1180	B biosolids—Denver and Los Angeles Hyperion Treatment Plant (UN-HABITAT, 2008)

WWTP, wastewater treatment plant; *n.r.*, not reported; *n.d.*, not detected or under quantification limits.

dissolved PAH concentrations rather than the total concentrations in order to better estimate their potential environmental availabilities.

For so-called "emerging" contaminants, scientific works appear and a global scheme is emerging since a decade. These contaminants are considered emerging because either the current analytical techniques finally allow their analyses in sludge, or the new industrial and domestic uses of certain products increase their concentrations in environmental matrices. Among these compounds are prominent pharmaceuticals, personal care products and residues, endocrine disruptors, and, more recently, nanoparticles and microplastics. In the excellent review of Clarke and Smith (2011), these authors ranked the following biosolid emerging contaminants (priority decreasing order): PFOS and PFOA, polychlorinated alkanes, polychlorinated naphthalenes, organotins, polybrominated diphenyl ethers, triclosan, triclocarban, benzothiazoles, antibiotics and pharmaceuticals, synthetic musks, bisphenol A, quaternary ammonium compounds, steroids, phthalate acid esters, and polydimethylsiloxanes.

The field of such OCs in sludge as well as their fates and behaviors following sludge land application are largely to be investigated to finally allow their comprehensive risk assessment and case-by-case sludge environmental (even ecotoxicological) assessments must be favored.

3.2 Pathogenic organisms

Sewage sludge contains biological agents that can be problematic for living organisms because some are pathogenic or may simply disturb natural ecosystems. Generally, four groups of pathogens can be found in sewage sludge: viruses, bacteria, parasites, and fungi. In fact, due to its very rich organic matter, sewage sludge can include many bacteria and fungi species in large quantities (Fijalkowski et al., 2017). Other organisms such as viruses and parasites are also regularly present in sewage sludge (Frąc et al., 2014). The concentration and type of pathogen depend on the type of WWTP, the source of wastewater, and some environmental factors (Romdhana et al., 2009). However, the majority of these pathogenic organisms are derived from human or animal feces (Bloem et al., 2017).

The microbial flora present in sewage sludge is very diverse and abundant due to the high content of organic matter. The majority of these bacteria are saprophytes; they are safe and play an important role in the process of wastewater treatment by forming flocs and degrading some contaminants (Tozzoli et al., 2017). However, some of these bacteria are pathogenic. Huang et al. (2018) identified 243 potentially pathogenic bacterial species in activated sludge, including six major pathogens (*Bacillus anthracis*, *Clostridium perfringens*, *Enterococcus faecalis*, *Escherichia coli*, *Pseudomonas aeruginosa*, and *Vibrio cholera*) that can reach abundances of 14% of the bacterial flora. Others pathogens such as *Salmonella*, *Shigella*, *Klebsiella*, *Serratia*, *Enterobacter*, or *Proteus* have also been identified (Korzeniewska, 2011). All these bacteria may cause various infections such as urinary tract infections (*E. coli*), pneumonia (*Klebsiella* and *Enterobacter*), blood infections (*Enterobacteriaceae*), and gastrointestinal infections (*E. coli*, *Salmonella*). These diseases can appear after contamination by gastrointestinal, respiratory, urinary, and biliary tracts (Korzeniewska, 2011).

Table 3 Total concentrations (average or range values) of selected PAH congeners in sewage sludge of different countries worldwide

	Concentrations of individual PAH (μg/kg of dry sludge solids except specific units for several values)								
	Na	**Ace**	**Ac**	**Fl**	**Phen**	**Ant**	**Fluo**	**Pyr**	**B[a]A**
China	16.23–180.95	20.08–289.86	3.11–131.14	17.37–91.77	48.37–466.41	22.46–214.44	138.40–658.33	120.47–317.31	170.58–2171.24
	140–16320	20–6570	**0**–3890	490–11940	**0–35100**	40–6140	620–9880	**0**–15680	180–4820
Czech Republic	n.r.	n.r.	n.r.	n.r.	17–3910	n.r.	12–877.9	9.5–2869.9	n.r.
France	n.r.	n.r.	n.r.	*n.r.*	n.r.	n.r.	40 **1930** **1070**	n.r.	n.r.
Italy	n.r.	0	0	0	0	0	0	0	0
	n.r.	n.r.	**1–184**	1–228	1–673	n.r.	2–844	1–1118	n.r.
Jordan	1.9 3.0 1.1	n.r.	**0.1** 0.3 0.4	0.5 1.9 3.5	3.0 4.8 **7.7**	0.4 3.9 0.5	**4.8** 3.3 1.4	4.2 6.2 4.8	2.0 **0** 0.7
Japan	n.r.	3.3	2.5	3.9	12.6	2.6	14.7	**15**	3.6
Malaysia	**4550–5380**	n.r.	n.r.	n.r.	**0**–4070	n.r.	**0–0**	n.r.	n.r.
Poland	**0**	**0**	**0**	**0**	68.7	17.9	**398.2**	**0**	84.8

Ch	B[b]F	B[k]F	B[a]P	D[ah]A	B[ghi]P	Ind	∑PAH* (μg/kg dry matter)	Comments and references
493.89–2958.66	572.71–**8514.09**	115.07–2138.07	226.73–6174.17	**0**–1134.62	**0**–1038.67	**0**–2682.70	2467.32–259723.79	Min and max six plants, Beijing *(16 US EPA) (Dai et al., 2007)
470–11200	**0**–1080	0–2170	230–7850	**0**–14050	**0**–6520	**0**–9370	33730–87500 13890–641200	Min and max 12 plants *(16 US EPA and 9 EC) (Hua et al., 2008)
n.r.	**434.4–3802.9§**	**434.4–3802.9§**	21.5–2468.4	n.r.	312.1–2724.6	305.3–2905.4	1481.3–17313.6	Min and max values of 45 samples of sludge (Vácha et al., 2005) In (Suciu et al., 2015)
n.r.	30 630 240	n.r.	**90** **800** 370	n.r.	n.r.	n.r.	n.r.	Min, max and mean in sludge collected along the treatment process of Seine Aval treatment plant (Blanchard et al., 2004)
0	0	0	0	0	0	0	n.r.	Sludge from the municipal WWTP of Brindisi (Nissim et al., 2018)
n.r.	2–**1511§**	**2–1511§**	1–1341	n.r.	1–1030	1–1310	11–3917	Min and max values from 35 WWTP in northern Italy 452 samples—survey of four years *(9 EC) (Suciu et al., 2015)
1.9 1.8 1.2	2.1 0.9 **0**	1.0 0.4 0.3	1.9 1.2 0.5	0.4 **0** **0**	4.7 **7.3** 6.3	1.4 1.2 **0**	34.6 39.3 28.7	WWTP sludge from an University complex or a municipal area and sludge from a raw wastewater and sludge disposal site, Karak * (∑16 with B[e]P instead of Ace) (Jiries et al., 2000)
3.6	4.4	3	4.3	1.6	2.8	**1.6**	82	Mix of six dewatered WWTP sludge: rural, urban and residential areas *(∑16 with B[e]P instead of Na) (Ozaki et al., 2017)
n.r.	n.r.	n.r.	**0–0**	n.r.	n.r.	n.r.	n.r.	Min and max from three plants, Johore (Ahmad et al., 2004)
0	214.5	**0**	134.7	**0**	**0**	88.2	2039.9	Min, max and mean of 15 municipal sewage treatment plants *(16 US EPA) (Baran and Oleszczuk, 2003)

Continued

Table 3 Continued

	Concentrations of individual PAH (μg/kg of dry sludge solids except specific units for several values)								
	Na	**Ace**	**Ac**	**Fl**	**Phen**	**Ant**	**Fluo**	**Pyr**	**B[a]A**
	0 **0** **0**	**7105.1** **2132.4** 0.0079	2946.5 886.6 0.0885	974.2 199.6 0.096	1149.2 526.7 0.1299	425.5 165.9 0.2822	5399.9 1937.8 **0.6060**	5050.5 1096.3 0.6115	2579.3 918.5 0.1295
	80.5	2.03	32.1	25.4	7.81	4.88	17.1	11.3	0.47
Portugal	**(ng/L)** 27–198 **0**–309	(ng/L) **0**–118 **0**–30	(ng/L) **0**–492 **0**–72	(ng/L) 28–704 77–909	(ng/L) 540–**2030** 250–**1760**	(ng/L) 34–234 **0**–292	(ng/L) 100–629 56–685	(ng/L) 277–702 112–706	(ng/L) 80–184 29–155
Romania	**0.575** **0.409**	n.r.	0.004 0.007	0.022 0.026	0.070 0.060	0.005 0.004	n.r.	0.024 0.000	0.009 0.004
Spain	n.r.	n.r.	**0**–300	**0**–750	20–**3000**	n.r.	33–570	**58**–731	n.r.
Turkey	n.r.	n.r.	n.r.	n.r.	**0**–1164.5	n.r.	12–**877.9**	9.5–2869.9	n.r.
UE countries	n.r.	n.r.	n.r.	n.r.	29.9–552.2	15.3–724.0	34.5–**3126.8**	47.2–2637.0	**9.1**–1832.6
	n.r.	n.r.	n.r.	n.r.	100% 29.9–552	84% n.r.	100% 34.5–3217	100% 47.2–2637	97% n.r.
UK	n.r.	n.r.	1700–6600	**3600**–8100	3200–**16000**	n.r.	1400–7400	2100–5600	n.r.

Less and more present congener reported values in a table line are in bold. Abbreviations: *n.r.*, not reported; *0*, not detected or under detection or quantification limits. *Na*, Naphthalene; *Ace*, Acenaphthylene; *Ac*, Acenaphtene; *Fl*, Fluorene; *Phen*, Phenanthrene; *Ant*, Anthracene; *Fluo*, Fluoranthene; *Pyr*, Pyrene; *B[a]A*, Benz[a]anthracene; *Ch*, Chrysene; *B[b]F*, Benzo[b] fluoranthene; *B[k]F*, Benzo[k] fluoranthene; *B[a]P*, Benzo[a]pyrene; *B[e]P*, Benzo[e]pyrene; *D[ah]A*, Dibenz[a,h] anthracene; *B[ghi]P*, Benzo[ghi]perylene; *Ind*, Indeno[1,2,3-cd]pyrene.

							∑PAH* (µg/kg dry matter)	Comments and references
Ch	**B[b]F**	**B[k]F**	**B[a]P**	**D[ah]A**	**B[ghi]P**	**Ind**		
1869.1 577.3 0.1543	7572.5 1857.3 0.0715	**0** **0** 0.0524	1786.0 610.7 0.0423	458.9 72.3 0.0822	835.1 262.6 0.0176	1395.6 368.8 0.1295	36034.1 11612.9 n.r.	Mean values in sludge from the treatment plant of Sokółka (Wołejko et al., 2018)
0.85	0.12	0.13	0.12	**0.012**	0.022	0.028	183.1	Freely dissolved PAHs in municipal WWTP sludge, Chełm *(16 US EPA) (Stefaniuk et al., 2018)
(ng/L) 79–312 13–283	(ng/L) 35–479 **0**–234	(ng/L) 11–289 **0**–95	(ng/L) 23–522 17–275	**(ng/L)** **0**–66 **0**–125	(ng/L) **0**–589 0–160	(ng/L) **0**–461 27–295	(ng/L) 2510–5520 1130–4120	Range values from two domestic and four industrial WWTP sludge *(16 US EPA) (Pérez et al., 2001)
0 **0**	0.005 0.002	0.001 **0**	0.003 0.001	0.009 0.003	0.032 0.023	0.022 0	n.r.	Averaged values in primary and in digested dehydrated sludge of Cluj-Napoca WWTP (Alhafez et al., 2012)
n.r.	**0**–242§	**0**–242§	17–100	n.r.	**0–60**	**0**–88	308–5118	Min and max values of 38 samples of sludge * (9 EC) (Sánchez-Brunete et al., 2007) In (Suciu et al., 2015)
n.r.	434.4–**3802.9§**	**434.4–3802.9§**	21.5–2468.4	n.r.	312.1–2724.6	305.3–2905.4	1481.3–17313.6	Min and max values of 14 samples of sludge * (9 EC) (Salihoglu et al., 2010) In (Suciu et al., 2015)
21.0–2020.5 94%	25.1–1919.4 91%	9.9–1048.0 100%	17.9–1475.5 100%	n.r.	n.r.	n.r.	*n.r.*	Range values and occurrence adapted from (Gawlik, 2012) In (Fijalkowski et al., 2017)
n.r.	**9.9–2967§§**	**9.9–2967§§**	17.9–1476	*n.r.*	29.7–1335	24.2–1401	*n.r.*	Min and max values of 32 samples of sludge from member states (JRC, 2012) In (Suciu et al., 2015)
n.r.	1800–11700§	1800–11700§	690–4000	n.r.	470–**2300**	**390**–2700	18000–50000	Min and max values of 14 samples of sludge * (9 EC) (Stevens et al., 2003) In (Suciu et al., 2015)

16 US EPA priority PAH: acenaphtene, acenaphthylene, anthracene, benzo[a]anthracene, benzo[b]fluoranthene, benzo[k]fluoranthene, benzo[ghi]perylene, benzo[a]pyrene, chrysene, dibenzo[a,h]anthracene, fluoranthene, fluorene, indeno[1,2,3-cd]pyrene, naphthalene, phenanthrene and pyrene.
9 CEC: EU proposed sum of PAHs (acenaphthene, fluorene, phenanthrene, fluoranthene, pyrene, benzo[b + j + k]fluoranthene, benzo[a]pyrene, indeno[1,2,3-cd]pyrene, benzo[ghi]perylene) should not exceed 6 mg/kg dry matter. in sludge for land application) (EC, 2000) ; § reported concentrations for benzo[b + j + k]fluoranthene; §§ reported concentrations for benzo[b + k]fluoranthene.

E. coli is already part of one of the quality criteria for sludge (European Directive 86/278/EEC) (EC, 1986). Also, *Salmonella* is one of the most-studied bacteria in WWTP sludge (Jr Krzyzanowski et al., 2016). These bacteria can survive once released into the environment in part through sludge spreading on agricultural plots (Jr Krzyzanowski et al., 2016; Bloem et al., 2017; Ellis et al., 2018). Thus, the consumption of food from these lands could be a way of contamination. It has been shown that even with low concentrations of *Salmonella* in sludge, some vegetables such as lettuce (Manios et al., 2013) and tomatoes (Asplund and Nurmi, 1991) may contain those bacteria in their tissues (Jr Krzyzanowski et al., 2016).

The risk of the presence of pathogenic bacteria could be aggravated by the presence of antibiotics in wastewater. This increases the number of antibiotic-resistant bacteria. Moreover, the high density of bacteria in WWTP reactors increases the probability of transfer of genetic material between bacteria (Turolla et al., 2018). In Austria, for example, Galler et al. (2018) isolated three multiresistant enterobacteria (extended-spectrum β-lactamase bacteria (ESBL)) from activated sludge: Gram-negative bacilli, methicillin-resistant *Staphylococcus aureus* (MRSA), and Vancomycin-resistant enterococci (VRE)). This could pose sanitary problems because of the dispersion of such antibiotic-resistant bacteria through trophic webs and in the environment (Reinthaler et al., 2013; Fijalkowski et al., 2017; Tozzoli et al., 2017).

The microflora of sewage sludge is also very rich in fungi (Frąc et al., 2014). Fungi play a crucial role in the treatment of wastewater by participating in the degradation of various contaminants (Tozzoli et al., 2017). Several are nevertheless pathogens for plants. For example, two common phytopathogens, *M. circinelloides* and *G. citri-aurantii*, are regularly observed and they affect crop yield by causing diseases in fruits and vegetables. In addition to this ecological and environmental/agronomic risk, with fungi being opportunistic organisms, they have potential pathogenic properties for humans and animals as well (Frąc et al., 2014). Frąc et al. (2017) have found in sewage sludge the fungus *Trichophyton* sp., which is responsible for dermatophytose.

Due to the origin of wastewater, sludge regularly contains viruses, especially of an intestinal origin. Schlindwein et al. (2010) highlighted the most common viruses in WWTP sludge samples from Brazil and tested their viability. The most common viruses were the adenovirus (AdV), the rotavirus (RV), the poliovirus (PV), and finally the hepatitis A virus (HAV). The viability of RV and HAV is around 15%–25% while that of AdV and PV is very high (100% and 90%, respectively), which shows that water and sludge treatment processes are not sufficient to inactivate viruses. This highlights the potential sanitary risks of the dispersal of sludge in the environment.

Furthermore, Bibby and Peccia (2013) found that the most abundant pathogenic viruses were herpes viruses in some US sludge samples. DNA viruses (adenovirus, herpes virus, papillomavirus, and bocavirus) are present in 90% of the samples, and RNA viruses (coronavirus, klassevirus, and rotavirus) are present in 80% of the sludge samples. These viruses can cause serious respiratory and gastrointestinal infections in humans and animals.

Like bacteria, viruses are able to survive once in the environment. Bloem et al. (2017) reported the persistence of enteric viruses for about 100 days in soils.

Works on sewage sludge also reported the presence of parasites such as nematodes and cestodes. Some of them are pathogens for humans and animals and are responsible for various diseases (Chaoua et al., 2017). Sludge frequently contains helminth eggs (*Ascaris*, *Trichuris*, *Toxocara*) (Da Rocha et al., 2016), which are among the most resistant organisms to sludge treatment. Their survival has already been observed for several years after the biosolid to soil application (Bloem et al., 2017).

Other parasites of the protozoan family are also present. Corrêa Medeiros and Antonio Daniel (2018) observed the presence of protozoan cysts in 100% of the samples they controlled and the presence of oocysts in more than half the same samples. A change in sludge treatment had no impact on the concentration and viability of these protozoan forms. Families with some pathogenic species for animals and humans have been observed such as *Cryptosporidium*, Giardia, and Entamoeba (Sabbahi et al., 2018; Khouja et al., 2010).

4 Conclusions and perspectives

Legislative pressure forces all countries to respect the common waste management hierarchy with prevention, reuse, recycling, and recovery the most preferable pathways while landfilling and disposal should be strictly limited (Rorat and Kacprzak, 2017). Authorities, communities, wastewater industries should therefore apply environmental assessments as decision-making tools, in addition to the economic and technical evaluation of each proposed solution. Life Cycle Assessment (LCA) is a tool that allows quantifying the environmental impact/cost of particular options for management of sewage sludge in order to choose the best suitable option for each stakeholder. The result of an LCA shall be understood to be an environmental profile of total and single lifecycle stages considering the use of resources, human health, and ecologic consequences; it does not show any economic or social factors (Cherubini et al., 2009). For instance, the impacts related to wastewater treatment could concern mainly: (1) energy consumption at different stages (global warming), (2) the presence of PTEs (toxicity) and (3) the content of chemical oxygen demand (COD), N and P (eutrophication) (Feijoo et al., 2018). In the case of sewage sludge, most studies examine environmental aspects related to sludge application through fuel requirements (transport on agricultural land), introduction of metals into soils, the reduced use of mineral fertilizers, greenhouse gas emissions, carbon storage, and nutrient leaching (Yoshida et al., 2013). Generally, land application is a contributor to global warming, eutrophication, and acidification while toxicity was considered to be related to the presence of Zn and Cu, according to (Yoshida et al., 2018). Lundin et al. (2004) have compared four different options for final disposal of sewage sludge: agricultural application, coincineration with waste, incineration combined with phosphorus recovery, and fractionation with phosphorus recovery. In most aspects, the agricultural land disposal was recognized as the least preferable from an environmental point of view while other options have good potential of sustainability. Lately, Turunen et al. (2018) have developed a multiattribute value theory (MAVT)-based decision support tool (DST) in order to supply the simple scoring method to count the environmental risks of particular scenarios. The constructed value tree helped to select pyrolysis above the other tested alternatives of composting and incineration.

It is worth noticing that no universal solution can be pointed to as the environmental cost depends on local conditions, which can be highly variable between regions/countries. Often, the decision-making tools omit the problems of the properties of soil, climate, fauna, and flora present in the environment that can also greatly change the final impact of the sewage sludge on the ecosystems. Unfortunately, the most important decisions considering the treatment of sewage sludge are still being made based on economic and political criteria.

References

ADEME, 1995. Les micropolluants métalliques dans les boues résiduaires des stations d'épuration urbaines. ADEME Editions, Paris, p. 209.

Ahmad, U., Ujang, Z., Woon, C., Indran, S., Mian, M., 2004. Development of extraction procedures for the analysis of polycyclic aromatic hydrocarbons and organochlorine pesticides in municipal sewage sludge. Water Sci. Technol. 50, 137–144.

Alhafez, L., Muntean, N., Muntean, E., Mihăiescu, T., Mihăiescu, R., Ristoiu, D., 2012. Polycyclic aromatic hydrocarbons in wastewater sewerage system from the Cluj-Napoca Area. Environ. Eng. Manag. J. 11.

Alvarez, E.A., Mochón, M.C., Sanchez, J.C.J., Rodríguez, M.T., 2002. Heavy metal extractable forms in sludge from wastewater treatment plants. Chemosphere 47, 765–775.

Andreoli, C.V., Von Sperling, M., Fernandes, F., Ronteltap, M., 2017. Sludge Treatment and Disposal. IWA Publishing.

Asplund, K., Nurmi, E., 1991. The growth of salmonellae in tomatoes. Int. J. Food Microbiol. 13, 177–181.

Atienza-Martínez, M., Ábrego, J., Mastral, J.F., Ceamanos, J., Gea, G., 2018. Energy and exergy analyses of sewage sludge thermochemical treatment. Energy 144, 723–735.

Awasthi, M.K., Wang, Q., Ren, X., Zhao, J., Huang, H., Awasthi, S.K., Lahori, A.H., Li, R., Zhou, L., Zhang, Z., 2016. Role of biochar amendment in mitigation of nitrogen loss and greenhouse gas emission during sewage sludge composting. Bioresour. Technol. 219, 270–280.

Babel, S., Del Mundo Dacera, D., 2006. Heavy metal removal from contaminated sludge for land application: a review. Waste Manag. 26, 988–1004.

Baran, S., Oleszczuk, P., 2003. The concentration of polycyclic aromatic hydrocarbons in sewage sludge in relation to the amount and origin of purified sewage. Pol. J. Environ. Stud. 12, 523–530.

Bastian, R.K., 1997. The biosolids (sludge) treatment, beneficial use, and disposal situation in the USA. Eur. Water Pollut. Control 7, 62–79.

Benmoussa, H., Tyagi, R., Campbell, P., 1997. Simultaneous sewage sludge digestion and metal leaching using an internal loop reactor. Water Res. 31, 2638–2654.

Bibby, K., Peccia, J., 2013. Identification of viral pathogen diversity in sewage sludge by metagenome analysis. Environ. Sci. Technol. 47, 1945–1951.

Blanchard, M., Teil, M., Ollivon, D., Legenti, L., Chevreuil, M., 2004. Polycyclic aromatic hydrocarbons and polychlorobiphenyls in wastewaters and sewage sludges from the Paris area (France). Environ. Res. 95, 184–197.

Bloem, E., Albihn, A., Elving, J., Hermann, L., Lehmann, L., Sarvi, M., Schaaf, T., Schick, J., Turtola, E., Ylivainio, K., 2017. Contamination of organic nutrient sources with potentially toxic elements, antibiotics and pathogen microorganisms in relation to P fertilizer potential and treatment options for the production of sustainable fertilizers: a review. Sci. Total Environ. 607, 225–242.

Bomboi, M., Hernandez, A., 1991. Hydrocarbons in urban runoff: their contribution to the wastewaters. Water Res. 25, 557–565.

Cano, R., Pérez-Elvira, S.I., Fdz-Polanco, F., 2015. Energy feasibility study of sludge pretreatments: a review. Appl. Energy 149, 176–185.

Cao, Y., Pawłowski, A., 2012. Sewage sludge-to-energy approaches based on anaerobic digestion and pyrolysis: brief overview and energy efficiency assessment. Renew. Sust. Energ. Rev. 16, 1657–1665.

CEC, C. O. T. E. C, 1999. The landfill directive (1999/31/EC). Off. J. Eur. Communities L. 182, 1–19.

Chaoua, S., Boussaa, S., Khadra, A., Boumezzough, A., 2017. Efficiency of two sewage treatment systems (activated sludge and natural lagoons) for helminth egg removal in Morocco. J. Infect. Public Health.

Cherubini, F., Bargigli, S., Ulgiati, S., 2009. Life cycle assessment (LCA) of waste management strategies: landfilling, sorting plant and incineration. Energy 34, 2116–2123.

Christodoulou, A., Stamatelatou, K., 2016. Overview of legislation on sewage sludge management in developed countries worldwide. Water Sci. Technol. 73, 453–462.

Clarke, B.O., Smith, S.R., 2011. Review of "emerging"organic contaminants in biosolids and assessment of international research priorities for the agricultural use of biosolids. Environ. Int. 37, 226–247.

Connor, R., Renata, A., Ortigara, C., Koncagül, E., Uhlenbrook, S., Lamizana-Diallo, B.M., Zadeh, S.M., Qadir, M., Kjellén, M., Sjödin, J., 2017. The United Nations World Water Development Report 2017. Wastewater: The Untapped Resource. The United Nations World Water Development Report.

Corrêa Medeiros, R., Antonio Daniel, L., 2018. Quantification and analysis of the viability of (oo) cysts of pathogenic protozoa in sewage sludge. Acta Scientiar. Technol. 40 https://doi.org/10.4025/actascitechnol.v40i1.28709.

Da Rocha, M.C.V., BarÉS, M.E., Braga, M.C.B., 2016. Quantification of viable helminth eggs in samples of sewage sludge. Water Res. 103, 245–255.

Dai, J., Xu, M., Chen, J., Yang, X., Ke, Z., 2007. PCDD/F, PAH and heavy metals in the sewage sludge from six wastewater treatment plants in Beijing, China. Chemosphere 66, 353–361.

Drescher-Kaden, U., Brüggeman, R., Matthes, B., Matthies, M., 1992. Contents of organic pollutants in German sewage sludges. In: Effects of Organic Contaminants in Sewage Sludge on Soil Fertility, Plants and Animals.pp. 14–34.

EC, 1986. 86/278/EEC: Council Directive of 12 June 1986 on the protection of the environment, and in particular of the soil, when sewage sludge is used in agriculture. Official J. Eur. Commun. L. 181, 04.

EC, 2000. European Commission working document on sludge. In: Third Draft, Brussels 27 April 2000, DG Environment.

EC, 2008. Environmental, economic and social impacts of the use of sewage sludge on land. Final report. Part I: overview report. In: Milieu Ltd, W. A. R. F. T. E. C, (Ed.), DG Environment Under Study Contract DG ENV.G.4/ETU/2008/0076R (ed.).

Ellis, S., Tyrrel, S., O'leary, E., Richards, K., Griffiths, B., Ritz, K., 2018. Proportion of sewage sludge to soil influences the durvival of Salmonella Dublin and *Escherichia coli*. Clean Soil Air Water 46, 1800042.

Eriksson, E., Christensen, N., Schmidt, J.E., Ledin, A., 2008. Potential priority pollutants in sewage sludge. Desalination 226, 371–388.

EU, 1999. 31/EC of 26 April 1999 on the landfill of waste. Official J. L. 182, 07.

Fang, W., Delapp, R.C., Kosson, D.S., Van Der Sloot, H.A., Liu, J., 2017. Release of heavy metals during long-term land application of sewage sludge compost: percolation leaching tests with repeated additions of compost. Chemosphere 169, 271–280.

Feijoo, G., Andrea, A., Moreira, M.T., 2018. Life cycle assessment of municipal wastewater and sewage sludge treatment. In: Life Cycle Assessment of Wastewater Treatment. CRC Press.

Feizi, M., Jalali, M., Renella, G., 2018. Nanoparticles and modified clays influenced distribution of heavy metals fractions in a light-textured soil amended with sewage sludges. J. Hazard. Mater. 343, 208–219.

Fijalkowski, K., Rorat, A., Grobelak, A., Kacprzak, M.J., 2017. The presence of contaminations in sewage sludge—the current situation. J. Environ. Manag. 203, 1126–1136.

Filali-Meknassi, Y., Tyagi, R., Narasiah, K., 2000. Simultaneous sewage sludge digestion and metal leaching: effect of aeration. Process Biochem. 36, 263–273.

Frąc, M., Jezierska-Tys, S., Oszust, K., Gryta, A., Pastor, M., 2017. Assessment of microbiological and biochemical properties of dairy sewage sludge. Int. J. Environ. Sci. Technol. 14, 679–688.

Frąc, M., Oszust, K., Lipiec, J., Jezierska-Tys, S., Nwaichi, E.O., 2014. Soil microbial functional and fungal diversity as influenced by municipal sewage sludge accumulation. Int. J. Environ. Res. Public Health 11, 8891–8908.

Frišták, V., Soja, G., 2015. Effect of wood-based biochar and sewage sludge amendments for soil phosphorus availability. Nova Biotechnologica et Chimica 14, 104.

Galler, H., Feierl, G., Petternel, C., Reinthaler, F.F., Haas, D., Habib, J., Kittinger, C., Luxner, J., Zarfel, G., 2018. Multiresistant bacteria isolated from activated sludge in Austria. Int. J. Environ. Res. Public Health 15, 479.

Garrido-Baserba, M., Molinos-Senante, M., Abelleira-Pereira, J.M., Fdez-Güelfo, L.A., Poch, M., Hernández-Sancho, F., 2015. Selecting sewage sludge treatment alternatives in modern wastewater treatment plants using environmental decision support systems. J. Clean. Prod. 107, 410–419.

Gawlik, B., 2012. Fate SEES results of a Pan-European Snapshot of randomly taken sewage sludge sample. In: Presented at the Proceedings of Workshop DG ENV and DG JrC2012.

Gianico, A., Braguglia, C.M., Mascolo, G., Mininni, G., 2013. Partitioning of nutrients and micropollutants along the sludge treatment line: a case study. Environ. Sci. Pollut. Res. 20, 6256–6265.

Grosser, A., 2018. Determination of methane potential of mixtures composed of sewage sludge, organic fraction of municipal waste and grease trap sludge using biochemical methane potential assays. A comparison of BMP tests and semi-continuous trial results. Energy 143, 488–499.

Harrison, E.Z., Oakes, S.R., Hysell, M., Hay, A., 2006. Organic chemicals in sewage sludges. Sci. Total Environ. 367, 481–497.

Healy, M.G., Fenton, O., Forrestal, P.J., Danaher, M., Brennan, R.B., Morrison, L., 2016. Metal concentrations in lime stabilised, thermally dried and anaerobically digested sewage sludges. Waste Manag. 48, 404–408.

Hua, L., Wu, W.-X., Liu, Y.-X., Tientchen, C.M., Chen, Y.-X., 2008. Heavy metals and PAHs in sewage sludge from twelve wastewater treatment plants in Zhejiang Province. Biomed. Environ. Sci. 21, 345–352.

Huang, K., Mao, Y., Zhao, F., Zhang, X.-X., Ju, F., Ye, L., Wang, Y., Li, B., Ren, H., Zhang, T., 2018. Free-living bacteria and potential bacterial pathogens in sewage treatment plants. Appl. Microbiol. Biotechnol. 102, 2455–2464.

ICON, 2001. Pollutants in urban wastewater and sewage sludge. Final Report. Office for Official Publications of the European Communities, Luxembourg.
Jiries, A., Hussain, H., Lintelmann, J., 2000. Determination of polycyclic aromatic hydrocarbons in wastewater, sediments, sludge and plants in Karak Province, Jordan. Water Air Soil Pollut. 121, 217–228.
Joseph, S., Lehmann, J., 2015. Biochar for environmental management: an introduction. In: Biochar for Environmental Management. Routledge.
Jr Krzyzanowski, F., De Souza Lauretto, M., Nardocci, A.C., Sato, M.I.Z., Razzolini, M.T.P., 2016. Assessing the probability of infection by Salmonella due to sewage sludge use in agriculture under several exposure scenarios for crops and soil ingestion. Sci. Total Environ. 568, 66–74.
JRC, 2012. Occurrence and Levels of Selected Compounds in European Sewage Sludge Samples. EUR 25598 EN.
Khouja, L.B.A., Cama, V., Xiao, L., 2010. Parasitic contamination in wastewater and sludge samples in Tunisia using three different detection techniques. Parasitol. Res. 107, 109–116.
Kończak, M., Oleszczuk, P., 2018. Application of biochar to sewage sludge reduces toxicity and improve organisms growth in sewage sludge-amended soil in long term field experiment. Sci. Total Environ. 625, 8–15.
Korzeniewska, E., 2011. Emission of bacteria and fungi in the air from wastewater treatment plants—a review. Front. Biosci. 3, 393–407.
Lazzari, L., Sperni, L., Bertin, P., Pavoni, B., 2000. Correlation between inorganic (heavy metals) and organic (PCBs and PAHs) micropollutant concentrations during sewage sludge composting processes. Chemosphere 41, 427–435.
Li, Z., Deng, H., Yang, L., Zhang, G., Li, Y., Ren, Y., 2018. Influence of potassium hydroxide activation on characteristics and environmental risk of heavy metals in chars derived from municipal sewage sludge. Bioresour. Technol. 256, 216–223.
Lundin, M., Olofsson, M., Pettersson, G.J., Zetterlund, H., 2004. Environmental and economic assessment of sewage sludge handling options. Resour. Conserv. Recycl. 41, 255–278.
Malińska, K., Golańska, M., Caceres, R., Rorat, A., Weisser, P., Ślęzak, E., 2017. Biochar amendment for integrated composting and vermicomposting of sewage sludge—the effect of biochar on the activity of Eisenia fetida and the obtained vermicompost. Bioresour. Technol. 225, 206–214.
Malińska, K., Zabochnicka-Świątek, M., Dach, J., 2014. Effects of biochar amendment on ammonia emission during composting of sewage sludge. Ecol. Eng. 71, 474–478.
Manios, S.G., Konstantinidis, N., Gounadaki, A.S., Skandamis, P.N., 2013. Dynamics of low (1–4 cells) vs high populations of *Listeria monocytogenes* and *Salmonella typhimurium* in fresh-cut salads and their sterile liquid or solidified extracts. Food Control 29, 318–327.
Maragkaki, A., Vasileiadis, I., Fountoulakis, M., Kyriakou, A., Lasaridi, K., Manios, T., 2018. Improving biogas production from anaerobic co-digestion of sewage sludge with a thermal dried mixture of food waste, cheese whey and olive mill wastewater. Waste Manag. 71, 644–651.
Matiasi, T.M., 2012. Wastewater Management and reuse for sustainable development. Scholar. J. Agri. Sci. 2 (11), 269–276.
Metcalf, E., Eddy, P., 1991. Wastewater Engineering: Treatment, Disposal, Reuse, 3rd ed. McGraw Hill, Singapore.
Metcalf, Inc, Tchobanoglous, G., Burton, F.L., Stensel, H.D., 2013. Metcalf, E. (Ed.), Wastewater Engineering: Treatment and Resource Recovery. McGraw-Hill Education.
Nikovski, G.N., Kalinichenko, K.V., 2014. Biotechnology of utilization of municipal wastewater sediments. Biotechnol Acta 7, 21–32.

Nissim, W., Cincinelli, A., Martellini, T., Alvisi, L., Palm, E., Mancuso, S., Azzarello, E., 2018. Phytoremediation of sewage sludge contaminated by trace elements and organic compounds. Environ. Res. 164, 356–366.

Nkoa, R., 2014. Agricultural benefits and environmental risks of soil fertilization with anaerobic digestates: a review. Agron. Sustain. Dev. 34, 473–492.

Novak, J.T., 2006. Dewatering of sewage sludge. Dry. Technol. 24, 1257–1262.

Opec, F.F.I.D., 2018. Wastewater Report 2018: The Reuse Opportunity. International Water Association: IWA.

Östman, M., Lindberg, R.H., Fick, J., Björn, E., Tysklind, M., 2017. Screening of biocides, metals and antibiotics in Swedish sewage sludge and wastewater. Water Res. 115, 318–328.

Ozaki, N., Nakazato, A., Nakashima, K., Kindaichi, T., Ohashi, A., 2017. Loading and removal of PAHS, fragrance compounds, triclosan and toxicity by composting process from sewage sludge. Sci. Total Environ. 605, 860–866.

Pathak, A., Dastidar, M., Sreekrishnan, T., 2009. Bioleaching of heavy metals from sewage sludge: a review. J. Environ. Manag. 90, 2343–2353.

Pathak, A., Dastidar, M.G., Sreekrishnan, T.R., 2008. Bioleaching of heavy metals from anaerobically digested sewage sludge. J. Environ. Sci. Health A. Tox. Hazard. Subst. Environ. Eng. 43, 402–411.

Peng, W., Pivato, A., Lavagnolo, M.C., Raga, R., 2018. Digestate application in landfill bioreactors to remove nitrogen of old landfill leachate. Waste Manag. 74, 335–346.

Pérez, S., Guillamón, M., Barceló, D., 2001. Quantitative analysis of polycyclic aromatic hydrocarbons in sewage sludge from wastewater treatment plants. J. Chromatogr. A 938, 57–65.

Puyol, D., Batstone, D.J., Hülsen, T., Astals, S., Peces, M., Krömer, J.O., 2016. Resource recovery from wastewater by biological technologies: opportunities, challenges, and prospects. Front. Microbiol. 7, 2106.

Reinthaler, F.F., Galler, H., Feierl, G., Haas, D., Leitner, E., Mascher, F., Melkes, A., Posch, J., Pertschy, B., Winter, I., 2013. Resistance patterns of Escherichia coli isolated from sewage sludge in comparison with those isolated from human patients in 2000 and 2009. J. Water Health 11, 13–20.

Romdhana, M.H., Lecomte, D., Ladevie, B., Sablayrolles, C., 2009. Monitoring of pathogenic microorganisms contamination during heat drying process of sewage sludge. Process Saf. Environ. Prot. 87, 377–386.

Rorat, A., Kacprzak, M., 2017. Eco-innovations in sustainable waste management strategies for smart cities. In: Brdulak, A., Brdulak, H. (Eds.), Happy City—How to Plan and Create the Best Livable Area for the People. Springer International Publishing, Cham.

Rorat, A., Suleiman, H., Grobelak, A., Grosser, A., Kacprzak, M., Płytycz, B., Vandenbulcke, F., 2015. Interactions between sewage sludge-amended soil and earthworms—comparison between *Eisenia fetida* and *Eisenia andrei* composting species. Environ. Sci. Pollut. Res. 23, 3026–3035.

Rorat, A., Wloka, D., Grobelak, A., Grosser, A., Sosnecka, A., Milczarek, M., Jelonek, P., Vandenbulcke, F., Kacprzak, M., 2017. Vermiremediation of polycyclic aromatic hydrocarbons and heavy metals in sewage sludge composting process. J. Environ. Manag. 187, 347–353.

Rouse, J.D., 2013. Sustainability of wastewater treatment and excess sludge handling practices in the federated states of Micronesia. Sustainability 5, 4183–4194.

Sabbahi, S., Trad, M., Ayed, L.B., Marzougui, N., 2018. Occurrence of intestinal parasites in sewage samples and efficiency of wastewater treatment systems in Tunisia. Water Qual. Res. J. 53 (2), 86–101.

Salihoglu, N.K., Salihoglu, G., Tasdemir, Y., Cindoruk, S.S., Yolsal, D., Ogulmus, R., Karaca, G., 2010. Comparison of polycyclic aromatic hydrocarbons levels in sludges from municipal and industrial wastewater treatment plants. Arch. Environ. Contam. Toxicol. 58, 523–534.

Samolada, M., Zabaniotou, A., 2014. Comparative assessment of municipal sewage sludge incineration, gasification and pyrolysis for a sustainable sludge-to-energy management in Greece. Waste Manag. 34, 411–420.

Sánchez-Brunete, C., Miguel, E., Tadeo, J.L., 2007. Analysis of 27 polycyclic aromatic hydrocarbons by matrix solid-phase dispersion and isotope dilution gas chromatography-mass spectrometry in sewage sludge from the Spanish area of Madrid. J. Chromatogr. A 1148, 219–227.

Sánchez-Monedero, M.A., Serramiá, N., Civantos, C.G.-O., Fernández-Hernández, A., Roig, A., 2010. Greenhouse gas emissions during composting of two-phase olive mill wastes with different agroindustrial by-products. Chemosphere 81, 18–25.

Schlindwein, A., Rigotto, C., Simões, C., Barardi, C., 2010. Detection of enteric viruses in sewage sludge and treated wastewater effluent. Water Sci. Technol. 61, 537–544.

Schowanek, D., David, H., Francaviglia, R., Hall, J., Kirchmann, H., Krogh, P.H., Schraepen, N., Smith, S., Wildemann, T., 2007. Probabilistic risk assessment for linear alkylbenzene sulfonate (LAS) in sewage sludge used on agricultural soil. Regul. Toxicol. Pharmacol. 49, 245–259.

Shamuyarira, K.K., Gumbo, J.R., 2014. Assessment of heavy metals in municipal sewage sludge: a case study of Limpopo Province, South Africa. Int. J. Environ. Res. Public Health 11, 2569–2579.

Singh, K.P., Mohan, D., Sinha, S., Dalwani, R., 2004. Impact assessment of treated/untreated wastewater toxicants discharged by sewage treatment plants on health, agricultural, and environmental quality in the wastewater disposal area. Chemosphere 55, 227–255.

Smith, S., 2009. Organic contaminants in sewage sludge (biosolids) and their significance for agricultural recycling. Philos. Trans. A Math. Phys. Eng. Sci. 367, 4005–4041.

Sogn, T.A., Dragicevic, I., Linjordet, R., Krogstad, T., Eijsink, V.G.H., Eich-Greatorex, S., 2018. Recycling of biogas digestates in plant production: NPK fertilizer value and risk of leaching. Int. J. Recycl. Org. Waste Agric. 7, 49–58.

Stefaniuk, M., Tsang, D.C., Ok, Y.S., Oleszczuk, P., 2018. A field study of bioavailable polycyclic aromatic hydrocarbons (PAHs) in sewage sludge and biochar amended soils. J. Hazard. Mater. 349, 27–34.

Stevens, J.L., Northcott, G.L., Stern, G.A., Tomy, G.T., Jones, K.C., 2003. PAHs, PCBs, PCNs, organochlorine pesticides, synthetic musks, and polychlorinated n-alkanes in U.K. sewage sludge: survey results and implications. Environ. Sci. Technol. 37, 462–467.

Suciu, N.A., Lamastra, L., Trevisan, M., 2015. PAHs content of sewage sludge in Europe and its use as soil fertilizer. Waste Manag. 41, 119–127.

Suleiman, H., Rorat, A., Grobelak, A., Grosser, A., Milczarek, M., Płytycz, B., Kacprzak, M., Vandenbulcke, F., 2017. Determination of the performance of vermicomposting process applied to sewage sludge by monitoring of the compost quality and immune responses in three earthworm species: *Eisenia fetida*, *Eisenia andrei* and *Dendrobaena veneta*. Bioresour. Technol. 241, 103–112.

Tozzoli, R., Di Bartolo, I., Gigliucci, F., Brambilla, G., Monini, M., Vignolo, E., Caprioli, A., Morabito, S., 2017. Pathogenic *Escherichia coli* and enteric viruses in biosolids and related top soil improvers in Italy. J. Appl. Microbiol. 122, 239–247.

Turolla, A., Cattaneo, M., Marazzi, F., Mezzanotte, V., Antonelli, M., 2018. Antibiotic resistant bacteria in urban sewage: role of full-scale wastewater treatment plants on environmental spreading. Chemosphere 191, 761–769.

Turunen, V., Sorvari, J., Mikola, A., 2018. A decision support tool for selecting the optimal sewage sludge treatment. Chemosphere 193, 521–529.

Tyagi, R., Couillard, D., Tran, F., 1988. Heavy metals removal from anaerobically digested sludge by chemical and microbiological methods. Environ. Pollut. 50, 295–316.

Un-Habitat, 2008. United Nations Human Settlements Programme, 2008, Global Atlas of Excreta, Wastewater Sludge, and Biosolids Management: Moving Forward the Sustainable and Welcome Uses of a Global Resource. ISBN 978-92-1-132009-1. 608 p.

Urbaniak, M., Wyrwicka, A., Tołoczko, W., Serwecińska, L., Zieliński, M., 2017. The effect of sewage sludge application on soil properties and willow (Salix sp.) cultivation. Sci. Total Environ. 586, 66–75.

Vácha, R., Horváthová, V., Vysloužilová, M., 2005. The application of sludge on agriculturally used soils and the problem of persistent organic pollutants. Plant Soil Environ. 51, 12–18.

Veeken, A., Hamelers, H., 1999. Removal of heavy metals from sewage sludge by extraction with organic acids. Water Sci. Technol. 40, 129–136.

Verheijen, F., Jeffery, S., Bastos, A., Van Der Velde, M., Diafas, I., 2010. Biochar application to soils. A critical scientific review of effects on soil properties, processes, and functions. European Commision, p. 162. 24099.

Wang, X., Li, C., Zhang, B., Lin, J., Chi, Q., Wang, Y., 2016. Migration and risk assessment of heavy metals in sewage sludge during hydrothermal treatment combined with pyrolysis. Bioresour. Technol. 221, 560–567.

Wilson, S., Duarte-Davidson, R.Z., Jones, K., 1996. Screening the environmental fate of organic contaminants in sewage sludges applied to agricultural soils: 1. The potential for downward movement to groundwaters. Sci. Total Environ. 185, 45–57.

Wołejko, E., Wydro, U., Jabłońska-Trypuć, A., Butarewicz, A., Łoboda, T., 2018. The effect of sewage sludge fertilization on the concentration of PAHs in urban soils. Environ. Pollut. 232, 347–357.

Wong, J.W.C., Selvam, A., 2006. Speciation of heavy metals during co-composting of sewage sludge with lime. Chemosphere 63, 980–986.

Xiang, L., Chan, L., Wong, J., 2000. Removal of heavy metals from anaerobically digested sewage sludge by isolated indigenous iron-oxidizing bacteria. Chemosphere 41, 283–287.

Yang, T., Huang, H.-J., Lai, F.-Y., 2017. Pollution hazards of heavy metals in sewage sludge from four wastewater treatment plants in Nanchang, China. Trans. Nonferrous Metals Soc. China 27, 2249–2259.

Yoshida, H., Christensen, T.H., Scheutz, C., 2013. Life cycle assessment of sewage sludge management: a review. Waste Manag. Res. 31, 1083–1101.

Yoshida, H., Ten Hoeve, M., Christensen, T.H., Bruun, S., Jensen, L.S., Scheutz, C., 2018. Life cycle assessment of sewage sludge management options including long-term impacts after land application. J. Clean. Prod. 174, 538–547.

Sludge activation, conditioning, and engineering

9

Ewa Wiśniowska
Department of Environmental Engineering, Czestochowa University of Technology, Czestochowa, Poland

1 Introduction

During both industrial and municipal wastewater treatment, sewage sludge is generated. This waste material can be separated from sewage during mechanical treatment by preliminary sedimentation or, if biological processes are applied, also in secondary clarifiers (secondary sludge, excess sludge). In municipal wastewater treatment plants, secondary sludge frequently contains chemical residues of phosphorus precipitation. In industrial wastewater plants, chemical sludge may be generated as a result of neutralization, precipitation of toxic compounds, or coagulation.

Quantities of sewage sludge and their quality depend on wastewater characteristics as well as the unit processes applied for their treatment. Usually, as the degree of wastewater increases, the quantity of sludge increases. Simultaneously, high-level requirements for the quality of effluent increase the level of sludge contamination (Sanin et al., 2011). A simplified scheme of municipal sludge classification is presented in Fig. 1. This scheme can be also used for industrial sludge.

Sewage sludge is waste that is hard to manage because of its properties. Solids removed during preliminary sedimentation contain a lot of water, high concentrations of organic matter (it makes it putrescible), and a high content of pathogenic microorganisms (Sanin et al., 2011; Bień, 2002). Primary sludge is vile, smelly, gray-brown, and not very susceptible for dewatering (should be dewatered mechanically) (Sanin et al., 2011).

Secondary sludge contains microorganisms that are quite resistant to hydrolysis. Not stabilized secondary sludge does not generate so many odors as primary. This type of sludge is yellow-brown, fluffy, biologically active and difficult to dewater (Sanin et al., 2011). In wastewater treatment plants where biological nutrient removal systems are applied secondary sludge contains high quantities of phosphorus in the form of polyphosphates. Under anaerobic conditions in methane digestion chambers, the phosphorus compounds accumulated in secondary sludge are released to the reject waters (Bien et al., 2004). A real problem in the case of sewage sludge is also its pollution by micropollutants such as heavy metals, polycyclic aromatic hydrocarbons, polychlorinated biphenyls, etc. The quality of both the primary and secondary sludge is very variable. Table 1 shows the ranges of the values of selected properties of sludge. The composition of sludge determines the methods of its stabilization, treatment, and disposal.

Industrial and Municipal Sludge. https://doi.org/10.1016/B978-0-12-815907-1.00009-X

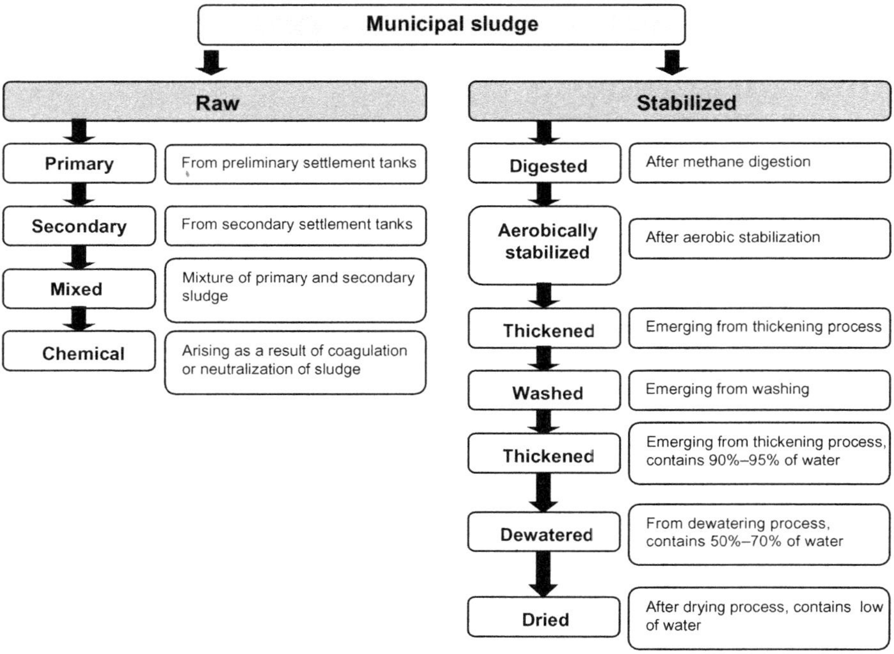

Fig. 1 Simplified scheme of sludge types in wastewater treatment plants.
Based on Kijo-Kleczkowska, A., Otwinowski, H., Środa, K., 2012. Properties and production of sewage sludge in Poland with reference to the methods of neutralizing. Arch. Waste Manage. Environ. Protect. 14(4), 59–78.

Table 1 Selected properties of municipal sewage sludge (Kijo-Kleczkowska et al., 2012)

Parameter	Unit	Raw sludge	Excess sludge	Well stabilized sewage sludge
pH	–	5.0–7.0	6.0–7.0	7.2–7.5
Alcalinity	$mg\,CaCO_3\,L^{-1}$	500–1000	55–1000	3000–4500
Volatile fatty acids	$mg\,CH_3COOH\,L^{-1}$	1800–3600	1800–3600	100–1000
Dry matter	%	4–12	0.3–5	4–12
Volatile solids	% d.m.	60–75	55–80	45–55
Total nitrogen	% d.m.	2–7	1.5–10	0.5–3.0
Total phosphorus	% d.m.	0.4–3	0.9–1.5	0.8–2.6
Potassium	% d.m.	0.1–0.7	0.1–0.8	0.1–0.3
Calorific value	$kJ\,g\,d.m.^{-1}$	16–20	15–21	10.5–15

Raw sewage sludge is rich in water (as a result high quantities of sludge must be processed) as well as in organic compounds. Because of the high content of volatile suspended solids in the sludge, it among others generates odors. Raw sludge also contains pathogens and because of this, it is sanitary dangerous. Because of this prior further management, sewage sludge from municipal wastewater treatment plants (WWTPs) must be stabilized and the water content should be decreased (by thickening, dewatering, and/or drying).

In practice, the biological processes of stabilization are usually applied, including mainly aerobic and anaerobic stabilization. After stabilization, the organic compounds in sludge decrease by several dozen percent (Bień, 2002). The digested sludge is black, musty, and produces biogas. Aerobically stabilized sewage sludge is yellow-brown and biologically active. The stabilized sludge is more susceptible for dewatering than raw sludge (Sanin et al., 2011).

The phase that limits the biological degradation of sludge is the hydrolysis of organic material. During hydrolysis, organic substrates such as proteins, fats, and carbohydrates are decomposed to more simple compounds, which can be assimilated by microorganisms. To increase the rate of biological processes, the conditioning of sludge may be applied. Chemical conditioning is also used to improve the dewatering processes.

2 Conditioning as a process of improving sludge dewatering

To improve sludge dewatering, primarily chemical processes are applied, but also thermal and freeze-thaw methods. By using these methods, sludge is prepared to separate water in for example, belt presses or centrifuges. The chemicals that are the most frequently used for sludge preparation are ferrous sulfate, ferric chloride, alum salts, and chlorinated copperas with or without lime (http://water.me.vccs.edu/courses/ENV149/conditioning.htm). The choice of the chemical is determined by local conditions such as the sludge characteristics, effectiveness, price, and legal requirements concerning further sludge treatment.

Chemical agents used as conditioners cause better filterability and dewaterability of sludge. The coagulation process consists of two main stages (Turowskiy and Mathai, 2006):

1. Neutralization of particles (colloids) charge: As a result of the introduction of a chemical agent into the sludge, neutralization of particle charges occurs. Because of this coagulants with charges opposite to those of colloids should be applied. Once the charge is neutralized, small suspended particles are capable of sticking together. As a result of this process, larger particles (called microflocs) are generated. The optimal pH for this process should be determined experimentally.
2. Binding individual particles into a floc structure. During this step, larger flocs are formed because the microflocs contact each other, which cause them to produce larger particles called macroflocs.

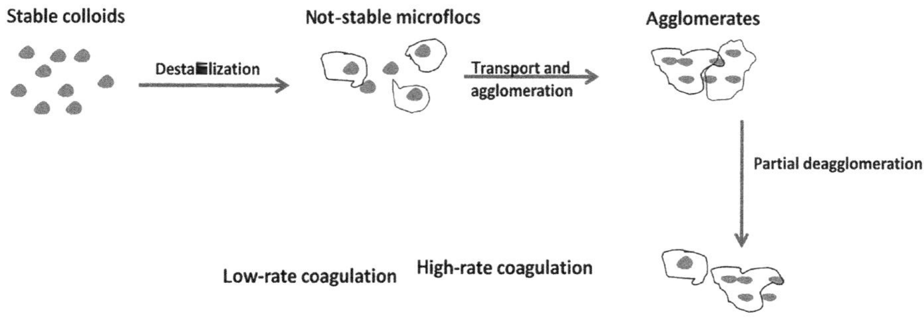

Fig. 2 Mechanism of coagulation.

The mechanism of the neutralization and binding process is presented in Fig. 2.

The dosages of iron salts used for coagulation are in the range of 3%–15% of dry matter and must be chosen experimentally. The polymer dosage is in the range of 3.0–8.0 kg/ton d.m.

A coagulation process with mineral salts is usually supported by using polymers. There are three basic types of polymers (Turowskiy and Mathai, 2006):

(a) Cationic (with positive charge) can be used with coagulants (e.g., aluminum sulfate).
(b) Anionic (with negative charge) are used with inorganic coagulants such as aluminum and iron salts.
(c) Nonionic (with equal amounts of positively and negatively charged groups in monomers).

The flocculants enhance the flocculation process because they have high molecular weights and electrically charged functional groups fixing destabilized particles to their chains. The formation of flocs releases the water from the sludge, improving the dewatering properties of this waste material. (https://www.snf.us/wp-content/uploads/2014/08/Sludge-Dewatering-E.pdf).

3 Conditioning as a process allowing to increase rate of biological decomposition of sewage sludge

More and more attention is being paid to increase the sludge hydrolysis rate, and as a result, to also increase the biogas volume generation. Simultaneously, the amount of volatile acids will also increase. The biogas can be used as an energy source, while volatile fatty acids can be applied as a source of organic carbon for denitrifiers. Preliminary conditioning of sludge also allows for a decrease of the dry matter content in the sludge (Cimochowicz-Rybicka, n.d.). The use of conditioning to destroy sludge structures is frequently called disintegration. Methods that are used for sludge conditioning can be divided into physical, chemical, thermal, and biological (Fig. 3). Some authors classify homogenization as a mechanical disintegration process.

The potential advantages of applying disintegration processes in WWTPs are listed in Table 2.

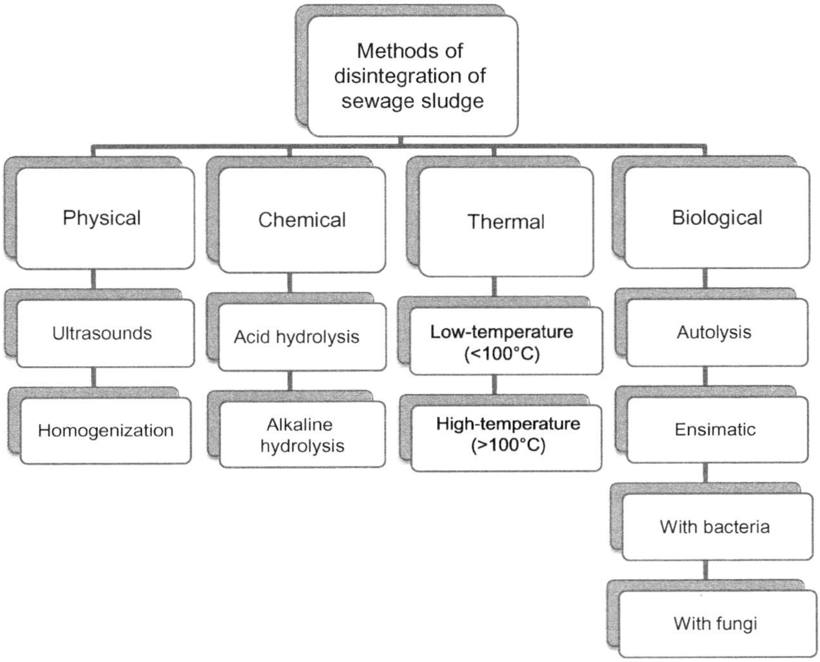

Fig. 3 Methods of sludge disintegration to improve the hydrolysis of organic compounds (Sincero and Sincero, 2003).

Table 2 Potential advantages of applying disintegration in WWTPs (Myszograj, 2017)

Field of use	Effect
Before or after thickening of raw sludge	Increasing easily degradable organic compound concentration in reject water. The supernatant can be used as a carbon source for denitrification and phosphorus removal.
	Enhancing the stabilization rate of raw sewage sludge.
Before secondary settlement tanks	Degasification and selective elimination of filamentous bacteria.
On recirculated sludge pipe	An additional source of carbon in the liquid phase.
Before and after thickening of excess sludge	Changes in the structure of activated sludge flocks, lysis of bacterial cells.
	Additional carbon source for denitrification or phosphorus removal.
	Speeding up hydrolysis of sludge in methane digesters.
Recirculation of sludge in digester	Reducing of foaming in digester, shortening hydrolysis, improving dewatering properties of sludge and its susceptibility to dewatering.
Before final dewatering	Improving dewatering properties of the sludge, sludge hygienization.

3.1 Thermal conditioning of sludge to increase the rate of hydrolysis

Thermal disintegration of sludge is a method that allows for the lysis of microorganisms and as a result, biogas production will increase while a COD value increase in reject water can be achieved (Topal and Arskan, 2009). The process temperature is usually in the range of 60–80°C; however it can reach even 250°C. Based on applied temperature, thermal methods can be divided into low-temperature ($T < 100$°C) and high-temperature processes ($T > 100$°C). The parameters that are the most frequently used are: temperature 150–200°C, pressure 1034–2068 kPa, time <1 h (Topal and Arskan, 2009; Lue-Hing et al., 1998). The minimal time of effective thermal disintegration is 10 min (Kepp et al., 2014). Thermal conditioning of sludge is often supported by the use of chemicals such as acids or lime (Topal and Arskan, 2009). When chemicals are used, usually the applied temperatures are higher compared to classical thermal conditioning. They are in the range of 175–360°C and under pressure 1279–11,377 kPa (Lue-Hing et al., 1998).

Thermal conditioning of sludge affects both biogas production and the composition of reject water. Thermal conditioning of excess sludge and mixed sludge at 70°C for 1 h allowed for increasing the biogas production by 10%–20% compared to non-disintegrated sludge (Davidsson and La Cour Jansen, 2006). Studies done by Zhang et al. (2014) obtained a 7%–11% increase of biogas production during methane digestion of sludge thermally conditioned at 170°C for 30 min Similar results (about 6% increase) were obtained by Janosz-Rajczyk et al. (2006) after 6 h preliminary conditioning of sludge at a temperature of 60°C. Use of combined methods, e.g., thermal-conditioning plus ultrasounds, thermal conditioning plus chemical disintegration allows obtaining a higher increase in biogas production than individual methods alone (Barański et al., 2012).

Thermal conditioning of sludge can also cause COD of reject water, however the data obtained by various authors are not clear, e.g., sludge disintegration at 70°C for 1 h increased COD value of reject water by just several percent (Davidsson and La Cour Jansen, 2006). This method, however, allowed to increase the content of soluble COD (chemical oxygen demand) in reject water by 56%–86% (Davidsson and La Cour Jansen, 2006). This is probably why this process of sludge disintegration is so effective in increasing the rate of the methane digestion process. According to Kepp et al., preliminary thermal digestion of sewage sludge allows for decreasing the required volume of digesters even by 50% compared to the classical methane digestion process.

3.2 Physical methods of sludge disintegration to increase the rate of hydrolysis

The most frequently used method of sludge conditioning is ultrasonic disintegration of its structure. Ultrasounds are sound vibrations at a frequency in the range from 20 kHz to 1 GHz. The use of ultrasounds for sludge conditioning has been widely described in the literature. For sludge conditioning, the frequently used ultrasounds are in the frequency 20–50 kHz (Davidsson and La Cour Jansen, 2006; Wang et al., 1999). Most researchers used the ones in a frequency of about 20 kHz (Xie et al., 2007; Salsabil et al., 2009; Bougrier et al., 2005; Erden and Filibeli, 2010). The results of the studies indicate that ultrasounds are especially effective for conditioning of excess sludge (Salsabil et al., 2009; Tiehm et al., 1997; Bougrier et al., 2005). Because of this, they are used mainly for conditioning of excess and mixed sludge, not the primary sludge (Schlafer et al., 2002; Bien et al., 2004). Ultrasounds effectively destroy bacterial cells and cause their lysis (Suschka and Gruebel, 2014).

As a result of sludge disintegration, an increase of produced biogas and COD in reject water occurs. Simultaneously, the degradation rate of organic solids in sludge is observed.

3.2.1 Enhanced biogas production by using ultrasound disintegration

The results obtained by Appels et al. (2008) indicate that the use of 25 kHz ultrasounds for several hours allowed for 15%–40% increase of biogas production during 8 days of methane digestion of the sludge. Similar results were obtained by Erden and Filibeli (2010) and Bien et al. (2004). Erden and Filibeli have observed a 44% increase in biogas production during mesophilic digestion of sludge disintegrated with 20 kHz ultrasounds. By using ultrasounds of the same frequency (180 W, 60 s), Bień et al. have observed a 24% increase in biogas production. In favorable conditions, even an 81%–>100% increase in biogas production can be achieved by using ultrasounds (Navaneethan, 2007; Tiehm et al., 1997; Schlafer et al., 2002). Just Tiehm et al. (1997) have stated that after 1.5 min of ultrasonic disintegration by 31 kHz ultrasounds, a more than 100% increase of biogas production can be achieved. Schlafer et al. (2002) have had conditioned the sludge with 20 kHz ultrasounds for 30 min and have obtained an increase in biogas production by 100% compared to not disintegrated sludge. The results cited above indicate that both ultrasound frequency and exposure time are important parameters affecting disintegration results while the third parameter is the type and characteristics of sewage sludge.

Ultrasonic disintegration of sludge was applied on a technical scale; however, the results were not so spectacular than in the laboratory scale. Under these conditions, increases of biogas production were in the range 15%–30% (Zafar, n.d.). According to the calculations of Appels et al. (2008) increase in biogas production at the level of at least 15%–40% is cost effective. One of the technologies that uses ultrasonic disintegration for increasing biogas production is the VTA GSD process (http://www.vta.cc/en/Plant-engineering/Disintegration-of-sewage-sludge). By using this method, up to a 30% increase of biogas production can be achieved. Simultaneously, it reduces sludge formation and reduces disposal costs as well as lowering the use of dewatering agents by 30%. In the method, the ultrasonic disintegration of sludge (25 kHz, solids retention time 60 min, 1 bar pressure) is performed in a flowthrough reactor.

3.2.2 COD in reject water and solids degradation rate increase by using ultrasound disintegration

The COD value increase in reject water by using ultrasound can be more than 30% (Tiehm et al., 1997). Davidsson and La Cour Jansen (2006) have stated that the COD of reject water conditioned by 50kHz ultrasounds can reach even 40,000 mgO_2/L (including 15,200 mgO_2/L of soluble fraction). It was stated that ultrasound disintegration is very specifically effective for the destruction of particles with a molecular weight of 40,000 and higher. The authors have observed that ultrasounds of a frequency higher than 100kHz are not as effective a disintegration agent. It explains why most authors use the frequency of 20–30kHz. The COD value increase in reject water is a result of the breakdown of organic polymers present in the sludge. They are released from the solid to water phase. Research by Tiehm et al. (1997) has shown that the degradation rate of volatile solids in sewage sludge treated by ultrasounds was about 50% higher compared to the one obtained during digestion of nonconditioned sludge. Ultrasound disintegration of sewage sludge before methane digestion is applied among others in wastewater treatment plants in Ahrensburg, Bamberg and Meldorf. Use of ultrasounds allowed for increasing the rate of suspended solids removal during methane digestion, by 25%–30% compared to the not conditioned sludge (Zafar, n.d.).

Ultrasounds can also be used in combination with other physical methods. For example, Neumann et al. (2018) evaluated the effect of combined ultrasound-thermal (55°C) pretreatment on the methane digestion process. Use of the disintegration method increased the methane yield by 30%. It was stated that the increase in biogas (methane) production was proportional to the effectiveness of the COD solubilization by the physical disintegration process. Pretreatment also improved the dewatering susceptibility of sludge.

3.3 Chemical sludge disintegration to increase the rate of hydrolysis

The chemical disintegration of sludge can also be considered a method of preliminary disintegration of sludge. For chemical disintegration of sludge, the following agents are usually used (Saphyr Process, n.d.; Barbusiński, 2010):

- Fenton's reagent, hydrogen peroxide, peracetic acid (CH_3COOOH), nitrites in an acid medium (Saphyr process).
- Sulfuric acid, hydrochloric acid.
- Sodium hydroxide.

Chemical disintegration of sludge can be used with a broader application than other disintegration methods. It can be applied for:

- Increasing the susceptibility of sludge to dewatering (Neyens et al., 2002).
- Supporting methane digestion and the aerobic sludge stabilization rate by speeding up the hydrolysis process (Ruiz-Hernando et al., 2013; Zhen et al., 2014; Dewil et al., 2005; Barbusiński and Filipek, 2000).
- Increasing the rate of organic and inorganic pollutants in the sludge (including micropollutants) (Wisniowska, 2008; Dewil et al., 2006).

The technological parameters of selected chemical disintegration methods are given in Table 3.

The doses of reagents used for sludge stabilization and disintegration vary a lot. They depend on the kind of sludge and the effect that should be achieved. When improvement of dewatering of sewage sludge is planned, high doses of reagents are usually high. For example, if we use Fenton's reagent they are in the range of 100–6000 mg of H_2O_2 per 1 L of sludge or up to 50,000 mgH_2O_2 per 1 g of d.m. (Zhen et al., 2014; Lu et al., 2003; Buyukkamaci, 2004; Neyens et al., 2003). If dewatering properties of sludge are the aim of disintegration, usually short process times are used that are not longer than 90 min. The use of Fenton's reagent allows decreasing the capillary suction time as well as the specific resistance to filtration by even more than 90% (Zhen et al., 2014; Lu et al., 2003; Buyukkamaci, 2004). Fenton's process is usually performed under acidic conditions (initial pH about 3.0) because they are optimal for the effective formation of hydroxyl radicals. This requires using additional reagents to acidify the sludge at the beginning as well as to neutralize after disintegration, prior to dewatering. pH is a very important parameter affecting the effectiveness of the chemical disintegration of sludge. Research work made by Lu et al. (2001) showed that changes in the rate of specific resistance to filtration depend on pH. A difference in the specific resistance to filtration rate of 90% was stated in favor of pH 2.6. pH affects not only process effectiveness but also has an effect on the fates of other pollutants. It should be emphasized that when we disintegrate the sludge under acidic conditions, the additional leaching of such micropollutants as heavy metals can occur (Janosz-Rajczyk and Wisniowska, 2005). The mechanism of the improvement of specific resistance to filtration has not been sufficiently explained yet. The most important processes are probably oxidation and destruction of extracellular polymers, which are components of activated sludge flocs. These polymers constitute the relevant cohesiveness of sludge floc and affect the way in which they bond with water (Barbusiński, 2010).

In the scientific literature, it is known that chemical conditioning/disintegration of sewage sludge affects not only the properties of dry matter but also the composition of reject water. Fenton's reagent can change COD of reject water. Studies made by Barbusiński and Filipek (2000), showed that low doses of Fenton's reagent decrease COD of supernatant. At low doses, Fenton's reagent does not oxidate and destroy the polymers present in the sludge. It only oxidizes organic compounds present in the supernatant or makes them more available for rapid degradation. At larger doses, it effectively oxidizes the sludge polymers (Mustranta and Viikari, 1993). In the studies by Barbusiński and Filipek (2000), relatively low doses of Fenton's reagent were used. At low doses, Fenton's reagent does not oxidate and destroy the polymers present in the sludge. It only oxidizes organic compounds present in the supernatant or makes them more available for rapid degradation. At larger doses, it effectively oxidizes the sludge polymers (Mustranta and Viikari, 1993). According to Mustranta and Viikari (1993), as a result of polymer degradation, biodegradable compounds of relatively low molecular weight are formed. They are released to the supernatant and increase the COD value. Simultaneously biodegradability of organic compounds

Table 3 Selected chemical parameters of chemical disintegration of sewage sludge

Effects of chemical disintegration	Chemical agent used for disintegration	Time of disintegration	Other parameters	Reference
Improving susceptibility of excess sludge to dewatering	Fenton's reagent	30 min	30°C, initial pH 3.0	Lu et al. (2003)
Improving susceptibility of excess sludge to dewatering	Fenton's reagent	5–45 min	pH <3.0	Buyukkamaci (2004)
Improving susceptibility of thickened sludge to dewatering, increasing the degree of organic compound degradation	Fenton's reagent	60–90 min	pH 3.0	Neyens et al. (2003)
Enhancing the rate of aerobic sludge stabilization	Fenton's reagent	20 days	Periodical dosage of Fentons's reagent	Barbusiński and Filipek (2000)
Increasing the hydrolysis rate of excess sludge	Peracetic acid	24 h	At room temperature	Zawieja and Wolski (2013)
Solubilization of sewage sludge	Sulfuric acid	5 h	121°C	Cited by Myszograj (2017)
Solubilization of sewage sludge	NaOH	1 h	175–200°C	
Improvement of sludge dewatering	NaOH H_2SO_4	1–20 min	60–90°C	
Increasing the biogas production	NaOH	0.5–24 h	20–40°C	

present in reject water increases. Good biodegradability of the organic fraction confirms the results of the biogas production rate during methane digestion confirms the thesis presented previously. For example, in the studies by Appels et al. (2008) 75% increase of biogas production was observed after conditioning of sewage sludge with Fenton's reagent. Very interesting results were obtained by Ping et al. (2018). They used calcium peroxide (CaO_2) for the disintegration of excess sludge. Disintegrated sludge was next digested at ambient, mesophilic, and thermophilic conditions. It was stated that disintegration of sludge with CaO_2 increased the production of short-chain fatty acids under ambient and mesophilic conditions. Under these conditions, addition of CaO_2 effectively released the polysaccharides and proteins present in the sludge into the reject water. CaO_3 addition was not effective under thermophilic conditions. It caused accumulation of ammonium nitrogen and humic-like substances. The results obtained by Ping et al. indicate that the effect of sludge disintegration should be analyzed not only by analyzing the parameters of the conditioning process, but also of the subsequent biological stabilization.

A significant increase of organic compound concentration in the supernatant is observed when sludge is treated with peracetic acid. It results from the chemistry of the CH_3COOOH action. During decomposition of this oxidizing agent, acetic acid is produced. Doses of peracetic acid used for sludge conditioning are usually not higher than $3.5\,cm^3/L$ of sludge (about $0.1\,cm^3/g$ d.m.) (Zawieja and Wolski, 2013; Wołczyński et al., 2009).

Chemical conditioning of sludge is a simple and flexible method of disintegration allowing for high efficiency of this waste material utilization. However, it also has some disadvantages. The main disadvantages are odors generated during disintegration and the corrosion of reactors caused by acids. Additional reagents are necessary to neutralize sludge, which brings costs and environmental risks (Ruiz-Hernando et al., 2013).

Acidic (usually with sulfuric acid) or alkaline (with sodium hydroxide) hydrolysis is also a disintegration method. Based on the literature data, it can be stated that alkaline hydrolysis is more effective in sludge conditioning than the acidic one (Rocher et al., 1999, 2001). A comparison of the effectiveness of acid and alkaline hydrolysis was done by Cassini et al. (2006). Comparable doses of H_2SO_4 and NaOH (20 mM/L) allowed for s the solubilization of 11% and 60% of organic compounds in sewage sludge, respectively. Doses of sodium hydroxide are rather high but comparable to oxidants such as hydrogen peroxide. For example, Banu et al. (2012) observed the effective hydrolysis of sludge at a dose of NaOH equal to 1600 mg/L. What was interesting was that the dose of NaOH was higher than the lime dose, which provided comparable results. The most effective conditioning method was using a mixture of NaOH with lime.

Simultaneous application of various methods described above was also considered in the literature, for example, the use of chemicals was supported by ultrasounds (Kim et al., 2010; Lu et al., 2018) or the use of chemicals was supported by heating of sludge (Penaud et al., 1999). Simultaneous use of oxidant and ultrasound disintegration allowed for the liquidation of about 70% of the organic compounds in the sludge

(Penaud et al., 1999). Simultaneous use of sodium hydroxide (20 g/L) and thermal conditioning (140°C, 30 min) allowed for the liquidation of 50% of the organic compounds. Results obtained by Lu et al. (2018) indicate that combined alkaline-ultrasound treatment enhanced sludge solubilization by producing low molecular weight substances.

4 Biological sludge disintegration to increase the rate of hydrolysis

Biological processes of sludge disintegration are mainly enzymatic hydrolysis of sludge. The problem is to speed up the rate of this process. Hydrolysis is intensified by using selected bacteria specimens such as *Acinetobacteria*, RutC-30, *Chrysosporium keratinophilum*, *Bacillus licheniformis,* or *Raoultella* (Guo and Xu, 2011). The specimens listed above secrete specific enzymes such as chitanases, celulases, keratinases, amylases, and proteases, respectively. Bacteria specimens can be replaced by enzymatic products (Parawira, 2012; Rashed et al., 2010). In trades, such commercial formulas are available as Alcalase, Carezyme, Cellulast, Termamyl, or Viscozyne (Termamyl, n.d.; Viscozyme, n.d.; Rashed et al., 2010). For example, the use of Viscozyme (at dose 0.1%) allowed for an increase of volatile solid removal by 16.3% compared to the control sample. Termamyl use at dose 0.5% allowed for the increase of volatile solid removal by 13.5%. Results obtained by Ayol (2005) indicate that the addition of enzymes (commercial product Enviro-Zyme 216, which contains protease, lipidase, *Apergillus oryzae* bacteria, and other hydrolytic enzymes) improved the dewaterability by degradation of compounds responsible for bed dewatering of sludge after methane digestion. Through enzyme use, a substantial reduction in sludge mass was also obtained.

The effectiveness of the process was connected with the composition of sludge, which was connected with the selectivity of enzyme action. The use of enzymatic preparations allowed also for improving the susceptibility of sewage sludge to dewatering (Cimochowicz-Rybicka, n.d.). The use of enzymatic agents is easy but involves huge operational costs (Cimochowicz-Rybicka, n.d.; Parawira, 2012).

5 Activation of sewage sludge

The increasing problem in sewage sludge treatment plants is micropollutants, including especially such ones as polycyclic aromatic hydrocarbons (PAHs), polychlorinated biphenyls (PCBs), pharmaceuticals, pesticides, etc. A part of them is removed from the sludge during biological stabilization processes (such as aerobic and anaerobic stabilization) (Sadecka et al., 2016; Barret et al., 2010). Enhanced removal of organic micropollutants from sludge can be achieved by using advanced oxidation methods for disintegration. So far, the effectiveness of Fenton's reagent in organic micropollutant removal has been confirmed (Wisniowska, 2008; Pignatello et al., 2007). Recently, however, new trends in the removal of micropollutants have developed, namely the preparation of activated carbons from sewage sludge by its activation (Kacan and Kutahyali, 2012; Kacan, 2016). Activated carbons obtained from sewage sludge are used as adsorbents. Both municipal and industrial sludge was used for this purpose. The activation of sewage sludge is necessary because direct application of dried sludge as an adsorbent is ineffective (Monsalvo et al., 2011). The results indicate that dried sewage sludge has a low surface area equal to about 5 m^2/g. Pyrolysis can increase the surface area of sewage sludge to 60–400 m^2/g and requires temperatures above 700°C (Monsalvo et al., 2011; Tu et al., 2012). Chemical activation of the sludge coupled with pyrolysis allows obtaining adsorbents of significantly higher surface areas (Chingombe et al., 2005). Prepyrolyzed material is usually used as a basis for sludge activation; however, chemical activation can also be the first step of activated carbon preparation before pyrolysis (Monsalvo et al., 2011). The temperature of pyrolysis affects the conversion of sewage sludge to carbon (Wang et al., 2017).

For activation of sewage sludge, various chemical reagents are used, including KOH, $ZnCl_2$, and H_3PO_4. As mentioned above, chemical activation is usually coupled with pyrolysis. Also, other chemical agents may be used to activate the sludge, including Fe^{2+}/Fe^{3+} systems (Gu et al., 2017; Monsalvo et al., 2011). Yu et al. have observed that acids modify char surface with more oxygen-containing groups (Wang et al., 2017; Yu et al., 2015).

Kacan (2016) has prepared textile sewage sludge activated carbon (TSSAC) from textile sludge by its activation with potassium hydroxide with pyrolization and carbonization. The optimal parameters of the process were as follows: pyrolyzed temperature 700°C, carbonization time 45 min, and chemical impregnation ratio 0.5. The maximum BET surface area of TSSAC was found to be 336 m^2/g. By changing the activation temperatures and impregnation ratio, different types of activated sludge can be prepared (Kacan and Kutahyali, 2012) with various sorption characteristics. The sorption capacities of the activated sludge obtained in this way depend on initial pH, temperature, and the kind of pollutant, among others.

Bedia et al. (2017) have used iron chloride ($FeCl_3$) for the activation of sewage sludge. In the experiment, they used aerobic granular sludge from industrial cosmetic wastewaters. The sludge was washed with distilled water to obtain constant conductivity. Then it was dried to constant weight (at 105°C). Dried solids were ground and fractions of 0.1–0.25 mm were collected. Dried biosolids were mixed with various amounts of $FeCl_3 \cdot 6H_2O$. The samples were next placed in the oven at 60°C (overnight) and pyrolyzed at 750°C for 2 h. Obtained preparations were cooled down to room temperature under continuous N_2 flow. Cooled samples were washed with 3M HCl and next with distilled water. Finally, they were dried at 60°C. As a result, activated carbon with 5.2%–7.3% content of iron was obtained. The carbon had well-developed porosity (BET surface area above 800 m^2/g). They were effective in TOC removal from water solutions. No iron leaching from samples was observed, which is evidence for not only good sorptive but also mechanical properties.

Activation of sludge can be also done by sequential electro-Fenton activation. Gu et al. (2017) have used municipal sludge to prepare Fe nanoparticle-supported magnetic carbon. In the electro-Fenton process, simultaneously oxidative radicals and ferric ions are produced. Iron is used as an anode and H_2O_2 is generated from dissolved oxygen. Fe^{2+} is produced by anode etching. A combination of Fe^{2+} and H_2O_2 produces strongly oxidative hydroxyl radicals (as in the classical Fenton's process). As indicates by Gu et al. (2017), during the electro-Fenton process, disintegration and stabilization of sludge flocs may occur.

6 Conclusions

Sewage sludge is generated both in municipal and industrial wastewater treatment plants in increasing amounts. It is waste material that is very difficult to manage because of the variability of its composition, pollution by pathogens and micropollutants, and high organics and water content. In practice, biological stabilization (aerobic and anaerobic digestion) is usually used as the first step of sludge treatment. Stabilization of the sludge allows for a decrease of the organic compound content in the sludge and improves the dewatering properties. The phase that limits the biological conversion of sludge is hydrolysis. Increasing the rate of hydrolysis increases the further biological mineralization of organic compounds. It simultaneously increases the volume of biogas as well as decreases the amount of sludge to be further treated. Because of this, disintegration technologies are also called minimization sludge technologies. Disintegration/conditioning of sludge is done by using various methods such as physical, chemical, thermal, or biological. Among the physical methods, ultrasound disintegration is the most frequently used. This method is well described in the laboratory scale; however, at present it is also used in wastewater treatment plants with good effects. At the technical scale, the main advantage of ultrasonic disintegration is an increase of biogas production by about 30% compared to the nondisintegrated sludge. The most frequently used ultrasounds are the ones of about 20 kHz. Ultrasounds are especially effective in waste-activated sludge conditioning/disintegration because they efficiently destroy bacterial cells. Thermal conditioning of sewage sludge for its disintegration is not as effective in the biogas rate increase but it effectively solubilizes the organic polymers present in sewage sludge. This can be performed as a low-temperature ($T < 100°C$) or a high-temperature process ($T > 100°C$). As a result of thermal disintegration, a significant increase of soluble COD content in reject water is observed. Because of this, thermal disintegration can be highly recommended as a method of generation of easily degradable organic compounds from primary sewage sludge. Organic compounds can be used by various heterotrophic bacteria, such as denitrifying ones. High-temperature thermal processes are also effective in sludge hygienization. The disadvantage connected with the use of thermal methods is the possible generation of refractory compounds, which can increase the toxicity of the sludge. Recently, more and more attention is being paid to chemical disintegration of sewage sludge. Chemical methods are simple to operate and very controllable. They simultaneously remove organic micropollutants from sludge, solubilize organic compounds, and increase biogas production. The most frequently considered chemical agents are Fenton's reagent, nitrites under acidic conditions, sulfuric acid, and sodium hydroxide. Doses and types of reagents must be set experimentally, depending on the aim of disintegration and taking into consideration local circumstances. The disadvantages of chemical disintegration methods use are high operational costs and generation of chemical sludge (as a result of oxidation, coagulation, and neutralization processes).

Recently, biological methods have also been placed on the market. Such commercial preparations as Alcalase, Carezyme, Cellulast, Termamyl, or Viscosine are available on the market. They contain enzymes of specific use (e.g., amylases, proteases, etc.). They are effective, but their use is still rather expensive. The preparation must be matched experimentally to individual sludge.

The increasing problem with contamination of sludge by toxic organic micropollutants resulted in another strategy for its management—the activation of sewage sludge. In this method, sewage sludge is converted into activated carbon and used as an adsorbent. Conversion in activated carbon is via pyrolysis of sludge. Additional activation by adding chemicals such as $FeCl_3$, KOH, and others allows for increasing the surface area of the adsorbent up to $>300\,m^2/g$. It provides good conditions for effective organic micropollutant removal. Also, electrochemical processes can be used for sewage sludge activation.

Based on the review of conditioning, disintegration, and activation methods, it can be stated that the future of these methods is minimizing sewage sludge generation by subsequently increasing biogas production. The interesting but not widely used method is the generation of a degradable source of organic compounds for deintrification and phosphorus removal. More and more attention is being paid to the methods that allow managing the pollution of sludge by micropollutants. Methods of sewage sludge activation are of greater and greater interest. These three directions–increasing the energy self efficiency of WWTPs by increasing the biogas production, decreasing the biomass of sludge, and managing micropollutants–seem to be at present the main trends in sewage sludge technology.

References

Appels, L., Devil, R., Baeyens, J., 2008. Ultrasonically enhanced anaerobic digestion of sludge. Int. J. Sustain. Eng. 1 (2), 94–104.

Ayol, A., 2005. Enzymatic treatment effects on dewaterability of anaerobically digested biosolids-I: performance evaluations. Process Biochem. 40 (7), 2427–2434.

Banu, J.R., Do, K.U., Adish Kumar, S., Ick-Tae, Y., Kaliappan, S., 2012. A novel method of sludge pretreatment using the combination of alkalis. J. Environ. Biol. 33 (2), 249–253.

Barański, M., Zawieja, I., Wolski, P., 2012. Effect of thermo-ultrasonic disintegration of excess sludge on the effectiveness of anaerobic stabilisation process. In: Paper read at ECOpole, at Opole.

Barbusiński, K., 2010. Zastosowanie odczynnika Fentona do wspomagania unieszkodliwiania osadów. In: Paper read at Nauka dla Praktyków. kierunki przeróbki i zagospodarowania osadów ściekowych, at Piaseczno.

Barbusiński, K., Filipek, K., 2000. Aerobic sludge digestion in the presence of chemical oxidizing agents. Part II. Fenton's reagent. Pol. J. Environ. Stud. 9 (3), 145–149.

Barret, M., Carrere, H., Delgadillo, L., Patureau, D., 2010. PAH fate during the anaerobic digestion of contaminated sludge: do bioavailability and/or cometabolism limit their biodegradation? Water Res. 44 (13), 3797–3806.

Bedia, J., Monsalvo, V.M., Rodriguez, J.J., Mohedano, A.F., 2017. Iron catalysts by chemical activation of sewage sludge with $FeCl_3$ for CWPO. Chem. Eng. J. 318, 224–230.

Bień, B., 2002. Osady ściekowe. Teoria i praktyka. Wydawnictwo Politechniki Częstochowskiej, Częstochowa.

Bien, J.B., Malina, G., Bien, J.D., Wolny, L., 2004. Enhancing anaerobic fermentation of sewage sludge for increasing biogas generation. J. Environ. Sci. Health Part A Tox. Hazard. Subst. Environ. Engr. 39 (4), 939–949.

Bougrier, C., Carrere, H., Delgenes, J.P., 2005. Solubilisation of waste-activated sludge by ultrasonic treatment. Chem. Eng. J. 106 (2), 163–169.

Buyukkamaci, N., 2004. Biological sludge conditioning by Fenton's reagent. Process Biochem. 39 (11), 1503–1506.

Cassini, S.T., Andrade, M.C.E., Abreu, T.A., Keller, R., Goncalves, R.F., 2006. Alkaline and acid hydrolytic processes in aerobic and anaerobic sludges: effect on total EPS and fractions. Water Sci. Technol. 53 (8), 51–58.

Chingombe, P., Saha, B., Wakeman, R.J., 2005. Surface modification and characterisation of a coal-based activated carbon. Carbon 43 (15), 3132–3143.

Cimochowicz-Rybicka, M., n.d. Minimization of Sewage Sludge Production—European Trends and Selected Technologies, https://www.kth.se/polopoly_fs/1.651120!/JPSU18P12.pdf.

Davidsson, A., La Cour Jansen, J., 2006. Pre-treatment of wastewater sludge before anaerobic digestion - hygenisation, ultrasonic treatment and enzyme dosing. Vatten 62, 335–340.

Dewil, R., Baeyens, J., Neyens, E., 2005. Fenton peroxidation improves the drying performance of waste activated sludge. J. Hazard. Mater. 117 (2–3), 161–170.

Dewil, R., Baeyens, J., Neyens, E., 2006. Reducing the heavy metal content of sewage sludge by advanced sludge treatment methods. Environ. Eng. Sci. 23 (6), 994–999.

Erden, G., Filibeli, A., 2010. Ultrasonic pre-treatment of biological sludge: consequences for disintegration, anaerobic biodegradability, and filterability. J. Chem. Technol. Biotechnol 85 (1), 145–150.

Gu, L., Li, C.X., Wen, H.F., et al., 2017. Facile synthesis of magnetic sludge-based carbons by using Electro-Fenton activation and its performance in dye degradation. Bioresour. Technol. 241, 391–396.

Guo, J., Xu, Y., 2011. Review of Enzymatic Sludge Hydrolysis. J. Bioremed. Biodegrad. https://doi.org/10.4172/2155-6199.10000130.

Janosz-Rajczyk, M., Wisniowska, E.W., 2005. Leaching of organic and inorganic micropollutants from chemically stabilized sewage sludge—OFMSW mixtures. Chem. Pap. 59 (6B), 453–457.

Janosz-Rajczyk, M., Dąbrowska, L., Rosińska, A., Płoszaj, J., Zakrzewska, E., 2006. Zmiany ilościowo-jakościowe PCB, WWA i metali ciężkich w kondycjonowanych osadach ściekowych stabilizowanych biochemicznie. Wydawnictwo Politechniki Częstochowskiej, Częstochowa.

Kacan, E., 2016. Optimum BET surface areas for activated carbon produced from textile sewage sludges and its application as dye removal. J. Environ. Manage. 166, 116–123.

Kacan, E., Kutahyali, C., 2012. Adsorption of strontium from aqueous solution using activated carbon produced from textile sewage sludges. J. Anal. Appl. Pyrolysis 97, 149–157.

Kepp, U., Machenbach, I., Weisz, N., Solheim, O.E., (cited 2014). Enhanced Stabilization of Sewage Sludge Through Thermal Hydrolysis—Three Years of Experience with Full Scale Plant.

Kijo-Kleczkowska, A., Otwinowski, H., Środa, K., 2012. Properties and production of sewage sludge in Poland with reference to the methods of neutralizing. Arch. Waste Manage. Environ. Protect. 14 (4), 59–78.

Kim, D.-H., Jeong, E., Sae-Eun, O., Shin, H.-S., 2010. Combined (alkaline plus ultrasonic) pretreatment effect on sewage sludge disintegration. Water Res. 44 (10), 3093–3100.

Lu, M.C., Lin, C.J., Liao, C.H., Ting, W.P., Huang, R.Y., 2001. Influence of pH on the dewatering of activated sludge by Fenton's reagent. Water Sci. Technol. 44 (10), 327–332.

Lu, M.C., Lin, C.J., Liao, C.H., Huang, R.Y., Ting, W.P., 2003. Dewatering of activated sludge by Fenton's reagent. Adv. Environ. Res. 7 (3), 667–670.

Lu, D., Xiao, K.K., Chen, Y., Soh, Y.N.A., Zhou, Y., 2018. Transformation of dissolved organic matters produced from alkaline-ultrasonic sludge pretreatment in anaerobic digestion: from macro to micro. Water Res. 142, 138–146.

Lue-Hing, C., Zenz, D.R., Tata, P., Kuchenrither, R., Malina, F.R., Sawyer, B., 1998. Municipal Sewage Sludge Management; A Reference Test on Processing, Utilization and Disposal. Water Quality Management Library, vol. 4. Technomic Publishing Company Inc., Lancaster.

Monsalvo, V., Mohedano, A., Rodriguez, J., 2011. Activated carbons from sewage sludge application to aqueous-phase adsorption of 4-chlorophenol. Desalination 277 (1–3), 377–382.

Mustranta, A., Viikari, L., 1993. Dewatering of activated-sludge by an oxidative treatment. Water Sci. Technol. 28 (1), 213–221.

Myszograj, S., 2017. Generation of biogas from excess sludge and organic wastes ultrasonically disintegrated. In: Publishing House of Institute of Environmental Engineering University of Zielona Góra, Zielona Góra (in Polish).

Navaneethan, N., 2007. Anaerobic Digestion of Waste Activated Sludge with Ultrasound Pretreatment. Asian Institute of Technology, School of Environment, Resources and Development, Thailand.

Neumann, P., Barriga, F., Alvarez, C., Gonzalez, Z., Vidal, G., 2018. Process performance assessment of advanced anaerobic digestion of sewage sludge including sequential ultrasound-thermal (55 degrees C) pre-treatment. Bioresour. Technol. 262, 42–51.

Neyens, E., Baeyens, J., Weemaes, M., De Heyder, B., 2002. Advanced biosolids treatment using H_2O_2-oxidation. Environ. Eng. Sci. 19 (1), 27–35.

Neyens, E., Baeyens, J., Weemaes, M., De Heyder, B., 2003. Pilot-scale peroxidation (H_2O_2) of sewage sludge. J. Hazard. Mater. 98 (1–3), 91–106.

Parawira, W., 2012. Enzyme research and applications in biotechnological intensification of biogas production. Crit. Rev. Biotechnol. 32 (2), 172–186.

Penaud, V., Delgenes, J.P., Moletta, R., 1999. Thermo-chemical pretreatment of a microbial biomass: influence of sodium hydroxide addition on solubilization and anaerobic biodegradability. Enzym. Microb. Technol. 25 (3–5), 258–263.

Pignatello, J.J., Oliveros, E., MacKay, A., 2007. Advanced oxidation processes for organic contaminant destruction based on the Fenton reaction and related chemistry (vol 36, pg 1, 2006). Crit. Rev. Environ. Sci. Technol. 37 (3), 273–275.

Ping, Q., Lu, X., Zheng, M., Li, Y.M., 2018. Effect of CaO_2 addition on anaerobic digestion of waste activated sludge at different temperatures and the promotion of valuable carbon source production under ambient condition. Bioresour. Technol. 265, 247–256.

Rashed, I.G.A.-A., Akunna, J., El-Halwany, M.M., Atiaa, A.F.F.A., 2010. Improvement in the efficiency of hydrolysis of anaerobic digestion in sewage sludge by the use of enzymes. Desalin. Water Treat. 21 (1–3), 280–285.

Rocher, M., Goma, G., Begue, A.P., Louvel, L., Rols, J.L., 1999. Towards a reduction in excess sludge production in activated sludge processes: biomass physicochemical treatment and biodegradation. Appl. Microbiol. Biotechnol. 51 (6), 883–890.

Rocher, M., Roux, G., Goma, G., Begue, A.P., Louvel, L., Rois, J.L., 2001. Excess sludge reduction in activated sludge processes by integrating biomass alkaline heat treatment. Water Sci. Technol. 44 (2–3), 437–444.

Ruiz-Hernando, M., Martinez-Elorza, G., Labanda, J., Llorens, J., 2013. Dewaterability of sewage sludge by ultrasonic, thermal and chemical treatments. Chem. Eng. J. 230, 102–110.

Sadecka, Z., Myszograj, S., Sieciechowicz, A., Pluciennik-Koropczuk, E., Wlodarczyk-Makula, M., 2016. Impact of selected insecticides on the anaerobic stabilization of municipal sewage sludge. Desalin. Water Treat. 57 (3), 1213–1222.

Salsabil, M.R., Prorot, A., Casellas, M., Dagot, C., 2009. Pre-treatment of activated sludge: effect of sonication on aerobic and anaerobic digestibility. Chem. Eng. J. 148 (2–3), 327–335.

Sanin, F., Clarkson, W., Vesilind, P.A., 2011. Sludge Engineering. The Treatment and Disposal of Wastewater Sludges. DEStech Publication Inc, Lancaster.

Saphyr Process, n.d., https://cms.esi.info/Media/documents/134247_1313503073922.pdf.

Schlafer, O., Onyeche, T., Bormann, H., Schroder, C., Sievers, M., 2002. Ultrasound stimulation of micro-organisms for enhanced biodegradation. Ultrasonics 40 (1–8), 25–29.

Sincero, A., Sincero, G., 2003. Physical-Chemical Treatment of Water and Wastewater. CRC Press, Boca Raton.

Suschka, J., Gruebel, K., 2014. Nitrogen in the process of waste activated sludge anaerobic digestion. Arch. Environ. Protect. 40 (2), 123–136.

Termamyl, n.d., http://www.ncbe.reading.ac.uk/materials/enzymes/termamyl.html.

Tiehm, A., Nickel, K., Neis, U., 1997. The use of ultrasound to accelerate the anaerobic digestion of sewage sludge. Water Sci. Technol. 36 (11), 121–128.

Topal, M., Arskan, E., 2009. Thermal conditioning of sludges. Saglik Bilim. Derg. 25 (1–2), 108–119.

Tu, Y.T., Tian, S.H., Kong, L.J., Xiong, Y., 2012. Co-catalytic effect of sewage sludge-derived char as the support of Fenton-like catalyst. Chem. Eng. J. 185, 44–51.

Turowskiy, I.S., Mathai, P.K., 2006. Wastewater Sludge Processing. John Wiley & Sons, Inc., Hoboken, NJ.

Viscozyme, n.d., https://www.sigmaaldrich.com/catalog/product/sigma/v2010?lang=pl&region=PL&gclid=EAIaIQobChMItNvnmrjX3gIVyYeyCh2EwgTPEAAYASAAEgIqvvD_BwE.

Wang, Q.H., Kuninobu, M., Kakimoto, K., Ogawa, H.I., Kato, Y., 1999. Upgrading of anaerobic digestion of waste activated sludge by ultrasonic pretreatment. Bioresour. Technol. 68 (3), 309–313.

Wang, X.P., Gu, L., Zhou, P., et al., 2017. Pyrolytic temperature dependent conversion of sewage sludge to carbon catalyst and their performance in persulfate degradation of 2-Naphthol. Chem. Eng. J. 324, 203–215.

Wisniowska, E., 2008. Effect of chemical stabilisation of sewage sludge on the fate of PAHs. Arch. Environ. Protect. 34 (3), 249–257.

Wołczyński, M., Wiśniowska, E., Janosz-Rajczyk, M., 2009. Wpływ kondycjonowania osadów ściekowych kwasem nadoctowym na zmiany ich właściwości fizyczno-chemicznych. Inży. Ochr. Śr. 12 (1), 51–63.

Xie, R., Xing, Y., Ghani, Y.A., Ooi, K.-E., Ng, S.-W., 2007. Full-scale demonstration of an ultrasonic disintegration technology in enhancing anaerobic digestion of mixed primary and thickened secondary sewage sludge. J. Environ. Eng. Sci. 6 (5), 533–541.

Yu, Y., Wei, H.Z., Yu, L., et al., 2015. Surface modification of sewage sludge derived carbonaceous catalyst for m-cresol catalytic wet peroxide oxidation and degradation mechanism. RSC Adv. 5 (52), 41867–41876.

Zafar, S., n.d. Ultrasonic Pretreatment in Anaerobic Digestion, https://www.bioenergyconsult.com/ultrasonic-pretreatment-ad-sewage/.

Zawieja, I., Wolski, P., 2013. Wpływ chemiczno-termicznej modyfikacji osadów nadmiernych na generowanie lotnych kwasów tłuszczowych w procesie fermentacji metanowej. Rocz. Ochr. Śr. 15, 2054–2070.

Zhang, L., Zhang, Y., Zhang, Q., et al., 2014. Sludge gas production capabilities under various operational conditions of the sludge thermal hydrolysis pretreatment process. J. Energy Inst. 87 (2), 121–126.

Zhen, G., Lu, X., Wang, B., et al., 2014. Enhanced dewatering characteristics of waste activated sludge with Fenton pretreatment: effectiveness and statistical optimization. Front. Environ. Sci. Eng. 8 (2), 267–276.

Sludge legislation-comparison between different countries

10

Ewa Wiśniowska, Anna Grobelak, Paulina Kokot, Małgorzata Kacprzak
Department of Environmental Engineering, Czestochowa University of Technology, Czestochowa, Poland

1 Introduction

Sewage sludge is produced during both municipal and industrial wastewater treatment. This type of waste material is defined in legal acts in many ways. The definitions are important because they limit the methods of sludge management. The most important definitions of sewage sludge are listed in Table 1.

Based on the definitions above, it can be stated that sewage sludge definitions include both municipal and industrial sludge; however the regulations concerning municipal sludge are more frequent. Municipal sludge is produced in municipal wastewater treatment plants (WWTPs) and it contains a lot of human excreta. In municipal WWTPs, typically, prior to thickening and dewatering, it consists of about 1% of wastewater inflow (Guidelines, 2004). Fecal excreta are also a component of fecal sludge. Sewage and fecal sludge contain organic matter and nutrients and because of this they can be used as a soil conditioner or a medium-grade soil fertilizer (Guidelines, 2004). The soil application of sewage sludge may improve the nutrients availability for plants for a longer period of time than artificial fertilizers (Grobelak et al., 2017). To the risks related to the use of sewage sludge in agriculture, the presence of pathogens, such as bacteria, viruses, helminthes, or protozoa as well as contamination by organic and inorganic micropollutants can be accounted (Guidelines, 2004). Generally, stabilized municipal sewage sludge includes 30%–55% organic matter, up to 3% total nitrogen, 0.7%–1.5% total phosphorus, 0.7% total potassium content, 10%–20% C/N ratio, and various levels of heavy metal ions. The pH of the sewage sludge is normally ranged 6.5–7.5. The heat value of the dry sludge is about 12.000–15.000 kJ/kg. Sludge also contains other elements such as potassium, magnesium, sulfur, and minor plant nutrients (Sewage Sludge, 2016; Kijo-Kleczkowska et al., 2012). Fecal sludge from septic tanks contains comparable levels of nutrient compounds; however, it usually has a higher content of suspended solids (about 3% versus 1%) and higher COD values (even >10,000 mg/L) than sludge produced in WWTPs (Niwagaba et al., 2014). Fecal sludge can contain high levels of fecal coliforms (1×10^5 cfu/100 mL) and helminthes eggs (up to 16,000 numbers/L) (Niwagaba et al., 2014). These kinds of sludge can be treated in a similar way. Industrial sludge has very diversified physicochemical properties, for example, it can contain toxic compounds and elements at very high concentrations.

Industrial and Municipal Sludge. https://doi.org/10.1016/B978-0-12-815907-1.00010-6

Table 1 Definitions of sewage sludge according to various legal acts

Legal act	Definition	Comments
Council Directive 91/271/EEC concerning urban wastewater treatment (Council Directive 91/271/EEC, 1991)	Residual sludge, whether treated or untreated, from urban wastewater treatment plants.	It concerns only urban wastewater, not industrial. It does not include fecal sludge.
Council Directive 86/278/EEC (Council Directive 86/278/EEC, 1986)	Residual sludge from: – Sewage plants treating domestic or urban wastewater and from other sewage plants treating wastewaters of a composition similar to domestic and urban wastewaters. – Residual sludge from septic tanks or other similar installations for the treatment of sewage. – Residual sludge from sewage plants other than mentioned above.	It is a wide definition that includes all types of sludge considered in the chapter.
USA Part 503 (40 CFR Part 503)	Sewage sludge is solid, semisolid, or liquid residue generated during the treatment of domestic sewage in a treatment works. Sewage sludge includes, but is not limited to, domestic septage; scum or solids removed in primary, secondary, or advanced wastewater treatment processes; and material derived from sewage sludge. Sewage sludge does not include ash generated during the firing of sewage sludge in a sewage sludge incinerator or grit and screenings generated during preliminary treatment of domestic sewage in a treatment works.	It concerns only the sludge produced during domestic wastewater treatment and septage sludge.

Table 1 Continued

Legal act	Definition	Comments
EPA definition 40 CFR 260 (40 CFR Part 260, n.d.)	*Sludge* means any solid, semisolid, or liquid waste generated from a municipal, commercial, or industrial wastewater treatment plant, water supply treatment plant, or air pollution control facility exclusive of the treated effluent from a wastewater treatment plant.	The definition is very wide and includes both industrial and municipal sludge, concerning even electroplating sludge.

Sludge management methods, which are regulated by law, can be recovery, disposal, recycling, or treatment. Directive 2008/98/EC of the European Parliament and of the Council (Directive 2008/98/EC, 2008) on Waste supplies the following definitions:

- *Recovery* means any operation the principal result of which is waste serving a useful purpose by replacing other materials that would otherwise have been used to fulfill a particular function, or waste being prepared to fulfill that function in the plant or in the wider economy.
- *Recycling* means any process of recovery in which waste materials are reprocessed into products, materials, or substances, whether for the original or other purposes. It includes the reprocessing of organic material, but does not include energy recovery and reprocessing into materials that are to be used as fuels or for backfilling operations.
- *Disposal* means any operation that is not recovery, even where the operation has a secondary consequence that is the recovery of substances or energy.
- *Treatment* means recovery or disposal operations, including preparation prior to recovery or disposal.

Countries differ widely in their approaches to sludge management. The aim of this study is to present and compare the legislation concerning sewage sludge in various regions of world.

2 European Union

The amount of sewage sludge generated in the EU is estimated at more than 10 million annually (Fytili and Zabaniotou, 2008). Management of such large quantities of sludge is a major challenge. In the EU, the management and treatment of sludge is regulated by legal acts that can be divided into the ones that give general principles (framework directives) of waste and sludge management and the ones that define the precise conditions of sludge processing. Legislation concerning sewage sludge in the EU treats this matter in a comprehensive way. The issue of sludge is considered not only in the acts directly dealing with waste management (such as the European Parliament and Council Directive 2008/98/EC of Nov. 19, 2008 (Directive 2008/98/EC, 2008), on sewage called the Waste Framework Directive), but also in, for example, the Water Framework Directive (Directive 2000/60/EC, 2000) that refers to achieving good qualitative status of water bodies, among others, by the control of materials applied to land or in Directive 2010/75/EC of Nov. 24, 2010 (Directive 2010/75/EC), on industrial emissions. This directive officially updates and combines other directives, including, among others, Directive 2008/1/EC on the integrated prevention of pollution and its control (IPPC) (Directive 2008/1/EC, 2008), and Directive 2010/75/EU on industrial emissions (Directive 2010/75/EC), which lays down the norms and rules for the incineration of waste as well as emission standards.

The Waste Framework Directive regulates waste recycling, including sewage sludge. The directive states that the prevention of waste production is the first priority with the next being the preparation of waste for reuse, recycling, or other forms of recovery, and finally waste disposal. It is not possible to prevent the production of sewage sludge. That is why other steps of dealing with this waste material are very important. That is preparation for reuse, understood as sludge reprocessing (including possible energy recovery or organic recycling). The Directive of the European Parliament and Council 2009/28/EC on the promotion of energy from renewable sources (Directive 2009/28/EC, 2009) amends, and as a consequence abates, Directive 2009/28/EC indicating compulsory energy levels to be derived from renewable energy sources. Another document, the Technical Report for End-of-Waste Criteria on Biodegradable Waste Subject to Biological Treatment (Technical Report, 2012), places sewage sludge on the positive waste list, allows clean sludge to be used as fertilizer, and gives way to qualify it as a waste product.

One of the most important EU legal acts directly concerning sewage sludge management is Directive 91/271/EEC of May 21, 1991 (Council Directive 91/271/EEC, 1991), concerning the treatment of municipal sewage. This operational directive obliges monitoring and reporting municipal sewage treatment and the final disposal of municipal sewage sludge for agglomerations. Article 14 of Directive 91/271/EEC refers to sludge produced in the course of sewage treatment and states that sewage has to be reused in every appropriate case, provided that adverse effects to the environment are prevented at all times. Limits regarding the storage of sludge were introduced by Council Directive 99/31/EC on the landfill of waste (Directive 99/31/EC, 1999). Based on the general rules given in the directives, individual countries have developed their guidelines and standards for sewage sludge treatment and disposal. Implementation of the above-mentioned EU directives in practice increased the stream of sewage sludge, but, on the other hand, it enabled methods of sludge reuse.

Taking into consideration various ways of sewage sludge treatment, it can be stated that in the EU, significant levels of sludge composting are in such countries as Austria, the Czech Republic, Finland, France, Germany, Hungary, Italy, Slovakia, and Sweden (Mininni and Dentel, 2013). German legislation promotes codigestion as a method of sludge preparation for use in agriculture (Christodoulou and Stamatelatou, 2016). In this country, phosphorus recovery from sludge/ashes is strongly promoted. Because of this, Germany could possibly became the first country to render this compound recovery obligatory (Christodoulou and Stamatelatou, 2016). In the United Kingdom, anaerobic digestion is a preferred method for sludge stabilization (Christodoulou and Stamatelatou, 2016). In the EU in 2001, a second draft of the document on the biological treatment of biowaste was published (Working Document, 2001). It established the requirements for the quality of waste (including sludge) that can be composted. As wastes suitable for biological treatment, the following sludges were qualified:

- Sludge from treatment of urban wastewater; septic tank sludge; sludge from washing, cleaning, peeling, centrifuging, and separation; green liquor sludge (from recovery of cooking liquor); deinking sludges from paper recycling; fiber rejects, fiber, filler, and coating sludges from mechanical separation; sludges in particular from onsite effluent treatment containing chromium; and sludges from biological treatment of industrial wastewater—if they fulfill the requirements of Directive 86/278/EEC for the use of sludge in agriculture.
- Sludge from washing and cleaning; and sludges from onsite effluent-treatment—if they fulfill the requirements of Directive 86/278/EEC for the use of sludge in agriculture and without prejudice to Directive 90/670/EEC on animal waste.

The general requirements for biological treatment plants were also pointed out. They included requirements for localization (among others, concerning distance from the boundary of the site to residential and recreation areas and the existence of surface and groundwater), wastewater and leachate management (with respect to the characteristics of the biowaste), control of odors, and minimizing nuisances and hazards. They have concerned both composting places and digesters. Moreover, general requirements for biogas use were established. When biogas is used as a fuel in internal combustion engines, its composition should comply with the emission limits: dust 50 mg/m^3, NO_x 500 mg/m^3, SO_2 500 mg/m^3, CO 650 mg/m^3, H_2S 5 mg/m^3, HCl 30 mg/m^3, and HF 5 mg/m^3. Biogas that cannot be used onsite or upgraded to natural gas quality shall be flared.

At present, the requirements for sludge stabilization plants in EU countries are regulated by best available techniques (BAT) standards. A reference document on BAT called BREF for waste treatment industries was published in 2006 (BREF, 2006). It includes, among others, composting installations in open air and composting in closed systems as well as digestion systems. Sewage sludge is, in this document, considered a type of biowaste. In the case of biowaste stabilization installations, special attention is paid for odor emissions, including measures that can be taken to reduce odor nuisance. The BREF also includes the solutions that would guarantee a reduction of emissions of NH_3 and dust, protection of groundwater and soil, reduction of noise, and a reduction of process residues and energy consumption. In the case of digestion plants, the techniques of storage and treatment of biogas are also presented. The technologies described as BAT were recommended based on a literature survey, the experiences of existing plants, and technical and economic studies.

When we consider the final management of sewage sludge, it can be agricultural use, incineration, or landfilling. According to the data from the European Commission, the agricultural application of sludge in 27 EU countries will reach in 2020 about 6,000,000 tons of dry solids per year (about 47% of total sludge solids production in EU) (SUSAN, 2008). Generally in Europe, the use of sewage sludge for agricultural purposes has increased significantly in recent years (Placek et al., 2017).

In the EU, the direct use of sewage sludge in agriculture and for land reclamation is regulated by Directive 86/278/EEC on the use of sludge in agriculture (Directive 86/278/EEC, 1986). According to the directive, sludge should be used for agricultural proposes in such ways as to prevent harmful effects on soil, vegetation, animals, and humans while encouraging its correct use. According to this document, because of valuable agronomic properties, it is justified to encourage the application of sewage sludge in agriculture. The application of sewage sludge must not impair the quality of soil and the harvest. It is, however, necessary to control concentrations of heavy metals in sewage sludge and soil. Sewage sludge use is prohibited when the concentration of heavy metals in sludge and soil exceeds limits given in the directive. The directive was implemented into the law systems of individual EU countries. A comparison between EU Directive 86/278/EEC and the legal systems of individual countries is given in Table 2. In EU countries, the agricultural use of sludge is widespread in the Czech Republic, France, Hungary, Portugal, Italy, Slovakia, Spain, and the United Kingdom. It is not popular in such countries as Austria, Belgium, Finland, Greece, Romania, Slovenia, and Switzerland (Mininni and Dentel, 2013). In countries such as Spain and Portugal, more than 65% of the produced sewage sludge is used as a fertilizer in agriculture. Regarding strict regulations on the reduction of the risk of soil and groundwater contamination by metals occurring in sewage sludge, their chemical composition is the key issue (Hemmat et al., 2010). In such countries as Germany (several lands) and the Netherlands, land application of sludge is banned (Christodoulou and Stamatelatou, 2016). However, federal regulations in Germany set the limits of contaminants in the case of sludge use for this purpose (Fytili and Zabaniotou, 2008).

The Fertilizer Law is in compliance with the 86/278/EEC Directive; however, organic micropollutants are also limited, not only heavy metals. Greece consequently aims to reduce significantly the quantities of sludge that are not landfilled (Christodoulou and Stamatelatou, 2016). Greece has implemented the 86/278/EEC Directive as a national framework, "Methods, Conditions, and Restrictions on the Use in Agriculture of Sludge Deriving from Treatment of Domestic and Municipal Wastewater" (MD 80568/4225/1991). It took over the higher limits of heavy metal concentrations from the EU legislation. In the United Kingdom, the agricultural use of sewage sludge is preferred (Christodoulou and Stamatelatou, 2016). A national legislative framework on the use of sludge in agriculture complies with the 86/278/EEC Directive because Great Britain is still an EU country. Heavy metal concentrations are limited. Moreover, the Safe Sludge Matrix was established to regulate the safe use of sludge in agriculture. It provided dual criteria for pathogen reduction similar to Part 503 Classes A and B (Christodoulou and Stamatelatou, 2016). They concern Escherichia coli and Salmonella Spp. The agricultural use of sludge in Sweden is regulated by the SNFS 1994/2 Act (SNFS, 1994). Similarly to the EU legislation, it limits the concentrations of metals in soil and sewage sludge. Permissible heavy metal concentrations in soil and sewage sludge are comparable to the lowest found in the EU and other countries. In some countries, the limits of organic micropollutants in sludge were also established (Table 3).

Table 2 Permissible concentrations of heavy metals in sewage sludge used in agriculture, mg/kg d.m. (Working Document, 2010; Mininni and Dentel, 2013)

Country	Cd	Cr	Cu	Hg	Ni	Pb	Zn
EU Directive 86/278/EEC	20–40	Not limited	1000–1750	16–25	300–400	750–1200	2500–4000
Austria	2–10	50–500	300–500	2–10	25–100	100–500	1500–200
Belgium (Flanders)	6	250	375	5	100	300	900
Belgium (Walloon)	10	500	600	10	100	500	2000
Bulgaria	30	500	1600	16	350	800	3000
Czech Republic	5	200	500	4	100	200	2500
Denmark	0.8	100	1000	0.8	30	120	4000
Finland	3	300	600	2	100	150	1500
France	20	1000	1000	10	200	800	3000
Germany	2	80	600	1,4	60	100	1500
Greece	20–40	500	1000–1750	16–25	300–400	750–1200	2500–4000
Hungary	10	1000 (CrVI)	1000	10	200	750	2500
Italy	20	Not limited	1000	10	300	750	2500
The Netherlands	1.25	75	75	0.75	30	100	300
Poland	20	500	1000	16	300	750	2500
	25	1000	1200	20	400	1000	3500
	50	2500	2000	25	500	1500	5000
Portugal	20	1000	1000	16	300	750	2500
Romania	10	500	500	5	100	300	2000
Slovenia	0.5	40	30	0.2	30	40	100
Spain	20–40	1000–1750	1000–1750	16–25	300–400	750–1200	2500–4000
Sweden	2	100	600	2.5	50	100	800
Working document on sludge (3rd draft)	10	1000	1000	10	300	750	2500
Proposal of directive 2010	10	1000	1000	10	300	500	2500

Table 3 Limits of organic micropollutants for sludge used in agriculture, mg/kg d.m. (Mininni and Dentel, 2013; Working Document, 2000; Global Atlas, 2008)

Organic pollutant	EU draft	Denmark	Sweden	Lower and upper Austria	Austria Carinthia	France	Germany	Germany proposed limits	Slovenia	Czech Republic
AOX	500	–	–	500	500	–	500	400	140	500
DEHP	100	50	–	–	–	–	–	–	–	–
LAS	2600–5000	1300	–	–	–	–	–	–	–	–
NP/NPE	50–450	10	50	–	–	–	–	–	–	–
PAH	6[a]	3[a]	3[a]	–	6[a]	Fluoranthene 4 Benzo(b) fluoranthene 2.5 Benzo(a) pyrene 1.5	–	Benzo(a)pyrene 1.0	–	–
PCB	0.8[b]	–	0.4[b]	0.2[c]	1	0.8[b]	0.2 for each congener	0.1 for each congener	<0.05	0.6
PCDD/F	10^{-4}	–	–	10^{-4}	0.5×10^{-4}	–	10^{-4}	0.3×10^{-4}	–	–
Other compounds	–	–	–	–	–	–	–	2-Mercaptobenzothiazole +2-hydroxybenzothiazole 0.6 Tonalid 15 Glalaxolide 10	–	–

[a]Sum of acenapthene, fluorene, phenanthrene, fluoranthene, pyrene, benzo(b+j+k)fluoranthene, benzo(a)pyrene, benzo(ghi)perylene, indeno(1,2,3-c,d)pyrene.
[b]Sum of seven congeners.
[c]Sum of six congeners.

In recent years, more and more attention has been paid to codigestion or cocomposting of the sludge with other waste materials. If sewage sludge contains organic and inorganic pollutants, it can worsen the quality of the final product. It should be emphasized that for compost use in agriculture, stricter limits are established compared to sewage sludge (Table 4). In the case of compost, not only limits on heavy metals are established but also limits on dosage or load restrictions (Annex 2, 2004). This includes direct load limitation ($g\,ha^{-1}\,y^{-1}$), in most cases calculated on a basis of 2–10 years, restrictions on the admissible dosage of dry matter compost per ha per year, or restrictions according to a maximum nutrient supply (phosphorus or nitrogen) of the agricultural crops. Not only heavy metals and organic micropollutants are considered to be a problem in the case of agricultural use of sewage sludge in Europe. In some countries, standards for pathogens present in sludge were set (Table 5). Because the directive on sewage sludge was established in 1986, a number of attempts were made to change the rules and adapt them to the current situation. Until now, they have not been implemented; however, they indicate the planned directions of changes in EU law. One of them was the Third Working Document On Sludge ENV.E.3/LM 27.04.2000; Brussels published in 2000 (Working Document, 2000). It changed the definition of sewage sludge and established concentration limits not only for heavy metals but also for selected organic micropollutants (Table 3): AOX (sum of halogenated organic compounds), LAS (linear alkylbenzene sulfonates), DEHP (di (2-ethylhexyl)phthalate), NPE (substances nonylphenol and nonylphenolethoxylates with 1 or 2 ethoxy groups), PAH (sum of the following polycyclic aromatic hydrocarbons: acenapthene, phenanthrene, fluorene, flouranthene, pyrene, benzo(b+j+k) fluoranthene, benzo(a)pyrene, benzo(ghi)perylene, indeno(1, 2, 3-c,d)pyrene), PCB (sum of the polychlorinated biphenyl's components number 28, 52, 101, 118, 138, 153, 180), and PCD/Fs (polychlorinated dibenzodioxins/dibenzofuranes). In this document, the precise conditions of sludge use in agriculture were also noted, including the establishment of yearly loads of heavy metals discharged with sewage sludge into the soil. Moreover, the sampling frequency, required analyses, and sampling procedures were set. Sludge treatment processes applicable for this waste material when used in agriculture and land reclamation were described, including both conventional and advanced. The document has established the obligation of sludge treatment in order to reduce the danger of the spread of pathogenic agents in the environment. Attention was also drawn to odors. Medium-term and long-term strategies to ensure the beneficial use of sludge were planned. They were based on reducing the quantities of potentially hazardous substances, materials, and elements that end up in sewage sludge; maximizing the share of sewage sludge suitable for beneficial use; and informing the consumers of the composition of products. In 2001, the second draft of the working document on the biological treatment of biowaste was published (Working Document, 2001). The aim of this document was to promote the biological treatment of biowaste, protect the soil, and ensure that the use of treated biowaste results in agricultural benefits. It concerned sewage sludge, among other biowastes. It promoted carrying out the activities aimed at the prevention and reduction of sludge as well as its contamination by pollutants. It prohibited the disposal of shredded biowaste to the sewer to avoid an unjustified increase in the quantity of produced sewage

Table 4 Heavy metals limits for European compost standards (Annex 2, 2004)

Country	Regulation	Type of standard	Cd	Cr	Cu	Hg	Ni	Pb	Zn	As
Austria	Comp. ord.: Class A+–B	Statutory decree	0.7–3.0	70–250	70–500	0.4–3.0	25–100	45–200	200–1800	Not limited
Belgium	Royal Decree, 1998	Statutory decree	1.5	70	90	1	20	120	300	Not limited
Denmark	Statutory Order Nr. 49; Compost, 2000	Statutory decree	0.8	Not limited	1000	0.8	30	120/60 for private gardens	4000	25
Finland	Decisions of the Ministry of Agriculture and Forestry (46/94)	Statutory decree	3	Not limited	600	2	100	150	1,5000	150
France	NF COMPOST URBAIN	Standard	3	Not limited	Not limited	8	200	800	Not limited	Not limited
Germany	Biowaste ordinance, class I–II	Statutory decree	1–1.5	700–100	70–100	0.7–1	35–50	100–150	300–400	Not limited
Greece	Specifications framework and general programs for solid waste management	Statutory decree	10	510 (including CrVI 10)	500	5	200	500	2000	15
Ireland	Licensing of treatment plants as agreed with EPA	Voluntary	1.5	100	100	1	50	150	350	15
Italy	Technical regulation, DCI 27/07/84 MSWC	Statutory decree	10	500 (including 10 CrVI)	600	10	200	500	2500	10
	Law on fertilizers (L748/84; and: 03/98) BWC/GC/SSC		1.5	0.5 (CrVI)	150	1.5	50	140	500	Not limited
Luxembourg	Licensing for plants	–	1.5	100	100	1	50	150	400	Not limited

Table 4 Continued

Country	Regulation	Type of standard	Cd	Cr	Cu	Hg	Ni	Pb	Zn	As
Netherlands	BOOM Very clean—clean compost	Statutory decree	0.7–1	50	25–60	0.2–0.3	10–20	65–100	75–200	5–15
Portugal	Decree on sludge (limit values utilized also for MSW)	Statutory decree	20	1000	1000	16	300	750	2500	Not limited
Spain	B.O.E.n'm.131.2 June 1998	Statutory decree	10	400	450	7	120	300	1100	Not limited
	Spanish draft on composting Class AA—class A		2–5	250–400	300–450	2–5	100–150	150–300	400–1100	
Sweden	Guideline values of QAS	Voluntary	1	100	100	1	50	100	300	Not limited
Switzerland	Verordnung über umweltgefährdende Stoffe Nr. 814.013; 9. Juni 1986; rev.. 28.12.01)	Statutory decree	1	100	100	1	30	120	400	Not limited
UK	UKROFS fertil.org.farming, "Composted household waste"	Statutory (EC Reg. 2092/91)	0.7	70 (including 0 CrVI)	70	0.4	25	45	200	Not limited
EU ECO Label	Soil improvers and growing media	Voluntary	1	100	100	1	50	100	300	10
EC reg. 2092/91)	Compost from source separated biowaste	Statutory	0.7	70	70	0.4	25	45	200	Not limited
Canada			20	Not limited	Not limited	5	180	500	1850	75
USA	EPA CFR40/503 Sludge Rule		39	No ceiling	1500	17	420	300	2800	41
New Zealand			15	1000	1000	10	200	600	2000	Not limited

Table 5 Standards for maximum pathogens content in sludge (Mininni and Dentel, 2013)

Country	Salmonella sp.	Other pathogens
Poland	If sludge contains Salmonella it cannot be used in agriculture	Fecal streptococci: <100/g
France	8MPN/10 g d.m.	Enteroviruses 3 MPCN/10 g d.m. Helminths eggs: 3/10 g d.m
Finland	Not detected in 25 g	*Escherichia coli* <000 cfu
Italy	1000 MPN/ g d.m.	–
Luxembourg	–	Enterobacteria: 100/g no eggs of worm likely to be contagious
Hungary	–	Fecal coli and fecal streptococci decrease below 10% of original number
Denmark (only for advanced treated sludge)	No occurrence	Fecal streptococci: <100/g

sludge. According to the second draft of the working document on biological treatment (Working Document, 2001) of biowaste, not only heavy metals should be limited in composts and digestate or stabilized biowaste. In stabilized biowaste, also PCB (0.4 mg/kg d.m.) and PAH (3 mg/kg d.m.) concentration limits are set. These set values should be in consistent with the Sewage Sludge Directive.

In 2010, the next version of the document on sludge and biowaste (Working Document, 2010) was published. Based on this document, the maximal dose of sewage sludge applied in agriculture was established at a level of three tons of d.m. per hectare per year. Limits for heavy metals were maintained, and organic pollutants were limited to PAHs. When we use sludge in agriculture, limits for maximum nitrogen levels also should be taken into consideration (Directive 91/676/EEC, 1991). Another document, the Technical Report for End-of-Waste Criteria on Biodegradable Waste Subject to Biological Treatment (Technical Report, 2012) places sewage sludge on the positive waste list, allows clean sludge to be used as fertilizer, and gives way to qualify it as a waste product. The working documents *have still not been ade quately implemented,* but they make clear that the steps to be taken are yet to come.

According to the data of the European Commission in 2020, in 27 EU countries about 3,500,000 tons of dry solids per year will be incinerated (about 27% of the sludge produced) (SUSAN, 2008). At present, in the EU more than 20% of sludge d.m. is incinerated only in such countries as Austria, Belgium, Denmark, Germany, Netherlands, Slovenia, Switzerland, and the United Kingdom (Mininni and Dentel, 2013). Thermal treatment is especially popular in densely populated countries; however this is not s rule. In Austria, Germany, Belgium Flanders, and the Netherlands, this way of sludge utilization is promoted, among others, by setting low metals limits for land application of sewage sludge (Mininni and Dentel, 2013). In such countries as Portugal, incineration is not widely used because of the lack of social acceptance

(Global Atlas, 2008). Germany is focused on the incineration of sludge (and other thermal methods). The requirements for thermal disposal of sludge are set out in the Federal Emission Control Act (Christodoulou and Stamatelatou, 2016).

At present in the EU, landfilling of organic waste that contains more than 5% by dry weight of organic carbon is banned. The sludge directed to landfills should also contain no more than 6000 MJ of energy per kilogram of dry weight. Before 2016, the countries in which landfilling was one of the most frequently used for sludge disposal were Bulgaria, Greece, Poland, Iceland, Malta, Romania, and Slovenia (Mininni and Dentel, 2013). At present, the quantities of sludge in these countries that are directed to landfills have significantly decreased. According to the data of the European Commission, still about 12% of sludge will be landfilled in 27 EU countries in 2020 (SUSAN, 2008).

3 North America

Countries differ widely in their approaches to sludge management within Europe and internationally. Per capita production also differs, with the mean value in the EU of ~55 g/(P.E. × d). The mean value in the United States is around 60 g/(P.E. × d) while in Canada, it is more than 70 g/(P.E. × d). When we consider the stabilization methods of sludge in North America, it should be pointed out that in the United States, about 21% of the sludge is composted (Mininni and Dentel, 2013). The national policy banes landfilling of organic wastes and promotes other methods of sludge final disposal, thus composting and stabilization methods are used frequently. In the United States, the requirements for composting are regulated in individual US states (State Compost Regulations, 2018), e.g., in Alberta standards for composting facilities were set in July 2007 (Standards Alberta, 2007). According to these standards, all composting facilities should be designed, constructed, operated, and closed in a manner that is protective for air, water, land, biodiversity, and human health and quality of life. All plants must meet land use planning requirements and not create risks for the local community. Produced compost should also meet quality requirements. Facility operations were regulated as well as construction specifications. In Michigan, the best management practices for commercial scale composting operations were introduced in March 2015 (Compost Operator Guidebook, 2015). They include, among others, design and operations, but also financing and management strategies.

In the case of anaerobic digestion in the United States, the installations must meet local, state, and federal regulatory and permitting requirements for air, solid waste, and water. The guidelines were set in 18 states. Also federal requirements are involved. Waste processing facilities in the United States must meet Resource Conservation and Recovery Act (RCRA, 1976) Subtitle D requirements (which cover nonhazardous solid wastes) and 40 CFR Part 258 (which covers landfills). Federal laws do not set solid waste permits for the anaerobic digestion of manure at farms. Federal regulations specify also requirement for, for example, ignition internal combustion engines (40 CFR Part 60, Subpart JJJJ) (Anaerobic Digester Guidelines, 2018).

Also, in Canada, the composting and methane digestion guidelines are regulated in provinces. Their own regulations have British Columbia, Alberta, Saskatchewan,

Manitoba, and Ontario (Composting Regulations, 2018). The regulations concern the quality of compost but also the maintaining of composting operations. They include both general requirements (e.g., the Environmental Protection Act in Ontario) and also detailed guidelines (e.g., Municipal Waste Facility Operator Certification Guidelines for Landfills and Composting Operations in Alberta).

In the United States, the major part of generated sludge is used in agriculture (Christodoulou and Stamatelatou, 2016). Agricultural use of sludge in the United States is regulated by "Standards for the Use or Disposal of Sewage Sludge," which was established in 1993 (40 CFR Part 503). The standard establishes general requirements, pollutant limits, management practices, and operational standards for the final use or disposal of sewage sludge. It is limited to the sludge generated during the treatment of domestic wastewater in WWTPs. The standards are set for sludge applied to the land, placed on a surface disposal site, or fired in sewage sludge incinerators.

For sewage sludge applied to the land or placed on a surface disposal site, the inorganic pollutant concentrations are limited (Table 6), but the standard also includes pathogens (Table 7) and alternative vector attraction reduction requirements.

Table 6 Ceiling concentrations (mg/kg d.m.) and cumulative pollutant loading rates (kg/hectare per 365 day period) established by 40 CFR 503 (United States) (40 CFR Part 503)

Compound	Ceiling concentration for sludge applied to the land	Annual pollutant loading rate[a]
Arsenic	75	2.0
Cadmium	85	1.9
Copper	4300	75
Lead	840	15
Mercury	57	0.85
Molybdenum	75	–
Nickel	420	21
Selenium	100	5.0
Zinc	7500	140

[a]If bulk sludge is applied to agricultural land, forests, a public contact site, or a reclamation site, to a lawn and home garden.

Table 7 Standards for maximum concentrations of pathogens in sludge (40 CFR Part 503)

	Microorganism	
United States	***Salmonella***	**Other pathogens**
Class A	<3 MPN/4 g DM	Fecal coliforms <1000 MPN/g DM Enteric viruses <1 PFU/4 g DM Viable helminth ova <1/4 g DM
Class B		Fecal coliforms <2,000,000 MPN/g DM

40 CFR 503 establishes also standards for domestic septage. The annual application rate for domestic septage applied to agricultural lands, forest or a reclamation site should not exceed the annual application rate calculated using the equation taking into consideration amount of nitrogen discharged annually. When domestic septage is applied to mentioned above matrices the person who applies its should also develop other information (for 5 years): the location, of the place where sludge is applied, area of the place, date of application, rate of application, certification statement (40 CFR Part 503). Also in Canada agricultural application and land reclamation is the most frequently used method of sludge management (Global Atlas, 2008).

In USA about 15% of sludge is incinerated, whereas in Canada about 33% (Mininni and Dentel, 2013; Global Atlas, 2008). Landfilling is practiced for almost 30% of sludge in USA (Mininni and Dentel, 2013). In Canada 28.8% of sludge was incinerated in 2015, whereas 12.1% was landfilled (Tessier, 2017). As indicates J. Tessier in Canada disposal by incineration is decreasing trend, whereas land application disposed sludge increases. It is connected both with regulatory influences and voluntary improvements in biosolids quality.

4 Latin America

In Latin America, a lot of small anaerobic digesters are used, including septic tanks. Larger-scale installations include mainly stabilization ponds. In many countries of Latin America, wastewater treatment coverage is not high and because of this, the generation of sewage sludge is also limited. For example, in Argentina about 4.1% of the total produced wastewater is treated while in Chile, it reaches 37% (Wastewater as a resource, 2017). In Latin America, little attention is paid to sludge management. Legal regulations usually quite simply adopt regulations from more developed and industrialized countries, making this a poor fit with local conditions (Spinosa, 2007). Moreover, 77% of the population of Latin America and the Caribbean lacks access to safe sanitation. Only about 28% of the estimated wastewater is collected by public sewer receivers, mainly in country areas (Wastewater as a Resource, 2017). More than 50% of sewage is served by onsite sanitation systems (Ingallinella et al., 2002). The most advanced countries in the area of wastewater treatment are Argentina, Brazil, and Mexico. In Argentina, where 89% of the population lives in cities, about 54% of the population is connected to sewer systems (Ingallinella et al., 2002). In this country, the only legislation at the statutory level that applies to fecal sludge disposal and use is the Law on Hazardous Waste (Ingallinella et al., 2002). In Argentina, the agricultural use of both sludge and fecal sludge is permitted, but the criteria are not legally regulated for most of the country. Only in the province of Argentina Santa Fé the quality criteria for agricultural use were set, and moreover at levels similar to classes A and B US-EPA standards (Ingallinella et al., 2002). In Brazil, legal regulations on land use set restrictions on the slope (>5%), the kinds of crop biosolids, and the time of year when application can be done (Global Atlas, 2008). Limits for heavy metals were also set at levels comparable to United States legislation (Global Atlas, 2008). Also, Mexico set limits for heavy metals in sludge used in agriculture at the same level as the United States (Global Atlas, 2008). Based on the metal content in sludge, two classes can be distinguished: excellent and good. In other Latin America countries such as Columbia, only 5%–10% of municipalities carry out wastewater treatment. In such countries, the situation is similar to other nondeveloped countries, for example in Africa. They are at present at the stage of developing regulations on biosolid use in agriculture (Global Atlas, 2008).

Regulations on sludge landfilling have been established among others in Argentina (Law of Hazardous Waste) (Ingallinella et al., 2002). According to this legal act, fecal sludge is accepted in sanitary landfills; however, it should be discharged in special cells. Sludge destined for landfills should meet the requirements such as pH, total solids, volatile solids, combustibility, cyanides, and sulfides (Ingallinella et al., 2002). Also, the leachate quality is limited (including 25 parameters).

5 Australia and New Zealand

According to Christodoulou and Stamatelatou (2016), in Australia and New Zealand sewage sludge is widely used in agriculture. The agricultural use of sludge in Australia is regulated by local guidelines. For example, in Western Australia this is regulated by the Western Australian Guidelines for Direct Land Application of Biosolids and Biosolid Products; however, it is not mandatory, and at present it is under review (Global Atlas, 2008). In the case of the agricultural use of sludge, both inorganic and some organic micropollutants are limited. Also, strict constraints for pathogens were established (Christodoulou and Stamatelatou, 2016). The regulations seem to be rather complicated (Christodoulou and Stamatelatou, 2016) and restricted, which is connected to the fact that sludge in Australia is treated as waste. Land users who wish to receive more than 1000 wet tons of biosolids over a 12-month period must apply for a license under the Environmental Protection Act 1986 (Global Atlas, 2008). In South Australia, the South Australian Biosolid Guidelines are applied (Global Atlas, 2008). They classify biosolids according to the concentration of heavy metals into grades A, B, and C. They also set the acceptable stabilization processes to achieve appropriate stabilization grades. Both in Western Australia and Victoria heavy metals (arsenic, cadmium, chromium, copper, lead, mercury, nickel, selenium, zinc) and also organic contaminants such, as: DDT and derivatives 0.5–1 mg/kg d.m., organiochlorine pesticides 0.05–0.5 mg/kg d.m. and PCBs 0.2–1 mg/kg d.d. (first value refer to the biosolids grade C1 and the second to grade C2) are limited (Global Atlas, 2008).

Sludge incineration in Australia is not considered a preferred method because of the high investment and operational costs (Christodoulou and Stamatelatou, 2016). In this country, only one incineration plant is operated (Molonglo Plant) (Global Atlas, 2008). Thus, agricultural use of sewage sludge is the most frequently used method of sludge management in Australia (Global Atlas, 2008). Landfilling is not considered a method of sludge management, despite the fact that a lot of land is available in Australia. Landfilling is accepted in New Zealand, however (Global Atlas, 2008).

6 Africa

In Africa a few sewage systems have been developed. The most frequently used systems are the decentralized ones such as latrines or pit latrines. They produce mainly fecal sludge. Management systems of the sludge are usually ineffective. Fecal sludge taken from pit latrines, under the African conditions, is more or less stabilized. It contains nutrients but also pathogens. In most cases, there is no treatment procedure for sludge removed from the facilities mentioned above. The situation differs a lot in South Africa, which has well-developed wastewater treatment systems. Most of the large urban centers have sludge management master plans that are in various phases of implementation.

In Africa, the public sewer systems are not very popular. For example, in such countries as Tanzania and Ghana, more than 85% of sewage is served by onsite sanitation systems (Ingallinella et al., 2002). The waste from pit latrines and other similar excreta disposal systems is mainly used in agriculture, causing helminths and nematodes infections in humans and greezing animals (Ingallinella et al., 2002). In some African countries, the classification criteria for sludge was set. In South Africa, a three-class system was established based on microbial, stability, and pollution criteria. Microbial classification is based on the content of fecal coliform and helminth ova. Stability classes are divided based on the processes involved in treatment. In the case of pollutants, heavy metal concentration is limited at the following levels: arsenic (class a to c) <40–>75 mg/kg, cadmium <40–>85 mg/kg, chromium <1200–>3000 mg/kg, copper <1500–>4300 mg/kg, lead <300–>840 mg/kg, mercury <15–>55 mg/kg, nickel <420–>420 mg/kg, and zinc <2800–>7500 mg/kg (Guidelines South Africa, 2006).

7 Selected Asian countries

Asian countries differ a lot in terms of sludge management approaches and legislation. In such Asian countries that can be defined as middle-income countries in the area of wastewater treatment, the agricultural use of sewage sludge is common. According to a report (Global Atlas, 2008), in China and Russia, agricultural use and land application are the most affordable methods, and they are preferable to landfilling. In China, a huge amount of sewage sludge is generated every year (Yang et al., 2015). According to the data given by Yang et al., sludge management in China is regulated by two laws (the Environmental Protection Law and the Environmental Pollution Prevention and Control Law of Solid Wastes), two administrative regulations, and more than 30 standards. The laws and regulations establish the base for the protection of the environment against pollution connected with sludge management methods. More detailed guidelines concerning sludge treatment and disposal are given in the standards. The standards are, however, disparate in various parts of the country (Yang et al., 2015). Also, other Asian countries have set limits for heavy metal concentrations in sludge used in agriculture and/or land reclamation (Table 8).

Table 8 Permissible concentrations of heavy metals in sewage sludge used in agriculture, mg/kg d.m. (Global Atlas, 2008; Waste Management Rules India, 2016)

Country	Cd	Cr	Cu	Hg	Ni	Pb	Zn	Other metals
China (for sludge applied to soil of pH >6.5 and <6.5	20/5	1000/600	1500/800	15/5	200/100	1000/300	3000/ 2000	As: 75/75 Ba: 150/150
Japan (Fertilizer Control Law)	5	500	Not limited	2	300	100	Not limited	As: 50
Russian Federation	15	500	750	7.5	200	250	1750	As: 10
India	5	50	300	0.15	Not limited	100	1000	As: 10

In China, where land application is one of the frequently used methods, the Pollutants Discharge Standard of Municipal Wastewater Treatment Plants established that sludge should be stabilized before being discharged from WWTPs (Global Atlas, 2008). Anaerobic digestion, aerobic digestion, and composting are acceptable. During aerobic and anaerobic digestion, the organics degradation rate should be higher than 40%. In the case of composting, the rate of organics degradation should be more than 50%, the moisture content $<65\%$, the mortality rate of worm egg $>95\%$, and feces coliform >0.01 (Global Atlas, 2008). China has also established organic micropollutant concentration limits in sewage sludge used in agriculture. The limit values for soil were set for: AOX—500 mg/kg d.m., PCBs—0.2 mg/kg d.m., benzo(a) pyrene—3 mg/kg d.m., and PCDD/Fs—100 ng/kg d.m. (Global Atlas, 2008). Quantities of sewage sludge that are land applied in China are rather low (2.4%) compared to other management methods (Yang et al., 2015). The major part of sludge (83.6%) in China is improperly dumped; the latest data concern 2013 (Yang et al., 2015).

In Russia, a federal law was introduced called "On Residues of Production and Consumption" (hereinafter referred to as the "Waste Act"). The act sets out such principles such as the maximal reduction of their volumes, the maximal recovery of valuable components and reuse; environmentally safe temporary storage or landfilling (if required); elaboration of standards, regulations, rules, etc.; certification of certain types of technological processes, equipment, products (including the wastes itself), and works that may represent a potential threat to humans and environment; and licensing for certain kinds of activities related to toxic wastes" (Kalyuzhnyi, 2007). Sewage sludge management is also regulated in Federal Law "Sanitary-epidemiological wellbeing of population" of the 1997 year and "The temporary rules of protection of environment from residues production and consumption" that was approved in 1994 year by the Russian Ministry of Nature (Kalyuzhnyi, 2007). The two documents mentioned above extended the normative, legislative, and methodological bases related to waste management. Russia is a federal state, which is why waste management systems depend on federal and provincial authorities. The existing waste law regulates the division of power between the regional, federal, and local government institutions. The division of power between federal, regional, and local government institutions is regulated in the waste law. Since 2005, the Federal Agency for Environmental Technology and Nuclear Supervision has been the executive body authorized for state regulation of waste management. With regard to sewage sludge, the most important legislative and normative documents in Russia are:

- SanPiN 2.1.7.573-96: "Hygienic requirements for wastewater and sewage sludge use for land irrigation and fertilization."
- "Typical technological protocols for using wastewater sludge as an organic fertilizer," approved by the Russian Ministry of Agriculture in 2000.
- GOST R 17.4.3.07-2001: "Requirements for wastewater sludge for its application as a fertilizer."

The mentioned documents contain information on acceptable standards for sludge components, in particular for heavy metals (Table 8). They also determine the maximum amount of sediment that can be introduced into the soil (Kalyuzhnyi, 2007). Middle-income countries such as China and Russia prefer other methods to landfilling, mainly agricultural use and land reclamation (https://esa.un.org/iys/docs/san_lib_docs/habitat2008.pdf). However, in China improper dumping is common and landfilling is still a frequent method of sludge disposal (60 %). In China, requirements on the disposal of sludge in MSW landfill sites were set after the year 2007 (Ministerial standard No CJ/T249-2007). In the sludge directed to the landfill, the pH, moisture content, mineral oils, volatile phenol, total cyanide, and heavy metals are limited. What is interesting is that the acceptable concentrations of heavy metals are set at rather low levels, for example, Cd < 20 mg/kg d.m., Hg < 25 mg/kg d.m., Cr < 1000 mg/kg d.m., and Zn < 4000 mg/kg d.m. (Global Atlas, 2008).

Another big Asian country is India. In India, the final of postprocessing of fecal sludge and sewage is regulated by solid waste management rules implemented in 2016. This applies for the disposal and treatment of these waste materials before and after processing, including use as for composting and landfilling. In 2016, a draft of this legal act was prepared (Waste Management Rules India, 2016). In India, standards for sewage sludge used in agriculture were also set (Table 8).

In middle-income countries such as China or Russia, incineration is not widely used because of the high costs (Global Atlas, 2008); for example, in China only about 3.5% of sludge is incinerated. This method is popular in Japan, however, which is a high-income country. So it can be concluded that the use of thermal methods in Asia is connected with the stage of development.

Sludge management technologies differ in well-developed Asian countries. Japanese legislation in the New Sewerage Law (1970) established the principle of sludge production minimization (Christodoulou and Stamatelatou, 2016). The preferred methods are the ones that treat the sludge as a resource (e.g., phosphorus) or a fuel. Final disposal of sludge is minimized (Christodoulou and Stamatelatou, 2016). Also, agricultural use and land application are not considered as beneficial (Christodoulou and Stamatelatou, 2016). In Japan, sludge management is focused on thermal processing (incineration, gasification, melting, drying, carbonization) (Christodoulou and Stamatelatou, 2016). The "feed-in-tariff" is a regulation that mandates that electric power companies purchase electricity from renewable energy sources, including sewage sludge (Christodoulou and Stamatelatou, 2016). Sewage management is regulated in Japan both by laws concerning protection of the environment against the waste products of sludge/waste incineration (e.g., Waste Management and Public Cleaning Law, Air Pollution Control Law) and also limits for contaminants in sewage sludge (e.g., the Soil Contamination Countermeasures Law establishes contaminant limits obtained in elution tests). The Fertilizer Regulation Act includes criteria of sludge use as a fertilizer (Christodoulou and Stamatelatou, 2016). General principles for use of sludge in agriculture and land reclamation are given in the Waste Management and Public Cleaning Law mentioned above. Other provisions of Japanese laws regulate the use of sludge in cement products. Incinerated sludge is often mixed with a special cement and disposed in coastal landfills. The sludge-cement mixture is examined by leachate tests to assure its environmental safety (Global Atlas, 2008). In Japan, standards for biogas quality were set by the High Pressure Gas Safety Law and local laws. The concentrations of moisture, sulfides, oxygen, siloxane, and odors are controlled (Global Atlas, 2008). In Asian well-developed countries such as Japan, landfilling is not a preferable method.

8 Conclusions

Based on the data presented in this chapter, it can be stated that the methods of sludge management in individual countries are affected by many factors, of which population density, the area of arable lands, economic calculations, and social acceptance are the most important. Well-developed countries in Europe and North America have extensive legislation systems concerning sewage sludge management. In well-developed countries, the minimization of waste production is preferred, followed by recycling (e.g., by land application if possible). Landfilling is not a preferable option. In less-developed countries, more sludge is landfilled or disposed on lands without control.

Sewage sludge management in the EU is regulated both at the international level (by directives) and at the national level. In general, there are three types of legal regulations regarding sewage sludge in EU member states: EU directives and other legal acts; laws of EU member states created to implement EU directives; and standards and norms from non-EU countries. The content of internal legal regulations of the member states includes common characteristics (restrictions on the use of sewage sludge, methods of pretreatment of sewage sludge such as stabilization and hygienization); the permissible content of impurities (heavy metals) in sediments and soils where sewage sludge is to be used; limiting the choice of plants and the type of soil on which sediments can be used; and control of legal (legislative) compliance.

Most of the provisions of the EU directives have been transposed into the national laws of member states. Because of this, each member can create its own sludge management politics within the frameworks established by the directives. In the EU, significant levels of composting sewage sludge are only seen in Austria, the Czech Republic, Finland, France, Germany, Hungary, Italy, Slovakia, and Sweden. In the US 21% of sludge is composted. Agricultural use is very scarce in many countries such as Austria, Belgium, Finland, Greece, Romania, Slovenia. and now in Switzerland. Agricultural use is common in the Czech Republic, France, Hungary, Portugal, Italy, Slovakia, Spain, the UK, and the United States. (where 55% of all biosolids are land applied). In some German lands, the use of sludge for agricultural purposes is banned. Germany is a pioneer in phosphorus recovery from sludge. Incineration is practiced for more than 20% of dry sludge production in Austria, Belgium, Denmark, Germany, the Netherlands, Slovenia, Switzerland, and the UK. There is a correlation with population density in some cases. In the United States, 15% of all sludge is incinerated. Landfilling is practiced for more than 50% of dry sludge production in Bulgaria, Greece, Iceland, Malta, Romania, and Slovenia; however this management method is at present not preferable. Landfilling is practiced for 28% of biosolids in the United States.

In most EU countries and US and also in some Asian countries heavy metals limits in sewage sludge used in agriculture were specified. In the United States, limits appear to be generally much higher than in Europe, even for exceptional quality biosolids. Requirements focus on stabilization practices via pathogen requirements and other indices such as VS reduction. Differing regulations and practices are understandable due to different contaminant and pathogen levels, land availabilities, political and environmental expectations, and economic capabilities. Nonetheless, the lack of uniformity is an obstacle to sharing optimized treatment methods or technologies and can prevent economies of scale in sludge management.

One of the most important questions regards policy orientations for sludge management in the future. Are they biological methods? They are environmental friendly and allows for recycling of nutrients to the environment, which is a very important issue. The problem is the availability of arable grounds and the quality of sludge. Measures and technologies to improve sludge quality are of high importance, for example, heavy metals leaching from this waste material. Also, the technologies of the circular economy, such as phosphorus recovery from sludge are a promising alternative to other management methods. An alternative for direct agricultural use is composting sewage sludge. This method guarantees the stability of sludge, the destruction of pathogens, and the reduction of sludge mass and water content. Apart from sludge disposal, an important aim of composting is the possibility of reusing sludge in the economy and the environment. Organic substances subjected to composting may be used in fertilizing, soil structure improvement, and recultivation. Thermal treatment of sewage sludge based on incineration and coincineration are reasonable methods, and according to statistics, they are becoming predominant in EU countries. It is also widely used in Japan. Thermal methods are cost-consuming ones and moreover, in some countries there is no social acceptance for burning wastes. So this method of sludge disposal is one of the most controversial and its application is affected by many local factors. The chance to develop this method in Europe is that sludge can be considered a neutral energy source in terms of CO_2 emission (sludge is defined as biomass). This makes it possible for, for example, cement plants to avoid additional costs for CO_2 emissions, on which the cement factories have limits.

References

40 CFR Part 260, n.d. Hazardous Waste Management System: General. https://www.law.cornell.edu/cfr/text/40/part-260.

Anaerobic Digester Guidelines, 2018. Permitting. https://www.epa.gov/agstar/guidelines-and-permitting-livestock-anaerobic-digesters#permitting.

Annex 2, 2004. Heavy Metals and Organic Compounds From Wastes Used as Organic Fertilisers. http://ec.europa.eu/environment/waste/compost/pdf/hm_annex2.pdf.

BREF, 2006. Reference Document on Best Available Techniques for the Waste Treatment Industries. http://eippcb.jrc.ec.europa.eu/reference/BREF/wt_bref_0806.pdf.

Christodoulou, A., Stamatelatou, K., 2016. Overview of legislation on sewage sludge management in developed countries worldwide. Wat. Sci.Technol. 73 (3), 453–462.

Compost Operator Guidebook, 2015. Best Management Practices for Commercial Scale Composting Operations. https://www.michigan.gov/documents/deq/deq-oea-compostoperatorguidebook_488399_7.pdf.

Composting Regulations and Guidelines Across Canada. http://www.compost.org/pdf/ccc.sw&r.legislation.PDF.

Council Directive 86/278/EEC, 1986. Council Directive 86/278/EEC of 12 June 1986 on the protection of the environment, and in particular of the soil, when sewage sludge is used in agriculture. Off. J. L 181, 6–12.

Council Directive 91/271/EEC, 1991. Council Directive 91/271/EEC concerning urban wastewater treatment. Off. J. L 135, 40–52.

Directive 2000/60/EC, 2000. Directive 2000/60/EC of the European Parliament and of the Council of 23 October 2000 Establishing a Framework for Community Action in the Field of Water Policy.

Directive 2008/1/EC, 2008. Directive 2008/1/EC of the European Parliament and of the Council of 15 January 2008 Concerning Integrated Pollution Prevention and Control. https://eur-lex.europa.eu/legal-content/EN/TXT/HTML/?uri=LEGISSUM:l28045&from=LV.

Directive 2008/98/EC, 2008. Directive 2008/98/EC of the European Parliament and of the Council of 19 November 2008 on waste and repealing certain Directives. Off. J. L 312, 3–30.

Directive 2009/28/EC, 2009. Directive 2009/28/EC of the European Parliament and of the Council of 23 April 2009 on the promotion of the use of energy from renewable sources and amending and subsequently repealing Directives 2001/77/EC and 2003/30/EC. Off. J. L 140, 16–62.

Directive 91/676/EEC, 1991. Directive 91/676/EEC of 12 December 1991 concerning the protection of waters against pollution caused by nitrates from agricultural sources. Off. J. L 375, 1–8.

Directive 99/31/EC, 1999. Council Directive 1999/31/EC of 26 April 1999 on the landfill of waste. Off. J. L 182, 1–19.

Fytili, D., Zabaniotou, A., 2008. Utilization of sewage sludge in EU application of old and new methods—a review. Renew. Sust. Energ. Rev. 12, 116–140.

Global Atlas, 2008. Global Atlas of Excreta, Wastewater Sludge, and Biosolids Management: Moving Forward the Sustainable and Welcome Uses of a Global Resource. https://esa.un.org/iys/docs/san_lib_docs/habitat2008.pdf.

Grobelak, A., Placek, A., Grosser, A., Singh, B.R., Almås, Å., Napora, A., Kacprzak, M., 2017. Effects of single sewage sludge application on soil phytoremediation. J. Clean. Product. 155, 189–197.

Guidelines for Sewerage Systems: Biosolids Management. Agriculture and Resource Management Council of Australia and New Zealand and the Australian and New Zealand Environment and Conservation Council. http://www.agriculture.gov.au/water/quality/nwqms/sewerage-systems-biosolids-management.

Guidelines for the Utilization and Disposal of Wastewater Sludge. Republic of South Africa. http://www.dwaf.gov.za/Dir_WQM/docs/SewageSludgeMar06vol2.pdf.

Hemmat, A., Aghilinategh, N., Rezainejad, Y., Sadeghi, M., 2010. Long-term impacts of municipal solid waste compost, sewage sludge and farmyard manure application on organic carbon, bulk density and consistency limits of a calcareous soil in central Iran. Soil Tillage Res. 108, 43–50.

Ingallinella, A.M., Sanguinetti, G., Koottatep, T., Montanegro, A., Strauss, M., 2002. The challenge of Faecal Sludge management in urban, areas—strategies, regulations and treatment options. Water Sci. Technol. 16 (10), 285–294.

Kalyuzhnyi, S.V., 2007. Wastewater sludge management in the Russian federation: the current status and perspectives. Water Pract. Technol. 73–80.

Kijo-Kleczkowska, A., Otwinowski, H., Środa, K., 2012. Properties and production of sewage sludge in Poland with reference to the methods of neutralizing. Arch. Waste Manag. Environ. Prot. 14 (4), 59–78.

Mininni, G., Dentel, S., 2013. Highlights of Current Legislation on Sludge and Bio-Waste in EU Member States and in the United States. https://www.google.com/search?q=Mininni+G.%2C+and+Dentel+S.+Highlights+of+current+++legislation+on+++sludge+and+bio-waste+in+EU+member+++states+and+in+the+United+States&ie=utf-8&oe=utf-8&client=firefox-b.

Niwagaba, C.B., Mbéguéré, M., Strande, L., 2014. Fecal Sludge Quantification, Characterisation and Treatment Objectives. https://www.un-ihe.org/sites/default/files/fsm_ch02.pdf. (Chapter 2).

Placek, A., Grobelak, A., Hiller, J., Stępień, W., Jelonek, P., Jaskulak, M., Kacprzak, M., 2017. The role of organic and onorganic amendments in carbon sequestration and immobilization of heavy metals in degraded soils. J. Sust. Develop. Water Environ. Syst. 5 (4), 509–517.

RCRA, 1976. Resource Conservation and Recovery Act. https://www.epa.gov/rcra/resource-conservation-and-recovery-act-rcra-overview.

Sewage Sludge: Operational and Environmental Issues. Foundation for Water Research, Marlow. (FR/R0001).

SNFS, 1994. https://www.naturvardsverket.se/Documents/foreskrifter/nfs1994/snfs-1994-2.pdf.

Spinosa, L., 2007. Status and Perspectives of Sludge Management. http://www.bvsde.ops-oms.org/bvsaar/cdlodos/pdf/statusandperspectives103.pdf.

Standards for Composting Facilities in Alberta. http://aep.alberta.ca/waste/waste-facilities/documents/StandardsCompostingFacilitiesAlberta-2007.pdf.

State Compost Regulations, 2018. https://compostingcouncil.org/state-compost-regulations-map/.

SUSAN (Final Report), 2008. Part III Sustainable and Safe Re-Use of Municipal Sewage Sludge for Nutrient Recovery. https://cordis.europa.eu/project/rcn/78544_en.html.

Technical Report for End-of-Waste Criteria on Biodegradable Waste Subject to Biological Treatment, 2012. Third Working Document.

Tessier, J., 2017. Biosolids Mangement and Disposal Practices in Canada. http://cwwa.ca/pdf_files/2017WOO/Tessier.pdf.

Waste Management Rules, 2016. India. http://www.moef.gov.in/sites/default/files/SWM%202016.pdf.

Wastewater as a Resource, Not as Waste: Eye on Latin America and the Caribbean at the World Water Week, 2017. https://www.iadb.org/en/news/announcements/2017-08-28/water-week-latin-america-and-the-caribbean%2C11874.html.

Working Document on Biological Treatment of Biowaste 2nd Draft, 2001. https://www.compost.it/www/pubblicazioni_on_line/biod.pdf.

Working Document on Sludge, 3rd Draft, 2000. Brussels. http://susproc.jrc.ec.europa.eu/activities/waste/documents/IPTS_EoW_Biodegradable_waste_3rd_working_document_wo_line_nr.pdf.

Working Document on Sludge and Biowaste DG ENV C2/BZ/tb. 2010European Commission, Brussels.

Yang, G., Zhang, G.M., Wang, H.C., 2015. Current state of sludge production, management, treatment and disposal in China. Water Res. 78, 60–73.

Further reading

43 CFR Part 503, Standards for the Use or Disposal of Sewage Sludge. https://www.law.cornell.edu/cfr/text/40/part-503.

Directive 2010/75/EU, 2010. Directive 2010/75/EU of the European Parliament and of the Council of 24 November 2010 on industrial emissions (integrated pollution prevention and control). Off. J. L 334, 17–119.

Rheological characterization of sludge

11

*Ludovico Spinosa**, *Azize Ayol*†
*ISO and CEN Expert, Bari, Italy, †Department of Environmental Engineering, Dokuz Eylül University, Izmir, Turkey

1 Introduction

Nowadays, sludge management is of major concern due to environmental pressure and economic considerations. Optimizing management processes is a great challenge, and an accurate estimation of sludge rheological properties has been demonstrated to be useful in a variety of practical contexts for the design of sludge treatment and disposal facilities.

The rheological properties have been, and still are, widely applied to the study of slurry flow, but the existence of some correlations between them and other material characteristics has been shown.

In particular, the pioneering work by Geinopolos and Katz (1964), Dick and Ewing (1967), and others pointed out the close association between certain aspects of rheological behavior and the structure, size, and cohesion of flocs of agglomerated particles in suspension operations. These works have also shown that rheological measurements could be of great interest to quantitatively estimate sludge characteristics.

Specifically, rheological measurements have been shown to be of possible use as indicators of biological changes and an operational guide or control technique in several treatments, such as anaerobic digestion (Ayol et al., 2006a,b; Dieudé-Fauvel et al., 2014), conditioning (Abu-Orf and Dentel, 1999; Dieudé-Fauvel and Dentel, 2011), or dewatering (Stickland, 2015). As far as the disposal operations are concerned, the selection of the most suitable equipment and procedures for storage and transportation systems and for land application is strongly correlated to sludge consistency, and that compacting sludge in a landfill or forming a pile in composting is dependent on sludge shear strength rather than on its solids concentration (Spinosa and Lotito, 1999).

In this chapter, the techniques for rheological measurements and their application in sludge management are discussed.

It is also noted that when longstanding papers are referenced, that's because they are recognized milestones in the subject.

Industrial and Municipal Sludge. https://doi.org/10.1016/B978-0-12-815907-1.00011-8

2 Theoretical aspects

2.1 General

Rheology studies the influence of externally acting forces on the flow and deformation of both liquid and solid materials. In particular, if:

(i) The space of two parallel plates of area A, separated by a distance dy, are filled with a fluid.
(ii) A force F is applied to the upper plate, thus resulting in a plate velocity du while the lower plate is kept fixed.
(iii) No slipping is occurring between plate and fluid, that is, the fluid adjacent to a plate has the same plate velocity.

A velocity gradient du/dy, from now on denoted by *shear rate* ($\dot{\gamma}$), is achieved in the fluid as the shearing force F is uniform across the distance dy. The F/A ratio is known as *shear stress* (τ).

Although the complete stress state in a fluid is characterized by three functions, that is, the tangential shear flow (viscosity function), and the two normal stress coefficients, only the viscosity function, that is, the shear stress to shear rate relationship, is generally considered because normal stress effects become generally significant at high solids concentration, as confirmed by Mouzaoui et al. (2018).

For simple liquids such as water or very diluted suspensions, at a given temperature and pressure, the rheological behavior is defined by a constant viscosity (μ, $ML^{-1}T^{-1}$), and the fluid is called Newtonian, that is, the shear stress (τ, $ML^{-1}T^{-2}$) is linearly related to the shear rate ($\dot{\gamma}$, T^{-1}):

$$\tau = \mu \dot{\gamma} \quad (1)$$

This behavior basically represents only 5% of the existing natural fluids while 95% of the fluids are called non-Newtonian, as viscosity depends on the shear rate and, possibly, the mechanical history of the material. In non-Newtonian fluids, when the shear rate is varied, the shear stress does not vary in the same quantity even in the same direction, so the viscosity changes and the measured viscosity is called "apparent viscosity" (Brookfield, 1999; Franzini and Finnemore, 1997; Haake, 2000).

For some fluids, the flow curve does not pass through the origin, thus an initial resistance τ_0 (yield stress) has to be overcome before movement could occur. Below the yield stress, materials behave as viscoelastic solids.

In addition to the yield stress, other specific characteristics can be evidenced for non-Newtonian fluids, such as (Ayol, 2005; Mezger, 2000):

- *Shear thinning*: the apparent viscosity decreases when the shear rate increases.
- *Shear thickening*: the apparent viscosity increases when the shear rate increases.
- *Thixotropy*: the viscosity decreases with time while subjected to constant shearing, due to a reduction in structural strength that is regenerated during the subsequent rest phase. This succession is a reversible process. In case of a not complete structural regeneration, the process is referred to as "incomplete" or "false" thixotropy.
- *Rheopexy*: basically the opposite of thixotropy, as the viscosity increases with time while sheared at a constant rate, due to an increase in structural strength that is completely regenerated during the subsequent rest phase. This succession is an absolutely reversible process.

2.1.1 Viscosity

A general constitutive rheological equation (viscosity function) is given by (Govier and Aziz, 1972):

$$\tau = \tau_0 + K\dot{\gamma}^n \tag{2}$$

where:

τ_0 is the yield stress (=0, or >0).

K is the fluid consistency index.

n is the flow behavior index (<1, =1, or >1).

Depending on the τ_0 and n, values the following rheological behavior classification can be obtained (Larson, 1999; Mezger, 2000; Ayol, 2005; Eshtiaghi et al., 2013):

a) Dilatant: $\tau_0 = 0$, $n > 1$.
b) Newtonian: $\tau_0 = 0$, $n = 1$.
c) Pseudoplastic (Ostwald): $\tau_0 = 0$, $n < 1$.
d) Yield Dilatant: $\tau_0 > 0$, $n > 1$.
e) Bingham plastic: $\tau_0 > 0$, $n = 1$.
f) Yield Pseudoplastic (Ostwald): $\tau_0 > 0$, $n < 1$.

The most recurrent models found in the literature to model the liquid-like regime result are (Slatter, 1997; Sozanski et al., 1997; Baudez, 2001; Guibaud et al., 2003):

- Herschel-Bulkley: $\tau = \tau_0 + K\dot{\gamma}^n$.
- Ostwald pseudoplastic: $\tau = K\dot{\gamma}^n$.
- Bingham plastic: $\tau = \tau_0 + m\dot{\gamma}$, which is a special case of the Herschel-Bulkley model where the flow behavior index is equal to 1. This model is more adapted for high shear rates and has the advantage of defining a unique viscosity-type coefficient, m, called plastic viscosity or rigidity coefficient, measured by the slope of the line plotting shear stress versus shear rate.

The most applied rheological models in the laminar flow region are shown in Fig. 1 (CEN, 2016).

2.1.2 Viscoelasticity

At values below the yield stress, when present, non-Newtonian fluids show solid-like characteristics defined by an elasticity function through the association of a pure elastic Hookean behavior and a dissipative purely viscous behavior.

For viscoelastic materials, the stress can be separated into two components: an elastic stress and a viscous stress. The elastic stress is a measure of the extent to which a material behaves as an elastic solid; the viscous stress, the degree to which a material behaves as an ideal fluid (Larson, 1999; Ayol, 2005). The Hookean behavior is characterized by instantaneous linear relations between the components of stress and strain (measure of deformation), and by elasticity. The shear modulus, also called complex modulus, is the stress to strain ratio (G*, Pa). As long as this modulus is constant, a material is said to be in the linear viscoelastic (LVE) range (Larson, 1999). Once a critical strain level (γ_c) is reached, the material's behavior is nonlinear, and the transition allows γ_c to be determined (Larson, 1999; Mezger, 2000; Ayol et al., 2006a).

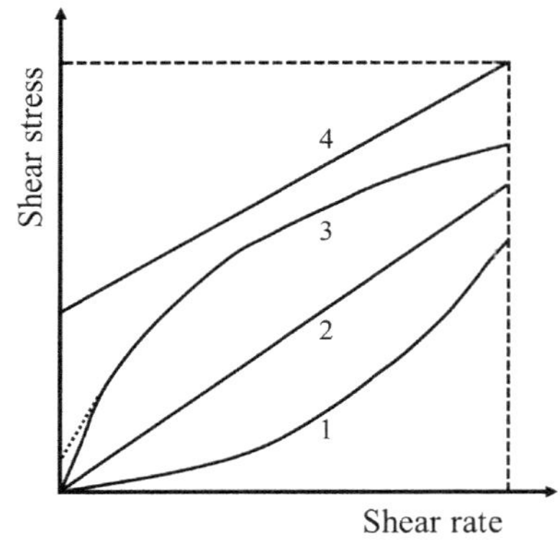

Fig. 1 Rheological models for liquids (*X*: shear rate; *Y*: shear stress). (1) Dilatant ($\tau_0=0$, $n>1$); (2) Newtonian ($\tau_0=0$, $n=1$); (3) Ostwald pseudoplastic ($\tau_0=0$ or >0, $n<1$); (4) Bingham plastic ($\tau_0>0$, $n=1$).
From CEN (European Committee for Standardization), 2016. Characterization of sludges—physical consistency—Part 1: determination of flowability—method by extrusion tube apparatus. EN 14742, Brussels.

At a large but finite strain or longer experimental times, the solid-like characteristics are dependent on both time and shear stress (Baudez et al., 2013), and viscoelastic models become more complex.

The storage modulus G′ (Pa), representing the storage of elastic energy, is the measure of the deformation energy stored in the sample during the shearing process. The loss modulus, G″ (Pa), represents the viscous dissipation of that energy, and is a measure of the deformation energy used up in the sample during the shear process and loss after that (Mezger, 2000; Larson, 1999). The ratio G″/G′, called the loss factor, is high ($\gg 1$) for the liquid-like materials while low ($\ll 1$) for the solid-like materials. A value equal to 1 is a significant point showing hardening or gel formation processes (Larson, 1999).

In Creep test mode, when a constant stress τ_e is applied to a perfectly elastic material, as represented by a spring, it responds instantaneously with a constant strain γ_e. According to Hooke's law, the degree of the strain is directly proportional to the stress due to neglecting all the internal effects (Larson, 1999; Wang et al., 2001). A Hookean spring recovers its original length completely when the constant stress is removed. Perfectly viscous behavior, as represented by a Newtonian dashpot, follows Newtonian law and the strain rate is proportional to the stress. The strain rate goes to zero and the fluid never recovers to its initial state when the constant stress is removed (Wang et al., 2001). Because the viscoelastic materials possess both viscous and elastic behavior, they can be modeled by a combination of springs and dashpots. The simplest model to define the viscoelastic solid properties is the Kelvin-Voigt model:

$$\frac{\gamma}{\tau}=\frac{1}{G}\left[1-\exp\left(-\frac{G}{\mu}t\right)\right] \tag{3}$$

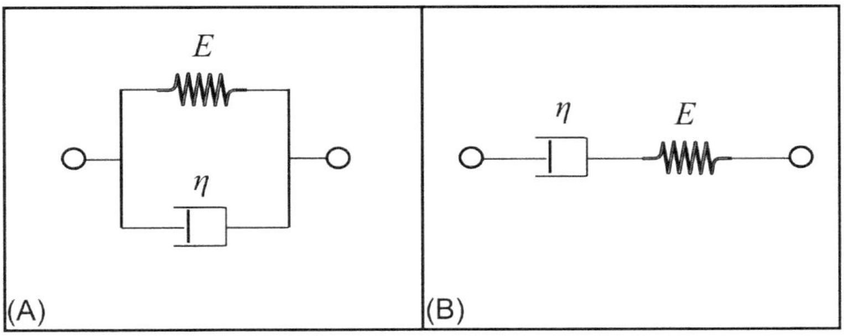

Fig. 2 Rheological models for solids: (A) Kelvin material; (B) Maxwell material.

where G is the elastic modulus and γ the shear strain.

This model is represented by a purely viscous damper and a purely elastic spring connected in parallel, as shown in Fig. 2A, and assumes the behavior, that is, the compliance (ratio $\frac{\gamma}{\tau}$), only depends of the time which is only valid at low strain and/or short experimental times (Baudez, 2008), as usually shown in dynamic rheological tests, which utilize an oscillatory input to the material (Larson, 1999).

The solid-like regime can be also modeled by the Burgers model, with the visco elastic material consisting of a Maxwell material having the properties of both elastic ity and viscosity (Fig. 2B), and a Kelvin material in series. Both highlight an instantaneous elasticity followed by viscous dissipation: the shear strain is an increasing function of time. More details on this model can be found elsewhere (e.g., Wang et al., 2001).

2.1.3 *Normal stress effects*

Normal forces usually appear for highly concentrated materials whose elastic component is significant and predominant on the viscous forces (Van Damme et al., 2002). These normal forces are related to nonlinear effects and occur due to the core microstructure that becomes anisotropic under flow conditions, thus inducing disturbing effects and measurement artifacts (such as cracks and fractures) that need to be controlled for accurate rheological characterization of condensed matter.

This is also observable for the cone and plate and coaxial cylinder measuring systems. On the rotation, the material will climb up the rotating inner rod or cylinder to an important level. This effect is known as the Weissenberg effect or die swell (Chanda and Roy, 2018). For accurate rheological measurements, this effect should also be considered.

2.1.4 *Thixotropy*

Thixotropy is referred to as the disintegration with time of the internal structure following application of shear stress (Eshtiaghi et al., 2013).

The complication of thixotropy arises because this reversible, microstructural change itself takes time to come about due to local spatial rearrangement of the components.

Fig. 3 Typical hysteresis loop.

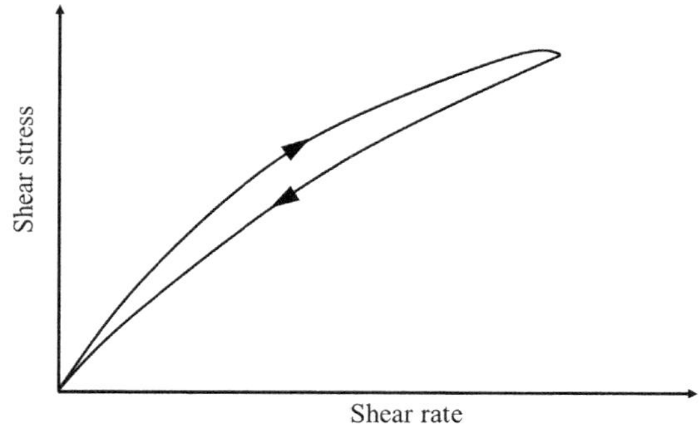

This frequently found time-response of a microstructure that is itself changing with time makes thixotropic, viscoelastic behavior one of the greatest challenges facing rheologists (Barnes, 1997).

Below a critical shear stress, colloidal forces tend to rebuild the solids structure and shearing forces tend to break the solids structure. When the critical shear is reached, the solids structure is completely collapsed and this can alter pipe transportation by producing a clog if the wall shear stress is not high enough to maintain a homogenous flow (Baudez, 2008).

Thixotropic materials are also dependent on the material mechanical history; regarding the rest period, the material undergoes partial or complete recovery of its internal structure.

A typical hysteresis loop for sludge is shown in Fig. 3.

2.2 *Sludge rheology*

From a biological point of view, sludge can have different forms (activated, physical-chemical, digested, etc.), but from a physical point of view, sludge can be "*simply*" seen as interacting particles in a suspending medium where bacteria form extracellular polymeric substances (EPSs) that present a highly hydrated three-dimensional gel-like biofilm matrix and are mainly responsible for the structural sludge aggregates (Wingender et al., 1999; Keiding et al., 2001; Sutherland, 2001).

Generally speaking, diluted sludge (i.e., generally less than 2% solids concentration) may be considered Newtonian while at higher solids concentration, the non-Newtonian characteristics appear and become predominant (Ratkovich et al., 2013). Sludge may exhibit a yield stress (Baudez and Coussot, 2001; Mori et al., 2006) that is shear-thinning (Chaari et al., 2003) or thixotropic (Tixier et al., 2003). In addition, sludge rheological behavior presents both thixotropic colloidal suspension characteristics at a high shear rate (Baudez, 2008) and polymeric behavior at a low shear strain (Baudez et al., 2011), on which time-temperature superposition applies.

However, a single parameter alone, for example, solids concentration, was shown to be unable to represent changes in the overall sludge characteristics, thus making it quite difficult to determine a priori the rheological behavior by using global macroscopic parameters. This was also because sludge continuously evolves due to microbiological activity, which is particularly strong in digesters where the temperature plays an important role (Monteiro, 1997; Spinosa and Wichmann, 2004).

Recently, it has been shown that temperature irreversibly modifies sludge rheology, as some solids are solubilized when the temperature increases and remain in this form even when the temperature decreases (Baudez et al., 2013; Farno et al., 2014).

Anyway, many studies have been carried out to examine the effect of several factors on sludge viscosity and yield stress. Some of those effects are summarized in the following.

2.2.1 Effect of solids concentration

For diluted suspensions whose behavior is still Newtonian, the relationship between viscosity and solids concentration can be described by the Einstein equation:

$$\eta = \eta_0 (1 + 2.5\varphi) \tag{4}$$

where η is the viscosity, η_0 that of the fluid phase, and φ the particle volume fraction. The equation assumes that suspended solids are spherical, noninteracting, insoluble, and rigid (Sanin, 2002).

As reported in Eshtiaghi et al. (2013), most authors have examined the effect of solids concentration on sludge yield stress. It is generally agreed that yield stress tends to increase as the solids concentration becomes higher.

The rheological properties of activated sludge with solids concentration ranging from 2.5 to 57.0 g/L have been examined by Mori et al. (2006), and the data fitted using the Herschel-Bulkley model; the yield stress of sludge was obtained through dynamic measurement and correlated with solids concentration by using an exponential law model.

A review of yield stress models used to describe various sludge types in different experimental conditions was carried out by Seyssiecq et al. (2003). In most cases, the Bingham model evidenced a yield stress also at a solids concentration equal to zero, which is unacceptable from a physical point of view because a minimum solids concentration is necessary to have a solids structure.

2.2.2 Effect of temperature

As summarized in Eshtiaghi et al. (2013), the effect of temperature on viscosity has been extensively studied and it is generally known that a decrease in sludge viscosity follows an increase in temperature.

A considerable influence of the thermal history, that is temperature variations with time, on sludge viscosity has been observed by Baudez et al. (2013). After heating and cooling, the Bingham viscosity of anaerobically digested sludge resulted higher probably due to the irreversible solubilization of some solids during heating. The correlation of yield stress to temperature is usually expressed by an exponential function.

2.2.3 Viscoelasticity

Following details already given, sludge exhibits viscoelasticity, which means that it behaves as elastic solids and viscous liquid and, when the applied stress reduces to zero, a partial elastic recovery is observed (Fig. 4). The partial recovery may be related to energy storage in interparticle bounds.

The viscoelastic properties are obtained through dynamic measurement by applying a sinusoidal deformation and measuring a sinusoidal stress (stress and viscous component) in response to deformation (Chhabra and Richardson, 2008).

2.2.4 Thixotropy

The sludge flow modification over time is very important in designing pipelines and pumping systems because a higher solids concentration involves worse flow conditions, as the shear stress is a power law function of solids concentration.

Further, the thixotropic behavior would result in the structural build-up of sludge over a long retention time in the mixing tank or reactors and form a stagnant region, thus requiring efficient stirring or mixing mechanisms to optimize the treatment process.

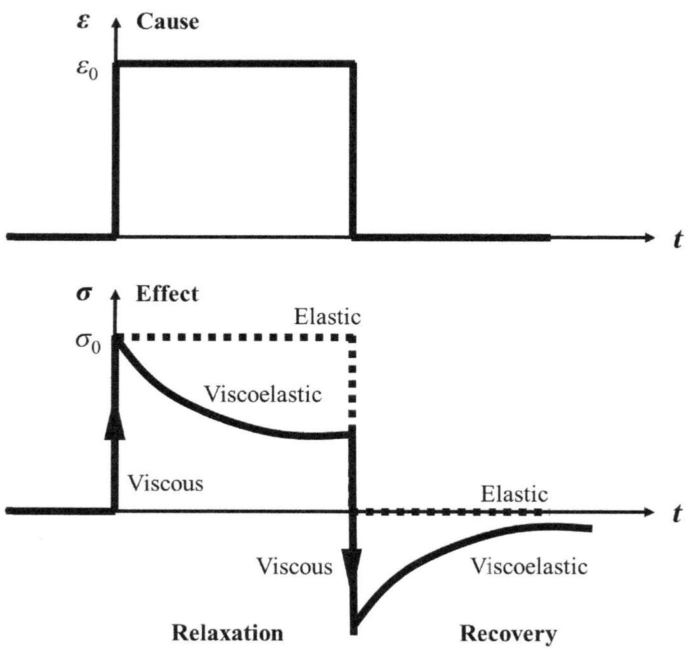

Fig. 4 Elastic and viscous response of a viscoelastic material to applied and removed deformation.
From Eshtiaghi, N., Markis, F., Yap, S.D., Baudez, J.C., Slatter, P., 2013. Rheological characterization of municipal sludge: a review. Water Res. 47, 5493–5510.

Some attempts to develop reliable models for characterizing thixotropy have been specifically addressed to sludge, but reliable correlations have not been presented. A comprehensive analysis of sludge thixotropy is reported in Eshtiaghi et al. (2013).

According to a new technique (Baudez, 2004, 2008), the sludge behavior can be modeled using a unique equation that consists of a solid and liquid component as well as a structural parameter (λ) measured as a function of time, which is a measure of the degree of structure in the suspension, ranging from 0 (fully broken) to 1 (fully structured). This structural parameter could be related to the rheological parameters of non-Newtonian fluids, but most of the models proposed still need to be experimentally validated.

3 Measurement systems

For non-Newtonian fluids, accurate flow curve determination implies the adoption of well-controlled experimental procedures to obtain reliable and reproducible results (Roberts and Barnes, 2001; Ahonguio et al., 2014). Problems are enhanced by the sludge time dependency (thixotropy) (Tabuteau et al., 2006).

Further, procedures must also be simple and easy when operated in the field. A review by Ratkovich et al. (2013) confirmed that the lack of measurement protocols involves great problems in comparing activated sludge rheological data.

3.1 Viscometers geometries

A lot of viscometers with different geometries are available. They generally fall into two general categories: (1) tube type and (2) rotating type (Fig. 5).

The use of a tube viscometer, which employs a pressure gradient to cause fluid to flow in a laminar region through a capillary tube of known diameter and length (Fig. 5A), seems a good approach. However, quite poor control of operative parameters is possible and the tube diameter must be large enough to prevent any clogging phenomenon, so the tube has to be quite long in order to obtain measurable head losses.

Problems can arise from the so-called end and wall effects. For a purely viscous fluid, this effect can be neglected if the length to diameter ratio of the capillary tube is of 100–120 (Nguyen et al., 2007) while for viscoelastic materials, higher values should be required.

Slatter et al. (1996) described a modified capillary device, called a Balanced Beam Tube Viscometer (BBTV), which consists of a transparent tube of various diameters that connects to two pressure vessels located at the two ends of the tube. The flow is obtained by compressed air at a known pressure and is not measured with a classical flow meter, but calculated by simple weighing.

For more simple and reliable measurements, an extrusion cell apparatus (Fig. 5B) can be used. It is based on the same principle as a tube viscometer and consists of a calibrated tube through which the sludge is extruded under gravity or applied pressure, thus allowing a quantitative measurement of the yield stress of the examined sludge to

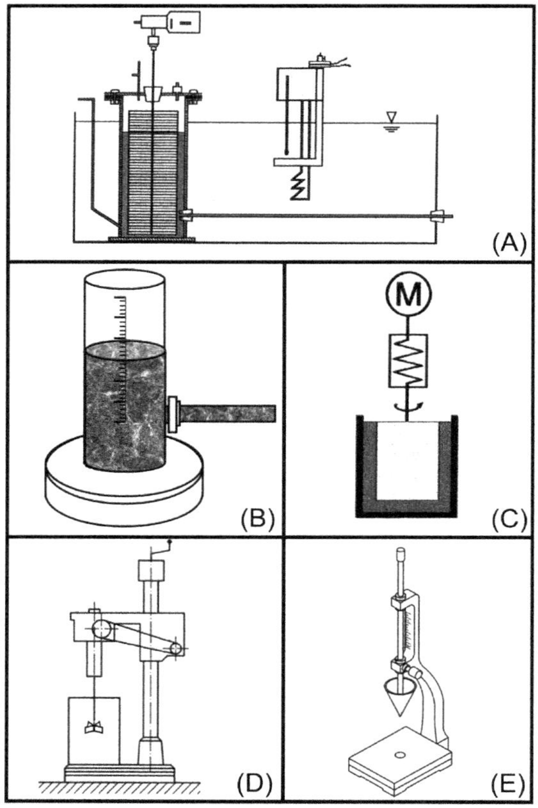

Fig. 5 Types of viscometers: (A) tube viscometer; (B) extrusion cell; (C) coaxial cylinders apparatus; (D) vane apparatus; (E) penetration device.

be obtained together with its degree of physical consistency. This device has been used to evaluate the sludge "*flowability*" and the experimental procedure standardized by CEN (2016). Details are given in Chapter 1 and mentioned in the following Section 8.

Rotating viscometers may have coaxial cylinders (also known as Couette geometry with a moving inner or outer cylinder), rotating blades, parallel plates, and cone-plate geometry. Gap size and roughness are two important characteristics that affect measurements. The gap size must be high enough to avoid jamming phenomena (Mori et al., 2006, 2008), but too large a gap size could produce turbulent effects leading to centrifugal actions.

In the coaxial cylinder viscometers, the torque is directly proportional to the rotational velocity for Newtonian fluids, so the rheometer's software approximates the shear rate with this assumption, even for non-Newtonian fluids, which is not correct unless for a very narrow gap.

Classical coaxial cylinder geometry (Fig. 5C) displays a small gap between the cylinders that is necessary for low concentrated sludge, but may be a drawback in the presence of coarse particles. A gap size at least 10 times greater than the particle size was recommended by Dick and Ewing (1967). This type of measuring system can be applicable for low-viscosity fluids that cannot flow out of the measuring system, even at high shear rate systems. If the Weissenberg effect occurs, the annular gap remains full as long as the homogeneous shear conditions, disposable cups, and bobs can be used (Mezger, 2000).

The possible separation of particles due to gravitational and centrifugal fields can be minimized with a diameter of inner to outer cylinder ratio larger than 0.99 (Chhabra and Richardson, 2008). The phenomenon of slipping occurring at the cylinder/liquid interface can be overcome by artificially increasing the roughness of the moving cylinder (Tabuteau et al., 2004). In case of slippage, measurements are not representative of bulk properties, but of interface conditions.

The effect of a rheometer's geometries on measurements has been demonstrated by Mori et al. (2006), who compared results obtained with a rotational rheometer with concentric cylinders (gap 1 mm) and with a double concentric cylinders (gaps 0.38 and 0.42 mm) for activated sludge mainly composed of macroflocs with mean diameter of 125 mm. They concluded that the concentric cylinder system is suitable for sludge characterization while the double concentric cylinder is too small with consequent blockage of flow.

In the case of viscometers with blades or vane apparatus (Fig. 5D), the velocity gradient is less well defined, so only a mean value based on the mechanical energy dissipated in the medium, calculated from measuring the drive torque of the mover, can be obtained.

Recent experiences have shown that, in practice, the methods based on vane apparatus are not fully reliable when applied to sludge; further studies are in progress (CEN, 2007).

A penetration device (Fig. 5E), where the measuring needle consists of a calibrated cone, can give a measure of the sludge plasticity by measuring the depth of penetration produced by applying a given load on the cone.

The parallel plates geometry consists of upper and lower concentric disks separated by a thin layer of the material to be tested. For representative results, the ratio of the gap between the disks to the radius of the discs must be much less than 1. Edge effects, such as cracks and fractures, and the phenomenon of slipping at the plate/liquid interface have to be carefully controlled.

Parallel plates systems allow (1) measuring dispersions holding large particles as well as samples with three-dimensional structures, (2) quickly and easily preparing high-viscosity samples in a relative wide gap, (3) varying the shear rate range by changing the gap distance, (4) performing tests over a wide temperature range, and (5) using profiled surfaces to counteract wall slipping effects (Mezger, 2000). The problem is that there is no constant shear gradient in the gap because the shear rate increases from 0 at the center of the plate to a maximum at the edge.

The cone-plate geometry consists of a horizontal plate surmounted by a cone-shaped plate. Both the lower plate and the upper cone have the same diameter while the cone makes an angle with the horizontal plan around 1–2 degrees, anycase lower 0.1 rad, depending on material and test type.

This system provides the constant strain and shear rate as well as accurate normal stress. Most systems have a truncated cone tip, and the gap distance between the cone and the plate should be set so that the virtual cone tip would touch the lower plate at one point. For this type, there is no abrasion on the surfaces and, therefore, no friction between them (Mezger, 2000), but the narrow gap (less than 0.1 mm) does not allow wastewater sludge rheology measurements.

Brookfield spindles, that is, the bobs in measuring systems that can take the shape of disks (with thickness and diameter like a coin), pins (installed in the direction of the rotational axis), or cylinders, can be found in many laboratories because they were the first commercially available rotational viscometers on the market (around 1945). They are also specified in many standards such as ASTM-2916 D (2015) for non-Newtonian fluids.

3.2 Measurement procedures

Different conclusion are often drawn when using different equipment and/or procedures, so the inaccuracy in sludge rheological measurements could be reduced by using two different geometries, which, in theory, should give identical results. By comparing the results obtained with them, it would be easy to check the validity of the experimental data and to highlight some potential disturbing effects that have not been taken into account.

Dick and Buck (1985) used a fabricated viscometer (gap 6.6 mm) and a commercial one (Haake Rotovisco RV3, gap of 2.6 mm). Compared to the fabricated viscometer, the commercial one had smoother and smaller cylinders, a narrower gap and the inner cylider rotating. Results provided contrasting rheological conclusions: the commercial viscometer appeared to be favorable for measuring apparent viscosity at high rates of shear while the fabricated one was superior for measuring yield strength or apparent viscosity at low shear rates.

Battistoni et al. (1990) suggested gently stirring the sludge sample inside the viscometer for stable and reproducible rheological results; however, different values of yield stress with or without preshear were noted.

Örmeci et al. (2004) published a laboratory protocol to measure the network strength of sludges while a comprehensive review of the use of a rotational rheometer for rheological characterization of sludge samples obtained from different stages or processes in sludge treatment has been reported in Eshtiaghi et al. (2013).

Dynamic measurements have proved to be a useful method to determine the elastic properties of sludge, thus providing a meaningful insight to technical matters such as mixing and pumping. A better understanding of the dynamic behavior may also improve the sludge process efficiency.

Ayol et al. (2006a) focused on two means of quantifying dynamic rheological test results: (1) the linear viscoelastic behavior (LVE) of sludge samples, as measured by the strain sweep analysis storage (G′) and loss moduli (G″), and (2) the flow curve, which presents the relationship between the shear viscosity and shear rate and determines the yield stress values of samples using the strain sweep test mode. It showed that these methods allow sludges to be more accurately characterized, and the behavior in treatment and transport processes better predicted.

4 Rheology and pumping

A special feature of sludge pumping systems relates mainly to the particular characteristics of the fluids to be pumped, whose characteristics affect all the design criteria that differ significantly from those used for clean water.

Laminar flow is generally more common for sludge. Turbulent flow may also occur, but at a much higher velocity than for water because the flow rate at which the flow regime in a pipe changes from being laminar to turbulent increases with viscosity.

4.1 Pipe flow

The major problem in pipe design is the estimation of the head loss due to pipe friction.

According to the old, but fundamental, well-known method developed by Darcy, the head loss (h_l) for laminar Newtonian fluid flow can be calculated for a Reynolds number less than 2000, using Eq. (5):

$$h_l = 4 f_F \,(L/D)\left(V^2/2g\right) \tag{5}$$

where:

- f_F is the Fanning friction factor given by $16/R = 16/(D*v*\rho/\mu)$, R being the Reynolds number.
- ρ is the fluid density (ML^{-3}).
- μ is the fluid viscosity ($ML^{-1}\,T^{-1}$).
- L is the pipe length (L).
- D is the pipe diameter (L).
- V is the average velocity ($L^3\,T^{-1}$).
- g the gravitational acceleration (LT^{-2}).

Metzner and Reed (1955) developed an expression for f_F valid for all types of fluids, Eq. (6):

$$f_F = 16/R' = 16\left(K\,8^{n-1}\right)/D^n * V^{2-n} * \rho \tag{6}$$

where K is the consistency index and n the flow behavior index.

Several approaches to estimate the flow parameters in turbulent flow conditions are found in the literature. Proposed procedures include graphical methods, for example, diagram plotting the Fanning friction factor versus the modified Reynolds number, practical methods based on the Hazen-Williams equation, and empirically derived graphs.

Again, values obtained with different procedures resulted in quite different results, thus requiring verification with experimental work.

In Table 1, the flow rate equation of sludge derived from the most-used non-Newtonian models is shown (Slatter, 2001, 2008, 2018).

Table 1 Flow rate for usual non-Newtonian models

Rheological model	Flow rate
Power law $\tau = k\dot{\gamma}^n$	$Q=\pi\left(\frac{\delta P}{2Lk}\right)^{1/n}\cdot\left(\frac{n}{3n+1}\right)\cdot R^{(3n+1)/n}$
Bingham $\tau=\tau_0+\mu\dot{\gamma}$	$Q=\left(\frac{\pi R^4\delta P}{8\mu L}\right)\cdot\left[1-\frac{4\tau_0}{3\tau_{max}}+\frac{1}{3}\left(\frac{\tau_0}{\tau_{max}}\right)^4\right]$
Herschel-Bulkley $\tau=\tau_0+k\dot{\gamma}^n$	$Q=\left(\frac{\pi R^3}{256}\right)\cdot\left(\frac{4n}{3n+1}\right)\cdot\left(\frac{\tau_{max}}{k}\right)^{1/n}\cdot\left(1-\frac{\tau_0}{\tau_{max}}\right)^{1/n}\cdot\left[1-\frac{\tau_0}{(2n+1)\tau_{max}}\left[1+\frac{2n}{n+1}\cdot\frac{\tau_0}{\tau_{max}}\cdot\left(1+\frac{n\tau_0}{\tau_{max}}\right)\right]\right]$

4.2 Optimizing system design

Excluding the extremely misleading techniques for the calculation of the head losses based on empirical factors used to multiply the corresponding losses for clean water, Castorani et al. (1985) developed a procedure to estimate these losses with more precision using Eq. (7):

$$i=\frac{8\lambda Q^2}{g\pi^2 D^5} \tag{7}$$

where i is the hydraulic gradient (dimensionless), λ the friction factor (dimensionless), Q the flow rate (L^3T^{-1}), g the gravitational acceleration (LT^{-2}), and D the pipe diameter (L).

Once the different kinds of rheological characterizations have been defined, the pump-pipeline system has to be sized in order to minimize costs. Given that total costs are determined by the sum of the energy cost (C_p), expressed as a function of the power, and of the linear cost of the pipeline, including ancillary equipment, connections, etc. (C_t), expressed as a function of pipe diameter, and supposing that the sludge is characterized using the power law model, the value of the optimal diameter can be obtained by Eq. (8):

$$D=\left[4.15\frac{C_p K(3n+1)^n}{\rho\alpha a(4n)^n}\left(\frac{32}{\pi}\right)^{n-1}\right]^{\frac{1}{\alpha+3n+1}} Q^{\frac{n+1}{\alpha+3n+1}} \tag{8}$$

In the case of turbulent flow, the friction factor can be correlated to the modified Reynolds number through the equation from Dodge and Metzner (Frost, 1982), and applying the same procedure as above, the optimal diameter can be obtained using Eq. (9):

$$D=\left[\frac{0.33Q^3\lambda C_p}{\alpha a}\left(\frac{c\cdot 0.434(3n-4)\sqrt{\lambda}}{0.5+c\cdot 0.434\left(1-\frac{n}{2}\right)}\right)+5\right]^{\alpha+5} \tag{9}$$

However, it must be pointed out that there is a possibility that the above procedures may lead to velocities below the minimum, thus causing possible solids deposition inside the pipe. This problem can be overcome by introducing a limiting velocity value as a constraint or operating under laminar flow conditions that are intermittently followed by turbulent flow to purge the system.

An interesting analysis on sludge pipeline design has been carried out by Slatter (2001) who compared the Bingham plastic model and the Metzner and Reed generalized approach.

By integrating twice the constitutive equation of the Bingham plastic reological model, the mean velocity is obtained by Eq. (10):

$$V = \frac{D}{2K\tau_x^3}(\tau_x - \tau_0)^2\left[\frac{(\tau_x - \tau_0)^2}{4} + \frac{2\tau_0(\tau_x - \tau_0)^2}{3} + \frac{2\tau_0{}^2}{2}\right] \tag{10}$$

where τ_x is the wall shear stress, that is, the stress that must exceed the yield stress in order for any flow to occur, and D the pipe diameter.

The application to the circular pipe flow of the dimensional analysis of this model allows three dimensionless groups to be obtained:

- The standard Fanning friction factor: $f = \tau_x/(1/2)\cdot\rho\cdot V^2$.
- The Bingham-Reynolds number: $\mathrm{Re_B} = (\rho\cdot V\cdot D)/K$.
- The Hedström number: $\mathrm{He} = (D^2\cdot\rho\cdot\tau_0)/K^2$.

In the approach by Slatter, the correlation applicable to large pipe flow, that is, $\mathrm{He} > 1.5\cdot 10^5$, yields a critical velocity, V_c, for the laminar/turbulent transition, which can be calculated directly from Eq. (11):

$$V_c = 26\cdot\sqrt{\tau_0/\rho} \tag{11}$$

The Metzner and Reed generalized approach sets a power law relationship between wall shear stress and bulk shear rate (Eq. 12):

$$\tau_x = \frac{D\cdot\Delta p}{4L} = K'\cdot\left(\frac{8V}{D}\right)^{n'} \tag{12}$$

where K' is the apparent fluid consistency index, n' the apparent flow behavior index, L the pipe length, and Δp the pressure drop across that pipe length.

This approach was extended to the development of a generalized Reynolds number, $\mathrm{Re_{MR}}$, which can be used for laminar/turbulent transition prediction at $\mathrm{Re_{MR}} = 2100$.

The two different methods resulted in significant differences when used to predict large pipe flow behavior, but the new method gave more accurate results, thus leading to improved design and efficiencies (Slatter, 2001, 2008).

4.3 Pump selection

Many types of pumps are available on the market for handling sludge. They can, in practice, be grouped into two categories: (1) rotodynamic (radial, mixed, vortex/torque flow, etc.) and (2) positive displacement (reciprocating piston/plunger, helical screw, diaphram, etc.) (Fig. 6).

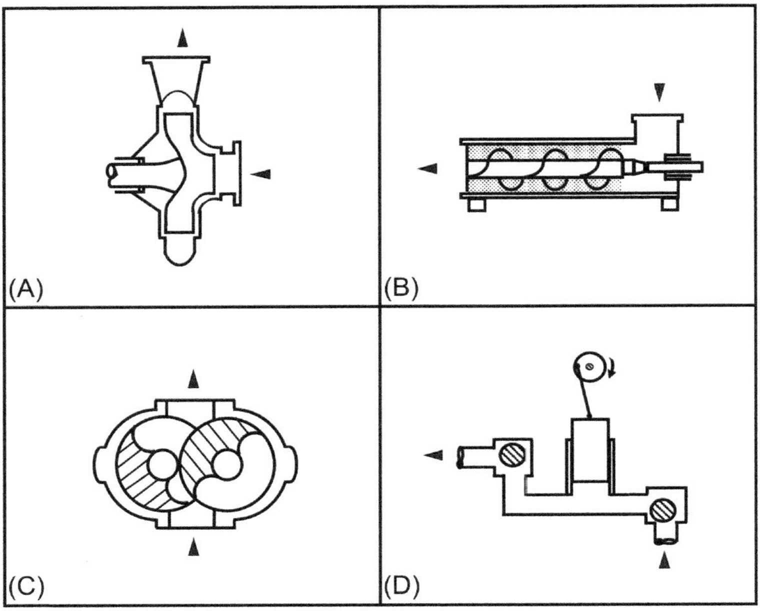

Fig. 6 Types of pumps. (A): Centrifugal; (B): Mono; (C): Rotary; (D) Plunger.

Rotodynamic pumps, for which it is possible to find operating conditions (flow rate and head)

corresponding to peak performance, are generally used for diluted sludge while positive displacement pumps are generally used for concentrated fluids and/or to obtain high heads.

Selection criteria based on various considerations, especially rheological ones, were defined by Frost (1981, 1982). The most important factors affecting pump selection are:

- Duty and flow regime, including the possible variations in solids concentration.
- Fluid properties, in particular solids concentration, viscosity, and the possible presence of fibrous materials.
- Energy consumptions.
- Costs that require the inclusion of proper safety factors to take into account unreliability and other factors difficult to be quantified.

From a general point of view, it appears that for solids concentrations ranging from 2% to 4% sludge behave almost similarly to water, whereas above 10% it is necessary to use volumetric pumps (single- or double-acting piston or helicoidal rotor) or rotodynamic pumps with radial, axial, or mixed flow. Highly concentrated sludge, such as a mechanically dewatered one with a concentration of 20%–45%, can be transported by eccentric helicoidally rotor pumps provided with a hopper and fed by a screw-drive.

5 Rheology versus processing

An improved understanding of the sludge flow properties has also become imperative because it is required in optimizing sludge management operations such as transporting, settling, conditioning, dewatering, and digestion (Ayol, 2005).

Seyssiecq et al. (2003) and Eshtiaghi et al. (2013) published a literature review on the state of the art in the rheological characterization of wastewater treatment sludge, and on the usefulness of rheological measurements in sludge treatment processes, respectively.

In the following, the possible use of rheological measurements in sludge management is more specifically discussed, depending on the type of operation (physical/chemical, biological, disposal/utilization) to which the sludge is submitted.

5.1 Physical/chemical operations

For many years, several correlations have been found between sludge rheological characteristics and physical/chemical operations.

In particular, a milestone is the work by Campbell and Crescuolo (1983), who showed that, in chemical conditioning, municipal sludges would develop a peak in the rheogram at a specific polymer dosage that was assessed to be the optimal one, thus opening to the automatic control of the conditioning process (Fig. 7).

Good results were obtained by correlating rheological parameters to both solids concentration and CST, thus allowing the evaluation of rheological parameters by

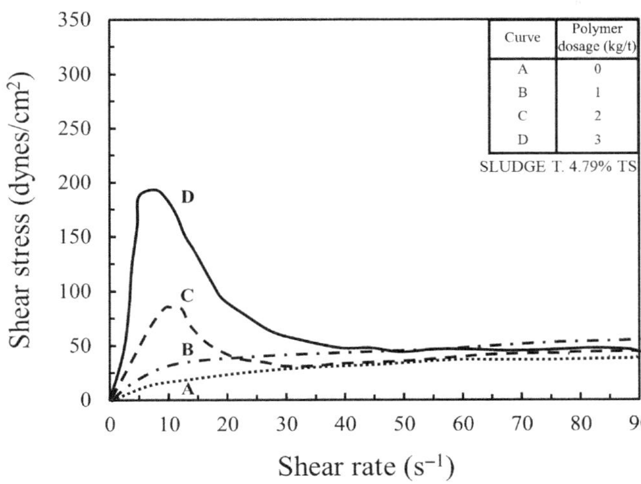

Fig. 7 Rheograms of conditioned sludge at different polymer dosages.
From Campbell, H.W., Crescuolo, P.J., 1983. Assessment of sludge conditionability using rheological properties. In: Casey, T.J., L'Hermite, P., Newman, P.J. (Eds.), Methods of Characterization of Sewage Sludge. D. Reidel Publ. Co., Dordrecht, The Netherlans, pp. 31–50.

means of simpler determinations and the use of rheograms for the correlation of variables involved in dewatering by filtration (Spinosa et al., 1989).

Other researchers (1) showed some relations between total solids, volatile solids, and the thixotropic nature of sewage and digested sludges (Battistoni, 1997); (2) investigated the effect of mixing on rheological characteristics of conditioned sludge, concluding that the yield stress of conditioned sludge increases with decreased mixing, thus increasing the initial floc strength (Abu-Orf and Dentel, 1997); (3) tested sewage sludge samples taken at different steps of treatment of different wastewater treatment plants, confirming that solids concentration is the main parameter affecting rheological behaviour (Lotito et al., 1997); (4) analyzed the viscosity of sludge in response to polymer conditioning and determined the optimum polymer dosage required for dewatering (Bache and Papavasilopoulos, 2000); and (5) evaluated the possibility of using rheological measurements to assess the dewaterability of inorganic water and organic activated sludge conditioned with fly ash and polymer, and concluded that the minimum viscosity values and rheogram peaks can be correlated to the dewaterability of inorganic water sludge but not for organic sludge (Hou and Li, 2003).

Further, Ward and Burd (2004) used a modified capillary rheometer to measure the viscosity of conditioned sludge at different pHs and solids concentrations at 100°C. Dentel et al. (2005), following the difficulties when attempting to characterize polymer-conditioned sludge with more traditional methods, proposed two methods for viscoelastic materials such as sludges: a strain amplitude sweep test and a creep test using either controlled shear rate or direct strain oscillatory mode. Ayol et al. (2006b), based on experimental results, compared the conditioning responses of thermophilic anaerobically, mesophilic anaerobically, and mesophilic aerobically digested sludge using rheological properties.

Other studies correlating rheological measurements to conditioning and dewatering have been carried out by Dieudé-Fauvel and Dentel (2011) and Stickland (2015).

The viscosity of municipal sludge, together with other physical properties such as particle size and bound water content, were significantly modified after thermal hydrolysis at 170°C for 60 min. The treated sludge was more like an inorganic particle system, and was shown to be accurately described by the Newtonian model (Feng et al., 2014).

Several studies have also been addressed to evaluate the influence on the rheological characteristics of sludge pretreatments for its reduction. In particular, Ruiz-Hernando et al. (2010) investigated the rheology of secondary sludge after alkaline pretreatment and hydrogen peroxide oxidation, and Jolis (2008) and Verma et al. (2007) demonstrated that after thermal hydrolysis pretreatment, the sludge viscosity is reduced and the fraction of soluble organic matter is increased. Also, Ayol and Dentel (2005) derived, from rheological measurements of anaerobic digested sludge after enzymatic treatment, parameters to possibly be used to characterize sludge drainability and filterability.

Dentel and Dursun (2009) investigated the shear sensitivity of digested sludge by two methods: a CST shear test and a strain sweep rheological method of measuring yield stress. They found that for all three polymer conditioners, longer shear times increased the dose requirements while the yield stress values were also significantly correlated with cake solids values and inversely correlated with filtrate solids.

Ayol et al. (2010) used, for the first time, an immobilization cell that determined the rheological properties during dewatering by filtration. One real sludge and one synthetic sludge, both conditioned with a high molecular weight cationic polyelectrolyte, were analyzed. The immobilization cell results agreed with the classical dewaterability test results, but provided more fundamental information.

5.2 Biological treatments/operations

Similarly to physical/chemical operations, for some decades correlations have been found between sludge rheological characteristics and biological ones.

In particular, historical papers by Dick and Ewing (1967) and Wood and Dick (1975) showed that the rheological characteristics of sludge consisting almost entirely of filamentous growths were remarkably different from those of nonfilamentous solids.

Other experiences showed that the removal of the solid substrate during the acidogenic phase produces a radical change in the rheological characteristics of the medium as it becomes fluid (Colin, 1982), and that the rheological characteristics of anaerobically digested and aerobically digested activated sludges are related to the surface charge carried by the component particles (Forster, 1982).

Moeller and Torres (1997) characterized primary and secondary sludges, treated both aerobically and anaerobically, in terms of physicochemical and flow properties. They found some correlations between both properties. The results showed that raw primary, raw secondary, and aerobically and anaerobically digested sludges had different shear rate versus shear stress profiles.

The apparent viscosity and yield stress of primary/secondary sludge mixtures was studied by Markis et al. (2016). It was evidenced that mixtures of primary and secondary sludge at 2.5%–7.0% solids concentration behave as non-Newtonian, shear thinning, yield stress materials, and that the apparent viscosity of primary/secondary sludge mixtures (with same total solids concentration) increases with increasing secondary sludge volume fraction. This would suggest that the weak flocs of primary sludge collapse such that the colloidal-like particles of primary sludge become trapped and entangled in the gel-like structure of secondary sludge. However, when dilute primary sludge is mixed with concentrated secondary sludge, or vice versa, the apparent viscosity and yield stress of the mixture increases with the increasing volume fraction of the concentrated sludge, regardless of sludge type. This is possibly due to the strengthening of the hydrodynamic and nonhydrodynamic interactions within concentrated sludge. This will help in designing and optimizing mixing and pumping systems in the anaerobic digestion of mixed sludge.

Other studies showed that rheological measurements are useful for assessing the effect of solids concentration on the mixing efficiency of the anaerobic digester (Eftekharzadeh et al., 2007), and confirmed that they can effectively be used as indicators of biological changes in anaerobic digestion (Dieudé-Fauvel et al., 2014).

Research by Liu et al. (2016) on the rheological properties of sewage sludge during enhanced anaerobic digestion with microwave-H_2O_2 pretreatment (MW-H_2O_2) showed the improvement of sludge flowability and the weakening of its viscoelastic properties.

6 Disposal/utilization operations

The most direct correlations in the field of sludge disposal/utilization were found to be with land spreading, including agricultural utilization and landfilling operations.

6.1 Land spreading

Sludge can be beneficially applied to land by spreading it onto the surface or incorporating it into the soil. It may also be applied to land in composted or dried forms, which can be more easily and longer stored with fewer environmental problems.

Liquid sludges may be spread without any problem by a common slurry spreader while paste-like or solid sludges have to be spread by a special manure spreader with special devices, such as a hydraulic bulkhead door and spreading table with horizontal propellers (Wiart, 1995).

It is therefore clear that all techniques are strongly dependent on the physical consistency of the material, so rheological characterization is useful in the selection of the best equipment to be used and the optimal procedures to be adopted.

6.2 Landfilling

The deposition capability of sewage sludge for disposal in sanitary landfills is often ascertained according to solids concentration. Values not less than 20%–25% are generally required, but in many cases, this is not sufficient because the corresponding physical consistency could be too low. This suggests that solids content should be replaced by a rheology-related parameter.

Results of tests carried out by a vane viscometer indicated that sludge is suitable for landfilling if its shear strength is at least higher than 10 kN/m^2, possibly around 20–25 kN/m^2, and that at a solids concentration of 35%, the vane shear strength could not be as high as 10 kN/m^2 (Loll, 1989).

The importance of rheological measurements regarding landfilling was also stressed by De Bekker and Van den Berg (1989), who showed that landfill stability decreases strongly if the sludge is present as a continuous horizontal layer. Stability improves when placing sludge packages perpendicular to the potential shearing line and applying a covering layer.

Further, landfilled sludge is subject to processes causing quite strong time-dependent changes in the chemical and structural properties; it follows the reduction of apparent viscosity, which can be measured by the vane shear strength. Experiences (Koehlhoff, 1990) showed that, after storage for 6 months, the vane shear strength of an anaerobic digested sludge at 40%–45% solids concentration had sunk from 20 to 15 kN/m^2.

At a depth of 1.5 m, where the sludge is completely waterlogged and could no longer be walked on, after a period of 2 years, the sludge exerted a vane shear strength of only 3.5 kN/m^2 in spite of the dry solids content, which instead increased.

7 Storage and transportation

An effective optimization of sludge management requires the correct planning of the operations linking the treatment steps to the utilization/disposal ones, that is, storage and transportation. These operations are particularly important in sludge agricultural use to overcome problems due to the variability of the time between sludge production and its use, which depends on the cultivation type and weather conditions. Further, rheological characteristics play an important role in the selection of the most appropriate systems for storage and transportation.

The most suitable ways of storing liquid sludge are in tanks/vessels and excavated lagoons/ponds while dumps, basins, and containers are used for paste-like materials. Rheological characteristics are important in deciding on the need to build containment walls or to form a pit instead of a platform, etc., thus obtaining a more or less favorable storage ratio in terms of m^3 of storage per m^2 of occupied area (Wiart, 1995). Within this framework, it must also be considered that the rheological parameters are decreasing during storage (Baudez and Coussot, 2001).

Lots of equipment can perform filling/emptying operations, so it is clear that the selection of the most suitable device to be used depends basically on whether the sludge is in the liquid state. Therefore, accurate knowledge of the physical consistency becomes of prime magnitude to determine what kind of installation to choose.

Sludge can be transported by pipeline, barge, rail, or truck. Transport by pipeline is a practice suitable for moving large quantities of sludge, especially at low solids concentrations. In this case, the knowledge of flow characteristics is an essential condition for designing and operating the system. The criteria for the design and optimization of sludge pumping systems have been already discussed.

Similarly, transport by rail and barge is more adapted and convenient for moving large quantities over long distances and needs adapted infrastructures while transport by truck is the most widespread method presently used because it requires relatively low investment costs and has a high degree of flexibility.

For all transport systems different from pipelines, the most suitable container/tanker, loading/unloading, and spreading devices must be chosen according to the physical state of the material, so the knowledge of sludge rheological characteristics and its physical consistency is necessary (Fig. 8).

8 Rheology and physical consistency

The physical consistency is a characteristic strictly linked to the rheological properties. It is of fundamental importance in sludge characterization because the selection of the most suitable equipment and procedure for land application, storage, and

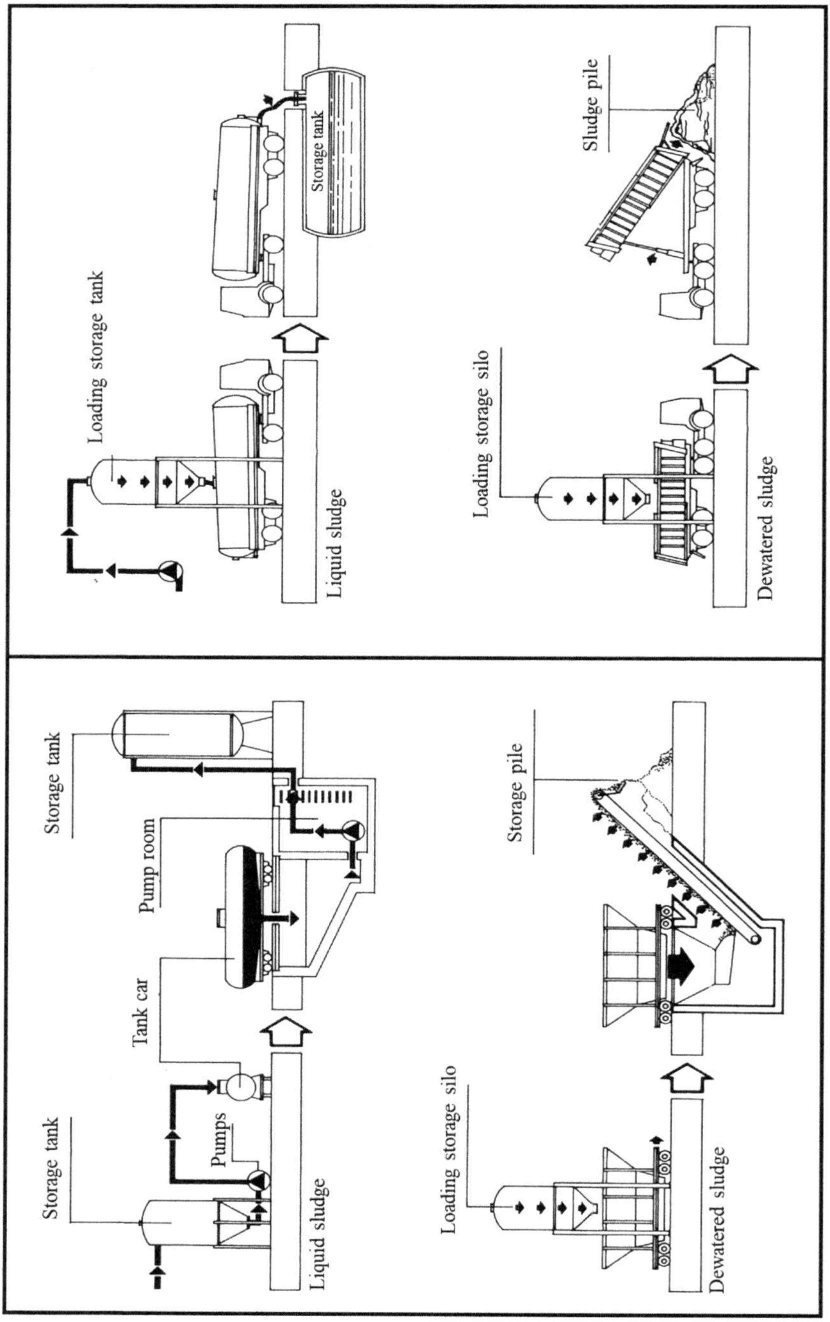

Fig. 8 Loading and unloading systems of railroad cars and trucks.

transportation is strongly connected to sludge consistency. Similarly, compacting sludge in a landfill or forming a pile in composting is dependent on sludge shear strength rather than on its solids concentration (Spinosa and Lotito, 1999).

Details on the evaluation of this property are given in Chapter 1, so only a general discussion is reported in the following.

According to CEN (2007), sludge shows three different behaviors: liquid, paste-like, and solid. Liquid is the state in which sludge is still able to flow under the effect of gravity or pressure below a certain threshold; solid is the state in which sludge shows a shear resistance above a certain threshold; and the paste-like physical state occurs when the sludge behavior is neither liquid nor solid.

The "*flowability*" parameter allows the boundary between the liquid and the paste-like physical state to be defined while the "*solidity*" parameter is between the solid and the paste-like physical state.

A simple procedure to evaluate sludge flowability has been standardized by CEN (2016) and is based on the use of an extrusion tube apparatus equipped with standardized discharge pipes at its bottom. Although the liquid physical state can be theoretically defined in the absence of yield stress, sludge can still be considered to behave as a liquid when the yield stress is below a certain threshold.

Solidity is the state in which a substance has no tendency to flow under moderate stress, resists shear stresses (such as compression) that tend to deform it, and retains a definite size and shape. According to Spinosa and Wichmann (2008), appropriate methods for evaluating solidity seem to be the laboratory vane shear apparatus and the pocket vane shear one. Penetrometers could also be used. However, more recent experiences have shown that, in practice, the above methods for the evaluation of solidity are not fully reliable when applied to sludge and further studies are in progress.

9 Conclusions

Rheological properties are widely applied to the study of water and slurry flows, but have also been shown to possibly be correlated with the performance of almost all sludge treatment operations.

Given that sludge always shows a non-Newtonian behavior, rheological measurements could be correlated to structure, size and cohesion of flocs of agglomerated particles in suspension and used as indicators of biological changes and operational guide or control technique in several treatment and disposal operations.

Rheology-type measurements would also allow quantitatively estimating the sludge physical consistency, which is a characteristic of fundamental importance in the selection of the most suitable equipment and/or procedure for sludge storage, transportation, land spreading, landfilling, composting, chemical conditioning before dewatering, etc.

In particular, sludge shows three different physical behaviors: (1) liquid, when sludge is still able to flow, (2) solid, when sludge shows a certain shear resistance, and (3) paste-like, which is an in-between state. The "flowability" parameter should allow the upper limit of the liquid physical state to be defined while the "solidity" parameter applies to the lower limit of the solid physical state.

Finally, the measurement of sludge rheological properties is a complex matter that requires the development of well-defined and, possibly, standardized procedures to allow comparable and reliable results to be obtained.

Acknowledgments

Thanks are due to Prof. Paul Slatter of Sheridan College in Perth (AU) for his valuable suggestions for the preparation of this chapter, and to Eng. Simone Visigalli of Politecnico di Milano (I) for his valuable assistance in editing this chapter.

References

Abu-Orf, M.M., Dentel, S.K., 1997. Effect of mixing on the rheological characteristics of conditioned sludge: full-scale studies. Water Sci. Technol. 36 (11), 51–60.

Abu-Orf, M.M., Dentel, S.K., 1999. Rheology as tool for polymer dose assessment and control. J. Environ. Eng. 125 (12), 1133–1141.

Ahonguio, F., Jossic, L., Magnin, A., 2014. Influence of surface properties on the flow of a yield stress fluid around spheres. J. Non-Newtonian Fluid Mech. 206, 57–70.

ASTM D2196-15, 2015. Standard Test Methods for Rheological Properties of Non-Newtonian Materials by Rotational Viscometer. ASTM International, West Conshohocken, PA. Available from:www.astm.org.

Ayol, A., 2005. Determination of rheological properties of sludges produced at different treatment stages. PhD thesisDokuz Eylul University, Izmir Turkey.

Ayol, A., Dentel, S.K., 2005. Enzymatic treatment effects on dewaterability of anaerobically digested biosolids II: laboratory characterizations of drainability and filterability. Process Biochem. 40 (7), 2435–2442.

Ayol, A., Dentel, S.K., Filibeli, A., 2006a. Toward efficient sludge processing using novel rheological parameters: dynamic rheological testing. Water Sci. Technol. 54 (5), 17–22.

Ayol, A., Dentel, S.K., Filibeli, A., 2006b. Evaluation of conditioning responses of thermophilic-mesophilic anaerobically and mesophilic aerobically digested biosolids using rheological properties. Water Sci. Technol. 54 (5), 23–31.

Ayol, A., Dentel, S.K., Filibeli, A., 2010. Rheological characterization of sludges during the belt filtration dewatering using an immobilization cell. J. Environ. Eng.-ASCE 136 (9), 992–999.

Bache, D.H., Papavasilopoulos, E.N., 2000. Viscous behaviour of sludge centrate in response to polymer conditioning. Water Res. 34 (1), 354–358.

Barnes, H.A., 1997. Thixotropy—a review. J. Non-Newtonian Fluid Mech. 70 (1-2), 1–33.

Battistoni, P., 1997. Pre-treatment, measurement execution procedure and waste characteristics in the rheology of sewage sludges and the digested organic fraction of municipal solid wastes. Water Sci. Technol. 36 (11), 33–41.

Battistoni, P., Fava, G., Ruello, M.L., 1990. Rheology of sewage sludge. Development of a standardized methodology for rheological properties measurement. Ingegneria Sanitaria 2, 10–19.

Baudez, J.C., 2001. Rheology and physico-chemistry of pasty sewage sludge in view of storing and spreading. PhD thesis (in French)ENGREF, Paris.

Baudez, J.C., 2004. New technique for reconstructing instantaneous velocity profiles from viscometric tests: application to pastry material. J. Rheol. 48 (1), 69–82.
Baudez, J.C., 2008. Physical aging and thixotropy in sludge rheology. Appl. Rheol. 18 (1), 13495-1–13495-8.
Baudez, J.C., Coussot, P., 2001. Rheology of aging, concentrated, polymeric suspensions—application to pasty sewage sludges. J. Rheol. 45 (5), 1123–1139.
Baudez, J.C., Markis, F., Eshtiaghi, N., Slatter, P., 2011. The rheological behaviour of anaerobic digested sludge. Water Res. 45 (17), 5675–5680.
Baudez, J.C., Slatter, P., Eshtiaghi, N., 2013. The impact of temperature on the rheological behavior of anaerobic digested sludge. Chem. Eng. J. 215–216, 182–187.
Brookfield, 1999. Technical Solutions. Brookfield Number D 1056-1014/11.99. Brookfield Engineering, Inc., USA.
Campbell, H.W., Crescuolo, P.J., 1983. Assessment of sludge conditionability using rheological properties. In: Casey, T.J., L'Hermite, P., Newman, P.J. (Eds.), Methods of Characterization of Sewage Sludge. D. Reidel Publ. Co., Dordrecht, The Netherlands, pp. 31–50.
Castorani, A., Spinosa, L., Troisi, S., 1985. Preliminary criteria for the design of sewage and sludge pumping systems. Phoenix Int. 3 (4), 22–29.
CEN (European Committee for Standardization), 2007. Characterization of sludges—physical consistency, Thixotropic behaviour and Piling behavior. Technical Report 15463, Brussels.
CEN (European Committee for Standardization), 2016. Characterization of sludges—physical consistency—Part 1: determination of flowability—method by extrusion tube apparatus. EN 14742, Brussels.
Chaari, F., Racineux, G., Poitou, A., 2003. Rheological behavior of sewage sludge and strain-induced dewatering. Rheol. Acta 42, 273–279.
Chanda, M., Roy, S.K., 2018. Plastics Technology Handbook, fifth ed CRC Press, Taylor and Francis. ISBN 9781315155876.
Chhabra, R.P., Richardson, J.F., 2008. Non-Newtonian Flow and Applied Rheology Engineering Applications, second ed Elsevier.
Colin, F., 1982. Characterization of the physical state of sludges. EEC Concerted Action Cost 68a. Working Party nr. 1Institute de Recherches Hydrologiques, Nancy, France.
De Bekker, P.H.A.M.J., Van den Berg, J.J., 1989. Landfilling with sewage sludge. In: Bruce, A.M., Colin, F., Newman, P.J. (Eds.), Treatment of Sewage Sludge. Elsevier Applied Science, London, pp. 72–79.
Dentel, S.K., Ayol, A., Filibeli, A., 2005. Modern rheometric characterization of sludges. J. Residual Sci. Technol. 2 (4), 233–240.
Dentel, S.K., Dursun, D., 2009. Shear sensitivity of digested sludge: comparison of methods and application in conditioning and dewatering. Water Res. 43 (18), 4617–4625.
Dick, R.I., Buck, J.H., 1985. Measurement of activated sludge rheology. In: O'Shoughnessy, J.C. (Ed.), Proc. E.E.D., Spec. Conference A.S.C.Epp. 539–545.
Dick, R.I., Ewing, B.B., 1967. The rheology of activated sludge. J. Water Pollut. Control Federation 39 (4), 543–560.
Dieudé-Fauvel, E., Dentel, S.K., 2011. Sludge conditioning: impact of polymers on floc structure. J. Residuals Sci. Technol. 8 (3), 101–108.
Dieudé-Fauvel, E., Héritier, P., Chanet, M., Girault, R., Pastorelli, D., Guibelin, E., Baudez, J.C., 2014. Modelling the rheological properties of sludge during anaerobic digestion in a batch reactor by using electrical measurements. Water Res. 51, 104–112.
Eftekharzadeh, S., Harrison, D., Marx, J., Wilson, T.E., 2007. Applying rheological techniques to upgrade anaerobic digesters and handle high solids concentrations. Water Pract. 1 (1), 1–9.

Eshtiaghi, N., Markis, F., Yap, S.D., Baudez, J.C., Slatter, P., 2013. Rheological characterization of municipal sludge: a review. Water Res. 47, 5493–5510.
Farno, E., Baudez, J.C., Parthasarathy, R., Eshtiaghi, N., 2014. Rheological characterization of thermally-treated anaerobic digested sludge: impact of temperature and thermal history. Water Res. 56, 156–161.
Feng, G., Tan, W., Zhong, N., Liu, L., 2014. Effects of thermal treatment on physical and expression dewatering characteristics of municipal sludge. Chem. Eng. J. 247, 223–230.
Forster, C.F., 1982. Sludge surfaces and their relation to the rheology of sewage sludge suspensions. J. Chem. Technol. Biotechnol. 32, 799–807.
Franzini, J.B., Finnemore, E.J., 1997. Fluid Mechanics with Engineering Applications. McGraw-Hill Int. Editions. ISBN 0-07-021914-1.
Frost, R.C., 1981. How to Design Sewage Sludge Pumping Systems. In: W.R.C. Process Engineering. 85-S.
Frost, R.C., 1982. A method of estimating viscosity and designing pumping systems for thickened heterogeneous sludges. In: 8th International Conference Hydraulic Transport Solids in Pipes, Johannesburg, SA, August 25–27.
Geinopolos, A., Katz, W.J., 1964. A study of the rotating cylinder sludge collector in the dissolved air flotation process. J. Water Pollut. Control Federation 36 (6), 712–721.
Govier, G.W., Aziz, A., 1972. The Flow of Complex Mixtures in Pipes. Van Nostrand Reinhold Co..
Guibaud, G., Dollet, P., Tixier, N., Dagot, C., Baudu, M., 2003. Characterization of the evolution of activated sludges using rheological measurements. Process Biochem. 76, 2.
Haake, 2000. Rheology seminar. Course Notes. January 18. Available from: http://haake.de.
Hou, C.H., Li, K.C., 2003. Assessment of sludge dewaterability using rheological properties. J. Chin. Inst. Eng. 26 (2), 221–226.
Jolis, D., 2008. High-solids anaerobic digestion of municipal sludge pretreated by thermal hydrolysis. Water Environ. Res. 80 (7), 654–662.
Keiding, K., Wybrandt, L., Nielsen, P., 2001. Remember the water—a comment on EPS colligative properties. Water Sci. Technol. 43 (6), 17–23.
Koehlhoff, D., 1990. Disposal of sewage sludge in landfills. In: IAWPRC Sludge Mgt. Conference, Loyola Marymount Univ., Los Angeles, January 8–12.
Larson, R.G., 1999. The Structure and Rheology of Complex Fluids. University of Michigan, Ann Arbor Oxford University Press.
Liu, J., Yu, D., Zhang, J., Yang, M., Wang, Y., Wei, Y., Tong, J., 2016. Rheological properties of sewage sludge during enhanced anaerobic digestion with microwave-H_2O_2 pretreatment. Water Res. 98, 98–108.
Loll, U., 1989. Measurement of laboratory vane shear strength as community standard or reference method for testing the suitability of dewatered sludges for use as landfill. In: Thomé-Kozmiensky, K.J. (Ed.), Recycling Int.In: vol. 2. pp. 855–860.
Lotito, V., Spinosa, L., Mininni, G., Antonacci, R., 1997. The rheology of sewage sludge at different steps of treatment. Water Sci. Technol. 36 (11), 79–85.
Markis, F., Baudez, J.C., Parthasarathy, R., Slatter, P., Eshtiaghi, N., 2016. The apparent viscosity and yield stress of mixtures of primary and secondary sludge: impact of volume fraction of secondary sludge and total solids concentration. Chem. Eng. J. 288, 577–587.
Metzner, A.B., Reed, J.C., 1955. Flow of non-Newtonian fluids—correlation of the laminar, transition, and turbulent-flow regions. AICHE J. 1 (4), 434–440.
Mezger, T., 2000. The Rheology Handbook. . ISBN 3-87870-567-0.
Moeller, G., Torres, L.G., 1997. Rheological characterization of primary and secondary sludges treated by both aerobic and anaerobic digestion. Bioresour. Technol. 61, 207–211.

Monteiro, P.S., 1997. The influence of the anaerobic digestion process on the sewage sludges rheological behavior. Water Sci. Technol. 36 (11), 61–67.
Mori, M., Seyssiecq, I., Roche, N., 2006. Rheological measurements of sewage sludge for various solids concentrations and geometry. Process Biochem. 41 (7), 1656–1662.
Mori, M., Seyssiecq, I., Roche, N., 2008. Effect of measuring geometries and of exocellular polymeric substances on the rheological characteristics of sewage sludge. Chem. Eng. Res. Des. 86 (6), 554–559.
Mouzaoui, M., Baudez, J.C., Sauceau, M., Arlabosse, P., 2018. Experimental rheological procedure adapted to dewatered sludge up to 45 % dry matter. Water Res. 133, 1–7.
Nguyen, T.P., Hankins, N.P., Hilal, N., 2007. A comparative study of the flocculation behaviour and final properties of synthetic and activated sludge in wastewater treatment. Desalination 204 (1-3), 277–295.
Örmeci, B., Cho, K., Abu-Orf, M.M., 2004. Development of a laboratory protocol to measure network strength of sludges. J. Residuals Sci. Technol. 1 (1), 35–44.
Ratkovich, N., Horn, W., Helmus, F.P., Rosenberger, S., Naessens, W., Nopens, I., Bentzen, T.R., 2013. Activated sludge rheology: a critical review on data collection and modelling. Water Res. 47, 463–482.
Roberts, G.P., Barnes, H.A., 2001. New measurements of the flow-curves for Carbopol dispersions without slip artefacts. Rheol. Acta 40 (5), 499–503.
Ruiz-Hernando, M., Labanda, J., Llorens, J., 2010. Effect of ultrasonic waves on the rheological features of secondary sludge. Biochem. Eng. J. 52 (2-3), 131–136.
Sanin, D.F., 2002. Effect of solution physical chemistry on the rheological properties of activated sludge. Water SA 28 (2), 207–212.
Seyssiecq, I., Ferrasse, J.H., Roche, N., 2003. State-of-the-art: rheological characterisation of wastewater treatment sludge. Biochem. Eng. J. 16 (1), 41–56.
Slatter, P., 1997. The rheological characterization of sludges. Water Sci. Technol. 36 (11), 9–18.
Slatter, P., 2001. Sludge pipeline design. Water Sci. Technol. 44 (5), 115–120.
Slatter, P., 2008. Pipe flow of highly concentrated sludge. J. Environ. Sci. Health Part A 43 (13), 1516–1520.
Slatter, P., 2018. Personal communication. Faculty of Mathematics, Physical Sciences and Life Sciences, Sheridan College, Perth, WA, Australia.
Slatter, P.T., Moodie, L., Petersen, F.W., 1996. Recent developments in balanced beam tube viscometer. In: British Hydromechanics Research Group 13th International Conference on Slurry Handling and Pipeline Transport: Hydrotransport 13, Johannesburg.
Sozanski, M.M., Kempa, E.S., Grocholski, K., Bien, J., 1997. The rheological experiment in sludge properties research. Water Sci. Technol. 36 (11), 69–78.
Spinosa, L., Lotito, V., 1999. Rheological measurements and their application in biosolids management. Water 21, 28–29.
Spinosa, L., Santori, M., Lotito, V., 1989. Rheological characterization of sewage sludges. In: Thomé-Kozmiensky, K.J. (Ed.), Recycling Int.In: vol. 2. pp. 847–854.
Spinosa, L., Wichmann, K., 2004. Sludge characterization: the role of physical consistency. Water Sci. Technol. 49 (10), 59–65.
Spinosa, L., Wichmann, K., 2008. Physical consistency: a useful tool for sludge characterization. Ingénieries Special 2008, 9–22.
Stickland, A.D., 2015. Compressional rheology: a tool for understanding compressibility effects in sludge dewatering. Water Res. 82, 37–46.
Sutherland, I.W., 2001. Exopolysaccharides in biofilms, flocs and related structures. Water Sci. Technol. 43, 77–86.
Tabuteau, H., Baudez, J.C., Bertrand, F., Coussot, P., 2004. Mechanical characteristics and origin of wall slip in pasty biosolids. Rheol. Acta 43 (2), 168–174.

Tabuteau, H., Coussot, P., Baudez, J.C., 2006. A new approach to the thixotropic character of sewage sludge. J. Residuals Sci. Technol. 3 (4), 233–240.

Tixier, N., Guibaud, G., Baudu, M., 2003. Towards a rheological parameter for activated sludge bulking characterization. Enzym. Microb. Technol. 33, 292–298.

Van Damme, H., Mansoutre, S., Colombet, P., Lesaffre, C., Picart, D., 2002. Pastes: lubricated and cohesive granular media. C. R. Physique 3, 229–238.

Verma, M., Brar, S.K., Riopel, A.R., Tyagi, R.D., Surampalli, R.Y., 2007. Pretreatment of wastewater sludge biodegradability and rheology study. Environ. Technol. 28 (3), 273–284.

Wang, H., Thompson, D.G., Schoonover, J.R., Aubuchon, S.R., Palmer, R.A., 2001. DMA-FTIR creep-recovery study of a poly(ester urethane) elastomer with molecular-level viscoelastic modeling. Macromolecules 34, 7084–7090.

Ward, O.P., Burd, H., 2004. Treatment of Sewage Sludge. Lystek International Inc., United States.

Wiart, J., 1995. Personal communication. Agence de l'Environnement et de la Maitrise de l'Energie, Angers, France.

Wingender, J., Neu, T.R., Flemming, H.C., 1999. Microbial Extracellular Polymeric Substances. Springer, Berlin.

Wood, R.F., Dick, R.I., 1975. Some effect of sludge characteristics on dissolved air flotation. Progr. Water Technol. 7 (2), 173–182.

Industrial sludge for ceramic products and its benefit for metal stabilization

12

*Yuanyuan Tang**, *Pengfei Wu**, *Kaimin Shih*[†], *Changzhong Liao*[†]
*School of Environmental Science and Engineering, Southern University of Science and Technology, Shenzhen, China, [†]Department of Civil Engineering, The University of Hong Kong, Hong Kong SAR, China

1 Introduction

Hazardous metal wastes are among the most environmentally dangerous types of wastes, as they can be highly mobile and are capable of leaching into water supplies, leading to bioaccumulation. Industrial activities have been regarded as the main source of heavy metal pollution in China (Cheng, 2003). Heavy metals commonly exist in the waste derived from electroplating, electronics, and metal cleaning industries (Katsou et al., 2011; Tang et al., 2010); these can cause various diseases and disorders when exceeding specific limits (Veli and Alyuz, 2007). Common strategies that are chosen to remove hazardous metals from wastewater include physicochemical processes, such as precipitation, coagulation, reduction, ion exchange, and membrane processes (Park et al., 2005). However, the treatments mentioned above always result in the production of large amounts of hazardous-metal bearing sludge, which requires additional treatment. Solidification/stabilization (S/S) is a widely accepted treatment process for hazardous metal immobilization, and this method achieves pollutant encapsulation through an interlocking framework of hydrated minerals (Chen et al., 2009; Sophia and Swaminathan, 2005; Zhou et al., 2006). Even when solidified and encapsulated in cement-based materials, metal-laden sludge still possesses enough long-term metal leachability for concern about its discharge because the main immobilization mechanism was simply due to the reduction of the available leaching surface.

A promising strategy has been reported to effectively incorporate metal-containing waste materials into a variety of ceramic products (bricks, tiles, and refractory aggregates) (Okuno et al., 2004; Shih and Leckie, 2007). Converting hazardous metal-containing waste sludge into ceramic products enables the reduction of the environmental impact of waste streams because of the significant reduction in metal leachability as a result of the irreversible phase change in mineral composition after thermal treatment. This strategy explores the possibility of a waste-to-resource technology that uses waste materials as part of the raw materials in the production of new ceramics. The stabilization of heavy metals can be successfully achieved from the ceramic sintering process.

Industrial and Municipal Sludge. https://doi.org/10.1016/B978-0-12-815907-1.00012-X

1.1 Metal pollution from industrial sludge

Hazardous-metal bearing sludge is the settleable solids separated from liquids during industrial processes (Cheremisinoff, 2002), and has become one of the most difficult problems for disposal. In China, the discarded electroplating sludge contains more than 100,000 tons of hazardous metals every year (Wang, 2006). The US Environmental Protection Agency (EPA) estimated about 1.3 million tons of wet electroplating sludge generated a year that needed to be disposed of in hazardous waste landfills. Copper and zinc commonly exist in the waste derived from the electroplating, electronics, and metal cleaning industries (Katsou et al., 2011; Tang et al., 2010). Millions of tons of chromium (Cr) slag worldwide have been dumped and spread over large areas without receiving appropriate treatment at industrial sites (Gao and Xia, 2011). The continuous leaching of the heavy metals can seriously contaminate the surrounding environment, such as soil, sediment, groundwater, and surface water. Environmental contamination by the heavy metals poses a serious threat to humans, plants, and animals. Therefore, an effective method of treating the industrial waste is urgently required to protect the environment and human health. Sludges generated from electroplating industries are highly complex due to the presence of metal ions and other inorganics, which pose a serious problem in regard to their disposal (Sophia and Swaminathan, 2005). It was pointed out that the extraction of copper, zinc, and chromium metals from sludge might not always be an economical operation, due to the influences of different facility scales and metal levels (Naja and Volesky, 2009).

1.2 Ceramic products

Ceramics show unique and favorable properties, such as good creep strength, high hardness, high wear resistance, chemical inertness, and resistance against high temperature corrosion. Ceramic raw materials can vary widely in nominal chemical and mineral composition, purity, physical and chemical structure, particle size, and price (Reed, 1995). For the production of refractories, structural ceramics, and electrical ceramics, minerals like silica, aluminosilicates, aluminum oxide, zirconia and the mixtures of them have been used for a long time. Conventional ceramic processing of engineering ceramics generally concerns fabrication from one or more ceramic materials, one or more liquids, and one or more special additives. The ceramic processing always includes the following steps. The starting materials or the batched system may benefit chemically and physically using operations such as crushing, milling, washing, chemical dissolving, settling, flotation, magnetic separation, dispersion, mixing, and spray drying. Mixing of the constituents is of increasing importance for composites, and it is used to minimize variations of raw materials, intermix different particle sizes, and to control microstructure (Binner, 1990; Rice, 1990). Drying moves some or all of the residual processing liquids. The finished material is then commonly heat treated to produce a sintered microstructure, and the sintered product may be a single component or a multicomponent composite structure. Sintering is used to describe the consolidation of the product during firing, and this process is usually the final and major determinant of the product microstructures.

1.3 Metal stabilization

The stabilization of heavy metals in industrial waste is usually conducted by fixing them into cement-based or pozzolan-based (fly ash, silica fume, cement kiln dust, etc.) systems (Dermatas and Meng, 2003). One such technology is the stabilization/solidification (S/S) technique, which adds cementitious binders such as lime and cement to form stable solids. The material used in the S/S process not only solidifies the hazardous waste by chemical means, but also insolubilizes, immobilizes, encapsulates, destroys, sorbs, or otherwise interacts with the selected waste components (Malviya and Chaudhary, 2006). Portland cement is the most commonly used media in S/S, and in practice, cement-based S/S products are still subject to disposal in secure landfills (Chen et al., 2009). However, the main disadvantages of the S/S method include the significant increase of waste volume and the possible release of hazardous metals over the medium to long term.

It has been reported that thermal treatment may be a promising approach for effective incorporation of metal-containing waste materials into a variety of ceramic products, such as bricks, tiles, and refractory aggregates (Okuno et al., 2004; Shih and Leckie, 2007). The feasibility of stabilizing nickel-laden sludge with commonly available ceramic precursors was investigated (Shih et al., 2006a,b) and the results indicated high nickel incorporation efficiency achieved via spinel formation during the sintering process. It was also previously reported that copper could be incorporated into the products sintered from clay materials, but the incorporation mechanism and phase transformation pathway have not been discussed in detail (Wei et al., 2001). Formation of zinc aluminate spinel ($ZnAl_2O_4$) has also been found in thermally treated sediments (Ndiba and Axe, 2009). It is thus anticipated that the incorporation of copper- and/or zinc-laden sludge into ceramic materials through thermal treatment may be a promising strategy for stabilizing hazardous metal wastes. With well-controlled thermal treatment schemes, the possibility of removing hazardous metals from the waste stream and making them reusable as part of raw materials for marketable ceramic products was preliminarily investigated in our previous works (Shih et al., 2006a,b; Tang et al., 2010). However, the potential of initiating metal spinel formation in the industrial short-sintering scheme of ceramic products (i.e., tiles, insulators, refractories) will require further investigation.

2 Experimental section

2.1 Chemicals and raw materials

The copper-containing electroplating sludge was collected from an industrial sludge treatment facility in Guangdong Province in China, and the collected sludge was dried at 105°C for further sintering experiments in this study. The sludge elemental compositions detected via XRF were normalized by their oxide forms (Fig. 1A) and showed copper as the predominant constituent. The XRD pattern of the 105°C dried electroplating sludge in Fig. 2 shows that the predominant crystalline copper-containing phase was detected to be posnjakite ($Cu_4(SO_4)(OH)_6(H_2O)$, PDF#83-1410). Other copper-containing crystalline phases were identified as malachite

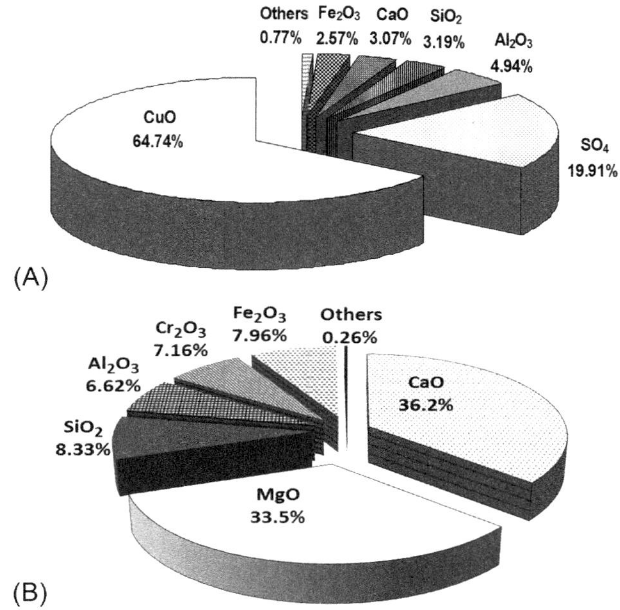

Fig. 1 Major compositions of the (A) dehydrated copper-bearing electroplating sludge and (B) 900°C sintered chromite ore processing residue (COPR) used in the study. The elemental compositions of the sludge were analyzed by XRF and expressed as the basic oxide forms.

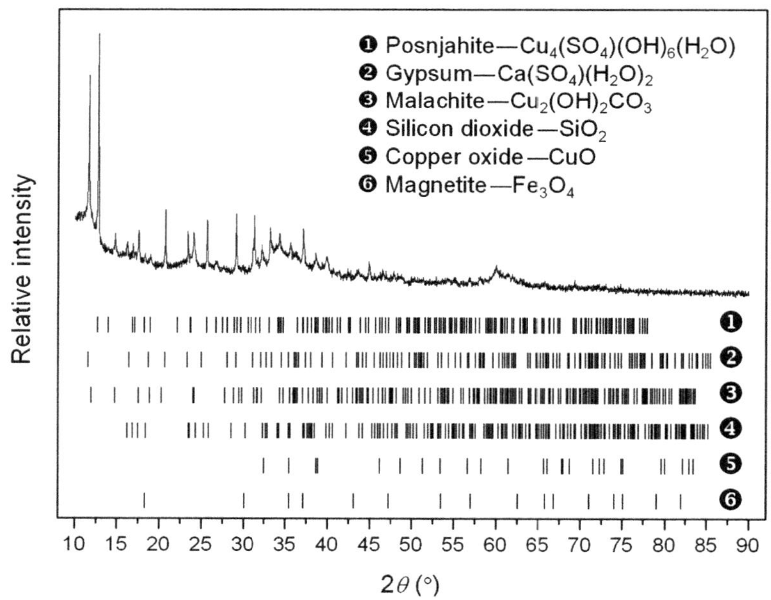

Fig. 2 XRD pattern of the dehydrated copper-bearing electroplating sludge. The crystalline phases are identified as: (1) posnjakite ($Cu_4(SO_4)(OH)_6(H_2O)$, PDF#83-1410); (2) gypsum ($Ca(SO_4)(H_2O)_2$, PDF#70-0982); (3) malachite ($Cu_2(OH)_2CO_3$, PDF#76-0660); (4) silicon oxide (SiO_2, PDF#82-1569); (5) copper oxide (CuO, PDF#80-1268); and (6) magnetite (Fe_3O_4, PDF#0866).

($Cu_2(OH)_2CO_3$, PDF#76-0660) and copper oxide (CuO, PDF#80-1268). γ-Al_2O_3 was prepared from HiQ-7223 alumina powder (Alcoa Corp.). The phase identification of HiQ-7223 alumina was confirmed by powder X-ray diffraction (XRD) as boehmite (AlOOH; PDF#74-1895), which converts to γ-Al_2O_3 upon thermal treatment at 650°C for 3h. Both CuO and ZnO were purchased from Sigma Aldrich. The sewage sludge was collected from Hong Kong and calcined at 900°C for further use.

The chromite ore processing residue (COPR) was collected from a landfill site in Guangdong Province, China. The sample was sintered at 900°C for 3h for X-ray fluorescence (XRF) analysis to measure the residual metal content (Fig. 1B). To produce glass ceramics, the raw materials were mixed on the basis of the obtained XRF results and the eutectic point of the $CaMgSi_2O_6$-$CaAl_2Si_2O_8$ pseudobinary phase diagram. The raw materials were mixed by mortar grinding and then melted at 1450°C for 3h in alumina crucibles. The melt was then poured into another crucible and cooled immediately to obtain the glass-ceramic products. To compare the effect of chromium incorporation in the glass-ceramic system, the simulated glass-ceramic samples were also prepared by mixing pure chemicals (MgO, CaO, SiO_2, Al_2O_3, and Cr_2O_3).

2.2 Sample preparation and characterization

A laboratory-scale porcelain jar is used for ball milling by partially filling it with zirconia short cylinders. The precursor and metal-containing samples were mixed to a certain weight at a fixed molar ratio together with deionized water for ball milling of 18h. The slurry samples were then dried and homogenized by mortar grinding. The pressing of samples is achieved by means of compressing a pellet die set, which was composed of a tool-steel sample chamber and carbide piston. The samples were shaped into 20mm pellets by 650MPa without any binding agent. After pressing, the pellets were fired in a top-hat furnace (LHT 02/16 LB, LBR, Nabertherm Inc.). After sintering, the pellets were quenched in air for further analyses, including being ground into randomly oriented powders for the observation of metal incorporation behavior.

After sintering, the samples were air quenched and ground into powders for XRD analysis to determine the formation of metal-bearing phase(s) in the products. Diffraction patterns were collected using a Bruker D8 diffractometer (Bruker Co. Ltd.) equipped with a Cu X-ray tube operated at 40kV and 40mA and a LynxEye detector. Scans were collected from 10 to 90 degree 2θ-angle, with a step size of 0.02 degrees and a counting time of 0.3 s $step^{-1}$. Phase identification was executed by matching the observed diffraction patterns with powder diffraction files (PDF) of the International Center for Diffraction Data (ICDD). To quantify the content of both crystalline and amorphous phases, the powder samples were mixed with 20wt.% CaF_2 (Merck, 449717-25G) as the internal standard. The quantitative XRD analysis was carried out with TOPAS 4.2 software (Bruker AXS, Karlsruhe, Germany).

The samples for TEM characterization were processed by mechanical thinning, followed by ion beam milling (Fischione Model 1010 Ion Beam Milling). TEM, selected area electron diffraction (SAED), and EDX were performed with an FEI Tecnai G220 S-TWIN operating at 200kV. For SEM characterization, the samples were polished with diamond paste and observed with a Hitachi S-3400N VP-SEM under variable-pressure mode. The Cr K-edge XANES spectra were collected in fluorescent mode using a Lytle detector at beamline 16A of the National Synchrotron Radiation Research Center (NSRRC), Hsinchu, Taiwan. The spectra of pure chemicals Cr_2O_3, $K_2Cr_2O_7$, and CrO_3 were also measured to serve as the references of Cr(III) and Cr(VI) oxidation states. The energy of the monochromator was calibrated using the Cr K-edge at 5989 eV. Processing and analysis of XANES spectra were carried out using ATHENA (Demeter software package version 0.9.18).

2.3 Leaching experiments

The Toxicity Characteristic Leaching Procedure (TCLP) is designed to determine the mobility of both organic and inorganic components present in liquid, solid, and multiphasic wastes (USEPA, 1992). The TCLP was intended to be a laboratory test designed to simulate leaching in a municipal landfill, and the solid phase is extracted with a mass of fluid equal to 20 times the mass of the solid phase. The agitation apparatus must be capable of rotating the extraction vessel in an end-over-end fashion at 30 ± 2 rpm for 18 ± 2 h. However, due to the TCLP's dependence on a single-point and short-term leaching result, its use in leaching prediction may result in the misclassification of waste, consequently leading to the underestimation of long-term leachability or unnecessary treatment costs (Kosson et al., 2002). In addition, the study of intrinsic leaching behavior in the metal-bearing phase is an important way to obtain valuable information about the chemical properties of the metals in stabilized products and their potential environmental risks (Li et al., 2001). A better understanding of the leaching and speciation mechanisms of individual metals is essential for predicting the long-term immobilization of metals exposed to different environmental conditions (Karamalidis and Voudrias, 2009).

As the purpose of this study was to predict long-term leaching behavior, the TCLP was modified to include an extraction period of up to 22 days. Each leaching vial was filled with 10 mL of TCLP extraction fluid and 0.5 g powder (unless otherwise specified), and the leaching vials were tumbled end over end at 60 rpm from 0.75 to up to 22 days. At the end of each agitation period, the leachates were filtered with 0.2-μm syringe filters. The pH was measured via a pH meter (Thermo Scientific) and the concentrations of all metals were derived from inductively coupled plasma atomic emission (ICP-AES, Perkin-Elmer Optima 3300 DV). In this study, extraction fluid #2 was used to evaluate the stabilization effect of the metals, and the preparation method was diluting 5.7 mL glacial acetic acid CH_3CH_2OOH with reagent water to a volume of 1 liter. When correctly prepared, the pH of this fluid will be 2.88 ± 0.05.

3 The potential of making industrial sludge into ceramic products

3.1 Copper-laden electroplating sludge

3.1.1 Copper transformation in aluminum-rich ceramic matrices

Samples for sintering experiments were prepared by mixing each precursor with the dried electroplating sludge for a total dry weight of 60 g at a Cu:Al molar ratio of 1:2.5 (with

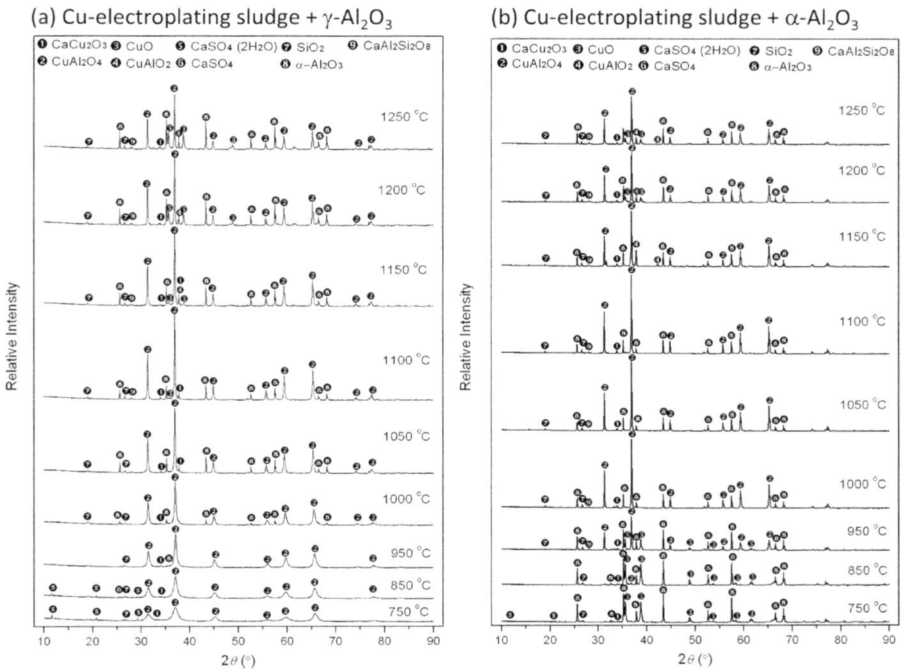

Fig. 3 XRD patterns of the (A) copper-electroplating sludge + γ-Al_2O_3 system and (B) copper-electroplating sludge + α-Al_2O_3 system show the reaction within the temperatures ranging from 750°C to 1250°C for 3h. The crystalline phases are identified as: (1) $CaCu_2O_3$ (PDF#34-0284), (2) $CuAl_2O_4$ (PDF#78-1605), (3) CuO (PDF#80-1268), (4) $CuAlO_2$ (PDF#75-1988), (5) gypsum ($CaSO_4 \cdot 2H_2O$, PDF#70-0982), (6) anhydrite ($CaSO_4$, PDF#37-1496), (7) quartz (SiO_2, PDF#79-1910), (8) α-Al_2O_3 (PDF#10-0173), and (9) anorthite (Ca ($Al_2Si_2O_8$), PDF#89-1473).

excess alumina available for copper incorporation). XRD patterns in Figs. 3 and 4 show the transformation of copper-containing crystalline phases using aluminum-rich precursors through a 3-h sintering scheme at temperatures ranging from 650°C to 1250°C. In Fig. 3A, the peaks of $CuAl_2O_4$ phase were clearly observed, together with the disappearance of the copper-containing phases (posnjakite, malachite and CuO) in the starting material when the mixture of sludge + γ-Al_2O_3 was sintered at 750°C for 3h. Another copper-containing phase of $CaCu_2O_3$ (PDF#34-0284) in the sintered products was also identified, but with the very low peak intensities. When the temperature was higher than 1100°C, CuO reappeared and was accompanied by another Cu-Al-O compound ($CuAlO_2$). When using α-Al_2O_3 as a precursor (Fig. 3B), the $CuAl_2O_4$ spinel phase was detected at sintering temperatures above 850°C, and the CuO was observed as the main copper-containing component in the α-Al_2O_3 system at lower temperatures. Similar to the system of using the γ-Al_2O_3 precursor, the $CaCu_2O_3$ phase existed as a minor copper-containing component, together with the formation of $CuAlO_2$ and the

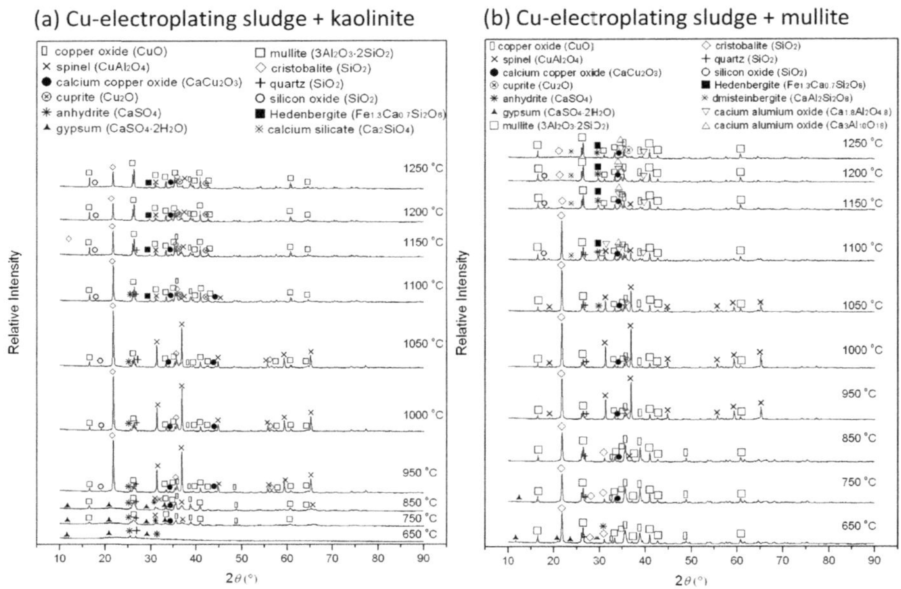

Fig. 4 XRD patterns of the (A) copper-electroplating sludge+kaolinite system and (B) Cu-electroplating sludge+mullite system show the reactions within the sintering temperatures ranging from 650°C to 1250°C for 3h. The crystalline phases were identified as: CuO (PDF#80-1268), $CuAl_2O_4$ (PDF#78-1605), $CaCu_2O_3$ (PDF#34-0284), Cu_2O (PDF#77-0199), $CaSO_4$ (PDF#37-1496), $CaSO_4{\cdot}2H_2O$ (PDF#70-0982) mullite (PDF#79-1455), cristobalite (PDF#76-0938), quartz (PDF#79-1910), SiO_2 (PDF#82-1563), hedenbergite (PDF#70-1876), Ca_2SiO_4 (PDF#33-0303), dmisteinbergite (PDF#31-0248), $Ca_{1.8}Al_2O_{4.8}$ (PDF#21-0129), and $Ca_3Al_{10}O_{18}$ (PDF#01-0572).

reappearance of CuO at temperatures higher than 1100°C. When comparing the XRD patterns of the 1000°C and 3-h sintered samples from sludge+γ-Al_2O_3 and sludge+α-Al_2O_3 systems (Fig. 3), the crystalline phases of $CuAl_2O_4$, α-Al_2O_3, quartz, and $CaCu_2O_3$ were identified. Peak intensities of the $CuAl_2O_4$ spinel phase were significantly higher than those derived from the other phases in the sintered products. Although more crystalline phases were found in the sintered products of sludge+α-Al_2O_3 system, the $CuAl_2O_4$ spinel also possesses the higher peak intensity compared to the other phases. After sintering, the generation of the predominant $CuAl_2O_4$ phase in both systems suggests that the copper in the electroplating sludge could be incorporated into the spinel structure.

XRD patterns in Fig. 4A show the phase transformation in the kaolinite system through the 3-h sintering process at temperatures ranging from 650°C to 1250°C. The 650°C and 3-h sintered sample contains significant amounts of amorphous phases, due to the formation of metakaolin during the kaolinite to mullite reaction series. The peaks of the $CuAl_2O_4$ phase were first detected in the 750°C sintered sample, together with the appearance of the mullite phase. A substantial increase in the

$CuAl_2O_4$ phase peak intensity was detected when the sample was sintered at 950°C. Meanwhile, the excess silica crystallized as cristobalite at this sintering temperature. However, when the temperature was increased to above 1100°C, the amount of $CuAl_2O_4$ spinel declined dramatically. This reaction process was accompanied with the reappearance of CuO and the generation of Cu_2O, as illustrated in the XRD patterns. At the same time, the aluminum decomposed from $CuAl_2O_4$ spinel reacted with the excess silica and formed the mullite phase again. XRD patterns in Fig. 4B show the phase transformation of samples of the mixture of copper-electroplating sludge +mullite through the 3-h sintering process at temperatures ranging from 650°C to 1250°C. The $CuAl_2O_4$ phase was first observed with small peaks in the 850°C sintered products, and there was a significant increase after sintering at 950°C for 3h. Similar to the phase changes in the kaolinite system, the $CuAl_2O_4$ spinel also decomposed at 1100°C. This reaction process was also accompanied with the reappearance of CuO and the generation of Cu_2O, as illustrated in the XRD patterns. The aluminum decomposed from the $CuAl_2O_4$ spinel was incorporated into the mullite phase again.

According to the standard $CuAl_2O_4$ XRD pattern (PDF#78-1605), its strongest peak is located at $2\theta = 36.868$ degrees, corresponding to the diffraction plane of (3 1 1). It is not overlapped by the major peaks of the other phases in the system. Therefore, the 2θ range of 36.6–37.6 degrees was selected to further observe the peak evolution, representing the generation of the $CuAl_2O_4$ spinel at elevated temperatures of the system (Fig. 5). Fig. 5A shows a distinguishable $CuAl_2O_4$ peak in the 750°C sintered sample, and the continuous growth of $CuAl_2O_4$ in the system was observed with the elevated temperatures. A substantial growth of the $CuAl_2O_4$ phase occurred from 1000°C to 1100°C, and its maximum peak intensity was observed at 1100°C. When the sintering temperature was higher than 1100°C, the peak intensity of the $CuAl_2O_4$ phase was found to decrease with the elevated temperatures. With α-Al_2O_3 as the sintering precursor, a small but observable $CuAl_2O_4$ peak was detected in the 850°C and 3-h sintered sample (Fig. 5B). With a further increase of 100°C, a substantial incorporation of copper was observed, as shown by the significant amplification of the $CuAl_2O_4$ peak intensity. At temperatures above 1100°C, the peak intensity of the $CuAl_2O_4$ phase was found to decrease. In both systems of kaolinite and mullite (Fig. 5C and D), the dramatic increase of the peak intensity occurred at 950°C, and the maximum $CuAl_2O_4$ peak intensity was observed at 1000°C. The decomposition of the $CuAl_2O_4$ spinel was observed with the decrease of $CuAl_2O_4$ peak intensity in both the kaolinite and mullite systems. However, the decomposition of $CuAl_2O_4$ in the kaolinite system was found much faster than that in mullite system.

3.1.2 Stabilization effect of copper in the ceramic products

3.1.2.1 pH value

The change of leachate pH during the prolonged leaching procedure is provided in Fig. 6. It was found that the leachates from the unsintered samples had the highest pH value (4.5–4.9). However, with γ-Al_2O_3 as the sintering precursors, the leachate pH decreased significantly, even when the sample was sintered at the lowest temperature (750°C) in the experiment (Fig. 6A). With the increase of sintering temperature,

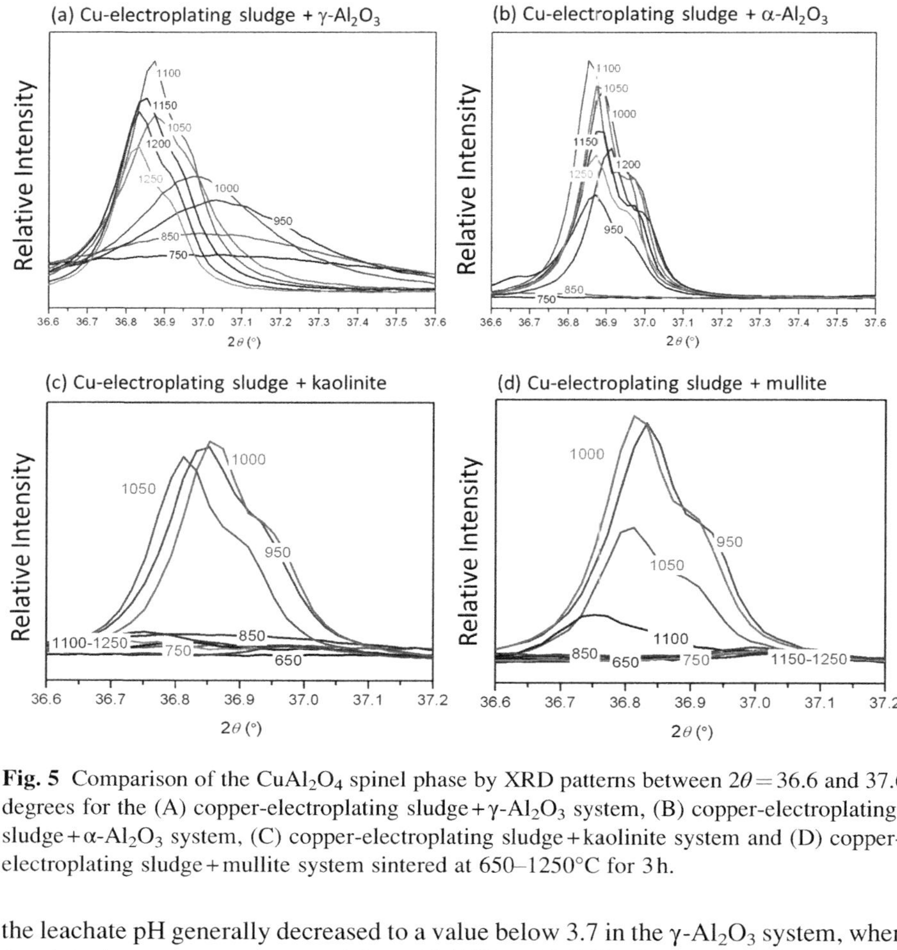

Fig. 5 Comparison of the $CuAl_2O_4$ spinel phase by XRD patterns between $2\theta = 36.6$ and 37.6 degrees for the (A) copper-electroplating sludge + γ-Al_2O_3 system, (B) copper-electroplating sludge + α-Al_2O_3 system, (C) copper-electroplating sludge + kaolinite system and (D) copper-electroplating sludge + mullite system sintered at 650–1250°C for 3 h.

the leachate pH generally decreased to a value below 3.7 in the γ-Al_2O_3 system, when the sintering temperature ranged from 1050°C to 1100°C. By using α-Al_2O_3 precursor, the leachate pH was found to reach the lowest value (around 3.3) at the similar temperature range. After being sintered at 650°C, the pH value decreased to 4.0 for the kaolinite system and 4.2–4.6 in the mullite system. With the further increase in sintering temperature, the leachate pH continuously decreased in both the kaolinite and mullite precursor systems, and reached the lowest value within a temperature range of 950–1000°C. Above this temperature range, the pH value first increased and then slightly decreased.

3.1.2.2 Copper and aluminum concentration in the leachates

The copper concentrations in the leachates of the γ-Al_2O_3 and α-Al_2O_3 systems are shown in Fig. 7A and B, and the copper leachability was continuously decreased with the increase of sintering temperature before 1100°C. When the mixture of sludge + γ-Al_2O_3 was sintered at 750°C for 3 h, the copper leachability could be

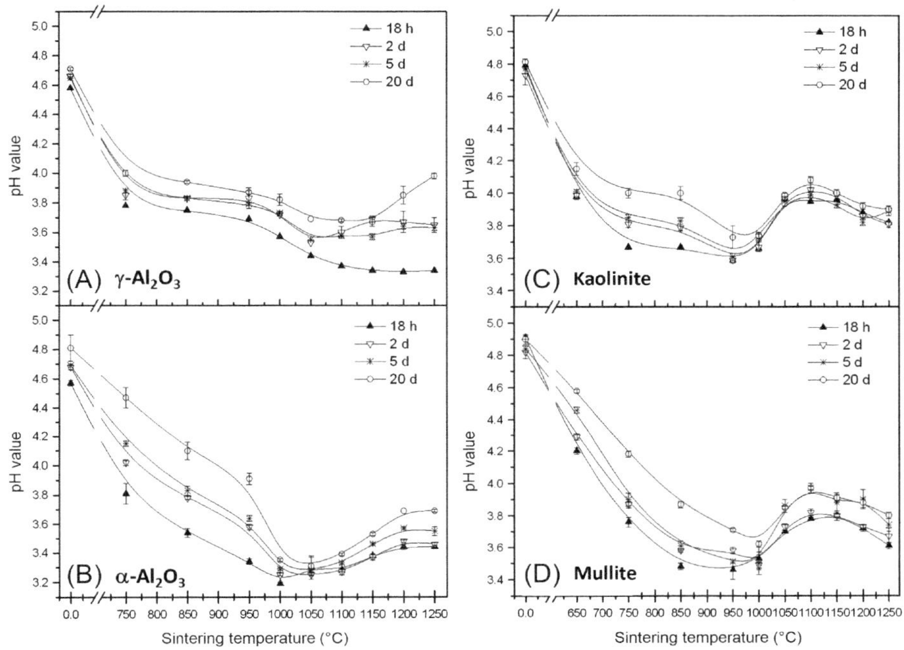

Fig. 6 Leachate pH values from (A) copper-electroplating sludge + γ-Al_2O_3 system, (B) copper-electroplating sludge + α-Al_2O_3 system, (C) copper-electroplating sludge + kaolinite system and (D) copper-electroplating sludge + mullite system before and after sintering at 650–1250°C for 3h. The leaching solution was TCLP extraction fluid no. 2 (acetic acid solution) with a pH of 2.9. Each leaching vial was filled with 5mL of extraction fluid and 0.25g of powder sample, and then rotated end over end between 0.75 and 22d.

reduced from 1500mgL^{-1} (without thermal treatment) to 190mgL^{-1}. The lowest copper concentration was found to be around 10mgL^{-1} at 1100°C, and such significant reduction (150 times) of the copper leachability clearly reflects the strong potential of stabilizing copper by thermal treatment. With α-Al_2O_3 precursor, copper concentrations decreased slightly at lower temperatures ($\leq$850°C), and declined dramatically to the lowest value (around 15mgL^{-1}) when the samples were sintered within the temperature range of 1000–1150°C. Therefore, the leachability of copper after such thermal treatment was also reduced to lower than the 1/100 level. When the leachability of aluminum in the sludge + γ-Al_2O_3 system was taken into consideration (Fig. 7C and D), it first increased after being sintered at 750–950°C for 3h. However, with the further increase of temperature, the concentration of aluminum decreased and reached the lowest value (<20mgL^{-1}) at 1050°C. Furthermore, the leachability of aluminum in the sample increased slightly again after being sintered at temperatures higher than 1050°C. The aluminum in the unheated sludge + α-Al_2O_3 samples had the lowest leachability, which was about 13mgL^{-1}. The aluminum concentrations kept continuously and slowly increasing with the elevated sintering temperature, and reached a maximum value when the sample was sintered at 1200°C.

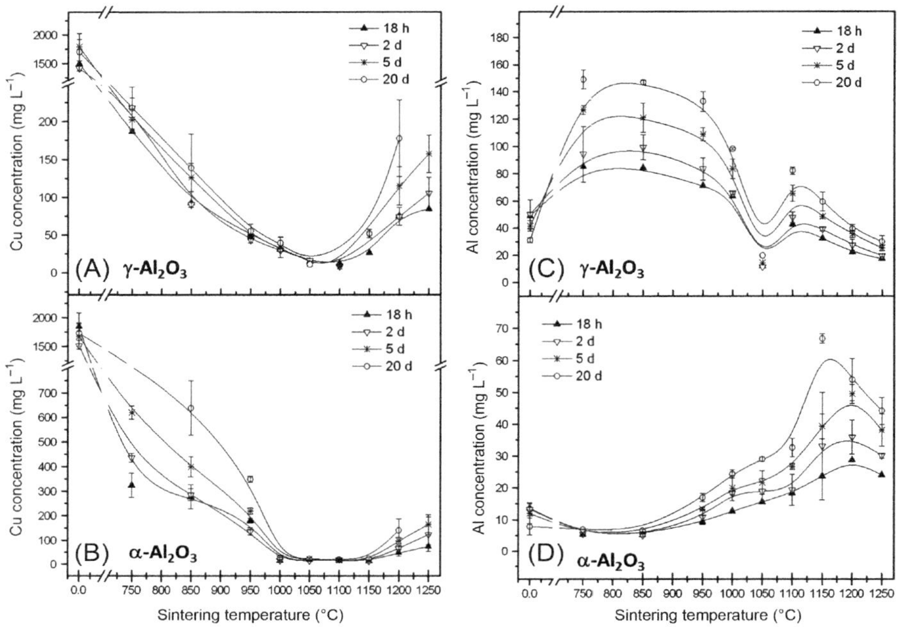

Fig. 7 Copper concentrations of the leachates from (A) copper-electroplating sludge + γ-Al_2O_3 system, and (B) copper-electroplating sludge + α-Al_2O_3 system; aluminum concentrations of the leachates from (C) copper-electroplating sludge + γ-Al_2O_3 system, and (D) copper-electroplating sludge + α-Al_2O_3 system. The leaching solution was TCLP extraction fluid no. 2 (acetic acid solution) with a pH of 2.9. Each leaching vial was filled with 5 mL of extraction fluid and 0.25 g of powder sample, and then rotated end over end between 0.75 and 22 d.

The variation of copper concentrations in the leachates of both the kaolinite and mullite systems was illustrated in Fig. 8A and B. Copper concentrations leached from the unsintered samples reached 2000 mg L^{-1} in both the kaolinite and mullite systems. A substantial decrease from 2000 to 500 mg L^{-1} was observed when the sample of copper-containing sludge + kaolinite was sintered at 650°C. However, the leached copper from the mullite system after being thermally treated at the same temperature was observed as high as 1700 mg L^{-1}. The minimum value of copper concentrations in the leachates was from samples sintered at 950–1000°C from both systems. Such values were about 200 mg L^{-1} for the sample in the kaolinite system and lower than 100 mg L^{-1} for the sample in the mullite system. To further investigate the leaching behavior of sintered products, the aluminum concentrations were also measured, as shown in Fig. 8C and D, respectively. The leached aluminum from the sample produced in the kaolinite system was decreased from 250 to 15 mg L^{-1} as the temperature increased. The leached aluminum slowly increased with the increase of temperature until a slight decrease occurred after 1100°C.

The variation of the copper concentration with the leaching time is also illustrated in Fig. 7. Generally, the copper concentrations in the leachates increased with a

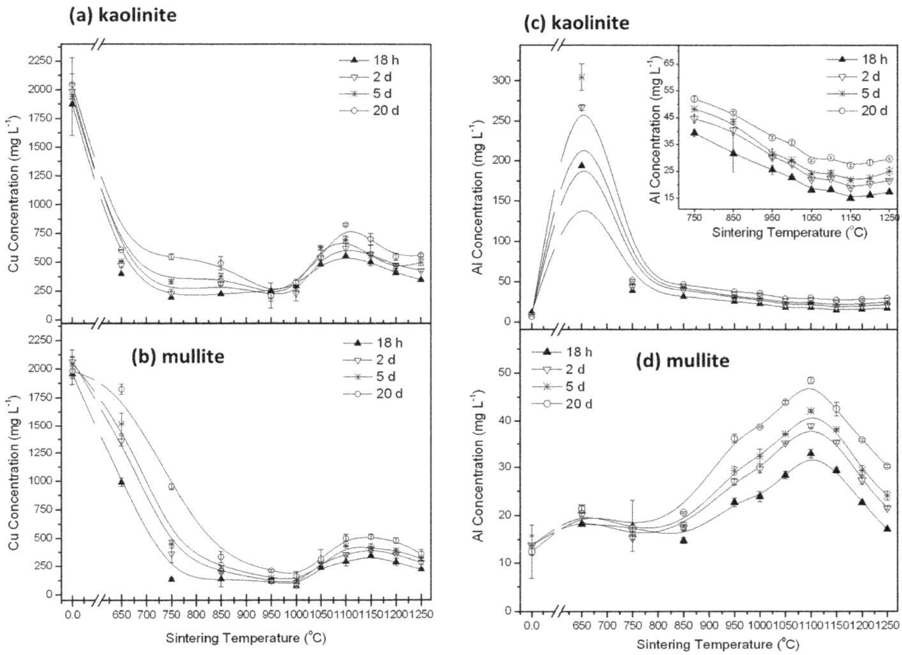

Fig. 8 Copper concentrations in the leachates from (A) copper-electroplating sludge + kaolinite system and (B) copper-electroplating sludge + mullite system; aluminum concentrations in the leachates from (C) copper-electroplating sludge + kaolinite system and (D) copper-electroplating sludge + mullite system. The leaching solution was TCLP extraction fluid no. 2 (acetic acid solution) with a pH of 2.9. Each leaching vial was filled with 5 mL of extraction fluid and 0.25 g of powder sample, and then rotated end over end between 0.75 and 22 d.

prolonged leaching period in both the γ-Al_2O_3 and α-Al_2O_3 systems. However, the copper concentrations in leachates increased differently for the γ-Al_2O_3 and α-Al_2O_3 systems. When the leaching period was prolonged from 18 h to 20 d, Fig. 7A illustrated a slow increase of the leached copper within a temperature range of 650–1100°C and a relatively rapid increment when the temperature was over 1100°C. While in Fig. 7B, the copper concentration in the leachates quickly increased at low temperatures with the prolonged leaching time, but such an increment slowed down dramatically when the temperature was above 1000°C. A slight increase of the leached copper was also found when the temperature was higher than 1150°C in the α-Al_2O_3 system. The change of the leached aluminum in the prolonged leaching period was investigated in Fig. 7C and D. The most significant increase of aluminum concentrations (from about 70 to 150 $mg\,L^{-1}$) was found when the sludge was thermally treated at a temperature range of 750–950°C with the γ-Al_2O_3 precursor. However, when the α-Al_2O_3 precursor was used to stabilize the electroplating sludge, the aluminum in the solid was relatively stable compared to that in the γ-Al_2O_3 system. The most significant increase of aluminum was observed with 20–60 $mg\,L^{-1}$.

3.1.2.3 Leaching behavior of the ceramic products

When the leaching of copper from the products is correlated with the thermal treatment process, it was found that the metal concentrations in leachates generally decreased with the formation of the $CuAl_2O_4$ spinel phase in the samples. For alumina matrices, the lowest copper concentration was observed within the temperature range of 1000–1100°C, which was detected as the most promising temperature range for the generation of the $CuAl_2O_4$ phase (Fig. 3). When the temperature was higher than 1150°C, a slight but continuous increase of the leached copper was observed. A small increase of copper in the leachate was likely due to the reappearance of CuO and the formation of the $CuAlO_2$ phase in the sample. Although copper can also be stabilized in the $CuAlO_2$ structure, as reported in our previous study, the resistance of the $CuAlO_2$ phase to the acidic attack was several times weaker than that of the $CuAl_2O_4$ phase. Furthermore, copper tends to leach out more at the prolonged leaching period from the $CuAlO_2$ phase in comparison to the $CuAl_2O_4$ phase. This can further explain the increment of copper concentrations with the prolonged leaching time at temperatures above 1150°C. Similarly, the lowest pH value was also consistent with the optimized sintering temperature range for $CuAl_2O_4$ formation. Moreover, the increase of pH with the extended leaching time for the sample sintered at $\geq$1100°C might also be due to the decomposition of the $CuAl_2O_4$ phase. However, with the leaching of major metals, the aluminum concentrations were much lower than the copper concentrations in the leachate.

For the kaolinite and mullite systems, when the leaching behavior was associated to the phase transformation occurring during the sintering process, the continuous decrease of the leached copper at higher temperatures was due to the development of $CuAl_2O_4$ spinel in products. The maximum values of pH change and copper concentration occurred at a temperature range of 950–1000°C, which was also confirmed as the optimized temperature range for the formation of the $CuAl_2O_4$ spinel. The increase of sample leachability was also associated with the decomposition of the $CuAl_2O_4$ spinel phase. However, the value further decreased when the temperature was higher than 1100°C in both systems, and this might be caused by the encapsulation effect of the silica melt. The stabilization of hazardous metals has been studied through the vitrification technique, and the metal can be stabilized through encapsulation into the glass matrix (Park and Heo, 2002; Scarinci et al., 2000). The impurities in samples can decrease the melting temperature of silicon oxide (O'Mara et al., 1990) and the sewage sludge ash with about 40% of silicon oxide observed to be melted at a temperature higher than 1050°C (Tsai et al., 2006). Therefore, in this system, the melting of silicon oxide at 1100°C can fix metals in the glass matrix and reduce their leachability.

3.2 Chromium ore processing residue for ceramic products

3.2.1 Phase compositions and morphology of the obtained glass ceramics

The results of phase identification are illustrated in Fig. 9, which shows the phase compositions in chromite ore processing residue (COPR) converted glass ceramics and products from two simulated systems containing 2% and 4% (weight percentage)

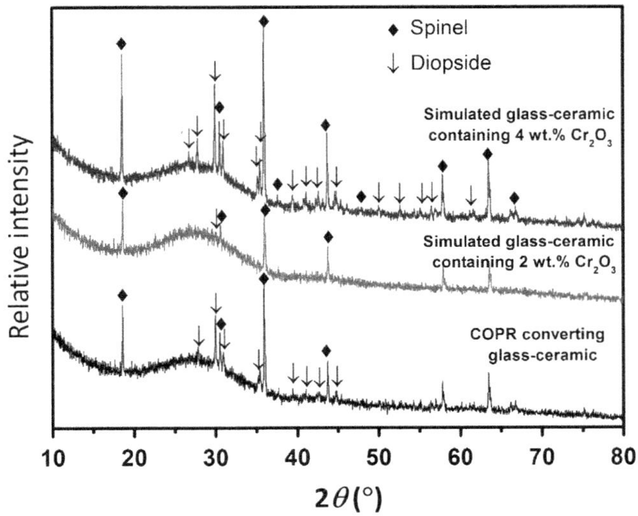

Fig. 9 Phase identification results for the COPR converting glass-ceramic and simulated glass-ceramic samples containing 2% and 4% Cr_2O_3.

of Cr_2O_3. Both spinel and diopside (ideal formula $CaMgSi_2O_6$) were observed in the three series, and the peak intensity of the diopside phase is higher in the simulated glass-ceramics system with 4% of Cr_2O_3 than that observed in the COPR converted glass ceramics. The higher peak intensity with higher Cr_2O_3 content indicates that more Cr_2O_3 loading may promote the crystallization of diopside in the CaO-MgO-SiO_2-Al_2O_3-Cr_2O_3-Fe_2O_3 system. An obvious hump was observed between 20–35 degrees in all XRD patterns of the three series due to the residual glass in the products, which further confirms that the obtained products contain not only crystalline phases but also glass components.

Bright-field backscattering SEM images (Fig. 10) show the morphology of the COPR converted glass-ceramic product, in which the spinel phase is presented in white contrast while the residual glass is in gray contrast. The diopside phase cannot be observed obviously due to the similarity in chemical compositions between diopside and residual glass. Fig. 10A shows that small spinel crystals homogeneously disperse in the residual glass and no obvious crack was observed between the crystal and glass components. Fig. 10B presents the SEM image for the sample without polishing, in which spinel crystal embeds in the residual glass. These microstructure results further indicate that the obtained COPR converted product is a homogeneous glass ceramic and crystals in this product are well encapsulated by the residual glass.

3.2.2 Detoxification of the chromite ore processing residue

About 25% of the chromium in COPR exists as Cr(VI) and is distributed in different minerals, such as hydrogarnet, hydrocalumite, and ettringite (Hillier et al., 2003). Because Cr(VI) is a toxic and mobile chromium species, the complete reduction of

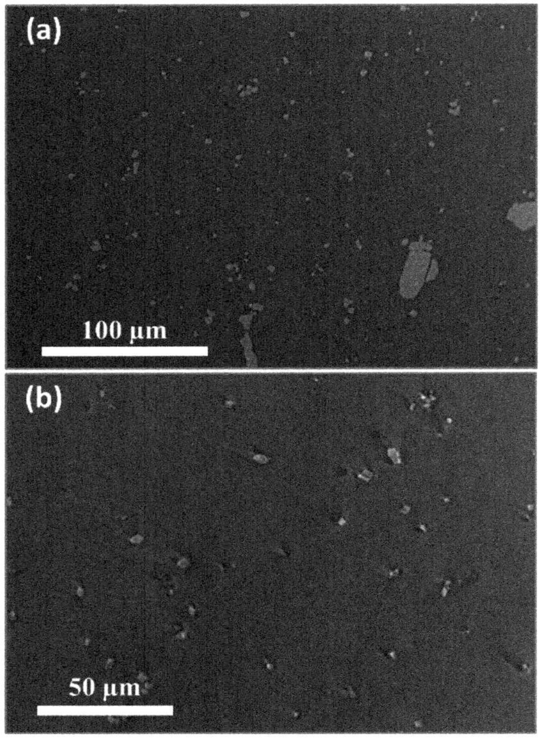

Fig. 10 Backscattered SEM images of the chromite ore processing residue (COPR) converting a glass-ceramic sample with polishing (A), and without polishing (B).

Cr(VI) is one of the most critical criteria for the safe disposal of COPR. To detoxify COPR, the oxidation states of chromium in the final product were analyzed by Cr k-edge XANES. In the Cr k-edge XANES, the Cr(VI) and Cr(III) can be distinguished with the pre-edge region. The pre-edge spectrum of Cr(VI) showed a strong absorption peak, whereas this phenomenon was absent in that of Cr(III) (Lin and Wang, 2012). Fig. 11 shows the Cr K-edge XANES spectra of COPR and COPR converted glass ceramic and those of the reference compounds containing Cr(III) and Cr(VI) (Cr_2O_3 for Cr(III), and CrO_3 and $K_2Cr_2O_7$ for Cr(VI)). Cr(VI) can be clearly observed in the COPR sample, but the pre-edge feature of Cr(VI) was not observed in the COPR converted glass-ceramic sample. As the detection limit of Cr(VI) by XANES was reported to be in ppm range (Huggins et al., 1999), the absence of this pre-edge feature indicates a nearly complete transferring all Cr(VI) into Cr(III) in the COPR converted glass-ceramic. This finding is consistent with those of other studies on thermal treatment of Cr(VI)-bearing materials, in which Cr(VI) was not detected after sintering at 1100°C (Wei et al., 2002, 2010). It was also pointed out by Mao et al. that CrO_4^{2-} can be reduced to CrO_2^{1-} and reacted with MgO to form $MgCr_2O_4$ when sintered at temperatures above 1000°C (Mao et al., 2015). By fabrication of COPR converted glass ceramic, the Cr(VI) bearing forms were detoxified into the CrO_2^{1-} form and captured by MgO to form the spinel phase or dissolved into the glassy network.

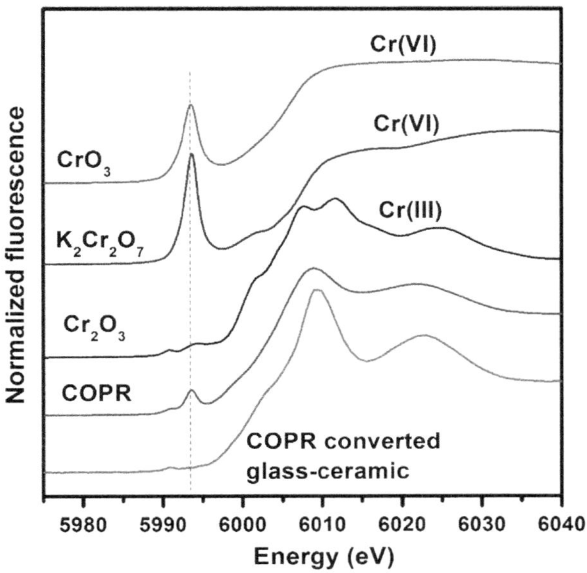

Fig. 11 Cr K-edge XANES spectra of COPR, COPR converted glass-ceramic, and reference compounds containing Cr(III) and Cr(VI) (Cr_2O_3 for Cr(III), and CrO_3 and $K_2Cr_2O_7$ for Cr(VI)).

3.2.3 Leachability of the COPR converted glass ceramics

To evaluate the immobilization effect of chromium in the obtained COPR converted glass ceramic, a prolonged TCLP experiment was carried out to observe the resistance to the acidic attack of the product compared to that of Cr_2O_3. Fig. 12A presents the concentrations of chromium in the leachates of both Cr_2O_3 and the COPR converted glass ceramic. The chromium leached from Cr_2O_3 is in general more than 20 times that from the COPR converted glass ceramic, which indicates that the obtained glass ceramic is much more resistant to acid attack than Cr_2O_3. After the first 18 h of leaching, the chromium concentration in the Cr_2O_3 leachate is about 25 ppm, whereas only about 0.5 ppm was detected in the leachate from the glass ceramic. Moreover, the overall chromium concentration in the leachate of the glass ceramic was less than 2 ppm throughout the 25-day leaching period, which is lower than the EPA regulatory limit of 5 ppm (USEPA, 1998).

In the obtained COPR converted glass ceramic, the spinel solid solution phase hosted most of the chromium ions (about 77%). This means the spinel phase acts as the basic barrier to resist the leaching of chromium, and the encapsulating glass matrix acts as an additional barrier. Because the glass matrix will first encounter the acidic attack, Fig. 12B presents the calcium, magnesium, aluminum, silicon, and iron concentrations, which were mainly leached from the glass matrix during the leaching experiments. Both calcium and magnesium concentrations increased continuously with the prolonged leaching time, whereas those of aluminum, silicon, and iron remained relatively stable during the entire leaching process. The relatively greater leaching of Mg^{2+} and Ca^{2+} may result from their roles as glass network modifiers; it is thus relatively easier to

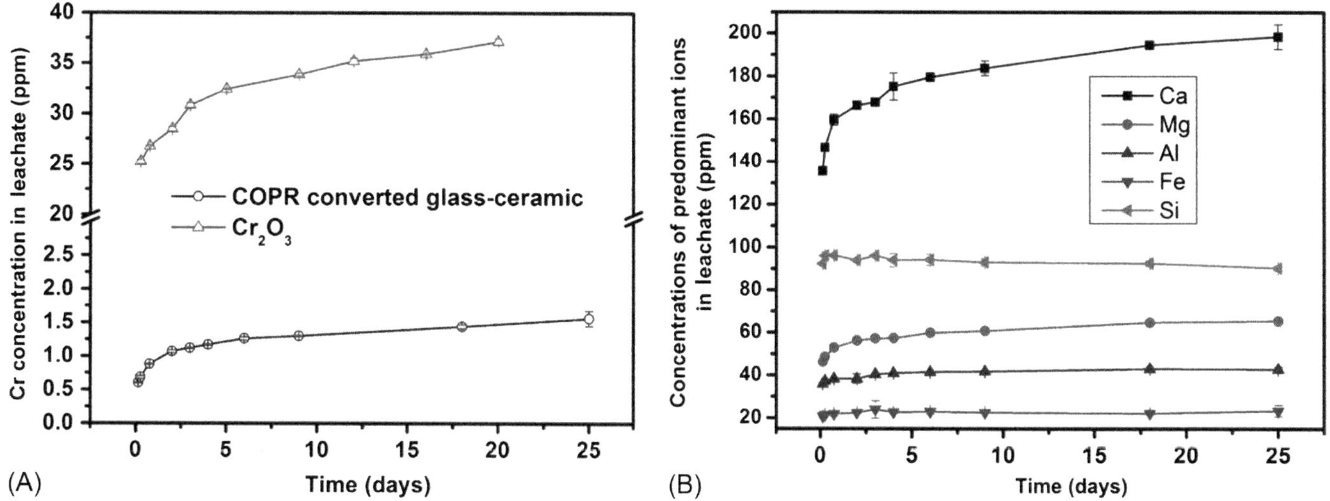

Fig. 12 Concentrations of (A) chromium in Cr_2O_3 and COPR converted glass-ceramic leachates, and (B) predominant elements (calcium, magnesium, aluminum, silicon, and iron) in COPR converted glass-ceramic leachate. Prolonged TCLP was tested with pH 2.9 acetic acid solution for agitation periods from 3 h to 25 d. Each leaching vial was filled with 5 mL of extraction fluid and 0.25 g of powder sample.

remove them from the glass networks (Edén et al., 2011). Aluminum, silicon, and iron usually act as the glass network formers (Leturcq et al., 1999) and are more stable in the glass networks. This result may show that the fundamental glass network structure is generally robust and effective in resisting acid attack to prevent the leaching of incorporated chromium.

4 Metal stabilization mechanisms in ceramic matrix

From the results in Section 3, it is clearly observed that the industrial waste can be used as raw materials for the fabrication of ceramic products. Moreover, during the ceramic sintering processes, the heavy metals such as copper and chromium can be successfully incorporated into the ceramic matrix. As the reaction pathway and incorporation behavior are regarded as the most critical mechanisms for the stabilization of heavy metals, this study further simulated the reaction series using metal oxides and quantified the incorporation efficiencies under a variety of sintering temperatures.

The X-ray diffraction (XRD) technique is a basic material characterization method and has been used for many decades with exceptional success to provide accurate information about the structure of materials. Nowadays, in studies of toxic metal solidification/stabilization, the XRD technique is often used to monitor and identify toxic metal phase(s) contained in samples as a qualitative technique (Young, 1995; Walenta and Fullmann, 2004). However, XRD can also be used to quantify phase compositions of natural or industrial materials, when combined with the Rietveld refinement method. It has been proved to be a reliable, precise, and very reproducible technique to quantify the relative phase abundances (Young, 1995; Walenta and Fullmann, 2004) such as in the cement industry where it is needed for accurate phase compositions. All crystalline phases in the products were subjected to quantitative analysis by TOPAS 4.2, a program that uses the Rietveld refinement method.

4.1 Mechanisms of copper stabilization in aluminum-rich ceramic matrix

4.1.1 Reaction pathways

The powder XRD patterns illustrated in Fig. 13 indicate that copper could be successfully incorporated into the $CuAl_2O_4$ structure by using γ-Al_2O_3 as the aluminum-rich ceramic precursors within a 3-h sintering mechanism. The chemical reactions for the formation of $CuAl_2O_4$ can be listed as follows:

$$CuO + \gamma - Al_2O_3 \rightarrow CuAl_2O_4 \quad (1)$$

The results in Fig. 13 reveal that the sample sintered at 750°C developed a distinguishable $CuAl_2O_4$ crystalline phase. The formation of the copper aluminate spinel has been observed in the equilibrium thermal experiment (Jacob and Alcock, 1975), which reported the spinel formation temperature starting at 612°C. However, when

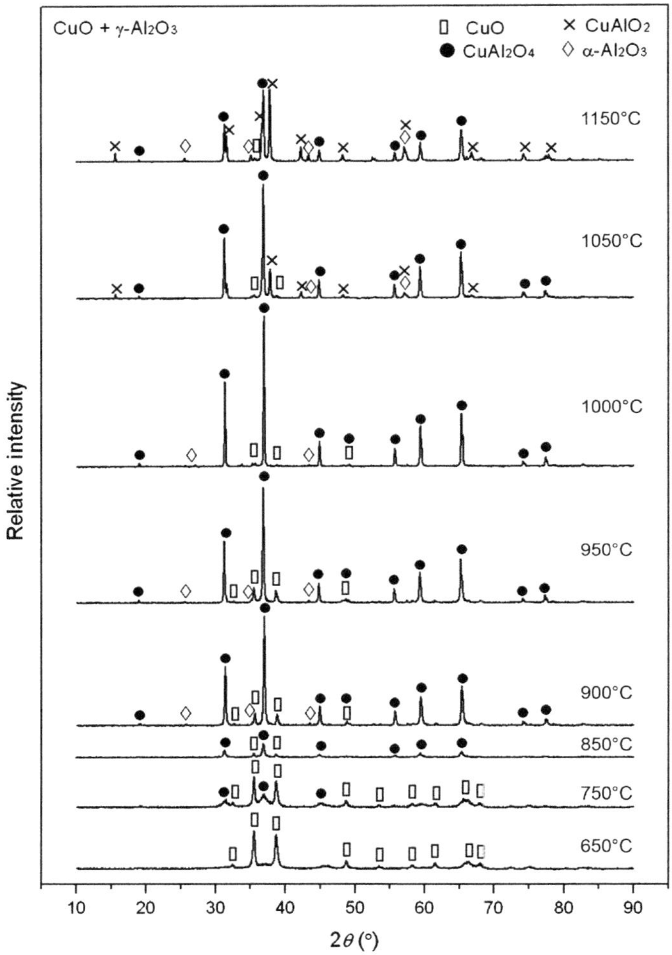

Fig. 13 XRD patterns of 650–1150°C and 3 h sintered CuO+γ-Al_2O_3 samples. The standard patterns include CuO (PDF#80-1268), $CuAl_2O_4$ (PDF#78-1605), $CuAlO_2$ (PDF#75-2356), and α-Al_2O_3 (PDF#10-0173).

comparing to the results observed in our short sintering experiment, it was found that an effective sintering temperature for a short sintering scheme to incorporate copper into $CuAl_2O_4$ in ceramic products should be at least above 750°C. Because the solid-state reaction is usually affected by both thermodynamic conditions and the diffusion process, this comparison may further suggest that spinel formation at temperatures below 750°C is largely limited by the prevailing slow diffusion, although it is thermodynamically feasible at temperatures above 612°C. Below 750°C, the $CuAl_2O_4$ phase formed by the short sintering scheme may only be limited at the grain boundary of the reactants, and the very small quantity of such a phase in the system was not reflected in the XRD results. Although there was a distinguishable $CuAl_2O_4$ phase in the 750°C sintered sample, the substantial growth of $CuAl_2O_4$ in the system was observed to be above 850°C, which may indicate the energy needed to overcome the major diffusion barrier in the system. Below 1000°C, the peak intensity of the $CuAl_2O_4$ phase increases as the temperature increases. However, at sintering temperatures higher than 1000°C, the peak intensity of the $CuAl_2O_4$ phase was found to decrease with elevated temperatures.

Furthermore, the XRD patterns in Fig. 13 also show that the decrease of $CuAl_2O_4$ at higher temperatures was due to the formation of another new Cu-Al oxide phase, cuprous aluminate ($CuAlO_2$). The potential reaction pathways are (Jacob and Alcock, 1975):

$$2CuAl_2O_4 \rightarrow 2CuAlO_2 + Al_2O_3 + 0.5O_2 \tag{2}$$

$$CuAl_2O_4 \rightarrow CuO + Al_2O_3 \tag{3}$$

$$2CuAl_2O_4 \rightarrow Cu_2O + 2Al_2O_3 + 0.5O_2 \tag{4}$$

$$2CuO \rightarrow Cu_2O + 0.5O_2 \tag{5}$$

$$Cu_2O + Al_2O_3 \rightarrow 2CuAlO_2 \tag{6}$$

4.1.2 Transformation behavior of copper in an aluminum-rich ceramic matrix

Four copper-containing crystalline phases were found in the products, including CuO (PDF#80-1268), $CuAl_2O_4$ (PDF#78-1605), and $CuAlO_2$ (PDF#75-2356). They were all subject to quantitative phase analysis together with the α-Al_2O_3 (PDF#10-0173) phase by using the Rietveld refinement method. Upon sintering the mixture of CuO and γ-Al_2O_3, the $CuAl_2O_4$ spinel was generally observed as the dominant product in this study (Fig. 14). Based on quantification results from the Rietveld refinement, the investigation of crystalline phase distribution was conducted by using a "transformation ratio (TR)" index to indicate metal incorporation efficiency by kaolinite and mullite precursors. The transformation ratio (TR, %) is defined as:

$$\mathrm{TR}\ (\%) = \frac{a_n \dfrac{\mathrm{wt\%\,of\,M}_n}{\mathrm{MW\,of\,M}_n}}{a_1 \dfrac{\mathrm{wt\%of\,M}_1}{\mathrm{MW\,of\,M}_1} + a_2 \dfrac{\mathrm{wt\%of\,M}_2}{\mathrm{MW\,of\,M}_2} + a_3 \dfrac{\mathrm{wt\%of\,M}_3}{\mathrm{MW\,of\,M}_3} + \cdots + a_n \dfrac{\mathrm{wt\%of\,M}_n}{\mathrm{MW\,of\,M}_n}} \tag{7}$$

where MW is the molecular weight; an ($n = 1, 2, 3, \ldots$) is the number of target metal atoms contained in the corresponding phase; and wt% is the weight fraction of its

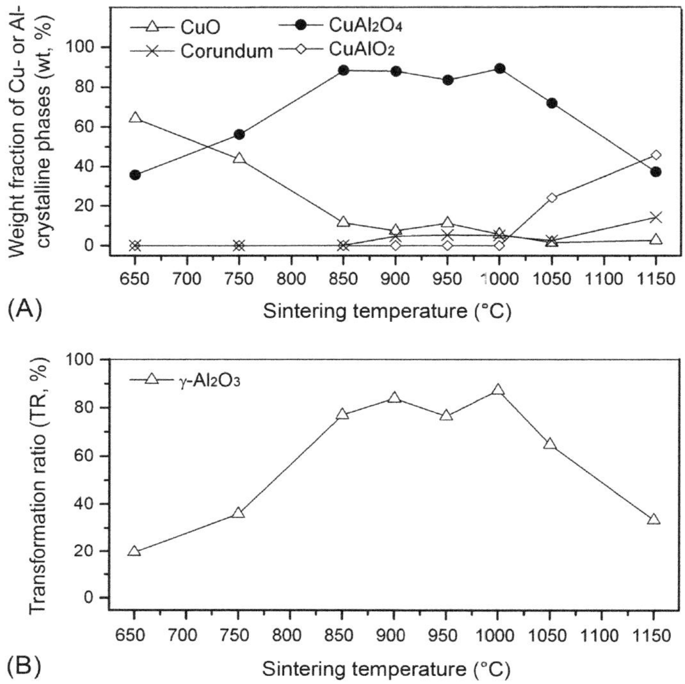

Fig. 14 (A) Variations of weight fractions of copper- and aluminum-containing crystalline phases obtained from the CuO sintering reactions with γ-Al_2O_3, and (B) the transformation ratio (TR, %) of copper into the $CuAl_2O_4$ spinel phase.

corresponding phase obtained via the Rietveld refinement. The total of metal fractions for all product phases should be 100%.

At 650–850°C, although the system was still dominated by the poorly crystalline γ-Al_2O_3 precursor, the growth of the $CuAl_2O_4$ was shown in the XRD result. When the mixture was sintered at 900°C for 3 h, the development of α-Al_2O_3 from the residual alumina content was observed in ~5% within the crystalline phases. This α-Al_2O_3 formation temperature was approximately 100°C lower than that of a previous study reported when only with alumina (Wolverton and Hass, 2001). Our result highlights the role of the CuO impurity in enhancing the thermal transformation behavior of alumina. For the product sintered at 650°C, the CuO and $CuAl_2O_4$ were refined with around 65% and 35%, respectively, in the crystalline phases, which indicates that about 20% of copper was transformed from CuO into $CuAl_2O_4$ (Fig. 14B). The maximum TR of the CuO + γ-Al_2O_3 samples reached 80% at temperatures between 850°C and 1000°C, and then declined at higher temperatures because of the formation of $CuAlO_2$. In addition, the weight fraction of CuO decreased dramatically from 65% to 8%, indicating the successful stabilization of copper within the temperature range of both $CuAl_2O_4$ and $CuAlO_2$ phase transformations.

4.2 Zinc stabilization in aluminum-rich ceramic matrix

Fig. 15 collates the XRD patterns of the sintered samples within the temperature range of 750–1450°C, and $ZnAl_2O_4$ was found as the predominant product phase after sintering. Therefore, the potential reactions between ZnO and the alumina precursor (γ-Al_2O_3) are expressed as follows:

$$ZnO + \gamma - Al_2O_3 \rightarrow ZnAl_2O_4 \quad (8)$$

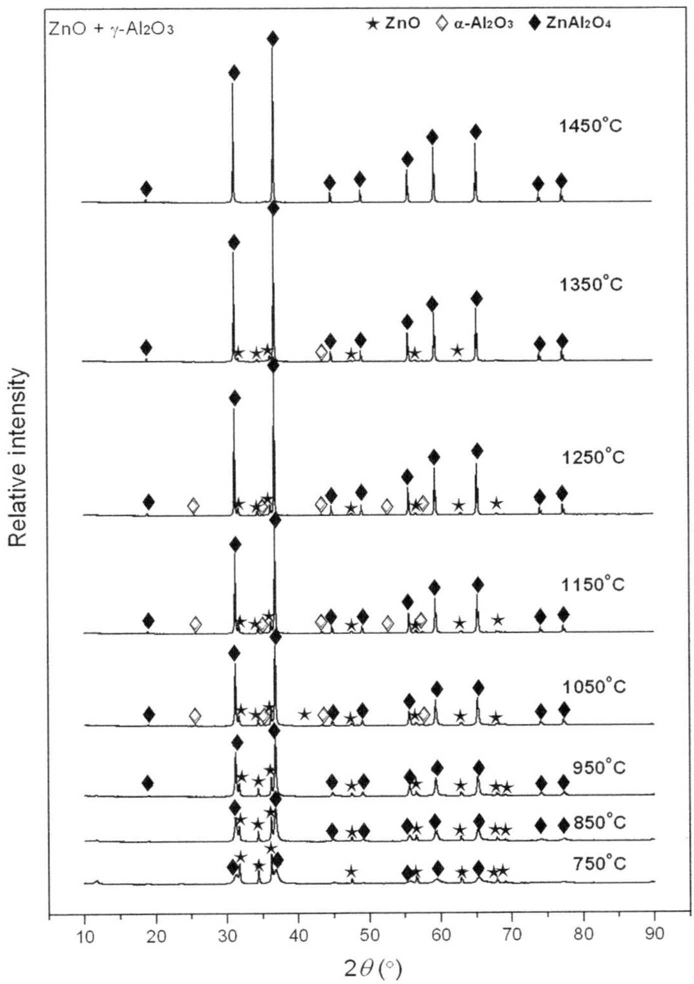

Fig. 15 XRD patterns of sintering ZnO+γ-Al_2O_3, at 750–1450°C for 3h. The standard patterns include ZnO (PDF#89-0511), $ZnAl_2O_4$ (PDF#82-1036), and α-Al_2O_3 (PDF#10-0173).

When γ-Al_2O_3 was used as a precursor, the $ZnAl_2O_4$ spinel phase was clearly initiated when sintered at 750°C for 3h. Once the sintering temperature was above the first appearance of spinel phases, the continuing growth of $ZnAl_2O_4$ was observed with an increased peak intensity. The signals of reactants eventually diminished after 3h of sintering at 1450°C and 1250°C for the $ZnO+\gamma$-Al_2O_3 system.

Two zinc-containing crystalline phases were found in the products—ZnO (PDF#89-0511) and $ZnAl_2O_4$ (PDF#82-1036). They were all subjected to quantitative phase analysis together with the α-Al_2O_3 (PDF#10-0173) phase, using Topas 4-2, which employs the Rietveld refinement method. To further quantify the variation of zinc incorporation efficiency in both precursor systems at different sintering temperatures, the distribution of crystalline phases in the products can be expressed by a transformation ratio (TR, %) (also calculated according to Eq. 8).

Fig. 16 summarizes the TR values of zinc incorporated into the $ZnAl_2O_4$ phase when $ZnO+\gamma$-Al_2O_3 samples were sintered over the temperature range of 750–1450°C. The reaction of zinc with the γ-Al_2O_3 precursor was strong and even reached more than 60% transformation at the lowest sintering temperature (750°C). The zinc transformation continued to increase with the elevation in sintering temperature and reached nearly full incorporation after sintering at 1450°C. The curve suggests that the ZnO and γ-Al_2O_3 interaction was dominant at a lower temperature. The driving forces for mass transfer during sintering are the differences in chemical potential (free enthalpy or molar Gibbs energy) between fired materials (Sarikaya et al., 2008). The value of ΔH_f° (standard enthalpy of formation) for γ-Al_2O_3

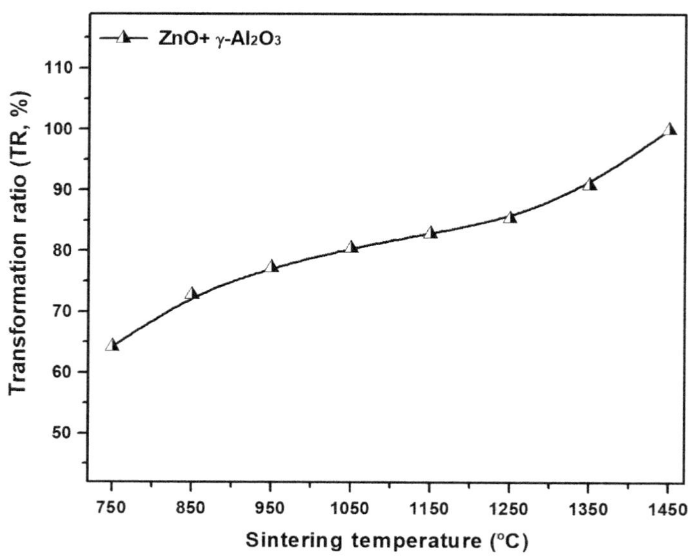

Fig. 16 The transformation ratio (TR, %) of zinc into the $ZnAl_2O_4$ spinel structure by γ-Al_2O_3 precursor. The incorporation of zinc reached 100% by the end of sintering processes.

is $+18.8\,kJ\,mol^{-1}$ (McHale et al., 1997), and thus the reaction between ZnO and γ-Al_2O_3 is energetically favored. This may explain the greater zinc incorporation efficiency by γ-Al_2O_3 at lower temperatures.

4.3 Mechanisms of chromium stabilization in glass ceramics produced by aluminum-rich precursors

4.3.1 Phase identification of the solid spinel solution

Glass-ceramic products were obtained in a simulated chromium-containing CaO-MgO-SiO_2-Al_2O_3-Cr_2O_3 slag system by high-temperature melting. The XRD results (Fig. 17A) revealed two crystalline phases in the glass-ceramic products: spinel and diopside (normal formula: $CaMgSi_2O_6$). However, the phase composition was found to differ according to Cr_2O_3 content. A crystalline phase with a spinel structure was clearly detected in the glass ceramics produced using 2% of Cr_2O_3, whereas the diopside signal for this system was very weak. With the continuous addition of Cr_2O_3 to the reaction system, both a spinel phase and a diopside phase were observed in the glass-ceramic products, with a clear increase in peak intensity in each case. The XRD patterns in Fig. 17B enable further comparison of the predominant peaks of diopside and spinel within the 2θ range of 29.0–31.0 and 35.0–37.0 degrees, respectively. The patterns clearly demonstrate that both diopside and spinel peak intensity increased with the weight percentage of Cr_2O_3.

In addition, as the Cr_2O_3 content increased, the 2θ position of the spinel peaks was observed to shift from a higher to a lower degree while smaller changes were observed in the diopside peaks (Fig. 17B). This shift in the spinel phase peaks may have been due to changes in the lattice parameters of the spinel structure. When the system contained various proportions of MgO, Cr_2O_3, and Al_2O_3, a solid spinel solution was formed during the melting process with a chemical formula of $MgCr_xAl_{2-x}O_4$ (Juhin et al., 2007). The value of x differed according to the chemical composition of the raw materials. Changes in x altered the lattice parameters of the spinel crystal, causing the diffraction peaks to shift. The Rietveld refinement was applied to the obtained XRD patterns. The lattice parameters of the spinel phase are shown in Table 1. The lattice parameters clearly increased with the Cr_2O_3 content of the starting mixture, indicating that chromium incorporation into the spinel phase was affected by the initial level of Cr_2O_3. To determine the chemical composition of the solid spinel solution (specifically the value of x) in the obtained glass-ceramic products, TEM-EDX was used to detect the signals from each element (Fig. 18). As shown in Fig. 18A, two key areas (areas 1 and 2) were selected near the edge of the crystal grains and the glass, respectively. The SAED pattern shown in Fig. 18B further demonstrates that area 1 was composed of spinel crystals. The chemical composition of the spinel crystals was also obtained in grain areas similar to area 1, located at the edge of the thin sample without overlap with the glass phase. Fig. 18C and D illustrate the significantly different element compositions of areas 1 and 2, as measured by TEM-EDX. Area 1 was composed mainly of magnesium, aluminum, chromium, and oxygen while silicon and calcium made up the majority of area 2. The results confirm that area 1 was

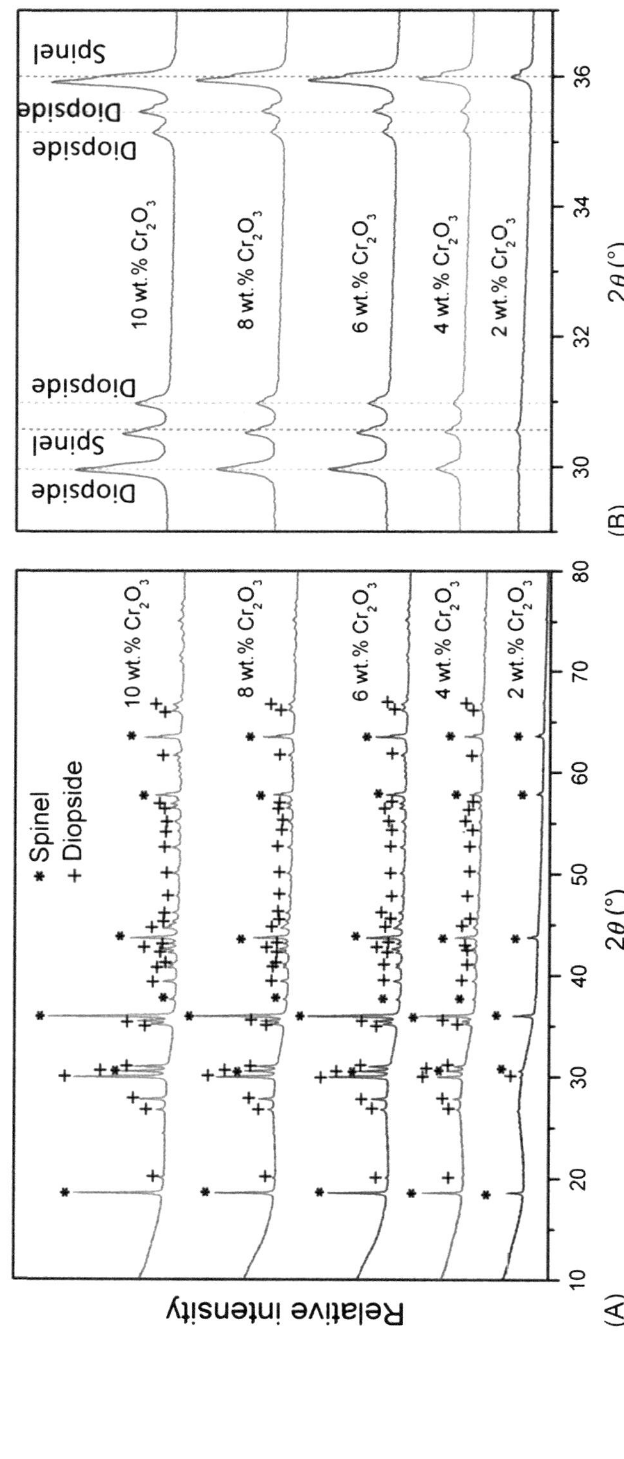

Fig. 17 (A) XRD patterns for glass-ceramics produced from starting mixtures containing different proportions of Cr_2O_3; (B) comparison of the major peaks for spinel and diopside, showing a peak shift for the spinel phase but no evidence of a peak shift for diopside.

Table 1 Lattice parameter and atomic ratio of the spinel phase based on the XRD patterns of the glass-ceramic products with different Cr_2O_3 contents

Sample no.	Cr_2O_3 content (wt.%)	Lattice parameter *a* (Å)	Atomic ratio of metal element (atomic, %)		
			Mg	Al	Cr
1	2	8.2773(1)	33.5	17.0	49.5
2	4	8.2828(1)	35.9	16.4	47.8
3	6	8.2861(1)	34.1	14.7	51.2
4	8	8.2896(1)	34.8	13.5	51.7
5	10	8.2912(1)	33.7	13.4	52.8

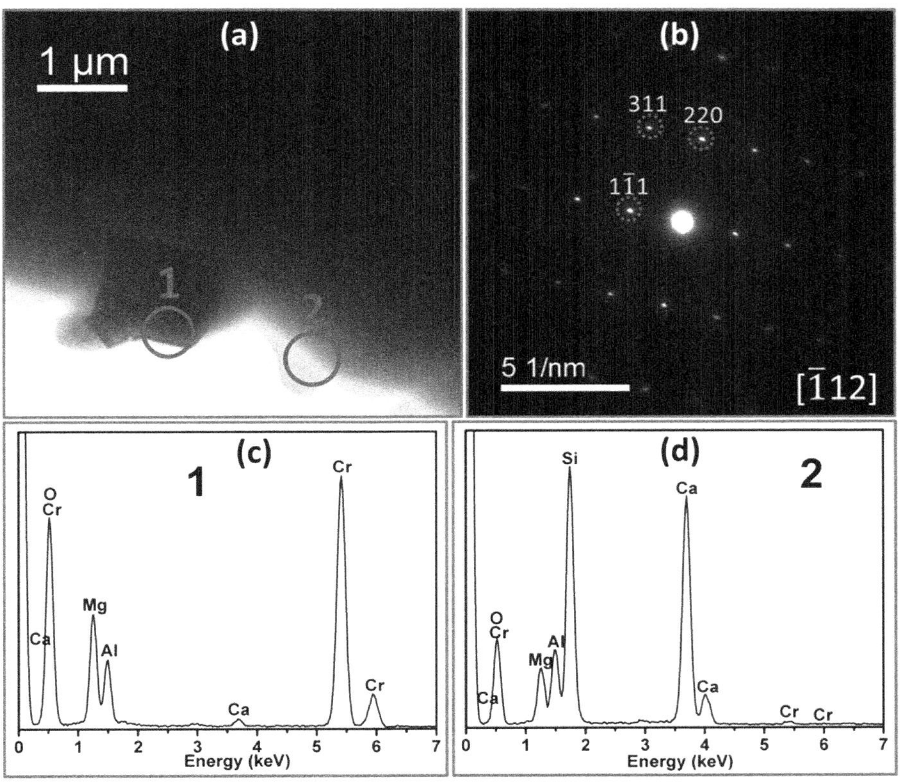

Fig. 18 (A) Bright-field TEM image of the glass-ceramics with a starting mixture containing 2% of Cr_2O_3; (B) SAED results to confirm the spinel crystal; TEM-EDX spectra for the crystal (area 1) (C) and the residual glass (area 2) (D).

located on a spinel grain with a chemical formula of $MgCr_xAl_{2-x}O_4$, whereas area 2 was located on the matrix, and was composed mainly of silicon- and calcium-containing minerals. The precise element composition of the spinel crystals was determined accordingly. The average values were obtained by statistically analyzing the composition of the elements detected in five spinel grains. The results are shown in Table 1. The atomic percentage of chromium in the solid solution $MgAl_{2-x}Cr_xO_4$ increased with the continuous addition of Cr_2O_3 to the starting raw materials, whereas those of aluminum decreased. As the radius of the chromium ions was larger than that of the aluminum ions, the lattice parameters of the spinel phase increased due to the continuous incorporation of Cr to substitute for aluminum.

4.3.2 Quantification of chromium-incorporation mechanisms

After the XRD patterns had been subjected to Rietveld refinement with standard reference materials, the weight fractions of the components (including crystalline and amorphous phases) of the glass-ceramic products were obtained. The results are shown in Fig. 19. When 2% of Cr_2O_3 was added to the system, the glass phase(s) made up more than 95% of the products. With a further increase in Cr_2O_3 content to 10%, the weight percentage of the glass phase(s) decreased linearly, finally reaching approximately 50%. In contrast, the crystalline phase(s) continued to grow, as illustrated in Fig. 19A. The weight percentage of the spinel phase increased from about 2% to around 10%, and that of the diopside phase ultimately exceeded 30% after adding 10% of Cr_2O_3 to the system. These results indicate that increasing the Cr_2O_3 content can enhance crystallization in a CaO-MgO-SiO_2-Al_2O_3-Cr_2O_3 system.

As spinel is the dominant chromium-hosting crystalline phase, it was crucial to study the distribution of chromium between the spinel and other phases in the glass-ceramics products. The distribution of chromium was examined using a "partitioning ratio (PR)" index, and it was defined as the ratio of the weight of Cr_2O_3 incorporated into the spinel phase to the total Cr_2O_3 used in the mixture. The equation can be described as follows:

$$PR = \frac{\text{wt.\% of } MgCr_xAl_{2-x}O_4 \times \dfrac{\text{MW of } Cr_xO_{1.5x}}{\text{MW of } MgCr_xAl_{2-x}O_4}}{\text{wt.\% of } Cr_2O_3 \text{ in sample}} \tag{9}$$

where the chemical formula of $MgCr_xAl_{2-x}O_4$ is determined by TEM-EDX, the weight percentage of $MgCr_xAl_{2-x}O_4$ is determined by XRD quantification analysis with a standard reference, and MW indicates molecular weight. A PR of 100% denotes the complete incorporation of chromium into the spinel phase; a PR of 0% indicates that no chromium has been incorporated into the spinel. As shown in Fig. 19B, the PR of chromium in the spinel phase continuously increased with the amount of Cr_2O_3 added to the system. When 2% of Cr_2O_3 was added to the glass-ceramics system, approximately 70% of the chromium was incorporated into the crystalline spinel phase while 30% remained in the residual glass matrix. However, when the content

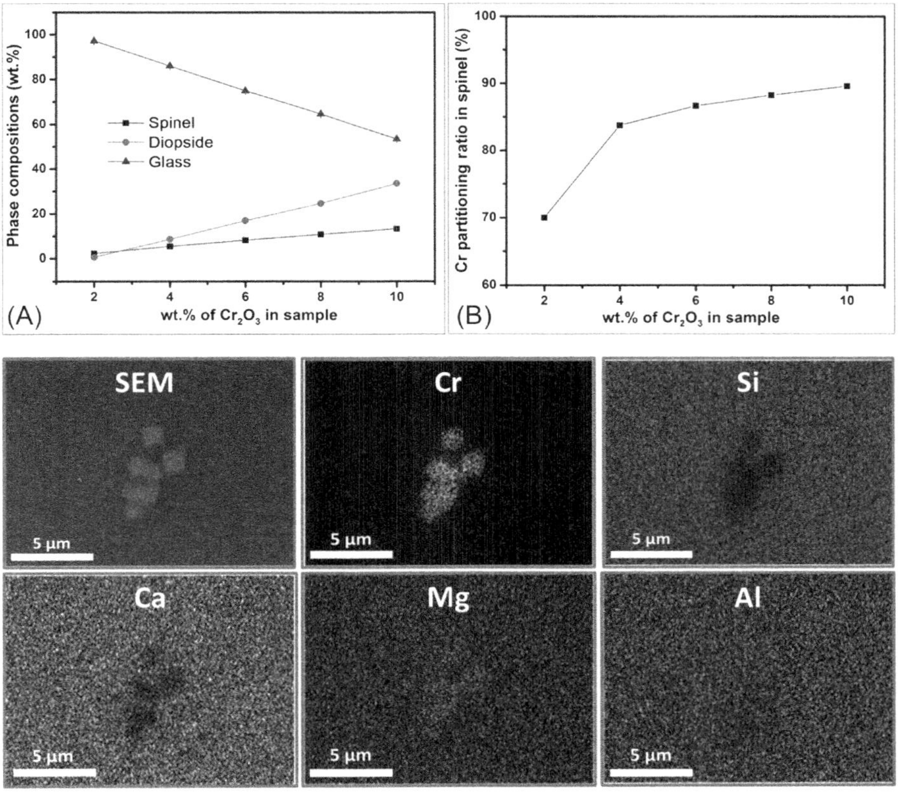

Fig. 19 (A) Quantitative composition of spinel, diopside, and glass in the as-formed glass-ceramic system with different percentages of Cr_2O_3, and (B) Partitioning ratio (PR) of chromium in the spinel phase, together with the SEM image and the corresponding elemental mapping (aluminum, calcium, chromium, magnesium, and silicon) of the sintered artificial clay+Cr_2O_3 (10%).

of Cr_2O_3 was 4%, the amount of chromium incorporated into the spinel phase increased significantly, from 70% to 85%. With a further increase in Cr_2O_3 content, the PR of chromium in the spinel continued to increase, but at a lower rate. PR reached its maximum value of approximately 90% when the largest proportion of Cr_2O_3 (10%) was added. This increase in PR may be related to the large octahedral site preferential energy of chromium in glass structures, leading to the formation of chromium-concentrated spinel (Jantzen and Brown, 2007; Tathavakar et al., 2005). These results indicate that most of the chromium element was incorporated into the spinel phase, with only a small fraction dispersed in the residual glass and/or diopside phase. In addition, the element-mapping profile for glass ceramics with 10% Cr_2O_3 clearly shows that the chromium element was concentrated in the spinel crystal and that the residual glass was rich in silicon, calcium, and aluminum.

5 The potential to use incinerated sewage sludge ash and coal fly ash for copper stabilization

5.1 Copper stabilization using sewage sludge ash

5.1.1 Characterization of sewage sludge and incinerated sludge ash

The sewage sludge was collected from Hong Kong (HKS), and experiments were carried out to further characterize the sewage sludge samples. Fig. 20A shows the weight loss of the dried sludge within the temperature range from 200°C to 1000°C. The weight of the sludge was observed to decrease dramatically by more than 80% before the heating temperature reached 500°C. Dried sludge was then fired at 900°C for 30 min to remove the organic content and then ground into powder for elemental composition analysis by X-ray fluorescence (XRF) spectroscopy (JEOL JSX-3201Z). Normalization into metal oxides (Table 2) shows that the top three elements were silicon, iron, and calcium. The XRD patterns in Fig. 20B further indicate complicated compositions in the calcined sludge. A strong hematite (Fe_2O_3) signal was detected, and aluminum mainly coexisted with calcium and/or silicon in the form of aluminosilicates. Due to the existence of iron and aluminum, it is possible that the sewage sludge ash from Hong Kong could be used beneficially as an iron- and aluminum-rich raw material to incorporate hazardous metals in waste materials.

5.1.2 Beneficial use of sewage sludge for copper stabilization

Samples prepared by mixing the sewage sludge incineration ash with CuO in a total dry weight of 60 g at a Cu:Fe molar ratio of 1:2 were examined by XRD, as shown in Fig. 21A. A clear signal reflected by the peaks of the $CuFe_2O_4$ spinel phase was observed when the sample was sintered at 750°C for 3 h. A further increase of sintering temperature increased the intensities of the spinel phase signal in the sintered samples. Besides the development of the spinel phase, the decrease of quartz and the other crystalline phases containing aluminum and/or silicon was generally very stable throughout the sintering process.

Due to the standard diffraction pattern of $CuFe_2O_4$ spinel (PDF#77-0010), the major peak at $2\theta \sim 35.5$ degrees was overlapped by the other phases' major peaks. Therefore, the second highest peak at $2\theta \sim 62.7$ degrees was selected for further observation of the $CuFe_2O_4$ spinel in the sintered products, as shown in Fig. 21B. The peak intensity of the $CuFe_2O_4$ spinel increased after its first appearance at 750°C and reached the maximum value at 950°C. The cubic $CuFe_2O_4$ remained relatively stable with a further increase of sintering temperature.

5.2 Copper stabilization by coal fly ash

5.2.1 Characterization of coal fly ash

The coal fly ash was collected from Castle Peak Power Station, the largest coal-fired power station in Hong Kong. The elemental compositions of the fly ash were detected

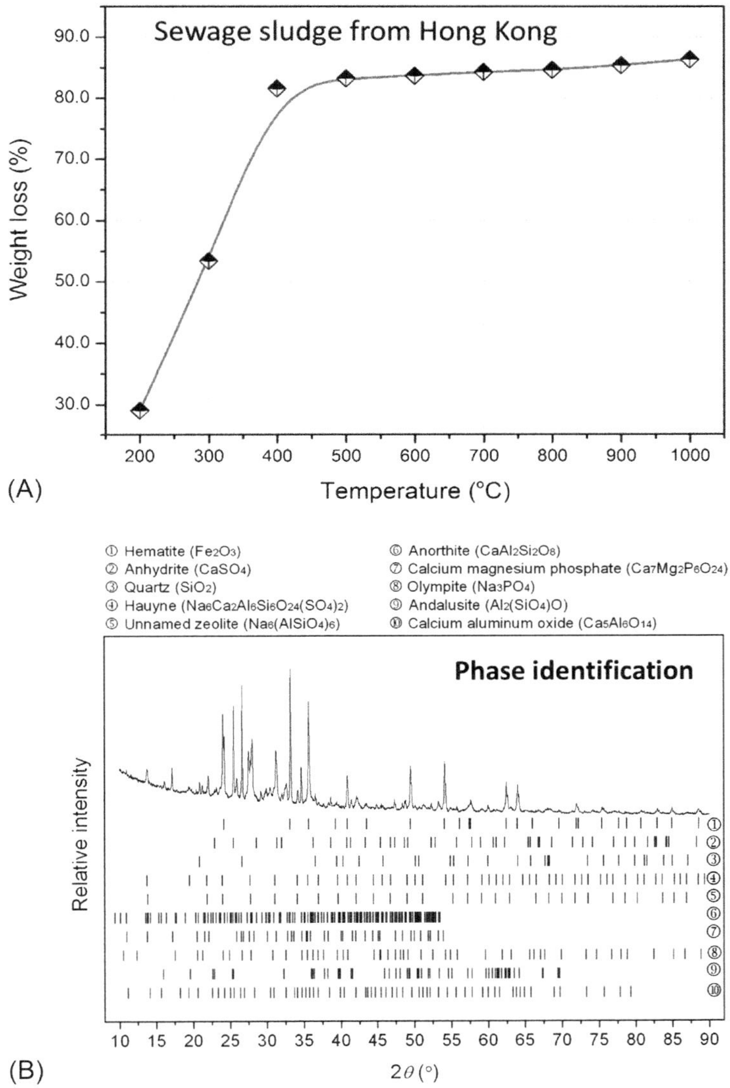

Fig. 20 (A) Weight losses of the dried sewage sludge collected from Hong Kong. (B) XRD pattern of sewage sludge calcined at 900°C for 30 min. The crystalline phases are identified as: (1) hematite (Fe_2O_3, PDF#85-0599); (2) anhydrite ($CaSO_4$, PDF#37-1496); (3) quartz (SiO_2, PDF#79-1910); (4) hauyne ($Na_6Ca_2Al_6Si_6O_{24}(SO_4)_2$, PDF#73-1920); (5) unnamed zeolite ($Na_6(AlSiO_4)_6$, PDF#42-0217); (6) anorthite ($Ca(Al_2Si_2O_8)$, PDF#89-1473); (7) calcium magnesium phosphate ($Ca_7Mg_2P_6O_{24}$, PDF#20-0348); (8) olympite (Na_3PO_4, PDF#33-1272); (9) andalusite ($Al_2(SiO_4)O$, PDF#39-0376); and (10) calcium aluminum oxide ($Ca_5Al_6O_{14}$, PDF#11-0357).

Table 2 Major compositions of the incinerated sewage sludge ashes

Composition	Fe_2O_3	CaO	SiO_2	P_2O_5	SO_4	Al_2O_3	K_2O	Others
Weight percentage (%)	18.62	14.57	24.35	8.78	10.52	11.43	1.22	10.51

via X-ray fluorescence (XRF) (JSX-3201Z, JEOL, Tokyo, Japan) and normalized by their oxide forms (Table 3). From the results, the fly ash can be classified as Class F by the American Society for Testing and Materials (ASTMs), with a combined SiO_2, Al_2O_3, and Fe_2O_3 content greater than 70%. The particle size distribution of the collected fly ash was measured by the laser scattering technique (Mastersizer 3000, Malvern), with the value of $D_v(10)$, $D_v(50)$, $D_v(90)$ as 37.5, 62.2, 91.1 μm, respectively (Fig. 22A). From the XRD pattern shown in Fig. 22B, quartz is one of the predominant crystal components contained in the coal fly ash. Moreover, mullite and hematite can also be clearly detected with relatively high peak intensities, which is consistent with the XRF results that aluminum, silicon, and iron are dominant elements contained in coal fly ash. Therefore, with the existence of aluminum- and iron-containing components, coal fly ash can be potentially used for copper stabilization.

5.2.2 Beneficial use of coal fly ash for copper stabilization

Fig. 23 exhibits XRD patterns of the copper-adsorbed fly ash before and after thermal treatment at temperatures ranging from 900°C to 1050°C. From the XRD pattern of the unsintered copper-adsorbed fly ash, the main copper-hosting phase was found to be gerhardtite ($Cu_4(NO_3)_2(OH)_6$; PDF#77-0148). When samples were sintered at 900°C for 3 h, the phase identification results showed that copper oxide (CuO; PDF#80-0076) was generated from gerhardtite and became the predominant copper-hosting phase. Besides the formation of CuO, copper iron oxide ($CuFe_2O_4$; PDF#77-0010) was also detected as another copper-hosting crystalline phase at 900°C. The generation of $CuFe_2O_4$ might be attributed to interactions between copper ions adsorbed on fly ash and iron ions originally contained in the fly ash. Reactions between copper and iron ions have also been reported previously in a solid CuO-Fe_2O_3 system followed by the formation of $CuFe_2O_4$ with a spinel structure (Tang et al., 2011). The copper leachability from the spinel decreased over two orders in comparison with that from copper oxides in acetic acid solutions (Tang et al., 2011), which confirmed the effective immobilization of copper ions in copper-adsorbed fly ash after sintering. With a further increase of the sintering temperature, the peaks of $CuFe_2O_4$ grew up eventually while the CuO signals reduced accordingly. At the highest temperatures (1050°C) of the sintering scheme in this experiment, peak intensities of the $CuFe_2O_4$ reached the maximum value while no further diffraction was observed from CuO.

Besides the crystallization of the $CuFe_2O_4$ spinel from reactions between copper and iron ions, aluminum-containing oxides such as kaolinite and alumina (Al_2O_3)

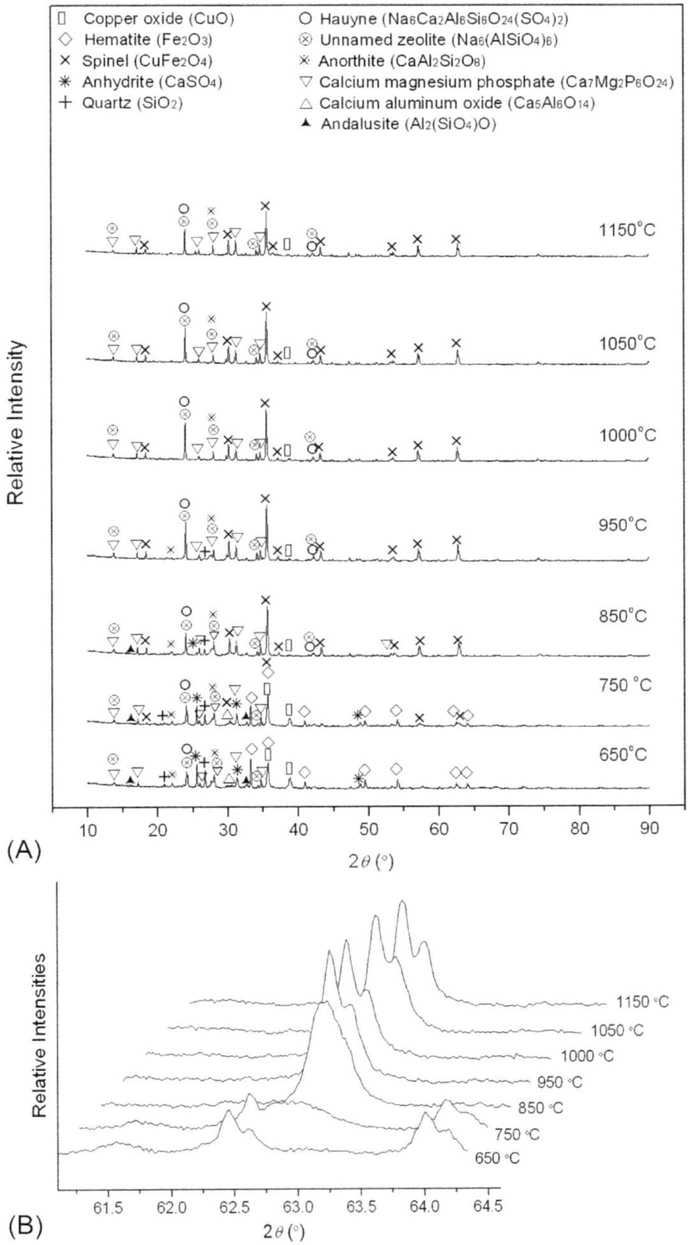

Fig. 21 (A) XRD patterns and (B) comparison of the $CuFe_2O_4$ spinel phase by XRD patterns between $2\theta = 61.5$ and 64.5 degrees of the temperature ranging from 650°C to 1150°C for sewage sludge ash + CuO sintered at 650–1150°C for 3 h. The crystalline phases were identified as: CuO (PDF#80-1268), Fe_2O_3 (PDF#85-0599), $CuFe_2O_4$ (PDF#77-0010), anhydrite ($CaSO_4$, PDF#37-1496), quartz (PDF#79-1910), hauyne ($Na_6Ca_2Al_6Si_6O_{24}(SO_4)_2$, PDF#73-1920), unnamed zeolite ($Na_6(AlSiO_4)_6$, PDF#42-0217), anorthite ($Ca(Al_2Si_2O_8)$, PDF#89-1473), calcium magnesium phosphate ($Ca_7Mg_2P_6O_{24}$, PDF#20-0348), calcium aluminum oxide ($Ca_5Al_6O_{14}$, PDF#11-0357), and andalusite ($Al_2(SiO_4)O$, PDF#39-0376).

Table 3 Chemical compositions (%) of the coal fly ash

Components	Weight percentage (wt%)
SiO_2	41.6
Fe_2O_3	17.3
Al_2O_3	16.3
CaO	15.1
MgO	3.9
P_2O_5	1.6
K_2O	1.4
SO_3	1.3
TiO_2	1.0

have also been reported as effective precursors to incorporate copper in the $CuAl_2O_4$ spinel (Tang et al., 2010). However, although aluminum was measured as another major element in fly ash, there was no $CuAl_2O_4$ detected by the XRD analysis. From the XRD patterns shown in Fig. 22B, aluminum exists in mullite ($3Al_2O_3{\cdot}2SiO_2$, PDF#79-1455) and other poor crystallized phases in the raw fly ash. However, the major aluminum-containing crystalline phase was identified as anorthite ($Ca(Al_2Si_2O_8)$; PDF#89-1462) when the copper-adsorbed fly ash was sintered at temperatures higher than 950°C. When the silicon-containing phase(s) were taken into consideration, the amount of quartz (SiO_2; PDF#89-1961) decreased significantly at temperatures higher than 950°C, which is attributed to the formation of anorthite as followed. The generation of anorthite indicates that aluminum ions were more likely to react with quartz rather than to compete with iron ions for the spinel structure. A further study was carried out to explore the detailed stabilization mechanisms of copper ions in aluminum and iron coexisting ceramic matrices such as the incinerated sewage sludge ash and coal fly ash.

5.3 Mechanisms of copper stabilization in aluminum- and iron-rich ceramic matrices

The immobilization mechanisms of copper ions in both aluminum- and iron- containing matrices were further studied in systems of $CuO+Fe_2O_3$+kaolinite and $CuO+Fe_2O_3+Al_2O_3$ with a molar ratio of Cu:Fe:Al as 1:2:2 for both mixtures, through a sintering scheme at 650–1250°C with a dwelling time of 3h. Fig. 24 collates the XRD patterns of both $CuO+Fe_2O_3$+kaolinite and $CuO+Fe_2O_3+Al_2O_3$ series before and after thermal treatment. In Fig. 24A, a new phase was observed with the same XRD patterns as spinel crystals when the mixture of $CuO+Fe_2O_3$+kaolinite was sintered at 850°C for 3h. Because the spinel formation might be due to potential reactions between the copper, iron, and aluminum contained in the system, a further peak comparison was carried out among $CuAl_2O_4$ (PDF#78-1605), $CuFe_2O_4$ (PDF#77-0010), and the as-formed spinel phase in this experiment. Results found that the peaks of the spinel phase in this experiment are located between the corresponding

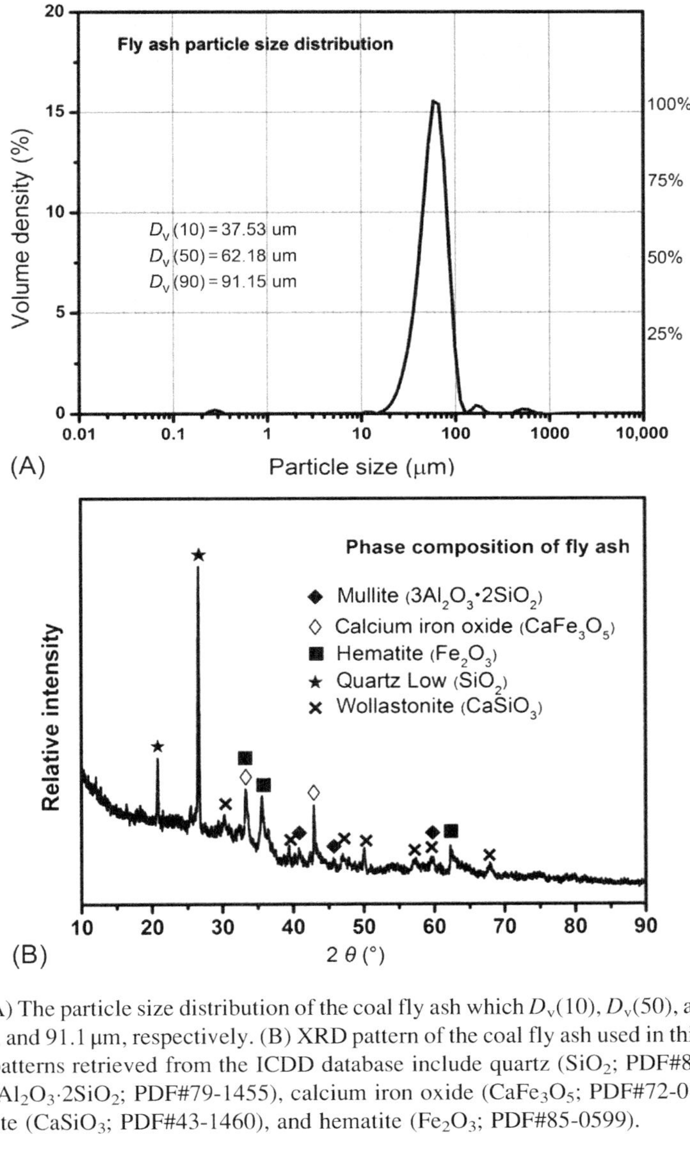

Fig. 22 (A) The particle size distribution of the coal fly ash which $D_v(10)$, $D_v(50)$, and $D_v(90)$ is 37.5, 62.2, and 91.1 μm, respectively. (B) XRD pattern of the coal fly ash used in this study. The standard patterns retrieved from the ICDD database include quartz (SiO_2; PDF#89-1961), mullite ($3Al_2O_3 \cdot 2SiO_2$; PDF#79-1455), calcium iron oxide ($CaFe_3O_5$; PDF#72-0890), wollastonite ($CaSiO_3$; PDF#43-1460), and hematite (Fe_2O_3; PDF#85-0599).

peaks of both $CuAl_2O_4$ and $CuFe_2O_4$ spinels, indicating the formation of a $CuAl_xFe_{2-x}O_4$ spinel solid solution. Peaks of the spinel solid solution kept increasing with the elevated temperatures until reaching the maximum value at 1150°C. With a further increase of the sintering temperature afterwards, the peak intensities of the $CuAl_xFe_{2-x}O_4$ spinel decreased due to the decomposition caused by the reduction of Cu(II) to Cu(I) at higher temperatures (>1150°C) (Andersson et al., 2011; Hanaor et al., 2014). In Fig. 24B, a similar spinel solid solution was also generated but at a lower temperature (750°C)

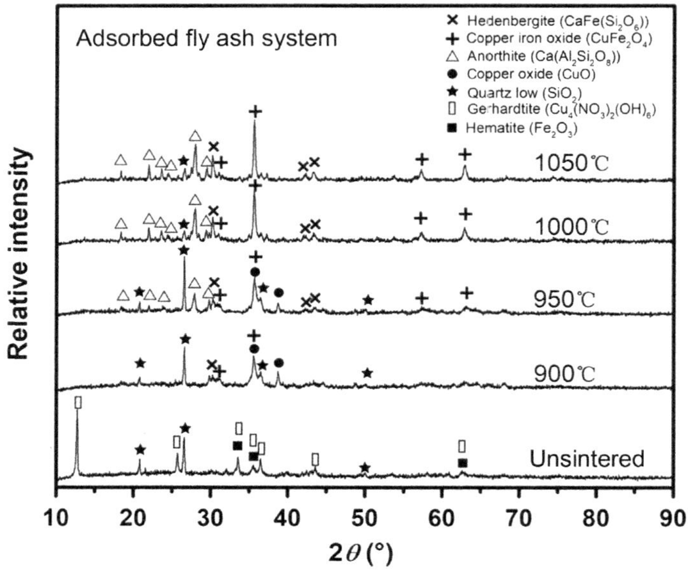

Fig. 23 XRD patterns of copper-absorbed fly ash before and after thermally treated at temperatures ranging from 900°C to 1050°C for 3h. The standard patterns retrieved from the ICDD database include quartz (SiO_2; PDF#89-1961), anorthite ($Ca(Al_2Si_2O_8)$; PDF#89-1462), hedenbergite ($CaFe(Si_2O_6)$; PDF#87-1695), copper oxide (CuO; PDF#80-0076), copper iron oxide ($CuFe_2O_4$; PDF#77-0010), and gerhardtite ($Cu_4(NO_3)_2(OH)_6$; PDF#77-0148).

in the thermally sintered $CuO+Al_2O_3+Fe_2O_3$ series. The spinel peaks also grew up continuously with the elevated sintering temperature, but peak shrinkage was not clearly detected at higher temperatures (>1150°C). Meanwhile, the signals of reactants such as CuO, Fe_2O_3, and aluminum-containing compounds were diminished eventually with the continuous formation of the $CuAl_xFe_{2-x}O_4$ spinel.

Fig. 24 also illustrates the amplification and shifting of the peaks of $CuAl_xFe_{2-x}O_4$ generated in the $CuO+Fe_2O_3$+kaolinite and $CuO+Al_2O_3+Fe_2O_3$ series, respectively. The peak intensity increased dramatically when the temperature rose from 850°C to 1050°C, and meanwhile, an obvious shifting of the peak was detected from lower to higher 2θ positions. The peak reached the maximum value at 1150°C and decreased at a higher temperature (1250°C), with a slight change in the 2θ position. In the $CuO+Al_2O_3+Fe_2O_3$ series, the peak intensity also increased significantly when the sintering temperature rose from 850°C to 950°C but kept a relatively steady increase afterward with the elevated temperature until reaching the maximum value at 1150°C. Although not obviously detected, a slight decrease in peak intensity was also found with a further 100°C increase in the sintering temperature to 1250°C. However, unlike the variation of peak positions in $CuO+Fe_2O_3$+kaolinite systems, the peak position of $CuAl_xFe_{2-x}O_4$ in the $CuO+Al_2O_3+Fe_2O_3$ series kept a steady increase from lower to higher 2θ value. According to Bragg's equation, the 2θ positions of the diffraction peaks are related to the interplanar spacing (d) of a

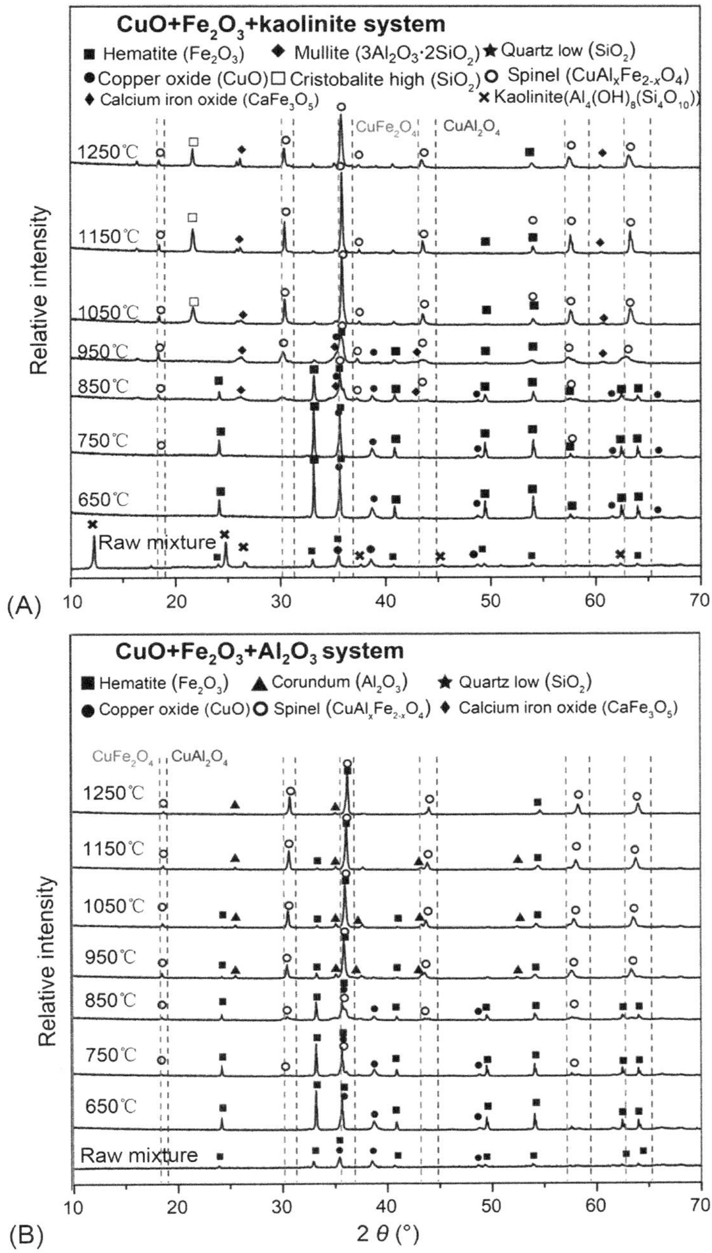

Fig. 24 XRD patterns of sintered (A) $CuO + Fe_2O_3$ + kaolinite system and (B) $CuO + Fe_2O_3 + Al_2O_3$ system show the thermal incorporation of copper at temperatures ranging from 650°C to 1250°C for 3h. The crystalline phases are identified as: copper oxide (CuO; PDF#80-0076), corundum (α-Al_2O_3; PDF#10-1073), hematite (Fe_2O_3; PDF#85-0599), spinel ($CuAl_xFe_{2-x}O_4$), mullite ($3Al_2O_3{\cdot}2SiO_2$; PDF#79-1455), and cristobalite (SiO_2; PDF#85-0621).

crystal structure. Because the radius of aluminum atoms is smaller than that of iron atoms, the diffraction peaks of the $CuAl_2O_4$ spinel locate at a higher 2θ value when compared with peaks from the same lattice plane in the $CuFe_2O_4$ crystals. Therefore, the peak shifting of $CuAl_xFe_{2-x}O_4$ is due to the shrinkage of distances between lattice planes, which is caused by the continuous enlargement of aluminum in the formation of the spinel solid solution. The spinel peak shifting from a lower to higher 2θ value indicates that iron ions are easier to be engaged in the formation of the spinel phase at lower temperatures but were outcompeted by aluminum ions at higher temperatures. Previous studies have also reported a lower sintering temperature for ferrite spinel when compared to aluminate spinel. The tendency of peak shifting with temperature increasing was further analyzed and the results were illustrated in Fig. 24. A perfect linear relationship was found between the 2θ value of the peak position and the sintering temperature in the $CuO+Al_2O_3+Fe_2O_3$ series, which can be explained by the continuous and steady engagement of aluminum ions in the formation of the $CuAl_xFe_{2-x}O_4$ spinel with elevated sintering temperatures. However, aluminum ions in $CuO+Fe_2O_3$+kaolinite were involved in mullite formation, as reflected from the increase of mullite peaks at temperatures higher than 850°C. Therefore, there are no obvious changes observed in the 2θ positions of the $CuAl_xFe_{2-x}O_4$ spinel. An overall higher 2θ value of the spinel peaks was found in the $CuO+Al_2O_3+Fe_2O_3$ series when compared with the $CuO+Fe_2O_3$+kaolinite systems, which shows a stronger competition between aluminum and iron in ceramic matrices without silicon.

6 Conclusion

This study reported the potential of blending industrial sludge for ceramic products, and further found a benefit for metal stabilization in the ceramic matrices. Ceramic products were obtained by sintering the copper-electroplating sludge or chromium ore processing residue with aluminum-rich precursors. The formation of a copper aluminate spinel ($CuAl_2O_4$) was observed when the copper-electroplating sludge was sintered with the alumina precursor. Moreover, the most critical crystalline phase for chromium immobilization was detected to be a spinel solid solution with the formula $MgCr_{1.32}Fe_{0.19}Al_{0.49}O_4$. The metal reaction pathways and the incorporation efficiencies for copper, zinc, and chromium were quantitatively provided by the application of the Rietveld refinement method. Results show that the maximum transformation ratio of copper reached 80% at temperatures between 850°C and 1000°C, while the partition ratio of chromium reached its maximum value of approximately 90% when the largest proportion of Cr_2O_3 (10%) was added. Furthermore, the potential to use coal fly ash and incinerated sewage sludge ash for copper stabilization was further studied, and the results show that copper can be well stabilized in the ceramic products from both aluminum- and iron-rich systems such as coal fly ash and incinerated sewage sludge ash. The mechanism study further identified the formation of the $CuAl_xFe_{2-x}O_4$ spinel solid solution as a critical phase for copper stabilization during the reaction processes.

References

Andersson, K.I., Eriksson, M., Norgren, M., 2011. Removal of lignin from wastewater generated by mechanical pulping using activated charcoal and fly ash: adsorption isotherms and thermodynamics. Ind. Eng. Chem. Res. 50 (13), 7722–7732.

Binner, J.G.P., 1990. Advanced Ceramic Processing and Technology. William Andrew Publishing/Noyes, Nottingham, UK.

Chen, Q.Y., Zhang, L.N., Ke, Y.J., Hills, C., Kang, Y.M., 2009. Influence of carbonation on the acid neutralization capacity of cements and cement-solidified/stabilized electroplating sludge. Chemosphere 74 (6), 758–764.

Cheng, S., 2003. Heavy metal pollution in China: origin, pattern and control. Environ. Sci. Pollut. Res. Int. 10 (3), 192–198.

Cheremisinoff, N.P., 2002. Handbook of Water and Wastewater Treatment Technologies. Butterworth-Heinemann, Boston.

Dermatas, D., Meng, X.G., 2003. Utilization of fly ash for stabilization/solidification of heavy metal contaminated soils. Eng. Geol. 70 (3–4), 377–394.

Edén, M., Sundberg, P., Stålhandske, C., 2011. The split network analysis for exploring composition–structure correlations in multi-component glasses: II. Multinuclear NMR studies of alumino-borosilicates and glass-wool fibers. J. Non-Cryst. Solids 357, 1587–1594.

Gao, Y., Xia, J., 2011. Chromium contamination accident in China: viewing environment policy of China. Environ. Sci. Technol. 45 (20), 8605–8606.

Hanaor, D.A.H., Ghadiri, M., Chrzanowski, W., Gan, Y., 2014. Scalable surface area characterization by electrokinetic analysis of complex anion adsorption. Langmuir 30, 15143–15152.

Hillier, S., Roe, M.J., Geelhoed, J.S., Fraser, A.R., Farmer, J.G., Paterson, E., 2003. Role of quantitative mineralogical analysis in the investigation of sites contaminated by chromite ore processing residue. Sci. Total Environ. 308, 195–210.

Huggins, F.E., Najih, M., Huffman, G.P., 1999. Direct speciation of chromium in coal combustion by-products by X-ray absorption fine-structure spectroscopy. Fuel 78, 233–242.

Jacob, K.T., Alcock, C.B., 1975. Thermodynamics of $CuAlO_2$ and $CuAl_2O_4$ and phase-equilibria in system Cu_2O-CuO-Al_2O_3. J. Am. Ceram. Soc. 58 (5–6), 192–195.

Jantzen, C.M., Brown, K.G., 2007. Predicting the spinel–nepheline liquidus for application to nuclear waste glass processing. Part I: primary phase analysis, liquidus measurement, and quasicrystalline approach. J. Am. Ceram. Soc. 90 (6), 1866–1879.

Juhin, A., Calas, G., Cabaret, D., Galoisy, L., Hazemann, J.L., 2007. Structural relaxation around substitutional Cr^{3+} in $MgAl_2O_4$. Phys. Rev. B. 76(5)054105.

Karamalidis, A.K., Voudrias, E.A., 2009. Leaching and immobilization behavior of Zn and Cr from cement-based stabilization/solidification of ash produced from incineration of refinery oily sludge. Environ. Eng. Sci. 26 (1), 81–96.

Katsou, E., Malamis, S., Haralambous, K.J., 2011. Industrial wastewater pre-treatment for heavy metal reduction by employing a sorbent-assisted ultrafiltration system. Chemosphere 82 (4), 557–564.

Kosson, D.S., van der Sloot, H.A., Sanchez, F., Garrabrants, A.C., 2002. An integrated framework for evaluating leaching in waste management and utilization of secondary materials. Environ. Eng. Sci. 19 (3), 159–204.

Leturcq, G., Berger, G., Advocat, T., Vernaz, E., 1999. Initial and long-term dissolution rates of aluminosilicate glasses enriched with Ti, Zr and Nd. Chem. Geol. 160, 39–62.

Li, X.D., Poon, C.S., Sun, H., Lo, I.M.C., Kirk, D.W., 2001. Heavy metal speciation and leaching behaviors in cement based solidified/stabilized waste materials. J. Hazard. Mater. 82 (3), 215–230.

Lin, Y.C., Wang, S.L., 2012. Chromium(VI) reactions of polysaccharide biopolymers. Chem. Eng. J. 181, 479–485.

Malviya, R., Chaudhary, R., 2006. Factors affecting hazardous waste solidification/stabilization: a review. J. Hazard. Mater. 137 (1), 267–276.

Mao, L., Gao, B., Deng, N., Zhai, J., Zhao, Y., Li, Q., Cui, H., 2015. The role of temperature on Cr(VI) formation and reduction during heating of chromium-containing sludge in the presence of CaO. Chemosphere 138, 197–204.

McHale, J.M., Navrotsky, A., Perrotta, A.J., 1997. Effects of increased surface area and chemisorbed H_2O on the relative stability of nanocrystalline gamma-Al_2O_3 and alpha-Al_2O_3. J. Phys. Chem. B 101 (4), 603–613.

Naja, G.M., Volesky, B., 2009. Treatment of metal-bearing effluents: removal and recovery. In: Wang, L.K., Chen, J.P., Hung, Y.T., Shammas, N.K. (Eds.), Handbook on Heavy Metals in the Environment. Taylor & Francis and CRC Press, Boca Raton, FL, pp. 247–291.

Ndiba, P.K., Axe, L., 2009. Sequential extraction of phosphate- and thermal-treated New York/New Jersey Harbor dredged sediments. Environ. Eng. Sci. 26 (12), 1755–1764.

O'Mara, W.C., Herring, R.B., Hunt, L.P., 1990. Handbook of Semiconductor Silicon Technology. Noyes Publications, Park Ridge, NJ.

Okuno, N., Ishikawa, Y., Shimizu, A., Yoshida, M., 2004. Utilization of sludge in building material. Water Sci. Technol. 49 (10), 225–232.

Park, Y.J., Heo, J., 2002. Vitrification of fly ash from municipal solid waste incinerator. J. Hazard. Mater. 91 (1–3), 83–93.

Park, D., Lee, D.S., Park, J.M., Chun, H.D., Park, S.K., Jitsuhara, I., Miki, O., Kato, T., 2005. Metal recovery from electroplating wastewater using acidophilic iron oxidizing bacteria: pilot-scale feasibility test. Ind. Eng. Chem. Res. 44 (6), 1854–1859.

Reed, J.R., 1995. Principles of Ceramics Processing, second ed. John Wiley & Sons, New York.

Rice, R.W., 1990. Ceramic processing—an overview. AICHE J. 36 (4), 481–510.

Sarikaya, Y., Ada, K., Onal, M., 2008. A model for initial-stage sintering thermodynamics of an alumina powder. Powder Technol. 188 (1), 9–12.

Scarinci, G., Brusatin, G., Barbieri, L., Corradi, A., Lancellotti, I., Colombo, P., Hreglich, S., Dall'Igna, R., 2000. Vitrification of industrial and natural wastes with production of glass fibres. J. Eur. Ceram. Soc. 20 (14–15), 2485–2490.

Shih, K., Leckie, J.O., 2007. Nickel aluminate spinel formation during sintering of simulated Ni-laden sludge and kaolinite. J. Eur. Ceram. Soc. 27 (1), 91–99.

Shih, K., White, T., Leckie, J.O., 2006a. Nickel stabilization efficiency of aluminate and ferrite spinels and their leaching behavior. Environ. Sci. Technol. 40 (17), 5520–5526.

Shih, K., White, T., Leckie, J.O., 2006b. Spinel formation for stabilizing simulated nickel-laden sludge with aluminum-rich ceramic precursors. Environ. Sci. Technol. 40 (16), 5077–5083.

Sophia, A.C., Swaminathan, K., 2005. Assessment of the mechanical stability and chemical leachability of immobilized electroplating waste. Chemosphere 58 (1), 75–82.

Tang, Y., Shih, K., Chan, K., 2010. Copper aluminate spinel in the stabilization and detoxification of simulated copper-laden sludge. Chemosphere 80 (4), 375–380.

Tang, Y., Chui, S.S.Y., Shih, K., Zhang, L., 2011. Copper stabilization via spinel ormation during the sintering of simulated copper-laden sludge with aluminum-rich ceramic precursors. Environ. Sci. Technol. 45, 7609–7610.

Tathavakar, V.D., Antony, M.P., Jha, A., 2005. The physical chemistry of thermal decomposition of South African chromite minerals. Metall. Mater. Trans. B 36 (1), 75–84.
Tsai, C.C., Wang, K.S., Chiou, I.J., 2006. Effect of SiO_2-Al_2O_3-flux ratio change on the bloating characteristics of lightweight aggregate material produced from recycled sewage sludge. J. Hazard. Mater. 134 (1–3), 87–93.
United States Environmental Protection Agency (USEPA), 1992. EPA Method 1311—Toxicity Characteristic Leaching Procedure. United States Environmental Protection Agency.
United States Environmental Protection Agency (USEPA), 1998. Method 1311, Toxicity Characteristic Leaching Procedure, SW-846: Test Methods for Evaluating Solid Waste Physical/Chemical Methods. United States Environmental Protection Agency (Placeholder Text).
Veli, S., Alyuz, B., 2007. Adsorption of copper and zinc from aqueous solutions by using natural clay. J. Hazard. Mater. 149 (1), 226–233.
Walenta, G., Fullmann, T., 2004. Advances in quantitative XRD analysis for clinker, cements, and cementitious additions. Powder Diffract. 19 (1), 40–44.
Wang, Q., 2006. Industrial Solid Waste Disposal and Recycling. China Environmental Science Press, Beijing.
Wei, Y., Yang, Y., Cheng, N., 2001. Study of thermally immobilized Cu in analogue minerals of contaminated soils. Environ. Sci. Technol. 35 (2), 416–421.
Wei, Y.L., Chiu, S.Y., Tsai, H.N., Yang, Y.W., Lee, J.F., 2002. Thermal stabilization of chromium(VI) in kaolin. Environ. Sci. Technol. 36, 4633–4641.
Wei, Y.L., Hsieh, H.F., Peng, Y.S., Yang, J.C., Wang, H., Lin, C.Y., Shih, W.L., Hsu, C.C., 2010. Thermal detoxification and bloating of chromium(VI) with bentonite. Nucl. Instrum. Methods Phys. Res. A: Accel. Spectrom. Detect. Assoc. Equip. 619, 108–111.
Wolverton, C., Hass, K.C., 2001. Phase stability and structure of spinel-based transition aluminas. Phys. Rev. B. 63(2)024102.
Young, R.A., 1995. The Rietveld Method. International Union of Crystallography/Oxford University Press, Chester, England, Oxford/New York.
Zhou, Q., Milestone, N.B., Hayes, M., 2006. An alternative to Portland cement for waste encapsulation—the calcium sulfoaluminate cement system. J. Hazard. Mater. 136 (1), 120–129.

Detoxification of organic sludge from water-treatment plants by active forms of Ca and Si

13

Vladimir Matichenkov*, John Campbell†
*Department of Ecology and Physiology of Autotrophic Organisms, Institute Basic Biological Problems, Pushchino, Russia, †Everse Limited, Hamilton, New Zealand

1 Introduction

Modern living systems, which provide water treatment facilities, produce large quantities of sewage sludge around the world. The amount of sewage sludge produced increases in line with the growing human population and sewage treatment advances. For example, in China, 30 million metric tons of sludge (at a moisture content of 80%) were produced in 2012, and in 2015, that increased to 34 million metric tons (Feng et al., 2015). It is estimated that Europe and the United States produce approximately 10 million tons of dry solids a year^{-1} (Genedebien et al., 1999). However, the utilization of sewage sludge is critically low, even in developed areas such as the United States and the European Union.

Historically, more than 80% of the sludge has not been treated and disposed of effectively and safely. This poses a great threat to the environment, particularly because of the ubiquitous use of combined treatment systems for municipal wastewater, industrial wastewater, and rainwater treatment (Feng et al., 2015). In 2013, there were about 2600 sludge treatment plants in China, but only about 60 plants had adopted anaerobic digestion processes and just 10 – 30 of them were actually operating, resulting in wasted infrastructure and treatment facilities (Feng et al., 2015). In India, only about 30% of the wastewater generated from major cities is being processed (Kapshe et al., 2013). Sometimes raw sewage is directly discharged into natural waters. This happens most often in small and less-developed countries. For instance, almost 30% of the sewage produced in the Federated States of Micronesia goes into the waters of the Pacific Ocean without prior purification (Rouse, 2013).

A large part of sewage sludge is transported out of the cities, naturally composted, and used in agriculture (Ju et al., 2005). Treated sewage sludge has been used in agriculture in the UK, Europe, and China for more than 80 years. However, there is increasing pressure in some countries to stop the practice of land application due to soil contamination and public outrage. The restrictions on using sewage sludge for agricultural purposes is related to the following problems:

(1) Contamination (Shao et al., 2015), including Al, because it is commonly used for organic matter precipitation (Xu et al., 2009).

Industrial and Municipal Sludge. https://doi.org/10.1016/B978-0-12-815907-1.00013-1

(2) Pathogenic microorganisms (Wéry et al., 2008).
(3) Unpleasant odor.

Quicklime (QL) and calcium hydroxide (hydrated lime) have been used to treat biological organic wastes for more than 100 years. Treatment of human wastewater sludges (i.e., biosolids) with lime is specifically prescribed in Environmental Protection Agency (EPA) regulations. This treatment provides a reduction or elimination of pathogenic microorganisms in sewage sludge due to high temperatures and high pH values (Zdybel et al., 2015).

The coagulation-flocculation step in water treatment plants, which requires the addition of alum (Aluminum sulfate) as a chemical coagulant, is the critical treatment that is required to remove colloidal solids. Alum is cheap and highly effective in the coagulation process and used in many countries (Hall and Dietrich, 2000). Alum is the common name for aluminum sulfate $Al_2(SO_4)_3{*}xH_2O$. Alum has been used for drinking water treatment since 500 BCE. It not only settles certain hazardous chemicals and suspended solids but also bacterial colonies. The use of alum to remove suspended solids first occurred in Egypt. During the 13th to 15th centuries BCE, Egyptians developed a spectacular sedimentation apparatus that used alum for drinking water treatment. Naturally occurring alum was first recognized in the deserts of Egypt (Hall and Dietrich, 2000). The sewage sludge that was obtained after alum treatment contains high levels of soluble Al, which is problematic for the utilization of such sludge in agriculture.

Aluminum is the third most abundant and ubiquitously distributed metallic element in the Earth's crust, comprising 7% of its mass after oxygen and silicon. Aluminum is one of the integral components of mineral soil, in which it is usually incorporated in the precipitated form as gibbsite or the harmless aluminosilicates form (Singh et al., 2017). The inception of soil acidification ($pH < 5.0$) initiates Al solubilization and the formation of such forms of Al as monomeric Al^{3+}, which also formulates the mononuclear species such as $AlOH^{2+}$, $Al(OH)_2^+$, $Al(OH)_3$, and $Al(OH)_4$. However, among all forms of aluminium, Al^{3+} is considered the most toxic one (Schmitt et al., 2016; Silva, 2012). It has been estimated that more than 50% of the world's potentially arable lands are acidic (Kochian et al., 2005); hence, Al toxicity is a very important worldwide limitation to crop production. Aluminum toxicity restrains a broad range of physical, cellular, and molecular processes. Aluminum toxicity occurs due to the strong and rapid interactions of Al with apoplasmic and symplasmic targets. Imposition of Al instigates multifarious detrimental effects in plants. Some of these responses could be located within seconds to minutes while the effects of others are noticeable only at prolonged exposure. Therefore, it is really enigmatic to decide a definite time for Al toxicity (Singh et al., 2017). As a result, the application of the alum-treated sewage sludge may have a negative impact on the growing of the cultivated plants, despite the availability of the macro- and micronutrients in this product (Dassanayake et al., 2015; Matichenkov and Bocharnikova, 2016).

Active forms of Si, such as monosilicic acid (MA), can remarkably reduce the mobility and toxicity of Al and HMs (Exley et al., 2002; Matichenkov et al., 2013). The Si action usually relies on the formation of insoluble or slightly soluble silicates of HMs (Matichenkov and Bocharnikova, 2001). Monosilicic acid under different concentrations is able to combine with heavy metals (Cd, Pb, Zn, Hg, and others), forming soluble complex compounds if the monosilicic acid concentration is slight (Schindler et al., 1976) and slightly soluble heavy metal silicates when the concentration of monosilicic acid is greater in the system (Lindsay, 1979; Ji et al., 2016).

$$2Cd^{2+} + H_4SiO_4 = CdSiO_4 + 4H^+$$

$$2Pb^{2+} + H_4SiO_4 = PbSiO_4 + 4H^+$$

Monosilicic acid possesses high chemical activity (Iler, 1979; Lindsay, 1979). Monosilicic acid can react with aluminum with the formation of slightly soluble silicates (Lumsdon and Farmer, 2010):

$$2Al^{3+} + 2H_4SiO_4 = Al_2Si_2O_5 + 2H^+ + 3H_2O$$

$$2Al^{3+} + 2H_4SiO_4 + H_2O = Al_2Si_2O_5(OH)_4 + 6H^+$$

It is importhat that monosilicic acid mostly react with the most toxic form of Al.

We supposed that the sludge treatment by Ca and Si compounds could be used in technology development for the utilization of water treatment plant waste with final receiving environmentally friendly product.

The main aim of this study was to test the efficacy of sewage sludge treatment with QL and MA with regard to detoxification of pathogens and immobilization of Al and HMs.

2 Material and methods

Three different greenhouse experiments were conducted for verification of the suggested technology:

(a) Experiment with aluminum sulfate $Al_2(SO_4)_3$, monosilicic acid, and quick lime.
(b) Experiment with sewage sludge and barley.
(c) Experiment with sewage sludge and tomatoes.

2.1 Experiment with aluminum sulfate $Al_2(SO_4)_3$

The first greenhouse experiment was carried out in pots of 10-cm diameter and 10-cm depth. The treatments included:

1. Control.
2. Alum application 1 (AL-1).
3. Alum application 2 (AL-2).
4. MA 1 (MA-1).
5. MA 2 (MA-2).
6. Alum application 1 + MA 1 (AL-1 + MA-1).
7. Alum application 1 + MA 2 (AL-1 + MA-2).

8. Alum application 2 + MA 1 (AL-2 + MA-1).
9. Alum application 2 + MA 2 (AL-2 + MA-2).
10. QL 1 (QL-1).
11. Alum application 1 + QL 1 (AL-1 + QL-1).
12. Alum application 2 + QL 1 (AL-2 + QL-1).

Barley (*Hordium vulgare* L. vr. Moscow 3) was grown in washed quartz sand. Alum was applied at rates of 0.05 and 0.1 g kg^{-1} of soil, which are equal to 50 and 100 kg ha^{-1}, accordingly. This amount of alum is similar for treatment of the SS and then SS application at the 1 and 2 ton/hectare. The MA acid ("Accor" LSD, Moscow region, 15% of total Si) was applied at the same rate. The application of the MA was conducted by using the diluted solution (1:500). Considering the large volume of the diluted solution, the application of the MA was conducted by irrigation of the barley during growing. Industrial powdered QL ("Musson" LSD, Ufa, Russia; purity of about 90%) was applied at the rate of 0.1 g kg^{-1} of soil, which is equal to 100 kg ha^{-1}. Alum and QL were applied before barley seed planting. The air temperature in the growth chamber was kept at 20 ± 2°C during the day and 12 ± 2°C during the night. The light period was 12 h; the light intensity was 950 μmol photons m^{-2} s^{-1}. The relative air humidity was 45% ± 5% during the day and 70% ± 5% during the night. One-month-old barley plants were harvested and the biomass of shoots was measured. The total shoot content of Al was analyzed using microwave digestion with subsequent testing by an ICP OES spectrometer Perkin Elmer Optima 5300 DV. Soil samples were collected from pots and analyzed for (0.1 n HCl)-extractable Al and pH. The concentrations of Al in filtrated extracts were measured by the ICP OES spectrometer Perkin Elmer Optima 5300 DV. Each treatment had five replications.

2.2 Experiment with sewage sludge and barley

Sewage sludge (SS) from Serpukhov (southern Moscow region, Russia) was used in the technology demonstration. The sludge contained 67% of C (on dry weight), pH = 7.6, and moisture = 25%. This sludge was formed using the $AlSO_4$-based technique. The basic physical-chemical characteristics of the original SS are presented in Table 1.

The original SS was treated with QL at a ratio of 10:1 and 10:2 (SS:QL, accordingly) with intensive mixing for the first 10 min and then mixing every hour for the next 10 h. The treatment was conducted in a heat-resistant plastic box at room temperature, then the QL-treated SS was cooled. The industrial powdered QL ("Musson" LSD, Ufa, Russia) had a purity of about 90%.

Part of the QL-treated SS was mixed with concentrated MA ("Accor" LSD, Moscow region, 15% of total Si) at a ratio of 100:2 and 100:5 (QL-treated SS: MA). Monosilicic acid was sprayed on continuously, agitating the QL-treated SS in the same plastic box being used for QL treatment. The MA-QL-treated SS was kept for one week with periodic agitation. After that, samples of untreated, QL-treated, and QL + MA-treated SS were dried at +65°C for 48 h. The dried material was ground, passed through a 0.2 mm sieve, and used in a greenhouse test.

The contents of the pathogenic microorganisms and eggs of the intestinal worms were measured in all samples of sludge using the serological methods of diagnostics

Table 1 Physicochemical characteristics of SS (mean; $n = 5 \pm SD$)

Parameters	**SS**	**QL-1**	**QL-2**	**QL-1+MA-1**	**QL-1+MA-2**	**QL-2+MA-1**	**QL-2+MA-2**
pH	7.4 ± 0.2	8.2 ± 0.2	8.5 ± 0.2	8.1 ± 0.2	8.1 ± 0.2	8.4 ± 0.2	8.3 ± 0.2
EC (mS cm $^{-1}$)	4.33 ± 0.32	6.54 ± 0.30	8.33 ± 0.32	7.21 ± 0.31	8.24 ± 0.30	9.33 ± 0.33	10.04 ± 0.34
Total organic matter, g kg^{-1}	46.5 ± 1.23	40.2 ± 1.23	34.4 ± 1.23	40.1 ± 1.23	40.4 ± 1.23	33.2 ± 1.23	34.5 ± 1.23
Total N, %	4.85 ± 0.25	4.23 ± 0.25	3.84 ± 0.22	4.20 ± 0.22	4.15 ± 0.25	3.82 ± 0.20	3.75 ± 0.27
Total P, %	5.86 ± 0.32	5.02 ± 0.35	4.32 ± 0.29	5.11 ± 0.25	5.12 ± 0.25	4.75 ± 0.30	4.86 ± 0.28
Total K, %	2.43 ± 0.11	1.95 ± 0.11	1.48 ± 0.10	1.97 ± 0.11	1.95 ± 0.10	1.43 ± 0.12	1.46 ± 0.11
Total Ca, %	7.56 ± 0.24	9.21 ± 0.21	10.3 ± 0.20	9.11 ± 0.11	9.04 ± 0.23	10.2 ± 0.11	10.2 ± 0.21
Total Si, %	4.33 ± 0.12	4.11 ± 0.11	4.01 ± 0.13	5.43 ± 0.11	6.12 ± 0.13	5.44 ± 0.14	6.15 ± 0.15
Total Al, %	0.45 ± 0.03	0.42 ± 0.03	0.34 ± 0.02	0.43 ± 0.03	0.40 ± 0.02	0.31 ± 0.03	0.30 ± 0.04
Total Mg, ppm	244 ± 24	210 ± 24	185 ± 24	213 ± 24	218 ± 24	189 ± 24	194 ± 24
Total Cd, ppm	3.1 ± 0.1	2.8 ± 0.1	2.4 ± 0.1	2.7 ± 0.1	2.7 ± 0.1	2.1 ± 0.1	2.2 ± 0.1
Total Pb, ppm	2.0 ± 0.1	1.7 ± 0.1	1.4 ± 0.1	1.8 ± 0.1	1.7 ± 0.1	1.3 ± 0.1	1.3 ± 0.1

(Vainionpaa et al., 2008). The standard analysis on the pathogenic microorganisms and worm eggs was conducted in the certified sanitary laboratory in Serpukhov (Russia).

The greenhouse experiment was carried out in pots of 10-cm diameter and 10-cm depth. The treatments included:

1. Control—untreated SS.
2. SS:QL of 10:1 (QL-1).
3. SS:QL of 10:2 (QL-2).
4. QL-1:MA of 100:2 (QL-1+MA-1).
5. QL-2:MA of 100:2 (QL-2+MA-1).
6. QL-1:MA of 100:5 (QL-1+MA-2).
7. QL-2:MA of 100:5 (QL-2+MA-2).

Barley (*Hordium vulgare* L. vr. Moscow 3) was grown in washed quartz sand. Untreated or treated SS was applied at rates of 1 and 2 g kg^{-1} of soil, which are equal to 1 and 2 t ha^{-1}, accordingly. The air temperature in the growth chamber was kept at 20 ± 2°C during the day and 12 ± 2°C during the night. The light period was 12 h; the light intensity was 950 μmol photons m^{-2} s^{-1}. The relative air humidity was $45 \pm 5\%$ during the day and $70\% \pm 5\%$ during the night. One-month-old barley plants were harvested and the biomass of shoots was measured. The total shoot content of Al, Ca, Cd, K, Mg, P, Pb, and Si was analyzed using microwave digestion with subsequent testing by the ICP OES spectrometer Perkin Elmer Optima 5300 DV. Total C and N in untreated and treated SS and total N in plant samples were measured by the 2400 CHN Elemental Analyzer (Perkin Elmer).

Soil samples were collected from pots and analyzed for pH, total C, acid (0.1 n HCl)-extractable Al, Ca, Cd, K, Mg, P, Pb, and Si. The concentrations of Al, Ca, Cd, K, Mg, P, Pb, and Si were measured by the ICP OES spectrometer Perkin Elmer Optima 5300 DV. Each treatment had five replications.

2.3 Experiment with sewage sludge and tomatoes

A greenhouse experiment with tomatoes was carried out in large pots of 30-cm diameter and 20-cm depth, which give the possibility of measuring the yield of the plants. The treatments included:

1. Control.
2. Untreated SS.
3. QL-1.
4. QL-2.
5. SS:QL of 10:1 (SS+QL-1).
6. SS:QL of 10:2 (SS+QL-2).
7. MA-1.
8. MA-2.
9. SS+QL-1:MA of 100:2 (SS+QL-1+MA-1).
10. SS+QL-2:MA of 100:2 (SS+QL-2+MA-1).
11. SS+QL-1:MA of 100:5 (SS+QL-1+MA-2).
12. SS+QL-2:MA of 100:5 (SS+QL-2+MA-2).

Tomatoes (*Solánum lycopérsicum* vr.Avrora) were grown in muck soil (Fasco, Garden Retail Service, Khimki, Russia). Untreated or treated SS was applied at rates of 1 and 2 g liter^{-1} of soil, which are equal to 1 and 2 t ha^{-1}, accordingly. The MA acid ("Accor" LSD, Moscow region, 15% of total Si) was applied at the rate of 0.05 and 0.1 g kg^{-1} of soil, which are equal to 50 and 100 t ha^{-1}, accordingly. The application of the MA was conducted by using the diluted solution (1:500). Considering the large volume of the diluted solution, the application of the MA was conducted by irrigation of the barley during growing. An industrial-powered QL ("Musson" LSD, Ufa, Russia; purity of about 90%) was applied at the rate 0.1 g and 0.2 kg^{-1} of soil, which are equal 100 and 200 kg ha^{-1}. Solid materials were applied before planting the tomato seeds. The air temperature in the growth chamber was kept at 20 ± 2°C during the day and 12 ± 2°C during the night. The light period was 12 h; the light intensity was 950 μmol photons m^{-2} s^{-1}. The relative air humidity was 45% ± 5% during the day and 70% ± 5% during the night. Tomatoes were harvested during three months and the biomass-harvested vegetables were measured. The vitamin C was measured by Tillman's method (PN-A-04019, 1998), total sugars was measured by Schoorl-Rogenhausen's method (Rutkowska, 1981) and NO_3^- was determined by ionometric method (Twin NO_3^-, Horiba, Japan). The total content of Al, Ca, Cd, K, P, Pb, and Si was analyzed in the tomatoes after drying and by using microwave digestion with subsequent testing by the ICP OES spectrometer Perkin Elmer Optima 5300 DV.

Soil samples were collected from pots and analyzed for pH, acid (0.1 n HCl)-extractable Al, Ca, Cd, K, Mg, P, Pb, and Si. The concentrations of Al, Ca, Cd, K, Mg, P, Pb, and Si were measured by the ICP OES spectrometer Perkin Elmer Optima 5300 DV. Each treatment had five replications.

3 Results

3.1 Experiment with aluminum sulfate $Al_2(SO_4)_3$

The application of the aluminum sulfate significantly reduced the biomass of the shoots and roots of the barley (Fig. 1). The QL application had no significant influence on the barley biomass while using the MA-1 or MA-2 significantly increased the biomass of the shoots and roots of the barley. The maximum effect was obtained for MA-2, and the fresh biomass of the roots and shoots was increased from 0.40 to 0.53 g plant^{-1} and from 0.38 to 5.9 g plant^{-1}, accordingly.

The combination of the alum and MA results in a significant increase of the barley biomass, which has demonstrated that monosilicic acid can neutralize the Al toxicity.

Using the QL also reduced the Al toxicity. The weight of the barley root and shoot under AL-1 application was 0.25 and 0.32 g plant^{-1}; the application of the AL-1 with QL increased the fresh weight of the barley root and the shoot was 0.32 and 0.34 g plant^{-1} accordingly. The effect on the AL-2 was more dramatic. This effect is related with pH changing (Table 2). The application of the aluminum sulfate to soil significantly reduced the pH soil from 6.2 to 6.0 and 5.7, according for application AL-1 and AL-2. The application of just MA slightly increased the pH only for the

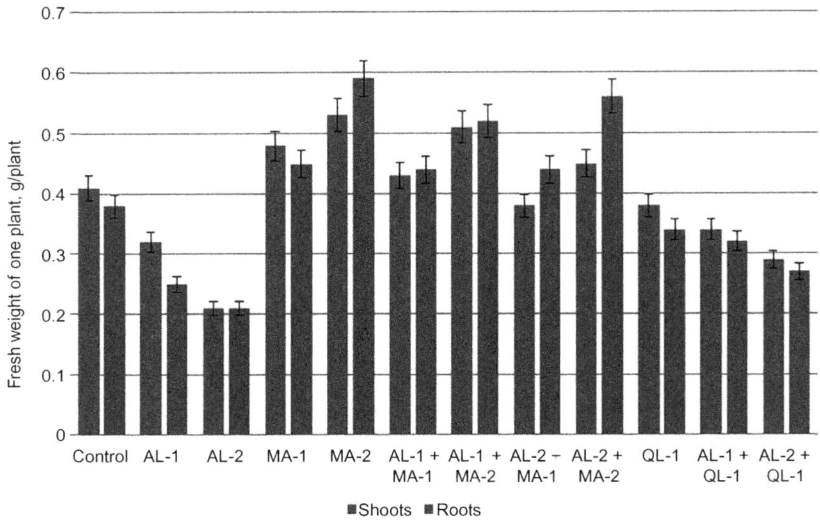

Fig. 1 Effect of alum (AL-1 and AL-2) and monosilicic acid (MA-1 and MA-2) on the fresh biomass of the barley.

Table 2 Effect of aluminum sulphate, MA, and QL on the soil pH, content of mobile Al in the soil, and total Al in the roots and shoots of the barley (mean; $n = 5 \pm SD$)

Treatments	**pH**	**Mobile Al in soil, mg kg^{-1}**	**Total Al in plants**	
			Roots	**Shoots**
			%	
Control	6.2 ± 0.1	0.1 ± 0.1	0.01 ± 0.01	0.01 ± 0.01
AL-1	6.0 ± 0.1	45.3 ± 5.1	0.29 ± 0.03	0.17 ± 0.03
AL-2	5.7 ± 0.1	89.4 ± 9.0	0.54 ± 0.05	0.23 ± 0.04
MA-1	6.1 ± 0.1	0.1 ± 0.1	0.01 ± 0.01	0.01 ± 0.01
MA-2	6.4 ± 0.1	0.1 ± 0.1	0.01 ± 0.01	0.01 ± 0.01
AL-1 + MA-1	6.0 ± 0.1	12.5 ± 0.3	0.05 ± 0.01	0.02 ± 0.01
AL-1 + MA-2	6.1 ± 0.1	8.5 ± 0.4	0.03 ± 0.01	0.01 ± 0.01
AL-2 + MA-1	6.0 ± 0.1	21.4 ± 0.3	0.07 ± 0.01	0.03 ± 0.01
AL-2 + MA-2	6.2 ± 0.1	14.2 ± 0.2	0.03 ± 0.01	0.02 ± 0.01
QL-1	8.2 ± 0.1	0.1 ± 0.1	0.01 ± 0.01	0.01 ± 0.01
AL-1 + QL-1	8.1 ± 0.1	32.8 ± 0.3	0.015 ± 0.03	0.11 ± 0.02
AL-2 + QL-1	7.9 ± 0.1	48.9 ± 0.6	0.21 ± 0.03	0.17 ± 0.03

MA-2 application. This can be related with the chemistry of the monosilicic acid, which has n equilibrated pH of about 7 (Iler, 1979):

$$H_4SiO_4 + OH^- = H_3SiO^- + H_20$$

The combination of the AL-1 or AL-2 with MA-1 or MA-3 results in slightly increasing the pH compared with only alum application. The application of the QL significantly increased the pH level up to 8.2. The soil treated by AL-1 and AL-2 with QL also is characterized by high pH level–0.1 and 7.9 for AL-1+QL-1 and AL-2+QL-1, respectively.

The original soil (sand) had no Al. The total content of Al in the barley roots and shoots also was not detected in plants from these plots. The alum application dramatically increased the content of mobile Al up to 45.3 ± 0.1 and 89.4 ± 0.1 mg kg^{-1} accordingly for AL-1 and AL-2 application. The application of the MA together with AL significantly reduced the mobile Al in the soil and the total content of Al in the barley roots and barley. The combined application of the alum and QL also reduced the Al mobility in the soil from 45.1 ± 5.1 to 32.8 ± 0.3 mg kg^{-1} and from 89.4 ± 9.0 to 48.9 ± 0.6 mg kg^{-1} accordingly for AL-1+QL-1 and AL-2+QL-1.

3.2 Experiment with sewage sludge and barley

The basic physical-chemical characteristics of untreated and treated SS are presented in Table 1. Quicklime provided a significant increase in pH from 7.4 in untreated SS to 8.2 and 8.5 in QL-1 and QL-2, accordingly. Monosilicic acid reduced the pH of QL-treated SS, probably due to the feature of MA usually having an equilibrium pH of about 7 (Iler, 1979). Although the concentrated MA is stabilized by NaOH and has a high pH, the pH of final diluted (1:500 and more) water solution can be stabilized with a pH of about 7.

The electrical conductivity of SS was increased as a result of QL treatment and then was increased more after mixing with MA. A higher rise in the electrical conductivity was detected in QL-2+MA-2.

In QL-treated SS, the total content of organic matter and elements, except Ca, significantly declined due to the dilution process. Monosilicic acid also reduced the content of organic matter and all tested elements, except Si.

Pathogen contamination of SS can be eliminated by treatment with QL in two ways: pH increase and temperature rise. When QL is used with water, an exothermic reaction occurs that releases heat. As a result, the temperature can increase up to 70°C, thereby providing effective pasteurization (Lopes et al., 2015). The data obtained showed that QL added at low and high rates eliminated the eggs of helminths and most of the tested pathogens (Table 2). However, QL-1 still contained some pathogens such as *Intestine baciluus*, *Trichomoniasis*, and *Clostridium perfringos*. The same pathogens were observed after MA treatment at a low rate, whereas a high rate of MA resulted in elimination of these pathogens. By this means, the combination of QL and MA showed higher efficacy of the pathogenic microorganism elimination than QL alone. Sewage sludge treatment by QL at a high rate completely eliminated colonies of pathogenic microorganisms.

Table 3 Availability of pathogenic microorganism colonies in the untreated and treated municipal sludge

Parameters	SS	QL-1	QL-2	QL-1 +MA-1	QL-1 +MA-2	QL-2 +MA-1	QL-2 +MA-2
Eggs of helminth	+++	–	–	–	–	–	–
Intestine bacillus	+++	+		+	–	–	–
Streptococcus	++	–	–	–	–	–	–
Micrococcus	+++	–	–	–	–	–	–
Diplococcic	+++	–	–	–	–	–	–
Staphylococci	++	–	–	–	–	–	–
Trichomoniasis	+++	+	–	+	–	–	–
Intestine amoeba	++	–	–	–	–	–	–
Clostridium perfringos	+++	+	–	+	–	–	–

+++, many colonies of microorganisms; ++, medium colonies of microorganisms.
+, low colonies; –, no colonies.

Greenhouse tests showed that untreated SS had no significant effect on the biomass of one-month-old barley shoots (Tables 3 and 4). However, some reduction in the plant biomass was observed. We suggest that this is related with the high level of Al in SS. Our previous tests have demonstrated the same negative influence of Al (Matichenkov et al., 2013). A negative influence of Al on the plant growth has been reported in some literature (Singh et al., 2017). The application of only QL- and QL+MA-treated SS increased the plant biomass significantly up to 60% and 95%, accordingly, at low and high rates of QL+MA-treated SS.

Total Al in the plant tissue was increased from 0.01% to 0.35% under the application of untreated SS. Quicklime- and MA-treated SS markedly reduced the total plant Al and a level of the reduction was positively correlated with rates of QL and MA. The action of QL and MA treatments is based on increasing soil pH by QL and on the neutralization of Al by MA:

$$2Al^{3+} + 2H_4SiO_4 = Al_2Si_2O_5 \downarrow + 2H^+ + 3H_2O$$

So, QL+MA treatment provided a reduction in the Al accumulation by barley and mitigated the Al toxic effect on plant growth.

Total Ca in the plant tissue was considerably increased under applications of all tested materials. Under application of QL-treated SS, the plants contained a maximum Ca of 0.57 and 0.65%, accordingly, at QL-1 and QL-2 application. The SS treatment with QL+MA slightly reduced the total Ca in the barley shoots compared to the plants grown with QL-treated SS. This fact perhaps is related to the MA impact on the absorption and redistribution of elements in plants (Snyder et al., 2006).

Total P in the plant tissue was remarkably increased by untreated and QL+MA-treated SS while QL-treated SS reduced the total P in the barley shoots in comparison

Table 4 Effect of tested materials applied at a rate of 1 t ha^{-1} on fresh biomass of barley shoots (g plant^{-1}) and total plant content of Al, N, P, K, Ca, Si, Mg, Cd, and Pb on dry weight (mean; $n = 5 \pm$ SD)

Parameter	Control	SS	QL-1	QL-2	QL-1 +MA-1	QL-1 +MA-2	QL-2 +MA-1	QL-2 +MA-2
Shoots, g plant^{-1}	0.40±0.05	0.39±0.03	0.42±0.01	0.40±0.05	0.49±0.03	0.53±0.02	0.59±0.01	0.64±0.03
Total N, %	0.85±0.11	1.23±0.14	1.32±0.14	1.22±0.13	1.35±0.13	1.38±0.14	1.30±0.15	1.34±0.15
Total P, %	0.28±0.01	0.44±0.03	0.41±0.03	0.34±0.02	0.54±0.03	0.67±0.03	0.55±0.03	0.68±0.04
Total K, %	0.23±0.11	0.58±0.11	0.59±0.06	0.59±0.14	0.64±0.13	0.67±0.12	0.66±0.11	0.69±0.13
Total Ca, %	0.24±0.10	0.38±0.10	0.49±0.11	0.57±0.13	0.48±0.11	0.49±0.13	0.50±0.13	0.54±0.12
Total Si, %	0.32±0.01	0.34±0.03	0.30±0.02	0.27±0.02	0.45±0.05	0.59±0.05	0.48±0.04	0.64±0.06
Total Al, %	0.01±0.01	0.35±0.01	0.21±0.01	0.16±0.01	0.07±0.01	0.03±0.01	0.05±0.01	0.02±0.01
Total Mg, ppm	n/d	0.87±0.07	0.34±0.05	0.12±0.03	0.96±0.09	1.24±0.09	0.86±0.08	0.95±0.09
Total Cd, ppm	n/d	0.25±0.01	0.12±0.01	0.06±0.01	0.05±0.01	0.04±0.01	0.03±0.01	0.01±0.01
Total Pb, ppm	n/d	0.23±0.01	0.14±0.01	0.04±0.01	0.05±0.01	0.02±0.01	0.03±0.01	0.01±0.01

n/d, not detectable level.

with the plants grown at untreated SS. A higher effect on the plant P was observed with using QL+MA-treated SS. The leaf P raised from 0.28% in the control plants to 0.68% and 0.82% in the plants under QL-2+MA-2 at low and high rates, accordingly. The mechanisms responsible for this increasing effect are related to the QL-induced precipitation of mobile forms of P with the formation of slightly soluble Ca phosphates (Matichenkov et al., 2017):

$$.½\,Ca^{2+} + H_2P_2O^-{}_4 = Ca(H_2PO_4)_2$$

$$Ca^{2+} + HPO_2{}^{-4} = CaHPO_4$$

However, MA can reverse this interaction and increase the soluble forms of P via the following reaction:

$$CaHPO_4 + Si(OH)_4 = CaSiO_3 + H_2O + H_3PO_4$$

Using standard thermodynamic data, the diagram of the $Ca(H_2PO_4)_2$ and $CaHPO_4$ dissolving for different pH levels was calculated.

3.3 $Ca(H_2PO_4)_2$ dissolving

$$Ca(H_2PO_4)_2 \leftrightarrow ½Ca^{2+} + H_2P_2O^-{}_4$$

The algorithm for calculation of the dissolved phosphate anion was the following:

$$\lg C_{PO4} = -1.50 - 0.5\lg C_{Ca} - 0.5\lg\alpha Ca^{2+} - \lg\alpha H_2PO_4{}^- - \lg\gamma H_2PO_4{}^- - 0.5\lg\gamma Ca^{2+}.$$

The competition agents for the formation of $\beta CaSiO_3$ and Ca_2SiO_4 were used:

	Log K
$\beta CaSiO_3 + 2H^+ + H_2O = Ca^{2+} + H_4SiO_4$	13.27
$Ca_2SiO_4 + 4H^+ = Ca^{2+} + H_4SiO_4$	37.82
For calculation of $\lg\alpha\ SiO_4^{4-}$ and $\lg\alpha\ SiO_3^{2-}$, the following constants of silicic acid dissociation were used:	
$H_4SiO_4 \leftrightarrow H_3SiO_4^- + H^+$	−9.71
$H_3SiO_4^- \leftrightarrow H_2SiO_4^{2-} + H^+$	−11.80
$H_2SiO_4^- \leftrightarrow HSiO_4^{3-} + H^+$	−12.00
$HSiO_4^- \leftrightarrow SiO_4^{4-} + H^+$	−12.00
$H_4SiO_4 + OH^- = HSiO_3^- + 2H_2O$	$K_1(H^+) = 10^{-9,8}$
$HSiO_3^- + OH^- = SiO_3^{2-} + 2H_2O$	$K_2(H^+) = 10^{-12,16}$
The ion force of the solution was $J = 0.01$.	
The calcium concentration in the calculations was equal to $C_{Ca} = 10^{-1}$ M or $\lg C_{Ca} = -1$.	

The monosilicic acid was equal to $C_{H4SiO4} = 0.01$ M, which is approximately the same as the concentration of monosilicic acid under dissolving of the amorphous silica.

The calculation for the competition reaction with the formation of $\beta CaSiO_3$ was realized by using the following algorithm:

$$\lg C_{PO4} = \lg C_{PO4} + \lg \alpha H_4SiO_{4,}$$

$$\lg \alpha H_4SiO_4 = \lg\left(1 + K_{\beta CaSiO3} C_{H4SiO4} \alpha SiO_3{}^{2-}\right),$$

$$\text{где } \alpha SiO_3{}^{2-} = K_1K_2 / \left([H^+]^2 + K_1[H^+] + K_1K_2\right).$$

The calculation for the competition reaction with the formation of Ca_2SiO_4 was realized by using the following algorithm:

$$\lg C_{PO4} = \lg C_{PO4} + \lg \alpha H_4SiO_{4,}$$

$$\lg \alpha H_4SiO_4 = \lg\left(1 + K_{Ca2SiO4} C_{H4SiO4} \alpha SiO_4{}^{4-}\right),$$

$$\text{где } \alpha SiO_4{}^{4-} = K_1K_2K_3K_4 / \left([H^+]^4 + K_1K_2[H^+]^3 + K_1K_2K_3[H^+]^2 + K_1K_2K_3K_4\right).$$

The obtained data (Fig. 2) has showed that during βCaSiO3 formation, the concentration of phosphate-anion in solution occurs only for pH 6 and higher, while for Ca_2SiO_4 formation, the phosphate-anion concentration can be increased from pH 3 and higher. By this means, for soil with neutral and high levels of pH, the competition reaction will increase the phosphate-anion concentration in the soil solution under increasing monosilicic acid concentrations.

3.4 Dicalcium phosphate $CaHPO_4$

$$CaHPO_4 \leftrightarrow Ca^{2+} + HPO^{2-}{}_4$$

The algorithm for calculation of the dissolved phosphate anion was the following:

$$\lg C_{PO4} = -6.66 - \lg C_{Ca} - \lg \alpha Ca^{2+} - \lg \alpha HPO_4{}^{2-} - \lg \gamma HPO_4{}^{2-} - \lg \gamma Ca^{2+}.$$

The competition agents for formation of $\beta CaSiO_3$ and Ca_2SiO_4 with all conditions were the same used for $Ca(H_2PO_4)_2$

The obtained data is presented in Fig. 3.

By this means, the availability of P in the soil can be remarkably increased. This process was described in our previous study (Bocharnikova et al., 2010).

Total N and K in the barley shoots was increased by all tested materials, accordingly to their application rate. There was no significant influence of QL or MA on the accumulation of these elements.

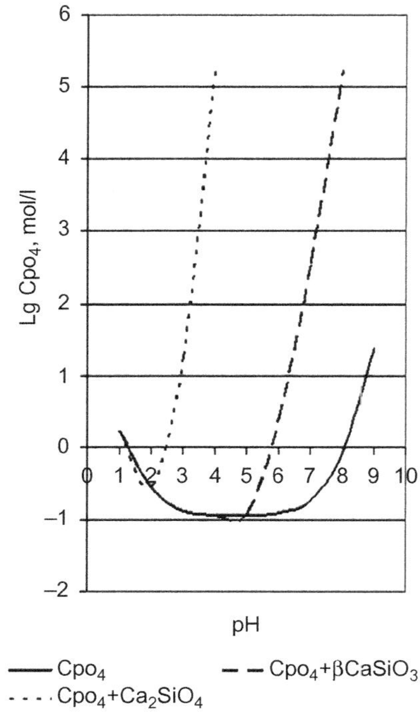

Fig. 2 The effect of competing reactions of the formation of $\beta CaSiO_3$ and Ca_2SiO_4 on the content of the phosphate anion in a saturated solution upon dissolution of $Ca(H_2PO_4)_2$.

Fig. 3 The effect of competing reactions of the formation of $\beta CaSiO_3$ and Ca_2SiO_4 on the content of the phosphate anion in a saturated solution upon dissolution of $CaHPO_4$.

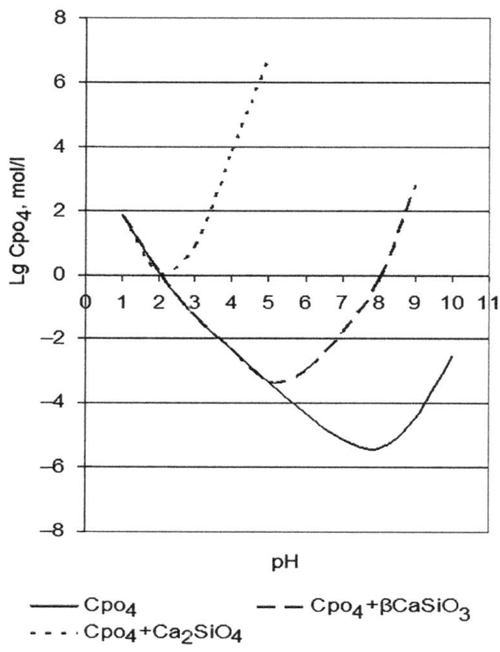

The shoot Mg was increased under application of untreated SS, whereas QL-treated SS significantly reduced the shoot Mg, possibly due to reduced Mg plant availability in the soil by high pH. This fact is confirmed by the data presented in Tables 5 and 6. The application of QL+MA-treated SS has led to increasing Mg uptake by plants, which may be caused by a decline in the pH level and by the ability of MA to optimize plant nutrient absorption (Hernandez-Apaolaza, 2014).

The total content of pollutants (Cd and Pb) in the barley shoots was increased under untreated SS addition. However, QL– or QL+MA-treated SS significantly reduced the total Cd and Pb in the barley shoots. The treatment with QL increased the pH of the soil, resulting in the reduced mobility of heavy metals (HMs) in the soil-plant system. The treatment of SS with QL+MA intensified the HM immobilization through the precipitation by silicate ions:

$$H_4SiO_4 + Cd^{2+} = CdSiO_3\downarrow + H_2O + H^+$$

$$H_4SiO_4 + Pb^{2+} = PbSiO_4\downarrow + 4H^+$$

The results of soil examinations after the greenhouse test are presented in Tables 5 and 6. Untreated SS increased the pH of sand from 6.8 to 7.1 and 7.2 at rates of 1 and 2 g kg^{-1}, accordingly. Quicklime-treated SS increased pH up to 8.6 under QL-2 application at a higher rate while QL+MA-treated SS significantly reduced pH compared to QL-treated SS alone. This phenomenon was described and explained above.

Untreated SS induced a considerable rise in the mobile Al in the sand, the concentration of which reached 169.3 ± 1.8 mg kg^{-1} of Al. On the contrary, the addition of QL-treated SS caused an almost 12-fold decline in the mobile Al in the sand. The reducing effect was strengthened by QL+MA-treated SS (Table 7).

The mobility of Ca in the sand was increased under the application of all types of SS-based products, with a higher influence demonstrated by QL-treated SS. The addition of MA to QL-treated SS significantly reduced the Ca mobility. However, this reduction was not as significant as that of the total Ca in the barley shoots. Consequently, it is possible to suggest that MA mostly impacted the Ca transport inside the plant, but not in the growing media.

The soil acid-extractable Si was significantly increased under the untreated SS application (up to 49.5 ± 0.4 mg kg^{-1} of Si). Quicklime-treated SS increased the acid-extractable Si up to 59.2 ± 0.3 mg kg^{-1} of Si. Finally, the maximum mobile Si was observed in the QL-2+MA-2-added sand (89.5 ± 0.9 mg/kg of Si).

The application of QL-treated SS markedly reduced the mobile forms of P and Mg in the sand compared to those at untreated SS, whereas QL+MA-treated SS significantly increased the mobility of these elements.

The content of the mobile K in the sand was positively related with the rate of untreated or treated SS. The mobility of both HMs (Cd and Pb) in the sand significantly depended on the material applied. The mobile forms of Cd and Pb were most enhanced by untreated SS at a higher rate (2.4 ± 0.4 and 2.0 ± 0.2, accordingly). The addition of QL or QL+MA remarkably reduced the mobility of these HMs. The mechanisms of these processes were described above.

Table 5 Effect of treated materials applied at a rate of 2 t ha^{-1} on fresh biomass of barley shoots (g $plant^{-1}$) and total plant content of Al, N, P, K, Ca, Si, Mg, Cd, and Pb on dry weight (mean; n = 5 ± SD)

Parameter	Control	SS	QL-1	QL-2	QL-1 +MA-1	QL-1 +MA-2	QL-2 +MA-1	QL-2 +MA-2
Shoots, g $plant^{-1}$	0.40 ± 0.05	0.38 ± 0.04	0.45 ± 0.04	0.42 ± 0.05	0.65 ± 0.04	0.73 ± 0.04	0.58 ± 0.03	0.78 ± 0.05
Total N, %	0.85 ± 0.11	1.43 ± 0.15	1.42 ± 0.14	1.42 ± 0.15	1.55 ± 0.12	1.68 ± 0.14	1.50 ± 0.15	1.64 ± 0.15
Total P, %	0.28 ± 0.01	0.59 ± 0.04	0.38 ± 0.03	0.32 ± 0.02	0.74 ± 0.03	0. 87 ± 0.03	0.75 ± 0.05	0.82 ± 0.06
Total K, %	0.23 ± 0.11	0.59 ± 0.11	0.69 ± 0.13	0.79 ± 0.17	0.69 ± 0.14	0.77 ± 0.15	0.77 ± 0.15	0.89 ± 0.14
Total Ca, %	0.24 ± 0.10	0.43 ± 0.10	0.59 ± 0.16	0.65 ± 0.17	0.54 ± 0.11	0.59 ± 0.16	0.53 ± 0.13	0.59 ± 0.12
Total S, %	0.32 ± 0.02	0.34 ± 0.02	0.31 ± 0.02	0.31 ± 0.03	0.49 ± 0.03	0.69 ± 0.05	0.63 ± 0.05	0.75 ± 0.06
Total Al, %	0.01 ± 0.01	0.65 ± 0.03	0.25 ± 0.02	0.18 ± 0.02	0.08 ± 0.01	0.04 ± 0.01	0.06 ± 0.01	0.02 ± 0.01
Total Mg, ppm	n/d	1.23 ± 0.12	0.44 ± 0.07	0.22 ± 0.03	1.45 ± 0.14	1.53 ± 0.14	1.24 ± 0.13	1.44 ± 0.11
Total Cd, ppm	n/d	0.35 ± 0.03	0.15 ± 0.01	0.08 ± 0.01	0.06 ± 0.01	0.05 ± 0.01	0.03 ± 0.01	0.01 ± 0.01
Total Pb, ppm	n/d	0.27 ± 0.02	0.16 ± 0.01	0.09 ± 0.01	0.06 ± 0.01	0.03 ± 0.01	0.03 ± 0.01	0.01 ± 0.01

n/d, not detectable level.

Table 6 Effect of tested materials applied at a rate of 1 t ha^{-1} on soil pH and acid-extractable Al, Si, P, K, Ca, Mg, Cd, and Pb in soil (mean; $n=5\pm SD$)

Parameter	Control	SS	QL-1	QL-2	QL-1+MA-1	QL-1+MA-2	QL-2 +MA-1	QL-2 +MA-2
pH	6.8±0.1	7.1±0.1	8.2±0.1	8.3±0.1	7.6±0.1	7.3±0.1	7.8±0.1	7.6±0.1
Al mg kg^{-1}	0.1±0.1	69.4±0.9	20.3±0.7	9.4±0.5	10.4±0.5	2.3±0.1	3.7±0.2	1.5±0.1
Si mg kg^{-1}	45.8±0.3	47.5±0.3	50.3±0.8	49.4±0.3	59.6±0.6	68.6±0.5	55.5±0.3	59.6±0.5
P mg kg^{-1}	12.2±0.2	38.4±0.3	24.3±0.7	12.6±0.5	45.6±0.5	50.4±0.4	43.8±0.5	47.5±0.5
K mg kg^{-1}	21.2±0.3	86.5±0.9	74.5±0.9	71.5±0.8	74.8±0.9	74.5±0.9	77.4±0.9	77.8±0.8
Ca mg kg^{-1}	0.1±0.1	21.5±0.2	44.8±0.6	56.7±0.6	38.4±0.4	37.8±0.4	52.3±0.6	50.4±0.6
Mg mg kg^{-1}	0.1±0.1	3.5±0.2	0.3±0.2	0.2±0.2	4.5±0.3	5.4±0.3	3.2±0.2	3.9±0.2
Cd mg kg^{-1}	n/d	1.3±0.2	0.4±0.1	0.2±0.1	0.1±0.1	0.1±0.1	0.1±0.1	0.1±0.1
Pb mg kg^{-1}	n/d	1.1±0.2	0.3±0.1	0.1±0.1	0.1±0.1	0.1±0.1	0.1±0.1	0.1±0.1

n/d, not detectable level.

Table 7 Effect of tested materials applied at a rate of $2\,t\,ha^{-1}$ on soil pH and acid-extractable Al, Si, P, K, Ca, Mg, Cd, and Pb in soil (mean; $n=5\pm SD$)

Parameter	Control	SS	QL-1	QL-2	QL-1+MA-1	QL-1+MA-2	QL-2+MA-1	QL-2 +MA-2
pH	6.8±0.1	7.2±0.1	8.5±0.1	8.6±0.1	7.7±0.1	7.6±0.1	7.5±0.1	7.6±0.1
Al mg kg^{-1}	0.1±0.1	169.3±1.8	40.7±0.8	14.4±0.6	10.8±0.5	2.8±0.1	4.3±0.2	1.7±0.1
Si mg kg^{-1}	45.8±0.3	49.5±0.4	50.4±0.8	59.2±0.3	69.5±0.5	78.4±0.6	65.2±0.4	89.5±0.9
P mg kg^{-1}	12.2±0.2	47.5±0.7	22.4±0.4	15.5±0.6	55.8±0.8	58.6±0.7	46.9±0.7	49.4±0.7
K mg kg^{-1}	21.2±0.3	96.2±0.9	84.2±1.2	83.4±0.8	84.2±0.9	81.5±0.9	84.1±1.3	89.4±0.9
Ca mg kg^{-1}	0.1±0.1	25.7±0.5	57.8±0.8	76.2±0.8	68.1±0.6	67.2±0.6	64.7±0.6	60.1±0.4
Mg mg kg^{-1}	0.1±0.1	5.8±0.5	0.4±0.2	0.4±0.2	6.7±0.6	6.7±0.5	4.4±0.3	4.9±0.4
Cd mg kg^{-1}	n/d	2.4±0.4	0.7±0.1	0.2±0.1	0.2±0.1	0.1±0.1	0.1±0.1	0.1±0.1
Pb mg kg^{-1}	n/d	2.0±0.2	0.5±0.1	0.2±0.1	0.2±0.1	0.1±0.1	0.1±0.1	0.1±0.1

n/d, not detectable level.

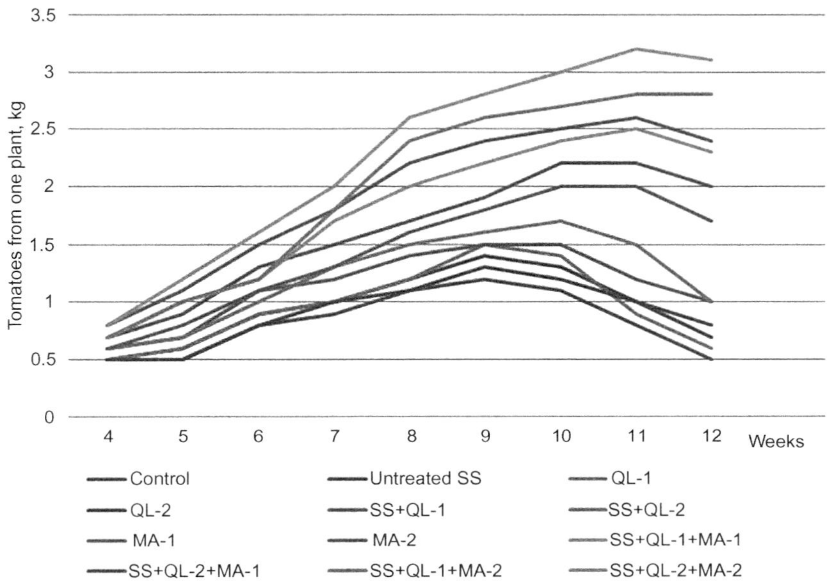

Fig. 4 The dynamic of the tomato harvest (tomatoes kg from one plant).

Generally, the sand content of mobile forms of tested elements agreed with their total content in the barley shoots.

3.5 Experiment with sewage sludge and tomatoes

The dynamic of the tomatoes weight is present in the Fig. 4. The harvesting of the tomatoes was started in the fourth week of the plant's growth and finished in the 12th week. For control plants (growing only on much soil, which was not treated by any substances) the maximum tomato production was obtained in the ninth week. The maximum production in the ninth week also was determined for tomatoes growing in the soil treated by untreated SS, QL-1, and QL-2. However, using SS treated by QL or by MA provides maximum production in the 10th or 11th week. But the amount of collected tomatoes from these plots had maximum during all time-period of the tomatoes growing. By this means using of the SS treated by QL or by MA increase the prolongation of the maximum tomatoes formation. It is important that using untreated SS has no negative influence on tomato production (Al toxicity). This is probably related to the cultivated plant, which had more resistance to Al toxicity compared with barley.

The application of QL, MA, and treated and untreated SS significantly increased the tomatoes productivity (Fig. 5). Using the untreated SS increased the tomato yield 16%. The application of only QL gives the possibility to increase the tomato yield by 16 and 10%, respectively, for QL-1 and QL-2. Using the SS treated by QL had more effect and the tomato yield increased by 37 and 47% accordingly for SS+QL-1 and

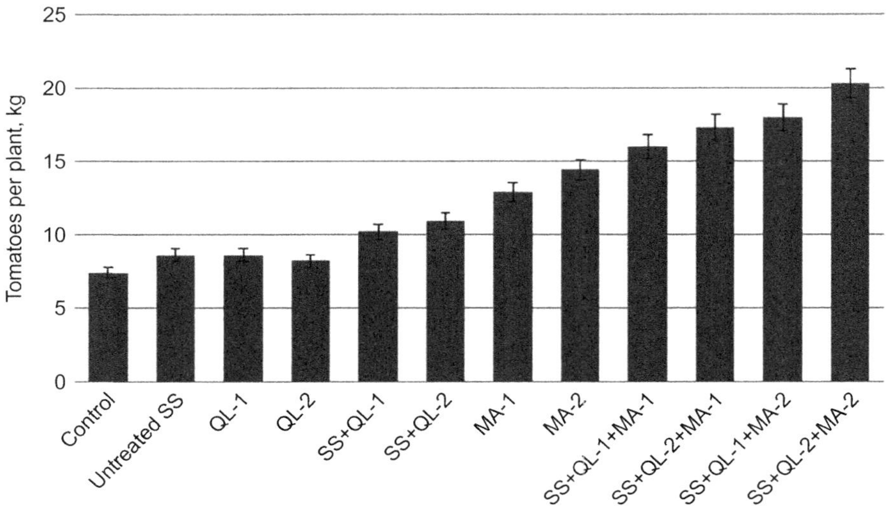

Fig. 5 Summarizing the yield of the tomatoes (tomatoes kg from one plant).

SS+QL-2. Using only MA-1 was more effective of 74% compare with using untreated SS and application of MA-2 was more effective than SS treated by QL on 94%. This data is correlated with our previous work and the literature data (Matichenkov and Bocharnikova, 2004; Toresano-Sánchez et al., 2012; Zhao et al., 2007).

The quality control on tomatoes (vitamin C, sugar content, and NO_3^-) is shown in Table 8. The application of all tested materials significantly increased the vitamin C content in the tomatoes. The maximum effect on vitamin C was obtained for plants grown in soil treated by SS+QL-2+MA-2 (increasing from 18.4 ± 0.5 to 22.4 ± 0.4 mg·100 g^{-1} fresh weight). The vegetables from this plot also were characterized by a maximum content of sugar (from 2.15 ± 0.04 to 2.85 ± 0.02 % of fresh weight). By this means, the complete treatment of the SS by QL and MA provided a better taste of the tomatoes.

The nitrate content in the vegetable is one of the most important quality parameters. Nitrate is a naturally occurring form of nitrogen and is an integral part of the nitrogen cycle in the environment. Nitrate is formed from fertilizers, decaying plants, manure, and other organic residues. A high content of NO_3^- is related to a high rate of nitrogen fertilizer. The presence of nitrate in vegetables, as in water and generally in other foods, is a serious threat to human health (Santamaria, 2006). The obtained data showed that the application of the SS significantly increased the nitrate content in the tomatoes from 54 ± 3 to 97 ± 8 mg kg^{-1} fresh weight. The application of only QL had no significant influence on the nitrate content in tomatoes. However, the application of the MA only significantly reduced the nitrate content in the vegetable—from 54 ± 3 to 23 ± 3 and 21 ± 2 mg kg^{-1} fresh weight, accordingly, for MA-1 and MA-2,

Table 8 The tomato quality of vitamin C, sugar content, and NO_3^-

Treatment	Vitamin C, mg·100 g^{-1} fresh weight	Sugar content, % fresh weight	NO_3^-, mg kg^{-1} fresh weight
Control	18.4±0.5	2.15±0.04	54±3
untreated SS	19.2±0.5	2.20±0.03	97±8
QL-1	19.0±0.5	2.17±0.04	53±5
QL-2	19.2±0.4	2.18±0.03	52±5
SS+QL-1	19.4±0.3	2.19±0.03	99±6
SS+QL-2	19.3±0.5	2.19±0.02	103±5
MA-1	20.4±0.5	2.32±0.04	23±3
MA-2	21.5±0.3	2.79±0.04	21±2
SS+QL-1+MA-1	20.5±0.3	2.59±0.03	45±4
SS+QL-2+MA-1	21.2±0.4	2.84±0.04	48±4
SS+QL-1+MA-2	20.6±0.5	2.65±0.02	38±3
SS+QL-2+MA-2	22.4±0.4	2.85±0.02	39±2

accordingly. The maximum nitrate content in tomatoes was observed for plants growing under application of the SS+QL-1 (99±6 mg kg^{-1} of fresh weight) and SS+QL-2 (103±5 mg kg^{-1} of fresh weight). The treatment of these types of SS by MA significantly reduced the nitrate content in the vegetable.

The selected element composition of the tomatoes is present in Table 8. The Al and heavy metals are not detected in the tomatoes, which were harvested from plants growing on control plots and plots treated by QL or MA only, because only SS is a source of the pollutants in the tested system. The maximum total content of Al (8.5±0.4 mg kg^{-1} of dry weight), Cd (2.5±0.1 mg kg^{-1} of dry weight), and Pb (1.3±0.1 mg kg^{-1} of dry weight) was determined in the tomatoes from plants grown in the soil treated by the original SS. Using only QL for SS treatment reduced the bioavailability of the Al, Cd, and Pb. However, the maximum effect was obtained for SS treated by QL and MA (Table 9).

The application of untreated or treated SS allowed a significant increase in the content of Ca, P, and Si in the tomatoes because SS is also a source of the macro and micronutrients. The maximum total content of Ca (944±87 mg kg^{-1} of dry weight) was obtained for plants growing in the plot with QK-2 application. However, these tomatoes also were characterized by the lowest total content of P (1845±18 mg kg^{-1} of dry weight). This is related with fixation of the mobile and potentially P by Ca (Leytem and Mikkelsen, 2005). The application of the MA makes it possible to increase the P mobility. The mechanism of this is described in the previous subchapter. Using MA alone or SS treated by MA dramatically increased the total content of Si in the tomatoes (from 52±8 mg kg^{-1} of dry weight in control to 105±9 mg kg^{-1} of dry weight for vegetables harvested from plants growing on a plot with SS+QL-2+MA-2 application).

Table 9 The selected element composition of the tomatoes

Treatment	Al	Ca	Cd	P	Pb	Si
	mg kg^{-1} of dry weight					
Control	n/d	534±45	n/d	2845±25	n/d	52±8
untreated SS	8.5±0.4	739±65	2.5±0.1	3485±34	1.3±0.1	83±8
QL-1	n/d	893±80	n/d	2107±37	n/d	64±5
QL-2	n/d	944±87	n/d	1845±18	n/d	61±6
SS+QL-1	4.5±0.3	869±67	1.2±0.1	2845±25	0.5±0.1	74±5
SS+QL-2	4.0±0.3	876±80	1.1±0.1	2104±20	0.4±0.1	72±5
MA-1	n/d	634±65	n/d	3024±31	n/d	89±8
MA-2	n/d	655±60	n/d	3110±30	n/d	95±7
SS+QL-1+MA-1	2.1±0.2	821±70	0.3±0.1	3496±35	0.2±0.1	85±8
SS+QL-2+MA-1	2.0±0.2	820±83	0.3±0.1	3674±40	0.2±0.1	89±9
SS+QL-1+MA-2	1.2±0.1	833±90	0.2±0.1	3794±40	0.1±0.1	96±9
SS+QL-2+MA-2	1.0±0.1	874±85	0.1±0.1	3822±45	0.1±0.1	105±9

n/d, not detected.

3.6 Discussion

The tested QL- and MA-based treatment could be a perspective for the utilization of municipal wastes and manufacturing a highly effective, environmentally friendly product that can be used in agriculture, the greenery of large cities, and bioremediation of contaminated areas. The technology includes several steps:

(a) Elimination of pathogenic microorganisms using QL at a ratio of 10:2 (SS:QL).
(b) Detoxification of Al and HMs by active MA.
(c) Drying and granulation of the final product.

The obtained data also showed that the SS treated by QL and by MA can be classified as a complex organo-Si fertilizer. The effect of complex and organic fertilizer is very good, as described in the scientific literature (Diacono and Montemurro, 2011; Kanagachandran and Jayaratne, 2006; Singh and Agrawal, 2008). However, the information about Si fertilizers is not common in agricultural and scientific societies. Therefore, it is critically important to understand what Si fertilizer is and the main effects of this type of fertilizer on the soil-plant system.

The first trial of silicon fertilizers was conducted in a greehouuse by the great chemist and agrochemist Justus von Liebig in Germany in 1840 (Liebig, 1840). He developed the mineral theory of plant nutrition and identified four chemical elements as essential to plant growth: N, P, K, and Si. As a matter of fact, the last element was forgotten more than 100 years ago, although some attempts to introduce silicon fertilizers into agriculture were made (see below). Thus, the first field trials with sodium silicate were started at the Rothamsted Station in 1856 (Rothamsted experimental Station, Rothamsted Experimental Station Guide the Classical Experiment, 1991). These studies initiated investigations and scientifically substantiated the practical implementation of silicon fertilizers. The positive impacts of silicon fertilizers were observed in different soils for numerous agricultural plants: rice, sugar cane, barley, wheat, oat, rye, sorghum, corn, sunflower, bean, soya, clover, alfalfa, millet, tomato, cucumber, squash, salad, tobacco, sugar beat, lemon, tangerine, grape, apple, melon, and others (Aleshin, 1982; Adatia and Besford, 1986; Bakhat et al., 2018; Bowen et al., 1992; Etesami and Jeong, 2018; Kudinova, 1974; Taranovskaya, 1939; Tubana et al., 2016; Vasanthi et al., 2015).

An important feature of silicon fertilizers is multifunctionality. Two types of silicon that impact cultivated plants can be distinguished as direct impact on plants and indirect impact through soil (Ma and Takahashi, 1991; Suntheim, 1986). Silicon was found to be taken by plants in a form of monosilicic acid, H_4SiO_4 (Yoshida, 1975). In the plant, monosilicic acid is polymerized and accumulated in the epidermal tissues of bark, leaves, and roots, or forms different phytoliths (Mann and Ozin, 1996). In the epidermal tissues, Si forms the double cuticle layer protecting and mechanically strengthening the plants (Yoshida, 1965, 1975) as well as preventing excess evaporation from leaves (Mann and Ozin, 1996; Efimova and Dokuchan, 1986). Optimized plant silicon nutrition results in increasing the weight of roots (Adatia and Besford, 1986; Emadian and Newton, 1989), their volume, and total and adsorbing surfaces (Kudinova, 1974; Yamaguchi et al., 1995). Silicon fertilizers improve root respiration (Matichenkov et al., 1999a,b). Our investigations of the silicon fertilizer impacts on citrus and fodder grasses have shown an Si-induced increase in the amounts of secondary and tertiary roots (Kudinova, 1975; Matichenkov et al., 1999a,b). Amorphous silicon dioxide is known to be one of the main components of coronary cells of root cap (Matichenkov et al., 1999a,b; Jones et al., 1963).

Monosilicic acid was found to increase the germination of cereals and citrus, accelerate the fruit formation of tomatoes and rice, mature corn and the growth of citrus trees, and increase sugar content in sugar beet and sugarcane (Jones and Handreck, 1967; Kudinova, 1975; Matichenkov et al., 1999a,b). Our research indicates that Si promotes increased cold tolerance of rice, sugar cane, and citrus trees as well as plant tolerance to salts, heavy metals, and petroleum hydrocarbons (Bocharnikova et al., 1999; Matichenkov, 1990a,b).

Some scientists have demonstrated that improved plant silicon nutrition results in enhanced stability of the DNA and RNA molecules (Aleshin, 1982; Bocharnikova et al., 2014; Voronkov et al., 1978). Silicon was detected in ribosomes, mitochondrions, chloroplasts, and microsomes (Reichert et al., 1994; Vaculík et al., 2015).

The main function of Si in plants is enhancing the resistance to unfavorable growth conditions due to thickening epidermal tissues (mechanical protection), accelerated growth and increased root system (physiological protection), and increased tolerance to abiotic stresses (biochemical protection). Considering that many plants were tested, we can confirm, that this function of Si is universal and both class of plants: Si-accumulative and—nonaccumulative plants have high respond.

The cultivation of agricultural plants disturbs the balance of mineral nutrients because of the annual removal with crops. Plants mostly take up Si in higher amounts than those of other nutrients (Anderson, 1991). For example, potato adsorbs 50 to 70 kg ha^{-1} Si and cereals, 100 to 300 kg ha^{-1} Si (Bazilevich et al., 1975). Maximal Si uptake is typical for sugar cane—up to 700 kg ha^{-1} Si annually (Anderson, 1991). Based on the available data on Si adsorption by different cultivated plants and crop yields, it was calculated that about 210 to 224 million tons of Si is removed irreversibly with crops worldwide every year (Bocharnikova and Matichenkov, 1994, 2012).

Cultivation of agricultural plants results in a fast decrease in the concentrations of monosilicic acid in the arable soil (Matichenkov et al., 1994, 1997). Because Si is not only a nutrient but also a constitutive soil element, growing Si deficiency leads to certain negative consequences. Monosilicic acid deficiency and decreased content of amorphous silicon dioxide initiate the destruction of the organo-mineral soil complex, accelerate organic matter decomposition, and deteriorate mineral composition (Karmin, 1986; Marsan and Torrent, 1989). Thus, silicon fertilizers are necessary to maintain the balance of monosilicic acid for providing sufficient plant silicon nutrition and preventing soil degradation. In addition, several specific mechanisms of the silicon fertilizer's influence on soil fertility have been suggested.

The trend and extent of the Si fertilizer's influence on physical soil properties depend on soil properties and the fertilizer used. Munk (1982)) has stated improved physical soil properties at the silica application rate of 200–800 kg ha^{-1} per year. Polysilicic acids are able to link soil particles. Soil structure is improved due to forming Si bridges between silt particles. That leads to improving structure, water-holding capacity, exchange capacity, and the buffer capacity of light soils (Matichenkov and Bocharnikova, 2001).

1. Silicon fertilizers benefit plant defense systems, including mechanical, physiological, and biochemical defenses.
2. Agricultural activity today result in reduction plant-available Si in soil, which reinforce the degradation of soil cover.
3. Silicon fertilizers optimize phosphorous plant nutrition due to transforming the plant-unavailable P into plant-available, reducing P leaching, and keeping P in a plant-available form.
4. Silicon fertilizers can be used for increasing the pH of acid soils.
5. Silicon fertilizers can optimize physical soil properties.
6. Silicon fertilizers are specific. They should be used in combination with conventional fertilizers. The multiformity of the silicon fertilizer effects on soils and plants indicates that the plant-available Si deficiency may be the main factor limiting the crop yield.

4 Conclusions

Tested organic sludge treatment technology is based on using concentrated MA as an active Si form and QL as a Ca compound. Quicklime eliminated pathogens present in sludge. Concentrated MA detoxified Al and HMs via their chemical precipitation with the formation of insoluble silicates. Sewage sludge treated with QL and MA could be safely used in agriculture, the greenery of large cities, covering mine tailings, and for other purposes. The QQ treated by QL and MA can be classified as a complex organo-Si fertilizer and improve the quality and quantity of the cultivated plant crop. However, a technology adaptation is required, depending on the type of SS that may differ in the content and composition of pathogenic microorganisms and pollutants.

References

Adatia, M.H., Besford, R.T., 1986. The effects of silicon on cucumber plants grown in recirculating nutrient solution. Ann. Bot. 58, 343–351.

Aleshin, N.E., 1982. The content of silicon in rice RNA. Doklady Vashnil. 6, 6–7.

Anderson, D.L., 1991. Soil and leaf nutrient interactions following application of calcium silicate slag to sugarcane. Fertilizer Res. 30, 9–18.

Bakhat, H.F., Bibi, N., Zia, Z., Abbas, S., Hammad, H.M., Fahad, S., Saeed, S., 2018. Silicon mitigates biotic stresses in crop plants: a review. Crop Prot. 104, 21–34.

Bazilevich, N.I., Rodin, L.E., Rozov, N.N., 1975. Biological production and cycling of the chemical elements in the plant associations. Resources Biosphere Leningrad 1975 (1), 5–33.

Bocharnikova, E.A., Loginov, S.V., Matychenkov, V.V., Storozhenko, P.A., 2010. Silicon fertilizer efficiency. Russ. Agric. Sci. 36 (6), 446–448.

Bocharnikova, E.A., Matichenkov, V.V., 1994. Silicon soil state and biogeochemical balance in forest and grass ecosystems. In: Sustainable Development: The View from the Less Industrialized Countries. UNED, San Jose, Costa Rica, pp. 453–466.

Bocharnikova, E.A., Matichenkov, V.V., 2012. Influence of plant associations on the silicon cycle in the soil-plant ecosystem. Appl. Ecol. Environ. Res. 10 (4), 547–560.

Bocharnikova, E.A., Matichenkov, V.V., Snyder, G.H., 1999. The management of heavy metal behavior and mobility in the soil-plant system. In: Procceeding 31th Mid-Atlantic Industrial and Hazardous Waste Conference, June 1999, pp. 614–622.

Bocharnikova, E.A., Pakhnenko, E.P., Matychenkov, V.V., Matychenkov, I.V., 2014. The effect of optimization of silicon nutrition on the stability of barley DNA. Moscow Univer. Soil Sci. Bull. 69 (2), 84–87.

Bowen, P., Menzies, J., Ehret, D., Samuel, L., Glass, A.D.M., 1992. Soluble silicon sprays inhibit powdery mildew development on grape leaves. J. Am. Soc. Hortic. Sci. 117 (6), 906–912.

Dassanayake, K.B., Jayasinghe, G.Y., Surapaneni, A., Hetherington, C., 2015. A review on alum sludge reuse with special reference to agricultural applications and future challenges. Waste Manag. 38, 321–335.

Diacono, M., Montemurro, F., 2011. Long-term effects of organic amendments on soil fertility. In: Sustainable Agriculture Volume 2. Springer, Dordrecht, pp. 761–786.

Efimova, G.V., Dokuchan, S.A., 1986. Anatom-morphological composition of rice leaves epidermal layer and increasing of protection function under silicon. Agri. Biol. 3, 57–61.

Emadian, S.F., Newton, R.J., 1989. Growth enhancement of loblolly pine (*Pinus taeda L.*) seedlings by silicon. J. Plant Physiol. 134 (1), 98–103.

Etesami, H., Jeong, B.R., 2018. Silicon (Si): review and future prospects on the action mechanisms in alleviating biotic and abiotic stresses in plants. Ecotoxicol. Environ. Saf. 147, 881–896.

Exley, C., Schneider, C., Doucet, F.J., 2002. The reaction of aluminum with silicic acid in acidic solution: an important mechanism in controlling the biological availability of aluminium? Coord. Chem. Rev. 228, 127–135.

Feng, L., Jingyang, L., Yinguang, C., 2015. Dilemma of sewage sludge treatment and disposal in China. Environ. Sci. Technol. 49, 4781–4782.

Genedebien, A., Carlton-Smith, C., Izzo, M., 1999. UK Sewage Sludge Survey: Regional Presentation. Environment Agency.

Hall, E.L., Dietrich, A.M., 2000. A brief history of drinking water. Opflow 26 (6), 46–49.

Hernandez-Apaolaza, L., 2014. Can silicon partially alleviate micronutrient deficiency in plants? A review. Planta 240 (3), 447–458.

Iler, R.K., 1979. The Chemistry of Silica. Wiley, New York.

Ji, X., Liu, S., Huang, J., Bocharnikova, E., Matichenkov, V., 2016. Monosilicic acid potential in phytoremediation of the contaminated areas. Chemosphere 157, 132–136.

Jones, L.H.P., Handreck, K.A., 1967. Silica in soil, plants and animals. Adv. Agron. 19, 107–149.

Jones, L.H.P., Milne, A.A., Wadham, S.M., 1963. Studies of silica in the oat plant: II distribution of silica in the plant. Plant Soil 18, 358–371.

Ju, X., Zhang, F., Bao, X., Romheld, V., Roelcke, M., 2005. Utilization and management of organic wastes in Chinese agriculture: past, present and perspectives. Sci. China C. Life Sci. 48 (2), 965–979. https://doi.org/10.1007/BF03187135.

Kanagachandran, K., Jayaratne, R., 2006. Utilization potential of brewery waste water sludge as an organic fertilizer. J. Inst. Brew. 112 (2), 92–96.

Kapshe, M., Kuriakose, P.N., Srivastava, G., Surjan, A., 2013. Analysing the co-benefits: case of municipal sewage management at Surat, India. J. Clean. Prod. 58, 51–60.

Karmin, Z., 1986. Formation of ferrihydrite by inhibition of grun rust structures in the presence of silicon. Soil Sci. Soc. Am. J. 50 (1), 247–254.

Kochian, L.V., Pineros, M.A., Hoekenga, O.A., 2005. The physiology, genetics and molecular biology of plant aluminum resistance and toxicity. In: Root Physiology: From Gene to Function. Springer, Dordrecht, pp. 175–195.

Kudinova, L.I., 1974. The effect of silicon on barley weight. Agrokhimiya 1, 142–144.

Kudinova, L.I., 1975. The effect of silicon on plant growth, leaves area size and adsorption capacity of plant roots. Agrokhimiya. 10, 117–120.

Leytem, A.B., Mikkelsen, R.L., 2005. The nature of phosphorus in calcareous soils. Better Crops 89 (2), 11–13.

Liebig, J., 1840. Organic Chemistry in Its Application to Agriculture and Physiology. Printed for Taylor and Walton, London.

Lindsay, W.L., 1979. Chemical Equilibria in Soil. John Wiley & Sons, New York.

Lopes, M., Leite, F.L., Valente, B.S., Heres, T., Dai Prá, M.A., Xavier, E.G., Roll, V.F.B., 2015. An assessment of the effectiveness of four in-house treatments to reduce the bacterial levels in poultry litter. Poult. Sci. 94 (9), 2094–2098.

Lumsdon, D.G., Farmer, V.C., 2010. Solubility characteristics of proto-imogolite sols: how silicic acid can de-toxify aluminium solutions. Eur. J. Soil Sci. 46 (2), 179–186.

Ma, I.F., Takahashi, E., 1991. Effect of silicate on phosphate availability for rice in a P-deficient soil. Plant Soil 133, 151–155.

Mann, S., Ozin, G.A., 1996. Synthesis of inorganic materials with complex form. Nature 382, 52–57.

Marsan, F.A., Torrent, J., 1989. Fragipan bonding by silica and iron oxides in a soil from Northwestern Italy. Soil Sci. Soc. Amer. J. 53 (4), 1140–1145.

Matichenkov, V., Bocharnikova, E., 2004. Si in horticultural industry. In: Production Practices and Quality Assessment of Food Crops. Springer, Dordrecht, pp. 217–228.

Matichenkov, V., Bocharnikova, E., 2016. Utilization of sludge as manure. In: Environmental Materials and Waste, pp. 213–220.
Matichenkov, V.V., 1990a. Amorphouse Silicon Dioxide in Soddy Podzolic Soil and Its Influence on Plants: Aftoref Candidate Dissertation. MSU, Moscow, p. 26.
Matichenkov, V.V., 1990b. The silicon fertilizer effect of root cell growth of barley. In: Abstracts of 5th Symposium International Society of Root Research Clemson, South Carolina, USAp. 110.
Matichenkov, V.V., Ammosova, Y.M., Bocharnikova, E.A., 1997. Method of analysis of plant-available silicon in soils. Agrokhimiya 1, 76–84.
Matichenkov, V.V., Bocharnikova, E.A., 2001. The relationship between silicon and soil physical and chemical properties. In: Studies in Plant Science. Vol. 8. Elsevier, pp. 209–219.
Matichenkov, V.V., Bocharnikova, E.A., Pakhnenko, E.P., Khomiakov, D.M., 2017. Effect of Si-rich substances on phosphorous adsorption by sandy soils. Environ. Sci. Pollut. Res. 24 (31), 24311–24317.
Matichenkov, V.V., Calvert, D.V., Snyder, G.S., 1999a. Prospective of silicon fertilization for citrus in Florida. Soil Crop Sci. Florida Proc. 59.
Matichenkov, V.V., Calvert, D.V., Snyder, G.S., 1999b. Silicon fertilizers for citrus in Florida. Proc. Flo. State Hort. Soc. 112, 5–8.
Matichenkov, V.V., Pinsky, D.L., Bocharnikova, E.A., 1994. The effect of mechanic density of soil on content plant-available silicon. Eur. Soil Sci. 11, 71–76.
Matichenkov, V.V., Wei, X., Liu, D., Bocharnikova, E.A., 2013. Theory, practice and prospection of Si fertilizer. Agric. Sci. Technol. 14 (3), 498–502.
Munk, H., 1982. Zur bedentung silikatisher stoffe bei der oungung landwirtschaftlecker rulturpflanzen. Landwirt Forsch. 34 (38), 264–277.
PN-A-04019, 1998. Oznaczanie zawartości witaminy C.
Reichert, J.M., Norton, L.D., Huang Chi-hua, L.D., 1994. Sealing, amendment, and rain intensity effects on erosion of high-clay soils. Soil Sci. Soc. Am. J. 58, 1199–1205.
Rothamsted Experimental Station Guide the Classical Experiment, 1991. Lawes Agricultural Trust. Rapide Printing, Watton, Norfolk.
Rouse, J.D., 2013. Sustainability of wastewater treatment and excess sludge handling practices in the federated states of Micronesia. Sustainability 5 (10), 4183–4194.
Rutkowska, U., 1981. Wybrane metody badań składu i wartości odżywczej żywności. PZWL, Warszawa.
Santamaria, P., 2006. Nitrate in vegetables: toxicity, content, intake and EC regulation. J. Sci. Food Agri. 86, 10–17.
Schindler, P.W., Furst, B., Dick, R., Wolf, P.O., 1976. Ligand properties of surface silanol groups. I. Surface complex formation with Fe^{3+},Cu^{2+}, Cd^{3+}, and Pb^{2+}. J. Colloid Interface Sci. 55 (2), 469–475.
Schmitt, M., Watanabe, T., Jansen, S., 2016. The effects of aluminum on plant growth in a temperate and deciduous aluminum accumulating species. AoB Plants. https://doi.org/10.1093/aobpla/plw065.
Shao, J., Yuan, X., Leng, L., Huang, H., Jiang, L., Wang, H., Zeng, G., 2015. The comparison of the migration and transformation behavior of heavy metals during pyrolysis and liquefaction of municipal sewage sludge, paper mill sludge, and slaughterhouse sludge. Bioresour. Technol. 198, 16–22.
Silva, S., 2012. Aluminum toxicity targets in plants. Aust. J. Bot. 2012,219462. https://doi.org/10.1155/2012/219462. 8 p.
Singh, R.P., Agrawal, M., 2008. Potential benefits and risks of land application of sewage sludge. Waste Manag. 28 (2), 347–358.

Singh, S., Tripathi, D.K., Singh, S., Sharma, S., Dubey, N.K., Chauhan, D.K., Vaculík, M., 2017. Toxicity of aluminum on various levels of plant cells and organism: a review. Environ. Exp. Bot. 137, 177–c 193.
Snyder, G.H., Matichenkov, V.V., Datnoff, L., 2006. Silicon. In: Barker, A.V., Pilbeam, D.J. (Eds.), Handbook of Plant Nutrition. Massachusetts University, pp. 551–568.
Suntheim, R.W.L., 1986. Significance of silicon in plant nutrition. Arch. Acker-Pflanzenbau Bodenkd. 30 (12), 737–744.
Taranovskaya, V.G., 1939. Silicification of subtropical nurseries and plantations. Soviet Subtropis. 7, 32–37.
Toresano-Sánchez, F., Valverde-García, A., Camacho-Ferre, F., 2012. Effect of the application of silicon hydroxide on yield and quality of cherry tomato. J. Plant Nutr. 35 (4), 567–590.
Tubana, B.S., Babu, T., Datnoff, L.E., 2016. A review of silicon in soils and plants and its role in US agriculture: history and future perspectives. Soil Sci. 181 (9/10), 393–411.
Vaculík, M., Pavlovič, A., Lux, A., 2015. Silicon alleviates cadmium toxicity by enhanced photosynthetic rate and modified bundle sheath's cell chloroplasts ultrastructure in maize. Ecotoxicol. Environ. Saf. 120, 66–73.
Vainionpaa, R., Waris, M., Leinikki, P., 2008. Diagnostic techniques: serological and molecular approaches. Encyclopedia of Virology 29–37.
Vasanthi, N., Saleena, L.M., Raj, S.A., 2015. Silicon in crop production and crop protection—a review. Agric. Rev. 35 (1), 14–23.
Voronkov, M.G., Zelchan, G.I., Lukevic, A.J., 1978. Silicon and life. Riga: Zinatne. 578.
Wéry, N., Lhoutellier, C., Ducray, F., Delgenès, J.P., Godon, J.J., 2008. Behavior of pathogenic and indicator bacteria during urban wastewater treatment and sludge composting, as revealed by quantitative PCR. Water Res. 42 (1), 53–62.
Xu, G.R., Yan, Z.C., Wang, Y.C., Wang, N., 2009. Recycle of alum recovered from water treatment sludge in chemically enhanced primary treatment. J. Hazard. Mater. 161 (2), 663–669.
Yamaguchi, T., Tsuno, Y., Nakano, J., Mano, P., 1995. Relationship between root respiration and silica: calcium ratio and ammonium concentration in bleeding sap from stem in rice plants during the ripening stage. Japan J. Crop Sci. 64 (3), 529–536.
Yoshida, S., 1965. Chemical aspects of the role of silicon in physiology of the rice plant. Bull. Nat. Inst. Agric. Sci. Ser. B 15.
Yoshida, S., 1975. The physiology of silicon in rice. Food Fert. Tech. Centr. Tech. Bull. Taipei, Taiwan. (4).
Zdybel, J., Cencek, T., Karamon, J., Kłapeć, T., 2015. Effectiveness of selected stages of wastewater treatment in elimination of eggs of intestinal parasites. Bull. Vet. Inst. Pulawy 59 (1), 51–57.
Zhao, F.L., Wen, C.B., Hou, H.E., Liu, X.R., 2007. The utilization efficiency of silicon-potash fertilizer on tomato. J. Henan Agri. Sci. 8, 035.

Sludge multifunctions in a phytobiome—Forest and plantation application including microbial aspects

14

Anna Grobelak, Marta Jaskulak
Department of Environmental Engineering, Czestochowa University of Technology, Czestochowa, Poland

1 Biological and chemical components applied to the soil with sewage sludge

Consistently increasing industrialization and urbanization are leading to a rise in the quantity of wastewater produced in developed countries (Ashworth and Alloway, 2004). These large volumes of sewage sludge have to be disposed of or treated in some manner. One way of sludge disposal is to use it as a fertilizer, which, in recent years, has become a more common practice (Mantovi et al., 2005). The application of sewage sludge on agricultural areas is also gaining interest in many countries around the world. It was shown before that this kind of action improves soil properties and increases crop productivity (Passuello et al., 2012). The beneficial application of organic waste such as sewage sludge in agricultural soils can allow maintaining and restoring the quality of previously degraded soils as well as reduce the need for application of synthetic fertilizers (Ojeda et al., 2003). Today, more and more lands have been chronically suffering from a severe decrease in organic matter caused mostly by overexploitation and droughts (Latare et al., 2014). Therefore, the reuse of biosolids such as sewage sludge is proposed as an alternative solution for improvement in organic matter content and soil quality at low cost (Passuello et al., 2012). In the past 20 years, the idea of reusing nutrients by applying sewage sludge to agricultural soil has become a more common practice. It is currently considered a viable alternative to other disposal methods such as incineration or storing and processing at landfills (Roig et al., 2012). Because urban sewage sludge contains vast quantities of organic matter and plant nutrients, it has been shown to be beneficial to soil productivity (Grobelak et al., 2017). The soil application of organic waste such as sewage sludge contributes to an increase in the level of soil organic matter content which tends to lower across almost all Europe mostly as a result of overproduction, industrialization, contamination with heavy metals (HM) and droughts (Fytili an Zabaniotou, 2008). The benefits of reusing sewage sludge as fertilizers are its high content of organic carbon and crucial nutrients such as nitrogen, phosphorus, and potassium

Industrial and Municipal Sludge. https://doi.org/10.1016/B978-0-12-815907-1.00014-3

(Zaman et al., 2004). For agricultural purposes, the urban sewage sludge that is mostly the residual semisolid fraction resulting from wastewater treatment is most commonly used (Liu et al., 2014). Usually, it is composed of organic matter (OM), macronutrients such as NPK, micronutrients, other inorganic and organic substances, and microorganisms (Petersen et al., 2003a). Fertilizing soils with this organic amendment was shown before to improve soil properties and crop yields (Liu et al., 2014; Grobelak et al., 2013).

1.1 Macro and micro nutrients

Sewage sludge is characterized by a high abundance of fertilizing components, including crucial macro- and micronutrients for crop production (Chitdeshwari et al., 2002). However, there are significant variations in the chemical composition of sewage sludge that require constant analytical control in the case of agricultural use (Grobelak et al., 2017). Taking into consideration the quantitative fluctuations of components, one must emphasize the vast source of resources contained in urban sewage sludge, which could be returned to the soil to some extent (Table 1) (Perez-Espinosa et al., 2000).

While applying sewage sludge to the soil, special attention should be paid to nitrogen (N), a component of great importance for the natural environment and subject to a substantial decrease in it while the soil is overexploited (Grobelak and Napora, 2015).

From the agricultural point of view, the content of magnesium (Mg) and sulfur (S) in sewage sludge is also essential, as a strong deficiency in those two elements is often observed in soils across all Europe (Fytili and Zabaniotou, 2008). The presence of calcium (Ca) should also be considered, not only as a possibility of soil enrichment with this component, especially in the case of calcium-stabilized sludges, but also with regard to the possibility of reducing the mobility of some forms of heavy metals in sewage sludge, which limits their direct availability to plants (Bozkurt and Yarilgac, 2003).

Table 1 An example of the content of organic matter, organic carbon, and macro/micronutrients in urban sewage sludge (Perez-Espinosa et al., 2000; Grobelak et al., 2017; Gasco et al., 2005)

Parameter	Average content in sewage sludge	Unit
Organic Matter (OM)	9.5–85	% of dry matter
Organic Carbon (OC)	12–63	% of dry matter
N	1.2–84	$g\,kg^{-1}$ dry weight
P	0.4–42	$g\,kg^{-1}$ dry weight
K	0.2–9	$g\,kg^{-1}$ dry weight
Ca	2.5–110	$g\,kg^{-1}$ dry weight
S	3–15	$g\,kg^{-1}$ dry weight
Mg	0.4–20	$g\,kg^{-1}$ dry weight

1.2 Organic matter

Soil OM plays a valuable role in the conservation and restoration of soil over the long term due to sustaining its fertility as well as the improvement of the chemical, physical, and biological properties (Parat et al., 2005). The content of organic matter is normally a result of the residues from plants, animals, and microorganisms and their decomposition through the mineralization of OM. These processes depend on proper temperature and moisture leading to the release and further cycling of the most crucial nutrients: nitrogen, potassium, and sulfur (Grobelak et al., 2017). An increase in soil organic matter content by application of sewage sludge can improve the soil potential for sequestration of organic carbon (Sobik-Szołtysek et al., 2017). It was shown that both the content of organic carbon and nitrogen increases immediately after application of sewage sludge (Grobelak et al., 2018). Overall, the rise in levels of organic matter in soils amended with sewage sludge leads to: (i) favoring of the establishment of vegetal cover, (ii) improving soil quality, (iii) improving water retention, and (iv) increasing the levels of organic carbon stored in the soil (Fig. 1) (Sanchez et al., 2004). Given these valuable benefits, the use of sewage sludge as an organic amendment is gaining more interest throughout the world (Pascual et al., 2008). Moreover, the addition of organic matter to soil can not only provide the soil with organic carbon and symbiotic microorganisms, but it also stimulates the soil microbial population, which increases the stability of the soil aggregates (Qureshi et al., 2003). It was shown before that the soil OM and microbiome are positively affected by sewage sludge. The same studies have shown that the microbiological properties are the most sensitive to sludge amendments (Mantovi et al., 2005). Soil enzyme activity that reflects the microbial activities of soil's overall "health" plays a crucial role in the mineralization of carbon and the stabilization of organic matter (Kizilkaya and Bayrakli, 2005).

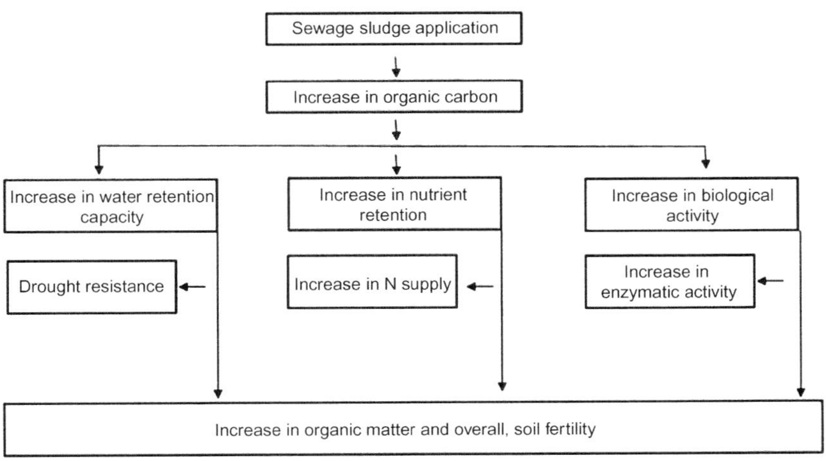

Fig. 1 Main benefits after sewage sludge application in soil (Sanchez Monedero et al., 2004; Grobelak et al., 2017).

1.3 Problems of heavy metals and microcontaminants

The idea of using sewage sludge as a fertilizer has a couple of important drawbacks that must be taken into consideration (Kiikkilä et al., 2002). Wrongly processed sewage sludge can have a wide variety of undesired traits and in consequence, adverse effects on a given environment (Singh et al., 2002). Due to this risk, it is crucial to gather knowledge about reducing the risks associated with using organic wastes such as sewage sludge as a soil amendment (Fijalkowski et al., 2017). Previously, the agricultural processes used to increase soil productivity have often led to adverse effects on soil fertility and quality due to the lack of regulations regarding contaminants in the sludge (Karathanasis et al., 2005). In the European Union, the reuse of sewage sludge in agriculture was regulated for the first time by Council Directive 86/278/EEC, which mostly dealt with the risks of sewage-borne metal contamination on human health (Fytili and Zabaniotou, 2008). In the meantime, countries such as Portugal have written their updated legislation, which introduced the criteria for pathogenic microorganisms and organic contaminants in sewage sludge applied to soils (Petersen et al., 2003b). Countries such as Portugal limit the drawbacks of sewage sludge application in their Decree-Law No. 276/2009, where they restrict the values of HM content and the occurrence of pathogenic microorganisms.

In order to prevent sludge-borne heavy metal contamination, wastewater treatment technologies are gaining more interest (Theodoratos et al., 2000). Composting, cocomposting, and anaerobic digesting are the most common methods used for sewage sludge treatment before application to soils as a fertilizer (Suleiman et al., 2017; Grosser, 2017). In anaerobic digesting, sludge is incubated without air for at least 14 days in temperatures between 35 and 55°C (Grosser, 2018). During the process of cocomposting, the sewage sludge is often mixed with a selected bulking agent: straw, wood chips, bark, and cereal hulls are typical at this stage. After that, the sewage sludge is allowed to decompose in aerobic conditions for several weeks (Rorat et al., 2016). All three of the mentioned processes—aerobic compositing, vermicomposting, and anaerobic digestion—make it possible to stabilize the sewage sludge, decrease the water content, reduce the number of potential pathogens, and reduce the availability of heavy metals, mostly due to their immobilization (Suleiman et al., 2017; Rorat et al., 2017).

The potential reuse of sewage sludge for soil fertilization faces a variety of technical problems, including the constant threat of finding harmful contaminants and even the simple fact that the sludge is produced year round and the application of fertilizers takes place only once or, at best, twice a year (Tsakou et al., 2003). The solution for this problem can be building facilities for storing and composting with other organic fractions, including agricultural or food production waste (Aggelides and Londra, 2000). Such cocomposting would help with overcoming some problems such as high water concentration, lack of stability, or contamination with human or livestock pathogens (Suleiman et al., 2017). The cocomposting is also a highly recommended procedure due to the high temperatures generated during such process, which destroy all parasites and pathogenic microorganisms (Rorat et al., 2017). It has to be taken into consideration that poor quality or wrongly processed sludge always poses a threat to the environment, which can result in toxic element accumulations (such as heavy metals and organic contaminants PAH and PCF) and leaching (Kunito et al., 2001). The transfer of metals from sewage sludge to soils and then groundwater and plants can be a potential risk to health and the environment because heavy metals are known to be highly persistent in soil and can stay there for thousands of years (Rorat et al., 2017). Metals added with such sewage sludge can be retained in soil due to their adsorption on clays, hydrous oxides, and organic matter (Kunito et al., 2001). Moreover, the soil pH is also a significant factor in metal retention. Because of that, $CaCO_3$ has been found to increase metal retention and overall, in soils with more basic pH, the heavy metal persistency is high (Gove et al., 2001). A vast number of previous studies dealt with this issue, and some of them show that the migration of metals is negligible and some have demonstrated a significant amount of Zn, Cr, and Cu leached from sewage sludge (Grobelak et al., 2013; Sobik-Szołtysek et al., 2017).

Urban sewage sludge can also consist of various contaminants of both chemical and biological origin (Aparicio et al., 2009). The significant risk associated with the application of sewage sludge to agricultural soil is the possibility of acidification and contamination of soil with heavy metals such as Cu, Zn, Cd, and Pb, which can have toxic effects on microorganisms, plants, and human health (Gondek, 2009; Jaskulak et al., 2018). However, many previous long-term studies reported that land application of sewage sludge did not lead to an increase in HM in soils (Hemmat et al., 2010). Consequently, the processes of soil fertilization using sewage sludge is possible without compromising soil as long as the ecotoxicological guidelines for sewage sludge safety are observed (Smith, 2009a). It was shown before that the use of complementary techniques, including toxicity bioassays and physical-chemical analysis, are valuable methods that lead to the minimization of risks associated with adding sewage sludge to agricultural soils (Placek et al., 2016).

2 Changes of physical and chemical parameters of soils after fertilization with sewage sludge

Application of organic wastes such as urban sewage sludge to soils is becoming consistently more popular due to the prospect of recycling crucial plant nutrients (Singh and Agrawal, 2008). Land application of sewage sludge could decrease or even eliminate the need for synthetic fertilizers (Sanchez et al., 2004). In many previous studies, the soil physical condition and quality have been improved after the application of organic waste (Pascual et al., 2008; Grobelak et al., 2017). The content of metals in sewage sludge depends on a couple of factors such as the sewage sludge treatment process and origin (Wang et al., 2012). Bioavailability of those metals is influenced mostly by the pH and redox potential (Morera et al., 2002). For this reason, soil pH is especially important in the case of metal presence in sewage sludge and/or soil (Aggelides and Londra, 2000). In past studies, both an increase and a decrease in pH after sewage sludge application have been noticed, which is related to the content of calcium carbonate present in sewage sludge

(Theodoratos et al., 2000; Liu et al., 2014). Organic matter added in the process resulted in a significant improvement in soil properties, including water retention capacity and aggregate stability (Petersen et al., 2003b). The total content of NPK has been shown to increase in sewage sludge-amended soil and relatively high doses of SS can also increase the cation exchange capacity that will cause a better retention of plant nutrients (Grobelak et al., 2018). In a study by Liu et al. (2014), soil amended with SS has CEC at around 22.8 cmol (p+/kg) with unamended only about 18 cmol (p+/kg). Moreover, the application of SS was also shown to increase the water-holding capacity and water retention (Roig et al., 2012). Studies by Pascual et al. (2009) tried to assess the risk of metal contamination caused by SS application in agricultural soil. It was found that in all doses, the content of OM significantly increased but total and available metals did not change. Because metal bioavailability depends on the form of OM, whether it is soluble (mostly fluvic acid) or insoluble (humic acid), the insoluble OM inhibits the bioavailability and uptake of metals. With stabilization of decomposition rates, the level of soluble OM decreases, which leads to a reduction in the bioavailability of metals (Pascual et al., 2009; Liu et al., 2014). In the Sukkariyah et al. (2005) studies, high organic matter addition with sewage sludge to soil contaminated by Cu led to the creation of higher Cu complexes and impaired the uptake by plants.

3 Changes in the population of microorganisms in soils fertilized with sewage sludge

The consistently rinsing needs for agricultural production as well as ensuring food safety require constant improvement of soil fertility (Hemmat et al., 2010). Organic waste as a byproduct can be used not only in the improvement of physical and chemical soil properties but also to increase the biological activity of soils (Viera and de Souza Silva, 2003). Over the years, there have been more and more studies that showed an improvement in soil biological activities after sewage sludge treatment, in comparison to unamended soils (Aggelides and Londra, 2000; Ojeda et al., 2003). Such effects mainly depend on the SS dose and quality (Smith, 2009b).

The improvement in soil quality and microbial activity is associated with an enrichment of the fertilized soil with sewage sludge microorganisms (Viera and de Souza Silva, 2003). It is well established that the microbial biomass and activity are increased after supplementation with sewage sludge (Hemmat et al., 2010). On the other hand, the adverse effects, including a rapid decrease in soil microflora, have also been reported when the sewage sludge used in soil application contained HM or organic contaminants or the dose of sewage sludge was too high (Theodoratos et al., 2000). Moreover, adding biosolids such as urban sewage sludge to soil was shown to provoke bioaugmentation and biostimulation of autochthonous microorganisms in the soil (Zaman et al., 2004). Degradation of carbohydrates and proteins by the soil microbiome can lead to the formation of other microbial products, depending on the ratio between carbon, nitrogen, phosphorus, and sulfur (Ojeda et al., 2003).

Correlation between the variation of a total number of bacteria and fungi in soil also confirms the positive effects of sewage sludge application on both groups (Wang et al., 2012). The biological activity in soils after sewage sludge amendment is currently an interest of many researchers because the potential presence of HM and other micropollutants can counterbalance all positive effects of increasing levels of organic matter, nutrients, and microorganisms in sludge-amended soils (Fijalkowski et al., 2017).

It was shown before that the soil microbiome is extremely sensitive and primarily affected when the physiochemical properties of soils are changed (Pascual et al., 2008). Therefore, soil microbial assays are used for the estimation of the effects due to various contaminants (heavy metals, fertilizers, pesticides, etc.) on the microbiome, soil fertility, and productivity (Liu et al., 2010). According to Pascual et al. (2008), microbial parameters such as biomass, activity, and enzyme activity can be useful indicators of changing conditions in the soil. Overall, soil respiration, fixation of atmospheric nitrogen, and overall microbial biomass are the most commonly used markers for assessing the soil microbiome (Kiikkilä et al., 2002).

In many studies, enzymatic activities are being estimated to establish soil fertility (Wang et al., 2012). The soil microbiome can act as a biocatalyst for many essential reactions to produce essential compounds for plants and other microorganisms (Zaman et al., 2004). Soil enzymatic activity can be significantly affected by heavy metal presence in sewage sludge. Due to that, the effects of sewage sludge application on the biological activity of soils can also be used as an indication of soil contamination (Pascual et al., 2008).

In studies by Viera and de Souza Silva (2003), the application of sewage sludge increased the activity of both urease and phosphatase. Enhancement of these enzymes was also noticed in different studies after sewage sludge application. This is often associated with an increase of organic matter and nutrients after fertilization, which leads to the stimulation of microbial activity (Qureshi et al., 2003).

In recent studies by Wang et al. (2012) concerning sewage sludge application and the biological activity of the soil, sewage sludge increased enzymatic activity, soil respiration, and microbial activity. A rapid and significant increase in enzymatic activities (mostly b-glucosidase, urease, alkaline phosphatase, and arylsulphatase) was noticed at all doses of sewage sludge, in comparison to unamended soil. Overall, the soils that after supplementation had the highest C/N ratio also had the highest activity of b-glucosidase. Moreover, in all doses, the microbial biomass remained higher than in unamended soils. This increase in enzyme activity was attributed to an enhancement of microbial activity that was provoked by an increase in nutrients and the organic matter of sewage-sludge amended soils (Wang et al., 2012).

It was also noticed by Wang et al. (2012) that the mineralization rates of organic carbon and nitrogen rapidly increased the activity of bacterial proteinase in the soil. The actinomycetes and some groups of fungi reached the maximum presence three weeks after fertilization, but remained at a stable level after that. Overall, the activity of dehydrogenase rose from 18% to 25% and phosphatase from 9% to 23% in comparison to unamended soil.

4 Case study—Long and short term microbial changes in soil after sewage sludge application

Microbiological activity of soils is very reactive to even small changes occurring in the fertilizing processes. Therefore, the assessment of such activity can be used to evaluate the effects of soil supplementation (Kunito et al., 2001). Microorganisms such as fungi, bacteria, actinomycete, and microalgae play a major role in the decomposition of organic matter and nutrient cycling (Rorat et al., 2017). When organic carbon is used for energy by microorganisms, it is then assimilated into their tissues, respirated as CO_2, or realized as other metabolic products (Pascual et al., 2008).

As an organic amendment, sewage sludge improves not only the content of OM and nutrients but also the soil microbiome (Latare et al., 2014). However, if wrongly processed, it can become a threat due to the potential presence of pollutants such as heavy metals and organic contaminants. Because of their persistence and potential toxicity, they can cause a risk to ecosystems, making it necessary to quality control sewage sludge before such an application (Ashworth and Alloway, 2004). Moreover, the use of sewage sludge can also have negative effects due to the possible presence of heavy metals or xenobiotics, which can severely damage the microbial communities in soils. Some of the contaminants such as polycyclic aromatic hydrocarbon (PAH) or, for example, polychlorinated biphenyls (PCB), are known to be strongly persistent in the soil, which creates more risks of their accumulation in soil (Gasco et al., 2005; Rorat et al., 2017).

The addition of organic fertilizers such as sewage sludge affects the soil for the short and long term (Table 2). Sewage sludge is shown to have beneficial effects on soil quality, such as the increase in organic matter, plant nutrients, the water holding capacity, and the cation exchange capacity (Qureshi et al., 2003). In soils fertilized by sewage sludge, autochthonous microbiological activity and growth are also often stimulated (Rahube et al., 2014). As a general rule, the kind and dose of sewage sludge added to the soil are the crucial factors determining the abundance of certain microbial groups and their activities (Wang et al., 2012).

Several long-lasting experiments have shown that soil biological properties, including basal respiration, enzymatic activities, and microbial biomass, are improved after sewage sludge application (Pourcher et al., 2007). Moreover, some studies have already demonstrated that sewage sludge application on soil can result in a longer availability of plant nutrients over an extended period (Grobelak et al., 2017). One study showed that fouryears after sewage sludge application, the soil still had 20% more microbial biomass in comparison to soil supplemented only by a synthetic fertilizer (Mantovi et al., 2005). Another study has shown that 17 years since the last application of sewage sludge, the soil organic matter content was 70% higher than before the first application (Kunito et al., 2001). The same authors also found a significant increase in the activities of phosphatase, urease, B-glucosidase, and o-diphenoloxidase, thus improving the soil microbiological activity and overall quality.

Table 2 Examples of long-term trials with positive results of sewage sludge on soil fertility

Site	Trial period	Crop species	Reference
Tokyo, Japan	23 years	*Zea maize* L. *Hordeum vulgare* L.	Zaman et al. (2004)
Pamplona, Spain	16 years	None	Roig et al. (2012)
Po Valley, Italy	17 years	*Triticum aestivum* L. *Beta vulgaris* L. *Zea mays* L.	Mantovi et al. (2005)
Linz, Austria	12 years	*Zea mays* L. *Hordeum* L. *Triticum* L.	Ros et al. (2006)
Bennett, USA	12 years	*Triticum* L. *Zea mays* L.	Barbarick and Ippolito (2007)
Murcia, Spain	17 years	None	Bastida et al. (2008)
Orange, USA	19 years	*Hordeum* L. *Raphanus sativus* *Lactuca sativa* L.	Sukkariyah et al. (2005)
Tokyo, Japan	19 years	*Zea mays* L. *Hordeum* L. *Oryza* L.	Kunito et al. (2001)
Fidenza, Italy	12 years	*Triticum* L. *Zea mays* L.	Saviozzi et al. (1999)

5 Case study of agricultural soils amended with sewage sludge-microbial changes

Sewage sludge is an inevitable byproduct of processes involved with wastewater treatment (Chitdeshwari et al., 2002). Along with growing industry and agriculture, the quantity of sewage sludge increases every year (Bozkurt and Yarilgac, 2003). In China, the amount of municipal wastewater produced per annum is more than 30 billion of m^3. Sewage sludge treatments and most importantly its disposal have seriously challenged the development of the industries involved in wastewater treatment and environmental protection of China (Liu et al., 2014). Recently, more and more research has been performed on the potential use of sewage sludge as a fertilizer for crops (Table 2) (Hemmat et al., 2010). The use of sewage sludge as a fertilizer in crop production provides essential nutrients for proper plant growth and development while at the same time reducing the economic and environmental considerations of SS incineration or storing in landfills (Parat et al., 2005). However, for safe production, a careful assessment of all characteristics of such soil and the quality of sewage sludge must be performed prior to land application (Morera et al., 2002).

In the EU, the quantity of sewage sludge production is estimated at more than 10 million tons (of dry solids) per year. Nearly 40% of this is applied to land for agricultural purposes (Fytili and Zabaniotou, 2008). Recycling of sewage sludge and its application to crop production vary significantly among different members of the EU (Placek et al., 2017). In Portugal and Spain, more than 65% of all sewage sludge produced is being used as a fertilizer in agriculture. Although this number can be similar across the EU, there are countries such as the Netherlands where strict policies are resulting in a ban on reusing sewage sludge in agriculture (Fytili and Zabaniotou, 2008).

The chemical composition of sewage sludge that is supposed to be used in land application is a crucial issue in developing strict regulations to reduce the risks of sludge-borne metal contamination in both soil and groundwater (Hemmat et al., 2010). In recent years, the application of sewage sludge for agricultural use has gained a significant increase among farmers in Europe (Placek et al., 2017). It appears that application of this biosolid on agricultural land is a logical and reasonable use due to its improvement in many soil properties, including the content of plant nutrients, organic matter, porosity, water retention, bulk density, and even pH. In Belgium, more than 57% of sewage sludge is applied to agricultural land (Fytili and Zabaniotou, 2008). At the same time, it needs to be taken into consideration that sewage sludge, among all nutrients such as nitrogen, phosphorus, and potassium, can also contain toxins such as heavy metals that can cause harm to not only the soil but also to groundwater and plants, posing a risk to human health (Zaman et al., 2004). This is the main reason why many countries regulate the quality and dose of sewage sludge that is appropriate to use in agriculture (Bastida et al., 2008).

6 Forest soils amended with sewage sludge-microbial changes

In the Mediterranean region, particularly semiarid lands, large quantities of soil are severely degraded as a result of past forest fires, inappropriate agricultural use, and overgrazing. Currently, in such areas, big restoration programs are implemented to try and stop further land degradation (Mantovi et al., 2005). Due to the dry and semiarid condition in forests of the Mediterranean area, soils often show an alarmingly low level of organic matter, which increases the degradation processes. In such areas, the application of sewage sludge that is high with organic matter can be considered a means to enhance the cycling of nutrients and also to help the establishment of new plants, which can prevent further degradation. Due to these advantages of sewage sludge application, it is more and more often used on barren soils (Placek et al., 2017).

In a study by Soudani et al. (2017), *Eucalyptus camaldulensis* was selected to grow on degraded soil with the addition of sewage sludge. The choice of plant species was made based on plant growth characteristics and its high economic value. Plants ware grown for 1 year, and the results demonstrated the outstanding fertilizing properties of sewage sludge. Overall, plants after fertilization with sewage sludge had 20% better growth and more than 400% more leaves in comparison to plants grown on unamended soil. No chlorosis and other diseases were observed after application.

Acknowledgments

The authors declare have no conflict of interest. Research funded from internal grant BS PB 401/304/11 and MNISW/2017/89 DiR/NN2.

References

Aggelides, S.M., Londra, P.A., 2000. Effect of compost produced from town wastes and sewage sludge on the physical properties. Bioresour. Technol. 71, 253–259.

Aparicio, I., Santos, J.L., Alonso, E., 2009. Limitation of the concentration of organic pollutants in sewage sludge for agricultural purposes: a case study in South Spain. Waste Manag. 29, 1747–1753.

Ashworth, D.J., Alloway, B.J., 2004. Soil mobility of sewage sludge-derived dissolved organic matter, copper, nickel and zinc. Environ. Pollut. 127, 137–144.

Barbarick, K.A., Ippolito, J.A., 2007. Nutrient assessment of a dryland wheat agroecosystem after 12 years of biosolids applications. Agron. J. 99, 715–722.

Bastida, F., Kandeler, E., Hernández, T., García, C., 2008. Long-term effect of municipal solid waste amendment on microbial abundance and humus-associated enzyme activities under semiarid conditions. Microb. Ecol. 55, 651–661.

Bozkurt, M.A., Yarilgac, T., 2003. The effects of sewage sludge applications on the yield, growth, nutrition and heavy metal accumulation in apple trees growing in dry condition. Turk. J. Agric. For. 27, 285–292.

Chitdeshwari, T., Savithri, P., Mahimairaja, S., 2002. Effect of sewage biosolid composts on the yield of crops. Indian J. Environ. Protect. 21 (10), 911–912.

Fijalkowski, K., Rorat, A., Grobelak, A., Kacprzak, M., 2017. The presence of contaminations in sewage sludge – the current situation. J. Environ. Manag. 203, 1126–1136.

Fytili, D., Zabaniotou, A., 2008. Utilization of sewage sludge in EU application of old and new methods – a review. Renew. Sust. Energ. Rev. 12, 116–140.

Gasco, G., Lobo, M.C., Guerrero, F., 2005. Land application of sewage sludge: a soil columns study. Water SA 31 (3), 309–318.

Gondek, K., 2009. Zinc content in maize (*Zea mays* L.) and soils fertilized with sewage sludge and sewage sludge mixed with peat. Polish J. Environ. Studies 18 (3), 359–368.

Gove, L., Cooke, C.M., Nicholson, F.A., Beck, A.J., 2001. Movement of water and heavy metals (Zn, Cu, Pb and Ni) through sand and sandy loam amended with biosolids under steady-state hydrological conditions. Bioresour. Technol. 78, 171–179.

Grobelak, A., Napora, A., 2015. The chemophytostabilisation process of heavy metal polluted soil. PLoS One 10 (6), e0129538.

Grobelak, A., Kacprzak, M., Grosser, A., Napora, A., 2013. Chemophytostabilisation of soil contaminated with cadmium, lead and zinc. Annuala Set Environ. Protect. 15 (2), 1982–2002.

Grobelak, A., Placek, A., Grosser, A., Singh, B.R., Almås, Å., Napora, A., Kacprzak, M., 2017. Effects of single sewage sludge application on soil phytoremediation. J. Clean. Prod. 155, 189–197.

Grobelak, A., Rorat, A., Kokot, P., Singh, B.R., Kacprzak, M., 2018. Chapter 28: Mine waste rehabilitation case studies from Poland. In: Prasad, M.N.V., de Campos Favas, P.J.,

Maiti, S.K. (Eds.), Bio-Geotechnologies for Mine Site Rehabilitation. first ed. Elsevier, pp. 515–529.
Grosser, A., 2017. The influence of decreased hydraulic retention time on the performance and stability of co-digestion of sewage sludge with grease trap sludge and organic fraction of municipal waste. J. Eviron. Manag. 203 (3), 1143–1157.
Grosser, A., 2018. Determination of methane potential of mixtures composed of sewage sludge, organic fraction of municipal waste and grease trap sludge using biochemical methane potential assays. A comparison of BMP tests and semi-continuous trial results. Energy 143, 488–499.
Hemmat, A., Aghilinategh, N., Rezainejad, Y., Sadeghi, M., 2010. Long-term impacts of municipal solid waste compost, sewage sludge and farmyard manure application on organic carbon, bulk density and consistency limits of a calcareous soil in Central Iran. Soil Till. Res. 108, 43–50.
Jaskulak, M., Rorat, A., Grobelak, A., Kacprzak, M., 2018. Antioxidative enzymes and expression of rbcL gene as tools to monitor heavy metal-related stress in plants. J. Environ. Manag. 218, 71–78. https://doi.org/10.1016/j.jenvman.2018.04.052.
Karathanasis, A.D., Johnson, D.M.C., Matocha, C.J., 2005. Biosolid colloid mediated transport of copper, zinc and lead in waste-amended soils. J. Environ. Qual. 34, 1153–1164.
Kiikkilä, O., Pennanen, T., Perkiömäki, J., Derome, J., Fritze, H., 2002. Organic material as a copper immobilising agent: a microcosm study on remediation. Basic Appl. Ecol. 3, 245–253.
Kizilkaya, R., Bayrakli, B., 2005. Effects of N-enriched sewage sludge on soil enzyme activities. Appl. Soil Ecol. 30 (3), 192–202.
Kunito, T., Saeki, K., Goto, S., Hayashi, H., Oyaizu, H., Matsumoto, S., 2001. Copper and zinc fractions affecting microorganisms in longterm sludge amended soils. Bioresour. Technol. 79, 135–146.
Latare, A.M., Omkar, K., Singh, S.K., Gupta, A., 2014. Direct and residual effect of sewage sludge on yield, heavy metals content and soil fertility under rice–wheat system. Ecol. Eng. 69, 17–24.
Liu, S., Song, F., Zhu, N., Yuan, H., Cheng, J., 2010. Chemical and microbial changes during autothermal thermophilic aerobic digestion (ATAD) of sewage sludge. Bioresour. Technol. 101, 9438–9444.
Liu, T., Liu, B., Zhang, W., 2014. Nutrients and heavy metals in biochar produced by sewage sludge pyrolysis: its application in soil amendment. Polish J. Environ. Stud. 23 (1), 271–275.
Mantovi, P., Baldoni, G., Toderi, G., 2005. Reuse of liquid, dewatered, and composted sewage sludge on agricultural land: effects of long-term application on soil and crop. Water Res. 39, 289–296.
Morera, M.T., Echeverria, J., Garrido, J., 2002. Bioavailability of heavy metals in soils amended with sewage sludge. Can. J. Soil Sci. 82, 433–438.
Ojeda, G., Alcaniz, J.M., Ortiz, O., 2003. Runoff and losses by erosion in soils amended with sewage sludge. Land Degrad. Dev. 14 (6), 563–573.
Parat, C., Chaussod, R., Leveque, J., Andreux, F., 2005. Long-term effects of metal-containing farmyard manure and sewage sludge on soil organic matter in a fluvisol. Soil Biol. Biochem. 37, 673–679.
Pascual, I., Avilés, M., Aguirreolea, J., Sánchez-Díaz, M., 2008. Effect of sanitized and non-sanitized sewage sludge on soil microbial community and the physiology of pepper plants. Plant Soil 310, 41–53.
Pascual, I., Azcona, I., Morales, F., Aguirreolea, J., Sánchez-Díaz, M., 2009. Growth, yield and physiology of Verticillium-inoculated pepper plants treated with ATAD and composted sewage sludge. Plant Soil 319, 291–306.

Passuello, A., Cadiach, O., Perez, Y., Schuhmacher, M., 2012. A spatial multicriteria decision making tool to define the best agricultural areas for sewage sludge amendment. Environ. Int. 38, 1–9.

Perez-Espinosa, A., Moreno-Caselles, J., Moral, R., Perez-Murcia, M.D., Gomez, I., 2000. Effects of sewage sludge application on salinity and physico-chemical properties of a calcareous soil. Arch. Agron. Soil Sci. 45, 51–56.

Petersen, S.O., Petersen, J., Rubæk, G.H., 2003a. Dynamics and plant uptake of nitrogen and phosphorus in soil amended with sewage sludge. Appl. Soil Ecol. 24, 187–196.

Petersen, S.O., Henriksen, K., Mortensen, G.K., Krogh, P.H., Brandt, K.K., Sorensen, J., Madsen, T., Petersen, J., Gron, C., 2003b. Recycling of sewage sludge and household compost to arable land: fate and effects of organic contaminants, and impact on soil fertility. Soil Till. Res. 72, 139–152.

Placek, A., Grobelak, A., Kacprzak, M., 2016. Improving the phytoremediation of heavy metals contaminated soil by use of sewage sludge. Int. J. Phytoremediat. 18 (6), 605–618.

Placek, A., Grobelak, A., Hiller, J., Stępień, W., Jelonek, P., Jaskulak, M., Kacprzak, M., 2017. The role of organic and inorganic amendments in carbon sequestration and immobilization of heavy metals in degraded soils, journal of sustainable development of energy. Water Environ. Syst. 5 (4), 509–517.

Pourcher, A.-M., Françoise, P.-B., Virginie, F., Agnieszka, G., Vasilica, S., Gérard, M., 2007. Survival of faecal indicators and enteroviruses in soil after land-spreading of municipal sewage sludge. Appl. Soil Ecol. 35, 473–479.

Qureshi, S., Richards, B., Khay, A.G., McBride, M.B., Baveye, P., Tsai, C.C., Steenhius, T.S., 2003. Effect of microbial activity on trace element release from sewage sludge. J. Environ. Qual. 37, 3361–3366.

Rahube, T.O., Marti, R., Scott, A., Tien, Y.C., Murray, R., Sabourin, L., Zhang, Y., Duenk, P., Lapen, D.R., Topp, E., 2014. Impact of fertilizing with raw or anaerobically-digested sewage sludge on the abundance of antibiotic-resistant coliforms, antibiotic resistance genes and pathogenic bacteria in soil, and on vegetables at harvest. Appl. Environ. Microbiol. 80 (22), 6898–6907.

Roig, N., Sierra, J., Martí, E., Nadal, M., Schuhmacher, M., Domingo, J.L., 2012. Long-term amendment of Spanish soils with sewage sludge: effects on soil functioning. Agric. Ecosyst. Environ. 158 (2012), 41–48.

Rorat, A., Suleiman, H., Grobelak, A., Grosser, A., Kacprzak, M., Płytycz, B., Vandenbulcke, F., 2016. Interactions between sewage sludge-amended soil and earthworms—comparison between *Eisenia fetida* and *Eisenia andrei* composting species. Environ. Sci. Pollut. Res. 23, 3026–3035.

Rorat, A., Wloka, D., Grobelak, A., Grosser, A., Sosnecka, A., Milczarek, M., Jelonek, P., Vandenbulcke, F., Kacprzak, M., 2017. Vermiremediation of polycyclic aromatic hydrocarbons and heavy metals in sewage sludge composting process. J. Environ. Manag. 187, 347–353.

Ros, M., Klammer, S., Knapp, B., Aichberger, K., Insam, H., 2006. Longterm effects of compost amendment of soil on functional and structural diversity and microbial activity. Soil Use Manage. 22, 209–218.

Sanchez Monedero, M.A., Mondini, C., De Nobili, M., Leita, L., Roig, A., 2004. Land applications of biosolids. Soil response to different stabilization degree or treated organic matter. Waste Manag. 24 (4), 325–332.

Sanchez, M.E., Estrada, I.B., Martinez, O., Martin-Villacorta, J., Aller, A., Moran, A., 2004. Influence of the application of sewage sludge on the degradation of pesticides in the soil. Chemosphere 57, 673–679.

Saviozzi, A., Biasci, A., Riffaldi, R., Levi-Minzi, R., 1999. Long-term effects of farmyard manure and sewage sludge on some soil biochemical characteristics. Biol. Fert. Soils 30, 100–106.

Singh, R.P., Agrawal, M., 2008. Potential benefits and risks of land application of sewage sludge. Waste Manage. Res. 28, 347–358.

Singh, M.K., Sinha, S.S., Singh, S.K., 2002. Effect of untreated and treated sewage on seed germination and seedling growth of Vicia Faba. Ind. J. Environ. Protect. 22 (4), 426–430.

Smith, S.R., 2009a. A critical review of the bioavailability and impacts of heavy metals in municipal solid waste composts compared to sewage sludge. Environ. Int. 35, 142–156.

Smith, S.R., 2009b. Organic contaminants in sewage sludge (biosolids) and their significance for agricultural recycling. Phil. Trans. R. Soc. A 367, 4005–4041.

Sobik-Szołtysek, J., Wystalska, K., Grobelak, A., 2017. Effect of addition of sewage sludge and coal sludge on bioavailability of selected metals in the waste from the zinc and lead industry. Environ. Res. 156, 588–596.

Soudani, L., Maatoug, M., Heilmeier, H., Kharytonov, M., Wiche, O., Moschner, C., Onyshchenkoc, E., Bouchenafa, N., 2017. Fertilization value of municipal sewage sludge for *Eucalyptus camaldulensis* plants. Biotechnology Reports, 13.

Sukkariyah, B.F., Evanylo, G., Zelazny, L., Chaney, R.L., 2005. Cadmium, copper, nickel, and zinc availability in a biosolids-amended piedmont soil years after application. J. Environ. Qual. 34, 2255–2262.

Suleiman, H., Rorat, A., Grobelak, A., Grosser, A., Milczarek, M., Płytycz, B., Kacprzak, M., Vandenbulcke, F., 2017. Determination of the performance of vermicomposting process applied to sewage sludge by monitoring of the compost quality and immune responses in three earthworm species: *Eisenia fetida*, *Eisenia andrei* and *Dendrobaena veneta*. Bioresour. Technol. 241, 103–112.

Theodoratos, P., Moirou, A., Xenidis, A., Paspaliaris, I., 2000. The use of municipal sewage sludge for the stabilization of soil contaminated by mining activities. J. Hazard. Mater. B77, 177–191.

Tsakou, A., Roulia, M., Christodoulakis, N.S., 2003. Growth parameters and heavy metal accumulation in poplar tree cultures (*Populus euramericana*) utilising water and sludge from a sewage treatment plant. Bull. Environ. Contam. Toxicol. 71, 330–337.

Viera, R.F., de Souza Silva, C.M.M., 2003. Soil amendment with sewage sludge and its impact on soil microflora. Braz. J. Microbiol. 34, 56–58.

Wang, R., Baldy, V., Périsso, C., 2012. Influence of plants on microbial activity in a vertical-downflow wetland system treating waste sewage sludge with high organic matter concentrations. J. Environ. Manag. 95, 158–164.

Zaman, M., Matsushima, M., Chang, S.X., Inubushi, K., Nguyen, L., Goto, S., Kaneko, F., Yoneyama, T., 2004. Nitrogen mineralization, N_2O production and soil microbiological properties as affected by longterm applications of sewage sludge composts. Biol. Fert. Soils 40, 101–109.

Co-composting of sewage sludge and wetland plant material from a constructed wetland treating domestic wastewater

15

Anna Kwarciak-Kozłowska
Faculty of Infrastructure and Environment, Institute of Environmental Engineering, Czestochowa University of Technology, Czestochowa, Poland

The increasing population and the use of modern, more effective technologies of wastewater treatment result in a substantial increase in the amount of produced sewage sludge. The costs involved in processing and management are also increasing. The best method to solve the problem of sewage sludge is by composting, which allows for processing the waste into a safe product (organic fertilizers, soil improvers, etc.) and safe use in the environment. This method is consistent with EU recommendations concerning waste management, which recommend waste recovery as the first method. However, due to substantial problems related to their composting, the attractive proposal is co-composting of sewage sludge and macrophytes from hydrobotanical treatment plants, termed constructed wetlands treatment (CW).

1 Legal regulations and directions of sewage waste management in the European Union

Sewage sludge is inherent in the wastewater treatment process. To date, no nonsludge technologies and solutions that would allow for complete elimination of sewage sludge from the environment have been developed (Miksch and Sikora, 2018; Bień et al., 2011). As high-performance methods of biological and chemical wastewater treatment methods are developed, the amount of sewage sludge increases, whereas its processing represents an integral part of technological processes in each treatment plant. According to the EU report, more than 10 million tons of dry mass of sewage sludge were produced in 2008 in 26 member states (Commission, 2008a). It is estimated that one person can generate from 10 to 15 kg of dry mass of sewage sludge per year (Biotenmare Kick Off Meeting, 2014; Eurostat, 2015). The biggest amounts of sewage sludge in the EU are generated in Germany, the UK, and France. A little less sewage sludge is generated by Italy and Spain (each country generates more than 700,000 t DM per year). Only these five countries generate nearly 75% of the total amount of sewage sludge in the EU (Kacprzak et al., 2017; Bourioug et al., 2015; Grobelak et al., 2018). According to Fytili and Zabaniotou (2008), sewage sludge generation in the

Industrial and Municipal Sludge. https://doi.org/10.1016/B978-0-12-815907-1.00015-5

EU has increased by 50% since 2005. Therefore, the optimization of sewage sludge management represents a key element in the sector of wastewater treatment (Uggetti et al., 2011).

Sewage sludge management is regulated by the Directive of the European Parliament and of the Council of 19 November 2008 (2008/98/CE) on waste (also called the Waste Framework Directive). According to the Waste Framework Directive, the following waste hierarchy shall apply as a priority order in waste prevention and management legislation and policy:

(1) Prevention.
(2) Preparing for reuse.
(3) Recycling.
(4) Other recovery processes.
(5) Disposal.

Preventing sewage sludge generation is, however, impossible, as it represents the type of waste that is unavoidable. Therefore, more importance is attached to the next priorities in the waste hierarchy, that is, preparing for reuse (including organic recycling and energy recovery) or final disposal (Kacprzak et al., 2017; Commission, 2008b; Bartkowska, 2017). The division of the industrial and nonindustrial methods of sewage sludge management is presented in Fig. 1. (Kosobucki et al., 2000)

The prohibition of disposal of waste to the sea and necessity of implementation of the 99/31/EC directive on the landfill of waste, which limits landfill of biodegradable waste (and, consequently, sewage sludge), the major directions of their management is the use in nature and thermal processing. Among these directions, the methods of disposal can be divided into two groups:

1. Organic recycling.
2. Energy and material recycling.

Legal regulations of the European Union concerning the problems of waste management include in particular the Directive of the Council 86/278/EEC of 12 June 1986 (the so-called Sewage Sludge Directive) on the protection of the environment, and in particular of the soil when sewage sludge is used in agriculture, which results in substantial limitations for the use of sewage sludge in agriculture and nature. The directive defined the maximal permissible levels of heavy metals in soil and sewage sludge and the maximal amounts of heavy metals that can be introduced into the soil. The shortest intervals between the use of sewage sludge at individual types of agricultural soils and the directions of the use of these soils were also specified.

Analysis of the directions of sewage sludge management in the EU countries in 2002–2013 reveals a strong tendency in the reduction of sewage sludge deposited in landfill sites, to the benefit of their use in agriculture or composition (Table 1). A mere 7% of sewage sludge generated in Poland is used in agriculture there. In some Western European countries, this index is much better, being, for example, 47% in Germany or 73% in France (Legroux and Truchot, 2009; Bourioug et al., 2015).

It is estimated that the costs connected with processing of sewage sludge and its subsequent management represent from 20% to 60% of the total expenditures connected with the functioning of wastewater treatment plants. This is especially

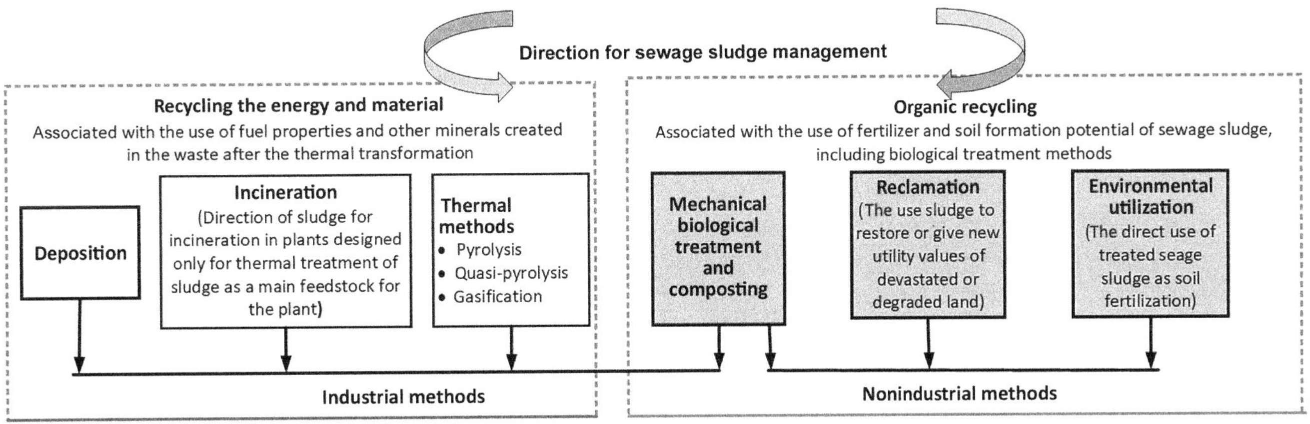

Fig. 1 Direction for sewage sludge management (Kacprzak et al., 2017; Kosobucki et al., 2000).

Table 1 Sewage sludge management in selected countries of the EU (Kacprzak et al., 2017; Bianchini et al., 2015)

UE country	Total disposal	Landfill	Incineration	Agricultural use	Compost and other application
	(10^3 t)				
	2013	2013	2013	2013	2013
Belgium	0	0 (2010)	113 (2010)	17 (2010)	n/a
Germany	1795	0	1034.77	491.33	264.4
Estonia	18.8	1.81	n/a	0.29	16.27
Greece	n/a	80 (2010)	36 (2011)	n/a	n/a
France	869.74	30.92	160.63	368.58	287.49
Latvia	20.74	0.24	0	7.48	2.3
Poland	540.3	31.4	72.9	105.4	32.6
Slovakia	57.43	6.64	5.01	0.52	35.21
United Kingdom	n/a	9 (2010)	260 (2010)	n/a	n/a
Norway	131.2	16.09	n/a	82/6	29.9
Switzerland	194.5	4	188.3	0	0

important in the case of wastewater treatment plants located in rural areas, which in practice are forced to transport sewage sludge to bigger plants instead of implementing their own methods of disposal (Uggetti et al., 2010). However, evaluation of the costs of each technology of sewage sludge treatment is difficult, whereas the most important factors that impact the total cost of their management include:

- Physical and chemical characterization of sewage sludge.
- Amount of sewage sludge.
- National and/or EU legal constraints.
- potential of sewage sludge for valorization and reuse.
- Other costs related to sewage sludge processing (e.g., consumption of electricity and fuel).

Fig. 2 presents the minimal and maximal costs of the management of 1 ton of dry sludge in the EU countries. Table 2 presents the advantages and disadvantages of the main methods of sludge disposal.

2 Division and properties of sewage sludge

With processes occurring in the facilities, sewage sludge that is generated in wastewater treatment plants forms the so-called wastewater and sludge line. The most popular division of sewage sludge results from the source of its origin:

- Raw sludge, originating directly from wastewater treatment processes (initial, secondary, secondary recirculated, secondary excess sludge, and chemical sludge).
- Stabilized sludge, subjected to treatment processes that reduce their susceptibility to putrefaction and the amount of organic substances (fermented sludge, aerobically stabilized

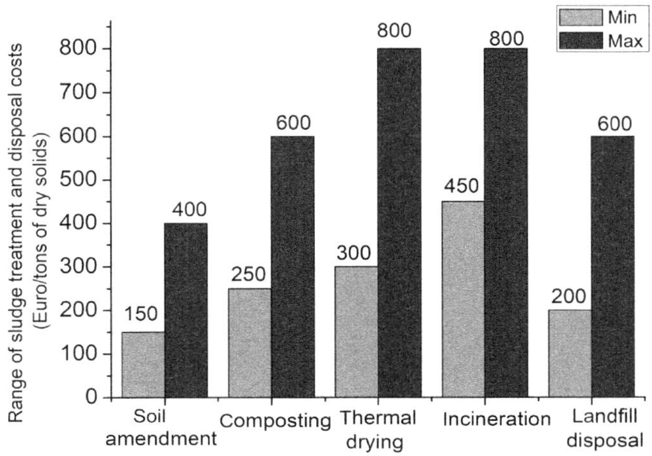

Fig. 2 Range of sludge treatment and disposal costs (Hall, 1999; Kalderis et al., 2010).

Table 2 Advantages and disadvantages of the main methods of sludge disposal (Bień et al., 2014, 2015; Janosz Rajczyk, 2014; Malczewska et al., 2017)

Disposal alternative	Advantages	Disadvantages
Incineration	• Drastic volume reduction • Sterilization • Energetic valorization of sludges • Low sensitivity to sludge composition • Minimization of odors, due to closed systems and high temperature	• High costs • Ash disposal • Atmospheric pollution
Landfill disposal	• Low cost	• Problems with locations near urban centers • Gas and landfill leachate production • Difficulty in reintegrating the area after decommissioning • Requirement of large areas • Requirement of special soil characteristics
Thermal drying	• Total destruction of organic matter and total mineralization of sewage sludge • The possibility of burning both stabilized and unstable sewage sludge	• Large investment cost, high operating costs • Energy consumption of the installation • The necessity of using predried, dehydrated sediments;

Continued

Table 2 Continued

Disposal alternative	**Advantages**	**Disadvantages**
	• Up to 10-fold reduction of sludge volume • Energy recovery • No odor emission	• Emission of dusts, gases (SO_x, NO_x, HCL, HF) • The need to store the resulting ashes
Composting	• The possibility of burning both stabilized and unstable sewage sludge • Sewage sludge mass and volume reduction • Sludge water content reduction • The product can be used as a fertilizer as it contains nitrogen, phosphorus, potassium, and microelements, and it improves soil properties • Easy operation of the installation	• Windrow and aerated static pile composting require relatively large areas, and odor control is a common problem • Ambient temperatures and weather conditions influence windrow and aerated static pile composting • In-vessel reactors have limited flexibility to handle changing conditions and are maintenance intensive
Agricultural reuse	• Large area availability • Potential as a fertilizer • Positive effects for the soil • Positive outcome for the crops • Long-term solution	• Contamination of the soil by metals • Odors • Limitations regarding composition and application rates • Food contamination with toxic elements and pathogenic

sludge, thickened sludge, dewatered sludge, hygienized sludge, and dried sludge) (Bień et al., 2015; Miksch and Sikora, 2018; Grosser, 2017).

Fig. 3 presents the places of sewage sludge generation in the technological line of a wastewater treatment plant with consideration for the method of their subsequent processing.

Although the sewage sludge does not exceed 3% wastewater, it contains more than half of the entire waste that is transferred in raw sludge (Miksch and Sikora, 2018; Kosicka-Dziechciarek et al., 2016; Grosser et al., 2013). It is estimated that each cubic meter of wastewater contains about 0.25 kg of sludge. Each treatment plant generates sludge with different physicochemical properties whose common features include:

- High water content (from 99% in the case of raw sewage sludge, 55%–80% for dewatered sludge and below 10% for thermally dried sludge).
- High content of organic compounds (from 75% d.m. in raw sludge to 45%–55% d.m. in stabilized sludge).
- High content of nitrogen compounds (2%–7% d.m.) at lower content of phosphorus and potassium.

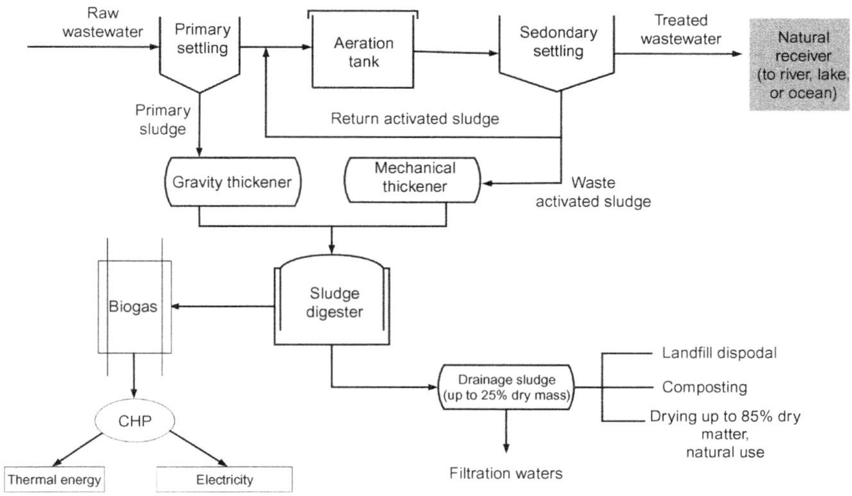

Fig. 3 Generation of sewage sludge in the technological process of a sewage treatment plant (Janosz Rajczyk, 2014).

Table 3 Comparison between raw and digested sludge (Kosicka-Dziechciarek et al., 2016; Bień et al., 2015)

Raw sewage sludge	**Digested sludge**
• Unstable organic matter (from 75% d.m.) • High water content (from 99%) • High biodegradable fraction in the organic matter • High potential in the generation of odors • High concentration of pathogens	• Stabilized organic matter (45%–55% d.m.) • Low proportion of the biodegradable fraction • Low potential in generation of odors • Concentration of pathogens lower than in raw sludge

Table 3 presents a comparison between raw and digested sludge.

A part of the contaminants in the wastewater reaching the treatment plant is not decomposed during treatment and is accumulated in sewage sludge. The level of these contaminants is usually not reduced in sludge treatment processes (Bever et al., 1997). These pollutants are potentially toxic and can include heavy metals, polychlorinated biphenyls (PCBs), polycyclic aromatic hydrocarbons (PAHs), adsorbable organic halides (AOXs), polychlorinated dibenzodioxins (dioxins, PCDDs), and polychlorinated dibenzofurans (Antonkiewicz and Jasiewicz, 2009; Fijalkowski et al., 2018; Grobelak et al., 2017).

Sewage sludge that was not treated in the hygienization processes can represent an epidemiological hazard due to the presence of various pathogenic organisms. The most frequent parasite eggs isolated from sewage sludge include *Ascaris lumbricoides* *Trichuris* sp. and *Toxocara* sp. (Kosicka-Dziechciarek et al., 2016; Kacprzak et al., 2015; Reimers et al., 1986; Romdhana et al., 2009; Zdybel et al., 2009). A substantial amount of bacteria is also present in sewage sludge, such as *Escherichia* sp., *Salmonella* sp., *Shigella* sp. *Bacillus anthracis*, *Pseudomonas aeruginosa*, *Listeria monocytogenes*, *Vibrio cholerae*, *Proteus vulgaris*, *Clostridium perfringens*, *Mycobacterium tuberculosis*, and *Streptococcus pyogenes*. Another numerous group of pathogenic microorganisms is viruses, mainly polyviruses that cause poliomyelitis, rotaviruses, and HIV and HCV viruses. However, the most frequent microorganisms isolated from sewage sludge are fungi: *Penicillium*, *Verticillium*, *Mortierelta*, *Fusarium*, *Aspergillus*, *Mucor*, *Geotrichum*, and *Trichoderma* (Pepper et al., 2006; Prażmo et al., 2003; Kosicka-Dziechciarek et al., 2016; Amin, 1988; Aubain et al., 2002; Stańczyk-Mazanek et al., 2012). Table 4 shows the characteristics of different types of sewage sludge (Kacprzak et al., 2017).

3 Constructed wetland

Wastewater treatment by using low-cost ecotechnology has gained importance in recent years (Rana et al., 2011; Gomesa et al., 2014; Scholz, 2011). The methods of environmental treatment based on phytotechnologies are gaining more and more supporters due to their ecological character, easy operation, effectiveness, and opportunity for conducting the wastewater treatment in situ. In Europe, such systems have been successfully used for municipal and industrial wastewater treatment and removal of heavy metals from road drainage systems or dewatering of sewage sludge (Manios et al., 2002; Vymazal and Kröpfelová, 2008; Herath and Vithanage, 2015; Brix, 1994; Manios, 2004; Obarska-Pempokowiak and Klimkowska, 1999). The first experiments with wastewater treatment using macrophytes were conducted in Germany in the early 1950s (Vymazal, 2011).

Removal of contaminants in these natural systems occurs by a combination of physical, chemical, and biological processes. The processes connected with the removal of contaminants is sedimentation, sorption, evapotranspiration, photooxidation, diffusion, and microbiological degradation (e.g., nitrification, denitrification, reduction of sulfates, and carbon metabolism (Herath and Vithanage, 2015).

A classification of constructed wetlands can be presented in three ways (Fig. 4):

- According to the type of plants used (submerged, floating, floating leaves, emergent).
- According to the direction of wastewater flow (horizontal, vertical, mixed).
- According to hydrological conditions (subsurface, surface, so-called free water).

The most popular constructed wetland systems are surface flow (SF) systems, horizontal subsurface flow (HSF), and vertical subsurface flow (VSF) systems.

- The surface flow systems (SF) are usually shallow ponds and ditches (often from 20 to 40cm) with water macrophytes. These systems are characterized by low investment expenditures and uncomplicated use. However, they require a large surface (up to 20 m^2/PE). The

Table 4 Characteristics of municipal sewage sludge (Kacprzak et al., 2017; Środa et al., 2012)

Parameter	Type of sludge				
	Untreated primary sludge	**Digested primary sludge (poorly)**	**Digested primary sludge (weakly)**	**Digested primary sludge (good)**	**Secondary sludge**
pH	5.0–7.0	6.5–7.0	6.8–7.3	7.3–7.8	6.5–8.0
Total dry solids—TS (%)	2.0–8.0	4–12	4–12	4–12	0.8–1.2
Volatile solids (% of TS)	60–80	55–80	55–80	30–55	55–80
Volatile fatty acids ($mgCH_3COOH/dm^3$)	1800–3600	2500–4000	1000–2500	3000–4000	1800–3600
Nitrogen (% of TS)	2–7	1–5	1–3.5	0.5–2.5	3–10
Phosphorus (% of TS)	0.4–3	0.9–1.5	0.8–2.6	0.8–2.6	0.9–1.5
Potassium (% of TS)	0.1–07	0.1–0.3	0.1–0.3	0.1–0.3	0.1–0.8
Filtration option (m/kg)	10^{11}–10^{13}	$51{\cdot}0^{11}$–$5{\cdot}10^{13}$	10^{11}–10^{12}	10^{12}–10^{13}	10^{10}–10^{11}
Heating value (kJ/g)	16–20	15–18	12.5–16	8–11	15–21

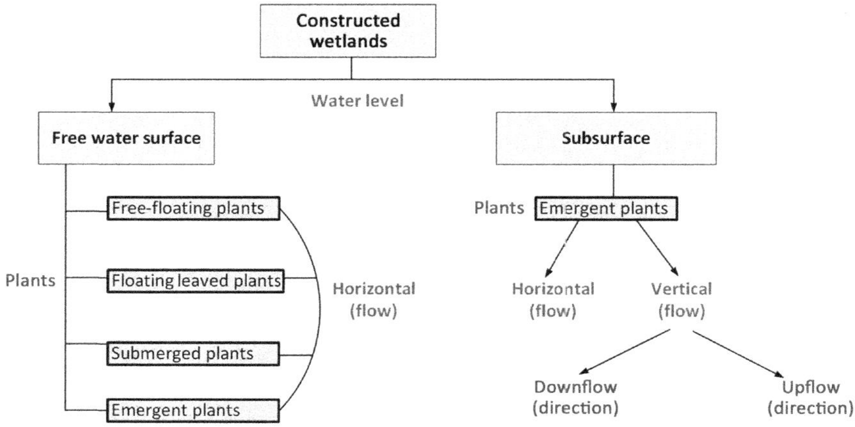

Fig. 4 Types of constructed wetlands for wastewater treatment (Vymazal, 2008; Herath and Vithanage, 2015).

problems occur with reduced performance in the period other than growing season and releasing an unpleasant smell and freezing.

- Horizontal subsurface flow (HSF) systems are cells filled with media (from 30 to 60 cm deep) in which aquatic vegetation is planted. This kind of wetland is saturated and the water column is not exposed to the atmosphere, usually remaining about 5–10 cm under the surface of the bed and therefore avoiding fouling odors and the proliferation of vectors. In HSF systems, the contaminants contained in sewage sludge are removed mainly during the anaerobic processes. The HSF systems require smaller surface (up to 5 m^2/PE) and are characterized by high resistance to freezing.
- Vertical subsurface flow systems (VSF), The subsurface VSF systems are cells filled with coarse sand or fine gravel, usually from 60 to 100 cm deep, and planted with aquatic vegetation. Due to the convenient conditions, quick and efficient removal of organic contaminants (nitrification capabilities) can be observed. Compared to other types of CW, they require the least surface, with 2–3 m^2/PE. (De la Verga et al., 2017; Zhang, 2012; Vymazal, 2010; Ramachandra et al., 2017).

The benefits of hydrophyte treatment plants include:

- Easy operation.
- Resistance to uneven flow of wastewater.
- Lower costs compared to conventional systems.
- Easy matching of the facilities with the present landscape.
- The lack of production of secondary wastewater treatment.
- Simultaneous removal of organic compounds, phosphorus, nitrogen, and heavy metals.

The main drawbacks are:

- Demand for large surface area.
- Difficulties connected with the adaptation of plants in mineral soil and the related long period that allows for a full development of the rhizosphere (Sadecka, 2005; Davis, 1995). Hydrobiological plants develop in wetland, which are oxygen-deficient. The very

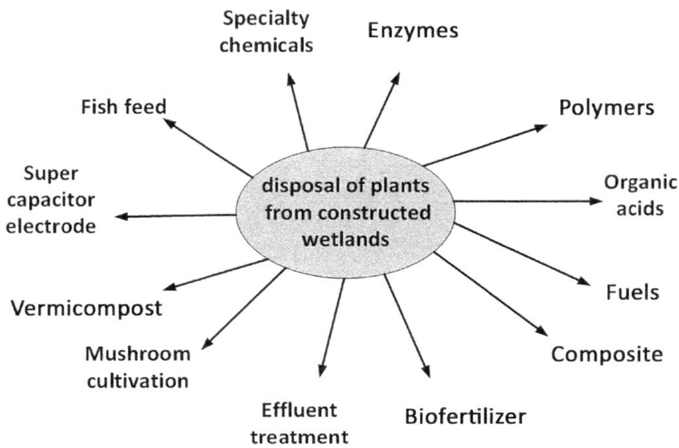

Fig. 5 Ways to dispose of plants from constructed wetlands (Sindu et al., 2017).

> good adaptation of a substantial part of macrophytes to these conditions is due to the spongy tissue termed aerenchyma. Aerenchyma is characterized by large air channels that represent containers for air necessary for the breathing process. They are present in the entire plant, thus facilitating oxygen transport from overground parts to underwater organs. This helps supply this life-giving element to the entire plant, which can grow in such unfavorable settings (Koutika and Rainey, 2015; Vymazal and Kröpfelová, 2008; Wetzel, 2001)

The costs of treatment of this type of vegetation represent a particularly serious challenge to small rural communities. Macrophytes are mostly annual aquatic vascular plants. Their life cycle ends in late autumn and their harvest should be planned for this season of the year. At the moment of becoming the waste, it also becomes a challenge for the managers of wastewater treatment plants. Among the methods of their disposal is incineration (e.g., *Salix viminalis*) or use for the production of enzymes, polymers, organic acids, fuel, and composting or vermicomposting (Fig. 5).

The characterization of the macrophytes used in CW presented in the Table 5 illustrates their potential as biomass, which can be successfully used in organic recycling. Wetzel (2001) gives the following values of ash (as the dry mass percentage): emergent species 12% (5%–25%), floating-leaved 16% (10%–25%), submerged 21% (9%–25%), average for all species 18%.

4 Co-composting of sewage sludge

Composting of sewage sludge is an aerobic process that ensures:

- Stabilization of sewage sludge.
- Destruction of pathogenic organisms.
- Reduction of mass and water content.

Table 5 Percentage of total biomass found in underground tissues of mature aquatic macrophytes (Vymazal and Kröpfelová, 2008; Wetzel, 2001)

Type	Species
% of total biomass (<10)	
Submerged	*Ceratophyllum demersum* (Coontail) *Elodea canadensis* (Common waterweed)
Emergent	*Cyperus fuscus* (Browb cyperius)
% of total biomass (10–20)	
Submerged	*Myriophyllum spicatum* (Eurasian watermilfoil) *Potamogeton pectinatus* (Sago pondweed)
Emergent	*Zizania aquantica* (Annual wildrice)
% of total biomass (>20)	
Submerged	*Potamogeton perfoliatus* (Read-head grass)—31%–51% *Vallisneria americana* (Wild celery)—48% *Isoetes lacustris* (Quillwort)—20%–52%
Floating	*Eichhornia crassipes* (Water hyacinth)—10%–56%
Floating-leaved	*Nuphar* spp. (Yellow water lily)—46%–80% *Nymphaea* spp. (Water lily)—48%–80%
Emergent	*Acorus calamus* (Sweet flag)—49%–66% *Alisma plantago-aquatica* (Water plantain)—40% *Carex lasiocarpa* (Hairy fruited sedge)—50%–78% *Cyperus popyrus* (Papyrus)—31% *Eleocharis rostellata* (Small-beaked spikerush)—47% *Equisetum fluviatile* (Water horsetail)—40%–83% *Typha angustifolia* (Broadleaf cattail)—32%–67% *Typha latifolia* (Broadleaf cattail)—29%–82%

The main process that occurs in the composted mass is biochemical decomposition of organic matter. In optimal conditions, composting occurs in four phases characterized by the different activities of specific groups of microorganisms (Bień and Wystalska, 2011; Jędrczak, 2008; Bień et al., 2014; Moretti et al., 2015):

- The phase of initial composting: the mesophilic phase or temperature rise phase that takes up to several days.
- Phase of intensive composting: thermophilic or high-temperature phase that can take from several days to several weeks (increase in temperature to 60–75°C). In this phase, easily biodegradable organic compounds are decomposed. The products include water, carbon dioxide, and ammonia.
- The phase of transitions, also termed main composting. It starts from weeks 3 to 5 and its duration is another 3–5 weeks (decline in temperature to 30–40°C). The characteristic features of this phase are: a decline of temperature, transformation of hardly decomposing compounds (e.g., lignin, fats, wax, resins) by mesophilic bacteria and fungi, and a noticeable reduction in the volume of waste).

- The compost maturation phase, also termed secondary composting. It leads to cooling down the material, with a stable compost fraction (humus) and the macrofauna is intensively generated. Duration of the phase may reach several months.

Distribution of organic substances (A) and the activity of microorganisms (B) during the process of composting are presented in Fig. 6.

The composting procedure depends on many parameters. Especially important are temperature and reaction. Furthermore, composting effects also depend on many other factors, which include the type and number of microorganisms, the humidity, the degree of substrate fragmentation, the content of organic substances, and the oxygen concentration in the process (Som et al., 2009; Bień et al., 2015).

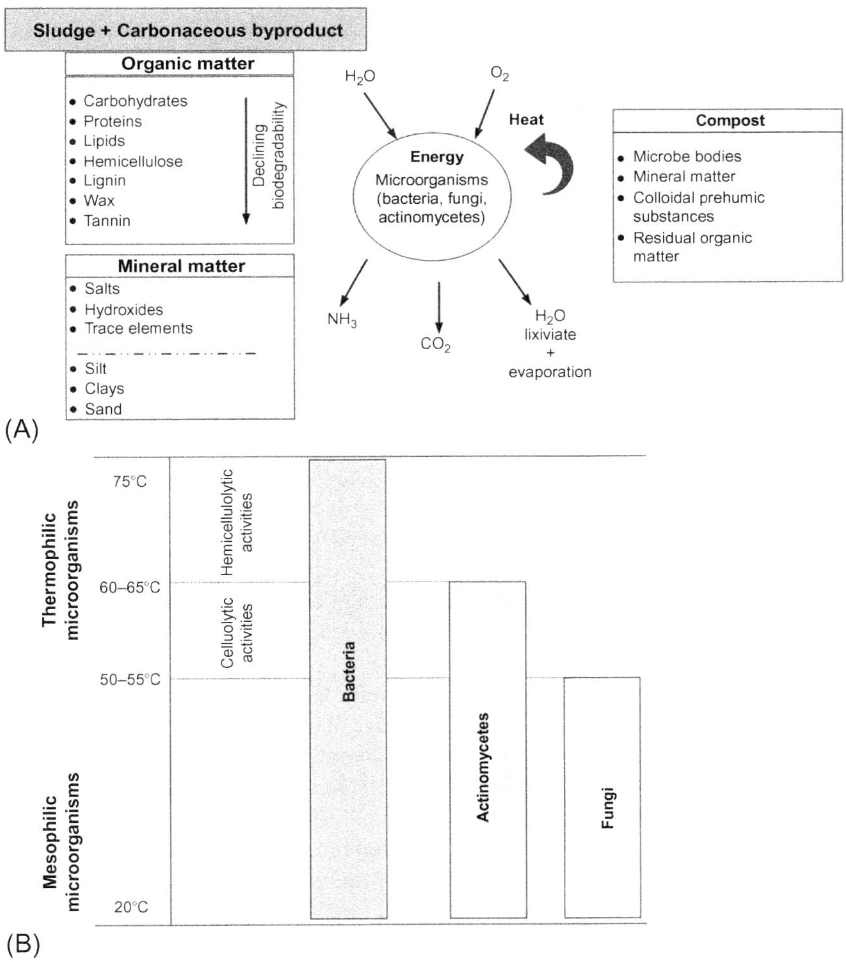

Fig. 6 Degradation of organic matter (A) and microbial activity (B) during the composting process (https://www.suezwaterhandbook.com).

After transformation into compost, the organic substance can be used as a fertilizer, structure reforming, or reclamation materials. The addition of compost impacts the change in the physical and chemical properties of the soil through improving the water and air conditions (e.g., aeration or increasing the capillary water capacity) and soil abundance in nutrients (i.e., nitrogen, phosphorus, or potassium). Microorganisms present in the compost mass enhance soil microflora and microfauna, thus intensifying the biological life of the soil and the soil-forming processes. High temperature of the composting process ensures compost safety in sanitary terms.

The two basic criteria of quality and efficiency of the compost are its stability and maturity. According to Som et al. (2009), compost maturity should be evaluated from the standpoint of several physicochemical parameters, such as:

- Moisture content.
- pH.
- Electrical conductivity (EC).
- C/N ratio.
- Organic matter content and biological, including:
- Biological activity indices (respiratory oxygen and CO_2 production).
- Dehydrogenase activity.
- Microbiological analysis.
- Compost phytotoxicity evaluation (Som et al., 2009; Tomati et al., 2010; Eggen and Vethe, 2001).

Composting can be used for unstabilized, stabilized, dewatered, and nondewatered sewage sludge. Fermented and dewatered sludge is safe in sanitary terms. However, the process of methane fermentation helps remove organic substances, which leads to the necessity of adding easily degradable organic waste. Watered sludge used for composting requires additional material in order to achieve the desired water content and create adequate aeration conditions. The advantage of raw sludge composting is a high content of organic substance, which is the source of organic carbon (18%–25% d.m.). However, this type of sludge can be contaminated with pathogenic microorganisms (Janosz Rajczyk, 2014).

From the standpoint of technology, raw sludge composting can be performed in heaps or reactors or with or without the additives of structure-reforming material (Fig. 7).

The choice of the sludge composting system depends on the degree of stabilization of sewage sludge and other types of waste used for composting. Nonstabilized sludge requires a closed and dynamic system at the first stage in order to limit the odor effect (Bień et al., 2014). Another option of this system is a static pile with forced suction, which also allows for control of cleaning of the polluted air. In the case of stabilized sludge, which ensures $C/N \approx 30$ and a good structure of the compost heap, the recommended method is windrow turning that limits the costs of the process (Bień et al., 2014; Wójtowicz et al., 2013). The reactor dynamic methods represent the first phase of composting of sewage sludge

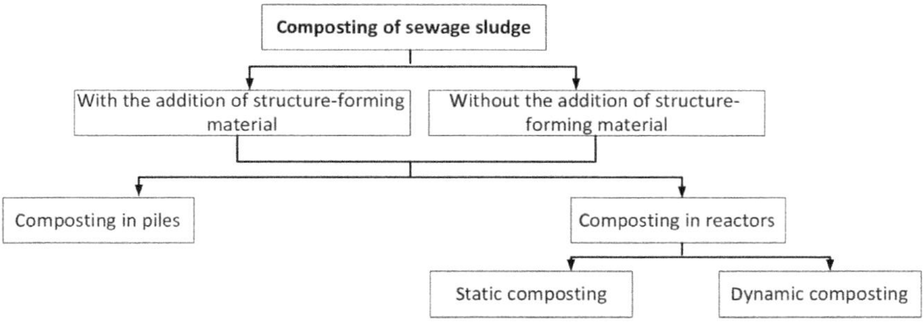

Fig. 7 Sewage sludge composting technologies (Janosz Rajczyk, 2014).

and nonstabilized waste. They reduce the time and help solve the odor problem during the initial phase and intensive decomposition. In the second phase, the second composting in the static or quasidynamic systems must be used because mycelium cannot be formed due to the cutting forces generated during the transfer of the composted material (fungi activity is necessary for a complete composting). The best solutions for sewage sludge processing are windrow composting and tunnel composting, with use and maintenance being relatively simple. Windrow composting is performed in the open air, on the surfaces covered with specific materials (e.g., asphalt) in order to protect the soil. This type of composting is used in small and medium-sized plants and can be also used for large wastewater treatment plants if a large composting surface is available (50–100 × 150–200 m) (Bień et al., 2014, Wójtowicz et. al. 2016, Sariola and Åkerman, 2012). Table 6 present the main sludge composting processes.

Numerous studies have shown that composting of only sewage sludge yields compost with poor quality due to a high humidity content and low air permeability. Furthermore, sewage sludge contains high contents of nitrogen compounds and heavy metals. In order to control the C:N ratio (from 25:1 to 50:1), it is necessary to mix sewage sludge with other components, such as rice straw, sawdust, grass, leaves, or plants from hydrobiological wastewater treatment plants (Petric and Selimbašić, 2008; Iqbal et al., 2010; Amir et al., 2008; Kouki et al., 2016). It is estimated that the mass percentage of sludge in the compost mass that ensures good product quality should not exceed 30%.

The conditions necessary for efficient co-composting of sewage sludge include (Siuta et al., 1996; Wasiak, 1994; Walker, 2001):

- High content of organic matter and specific ratio of sludge to additives (40%–50% d.m.).
- C:N ratio = 25–65, (the best C:N = 30:1).
- Composting temperature 55–60°C.
- Humidity in the composting mixture 50%–60%.

Table 6 Main sludge composting processes (https:/www.suezwaterhandbook.com; Sperling, 2007)

Composting method	System	Aeration	Mechan. turing over	Temp./ ventilation control	Odor control	Advantages	Disadvantages
Windrows	Open	Natura	Possible	–	–	• Low investment cost • Low operation costs	• High composting period • Possible mixing problem
		Forced	Possible	Possible	–		
	Confined	Forced	–	Possible	+		
Channels	Confined	Forced	+	+	+	• Batter odor control • Lower reaction time	• Investment required for the aeration system • Moderate operation and maintenance costs
Tunnels	Confined	Forced	–	+	+		
Bins	Confined or open	Forced	Possible	+	+		
Biological reactor	Confined	Forced	+	+	+	• Small land requirement • High degree of process control • Easiness in controlling temp. and odors	• Higher investment and operation and maintenance costs • Economically feasible only for large scales

- Aeration (mean from 90 to 160 $m^3/t_{d.m}$ h and 300 $m^3/t_{d.m.}$ h during the highest activity at the temperature of 60°C).
- The most beneficial composting time: minimum 21 days.
- Length of maturity period: 10–30 days.

There are many examples of the use of such green waste as additives for co-composting with sewage sludge. As mentioned before, sewage sludge cannot be composted without supplying additional carbon. Mixing it with macrophytes can lead to a higher C/N ratio and ensure adequate aerobic conditions for composting. The process of plant processing of sewage sludge uses, for example, *Phragmites australis*, *Typha latifolia*, *Salix viminalis*, or aquatic plants such as *Lemna minor* or the water hyacinth (Sobczyk and Sykuła, 2011; Kouki et al., 2016; Singh and Kalamdhad, 2012; Singht and Kalamdhad, 2014).

One of the worst and most invasive macrophytes used in CW is water hyacinth called "blue devil." It grows very fast in the form of a green mat in water reservoirs in tropical and subtropical regions of the world. It is abundant in all seasons of the year, which makes it an easily available material for co-composting with sewage sludge or municipal waste. (Gupta, 1987; Wafo et al., 2016; Dienye et al., 2017). Fresh water hyacinth is characterized by high water content (92.8%) and C:N ratio 36:1 (mean total organic carbon content: 338 g/kg, total nitrogen content: 9.5 g/kg). Due to high water content, initial processing of this macrophyte is recommended before it is used for composting. According to the literature, one of the methods is to use the methane fermentation process (Sankar Ganesh et al., 2008). Co-composting of fermented hyacinth with sewage sludge (hyacinth content from 25% to 50% vol. in the mixture) yields a stable and mature product of B class in terms of pathogen content.

The attempt was also made to increase the content of nutrients in the final water hyacinth compost product by additional inoculation of fungi and actinobacteria (Cellulolytic fungi: *Paecilomyces variotti* and *Chaetomium globosum*, ligninolytic fungi: *Pleurotus florida* and *Tramates versicolor* and actinobacteria: *Streptomyces lavendulae* and *Thermobifida fusca*). Inoculation of specific microflora at various stages of composting accelerates and increases the degree of carbon and cellulose removal, even by 20%. The time of the process can be substantially reduced whereas the product obtained is of high quality (Taiwo and Oso, 2004). Water hyacinth can be successfully used as a material for vermicomposting (Najar and Khan, 2013). The benefit of this strategy is that the vermicompost can be produced in 45 days, which represent half the duration of the conventional process and the product contains 30% water.

Fig. 8 presents a schematic diagram of the composting process sewage sludge with macrophyte.

Table 7 presents SWOT analysis of composting of sewage sludge with macrophytes.

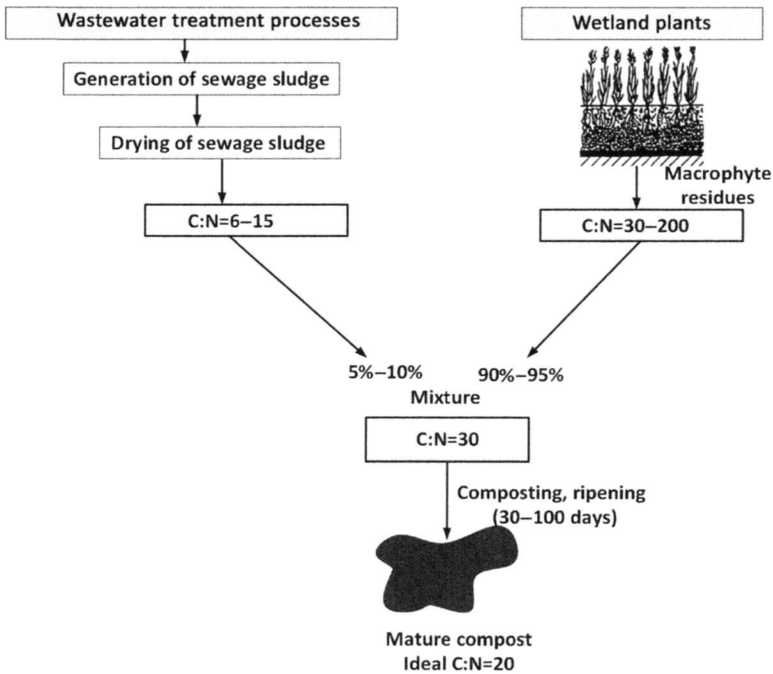

Fig. 8 Schematic diagram of the composting process sewage sludge with macrophyte (Kouki et al., 2016).

Table 7 SWOT analysis (Malczewska et al., 2017)

SWOT analysis of composting of sewage sludge with macrophytes	
Advantages/strengths	*Disadvantages/weaknesses*
• Opportunities for using both stabilized and nonstabilized sewage sludge • Opportunities for co-composting cheap biomass in the form of macrophytes • Reduction in N and P from wetland • Good sludge stabilization and hygienization • Sewage sludge mass and volume reduction • Sludge water content reduction • The obtained product is biochemically stable, with good physical properties and does not produce burdensome odors	• High level of investments (250–600 euro/tons of dry solids) • Necessity to use structure-forming materials • Energy consumption for compost aeration • Long-term process • Demand for the surface for compost maturing • Potential problems with selling the final product • Likelihood of dust emissions • Transport, sieving, and windrow throwing

Table 7 Continued

SWOT analysis of composting of sewage sludge with macrophytes	
• The product can be used as a fertilizer as it contains nitrogen, phosphorus, potassium, and microelements, and it improves soil properties • Easy operation of the installation	• Final product can be contaminated by pathogenic microorganisms • Required initial macrophyte processing due to high water content
Opportunities	*Threats*
• Solving the problems of sewage sludge management • Cheap methods to manage invasive macrophytes • Opportunities for financial benefits on selling the compost • Reduction in the amount of stored biodegradable waste in landfills • Opportunity for composting plant material from hydrobiological treatment plants • Opportunities for reaching the limits defined for reduction of • Biodegradable waste designed for storage • Intensification of the process through inoculation with fungi and actinobacteria	• Risk of local contamination of surface waters caused by liquor from windrow composting • Negative attitudes of local communities (odor problems)

References

Amin, O.M., 1988. Pathogenic micro-organisms and helminths in sewage products. Arabian Gulf, Country of Bahrain. Am. J. Public Health 78 (3), 314–315.

Amir, S., Merlina, G., Pinelli, E., Winterton, J., Revel, M., Hafidi, 2008. Microbial community dynamics during composting of sewage sludge and straw studies through phosholipid and neutral lipid analysis. J. Hazard. Mater. 159, 593–601.

Antonkiewicz, J., Jasiewicz, C., 2009. Zanieczyszczenia organiczne w osadach ściekowych. Zesz. Probl. Postep. Nauk. Rol. 537, 15–23. (In Polish).

Aubain, P., Gazzo, A., le Moux, J., Mugnier, E., 2002. Disposal and Recycling Routes for Sewage Sludge. Synthesis Report 22 February 2002. Arthur Andersen, EC DG Environment—B/2. http://ec.europa.eu/environment/waste/sludge/pdf/synthesisreport020222.pdf.

Bartkowska, I., 2017. Autotermiczna termofilna stabilizacja osadów ściekowych. Wydawnictwo Seidel-Przywecki. (In Polish).

Bever, J., Stein, A., Teichmann, H., 1997. Zaawansowane metody oczyszczania ścieków. Wyd. Projprzem-EKO, Bydgoszcz. (In Polish).

Bianchini, A., Bonfiglioli, L., Pellegrini, M., Saccani, C., 2015. Sewage sludge drying process integration with a waste-to-energy power plant. Waste Manag. 42, 159–165.

Bień, J., Wystalska, K., 2011. Osady ściekowe. Teoria i praktyka, Wydawnictwo Politechniki Częstochowskiej. (In Polish).

Bień, J., Neczaj, E., Worwąg, M., Grosser, A., Nowak, D., Milczarek, M., Janik, M., 2011. Kierunki zagospodarowania osadów w Polsce po roku 2013. Inżynieria. Ochr. Śr. 14 (4), 375–384. In Polish.

Bień, J., Górski, M., Gromiec, M., Kacprzak, M., Kamizela, T., Kowalczyk, M., Neczaj, E., Pająk, T., Wystalska, K., 2014. Ekspertyza, która będzie stanowić materiał bazowy do opracowania strategii postępowania z komunalnymi osadami ściekowymi na lata 2014-2020. Projekt współfinansowany ze środków Unii Europejskiej w ramach Programu Operacyjnego Pomoc Techniczna. (In Polish).

Bień, J., Kacprzak, M., Kamizela, T., Kowalczyk, M., Neczaj, E., Pająk, T., Wystalska, K., 2015. Komunalne osady ściekowe zagospodarowanie energetyczne i przyrodnicze. Wydawnictwo Politechniki Częstochowskiej. (In Polish).

Biotenmare Kick Off Meeting, 2014. Norway Grants in the Polish-Norwegian Research Programme operated by the National Centre for Research and Development "Innovation in recycling technologies of sewage sludge and other biowaste—energy and matter recovery" [POL-NOR/201734/76/13]. www.biotenmare.com. (In Polish).

Bourioug, M., Gimbert, F., Alaoui-Sehmer, L., Benbrahim, M., Aleya, L., Alaoui-Sossé, B., 2015. Sewage sludge application in a plantation: effects on tracemetal transfer in soil–plant–snail continuum. Sci. Total Environ. 502, 309–314.

Brix, H., 1994. Function of macrophytes in constructed wetlands. Water Sci. Technol. 29, 71–78.

Commission, E., 2008a. Directive 2008/98/EC of the European Parliament and of the Council of 19 November 2008 on Waste and Repealing Certain Directives (Waste Framework Directive, R1 Formula in Footnote of Attachment II).

Commission, E., 2008b. Environmental, Economic and Social Impacts of the Use of Sewage Sludge on Land. Final Report; Part III: Project Interim Reports. W.a. R. f. t. E. C. Milieu Ltd. DG Environment Under Study Contract DG ENV.G.4/ETU/2008/0076r.

Davis, L., 1995. A Handbook of Constructed Wetlands. vol. 1. U.S. Government Printing Office Superintendent of Document Mail Stop, Washington, DC.

De la Verga, D., Soto, M., Arias, C.A., van Oirschot, D., Kilian, R., Pascual, A., Alvarez, J., 2017. Constructed wetlands for industry wastewater treatment and removal of nutrients. In: Val de rio, A., Campos Gomez, J.L., Corral, A.M. (Eds.), Technologies for the Treatment and Recovery of Nutrients From Industral Wastewater. IGI. (Chapter 8).

Dienye, H.E., Olopade, O.A., Ikwuemesi, J.C., 2017. Macrophytes in Niger Delta Inland waters. Int. J. Horticult. Agr. Food Sci. 1, 7–14.

Eggen, T., Vethe, O., 2001. Stability indices for different composts. Compost. Sci. Utilisat. 9 (2), 27–37.

Eurostat, 2015. Sewage Sludge Production and Disposal From Urban Wastewater in Dry Substance (d.s).

Fijalkowski, K., Rosikon, K., Grobelak, A., Hutchison, D., Kacprzak, M.J., 2018. Modification of properties of energy crops under polish condition as an effect of sewage sludge application onto degraded soil. J. Environ. Manag. 217, 509–519.

Fytili, D., Zabaniotou, A., 2008. Utilization of sewage sludge in EU application of old and new methods—a review. Renew. Sust. Energ. Rev. 12, 116–140.

Gomesa, M.V.T., Rodriguesde-Souza, R., Teles, V.S., Mendes, E.A., 2014. Phytoremediation of water contaminated with mercury using Typha domingensis in constructed wetland. Chemosphere 103, 228–233.

Grobelak, A., Placek, A., Grosser, A., Singh, B.R., Almås, Å., Napora, A., Kacprzak, M., 2017. Effects of single sewage sludge application on soil phytoremediation. J. Clean. Prod. 155, 189–197.

Grobelak, A., Rorat, A., Kokot, P., Singh, B.R., Kacprzak, M., 2018. Mine waste rehabilitation case ctudies from Poland. In: Prasad, M.N.V., de Campos, P.J., Maiti, F.S.K. (Eds.), Bio-Geotechnologies for Mine Site Rehabilitation. first ed. Elsevier, pp. 515–529 (Chapter 28).

Grosser, A., 2017. The influence of decreased hydraulic retention time on the performance and stability of codigestion of sewage sludge with grease trap sludge and organic fraction of municipal waste. J. Environ. Manag. 203, 1143–1157.

Grosser, A., Worwag, M., Neczaj, E., Kamizela, T., 2013. Codigestion of organic fraction of municipal solid waste with different organic wastes: a review. In: Environmental Engineering IV—Proceedings of the Conference on Environmental Engineering IV.

Gupta, O.P., 1987. Aquatic Weed Management a Text Book and Manual. Today and Tomorrow Printers and Publishers, New Delhi.

Hall, J., 1999. Ecological and economical balance for sludge management option. In: Proceedings of the Workshop on Problems Around Sludge, Stresa, Italy Available from: http://ec.europa.eu/environment/waste/sludge/problems.

Herath, I., Vithanage, M., 2015. Phytoremediation in constructed wetlands. Phytoremediation 243–263.

Iqbal, M.K., Shafiq, T., Ahmed, K., 2010. Characterization of bulking agents and its effects on physical properties of compost. Bioresour. Technol. 101 (6), 1913–1919.

Janosz Rajczyk, M., 2014. Komunalne osady ściekowe podział, kierunki zastosowań oraz technologie przetwarzania, odzysku i unieszkodliwiania, Częstochowa. (In Polish).

Jędrczak, A., 2008. Biologiczne przetwarzanie odpadów Wydawnictwo Naukowe PWN, Warszawa. (In Polish).

Kacprzak, M., Fijałkowski, K., Grobelak, A., Rosikoń, K., Rorat, A., 2015. *Escherichia coli* and *salmonella* spp. early diagnosis and seasonal monitoring in the sewage treatment process by EMA-qPCR method. Pol. J. Microbiol. 64 (2), 143–148.

Kacprzak, M., Neczaj, E., Fijałkowski, K., Grobelak, A., Grosser, A., Worwag, M., Rorat, A., Brattebo, H., Almås, A., Singh, B.R., 2017. Sewage sludge disposal strategies for sustainable development. Environ. Res. 156, 39–46.

Kalderis, D., Aivalioti, M., Gidarakos, E., 2010. Options for sustainable sewage sludge management in small wastewater treatment plants on islands: the case of Crete. Desalination 260, 211–217.

Kosicka-Dziechciarek, D., Mazurkiewicz, J., Mazur, R., Wolna-Maruwka, A., 2016. Kompostowanie osadów ściekowych komunalnych i przydomowych. Technologia Wody 2 (46), 56–62. (In Polish).

Kosobucki, P., Chmarzyński, A., Buszewski, B., 2000. Sewage sludge composting. Pol. J. Environ. Stud. 9 (4), 243–248.

Kouki, S., Saidi, N., M'hiri, F., Hafiane, A., Hassen, A., 2016. Co-composting of macrophyte biomass and sludge as an alternative for sustainable management of constructed wetland by-products. Clean: Soil, Air, Water 44 (6), 694–702.

Koutika, L.-S., Rainey, H.J., 2015. A review of the invasive, biological and beneficial characteristics of aquatic species *Eichhornia Crassipes* and *Salvinia Molesta*. Appl. Ecol. Environ. Res. 13 (1), 263–275.

Legroux, J.P., Truchot, C., 2009. Bilan de dix années d'application de la réglementation relative à l'épandage des boues issues du traitement des eaux usées. Rapport n°771CGAAER du ministère de l'écologie, du développement et de l'aménagement durables (direction de l'eau) et du ministère de l'agriculture et de la pèche (direction générale de la forêt et des affaires rurales) (In French).

Malczewska, B., Woźniak, S., Jawecki, B., 2017. Zalety i wady kompostowania osadów ściekowych w porównaniu z termicznym ich spalaniem—studium przypadku. Sci. Rev. Energy Environ. Sci. 26 (1), 125–135. (In Polish).

Manios, T., 2004. The composting potential of different organic solid wastes: experience from the island of Crete. Environ. Int. 29, 1079–1089.

Manios, T., Stentiford, E.I., Millner, P.A., 2002. The removal of NH_3-n from primary treated wastewater in subsurface reed beds using different substrates. J. Environ. Sci. Health Part A A37 (3), 297–308.

Miksch, K., Sikora, 2018. Biotechnologia ścieków, Warszawa. (In Polish).

Moretti, S., Bertoncini, E., Abreu-Junior, C., 2015. Composting sewage sludge with green waste from tree pruning. Sci. Agric. 72 (5), 432–439.

Najar, I.A., Khan, A.B., 2013. Management of fresh water weeds (macrophytes) by vermicomposting using Eisenia fetida. Environ. Sci. Pollut. Res. 20, 6406–6417.

Obarska-Pempokowiak, H., Klimkowska, K., 1999. Distribution of nutrients and heavy metals in a constructed wetland system. Chemosphere 39, 303–312.

Pepper, I.L., Brooks, J.P., Gerba, C.P., 2006. Pathogens in biosoilds. Adv. Agr. Els. 90, 1–41.

Petric, I., Selimbašić, V., 2008. Development and validation of mathematical model for aerobic composting process. Chem. Energy J. 139 (2), 304–317.

Prażmo, Z., Krysińska-Traczyk, E., Skórska, C., Siatkowska, J., Cholewa, G., Dutkiewicz, J., 2003. Exposure to bioaerosol In a minicipial sewange treatment plant. Ann. Agric. Envirion. Med. 10 (2), 241–248.

Ramachandra, T.V., Bhat, S., Shivamurthy, V., 2017. Constructed wetlands for tertiary treatment of wastewater. ENVIS Technical Report, 124.

Rana, S., Jana, J., Bag, S.K., Mukherjee, S., Biswas, J.K., Ganguly, S., Sarkar, D., Jana, B.B., 2011. Performance of constructed wetlands in the reduction of cadmium in a sewage treatment cum fish farm at Kalyani, West Bengal, India. Ecol. Eng. 37, 2096–2100.

Reimers, R.S., McDonell, D.B., Little, M.D., Bowman, D.D., Englande, A.J., Henriques, W.D., 1986. Effectiveness of wastewater sludge treatment processes to inactivate parasites. Water Sci. Technol. 18, 387–404.

Romdhana, M.H., Lecomte, D., Ladevie, B., Ladevie, B., Sablayrolles, C., 2009. Monitoring of pathogenic microorganisms contamination during heat drying process of sewage sludge. Process. Saf. Environ. Prot. 87, 377–386.

Sadecka, Z., 2005. Zalety i wady oczyszczalni hydrobotanicznych. Ekotechnika 4, 24–27. (In Polish).

Sankar Ganesh, P., Sanjeevi, R., Gajalakshmi, S., Ramasamy, E.V., Abbasi, S.A., 2008. Recovery of methane-rich gas from solid-feed anaerobic digestion of ipomoea (Ipomoea carnea). Bioresour. Technol. 99 (4), 812–818.

Sariola, S., Åkerman, M., 2012. Projekt PURE, Dobre praktyki związane z gospodarką osadami ściekowymi, Komisja Środowiska Naturalnego Związku Miast Bałtyckich, Vanha Suurtori, FIN-20500 Turku. (In Finlandia).

Scholz, M., 2011. Wetland System, Storm Water Management Control. 2011,Springer-Verlag, London Part of the Green Energy and Technology book series (GREEN).

Sindu, R., Binod, P., Pandey, A., Madhavan, A., Alphonsa, J.A., Vivrk, N., Gnansounou, E., Castro, E., Faraco, V., 2017. Water hyacinth a potential source for value addition: an overview. Bioresour. Technol. 230, 152–162.

Singh, W.R., Kalamdhad, A.S., 2012. Concentration and speciation of heavy metals during water hyacinth composting. Bioresuor. Technol. 124, 169–179.

Singht, W.R., Kalamdhad, A.S., 2014. Potental for comosting of green Phumdi biomass of Loktak lake. Ecol. Eng. 67, 119–126.

Siuta, J., Wasiak, G., Chłopecki, K., Kaźmieczuk, M., Jońca, M., Mamelka-Sułek, S., 1996. Przyrodniczo-techniczne przetwarzanie osadów ściekowych na kompost. IOŚ ss, Warszawa, p. 40. (In Polish).

Sobczyk, R., Sykuła, M., 2011. Wykorzystanie makrofitów do przetwarzania osadów ściekowych na mursz. Forum Eksploatatora 4, 46–48. (In Polish).

Som, M.-P., Lemee, L., Ambles, A., 2009. Stability and maturity of a green waste and biowaste compost assessed on the basis of a molecular study using spectroscopy, thermal analysis, thermodesorption and thermochemolysis. Bioresour. Technol. 100 (19), 4404–4416.

Sperling, M., 2007. Wastewater Characteristics. In: Treatment and Disposal. IWA Publishing.

Środa, K., Kijo-Kleczkowska, A., Otwinowski, H., 2012. Termiczne unieszkodliwianie osadów ściekowych. Inżynieria Ekologiczna 28, 69–81 (In Polish).

Stańczyk-Mazanek, E., Nalewajek, T., Zabochnicka-Świątek, M., 2012. Drug-resistant microorganisms in soils fertilized with sewage sludge. Arch. Environ. Prot. 38 (1), 97–102.

Taiwo, L.B., Oso, B.A., 2004. Influence of composting techniques on microbial succession, temperature and pH in a composting municipal solid waste. Afr. J. Biotechnol. 3, 239–243.

Tomati, U., Madejon, E., Galli, E., 2010. Evolution of humic acid molecular weight as an index of compost stability. Compost. Sci. Util. 8 (2), 108–115.

Uggetti, E., Ferrer, I., Llorens, E., García, J., 2010. Sludge treatment wetlands: a review on the state of the art. Bioresour. Technol. 101, 2905–2912.

Uggetti, E., Ferrer, I., Molist, J., Garci, J., 2011. Technical, economic and environmental assessment of sludge treatment wetlands. Water Res. 45, 573–582.

Vymazal, J., 2008. Constructed wetlands for wastewater treatmen: a review. In: Sengupta, M., Dalwani, R. (Eds.), Proceedings of Taal 2007: The 12th Work Lake Conference. pp. 965–980.

Vymazal, J., 2010. Constructed wetlands for wastewater treatment. Water 2: 530–549.

Vymazal, J., 2011. Constructed wetlands for wastewater treatment: five decades of exerience. Environ. Sci. Technol. 45 (1), 65–69.

Vymazal, J., Kröpfelová, L., 2008. Wastewater Treatment in Constructed Wetlands With Horizontal Sub-Surface Flow. Springer Science + Business Media B.V. (Chapter 3).

Wafo, G.V.D., Nguemte, P.M., Nzouebet, W.A.L., Djocgoue, P.F., Kengne, I.M., 2016. Co-composting of sewage sludge and *Echinochloa pyramidalis* (Lam.) Hitchc. & Chase plant material from a constructed wetland system treating domestic wastewater in Cameroon. Afr. J. Environ. Sci. Technol. 10 (9), 272–282.

Walker, J.M., 2001. U.S. environmental protection agency regulations governing compost production and use. (Chapter 18), In: Stofella, P., Kahn, B. (Eds.), Compost Utilizaton in Horticultural Systems. CRC Press, Boca Raton, FL.

Wasiak, G., 1994. Wytwarzanie, właściwości i gospodarka osadami ściekowymi w Polsce na tle Zachodniej Europy i USA. W: Przyrodnicze użytkowanie osadów ściekowych Mater. semin. nauk.-tech. Warszawa, październik 1994. IOŚ, AG-CHEM EQ CO. INC s, Warszawa, pp. 11–23. (In Polish).

Wetzel, R.G., 2001. Limnology: Lake and River Ecosystems. Academic, San Diego, CA.

Wójtowicz, A., Jędrzejewski, C., Bieniowski, M., Darul, H., 2013. Modelowe rozwiązania w gospodarce osadowej. In: Izba Gospodarcza "Wodociągi Polskie". Stowarzyszenie Eksploatatorów Obiektów Gospodarki Wodno-Ściekowej. (In Polish).

Zdybel, J., Karamon, J., Cencek, T., 2009. Występowanie jaj nicieni pasożytniczych z rodzaju Ascaris, Trichuris i Toxocara w nawozach organicznych i organiczno-mineralnych oraz osadach ściekowych. Życie Wet. 84 (12), 992–996. (In Polish).

Zhang, Y., 2012. Design of a Constructed Wetland for Wastewater Treatment and Reuse in Mount Pleasant. Utah state University, Utah.

Further reading

Directive, C., 1999. Council directive 99/31/EC of 26 April 1999 on the landfill of waste (landfill directive). Off. J. Eur. Commun. L 182 (16) (Composting of sewage sludge with solid fraction of digested pulp from agricultural biogas plant).

Kannepallia, S., Ravita, B., Stroma, P.F., 2016. Composting of aged reed bed biosolids for beneficial reuse: a case study in New Jersey, USA. Compost Sci. Util. 24 (4), 281–290.

Coprocessing of sewage sludge in cement kiln

16

Jolanta Sobik-Szołtysek, Katarzyna Wystalska
Institute of Environmental Engineering, Czestochowa University of Technology, Częstochowa, Poland

1 Introduction

The gradual technological advances and optimization of the currently used technologies of wastewater treatment have led to an increase in the efficiency of the treatment process and, consequently, a continuous increase in the amount of generated sewage sludge (Chen et al., 2012; Grosser, 2017). Sludge is generated at different stages of wastewater treatment, whereas the amount and physical and chemical properties of sludge depend on the composition of the wastewater and treatment methods. It is estimated that sewage sludge is generated with the amount of about 3% of the wastewater subjected to treatment (Wzorek, 2012). In the past, sludge was mostly stored (including at sea) or used as an agricultural fertilizer. Nowadays, with the continuous increase in the mass of the sludge and the extension of the ban for storage, an intensive development of techniques of processing and treatment of this waste is being seen. The problem of using these methods and their environmental impact has been widely discussed (Li et al., 2009, 2010; Xiao et al., 2009, 2010; Magdziarz et al., 2011). The basic aim of efficient waste management is to search and use technologies that allow for maximal utilization of the potential of resources and energy. This allows for more efficient and environmentally friendly treatment integrated with the protection of natural resources, which is consistent with the principles of sustainable development and the introduction of closed matter circulation. Current tendencies for sewage sludge management include (Peters and Rowley, 2009; Murakami et al., 2009; Han et al., 2015; Chen et al., 2014; Zhou et al., 2014; Grosser, 2018; Kacprzak et al., 2017; Grobelak et al., 2017):

- Agricultural use as a fertilizer.
- Use for reclamation of industrial areas and waste landfills.
- Production of compost.
- Energy generation in the process of aerobic fermentation (biogas) and using it for intensification of biogas production in landfills.
- Thermal treatment.
- Storage within current legal regulations.

Burning belongs to the most expensive methods of sewage sludge utilization. However, in light of legal and technological limitations concerning storage and use of sludge in nature, thermal methods become especially important. This represents

Industrial and Municipal Sludge. https://doi.org/10.1016/B978-0-12-815907-1.00016-7

one of the most efficient methods of treatment of sludge containing organic substances, for example, the degree of destruction of organic matter during combustion in cement kilns reaches 99.999% (Genon and Brizio, 2008; Conesa et al., 2008; Arshed et al., 2014). Thermal technologies include combustion in furnaces and fluidized beds, cocombustion with other fuels, and alternative methods such as gasification, pyrolysis and wet oxidation (Werle and Wilk, 2010; Werle, 2012). Fig. 1 presents the methods used for thermal treatment of sewage sludge.

Combustion in conventional incineration plants leads to emissions of substantial amounts of gas contaminants (CO_2, CO, SO_2, NO_x), WWA, dioxins, furans, ash, and dust from dust-removing equipment or sludge from flue gas treatment installations. This leads to serious environmental threats (Liu et al., 2010a,b,c). Due to their composition, the ashes from burning sewage sludge should not be used directly in the environment. They can be used as an addition to cement or granular material for concrete structures (Chen et al., 2013; Cyr et al., 2007). Waste from combustion of sewage sludge modified in the process of vitrification can also be used as a granular material (Wystalska et al., 2013), which ensures opportunities for reduction of the sludge

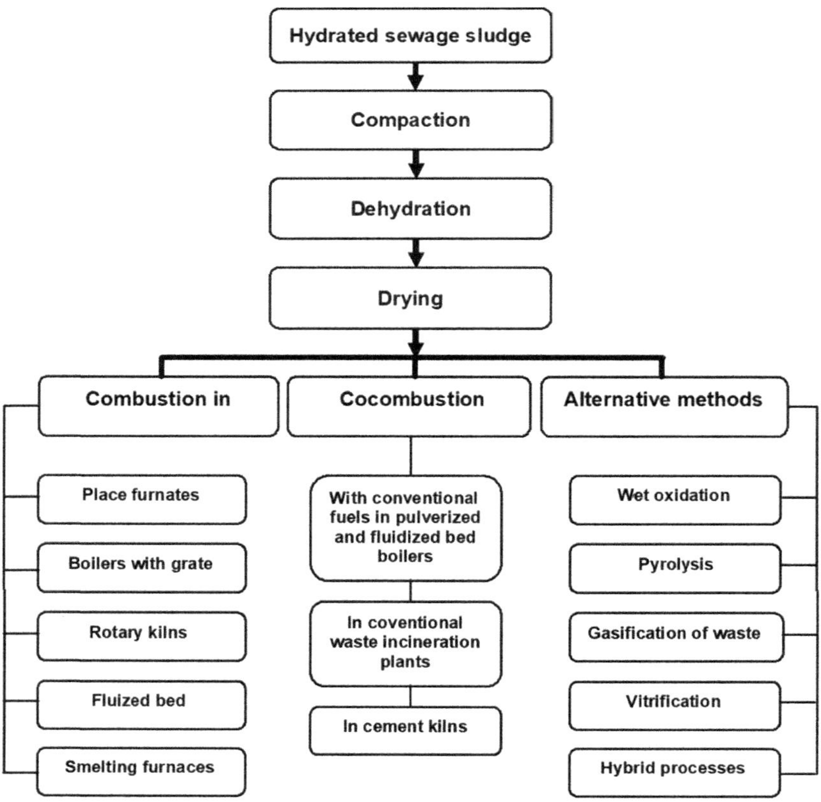

Fig. 1 Methods used for thermal treatment of sewage sludge.

volume to about 95%, integration of inorganic (also toxic) components in the glass structure, and a high degree of material conversion. The vitrification process is particularly useful for heavily polluted industrial sewage sludge, for example, from the tannery industry (Celary and Sobik-Szołtysek, 2014).

Combustion of sewage sludge in conventional incineration plants also leads to social protests due to the fear of a negative impact of thermal installations on the environment. Building an independent thermal plant for sewage sludge treatment generates a high investment expenditure and is economical only in the case of big treatment plants. The insufficient availability of many sludge incineration plants and the high costs of utilization suggest that the most economic and environmentally friendly technology is cocombustion of municipal waste with different types of fuels in existing facilities, for example, incineration plants for municipal waste, power plants, combined heat and power plants, or cement plants (Stasta et al., 2006; Stelmach and Wasilewski, 2008; Werle and Wilk, 2010; Lin et al., 2012). Due to the scale of production and substantial demand for fuels and raw materials, it is especially preferred to utilize cement kilns for this purpose. Large quantity of gas contaminants are generated during the combustion of conventional fuels, for example, CO_2, SO_2, and NO_x (Ali et al., 2011; Karagiannidis, 2012). In China, for instance, the cement industry represents the third-biggest source of emissions of these gases (Fu et al., 2015; Li et al., 2014), which intensify the greenhouse effect and the phenomenon of acid rains (Zhang et al., 2011). According to the most recent studies (Mikulcic et al., 2016; Teklay et al., 2016; Olivier et al., 2015), the cement industry is responsible for 8% of global anthropogenic emissions of CO_2 and 7% of consumption of conventional fuels (Chen et al., 2010a,b). Cement production processes lead to the release of 0.83 mg CO_2 per 1 mg of product. The use of an alternative fuel in the form of, for example, sewage sludge can substantially reduce these problems. Disposal of waste in cement kilns is cost effective and ecological because most of the contaminants contained in the waste are effectively destroyed or solidified in the cement clinker (Strazza et al., 2011; Rodríguez et al., 2013; Chatziaras et al., 2014; Pavlik et al., 2016). Other benefits for cement plants include reduction of costs of fuel and limitation of CO_2 emissions (Stehlik, 2009; Murakami et al., 2009; Lin et al., 2012; Li et al., 2012a,b). Furthermore, reduction of the use of nonrenewable fossil fuels translates into protection of natural energy resources and limitation of environmental damage caused by the utilization and incineration of fossil fuels. It also allows for the achievement of specific objectives contained in the Johannesburg Declaration on sustainable development and the Millennium Development Goals, contained in the United Nations Millennium Project (Georgiopoulou and Lyberatos, 2018).

2 Cement production process

The basic component in the process of cement production is a cement clinker that typically contains 68% CaO, 22% SiO_2, 6% Al_2O_3, and 3% Fe_2O_3. The raw materials for its production include limestone, marl, and clay that contain oxides of Si, Al, and Fe. Correction of the raw material mixture composition is provided by addition of sand, iron ore, or bauxite. The substitutes of natural resources and additives to cement are waste materials such as fly ash from the energy sector, blast furnace slag, silica dust, noncoal bearing shales, synthetic gypsum from flue gas desulfurization, or sewage sludge stabilized with calcium. The use of additives with a waste character at specific proportions does not have a negative effect on the quality of produced cement (Xu et al., 2014).

Two methods have been used in the cement industry to manufacture cement: dry and wet. In the dry method, milling materials and drying with hot waste gas occur in ball mills or cylinder and bowl mills. In the wet method, milling occurs in ball mills for wet milling, whereas raw material in the form of sludge with a water content of 32%–38% is collected in swimming pools, mixed, and, if the need arises, subjected to correction of chemical composition. Clinker burning is performed in rotational, pipe cement kilns with an inclined and rotating drum with a diameter of several meters and a length from several tens to even 200 m; the use of primary energy in the furnace accounts mostly for 70%–74% of the total demand for energy (Worrell, 2004). During high-temperature burning, the components of the raw material mixture are decomposed and gas products of decomposition are released: CO_2 from limestone materials, SO_2, and water vapor. The products of raw material mixture decomposition react with each other and form basic clinker materials in the form of silicates, aluminates, and calcium aluminoferrites. They are characterized by the ability to react with water, forming a hard concrete structure during crystallization.

Fuel supplied to the furnace is burnt in the lower part of the drum, whereas the stream of generated flue gases heats the burning material, moving according to furnace axis inclination. The so-called calcinators are also used, allowing for combustion of part of the fuel outside the drum before heat exchangers. The ash from fuel combustion is mixed with the burnt material and included in clinker composition. Flue gas, mixed with technological gases, is emitted to the atmosphere as exhaust gases. Burnt and powdery clinker is moved to the furnace (clinker cooler), where it is cooled down to about 100°C. Time of passage of the material through the pipe furnace is about 1 h in the dry method and 3–4 h in the wet method. The maximal incineration temperature in the rotary furnace reaches 2000°C, ensuring an intensive heat exchange that translates into furnace efficiency. Cement processes are characterized by a substantial surface of exchange of mass and 1200–3200 m^2 in the wet method and up to 80,000 m^2 in the dry method. This ensures a full reaction of volatile products of waste decomposition (HCl, HF, Cl, CN, NH_3, SO_2) with the components of raw material charge or their adsorption on the surface of the distributed solid phase. Consequently, these contaminants do not leave the system of the cement kiln. The material for cement production is heated in the furnace to about 1450°C and remains there for about 0.5 h.

The demand for heat energy in the process of cement clinker production includes the energy required by chemical/mineralogical reactions in the process of clinker burning (1700–1800 MJ/mg clinker) and heat energy needed for drying and heating raw materials, depending on the water content in the raw materials (BAT, 2013). Heat consumption during cement clinker burning also depends on the type of cement furnace and accounts for 3–4 GJ/mg product in the dry method and 5–6.4 GJ/mg product in the wet method (Alsop et al., 2007; Habert et al., 2010). The basic fuels in the cement industry include hard and brown coal and mazut. Due to high energy consumption and constant increases in prices of conventional fuels, this industry searches for opportunities to increase the percentage of alternative fuels in fuel mixtures used in rotary furnaces.

3 Sewage sludge as alternative fuels in cement kilns

Alternative fuels are obtained by mixing conventional fuels with waste that is not hazardous and should be, according to Lemarchand (2000), produced meeting the following conditions:

- Chemical quality of the fuel should meet legal standards in terms of environmental protection.
- Calorific value should ensure stability and continuity of the combustion process.
- Physical form of fuel should ensure ease of transport, storage, and supply to the incineration installation.

Due to the economics and efficiency of the clinker burning process, the components of alternative fuels include all the substances with waste character that have adequate energy properties (contain C and H in their composition) and chemical compounds in the amounts that can lead to deterioration of the quality of cement and components which are especially dangerous for the environment for example, radioactive substances. The European Committee for Standardization (CEN/TC 343, n.d.) specifies the parameters of the substitute fuel for the cement industry:

- Calorific value of 16–18 MJ/kg.
- Content of Cl $< 0.5\%$.
- Content Hg < 10 mg/kg d.m.

Processing of waste in the cocombustion process in cement kilns does not require an additional source of heat. By reducing the amount of greenhouse gases released, co-firing affects the reduction of global warming and the reduction of the potential for acidification of the environment (Garg et al., 2007). The only nuisance can be caused by odorants, which, in this case, require a specific modernization of the circulation of the used exhaust gases (Czechowska-Kosacka et al., 2015).

Individual fuels from waste are substantially varied in terms of chemical properties. Changes in the chemical composition of alternative fuels do not impact the emission of contaminants, the efficiency of processes occurring in cement kilns, or the product quality (Parlikar et al., 2016). It is necessary to conduct proper and continuous monitoring of the charge and cocombustion process conditions.

Adequate thermal conditions in the incineration zone are ensured by the calorific value of the fuel introduced to the main burner at the level of min. 22–24 MJ/kg. The calorific values of alternative fuels are usually lower. Therefore, their proportions in the fuel cofired in the main burner of the furnace have to ensure calorific value at a specific level. A lower calorific value of alternative fuels compared to conventional fuels and high water content lead to the reduction of flame temperature. It is estimated that the decline of the flame temperature by 100°C reduces furnace performance by

10% at a simultaneous increase in heat consumption by about 5%. The increase in the proportion of fuels from waste in the fuel mixture causes additional heat demand that is (Klein and Koenig, 2006):

- with participation to 40% fuels from waste: up to 100 MJ/mg of clinker.
- with participation from 40% to 70% fuels from waste: up to 210 MJ/mg of clinker.
- with participation from 70% to 100% fuels from waste: up to 330 MJ/mg of clinker.

The content of components that impact negatively on the product obtained in cement installations has to be constantly monitored, whereas the expected parameters of the alternative fuel require individual choice. For this reason, one should adequately choose the composition of the alternative fuels in the proportioning process, that is, the method of preparation of the substitute fuel with the waste material, in order to ensure adequate calorific value and adequate chemical composition of the clinker.

Apart from the economic benefits, the use of alternative fuels in the process of clinker burning stimulates the construction of more complex cogeneration installations (Cheung et al., 2006; Genon and Brizio, 2008). One inconvenience of the substitution of secondary fuels results from their low density compared to conventional fuels. With higher costs of transport and storage, the costs of alternative fuels calculated per unit of generated heat are higher than the costs of a conventional fuel (coke or carbon).

The chemical composition of sewage sludge (Table 1) confirms the opportunities for its use as a component of alternative fuels.

Potential emissions of contaminants and composition of the final product in the cement process are also affected by the presence of other contaminants in the sludge, especially heavy metals (Chen et al., 2012) whose content, according to various authors, is presented in Table 2.

The presence of heavy metals in sewage sludge can lead to their transfer to the clinker. With varied content of metals, the level of clinker contamination after substitution of the conventional fuel can increase or be reduced.

According to the terminology used in the European Union, sewage sludge is numbered among solid recovered fuels (SFR) (Van Tubergen et al., 2005; Van Tubergen, 2006), whereas the most frequent method of disposal is incineration. Cocombustion of sewage sludge is performed in many cement plants in Europe, mainly in Germany, Switzerland, Poland, Belgium, and the Netherlands as well as China, Japan, and the United States (Li et al., 2012a,b; Fang et al., 2015; Rahman et al., 2015). Substitution of natural fuels by combustible waste in cement kilns is on average 15%–30%, in individual cases exceeding 80%.

Unlike conventional fuels, sewage sludge is characterized by a high humidity and ash content, a low calorific value, and a greater content of volatile substances connected with a high content of elements, for example, N and S (Vainio et al., 2013). In this situation, the prerequisite for the combustion or cocombustion process is maximal thickening of the sludge. The basic parameter that determines the conditions of thermal use of sewage sludge is calorific value in the working state (Table 3.). Its value increases with the increase in the amount of organic substances and content of dry mass.

The high and stable content of organic substances in the dry matter of sludge ensures a positive energy balance and stable combustion heat. In energy terms, dried

Table 1 Content of basic elements in sewage sludge, according to various authors

Reference	Contents of main elements (% of dry matter)					
	C	H	O	N	S	Cl
Otero et al. (2008)	22.70	3.30	15.50	3.10	1.60	–
Gross et al. (2008)	30.10	4.12	17.84	3.79	0.85	–
Zabaniotou and Theofilou (2008)	49–53	–	–	4.5–7.5	–	0.8
Llorente and Garcia (2008)	38.4	5.9	15.8	5.2	0.8	0.08
Wzorek and Król (2009)	23.75–27.24	3.37–3.72	26,57–27.90	3.98–4.37	1.11–1.18	0.17–0.85
Wzorek (2012)	24.99	3.73	27.18	4.20	1.07	0.13
Hai and Xiaoqian (2012)	36,4	7,38	39,7	4,5	0,47	0,06
Kijo-Kleczkowska et al. (2013)	30.77	3.92	18.23	4.26	1.44	–
Werle (2013)	28–32	3.81–4.36	14–15	3.59–4.86	1.67–1.81	0.033–0.217
Huang et al. (2016)	34.04	5.03	23.48	6.09	1.67	–

Table 2 Content of certain heavy metals in sewage sludge, according to various authors

Reference	Contents of metals (ppm)										
	As	**Cr**	**Sn**	**Zn**	**Co**	**Cu**	**Ni**	**Pb**	**Hg**	**Fe**	**Cd**
Shih et al. (2005)	–	66-2021	–	354–640	–	80–2300	37–179	26–465	–	–	2.3–10.0
Otero et al. (2008)	6.2–15.3	10.,0–380.0	23.1–2.,1	2432–6100	10.9–40.0	80–800	16.0–50.0	20.0–49.5	1.99–2.50	23,586–26,000	–
Fytili and Zabaniotou (2008)	1.1–230	10–999,000	1.7–17.2	101–49,000	11.3–2490	204–1337	23.2–36.5	13–223	0.6–56	1000–154,000	1.0–3.41
Zabaniotou and Theofilou (2008)	–	3.8–31.7	–	804.4–2640	–	4.5–139.0	25.8–120	18.8–24.5	0.4–0.65	15,000–35,813	1.0–1.6
Wzorek and Król (2009)	2.77–6.72	2.,0–53.56	0.06–1.13	1931–3503	2.88–9.30	104–193.5	1.96–33.39	36.34–61.89	0.91–2.57	1236–57,994	1.03–3.09
Xu et al. (2014)	29.52	125.45	–	523.59	–	202.76	30.33	41.34	–	–	1.21

Table 3 Calorific value of sewage sludge, according to various authors

Reference	Calorific value (MJ/kg)
Werle and Wilk (2010) and Houdková et al. (2008)	6.7–12.0 (digested sludge) 13.3–17.5 (crude sludge) 15.0–17.0 (excess sludge)
Kijo-Kleczkowska et al. (2013)	12.574
Werle (2013)	10.7–13.0
Czechowska-Kosacka et al. (2015)	14.43–14.57 (dried sludge)
Fytili and Zabaniotou (2008)	16–23 (activated sludge) 9–13 (digested sludge)
Vamvuka and Sfakiotakis (2011)	14.8
Jiang et al. (2014)	15.59

sludge can be compared to wood, whose calorific value ranges from 14 to 19 MJ/kg (Czechowska-Kosacka et al., 2015). After initial processing on filtration presses, typical sewage sludge is characterized by a water content of about 80%, which translates into a calorific value of 0.5 MJ/kg. After a reduction in water content (in, for example, centrifuges) to 60%, calorific value increases to the level of 3–6 MJ/kg. A further increase in the calorific value of sludge can be obtained in the process of drying. The granulate obtained from drying, with a dry matter content of more than 80% and a calorific value of about 12–18 MJ/kg d.m., does not contain pathogenic organisms and is resistant to biodegradation. According to Stasta et al. (2006), sludge for cocombustion in cement kilns should be dewatered to a maximal 20% of water content and have a calorific value of minimal 9 MJ/kg. Excessive water content can lead to a decline in temperature in the furnace to the lower boundary value for the process of clinker burning and a reduction in product quality. According to Trezza and Scian (2005), alternative fuels for incineration in cement kilns should have the following parameters:

- Calorific value >14 MJ/kg.
- Water content <15%.
- Chlorine content <0.8% or <0.2%.
- Sulfur content <2.5%.

The combustion of the sediment carbonization is much slower than the combustion of hard coal, and 80% of the combustion products go to the raw material. A complete conversion of sewage sludge can be obtained through increased fragmentation, reduced water content, and a choice of adequate parameters of air for combustion (Nørskov, 2012). Furthermore, the composition of sewage sludge causes an intensification of the combustion process, resulting in a reduction in the ignition temperature for the fuel in the furnace (Kijo-Kleczkowska et al., 2013). Energy recovery from waste burnt in cement kilns is usually more effective compared to conventional incineration plants with energy generation systems. This is ensured by a more efficient use of waste heat, for example, for drying hydrated sewage sludge before cocombustion in the furnace (Damtoft et al., 2008; Dornburg et al., 2006).

Adjustment of cement kilns for cocombustion of sludge does not require building an additional combustion chamber. The dried sludge has a powdery form. Therefore, it can be supplied to the furnace by means of the equipment used for the transport of conventional fuels.

4 Sewage sludge as a raw material for cement production

Because sewage sludge contains SiO_2, CaO, Al_2O_3, Fe_3O_4, and other components of cement, it can be also used as a raw material for cement production (Yan et al., 2014). Lin et al. (2012) examined the effect of the addition of dried sewage sludge to raw clinker powder on properties of cement obtained in the process of clinker burning. Sewage sludge was added with the amount of 0.5%–15.0%, whereas the obtained mixtures were burnt at 1450°C for 2h. The major components of ecocement (cement obtained from sewage sludge) were similar to components contained in normal Portland cement. It was found that with the increase in sewage sludge, the amount of the C_2S phase increases in the raw powder. All ecocements were characterized by initial and final binding time compared to conventional cement. Binding time as extended with the increase in the content of sewage sludge in raw clinker powder. Ecocements also had lower early bending strength, increasing with the ratio of sewage sludge. An insignificant decline in compression strength was observed, yet without a significant effect on the final properties of the mature cement. The content of certain heavy metals in the clinker (Ti, Ba, Zn, Cr, Cu, Ni, and Pb) were increasing with the content of sewage sludge in the raw powder, whereas the contents of Mg, Sr, Al, Fe, and Mn were declining. The standards concerning leaching were met for all types of ecocement.

Rodríguez et al. (2013) analyzed opportunities for the use of ash from the combustion of sewage sludge as a component of raw material powder. Ash can be introduced directly to the molten clinker and, depending on their composition, the amount of natural raw materials needed for cement production can be substantially limited. This method of managing ash is possible despite the fact that it has a very small or even negligible pozzolan activity, even at substantial fragmentation (Donatello et al., 2010a,b). However, the presence of phosphates derived from ash in raw material powder increases the content of free calcium (Kwon et al., 2005). Therefore, the addition of this ash to powder is limited by the content of phosphates to the level above which the time of cement binding is increased and the binding force is decreased (Lin et al., 2009).

Around 1.6mg of raw material powder is needed for the production of 1mg of clinker (Huntzinger and Eatmon, 2009), of which 0.224mg can be dried sewage sludge. According to Rodríguez et al., 2013this technology allows for about a 14% decline in the demand for natural cement raw materials while reducing the content of free calcium in the clinker.

The biggest cement producer worldwide is China, where 2400 million mg of cement was produced in 2017, that is, more than 58.5% of the total production, followed by India (6.83%). The amount of cement produced worldwide is presented in Fig. 2.

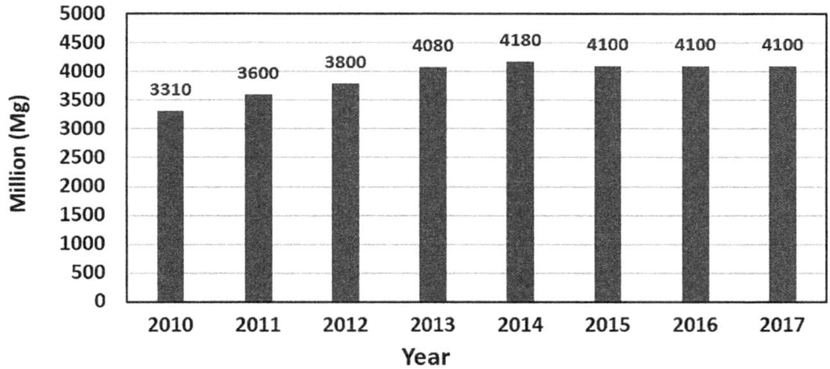

Fig. 2 World production of cement based on www.statista.com.

The mean consumption of raw materials for cement production is about 1.27 mg/mg cement (BAT, 2013). With annual cement production of 4100 million mg and assumption that 14% of raw clinker powder can be composed of dried sewage sludge, it is possible to use more than 570 million mg of this sludge as a substitute for natural resources. However, with the scale of production in the cement industry, the problem can arise with obtaining a sufficient amount of dried sewage sludge at a technologically and economically justified transport distance from a cement plant. According to the facts presented, sewage sludge can be used as both fuel and raw material. Fig. 3. presents a diagram of the cement production process using sewage sludge.

5 Effect of sewage sludge on process emissions and cement composition

In the process of cocombustion, sewage sludge is neutral in terms of CO_2 emissions. This is especially important because the cement industry emits huge amounts of this gas, especially during decarbonization of $CaCO_3$ (process emissions) and fuel combustion (fuel emissions). This accounts for 50% and 40%, respectively, of total CO_2 emissions. Therefore, it is important to use the solutions that limit the level of these emissions. The amount of carbon dioxide generated during clinker production is proportional to the content of CaO in carbonates from feed material. Process emissions are reduced if raw materials with a lower content of carbonates (fly ash or blast furnace slag) are used. Furthermore, fuel emissions can be reduced by increasing the cocombustion of alternative fuels with sewage sludge.

Cement production combined with thermal transformation of sewage sludge allows for a substantial limitation of the total load of contaminants released to the atmosphere compared to total emissions, which would be observed in the case of an independent and separate process of cement production and combustion of sludge in incineration plants. Emissions of gas contaminants during cocombustion of alternative waste fuels with conventional fuels is very similar to emissions observed during combustion of

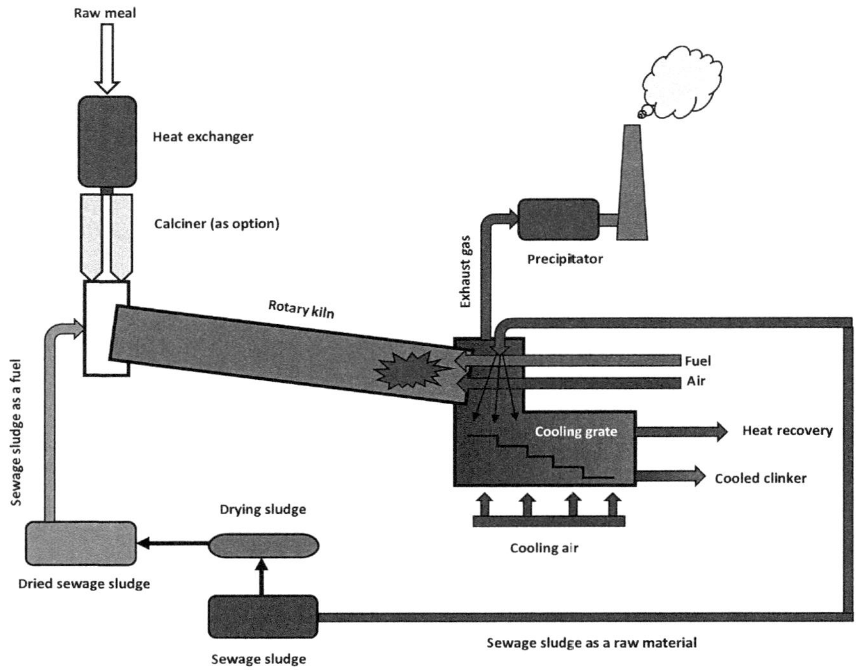

Fig. 3 Diagram of the cement production process using sewage sludge.

only conventional fuels (Conesa et al., 2008). In addition to gases, emissions from cement plants to the atmosphere also include other pollutants such as semivolatile organic compounds, heavy metals (Isikli et al., 2006; Marengo et al., 2006; Al-Khashman and Shawabkeh, 2006; Reijnders, 2007; Yatkin and Bayram, 2008), and persistent organic pollutants (POPs) in the form of polychlorinated dibenzo-*p*-dioxins and dibenzofurans (PCDD/F) (Karstensen, 2008; Liu et al., 2009; 2010a, b,c; Hu et al., 2013, Rivera-Austrui et al., 2014). These pollutants can be emitted in the form of gas or adsorbed on solids. The high and stable temperature reached in cement kilns ensures the total decomposition of organic substances and a very high degree of decomposition of toxic compounds (Conesa et al., 2011). Mineral solid components formed during combustion and heavy metals contained in the sludge are almost completely integrated in the clinker structure, which, while maintaining specific limitations in terms of the amount of the alternative fuel used, does not lose its beneficial properties (Pipilikaki et al., 2005; Shih et al., 2005; Trezza and Scian, 2005). The major source of PCDD/PCDF in gases is reactions in the gaseous phase and at the contact between gaseous and solid phases in the heater, precalcinator, or cyclone heat exchanges as well as the processes of adsorption and desorption (Karstensen, 2008). The in-depth analysis of the mechanisms of PCDD/F formation in the cement kilns for cocombustion of municipal solid waste and sewage sludge was published by Zhao et al. (2017). The study emphasized, for example, the importance of the chlorination reaction in the formation of PCDD/F.

Process emissions also depend on the presence of nitrogen, sulfur, and chlorine in the alternative fuel. Formation of NO_x depends on the amount of nitrogen in the fuel, the process temperature in the cement kiln, the type of burner, and the time of raw materials remaining in the furnace (Genon and Brizio, 2008). The low content of nitrogen in alternative fuels (0.3%–0.5%) translates into limitation of these pollutants. At high temperatures, sewage sludge releases ammonia very quickly (Cao et al., 2013). Around 70% of compounds that contain nitrogen from sewage sludge are transformed into ammonia (Liu et al., 2015). The study by Lv et al. (2016) found that the use of the addition of sewage sludge in the cement kiln causes a reduction in NO_x emissions, whose level depends mainly on the amount of NH_3 released from the sludge. Fang et al. (2015) examined the impact of the location and rate of dosing sewage sludge on the cocombustion process, with particular focus on the efficiency of NO_x removal. The results of the experiment demonstrated that the use of sewage sludge as an alternative fuel and denitration agent is conducive to a reduction in NO_x, which mainly depends on the speed of material transport in the furnace and the point of charge application. With the speed of sludge supply of 9 mg/h in the combustion zone of the precalcinator and combustion temperature from 850°C to 1000°C, NO_x removal was 75.82%. It was also found that the speed and location of sludge supply in the furnace can substantially impact the reduction of CH_4, HCN, and NH_3.

Sulfur content in the dry mass of sewage sludge is substantial and comparable to the content of this element in hard coal of low quality. Effective absorption of sulfur dioxide in the cement kiln results in its minimal emissions to the atmosphere, but the SO_2 emissions are substantially affected by the content of volatile sulfur in raw materials for cement production rather than fuel.

Examinations by Conesa et al. (2008) confirmed the lack of relationships between the amount of chlorine introduced to the production process with waste fuels and HCl emissions. The presence of calcium oxide also allows for a full neutralization of gaseous organic fluorine compounds (freons), which are decomposed in the furnace temperature and release hydrogen fluoride. This compound reacts with CaO and forms hardly water-soluble calcium fluorides that are transferred to the cement mass and fixed.

Emissions of mercury from clinker burning in the EU are at the level of 0.01–0.05 mg/m^3, with mean of 0.02 mg/m^3 (BAT, 2013). In a rotary furnace, mercury comes from the raw material for clinker production and fuel. Some authors (Schreiber et al., 2005) have linked its presence to the presence of stable mercury silicates (for example, $HgSiO_3$, $Hg_6Si_2O_7$). Mean concentration of this element in the clinker is 5.2 ppb (Mlakar et al., 2010). The discharge of mercury from the combustion system occurs with exhaust gases and the clinker. The mercury content in sewage sludge is estimated at 0.31–1.45 ppm d.m. (CEMBUREAU-CSI, 2009). Zabaniotou and Theofilou (2008) presented the results of examinations concerning cocombustion of humid sewage sludge in the Vassiliko cement plant in Cyprus. The analysis of Hg emissions revealed that the use of sewage sludge did not lead to exceeding the standards of emissions of harmful substances due to high temperature, reaching up to 1800°C, and a long time of gas remaining in the combustion chamber. Furthermore, in the alkaline environment, acid components of combustion gases were subjected to chemical binding.

Yu et al. (2017) examined the effect of chlorine on the volatilization of heavy metals during cocombustion of sewage sludge. These researchers analyzed the effect of the addition of chlorides on the volatilization of Cu, Ni, and Pb. With the increase in the content of $CaCl_2$ from 0% to 1.2%, the volatilization of Cu and Ni increased by 30%. This pattern was not observed with respect to Zn. Compared to NaCl, a greater efficiency of removal of heavy metals was found for other chlorine salts, for example, $AlCl_3 \cdot 6H_2O$, $FeCl_3 \cdot 6H_2O$, and $CaCl_2$. Maximal coefficient of volatilization of individual heavy metals at 1450°C was from 60.8% (for Cu) to 83.0% (for Pb). The increase in process temperature in the cement kiln accelerates the formation of volatile chlorides of heavy metals and, consequently, leads to a rapid increase in their volatilization with the flue gases. With the temperature increase from 950°C to 1050°C, Pb volatilization increased by nearly 30%. Intensity of heavy metal removal also depends on the energy of chloride binding (Luan et al., 2013), with greater efficiency observed for organic chlorides compared to inorganic (Li et al., 2015).

The limitation to the processes of cocombustion of sewage sludge in cement kilns is to obtain the expected composition of the product. Cement clinker contains elements and minerals that are present in the fuel as contaminants. During the burning process, these components will be built into the crystalline network of clinker materials, whereas some of them may cause deterioration of clinker properties. An example is ZnO, which, even with insignificant amounts, substantially reduces the cement strength and significantly extends binding time. Some chemical components contained in sludge, for example, Zn and Pb, can delay the reaction of clinker hydration (Liu et al., 2010a,b,c). This also concerns the content of compounds of Na, K, Cr, Pb, and other metals, chlorides, and sulfates. Excessive content of certain heavy metals in alternative fuels can be the cause of disturbances in the process of clinker formation. Furthermore, an elevated concentration of heavy metals in cement and cement dust increases the risk of their leaching into the environment. Alkali (K_2O and Na_2O) are introduced with feed material and fuel and, due to sublimation at high temperatures, form high concentrations in the gaseous phase. This can lead to formation of the build-up layers in cyclone exchangers and at the inlet to the furnace as well as the creation of lumps and rings inside the furnace.

Excessive content of chlorine can initiate the synthesis of dioxins, reduce the quality of produced cement, and lead to the destruction of the ceramic lining of the cement kiln as a result of corrosion caused by concentration of chlorides in the external and internal circulation of the rotary furnace. This leads to the formation of build-ups on cooler parts of linings and exchangers that heat raw material (Kääntee et al., 2004). Solid material can accumulate in cyclones, piping, and heater, disturbing a stable flow of process gases and other streams of matter in the furnace (Chatterjee, 2011). The presence of chlorine in fuel has an effect on the emissions of acid gases. Chlorine formed during combustion reacts with alkaline metals of raw material (Na and K), whereas salts evaporate at furnace temperature and are discharged with waste gases.

Increased presence of P_2O_5 in raw material powder or fuel can be the cause of a reduction in the early reactivity of cement, its strength, and clinkerization potential (Li and Zhang, 2011; Strigáč, 2015; Yang et al., 2016).

Sewage sludge combustion as an alternative fuel does not have a negative effect on clinker quality. The physical properties and mineral composition of cement clinkers produced with and without the use of sewage sludge are almost identical.

The use of solid waste, including sewage sludge, as a supplementing fuel with adequate calorific value in furnaces for cement production is one of the best available technologies for complete and safe treatment, combining the benefits of waste treatment and the production of energy (Conesa et al., 2011; Lin et al., 2012; Valderrama et al., 2013). Cocombustion of sewage sludge is not life threatening, not hazardous to health, and does not lead to negative changes in the environment (Schuhmacher et al., 2009; Nadal et al., 2009; Rovira et al., 2010; Hong and Li, 2011). The benefits of cocombustion of sewage sludge include:

- Reduction in the volume of generated sludge.
- Low sensitivity to the variability of fuel composition.
- Minimization of environmental odors.
- Opportunity for using secondary combustion processes.
- Long time of material and exhaust gases remaining at high temperatures.
- Very high thermal capacity of the furnace, ensuring constant temperature and process continuity, even in the case of an emergency stopping the combustion process.
- Combustion occurring in an alkali environment, allowing for neutralization of acid forming gases.
- Waste-free fuel combustion because ash after incineration is mixed with the burnt material whereas its components are included in the cement clinker.

Acknowledgment

This research was supported by BS/PB-401-304/11 in the Institute of Environmental Engineering, Częstochowa University of Technology.

References

Ali, M.B., Saidur, R., Hossain, M.S., 2011. A review on emission analysis in cement industries. Renew. Sust. Energ. Rev. 15 (5), 2252–2261.

Al-Khashman, O.A., Shawabkeh, R.A., 2006. Metals distribution in soils around the cement factory in southern. Jordan. Environ. Pollut. 140, 387–394.

Alsop, P.A., Chen, H., Tseng, H., 2007. The Cement Plant Operations Handbook, fifth ed. Tradeship Publications, Surrey, UK.

Arshed, S.F., Raza, S.A., Akhtar, A.B.T., Abdullah, Y., 2014. Comparative analysis of air emission from cement plant using TDF as partial substitute for coal. Environmental research. Eng. Manage. 68 (2), 61–68.

BAT—Best Available Techniques, 2013. Reference Document for the Production of Cement, Lime and Magnesium Oxide. http:/eippcb.jrc.ec.europa.eu/reference/BREF/CLM_30042013_ DEF.pdf.

Cao, J.P., Li, L.Y., Morishita, K., Xiao, X.B., Zhao, X.Y., Wei, X.Y., Takarada, T., 2013. Nitrogen transformations during fast pyrolysis of sewage sludge. Fuel 104, 1–6.

Celary, P., Sobik-Szołtysek, J., 2014. Vitrification as an alternative to landfilling of tannery sewage sludge. Waste Manag. 34, 2520–2527.

CEMBUREAU-CSI, 2009. Mercury in the Cement Industry. Universite de Liege. April 2009, http:/www.unep.org/hazardoussubstances/Portals/9/Mercury/A_Inventories20Industry% 20-%20 Hg%20report%20CEMBUREAU%20April%202010.pdf. (Accessed 21 March 2012).

CEN/TC 343, n.d. Solid Recovered Fuels.

Chatterjee, A.K., 2011. Chemistry and engineering of the clinkerization process—incremental advances and lack of breakthroughs. Cem. Concr. Res. 41, 624–641.

Chatziaras, N., Psomopoulos, C.S., Themelis, N.J., 2014. Use of alternative fuels in cement industry. In: Proceedings of the 12th International Conference on Protection and Restoration of the Environment. ISBN: 978-960-88490-6-8.

Chen, C., Habert, G., Bouzidi, Y., Jullienc, A., Venturac, A., 2010a. LCA allocation procedure used as an incitative method for waste recycling: an application to mineral additions in concrete. Resour. Conserv. Recycl. 54, 1231–1240.

Chen, C., Habert, G., Bouzidi, Y., Jullienc, A., 2010b. Environmental impact of cement production: detail of the different processes and cement plant variability evaluation. J. Clean. Prod. 18, 478–485.

Chen, H., Yan, S.H., Ye, Z.L., Meng, H.J., Zhu, Y.G., 2012. Utilization of urban sewage sludge: Chinese perspectives. Environ. Sci. Pollut. Res. 19, 1454–1463.

Chen, M., Blanc, D., Gautier, M., Mehu, J., Gourdon, R., 2013. Environmental and technical assessments of the potential utilization of sewage sludge ashes (SSAs) as secondary raw materials in construction. Waste Manag. 33, 1268–1275.

Chen, Y.Q., Yu, F., Liang, S.W., Wang, Z.P., Liu, Z.Z., Xiong, Y., 2014. Utilization of solar energy in sewage sludge composting: fertilizer effect and application. Waste Manag. 34, 2014–2021.

Cheung, W.H., Choy, K.K.H., Hui, D.C.W., Porter, J.F., McKay, G., 2006. Use of municipal solid waste for integrated cement production. Dev. Chem. Eng. Miner. Process. 14 (1/2), 193–202.

Conesa, J.A., Gálvez, A., Mateos, F., Martín-Gullón, I., Font, R., 2008. Organic and inorganic pollutants from cement kiln stack feeding alternative fuels. J. Hazard. Mater. 158, 585–592.

Conesa, J.A., Rey, L., Egea, S., Rey, M.D., 2011. Pollutant formation and emissions from cement kiln stack using a solid recovered fuel from municipal solid waste. Environ. Sci. Technol. 45, 5878–5884.

Cyr, M., Coutand, M., Clastres, P., 2007. Technological and environmental behavior of sewage sludge ash (SSA) in cement-based materials. Cem. Concr. Res. 37, 1278–1289.

Czechowska-Kosacka, A., Cel, W., Kujawska, J., Wróbel, K., 2015. Alternative fuel production based on sewage sludge generated in the municipal wastewater treatment. Annu. Set Environ. Prot. 17, 246–255.

Damtoft, J.S., Lukasik, J., Herfort, D., Sorrentino, D., Gartner, E.M., 2008. Sustainable development and climate change initiatives. Cem. Concr. Res. 38, 115–127.

Donatello, S., Tyrer, M., Cheeseman, C.R., 2010a. Comparison of test methods to assess pozzolanic activity. Cem. Concr. Compos. 32, 121–127.

Donatello, S., Freeman-Pask, A., Tyrer, M., Cheeseman, C.R., 2010b. Effect of milling and acid washing on the pozzolanic activity of incinerator sewage sludge ash. Cem. Concr. Compos. 32, 54–61.

Dornburg, V., Faaij, A.P.C., Meuleman, B., 2006. Optimizing waste treatment systems. Part A: methodology and technological data for optimizing energy production and economic performance. Resour. Conserv. Recycl. 49, 68–88.

Fang, P., Tang, Z.J., Huang, J.H., Cen, C.P., Tang, Z.X., Chen, X.B., 2015. Using sewage sludge as a denitration agent and secondary fuel in a cement plant: a case study. Fuel Process. Technol. 137, 1–7.

Fu, S.L., Song, Q., Yao, Q., 2015. Study on the catalysis of $CaCO_3$ in the SNCR $deNO_X$ process for cement kilns. Chem. Eng. J. 262, 9–17.

Fytili, D., Zabaniotou, A., 2008. Utilization of sewage sludge in EU application of old and new methods—a review. Renew. Sust. Energ. Rev. 12, 116–140.

Garg, A., Smith, R., Hill, D., Simms, N., Pollard, S., 2007. Wastes as co-fuels: the policy framework for solid recovered fuel (SRF) in Europe, with UK implications. Environ. Sci. Technol. 41 (14), 4868–4874.

Genon, G., Brizio, E., 2008. Perspectives and limits for cement kilns as a destination for RDF. Waste Manag. 28, 2375–2385.

Georgiopoulou, M., Lyberatos, G., 2018. Life cycle assessment of the use of alternative fuels in cement kilns: a case study. J. Environ. Manag. 216, 224–234.

Grobelak, A., Placek, A., Grosser, A., Singh, B.R., Almås, Å.R., Napora, A., Kacprzak, M., 2017. Effects of single sewage sludge application on soil phytoremediation. J. Clean. Prod. 155, 189–197.

Gross, B., Eder, C., Grziwa, P., Horst, J., Kimmerle, K., 2008. Energy recovery from sewage sludge by mean of fluidized bed gasification. Waste Manag. 28, 1819–1826.

Grosser, A., 2017. The influence of decreased hydraulic retention time on the performance and stability of co-digestion of sewage sludge with grease trap sludge and organic fraction of municipal waste. J. Environ. Manag. 203, 1143–1157.

Grosser, A., 2018. Determination of methane potential of mixtures composed of sewage sludge, organic fraction of municipal waste and grease trap sludge using biochemical methane potential assays. A comparison of BMP tests and semi-continuous trial results. Energy 143, 488–499.

Habert, G., Billard, C., Rossi, P., Chen, C., Roussel, N., 2010. Cement production technology improvement compared to factor 4 objectives. Cem. Concr. Res. 40, 820–826.

Hai, L., Xiaoqian, M., 2012. Simulation of co-incineration of sewage sludge with municipal solid waste in a grate furnace incinerator. Waste Manag. 32, 561–567.

Han, Y., Hwang, G., Kim, D., Park, S., Kim, H., 2015. Porous Ca-based bead sorbents for simultaneous removal of SO_2, fine particulate matters, and heavy metals from pilot plant sewage sludge incineration. J. Hazard. Mater. 283, 44–52.

Hong, J.L., Li, X.Z., 2011. Environmental assessment of sewage sludge as secondary raw material in cement production—a case study in China. Waste Manag. 31, 1364–1371.

Houdková, L., Boráan, J., Ucekaj, V., Elsäßer, T., Stehlík, P., 2008. Thermal processing of sewage sludge—II. Appl. Therm. Eng. 28, 2083–2088.

Hu, J.C., Zheng, M.H., Nie, Z.Q., Liu, W.B., Liu, G.R., Zhang, B., Xiao, K., 2013. Polychlorinated dibenzo-p-dioxin and dibenzofuran and polychlorinated biphenyl emissions from different smelting stages in secondary copper metallurgy. Chemosphere 90, 89–94.

Huang, L., Liu, J., He, Y., Sun, S., Chen, J., Sun, J., Chang, K., Kuo, J., Ning, X., 2016. Thermodynamics and kinetics parameters of co-combustion between sewage sludge and water hyacinth in CO_2/O_2 atmosphere as biomass to solid biofuel. Bioresour. Technol. 218, 631–642.

Huntzinger, D.N., Eatmon, T.D., 2009. A life-cycle assessment of Portland cement manufacturing: comparing the traditional process with alternative technologies. J. Clean. Prod. 17, 668–675.

Isikli, B., Demir, T.A., Akar, T., Berber, A., Urer, S.M., Kalyoncu, C., Canbek, M., 2006. Cadmium exposure from the cement dust emissions: a field study in a rural residence. Chemosphere 63, 1546–1552.

Jiang, L., Liang, J., Yuan, X., Li, H., Li, C., Xiao, Z., Huang, H., Wang, H., Guangming Zeng, G., 2014. Copelletization of sewage sludge and biomass: the density and hardness of pellet. Bioresour. Technol. 166, 435–443.

Kääntee, U., Zevenhoven, R., Backman, R., Hupa, M., 2004. Cement manufacturing using alternative fuels and the advantages of process modeling. Fuel Process. Technol. 85, 293–301.

Kacprzak, M., Neczaj, E., Fijałkowski, K., Grobelak, A., Grosser, A., Worwag, M., Rorat, A., Brattebo, H., Almås, Å., Singh, B.R., 2017. Sewage sludge disposal strategies for sustainable development. Environ. Res. 156, 39–46.

Karagiannidis, A., 2012. Waste to Energy: Opportunities and Challenges for Developing and Transition Economies. Springer Verlag, London.

Karstensen, K.H., 2008. Formation, release and control of dioxins in cement kilns. Chemosphere 70, 543–560.

Kijo-Kleczkowska, A., Środa, K., Otwinowski, H., 2013. Study into combustion of sewage sludge as energetic fuel. Arch. Min. Sci. 58 (4), 1085–1110.

Klein, H., Koenig, V., 2006. Model calculations of the fuel energy requirement for the clinker burning process. Cement Int. 4 (3), 44–63.

Kwon, W.-T., Kim, Y.-H., Chu, Y.-S., Lee, J.-K., Kim, I.-S., Kim, S.-R., 2005. Effect of P_2O_5 and chloride on clinkering reaction. J. Mat. Online 1, 1–8.

Lemarchand, D., 2000. Burning issues. Int. Cement Rev. 2, 65–67.

Li, F., Zhang, W., 2011. Combustion of sewage sludge as alternative fuel for cement industry. J. Wuhan Univ. Technol. 26 (3), 556–560.

Li, P.-S., Hu, Y., Yu, W., Yue, Y.-N., Xu, Q., Hu, S., Hu, N.-S., Yang, J., 2009. Investigation of sulfur forms and transformation during the co-combustion of sewage sludge and coal X-ray photoelectron spectroscopy. J. Hazard. Mater. 167, 1126–1132.

Li, W., Li, W., Liu, H., 2010. Effect of sewage sludge on rheological characteristics of coal-water slurry. Fuel 89, 2505–2510.

Li, Y., Wang, H., Zhang, J., Wang, J., 2012a. Disposal of obsolete pesticides including DDT in a Chinese cement plant as blueprint for future environmentally sound co-processing of hazardous waste including POPS in the cement industry. Proc. Environ. Sci. 16, 624–627.

Li, Y.Q., Wang, H.Z., Zhang, J., Wang, J.J., Ouyang, L., 2012b. The industrial practice of coprocessing sewage sludge in cement kiln. Proc. Environ. Sci. 16, 628–632.

Li, C., Nie, Z.R., Cui, S.P., Gong, X.Z., Wang, Z.H., Meng, X., 2014. The life cycle inventory study of cement manufacture in China. J. Clean. Prod. 72, 204–211.

Li, R., Zhao, W., Li, Y., Wang, W., Zhu, X., 2015. Heavy metal removal and speciation transformation through the calcination treatment of phosphorus-enriched sewage sludge ash. J. Hazard. Mater. 283, 423–431.

Lin, K.L., Lin, D.F., Luo, H.L., 2009. Influence of phosphate of the waste sludge on the hydration characteristics of eco-cement. J. Hazard. Mater. 168, 1105–1110.

Lin, Y.M., Zhou, S.Q., Li, F.Z., Lin, Y.X., 2012. Utilization of municipal sewage sludge as additives for the production of eco-cement. J. Hazard. Mater. 213–214, 457–465.

Liu, G.R., Zheng, M.H., Liu, W.B., Wang, C.Z., Zhang, B., Gao, L.R., Su, G.J., Xiao, K., Lv, P., 2009. Atmospheric emission of PCDD/Fs, PCBs, hexachlorobenzene, and pentachlorobenzene from the coking industry. Environ. Sci. Technol. 43, 9196–9201.

Liu, G.R., Zheng, M.H., Lv, P., Liu, W.B., Wang, C.Z., Zhang, B., Xiao, K., 2010a. Estimation and characterization of polychlorinated naphthalene emission from coking Industries. Environ. Sci. Technol. 44, 8156–8161.

Liu, G.X., Li, Y.Q., Zeng, F., Fan, W., Guo, L.X., 2010b. Experimental study on sludge disposal of wastewater treatment plants utilizing cement kilns. J. Changchun Univ. Sci. Technol. (Nat. Sci. Ed.) 33, 134–135.

Liu, Z.Z., Qian, G.R., Sun, Y., Xu, R., Zhou, J.Z., Xu, Y.F., 2010c. Speciation evolutions of heavy metals during the sewage sludge incineration in a laboratory scale incinerator. Energy Fuel 24, 2470–2478.

Liu, W., Xu, J., Liu, J., Cao, H., Huang, X.F., Li, G., 2015. Characteristics of ammonia emission during thermal drying of lime sludge for co-combustion in cement kilns. Environ. Technol. 36, 226–236.

Llorente, M.J.F., Garcia, J.E.C., 2008. Suitability of thermo-chemical correlations for determining gross caloric value in biomass. Thermochim. Acta 468, 101–107.

Luan, J., Li, R., Zhang, Z., Li, Y., Zhao, Y., 2013. Influence of chlorine, sulfur and phosphorus on the volatilization behavior of heavy metals during sewage sludge thermal treatment. Waste Manag. Res. 31, 1012–1018.

Lv, D., Zhu, T., Liu, R., Lv, Q., Sun, Y., Wang, H., Liu, Y., Zhang, F., 2016. Effects of co-processing sewage sludge in cement kiln on NO_x, NH_3 and PAHs emissions. Chemosphere 159, 595–601.

Magdziarz, A., Wilk, M., Kosturkiewicz, B., 2011. Investigation of sewage sludge preparation for combustion process. Chem. Process Eng. Inz. 32 (4), 299–309.

Marengo, E., Bobba, M., Robotti, E., Liparot, M.C., 2006. Modeling of the polluting emissions from a cement production plant by partial least-squares, principal component regression, and artificial neural networks. Environ. Sci. Technol. 40, 272–280.

Mikulcic, H., Klimes, J.J., Vujanovic, M., Urbaniec, K., Duic, N., 2016. Reducing greenhouse gasses emissions by fostering the deployment of alternative raw materials and energy sources in the cleaner cement. J. Clean. Prod. 136, 119–132.

Mlakar, T.L., Horvat, M., Vuk, T., Stergaršek, A., Kotnik, J., Tratnik, J., Fajon, V., 2010. Mercury species, mass flows and processes in a cement plant. Fuel 89, 1936–1945.

Murakami, T., Suzuki, Y., Nagasawa, H., Yamamoto, T., Koseki, T., Hirose, H., Okamoto, S., 2009. Combustion characteristics of sewage sludge in an incineration plant for energy recovery. Fuel Process. Technol. 90, 778–783.

Nadal, M., Schuhmacher, M., Domingo, J.L., 2009. Cost–benefit analysis of using sewage sludge as alternative fuel in a cement plant: a case study. Environ. Sci. Pollut. Res. 16, 322–328.

Nørskov, L.K., 2012. Combustion of Solid Alternative Fuels in the Cement Kiln Burner. (Ph.D. Thesis). Technical University of Denmark. http:/orbit.dtu.dk/files/74227236/PhD.thesis_Linda_N_rskov_31072012..PDF

Olivier, J.G.J., Janssens-Maenhout, G., Muntean, M., Peters, J., 2015. Trends in Global CO_2 Emissions. (Report). PBL Netherlands Environmental Assessment Agency/European Commission, Joint Research Centre, Ispra/The Hague.

Otero, M., Gomez, X., Garcia, A.I., Moran, A., 2008. Non-isothermal thermogravimetric analysis of the combustion of two different carbonaceous materials. J. Therm. Anal. Calorim. 93, 619–626.

Parlikar, U., Bundela, P.S., Baidya, R., Ghosh Sannidhya, K., Ghosh Sadhan, K., 2016. Effect of variation in the chemical constituents of wastes on the co-processing performance of the cement Kilns. Proc. Environ. Sci. 35, 506–512.

Pavlik, Z., Fort, J., Zaleska, M., Pavlikova, M., Trnik, A., Medved, I., Keppert, M., Koutsoukos, P.G., Cerny, R., 2016. Energy-efficient thermal treatment of sewage sludge for its application in blended cements. J. Clean. Prod. 112, 409–419.

Peters, G.M., Rowley, H.V., 2009. Environmental comparison of biosolids management systems using life cycle assessment. Environ. Sci. Technol. 43, 2674–2679.

Pipilikaki, P., Katsioti, M., Papageorgiou, D., Fragoulis, D., Chaniotakis, E., 2005. Use of tire derived fuel in clinker burning. Cement Concr. Compos. 27, 843–847.

Rahman, A., Rasul, M.G., Khan, M.M.K., Sharma, S., 2015. Recent development on the uses of alternative fuels in cement manufacturing process. Fuel 145, 84–99.

Reijnders, L., 2007. The cement industry as a scavenger in industrial ecology and the management of hazardous substances. J. Ind. Ecol. 11, 15–25.

Rivera-Austrui, J., Martinez, K., Marco-Almagro, L., Abalos, M., Abad, E., 2014. Long term sampling of dioxin-like substances from a clinker kiln stack using alternative fuels. Sci. Total Environ. 485, 528–533.

Rodríguez, N.H., Martínez-Ramírez, S., Blanco-Varela, M.T., Donatello, S., Guillem, M., Puig, J., Fos, C., Larrotcha, E., Flores, J., 2013. The effect of using thermally dried sewage sludge as an alternative fuel on Portland cement clinker production. J. Clean. Prod. 52, 94–102.

Rovira, J., Mari, M., Nadal, M., Schuhmacher, M., Domingo, J.L., 2010. Partial replacement of fossil fuel in a cement plant: risk assessment for the population living in the neighborhood. Sci. Total Environ. 8, 5372–5380.

Schreiber, R.J., Kellett, C.D., Joshi, N., Skokie, I.L., 2005. Inherent Mercury Controls Within the Portland Cement Kiln System. Research & Development Information, Skokie, IL. (Portland Cement Association. Serial No. 2841).

Schuhmacher, M., Nadal, M., Domingo, J.L., 2009. Environmental monitoring of PCDD/Fs and metals in the vicinity of a cement plant after using sewage sludge as a secondary fuel. Chemosphere 74, 1502–1508.

Shih, P.H., Chang, Lu, J.E., Chinag, L., 2005. Reuse of heavy metal-containing sludges in cement industry. Cem. Concr. Res. 35, 2110–2115.

Stasta, P., Boran, J., Bebar, L., Stehlik, P., Oral, J., 2006. Thermal processing of sewage sludge. Appl. Therm. Eng. 26, 1420–1426.

Stehlik, P., 2009. Contribution to advances in waste-to-energy technologies. J. Clean. Prod. 17, 919–931.

Stelmach, S., Wasilewski, R., 2008. Co-combustion of dried sewage sludge and coal in pulverized coal boiler. J. Mater. Cycles Waste 10, 110–115.

Strazza, C., Del Borghi, A., Gallo, M., Del Borghi, M., 2011. Resource productivity enhancement as means for promoting cleaner production: analysis of co-incineration in cement plants through a life cycle approach. J. Clean. Prod. 19, 1615–1621.

Strigáč, J., 2015. Effect of selected alternative fuels and raw materials on the cement clinker quality SSP. J. Civ. Eng. 10 (2), 81–92.

Teklay, A., Yin, C., Rosendahl, L., 2016. Flash calcination of kaolinite rich clay and impact of process conditions on the quality of the calcines: a way to reduce CO_2 footprint from cement industry. Appl. Energy 162, 1218–1224.

Trezza, M.A., Scian, A.N., 2005. Waste fuels: their effect on Portland cement clinker. Cem. Concr. Res. 35, 438–444.

Vainio, E., Yrjas, P., Zevenhoven, M., Brink, A., Laurén, T., Hupa, M., Kajolinna, T., Vesala, H., 2013. The fate of chlorine, sulfur, and potassium during co-combustion of bark, sludge, and solid recovered fuel in an industrial scale BFB boiler. Fuel Process. Technol. 105, 59–68.

Valderrama, C., Granados, R., Cortina, J.L., 2013. Stabilization of dewatered domestic sewage sludge by lime addition as raw material for the cement industry: understanding process and reactor performance. Chem. Eng. J. 232, 458–467.

Vamvuka, D., Sfakiotakis, S., 2011. Combustion behavior of biomass fuels and their blends with lignite. Thermochim. Acta 526, 192–199.

Van Tubergen, J., 2006. SRF: an important contribution to achieving environmental and energy-related goals. In: Waste Management and Solid Recovered Fuels Potential in the Enlarged European Union, Workshop Proceedings, Larnaca (Cyprus), June 20 – 23. ERFO.

Van Tubergen, J., Glorius, T., Waeyenbergh, E., 2005. Classification of Solid Recovered Fuels. ERFO.

Werle, S., 2012. Analysis of the possibility of the sewage sludge thermal treatment. Ecol. Chem. Eng. A 19, 137–144.

Werle, S., 2013. Potential and properties of the granular sewage sludge as a renewable energy source. J. Ecol. Eng. 1, 17–21.

Werle, S., Wilk, R.K., 2010. A review of methods for the thermal utilization of sewage sludge: the Polish perspective. Renew. Energy 35, 1914–1919.

Worrell, E., 2004. Cement and energy. In: Encyclopedia of Energy. 1, pp. 307–315.

Wystalska, K., Sobik-Szołtysek, J., Bień, J.B., 2013. Vitrification and devitrification of ash after sewage sludge combustion. Annu. Set Environ. Prot. 15, 181–191.

Wzorek, M., 2012. Characterization of the properties of alternative fuels containing sewage sludge. Fuel Process. Technol. 104, 80–89.

Wzorek, M., Król, A., 2009. Analysis of selected properties of sewage sludge within the consideration of its energetic application in forming alternative fuels. Pol. J. Environ. Stud. 6, 132–136.

Xiao, H., Ma, X., Lai, Z., 2009. Isoconversional kinetic analysis of co-combustion of sewage sludge with straw and coal. Appl. Energy 86, 1741–1745.

Xiao, H., Ma, X., Liu, K., 2010. Co-combustion kinetics of sewage sludge with coal and coal gangu dunder different atmospheres. Energ. Convers. Manage. 51, 1976–1980.

Xu, W., Xu, J., Liu, J., Li, H., Cao, B., Huang, X., Li, G., 2014. The utilization of lime-dried sludge as resource for producing cement. J. Clean. Prod. 83, 286–293.

Yan, D., Peng, Z., Karstensen, K.H., Ding, Q., Wang, K., Wang, Z., 2014. Destruction of DDT wastes in two preheater/precalciner cement kilns in China. Sci. Total Environ. 476–477, 250–257.

Yang, Z., Zhang, Y., Liu, L., Seetharaman, S., Wang, X., Zhang, Z., 2016. Integrated utilization of sewage sludge and coal Gangue for cement Clinker products: promoting tricalcium silicate formation and trace elements immobilization. Materials 9 (4), 275–287.

Yatkin, S., Bayram, A., 2008. Determination of major natural and anthropogenic source profiles for particulate matter and trace elements in Izmir, Turkey. Chemosphere 71, 685–696.

Yu, S., Zhang, B., Wei, J., Zhang, T., Yu, Q., Zhang, W., 2017. Effects of chlorine on the volatilization of heavy metals during the co-combustion of sewage sludge. Waste Manag. 62, 204–210.

Zabaniotou, A., Theofilou, C., 2008. Green energy at cement kiln in Cyprus—use of sewage sludge as a conventional fuel substitute. Renew. Sust. Energ. Rev. 12, 531–541.

Zhang, Y., Cao, S., Shao, S., Chen, Y., Liu, S., Zhang, S., 2011. Aspen plus-based simulation of a cement calciner and optimization analysis of air pollutants emission. Clean Techn. Environ. Policy 13 (3), 459–468.

Zhao, Y., Zhan, J., Liu, G., Ren, Z., Zheng, M., Jin, R., Yang, L., Wang, M., Jiang, X., Zhang, X., 2017. Field study and theoretical evidence for the profiles and underlying mechanisms of PCDD/F formation in cement kilns co-incinerating municipal solid waste and sewage sludge. Waste Manag. 61, 337–344.

Zhou, H.B., Ma, C., Gao, D., Chen, T.B., Zheng, G.D., Chen, J., Pan, T.H., 2014. Application of a recyclable plastic bulking agent for sewage sludge composting. Bioresour. Technol. 152, 329–336.

Section C

Contaminants in sludge and treatment strategies

Enhanced sludge degradation process using a microbial electrolysis cell

17

Iwona Zawieja
Faculty of Infrastructure and Environment, Institute of Environmental Engineering, Czestochowa University of Technology, Czestochowa, Poland

1 The idea of biohydrogen generation

Currently, one of the main research issues undertaken by many scientists in connection with growing environmental pollution is the modernization of existing wastewater treatment technologies and the creation of new, more effective solutions. Therefore, a number of studies are being carried out to improve the anaerobic stabilization process, taking into account the energy aspect of the process, that is, the path for obtaining biohydrogen. In the era of development of unconventional energy sources and the search for new technologies, a number of tests have been carried out to intensify biogas production (Grosser et al., 2013, 2017). In connection with the assumptions of the Kyoto Protocol, which entered into force on February 16, 2005, a common tendency is the attempt to replace energy from traditional sources with renewable energy. In addition, due to the current trend in the European Union regarding waste disposal with simultaneous energy acquisition, the principle of maximizing waste recovery and limiting storage in force is being sought.

The production of sludge increases with the use of highly effective methods of biological and chemical wastewater treatment. The amount of raw sludge formed (preliminary and excess) in municipal wastewater treatment plants depends on the composition of wastewater and the technology used for its treatment. Excess sludge formed in wastewater treatment plants where technologically advanced activated sludge methods are carried out shows a significantly limited susceptibility to the biochemical degradation process under anaerobic conditions. Excess sludge distinguished by a characteristic flocculating structure is susceptible to breaking up this structure in disintegration methods. It is advisable to subject excess sludge to disintegration before its mesophilic anaerobic stabilization, which significantly increases the susceptibility of such modified sludge to biodegradation. Waste-activated sludge (WAS) is a widely used substrate in the process of anaerobic stabilization for the production of biogas and nowadays hydrogen is being produced more and more.

Biohydrogen is a transient byproduct of many biochemical reactions taking place with the participation of microorganisms. Among others, it is a natural product generated during anaerobic digestion. There is also the possibility of generating

Industrial and Municipal Sludge. https://doi.org/10.1016/B978-0-12-815907-1.00017-9

biohydrogen from water with the help of specialized groups of microorganisms in the presence of sunlight. In addition, this gas can be formed as the result of anaerobic decomposition of monosaccharides, disaccharides, starch, and hemicellulose (Jadhav et al., 2017). Therefore, according to Gürtekin (2014) in the case of biological methods of biohydrogen production, direct biophotolysis, indirect biophotolysis, photo fermentation, dark fermentation, a combination of dark and photo fermentation, and biocatalyzed electrolysis can be distinguished. Fig. 1 presents the methods of biohydrogen production.

Nowadays, the acquisition of anaerobic stabilization of biohydrogen, as a source of energy that can be used to power hydrogen cells, is of great interest. In the hydrolysis process, which is the first stage in the decomposition of organic compounds and is carried out by saprophytic bacteria that act on nondissolved compounds with extracellular enzymes, larger particles are transformed into smaller ones, suitable for direct decomposition. The hydrolytic phase of the anaerobic stabilization process occurs slowly, which is why it is the phase determining the rate of transformation of the fermentation process. The acid-forming phase consists of the decomposition of products formed in the hydrolysis process, mainly to volatile fatty acids (formic, propionic,

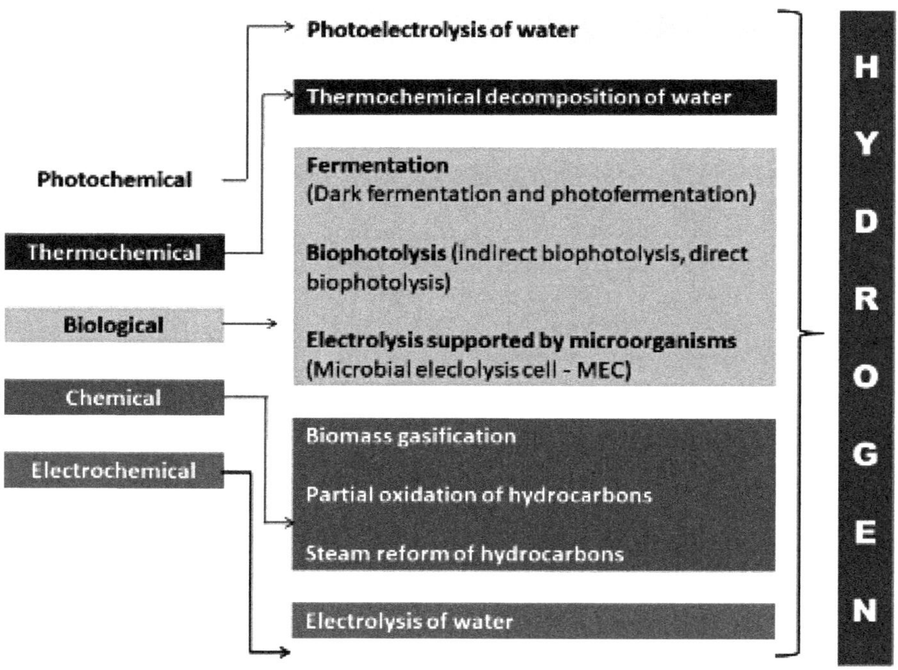

Fig. 1 The methods of hydrogen production.
Developed based on Moritz, M., 2012. Biological methods of hydrogen production. Chemist. 66 (8), 827–834; Jadhav, P., Sarkar, J., Patil, R., Vinchurkar, S., 2017. Microbial disintegration of bio-waste for hydrogen generation for application in fuel cell. Int. Res. J. Eng. Technol. 4 (3), 229–232.

butyric), alcohols, aldehydes, and gaseous products such as CO_2 and H_2, and, to a lesser extent, to acetates. It is known that in standard fermentation, with the ratio of acetate to butyrate as 3 to 4 and pH5.5, we observe the highest intensity of biohydrogen production. Acid-forming bacteria include, among others, *Clostridium*, *Bacteroides*, *Ruminococcus*, *Butyryvibrio*, *Escherichia coli*, *Bacillus*, *Bifidobacterium*, *Aerobacter*, *Streptococcus*, and *Micrococcus*, which are relatively tolerant to changes in pH and temperature and multiply much faster than methane bacteria. Acetogenic bacteria (*Syntrophobacter*, *Syntrophomonas*) use products derived from the acid fermentation process. They convert these products into acetates, CO_2, and H_2, which in turn are used for methanogenic bacteria. Acetogenic bacteria are sensitive to the presence of H_2 in the environment. At a slow rate of H_2 removal, they can be displaced, which may cause the accumulation of volatile organic acids in the environment. That in turn may reduce the pH and then inhibit the growth of methanogenic bacteria. There are various methods for the production of biohydrogen, for example, as a result of modification of the fermentation process through sediment conditioning, fermentation division into two stages—acid and methane fermentation, which differ in the kinetics of biochemical transformations, different degradation products and the species composition of the bacteria involved in the processes.

According to Kadier et al. (2016) the main challenges of MEC technology are enhancing the hydrogen production rate and lowering the energy input. The hydrogen gas can be an environmentally acceptable energy carrier for vehicles.

WAS during the fermentation process is characterized by low hydrogen yield because it contains limited amounts of carbohydrates suitable for use by hydrogen-producing bacteria (Lu et al., 2012).

The main obstacles in the production of biohydrogen are the low level of soluble carbohydrate and the occurrence of proteins and carbohydrates in the form of flocs made up of microbial cells and extracellular polymeric substances. In addition, hydrogen is used in the methane fermentation process by homoacetogenic and methanogenic microorganisms (Cai et al., 2004).

The combination of alkaline pretreatment and high initial pH could maintain a suitable pH range for the growth of dominant hydrogen-producing anaerobes, but it could also inhibit the growth of hydrogen-consuming anaerobes (Cai et al., 2004; Liu et al., 2012).

Almost all protein and carbohydrates are enveloped in flocs made up of microbial cells and extracellular polymeric substances. The substrates for electrohydrogenesis are not only carbohydrates, but protein and its acidification products such as volatile fatty acids are also responsible for a portion of H_2 generation in microbial electrolysis cells (MECs) (Lu et al., 2012).

Results received by Cai et al. (2004) showed that protein is the major substrate for biohydrogen production from sewage sludge and that *Eubacterium multiforme* and *Paenibacillus polymyxa* were the dominant bacteria in biohydrogen production from sludge.

The implementation of the process using MEC for the production of biohydrogen must be preceded by studies on the optimization of the MEC configuration with the determination of the most favorable conditions for pretreatment of WAS (Lu et al., 2012).

2 Pretreatment methods

Optimizing MEC configurations and operation conditions of the WAS pretreatment processes is necessary to implement this technology on an industrial scale.

In MECs, the short-chain fatty acids (SCFAs) accumulated in WAS fermentation should be adopted as an alternative extra carbon source for biohydrogen generation (Liu et al., 2012). Therefore, it is advisable to subject sludge to disintegration using methods of both a chemical nature (alkalization, acidification) and a physical (ultrasonic field, freezing/thawing) one. These methods, acting in support of the hydrogenic fermentation process, contribute to the inactivation of metanogenic microorganisms that are competing for the microorganisms responsible for the generation of biohydrogen (Lu et al., 2012; Cai et al., 2004)

The SCFAs accumulated in WAS fermentation can be adopted as an alternative extra carbon source for biohydrogen production in MECs by, for example, bifrequency ultrasonic. During the third day of the process conducted by Liu et al. (2012), the highest SCFAs were observed and the total production of biohydrogen obtained during the research was 120 mL H_2/gVSS/d. In the case of disintegration using an ultrasonic field, the vibration frequency was 28–40 kHz, an ultrasonic energy density of 0.5 kW/L, and a pretreatment time of 10 min. After the disintegration process, the SCOD value increased to 6040 mg/L, consisting of acetic acid over 40%.

The process of MECs might be intensified by alkaline conditioning. Nearly doubled H_2 yields were reported in the case of alkaline pretreated WAS. In the case of two-chamber MECs (TMECs) from alkaline pretreatment WAS, H_2 yields of 15.08 ± 1.41 mg H_2/gVSS were obtained and 5.67 ± 0.61 mg H_2/gVSS from raw waste activated sludge. This was several times higher than yields obtained during conventional fermentation (Lu et al., 2012).

3 Construction and principles of operation of MECs

As reported by Kadier et al. (2016), MECs were discovered in 2005 by two independent research groups, one at Penn State University in the United States and the second at Wageningen University in the Netherlands. The microbial fuel cell consists of two compartments, the anode and cathode, separated by a semipermeable proton exchange membrane (PEM) that provides anaerobic conditions in the naval compartment (Neczaj, 2016; Zhang et al., 2010). The first of the constructions designed in 2005 based on the bioelectrohydrogenesis process was a two-chamber MEC. In this case, the anode and cathode chambers were separated by an ion (proton) exchange membrane (PEM) (Liu et al., 2005). In the case of a two-chamber tank with MEC, the total hydrogen yield is only 60%–73%. The reason for this process efficiency is the loss in the generation of biohydrogen occurring in the unwanted processes within the MEC, such as biomass production, conversion of substrates into polymers, and methanogenesis from hydrogen and acetate. A membrane was used to administer the hydrogen diffusion into the anode chamber. The introduction of the membrane, however, forces the process costs to increase and affects the lack of compactness of the construction itself, which is an important feature of single-chamber tanks (Karthikeyan et al., 2017). As reported by Cheng and Logan (2011) in the case of two-chamber MECs, hydrogen production rates ranging from 0.01 to 6.3 $m^3/m^3/d$. The construction of the single-chamber MECs is based on placing both the anode and cathode in one chamber, which ensures the compactness of the solution. Single-chamber MECs are characterized by a lower internal resistance. In addition, according to Call and Logan (2008), in the case of single-chamber MEC, the biofuel production efficiency was twice as high (3.12 $m^3/m^3/d$ at an applied voltage of 0.8 V) than in the case of the two-chamber MEC construction while maintaining identical operating conditions. Fig. 2 shows the construction of a single-chamber microbial electrolysis cell (MEC).

In the case of MEC technology, the essence of the process is the oxidation of organic matter by electrochemically active bacteria, which results in the production of CO_2, electrons, and protons. With the participation of bacteria, transports the electrons to the anode and releases the protons to the solution take place. The electrons flow through a wire to a cathode and combine in solution with the free protons. This process requires initialization. MEC reactors require an externally supplied voltage that is ≥ 0.2 V and a biologically assisted condition (pH = 7, $T = 30\,°C$, $P = 1$ atm). During fermentation, a maximum of four moles of hydrogen per mole of glucose can be obtained. The process products (acetates) produced in a concentially conducted fermentation do not show susceptibility to transformation by anaerobic bacteria into hydrogen. This biochemical limitation can be offset by biohydrogen production from the acetate using a completely anaerobic microbial fuel cell (MFC). The improvement of biohydrogen production efficiency is possible by increasing the electrochemical potential achieved by bacteria in the MFC caused by an additional voltage of 250 mV or more. The increase in the value of the electrochemical potential enables the production of hydrogen at the cathode directly from the oxidized organic matter. Under these conditions, it keeps an overall Coulombic efficiency of 60%–78%, and >90% of the protons and electrons produced by biohydrogen. Theoretically, any biodegradable dissolved organic matter can be used to generate hydrogen from the complete oxidation of organic matter (Liu et al., 2005; Kadier et al., 2016)

Results obtained by Liu et al. (2012) showed that >90% of acetate and ~90% of propionate were effectively converted to hydrogen, and next were n-butyrate and n-valerate (at dilutions), but <20% of isobutyrate and isovalerate were converted.

The reduction of metal ions, for example, manganese, iron, uranium, and copper, with simultaneous oxidation during the redox reaction of organic substrates occurs with the participation of various types of bacteria, including Escherichia, Shewanella, Closridium, and Desulfovibrio (Neczaj, 2016; Du et al., 2007).

Fig. 3 shows the essence of the two-chamber MEC operation.

As a result of covering the anode with a bacterial film, its potential decreases. The decrease in the potential value is related to the concentration gradient of substrates and products within the bacterial biospectrum, the intracellular transport of electrons, and the extracellular transport of electrons to the surface of the anode (Lee et al., 2010). These processes, leading to the reduction of voltage, allow the electrolysis process to be carried out. A biological system containing an acetate acetate anion of 0.8 g/dm^3 allows obtaining a bioanode potential of approximately −0.300 V, and the value of the

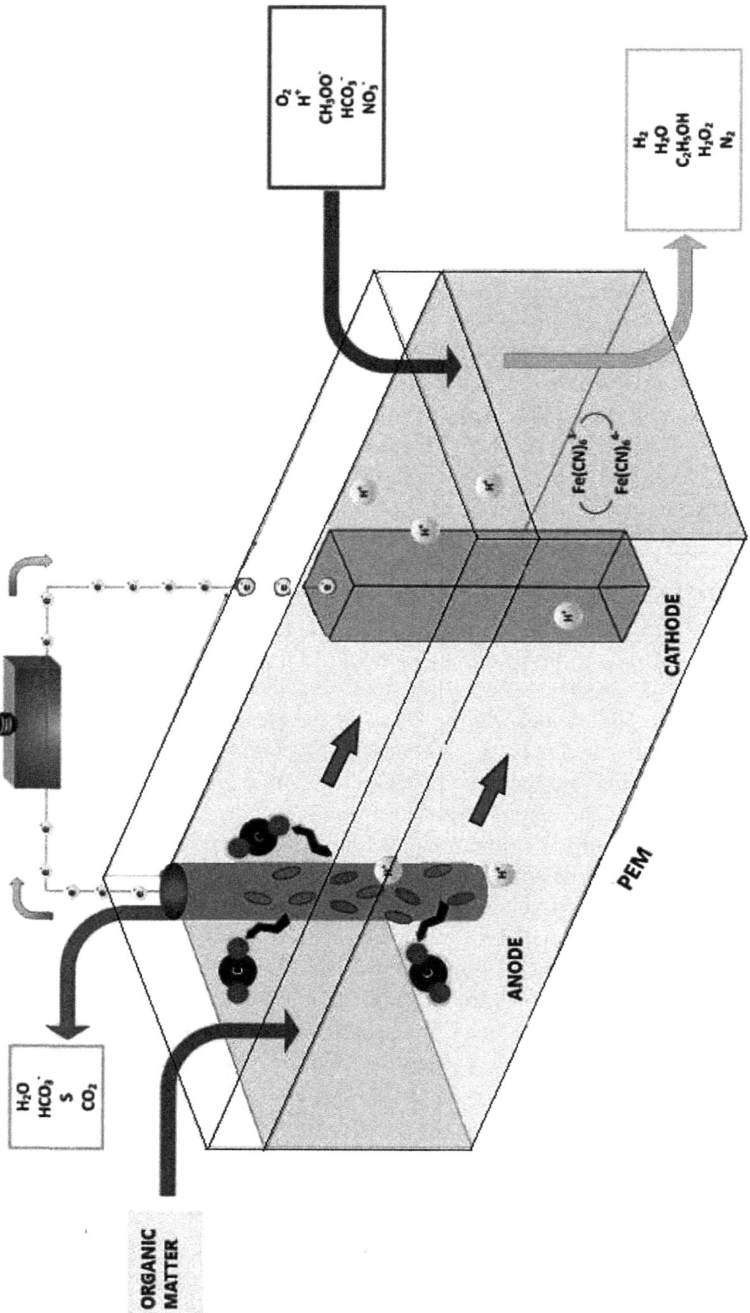

Fig. 2 The construction of a single-chamber microbial electrolysis cell (MEC).
Developed based on Lu, L., Ren, Z.J., 2016. Microbial electrolysis cell for waste biorefinery: a state of the art review. Bioresour. Technol. 215, 254–264; Neczaj, E., 2016. Methods of Intensification of Biogas Production from Municipal Sewage Sludge. Publisher of Czestochowa University of Technology Czestochowa.

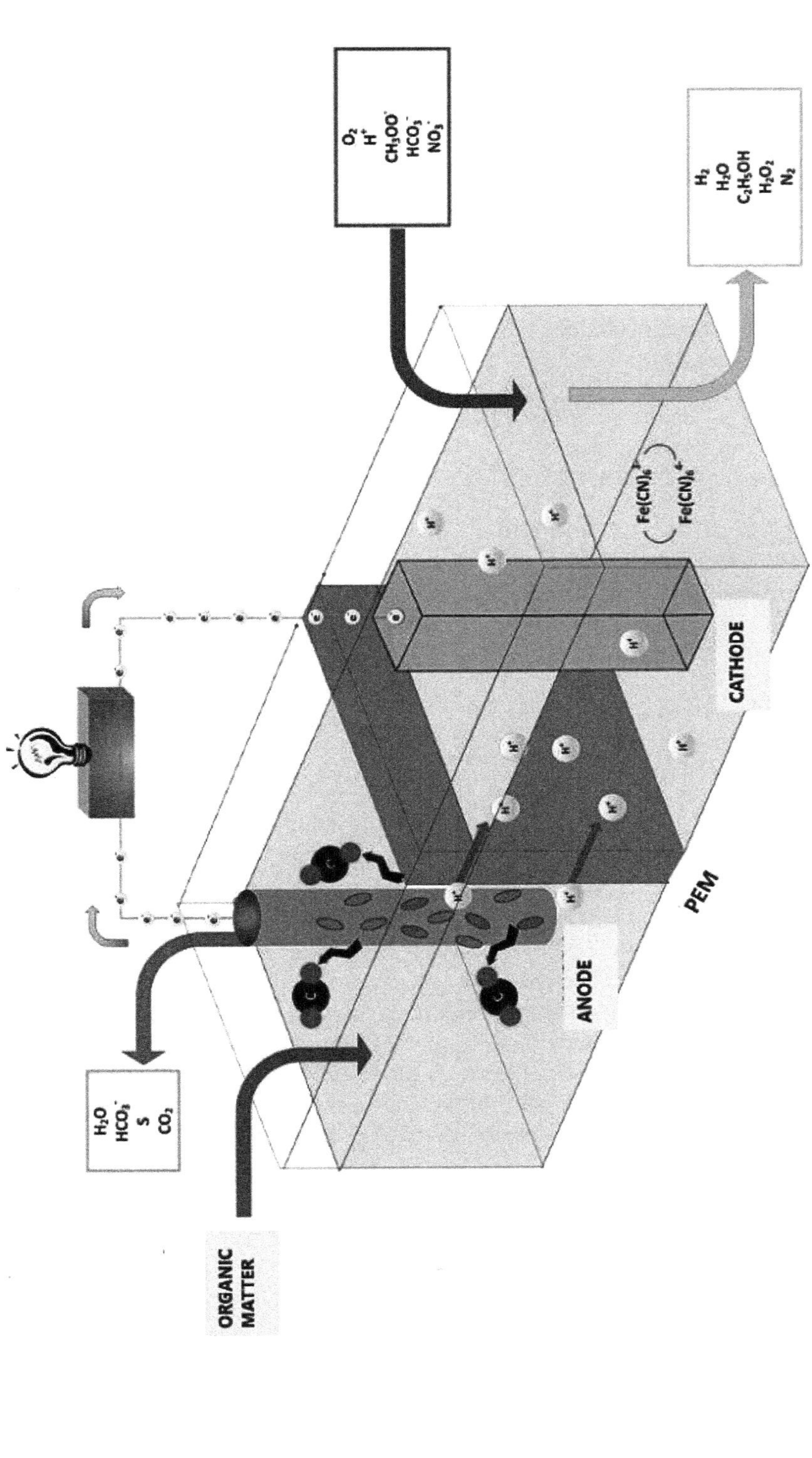

Fig. 3 Construction and operation of two-chamber MEC.
Developed based on Neczaj, E., 2016. Methods of Intensification of Biogas Production from Municipal Sewage Sludge. Publisher of Czestochowa University of Technology Czestochowa.

potential necessary to reduce protons is −0.414 V. This means that applying an external voltage correspondingly greater than the potential difference of individual half-cells (0.114 V) will initiate the electrolysis process. In practice, the voltage applied to the terminals of the bioelectrolyzer must, however, be slightly higher, on the order of 0.6–1.2 V (Moritz, 2012).

The reducing power obtained using a bioanode is not sufficient to initiate the hydrogen evolution reaction (HER) at the cathode. Supplying the system via a small voltage (normally ranging from 0.2 V to 1.0 V) causes the cathodic HER to occur (Reactions 1 and 2). In the case of conventional water electrolysis, a higher voltage ($E_0 > 1.2$ V) is required (Reactions 3 and 4). As a result, MEC technology is considered an energy-efficient solution (Logan et al., 2008).

Microbial electrolysis:

$$CH_3COO^- + 4H_2O \rightarrow 2HCO^- + 9H^+ + 8e^- E_{anode} = -0.297\ V \quad (1)$$

$$2H^+ + 2e^- \rightarrow H_2 E_{cathode} = -0.414\ V \quad (2)$$

$$E^0 = E_{cathode} - E_{anode} = -0.135\ V$$

Water electrolysis:

$$2H_2O \rightarrow O_2 + 4H^+ + 4e^- - E_{anode} = 0.82\ V \quad (3)$$

$$2H^+ + 2e^- \rightarrow H_2 E_{cathode} = -0.414\ V \quad (4)$$

$$E^0 = E_{cathode} - E_{anode} = -1.22\ V$$

According to Karthikeyan et al. (2017), the energy requirement for MECs is only about 0.6 kWh m^3 (0.2 mol H_2 energy/mol H_2 produced), while in the case of water electrolysis 4.5–5 kWh m^3 is necessary (1.5–1.7 mol H_2 energy/mol H_2 produced).

The economic viability of the process is obtained in the case of obtaining hydrogen from organic substances in 60–80% (Łuczyński and Gołaszewski, 2015).

In the MEC operation, different constructions are used, for example, the MEC chamber is modified by removing the membranes and creating a one-chamber structure. There are cells with a flat, cylindrical, tubular configuration or associated in systems ensuring the continuity of the process of obtaining biohydrogen. Due to the method of dispensing the subtreasure to the anode compartment, there are designs powered periodically or in a flowthrough manner (Markowska et al., 2013).

4 Effect of electrohydrogenesis on biomass

The calculation of electricity conversion efficiency in MEC technology can be made on the basis of the change in the analyte saturation that results from the release of electricity or the production of hydrogen (Łuczyński and Gołaszewski, 2015). When using MEC, the biomass of various origins can theoretically be transformed into

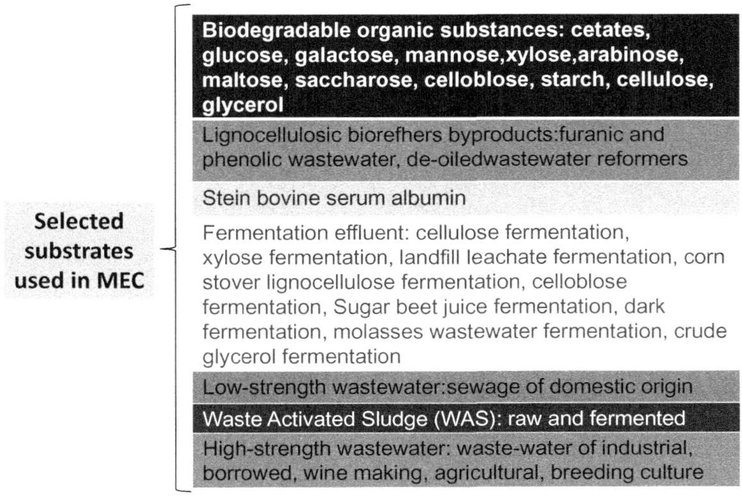

Fig. 4 Selected substances used in MEC and their sources of origin.
Based on Lu, L., Ren, Z.J., 2016. Microbial electrolysis cell for waste biorefinery: a state of the art review. Bioresour. Technol. 215, 254–264.

biohydrogen. Fig. 4 shows the selected substrate used in MEC and the sources of their origin based on Lu and Ren (2016).

Acetates are used by the bacteria to produce energy through its oxidation in the lemon cycle. The redox potential of this reaction is −0.288 V. Oxygen being an external electron acceptor in MEC cells provides bacteria with the highest energy yield. Oxygen reduction on solid electrodes runs with very high overpotential. The overpotential of oxygen reduction results from the strong adsorption of oxygen and hydroxides on the surface of the electrode. The energy associated with an overvoltage phenomenon is not available for bacteria. There is a loss of energy that affects the operation of cells, especially the value of generated voltage and power. The degree of biomass to biohydrogen conversion for the MEC technology used is a parameter affecting the efficiency of biohydrogen generation (Łuczyński and Gołaszewski, 2015).

In the case of the MEC operation, the hydrogen yield is an important technological indicator. This indicator is directly related to the system and speed of production. The energy efficiency is defined as the input electrical energy required. This efficiency can be used to evaluate the course of the microbial electrolysis process (Lu and Ren, 2016). According to Logan et al. (2008), the substrates used in biorefineries for the production of hydrogen using MEC are "free" and obtained chemical energy is not included in the calculation for energy efficiency. Table 1 shows the efficiency of selected installations based on MEC technology.

As a result of MEC technology, high H_2 yields were obtained for the tested substrates, especially with regard to the obtained effects for fermentation. In the case of fermentation, the theoretical value of H_2 yield equal 33% (4 mole H_2 per mole hexose) is the limit value of hydrogen production due to incomplete decomposition of organic matter. For pure nonfermentable organics such as VFAs and alcohols,

Table 1 The efficiency of selected installations as a result of obtaining hydrogen was developed on the basis of Lu and Ren (2016)

Technological conditions of the installation's operation	The obtained effects	Representative studies
The 40 mL two-chamber reactor, 1 V voltage, and new Ni-foam cathode, substrate: acetate	a maximum H_2 rate of 50 m^3/m^3/day with a current density 5704 A/m^3	Jeremiasse et al. (2010)
The several pilot scale systems: 100–1000 L, substrate: domestic wastewater	H_2 production rates <0.02 m^3/m^3/day	Heidrich et al. (2013)
The 1000 L, single chamber MEC, operating time 100 days, substrate: high-strength wastewater	Maximum current density of 7.4 A/m^3, H_2 production rates of 0.07 m^3/m^3/day with a decent COD removal of 62%.	Cusick et al. (2011)
The reactor volumes ranged from 10 to 50 mL and applied voltage of 1–1.5 V. Substrate: acetate	H_2 production rates of 6.3–50 m^3/m^3/day	Cheng and Logan (2011)

respectively, >80% H_2 recovery and >90% recovery was achieved. For mixed wastewater, including fermentation effluents, wastewater streams, and lignocellulosic biorefinery byproducts achieved >30% of recovery. H_2 yield from proteins and waste sludge ranges from 3% to 72%, which is due to less susceptibility to degradation and loss of electrons to competing mechanisms (Lu and Ren, 2016).

5 Advantages and possibilities of using MEC for the production of biohydrogen

In counteracting the energy crisis and environmental pollution, it seems reasonable to use hydrogen as a biofuel, which is a universal energy carrier (Moritz, 2012).

The use of biohydrogen as a potential source of energy is determined not only by the high heat of combustion, but also by the absence of toxic gases and dust during the process. The value of the heat of combustion for hydrogen is about 142 kJ/kg, this is about three times higher than gasoline and about four times higher than biodiesel. A problem requiring a solution is storing of the liquefied hydrogen in low temperature (−252.9°C) with all precautions, because the flammability range of hydrogen in the mixture with air is 4%–75% vol. (Midilli et al., 2005).

As reported by Nagai et al. (2003), bioelectrochemical production of hydrogen can be carried out using fermentation products or organic pollutants contained in wastewater. According to Karthikeyan et al. (2017), compared to abiotic water electrolysis, a much lower electrical voltage (~0.2 V) is required for hydrogen production in MECs.

The increase of hydrogen production from WAS is caused by the cascade utilization of organic matter in MECs. The advantages of MECs are high hydrogen yield from WAS, fewer H_2 sinks, and insensitivity to temperature. According to literature data (Neczaj 2016), the efficiency of microbial fuel cells depends on many parameters such as pH, temperature, type of substrates, type of microorganisms, and the structure and nature of electrodes. The first studies on the use of MEC using sewage sludge were conducted by Dentel et al. (2004), who obtained a current of 60 μA and a voltage of several thousand mV. It must be stated that MECs integrated with activated sludge disposal are valuable energy sources by recovering hydrogen gas.

The useful electrical power of MEC depends on many factors, including the efficiency of electron transfer from microorganisms to the anode, the fuel type, the electrode surface, internal fuel resistance, the cathode kinetic reduction reaction, and the operating conditions of imposed cells such as temperature, fuel transport to the anode, and oxygen to cathode. MEC technology can be used in the future in small wastewater treatment plants for the treatment of wastewater from animal farms and food processing plants. In the future, MEC technology can be used in small sewage treatment plants, for cleaning wastewater from animal farms, food processing plants, removing single impurities such as nitrates and p-nitrophenols. MEC technology can also be used to produce biosensors for measuring the biological oxygen demand in wastewater (Łuczyński and Gołaszewski, 2015).

Hydrogen stores a large amount of chemical energy per unit of mass (142 MJ/kg), which can be released without the emission of combustion byproducts into the atmosphere. The great interest in obtaining energy from hydrogen is associated primarily with Krawiec (2018):

- Reducing the use of fossil fuels as the primary energy source and greenhouse gas emissions.
- Increasing the share of renewable raw materials in the total raw material balance—increasing energy security.
- Lack of unfavorable combustion products (the final product of combustion is water).
- Using biofuels in accordance with the policy of sustainable development, high efficiency of bioconversion of hydrogen in fuel cells (at the level of 45%–60%).
- Processing of organic waste into a product that is less burdensome for the environment.
- The use of ecological recycling.

The ability of microorganisms dwelling in the MEC to degrade a broad spectrum of environmental pollutants is as important as the production of energy and biohydrogen, especially in systems using the analyzed technology for the treatment of sewage and sludge in situ (Franks and Nevin, 2010).

The alternative to the diffusional electron exchange via small molecules is the discovery of direct extracellular electron transfer. The establishment of electronic connections between electron donors and acceptors in microbial communities is crucial for electron transfer. Various microbial aggregates allow this type of connection. However, the examination of conductivity in conceptual and functional models is still being developed and debated (Cheng and Logan, 2011).

6 Summary

The essence of MECs is a bioelectrochemical pathway known as "bioelectrohydrogenesis." The mechanics of this pathway are not fully known and there are a number of barriers that need to be overcome in order to optimize the hydrogen yield and purity (Karthikeyan et al., 2017).

With reference to the concept of the development of a sustainable energy system, the biological methods of obtaining hydrogen, especially microbial electrolysis cells (MECs), are the current technological solution.

The choice of hydrogen as an energy carrier depends, inter alia, on the value of the heat of combustion amounting to approximately 142 kJ/kg. The value of this parameter is three times higher than the gasoline combustion heat of about 47.4 kJ/kg and four times higher than the biodiesel heat value of 37.0 kJ/kg (Moritz, 2012). In the process of biohydrogen combustion only water is released. The production of biohydrogen in the MEC process is economically viable when the waste biomass is the substrate for its production.

The essence of the MEC operation is the transformation to electricity of organic substances contained in biomass by microorganisms. An important advantage of the MEC operation is the possibility of obtaining energy from the moisture of organic substances susceptible to biodegradation such as pure compounds (glucose) or waste biomass, for example of agricultural origin or municipal or animal waste (Markowska et al., 2013).

Acknowledgments

The research was funded by the Ministry of Science and Higher Education by the project No. BS-PB 401/301/11.

References

Cai, M., Liu, J., Wei, Y., 2004. Enhanced biohydrogen production from sewage sludge with alkaline pretreatment. Environ. Sci. Technol. 38 (11), 3195–3202.

Call, D., Logan, B.E., 2008. Hydrogen production in a single chamber microbial electrolysis cell lacking a membrane. Environ. Sci. Technol. 42, 3401–3406.

Cheng, S.A., Logan, B.E., 2011. High hydrogen production rate of microbial electrolysis cell (MEC) with reduced electrode spacing. Bioresour. Technol. 102, 3571–3574.

Cusick, R.D., Bryan, B., Parker, D.S., Merrill, M.D., Mehanna, M., Kiely, P.D., Liu, G.L., Logan, B.E., 2011. Performance of a pilot-scale continuous flow microbial electrolysis cell fed winery wastewater. Appl. Microbiol. Biotechnol. 89, 2053–2063.

Dentel, S.K., Strogen, B., Chiu, P., 2004. Direct generation of electricity from sludges and other liquid wastes. Water Sci. Technol. 50 (9), 161–168.

Du, Z., Li, H., Gu, T., 2007. A state of the art review on microbial fuel cells: a promising technology for wastewater treatment and bioenergy. Biotechnol. Adv. 25 (5), 464–482.

Franks, A.E., Nevin, K.P., 2010. Microbial fuel cells, a current review. Energies 3, 899–919.

Grosser, A., Worwag, M., Neczaj, E., Kamizela, T., 2013. Co-digestion of organic fraction of municipal solid waste with different organic wastes: a review. Environ. Eng. 4, 231–242.

Grosser, A., Neczaj, E., Singh, B.R., Almas, A.R., Brattebø, H., Kacprzak, M., 2017. Anaerobic digestion of sewage sludge with grease trap sludge and municipal solid waste as co substrates. Environ. Res. 155, 249–260.

Gürtekin, E., 2014. Biological Hydrogen Production Methods. Department of Environmental Engineering Firat University, Turkey. Akademic Platform.

Heidrich, E., Dolfing, J., Scott, K., Edwards, S., Jones, C., Curtis, T., 2013. Production of hydrogen from domestic wastewater in a pilot-scale microbial electrolysis cell. Appl. Microbiol. Biotechnol. 97, 6979–6989.

Jadhav, P., Sarkar, J., Patil, R., Vinchurkar, S., 2017. Microbial disintegration of bio-waste for hydrogen generation for application in fuel cell. Int. Res. J. Eng. Technol. 4 (3), 229–232.

Jeremiasse, A.W., Hamelers, H.V., Saakes, M., Buisman, C.J., 2010. Ni foam cathode enables high volumetric H_2 production in a microbial electrolysis cell. Int. J. Hydrog. Energy 35, 12716–12723.

Kadier, A., Simayi, Y., Abdeshahian, P., Azman, N.F., Chandrasekhar, K., Kalil, M.S., 2016. A comprehensive review of microbial electrolysis cells (MEC) reactor designs and configurations for sustainable hydrogen gas production. Alexandria Eng. J. 55 (1), 427–443.

Karthikeyan, R., Cheng, K.Y., Selvama, A., Bosed, A., Wong, J., 2017. Bioelectrohydrogenesis and inhibition of methanogenic activity in microbial electrolysis cells – a review. Biotechnol. Adv. 35, 758–771.

Krawiec, M., 2018. Biofuels. https://prezi.com/3hqnoptidnok/biopaliwa.

Lee, H.-S., Vermaas, W.F.J., Rittmann, B.E., 2010. Biological hydrogen production: prospects and challenges. Trends Biotechnol. 28, 262–271.

Liu, H., Grot, S., Logan, B.E., 2005. Electrochemically assisted microbial production of hydrogen from acetate. Environ. Sci. Technol. 39, 4317–4320.

Liu, W., Huang, S., Zhou, A., Zhou, G., Rena, N., Wang, A., Zhuang, G., 2012. Hydrogen generation in microbial electrolysis cell feeding with fermentation liquid of waste activated sludge. Int. J. Hydrogen Energ. 37 (18), 13859–13864.

Logan, B.E., Call, D., Cheng, S., Hamelers, H.V.M., Sleutels, T., Jeremiasse, A.W., Rozendal, R.A., 2008. Microbial electrolysis cells for high yield hydrogen gas production from organic matter. Environ. Sci. Technol. 42, 8630–8640.

Lu, L., Ren, Z.J., 2016. Microbial electrolysis cell for waste biorefinery: a state of the art review. Bioresour. Technol. 215, 254–264.

Lu, L., Xing, D., Liu, B., Ren, N., 2012. Enhanced hydrogen production from waste activated sludge by cascade utilization of organic matter in microbial electrolysis cells. Water Res. 46(4), 1015–1026.

Łuczyński, M.K., Gołaszewski, J., 2015. Biorefinery Production of Alternative Fuels – Process Conditions. Publisher of the University of Warmia and Mazury, Olsztyn.

Markowska, K., Grudniak, A.M., Wolska, K.I., 2013. Microbial fuel cells: Technology basics, its limitations and potential applications. Adv. Microbiol. 52 (1), 29–40.

Midilli, A., Ay, M., Dincer, I., Rosen, M.A., 2005. On hydrogen and hydrogen energy strategies: current status and needs. Renew. Sust. Energ. Rev. 9, 255–271.

Moritz, M., 2012. Biological methods of hydrogen production. Chemist 66 (8), 827–834.

Nagai, N., Takeuchi, M., Kimura, T., Oka, T., 2003. Existence of optimum space between electrodes on hydrogen production by water electrolysis. Int. J. Hydrogen Energ. 28, 35–41.

Neczaj, E., 2016. Methods of Intensification of Biogas Production from Municipal Sewage Sludge. Publisher of Czestochowa University of Technology, Czestochowa.

Zhang, X., Cheng, S., Huang, X., Logan, B.E., 2010. Improved performance of single-chamber microbal fuel cell through control of membrane deformation. Biosens. Bioelectron. 25 (7), 1825–1828.

Further reading

Li, C., Lesnik, K.L., Liu, H., 2017. Stay connected: Electrical conductivity of microbial aggregates. Biotechnol. Adv. 35 (6), 669–680.

Sludge dewatering: Processes for enhanced performance

18

Tomasz Kamizela, Mariusz Kowalczyk
Institute of Environmental Engineering, Czestochowa University of Technology, Częstochowa, Poland

1 Introduction

The reduction in the volume of sewage sludge is one of the most basic tasks of sludge processing. It consists in decreasing water content in the sludge and is performed through the processes of thickening, dewatering, and drying. In the process of gravitational thickening, free water is mainly removed from the sludge (not bound to sewage sludge solids), whereas capillary water (bound by capillary forces) is removed in the process of mechanical dewatering or in natural conditions. The bound water is removed in the drying processes (Fig. 1) (Sozański and Dymaczewski, 2002). Fig. 1 shows the changes in the volume of sewage sludge with increasing content of dry matter.

Substantial change in sludge volume during water removal forces designers to pay particular attention to the thickening and dewatering processes because these processes and their efficiency will lead to serious economic effects (Fukas-Płonka et al., 2006). Various examinations demonstrated that the type of sludge and the equipment used for dewatering substantially impact the dewatering efficiency (Chu and Lee, 1999; Rehmat et al., 1997; Nellenschulte and Kayser, 1997).

Filtration presses and centrifuges are used most frequently for dewatering sewage sludge. Rehmat et al. (1997) compared the operations of chamber and belt filtration presses for sludge dewatering and found that the content of dry matter of the sludge in the filtration cake in both devices was very similar. In order to improve the sewage sludge dewatering effect, many researchers have focused on choosing polymers and their doses and ignored a key role of dewatering equipment and its effect on the dewatering parameters.

Sludge processing is characterized by changes in the volume and content of dry matter, which decreases if hygienization with calcium is not conducted (Oleszkiewicz, 1998). It is possible to obtain a reduction in water content in the process of dewatering of municipal sewage sludge to a final hydration of 70%–88%. It is important during this process that sewage sludge increases its volume more than three times compared with the volume of thickened sludge (Fig. 1) (Bień, 2007). Wastewater dewatering is of substantial importance to sludge management. This is connected with the reduced volume of sludge and, therefore, the costs of transport. Furthermore, dewatered sludge is much easier for further processing. Nevertheless, sludge

Industrial and Municipal Sludge. https://doi.org/10.1016/B978-0-12-815907-1.00018-0

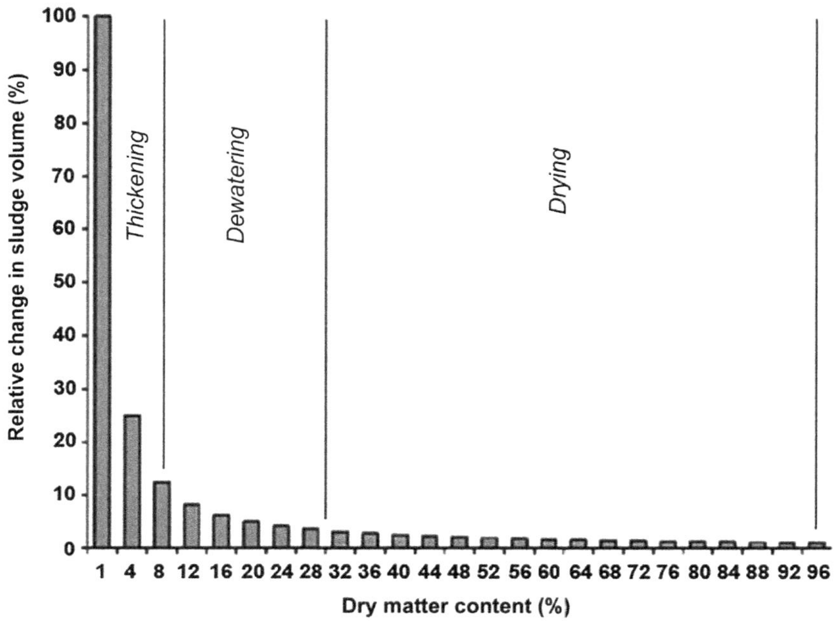

Fig. 1 Changes in sewage sludge volume during processing.

dewatering continues to be one of the most expensive and most demanding processes in sewage treatment plants (Bruus et al., 1992). The choice of dewatering methodology is affected by the sewage treatment plant location, investor capabilities, sludge designation, and the results of examinations of sludge characterization (e.g., analytical examinations such as density, specific resistance, zeta potential, particle size, colloid percentage, and many others) (Mizera, 2002; Kowalczyk, 2012).

2 The impact of conditioning methods on the technological parameters of sewage sludge dewatering

In order to improve the dewatering, sewage sludge is subjected to chemical conditioning. Chemical substances are added such as aluminum sulfate, iron chloride (III), iron sulfate (II), and polyelectrolytes. Nowadays, the initial processing of the sludge before the dewatering processes uses polyelectrolytes. However, the use of polyelectrolytes leads to higher costs of sludge processing and can be the cause of secondary environmental pollution. Therefore, various alternative methods of sewage sludge conditions, including dosing acids and surface active means (Chen et al., 2001), Fenton's reagents (deepened oxidation) (Lu et al., 2003; Chang et al., 2001; Parkitna et al., 2013a,b) and ultrasound field (Na et al., 2007; Parkitna et al., 2013a,b) and ultra-quick freezing (Parker and Collins, 1999) are proposed. Among these methods, ultrasound is considered a conditioning factor that is effective in the disintegration of sewage sludge for improvement in biodegradation (Tiehm et al., 1997). It is also regarded as a tool that leads to an increase in the efficiency of mechanical sewage sludge dewatering, with a negligible effect on the environment (Bougrier et al., 2006). Examinations conducted by different authors demonstrated that conditioning of sewage sludge with ultrasound can have both a positive and negative effect on sludge dewatering (Bień et al., 2009a). Na et al. (2007) found that the use of sonication for preparation of municipal sewage sludge before dewatering improved dewaterability, which is reflected by the decline in capillary suction time (CST) with the increase in the dose of ultrasound energy. CSK declined from 53 s for the raw control sample to <14 s for sonicated sludge with the ultrasound energy of 2000 kJ/cm^3. Yin et al. (2006) demonstrated that the use of initial ultrasound processing of sludge from petrochemical wastewater treatment subjected to dewatering in the filtration process led to a reduction in sludge water content from 99% to 80%. The initial sonication of sludge with the addition of a flocculant led to a reduction in specific filtration resistance from 3.59×10^{12} to 1.18×10^{12} m/kg. The flocculant dose declined by about 25%–50%. The content of water bound in the sludge was reduced from 16.7 to about 2.0 g/g d.m. after 2–4 min exposure in the ultrasound field with a power of 400 W/m^2. Kim and Kim (2003) also found that sludge sonication leads to a substantial increase in its susceptibility to dewatering. The efficiency of dewatering was varied and depended on ultrasound energy and sonication duration. Furthermore, Wang et al. (2006) demonstrated that dewatering capacity of sludge was reduced after ultrasonic disintegration because the increase in filtration resistance and CST was found during dewatering. After the use of ultrasound with a power density of 0.528 W/cm^3 for 5 min, the filtration resistance and CST reached values of 1.33×10^{14} m/kg and 344 s, respectively. These values were substantially greater than those obtained for raw sewage sludge (filtration resistance of 1.67×10^{12} m/kg, CSK 82 s). Similarly, Dewil et al. (2006) found that the increase in ultrasound energy at sludge conditioning leads to a decline of efficiency of sewage sludge dewatering. Both CST and filtration parameters increase. With ultrasonic energy increased from 7500 to 20,000 kJ/kg d.m., the polyelectrolyte dose was doubled, which was needed for the achievement of the same content of dry matter in dewatered sludge and nonsonicated sludge (Bień et al., 2007). In their study, Feng et al. (2008) dewatered sonicated sewage sludge while supplying varied ultrasonic energy (Kowalczyk, 2012; Bień et al., 2010). Sludge dewatering effects were evaluated based on CSK and filtration resistance versus specific energy used. They found that the values of CST and resistance reduced from 94.2 s and 2.35×10^{10} m/kg for the sludge not subjected to sonication to 83.1 s and 1.30×10^{10} m/kg for sonicated sludge using the specific energy of 800 kJ/kg d.m. When the ultrasound energy was increased to 35,000 kJ/kg d.m., an increase in CST and filtration resistance was increased to the values of 673.4 s and 3.81×10^{10} m/kg. The improved sludge dewatering effectiveness was obtained using the energy of ultrasounds of 800 and 2200 kJ/kg d.m. An insignificant reduction in the size of sludge flocs and a small increase in the concentration of extracellular polymeric substances (Zhang et al., 2007) can explain the observed improvement in sludge dewatering capacity. Supplying ultrasonic energy of more than 4400 kJ/kg d.m. leads to the deterioration of the dewatering process. These findings were consistent with the dewatering effects using ultrasound obtained by Chu et al. (2001), who found that exposure to ultrasound with low power density and a short time of exposure to the ultrasonic field led to a reduction in the CST of sewage sludge. However, these results are not consistent with the findings of other studies (Dewil et al., 2006; Wang et al., 2006), which found that the use of an ultrasonic field, regardless of the supplied energy, leads to reduced effects of sewage sludge dewatering. Nevertheless, Na et al. (2007) presented other conclusions. They found that CST was reduced following the use of an energy density higher than 2000 kJ/dm^3. These contradictory conclusions are difficult to explain. Usually high doses of energy lead to a destruction in the structure of flocs and cell membranes and also reduce floc size and increase concentration of extracellular polymeric substances. All these processes should have an effect on the reduction of dewatering of sewage sludge.

Because municipal sewage sludge is incompressible, excessive pressure in the pressure filtration does not always lead to increased dewatering effects and filtration efficiency. Laheij et al. (1996) found a high content of dry matter by weight (35%–0%), which can be achieved using the pressure of 300–400 kPa during filtration. Zhao and Bache (2002) investigated the process of pressure filtration while taking into account pressure and filtration time and doses of polymers in order to examine the integrated effect on dewatering of aluminum sludge. The study results demonstrated that dewatering effects can be magnified by increased pressure and filtration time. After achievement of a specific optimal point, a further increase in pressure and extension of the filtration time yield insignificant improvements. Wu et al. (2001) found experimentally that there was a pressure threshold at which the filtration process substantially deteriorated. The use of polyelectrolyte or freezing/defrosting for sewage sludge conditioning leads to a structure that allows for easier water release compared to the primary structure of the sludge. Based on analogue computations, Sorensen et al. (1996) demonstrated that high compressibility substantially deteriorates the sludge dewatering effects in the pressure filtration process and even inhibits the process. This means that in the case of sludge with very high compressibility, nearly unlimited time is needed for a substantial increase in dry matter content during dewatering while the filtration rate is becoming independent of the pressure used. In practice, it can explain why dewatering of biological waste is becoming time consuming and leads to low content of dry matter in dewatered sludge, such as 20%–25%. A reduction in the compressibility of sewage sludge represents a substantial technological challenge for researchers. Finding a way to reduce the compressibility of biological waste would lead to a high increase in dry matter concentration in dewatered sludge. Chu and Lee (2001) examined the pattern of sludge centrifuging subjected to cation conditioning with polyelectrolytes in order to reduce the water content in the excess sludge. They found that there was an optimum rotational speed, centrifuging time, and polyelectrolyte dose at which the final water content is maximal. Davis and Mafia (1995) demonstrated that collagen doses of from 1% to 2% d.m. to sewage sludge improve the sedimentation speed and reduce the specific resistance of sludge after filtration. Hwa and Jeyaseelan (1997) compared the results of examinations obtained during dosing calcium and aluminum salts and found that aluminum salts lead to better effects in the form of reduced CST and reduced specific resistance during filtration compared to calcium doses. Ozacar and Sengil (2000) found that tannins (organic compounds made of many particles of the gallic acid (3,4,5-trihydroxybenzoic acid) and D-glucose) and $Al_2(SO_4)_3$ dosed together led to a substantial improvement in the dewatering degree during sludge dewatering compared to dosing only $Al_2(SO_4)_3$. Omerci and Vesilind (2001) argued that sludge conditioning by cold treatment can reduce the content of proteins, carbohydrates, and cations in sludge before dewatering. The use of conventional means of conditioning of excess sludge demonstrated that sludge can be better dewatered after a reduction of the content of proteins, carbohydrates, and cations. The obtained agglomerates of sludge flocs were thicker. Parker and Collins (1999) also found that the use of ultrafast sludge freezing can increase the susceptibility to dewatering. The biggest freezing rate was 37.5 mm/h and, after this sludge processing, the CST could be reduced from 50–200 s to 10 s. Chang et al. (1998) examined sludge conditioned with kaolin combined with various cation polyelectrolytes. The final dewatering of sludge cakes depended on the molecular weight of the polyelectrolyte as well as the concentration and dose size. Zhao and Bache (2002) demonstrated that gypsum can be used as a material for building a skeleton of aluminum sludge in order to reduce compressibility and, consequently, improve dewatering in the filtration process.

Analysis of the obtained results and conclusions from the presented examinations are unequivocal and often controversial. Therefore, further examinations are needed to establish whether ultrasound actually improves the effects of sewage sludge dewatering and which conditions should be ensured during conditioning with an ultrasonic field and chemical substances and which beneficial aspects of this conditioning technology can justify its use on a technological scale (Kowalczyk, 2012).

3 Sludge conditioning for support of dewatering processes

Sewage sludge conditioning is aimed at changing the structure and physico-chemical properties of sewage sludge. With this process, part of water from bound forms is transformed into free water of semibound water, which is easier to be removed during dewatering at lower energy expenditure. Conditioning can be conducted at various stages of the technological line in wastewater treatment plants (Fig. 2) through:

- Chemical methods.
- Thermal methods.
- Mechanical methods.
- Unconventional methods.

Conditioning is aimed at increasing the susceptibility of the sludge for its thickening and dewatering. It is most often used before mechanical dewatering and less frequently before gravitational thickening (Kowal and Świderska-Bróż, 2000; Bień, 2007). Sludge conditioning can be performed in various manners. Depending on

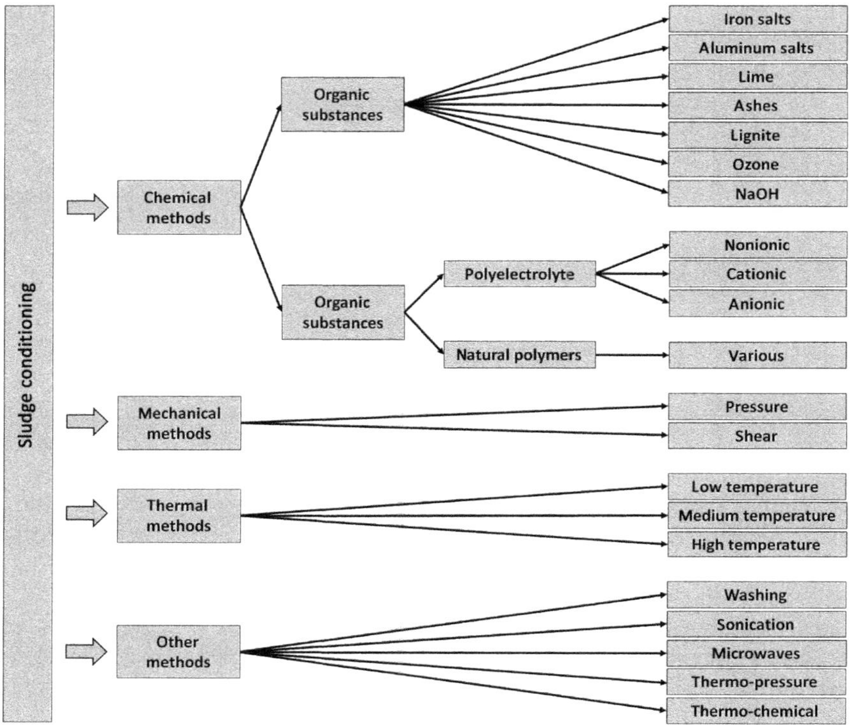

Fig. 2 Overview of the methods of sewage sludge conditioning (Kowalczyk, 2012).

the selected method of sludge conditioning, different final effects are achieved using the same dewatering devices. For this reason, it is justified to analyze the drawbacks and benefits of the use of individual processing methods.

The following effects can be observed:

- Striving to obtain clear liquor.
- Striving for an increase in the efficiency of dewatering devices.
- Obtaining a higher degree of phase separation (separation of suspended substances from free water).
- Minimization of byproducts of sludge processing that have a negative effect on treatment and processes and represent a burden to the environment (Bień, 2007).

3.1 Conventional methods

Increased dewatering of sewage sludge can be obtained by adding polymers or, less frequently, ash, calcium, gypsum, and diatomaceous earth as well as freezing and defrosting or heating the dewatered substances.

3.1.1 Chemical conditioning

In technologies of mechanical sludge dewatering, the initial preparation of sewage sludge, apart from conventional coagulants, is based on polyelectrolytes, which are especially useful in the case of hardly thickening and dewatered fermented sludge. Connecting small particles into bigger agglomerations (flocs) in the process of flocculation leads to faster and easier sedimentation (Kosmala-Szczurek and Tal-Figiel, 1998; Gawroński, 1996). During the examinations of dewatering of sludge conditioned with polyelectrolytes in the pressure filtration process, polymer doses are chosen based on a specific filtration resistance by comparison to the results of capillary suction time. Christensen et al. (1993) dewatered sludge containing oil, ash, calcium, and aluminum salts conditioned with various polymers and observed that the correlation between capillary suction time (CST) and specific resistance was similar. The above findings were confirmed by the tests performed by Hwa and Jeyaseelan (1997) and Tay and Jeyaseelan (1997). However, these researchers did not find a significant correlation between the listed indices of sludge dewatering capacity and their physical properties such as particle size. However, there is a view (Chang et al., 2001) that this correlation does exist because an increase in CST was observed with the increase in particle size. After examination of the sludge conditioned with organic polymers, it was demonstrated that the doses evaluated based on specific resistance were substantially higher than those that were determined based on CST. Other researchers (Wu et al., 1997; Papavasilopoulos, 1997; Papavasilopoulos and Bache, 1998; Pan et al., 2003) emphasized that with sludge conditioning with polyelectrolytes, the optimal values of polymer doses should be determined based on the capillary suction time (CSK) test. Doses of polyelectrolyte and polymers are insignificant and usually not higher than 15 mg/g d.m sludge. Polymer doses are usually $(1\text{–}20) \times 10^{-3}$ g polymer/g d.m of sludge or 0.05%–0.5% d.m. (Wu et al., 1997). Overdosing leads to a significant deterioration of thickening effects and sludge dewatering. Small doses of polymers do not increase the sludge mass through their conditioning, which represents a substantial benefit of these reagents compared to conventional coagulants, mainly hydrolyzing iron, calcium, and aluminum salts (Florkiewicz et al., 1999; Kowalczyk, 2012). The reagents have to be well mixed with the sludge. Sludge and washings from water treatment are conditioned nearly only by means of chemical methods using polyelectrolytes (polymers). Polymers used in sludge conditioning are water-soluble, high-molecular synthetic or natural compounds, whose very high coagulation or flocculation efficiency results mainly from their structure (mostly linear), high molecular weight (up to several million), presence of functional groups (e.g., —OH, —COOH, —NH_2, —SO_3H), and adsorptive activity. Polyelectrolytes are usually more effective coagulants and flocculants compared to nonionic polymers and, depending on the electric charge, are divided into cationic (polycation, e.g., vinylpyridine), anionic (polyanion, e.g., sodium polyacrylate), and cationic-anionic polyelectrolytes, which during dissociation form compounds with both an acidic and an alkaline character (e.g., polyacrylonitrile and polyacrylamide). Types and doses of polyelectrolytes for sewage sludge conditioning are chosen during experimental examinations.

The mechanisms of polymer and polyelectrolyte effects in sludge conditioning have not been fully explained to date. With respect to sewage sludge, the flocculation effect is explained by the formation of the hydrogen and electrostatic bonds, less frequently ionic bonds, between their chains and the surface of sludge particles.

3.1.2 Physical and physicochemical methods

Conditioning can be also carried out by means of adding structure-forming substances to the sludge, for example, ash, carbon, or diatomaceous earth preparations. The addition of these substances forming a specific scaffold in the sludge significantly improves the susceptibility of sludge to dewatering. Previous studies carried out in Poland and abroad have demonstrated that the use of polyelectrolytes, ash, or gypsum in sludge conditioning improves the filtration properties, thus leading to changes in the sludge structure (Kowalczyk, 2012; Bień, 2007; Wu et al., 1997; Lee and Liu, 2001; Zhao, 2003). A significant improvement in sludge dewatering was obtained when using a carbon suspension. It is used due to a higher density of carbon particles, which ensures higher resistance to shear strain and compression during mechanical dewatering in filtration presses. Sander et al. (1989), Broeckel et al. (1996), and Thapa et al. (2009) have used carbon dust with polyelectrolytes in order to improve the efficiency of sewage sludge dewatering. Further, Smollen and Kafaar (1997) used powder carbon for the methods combined with polyelectrolytes in order to improve dewatering. They found that adding powder carbon to loose sludge flocs that were formed during flocculation results in the formation of a porous, permeable, and rigid structure of the net. Such materials as fly ash, cement dust and other materials have been also used as "skeleton-forming" materials (Nelson and Brattlof, 1979; Benitez et al., 1994; Lin et al., 2001). These substances are claimed to form a permeable and rigid structure of a net, which is likely to remain porous during mechanical dewatering in the filtration process. This intensifies the effect of sludge dewatering. In order to improve the effect of dewatering, Zhao (2002) and Bień et al. (2008) used skeleton-forming gypsum connected with polyelectrolytes to magnify the dewatering effects. The addition of brown coal to sewage sludge (Durie, 1991) causes an increase in the caloric value of the sludge, which, after dewatering, can be combusted without supplying additional fuel. Sludge susceptibility to dewatering as a result of conditioning and the amount of conditioning agents to be used are determined in an experimental manner. Sludge dewatering is assessed after adding scaffold-building substances, using a filtration resistance test. Good dewatering effects are typically observed when specific filtration resistance does not exceed the level of from 2 to 3×10^{12} m/kg. Filtration resistance decreases with the higher doses of additional substance. In some cases, sludge conditioning with these methods does not produce the expected effects or high amounts of the compound are needed. The combined methods of the previous addition of conditioning chemicals are recommended in such cases (Bień, 2007; Zielewicz-Madej and Fukas-Płonka, 1998).

3.1.3 Conditioning with thermal methods

Thermal methods of sewage sludge conditioning before dewatering use changes in sludge structure obtained through supply or removal of heat. A very effective method is cold treatment of the sludge. Huang et al. (2001) demonstrated that the use of low temperatures in sludge conditioning (−4°C) leads to more effective thickening and dewatering. Omerci and Vesilind (2001) conditioned postcoagulation sludge and activated sludge using freezing and defrosting. They found that the conditioning process occurs more effectively for postcoagulation sludge with a low content of organic substances. The process of cold treatment of sewage sludge is composed of two phases: slow freezing and defrosting (Bień, 2007). Freezing and defrosting of postcoagulation sewage sludge destroys the flocculant structure and releases free water. Freezing also leads to molecule agglomeration and the formation of greater flocs. After defrosting, flocs are separated from water and sediment, leading to a reduction of sludge volume (Martel et al., 1998; Chen et al., 2001; Hong et al., 1995; Hung et al., 1996). In natural conditions, cold treatment of sludge with water content of 98% may ensure a reduction in this content to 80%. On an industrial scale, this method is recommended if it can be performed in natural conditions. Attempts to perform artificial cold treatment demonstrated that the efficiency of the installations used is insufficient for the purpose, whereas the process is energy-intensive. The attempts at sludge conditioning by heating demonstrated that the method is less efficient than cold treatment and also more energy consuming. Therefore, sludge heating has not been used in preparation for dewatering.

3.1.4 Unconventional methods

Unconventional methods include the effect of aultrasonic field, a magnetic field, and microwaves.

Transition of the ultrasonic wave through a liquid is accompanied by, for example, ultrasonic coagulation and electrokinetic phenomena; the effects depend on the sludge particle size and sound intensity. A very short sonication time is sufficient for the observation of changes in sludge properties. The effect of an ultrasonic field enhances the spatial packing of particles, which improves the conditions of free water removal during vacuum filtration; thickening of the sludge treated with ultrasound also occurs faster. It seems very efficient to add polyelectrolytes before the exposure of the sludge to the effect of an ultrasonic wave. The effect of ultrasonic waves substantially impacts the improvement of the sludge dewatering conditions in the filtration process (Bień, 2007). Using an ultrasound field in sewage sludge conditioning was aimed at intensification of the dewatering effect and limitation in the consumption of chemical reactants (polyelectrolytes) (Bień and Wolny, 2003). Examinations of the use of polyelectrolytes prepared with an ultrasound field have also been performed. The energy of the ultrasound field affects the polyelectrolyte structure so that it is changed by tearing the polyelectrolyte chains without changing its molecular weight.

An electromagnetic field is a state of space where forces act on electric charges and magnetic poles (Strzelczyk et al., 2002). This field is characterized by applying four vectors in each point: electric field intensity, *E*; electric induction, *D*. magnetic induction, *B*. and magnetic field intensity, *H*. The use of the magnetic field for raw sludge conditioning has a positive effect on CST test results, where capillary suction time was shortened with the increase in the exposure time. The effect of the magnetic field on raw sludge also has an effect on the reduction in final water content. Preparation of sewage sludge using a magnetic field and polyelectrolytes simultaneously yields substantially lower values of specific filtration resistance compared to the use of only polyelectrolytes (Bień, 2007).

Microwaves belong to electromagnetic waves with a wave length of from 1 mm to 1 m, which corresponds to the frequency from 300 MHz to 300 GHz. Specific phenomena occur with the medium with substantial water content placed in the microwave field, with part of them closely related to the increase in medium temperature (thermal effects), whereas others are characteristic for electromagnetic radiation and can be achieved through heating (nonthermal effects) (Thostenson and Chou, 1999; Wojciechowska and Kowalik, 2003). Volumetric heating is observed in the layer of the material where radiation penetration occurred. Microwave radiation energy is transformed into the mechanical energy of particle motion directly inside the irradiated material through molecular interactions with the electromagnetic field. The local electric charge (e.g., dipoles) is polarized and changes orientation according to the direction of the applied field. Resistance forces counteract the change in dipole alignment, leading to energy dissipation and heat release. In the layer beyond the microwave's reach, heating may occur through diffusion Wojciechowska and Kowalik, 2003).

4 Sewage sludge dewatering

Sewage sludge dewatering consists in a reduction of water content in the sludge such that its final water content ranges from 50% to 88%. This process is accompanied by a reduction in sludge volume (Cywiński et al., 1972). Processes of sludge dewatering include thickening, main dewatering, and drying. Depending on the type of water contained in the sludge, it is removed using various methodologies (Table 1).

4.1 Sludge thickening

Sludge thickening is understood to mean processes where, through separation of the solid phase from the liquid, the sludge reduces its volume without losing its liquid consistency (Bień et al., 2009b). Thickening leads to a reduction in water content to about 92%–94%. The process allows only for the removal of free water (primary water or water released after the conditioning process). Thickening is the first process of sludge processing. Its use reduces the investment and functional costs of individual stages of the treatment line and impacts the functioning of the entire wastewater treatment plant.

Table 1 Types of water contained in sludge and methods to remove them (Bień and Wystalska, 2011; Sorensen et al., 1996; Nellenschulte and Kayser, 1997)

Type of water contained in sludge		Water removal process
Free water	Water not bound to sludge solid particles	Gravitational thickening (sedimentation process)
Water bound in colloids	Is bound with surface tension and can be removed only after destruction of the sludge floc structure	Mechanical dewatering
Capillary water	Is bound with capillary forces, composed of adhesion and cohesion forces	Partially removed during mechanical dewatering
Bound water	Intracellular water that is present only in living organisms, chemically bound water	Partially removed through mechanical dewatering supported by modern methodologies of sludge conditioning, thermal processes

The following types of thickening can be used, depending on the process used:

- Gravitational.
- Mechanical.
- Floatation.

4.2 Gravitational thickening

Gravitational thickening occurs through the sedimentation of sludge particles and their compression with gravity (Cywiński et al., 1972). The sedimentation of sludge particles occurs due to the difference in specific weight of liquids and sludge particles, whose specific weight is from 1.1 to 1.3 g/cm^3. This occurs in the primary and secondary settling tank hoppers and in dedicated devices called gravitational thickeners. Gravitational thickening is performed periodically or continuously with a constant supply or discharge of the sludge and a discharge of the sludge liquor. With periodical thickening, sludge particles fall down gradually, whereas the sludge liquor is transported upward with the total amount of liquid remaining the same. Contrary to periodical thickening, continuous thickening is characterized by a constant supply of sludge and a simultaneous discharge of the sludge liquor by an overflow channel and the thickened sludge from the tank bottom. The design of gravitational thickeners is similar to that of vertical and horizontal radial settling tanks. The process is intensified by slow-speed rod mixers used to cut and regroup the sludge structure in order to release the water contained in the spaces between sludge particles (Podedworna and Umiejewska, 2008). Prolonged storage of the sludge in the gravitational thickeners can lead to the decomposition of organic compounds and the release of gas bubbles, which consequently leads to sludge flotation and the emission of unpleasant odors (Bień and Wystalska, 2011). Gravitational thickening is used for thickening of the initial sludge, reaching the content of dry matter of about 4%–6 %.

4.3 Mechanical thickening

Mechanical thickening is used for excess sludge that is poorly thickened gravitationally. Therefore, the introduction of an additional force (e.g., centrifugal force) in the mechanical devices accelerates the phase separation. Mechanical thickening of the sludge is performed by means of:

- Belt thickeners.
- Drum thickeners.
- Thickening centrifuges (Wójtowicz et al., 2013).

The efficiency of the equipment for mechanical sludge thickening is 2–150 m^3/h. In most of the devices except centrifuges, the effective level of phase separation and the thickening effect need an initial flocculation of the sludge with polyelectrolytes. Polymer consumption ranges from 2 to 8 g/kg d.m. The degree of phase separation is substantially higher for devices with filtration screens. Properly chosen types and doses of polyelectrolytes allow obtaining a degree of separation of 99% with these devices, at a dry matter content in the liquor of 50–150 mg/dm^3 (Zielewicz-Madej and Fukas-Płonka, 1998).

Belt thickeners act as sieves. A belt is stretched on the rotating rolls and the well-flocculated sludge is supplied on the belt. During belt movement, water is filtered partially through the sieve mesh (Wójtowicz et al., 2013).

Drum thickeners also act using the same principle, but the sieve has the shape of a drum supported on the load-carrying frame by means of the central shaft (Wójtowicz et al., 2013). The drum rotates on the rolls attached to the external frame. The sludge is supplied to the drum interior from one side and gradually moved to its other end. Water is removed outside through the sieve. The drum can be equipped in the devices that facilitate sludge motion (e.g., spiral, chicanes) (Wójtowicz et al., 2013).

Similar to dewatering centrifuges, the thickening centrifuges (slightly different in their design) operate based on the centrifugal forces. Sludge is typically supplied with the pipe to the interior of the rotating centrifuge drum. With the effect of the centrifugal force, sludge is pressed to the internal wall of the drum. The auger conveyor usually rotates slightly faster than the drum and transports the sludge to the conical part of the drum, then removes the sludge from the device. The thickening centrifuges allow for obtaining dry matter content of 5%–7% without the addition of the flocculant and 6%–8% with its addition (Podedworna and Umiejewska, 2008).

4.4 Floatation thickening

Floatation thickening leads to the flow of sludge up in the form of a floating skin, which is next removed by means of dedicated sweeps (Bień et al., 2004). The following types of floatation can be observed, depending on the floatation mechanism (Podedworna and Umiejewska, 2008; Bień and Wystalska, 2011):

- Biological floatation, which consists of the use of sludge from biogas suspension. It is used relatively rarely and occurs at the temperature of 35°C for 180 h, with CO_2 released during the process that leads to the transport of the sludge upward. Sludge is thickened to the content of 8% d.m.

- Chemical floatation, caused by the use of specific chemical agents. These agents, usually with acid reaction, react with the alkaline substances of the sludge. Gas is usually released, mainly CO_2, causing the transport of the sludge liquor particles to the surface.
- Air floatation, which, depending on the methods to introduce air to the sludge, can be mechanical (with fragmented air) and pressurized (with solved air). Floatation with fragmented air consists of the dispersion of the compressed air supplied to the floatation chamber by porous or perforated material. However, pressure floatation with solved air is most frequently used. The process consists of the exposure of the mixture of sludge and air to the effect of high pressure, leading to solving the air in the sludge liquor. With a sudden reduction of the pressure, the solubility of air is reduced while its excess is released in the form of very tiny bubbles. Air bubbles "catch" sludge particles contained in the liquor and transport them toward the surface.

4.5 Main dewatering

Sewage sludge dewatering can be theoretically performed using two methods: dewatering in natural conditions and dewatering in mechanical devices. Dewatering in natural conditions is not recommended. Therefore, only mechanical dewatering methods will be characterized, allowing for obtaining the content of dry matter in the sludge of 19%–40%.

4.5.1 Filtration presses

Presses are commonly used among mechanical devices, and their operation consists of pressing the water from the sludge, usually using filtration. The process requires good sludge preparation in terms of forming the compression resistance. Sludge dewatering is usually performed using the following filtration presses: belt presses, frame presses, and membrane presses (Bień and Wystalska, 2011).

4.5.2 Belt presses

The most popular devices used in Poland for sludge dewatering are belt filtration presses, which dewater sludge with a content of dry matter of 30%–35% (initial sludge), 24%–30% (mixture of initial and excess sludge), 15%–22% (aerobically stabilized sludge), and 20%–28% (fermented sludge). Filtration pressure in belt presses is obtained using rotary rolls pressed against the filtration fabric. Consequently, the consumption of energy in such presses is substantially lower compared to filtration presses. The drawback of such devices is the substantial surface of the filtration fabric and the much faster wear of the fabric. Belt filtration presses are very complex in technological terms. Sludge dewatered on belt filtration presses requires conditioning similar to that in filtration presses. The load of dry matter of the sludge to the press per 1 m of the fabric width depends on many parameters, the most important of which include belt speed, time of storage in the gravitational dewatering zone, and the water content of the supplied sludge. For the sludge generated during municipal waste treatment, this

Table 2 Technological parameters of belt presses (Dymaczewski, 1997; Podedworna and Umiejewska, 2008)

Sludge type	Sludge load (kg d.m./(m h))	Content of solids (%)		Polymer dose (kg/t d.m.)
		Feed material	Cake	
Raw sludge				
Initial	260–680	3–10	24–44	1–5
Excess activated sludge	40–230	0.5–4.0	20–35	1–10
Mixed initial and excess sludge	180–590	3–6	20–40	2–8
Mixed initial and secondary sludge from biological beds	180–590	3–6	20–40	2–8
Fermented sludge				
Initial	360–680	3–10	25–36	1–5
Excess activated sludge	40–136	3–4	12–22	2–10
Mixed initial and excess sludge	180–680	3–9	18–44	2–8
Sludge after aerobic stabilization				
Mixed initial and excess sludge	90–230	1–3	12–20	2–8

load ranges from 45–550 kg d.m./m h (Tchobanoglous et al., 2003). It is important to ensure even sludge distribution over the belt width. Increasing belt speed leads to a reduction in the sludge dewatering degree (Roman, 1986).

The cycle of belt filtration press operation has three filtration phases:

- Phase of gravitational dewatering: the initially conditioned sludge is supplied to the filtration belt, where, with gravitational forces, the liquor released during coagulation is filtered.
- The wedge phase: dewatering at low pressure in the wedge zone, and the initially formed cake is subjected to the effect of the pressure belt.
- The phase of high-pressure dewatering: the sludge is contained between the two belts and moved to the high-pressure zone, where the main sludge dewatering occurs. This zone is a set of rolls with various diameters where belts are moved (Bień and Wystalska, 2011; Oleszkiewicz, 1998).

Technological parameters that characterize the operation of belt presses are shown in Table 2 (Dymaczewski 1997).

4.5.3 *Chamber and membrane filtration presses*

The operation of filtration presses is based on the same processes as in belt presses. The difference lies in the methods of obtaining filtration pressure and continuous belt movement (in the case of belt presses) (Bień and Wystalska, 2011).

Chamber presses can ensure a sludge dewatering degree of even 45% at efficient conditioning, whereas membrane presses yield the content of dry matter higher by 2%–3%. However, filtration presses are very expensive and therefore are rarely used.

Dewatering in chamber presses occurs through the introduction of the sludge to the interior of the press and pressing a set of filtration plates equipped in the filtration fabric. The plates have a shape such that after pressing against each other, they form chambers between each other. The sludge is supplied to each chamber. The chamber is separated from its walls, formed by two neighboring plates from the filtration fabric. The fabric rests on grooves in the plate used to supply the filtrate to the channel separately for each plate or a common channel. The chamber press can operate as a low-pressure or a high-pressure chamber. The use of stepped pressure, for example with 400 kPa steps, improves the filtration effects (Bień and Wystalska, 2011). The dewatering effect using chamber presses depends on the type and properties of the dewatered sludge. Chamber press dewatering of the primary and the mixture of the primary and excess sludge conditioned using lime or iron salts may lead to obtaining the content of dry matter of 45%, whereas this value can reach 35% d.m for aerobically stabilized sludge and 40% d.m. for the fermented sludge. The method of dewatering on chamber presses is especially useful in the case of sludge with a relatively high content of inorganic substances. Therefore, it is suitable for dewatering the sludge containing a substantial percentage of the sludge from chemical phosphorus precipitation.

The membrane filtration press differs from the conventional chamber press mainly in the method of sludge dewatering. The sludge is dewatered at two stages: through filtration and compression of the membrane at pressure. After stopping the filling, the air pressure to the membrane is switched on, additionally pressing the water from the already formed cake. This additional pressure is 1500–2000 kPa (Bień and Wystalska, 2011). The membrane press represents an improved version of the chamber press and allows obtaining 2%–3% higher content of dry matter. Unfortunately, this leads to higher investment costs.

4.5.4 Centrifuges

At present, belt presses are gradually being ousted by centrifuges that ensure a higher dewatering degree. However, this involves slightly higher investment costs. Centrifuges allow obtaining dry matter content of 32%–40% in the dewatered initial sludge, 26%–32% in the dewatered mixture of initial and excess sludge, 18%–24% in dewatered aerobically stabilized sludge, and 22%–30% in dewatered fermented sludge.

Centrifuging is the process of accelerated sedimentation through the effect of the centrifugal force on the small surface at an acceleration 1500–3000 times higher than gravitational acceleration. Centrifuges need less space, operate in a closed system while isolating the operator from contact with the sludge, require less work and control of the operator, and are fully automated (Oleszkiewicz, 1998). The basic benefit of the centrifuges is low demand for the surface area, whereas their drawbacks are high functional costs, being a burden to the environment (noise), and the formation of the concentrated sewage sludge liquor (Roman, 1986). The effect of centrifuge operation depends on mechanical factors (drum design, rotational speed), sludge properties (water content, temperature, viscosity, particle size), accurate choice of the polymer, and its dose. The most frequently used polymers are cation polymers with doses of 3–8 g/kg d.m.

The main components of the decanter centrifuge include the cylindrical-conical drum and the auger conveyor contained inside the drum. The sludge is typically supplied with the channel put into the interior of the empty shaft of the auger. There are holes in the central part of the auger that supply the sludge to the centrifuge drum. With the centrifugal force, the sludge is accumulated at the drum wall. The drum and auger rotate in the same direction, but at different speeds (auger by 1%–3% faster than the drum). The difference in the speeds, termed differential speed, determines the degree of sludge dewatering (Oleszkiewicz, 1998). Sludge is moved by means of the auger toward the conical part of the drum, whereas water that is separated from the sludge flows to the other side (the cover of the cylindrical part of the drum). Water flows out from the drum through the holes in the cover, whereas dewatered sludge is transported from the drum with the holes in the conical part. Centrifuges mostly operate in the counterflow system, but they can also operate in the parallel flow system where the inflow of the dewatered sludge is oriented to the same direction as the outflow of the sludge. The counterflow centrifuge operates at higher accelerations and offers a drier cake (Oleszkiewicz, 1998).

Centrifuges have been mostly used for dewatering fermented sludge and aerobically nonstabilized sludge. As mentioned above, they ensure a higher degree of dewatering compared to belt presses. However, this also depends on the type of dewatered sludge and the used polymer dose (Table 3).

4.5.5 Piston presses

In recent years, reports have been published concerning the use of piston presses to ensure very good dewatering effects with sewage sludge dewatering. Piston presses proposed by Bucher have been operating in the municipal industry for about 10 years. The main components of the press include the piston and cylinder to which dewatered sludge is supplied. There are special drains located between these components to discharge the liquor. The principle of press operation is based on the cyclic phases of compression and decompression with simultaneous rotational motion of the piston with the cylinder. During pressing, the drains bend between the piston and the sludge and discharge the liquor (Proffico, n.d.). Piston presses operate in installations in Germany, Switzerland, Austria, Sweden, and Norway. The biggest installation constructed so far is used in Stockholm (Kappala treatment plant). In this plant, four Bucher piston presses allow for processing the sludge from 700,000 PE while dewatering about 10,000 mg d.m./year at a maximal dry matter content of up to 50% (Proffico, n.d.). The authors (Wójtowicz et al., 2013) argue that at typical polymer doses, the dewatering degree is at the level of 35% d.m.

Table 3 Effects of dewatering in centrifuges (Podedworna and Umiejewska, 2008; Dymaczewski, 1997)

Sludge type	Content of solids in the cake (%)	Separation factor (%)	Polymer dose (g/kg d.m.)
Initial raw sludge	28–34	90–95	2–3
Initial fermented sludge	26–34	90–95	2–3
Excess raw sludge	14–18	90–95	6–10
Excess fermented sludge	14–18	90–95	6–10
Mixed raw sludge	18–25	90–95	3–7
Mixed fermented sludge	17–24	90–95	3–8
After aerobic stabilization	12–16	90–95	6–10

4.5.6 Bagging machines

The most often used solution in smaller wastewater treatment plants is based on sludge dewatering in bag filters. Such devices, known as DRAIMAD, were developed by an Italian manufacturer and are designed to dewater sludge from the treatment of domestic and industrial waste. Sludge dewatering occurs in the bags made of a special hydrophobic fabric. The bags are placed on the supporting structure made of stainless steel that ensures adequate distribution of the sludge and discharge of the filtrate. Cycles of filling, dewatering, and replenishment are controlled electronically if the automatic control version is used. Before dewatering, the sludge should be mixed with polyelectrolyte, with the amount of 2–4 g per 1 kg of dry matter of the sludge. The choice of the type of polyelectrolyte and its dose is made during start-up of the treatment plant. During the working cycle, the sludge is pumped to the closed (in the M-open version) distribution tank, with the terminals in the bottom on which the bags are hung. The sludge is poured to the bags and water is filtrated to the outside, whereas sludge solids remain inside the bags. Depending on the sludge type, its water content is reduced to 80%–85% over a several-hour work cycle, whereas for certain types, this value is even lower. Completion of the first dewatering phase in DRAIMAD occurs after 10–24 h. After this time, the bags, containing up to 15 kg of dry matter of the sludge, are closed and transported using a special cart and stored in the open air or under a roofed shelter. The sludge continues to reduce its volume through natural evaporization. This process is independent of the weather conditions because the bags are made of a hydrophobic material and do not allow rainwater to permeate. After storage for 2–3 months, the dry matter content reaches 50%–70%, which corresponds to a water content of 30%–50%. These devices, characterized by relatively low prices, are very useful wherever daily or monthly volumes of sludge are very low. In some cases, this can represent a technically and economically justified solution to the problems with small volumes of sludge (Bień and Wystalska, 2011).

5 Sewage sludge drying

Sewage sludge drying allows for mass and volume reduction and obtaining a product with specific parameters that facilitate the final use of the sludge. The drying process does not modify the chemical composition of the sludge, thus the fertilizing properties remain the same whereas the calorific value of the sludge increases through a reduction of water content. Sludge drying is used as an indirect process to prepare sludge for further processing, especially burning. The process also allows for storing and transporting sludge. Dried sewage sludge can be incinerated in monoincinerators or coburnt in cement plants and waste incineration plants. It can be also used in nature, obviously after meeting specific legal requirements (Bień and Wystalska, 2011). Sewage sludge drying can be performed using mechanical devices (various types of dryers), thermal drying, and solar energy (solar drying).

5.1 Thermal drying

Installations for thermal drying of sewage sludge are based on various types of dryers. Sewage sludge drying can occur using direct or indirect systems or their combination. In direct installations (convection), the factor that transports heat remains in contact with the dried material, whereas in indirect (contact) installations, the heat-transporting medium is separated with the heated surface from the dried material (Korczak-Niedzielska and Gromiec, 1998). The process of drying can be performed at low temperature (about 60°C), middle temperature (60–130°C), or high temperature (130–500°C), with low-temperature installations characterized by substantially higher energy consumption than middle-temperature and high-temperature installations (Fukas-Płonka and Janik, 2008). Radiant drying is also used, utilizing the process of heat transfer without using fuel, by means of electromagnetic radiation or infrared radiation. Thermal drying allows for obtaining a high content of dry matter in the dried sludge, reaching about 90%–95% (Wójtowicz et al., 2013). The convection drying process is most often performed using belt dryers, drum dryers, air dryers, and fluidized-bed dryers (Wójtowicz et al., 2013). In the case of belt dryers, sewage sludge is distributed evenly over a special belt. The stream of hot air, flowing from the top downward, takes over the water from the sludge, which is next condensed (in a condenser) using water supplied from the treated outflow from the treatment plant. One modification of the belt dryers is air dryers, which utilize natural drying in the non-heated air. Substantial volumes of air supplied with fans or blowers are required (Wójtowicz et al., 2013; Krevox, n.d.). In drum dryers, sludge granulated in the mixer is transported by means of the hot air through the drum composed of three cylinders, rotating around its axis. Part of the sludge is recirculated. The final product of the process is dust-free granules, fully hygienized, with a dry matter content of 90%–95%. *Fluidized dryers can* be composed of two chambers, placed one over another and separated from each other with sieves. The sludge, dewatered to a minimum of 20% d.m., is supplied to the top chamber and spread on the sieve surface. The chamber is equipped in the mixers, supporting proper movement of the dried sludge in the chamber.

The hot air ($t = 135°C$) is pressed to the bottom chamber, from which it is transported to the chamber with the sludge. The air "washes" the sludge and, due to the contact of hot air with wet sludge, the drying process is performed, during which the drier and lighter sludge floats and forms the fluidized layer. Wet and cooled air is supplied from the drier and, after moving through a system of dust collectors, exchangers, and heaters, is returned to the dryer. Dried and hot sludge from the top chamber of the dryer is discharged to the postdryer to which an insignificant stream of hot air is supplied. Sludge is hygienized through the contact of the dried and hot sludge with air with an even higher temperature (140°C) (Bień and Wystalska, 2011).

The contact driers used most frequently are thin-layer dryers, disc dryers, and rotational dryers with a pipe heat exchanger (Wójtowicz et al., 2013). The process of drying in thin-layer dryers consists of drying the sludge in the form of a thin layer with high turbulence, moving on the internal wall of the static drying module, with the sludge motion being forced by the rotary motion of the rotor (Bień and Wystalska, 2011). Disc dryers can be used for predrying of sludge (to about 65%–55%) or complete drying (90%–98% d.m.). During the drying process, the dewatered sludge is supplied to the dryer using the auger pump. In the complete drying mode, the sludge is supplied to the mixer, where it is mixed with the dried product (recirculation of the dry product with dry matter content of about 90%–98%) so that the obtained mixture has a water content of less than 30%. This material, characterized by low adhesiveness, is supplied to the dryer drum. Water steam is supplied as a drying factor by means of a special rotary head to the hollow shaft with the discs, located inside the dryer drum. The steam is condensed on the internal discs of the walls and transfers the heat to the dried material. The heating surface is around $100\,m^2$. However, it depends on the efficiency, which ensures a high speed and efficiency of energy transfer. The dried sludge is discharged from the device through cooled auger conveyors. Temperature in the dryer drum is about 170°C. This allows for a full hygienization of the sludge (Klimapol, n.d.). In rotational driers with a pipe heat exchanger, heating is performed by means of a heat exchanger installed in the rotational drum, which rotates, mixing the sludge and throwing it on the pipe exchanger. The heating medium is saturated water steam. After sieving and separation of the dust, granulate with the content of dry matter of about 90%–95% is obtained (Wójtowicz et al., 2013).

Sludge-drying installations mostly ensure its granulation, whereas dried sludge is characterized (Zarzycki and Wielgosiński, 2001) by a spherical shape (3–4 mm), high hardness, mechanical resistance, and high bacteriological quality. Depending on the technology used, dried sludge can be in the form of:

Circular, regular, and hard granules with diameter of 2–6 mm, not washable by water, nearly odorless.

Irregularly cut hard "pasta" with diameter of 2–6 mm, length of 1–2 cm, not washable by water, nearly odorless.

Irregular pieces, nearly odorless but with lower hardness than those described above, with a tendency for formation of the dust during emptying or transport (Gromiec and Korczak-Niedzielska, 2000).

Thermally dried sludge is characterized by a mean content of dry matter of 85%–92%; dust content of $<100\,\mu m$, 1%; bulk weight of $650–750\,kg/m^3$; and a calorific value from 11,000 kJ/kg d.m. (anaerobically stabilized sludge, mixed initial and secondary sludge) to 20,000 kJ/kg d.m. (raw mixed sludge, initial sludge + secondary sludge) (Gromiec and Korczak-Niedzielska, 2000). The chemical composition of the sludge is usually unchanged during drying. Changes occur only if the temperature at which the sludge is dried exceeds 85°C, and the organic substances are carbonized. Chemical reaction of dry sludge is pH 6.5. During drying, sludge is partially pasteurized (temp. 85°C for about 30 min). Furthermore, the sludge heated during the contact method up to 100–140°C is even subjected to partial sterilization (Gromiec and Korczak-Niedzielska, 2000). Sludge drying is an expensive process due to the large amounts of energy needed for the process. The more watered the sludge, the higher the demand for energy (Bień and Wystalska, 2011).

The processes of thermal drying used in Polish wastewater treatment plants will be further discussed in individual chapters of this study.

5.2 Solar drying

Solar drying is considered a cheaper method of removing water from sludge. Solar drying facilities operate based on the use of the greenhouse effect that releases substantial amounts of energy used for the intensification of water dewatering. Drying sewage sludge can be done using only the greenhouse effect or by supporting solar drying with an additional heat source in so-called hybrid dryers. Sewage sludge drying in solar dryers mainly depends on:

- Intensity of the solar radiation (amount of heat supplied to the sewage sludge plate).
- Drying air temperature and its relative humidity.
- Flow rate of the drying air stream over the sludge layer.
- Bed surface area (sewage sludge cake).
- Thickness of sewage sludge cake.
- Temperature and water content in sewage sludge.
- Physical and chemical properties of sewage sludge.

Using the radiation energy, one can remove about 850 kg water per 1 m^2 of the active surface of the dryer in the period of spring, summer, and fall. In order to reduce the active surface of the dryer and stabilize the drying process, the solar energy can be supported with heat from the external source. In such cases, the amount of water removed from 1 m^2 of the dryer can reach about 3500 kg H_2O/year, allowing for nearly a fourfold reduction of the surface area of drying facilities. Opportunities for solar drying of sewage sludge depend on weather conditions, with particular focus on the "dose" of solar radiation, the air temperature, and the relative humidity. It is generally accepted that Polish drying facilities have to operate effectively from April to October.

Solar drying of sewage sludge is conducted in specifically designed and equipped drying rooms as part of solar drying facilities. The basic components of the solar dryer are:

- Drying rooms.
- Shakers for dried sludge.

- Gravitational and mechanical ventilation systems.
- Measurement, control, and monitoring systems that allow for automated operation of the drying plant.
- System of supply of wet sludge and discharge of dried sludge from the drying room.

Acknowledgment

The authors declare have no conflict of interest. Research funded from internal Grant BS PB 407/302/11 and BS/PB 401/301/11.

References

Benitez, J., Rodriguez, A., Suarez, A., 1994. Optimisation technique for sewage sludge conditioning with polymer and skeleton builders. Water Res. 28 (10), 2067–2073.

Bień, J.B., 2007. Sewage Sludge Theory and Practice. Wydawnictwo Politechniki Częstochowskiej, Częstochowa.

Bień, J., Wolny, L., 2003. Investigation of changes in some properties of sewage sludge in the ultrasonic field conditioning process. Inż. Ochro. Śr. 2, 11–16. (in Polish).

Bień, J., Wystalska, K., 2011. Sewage Sludge, Theory and Practice. Wydawnictwo Politechniki Częstochowskiej, Częstochowa.

Bień, J., Kamizela, T., Kowalczyk, M., Stępniak, L., 2004. The analysis of sedimentation and thickening of sewage sludges after ultrasonic disintegration. Environ. Prot. Eng. 30 (4), 24–28.

Bień, J., Kamizela, T., Kowalczyk, M., 2007. The contamination of supernatant after sedimentation of sonicated activated sludge. Environ. Prot. Eng. 2, 53–60.

Bień, J., Kowalczyk, M., Kamizela, T., 2008. The effectiveness of dewatering of sludge conditioned with chemical substances in methods combined with ultrasounds. Inż. Ochro. Śr. 1, 65–72. (in Polish).

Bień, J., Kowalczyk, M., Kamizela, T., 2009a. Influence of parameters of the filtration layer and the method of preparation of sewage sludge on the effects of their dewatering. In: Szymański, K. (Ed.), Gospodarka Opadami Komunalnymi, Komitet chemii analitycznej PAN. Tom V, Koszalin, pp. 129–138. (in Polish).

Bień, J., Kamizela, T., Kowalczyk, M., Mrowiec, M., 2009b. Possibilities of gravitational and mechanical separation of sonicated activated sludge suspensions. Environ. Prot. Eng. 35 (2), 67.

Bień, J.B., Kowalczyk, M., Kamizela, T., Mrowiec, M., 2010. The influence of ultrasonic disintegration aided with chemicals on the efficiency of sewage sludge centrifugation. Environ. Prot. Eng. 36 (1), 35–43.

Bougrier, C., Albasi, C., Delgenes, J.P., Carrere, H., 2006. Effect of ultrasonic, thermal and ozone pre-treatments on waste activated sludge solubilisation and anaerobic biodegradability. Chem. Eng. Process. 45, 711–718.

Broeckel, U., Roemer, R., Meyer, J., Mittelstrass, A., Witt, R., Schubert, E., Blei, P., Fessler, J., 1996. Method for Filtering and Dewatering of Sewage Sludge, EP, BASF A.-G., Germany, Patent No. 690.

Bruus, J.H., Nielsen, P.H., Keiding, K., 1992. On the stability of activated sludge flocs with implications to dewatering. Water Res. 26, 1597–1604.

Chang, M.-C., Lin, H.-L., Huan, C.-L., Wang, Y.-Y., Wan, C.-C., 1998. Conditioning characteristics of kaolin sludge with different cationic polyelectrolytes. Colloids Surf. A 139, 75–80.

Chang, G.R., Liu, J.C., Lee, D.J., 2001. Co-conditioning and dewatering of chemical sludge and waste activated sludge. Water Res. 35, 786–794.

Chen, Y.G., Yang, H.Z., Gu, G.W., 2001. Effect of acid and surfactant treatment on activated sludge dewatering and settling. Water Res. 35, 2615–2620.

Christensen, J.R., Sorensen, P.B., Christensen, G.L., Hansen, J.A., 1993. Mechanisms for overdosing in sludge conditioning. J. Environ. Eng. 119, 159–171.

Chu, C.P., Lee, D.J., 1999. Three stages of consolidation dewatering of sludge. J. Environ. Eng. 125, 959–965.

Chu, C.P., Lee, D.J., 2001. Experimental analysis of centrifugal dewatering process of polyelectrolyte flocculated waste activated sludge. Water Res. 35 (10), 2377–2384.

Chu, C.P., Chang, B.-V., Liao, G.S., Jean, D.S., Lee, D.J., 2001. Observations on changes in ultrasonically treated waste activated sludge. Water Res. 35 (2), 1038–1046.

Cywiński, I., Gdula, S., Kempa, E., Kurbiel, J., Płoszański, H., 1972. Treatment of Urban Wastewater. Arkady, Warszawa. (in Polish).

Davis, J.F., Mafia, G.J., 1995. Collagen dispersions for liquid-solid separation in water treatment and sludge dewatering. Sep. Technol. 5, 147–152.

Dewil, R., Baeyens, J., Goutvrind, R., 2006. Ultrasonic treatment of waste activated sludge. Environ. Prog. 25, 121–128.

Durie, R.A., 1991. The Science of Victorial Brown Coal: Structure, Properties and Consequences for Utilisation. Butterworth-Heinemann Ltd.

Dymaczewski, Z. (Ed.), 1997. Guide to the Operator of the Sewage Treatment Plant. PZITS, Poznań. (in Polish).

Feng, X., Deng, J., Lei, H., Bai, T., Fan, Q., Li, Z., 2008. Dewaterability of waste activated sludge with ultrasound conditioning. Bioresour. Technol. 100, 1074–1081.

Florkiewicz, A., Krajewski, P., Leszczyńska, M., Sozański, M.M., 1999. Technology of Removing and Neutralizing Sediments from Water Treatment. Wydawnictwo Politechniki Poznańskiej, Poznań. (in Polish).

Fukas-Płonka, Ł., Janik, M., 2008. Pros and cons of drying sewage sludge. Forum Eksploatatora 5, 25–27. (in Polish).

Fukas-Płonka, Ł., Kuś, K., Janik, M., 2006. Characterization of physical properties of mechanically thickened sludge. In: Sympozjum Ogólnokrajowe: Hydroprezentacje, NOT, Ustroń. (in Polish).

Gawroński, R., 1996. Liquid Purification Processes. Oficyna Wydawnicza Politechniki Warszawskiej, Warszawa. (in Polish).

Gromiec, M.J., Korczak-Niedzielska, M., 2000. The use of thermally dried municipal sewage sludge. Inż. Ochro. Śr. 1-2, 211–218. (in Polish).

Hong, S.G., Chen, G.W., Chang, I.L., Hung, W.T., Lee, D.J., 1995. Freeze/thaw treatment on waste activated sludge: an FTIR spectroscopic study. J. Environ. Sci. Health A 30, 1717–1724.

Hung, W.T., Chang, I.L., Lin, W.W., Lee, D.J., 1996. Uni-directional freezing of waste activated sludges: effects of freezing speed. Environ. Sci. Technol. 30, 2391–2396.

Huang, Ch, Pan, J.R., Pang, C.-H., Li, K.-Ch., Lin, Ch.-F., 2001. Effect of low temperature treatment on dewatering characteristics of polyelectrolyte—conditioned fresh water sludge. In: Lee, D.J. (Ed.), Sludge Management Entering the 3rd Millennium: Conference Proceedings of the IWA International Specialized Conference "Sludge Management Entering the 3rd Millennium—Industrial, Combined, Water and Wastewater Residues", Taipei, Chinese Taiwan, 25–28 March 2001, 496–503.

Hwa, T.J., Jeyaseelan, S., 1997. Comparison of lime and alum as oily sludge conditioners. Water Sci. Technol. 36 (12), 117–124.

Kim, Y.U., Kim, B.I., 2003. Effect of ultrasound on dewaterability of sewage sludge. Jpn. J. Appl. Phys. Part 1 (42), 5898–5899.

Klimapol, n.d. http://www.klimapol.pl (Accessed 10.06.2017).

Korczak-Niedzielska, M., Gromiec, M.J., 1998. Drying of sewage sludge. In: Mat. Konf. Nauk. -Techn. nt. Osady ściekowe w praktyce. Wydawnictwo Politechniki Częstochowskiej, Częstochowa-Ustroń. (in Polish).

Kosmala-Szczurek, B., Tal-Figiel, B., 1998. The effect of the combined action of ultrasounds and polyelectrolytes on the rate of precipitation of the suspension particles. Inż. Ap. Chem. 4, 21–24. (in Polish).

Kowal, A.L., Świderska-Bróż, M., 2000. Purification of Water. PWN, Warszawa. (in Polish).

Kowalczyk, M., 2012. Enhancement of Processes for Sewage Sludge Dewatering. Wydawnictwo Politechniki Częstochowskiej, Częstochowa. (in Polish).

Krevox, n.d. http://www.krevox.com.pl (Accessed 02.03.2018).

Laheij, E.J., Kerkhof, P.J.A.M., Hewijn, A.J.M., Coumans, W.J., 1996. Fundamental aspects of sludge filtration and expression. Water Res. 30 (3), 697–703.

Lee, C.H., Liu, J.C., 2001. Sludge dewaterability and floc structure in dual polymer conditioning. Adv. Environ. Res. 5, 129–136.

Lin, Y.-F., Jing, S.-R., Lee, D.-Y., 2001. Recycling of wood chips and wheat dregs for sludge processing. Bioresour. Technol. 76 (2), 161–163.

Lu, M.C., Lin, C.J., Liao, C.H., Huang, R.Y., Ting, W.P., 2003. Dewatering of activated sludge by Fenton's reagent. Adv. Environ. Res. 7, 667–670.

Martel, J.C., Affleck, R., Yushak, M., 1998. Operational parameters for mechanical freezing of alum sludge. Water Res. 9, 2646–2654.

Mizera, A., 2002. Sewage Sludge (Un)safe Waste. GreenWorld—Ochrona Środowiska, Ekologia. (in Polish).

Na, S., Kim, Y.-U., Khim, J., 2007. Physiochemical properties of digested sludge with ultrasonic treatment. Ultrason. Sonochem. 14, 281–285.

Nellenschulte, T., Kayser, R., 1997. Change of particle structure of sewage sludges during mechanical and biological processes with regard to the dewatering result. Water Sci. Technol. 36 (4), 293–306.

Nelson, R.F., Brattlof, B.D., 1979. Sludge pressure filtration with fly ash addition. J. Water Pollut. Control Fed. 51 (5), 1024–1031.

Oleszkiewicz, J., 1998. Sewage Sludge Management, Decider's Guide. LEM, Kraków. (in Polish).

Omerci, B., Vesilind, P.A., 2001. Effect of dissolved organic material and cations on freeze-thaw conditioning of activated and alum sludges. Water Res. 35 (18), 4299–4306.

Ozacar, M., Sengil, A., 2000. Effectiveness of tannins obtained from valonia as a coagulant aid for dewatering of sludge. Water Res. 34 (4), 1407–1412.

Pan, J.R., Huang, C., Cherng, M., Li, K., Lin, C., 2003. Correlation between dewatering index and dewatering performance of three mechanical dewatering devices. Adv. Environ. Res. 7, 599–602.

Papavasilopoulos, E.N., 1997. Viscosity as a criterion for optimum dosing of polymers in waterworks sludge. J. CIWEM 11, 217–224.

Papavasilopoulos, E.N., Bache, D.H., 1998. On the role of aluminum hydroxide in the conditioning of an alum sludge. Water Sci. Technol. 38 (2), 33–40.

Parker, P.J., Collins, A.G., 1999. Ultra-rapid freezing of water treatment residuals. Water Res. 33 (10), 2239–2246.

Parkitna, K., Kowalczyk, M., Kamizela, T., Milczarek, M., 2013a. Dewatering of sewage sludge conditioned by means of the combined method of using ultrasound field, fenton's reaction and gypsum. In: Environmental Engineering IV. CRC Press. (in Polish).

Parkitna, K., Kowalczyk, M., Krzemińska, D., 2013b. Change in the amount of ultrasound energy introduced to sewage sludge depending on the content of dry matter. Rocz. Ochr. Śr. 15, 2039–2053. (in Polish).

Podedworna, J., Umiejewska, K., 2008. Sewage Sludge Technology. Oficyna Wydawnicza Politechniki Warszawskiej, Warszawa. (in Polish).

Proffico, n.d. http://www.proffico.com (Accessed 24.09.2017).

Rehmat, T., Branion, R., Duff, S., Groves, M., 1997. A laboratory sludge press characterizing sludge dewatering. Water Sci. Technol. 35 (2–3), 189–196.

Roman, M. (Ed.), 1986. Sewerage—Wastewater Treatment. Arkady, Warszawa. (in Polish).

Sander, B., Lauer, H., Neuwirth, M., 1989. Process for Producing Combustible Sewage Sludge Filter Cakes in Filter Presses, US, BASF Aktiengesellschaft, Fed. Rep. Germany, Patent No. 4, 840, 736.

Smollen, M., Kafaar, A., 1997. Investigation into alternative sludge conditioning prior to dewatering. Water Science and Technology. In: Proceedings of the 1997 International Specialized Conference on Sludge Management vol. 38 (11), pp. 115–119.

Sorensen, P.B., Moldrup, P., Hansen, J.A.A., 1996. Filtration and expression of compressible cakes. Chem. Eng. Sci. 51 (6), 967–979.

Sozański, M.M., Dymaczewski, Z., 2002. Water supply and sewage system in Poland. In: Tradition and Modernity. Polska Fundacja Ochrony Zasobów Wodnych, Poznań/Bydgoszcz. (in Polish).

Strzelczyk, M., Kamizela, T., Wolny, L., 2002. Determination of permanent, electromagnetic field influence on sewage sludges conditioning. Environ. Prot. Eng. 28 (1), 49–53.

Tay, J.H., Jeyaseelan, S., 1997. Conditioning of oily sludges with municipal solid wastes incinerator fly ash. Water Sci.Technol. 35 (8), 231–238.

Tchobanoglous, G., Burton, F., Stensel, H.D., 2003. Wastewater Engineering. Treatment, Disposal, Reuse, Matcalf & Eddy fourth ed. Mc Gaw-Hill Inc., New York.

Thapa, K.B., Qi, Y., Hoadley, A.F.A., 2009. Interaction of polyelectrolyte with digested sewage sludge and lignite in sludge dewatering. Colloids Surf. A: Physicochem. Eng Asp. 334, 66–73.

Thostenson, E.T., Chou, T., 1999. Microwave processing: fundamentals and applications. Composites 30, 1055–1071.

Tiehm, A., Nickel, K., Neis, U., 1997. The use of ultrasound to accelerate the anaerobic digestion of sewage sludge. Water Sci. Technol. 36, 121–128.

Wang, F., Ji, M., Lu, S., 2006. Influence of ultrasonic disintegration on the dewaterability of waste activated sludge. Environ. Prog. 25, 257–260.

Wojciechowska, E., Kowalik, P., 2003. The use of microwave radiation for the conditioning of sewage sludge. Inż. Ochr. Śr. 2, 167–178. (in Polish).

Wójtowicz, A., Jędrzejewski, C., Bieniowski, M., Darul, H., 2013. Model solutions in the sludge management. In: Izba Gospodarcza "Wodociągi Polskie". Stowarzyszenie Eksploatatorów Obiektów Gospodarki, Wodno-Ściekowej. (in Polish).

Wu, C., Huang, C., Lee, D.J., 1997. Effects of polymer dosage on alum sludge dewatering characteristics and physical properties. Colloids Surf. Physicochem. Eng. Asp. 122, 89–96.

Wu, R.M., Lee, D.J., Wang, C.H., Chen, J.P., Tan, R.B.H., 2001. Novel cake characteristics of waste-activated sludge. Water Res. 35 (5), 1358–1362.

Yin, X., Lu, X.P., Han, P.F., Wang, Y.R., 2006. Ultrasonic treatment on activated sewage sludge from petro-plant for reduction. Ultrasonics 44, 397–399.

Zarzycki, R., Wielgosiński, G., 2001. Problems of sewage sludge management. In: Zarzycki, R. (Ed.), Gospodarka komunalna w miastach. Polska Akademia Nauk Oddział w Łodzi, Łódź, pp. 215–260. (In Polish).
Zhang, P.Y., Zhang, G.M., Wang, W., 2007. Ultrasonic treatment of biological sludge: floc disintegration, cell lysis and inactivation. Bioresour. Technol. 98, 207–210.
Zhao, Y.Q., 2002. Enhancement of alum sludge dewatering capacity by using gypsum as skeleton builder. Colloids Surf. A Physicochem. Eng. Asp. 211, 205–212.
Zhao, Y.Q., 2003. Correlations between floc physical properties and optimum polymer dosage in alum sludge conditioning and dewatering. Chem. Eng. J. 92, 227–235.
Zhao, Y.Q., Bache, D.H., 2002. Integrated effects of applied pressure, time, and polymer doses on alum sludge dewatering behaviour. Waste Manag. 22, 813–819.
Zielewicz-Madej, E., Fukas-Płonka, Ł., 1998. Thickening and Conditioning of Sludge. Podstawy oraz praktyka przeróbki i zagospodarowania osadów. LEM, Kraków. (in Polish).

Further reading

http://www.swb.com.pl (Accessed 15.07.2017).

Traditional contaminants in sludge

19

Agata Rosińska
Faculty of Infrastructure and Environment, Czestochowa University of Technology, Częstochowa, Poland

Abbreviations

AOC	absorbable organic halogens
BTBPE	1,2-bis-(2,4,6-tribromophenoxy)ethane
DCF	dichlorofluorescein
DDD	dichlorodiphenyldichloroethane
DDE	dichlorodiphenyldichloroethylene
DDT	dichlorodiphenyltrichloroethane
DM	dry mass
DMDT	methoxy chloride
E1	estrone
E2	17b—estradiol
EE2	17a—ethinylestradiol
EH-TBB	ethyl-hexyl tetrabromobenzoate 2-ethyl-1-hexyl-2,3,4,5-tetrabromobenzoate
EPA	Environmental Protection Agency
HBCD	hexabromocyclododecane
HCB	hexachlorobenzene
HCH	hexachlorocyclohexane
HMW	high molecular weight
IPB	ibuprofen (iso-butyl-propanoic-phenolic acid)
LMW	low molecular weight (LMW)
LODs	maximum limits of detection
LOQs	limits of qualification
OCP	organochlorine pesticides
OMPs	organic micropollutants
PAHs	polycyclic aromatic hydrocarbons
PBBs	polybrominated biphenyls
PBDEs	polybrominated diphenylethers
PCB 126	3,3′4,4′,5-pentachlorobiphenyl
PCB 169	3,3′,4,4′,5,5′-hexachlorobiphenyl
PCB 77	3,3′4,4′-tetrachlorobiphenyl
PCB 81	3,4,4′,5-tetrachlorobiphenyl
PCBs	polychlorinated biphenyls
PCDDs	polychlorinated dibenzo-p-dioxins
PCDFs	polychlorinated dibenzofurans
PCT	pentachlorophenol
PFBS	perfluorobutanesulfonate
PFCA	perfluorinated carboxylic acid

Industrial and Municipal Sludge. https://doi.org/10.1016/B978-0-12-815907-1.00019-2

PFCs	perfluorinated compounds
PFDA	perfluorodecanoic acid
PFDoA	perfluorododecanoic acid
PFDS	perfluorodecanesulfonate
PFHpA	perfluoroheptanoic acid
PFHpS	perfluoroheptanesulfonate
PFHxS	perfluorohexanesulphonic acid
PFNA	perfluorononanoic acid
PFOA	perfluorooctanoic acid
PFOS	perfluorooctane sulfonate
PFOSA	perfluorooctane sulphonamide
PFPeA	perfluoropentanoic acid
PFTeDA	perfluorotetradecanoic acid
PFTrDA	perfluorotetranoic acid
PPFUdA	perfluoroundecanoic acid
TBBPA	tetrabromobisphenol A
TBPH	bis(2-ethyl-1-hexyl) tetrabromophthalate
TCDD	2,3,7,8-tetradibenzo-p-dioxin
WWTPs	Wastewater Treatment Plants

1 Introduction

In recent years, significant changes have been observed in the approach to problems related to the development of sludge management in wastewater treatment plants (Hara et al., 2016; Kacprzak et al., 2017). Particular actions were undertaken toward:

- Limiting the content of harmful and toxic substances both in sewage wastewater entering wastewater treatment plants and in sewage sludge.
- Recovery of energy.
- Rational and safe management of sludge.

The content of organic substances, including organic micropollutants (OMPs), in sludge depends on the type of wastewater and processes applied during sludge treatment. Raw sludge contains $75 \div 85\%$ of organic substances in dry matter while stabilized sludge contains approximately 50% of organic substances in dry mass (DM). The most important OMPs occurring in sewage sludge include PAH, PCDDs, PCDFs, PCBs, PBBs, PBDEs, PFCs, and pesticides (e.g., organochlorine pesticides OCPs).

2 Polycyclic aromatic hydrocarbons (PAHs)

PAHs are a ubiquitous and very diverse group of pollutants, occurring both in the environment and in food products (Boström et al., 2002; Haritash and Kaushik, 2009; Lawal, 2017; Masih and Lal, 2014; Nguyen et al., 2014; Sharma, 2014). PAHs constitute a group of compounds containing from two to >10 benzene rings in the molecule, with any number of alkyl substituents. There are >200 known PAHs that occur in the human environment, 33 of which have been recognized by the SCF (Scientific

Committee on Food) as particularly toxic (Neff et al., 2005). The best-studied PAH hydrocarbon is benzo(a)pyrene, which, due to the strength of the carcinogenic effect and the prevalence in the environment,is considered an indicator of the entire PAH group (Boström et al., 2002; Kim et al., 2013; Lawal, 2017).

In the pure state, PAHs occur in the form of colorless, white, light-yellow, or light-green crystals that are characterized by a low vapor pressure and a high melting point. They demonstrate sensitivity to changes in the temperature and pH of the environment in which they are found as well as to the effects of UV radiation and the presence of surfactants, pesticides, and oxidizing agents such as oxygen and ozone. As nonpolar lipophilic macromolecular compounds, PAHs dissolve poorly in water (Pavlova and Ivanova, 2003), but well in organic solvents such as benzene, cyclohexane, dichloromethane, and toluene. The solubility of PAH in water decreases with the increase of molecular weight (Stogiannidis and Laane, 2015). For example, the solubility of naphthalene (128u) in water equals 32 mg/dm^3, benzo(a)pyrene (252)–0.003 mg/L. The physicochemical properties of selected parent PAHs are shown in Table 1.

Table 1 The physicochemical properties of selected parent PAH (Stogiannidis and Laane, 2015)

PAH	Abbreviations	MW	RN	S (mg/L)	P (Pa)	K_{ow}	Bp (°C)
Nafthalene	N	128	2	32	11	3.37	218
Acenaphtylene	Acy	152	3	3.9	$9.0 \cdot 10^{-1}$	4.1	280
Acenahtylene	Ace	152	3	3.9	$3.0 \cdot 10^{-2}$	3.9	279
Fluorene	F	166	3	1.9	$9.0 \cdot 10^{-2}$	4.18	295
Phenanthrene	P	178	3	1.1	$2.0 \cdot 10^{-2}$	4.57	340
Anthracene	Ant	178	3	0.05	$1.0 \cdot 10^{-3}$	4.54	342
Fluoranthene	Fl	202	4	0.26	$1.2 \cdot 10^{-3}$	5.22	375
Pyrene	Pyr	202	4	0.13	$6.0 \cdot 10^{-4}$	5.18	393
Benz(*a*) anthracene	BaAnt	228	4	0.009–0.014	$2.8 \cdot 10^{-5}$	5.6	400
Chrysene	Chr	228	4	0.002	$1.4 \cdot 10^{-6}$	5.86	448
Benzo(*b*) fluoranthene	BbFl	252	4	0.0014	$6.7 \cdot 10^{-5}$	5.8	481
Benzo(*k*) fluoranthene	BkFl	252	5	0.0007–0.008	$5.2 \cdot 10^{-8}$	6	480
Benzo(*a*) pyrene	BaPyr	252	5	0.003	$6.0 \cdot 10^{-7}$	6.0	496
Benzo(*ghi*) perylene	BghiPer	276	6	0.00026	$1.4 \cdot 10^{-8}$	7.1	550
Indeno(*1,2,3-cd*)pyrene	IPyr	276	6	0.00019	$1.3 \cdot 10^{-8}$	6.6	536
Dibenz(*ah*) anthracene	DBahAnt	278	5	0.0005	$3.7 \cdot 10^{-8}$	6.5	524

MW, molecular weight; *RN*, ring numer; *S*, aqueos solubility (25°C); K_{ow}, the logarithm of the octanol-water partition coefficient; *Bp*, boiling point.

PAHs demonstrate high affinity for the surface of solids; therefore they occur mainly in a form adsorbed on particles that form suspended solids (Bertilsson and Widenfalk, 2002). Strongly bound to suspended solids and the surface of solid substances are indeno(1,2,3-c,d)pyrene, benzo(g,h,i)perylene, dibenz(a,h)anthracene, and benzo(k)fluoranthene. Naphthalene and acenaphthylene are unstably bound to suspended solids, and the remaining hydrocarbons constitute a group of compounds that can occur both in a dissolved form and adsorbed on the surface of solids (Abdel-Shafy and Mansour, 2016; Lawal, 2017).

Many PAHs have carcinogenic and mutagenic properties as well as being genotoxic and embryotoxic (Boehm et al. 2007; Guo et al., 2007; Swietlik et al., 2002). These compounds have been included in the list of priority organic compounds of the US Environmental Protection Agency (USEPA) (USEPA, 2011), and have become compounds that are subject to frequent monitoring (Lawal, 2017; Lung et al., 2004). Since 2005, the European Union (EU) has adopted a list of PAHs that also have to be monitored. Benzo(a)pyrene (BaP) is most likely the most-studied PAH in environmental matrices. In 1987, the International Agency for Research in Cancer (IARC) described BaP as the main human carcinogen. Therefore, a BaP assay is common in environmental analyses as a marker of the total PAH content (Lawal, 2017).

Polycyclic aromatic hydrocarbons occurring in the environment originate mainly from anthropogenic sources (Bertilsson and Widenfalk 2002; Morillo et al., 2008). The amount of PAHs produced from natural sources is small compared to the amount resulting from human activity (Abdel-Shafy and Mansour, 2016; Jiao et al., 2009). Combustion of anthropogenic substances produces >90% of the total amount of PAHs in the surrounding environment. PAHs are formed during incomplete combustion of organic matter. Polycyclic aromatic hydrocarbons occur also in industrial raw materials and in their treatment products (Kim et al., 2008; Morillo et al., 2008; Yan et al., 2006). Smoking tobacco, both passive and active, is a separate source of PAHs (Lung, et al., 2004).

The composition and amount of mixtures of PAHs emitted to the environment depend on the type of combusted substance, the method of combustion, and the use of filters and other devices protecting against their emission. Most PAHs in the air occur in the form of vapors and aerosols adsorbed on dust with a diameter of approximately 0.5 nm. Thus, adsorbed PAH molecules can be transferred by wind away from the places of origin, contaminating soil, water, and plants.

The presence of PAHs in sewage sludge results from the discharge of rainwater to the treatment plant (Perez et al., 2001a,b). These waters contain not only the pollution of atmospheric air, but also that rinsed from the roads, the products of asphalt surface abrasion, and the wear of car components, which are an important source of wastewater and sewage sludge contamination by PAHs (Lawal, 2017).

It was found that during intensive rainfall and thaw, the concentration of hydrocarbons in urban wastewater may increase by 100 times, and the concentration of benzo(a)pyrene may significantly exceed its permissible amount in wastewater. The main source of PAHs in sewage sludge is industrial wastewater and domestic wastewater. The character of industrial wastewater depends mainly on the production profile of a given plant (coal and crude oil processing, coking plants, power plants), which simultaneously affects the hydrocarbon concentration.

Sewage sludge is separated or/and generated in particular stages of wastewater treatment. PAHs, condensed in sewage sludge, occur in concentrations several orders larger than in raw wastewater (Fernández-Luqueño et al., 2016).

The research carried out by Perez et al. (2001a) concerned changes in PAH content in sewage sludge originating from two treatment plants, the first of which collected only domestic and commercial wastewater while the second collected selected industrial wastewaters. The authors found that higher total concentrations of the analyzed compounds occurred in sewage sludge after treatment of domestic wastewater. Phenanthrene dominated among the assayed hydrocarbons, it constituted from 24% to 37% in reference to aggregated PAH concentration. The degree of PAH load in sewage sludge may also result from specific conditions of a given region, such as the location of the treatment plant in the urban agglomeration, a significant amount of industrial wastewater, and the existence of various types of sewage systems (combined, separate). Sludge from highly industrialized areas usually shows higher concentrations of PAHs as compared to the sludge from small wastewater treatment plants, most often operating in an area with a separate sewage system (Perez et al., 2001a,b). Studies of PAH content in various sewage sludges originating from municipal wastewater treatment plants conducted by many authors showed that the highest concentrations of the discussed compounds were observed in digested sludge (25.8–202.0 mg/kg DM), raw sludge (2.4–97.6 mg/kg DM), and collected on sludge beds (0.3–54.2 mg/kg DM).

3 Dioxins and furans

Of the many chemical compounds present in the natural environment, many are released from industries and households. Only a small number of chemical pollutants are regularly analyzed and monitored. This group includes organochlorine compounds, referred to as dioxins. These compounds appeared in the human environment as unintentional trace contaminants, a product of transformations occurring in technological processes and side reactions in the environment. Dioxins were never produced commercially and no practical use for them was found (Schiavon et al., 2016).

Dioxins constitute a group of aromatic organochlorine compounds including polychlorinated dibenzo-p-dioxins (PCDDs) and dibenzo-p-furans (PCDFs) (Fig. 1). These compounds are composed of two benzene rings connected by one or two oxygen bridges.

With respect to the amount and place of chlorine substitution, 75 isomers of PCDDs and 135 isomers of PCDFs (congeners) are distinguished. There are 210 PCDD/Fs

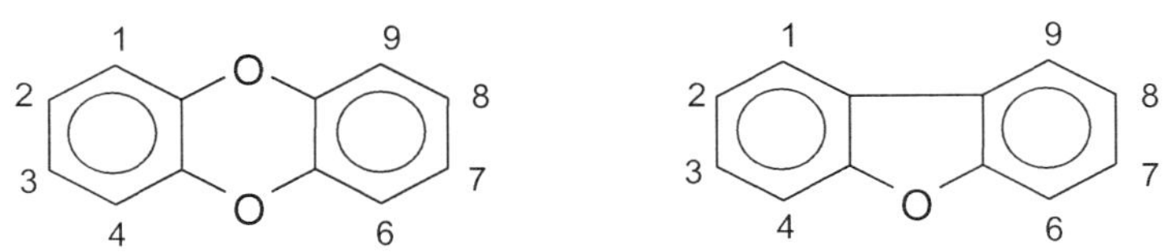

Fig. 1 General formula: (A) dibenzo-p-dioxins and (B) dibenzo-p-furan with carbon numeration.

congeners known and almost all occur in environmental samples. Due to the toxicity to humans, only 17 congeners (seven dioxins and 10 furans) are considered the most dangerous, and usually only those 17 are determined in environmental samples. Therefore, the term "dioxins" refers only to these 17 compounds. The most toxic PCDD congener is 2,3,7,8-tetradibenzo-p-dioxin that is 2,3,7,8-TCDD. TCDD is the most potent synthetic toxin, second only to venoms, that is, biospherical toxins. Dibenzo-p-dioxin and dibenzofuran can also be substituted with other halogens, that is, fluorine, bromine and iodine. As a result, there are 75 dioxin congeners and 135 dibenzofuran congeners (Schiavon et al., 2016).

Polychlorinated dibenzo-p-dioxins and dibenzofurans are colorless solids with melting points above 100°C, low vapor pressures, and high boiling points. They are poorly soluble in water and relatively well soluble in apolar solvents, but dissolve best in fats. For this reason, in living organisms they accumulate in adipose tissue and bioaccumulate in the trophic cycle. Dioxins are characterized by long half-lives ($T_{1/2}$). It is usually estimated that the half-life of PCDD/Fs in the human body equals on average approximately 8 years; however, data in the literature give longer half-lives in the range of 12–132 years (Geyer et al., 2002).

The result of lipophilic properties of dioxins and low saturated vapor pressure is their sorption on apolar surfaces of solids, for example, activated carbon, as well as on human skin, clothing, etc.

PCDD/Fs are characterized by exceptional durability, thermal stability, and chemical resistance to oxidation and processes of biological degradation. Particular resistance to biochemical degradation is demonstrated by dioxin congeners containing chlorine at 2,3,7,8 positions, whereas the factor that can cause degradation of PCDD and PCDF is ultraviolet radiation, affecting especially compounds dissolved in organic solvents.

Dioxins undergo thermal decomposition at temperatures above 850°C. This process is used for the thermal destruction of waste, where high temperature and excess oxygen advance the reduction of dioxin emissions to the atmosphere.

Dioxins are classified as substances of high toxicity; however, not all PCDD and PCDF congeners are equally toxic. The degree of toxicity of congeners depends on the amount and location of chlorine substituents. It is assumed that congeners with 1–3 substituents do not show toxicity, or that it is is negligible. As the most toxic are considered congeners with substituents in positions 2, 3, 7, and 8. The above compounds were given values that were described as toxicity equivalent factors—TEFs. The normative content of these compounds is stated as the toxicity equivalent quantity—TEQ.

Sources of dioxins can be divided into primary and secondary (reservoirs). Primary sources include processes and/or reactions that result in dioxins being formed from elements or precursors (Correa et al., 2006). Secondary sources, called secondary environmental pollution, are products and substances containing PCDD/Fs at a trace level (Everaert and Baeyens, 2002; Gworek et al., 2013; Ren and Zheng, 2009; Wielgosinski, 2011; Yu et al., 2006). Dioxins are formed as byproducts during uncontrolled combustion processes such as, for example, household waste incineration. The sources of organochlorine compounds are also facilities for processing secondary raw materials, landfills from which hazardous effluents may escape, combustion processes, and landfilling of industrial waste. The presence of dioxins in the environment may be a result of fires and industrial processes, for example, cellulose and paper production, secondary processing of nonferrous metals, and the production of copper, magnesium, and steel. PCDDs and PCDFs are also formed in photochemical reactions and metabolic processes occurring in nature, mainly in soil where the precursors may be, for example, residues of chlorinated plant protection products and others (Carroll, 2001; Domingo et al., 2001; Grochowalski and Konieczyński, 2008; Gworek et al., 2013; Oh et al., 2006; Rahman et al., 2014; Smith et al., 2009; Vassiliadou et al., 2009; Wu et al., 2010).

The presence of PCDDs and PCDFs in sewage sludge can originate from the atmosphere through dry and wet deposition, inflows from wastewater treatment plants, and processes occurring during wastewater treatment in treatment plants. Atmospheric deposition can have an impact on the level of PCDD/Fs in the sewage sludge itself (during drying or thermal stabilization processes) and on the inflow of these compounds to municipal wastewater through storm runoffs as well as runoff from streets and roofs. The source of atmospheric deposition is PCDD/Fs emission from thermal processes. In cities, local sources, including car traffic, may also be important. Moreover, sewage sludge itself can be a source of PCDD/Fs. In wastewater treatment plants, PCDD/F formation is possible, for example, as a result of enzymatic reactions and dechlorination reactions of higher-substituted congeners. It was found that PCDD/Fs may be formed as a result of oxidation of chlorophenols, catalyzed by peroxidase. Hepta- and octachlorinated dioxins can be formed from pentachlorophenol (PCT). It is believed that approximately 2.4 kg of PCDD/Fs per year in sewage sludge could be derived from enzymatic reactions from such precursors as chlorophenols or PCP to chlorate (I). It has been shown that the concentration of these compounds increases during the thermal stabilization processes of sludge.

The range of PCDD/F concentration in sewage sludge in various countries is diverse (Eljarrat et al., 2003; Fuentes et al., 2007; Harrison et al., 2006). However, regardless of the country in which the research was conducted, it was observed that in sewage sludge, the total PCDD concentration is higher compared to the PCDF concentration and the highest chlorinated dioxins predominate in sludge. At the end of the 1980s, in industrialized countries a decrease in PCDD/F concentration in sewage sludge was observed, which was the result of a ban on PCP usage. However, still high concentrations in sludge in many countries may indicate that PCPs used in textile materials are still a significant factor affecting the PCDD/F level in sewage sludge. Lower chlorinated PCDD/Fs are transferred to sewage sludge from surface runoff contaminated by atmospheric deposition.

4 Polychlorinated biphenyls

Polychlorinated biphenyls (PCBs) are harmful chemical compounds while the degree of toxicity depends on the number and place of substitution of chlorine atoms in the biphenyl molecule. These contaminants enter the human body via the alimentary and respiratory routes as well as through the skin. PCBs demonstrate a variety of toxic

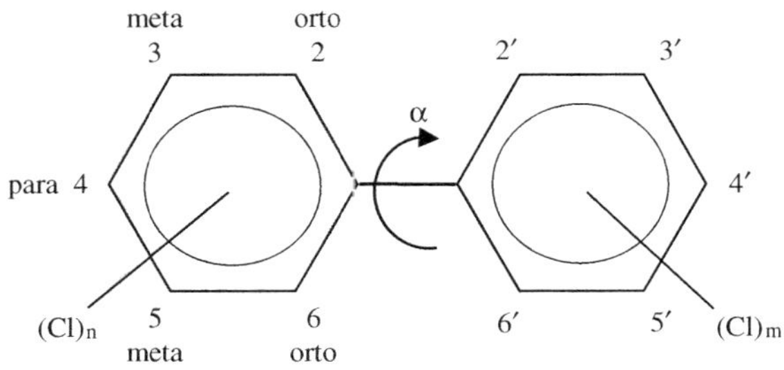

Fig. 2 Chemical structure and numbering of carbon atoms in biphenyl molecule.

effects. They concentrate in the subsequent stages of movement in the food chain and become harmful to humans and animals. In these organisms, PCBs quickly penetrate into the circulatory system and then accumulate in most tissues. The largest amounts of polychlorinated biphenyls were detected in adipose tissue (Robertson and Hansen, 2015).

Polychlorinated biphenyls usually occur as mixtures of congeners. On the technical scale, these compounds were obtained from the direct reaction of biphenyl with chlorine, resulting in mixtures of congeners with a composition depending on the proportion of chlorine and biphenyl and the conditions of the synthesis. The structural formula with carbon atom numeration in the biphenyl molecule is presented in Fig. 2. In terms of chemical structure, the chlorobiphenyl molecule is composed of two benzene rings connected by a bridge (biphenyl) in which one, several, or all the hydrogen atoms can be substituted with chlorine atoms. There are 209 PCB isomers that can be divided into 10 homologous groups (from mono to decachlorobiphenyl) (Robertson and Hansen, 2015). Depending on the place of substitution and the number of chlorine atoms, both rings form a different spatial configuration. Chlorobiphenyls with a flat spatial structure are called planar, coplanar PCBs, or nonortho due to the lack of chlorine atoms in the ortho position. There are four nonortho PCB congeners: 3,3′4,4′-tetrachlorobiphenyl (PCB 77), 3,4,4′,5-tetrachlorobiphenyl (PCB 81), 3,3′4,4′,5-pentachlorobiphenyl (PCB 126), and 3,3′,4,4′,5,5′-hexachlorobiphenyl (PCB 169). In the environmental samples, very often seven indicator PCB congeners (I-PCBs) are included with the following IUPAC codes: 28, 52, 101, 118, 138, 153, and 180, which identification is recommended by the USEPA. To determine the physical and chemical properties of PCBs, the level of isomeric groups was chosen. The place of substitution and the number of chlorine atoms in the chlorobiphenyl molecule determine the state of aggregation of a PCB congener, its behavior in the environment, and its toxicity.

PCBs are characterized by high thermal stability, a high ignition temperature of 170÷380°C, good thermal conductivity, high dielectric constant, and high density. These properties intensify with an increasing number of chlorine atoms in the molecule (Robertson and Hansen, 2015).

PCBs as nonionic, hydrophobic organic compounds are very slightly soluble in water and well soluble in fats, oils, and nonpolar solvents (hexane, isooctane) (Urbaniak, 2007). The lipophilicity of PCBs increases with the increasing number of chlorine atoms in the biphenyl molecule. The solubility of PCB congeners in water

Table 2 Selected physicochemical properties of PCBs

PCB	Mp (°C)	Bp (°C)	S (μg/L)	log K_{ow}	P (Pa)	BCF
Bifenyl	71	256	9.3	4.3	4.9	1000
$C_{12}H_9Cl$	25–77.9	285	$1.3·10^3$–$7·10^3$	4.63–4.66	1.1	2500
$C_{12}H_8Cl_2$	24.4–149	312	$0.6·10^2$–$7.9·10^2$	4.72–5.23	0.24	6300
$C_{12}H_7Cl_3$	28–87	337	$0.1·10^2$–$6.4·10^2$	5.04–5.71	0.054	$1.6·10^4$
$C_{12}H_6Cl_4$	47–180	369	$0.2·10^2$–$1.7·10^2$	4.84–6.32	0.012	$4.0·10^4$
$C_{12}H_5Cl_5$	76.5–124	381	4.5–0.9	5.60–6.92	$2.6·10^{-3}$	$1.0·10^5$
$C_{12}H_4Cl_6$	77–150	400	0.4–0.9	6.20–7.30	$5.8·10^{-4}$	$2.5·10^5$
$C_{12}H_3Cl_7$	122.4–149	417	0.5	6.55–7.72	$1.3·10^{-4}$	$6.3·10^5$
$C_{12}H_2Cl_8$	159–162	432	0.2–0.3	7.21–7.62	$2.8·10^{-5}$	$1.6·10^6$
$C_{12}HCl_9$	182.8–206	445	0.1	7.88–7.94	$6.3·10^{-6}$	$4.0·10^6$
$C_{12}Cl_{10}$	305.9	456	0.02	8.3	$1.4·10^{-6}$	$1.0·10^6$

Mp, melting point; *Bp*, boiling point; *BCF*, bioconcentration factor in fishes.

is varied and ranges from $1{,}3·10^3$–$7·10^3$ μg/L for monochlorobiphenyl to 0.02 μg/L for decachlorobiphenyl (Table 2).

The solubility of PCBs in water and organic solvents affects their transport and durability in the environment and plays an important role in degradation. PCBs with relatively good solubility in water can be more easily decomposed by microorganisms than those with low solubility (Urbaniak 2007). The lipophilic properties and the low saturated vapor pressure of PCBs determine sorption on polar surfaces of solids that are particularly rich in organic compounds (Durjava et al., 2007; (Zhou et al. 2005). The sorption process intensifies with the increase in the number of chlorine atoms in the biphenyl molecule. This indicates that the higher chlorinated biphenyls sorb better than the lower chlorinated ones, with the coplanar congeners containing four to six chlorine atoms sorb more strongly than the higher chlorinated biphenyls due to their flat spatial structure.

Due to dielectric properties, nonflammability and chemical resistance of PCBs, they are used as: insulating materials for filling transformers and capacitors; hydraulic fluids for energy transmission (mining, aviation); liquids for vacuum pumps; lubricants in gas turbines; cooling and lubricating liquids during machining; in the building materials industry (Robson et al., 2010); additives for paints, varnishes, plasticisers, thermosetting plastics and copying paper, plant protection products as well as preservatives and impregnating agents.

The wide usage of PCBs in products with long periods of practical use has led to the spread of these compounds in all elements of the natural human environment. PCBs penetrate into the environment as a result of evaporation (open systems), spills of oils contaminated with PCBs, during the storage and incineration of hazardous and medical waste (Yoshida et al., 2009), and from treatment of industrial wastewater.

The sources of PCBs in sewage sludge may be wastewater flowing into the treatment plant, dry and wet deposition of these compounds in the atmospheric air, and processes occurring during wastewater treatment in the treatment plant. The PCB content in sewage sludge is closely related to the occurrence of polychlorinated biphenyls in wastewater and depends on their type (Table 3) and share of industrial and rainwater wastewater. The values of PCB concentrations in sewage sludge in the literature vary significantly. In 2001–2003, Abad et al. (2005) determined the content of micropollutants in sewage sludge for more than a dozen treatment plants in Spain. The content of the analyzed PCBs ranged from 0.003 to 0.06 mg/kg. Similar PCB concentrations were determined by Blanchard et al. (2004) in sludge (0.07–0.65 mg/kg DM) from the Parisian wastewater treatment plant. In Australia in 1995–2006, sewage sludge collected from selected treatment plants was analyzed. PCBs were detected in only 8% of the tested samples (Clarke et al., 2008). No significant difference was demonstrated in indicator PCB concentrations in sewage sludge sampled from municipal and industrial wastewater treatment plants.

5 Polybrominated biphenyls and polybrominated diphenyl ethers

Polybromide derivatives of biphenyls have found widespread use as retardants in the incineration of plastics. The currently used flame retardants can be classified into four groups:

1. Inorganic (inorganic salts) antimony compounds, aluminium hydroxide, borax.
2. Phosphorous derivatives (phosphorus compounds) organophosphorus compounds.
3. Nitric (nitrogen-based compounds) melamine and melamine derivatives (melamine and melamine derivatives).
4. Chlorinated derivatives and halogenated compounds (paraffins, chlorinated alicyclic compounds, and brominated aromatic compounds) (Morosöe, 2006).

In terms of chemical structure, polybromide biphenyl derivatives can be divided into polybrominated biphenyls, polybrominated ethers, biphenyl ethers, hexabromocyclododecane (HBCD), tetrabromobisphenol-A (TBBPA), and recently introduced modern agents such as bis (2,4,6-triphenoxy) ethane, 1,2-bis-(2,4,6-tribromophenoxy)ethane (BTBPE), and ethyl-hexyl tetrabromobenzoate 2-ethyl-1-hexyl-2,3,4,5-tetrabromobenzoate (EH-TBB).

Halogen derivatives are among the most commonly used retardants and their consumption accounts for about 21% of the total demand for flame retardants. The effect of halogen retardants is based on the uptake of free radicals, which are responsible for the spread of fire. The effectiveness of their function is related to the type of halogen atom. Due to the high efficiency of free radical uptake, brominated retardants are most commonly used.

Table 3 PCBs concentration range in wastewater and sewage sludge (Aparicio et al., 2009; Clarke et al., 2008; de Souza Pereira and Kuch, 2005; Eljarrat et al., 2003; Katsoyiannis and Samara, 2005)

Congener	Sewage (ng/L)	Preliminary sludge (ng/L)	Excessive sludge (ng/g DM)	Active sludge (ng/g DM)	Sewage sludge after anaerobic stabilization (ng/g DM)	Dewatering sludge (pg/g DM)
PCB 28	4.5	4.8	5.6	4.2–11	<0.05–6.8	1.80–5.88
PCB 52	220	390	115	4.7–300	<0.05–160	2.71–8.92
PCB 101	130	260	150	1.1–120	4.0–91	3.22–15.10
PCB 118	12	15	18	1.9–13	3.5–30	1.98–10.70
PCB 138	4	11	15	2.0–10	1.4–22	4.26–12.7
PCB 153	10	14	17	8.3–9.8	4.3–22	3.38–17.10
PCB 180	250	340	140	4.4–150	4.3–210	2.89–9.55
∑PCB	630.5	1034.8	460.6	26.6–620	17.5–550	20.24–79.95

Polybrominated diphenyl ethers (PBDE) are used in consumer electronic devices such as computers and televisions as well as upholstery and carpets. Literature data show that PBDE, like other brominated retardants, undergo bioaccumulation, causing global environmental contamination; they have been identified in various environmental elements (Covaci et al., 2006; Frederiksen et al., 2009; Kolic et al., 2009; Kryłów and Rezka, 2017; Law et al., 2008; Marvin et al., 2011; Raab et al., 2008; Rice et al., 2002; Stapleton et al., 2008; Toms et al., 2009).

Hexabromocyclododecane (HBCD) is used in the production of foamed polystyrene insulation boards and fabrics. TBBPA is used as a reactive flame retardant in electrical devices (e.g., printed circuit boards) (Rosenberg et al., 2011).

The presence of newly introduced brominated flame retardants such as EH-TBB and bis(2-ethyl-1-hexyl) tetrabromophthalate (TBPH) was found at the recycling points for electrical and electronic waste and in the air and marine fauna in Asia as well as in the Arctic (Möller et al., 2011).

The main sources of bromine retardant emissions are plants producing technical mixtures of PBDEs as well as plastic and textile plants.

The processes of neutralizing electrical and electronic waste are important sources of brominated retardants. The direct effect of pollutant emission is the health risk of the inhabitants of developing countries such as China (Chan et al., 2007; Huo et al., 2007) and India (Eguchi et al., 2012).

Increasing environmental contamination has resulted in countermeasures in many countries, efforts that are aimed at limiting the production and use of brominated retardants. It should be noted that, at the same time, there was an increase in the production and use of other bromide derivatives.

For the first time, PBDE was detected in sewage sludge samples in 1979 in the United States. In 1992, the content of two congeners, BDE-47 and BDE-99, was assayed, with concentrations of 15 and 19 μg/kg, respectively (Eguchi et al., 2012). Similar BDE content was determined in municipal sewage sludge sampled from German wastewater treatment plants (Knoth et al., 2007). In subsequent years, there was an increase in the consumption of preparations containing BDE-209 in comparison with preparations containing congeners BDE-47 and BDE-99. In sewage sludge collected from municipal wastewater treatment plants in Switzerland, the BDE-209 concentrations were determined at 1100 μg/kg, which constituted more than a fivefold increase in the concentration of these compounds compared to the results obtained in 1993. In sludge samples collected from 11 municipal wastewater treatment plants in Germany from 2002 to 2003, a significant variation in concentrations of BDE-209 was observed (Table 4).

6 Perfluorinated aliphatic compounds

.0Perfluorinated aliphatic compounds (PFCs) are derivatives of hydrocarbons in which all or only part of the hydrogen atom is replaced with fluorine atoms. This group of compounds includes carboxylic derivatives (e.g., perfluorooctanoic acid—PFOA), sulfonates (perfluorooctane sulfonate—PFOS), and sulfonamides (e.g., perfluorooctane

Table 4 Concentrations of significant PBDEs in sewage sludge from different stages of the wastewater treatment process (Knoth et al., 2007; Kryłów and Rezka, 2017)

Concentration (ng/g DM)	Primary sludge	Secondary excess sludge	Digested sludge	Dewatered digested sludge
WWTP/population equivalent 48,000 (1)				
DeBDE 209	256	341	690	556
$\sum$Congeners 28–209	494	569	960	781
63,500 (1)				
DeBDE 209	239	234	605	417
$\sum$Congeners 28–209	328	321	699	523
WWTP/population equivalent 75,000 (2)				
DeBDE 209	169–217	182	411–1141	204–340
$\sum$Congeners 28–209	182–266	294	520–1261	280–452
WWTP/population equivalent 240,000 (1)				
DeBDE 209	199	334–486	393	–
$\sum$Congeners 28–209	234	466–640	537	–
WWTP/population equivalent 350,000 (1)				
DeBDE 209	97.1	206–220	217	–
$\sum$Congeners 28–209	142	352–437	266	–
WWTP/population equivalent 1,820,000 (1)				
DeBDE 209	209	182	–	133
$\sum$Congeners 28–209	305	264	–	186

sulphonamide—PFOSA) as well as their esters, salts, and fluorides. The construction of these compounds is characterized by a strong carbon-fluorine bond. The consequence of the substitution of hydrogen atoms in an aliphatic chain with fluorine atoms (which are about 20 times heavier) are the characteristic properties of perfluorinated compounds. These properties include, for example, a much greater rigidity of the aliphatic chain compared to nonfluorinated compounds, a higher C–F bonding energy (increasing with an increase in the number of fluorine substituents) than the C–H bonding energy. The C—F bonding is a very strong bond due to the high electronegativity of fluorine (Arvaniti and Stasinakis, 2015; Vecitis et al., 2009). The result is the thermal and chemical stability of perfluorinated compounds, which is why these compounds are poorly biodegradable in the natural environment (Ahrens, 2011; Prevedouros et al., 2006).

PFCs are characterized by higher density (about two times higher than water) and greater hydrophobicity compared to the corresponding nonfluorinated compounds. The above-mentioned features have resulted in PFCs being used commercially and industrially. They are most commonly used as high efficiency extinguishing foams, hydrogels applied on open wounds, protective creams, lubricants in metallurgy, impregnates for textiles (e.g., waterproof clothing), and in Teflon products (e.g., in pans and pots) (Giesy and Kannan, 2002). Perfluorinated compounds are also widely used as ingredients in fat-resistant food packaging as well as in a wide range of cleaning products and personal care products, including shampoos, threads and other dental agents, floor cleaners, waxes, footwear, and clothing impregnates.

Perfluorinated compounds are very stable compounds, which is why even if their production were to be terminated today, the levels of these compounds in the environment would still increase for many years. This is related to the fact that everyday objects are often used in households for a long time, and the release process itself is very slow. The studies pay special attention to the toxicity of PFCs and, consequently, to real threats to human health as well as their impact on the development of cancers. They also pay attention to the ability of perfluorinated compounds to accumulate in the human body as well as the mechanisms of their excretion.

The presence of perfluorinated compounds (in particular PFOA and PFOS and their salts) was confirmed in air and industrial wastewater, but also in surface waters, household dust samples, animal tissue samples, and human-derived material. Perfluorinated compounds were detected in samples originating from different regions of the world, from areas with varying degrees of population density and industrialization and in various elements of the environment. In most studies, the maximum limits of detection (LODs) and the limits of qualification (LOQs) were in the range of several ng/L and several ng/g. Slightly higher LOQ values were demonstrated by Huset et al. (2008) and Sinclair and Kannan (2006) with regard to wastewater and sewage sludge, respectively, up to 31.1 ng/L and up to 25.0 ng/g.

Most of the data on the occurrence of PFCs in sewage sludge have been published in the last 10 years (Arvaniti and Stasinakis, 2015). Most of the information comes from the United States (Arvaniti and Stasinakis, 2015; Boulanger et al., 2005; Loganathan et al., 2007; Schultz et al., 2006; Sinclair and Kannan, 2006) Asia (Guo et al., 2010; Ma and Shih, 2010; Murakami et al., 2009) and North European countries (Ahrens et al., 2009; Alder and van der Voet, 2015; Becker et al., 2008; Bossi et al., 2008; Huset et al., 2008) while limited information was available until recently for the Mediterranean area (Arvaniti et al., 2012; Campo et al., 2014; Stasinakis et al., 2013). To the best of our knowledge, only one study is available in Australia (Thompson et al., 2011). Data for African countries are not available. Most of this research focuses mainly on PFOS and PFOA while there is limited information about PFCA (Arvaniti et al., 2012; Campo et al., 2014, Sinclair and Kannan, 2006; Stasinakis et al., 2013). So far, most PFC monitoring studies have focused exclusively on the dissolved phase of wastewater (Guo et al., 2010; Ma and Shih, 2010). Literature data indicate a lack of seasonal variability of PSCs in wastewater and sewage sludge over the year. In untreated wastewater, the highest concentrations were demonstrated for PFOS, PFOA, and PFOSA, which amounted to 465.4 ng/L, 638.2 ng/L, and 615.0 ng/L, respectively. For PFPeA, PFHxA, PFNA, PFDA, PFTrDA, and PFDS, concentrations did not exceed

453.0 ng/L (PFTrDA). It was demonstrated that PFOS dominated in sewage sludge (up to 7304.9 ng/g DW). For PFOA, the highest determined concentration was equal to 241 ng/g DW, whereas PFCA concentration did not exceed 3209 ng/g DW (PFUdA). Higher concentrations of PFOA in wastewater or sewage sludge were detected in some wastewater treatment plants (WWTPs) in Singapore, Korea, Taiwan, Greece, and the United States. The elevated concentrations observed in these WWTPs may be related to population density as well as to differences in the human lifestyle and in the type of wastewater that each WWTP receives.

7 Residues of plant protection products

Pesticides are a group of synthetic compounds that are introduced into the environment as a result of an intended human decision. The basic role of pesticides is to increase the quantity and improve the quality of crops as well as for weed control (herbicides), fungi control (fungicides), and deterring or combating pests and pathogenic organisms (insecticides). Therefore, several classifications of pesticides can be distinguished. The most commonly used is the division of pesticides into zoocides, herbicides, and fungicides. Among zoocides, the most numerous subgroup is insecticides, in which the active substances belong to the following combinations: chlorinated hydrocarbons, organophosphorus compounds, carminic acid derivatives, and synthetic pyrethrine derivatives. Organochlorine pesticides are halogen derivatives of several-ringed cycloparaffins and combinations of diene terpenes, benzene, etc. This group of pesticides includes primarily DDT and its analogues: lindane, methoxy chloride (DMDT), aldrin, dieldrin, chlordane, endrin, heptachlor, and toxaphene. Compounds belonging to this group are characterized by a different chemical structure, which causes the chemical, physical, and toxicological properties of individual preparations to differ sometimes in principle. Organochlorine pesticides also have common features that include good solubility in lipids and the ability to bioaccumulate in tissues of living organisms. They are characterized by high resistance to detoxification agents and durability in the environment. Their mechanisms of toxic action rely on damaging cellular structures of the nervous tissue and disturbing their basic functions.

The range of properties of pesticides is extremely high, and the intensification of food production has caused their usage on a large scale. Huge amounts of food are stored in warehouses and must be properly secured to be kept fresh and safe from various insects. However, the use of pesticides has negative effects. The side effect of chemization of agriculture is the migration of pesticides to various elements of the natural environment. These compounds are found in air, soil, water, and living organisms. As a result, this leads to various types of diseases and food poisoning of people and animals. Despite the fact that the dangerous side effects of pesticides are known, it is impossible to eliminate the use of pesticides.

The main source of release of pesticides into the environment is the intentional use of these compounds in agriculture and industry. In addition, pesticides leak into the environment as a result of illegal storage of expired products containing these compounds (Wang et al., 2010). A major difference in sewage systems between nations can be the incorporation of stormwater drains. In countries such as Australia, the sewer system is closed. Therefore, the movement of OCPs directly into the sewage system is less likely than in European systems that also catch and treat rainwater (Clarke et al., 2010). In the literature, there are several examples of research on the fate of pesticides in sewage sludge. Studies have shown that lindane can be effectively removed (67 ± 10%) during wastewater treatment (Kipopoulou et al., 2004), which can be explained by the better solubility of lindane in water compared to other pesticides, and therefore its easier biodegradation. Lindane also has a higher vapor pressure so it could be volatilized, particularly in an aeration basin of an activated sludge wastewater treatment. The rate of pesticide degradation in the environment depends on the chemical structure of the compounds, but also to a large extent on their physical, chemical, and biological properties. There are general regularities characterizing transformations of pesticides in the environment, that is, polar compounds are degraded faster than nonpolar compounds. The above rule is the result of different solubilities of pesticides in water and their more efficient sorbing by organic substances.

Migration of pesticides in the environment has caused these compounds to enter wastewater along with pollutants, and then to sewage sludge (Sadecka, 2007). The most commonly detected in sewage sludge are pesticides such as hexachlorobenzene (HCB), lindane, DDT, and its DDE metabolite. The concentration range of the above insecticides was from 0.1 to 10 mg/kg DM for HCB, from 0.01 to 8.0 mg/kg DM for lindane, and from 0.01 to 0.15 mg/kg DM for DDT and DDE. For the first time, the presence of pesticides in British sewage sludge was demonstrated in 1982; the sludge contained DDE, dieldrin, and lindane. The lowest concentrations were found for DDE (10–40 μg/kg) and the highest for dieldrin (10–1260 μg/kg). In 1996, studies of Canadian sewage sludge showed that concentrations of aldrin, chlordane, heptachlor, and DDT were below the detection limit, that is, below 10 μg/kg. DDE and HCB were detected in some samples in the range of 10–130 μg/kg and 10–330 μg/kg, respectively. Also, in Sweden, concentrations of DDT, HCB, and lindane in sewage sludge were below 10 μg/kg. In Italy, the level of HCB in sewage sludge ranged from 10 to 310 μg/kg and DDE from 20 to 90 μg/kg (Clarke et al., 2010).

By the new millennium, a decrease over time in the concentration of pesticides in sewage sludge was observed. According to Wang et al. (2010), in 1998–1999, the average concentration of HCH in sewage sludge in China amounted to 584 ng/kg DM, and in 2005 it decreased to 145 ng/kg DM. Nevertheless, some researchers have decided to continue monitoring the content of pesticides in sewage sludge. Research on digested sludge from UK WWTPs showed that concentrations of HCHs, HCB, endosulphan, DDT, DDD, DDE, and chlordane were below the detectable level in most of the samples. Only concentrations of two compounds were above the limit of detection, that is, HCB (median 22 μg/kg DS) and DDE (median 13 μg/kg DS) (Stevens et al., 2003). It was demonstrated that the concentration of pesticides is decreasing in countries where government restrictions have been applied. DDE and dieldrin have a much longer half-life; therefore, they are still regularly detected in sewage sludge even 15 years after being banned.

8 Influence of sewage sludge treatment processes on OMP changes

The presence of OMPs in sewage sludge wasn't noticed until the 1980s. In the majority of European countries, for many years, the content of the abovementioned organic compounds in sewage sludge intended for natural, including agricultural, usage, has not been limited. This approach was based on the assumption that these compounds are not present in concentrations that may pose a threat to the environment and to human health. In the 1990s, acceptable levels of compounds such as, for example, dioxins, furans, PCBs, and AOX, were established in Germany, Austria, and Switzerland. In France in 1998, the content of the so-called trace organic compounds in sewage sludge was already specified as well as their permissible accumulation in soil within 10 years. It was emphasized that the sludge cannot be used agriculturally if the content of one of these compounds exceeds the permissible value. The current European regulation on the use of sewage sludge in agriculture (Directive 86/278/EWG) does not take into account the presence of most OMPs. Only seven EU countries have constituted limits for some harmful organic micropollutants in sewage sludge in their national legislation, although maximum values and target compounds (usually halogenated organic compounds, linear alkyl benzene sulphonates, polychlorinated biphenyls, dibenzodioxins/dibenzofurans, phthalates, NPs, and polycyclic aromatic hydrocarbons) vary depending on the country (Gonzalez-Gil et al., 2016; Kelessidis and Stasinakis, 2012). The problem of OMP presence in sewage sludge is particularly important due to the accumulation of micropollutants in biosolid-amended soil and their leaching into the groundwater after rainfall (Barron et al., 2010). The upcoming regulatory trends regarding the land application of biosolids (Inglezakis et al., 2014) plan to fill this gap. PFCs are an example of this. Although they are highly significant contaminants, so far they have not been regulated but that will likely change in the future. The proposed concentration limit for PFCs in sewage sludge used as agricultural fertilizer in some regions of Germany (e.g., North Rhine-Westphalia and Baden-Württemberg) would amount to 100 ng/g; PFCs are also included in the German Use of Fertilizers Ordinance (Gómez-Canela et al., 2012; German Federal Environmental Agency, UBA, 2009). While the literature gives many examples of research on the content and fate of dioxins, furans (Li et al., 2011), PCBs (Bertin et al., 2007), and PAH (Zhai et al., 2011) in sewage sludge, little is known about PFCs regarding this context.

In order to limit OMPs in sewage sludge, a number of studies have been carried out to intensify the treatment methods so that the final management of sewage sludge is completely safe and meets the requirements of environmental protection (Pedrazzani, et al., 2016). Therefore, sludge formed in the treatment plant must be processed into harmless final products and removed from the treatment plant area. During the stabilization of sludge, the amount of organic substances and pathogenic organisms is reduced and hydration is decreased. The most commonly used processes are biochemical aerobic or anaerobic stabilization. Currently, the activated sludge process is the most-used technology for biological treatment. In such cases, the aerobic treatment for nitrification is often supplemented by an anoxic denitrification, representing the state of the art and favoring the removal of organic pollutants. However, several biodegradation reactions require strictly anaerobic conditions, such as reductive dehalogenation (Bhatt et al., 2007; Radke and Maier, 2014; Völker et al., 2017), the reduction of nitro groups, and the demethylation of methoxy groups (Gasser et al., 2012). By improving the anaerobic stabilization, the biodegradation of micropollutants can be improved, thus reducing the toxicity of sewage sludge. Some studies have confirmed that anaerobic stabilization affects the transformation of OMPs (Carballa et al., 2007; Malmborg and Magner, 2015; Paterakis et al., 2012; Samaras et al., 2014) but the results are not always conclusive. It has been shown that for some OMPs (e.g., nonylphenol and their ethoxylates) the influence of temperature (mesophilic and thermophilic conditions) is not significant whereas for other OMPs (e.g., PCBs, dioxins) better removal effects were obtained at higher temperatures, that is, during thermophilic anaerobic stabilization (Dąbrowska and Rosińska, 2012). In general, the effectiveness of other OMP removal is quite diverse. Some studies have shown that during the treatment processes of sewage sludge, the efficiency of hormone (E1, E2, and EE2) removal ranges from 0 to <25% (des Mes et al., 2008; Malmborg and Magner, 2015). Similar results were obtained for diclofenac (Malmborg and Magnér, 2015; Narumiya et al., 2013), ibuprofen (Malmborg and Magnér, 2015; Narumiya et al., 2013), while other authors obtained opposite results. For example, Carballa et al. (2007) found that fragrances and hormones were removed to 95% and 70%, respectively. Also Samaras et al. (2014) demonstrated an elimination above 90% for diclofenac and ibuprofen, while for triclosan between 60% and 80%. The reasons for these discrepancies remain unclear.

It was demonstrated that during biochemical aerobic or anaerobic stabilization, changes in PCDD, PCDF, PCB, and PAH content occur in sewage sludge; this is a result of PCB transformations. These transformations include both biodegradation and processes without the participation of microorganisms. There are a few examples in the literature regarding the changes in PCB content during sewage sludge anaerobic stabilization. Transformations of these compounds take place under both aerobic and anaerobic conditions. It is believed that the process of biological degradation depends, on one hand, on the properties of hydrocarbons and, on the other hand, on the process conditions to which sewage sludge is subjected. Research results by Chang et al. (2005) showed that the rate of biodegradation under the conditions characteristic of the denitrification process was lower compared to the loss of PAH under conditions suitable for methane generation or sulfate reduction. Man et al. (2017) demonstrated that removal of PAHs mainly relied on the adsorption process in the primary treatment tank. Studies have shown the possibility of degradation of high molecular weight (HMW) PAHs (fluoranthene-dibenzo(a,h) anthracene) to low molecular weight (LMW) ones (naphthalene-anthracene). Zhang et al. (2016) showed that transformation reactions among PAH monomers can occur in the treatment process; these are closely related to temperature. It was observed that the release of PAH occurred mainly at 300–750°C (>90% of the total amount of PAH released from sewage sludge). Naphthalene may form indeno (1, 2, 3-cd) pyrene at 300°C. In the initial sewage sludge, 39% was indeno (1, 2, 3-cd) pyrene, which was formed during the transformation of other PAH monomers at 300°C. These studies indicate that during the process of thermal treatment of sewage sludge, reactions between PAH monomers occur, generating formation of other types of these compounds.

The described research mainly consisted in the determination of indicator PCB changes during mesophilic fermentation. It was demonstrated that during methane fermentation of sewage sludge, PCB content decreased on average to 33%–40% (Aparicio et al., 2009; Bertin et al., 2007; Patureau and Trably, 2006). According to Bertin et al. (2007), extending the fermentation process to 10 months enables further PCB loss and complete elimination of tetrachlorobiphenyls. Benabdallah El-Hadj et al. (2007) described the changes of indicator PCBs during mesophilic (35°C) and thermophilic (55°C) fermentation of sewage sludge. They showed that after mesophilic fermentation, the PCB decrease amounted to 33.0%–58.0%, and after thermophilic, it amounted to 59.4%–83.5%. The higher efficiency of PCB removal during sewage sludge thermophilic fermentation is confirmed by studies by Bertin et al. (2011). The authors reported that after mesophilic fermentation, the PCB decrease was equal to 47%, and after thermophilic, it was equal to 57%. According to many authors, the efficiency of PCB removal during methane fermentation is influenced by the type of sewage sludge, the concentration and type of PCB congener, the process temperature, the pH, the oxidation-reduction conditions, and the occurrence of an additional source of organic carbon and other substances (Barret et al., 2010a,b; Nollet et al., 2005). However, the process of methane digestion is very sensitive to any changes in the environment. In the group of compounds exerting a toxic effect on the process, special notice should be taken of pesticides (Sadecka, 2007).

Knoth et al. (2007), when examining changes in concentrations of lower BDE congeners (BDE-28, -47, -99, -153, -154, -183), did not demonstrate changes in the mutual rate of concentrations of these congeners at particular stages of sludge processing (initial sludge, excessive sludge, dehydrated, fermented), which suggests that in sewage sludge, no degradation of BDE-209 and other higher BDE congeners occurs (Table 4).

Research on deca-BDE degradation in activated sludge under anaerobic conditions (in the mesophilic process) has shown the presence of debromination products in the form of octa- and nona- diphenyl bromo ethers (congeners BDE-206, BDE-207, BDE-208). Debromination occurred mainly in the *meta* and *para* positions (Gerecke et al., 2005).

Research on debromination of the most frequently occurring congeners of polybrominated diphenylethers (47, 99, 100, 138, 153, 154, 183, 209) carried out in laboratory conditions and in pilot scale flow conditions in a fermenter demonstrated that in the laboratory scale, BDE 100 and BDE 209 congeners were decomposed to a small extent, and the remaining congeners were not degraded. In the pilot scale, the decrease in concentration of congeners was within the range of 20%–60%, with BDE 209 and BDE 138 congeners decomposing at 30% and 60%, respectively. In contrast to studies on the debromination of polybrominated biphenyls, there is a lack of reliable experimental data regarding the order of cleavage of substituents in the case of polybrominated diphenylethers and the effect of the position of bromine atoms in the diphenylether molecule on the rate of the dehalogenation (debromination) reaction.

It is also not possible to compare the velocity of debromination of corresponding polybrominated biphenyls and diphenyl ethers, which would allow determining the effect of the heteroatom (oxygen) on the reaction rate.

While monitoring the content of PFCs in WWTPs, the analysis of both the dissolved phase and of the solids in raw wastewater is of great importance because, depending on the PFCs type, it was found that PFDoA, PFTeDA, PFHpS, PFDS, and PFOSA occur mainly in the particulate phase while PFHpA, PFOA, PFNA, and PFHxS were detected primarily in the dissolved phase (Arvaniti and Stasinakis, 2015, Arvaniti et al., 2012; Stasinakis et al., 2013). This approach indicates the importance of analysis of the dissolved phase and particulate matter in raw wastewater and allows avoiding underestimating PFC levels in WWTPs. In the case of PFCs, it was shown that mainly PFBS predominates in sewage sludge. PFBS concentrations were in anaerobic digested sludge and centrifuged sludge and suggest degradation under anaerobic conditions (Gómez-Canela et al., 2012; Parsons et al., 2008). The presence of PFOS and PFNA in anaerobic digested sludge and centrifuged sludge compared to primary sludge may be caused by the degradation of 2-(N-ethylperfluorooctane sulfonamide)acetic acid, which could generate PFOS during sludge treatment (Higgins et al., 2005). Wherein PFBS, PFNA, and PFOS are the major components of ΣPFCs and show changes in the concentration of this group of compounds and define the level of wastewater contamination with perfluorinated compounds. Different concentrations of PFCs in anaerobic digested sludge and centrifuge sludge are influenced by the following factors: hydraulic residence times, the duration of digestion in the bioreactor, and the dewatering technique. The difference between PFC concentrations in primary sewage and anaerobic digested sludge is a weak indicator of their degradation unless the sludge samples are collected at high frequency for several weeks. The research results indicate the possibility that PFCs persist within sludge treatment pathways and are not removed during sludge treatment (Gómez-Canela et al., 2012). On the contrary, an increase toward the end of the treatment pathway may result from degradation of precursor materials, which is why it is necessary to develop new methods of PFC removal during sewage sludge treatment. Research in this direction should also pay attention to higher-frequency monitoring of PFCs in raw sludge being introduced into the treatment plant in order to solve the differences in concentrations between individual stages of sewage sludge treatment (Gómez-Canela et al., 2012).

9 Conclusion

Currently, it is believed that the combination of chemical and biological methods leads to the intensification of OMP degradation in sewage sludge and the removal of estrogenic and genotoxic activities during the mesophilic and thermophilic sludge digestion, at environmentally relevant concentrations. However, it is still necessary to deepen the knowledge on integrated assessment of the biotransformation of OMPs and these specific toxicities in order to assess the effectiveness of sewage sludge treatment processes in the removal of organic micropollutants (Gonzalez-Gil et al., 2016).

According to Völker et al. (2016), improved processes of sludge treatment significantly improve OMP removal. However, with regard to the heterogeneity of micropollutants (Schwarzenbach et al., 2006) as well as countless potential modes of action (Stamm et al., 2016), the evaluation of the efficiency of the sewage sludge treatment process cannot be based only on the removal of a specific group of organic micropollutants.

References

Abad, E., Martinez, K., Planas, O., Palacios, J., Caixach, J., Rivera, J., 2005. Priority organic pollutant assessment of sludges for agricultural purposes. Chemosphere 61, 1358–1369.

Abdel-Shafy, H.I., Mansour, M.S.M., 2016. A review on polycyclic aromatic hydrocarbons: source, environmental impact, effect on human health and remediation. Egypt. J. Pet. 25 (1), 107–123.

Ahrens, L., 2011. Polyfluoroalkyl compounds in the aquatic environment: a review of their occurrence and fate. J. Environ. Monit. 13 (1), 20–31.

Ahrens, L., Felizeter, S., Sturm, R., Xie, Z., Ebinghaus, R., 2009. Polyfluorinated compounds in waste water treatment plant effluents and surface waters along the river Elbe, Germany. Mar. Pollut. Bull. 58 (9), 1326–1333.

Alder, A.C., van der Voet, J., 2015. Occurrence and point source characterization of perfluoroalkyl acids in sewage sludge. Chemosphere 129, 62–73.

Aparicio, I., Santos, J.I., Alonso, E., 2009. Limitation of the concentration of pollutants in sewage sludge for agricultural purposes: a case study in South Spain. Waste Manag. 29 (5), 1747–1753.

Arvaniti, O.S., Stasinakis, A.S., 2015. Review on the occurrence, fate and removal of perfluorinated compounds during wastewater treatment. Sci. Total Environ. 524–525, 81–92.

Arvaniti, O.S., Ventouri, E.I., Stasinakis, A.S., Thomaidis, N.S., 2012. Occurrence of different classes of perfluorinated compounds in Greek wastewater treatment plants and determination of their solid-water distribution coefficients. J. Hazard. Mater. 24–31.

Barret, M., Barcia, C.A., Carrère, G.A., Patureau, D., 2010a. Influence of feed characteristics on the removal of micropollutants during the anaerobic digestion of contaminated sludge. J. Hazard. Mater. 181, 241–247.

Barret, M., Carrère, H., Delgadillo, L., Patureau, D., 2010b. PAH fate during the anaerobic digestion of contaminated sludge: do bioavailability and/or cometabolism limit their biodegradation? Water Res. 44, 3797–3806.

Barron, L., Nesterenko, E., Hart, K., Power, E., Quinn, B., Kelleher, B., Paull, B., 2010. Holistic visualisation of the multimodal transport and fate of twelve pharmaceuticals in biosolid enriched topsoils. Anal. Bioanal. Chem. 397, 287–296.

Becker, A., Gerstmann, S., Frank, H., 2008. Perfluorooctane surfactants inwaste waters, the major source of river pollution. Chemosphere 72, 115–121.

Benabdallah El-Hadj, T., Dosta, J., Torres, R., Mata-Alvarez, J., 2007. PCB and AOX removal in mesophilic and termophilic sewage sludge digestion. Biochem. Eng. J. 36, 281–287.

Bertilsson, S., Widenfalk, A., 2002. Photochemical degradation of PAHs in freshwaters and their impact on bacterial growth – influence of water chemistry. Hydrobiologia 469, 23–32.

Bertin, L., Capodicasa, S., Occulti, F., Girotti, S., Marchetti, L., Fava, F., 2007. Microbial processes associated to the decontamination and detoxification of a polluted activated sludge during its anaerobic stabilization. Water Res. 41, 2407–2416.

Bertin, L., Capodicasa, S., Fedi, S., Zannoni, D., Marchetti, L., Fava, F., 2011. Biotransformation of highly chlorinated PCB mixture in an activated sludge collected from a membrane biological reactor (MBR) subjected to anaerobic digestion. J. Hazard. Mater. 186, 2060–2067.

Bhatt, P., Kumar, M.S., Mudliar, S., Chakrabarti, T., 2007. Biodegradation of chlorinated compounds – a review. Crit. Rev. Environ. Sci. Technol. 37 (2), 165–198.

Blanchard, M., Teil, M.J., Ollivion, D., Legenti, L., Chevreuil, M., 2004. Polycyclic aromatic hydrocarbons and polychlorinated biphenyls in wasterwaters and sewage sludges from the Paris area (France). Environ. Res. 95, 184–194.

Boehm, P.D., Neff, J.M., Page, D.S., 2007. Assessment of polycyclic aromatic hydrocarbon exposure in the waters of Prince William sound after Exxon Valdez oil spill: 1989–2005. Mar. Pollut. Bull. 54, 339–367.

Bossi, R., Strand, J., Sortkjær, O., Larsen, M., 2008. Perfluoroalkyl compounds in Danish wastewater treatment plants and aquatic environments. Environ. Int. 34, 443–450.

Boström, C.-E., Gerde, P., Hanberg, A., Jernström, B., Johansson, C., Kyrklund, T., Rannug, A., Törnqvist, M., Victorin, K., Westerholm, R., 2002. Cancer risk assessment, indicators and guidelines for polycyclic aromatic hydrocarbons in ambient air. Environ. Health Persp. 110, 451–489.

Boulanger, B., Vargo, J., Schnoor, J., Hornbuckle, K., 2005. Evaluation of perfluorooctane surfactants in a wastewater treatment systemand in a commercial surface protection product. Environ. Sci. Technol. 39, 5524–5530.

Campo, J., Masiá, A., Picó, Y., Farré, M., Barceló, D., 2014. Distribution and fate of perfluoroalkyl substances in Mediterranean Spanish sewage treatment plants. Sci. Total Environ. 472, 912–922.

Carballa, M., Omil, F., Ternes, T., Lema, J.M., 2007. Fate of pharmaceutical and personal care products (PPCPs) during anaerobic digestion of sewage sludge. Water Res. 41, 2139–2150.

Carroll, W.F., 2001. The relative contribution of wood and poly(vinyl chloride) to emissions of PCDD and PCDF from house fires. Chemosphere 45 (8), 1173–1180.

Chan, J.K.Y., Xing, G.H., Xu, Y., Liang, Y., Chen, L.X., Wu, S.C., Wong, C.K.C., Leung, C.K.M., Wong, M.H., 2007. Body loadings and health risk assessment of polychlorinated dibenzo-p-dioxins and dibenzofurans at an intensive electronic waste recycling site in China. Environ. Sci. Technol. 41, 7668–7674.

Chang, V.B., Liao, C.S., Yuan, S.Y., 2005. Anaerobic degradation of diethyl phthalate, di*n*-butyl phthalate, and di-(2-ethylhexyl) phthalate from river sediment in Taiwan. Chemosphere 58 (11), 1601–1607.

Clarke, B., Porter, N., Symons, R., Blackbeard, J., Ades, P., Marriott, A., 2008. Dioxin-like compounds in Australian sewage sludge-review and national survey. Chemosphere 72, 1215–1228.

Clarke, B.O., Porter, N.A., Marriott, F.J., Blackbeard, J.R., 2010. Investigating the levels and trends of organochlorine pesticides and polychlorinated biphenyl in sewage sludge. Eniron. Int. 36 (4), 323–329.

Correa, O., Raun, L., Rifai, H., Suarez, M., Holsen, T., Koenig, L., 2006. Depositional flux of polychlorinated dibenzo-p-dioxins and polychlorinated dibenzofurans in an urban setting. Chemosphere 64 (9), 1550–1561.

Covaci, A., Gerecke, A.C., Law, R.J., Voorspoels, S., Kohler, M., Heeb, N.V., 2006. Hexabromocyclododeca-nes (HBCDs) in the environment and humans: a review. Environ. Sci. Technol. 40, 3679–3688.

Dąbrowska, L., Rosińska, A., 2012. Change of PCBs and forms heavy metals in sewage sludge during thermophilic anaerobic digestion. Chemosphere 88, 168–173.

de Souza Pereira, M., Kuch, B., 2005. Heavy metals, PCDD/F in sewage sludge samples from two wastewater treatment facilities in Rio de Janeiro state, Brazil. Chemosphere 60, 844–853.

des Mes, T.Z.D., Kujawa-Roeleveld, K., Zeeman, G., Lettinga, G., 2008. Anaerobic biodegradation of estrogens e hard to digest. Water Sci. Technol. 57, 1177–1182.

Domingo, J.L., Schuhmacher, M., Llobet, J.M., Műller, L., 2001. PCDD/F concentration in soil and vegetation of municipal waste incinerator after a prounced decrase in the emissions of PCDD/F from the facility. Chemosphere 43 (2), 217–226.

Durjava, M.K., ter Laak, T.L., Joop, L.M., Hermens, J.L.M., Jaap Struijs, J., 2007. Distribution of PAHs and PCBs to dissolved organic matter: high distribution coefficients with consequences for environmental fate modeling. Chemosphere 67, 990–997.

Eguchi, A., Nomiyama, K., Devanathan, G., Subramanian, A., Bulbule, K.A., Parthasarathy, P., Takahashi, S., Tanabe, S., 2012. Different profiles of anthropogenic and naturally produced organohalogen compounds in serum from residents living near a coastal area and e-waste recycling workers in India. Environ. Int. 47, 8–16.

Eljarrat, E., Caixach, J., Rivera, J., 2003. A comparison of TEQ contributions from PCDDs, PCDFs and dioxin-like PCBs in sewage sludges from Catalonia, Spain. Chemosphere 51 (7), 595–601.

Everaert, K., Baeyens, J., 2002. The formation and emission of dioxins in large scale thermal processes. Chemosphere 46, 439–448.

Fernández-Luqueño, F., López-Valdez, F., Dendooven, L., LunaSuárez, S., Ceballos-Ramírez, J.M., 2016. Why wastewater sludge stimulates and accelerates removal of PAHs in polluted soils? Appl. Soil Ecol. 101(1–4).

Frederiksen, M., Vorkamp, K., Thomsen, M., Knudsen, L.E., 2009. Human internal and external exposure to PBDEs—a review of levels and sources. Int. J. Hyg. Environ. Health 212, 109–134.

Fuentes, M.J., Font, R., Gómez-Rico, M.F., Moltó, J., 2007. Multivariant statistical analysis of PCDD/FS in sewage sludges from different areas of the Valencian community (Spain). Chemosphere 67 (7), 1423–1433.

Gasser, G., Pankratov, I., Elhanany, S., Werner, P., Gun, J., Gelman, F., Lev, O., 2012. Field and laboratory studies of the fate and enantiomeric enrichment of venlafaxine and O-desmethylvenlafaxine under aerobic and anaerobic conditions. Chemosphere 88 (1), 98–105.

Gerecke, A.C., Hartmann, P.C., Heeb, N.V., Kohler, H.P., Giger, W., Schmid, P., Zennegg, M., Kohler, M., 2005. Anaerobic degradation of decabromodiphenyl ether. Environ. Sci. Technol. 39, 1078–1083.

German Federal Environmental Agency (UBA), 2009. Do without perand polyfluorinated chemicals and prevent their discharge into the environment. In: German Use of Fertilizers Ordinance.pp. 1–17.

Geyer, H.J., Schramm, K.-W., Feight, E.A., Behecht, A., Steinberg, C., Bruggemann, R., Poiger, H., Henkelmann, B., Kettrup, A., 2002. Half-lives of tetra-, penta-, hexa-, hepta-, and octachlorodibenzo-p-dioxin in rats, monkeys, and humans – a critical review. Chemosphere 48, 631–644.

Giesy, J.P., Kannan, K., 2002. Perfluorochemical surfactants in the environment. Environ Sci. Technol. 36, 146A–152A.

Gómez-Canela, C., Barth, J.A.C., Lacorte, S., 2012. Occurrence and fate of perfluorinated compounds in sewage sludge from Spain and Germany. Environ. Sci. Pollut. Res. 19, 4109–4119.

Gonzalez-Gil, L., Papa, M., Feretti, M., Ceretti, D.E., Mazzoleni, G., Steimberg, N., 2016. Is anaerobic digestion effective for the removal of organic micropollutants and biological activities from sewage sludge? Water Res. 102, 211–220.

Grochowalski, A., Konieczyński, J., 2008. PCDDs/PCDFs, dl-PCBs and HCB in the flue gas from coal fired CFB boilers. Chemosphere 73, 97–103.

Guo, Z., Lin, T., Zhang, G., Zheng, M., Zhang, Z., Hao, Y., Fang, M., 2007. The sedimentary fluxes of polycyclic aromatic hydrocarbons in the Yangtze River estuary coastal sea for the past century. Sci. Total Environ. 386, 33–41.

Guo, R., Sim, W.J., Lee, E.S., Lee, J.H., Oh, J.E., 2010. Evaluation of the fate of perfluoroalkyl compounds in wastewater treatment plants. Water Res. 44, 3476–3486.

Gworek, B., Hajduk, A., Koda, E., Grochowalski, A., Jeske, A., 2013. Influence of a municipal waste landfill on the spatial distribution of polychlorinated dibenzo-p-dioxins and dibenzofurans (PCDDs/fs) in the natural environment. Chemosphere 92, 753–759.

Hara, K., Kuroda, M., Yabar, H., Kimura, M., Uwasu, M., 2016. Historical development of wastewater and sewage sludge treatment technologies in Japan – an analysis of patent data from the past 50 years. Environ. Dev. 19, 59–69.

Haritash, A.K., Kaushik, C.P., 2009. Biodegradation aspects of polycyclic aromatic hydrocarbons (PAHs): a review. J. Hazard. Mater. 169, 1–15.

Harrison, E.Z., Oakes, S.R., Hysell, M., Hay, A., 2006. Organic chemicals in sewage sludges. Sci. Total Environ. 367 (2–3), 481–497.

Higgins, C.P., Field, J.A., Criddle, C.S., Luthy, R.G., 2005. Quantitative determination of perfluorochemicals in sediments and domestic sludge. Environ. Sci. Technol. 39, 3946–3956.

Huo, X., Peng, L., Xu, X., Zheng, L., Qui, B., Qi, Z., Zhang, B., Han, D., Piao, Z., 2007. Elevated blood lead levels of children in Guiyu, an electronic waste recycling town in China. Environ. Health Persp. 115, 1113–1117.

Huset, C., Chiaia, A.C., Barofsky, D.F., Jonkers, N., Kohler, H.P.E., Ort, C., Giger, W., Field, J.A., 2008. Occurrence and mass flows of fluorochemicals in the Glatt Valley watershed, Switzerland. Environ. Sci. Technol. 42, 6369–6377.

Inglezakis, V., Zorpas, A.A., Karagiannidis, A., Samaras, P., Voukkali, I., Sklari, S., 2014. European Union legislation on sewage sludge management. Fresenius Environ. Bull. 23, 635–639.

Jiao, L., Zheng, G.J., Minh, T.B., Richardson, B., Chen, L., Zhang, Y., Yeung, L.W., Lam, J.C.W., Yang, X., Lam, P.K.S., Wong, M.H., 2009. Persistent toxic substances in remote lake and coastal sediments from Svalbard, Norwegian Arctic: levels, sources and fluxes. Environ. Pollut. 157, 1342–1351.

Kacprzak, M., Neczaj, E., Fijałkowski, K., Grobelak, A., Grosser, A., Worwag, M., Rorat, A., Brattebo, H., Almås, A., Singh, B.R., 2017. Sewage sludge disposal strategies for sustainable development. Environ. Res. 156, 39–46.

Katsoyiannis, A., Samara, C., 2005. Persistent organic pollutants (POPs) in the conventional activated sludge treatment process: fate and mass balance. Environ. Res. 97, 245–257.

Kelessidis, A., Stasinakis, A.S., 2012. Comparative study of the methods used for treatment and final disposal of sewage sludge in European countries. Waste Manag. 32, 1186–1195.

Kim, M., Kennicutt, M.C., Yaorong, Q., 2008. Source characterization using compound composition and stable carbon isotope ratio of PAHs in sediments from lakes, harbor and shipping waterway. Sci. Total Environ. 389, 367–377.

Kim, K.-H., Jahan, S.A., Kabir, E., Brown, R.J.C., 2013. A review of airborne polycyclic aromatic hydrocarbons (PAHs) and their human health effects. Environ. Int. 60, 71–80.

Kipopoulou, A.M., Zouboulis, A., Samara, C., Kouimtzis, T., 2004. The fate of lindane in the conventional activated sludge treatment process. Chemosphere 55 (1), 81–91.

Knoth, W., Mann, W., Meyer, R., Nebhuth, J., 2007. Polybrominated diphenyl ether in sewage sludge in Germany. Chemosphere 67, 1831–1837.

Kolic, T.M., Shen, L., Mac Pherson, K., Fayez, L., Gobran, T., Helm, P.A., et al., 2009. The analysis of halogenated flame retardants by GC–HRMS in environmental samples. J. Chromatogr. Sci. 47, 83–91.

Kryłów, M., Rezka, P., 2017. Polybrominated flame retardants in sewage sludge and sediments (review). Environ. Eng. 3, 153–166.

Law, R.J., Herzke, D., Harrad, S., Morris, S., Bersuder, P., Allchin, C.R., 2008. Levels and trends of HBCD and BDEs in the European and Asian environments, with some information for other BFRs. Chemosphere 73, 223–241.

Lawal, A.T., 2017. Polycyclic aromatic hydrocarbons. A review. Cogent Environ. Sci. 3, 1–86.

Li, X., Ke, Z., Dong, J., 2011. PCDDs and PCDFs in sewage sludges from two wastewater treatment plants in Beijing, China. Chemosphere 82, 635–638.

Loganathan, B., Sajwan, K., Sinclair, E., Kumar, K., Kannan, K., 2007. Perfluoroalkyl sulfonates and perfluorocarboxylates in two wastewater treatment facilities in Kentucky and Georgia. Water Res. 41, 4611–4620.

Lung, S.C.C., Wu, M.J., Lin, C.C., 2004. Customers' exposure to PM 2.5 and polycyclic aromatic hydrocarbons in smoking/nonsmoking sections of 24-h coffee shops in Taiwan. J. Expo. Anal. Environ. Epidemiol. 14, 529–535.

Ma, R., Shih, K., 2010. Perfluorochemicals in wastewater treatment plants and sediments in Hong Kong. Environ. Pollut. 158, 1354–1362.

Malmborg, J., Magnér, J., 2015. Pharmaceutical residues in sewage sludge: Effect of sanitization and anaerobic digestion. J. Environ. Manag. 153, 1–10.

Man, Y.B., Chow, K.L., Zhang Cheng, Z., Mo, W.Y., Chan, Y.C., Lam, J.C.W., Lau, F.T.K., Fung, W.C., Wong, M.H., 2017. Profiles and removal efficiency of polycyclic aromatic hydrocarbons by two different types of sewage treatment plants in Hong Kong. J. Environ. Sci. 53, 196–206.

Marvin, C.H., Tomy, G.T., Armitage, J.M., Arnot, J.A., Mc Carty, L., Covaci, A., et al., 2011. Hexabromocyclododecane: current understanding of chemistry, environmental fate and toxicology and implications for global management. Environ. Sci. Technol. 45, 613–623.

Masih, A., Lal, J.K., 2014. Concentrations and carcinogenic profiles of polycyclic aromatic hydrocarbons (PAHs) in groundwater of an urban site at a terai belt of North India. IJAER 9, 1–7.

Möller, A., Xie, Z., Cai, M., Zhong, G., Huang, P., Cai, M., et al., 2011. Polybrominated diphenyl ethers vs alternate brominated flame retardants and Dechloranes from East Asia to the Arctic. Environ. Sci. Technol. 45, 6793–6799.

Morillo, E., Romero, A.S., Madrid, L., Villaverde, J., Maqueda, C., 2008. Characterization and sources of PAHs and potentially toxic metals in urban environments of Sevilla (southern Spain). Water Air Soil Pollut. 187, 41–51.

Morosöe, G., 2006. An Overview of Alternatives to Tetrabromobisphenol A (TBBPA) and Hexabromocyclododecane (HBCD). Report prepared for the Lowell Center For Sustainable Production, USA.

Murakami, M., Shinohara, H., Hideshige, H., 2009. Evaluation of wastewater and street runoff as sources of perfluorinated surfactants (PFSs). Chemosphere 74, 487–493.

Narumiya, M., Nakada, N., Yamashita, N., Tanaka, H., 2013. Phase distribution and removal of pharmaceuticals and personal care products during anaerobic sludge digestion. J. Hazard. Mater. 260, 305–312.

Neff, J.M., Stout, S.A., Gunster, D.G., 2005. Ecological risk assessment of polycyclic aromatic hydrocarbons in sediments: identifying sources and ecological hazard. Integr. Environ. Assess. Manag. 1, 22–33.

Nguyen, T.C., Loganathan, P., Nguyen, T.V., Vigneswaran, S., Kandasamy, J., Slee, D., Stevenson, G., Naidu, R., 2014. Polycyclic aromatic hydrocarbons in road-deposited sediments, water sediments, and soils in Sydney, Australia: Comparisons of concentration distribution, sources and potential toxicity. Ecotox. Environ. Safe. 104, 339–348.
Nollet, H., van de Putte, I., Raskin, L., Verstraete, W., 2005. Carbon/electron source dependence of polychlorinated biphenyls dechlorination pathways for anaerobic granules. Chemosphere 58, 299–310.
Oh, J.-H., Choi, S.-D., Lee, S.-J., Chang, Y.-S., 2006. Influence of a municipal solid waste incinerator on ambient air and soil PCDD/Fs levels. Chemosphere 64, 579–587.
Parsons, J.R., Saez, M., Dolfing, J., de Voogt, P., 2008. Biodegradation of perfluorinated compounds. Rev. Environ. Contam. Toxicol. 196, 53–71.
Paterakis, N., Chiu, T.Y., Koh, Y.K.K., Lester, J.N., McAdam, E.J., Scrimshaw, M.D., Soares, A., Cartmell, E., 2012. The effectiveness of anaerobic digestion in removing estrogens and nonylphenol ethoxylates. J. Hazard. Mater. 199-200, 88–95.
Patureau, D., Trably, E., 2006. Impact of anaerobic and aerobic processes on polychlorobiphenyl removal in contaminated sewage sludge. Biodegradation 17, 9–17.
Pavlova, A., Ivanova, R., 2003. Determination of petroleum hydrocarbons and polycyclic aromatic hydrocarbons in sludge from wastewater treatment basins. J. Environ. Monit. 5, 319–323.
Pedrazzani, R., Bertanza, G., Lema, J.M., Carballa, M., 2016. Is anaerobic digestion effective for the removal of organic micropollutants and biological activities from sewage sludge? Water Res. 102, 211–220.
Perez, S., Guillamon, M., Barcelo, D., 2001a. Quantitative analysis polycyclic aromatic hydrocarbons in sewage sludge from wastewater treatment plants. J. Chromatogr. 938, 57–65.
Perez, S., La Farre, M., Garcia, M.J., Barcelao, D., 2001b. Occurrence of polycyclic aromatic hydrocarbons in sewage sludge and their contribution to its toxicity in the ToxAlert® 100 bioassay. Chemosphere 45, 705–712.
Prevedouros, K., Cousins, I.T., Buck, R.C., Korzeniowski, S.H., 2006. Sources, fate and transport of perfluorocarboxylates. Environ. Sci. Technol. 40, 32–44.
Raab, U., Preiss, U., Albrecht, M., Shahin, N., Parlar, H., Fromm, H., 2008. Concentrations of polybrominated diphenyl ethers, organochlorine compounds and nitro musks in mother's milk from Germany (Bavaria). Chemosphere 72, 87–94.
Radke, M., Maier, M.P., 2014. Lessons learned from water/sediment-testing of pharmaceuticals. Water Res 55, 63–73.
Rahman, M.M., Kim, K.-H., Brown, R.J.C., Bae, I.S., Park, C.G., 2014. PCDD and PCDF concentrations in a traffic tunnel environment. Sci. Total Environ. 493, 773–780.
Ren, Z., Zheng, M., 2009. Impacts of human activities on dioxins emissions at national scale. Chemosphere 76, 853–859.
Rice, C.P., Chernyak, S.M., Begnoche, L., Qunital, R., Hickley, J., 2002. Comparisons of PBDE composition and concentration in fish collected from Detroit River MI and Des Plaines River IL. Chemosphere 49, 731–737.
Robertson, L.W., Hansen, L.G. (Eds.), 2015. PCBs: Recent Advances in Environmental Toxicology and Health Effects. University Press of Kentucky.
Robson, M., Melymuk, L., Csiszar, S.A., Giang, A., Diamond, M.L., Helm, P.A., 2010. Continuing sources of PCBs: the significance of building sealants. Environ. Int. 36, 506–513.
Rosenberg, C., Meila, M., Tornaeus, J., Säkkinen, K., Puttonen, K., Korpi, A., et al., 2011. Exposure to flame retardants in electronics recycling sites. Ann. Occup. Hyg. 55, 658–665.
Sadecka, Z., 2007. Impact of pyrethroids on methane digestion. CEER 2, 59–69.

Samaras, V.G., Stasinakis, A.S., Thomaidis, N.S., Mamais, D., Lekkas, T.D., 2014. Fate of selected emerging micropollutants during mesophilic, thermophilic and temperature co-phased anaerobic digestion of sewage sludge. Bioresour. Technol. 162, 365–372.

Schiavon, M., Torretta, V., Rada, E.C., Ragazz, M., 2016. State of the art and advances in the impact assessment of dioxins and dioxin-like compounds. Water Res. 102, 211–220.

Schultz, M.M., Barofsky, D., Field, J., 2006. Quantitative determination of fluorinated alkyl substances by large-volume-injection liquid chromatography tandem mass spectrometry characterization of municipal wastewaters. Environ. Sci. Technol. 40 (1), 289–295.

Schwarzenbach, R.P., Escher, B.I., Fenner, K., Hofstetter, T.B., Johnson, C.A., von Gunten, U., Wehrli, B., 2006. The challenge of micropollutants in aquatic systems. Science 313 (5790), 1072–1077.

Sharma, T., 2014. In silico investigation of polycyclic aromatic hydrocarbons against bacterial 1-2 dioxygenase. J. Chem. Pharm. Res. 6, 873–877.

Sinclair, E., Kannan, K., 2006. Mass loading and fate of perfluoroalkyl surfactants in wastewater treatment plants. Environ. Sci. Technol. 40, 1408–1414.

Smith, B., Wong, C.A., Smith, T.C., Boyko, E.J., Gackstetter, G.D., Ryan, M.A.K., Millennium Cohort Study, T., 2009. Newly reported respiratory symptoms and conditions among military personnel deployed to Iraq and Afghanistan: a prospective population-based study. Am. J. Epidemiol. 170, 1433–1442.

Stamm, C., Räsänen, K., Burdon, F.J., Altermatt, F., Jokela, J., Joss, A., Ackermann, M., Eggen, R.I.L., 2016. Unravelling the impacts of micropollutants in aquatic ecosystems. Adv. Ecol. Res. 55, 183–223.

Stapleton, H.M., Allen, J.G., Kelly, S.M., Konstantinov, A., Klosterhaus, S., Watkins, D., et al., 2008. Alternate and new brominated flame retardants detected in U.S. house dust. Environ. Sci. Technol. 42, 6910–6916.

Stasinakis, A.S., Thomaidis, N.S., Arvaniti, O.S., Asimakopoulos, A.G., Samaras, V.G., Ajibola, A., Mamais, D., Lekkas, T.D., 2013. Contribution of primary and secondary treatment on the removal of benzothiazoles, benzotriazoles, endocrine disruptors, pharmaceuticals and perfluorinated compounds in a sewage treatment plant. Sci. Total Environ. 463-464, 1067–1075.

Stevens, J.L., Northcott, G.L., Stern, G.A., Tomy, G.T., Kevin, C., Jones, K.C., 2003. PAHs, PCBs, PCNs, organochlorine pesticides, synthetic Musks, and polychlorinated n-alkanes in U.K. sewage sludge: survey results and implications. Environ. Sci. Technol. 37 (3), 462–467.

Stogiannidis, E., Laane, R., 2015. Source characterization of polycyclic aromatic hydrocarbons by using their molecular indices: an overview of possibilities. Rev. Environ. Contam. Toxicol. 234, 49–133.

Swietlik, R., Kowalczyk, D., Dojlido, J., 2002. Influence of selected physicochemical factors on the degradation of PAHs in water. Pol. J. Environ. Stud. 11, 165–169.

Thompson, J., Eaglesham, G., Reungoat, J., Poussade, Y., Bartkow, M., Lawrence, M., Mueller, J., 2011. Removal of PFOS, PFOA and other perfluoroalkyl acids at water reclamation plants in south East Queensland, Australia. Chemosphere 82, 9–17.

Toms, L.M., Hearn, L., Kennedy, K., Harden, F., Bartkow, M., Temme, C., et al., 2009. Concentrations of polybrominated diphenyl ethers (PBDEs) in matched samples of human milk, dust and indoor air. Environ. Int. 35, 864–869.

Urbaniak, M., 2007. Polychlorinated biphenyls: Sources, distribution and transformation in the environment-a literature review. Act. Toxicologica 15 (12), 83–93.

US-EPA United States Environmental Protection Agency (US-EPA), 2011. January. Available on: www.epa.gov/waterscience/methods/pollutants.htm.

Vassiliadou, I., Papadopoulos, A., Costopoulou, D., Vasiliadou, S., Christoforou, S., Leondiadis, L., 2009. Dioxin contamination after an accidental fire in the municipal landfill of Tagarades, Thessaloniki, Greece. Chemosphere 74, 879–884.

Vecitis, C.D., Park, H., Cheng, J., Mader, B.T., Hoffmann, M.R., 2009. Treatment technologies for aqueous perfluorooctanesulfonate (PFOS) and perfluorooctanoate (PFOA). Front. Environ. Sci. Eng. China 3, 129–151.

Völker, J., Castronovo, S., Wick, A., Ternes, T.A., Joss, A., Oehlmann, J., Wagner, M., 2016. Advancing biological wastewater treatment: extended anaerobic conditions enhance the removal of endocrine and dioxin-like activities. Environ. Sci.Technol. 50 (19), 10606–10615.

Völker, J., Vogt, T., Castronovo, S., Wick, A., Ternes, T.A., Joss, A., Oehlmann, J., Wagner, M., 2017. Extended anaerobic conditions in the biological wastewater treatment: higher reduction of toxicity compared to target organic micropollutants. Water Res. 116, 220–230.

Wang, G., Yonglong, L.Y., Han, J., Luo, W., Shi, Y., Wang, T., Sun, Y., 2010. Hexachlorobenzene sources, levels and human exposure in the environment of China. Environ. Int. 36, 122–130.

Wielgosinski, G., 2011. The reduction of dioxin emissions from the processes of heat and power generation. J. Air Waste Manag. Assoc. 61, 511–526.

Wu, T., Lin, Y.-C., Chen, Y.-M., Chang-Chien, G.-P., 2010. Impacts of dust storms on PCDD/Fs and regulated harmful matter in a basin area. Atmos. Res. 97, 279–293.

Yan, B., Abrajano, T.A., Bopp, R.F., Benedict, L.A., Chaky, D.A., Perry, E., Song, J., Keane, D.P., 2006. Combined application of δ13C and molecular ratios in sediment cores for PAH source apportionment in the New York/New Jersey harbor complex. Org. Geochem. 37, 674–687.

Yoshida, H., Takahashi, K., Takeda, N., Sakai, S., 2009. Japan's waste managment policies for dioxins and polychlorinated biphenyls. J. Mater Cycles Waste Manag. 11, 229–243.

Yu, B.-W., Jin, G.Z., Moon, J.Y.H., Kim, M.K., Kyoung, J.D., Chang, Y.S., 2006. Emission of PCDD/Fs and dioxin-like PCBs from metallurgy industries in S. Korea. Chemosphere 62, 494–501.

Zhai, J., Tian, W., Liu, K., 2011. Quantitative assessment of polycyclic aromatic hydrocarbons in sewage sludge from wastewater treatment plants in Qingdao, China. Environ. Monit. Assess. 180, 303–311.

Zhang, H., Le Xu, L., Yifei Zhang, Y., Jiang, M., 2016. The transformation of PAHs in the sewage sludge incineration treatment. Front. Environ. Sci. Eng. 10 (2), 336–340.

Zhou, W., Zhai, Z., Wang, Z., Wang, L., 2005. Estimation of n-octanol/water partition coefficients (K_{ow}) of all PCB congeners by density functional theory. J. Mol. Struct. (THEOCHEM) 755, 137–145.

Further reading

Carballa, M., Fink, G., Omil, F., Lema, J.M., Ternes, T., 2008. Determination of the solid-water distribution coefficient (Kd) for pharmaceuticals, estrogens and musk fragrances in digested sludge. Water Res. 42, 287–295.

DanonSchaffer, M., Grace, J., Wenning, R., Ikonomou, M., Lukemburg, W., 2006. PBDEs in landfill leachate and potential for transfer from electronic waste. Organohalogen Compd. 68, 1759–1762.

de Wit, C.A., Alaee, M., Muir, D.C., 2006. Levels and trends of brominated flame retardants in the Arctic. Chemosphere 64, 209–233.
Qu, M., Kim, Y.-J., Sakai, S.-i., 2004. Leaching of brominated flame retardants in leachate from landfills in Japan. Chemosphere 57, 1571–1579.
Qu, W., Bi, X., Sheng, G., Lu, S., Fu, J., Yuan, J., Li, L., 2007. Exposure to polybrominated diphenyl ethers among workers at an electronic waste dismantling region in Guangdong, China. Environ. Int. 33, 1029–1034.

Emerging contaminants in sludge (endocrine disruptors, pesticides, and pharmaceutical residues, including illicit drugs/controlled substances, etc.)

20

Krzysztof Fijalkowski
Institute of Environmental Engineering, Czestochowa University of Technology, Czestochowa, Poland

1 Introduction

In recent decades, studies on the characteristics of not only wastewater but also sewage sludge have paid attention to the environmental occurrence of new compounds of anthropogenic origin. The presence of such compounds in trace concentrations has been classified as "emerging contaminants" (EC) and their impact on various life forms such as aquatic and terrestrial organisms, including human health, is now a subject of concern among scientists, engineers, and the general public (Gogoi et al., 2018). Due to the development of analytical techniques, unregulated organic trace micropollutants, known as emerging microcontaminants, have also become of interest (Richardson, 2007).

There is clear evidence that concentrations of heavy metals (main chemical pollution compound) have decreased since the mid-1980s due to regulatory controls and legal changes regarding the use and release of hazardous substances. These solutions have led to the termination or reduction of emissions of these heavy metals to the environment (Ivanová et al., 2018; Bloem et al., 2017).

Emerging contaminants (EC) are substances that show understandable, plausible, or real risks to human health and the environment; they are also usually present in low concentrations (range $ng\,L^{-1}$ to $mg\,L^{-1}$) (Archer et al., 2017). Generally, these relationships can be divided into three main categories (Tran et al., 2018; Gogoi et al., 2018):

- Pharmaceutically active compounds (PhACs).
- Personal care products (PCP).
- Endocrine disrupting compounds (ECD).

It should be noted that these are general categories of EC; knowledge about occurrences in the environment and the fate of many EC classes has been documented in numerous studies and a summary of this is shown in Fig. 1.

Industrial and Municipal Sludge. https://doi.org/10.1016/B978-0-12-815907-1.00020-9

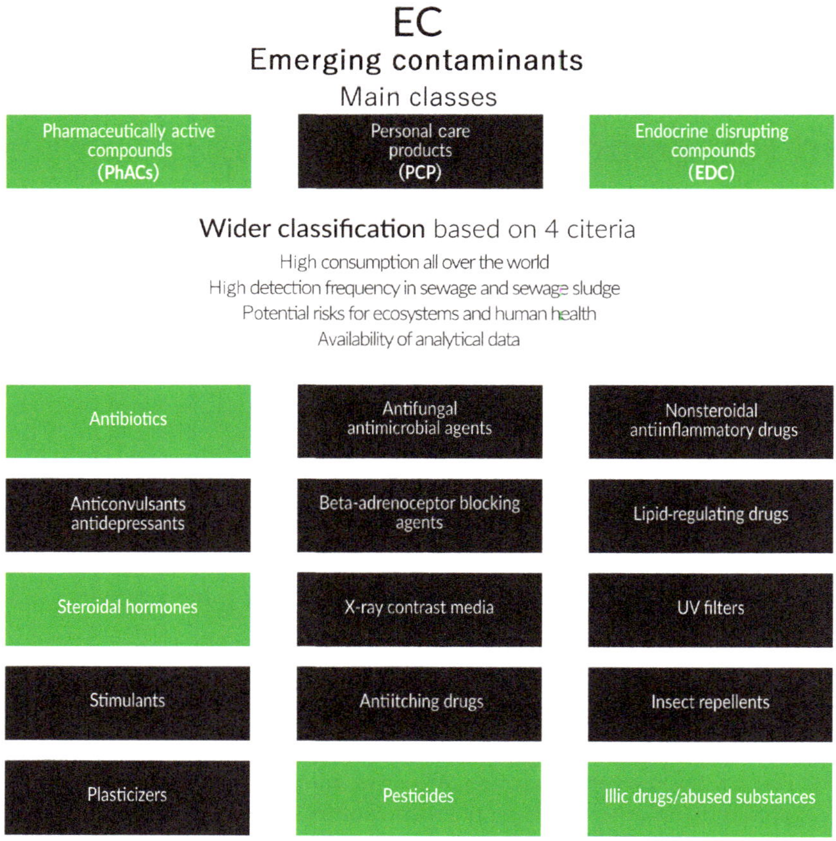

Fig. 1 General categories of ECs (Tran et al., 2018; Gogoi et al., 2018).

Due to the very wide range of EC occurrence in sewage sludge and the development of analytical tools, this article is limited to the discussion of only the selected EC pollution classes (marked in *greenor lighter gray*) in Fig. 1.

According to the latest point of view, sewage sludge is often referred to as a biosolid or biowaste produced in wastewater treatment plants (WWTP), where it is properly processed and stabilized. Stabilized sewage sludge is considered to be the most important waste stream of WWTP and its treatment is the most complex problem to be faced by sanitary engineers (Ivanová et al., 2018).

It is assumed that about 11 million tons of sewage sludge dry matter per year are generated in the EU, and in the future, that is expected to increase to 15 million tons. According to estimates, almost 50% of sewage sludge produced in EU countries is used for agricultural purposes while about 12% is composted and used in the fertilization of forests and gardens. These data are different for almost every country. For example, in Denmark, Ireland, France, Spain, and Great Britain, more than 50% of all biosolids are used in agriculture. In countries such as: Malta, Netherlands, Slovakia, and Slovenia, the biosolids are not used in agriculture at all, while in the not mentioned in here EU countries in around 10%.(Fijalkowski et al., 2018; Grobelak, 2016; Ivanová et al., 2018; Fijalkowski et al., 2017).

It should be noted that sewage sludge has a very high tendency to absorb many potential contaminants from wastewater. Therefore, substances not degraded into simple compounds in the treatment process at WWTP will mostly be adsorbed on sewage sludge (Ivanová et al., 2017; Ivanová et al., 2018).

2 Pharmaceutically active compounds (PhACs)

Many chemicals, through increasing the sensitivity of the analytical methods, can be determined in microquantities. Within them, pharmaceutically active compounds (PhACs) arouse great interest not only because of their toxic properties in the ecosystem, but above all, because of their continuous emission to the environment. Therefore, in 2013 and 2015, the European Council Directive and the EU Decision included such PhACs as azithromycin, clarithromycin, erythromycin, diclofenac, estrone, and 17β-ethinylestradiol, in monitoring wastewater and 17α-estradiol. However, there is no regulation of content of these compounds in sewage sludge, although this may change in the coming years due to current scientific reports (Ivanová et al., 2018).

There is not much data on the PhAC occurrence in sewage sludge/biosolids as final products for use on soils. With regard to the stabilized sewage sludge as solid waste from WWTP, the most studied classes of pharmaceuticals are analgesics/antiinflammatory drugs, psychotropic drugs, and antibiotics (Verlicchi and Zambello, 2015; Fernandez-Martinez et al., 2013; Ivanová et al., 2017, 2018; Fijalkowski et al., 2017). The categories of PhACs in sewage sludge are shown in Fig. 2.

The PhAC compound group is mainly psychiatric drugs affecting the nervous system, which have a mostly lipophilic character and thus strongly bind to solid bodies. Therefore, a large amount of this type of compound is detected in sewage sludge, mainly SSRI (selective serotonin reuptake inhibitors) (Malmborg and Magnér, 2015; Tran et al., 2018; Bergersen et al., 2012; Lajeunesse et al., 2013).

Another group of pharmaceuticals detected in sewage sludge is antibiotics (ciprofloxacin and tetracycline), which are attracted electrostatically by negatively charged particles of sewage sludge. These behawior also occurs for compounds such as anticonvulsants (carbamazepine) and beta blockers (metoprolol and propanolol). A specific group of compounds is nonsteroidal antiinflammatory drugs (NSAIDs), which, though having a low tendency toward sorption on solid particles, are detected in large amounts in sewage sludge due to the large mass load in WWTP. This applies to diclofenac and ibuprofen. Large amounts of a stimulant—caffeine—were also detected (Malmborg and Magnér, 2015; Tran et al., 2018; Carballa et al., 2007).

It should be noted that EC (PhACs are included in it) concentration in sewage sludge and biosolids also depends on environmental conditions and biodegradability. In some cases, a higher concentration of these substances is observed in secondary sewage sludge compared to primary ones. It is believed that this is due to the fact that a higher hydrolysis of conjugates and a higher content of organic substances occurs in secondary treatments in WWTP (Fijalkowski et al., 2017; Tran et al., 2018; Urase and Kikuta, 2005).

PhACs

Pharmaceutically active compounds

present in sewage sludge

Antibiotics	Amoxicilin Azithromycin Chlorampenicol Chlortetracycline Ciproflaxacin	Enrofloxacin Erythromycin Lincomycin Minocycline Ofloxacin	Oxytetracycline Sulfamethazine Sulfamethoxazole Tetracycline	Trimethoprim Tylosin Vancomycin Metronidazole
Antimicrobial /antifungal	Miconazole Thiabendazole Triclocarban Triclosan Ketoconazole	Clotrimazole Terbinafine Fluconazole Econazole Miconazole	Propiconazole	
NSAIDs Nonsteroidal antiinflammatory drugs	Acetaminophen Codeine Diclofenac Fenoprofen Ibuprofen	Indomethacin Ketoprofen Naproxen Salicylic acid		
Beta-blockers	Atenolol Metoprolol Propranolol			
Anticonvulsants	Carbamazepine			
Lipid regulators	Bezafibrate Clobric acid Gemfibrozil			
Hormones	Estrone Estriol 17a-ethinyestradiol			
Psychoactive	Citalopram Sertraline Norsertraline Carbamazepine Venlafaxine	Sulfapyridine Amitryptyline Caffeine Mianserin Clomipramine	Codeine Mirtazapine Methadone Diphenhyramine	Haloperidol Oxazepam Clonazepam Mephedrone

Fig. 2 Categories of PhACs in sewage sludge (Tran et al., 2018; Östman et al., 2017).

The concentrations of selected PhACs in sewage sludge with its removal degree is shown in Table 1.

2.1 PhAC removal treatments

Table 1 shows the PhAC removal in standard and common treatment processes in WWTP. Despite this, new and effective solutions are currently being sought. Salgado et al. (2012) indicated that the most common mechanism of PhAC removal was biological (45%), followed by adsorption (33%) and UV radiation (22%). In most

Table 1 Selected PhAC concentrations in sewage sludge with removal range (Petrie et al., 2016; Tran et al., 2018; Ivanová et al., 2018; Östman et al., 2017; Archer et al., 2017; Silva et al., 2015; Kim et al., 2014; Gracia-Lor et al., 2012)

Selected compound (PhACs in EC)	Sludge from WWTPs (ng g^{-1} dw)	Digested sludge (ng g^{-1} dw)	Removal range in full scale WWTPs (%)
Antibiotics			
Amoxicilin	<MQL	–	69.9–99.7
Azithromycin	<MQL–666	–	<0–99
Chloramphenicol	<MQL	–	11.8–73.8
Chlortetracycline	<MQL–1908	–	31.4–97.8
Ciproflaxin	1400–4800	–	<0–100
Enrofloxacin	7.6–11,560	–	<0–99
Erythromycin	6–79	–	<0–88.9
Lincomycin	<MQL–4967	–	0–67
Minocycline	1000–2000	–	26.6–77.7
Ofloxacin	1480–5760	–	<0–100
Oxytetracycline	<MQL–3790	–	44.8–86.9
Sulfamethazine	<MQL–54.58	<MQL	<0–99
Sulfamethoxazole	<MQL–84.4	1.8	54.6–96.3
Tetracycline	<MQL–466	–	<0–96.2
Trimethoprim	<MQL–13	21.5	<0–99
Tylosin	31–139	–	34–97
Vancomycin	<MQL	–	<0–97
Metronidazole	<MQL	–	96.6–99.9
Antymicrobial/antifungal			
Miconazole	<MQL–2609	–	94–99
Thiabendazole	<MQL–226	–	8–88
Triclocarban	362–8460	–	92–99
Triclosan	354–15,600	<MQL–99	<0–100
Ketoconazole	–	<MQL–10,000	–
Clotrimazole	–	73–590	–
Terbinafine	–	<MQL–300	–
Fluconazole	–	<MQL	–
Econazole	–	73–660	–
Miconazole	–	<MQL–700	–
Propiconazole	–	<MQL–120	–
NSAIDs			
Acetaminophen	<MQL–586	10.3	<0–100
Codeine	<MQL–79	–	<0–98
Diclofenac	<MQL–133	23.5	<0–98
Fenoprofen	<MQL	–	98.6–100
Ibuprofen	<MQL–3988	174	<0–99.8

Continued

Table 1 Continued

Selected compound (PhACs in EC)	**Sludge from WWTPs ($ng\,g^{-1}$ dw)**	**Digested sludge ($ng\,g^{-1}$ dw)**	**Removal range in full scale WWTPs (%)**
Indomethacin	<MQL–77	–	7–98.6
Ketoprofen	<MQL–58.4	<MQL	51.5–91.9
Naproxen	<MQL–1022	39.8	<0–99.3
Salicylic acid	<MQL–13,743	–	9–95.4
Beta-blockers			
Atenolol	<MQL–86	26.8	<0–96
Metoprolol	<MQL–226	1.0	<0–58.7
Propanolol	<MQL–849	192	<0
Anticonvulsants			
Carbamazepine	<MQL–50	–	<0–83
Lipid regulators			
Bezafibrate	17–64	<MQL	48.4–95.8
Clobric acid	<MQL	–	27.7–71.8
Gemfibrozil	<MQL–1192		0–100
Hormones			
Estrone	<MQL–17.5	–	0–100
Estriol	<MQL–49	–	18–100
17a-ethinyestradiol	<MQL–17	–	33–100
Psychoactive			
Citalopram	580	657	78
Sertraline	490	1138	100
Norsertraline	110	–	–
Carbamazepine	86	–	29–99
Venlafaxine	44	37.9	75
Sulfapyridine	33	–	–
Amitryptyline	29	–	85
Caffeine	26	–	100
Mianserin	23	–	–
Clomipramine	23	–	–
Codeine	16	–	99
Mirtazapine	15	66.1	–
Methadone	15	–	–
Diphenhydramine	7.8	–	–
Haloperidol	7.4	–	–
Oxazepam	6.6	–	–
Clonazepam	5.5	–	–
Mephedrone	3.7	–	–

MQL, method quantitation limit.

cases, the WWTO achieved >75% removal of the most detected PhACs with the exception of diclofenac. The results presented by Zhou et al. (2018) showed that the integrated modern process—MEADRD (mixing enzymes anaerobic digestion)—in both mesophilic and thermophilic conditions was suitable for the effective removal of PhACs from sewage sludge.

3 Endocrine-disrupting chemicals (EDC)

Various environmental chemicals have been identified as endocrine disrupters and have become serious environmental problems. Natural and synthetic steroid hormones are considered to be the strongest endocrine-disrupting compounds, even at a very low level (Amoah et al., 2017).

The natural steroid hormones groups include progestrogens, glucocorticoids, mineralocorticoids, androgens, and estrogens secreted by the hormonal systems of humans and animals. Synthetic steroids compounds are 17α-ethinylestradiol (EE2) and mestranol (MeEE2), commonly used in oral contraceptives and hormone replacement therapy. Natural steroid estrogens compounds are estrone (E1), 17β-estradiol (E2), and estriol (E3). Synthetic androgens and progestins are also used as growth promoters in animal farms (Fent, 2015; Ting and Praveena, 2017).

There have been many studies on the occurrence and effects of natural and synthetic steroidal estrogens (ethinyl estradiol, estradiol, estrone, and estriol) and estrogen-like compounds, and their disturbing effects have been well documented. The classification of biologically active EDCs is shown in Fig. 3.

Steroids with progestogenic activity are called gestagens, proestagens, or progestagens. In the case of gestagens present in the environment, the term "environmental gestagens" and the term progestin in the case of synthetic progestagens/gestagens are often used. In practice, synthetic progestins have endocrine activity, similar to EE2. Together with synthetic estrogen steroids, progestins belong to the most important group of interest of environmental drugs, but unlike estrogens, progestins are not well studied and their environmental hazards are not fully known (Fent, 2015).

The main sources of steroid estrogen are municipal, pharmaceutical, and hospitals wastewaters. However, the presence of steroidal estrogens in WWTPs comes mainly from municipal sources of human excretion as inactive glucuronides and sulfate conjugates or free forms (Sim et al., 2011).

It is also worth noting that the point source of EDC is animal breeding, where these substances are released from the urine and slurry. Also, spawning fish increase the concentration of steroid estrogens in the river receiving wastewater (Sarmah et al., 2006). WWTPs have limited ability to completely remove steroidal estrogens, and partially removed steroidal estrogens during wastewater treatment are, at the end, released into the environment (Ting and Praveena, 2017; Silva et al., 2015). Sorption and biodegradation are the main mechanisms of steroidal estrogen removal that enter the WWTP, and the effectiveness of that process depends on the type of wastewater treatment system. EDCs in the primary treatment are removed by sorption into the

EDC

Endocrine disrupting compounds

Class	Group	Group representatives	Concentration in sewage sludge (ng g^{-1} dm) *Swine farms	Removal efficiency (%)
Gestagens	Progesterone	Progesterone–P4	<MQL–56	100
	Progesterone derivatives	Dydrogesterone	27–98*	100
		Medrogestone	–	–
	Pregnane derivatives	Chlormadinone acetate	–	–
		Cyproterone acetate	–	–
		Merdoxyprogesterone acetate	–	–
	Norpregnane derivatives	Promegestone	xxx	xxx
		Nomegestrol acetate	xxxx	xxxx
	Estranes	Norethisterone	<MQL–106	–
		Lynestrenol	–	–
		Tibolone	–	–
	Gonanes (ethinylated deri.)	Levonorgestrel	<MQL–53	–
		Norgestimate	–	–
		Gestodene	–	–
	Gonanes (nonethinylated deri.)	Dienogest	–	–
Estrogens	Estrogens	Estrone–E1	<MQL-61.4	0–100
		17(Beta)-estradiol–E2	<MQL–50	62–100
		Estriol–E3	<MQL–210	18–100
		17(Alpha)-ethinylestradiol–EE2	<MQL–178	33–100

Fig. 3 Classification of biologically active EDCs (Besse and Garric, 2009; Komesli et al., 2015; Huber et al., 2016; Tran et al., 2018; Petrie et al., 2016; Esperanza et al., 2007; Liu et al., 2011, 2015).

primary sludge, followed by sorption on micropellets and biodegradation by microorganisms in the secondary treatment (Hamid and Eskicioglu, 2012).

Based on research by Johnson et al. (2000), it was estimated that men excreted 3.9 μg of E1, 1.6 μg of E2, and 1.5 μg of E3 of urine per day; for comparison, in pregnant women the excretion of steroidal estrogen was 600 μg of E1, 259 μg of E2, and 6000 μg of E3. The menstruating women excreted two orders of magnitude lower. It is very interesting that EDCs coming directly from birth control pills (pharmaceutical production) are difficult to degrade due to their complexity, in contrast to EDCs excreted by living organisms (Johnson et al., 2000; Cui et al., 2006; Ting and Praveena, 2017).

3.1 EDSs removing treatments

According to many researchers, sorption and biodegradation processes are the main EDC-removing mechanisms in WWTPs. In addition, these organic compounds have a hydrophobic character and their low water solubility promotes sorption on sewage sludge (Ting and Praveena, 2017; Silva et al., 2012; Pessoa et al., 2014; Lai et al., 2000). The presence of steroidal estrogens in an aqueous, solid, and mixed solution indicates that the removal took place on solids and then by biodegradation, which is the basic mechanism of removing steroidal estrogen. Bacteria present in wastewater and sewage sludge (mainly fecal forms from the *Coli* group) have the ability to completely biodegrade EDC to harmless products by deconjugation of sulfates and glucuronides (Khanal et al., 2006). For example, natural estrogen E2 is readily biodegradable while synthetic EE2 is not readily removed by biological processes. It is also disclosed that E1 and E2 as well as EE2 are biodegradable in all three redox states, where the aerobic state is most preferred and the elevated temperature promotes biodegradation by increasing microbial activity. Many studies show that both sludge retention (SRT) and HRT seem to be particularly important parameters in removing EDCs from secondary treatment systems. It has also been shown that at least a 10-day retention time is required for microbial growth to break down these compounds (Andersen et al., 2003; Grosser, 2017; Hamid and Eskicioglu, 2012; Khanal et al., 2006; Vymazal et al., 2015). The removal of EDC from sewage sludge, as for other substances from the EC group, is most effective when using anaerobic digestion technology, where separation into mesophilic or thermophilic is irrelevant. Thermal hydrolysis effectively removes EDC from sewage sludge and advanced oxidation processes using Fenton's reaction remove the more stable and remaining forms (Grosser, 2018; Grosser et al., 2013; Malmborg and Magnér, 2015).

4 Illicit drugs and controlled substances

Among ECs, illegal drugs are one of the newest groups of compounds identified in wastewater. These include synthetic or plant-derived pharmaceutical compounds for which nonmedical use is prohibited by national or international law (Chiavola et al., 2017; Baker and Kasprzyk-Hordern, 2011; Daughton, 2011).

Drugs are excreted in the urine and feces (also with saliva and sweat), and then transported to the sewage treatment plant in many ways. A significant number of analytical methods have been developed for the analysis of nonauthorized substances. Most of these studies consider only the amount of the compounds dissolved in the aqueous phase while only a few have been marked as bound to solid particles such as sewage sludge. For most of the tested drugs or abuse drugs, their share in the total amount of the detected substance is below 5%, but methadone (METH) and its metabolite (EDDP) amounted to more than 10% (Mastroianni et al., 2013; Castiglioni and Zuccato, 2011; Pal et al., 2013; Baker and Kasprzyk-Hordern, 2011; Metcalfe et al., 2010; Senta et al., 2013). Table 2 presents the list of illegal drugs tested in sewage sludge.

There is a lack of data on the occurrence of illegal and abused drugs as well as their metabolites in sewage sludge, but on the basis of the available data, it can be concluded that concentrations below 100 ng g^{-1} have been detected for MA, AM, MDMA, COC, BE, and DIA. However, concentrations up to 300 and 125 ng g^{-1} were detected for AM and EDDP. Drugs and abused drugs that are hydrophobic (log $K_{ow} > 3$) such as METH and EDDP, can be expected at high concentrations in sewage sludge. However, it is a complex process that does not depend only on one factor, thus assessing the migration in the environment and on solid matrices is still dependent on chemical analysis (Senta et al., 2013; Kaleta et al., 2006; Langford et al., 2011; Jones-Lepp and Stevens, 2007; Jelić et al., 2009; Peysson and Vulliet, 2013).

In the Mastroianni et al. study (Mastroianni et al., 2013), which tested WWTPs along four Spanish basins, METH, EDDP, and THC were present in all samples; COC, EPH, and CBD were present in at least 80% of the samples; and MOR, LSD, and OH-LSD were not detected at all. The benzodiazepines ALP and DIA were identified positively in 60% of 73% of the tested samples, but the levels did not exceed 2.5 ng g^{-1} d. in any case. The total amount of target illicit and abused drugs was in the range from 138 to 1560 ng g^{-1} d.

4.1 Illicit drugs and controlled substances removal treatments

In sewage treatment plants, drugs are subject to such processes as oxidation, adsorption, and biodegradation, depending on their physicochemical properties, wastewater specification (e.g., temperature, pH, redox environment), and operational parameters (e.g., retention time, hydraulic retention time). In most cases, biological wastewater treatment processes are not able to sufficiently reduce the load of these compounds. There are no data indicating the removal of illegal drugs from sewage sludge itself, and even high wastewater treatment efficiency does not guarantee their absence in sewage sludge. However, it can be assumed that, like all organic compounds, they are quite well removed during aerobic or anaerobic processes (Chiavola et al., 2017; Mastroianni et al., 2013; Petrie et al., 2016; Malmborg and Magnér, 2015).

5 Pesticides

Pesticides have been recognized as significant environmental pollution due to the their necessity and widespread use (Ponce-Robles et al., 2017). At present, the main problems are compounds such as aniline derivative, carbamate, chlorophenoxy compounds, organochlorines, organophosphorus, pyridine, pyrimidine, triazine, and urea (Table 3).

There is evidence that some pesticides, including resmethrine, cypermethrin, bifenthrin, tolylfluanid, and methoxychlor, play a key role in hormonal disturbances,

Table 2 Characteristics of illegal drugs and drugs abused also with the concentration in sewage sludge, wastewater and digested sludge (Mastroianni et al., 2013; Devault et al., 2016; Ivanová et al., 2018; Petrie et al., 2016)

Compound name	Abbreviation	Concentration range ($ng\,g^{-1}$ d.w.)	Removal efficiency range from wastewater (%)	Concentration range in digested sludge ($ng\,g^{-1}$ d.w.)
Cocaine	COC	0.7–23.0	34–100	3.1
Benzoylecgonine	BE	0.8–3.2	27–100	–
Cocaine coccine metabolite	CE	–	43 - 100	–
Ephedrine	EPH	0.6–44.1	–	–
Amphetamine	AM	2.1–13.9	50–100	7.5
Methamphetamine	MA	19	52–100	<MQL
3,4-Methylenedioxymethamphetamine	MDMA	0.4–2.7	3.95–69	10.3
Morphine	MOR	2.2–19.1	72.8–100	277
Methadone	METH	6.7–111.0	24–80.6	24.6
Methadone metabolite	EDDP	8.7–279.0	–	118
Alprazolam	ALP	0.5–1.9	–	–
Diazepam	DIA	0.6–2.5	–	–
THC cannabinoids	THC	29.9–579.0	48–100	–
Cannabinoid	CBD	32.2–479.0	–	–
Cannabidiol	CBN	7.1–188.0	–	–
Lysergic acid diethylamide metabolite	OH-THC	10.3–170.0	–	–

Table 3 Concentration ranges of pesticides in sewage sludge samples from industrial and urban WWTP (Campo et al., 2013; Leitgib et al., 2007; Masiá et al., 2015; Tadeo et al., 2010)

Compound name	Concentration range (ng g^{-1} d.w.)
Anilide	–
Azole	0.10–1166
Benzimidazole	0.10–1166
Carbamate	–
Carboxamide	1.41–7.31
Chloroacetanilide	2.24
Neonicotinoid	–
Organophosphorus	0.05–77.88
Triazine	2.59–10.61
Triazole	–
Urea	–
Acetamipiryd	<MQL–144
Carbendazim	–
Dimethoate	–
Imidacloprid	–
Myclobutanil	<MQL–34
Pirimicarb	<MQL–0.9
Prochloraz	<MQL–163
Propamocarb	<MQL–29
Pyrimethanil	<MQL–80,990
Thiabendazole	<MQL–3033
Glyphosate	200–2900
Aminomethylphosphonic acid	2300–33,300
Chlorpyrifos	0.45–703
Diazinon	9.59–20.1
Ethion	0.18–0.78
Hexythiazox	2.21–4.84
Imazalil	395–1,082
Pyriproxyfen	22.2–25.5
Thiabendazole	11.4–186

affecting the glucocorticoid hormone receptor that is a critical factor in the endocrine system. Therefore, this can cause a number of adverse health effects. As is commonly known, pesticide pollution results mainly from extensive cultivation operations, discharge of agricultural rainwater, and irrigation of arable land. Polar and highly soluble pesticides are often detected mainly in wastewater in amounts of ng L^{-1} (Campo et al., 2013). Although agricultural practices are considered to be the main source of pesticide contamination of waters, in recent years attention has been paid to nonagricultural sources such as municipal WWTPs, where the pesticides accumulate in sewage sludge. Therefore, the common practice of agricultural use of sewage sludge is an additional source of soil pesticide pollution (Ponce-Robles et al., 2017; Masiá et al., 2015; Olofsson et al., 2013). The agrifood industry has also been recognized as an important source of contamination with pesticides. For example, food processing plants use fungicides to neutralize fungal infections during storage of products by washing them, which generates very large amounts of wastewater rich in these compounds (Karas et al., 2015).

As shown in the studies of Masiá et al. (2015) chlorpyriphos was the most frequently detected pesticide in three types of samples at concentrations up to 65.3 ng g^{-1} in soil (83% of samples), and 560 and 703 ng g^{-1} in sewage sludge (90%). Thiabendazole, imazalil, diazinon, pyriproxyfen, and hexytioxazox were also commonly detected in sewage sludge. This is fairly consistent because imazalil and thiabendazole are used as postharvest fungicides while others are widely used in urban areas and are often reported in WWTPs. Other pesticides such as carbofuran, isoprotone, and two triazine metabolites—terbumetone-deethyl and terbuthylazine-deethyl—were detected in sewage sludge. These compounds are widely used as insecticides and herbicides. They can form sediments in the soil and usually reach the river, where they can accumulate on solid forms. The soil analysis showed the presence of pesticides (chlorpyrifos, thiabendazol, and pyriproxyfen) that were detected mainly in sewage sludge samples. This may indicate fertilizing the soil by sewage sludge (a common practice in some countries). In addition, some herbicides that are commonly sprayed near the soil, such as terbuthylazine and metolachlor, have also been found (Masiá et al., 2015).

5.1 Pesticide removal treatments

Due to the lack of literature data pertaining to the removal of pesticides from sewage sludge, we should focus on the possible techniques of their elimination from wastewater, which is a direct source of these compounds in sewage sludge.

Campo et al. (2013) state that most pesticides are only partially eliminated during secondary treatment in WWTPs (in about 50%). Therefore, modern and hybrid sewage sludge treatment techniques are increasingly being sought. The removal of glyphosate was investigated by means of a complex adsorption process and advanced oxidation processes. By using nanometal/resin complexes in advanced oxidation processes, the maximum rate of glyphosate degradation was increased to 60.5% (Laskowski et al. 2017; Zhang et al., 2011). The removal of glyphosate and aminomethylphosphonic acid (AMPA) with synthetic wastewater was carried out on a laboratory scale on a nanofiltration apparatus. Nanofiltration technology has proven to be a good approach to solving the problem of micropolicity of pesticides (Yuan et al., 2018). Benzimidazole is an important intermediate product in industry and is usually difficult to degrade with the help of many machining technologies. Based on the structure of contaminants, a microcell is now being used in combination with the Fenton technique to degrade industrial benzimidazole wastewater. Spectrophotometric studies in ultraviolet, Fourier transform infrared spectroscopy, and mass spectrometry with liquid chromatography showed that both 2-(α-hydroxyethyl)benzimidazole and 2-acetylbenzimidazole were decomposed to isopropanolamine and aniline after that treatment; the intermediates were then oxidized to oxalic acid after the Fenton reaction (Li et al., 2018). Organophosphorus pesticides, including chlorotriphene, fenitrotione, diazinon, methamidophos, edifenphos, mevinphos, fention, and acephate, present in agricultural wastewater can be effectively removed by extraction with supercritical carbon dioxide (SC-CO_2). Pesticides in Fenton's solution are completely degraded within one hour after collection. The combination of SC-CO_2 extraction and subsequent degradation with Fenton's reagent can provide an alternative strategy for treating agricultural wastewater with organophosphorus pesticide pollution (Krzywicka and Kwarciak-Kozłowska, 2014; Kwarciak-Kozłowska, 2015; Kwarciak-Kozłowska and Krzywicka, 2015; Yu, 2002).

The studies of Kahle et al. (2008) indicate that azole compounds (fluconazole, propiconazole, and tebuconazole) are largely insensitive to wastewater, but clotrimazole is effectively eliminated (>80%). Incubation studies with activated sludge did not show degradation of fluconazole and clotrimazole within 24 h, but strong sorption of clotrimazole into the activated sludge. Slow degradation and some sorption were observed for tebuconazole and propiconazole (half-lives in the decay phase, 2–3 days). Peng et al. (2012) states that clotrimazole, econazole, and miconazole are more susceptible to adsorption and persist in sewage sludge. Lipophilicity plays a major role in adsorption. The highest concentrations in raw sewage were observed in winter for azole pharmaceuticals, with the exception of fluconazole.

In the case of fungicides, Stamatis et al. (2010) indicate that with the exception of triadimhone, all azole fungicides and pyrimethanil showed a relatively low removal efficiency after secondary and tertiary treatment of wastewater. The average removal rate of fungicides after secondary treatment ranged from 31% for pyrimethanil to 65% for triadimefon. The average overall removal efficacy after tertiary treatment was from 46% for pyrimethanil to 93% for triadimefon.

6 Summary

The range of pollutants found in sewage sludge samples reflects what is used in municipal society. The behavior of emerging contaminants in sewage sludge requires further research because the available data give only a general overview. Sewage sludge tends to absorb many potential pollutants from wastewater such as pesticides, pharmaceuticals, illegal drugs, and other organic chemical residues.

Currently, in regard to research activities contributing to this area, new and improved methods are emerging. Currently, there are several advanced analytical methods for detecting and quantifying emerging contaminants, but few of them can be applied to sewage sludge.

The sample preparation procedure plays a fundamental role in the development of an analytical methodology for complex matrices such as sewage sludge. There are many publications describing techniques that provide quick, safe, and easy preparation procedures and use smaller amounts of samples and solvents.

The use of advanced technologies such as LC-MS/MS in environmental analysis has opened a range of compounds that can be analyzed, thus allowing a more comprehensive assessment of environmental pollution. Concentrations of selected pollutants determined in sewage sludge indicate that precise control and monitoring of organic pollutants is necessary in order to ensure the safe management of sewage sludge, which is estimated to be used 45% of the time in agriculture, 18% of the time in landfills, and incinerated 17% of the time.

In terms of removal, currently the best effects for a wide group of ECs are combining chemical and biological methods during thermophilic and mesophilic anaerobic digestion, which really promote safer removal of these compounds from sewage sludge.

Some countries, for example, Sweden, have introduced obligatory sanitization of sewage sludge before it can be released into the environment through the use of new technologies, which include a range of chemical and thermal treatments such as pasteurization, thermal hydrolysis, advanced oxidation reaction using Fenton's reaction, ammonia treatment, and thermophilic/mesophilic anaerobic digestion.

References

Amoah, I.D., et al., 2017. Detection and quantification of soil-transmitted helminths in environmental samples: a review of current state-of-the-art and future perspectives. Acta Trop. 169, 187–201.

Andersen, H., et al., 2003. Fate of estrogens in a municipal sewage treatment plant. Environ. Sci. Technol. 37 (18), 4021–4026.

Archer, E., et al., 2017. The fate of pharmaceuticals and personal care products (PPCPs), endocrine disrupting contaminants (EDCs), metabolites and illicit drugs in a WWTW and environmental waters. Chemosphere 174, 437–446.

Baker, D.R., Kasprzyk-Hordern, B., 2011. Multi-residue determination of the sorption of illicit drugs and pharmaceuticals to wastewater suspended particulate matter using pressurised liquid extraction, solid phase extraction and liquid chromatography coupled with tandem mass spectrometry. J. Chromatogr. A 1218 (44), 7901–7913.

Bergersen, O., Hanssen, K.Ø., Vasskog, T., 2012. Anaerobic treatment of sewage sludge containing selective serotonin reuptake inhibitors. Bioresour. Technol. 117, 325–332.

Besse, J.-P., Garric, J., 2009. Progestagens for human use, exposure and hazard assessment for the aquatic environment. Environ. Pollut. 157 (12), 3485–3494.

Bloem, E., et al., 2017. Contamination of organic nutrient sources with potentially toxic elements, antibiotics and pathogen microorganisms in relation to P fertilizer potential and treatment options for the production of sustainable fertilizers: A review. Sci. Total Environ. 607–608, 225–242.

Campo, J., et al., 2013. Occurrence and removal efficiency of pesticides in sewage treatment plants of four Mediterranean River Basins. J. Hazard. Mater. 263, 146–157.

Carballa, M., et al., 2007. Fate of pharmaceutical and personal care products (PPCPs) during anaerobic digestion of sewage sludge. Water Res. 41 (10), 2139–2150.

Castiglioni, S., Zuccato, E., 2011. Occurrence of illicit drugs in wastewater and surface water in Italy. In: Illicit Drugs in the Environment. Occurrence, Analysis, and Fate Using Mass Spectrometry. John Wiley & Sons, Inc., Hoboken, NJ, pp. 137–151.

Chiavola, A., Tedesco, P., Boni, M.R., 2017. Fate of selected drugs in the wastewater treatment plants (WWTPs) for domestic sewage. Environ. Sci. Pollut. Res. 1216, 1–11.

Cui, C.W., Ji, S.L., Ren, H.Y., 2006. Determination of steroid estrogens in wastewater treatment plant of a controceptives producing factory. Environ. Monit. Assess. 121 (1–3), 409–419.

Daughton, C.G., 2011. Illicit drugs: contaminants in the environment and utility in forensic epidemiology. In: Reviews of Environmental Contamination and Toxicology. vol. 210. Springer, New York, NY, pp. 59–110.

Devault, D.A., et al., 2016. The removal of illicit drugs and morphine in two waste water treatment plants (WWTPs) under tropical conditions. Environ. Sci. Pollut. Res. 24 (33), 25645–25655.

Esperanza, M., et al., 2007. Fate of sex hormones in two pilot-scale municipal wastewater treatment plants: conventional treatment. Chemosphere 66 (8), 1535–1544.

Fent, K., 2015. Progestins as endocrine disrupters in aquatic ecosystems: concentrations, effects and risk assessment. Environ. Int. 84 (C), 115–130.

Fernandez-Martinez, J., et al., 2013. Effect of environmental stress factors on ecophysiological traits and susceptibility to pathogens of five Populus clones throughout the growing season. Tree Physiol. 33 (6), 618–627.

Fijalkowski, K., Rorat, A., Grobelak, A., Kacprzak, M.J., 2017. The presence of contaminations in sewage sludge—the current situation. J. Environ. Manag. 203 (Pt 3), 1126–1136.

Fijalkowski, K., et al., 2018. Modification of properties of energy crops under Polish condition as an effect of sewage sludge application onto degraded soil. J. Environ. Manag. 217, 509–519.

Gogoi, A., et al., 2018. Occurrence and fate of emerging contaminants in water environment—a review. Groundwater Sustain. Dev. 6, 169–180.

Gracia-Lor, E., et al., 2012. Occurrence and removal of pharmaceuticals in wastewater treatment plants at the Spanish Mediterranean area of Valencia. Chemosphere 87 (5), 453–462.

Grobelak, A., 2016. Organic soil amendments in the phytoremediation process. In: Phytoremediation. Springer International Publishing, Cham, pp. 21–39.

Grosser, A., 2017. The influence of decreased hydraulic retention time on the performance and stability of co-digestion of sewage sludge with grease trap sludge and organic fraction of municipal waste. J. Environ. Manag. 203, 1143–1157.

Grosser, A., 2018. Determination of methane potential of mixtures composed of sewage sludge, organic fraction of municipal waste and grease trap sludge using biochemical methane potential assays. A comparison of BMP tests and semi-continuous trial results. Energy 143, 488–499.

Grosser, A., et al., 2013. Co-digestion of organic fraction of municipal solid waste with different organic wastes. In: Environmental Engineering IV. A Review. CRC Press, pp. 231–242.

Hamid, H., Eskicioglu, C., 2012. Fate of estrogenic hormones in wastewater and sludge treatment: a review of properties and analytical detection techniques in sludge matrix. Water Res. 46 (18), 5813–5833.

Huber, S., et al., 2016. A first screening and risk assessment of pharmaceuticals and additives in personal care products in waste water, sludge, recipient water and sediment from Faroe Islands, Iceland and Greenland. Sci. Total Environ. 562, 13–25.

Ivanová, L., et al., 2017. Estimation of amount of selected pharmaceuticals sorbed onto digested sludge from wastewater treatment plant Bratislava-Petržalka. Environ. Res. 155, 31–35.

Ivanová, L., et al., 2018. Pharmaceuticals and illicit drugs—a new threat to the application of sewage sludge in agriculture. Sci. Total Environ. 634, 1–10.

Jelić, A., Petrović, M., Barceló, D., 2009. Multi-residue method for trace level determination of pharmaceuticals in solid samples using pressurized liquid extraction followed by liquid chromatography/quadrupole-linear ion trap mass spectrometry. Talanta 80 (1), 363–371.

Johnson, A.C., Belfroid, A., Di Corcia, A., 2000. Estimating steroid oestrogen inputs into activated sludge treatment works and observations on their removal from the effluent. Sci. Total Environ. 256 (2–3), 163–173.

Jones-Lepp, T.L., Stevens, R., 2007. Pharmaceuticals and personal care products in biosolids/sewage sludge: the interface between analytical chemistry and regulation. Anal. Bioanal. Chem. 387 (4), 1173–1183.

Kahle, M., et al., 2008. Azole fungicides: occurrence and fate in wastewater and surface waters. Environ. Sci. Technol. 42 (19), 7193–7200.
Kaleta, A., Ferdig, M., Buchberger, W., 2006. Semiquantitative determination of residues of amphetamine in sewage sludge samples. J. Sep. Sci. 29 (11), 1662–1666.
Karas, P., et al., 2015. Dissipation, metabolism and sorption of pesticides used in fruit-packaging plants: towards an optimized depuration of their pesticide-contaminated agro-industrial effluents. Sci. Total Environ. 530–531, 129–139.
Khanal, S.K., et al., 2006. Fate, transport, and biodegradation of natural estrogens in the environment and engineered systems. Environ. Sci. Technol. 40 (21), 6537–6546.
Kim, M., et al., 2014. Removal of pharmaceuticals and personal care products in a membrane bioreactor wastewater treatment plant. Water Sci. Technol. 69 (11), 2221–2229.
Komesli, O.T., et al., 2015. Occurrence, fate and removal of endocrine disrupting compounds (EDCs) in Turkish wastewater treatment plants. Chem. Eng. J. 277, 202–208.
Krzywicka, A., Kwarciak-Kozłowska, A., 2014. Advanced oxidation processes with coke plant wastewater treatment. Water Sci. Technol. 69 (9), 1875–1878.
Kwarciak-Kozłowska, A., 2015. Fenton process supported with ultrasonic field in coke wastewater treatment Wspomaganie procesu Fentona polem ultradźwiękowym w oczyszczaniu ścieków koksowniczych. Przem. Chem. 1 (9), 93–95.
Kwarciak-Kozłowska, A., Krzywicka, A., 2015. The comparison of efficiency of Fenton and photo-Fenton treatment of stabilised landfill leachate/Porównanie efektywności oczyszczania odcieków z ustabilizowanego składowiska odpadów komunalnych w procesie Fentona i foto-Fentona. Ochr. Śr. Zasobów Nat. 26 (3), 49–53.
Lai, K.M., et al., 2000. Binding of waterborne steroid estrogens to solid phases in river and estuarine systems. Environ. Sci. Technol. 34 (18), 3890–3894.
Lajeunesse, A., et al., 2013. Ozone oxidation of antidepressants in wastewater—treatment evaluation and characterization of new by-products by LC-QToFMS. Chem. Cent. J. 7 (1), 15.
Langford, K.H., Reid, M., Thomas, K.V., 2011. Multi-residue screening of prioritised human pharmaceuticals, illicit drugs and bactericides in sediments and sludge. J. Environ. Monit. 13 (8), 2284–2291.
Laskowski, L., et al., 2017. New class of antimicrobial agents: SBA-15 silica containing anchored copper ions. J. Nanomater 2017, 1–12.
Leitgib, L., Kálmán, J., Gruiz, K., 2007. Comparison of bioassays by testing whole soil and their water extract from contaminated sites. Chemosphere 66 (3), 428–434.
Li, C., et al., 2018. Process design of high-concentration benzimidazole wastewater treatment based on the molecular structure of contaminants. Environ. Technol. 39 (8), 1007–1016.
Liu, Z.-H., et al., 2011. Occurrence, fate and removal of synthetic oral contraceptives (SOCs) in the natural environment: a review. Sci. Total Environ. 409 (24), 5149–5161.
Liu, S.-S., et al., 2015. Occurrence and removal of progestagens in two representative swine farms: effectiveness of lagoon and digester treatment. Water Res. 77, 146–154.
Malmborg, J., Magnér, J., 2015. Pharmaceutical residues in sewage sludge: effect of sanitization and anaerobic digestion. J. Environ. Manage. 153 (C), 1–10.
Masiá, A., Vásquez, K., Campo, J., Picó, Y., 2015. Assessment of two extraction methods to determine pesticides in soils, sediments and sludges. Application to the Túria River Basin. J. Chromatogr. A 1378, 19–31.
Mastroianni, N., et al., 2013. Illicit and abused drugs in sewage sludge: method optimization and occurrence. J. Chromatogr. A 1322, 29–37.
Metcalfe, C., et al., 2010. Illicit drugs in Canadian municipal wastewater and estimates of community drug use. Environ. Pollut. 158 (10), 3179–3185.

Olofsson, U., et al., 2013. Comprehensive mass flow analysis of Swedish sludge contaminants. Chemosphere 90 (1), 28–35.

Östman, M., et al., 2017. Screening of biocides, metals and antibiotics in Swedish sewage sludge and wastewater. Water Res. 115, 318–328.

Pal, R., et al., 2013. Illicit drugs and the environment—a review. Sci. Total Environ. 463–464, 1079–1092.

Peng, X., et al., 2012. Distribution, behavior and fate of azole antifungals during mechanical, biological, and chemical treatments in sewage treatment plants in China. Sci. Total Environ. 426, 311–317.

Pessoa, G.P., et al., 2014. Occurrence and removal of estrogens in Brazilian wastewater treatment plants. Sci. Total Environ. 490, 288–295.

Petrie, B., et al., 2016. Multi-residue analysis of 90 emerging contaminants in liquid and solid environmental matrices by ultra-high-performance liquid chromatography tandem mass spectrometry. J. Chromatogr. A 1431, 64–78.

Peysson, W., Vulliet, E., 2013. Determination of 136 pharmaceuticals and hormones in sewage sludge using quick, easy, cheap, effective, rugged and safe extraction followed by analysis with liquid chromatography-time-of-flight-mass spectrometry. J. Chromatogr. A 1290, 46–61.

Ponce-Robles, L., et al., 2017. Determination of pesticides in sewage sludge from an agro-food industry using QuEChERS extraction followed by analysis with liquid chromatography-tandem mass spectrometry. Anal. Bioanal. Chem. 409, 6181–6193.

Richardson, S.D., 2007. Water analysis: emerging contaminants and current issues. Anal. Chem. 79 (12), 4295–4324.

Salgado, R., et al., 2012. Assessing the removal of pharmaceuticals and personal care products in a full-scale activated sludge plant. Environ. Sci. Pollut. Res. Int. 19 (5), 1818–1827.

Sarmah, A.K., et al., 2006. A survey of endocrine disrupting chemicals (EDCs) in municipal sewage and animal waste effluents in the Waikato region of New Zealand. Sci. Total Environ. 355 (1–3), 135–144.

Senta, I., et al., 2013. Integrated procedure for multiresidue analysis of dissolved and particulate drugs in municipal wastewater by liquid chromatography-tandem mass spectrometry. Anal. Bioanal. Chem. 405 (10), 3255–3268.

Silva, C.P., Otero, M., Esteves, V., 2012. Processes for the elimination of estrogenic steroid hormones from water: a review. Environ. Pollut. 165, 38–58.

Silva, B., et al., 2015. Removal of psychiatric drugs in water treatment systems. In: Psychiatric Pharmaceuticals as Emerging Contaminants in Wastewater, Springer Briefs in Molecular Science. Springer, Cham, Switzerland, pp. 47–71.

Sim, W.-J., et al., 2011. Assessment of fates of estrogens in wastewater and sludge from various types of wastewater treatment plants. Chemosphere 82 (10), 1448–1453.

Stamatis, N., Hela, D., Konstantinou, I., 2010. Occurrence and removal of fungicides in municipal sewage treatment plant. J. Hazard. Mater. 175 (1–3), 829–835.

Tadeo, J.L., et al., 2010. Determination of pesticide residues in sewage sludge: a review. J. AOAC Int. 93 (6), 1692–1702.

Ting, Y.F., Praveena, S.M., 2017. Sources, mechanisms, and fate of steroid estrogens in wastewater treatment plants: a mini review. Environ. Monit. Assess 189, 139.

Tran, N.H., Reinhard, M., Gin, K.Y.-H., 2018. Occurrence and fate of emerging contaminants in municipal wastewater treatment plants from different geographical regions-a review. Water Res. 133, 182–207.

Urase, T., Kikuta, T., 2005. Separate estimation of adsorption and degradation of pharmaceutical substances and estrogens in the activated sludge process. Water Res. 39 (7), 1289–1300.

Verlicchi, P., Zambello, E., 2015. Pharmaceuticals and personal care products in untreated and treated sewage sludge: occurrence and environmental risk in the case of application on soil—a critical review. Sci. Total Environ. 538, 750–767.

Vymazal, J., Březinová, T., Koželuh, M., 2015. Occurrence and removal of estrogens, progesterone and testosterone in three constructed wetlands treating municipal sewage in the Czech Republic. Sci. Total Environ. 536 (C), 625–631.

Yu, J.-J., 2002. Removal of organophosphate pesticides from wastewater by supercritical carbon dioxide extraction. Water Res. 36 (4), 1095–1101.

Yuan, J., et al., 2018. Removal of glyphosate and aminomethylphosphonic acid from synthetic water by nanofiltration. Environ. Technol. 39 (11), 1384–1392.

Zhang, M.D., et al., 2011. Glyphosate degradation with industrial wastewater effluent by combined adsorption treatment and advanced oxidation processes. Adv. Mater. Res. 233–235 369–372.

Zhou, H., et al., 2018. Fate of pharmaceutically active compounds in sewage sludge during anaerobic digestions integrated with enzymes and physicochemical treatments. Waste Manage. 78, 911–916.

Carbonization of sewage sludge as an adsorbent for organic pollutants

21

Lingjun Kong[*,†], *Minhua Su*[*], *Kaimin Shih*[†], *Diyun Chen*[*]
[*]Guangdong Provincial Key Laboratory of Radionuclides Pollution Control and Resources, School of Environmental Science and Engineering, Guangzhou University, Guangzhou, China, [†]Department of Civil Engineering, The University of Hong Kong, Hong Kong SAR, China

1 Introduction

Sludge is a solid byproduct generated from water and the wastewater treatment process. Generally, it is composed of organic compounds, heavy metals, and inorganic compounds. It is classified into sewage sludge, industrial sludge, and water-working sludge (Xu et al., 2015). The characteristics of sludge are different from their sources. Sewage sludge is a mixture of exhausted biomass generated from aerobic and anaerobic microorganisms during municipal sewage treatment and inorganic materials, including sand and chemical reagents. The quantity of sludge production has increased vastly due to the rapid development of urbanization in China. Many toxic substances, including pathogens, organic containments, and heavy metals, are found in sludge (Dong et al., 2013). Public concern regarding sludge disposal has grown greatly in the last decades. Thus, it is a great challenge to manage and treat sludge efficiently.

The common options for sludge disposal are landfill, land application, and incineration, but each option has important limitations with increasing environmental and legislative constraints. Although landfill is the most widely used option due to its advantages in operation and low cost, it is not a development strategy because a large amount of land is required and it does not include material recovery (Hadi et al., 2015). In addition, land application as a fertilizer is also limited because pathogens, metals, and low concentrations of antibiotics are found in the sludge. Considering incineration, a considerable reduction in the volume and high-energy output were obtained (Gomez-Pacheco et al., 2012). However, sulfur and nitrogen oxides, dioxins, and heavy metals were emitted from the incineration plants, thus resulting in public unacceptability (Bennett and Knapp, 1982). Therefore, these challenges require sustainable development techniques for sludge disposal. One of the most promising options is resource recycling.

Because sludge contains a lot of organic compounds, it could be a potential precursor of carbonaceous adsorbents. Carbonization of waste sludge for biochar has

Industrial and Municipal Sludge. https://doi.org/10.1016/B978-0-12-815907-1.00021-0

received much attention in recent years because the sludge could be stabilized and reduced during carbonization. During the carbonization process, the sludge was heated to 500–800°C under an inert atmosphere; the noncarbon elements were decomposed. Also, it leads to the grouping together of the free atoms of elementary carbon as elementary graphitic crystallites. A rudimentary pore structure were observed, meanwhile, many pores were blocked by the tarry substance during carbonization process. To enhance the value of sludge carbon as an efficient adsorbent, activation processes were considered to improve its porosity. Physical activation and chemical activation are widely considered at the present.

Therefore, conversion of sludge into adsorbent for removing pollutants from environments is being concerned because adsorption is one of the most concerned decontamination technologies (Smith et al., 2009). In the last decades, organic pollutants have been widely discharged in many industrial processes. Thus, this work addresses the important view of conversion of sludge into adsorbents as a substitute for pollutant removal from aqueous solutions and effluent gases.

2 Materials and method

2.1 Materials

The raw sewage sludge was obtained from a sewage treatment plant. It was a byproduct of sewage disposal. Chemical activation reagents, including zinc chloride ($ZnCl_2$), potassium hydroxide (KOH), sodium hydroxide (NaOH), and sulfuric acid, (H_2SO_4) were analytical grade. Physical activation reagents were CO_2 and steam.

2.2 Preparation of sewage sludge derived adsorbents

Sludge-derived carbon adsorbents were prepared by a carbonization process coupled with an activation process. The dried sludge in the presence of activation reagents was carbonized by an electrical furnace set at a determined temperature range from 500 to 900°C. During the carbonization process, the atmosphere is free of oxygen. Generally, nitrogen was conducted to sweep the air in the furnace. The carbonized samples were washed by deionized water and an acid solution to remove the activation reagents and residual minerals in the sludge carbon.

2.3 Analytical methods

2.3.1 XRF

X-ray fluorescence was applied to analyze the mineral content in the sludge-derived adsorbents. The instrument has a titanium target X-ray tube and a high-resolution detector. The samples were studied in a solid phase after grinding and sieving to use the matrices with similar physical properties (Bandosz and Block, 2006).

2.3.2 XPS

The surface characteristics of the sludge carbon were analyzed by X-ray photoelectron spectroscopy (XPS) using an X-ray photoelectron spectrometer. The binding energies of all core levels were referenced to the C 1s C-C bond at 284.6 eV (Wang et al., 2017).

2.3.3 XRD

X-ray diffraction measurements were conducted by the standard powder diffraction procedure. The sludge carbon was ground in the presence of methanol using a small agate mortar. The sludge carbon was ground by hand, ensuring particle sizes between 5 and 10 μm, which prevents line broadening in diffraction peaks. The mixtures were smear mounted onto the zero background quartz window of a Phillips specimen holder and allow to air dry. Samples were analyzed by Cu KR radiation generated in an X-ray diffractometer. A quartz standard slide was run to check for instrument wander and to obtain the accurate location of 2θ peaks.

2.3.4 Nitrogen adsorption-desorption

The sorption and desorption of nitrogen was carried out using an auto physical adsorption analyzer. Before the experiments, the samples were dried and outgassed at 120°C for 2 h to constant vacuum (10^{-6} Torr). From the isotherms, the surface areas (BET method), total pore volume (from the last point of isotherm at relative pressure equal to 0.99), micropore volume, mesopore volume, and pore size distributions were calculated using corresponding adsorption models.

2.3.5 SEM

The morphologies of the sludge carbon were observed using a field emission scanning electron microscope (SEM). The element compositions were acquired with an energy dispersive x-ray spectrometer (EDS) (Inca300, Oxford). The element mapping was analyzed by SEM-mapping analysis.

2.3.6 FTIR

The FTIR spectra were observed through a MAGNA 560 IR Spectrometer (Nicolet, United States). The number of scans and the scanning resolution were set to 250 cm^{-1} and 4 cm^{-1}, respectively. A background measurement was made of the air within the sample chamber to correct the subsequent sample spectrum by the FTIR software before analysis. Prior to analysis, it was necessary to mount the samples in a KBr medium. Thus, samples were dried, mixed with KBr (Spectrosol, VWR International Ltd., Poole, UK) at a ratio of *ca.* 1:150 (w/w), and crushed using an agate pestle and mortar. Approximately 150 mg of the sludge carbon–KBr mixture was loaded into a pneumatic, 13 mm die press (Specac, UK), and 10 metric tons of pressure was applied for 10 min to form a disk. The disk was conducted for further analysis.

2.3.7 Thermogravimetric analysis

Thermalgravimetric analysis was carried out using a TG Instrument thermal analyzer. The instrument settings were of a determined heating rate and a nitrogen atmosphere. For each measurement, about 25 mg of a ground adsorbent sample was used. The gravimetric results were recorded during the heating process.

2.4 Adsorption experiments

Batch adsorption experiments were conducted to investigate the adsorption behavior of organic pollutants from aqueous. For each test, a determined sludge carbon was thrown into organic pollutants containing solution. The isotherm adsorption experiments were conducted at a determined temperature controlled by a water bath. Langmuir and Freundlich models were evolved to fit the experimental results. For the kinetic adsorption experiments, the solution was drawn at determined time intervals and the residual concentration of adsorbates was analyzed. Pseudo first-order and pseudo second-order models were conducted to fit the adsorption kinetics. The equilibrium adsorption capacity (q_e, mg g^{-1}) and the adsorption capacity at different time t (q_t, mg g^{-1}) were calculated as follows:

$$q_e = \frac{(C_0 - C_e)V}{m} \tag{1}$$

$$q_t = \frac{(C_0 - C_t)V}{m} \tag{2}$$

where C_0 (mg L^{-1}), C_e (mg L^{-1}), and C_t (mg L^{-1}) represented the initial concentration, the equilibrium concentration, and the concentration at time t, respectively. V (L) represents the volume of the solution, and m (g) represents the adsorbent mass.

The adsorption experiments to VOC were conducted in a fixed-bed reactor. The adsorption experimental setup consisted of a VOCs vapor generator, a glass adsorption column filled with adsorbents, and a gas analysis system, as shown in Fig. 1. First, the nitrogen steam from a compressor was divided into two steams using two mass flow controllers. One was conducted to the bubble saturator containing pure liquid pollutant. Another was conducted to adjust the VOC concentration. Then the mixed steams were passed through the adsorption column filled with adsorbents.

The VOC concentration was analyzed with a gas chromatograph (GC, Shimadzu GC-2014) equipped with a flame ionization detector (FID). The residual concentration of VOC at the outlet was measured by a GC-FID at determined time intervals. The amount of the adsorbed VOC at equilibrium was calculated as follows:

$$w = \frac{F}{m}\left(c_0 - \int_0^t c_i dt\right) \tag{3}$$

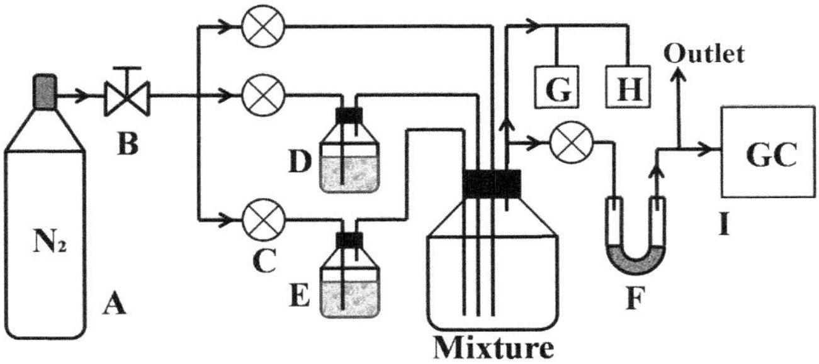

Fig. 1 The experimental diagram of dynamic adsorption of toluene (A: N_2; B: wave; C: flow meter; D: toluene; E: water; F: adsorbents; G: detector of flow pressure; H: absorption for off-gas; I: GC-FID).

where w is the adsorption amount (mg/g); F is the volume flow rate (mL min^{-1}); c_i and c_0 are the outlet and inlet concentration of adsorbate (g mL^{-1}), respectively; m is the mass of the adsorbent (g); and t is the adsorption equilibrium time (min).

3 Results and discussion

3.1 Effect of activation process on pore characteristics

3.1.1 Physical activation

A two-step process was carried out in physical activation. In the first step, the precursor was pyrolyzed at low temperature in the absence of air. The resulted char had low porosity. At the second step, it evolved the selective carbon atom burn-off under controlled gasification atmosphere. In this process, carbon dioxide and steam are the most addressed physical activation reagents. To achieve the burn-off process, the carbon structure is exposed to a carbon dioxide and steam atmosphere at about 700–1100°C. The tarry substance on the carbon surface was partly oxidized, it was served to fabricate pores. In addition, steam activation could fabricate microporosity and mesoporosity. The mechanism could be ascribed to the coupling effect of devolatilization and loss of fixed carbon during the reaction of carbon with steam: $C+H_2O \rightarrow CO+H_2$. The porosity and BET surface area increased with activation temperatures. However, more particles were burned out with the temperature increased higher than 850°C, leading to the collapse of the pore network. Thus, the BET surface area and porosity declined (González et al., 2009).

Carbon dioxide is also active for enhancing the porosity of carbon due to the gasification of the carbon atom in the interior of the particle. It is more competitive than steam activation if the temperature exceeds the maximum temperature of 800°C. A high activation temperature and a long dwell time are required. It is reported that the expansion of opened micropores and the opening of closed micropores were both observed during carbon dioxide activation. The functional groups, including —OH,

—NH and C=O functionalities, were decreased with increased pyrolysis temperature, indicating a decrease in the acidity of the char (Jindarom et al., 2007).

To further increase the porosity of the adsorbent, a washing process before and after carbonization was conducted to remove the inorganic fraction. HCl washing prior to CO_2 activation produced an adsorbent with a BET surface area of 269 m^2/g. It is much higher than 7 m^2/g for that without washing (Ros et al., 2006). Compared to the results obtained by HCl postwashing, the increase in BET surface area is less effective than that prior to activation (Fitzmorris et al., 2006). The improvement of the BET surface area in the presence of the washing process is ascribed to the reduction of inorganic matter. Thus, it is not hard to understand that HCl washing prior to activation is more effective than postactivation washing because the inorganic fraction could be converted into mineral-like compounds. Therefore, the minerals could be encapsulated into the carbon phase, increasing the stability of the inorganic fraction (Bagreev et al., 2001). Therefore, it is evident that improved washing efficiency for removing the inorganic fraction is necessary.

Alvarez et al. (2016) reported that CO_2 activation for preparing sewage sludge char could increase its surface area. A new washing process in the presence of Na_2CO_3 solution was conducted to enhance the silica extraction. The ash content of sludge carbon decreased from 44.7% to 28.0% after washing. Thus, the values of the BET surface area increased from 43 to 235 m^2/g and 311 m^2/g after postwashing in the presence of HCl and HCl coupled with Na_2CO_3 solution, respectively.

3.1.2 Chemical activation

Chemical activation is another technique for enhancing the BET surface area. Table 1 summarizes the different chemical reagents with various efficiencies on enhancing the BET surface area. Compared to the results in the presence of physical activation, a higher surface area and a lower activation temperature were required by chemical activation. The details in the chemical activation process are discussed as follows.

KOH

Potassium hydroxide is one of the most widely concerned activation reagents. It clearly can be seen from Table 1 that potassium hydroxide activation has the most significant effect on enhancing the surface area. The obtained surface area values are different from the activation temperatures, the source of sludge, the mass ratio of potassium hydroxide to sludge. Three kinds of sludge were reported by Lillo-Ródenas et al. (2008) to prepare activated carbon in the presence of potassium hydroxide. During these processes, the tested sludge was carbonized, and further activated by potassium hydroxide. In the absence of the activation process, the surface areas of three chars were all not higher than 50 m^2/g. Once the activation process was presented, the surface area was highly increased to more than 1000 m^2/g, but the values were different. It is not hard to understand that the difference is ascribed to the properties of chars. The surface area depends on the ash content of the char. The lower the ash content of char, the higher the surface area is. The activation reaction occurred between the char and the activation reagent. The mass ratio of potassium hydroxide

Table 1 Effect of chemical activation on S_{BET} of sludge-derived adsorbents

References	Chemical reagent	Mass ration	Temperature	S_{BET} (m^2/g)	Post treatment
Lillo-Ródenas et al. (2008)	KOH[a]	1:1	700°C	1058	HCl
Lillo-Ródenas et al. (2008)	KOH[1]	1:1	700°C	1301	HCl
Monsalvo et al. (2011)	KOH	1:1	750°C	950	HCl
Monsalvo et al. (2011)	KOH	3:1	750°C	1832	
Hunsom and Autthanit (2013)	KOH		800°C	853	HCl
Kong et al. (2013b)	KOH	1:1	800°C	853	HCl-CA
Hunsom and Autthanit (2013)	K_2CO_3		800°C	810	HCl
Ros et al. (2007)	NaOH		700°C	725	HCl
Ros et al. (2006)	NaOH		700°C	689	HCl
Zhang et al. (2005)	$ZnCl_2$		650°C	555	HCl
Chen et al. (2002)	$ZnCl_2$		500°C	647.4	HCl
Kong et al. (2013a,b)	$ZnCl_2$		500°C	517.4	CA-HCl
Martin et al. (2003)	H_2SO_4		700°C	253	NaOH
Zhang et al. (2005)	H_2SO_4		650°C	408	NaOH
Zhang et al. (2005)	H_3PO_4		650°C	289	NaOH

[a]With different sludge precursor.

to sludge was also a key factor determining the value of the surface area. Although a high ratio of potassium hydroxide resulted in a high surface area, the yield is low in the presence of a high mass ratio. During potassium hydroxide activation, a high temperature of 700°C is required.

Generally, carbonization processes was evolved as one step and two steps to prepare the activation carbon. For the one-step process, it is steam from the dehydration that KOH could react directly with organic compounds during carbonization, especially react with oxygen containing functional groups, favoring the development of porosity. It is proposed that dehydration occurred by reacting with KOH and NaOH, leading to the rise in the C:H and C:O ratios (Hsu and Teng, 2000).

To remove the residual chemical activation reagent, postwashing was required. The traditional washing solution was HCl solution. However, the surface area depends on the inorganic compound content in the activated char, so it is important to enhance the efficiency to remove inorganic compounds. Citric acid coupled with the HCl washing process was conducted to enhance the postremoval efficiency (Kong et al., 2013a). Obviously, postwashing could remove the minerals from the sludge-derived activated carbon. Only water dissolvable ions such as K^+ and Na^+ ions were removed by deionized water washing. HCl solution with a pH value of 1 was used to wash SC by protonation besides water dissolution. Parts of water-insoluble ions could be removed in the presence of the HCl solution. However, the removal efficiency is unfavorable. K −, Na −, and parts of Ca −, Mg −, and Mn − containing minerals could be removed by HCl washing while the others could not be removed easily due to the formation of stable minerals such as oxides, phyllosilicates, hydrate, or stable crystals fixed on SC. Interestingly, besides K, Na, Ca, Mg, and Mn, more Fe and Al-containing minerals in SC were removed by CA washing comparing to HCl washing. The enhanced removal efficiencies of Fe, Al, and Mn were ascribed to chelation besides protonation. A mixed solution of HCl-CA with a pH value of 1 was conducted to wash the SC. Results showed that HCl-CA-SC had the highest demineralization efficiency. The removal efficiencies and removal rates of Ca, Mg, Fe, Mn, and Al washed by HCl-CA were the highest compared to other washing solutions. The promoting demineralization mechanisms of HCl-CA washing could be concluded as: (a) enhancing the dissolution rate far from equilibrium through protonation; and (b) forming soluble metals complexes with organic ligands in the presence of CA.

Thus, the resulting DW-SC, HCl-SC, CA-SC, and HCl-CA-SC washed by deionized water, HCl, citric acid, and a mixture solution of HCl-CA had different ash contents that were related to the mineral removal efficiencies. In addition, the values of S_{BET} were in agreement with their ash content, as shown in Table 2. The HCl-CA-SC with the lowest ash content had the highest S_{BET}. Considering the total pore volume and micropore volume after postwashing, it is not hard to see that the micropore volume could be increased but the postwashing process mainly resulted in an increase in the total pore volume. Thus, the HCl-CA-SC had the lowest ratio of V_{micro}/V_T. All these results indicated that postwashing could influence the mineral content, further enhancing the S_{BET} and pore volume.

Table 2 Pore structure parameters of SC, DW-SC, HCl-SC, CA-SC, and HCl-CA-SC

			Pore structure			
Adsorbents	**Yield (%)**	**Ash (%)**	**S_{BET} (m^2g^{-1})**	**V_t (m^3g^{-1})**	**V_{micro} (m^3g^{-1})**	**V_{micro}/V_T**
SC	100	75	116.8	0.09	0.05	0.602
DW-SC	71.26	64	318.5	0.20	0.09	0.443
HCl-SC	57.74	57	325.4	0.25	0.09	0.362
CA-SC	45.52	46	385.8	0.28	0.10	0.339
HCl-CA-SC	37.46	32	605.2	0.46	0.15	0.316

NaOH

NaOH was another activation reagent. Its effect on generating a high surface area was obvious. The activation mechanism was ascribed to the dehydration. Also, post-washing was required to remove the residual activation reagent. Considering the result of the surface area, the resulted S_{BET} obtained from NaOH activation is not as high as KOH activation. But it is more effective in the activation of coal pitch-derived carbon fibers in the presence of NaOH (Maciá-Agulló et al., 2004). More attention may be paid on synthesis of sludge-derived activated carbon in future.

$ZnCl_2$

Zinc chloride activation for preparing activated carbon from paper mill sludge was reported by Khalili et al. (2000). The increase in the mass ratio of $ZnCl_2$ to sludge from 0.75 to 2.5 resulted in a 600% increase in the mesopore volume. A mass ratio of $ZnCl_2$ to sludge less than 1.0 and greater than 1.5 resulted in the production of micro- and mesoporous carbons, respectively. After that, the activation of sewage sludge by $ZnCl_2$ at 650°C was reported for the activation of anaerobically digested and undigested sewage sludge. The undigested sludge with higher carbon content and lower ash content had a higher BET surface area and pore volume (Tay et al., 2001).

To further enhance the S_{BET}, a coupling template and $ZnCl_2$ were evolved to prepare sludge-derived porous carbon in our previous study. The fabrication of hierarchical pores was reported in two steps. For the one-step process, 40.0 g sludge was first impregnated into 0.25 mol citric acid solution, being dried and carbonized in a programmable tube electric furnace (SKF-210, Hangzhou Lantian Instrument Co., China) at a rate of 20°C min^{-1} to 500°C in the presence of N_2 and holding for 1 h. The obtained sample named SC_{CA} was further impregnated into 0.5 mol $ZnCl_2$ solution overnight and activated for 1 h at 500°C. After cooling to room temperature, the product was ground to less than 80 mesh, washed with citric acid-hydrochloric acid solution to remove metal ions and then rinsed with deionized water until the pH reached 7 (Kong et al., 2013b). The obtained product was dried at 105°C overnight, and named $SC_{CA/Zn}$. The bulk density of SC_{CA} was 0.58 kg dm^{-3} and the bulk density of $SC_{CA/Zn}$ was 0.36 kg dm^{-3}, which were lower than that (1.02 kg dm^{-3}) of SC. In addition, the S_{BET} of SC_{CA} and $SC_{CA/Zn}$ increased to 117.6 and 867.6 $m^2\ g^{-1}$, respectively. Obviously, the S_{BET} of $SC_{CA/Zn}$ was higher than that of SC activated by $ZnCl_2$, indicating that the coupling use of the template and activation reagent caused a synergistic effect on fabricating pores. Pores in a wide size distribution can be seen on the surface of $SC_{CA/Zn}$. Significantly, a lot of micropores were located on the surface of macropores, leading to an appearance of "pores in pores" as obtained by SEM analysis. Fabrication of hierarchical pores can be described as follows: First, CA is decomposed into plenty of CO_2 and H_2O vapors in pyrolysis. The released CO_2 and H_2O vapors favored fabricating abundantly hierarchical macropores, providing more interspace and macropores for $ZnCl_2$ activation to fabricate abundant micropores and mesopores, as shown in Fig. 2.

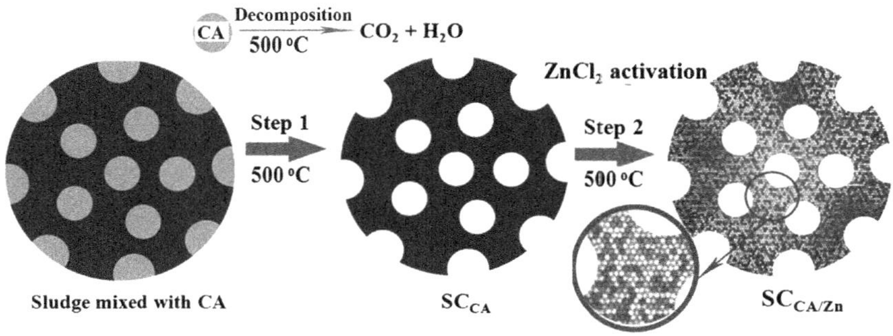

Fig. 2 The diagram of pore fabrication process coupling CA and zinc chloride activation.

Besides, the one-step process coupling template and activation reagent was conducted to prepare the hierarchical porous carbon. A total of 40.0 g of sludge was impregnated into a 100 ml mixed solution containing 0–2 mol CA and 0.5 mol $ZnCl_2$, then stirred overnight to form homogenous sol-gels. The sol-gels were dried and carbonized, subsequently washed with hydrochloric acid solution to remove metal ions, then rinsed with deionized water until the pH reached 7, and finally dried at 105°C overnight. The product with the maximum BET surface area was named $SC_{CA\text{-}Zn}$. The maximum BET surface area of $SC_{CA\text{-}Zn}$ was 792.4 $m^2\,g^{-1}$. As being similar to the results described above, the resulting $SC_{CA\text{-}Zn}$ possessed hierarchical pores, and lots of well-developed pores with various sizes were observed on the surface of $SC_{CA\text{-}Zn}$. Coupling the template CA and $ZnCl_2$ by one step and two steps could be an efficient method for fabricating hierarchical pores with great surface area.

For the sludge char powers, it is difficult to separate them from the aqueous solution in application. The granular sludge chars were easily separated from aqueous solution, which could address the above issues. However, the interior surface of these granular char acted less because adsorption only occurred on the surfaces of granular char. Therefore, the development of a new sludge char with high availability and easy separation was highly concerned.

Hollow-like spherical sludge chars were prepared by coating the sludge pastes impregnated with $ZnCl_2$ in different mass ratios onto a cotton ball to form a kind of cotton/sludge sphere. The diagram is described in Fig. 3. The layer of the sludge paste was about 1 mm, and then the sludge-coated spheres were dried at 105°C. Finally, the

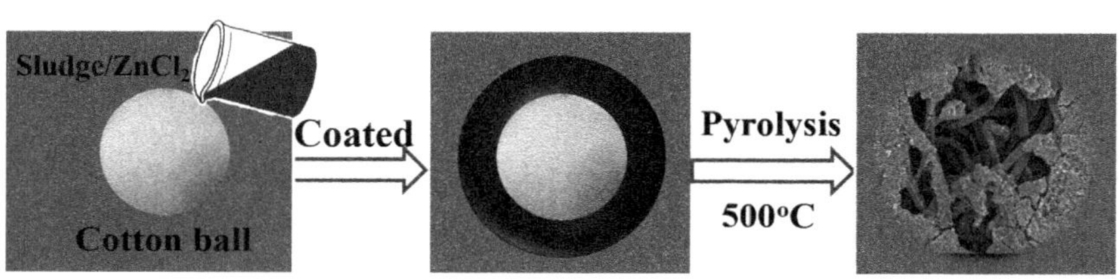

Fig. 3 The diagram of preparing hollow-like sludge char sphere.

obtained spheres were put into a ceramic ark and heated in a programmable tube electric furnace at a rate of 20°C min^{-1} to 500°C in the presence of N_2 holding for 2 h (Rozada et al., 2008). After cooling to room temperature, the product was washed with hydrochloric acid solution to remove inorganic compounds and then rinsed with deionized water until the pH value reached 7. The resulting granular char was named HSC. The S_{BET} of HSC was 1008.4 m^2/g. Generally, the S_{BET} of HSC could be equal to the sum of the S_{BET} of cotton carbon fiber and sludge char in the HSC according to their mass percentage. Considering that the mass percentage of the sludge char and cotton carbon fiber in HSC were 87.8% and 12.2%, respectively, while the S_{BET} of cotton carbon fiber was 2150.6 m^2/g, the S_{BET} of HSC can be calculated as 675.1 m^2/g. The calculated S_{BET} of HSC is clearly less than the experimental S_{BET} of HSC. The fact shows that hollow-like spheres possessed the advantage of great surface areas. The apparent density of HSC was 0.1008 g/cm^3. It was only 15% of that of SC. The low apparent density resulted in low resistant loss in the filling absorbent column, as shown by a pressure drop of 38.8 mm H_2O column (Wu et al., 2014).

Further, we report a process for preparing hollow spherical sludge carbon via using different sized polystyrene (PS) foam waste as the sacrificial hard templates and as an inner core, and the sludge paste impregnated with $ZnCl_2$ was coated onto the PS surface. The obtained PS@sewage sludge was pyrolyzed. During pyrolyzing, the PS could be decomposed, leaving a hollow structure, and the sludge was converted into carbon, forming a hollow spherical sludge carbon. $ZnCl_2$ acted as the activation reagent to fabricate micropores. Thus, the apparent density of hollow sludge carbon with a shell thickness of 0.2 mm was 0.0905 g/cm^3, only yielding 27.0% that of the solid sludge carbon. The low apparent density resulted in a less resistant loss in the adsorbent filling column and a large external surface. This is the advantage of the hollow spherical sludge carbon as described above.

H_2SO_4

Chemical activation in the presence of H_2SO_4 was another attempt widely concerned for increasing the S_{BET} of sludge-derived carbon. The S_{BET} of sludge carbon increased from 137 to 408 m^2/g while the micropore and total pore volume increased from 0.016 and 0.227 to 0.026 and 0.523 cm^3/g, respectively, in the presence of H_2SO_4 (Zhang et al., 2005). However, the S_{BET} and pore volume of sludge carbon in the presence of H_2SO_4 were lower than those in the presence of $ZnCl_2$, indicating that $ZnCl_2$ activation was more efficient than H_2SO_4 activation. Besides, the resulting sludge carbon activated by H_2SO_4 was washed by NaOH solution. An experimental design methodology was used to optimize the experimental conditions of sorbent preparation from sewage sludge by chemical activation with sulfuric acid at 700°C. The optimal S_{BET} of sludge carbon was 353 m^2/g (Rio et al., 2005). Notably, sulfuric acid concentration is not higher than 3 M because destruction of the material structure was observed when the concentration was higher than 3 M during thermal treatment (Rio et al., 2003). Pan et al. (2011) reported that the micropores of the resulting carbon were destroyed and enlarged during H_2SO_4 activation. Thus, the pores in these absorbents were mainly composed of mesopores and macropores, even supermacropores.

3.2 Adsorption of organic pollutants from aqueous

3.2.1 Adsorption of dyes from aqueous solution

Because the activation reagents and minerals are founded in the sludge-derived adsorbents, postwashing played an important role in demineralization and further enhancing S_{BET}. Thus, the effect of postwashing on the adsorption capacities of sludge-derived adsorbents was investigated. Methylene blue and acid orange II were conducted as adsorbates. The effect of contact time on the adsorption capacities of acid orange II and methylene blue was studied by a kinetic experiment. The initial dye concentration was $500\,mg\,L^{-1}$. The pH value of acid orange II and methylene blue in solution was 5.85 and 3.93, respectively, without adjusting and the additive dosage of produced activated carbon was $1.5\,g\,L^{-1}$. The adsorption capacities of acid orange II and methylene blue on sludge carbon and demineralized sludge-derived carbons were increased rapidly in the first 30 min, and then slowed down toward equilibrium within 180 min. The maximum adsorption capacities of sludge carbons, deionized water-washed sludge carbon (DW-SC), HCl-washed sludge carbon (HCl-SC), citric acid-washed sludge carbon (CA-SC), and HCl and citric acid mixtures washed carbon (HCl-CA-SC) were 116, 147, 182, 239, and $319\,mg\,g^{-1}$, respectively, for acid orange II, and 175, 177, 192, 224, and $250\,mg\,g^{-1}$, respectively, for methylene blue, as shown in Fig. 4. They both increased with demineralization efficiency and S_{BET}. This could be explained by the increased specific surface area and the total pore volume, especially for mesopores and macropores, because opening and widening of micropores, mesopores, and interstices channels in sludge carbon by demineralization could provide more adsorption sites for removing acid orange II and methylene blue (Malik et al., 2007; Rodríguez et al., 2009). Thus, the demineralization of SC played an important role in increasing their adsorption capacities to acid orange II and methylene blue.

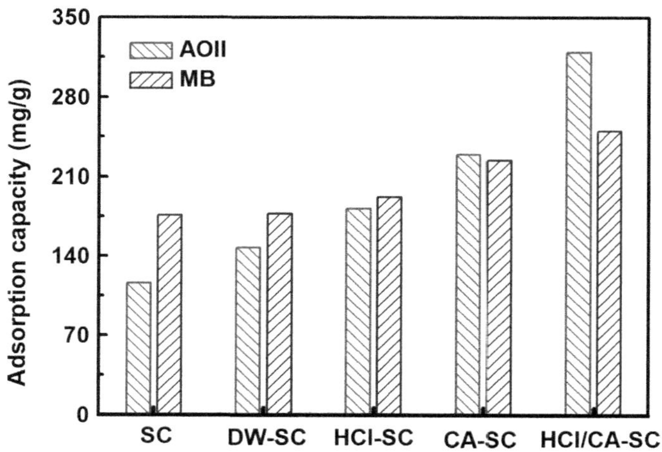

Fig. 4 Comparison of adsorption capacities of AOII and MB onto SC derived adsorbents washed by various washing reagents.

Comparing the adsorption capacities of sludge carbonaceous adsorbents, SC and DW-SC had higher adsorption abilities to cationic dye (methylene blue) than to anionic dye (acid orange II) while HCl-CA-SC had higher adsorption abilities to anionic dye than to cationic dye. It can be explained by the fact that SC and DW-SC had a favorable ability to uptake H^+ on their surfaces due to their basic surfaces (pH^a of 10.17 and 8.12). Then, SC and DW-SC were negatively charged when the pH value was higher than the pH_{zpc} (pH at which the surface has a net zero charge) due to their basic pH^a, and therefore SC and DW-SC had higher affinity to cationic dyes than to anionic dyes due to electrostatic attracting interaction between cationic dyes and SC or DW-SC. The surfaces of HCl-SC and CA-SC were changed to neutral and the content of minerals decreased after washing with HCl and CA, then HCl-SC and CA-SC had similar adsorption amounts and adsorption kinetics for methylene blue and acid orange II due to the neutral surfaces and the weakened ion exchange reaction. In turn, the surface of HCl-CA-SC ($pH^a = 4.26$) was protonated by acid washing and possessed a positive charge below pH_{zpc} by dissociation of OH^-, presenting higher adsorption capacities of $319\,mg\,g^{-1}$ to acid orange II than $250\,mg\,g^{-1}$ to methylene blue. This was consistent with the results of a previous study. Thus, postwashing with HCl-CA was effective for demineralization to enhance pore structures, and further significantly increased the adsorption capacities for the removal of anionic dyes.

The isotherm adsorption experimental data was fitted by Langmuir and Freundlich models to evaluate the adsorption characteristics. The correlation coefficients indicated almost perfect agreement with the experimental data fitted by the Langmuir model and the Freundlich model. However, the calculated maximum adsorption capacities of acid orange II and methylene blue from the Langmuir model were larger than the experimental adsorption capacities, suggesting that the isotherm adsorptions were not fitted well by the Langmuir model but were fitted well by the Freundlich model.

The filling column was conducted to investigate the adsorption performances of the resulted hollow-like granular sludge derived adsorbent (HSC). Methylene blue was conducted as a pollutant, and granular activated carbon (GAC) and sludge carbon (SC) were evolved as comparative adsorbents to investigate the adsorption behavior, as shown in Fig. 5. Adsorption breakthrough curves obtained under various experimental conditions were presented. The half breakthrough time (when the C_t/C_0 was 0.5) of GAC, SC, and HSC was 13.9, 10.6, and 20.2 h, respectively. The calculated adsorption capacity of HSC was 186.42 mg/g. It was about 8.4 and 12.2 times those of GAC and SC, although the S_{BET} of GAC was equal to HSC. Such great disparity between the adsorption capacities of HSC and those of GAC and SC to methylene blue can be contributed to the hollow-like structure of HSC because the hollow-like structure took the advantage in penetrability, as discussed above.

First, the Thomas model was described in Eq. (4) to evaluate the adsorption dynamic parameter to further understand the adsorption performance of HSC.

$$\ln\left(\frac{C_0}{C_t}-1\right)=\frac{k_{Th}q_0w}{Q}-k_{Th}C_0t \tag{4}$$

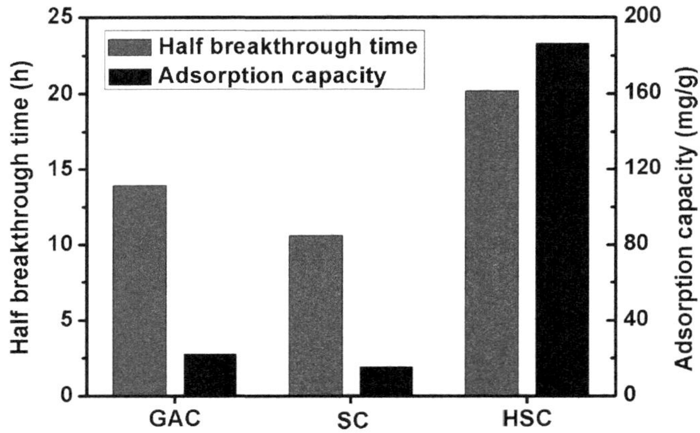

Fig. 5 Comparison of half breakthrough time and adsorption capacity of MB on GAC, SC, and HSC.

Table 3 Dynamic adsorption model parameters of methylene blue on varied adsorbents

	Thomas model		
Adsorbents	**k_{Th} (mL/min mg)**	**q_0 (mg/g)**	**R^2**
GAC	1.253	22.16	0.983
SC	1.227	14.67	0.979
HSC	1.547	169.40	0.985

where k_{Th} is the Thomas rate constant (mL/h mg), q_0 is the equilibrium adsorption capacity (mg/g), and C_0 and C_t are the initial concentration and outlet concentration(mg/L) of MB at time t (h), respectively. w is the adsorbent mass (g) and Q is the flow rate (mL/h). The values of k_{Th} and q_0 of the tested absorbents obtained by fitting the Thomas model are shown in Table 3. The calculated q_0 of GAC, SC, and HSC were 22.16, 14.67, and 169.40 mg/g, respectively. They were all consistent with their experimental values. The fact shows that the adsorption breakthrough curves of HSC to MB could be well described by the Thomas model (Zümriye and Ferda, 2004; Theodoridis et al., 2014). It is well known that the Thomas model is suitable for adsorption processes where the external and internal diffusions are absent. Therefore, the fitting result indicated that the adsorption of HSC to MB was not limited by diffusion. The result is consistent with the high surface permeability of hollow-like spherical structures for HSC.

3.2.2 Adsorption of benzene derivatives from aqueous solution

Because a kind of sludge carbon with hierarchical pores ($SC_{CA\text{-}Zn}$) was prepared by coupling template and activation reagent, the resulted sludge carbon ($SC_{CA\text{-}Zn}$) was conducted as an adsorbent for removing benzenes in aqueous solution. Fig. 6 presents

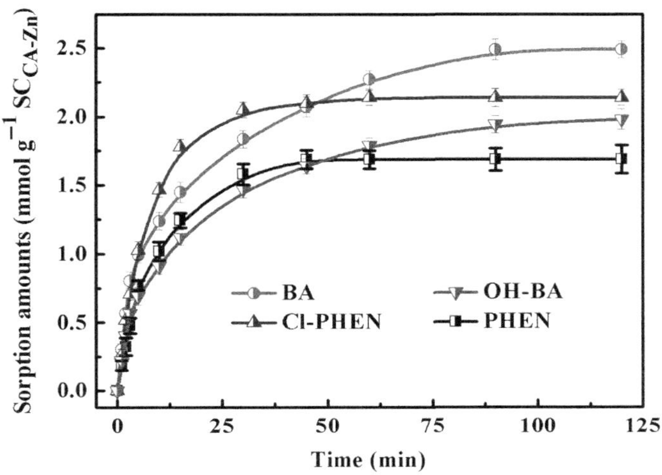

Fig. 6 Sorption kinetics of OH-BA, PHEN, BA, and Cl-PHEN on SC_{CA-Zn}.

the sorption amounts of four benzene derivatives on SC_{CA-Zn} at pH 4 as a function of contact time. Obviously, the sorption equilibrium of PHEN and Cl-PHEN was achieved at the first 60 min, but sorption of BA and OH-BA was continued after 60 min. The discrepancy was ascribed to the different sorption mechanism.

Sorption isotherms of four benzene derivatives on SC_{CA-Zn} at pH 4 and 25°C are investigated in Fig. 7 to understand the mechanism. Apparent two-step stepwise adsorption curves were observed. First, a rapid nonlinear increase in the low

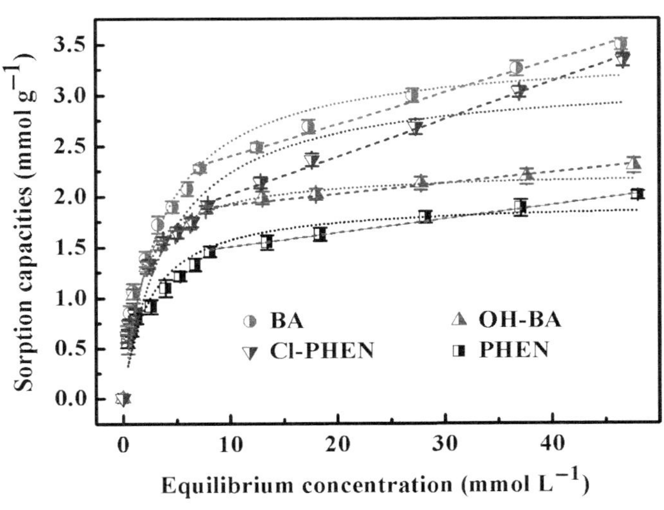

Fig. 7 Sorption isotherms of four benzene derivatives on SC_{CA-Zn} (Kong et al., 2014).

equilibrium concentration range occurred, following by a slow linear increase in the high equilibrium concentration range. The phenomena were also found by other studies (Mohamed et al., 2011; Chen et al., 2012a,b; Yuan et al., 2004; Chen et al., 2008). A dual reaction domain mode could be conducted to describe the sorption performances quantitatively (Chen et al., 2008):

$$Q_T = Q_A + Q_P \tag{5}$$

where Q_T is the total sorption amount (mmol g^{-1}), Q_A is the adsorption amount, and Q_P is the partition sorption amount (mmol g^{-1}). In the range of high solution concentration, the interaction between benzene derivatives and $SC_{CA\text{-}Zn}$ was mainly ascribed to partition sorption because the surface adsorption was preferentially saturated (And and Kile, 1998; Cornelissen and Gustafsson, 2004). In this case, the dual reaction domain model (5) can be transformed to the following equation:

$$Q_T = Q_A + Q_P = Q_A + K_P C_e \tag{6}$$

where Q_A is the saturated adsorption capacity, K_P is the partition sorption coefficient, and C_e is the solute equilibrium concentration. The sorption isotherms in high concentration can be fitted well with Eq. (6) and the regression parameters are presented in Table 4 (Kong et al., 2014).

The fitted results illustrated that partition coefficients (K_P) of the tested benzene derivatives to $SC_{CA\text{-}Zn}$ followed the order of Cl-PHEN > BA >PHEN > OH-BA. The order was consistent with their octanol-water partition coefficients (log K_{ow}), which indicated the hydrophobic. A higher value of log K_{ow} suggested hydrophobic adsorbates. More important, the values of K_P and log K_{ow} were linear related ($K_P = 0.0213 \log K_{ow} - 0.018$, $R^2 = 0.942$), as shown in Fig. 8. The above results indicated that sorption at a high solute concentration range was ascribed to partition adsorption, which was dependent on hydrophobic interactions between these benzene derivatives and the highly carbonized aromatic matrix in the $SC_{CA\text{-}Zn}$.

Except from the partition adsorption, the other factors contributed to sorption should be considered to understand the sorption mechanism. Therefore, it is necessary to investigate the corresponding saturated adsorption (Q_A). According to Eq. (6), the

Table 4 Partition sorption coefficients and saturated adsorption of benzene derivatives on the sludge-derived biochars from aqueous solution

Adsorbate	Log K_{ow}	K_P(L g^{-1})	Q_A (mmol g^{-1})	R^2	[a]Q_T (mmol g^{-1})
OH-BA	1.26	0.010	1.831	0.977	2.90
PHEN	1.46	0.014	1.367	0.988	2.01
BA	1.95	0.030	2.138	0.998	3.49
Cl-PHEN	2.45	0.036	1.672	0.997	3.34

[a]Total sorption amounts at initial concentration of 50 mmol L^{-1}.

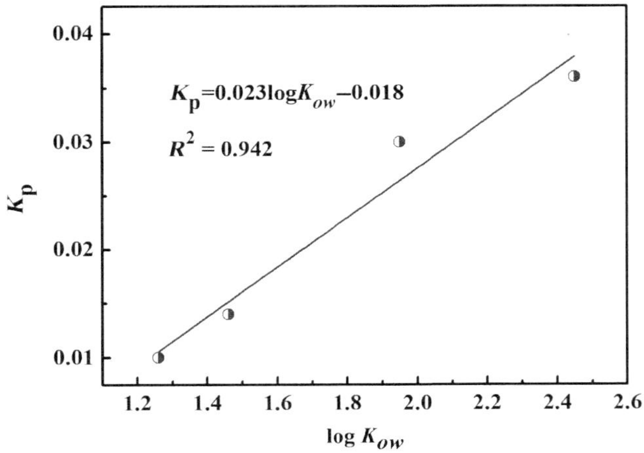

Fig. 8 Liner fitting of partition coefficient K_P plotted as log K_{ow}.

adsorption apart from partition sorption contributed to the whole isotherm can be calculated as Eq. (7):

$$Q_A = Q_T - Q_P = Q_T - K_P Ce \quad (7)$$

The saturated adsorption capacities (Q_A) of the tested benzene derivatives on $SC_{CA\text{-}Zn}$ were calculated from Eq. (7). The nonlinear adsorption isotherms excluding partition sorption can be fitted well with the Langmuir equation, as shown in Fig. 9 and Table 5.

Obviously, the Q_A at plateaus (Q_A^{max}) calculated from the Langmuir equation was nearly similar to the Q_A calculated from Eq. (7). They followed the order of BA > OH-BA > Cl-PHEN > PHEN, and the order was different from the K_P and log K_{ow}. Although pore-filling adsorption was widely known due to its rich pores, the discrepancy of Q_A suggested that another mechanism could contribute to the sorption of these benzene derivatives besides hydrophobic interaction and pore filling. It was interesting to find that the Q_A of the benzene derivatives with carboxyl were greater than that of the benzene derivatives without carboxyl, no matter how many the values of their log K_{ow} were. This fact indicated that the carboxyl groups may play an important role in the adsorption of these benzene derivatives on $SC_{CA\text{-}Zn}$.

Considering the fact that the elemental carbon and highly carbonized aromatic matrix contribute to the hydrophobicity (Chen et al., 2012a,b), thus, the non-hydrophobic interaction was supposed to drive the adsorption by inorganic ash in $SC_{CA\text{-}Zn}$. To confirm the assumption, the inorganic ash of $SC_{CA\text{-}Zn}$ was obtained by burning $SC_{CA\text{-}Zn}$ at atmosphere (650°C), and used to adsorb four derivatives benzenes. The sorption capacities of $SC_{CA\text{-}Zn}$ ash for benzoic acid, p-hydroxybenzoic acid, p-chlorophenol, and phenol were 0.052, 0.058, 0.025, and 0.023 mmol g^{-1}, respectively. The order of adsorption capacities was not in agreement with their log K_{ow} as well. However, the adsorption capacities to benzoic acid and p-hydroxybenzoic acid with the carboxyl group were obviously greater than those

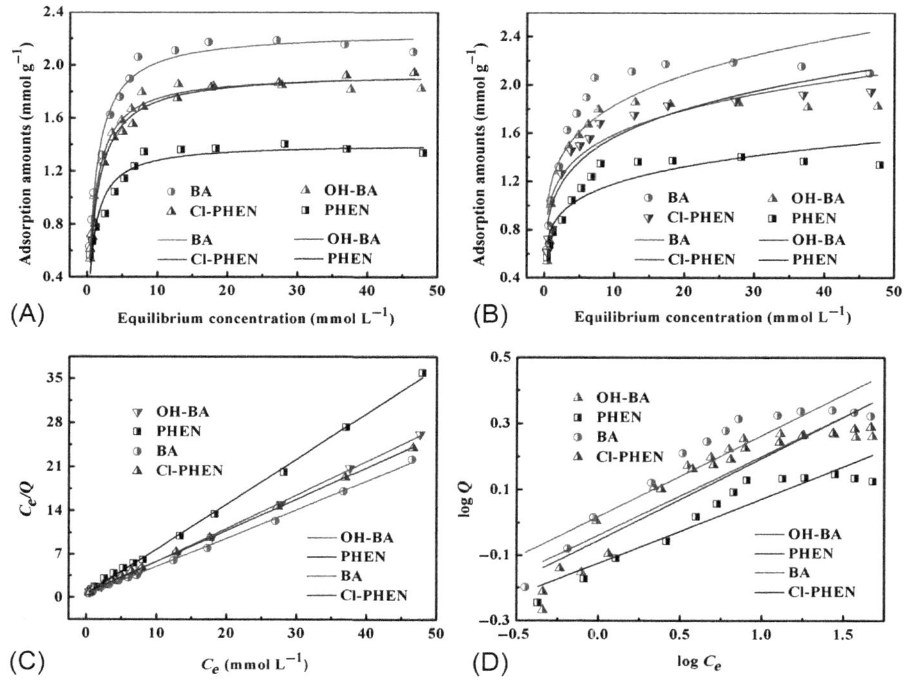

Fig. 9 Sorption isotherms of OH-BA, PHEN, BA, and Cl-PHEN onto $SC_{CA\text{-}Zn}$ except partition sorption: (A) Langmuir model, (B) Freundlich model, (C) linear fitting of Langmuir model, and (D) linear fitting of Freundlich model.

Table 5 Linear regression parameters of sorption isotherms of organic pollutants to $SC_{CA\text{-}Zn}$ from aqueous solution

	Langmuir			Freundlich		
Adsorbates	Q_A^{max}	k	R^2	n	k_f	R^2
OH-BA	1.868	1.777	0.999	4.016	0.882	0.899
PHEN	1.388	0.795	0.999	5.094	0.750	0.885
BA	2.185	0.846	0.999	4.072	1.042	0.861
Cl-PHEN	1.993	1.487	0.999	4.184	0.914	0.787

to phenol and p-chlorophenol without the carboxyl group. The preferable adsorption of carboxyl-containing benzene derivatives was ascribed to the inorganic ash in the $SC_{CA\text{-}Zn}$.

Because the inorganic ash was mainly composed of SiO_2 (82% wt.%), the effect of the amounts of SiO_2 on the sorption isotherms of the sludge char should be investigated. However, the S_{BET} for the adsorbents in various SiO_2 contents was different;

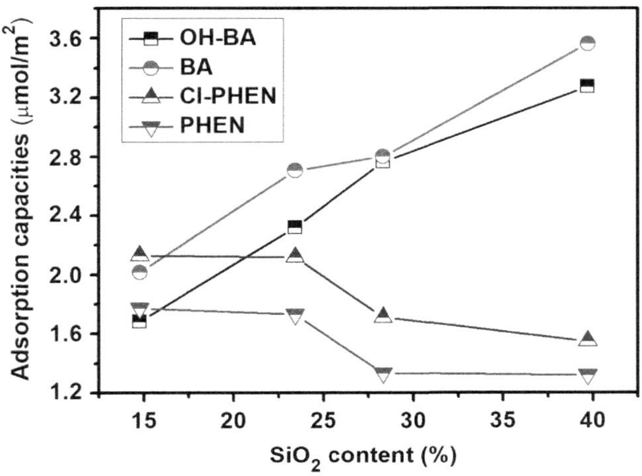

Fig. 10 BET surface area normalized adsorption capacity (Q_{SA}) of the benzene derivatives as a function of SiO_2 content.

the S_{BET} normalized adsorption capacities as a function of SiO_2 content were investigated in Fig. 10 to further understand the adsorption mechanism. The Q_{SA} of benzene derivatives without the carboxyl group changed slightly with SiO_2 content, but the Q_{SA} of benzene derivatives with the carboxyl group increased obviously with SiO_2 content. The positive effect of SiO_2 on the adsorption could be ascribed to the important interaction between carboxyl and SiO_2 in $SC_{CA\text{-}Zn}$.

A quantum chemistry calculation equipped with density functional theory and $DMol^3$ code embedded in the materials studio software was conducted to understand the interaction between carboxyl and SiO_2 in $SC_{CA\text{-}Zn}$ (Giannakas et al., 2013). The benzoic acid as a representative carboxyl-containing pollutant and one of the crystal faces of SiO_2 (100) as an interacting surface were presented in this work to investigate the interaction. Adsorption energy E_b, defined as $E_{(SiO2\ (100)+benzoic\ acid)} - [E_{SiO2\ (100)} + E_{(benzoic\ acid)}]$, was evolved to investigate the interaction, where $E_{(SiO2\ (100)+benzoic\ acid)}$ is the total energy of the benzoic acid-adsorbed surface SiO_2 (100), $E_{SiO2\ (100)}$ is the energy of pristine SiO_2 (100), and $E_{(benzoic\ acid)}$ is the energy of the isolated benzoic acid. The E_b value below 0 indicated exothermic adsorption, leading to stable minimal energy configurations.

Four possible interaction modes were observed between benzoic acid and SiO_2 (100), as shown in Fig. 11. Comparing the possible adsorption configurations, the E_b calculated from mode (a) and (b) is higher than that calculated from mode (c) and (d), indicating that the high E_b favors adsorption. Clearly, the preferential adsorption between the benzoic acid and SiO_2 (100) surface followed the rule: mode (a) and mode (b) > mode (c) and mode (d). A Si—O1 bond (1.83 Å for configuration (a) and 1.87 Å for configuration (b)) and O3-H hydrogen bond (1.61 Å for

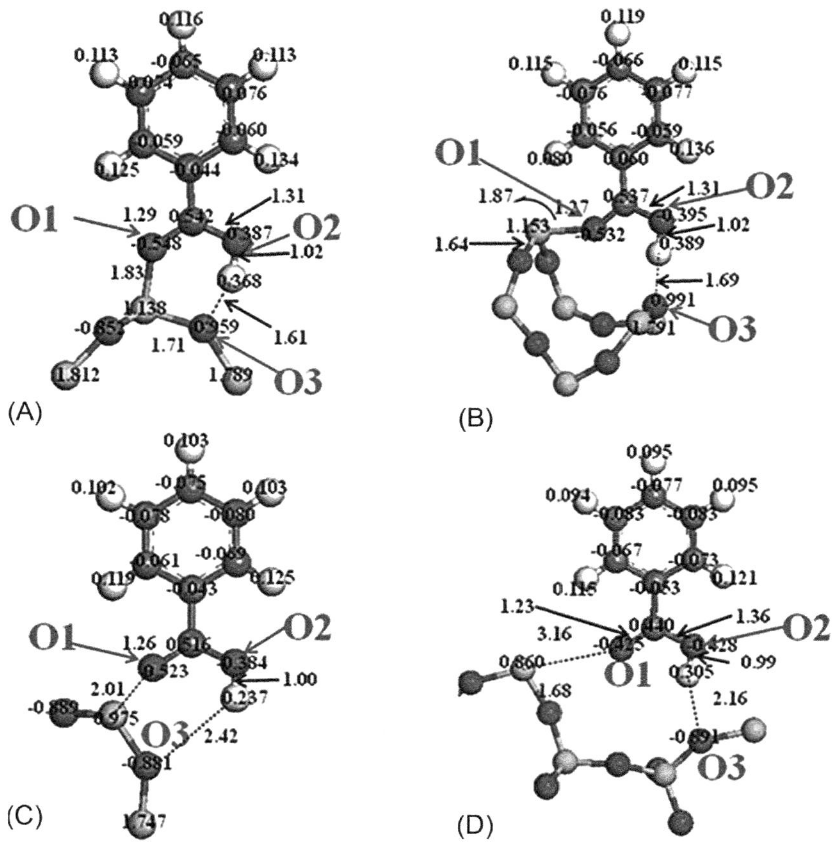

Fig. 11 The charge distributions of adsorption configurations of benzoic acid on SiO_2_quartz_beta (100) surface: Mode (a), Mode (b), Mode (c), Mode (d).

configuration (a) and 1.69 Å for configuration (b)) between the benzoic acid and the surface were formed, corresponding to the elongation of the neighboring C—O1 bond to 1.29 Å for mode (a) and 1.27 Å for mode (b), both from 1.22 Å of origin molecule, and the elongation of H-O2 bond from 0.98 to 1.02 Å for mode (a) and 1.01 Å for mode (b).

The Si atom on the active surface was two-fold coordination, unlike four-fold coordinated Si in the bulk structure. As shown in Fig. 11, the O atom of benzoic acid could attach Si for forming chemical bond Si—O. In addition, it is easy to form a hydrogen bond between the O and H atoms due to the large electronegativity of the O atom. Obviously, the chemical bond was much stronger than the hydrogen bond. Thus, chemisorption and physisorption contributed to the sorption of benzoic acid on the SiO_2 (100) surface; the chemisorption configuration with one Si-O bond and a hydrogen bond formed as mode (a) and mode (b) was favorable.

3.3 Adsorption of volatile organic pollutants

The sludge-derived adsorbent was also evolved for removing volatile organic pollutants. $SC_{CA/Zn}$ was first conducted as an adsorbent to investigate the adsorption breakthrough to toluene performed in dry air (0% relative humidity) at 298 K, in which other SC, SC_{Zn}, SC_{CA}, and ACF were evolved to investigate the adsorption behavior comparatively. The breakthrough time for SC, SC_{Zn}, SC_{CA}, and ACF were 3, 10, 8, and 35 min, respectively, as shown in Fig. 12. Fortunately, the breakthrough time for $SC_{CA/Zn}$ was 42 min, which was much greater than those of other chars. Importantly, it was about 133% longer than the sum of the breakthrough times of SC_{Zn} and SC_{CA}, indicating that $SC_{CA/Zn}$ displayed the greatest toluene adsorption ability among the test adsorbents.

The linear regression analysis of adsorption breakthrough curves by the Yoon-Nelson model $\ln[c/(c_0 - c)]$ *vs.* t was performed (Long et al., 2011) to describe the adsorption breakthrough curves, where c_0 and c were the inlet and outlet concentration (mg L^{-1}), respectively. The adsorption rate constant (k') was calculated from the slope of the $\ln[c/(c_0 - c)]$ versus time (t) plot. They were 0.065, 0.075, 0.131, 0.183, and 0.203 min^{-1} for SC, SC_{Zn}, SC_{CA}, $SC_{CA/Zn}$, and ACF, respectively. As expected, $SC_{CA/Zn}$ possessed the fastest adsorption rate. However, it was unexpected that the k' value of SC_{CA} was greater than that of SC_{Zn}, although SC_{CA} had a smaller S_{BET} and S_{micro} than SC_{Zn}. The unexpected result is possibly ascribed to more macropores of SC_{CA}, and these macropores could provide many inner passageways for toluene transportation, which can speed their adsorption.

The adsorption capacities of SC, SC_{CA} SC_{Zn}, $SC_{CA/Zn}$, and ACF to toluene calculated from their breakthrough curves were 0.25, 0.32, 0.49, 0.83, and 0.62 g g^{-1}, respectively, as shown in Fig. 12. They followed the order: $SC_{CA/Zn}$ > ACF > SC_{Zn} > SC_{CA} > SC. However, the adsorption capacities of $SC_{CA/Zn}$ and ACF are not

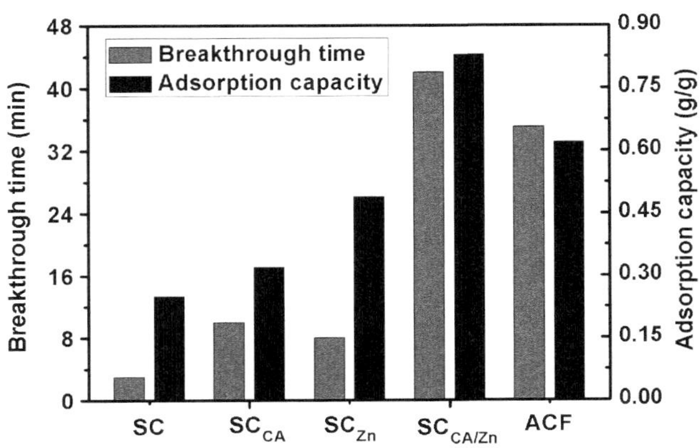

Fig. 12 Comparison of adsorption breakthrough time and adsorption capacities of toluene on sludge derived adsorbents with ACF (initial concentration = 70 mg L^{-1}; flow rate = 200 mL min^{-1}; T = 298 K).

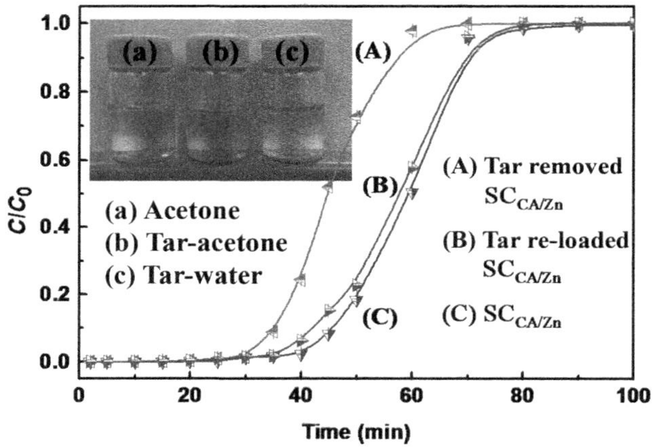

Fig. 13 Adsorption breakthrough curves of the adsorbents for toluene (initial concentration: 70 mg L^{-1}, flow rate: 200 mL min^{-1}, $T = 298$ K).

consistent with the order of their surface areas. The adsorption capacity of the $SC_{CA/Zn}$ was 34% greater than that of the ACF, although the surface area of $SC_{CA/Zn}$ was 867.6 m^2 g^{-1}, which was lower than that of ACF. The divergence denotes that other factors besides pore filling contribute to adsorption.

Compared to ACF, the sludge char was not completely carbonized, and it still contained some solid tar oil (Bernardoa et al., 2012; Keiluweit et al., 2010). First, the char was washed by acetone for 24h to remove the tar oil on the surface of $SC_{CA/Zn}$ (Bernardoa et al., 2012). The breakthrough curve of $SC_{CA/Zn}$ partly removed the tar oil and the pictures of the removed tar and acetone are shown in Fig. 13. The adsorption capacity of the oil-removed $SC_{CA/Zn}$ was 0.52 g g^{-1}. It deceased by 37% compared with the adsorption capacity of $SC_{CA/Zn}$. Interestingly, the adsorption capacity of tar oil-removed $SC_{CA/Zn}$ was recovered to 0.81 g g^{-1} once the oil-removed $SC_{CA/Zn}$ was tar oil reloaded by dipping. The adsorption capacity of the reloaded $SC_{CA/Zn}$ was restored to 96%, confirming that the adsorption of toluene on $SC_{CA/Zn}$ was solid tar oil dependent. Because the tar oil contains large amounts of multiple aromatic hydrocarbons and a few types of aliphatics (Sánchez et al., 2009), the surface of $SC_{CA/Zn}$ is hydrophobic. The surface of $SC_{CA/Zn}$ possessed a strong hydrophobic property, being confirmed by contact angles of 136.5 degrees, being greater than the contact angles of ACF (76.0 degrees), as shown in Fig. 14. The favorable adsorption of toluene on $SC_{CA/Zn}$ is partly dependent on the tar oil, based on the principle of similarity and intermiscibility (Bernardoa et al., 2012; Li et al., 2010a,b; Pan and Xing, 2008; Sánchez et al., 2009). Thus, the adsorption of toluene on $SC_{CA/Zn}$ could be ascribed to not only pore adsorption but also hydrophobic partition adsorption. The solid tar oil not only locates in these hierarchical pores but also covers the non-microporous surfaces.

Besides, the sludge-derived carbonaceous adsorbent was prepared from pyrolyzed sewage sludge, following by acid washing and alkaline hydroxide activation for VOC

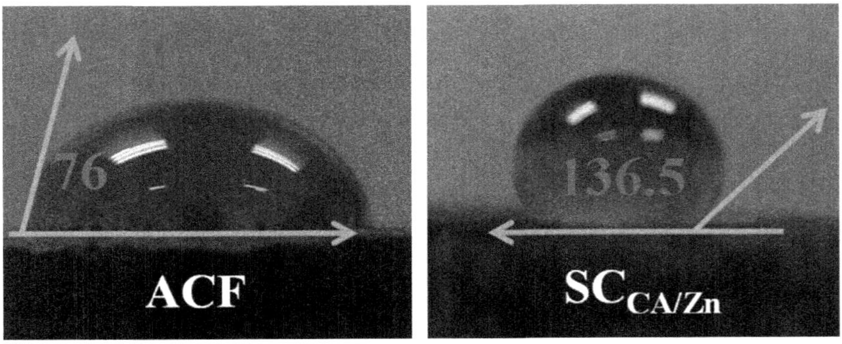

Fig. 14 Water drop shapes on ACF and $SC_{CA/Zn}$.

removal. Air streams with low concentrations (<100 ppm, v/v) of toluene, methyl ethyl ketone, and limonene were used in dynamic adsorption/desorption experiments. Adsorption capacities to toluene, methyl ethyl ketone, and limonene were 350, 220, and 640 mg per g of alkaline-activated sludge, respectively. The adsorption efficiency and rate were favorable due to the presence of mesopores for sludge-based adsorbents. The sludge adsorbent favored the adsorption to polar VOCs such as methyl ethyl ketone. Adsorption of toluene and limonene is controlled by the size of the molecules and their ability to access the inner (micro) pores. Toluene adsorption was found to be dependent on the narrow micropore (<0.7 nm) volumes while limonene adsorption was determined by the total pore volume. The VOC adsorption process is not fully reversible.

4 Conclusion

Sewage sludge could be converted into carbonaceous adsorbents by carbonization and activation. Acid washing was evolved to remove the inorganic compounds, and further to increase the S_{BET}. Physical activation and chemical activation were widely paid more attention. Carbon dioxide and steam were the most widely concerned physical activation reagents. Potassium hydroxide, sodium hydroxide, and zinc chloride were the highly concerned chemical activation reagents. Activation by potassium hydroxide and sodium hydroxide was usually conducted at about 800°C while activation by zinc chloride was carbonized at about 500°C. The resulting sludge-derived adsorbents were considered to be efficient adsorbents for removing organic pollutants from aqueous and atmosphere. Physical adsorption driven by pore filling and chemisorption being ascribed to complexation were the main adsorption mechanisms. The high mineral content in the sludge-derived adsorbent led to the S_{BET}, further limiting the wide application of sludge-derived adsorbents in pollutant decontamination. Therefore, the sludge could be converted into carbonaceous adsorbents in the view of resource recovery by carbonization, but it is a challenge to the wide application of sludge resource utilization as adsorbents.

Acknowledgments

This research was supported by the Nature Science Foundations of China (Nos. 51508116, 21407155), the Nature Science Foundations of Guangdong Province (2016A030310265), the Science and Technology Research Programs of Guangzhou City (201607010311), the Project of Guangdong Provincial Key Laboratory of radioactive contamination control and resources (2017B030314182), and the Hong Kong Scholarship (XJ2016037).

References

Alvarez, J., Lopez, G., Amutio, M., Bilbao, J., Olazar, M., 2016. Preparation of adsorbents from sewage sludge pyrolytic char by carbon dioxide activation. Process Saf. Environ. 103, 76–86.

And, C.T.C., Kile, D.E., 1998. Deviations from sorption linearity on soils of polar and nonpolar organic compounds at low relative concentrations. Environ. Sci. Technol. 32 (3), 338–343.

Bagreev, A., Bashkova, S., Locke, D.C., Bandosz, T.J., 2001. Sewage sludge-derived materials as efficient adsorbents forremoval of hydrogen sulfide. Environ. Sci. Technol. 35 (7), 1537–1543.

Bandosz, T.J., Block, K., 2006. Municipal sludge-industrial sludge composite desulfurization adsorbents: synergy enhancing the catalytic properties. Environ. Sci. Technol. 40 (10), 3378–3383.

Bennett, R.L., Knapp, K.T., 1982. Characterization of particulate emissions from municipal wastewater sludge incinerators. Environ. Sci. Technol. 16 (12), 831–836.

Bernardoa, M., Lapaa, N., Gonçalves, M., Mendes, B., Pinto, F., Fonseca, I., Lopes, H., 2012. Physico-chemical properties of chars obtained in the co-pyrolysis of waste mixtures. J. Hazard. Mater. 219–220, 196–202.

Chen, B., Zhou, D., Zhu, L., 2008. Transitional adsorption and partition of nonpolar and polar aromatic contaminants by biochars of pine needles with different pyrolytic temperatures. Environ. Sci. Technol. 42 (14), 5137–5143.

Chen, X., Jeyaseelan, S., Graham, N., 2002. Physical and chemical properties study of the activated carbon made from sewage sludge. Waste Manag. 22 (7), 755–760.

Chen, Z., Chen, B., Zhou, D., Chen, W., 2012b. Bisolute sorption and thermodynamic behavior of organic pollutants to biomass-derived biochars at two pyrolytic temperatures. Environ. Sci. Technol. 46 (22), 12476–12483.

Chen, Z.M., Chen, B.L., Chiou, C.T., 2012a. Fast and slow rates of naphthalene sorption to biochars produced at different temperatures. Environ. Sci. Technol. 46 (20), 11104–11111.

Cornelissen, G., Gustafsson, O., 2004. Sorption of phenanthrene to environmental black carbon in sediment with and without organic matter and native sorbates. Environ. Sci. Technol. 38 (1), 148–155.

Dong, B., Liu, X., Dai, L., Dai, X., 2013. Changes of heavy metalspeciation during high-solid anaerobic digestion of sewagesludge. Bioresour. Technol. 131, 152–158.

Fitzmorris, K.B., Lima, I.M., Marshall, W.E., Reimers, R.S., 2006. Anion and cation removal from solution using activatedcarbons from municipal sludge and poultry manure. J. Residuals Sci. Technol. 3 (3), 161–167.

Giannakas, A.E., Antonopoulou, M., Deligiannakis, Y., Konstantinou, I., 2013. Preparation, characterization of N–I co-doped TiO_2 and catalytic performance toward simultaneous Cr(VI) reduction and benzoic acid oxidation. Appl. Catal. B Environ. 140–141 (8), 636–645.

Gomez-Pacheco, C.V., Rivera-Utrilla, J., Sánchez-Polo, M., Lopez-Penalver, J.J., 2012. Optimizationof the preparation process of biological sludge adsorbents for application in water treatment. J. Hazard. Mater. 217–218, 76–84.

González, J.F., Román, S., Encinar, J.M., Martínez, G., Gonzalezperez, J.A., Gonzalezvila, F.J., et al., 2009. Pyrolysis of various biomass residues and char utilization for the production of activated carbons. J. Anal. Appl. Pyrolysis 85 (1), 134–141.

Hadi, P., Xu, M., Ning, C., Lin, C.S.K., Mckay, G., 2015. A critical review on preparation, characterization and utilization of sludge-derived activated carbons for wastewater treatment. Chem. Eng. J. 260, 895–906.

Hsu, L.Y., Teng, H., 2000. Influence of different chemical reagents on the preparation of activated carbons from bituminous coal. Fuel Process. Technol. 64 (1–3), 155–166.

Hunsom, M., Autthanit, C., 2013. Adsorptive purification of crude glycerol by sewage sludge-derived activated carbon prepared by chemical activation with H_3PO_4, K_2CO_3 and KOH. Chem. Eng. J. 229 (4), 334–343.

Jindarom, C., Meeyoo, V., Kitiyanan, B., Rirksomboon, T., Rangsunvigit, P., 2007. Surface characterization and dye adsorptive capacities of char obtained from pyrolysis/gasification of sewage sludge. Chem. Eng. J. 133 (1-3), 239–246.

Keiluweit, M., Nico, P.S., Johnson, M.G., Kleber, M., 2010. Dynamic molecular structure of plant biomass-derived black carbon (biochar). Environ. Sci. Technol. 44, 1247–1253.

Khalili, N.R., Campbell, M., Sandi, G., Golaś, J., 2000. Production of micro- and mesoporous activated carbon from paper mill sludge: I effect of zinc chloride activation. Carbon 38 (14), 1905–1915.

Kong, L., Tian, S.H., Luo, R., Wei, L., Tu, Y.T., Xiong, Y., 2013a. Demineralization of sludge-based adsorbent by post-washing for development of porosity and removal of dyes. J. Chem. Technol. Biotechnol. 88 (8), 1473–1480.

Kong, L., Xiong, Y., Sun, L., Tian, S., Xu, X., Zhao, C., et al., 2014. Sorption performance and mechanism of a sludge-derived char as porous carbon-based hybrid adsorbent for benzene derivatives in aqueous solution. J. Hazard. Mater. 274 (12), 205–211.

Kong, L., Xiong, Y., Tian, S., Luo, R., He, C., Huang, H., 2013b. Preparation and characterization of a hierarchical porous char from sewage sludge with superior adsorption capacity for toluene by a new two-step pore-fabricating process. Bioresour. Technol. 146 (146C), 457–462.

Li, S.D., Tian, S.H., Du, C.M., He, C., Cen, C.P., Xiong, Y., 2010a. Vaseline-loaded expanded graphite as a new adsorbent for toluene. Chem. Eng. J. 162, 546–551.

Li, S.D., Tian, S.H., Feng, Y.F., Lei, J., Wang, P., Xiong, Y., 2010b. A comparative investigation on absorption performances of three expanded graphite-based complex materials for toluene. J. Hazard. Mater. 183, 506–511.

Lillo-Ródenas, M.A., Ros, A., Fuente, E., Montes-Morán, M.A., Martin, M.J., Linares-Solano, A., 2008. Further insights into the activation process of sewage sludge-based precursors by alkaline hydroxides. Chem. Eng. J. 142 (2), 168–174.

Long, C., Liu, P., Li, Y., Li, A.M., Zhang, Q.X., 2011. Characterization of hydrophobic hypercrosslinked polymer as an adsorbent for removal of chlorinated volatile organic compounds. Environ. Sci. Technol. 45, 4506–4512.

Maciá-Agulló, J.A., Moore, B.C., Cazorla-Amorós, D., Linares-Solano, A., 2004. Activation of coal tar pitch carbon fibres: physical activation vs. chemical activation. Carbon 42 (7), 1367–1370.

Malik, R., Ramteke, D.S., Wate, S.R., 2007. Adsorption of malachite green on groundnut shell waste based powdered activated carbon. Waste Manag. 27 (9), 1129–1138.

Martin, M.J., Artola, A., Balaguer, M.D., Rigola, M., 2003. Activated carbons developed from surplus sewage sludge for the removal of dyes from dilute aqueous solutions. Chem. Eng. J. 94 (3), 231–239.

Mohamed, E.F., Andriantsiferana, C., Wilhelm, A.M., Delmas, H., 2011. Competitive adsorption of phenolic compounds from aqueous solution using sludge-based activated carbon. Environ. Technol. 32 (12), 1325–1336.

Monsalvo, V.M., Mohedano, A.F., Rodriguez, J.J., 2011. Activated carbons from sewage sludge: application to aqueous-phase adsorption of 4-chlorophenol. Desalination 277 (1), 377–382.

Pan, B., Xing, B.S., 2008. Adsorption mechanisms of organic chemicals on carbon nanotubes. Environ. Sci. Technol. 24, 9005–9013.

Pan, Z.H., Tian, J.Y., Xu, G.R., Li, J.J., Li, G.B., 2011. Characteristics of adsorbents made from biological, chemical and hybrid sludges and their effect on organics removal in wastewater treatment. Water Res. 45 (2), 819–827.

Rio, S., Faurbrasquet, C., Le, C.L., Courcoux, P., Le, C.P., 2005. Experimental design methodology for the preparation of carbonaceous sorbents from sewage sludge by chemical activation—application to air and water treatments. Chemosphere 58 (4), 423–437.

Rio, S., Faur-Brasquet, C., Le Coq, L., Le Cloirec, P., 2003. Structure characterization of porous carbonaceous materials from sewage sludge. In: Proceedings of Carbon'03, Oviedo.

Rodríguez, A., García, J., Ovejero, G., Mestanza, M., 2009. Adsorption of anionic and cationic dyes on activated carbon from aqueous solutions: equilibrium and kinetics. J. Hazard. Mater. 172 (2), 1311–1320.

Ros, A., Lilloródenas, M.A., Canalsbatlle, C., Fuente, E., Montesmorán, M.A., Martin, M.J., et al., 2007. A new generation of sludge-based adsorbents for h2s abatement at room temperature. Environ. Sci. Technol. 41 (12), 4375–4381.

Ros, A., Lillo-Ródenas, M.A., Fuente, E., Montes-Morán, M.A., Martín, M.J., Linares-Solano, A., 2006. High surface area materials prepared from sewage sludge-based precursors. Chemosphere 65 (1), 132–140.

Rozada, F., Otero, M., Moran, A., Garcia, A., 2008. Adsorption of heavy metals onto sewage sludge-derived materials. Bioresour. Technol. 99 (14), 6332–6338.

Sánchez, M.E., Menéndez, J.A., Domínguez, A., Pis, J.J., Martínez, O., Calvo, L.F., 2009. Effect of pyrolysis temperature on the composition of the oils obtained from sewage sludge. Biomass Bioenergy 33, 933–940.

Smith, K.M., Fowler, G.D., Pullket, S., Graham, N.J.D., 2009. Sewage sludge-based adsorbents: a review of their production, properties and use in water treatment applications. Water Res. 43 (10), 2569–2594.

Tay, J.H., Chen, X.G., Jeyaseelan, S., Graham, N., 2001. A comparative study of anaerobically digested and undigested sewage sludges in preparation of activated carbons. Chemosphere 44 (1), 53–57.

Theodoridis, A., Giescke, W.H., Du, T.I., 2014. Removal of Pb(II) ions from aqueous solution using water hyacinth root by fixed-bed column and ANN modeling. J. Hazard. Mater. 273 (3), 94–103.

Wang, J., Liu, T.L., Huang, Q.X., Ma, Z.Y., Chi, Y., Yan, J.H., 2017. Production and characterization of high quality activated carbon from oily sludge. Fuel Process. Technol. 162, 13–19.

Wu, Z., Xiong, Y., Guan, G., Kong, L., Tian, S., 2014. Preparation and adsorption behavior of new hollow-like spherical sludge chars for methylene blue. RSC Adv. 4 (98), 55256–55262.

Xu, G., Yang, X., Spinosa, L., 2015. Development of sludge-based adsorbents: preparation, characterization, utilization and its feasibility assessment. J. Environ. Manag. 151, 221–232.
Yuan, C., Sheng, G., Chiou, C.T., Xing, B., 2004. Compositions and sorptive properties of crop residue-derived chars. Environ. Sci. Technol. 38 (17), 4649–4655.
Zhang, F.S., Nriagu, J.O., Itoh, H., 2005. Mercury removal from water using activated carbons derived from organic sewage sludge. Water Res. 39 (2–3), 389–395.
Zümriye, A., Ferda, G., 2004. Biosorption of phenol by immobilized activated sludge in a continuous packed bed: prediction of breakthrough curves. Process Biochem. 39 (5), 599–613.

Further reading

Linares-Solano, A., 2006. High surface areamaterials prepared from sewage sludge-based precursors. Chemosphere 65 (1), 132–140.
Ros, A., Lillo-Rodenas, M.A., Fuente, E., Montes-Moran, M.A., Martin, M.J., Smith, K.M., Fowler, G.D., Pullket, S., Graham, N.J.D., 2009. Sewage sludge-based adsorbents: a review of their production, properties and use in water treatment applications. Water Res. 43 (10), 2569–2594.

Stabilization of cadmium in industrial sludge—Generation of crystalline products

22

Minhua Su, Lingjun Kong*,‡, Changzhong Liao†,‡, Diyun Chen*, Kaimin Shih‡*
*Guangdong Provincial Key Laboratory of Radionuclides Pollution Control and Resources, School of Environmental Science and Engineering, Guangzhou University, Guangzhou, China, †Guangdong Key Laboratory of Integrated Agro-Environmental Pollution Control and Management, Guangdong Institute of Eco-Environmental Science & Technology, Guangzhou, China, ‡Department of Civil Engineering, The University of Hong Kong, Hong Kong SAR, China

1 Introduction

In many parts of the world, large quantities of industrial wastewater containing a considerable amount of toxic metal ions and metal compounds are increasingly being produced through numerous industrial activities, such as coal and ore mining, electroplating, tanning, smelting, and metal refining (Allioux et al., 2017; Nair et al., 2008). These intensive industrial activities have resulted in the increased release of toxic heavy metals, such as cadmium (Cd), chromium (Cr), copper (Cu), lead (Pb), zinc (Zn), nickel (Ni), and mercury (Hg), into the environment (Ahmaruzzaman, 2011; Allioux et al., 2017). Unlike organic pollutants, heavy metals are nonbiodegradable (Badruddoza et al., 2013). Therefore, they are very stable and persist in nature. Because of their toxicity, abundance, and persistence, water and soil contamination by heavy metals has become a worldwide environmental concern (Järup, 2003). Heavy metals and their impacts on human health are reviewed regularly by some international organizations, such as the World Health Organization. After accumulating in living organisms in the human body via food chains, heavy metals may cause various fatal diseases, such kidney damage and cancers (Balsamo et al., 2011; Järup, 2003). Inappropriate metal-laden sludge management may lead to the release of high levels of heavy metals, adversely affecting both public health and the environment through atmospheric, soil, and water pollution. Therefore, waste streams containing heavy metals must be decontaminated or stabilized before being discharged into the environment (Islam et al., 2017).

1.1 Hazards of industrial sludge

Globally, wastewater treatment plants generate millions of tons of residual sludge every year; this quantity is expected to increase in the future because of improved sanitation and increasing numbers of domestic and industrial wastewater treatment plants

Industrial and Municipal Sludge. https://doi.org/10.1016/B978-0-12-815907-1.00022-2

(Krüger et al., 2014). Sludge, the pollutant sink of wastewater treatment, is the residue generated from various wastewater treatment processes; its disposal is becoming a critical concern (Edwards et al., 2017; Zhang et al., 2017). Industrial sludge—a residue from industrial wastewater treatment, generally containing undesirable levels of heavy metals or priority chemicals—is considered hazardous waste. Moreover, in many developing countries, industrial and municipal wastewaters are commonly mixed and treated together in wastewater treatment plants, thus increasing heavy metal accumulation in the sludge (Chen et al., 2008; Wang et al., 2015). A high level of heavy metal residue in the sludge is a major obstacle during sludge disposal or utilization. When these heavy metals are released into the soil or water bodies, secondary environmental pollution occurs, posing serious risks to the environment and human health (Fig. 1). Furthermore, the introduction of excessive levels of heavy metals into the soil through sludge can lead to increased heavy metal uptake by plants, resulting in damage to plants and significant impact on human health through the consumption of crops grown on contaminated soil. Metal-laden industrial sludge disposal is difficult because heavy metals are nonbiodegradable and can bioaccumulate in living tissues via food chains (Wang et al., 2015; Wu et al., 2015). Therefore, contamination of the environment by heavy metals from untreated industrial sludge is a serious environmental problem.

A commonly used approach for sludge disposal (particularly sewage sludge) is land application involving recycling the valuable components in the sludge. However, industrial sludge produced by battery factories, metal-plating facilities, and tanneries generally contains high levels of heavy metals. In particular, several metals, such as cadmium (Cd), chromium (Cr), copper (Cu), lead (Pb), and mercury (Hg), are harmful and toxic to the environment and humans (Wu et al., 2015). Thus, reducing the availability of heavy metals in sludge is a major concern in sustainable industrial sludge management (Nair et al., 2008; Wu et al., 2015). Furthermore, with stricter legislation and environmental awareness, the importance of implementing waste minimization and management solutions has increased considerably (Allioux et al., 2017). Therefore, effective removal and regeneration of heavy metals from sludge is quite imperative (Wu et al., 2015).

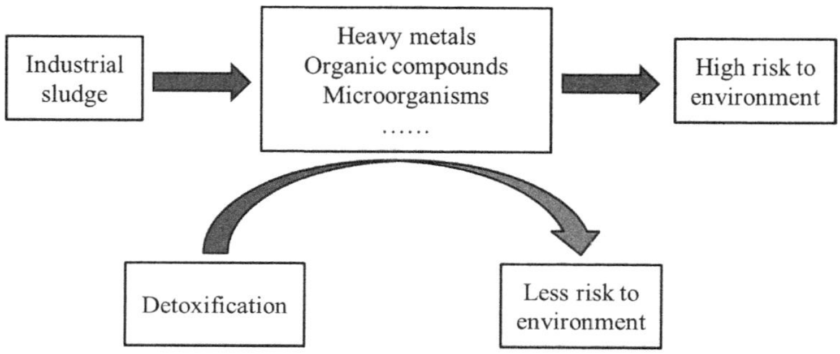

Fig. 1 Potential risks of untreated and treated industrial sludge to the environment.

1.2 Cadmium and cadmium pollution

Industries producing large amounts of waste are increasingly concerned about problems pertaining to the environment (Su et al., 2018). The waste sludge generated during various industrial activities often contains toxins, such as heavy metals (e.g., Cd, Ni, and Hg), at large concentrations (Ahmaruzzaman, 2011; Jabłońska and Siedlecka, 2015; Sun et al., 2016; Hua et al., 2012). Wu et al. (2015) indicated the presence of large amounts of cadmium and nickel in the industrial sludge they examined. Rivoallan et al. (1994) studied the feasibility of eliminating and recycling toxic metals from two industrial sludge types; of these, one type contained approximately 16% Cd and 25% Ni. As a priority environmental pollutant, cadmium has received more attention because of its high mobility and biological enrichment (Bhatnagar and Minocha, 2009). Cadmium, a nonessential trace metal, is highly toxic and carcinogenic; it can occur as a food contaminant and global pollutant (Khan et al., 2017; Liu et al., 2017). Long-term occupational exposure to Cd may contribute to the development of lung cancer, and high Cd exposure may lead to kidney and bone damage. Since the appearance of the *Itai-Itai* disease in Jinzu Valley, Japan, Cd in the environment has been an object of significant societal concern (Liu et al., 2017). Furthermore, for their high toxicity and long retention time in organisms after bioaccumulation, Cd and its compounds have been classified as Category I carcinogens by the International Agency for Research on Cancer and Group-B1 carcinogens by the US Environmental Protection Agency (Järup, 2003; NTP, 2014; USEPA, 1999). According to the British Geological Survey, the total worldwide Cd production in 2015 was approximately 24,900 metric tons (Khan et al., 2017). In 2016, the worldwide production of Cd, excluding that in the United States, was 23,000 metric tons (USGS, 2016). Cadmium is widely used for manufacturing several products, including Ni-Cd batteries, solders, pigments, coatings, alloys, polyvinyl chloride plastics, solar cells, and stabilizers (Khan et al., 2017; USGS, 2016). A large amount of cadmium (approximately 67%) is used for producing the Cd electroplates in nickel-cadmium batteries (Khan et al., 2017). In 2009 in China, the total estimated emission of Cd was 743 metric tons, of which 57% was from industrial processes (Cheng et al., 2014). During the manufacturing processes of Cd-containing products, high levels of cadmium are often introduced into aqueous media, which finally accumulate in the sludge as hydroxides through precipitation (Hossain and Mukherjee, 2012). Uncontrolled and improper waste disposal practices have significantly increased environmental Cd levels. Without treatment, such hazardous sludge would leach cadmium, posing great environmental threat. Currently, under the pressure of severe legislation, industries must explore a reliable and effective method for treating these hazardous wastes before disposal. The development of an effective and reliable technique for environmentally friendly and safe disposal of cadmium-laden sludge, which is also technically feasible and cost-effective, is often challenging.

1.3 Solidification/stabilization

Despite being two distinct techniques, solidification and stabilization are frequently mentioned together as solidification/stabilization (S/S). S/S is widely used to control the migration and exposure of toxins from contaminated soil, sludge, or sediment.

The S/S process was first introduced for road construction and has since been used for similar purposes, including harbor protection (Barth, 1990). The application of S/S to waste treatment began in the 1950s as a process for transforming liquid low-level radioactive waste into solid forms by using urea formaldehyde and asphalt systems to facilitate their easy transportation and better disposal (Conner, 1990). However, practices for the managing hazardous wastes were generally mandated after the 1970s, and S/S techniques were regulated in the 1980s as amendments to the Solid Waste Disposal Act, the Resource Conservation and Recovery Act, and the Hazardous and Solid Waste Amendments (Barth, 1990).

During the S/S process, waste material is generally converted to solid forms through interactions with cementitious materials. A few chemical reactions (e.g., sorption or cementation mechanisms) are usually involved; these reduce the mobility or solubility of the hazardous components in the wastes. Through the combination of solidification and stabilization processes, the contaminated waste materials can be completely mixed with cementitious materials, resulting in the physical and chemical immobilization of the hazardous components. The S/S technique is a nondestructive approach to eradicate or inhibit the mobility of contaminants in waste materials.

The ultimate goal of the S/S approach is to completely convert toxic waste components into their nontoxic forms; however, this process may not reliably prevent metals from leaching into acidic environments. In addition, the need for additives in the S/S process increases the treatment cost and product volume. Thus, the formation of new compounds through chemical transformation may be effective in treating hazardous waste (Hasegawa et al., 2016).

1.4 Thermal stabilization

A practicable and effective approach for controlling and utilizing metal-bearing wastes is urgently needed. A widely adopted approach to eliminate the hazards of wastes bearing heavy metals involves thermally converting them into nonhazardous products (Fig. 2). With this process, the metals in the waste can be transformed into crystal structures through thermal reactions by using low-cost and attainable ceramic raw materials, such as alumina, silica, and kaolinite (Shih et al., 2006; Tang et al., 2010). The metals in the wastes can be successfully converted into extremely stable crystalline products, thus virtually eliminating the leaching of metals into acidic

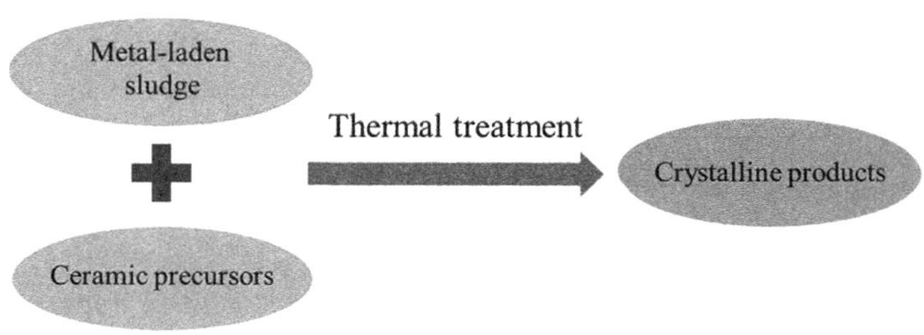

Fig. 2 Formation of crystalline products through thermal treatment.

environments. The strength and effectiveness of this thermal stabilization technique has been verified through stringent leaching tests, such as the toxicity characteristic leaching procedure (TCLP) and the constant pH leaching test (CPLT) (Shih et al., 2006; Su et al., 2015). During thermal treatment, most cadmium in the sludge converts to CdO, which is highly soluble in acid; therefore, a more acid-resistant cadmium-hosting phase is required. Both aluminum and iron oxides are common industrial materials used for manufacturing ceramics and other construction products; moreover, they are abundant in nature and locally available (Cornell and Schwertmann, 2006; Wefers and Misra, 1987). A few equilibrium diagrams have reported that CdO can thermally react with γ-Al_2O_3 and α-Fe_2O_3 through a solid-state reaction to form aluminate (Colin, 1968) and ferrite (Chinnasamy et al., 2001a,b; Mahmoud et al., 2003), respectively. However, to use a reaction mechanism in a feasible treatment technique, the processing parameters of effectively incorporating cadmium into crystalline products by using aluminum and iron oxides require investigation. Furthermore, the robustness of the products to acids should be examined with leaching tests for evaluating the metal stabilization effects (Su et al., 2015).

1.5 Evaluation of metal stabilization effect with leaching test

Environmental contamination caused by heavy metal leaching from hazardous waste is a major issue. After thermal stabilization, the chemical durability of products should be examined through leaching experiments to verify their metal stabilization effect. Leaching is a complex reaction because many factors can affect the release rate and behavior of the constituents from the waste under acidic or alkaline conditions over a period (Van der Sloot et al., 1996). The main factors are waste composition, pH, redox potential, complexation, liquid-to-solid ratio, and contact time (Van der Sloot et al., 1996).

ASTM D 4874 and the Dutch column test (NEN 7343) are the two major standards for column leaching systems for solid wastes (ASTM-D-4874, 2001; NEN-7343, 1995; Pendowski, 2003). However, these systems are considered for nonconstant pH leaching tests (Jackson et al., 1984). When the leaching fluid passes through the column to leach the sample, the pH cannot be monitored in time. Therefore, the column leaching systems cannot maintain a constant pH for leaching solid samples. The CPLT employs continuous stirring to maintain the homogeneity of the mixture, and the measured pH value can better reflect the actual pH between the solution and the solid surface at any given time. Similarly, the pH can be adjusted to provide sufficient protons for surface reactions immediately.

The CPLT is useful for assessing the leachability of products because it maintains pH at a steady level, unlike most of the other leaching tests (e.g., TCLP and the column leaching test) (Al-Abed et al., 2007; Jackson et al., 1984; Islam et al., 2004). Moreover, by using the CPLT, the leaching behaviors of various solids can be compared. During the CPLT, no buffer solution is added, thus avoiding any possible heavy metal complexation, which can abnormally increase leached metal concentration (Cappuyns and Swennen, 2008).

1.6 Potential precursors for cadmium stabilization

A few oxides (e.g., SiO_2, γ-Al_2O_3, and α-Fe_2O_3) can thermally react with CdO to produce different Cd-hosting crystalline product phases. For the reaction between CdO and SiO_2, an equilibrium diagram of the Cd-B-Si-O system has been reported (Lonsdale and Whitaker, 1978). This phase diagram illustrates that three types of cadmium silicates are generated after the thermal treatment of CdO and SiO_2 mixtures at various Cd/Si molar ratios. Furthermore, the phase diagram of the Cd-Al-O system indicates that cadmium aluminates (e.g., $CdAl_2O_4$, $CdAl_4O_7$, and $CdAl_{12}O_{19}$) may be formed when sintering CdO with alumina, which may aid in stabilizing cadmium in aluminas. According to Kurihara and Suib (1993), $CdAl_2O_4$ can be obtained through a two-step synthetic process involving sol-gel and sintering. $CdAl_4O_7$ can be prepared through conventional solid-state reactions by sintering CdO and alumina mixtures (Colin, 1968). Although little is known about the formation of $CdAl_{12}O_{19}$, Colin (1968) stated that $CdAl_{12}O_{19}$ may be formed through analogy with products from the reactions between β-Al_2O_3 and oxides (e.g., CaO, BaO, and PbO) under thermal conditions. Furthermore, $CdFe_2O_4$ can be formed in the Cd-Fe-O system by sintering CdO with α-Fe_2O_3 (Bashkirov and Kornilova, 1980). Al- and Si-rich materials—essential construction and ceramics materials—have a high potential for reuse. In addition, iron (Fe) is a main impurity widely existing in such materials. Therefore, the use of these naturally occurring materials as precursors for metal stabilization is of great interest.

1.6.1 SiO_2

Silicon (Si) is the second most abundant element (after oxygen) in Earth's crust (Chan, 1989; McLaughlin et al., 1997). Silica (SiO_2), the oxide form of Si, has many scientific and engineering uses because it is chemically inert, thermodynamically stable, nontoxic, and inexpensive (Nandiyanto et al., 2009). Silica has two forms: crystalline and noncrystalline (amorphous). Si and O atoms are arranged in a geometric structure in crystalline silica (e.g., quartz, cristobalite, and tridymite); by contrast, Si and O atoms demonstrate no spatial ordering in amorphous silica (e.g., naturally occurring silica, silica formed under uncontrolled conditions, and synthetic silica) (Merget et al., 2002). Silica fume with very fine amorphous silica particles is a byproduct of the manufacture of silicon and ferrosilicon (Rodella et al., 2017). Because of its large specific surface area and high amorphous silica content, silica fume can fix heavy metals (Huang and Huang, 2008; Khan and Siddique, 2011; Li et al., 2014). Notably, numerous forms of amorphous and crystalline silica are commonly present in incinerated sewage sludge ash (Nowak et al., 2013). Most silicates have good physical and chemical stability and thus have numerous industrial applications (Qu et al., 2009). Silicates can be prepared using various methods, including sol-gel, sintering, and solid-state diffusion methods. A few studies have indicated that Cd can react with silicates through solid-state reactions and form various silicate products; thus, the use of silicates to stabilize Cd may also be a promising strategy (Glasser, 1965; Lonsdale and Whitaker, 1978).

1.6.2 Al_2O_3

Aluminum (Al) is the third most abundant element in Earth's crust, accounting for 8 wt.% of its solid surface (Glasser, 1965; Lonsdale and Whitaker, 1978; Zuckerman et al., 2011). Alumina (Al_2O_3) is the oxide form of Al. The common forms of alumina are α, γ, η, δ, θ, κ, and χ forms. α-Al_2O_3 is the most stable phase, and the other forms can be obtained by the thermal decomposition of aluminum hydroxides or oxyhydroxides (Wolverton and Hass, 2000). Aluminas are widely used as raw materials in the manufacture of various functional and conventional ceramic products. γ-Al_2O_3 is the most common oxide form of Al in nature and has great application potential in many fields because of its unique crystal structure, large surface area, highly reactive nature, low cost, and attainability (Wang and Lu, 1999; Zhang et al., 2002). Although γ-Al_2O_3 has a cubic structure, its crystal structure is often defective, which may promote the incorporation of metals (Gutiérrez et al., 2001; Paglia et al., 2003, 2004).

1.6.3 Fe_2O_3 and Fe_3O_4

Fe is one of the most abundant elements on Earth, and it is common in the raw materials used for ceramic manufacturing (Schwertmann and Cornell, 2008; Shih et al., 2006). Several iron oxides exist, including hematite (α-Fe_2O_3), magnetite (Fe_3O_4), maghemite (γ-Fe_2O_3), β-Fe_2O_3, ε-Fe_2O_3, and wustite (FeO). Both β-Fe_2O_3 and ε-Fe_2O_3 are uncommon in nature (Cornell and Schwertmann, 2006). Hematite and magnetite are the two most widespread iron oxides in nature (Cornell and Schwertmann, 2006; Schwertmann and Cornell, 2008). Hematite has a corundum structure, and it is the most thermodynamically stable of all iron oxides. Because of the close-packed arrangement of Fe and O atoms, hematite has no charge excess or deficit. Magnetite, an inverse spinel, can be easily oxidized to maghemite and finally to hematite (Cornell and Schwertmann, 2006; Schwertmann and Cornell, 2008; Sharma et al., 2013). Few studies have revealed the effects of different iron oxides on $CdFe_2O_4$ formation. Hence, investigating the reactions of different iron oxides with Cd to form ceramic products is of great interest.

1.6.4 Incineration sewage sludge ash

Numerous sewage sludges are generated from sewage treatment plants nowadays (Xu et al., 2010; Tyagi and Lo, 2013). The disposal of inadequately treated sludge poses hazards to the environment because most sludge contains various heavy metals, pathogens, and organic and inorganic compounds (Han et al., 2012). However, sludge can be used to make ceramic products. Therefore, the transformation of sludge into useful products provides a potential solution to the sludge disposal problem (Shih et al., 2006; Tang et al., 2011a,b; Zou et al., 2009). To avoid the use of massive amounts of clay and develop a waste-to-resource strategy, ceramic products can be fabricated using ashes from sewage sludge. The exploration of other potential materials to replace clay in the manufacture of ceramics should be encouraged.

The use of sewage sludge as a component for ceramic production is significant from the viewpoint of waste recycling and reuse, waste volume reduction, and organic pollutant and pathogen removal (Hartman et al., 2005; Xu et al., 2010; Tyagi and Lo, 2013). Such conversion can transform waste sludge into low-cost raw materials for construction. The treated industrial sludge is usually rich in Al/Fe-containing compounds because Al/Fe salts are commonly used to enhance the removal efficiency of small particles from wastewater through coagulation or flocculation. Such sludge can be sintered at approximately 1000°C to produce ceramic products and stabilize heavy metals (e.g., Zn, Ni, Cu, Cr, Cd, and Pb) (Xu et al., 2009; Tang et al., 2011a,b).

In this chapter, we developed and employed an environmentally friendly and safe strategy to detoxify and stabilize simulated cadmium-laden sludge. We quantified the cadmium incorporation efficiency through the quantitative X-ray diffraction (XRD) technique, resulting in various crystallized cadmium-hosting product phases with desired acid-resistant capability. The leaching behavior of these products was compared and discussed by performing the CPLT to further assess the benefits of adopting this strategy.

2 Materials and methods

2.1 Chemicals and sample preparation

Gamma-alumina (γ-Al_2O_3), silica fume (amorphous SiO_2 type), hematite (α-Fe_2O_3), and magnetite (Fe_3O_4) were used as the precursors for cadmium incorporation. All chemicals used in this study were of reagent grade. Cadmium oxide powder, a cubic phase, was used as the cadmium source, representing its major form in thermally treated industrial sludge. The CdO powder was purchased from Fisher Scientific (United States). The alumina powder was obtained from Sasol. γ-Al_2O_3 was prepared by sintering the as-received PURAL SB alumina powder, which was identified as boehmite (AlOOH), at 650°C for 3 h. Hematite (α-Fe_2O_3) and magnetite (Fe_3O_4), used as Fe-rich precursors, were purchased from Sigma-Aldrich (United States) and Beijing HWRK Chem Co., Ltd. (China), respectively. Fresh dewatered sewage sludge samples were collected from a chemically enhanced primary sewage treatment works (CEPSTW). The dewatered sewage sludge samples were sintered at 900°C for 30 min to effectively remove all organic compounds from the sludge and generate incineration sewage sludge ash (ISSA). ISSA was characterized through X-ray fluorescence (XRF) and XRD to determine its elemental composition and mineral phases. To immobilize Cd, CdO was alternatively mixed with silica fume, γ-Al_2O_3, α-Fe_2O_3, and Fe_3O_4 at various Cd/Si, Cd/Al, and Cd/Fe molar ratios, as applicable. The mixtures were entirely blended through wet milling and then dried in a vacuum oven at 105°C for 24 h. After drying, the mixtures were ground in a mortar for further homogenization. The homogenized mixtures were then pressed into Φ 20-mm pellets at 250 MPa, followed by a thermal treatment scheme at temperatures of 600–1000°C with a 3 h retention time.

2.2 Product microstructure characterization and XRD analysis

The fired pellets were air cooled to room temperature. The as-obtained pellets were ground into powder for XRD analysis. XRD patterns were obtained using a diffractometer (Bruker D8 Advance) equipped with a LynxEye detector. The 2θ range was from 10 to 80 degrees with Cu $K\alpha_{1,2}$ radiation (1.54 Å) generated at 40 kV and 40 mA. The scanning step size was 0.02 degrees, and the scanning speed was 0.5 s/step. To monitor the phase transformation, the collected XRD patterns were indexed with those derived from the standard powder diffraction database of the International Center for Diffraction Data (ICDD PDF-2 Release 2008).

2.3 CPLT

A CPLT was designed to assess treatment effectiveness. Single phases of Cd-hosting products (i.e., $CdSiO_3$, Cd_2SiO_4, Cd_3SiO_5, $CdAl_4O_7$, and $CdFe_2O_4$) compared with CdO were leached using the CPLT under acidic conditions. A nitric acid (HNO_3; BDH Chemicals) aqueous solution (pH 4.0) was used as the initial leachant. The pH, continuously monitored using a pH meter, was maintained at 4.0 ± 0.2 by adding a few drops of 1 M HNO_3 aqueous solution (less than 20 μL for each adjustment). The CPLT was conducted in a jar filled with 500 mL of leachant and 0.5 g of the powders, with mechanical stirring (200 rpm). For each test, 13 aliquots with 5 mL of leachate together with suspended solids were taken over 0–120 min. The extracts were filtered through 0.2-μm syringe filters and stored in sealed plastic tubes. The leached metal concentrations in the CPLT leachate were analyzed to correlate them with the mobility results with the metal stabilization effect. The metal concentrations were analyzed using an inductively coupled plasma optical emission spectrometer (Perkin Elmer 800). On sampling, three independent sets of Cd, Si, Al, and Fe surrogate standards were run, providing a satisfactory calibration curve in the detection range of 1–2000 ppb. All aforementioned experiments were repeated in triplicate for each sample to ensure the reproducibility of the instrument portion as well as the CPLT itself and to justify its use for investigating the metal stabilization effect.

3 Results and discussion

3.1 Formation of cadmium silicates

Fig. 3 illustrates the XRD patterns of the sintered samples of CdO + amorphous SiO_2 at 900°C for 3 h. The Cd/Si molar ratios of the mixture were 1/1, 2/1, and 3/1. Two types of Cd-hosting product phases, namely orthorhombic Cd_2SiO_4 (space group *Fddd*, no. 70) and tetragonal Cd_3SiO_5 (space group *P4/nmm*, no. 129), were formed. Fig. 4 shows the XRD patterns of the sintered sample of CdO + amorphous SiO_2 at 950°C for 3 h. When the Cd/Si molar ratio was 1.0, in addition to the diffraction signals of orthorhombic Cd_2SiO_4, the diffraction signals of monoclinic $CdSiO_3$ (space group *P21/c*, no. 14) were noted in the XRD pattern. In this system, Cd_2SiO_4 was the predominant Cd-hosting product phase. When the Cd/Si molar ratio was 2.0, Cd_2SiO_4 and Cd_3SiO_5 were the predominant products; this phenomenon also appeared in the system of CdO + amorphous SiO_2, with the Cd/Si molar ratio of 3/1. No diffraction

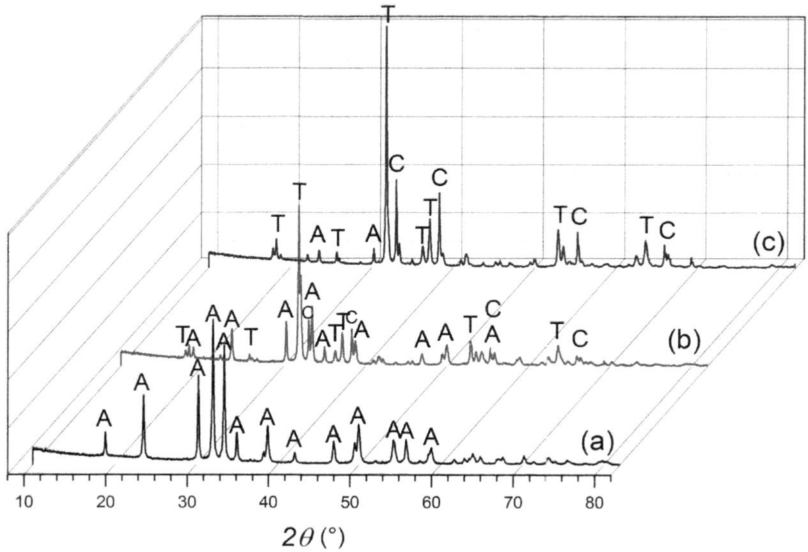

Fig. 3 XRD patterns of sintered CdO + amorphous SiO_2 [Cd/Si molar ratios = (a) 1/1, (b) 2/1, (c) 3/1] at 900°C for 3 h, showing the formation of Cd silicates. C denotes CdO (ICDD PDF no. 75-0592), A denotes Cd_2SiO_4 (ICDD PDF no. 89-0218), and T denotes Cd_3SiO_5 (ICDD PDF no. 85-0575).

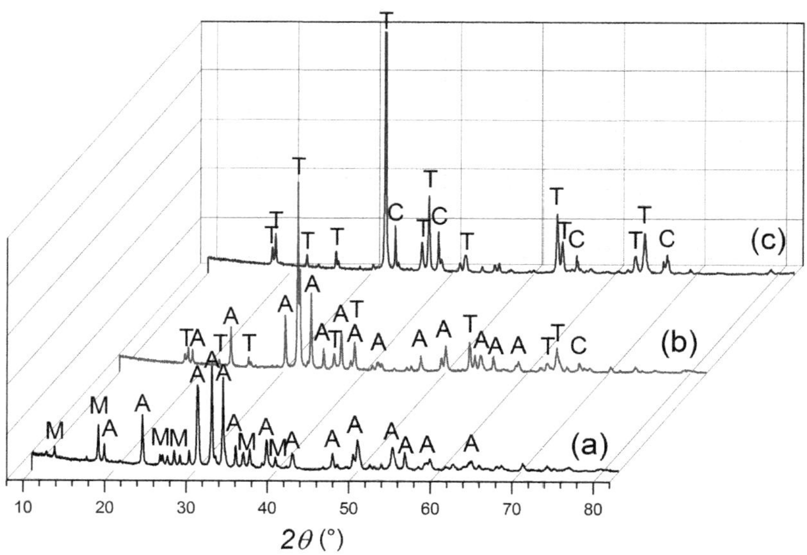

Fig. 4 XRD patterns of sintered CdO + amorphous SiO_2 [Cd/Si molar ratios = (a) 1/1, (b) 2/1, (c) 3/1] at 950°C for 3 h, showing the formation of Cd silicates. C denotes CdO (ICDD PDF no. 75-0592), M denotes $CdSiO_3$ (ICDD PDF no. 35-0810), A denotes Cd_2SiO_4 (ICDD PDF no. 89-0218), and T denotes Cd_3SiO_5 (ICDD PDF no. 85-0575).

signal of $CdSiO_3$ was observed in both systems with Cd/Si molar ratios of 2/1 and 3/1. Notably, in all systems, no diffraction signal related to cadmium silicate solid solutions was observed. Eqs. (1)–(3) demonstrated the crystallization processes of $CdSiO_3$, Cd_2SiO_4, and Cd_3SiO_5 by sintering CdO with amorphous SiO_2:

$$CdO + SiO_2 \rightarrow CdSiO_3 \tag{1}$$

$$2CdO + SiO_2 \rightarrow Cd_2SiO_4 \tag{2}$$

$$3CdO + SiO_2 \rightarrow Cd_3SiO_5 \tag{3}$$

Cadmium was effectively incorporated into cadmium silicates when sintered with amorphous SiO_2 at low Cd/Si molar ratios (1/1 and 2/1) and temperatures lower than 950°C. At high Cd/Si molar ratios, although small amounts of cadmium oxide were noted, the incorporation efficiency was quite high because the initial loading of cadmium was considerably high. The Cd incorporation reactions for various molar ratios of Cd/Si may guide the stabilization of cadmium wastes by using Si-rich precursors.

3.2 Formation of cadmium aluminate

Fig. 5 depicts the XRD patterns of sintered samples of CdO + γ-Al_2O_3 at 850–950°C for 3 h. Only the $CdAl_4O_7$ monoclinic structure was formed as the crystalline Cd-hosting product phase. Regarding solid-state reactions, both thermodynamic

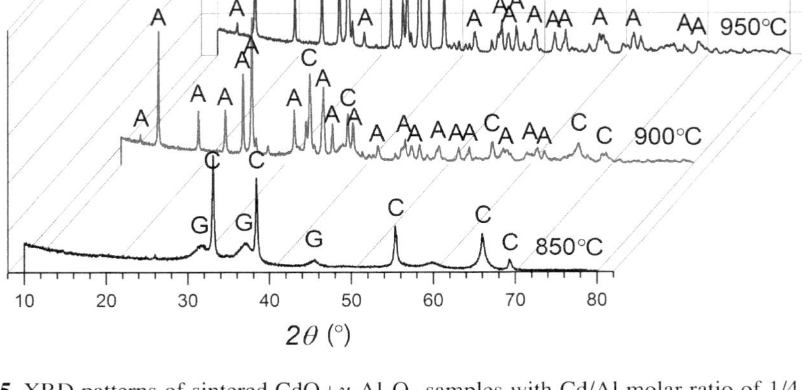

Fig. 5 XRD patterns of sintered CdO + γ-Al_2O_3 samples with Cd/Al molar ratio of 1/4 at 850–950°C for 3 h. A denotes $CdAl_4O_7$ (ICDD PDF no. 22-1061), C denotes CdO (ICDD PDF no. 75-0594), and G denotes γ-Al_2O_3 (ICDD PDF no. 50-0741).

constraints and kinetic process are two important parameters (Schmalzried, 2008). Here, the formation of $CdAl_4O_7$ was insignificant at 850°C, most likely because the cadmium incorporation reaction was constrained at the grain boundary among reactants within a short sintering scheme. Thus, this small quantity of crystalline Cd-hosting product phase in the sintered samples cannot be noticeably reflected in the XRD pattern. By sintering the $CdO+\gamma\text{-}Al_2O_3$ mixture at 900°C for 3h, the diffraction signals of $CdAl_4O_7$ appeared in the XRD pattern. At this treatment temperature, the intensity of the diffraction peaks of CdO decreased. At 950°C for 3h, more and stronger diffraction signals of $CdAl_4O_7$ were noted in the XRD patterns, indicating that Cd was significantly incorporated into $CdAl_4O_7$. Sufficient energy input could aid in overcoming the reaction barrier and more effective incorporation (Su et al., 2015). The reaction for cadmium incorporation into monoclinic $CdAl_4O_7$ by $\gamma\text{-}Al_2O_3$ could be described by Eq. (4):

$$CdO + 2\gamma\text{-}Al_2O_3 \rightarrow CdAl_4O_7 \tag{4}$$

3.3 Formation of cadmium ferrite

The XRD patterns of the $CdO+\alpha\text{-}Fe_2O_3$ samples treated with a Cd/Fe molar ratio of 1/2 at 600–800°C for 3h are shown in Fig. 6. Cd could be incorporated into $CdFe_2O_4$ spinel after it was sintered with $\alpha\text{-}Fe_2O_3$. The $CdFe_2O_4$ product phase crystallized into a cubic spinel structure (space group $Fd\overline{3}m$ (no. 227)). In the reactive system of $CdO+Fe_3O_4$ as well (Fig. 7), the crystalline Cd-hosting product phase was $CdFe_2O_4$ spinel. However, this product phase is formed significantly at temperatures as low as 600°C, indicating that at low temperatures, the cadmium incorporation reaction was more reactive with Fe_3O_4 than with $\alpha\text{-}Fe_2O_3$. Furthermore, Fe_3O_4 sintered at 600°C yielded a smaller XRD crystallite size than that yielded by the as-received $\alpha\text{-}Fe_2O_3$ (Su et al., 2017). Reactants with poor crystalline structure and small crystallite size often have high reactivity and thus fast reaction kinetics. By elevating the temperature to 800°C, cadmium was completely incorporated into $CdFe_2O_4$ by $\alpha\text{-}Fe_2O_3$, but incompletely by Fe_3O_4, as reflected by the residual CdO in the system with Fe_3O_4 as the precursor. It was mainly because of the phase transformation of Fe_3O_4 into $\alpha\text{-}Fe_2O_3$ with increasing sintering temperature. If this transformation rate exceeds the formation rate of $CdFe_2O_4$ spinel, the predominant formation mechanism of $CdFe_2O_4$ is the interaction between $\alpha\text{-}Fe_2O_3$ and CdO. The formation of $CdFe_2O_4$ through sintering CdO with $\alpha\text{-}Fe_2O_3$ and Fe_3O_4 precursors can be described as follows:

$$CdO + \alpha\text{-}Fe_2O_3 \rightarrow CdFe_2O_4 \tag{5}$$

$$6CdO + 4Fe_3O_4 + O_2 \rightarrow 6CdFe_2O_4 \tag{6}$$

The successful stabilization of Cd by Fe-rich precursors provides a promising strategy to safely and reliably immobilize Cd in the $CdFe_2O_4$ spinel structure as well (Figs. 6–7).

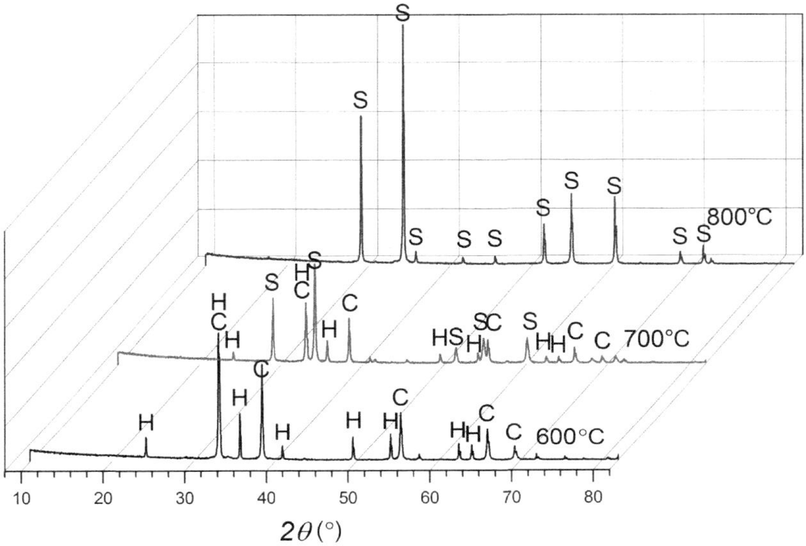

Fig. 6 XRD patterns of sintered CdO+α-Fe_2O_3 samples with a Cd/Fe molar ratio of 1/2 at 600–800°C for 3h. C denotes CdO (ICDD PDF no. 75-0594), H denotes α-Fe_2O_3 (ICDD PDF no. 87-1166), and S denotes $CdFe_2O_4$ (ICDD PDF no. 22-1063).

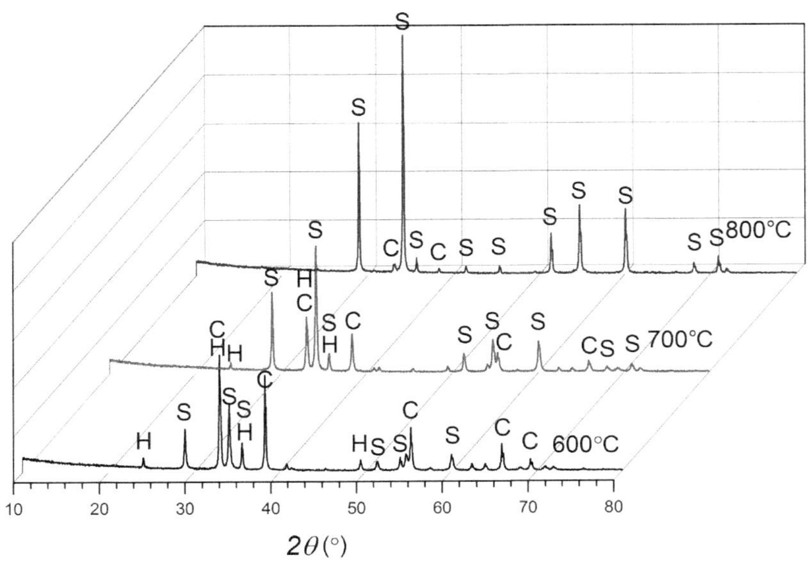

Fig. 7 XRD patterns of sintered CdO+Fe_3O_4 samples with a Cd/Fe molar ratio of 1/2 at 600–800°C for 3h. C denotes CdO (ICDD PDF no. 75-0594), H denotes α-Fe_2O_3 (ICDD PDF no. 87-1166), and S denotes $CdFe_2O_4$ (ICDD PDF no. 22-1063).

3.4 Use of ISSA for cadmium stabilization

A previous study indicated that CEPSTW ISSA is enriched with Si and Fe (Su and Shih, 2017). Fig. 8 illustrates the XRD patterns of the sintered CdO + CEPSTW ISSA samples with Cd/Si of 2/1 and 3/1 at 950°C for 3 h. When CEPSTW ISSA was sintered with CdO at a Cd/Si molar ratio of 2/1, cadmium was completely incorporated into both the Cd_3SiO_5 and $CdFe_2O_4$ phases. As the Cd/Si molar ratio increased to 3/1, the predominant Cd-hosting product phases in the sintered sample were Cd_3SiO_5 and $CdFe_2O_4$; however, significant diffraction signals of CdO were still observed, implying that cadmium could not be completely incorporated owing to a high loading of cadmium in the sample. As mentioned, at temperatures lower than 900°C, the interactions among CdO and Si-/Fe-rich precursors were much stronger than those between CdO and Al_2O_3. $CdFe_2O_4$ formation was initiated at 600°C during the sintering of the CdO-α-Fe_2O_3 mixture. The formation temperature of cadmium silicates was as low as 700°C. Different formation temperatures of these product phases suggest that the driving forces for mass transfer among their solid-state reactions are different, as summarized in Table 1. Different cadmium incorporation reactions occurred when ISSA with several elements was mixed and sintered with cadmium-containing waste. Thus, the ISSA enriched with Si/Fe can be beneficially reused as a matrix for cadmium stabilization, offering a promising waste management strategy for treating cadmium-containing wastes.

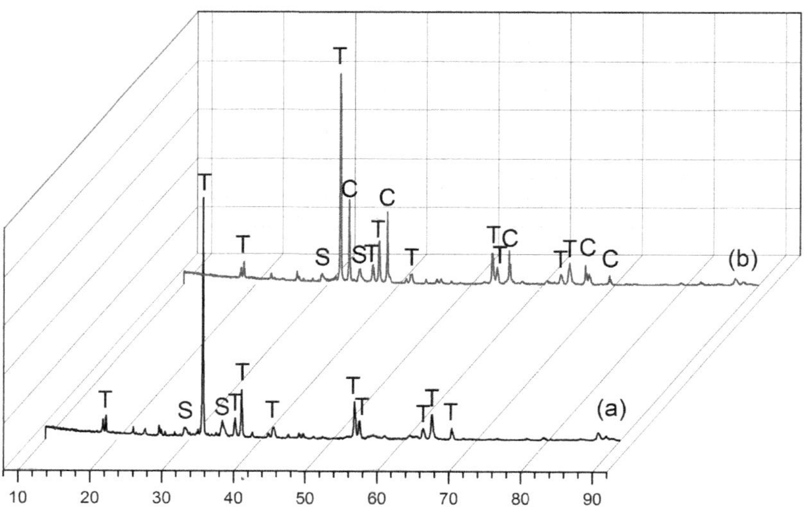

Fig. 8 XRD patterns of sintered CdO + CEPSTW ISSA samples with Cd/Si of (a) 2/1 and (b) 3/1 at 950°C for 3 h, showing beneficial use of ISSA for cadmium stabilization. C denotes CdO (ICDD PDF no. 75-0594), S denotes $CdFe_2O_4$ (ICDD PDF no. 22-1063), and T denotes Cd_3SiO_5 (ICDD PDF no. 85-0575).

Table 1 Standard thermodynamic values at 298 K

Formula	Enthalpy ΔH_f^0 (kJ/mol)	Entropy S^0 (J mol/K)	Gibbs free energy ΔG_f^0 (kJ/mol)
CdO	−258.15	54.81	−228.45
γ-Al_2O_3	−1656.86	59.83	−1562.72
Amorphous SiO_2	−903.49	46.86	−850.73
Quartz	−910.94	41.84	−856.67
α-Fe_2O_3	−824.25	87.40	−742.24
Fe_3O_4	−1118.38	146.44	−1015.46
$CdSiO_3$	−1189.09	97.49	−1105.41
$CdFe_2O_4$*	−1065.00	168.00	−1115.06

Notes:
(1) All these matters are solid.
(2) The data for $CdFe_2O_4$* derived from Knacke, O., Zimmermann, E., 1993. Thermal dissociation of CdO and $CdFe_2O_4$. Z. Anorg. Allg. Chem. 619(10), 1790–1792.
(3) All other data are from Dean, J.A., 1979. Lange's Handbook of Chemistry, twelfth ed. McGraw-Hill, New York, pp. 9-4–9-94.

3.5 Leachability of products

The CPLT was used to evaluate the leachability of $CdSiO_3$, Cd_2SiO_4, Cd_3SiO_5, $CdAl_4O_7$, and $CdFe_2O_4$. To compare it with those of the cadmium-hosting product phases, CdO used in the incorporation reaction was leached with the CPLT as well. The theoretical solubility constants of a few of these solid phases at equilibrium are known (Boyanov and Peltekov, 2014; Brezonik and Arnold, 2011); however, knowledge regarding their leachability and dissolution kinetics under unfavorable conditions is essential. Because the focus of the present study was on pure phase comparisons, single-phase samples were prepared. Single-phase monoclinic $CdSiO_3$ mixtures in a pellet with a Cd/Si molar ratio of 1/1 were presintered at 950°C for 3 h and then sintered at 1050°C and 1075°C for 6–12 h. Single-phase orthorhombic Cd_2SiO_4 was prepared through ceramic sintering at 950°C for 3 h and then at 1050°C and 1075°C for 6–12 h. Single-phase tetragonal Cd_3SiO_5 was prepared through two iterations of ceramic sintering at 950°C for 3 h and then at 1000°C and 1050°C for 3 h. Single-phase $CdAl_4O_7$ and $CdFe_2O_4$ samples were prepared by sintering the mixtures with Cd/Al and Cd/Fe molar ratios of 1/4 and 1/2, respectively, at 950°C and 850°C, respectively. An identical preparation process of blending, pelletizing, and sintering was repeated to ensure the complete transformation and homogeneity of the products. The final sintered products were further ball-milled into powder forms for the leaching test.

On the basis of the dissolution mechanism in acidic circumstances, congruent dissolution of CdO, $CdSiO_3$, Cd_2SiO_4, Cd_3SiO_5, $CdAl_4O_7$, and $CdFe_2O_4$ in an acidic solution can be expressed as follows:

$$CdO_{(s)} + 2H^{+}_{(eq)} \rightarrow Cd^{2+}_{(eq)} + H_2O \tag{7}$$

$$CdAl_4O_{7(s)} + 14H^+_{(eq)} \rightarrow Cd^{2+}_{(eq)} + 4Al^{3+}_{(eq)} + 7H_2O \tag{8}$$

$$CdFe_2O_{4(s)} + 8H^+_{(eq)} \rightarrow Cd^{2+}_{(eq)} + 2Fe^{3+}_{(eq)} + 4H_2O \tag{9}$$

$$CdSiO_{3(s)} + 2H^+_{(eq)} + H_2O \rightarrow Cd^{2+}_{(eq)} + Si(OH)_{4(eq)} \tag{10}$$

$$Cd_2SiO_{4(s)} + 4H^+_{(eq)} \rightarrow 2Cd^{2+}_{(eq)} + Si(OH)_{4(eq)} \tag{11}$$

$$Cd_3SiO_{5(s)} + 6H^+_{(eq)} \rightarrow 3Cd^{2+}_{(eq)} + Si(OH)_{4(eq)} + H_2O \tag{12}$$

Fig. 9 illustrates the leached Cd concentrations in the leachates of the tested samples, namely CdO, $CdSiO_3$, Cd_2SiO_4, Cd_3SiO_5, $CdAl_4O_7$, and $CdFe_2O_4$, at a constant pH of 4.0. Leached Cd concentrations in the $CdSiO_3$, Cd_2SiO_4, Cd_3SiO_5, $CdAl_4O_7$, and $CdFe_2O_4$ leachates were 2.6, 64.0, 68.9, 11.7, and 1.5 mg/L, respectively, all of which are lower than that in the CdO leachate (691.2 mg/L). Thus, superior stabilization and detoxification could be achieved by the formation of silicates, aluminate, and spinel crystal structures. Notably, leached Cd concentrations in the Cd_2SiO_4 and Cd_3SiO_5 leachates were considerably higher than in the $CdSiO_3$, $CdAl_4O_7$, and $CdFe_2O_4$ leachates. The leaching behavior of the Cd_2SiO_4 and Cd_3SiO_5 phases tends toward near-congruent dissolution, whereas that of the $CdSiO_3$, $CdAl_4O_7$, and $CdFe_2O_4$ phases tends toward incongruent dissolution. Furthermore, the high Cd content (more than 70 wt.%) in Cd_2SiO_4 and Cd_3SiO_5 implies that Cd atoms may be more exposed to the leaching fluid, possibly leading to high dissolution of Cd from these two cadmium silicate phases. Thus, the CPLT results demonstrated that the phase transformations to $CdSiO_3$, $CdAl_4O_7$, and $CdFe_2O_4$ spinel could more reliably enhance the inherent robustness of cadmium-hosting products to acids than could those to the CdO and other Cd-hosting product phases.

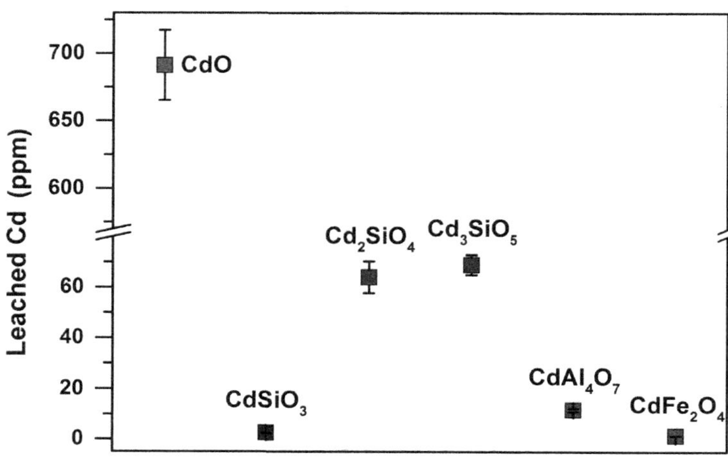

Fig. 9 Leached Cd concentrations in the leachates of Cd-hosting phases at the end of the CPLT (120 min).

4 Conclusion

This chapter is mainly concerned with the feasibility of stabilization by using different ceramic precursors to incorporate simulated cadmium-laden industrial sludge into durable crystal structures under thermal conditions. A well-controlled thermal treatment process is a cost-effective strategy to reduce heavy metal contamination through the generation of preferable crystalline phases. Silica fume (a typical Si-rich material), γ-Al_2O_3 (a typical Al-rich material), as well as α-Fe_2O_3 and Fe_3O_4 (two typical Fe-rich materials) could incorporate cadmium into crystal structures, including $CdSiO_3$, Cd_2SiO_4, Cd_3SiO_5, $CdAl_4O_7$, and $CdFe_2O_4$ under thermal conditions. In the reactive system of $CdO + SiO_2$, the product phases and cadmium silicate formation were highly affected by the sintering temperature and Cd/Si molar ratios (i.e., 1/1, 2/1, and 3/1). Monoclinic $CdSiO_3$ was only observed in the systems with the Cd/Si molar ratio of 1/1 treated at temperatures of more than 950°C for 3 h. Orthorhombic Cd_2SiO_4 was found commonly in all systems. With high amounts (Cd/Si molar ratio = 3/1) of CdO, significant amounts of tetragonal Cd_3SiO_5 were formed, and it was the dominant Cd-hosting product phase. The CPLT results showed that $CdSiO_3$ possessed the highest acid resistance. The leachability of Cd_2SiO_4 and Cd_3SiO_5 was similar. Furthermore, 950°C is the preferred treatment temperature for mixtures with a Cd/Al molar ratio of 1/4 because the Cd incorporation was nearly complete after a 3 h treatment scheme. A substantial reduction in Cd leachability could be achieved through formation of the monoclinic $CdAl_4O_7$ structure. $CdFe_2O_4$ spinel was the only Cd-hosting product phase in the sintered $CdO + \alpha$-Fe_2O_3 and $CdO + Fe_3O_4$ systems. Compared with hematite, magnetite facilitated more vigorous crystallization of the $CdFe_2O_4$ spinel at low temperatures. Both the tested Fe-rich precursors showed fairly good capability in terms of incorporation of Cd into the $CdFe_2O_4$ spinel. The CPLT results revealed that a very small amount of Cd leached out from the $CdFe_2O_4$ spinel. ISSA could effectively incorporate cadmium upon treatment at 950°C for 3 h. After sintering, the product phases could be Cd_2SiO_3, Cd_3SiO_5, and $CdFe_2O_4$, depending on the type of sludge ash used and the amount of cadmium doped in the systems.

Acknowledgments

This work was supported by National Natural Science Foundations of China (51708143), the Science and Technology Program of Guangzhou, China (No. 201804010366), Scientific Research Foundation for the Returned High-level Overseas Talents (2018), Guangzhou University's Training Program for Excellent New-recruited Doctors (YB201710), the Research Grants Council of Hong Kong (17212015, 17257616, C7044-14G, and T21-771/16R).

References

Ahmaruzzaman, M., 2011. Industrial wastes as low-cost potential adsorbents for the treatment of wastewater laden with heavy metals. Adv. Colloid Interface Sci. 166 (1–2), 36–59.

Al-Abed, S.R., Jegadeesan, G., Purandare, J., Allen, D., 2007. Arsenic release from iron rich mineral processing waste: influence of pH and redox potential. Chemosphere 66 (4), 775–782.

Allioux, F.-M., Kapruwan, P., Milne, N., Kong, L., Fattaccioli, J., Chen, Y., Dumée, L.F., 2017. Electro-capture of heavy metal ions with carbon cloth integrated microfluidic devices. Sep. Purif. Technol. 194 (3), 26–32.

ASTM-D-4874, 2001. ASTM D4874-95(R2001) Standard Test Method for Leaching Solid Material in a Column Apparatus. ASTM International, West Conshohocken, PA.

Badruddoza, A.Z.M., Shawon, Z.B.Z., Tay, W.J.D., Hidajat, K., Uddin, M.S., 2013. Fe_3O_4/cyclodextrin polymer nanocomposites for selective heavy metals removal from industrial wastewater. Carbohydr. Polym. 91 (1), 322–332.

Balsamo, M., Di Natale, F., Erto, A., Lancia, A., Montagnaro, F., Santoro, L., 2011. Cadmium adsorption by coal combustion ashes-based sorbents—relationship between sorbent properties and adsorption capacity. J. Hazard. Mater. 187 (1–3), 371–378.

Barth, E.F., 1990. An overview of the history, present status, and future direction of solidification/stabilization technologies for hazardous waste treatment. J. Hazard. Mater. 24 (2), 103–109.

Bashkirov, L.A., Kornilova, N.V., 1980. Equilibre de phases dans le systeme MgO-CdO-Fe_2O_3 dans la partie contenant Au moins 50 moles% de Fe_2O_3. Izvest. Akad. Nauk SSSR, Neorg. Mater. 16 (7), 1229–1233.

Bhatnagar, A., Minocha, A.K., 2009. Utilization of industrial waste for cadmium removal from water and immobilization in cement. Chem. Eng. J. 150 (1), 145–151.

Boyanov, B., Peltekov, A., 2014. Solubility of $ZnFe_2O_4$, $CdFe_2O_4$ and $CuFe_2O_4$ in solutions of H_2SO_4, HCl and HNO_3. In: The 46th International October Conference on Mining and Metallurgy.

Brezonik, P., Arnold, W., 2011. Water Chemistry: An Introduction to the Chemistry of Natural and Engineered Aquatic Systems. Oxford University Press, New York.

Cappuyns, V., Swennen, R., 2008. The application of pH_{stat} leaching tests to assess the pH-dependent release of trace metals from soils, sediments and waste materials. J. Hazard. Mater. 158 (1), 185–195.

Chan, S.H., 1989. A review on solubility and polymerization of silica. Geothermics 18 (1–2), 49–56.

Chen, M., Li, X.-M., Yang, Q., Zeng, G.-M., Zhang, Y., Liao, D.-X., Liu, J.-J., Hu, J.-M., Guo, L., 2008. Total concentrations and speciation of heavy metals in municipal sludge from Changsha, Zhuzhou and Xiangtan in middle-south region of China. J. Hazard. Mater. 160 (2), 324–329.

Cheng, K., Tian, H.Z., Zhao, D., Lu, L., Wang, Y., Chen, J., Liu, X.G., Jia, W.X., Huang, Z., 2014. Atmospheric emission inventory of cadmium from anthropogenic sources. Int. J. Environ. Sci. Technol. 11 (3), 605–616.

Chinnasamy, C., Narayanasamy, A., Ponpandian, N., Joseyphus, R.J., Chattopadhyay, K., Shinoda, K., Jeyadevan, B., Tohji, K., Nakatsuka, K., Greneche, J.M., 2001a. Ferrimagnetic ordering in nanostructured $CdFe_2O_4$ spinel. J. Appl. Phys. 90 (1), 527–529.

Chinnasamy, C., Narayanasamy, A., Ponpandian, N., Justin Joseyphus, R., Chattopadhyay, K., Shinoda, K., Jeyadevan, B., Tohji, K., Nakatsuka, K., Guerault, H., Greneche, J.M., 2001b. Structure and magnetic properties of nanocrystalline ferrimagnetic $CdFe_2O_4$ spinel. Scr. Mater. 44 (8–9), 1411–1415.

Colin, F., 1968. Des phases formees au cours de la reduction de certains oxydes mixtes $nAl_2O_3{\cdot}MO$ (phases obtained during the reduction of some $nAl_2O_3{\cdot}MO$ oxides). Rev. Int. Hautes Temp. Refract. 5 (4), 267–283.

Conner, J.R., 1990. Chemical Fixation and Solidification of Hazardous Wastes. Van Nostrand Reinhold/Chapman and Hall, New York, NY/London, ISBN 0-442-20511-2, p. 692.

Cornell, R.M., Schwertmann, U., 2006. The Iron Oxides: Structure, Properties, Reactions, Occurrences and Uses. Wiley-VCH Verlag GmbH & Co. KGaA, Weinheim.

Edwards, J., Othman, M., Crossin, E., Burn, S., 2017. Anaerobic co-digestion of municipal food waste and sewage sludge: a comparative life cycle assessment in the context of a waste service provision. Bioresour. Technol. 223 (Suppl. C), 237–249.

Glasser, L., 1965. Silicates M_3SiO_5. II. Relationships between Sr_3SiO_5, Cd_3SiO_5, and Ca_3SiO_5. Acta Crystallogr. 18 (3), 455–457.

Gutiérrez, G., Taga, A., Johansson, B., 2001. Theoretical structure determination of γ-Al_2O_3. Phys. Rev. B 65 (1), 012101.

Han, X., Niu, M., Jiang, X., Liu, J., 2012. Combustion characteristics of sewage sludge in a fluidized bed. Ind. Eng. Chem. Res. 51 (32), 10565–10570.

Hartman, M., Svoboda, K., Pohořelý, M., Trnka, O., 2005. Combustion of dried sewage sludge in a fluidized-bed reactor. Ind. Eng. Chem. Res. 44 (10), 3432–3441.

Hasegawa, H., Rahman, I.M.M., Rahman, M.A., 2016. Environmental remediation technologies for metal-contaminated soils. Springer, Japan.

Hossain, S.T., Mukherjee, S.K., 2012. CdO nanoparticle toxicity on growth, morphology and cell division in *Escherichia coli*. Langmuir 28 (48), 16614–16622.

Hua, M., Zhang, S., Pan, B., Zhang, W., Lv, L., Zhang, Q., 2012. Heavy metal removal from water/wastewater by nanosized metal oxides: a review. J. Hazard. Mater. 211–212, 317–331.

Huang, W.-J., Huang, H.-S., 2008. Using fume silica as heavy metals' stabilizer for high alkali and porous MSWI baghouse ash. J. Hazard. Mater. 152 (1), 176–182.

Islam, M.Z., Catalan, L.J., Yanful, E.K., 2004. Effect of remineralization on heavy-metal leaching from cement-stabilized/solidified waste. Environ. Sci. Technol. 38 (5), 1561–1568.

Islam, M.S., Ahmed, M.K., Raknuzzaman, M., Habibullah-Al-Mamun, M., Kundu, G.K., 2017. Heavy metals in the industrial sludge and their ecological risk: a case study for a developing country. J. Geochem. Explor. 172 (Suppl. C), 41–49.

Jabłońska, B., Siedlecka, E., 2015. Removing heavy metals from wastewaters with use of shales accompanying the coal beds. J. Environ. Manage. 155, 58–66.

Jackson, D.R., Garrett, B.C., Bishop, T.A., 1984. Comparison of batch and column methods for assessing leachability of hazardous waste. Environ. Sci. Technol. 18 (9), 668–673.

Järup, L., 2003. Hazards of heavy metal contamination. Br. Med. Bull. 68 (1), 167–182.

Khan, M.I., Siddique, R., 2011. Utilization of silica fume in concrete: review of durability properties. Resour. Conserv. Recycl. 57, 30–35.

Khan, M.A., Khan, S., Khan, A., Alam, M., 2017. Soil contamination with cadmium, consequences and remediation using organic amendments. Sci. Total Environ. 601–602 (Suppl. C), 1591–1605.

Krüger, O., Grabner, A., Adam, C., 2014. Complete survey of German sewage sludge ash. Environ. Sci. Technol. 48 (20), 11811–11818.

Kurihara, L.K., Suib, S.L., 1993. Sol-gel synthesis of ternary metal oxides. 1. Synthesis and characterization of MAl_2O_4 (M = Mg, Ni, Co, Cu, Fe, Zn, Mn, Cd, Ca, Hg, Sr, and Ba) and lead aluminum oxide ($Pb_2Al_2O_5$). Chem. Mater. 5 (5), 609–613.

Li, X., Chen, Q., Zhou, Y., Tyrer, M., Yu, Y., 2014. Stabilization of heavy metals in MSWI fly ash using silica fume. Waste Manage. 34 (12), 2494–2504.

Liu, Y., Xiao, T., Perkins, R.B., Zhu, J., Zhu, Z., Xiong, Y., Ning, Z., 2017. Geogenic cadmium pollution and potential health risks, with emphasis on black shale. J. Geochem. Explor. 176 (Suppl. C), 42–49.

Lonsdale, T.D., Whitaker, A., 1978. Phase equilibria in the system CdO-B_2O_3-SiO_2 at 800°C. J. Mater. Sci. 13 (7), 1503–1508.

Mahmoud, M.H., Abdallas, A.M., Hamdeh, H.H., Hikal, W.M., Taher, S.M., Ho, J.C., 2003. Mössbauer spectroscopic evaluation of high-energy ball-milled $CdFe_2O_4$. J. Magn. Magn. Mater. 263 (3), 269–274.

McLaughlin, J.K., Chow, W.H., Levy, L.S., 1997. Amorphous silica: a review of health effects from inhalation exposure with particular reference to cancer. J. Toxicol. Environ. Health A 50 (6), 553–566.

Merget, R., Bauer, T., Küpper, H., Philippou, S., Bauer, H., Breitstadt, R., Bruening, T., 2002. Health hazards due to the inhalation of amorphous silica. Arch. Toxicol. 75 (11–12), 625–634.

Nair, A., Juwarkar, A.A., Devotta, S., 2008. Study of speciation of metals in an industrial sludge and evaluation of metal chelators for their removal. J. Hazard. Mater. 152 (2), 545–553.

Nandiyanto, A.B.D., Kim, S.-G., Iskandar, F., Okuyama, K., 2009. Synthesis of spherical mesoporous silica nanoparticles with nanometer-size controllable pores and outer diameters. Microporous Mesoporous Mater. 120 (3), 447–453.

NEN-7343, 1995. Leaching Characteristics of Solid Earth and Stony Building and Waste Materials. Leaching Tests. Determination of the Leaching of Inorganic Components From Granular Materials with the Column Test. Nederlandse Normalisatie-instituut, Delft, Netherlands.

Nowak, B., Aschenbrenner, P., Winter, F., 2013. Heavy metal removal from sewage sludge ash and municipal solid waste fly ash—a comparison. Fuel Process. Technol. 105 (0), 195–201.

NTP (National Toxicology Program), 2014. Report on Carcinogens, thirteenth ed. U.S. Department of Health and Human Services, Public Health Service, Research Triangle Park, NC.

Paglia, G., Buckley, C., Rohl, A., Hunter, B., Hart, R., Hanna, J., Byrne, L., 2003. Tetragonal structure model for boehmite-derived γ-alumina. Phys. Rev. B 68 (14), 144110.

Paglia, G., Buckley, C.E., Udovic, T.J., Rohl, A.L., Jones, F., Maitland, C.F., Connolly, J., 2004. Boehmite-derived γ-alumina system. 2. Consideration of hydrogen and surface effects. Chem. Mater. 16 (10), 1914–1923.

Pendowski, J.J., 2003. An Assessment of Laboratory Leaching Tests for Predicting the Impacts of Fill Material on Ground Water and Surface Water Quality. (Publication No. 03-09-107).

Qu, X., Cao, L., Liu, W., Su, G., Qu, H., Xu, C., 2009. Synthesis of $CdSiO_3$ powder by sol–gel method. J. Alloys Compd. 484 (1–2), 641–644.

Rivoallan, A., Calmettes, M., Cohendet, M., 1994. Recovery and recycling of heavy metals in industrial sludges. Resour. Conserv. Recycl. 10 (1), 177–184.

Rodella, N., Bosio, A., Dalipi, R., Zacco, A., Borgese, L., Depero, L.E., Bontempi, E., 2017. Waste silica sources as heavy metal stabilizers for municipal solid waste incineration fly ash. Arab. J. Chem. 10 (Suppl. 2), s3676–s3681.

Schmalzried, H., 2008. Chemical Kinetics of Solids. VCH Verlagsgesellsehaft/VCH Publishers, Weinheim, Germany/New York.

Schwertmann, U., Cornell, R.M., 2008. Iron Oxides in the Laboratory. Wiley-VCH Verlag GmbH, Weinheim.

Sharma, V.K., Klingelhofer, G., Nishida, T., 2013. Mossbauer Spectroscopy: Applications in Chemistry, Biology, Industry, and Nanotechnology. John Wiley & Sons, Inc., Hoboken, NJ.

Shih, K., White, T., Leckie, J.O., 2006. Nickel stabilization efficiency of aluminate and ferrite spinels and their leaching behavior. Environ. Sci. Technol. 40 (17), 5520–5526.

Su, M., Shih, K., 2017. Utilisation of incinerated sewage sludge ash as a matrix for cadmium stabilisation. HKIE Trans. 24 (1), 35–41.

Su, M., Liao, C., Chuang, K.-H., Wey, M.-Y., Shih, K., 2015. Cadmium stabilization efficiency and leachability by $CdAl_4O_7$ monoclinic structure. Environ. Sci. Technol. 49 (24), 7.

Su, M., Liao, C., Lee, P.-H., Li, H., Shih, K., 2017. Formation and leaching behavior of ferrite spinel for cadmium stabilization. Chem. Eng. Sci. 158 (Suppl. C), 287–293.

Su, M., Liao, C., Chan, T.-S., Shih, K., Xiao, T., Chen, D., Kong, L., Song, G., 2018. Incorporation of cadmium and nickel into ferrite spinel solid solution: X-ray diffraction and X-ray absorption fine structure analyses. Environ. Sci. Technol. 52, 775–782.

Sun, Y., Sun, G., Xu, Y., Liu, W., Liang, X., Wang, L., 2016. Evaluation of the effectiveness of sepiolite, bentonite, and phosphate amendments on the stabilization remediation of cadmium-contaminated soils. J. Environ. Manage. 166, 204–210.

Tang, Y., Shih, K., Chan, K., 2010. Copper aluminate spinel in the stabilization and detoxification of simulated copper-laden sludge. Chemosphere 80 (4), 375–380.

Tang, Y., Chui, S.S.-Y., Shih, K., Zhang, L., 2011a. Copper stabilization via spinel formation during the sintering of simulated copper-laden sludge with aluminum-rich ceramic precursors. Environ. Sci. Technol. 45 (8), 3598–3604.

Tang, Y., Shih, K., Wang, Y., Chong, T.C., 2011b. Zinc stabilization efficiency of aluminate spinel structure and its leaching behavior. Environ. Sci. Technol. 45 (24), 10544–10550.

Tyagi, V.K., Lo, S.-L., 2013. Sludge: a waste or renewable source for energy and resources recovery? Renew. Sust. Energ. Rev. 25 (0), 708–728.

US Environmental Protection Agency, 1999. Integrated Risk Information System (IRIS) on Cadmium. National Center for Environmental Assessment. Office of Research and Development, Washington, DC.

USGS (U.S. Geological Survey), 2016. Mineral Commodity Summaries 2016: Cadmium. U.S. Geological Survey, Reston, VA.

Van der Sloot, H., Comans, R., Hjelmar, O., 1996. Similarities in the leaching behaviour of trace contaminants from waste, stabilized waste, construction materials and soils. Sci. Total Environ. 178 (1), 111–126.

Wang, S., Lu, G., 1999. A comprehensive study on carbon dioxide reforming of methane over $Ni/\gamma\text{-}Al_2O_3$ catalysts. Ind. Eng. Chem. Res. 38 (7), 2615–2625.

Wang, X., Chen, J., Yan, X., Wang, X., Zhang, J., Huang, J., Zhao, J., 2015. Heavy metal chemical extraction from industrial and municipal mixed sludge by ultrasound-assisted citric acid. J. Ind. Eng. Chem. 27 (Suppl. C), 368–372.

Wefers, K., Misra, C., 1987. Oxides and Hydroxides of Aluminum. Alcoa Laboratories, Pittsburgh, PA.

Wolverton, C., Hass, K., 2000. Phase stability and structure of spinel-based transition aluminas. Phys. Rev. B 63 (2), 024102.

Wu, Q., Cui, Y., Li, Q., Sun, J., 2015. Effective removal of heavy metals from industrial sludge with the aid of a biodegradable chelating ligand GLDA. J. Hazard. Mater. 283 (Suppl. C), 748–754.

Xu, G., Zou, J., Li, G., 2009. Stabilization/solidification of heavy metals in sludge ceramsite and leachability affected by oxide substances. Environ. Sci. Technol. 43 (15), 5902–5907.

Xu, G., Zou, J., Li, G., 2010. Stabilization of heavy metals in sludge ceramsite. Water Res. 44 (9), 2930–2938.

Zhang, Z., Hicks, R.W., Pauly, T.R., Pinnavaia, T.J., 2002. Mesostructured forms of $\gamma\text{-}Al_2O_3$. J. Am. Chem. Soc. 124 (8), 1592–1593.

Zhang, Q., Hu, J., Lee, D.-J., Chang, Y., Lee, Y.-J., 2017. Sludge treatment: current research trends. Bioresour. Technol. 243 (Suppl. C), 1159–1172.

Zou, J.L., Xu, G.R., Li, G.B., 2009. Ceramsite obtained from water and wastewater sludge and its characteristics affected by Fe_2O_3, CaO, and MgO. J. Hazard. Mater. 165 (1–3), 995–1001.

Zuckerman, B., Koester, D., Dufour, P., Melis, C., Klein, B., Jura, M., 2011. An aluminum/calcium-rich, iron-poor, white dwarf star: evidence for an extrasolar planetary lithosphere? Astrophys. J. 739 (2), 101.

Bioleaching for extracting heavy metals from electronic waste sludge

23

Jianfeng Bai[*,†], *Weihua Gu*[*,†,‡], *Changzhong Liao*[§], *Wenyi Yuan*[*,†], *Chenglong Zhang*[*,†], *Jingwei Wang*[*,†], *Bin Dong*[¶], *Kaimin Shih*[§]
[*]WEEE Research Center of Shanghai Polytechnic University, Shanghai, China, [†]Shanghai Collaborative Innovation Centre for WEEE Recycling, Shanghai, China, [‡]College of Environment, Zhejiang University of Technology, Hangzhou, China, [§]Department of Civil Engineering, The University of Hong Kong, Hong Kong SAR, China, [¶]Department of Environmental Science and Engineering, Tongji University, Shanghai, China

1 Introduction

Electronic waste (E-waste) refers to electrical appliances or electronic devices that are discarded by people, mainly including refrigerators, washing machines, televisions, air conditioners, and computers. Cell phones, monitors, printers, and copiers that are considered obsolete by their owners are also e-waste (Afroz et al., 2013). In recent years, with the rapid development of science and technology, the lifespan of electrical and electronic equipment has continuously shortened, thus lots of electrical and electronic equipment waste are being generated. According to the latest statistics derived from a research group at continuously United Nations University, the production of global e-waste was 47 million tons in 2016 with an increasing rate of 4.6% per year, and is expected to reach approximately 52.2 million tons in 2021 (Baldé et al., 2017). Due to the enormous production of e-waste, many approaches are being executed to solve this problem (Tuncuk et al., 2012; Khanna et al., 2014; Kaya, 2016; Sahni et al., 2016). Particularly, the hydrometallurgy method will produce waste water simultaneously, and finally forms the sludge, usually it's called e-waste sludge. This kind of sludge is different than sewage sludge, as e-waste sludge contains a variety of heavy metals while sewage sludge involves plenty of organic substances. To date, considering environmental protection, the physical method has been widely adopted by some regular enterprises to dismantle e-waste. To pursue economic benefits, some acid and alkali reagents are used by family dismantling workshops to extract metals from e-waste. Besides, many researchers worldwide are focusing on hydrometallurgy methods to dispose of e-waste (Habib, 2013; Wang et al., 2016). All these behaviors will eventually produce e-waste sludge. When some improper measures are adopted to treat that e-waste sludge, "secondary pollution" will be generated. That can pollute soil and groundwater, even endangering human health through the enrichment of continuously food chain (He and Xu, 2014). On the other hand, there are a variety of

Industrial and Municipal Sludge. https://doi.org/10.1016/B978-0-12-815907-1.00023-4

metals in e-waste sludge. If these metals can be recycled with some suitable methods, that can alleviate the shortage of metal resources around the world (Zeng et al., 2017). Generally speaking, due to the composition of heavy metals in e-waste sludge being mostly the same as that of heavy metals in sewage sludge, some methods that are used to dispose of sewage sludge can be applied equally to the treatment of e-waste sludge (Harikrushnan et al., 2016; Darezereshki et al., 2018). Considering environmental protection and resource circulation simultaneously, usually the disposal of e-waste sludge contains two methods: harmless treatment technology and recycling technology. The commonly used technologies for e-waste sludge are shown in Fig. 1. For a further segmentation, harmless treatment technology includes solidification and stabilization, incineration, and landfill (Cui and Zhang, 2008; Ogunniyi and Vermaak, 2009; Das et al., 2009). Recycling technology contains pyrometallurgy, hydrometallurgy, and materialization (Wang et al., 2016; Gu et al., 2017a,b). The solidification and stabilization technology refers to that using curing agents to solidify the e-waste sludge, making heavy metal solidification inside the mixture substances, or adding stabilizer to stabilize the e-waste sludge, and producing harmless substances. The incineration technology is an advanced oxidation and fusion process. Through the thermal treatment, the toxic substances inside the e-waste sludge will be eliminated. Usually, For preventing the toxic substances from being released, solidification and stabilization technology will be used to treat e-waste sludge. Considering the e-waste sludge's sanitary criteria, the index of heavy metals in e-waste sludge can't meet the standards for farm application, so landfill technology is widely adopted in China (Das et al., 2009). Compared to the pyrometallurgy technology, the hydrometallurgy technology is broadly utilized because of its lower cost and less secondary pollution.

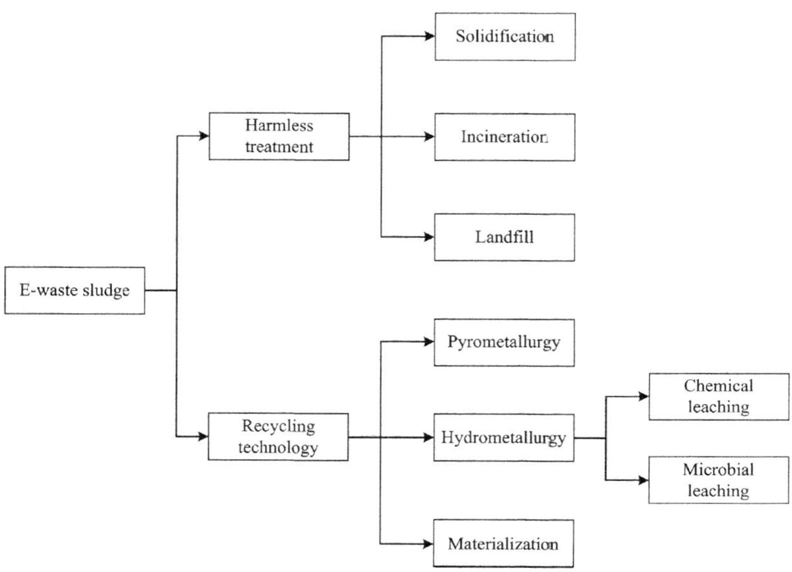

Fig. 1 The commonly used technologies for e-waste sludge.

The hydrometallurgy method contains chemical leaching and microbial leaching. Obviously, due to its high pollution to the surrounding environment, chemical leaching isn't a suitable method to apply in industrial application. Furthermore, it can generate more sludge after the chemical treatment. Microbial leaching refers to the use of direct and/or indirect effects generated by *Acidthiobacillus ferrooxidans* (*A. ferrooxidans*) and *Acidthiobacillus thiooxidans* (*A.thiooxidans*) to extract metals from sludge (Wang et al., 2009; Ilyas and Lee, 2014). The reactions include oxidation, reduction, exchange, adsorption, and dissolution, thus the metals inside the sludge can be leached. This method belongs to an emerging technology around the world, and is widely applied in sewage sludge, e-waste sludge, and mining waste fields. Materialization technology can use the function of some specific metals in sludge to develop new materials. For instance, a chromium-containing sludge can be manufactured to a dope. As of 2017, the specialized sludge disposal plants are distributed all over the world, especially in some developed countries, such as the United States, Japan, and Germany (Zhou and Xu, 2012). The US Environmental Protection Agency proposed a "three steps" pattern to dispose of sludge, particularly for hazardous sludge. The German government actively promotes the minimization and recycling of sludge, and the combinations of technologies above are adopted. The comprehensive utilization technology of sludge is immature in China. Usually, solidification and landfill technologies are commonly used. Many researchers have attempted the thermal treatment and materialization of sludge under laboratory conditions (Das et al., 2009; Wang et al., 2009). As for the treatment of e-waste sludge, there aren't many references focusing on it. This chapter focuses on introducing the bioleaching method for extracting heavy metals from e-waste sludge. Three parts developed by our group will be introduced. First, the sources of different e-waste sludge and the species and content of heavy metals in e-waste sludge were analyzed. Second, the species of microorganisms that were selected to do the bioleaching process were introduced, and *A. ferrooxidans* was chosen to bioleach heavy metals from e-waste sludge. Third, the optimized conditions, bioleaching mechanisms, and effects were also discussed.

2 Sources of e-waste sludge

In principle, hydrometallurgy technology for disposing of e-waste can produce wastewater that contains heavy metals that eventually transfer into the sludge, also known as e-waste sludge. As said above, the informal disposal of e-waste in family dismantling workshops and the hydrometallurgic treatment of e-waste under laboratory conditions can both produce e-waste sludge. Usually, as an e-waste, printed circuit boards (PCBs), liquid crystal displays (LCDs), cathode ray tubes (CRTs), and plastic are widely investigated under laboratory conditions. After the treatment of these e-wastes, it forms the residue, respectively, often we call these residues as PCB sludge, LCD sludge, and CRT sludge. Due to the different kinds of heavy metals among the PCBs, LCDs, and CRTs, the disposal technology is also different, and the composition of heavy metals in the PCB, LCD, and CRT sludge is quite different accordingly.

E-waste sludge contains electroplating sludge because of its composition. Many electronic products are constituted with some components, and the electroplating method is partly used to make the combination of these two parts. Furthermore, some metals are covered on the surface of electronic products through the electroplating technology. Generally speaking, the electroplating sludge contains Cr, Fe, Ni, Cu, and Zn. According to research, the compositions of electroplating sludge are 75%–80% moisture, 1.5%–3.0% heavy metals, 8%–10% $CaCO_3$, and 5% impurities (Ding et al., 2013).

For improving the properties of electronic products and satisfying people's interests of various electronic products, stainless steel is also adopted to manufacture electronic products. Stainless steel sludge can be produced among these processes. Heavy metals such as Cr, Ni, Fe, and Ca are included in stainless steel sludge from the analysis of stainless steel sludge by researchers (Fei et al., 2015). Besides, the leftover materials that come from electronic product manufacturing processes will also produce sludge. All the sludge mentioned above is collectively called e-waste sludge in this chapter, and more detailed presentations about e-waste sludge are as follows.

3 Species and contentsof heavy metals in e-waste sludge

This chapter mainly talks about e-waste sludge that contains stainless steel sludge, PCB sludge, and electroplating sludge. The species and content of heavy metals in e-waste sludge are shown in Table 1. As seen from this table, obviously, stainless steel sludge contains Cu, Ni, Zn, and Cr, and different sources of stainless steel sludge have different content of heavy metals, although they have the same species of heavy metals. PCB sludge has the same phenomena compared to stainless steel sludge, but the species of heavy metals are quite different between stainless steel sludge and PCB sludge, as PCB sludge mainly includes Cu, Ni, Zn, Mn, and Pb. The content of heavy metals in electroplating sludge derived from different plants in Guangdong province are also different. The electroplating sludge mainly contains Cu, Ni, Zn, and Mn. Moreover, part of the electroplating sludge also contains Sn or Pb. In general, these three kinds of e-waste sludge all contain Cu, Ni, and Zn.

4 Species of microorganisms selected for bioleaching metals from e-waste sludge

4.1 Research status of the bioleaching technology in e-waste sludge

The bioleaching technology was first applied in mineral metallurgy areas, and has received more and more worldwide attention because of its mild reaction conditions, low consumption of acids, and lower cost (Harikrushnan et al., 2016). To date, bioleaching technology has already been adopted for the extraction of metals from mines and sludge, and its industrialized application has also been achieved (Wang et al., 2016). In the mid-1980s, the bioleaching technology that was applied in low-grade copper mines was first developed on an industrialized application. In 1992, copper production was 1.5 million tons in Chile, and 20% of that was extracted by bioleaching technology.

Table 1 Species and contents of heavy metals in e-waste sludge

Unit: %	Sources	Cu	Ni	Zn	Cr	Mn	Sn	Pb
Stainless steel sludge 1	Guangdong	0.10	0.68	0.02	1.43	–	–	–
Stainless steel sludge 2	Jiangsu	0.00	0.02	7.04	0.09	–	–	–
PCB sludge 1	Guangdong	13.24	0.02	0.38	–	0.14	–	0.08
PCB sludge 2	Guangdong	15.75	0.87	0.12	–	0.22	–	0.10
Electroplating sludge 1	Guangdong	6.61	0.03	0.02	–	0.04	0.02	–
Electroplating sludge 2	Guangdong	5.55	0.22	0.24	–	0.01	0.74	–
Electroplating sludge 3	Guangdong	5.87	0.02	0.02	–	0.39	0.33	–
Electroplating sludge 4	Guangdong	2.67	–	0.01	–	0.07	0.41	–
Electroplating sludge 5	Guangdong	4.52	0.03	0.06	–	0.23	–	0.04
Electroplating sludge 6	Guangdong	0.82	7.78	0.41	–	0.16	–	0.05
Electroplating sludge 7	Guangdong	1.41	1.44	2.30	–	0.04	–	0.03
Electroplating sludge 8	Guangdong	0.64	2.64	5.88	–	0.11	–	0.03

Note: "–" means not be detected.

Besides, bioleaching technology produced $0.35 billion of copper from copper mine in America. Furthermore, 18% of the newly increased production of copper every year came from bioleaching technology (Zhou and Xu, 2012; Darezereshki et al., 2018). The applications of bioleaching technology can also be found in Australia and Brazil. In recent years, as an emerging technology, bioleaching has been applied in the recovery of metals from e-waste, sewage sludge, and mining waste, but this direction of research is uncommon in China. Generally speaking, microorganisms used in bioleaching can be divided into two types: auxoautotrophs and auxohetertrophs (Nakade, 2013; Gu et al., 2014). Auxoautotrophs include *A. ferrooxidans*, *A. thiooxidans*, *Leptospirillum ferrooxidans* (*L. ferrooxidans*), *Sulfobacillus*, *Acidianus*, and *Acidiphilium*, of which the *A. ferrooxidans*, *A. thiooxidans*, and *L. ferrooxidans* are the most widely used in practical application. Auxohetertrophs can bioleach heavy metals from the solid phase into the liquid phase through a function that can use organic substrates to produce acidic metabolites, and the mechanisms mainly contain acidolysis, complexation, biological reduction, and alkalinization. Usually, auxohetertrophs include bacteria and fungi. Most of the bacteria are *Bacillus* and *Pseudomonas* while their metabolites are carbonic acid, nitric acid, sulfuric acid, acetic acid, and formic acid. *Aspergillus* and *Penicillium* are the most commonly used fungi because they can produce metabolites such as oxalic acid, citric acid, and gluconic acid (Yang et al., 2009). As of 2017, a research group developed a bioleaching technology applied in electroplating sludge. Two kinds of *Thiobacillus* were adopted in their studies, and this indicated that the bioleaching rates of Cu and Ni were 96% and 91% after eight days of reaction, respectively (Tezcan et al., 2017). Schippers uses 11 kinds of auxohetertrophs to bioleach metals from alkaline slag, of which the consequences of fungi and saccharomycetes were the worst while the effect of corynebacterium was the best. The bioleaching rates of each metal were as follows, according to the tests: 38% Mn, 46% Mg, 68% Ca, 27% Zn, 15% Fe, 26% Ni, 40% Co, and 8% Pb (Schipper et al., 2002). The above studies mostly focused on the low solid-liquid ratio condition. Due to the inhibiting effect of high concentration of heavy metals to the normal growth of microorganisms, using microorganisms to bioleaching heavy metals under high concentration conditions is basically less. Supposing that microorganisms can bioleach heavy metals effectively under high solid-liquid ratio conditions of the sludge and culture medium, the disposal costs can be reduced greatly by converting the hazardous waste into ordinary waste. To date, in the authors' laboratory, PCB sludge, electroplating sludge, and stainless steel sludge extracted by *A. ferrooxidans* were investigated. Also, the conditions of optimization, bioleaching mechanisms, and harmless and recycling technologies were also discussed.

4.2 Harmless research of PCB sludge by bioleaching technology

4.2.1 Materials and methods

The sources of experimental sludge are shown in Table 1. From this table, the concentrations of heavy metals in different kinds of sludge are also known. Prior to the next experiment, the PCB sludge was naturally dried and crushed uniformly, then sieved by 60 mesh of sifter.

The microorganisms used in this chapter are isolated from the Dexing copper mine at Jiangxi Province, named *A. ferrooxidans* (Wang et al., 2009). Usually, *A. ferrooxidans* is used to bioleach heavy metals from waste PCBs in the authors' laboratory, and it has a good effect (Gu et al., 2017c). A bacteria which can tolerate high concentrations of heavy metals is selected as an initially inoculated solution on the process of bioleaching heavy metals from waste printed circuit board. Taking 5 mL bacteria solution to inoculate into 100 mL 9K culture medium, then cultivated five days in a shaker with its temperature of 30°C. Finally, the bacteria solution is inoculated into a fresh culture medium. Prior to the next experiment, the above processes are repeated three times.

The sludge property, extraction time, solid-liquid ratio, initial pH, and concentration of $FeSO_4 \cdot 7H_2O$ will affect the bioleaching effect of heavy metals from sludge. In order to explore the effects of bioleaching heavy metals from sludge, it is necessary to investigate the influence of these factors on the bioleaching process. 0.5 g, 1.0 g, 2.5 g, and 5.0 g dried PCB sludge was added into a 100 mL conical flask with 45 mL 9K culture medium, respectively. Five mL bacteria solution is inoculated, and the total solution is cultured in a shaker with 30°C and 160 rpm of conditions. Sampling and detecting pH processes are taken every other day. Taking 0.5 mL supernate to dilute into 50 mL by deionized water, then the heavy metals in the solution is detected by Inductively Coupled Plasma Optical Emission Spectrometer (ICP-OES). A fresh culture medium is added after every sampling process. No inoculation group is set as the control group, each group is paralleling with 3 times.

Five grams of dried PCB sludge are added into a 100 mL conical flask with 45 mL 9K culture medium, and a 10% bacteria solution is also added into the culture medium, then cultured in a shaker at 30°C and 160 rpm of conditions. The same method is adopted continuously four times, and the inoculated solution in the second generation is the mixture of that in the first generation. The variation of pH and the bioleaching rates of Cu, Zn, Ni, and Cr after six days of reaction are all detected. The optimization of conditions that can affect the reaction are performed. For instance, 1/10 of solid-liquid ratio, the fifth generation of inoculated solution, 0.5 of initial pH, and the effect of $FeSO_4 \cdot 7H_2O$ to the bioleaching rates of heavy metals from PCB sludge are investigated. Besides, the initially untreated PCB sludge is digested by aqua regia under a water bath with the conditions of 90°C and 3 h, then diluted and analyzed by deionized water and ICP-OES.

4.2.2 Results and discussion

4.2.2.1 Effect of solid-liquid ratio on the bioleaching rate

Recently, research indicated that using *A. ferrooxidans* to dispose of copper-containing sludge with its 3% solid-containing content, the bioleaching rates of Ni, Cr, and Cu after 15 days of reaction were 92.8%, 85.0%, and 96.8%, respectively, but the main reason to restrict its large-scale application is long time-consuming (Cui and Zhang, 2008). Considering the efficiency of industrial application, six days of reaction were selected as a control point in this experiment, and the effects of high concentration of heavy metals to the bioleaching rates were also investigated.

The variations of different heavy metals' bioleaching rates along with the solid-liquid ratio are shown in Fig. 2. Reflected from Fig. 2, the bioleaching rates of Cu and Ni were changed from 90% to 10% while the bioleaching rates of Zn and Cr were changed from 90% to 60%, and 50% to 10%, respectively. These phenomena indicated that an obvious inhibition to the normal growth of *A. ferrooxidans* existed when the solid-liquid ratio was increased gradually. Generally speaking, heavy metals are extracted through the direct and/or indirect role of *A. ferrooxidans*. The possibly presented mechanisms of the bioleaching process are listed as follows. Seen from Eqs. (1), (2), (3) (Liu et al., 1998). "M" represents any divalent metal cations.

$$MS + 2O_2 \rightarrow M^{2+} + SO_4^{2-} \quad (1)$$

$$4Fe^{2+} + O_2 + 4H^+ \rightarrow 4Fe^{3+} + 2H_2O \quad (2)$$

$$4Fe^{3+} + MS + 2H_2O + O_2 \rightarrow MSO_4 + 4Fe^{2+} + 4H^+ \quad (3)$$

A direct role means using *A. ferrooxidans* to bioleach heavy metals from sludge directly through direct oxidation. An indirect role shows *A. ferrooxidans* oxidizes Fe^{2+} into Fe^{3+} by using its metabolism, then metallic sulfide is oxidized to metallic sulfate by Fe^{3+}, and H^+ is produced simultaneously. Besides, the hydrolysis of Fe^{3+} can also produce H^+, leading to the reduction of pH in the solution, thus speeding up the reaction. According to the mechanism of the indirect role, when high concentrations of heavy metals produce inhibition to the metabolism of *A. ferrooxidans*, reaction (2) can be restrained, the oxidation rate of Fe^{2+} can also be reduced, the

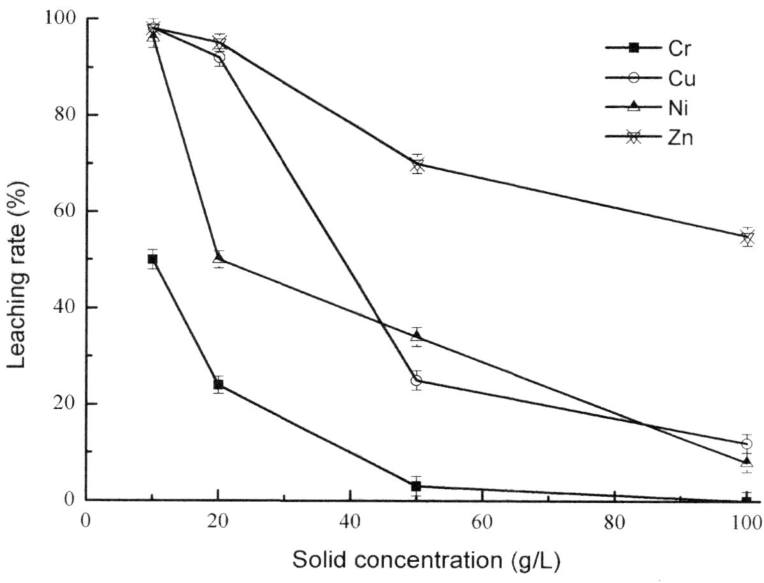

Fig. 2 Leaching rates of heavy metals under different solid-liquid ratios.

descent rate of pH slows down, even be rose, leading to the decreasing of bioleaching rates of heavy metals from PCB sludge.

Another study indicated that when the concentrations of Ni^{2+}, Cr^{3+}, Cu^{2+}, and Zn^{2+} are 30, 0.1, 2.5, and 30 g L^{-1}, respectively, they can inhibit the normal growth of *A. ferrooxidans* (Xie et al., 2013). Through the authors' investigation, when the solid-liquid ratio is 1/100 in the solution, the concentration of Cu^{2+} reaches 3.5 g L^{-1} after six days of reaction. Increasing the solid-liquid ratio can inhibit the bioleaching rates of heavy metals from PCB sludge.

4.2.2.2 Effect of the acclimation period on the bioleaching rate

A preliminary study investigated whether high concentrations of heavy metals can inhibit the normal growth of microorganisms. Therefore, a series of continuously domesticated processes were performed, and about five cycles were adopted at 1/10 of the solid-liquid ratio condition. The domestication effects of highly tolerated microorganisms were investigated.

A. ferrooxidans, which can tolerate acid and heavy metal surroundings, is selected through domestication processes. On one hand, the domesticated processes can increase the concentration of *A. ferrooxidans*. On the other hand, they can also shorten the lag phase of *A. ferrooxidans* among its growth process. Both those actions can accelerate the metabolic rate of *A. ferrooxidans*. According to the indirect mechanism, the metabolism of *A. ferrooxidans* can increase the concentration of H^+, thus the pH value in the solution can be reduced gradually.

Fig. 3 reflects the variations of heavy metals in the solution among five acclimation periods. The results indicated that the bioleaching rates of Cr, Cu, Ni, and Zn were increased along with the time. After five cycles' continuous acclimation, the bioleaching rates of Cr, Cu, Ni, and Zn in PCB sludge were increased from 3%, 17%, 34%, and 12% to 8%, 71%, 65%, and 78%, respectively. The increasing bioleaching rates also reflect the increasing metabolism of the microorganisms. Along with the acclimation cycles, the bioleaching rates of heavy metals in the solution were increased continuously. These might be the results of the domestication processes, the highly tolerated microorganisms were selected, and the time of microorganism's lag phase was shortened, thus the phenomena described above were emerged.

4.2.2.3 Effect of $FeSO_4 \cdot 7H_2O$ concentration on the bioleaching rate

The variation of pH in the solution is the result of the microorganism's metabolism and chemical reaction. Due to the alkalinity of the PCB sludge itself, the microbial reaction and acid-base neutralization happen simultaneously. On the one hand, the indirect action of the metabolism produces H^+, but on the other hand, the production and release of some alkaline substances have the buffering of the pH in the solution. When the concentration of $FeSO_4 \cdot 7H_2O$ is low, which means the substrate is insufficient, the metabolism of the microorganism is slow, and the chemical reaction has the dominant position, the pH value is rose slowly. When the substrate is sufficient, which means the microbial reaction has the dominant position, the pH value is decreased definitely.

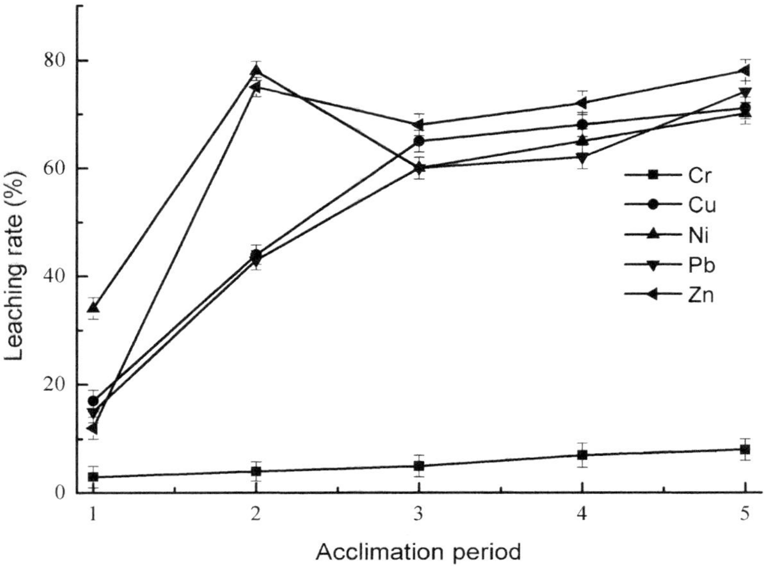

Fig. 3 Leaching rates of heavy metals under different acclimation periods.

Fig. 4 reflects the variation of bioleaching rates of heavy metals under different concentrations of $FeSO_4{\cdot}7H_2O$. Through the observation of bioleaching rates after six days of reaction, when the concentration of $FeSO_4{\cdot}7H_2O$ is less than 60 g L^{-1}, the bioleaching rates of Cu, Ni, and Zn are increased along with the Fe^{2+} increasing process. To some extent, when the concentration of Fe^{2+} is increased continuously, the bioleaching rates of Cu and Zn remain unchanged, and the bioleaching rate of Ni decline. Furthermore, the bioleaching rate of Cr remains less than 10%. All these phenomena indicate that a high concentration of Fe^{2+} can inhibit the bioleaching rate of Ni, so 60 g L^{-1} of $FeSO_4{\cdot}7H_2O$ is selected as the optimization condition.

When the substrate is sufficient, the microbial reaction has the dominant position, leading to the decreasing of pH, and the acidic environment accelerates the bioleaching rates of heavy metals from PCB sludge. Besides, the superfluous Fe^{2+} will be reacted with K^+ and $NH_4SO_4^-$ in the solution, finally forming a precipitate called jarosite. Unluckily, the precipitate will be adsorbed on the surface of PCB sludge, then inhibit the bioleaching effects of heavy metals. To some extent, on the process of turning Fe^{2+} into Fe^{3+}, the flocculation increases the effects of adsorption and coprecipitation, making the dissolved heavy metals return to the solid phase again.

4.2.2.4 Effect of initial pH of the culture medium on the bioleaching rate

Due to the alkalinity of the PCB sludge itself, the H^+ in the culture medium has been reacted with the alkaline substances that are released from the PCB sludge at the initial reaction stage. Increasing the production of acid can neutralize the alkaline substances in the culture medium, giving the culture medium a faintly acid environment and thus

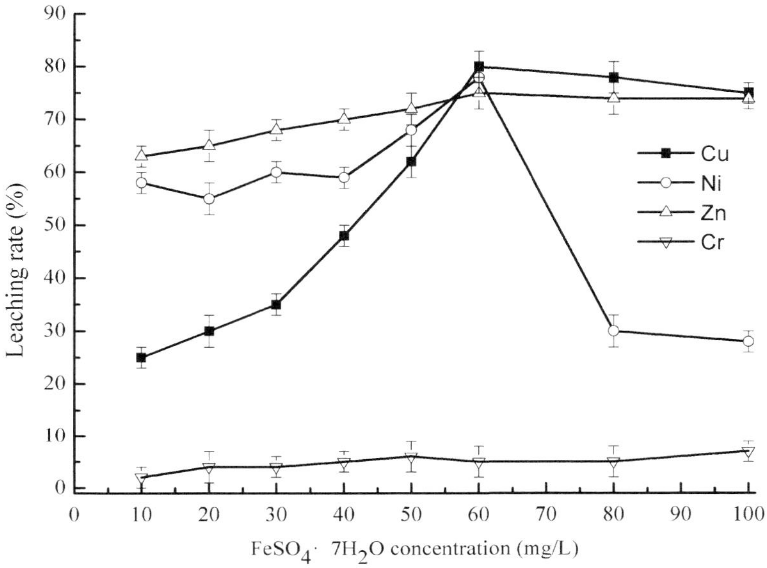

Fig. 4 Leaching rates of heavy metals under different concentration of $FeSO_4 \cdot 7H_2O$.

accelerating the reproduction of *A. ferrooxidans*. At the same time, the direct and indirect actions of *A. ferrooxidans* can bioleach heavy metals from PCB sludge. When the acidic substances are insufficient in the culture medium, the initial pH of the culture medium reflects neutral or alkaline, and the metabolism of *A. ferrooxidans* is slow, leading to the low bioleaching rates of heavy metals. Besides, the leached heavy metals may have the redeposition processes, and the bioleaching rates of heavy metals reflect the tendency of first rise and then fall.

The variations of heavy metal bioleaching rates under different initial pH (0.25, 0.5) of the 9K culture medium are shown in Fig. 5. The bioleaching rates of Cr, Cu, Ni, and Zn are 65%, 76%, 80%, and 77% in the culture medium with its initial pH 0.25 after six days of reaction. At the initial pH 0.5 of the culture medium, the bioleaching rates of Cr, Cu, Ni, and Zn are 2%, 78%, 53%, and 74%, respectively. The dissolution of heavy metals is mainly dominated by the variation of pH in the culture medium, and different heavy metals suffer from different levels of impacts. The difference of the bioleaching rate of Cr is the result of the initial pH in the culture medium, which doesn't reach the bioleaching limit of Cr. The increase of the pH on the bioleaching process results in the precipitation of Fe^{3+}, making the difference of the bioleaching rate of Fe. When the initial pH is 0.5 in the culture medium, compared to that of the initial pH 0.25, it may extract heavy metals selectively. This pH value is favorable in the subsequent recovery processes.

When the solid-liquid ratio is higher than 1/20, the high concentration of heavy metals in solution has an obvious inhibition to the microbial bioleaching process.

The bioleaching rates of Cu, Ni, and Zn, which are extracted by domesticated bacteria, are higher than that of bacteria that isn't domesticated. Through the continuous

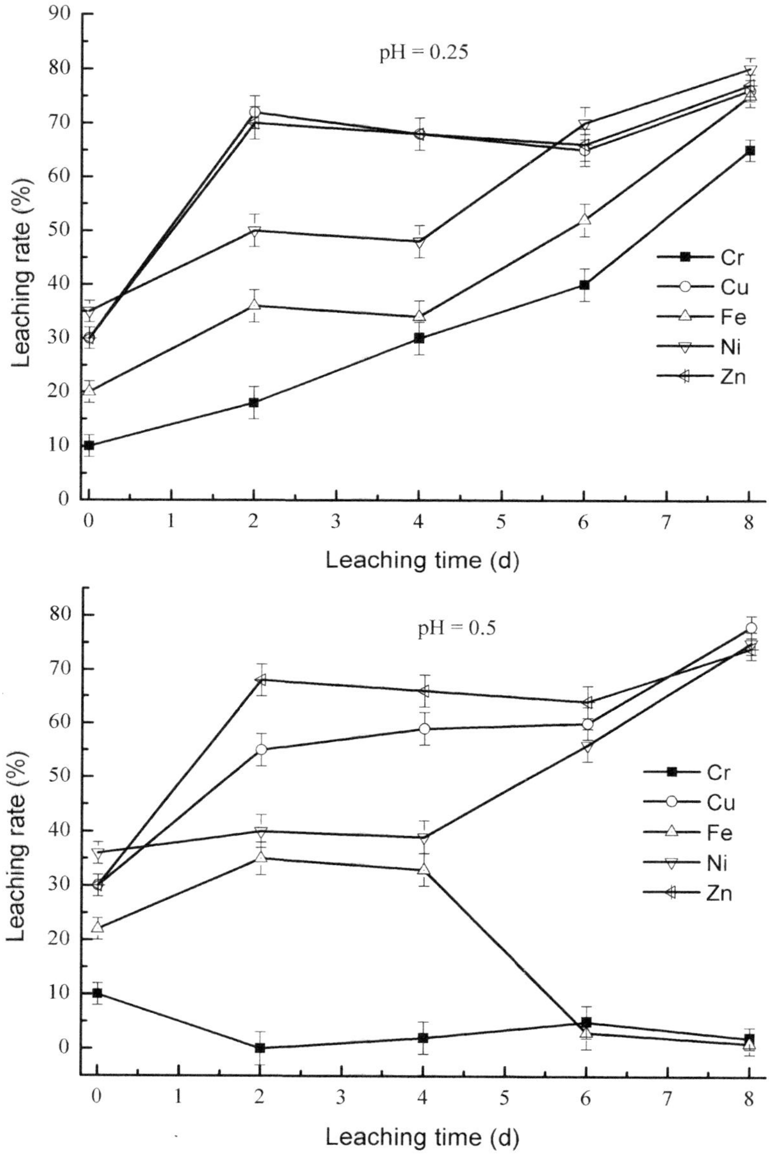

Fig. 5 Leaching rates of heavy metals under different initial pH of culture medium.

domestication, the tolerance of *A. ferrooxidans* to high concentrations of heavy metals is enhanced. Besides, the bioleaching rate of Cr from PCB sludge is not good because of its toxicity to the normal growth of *A. ferrooxidans*.

High bioleaching rates and fast starting speed are the results of choosing the initial pH of 0.25 or 0.5 in the 9K culture medium. When the initial pH is 0.5, Cu, Ni, Zn, and Fe can be leached selectively. Compared to the initial pH of 0.25, the consumption of acid is less. The initial pH of 0.5 is more suitable for the next industrial application.

Using *A. ferrooxidans* to bioleaching heavy metals from PCB sludge, the optimization condition is $FeSO_4{\cdot}7H_2O$ 60 g L^{-1}, initial pH 0.5, reaction time six days, and the bioleaching rates of Cu, Ni, and Zn are 76%, 74%, and 72%, respectively.

4.3 Harmless research of electroplating sludge by bioleaching technology

4.3.1 Materials and methods

The sources of electroplating sludge are also shown in Table 1. From this table, the concentrations of heavy metals in eight kinds of electroplating sludge are known. Besides, the moisture content of the selected electroplating sludge is 46.9%, and the appearance of this sludge is bronzing. Prior to the next experiment, the electroplating sludge was also naturally dried and crushed uniformly, then finally sieved by 60 mesh of sifter.

The microorganism that is used in this experiment is also isolated from the Dexing copper mine at Jiangxi Province, and is named *A. ferrooxidans* (Wang et al., 2009). The cultivation and domestication processes for the *A. ferrooxidans* that is adopted to extract heavy metals from electroplating sludge are the same as the reaction processes adopted between *A. ferrooxidans* and PCB sludge. Prior to the next experiment, the domestication processes are also repeated three times.

The influencing factors such as sludge property, extraction time, solid-liquid ratio, initial pH, and concentration of $FeSO_4{\cdot}7H_2O$ are almost the same as the reaction between *A. ferrooxidans* and PCB sludge. Therefore, the operation steps are almost the same as the bioleaching reactions between *A. ferrooxidans* and PCB sludge, and the only difference is the sample used in this experiment is the electroplating sludge. First, use ICP-OES to analyze the concentrations of heavy metals in the electroplating sludge. Then, use *A. ferrooxidans* to extract the heavy metals from the electroplating sludge.

4.3.2 Results and discussion

4.3.2.1 Effect of solid-liquid ratio on the bioleaching rate

As previously said, when the concentrations of Ni^{2+}, Cr^{3+}, Cu^{2+}, and Zn^{2+} are higher than 30 g L^{-1}, 0.1g L^{-1}, 2.5 g L^{-1}, and 30 g L^{-1}, respectively, there is an inhibition to the activity of *A. ferrooxidans* (Xie et al., 2013). Therefore, a high solid-liquid ratio can also affect the bioleaching rates of heavy metals from electroplating sludge by *A. ferrooxidans*. Using six days of reaction as a control point, the effects of the solid-liquid ratio to the bioleaching reaction are investigated. Fig. 6 shows the

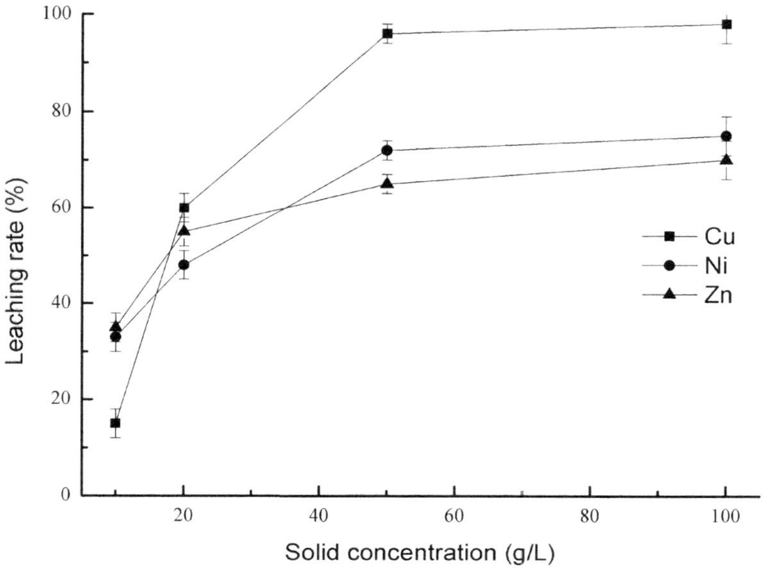

Fig. 6 Leaching rates of heavy metals under different solid-liquid ratios.

variations of heavy metal bioleaching rates under the different solid-liquid ratio conditions, of which the bioleaching rate of Cu declineds from 90% to 20%, and the bioleaching rates of Ni and Zn decline from 70% to 40%. The results indicate that when the solid-liquid ratio is higher than 1/50, the bioleaching rates of heavy metals from electroplating sludge are inhibited.

4.3.2.2 Effect of acclimation period on the bioleaching rate

The above experiment indicates that a high solid-liquid ratio can inhibit the bioleaching effect of electroplating sludge by *A.ferrooxidans*. This part of the study shows the effect of domesticated bacteria to bioleaching heavy metals from electroplating sludge on the continuously domesticated processes. The bioleaching rates of heavy metals from electroplating sludge along with the variations of the domestication period are shown in Fig. 7. Compared with the bioleaching rates of Cu, Ni, and Zn between the first domestication period and the fifth domestication period, the bioleaching rates of Cu, Ni, and Zn are improved from 10.16%, 40.20%, and 10.35% to 82.32%, 51.32%, and 70.32%, respectively. The improving of bioleaching rates reflects the enhancing of *A. ferrooxidans*'s metabolism. The domestication processes have already enhanced the bioleaching rates of heavy metals from electroplating sludge.

4.3.2.3 Effect of $FeSO_4 \cdot 7H_2O$ concentration on the bioleaching rate

According to the mechanism of the bioleaching process, two parts of action also exist in this process, including chemical and biological process. Due to the alkalinity of the electroplating sludge itself, there is an acid-base neutralization reaction existing in the

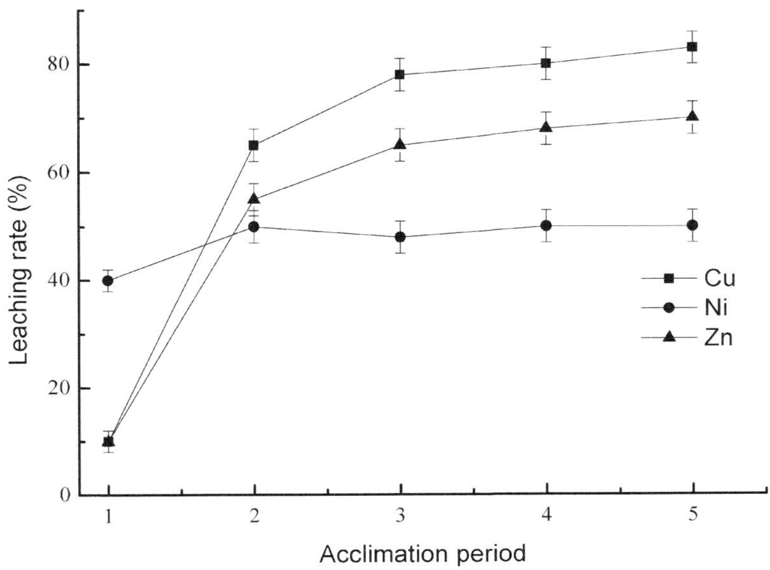

Fig. 7 Leaching rates of heavy metals under different acclimation periods.

bioleaching process. When the action of the microorganism is weak, less H^+ is produced, and the release of alkaline substances plays a dominant role in the culture medium, leading to the rise or invariability of pH in the solution. When the substrate is adequate, the metabolism of the microorganism has the dominant role, the pH in the solution declines, and the heavy metals in the electroplating sludge will be leached quickly. Therefore, the decline of pH can reflect the metabolism situation of the microorganism in the culture medium to some extent.

The variations of heavy metal bioleaching rates alone with the dosage of $FeSO_4{\cdot}7H_2O$ are shown in Fig. 8. When the concentrations of $FeSO_4{\cdot}7H_2O$ are $20\,g\,L^{-1}$, $40\,g\,L^{-1}$, and $60\,g\,L^{-1}$, the results show that the bioleaching rates of heavy metals are increased gradually, along with the increasing $FeSO_4{\cdot}7H_2O$ concentration. When the concentration of $FeSO_4{\cdot}7H_2O$ is $80\,g\,L^{-1}$, the bioleaching rate of Cu remains unchanged, and the bioleaching rates of Ni and Zn both decline. These investigations demonstrate that enough Fe^{2+} can enhance the bioleaching rates of heavy metals in electroplating sludge, but excess Fe^{2+} will inhibit the bioleaching rates of some heavy metals. In other words, when the solid-liquid ratio is 1/10 and the concentration of $FeSO_4{\cdot}7H_2O$ is less than $60\,g\,L^{-1}$, the bioleaching rates of heavy metals can be enhanced by improving the dosage of Fe^{2+}. While the concentration of $FeSO_4{\cdot}7H_2O$ exceeds $60\,g\,L^{-1}$, it will inhibit the bioleaching effects of some heavy metals. Therefore, under 1/10 of solid-liquid ratio, using $60\,g\,L^{-1}$ of $FeSO_4{\cdot}7H_2O$ as a control point is the most suitable condition in bioleaching heavy metals from electroplating sludge.

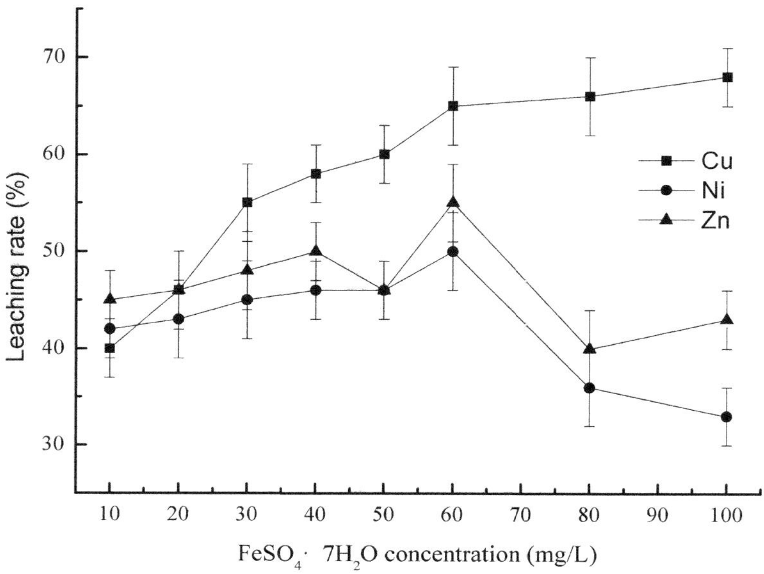

Fig. 8 Leaching rates of heavy metals under different concentrations of $FeSO_4{\cdot}7H_2O$.

4.3.2.4 Effect of initial pH of culture medium on the bioleaching rate

The variations of heavy metal bioleaching rates in the different initial pH of the 9K culture medium are shown in Figs. 9–11. When the initial pH is 2.5, the bioleaching rate of Zn decline from 32% to 0%, and when the initial pH is 1.0, the bioleaching rate of Zn also declinedfrom 32% to 24%, then rises to 35% gradually after two days of reaction. At the beginning of the reaction, the bioleaching rates of heavy metals rise continuously among the culture medium when the initial pH is 0.75, 0.5, and 0.25, respectively. When the initial pH is 0.5 or 0.25, the bioleaching rate of Zn reaches 80% after eight days of reaction. Due to the release of alkaline substances into the solution at the initial time of reaction, the pH is increased, leading to the dissolved heavy metals' redeposition in the solution. A low initial pH of the culture medium can slow down the increasing of the pH, which is affected by the release of alkaline substances. On the other hand, a pH value between 2.0 and 3.0 is suitable for the normal growth of *A. ferrooxidans*, and accelerates the bioleaching rates of heavy metals from electroplating sludge. The variations of the bioleaching rates of Cu and Ni are similar with that of Zn's. These phenomena indicate that using a low initial pH in the 9K culture medium can speed up the bioleaching rates of heavy metals. Considering the economic cost, the initial pH of 0.5 is selected as the control point under 1/10 of the solid-liquid ratio in bioleaching heavy metals from electroplating sludge.

When the solid-liquid ratio is higher than 1/50, the high concentration of electroplating sludge can inhibit the bioleaching process. Through continuous domestication processes, the tolerance of *A. ferrooxidans* can be enhanced. Compared to the nondomestication of *A. ferrooxidans*, the bioleaching rates of Cu, Ni, and Zn are enhanced 72.16%, 12.12%, and 59.97%, respectively. Besides, the bioleaching rate

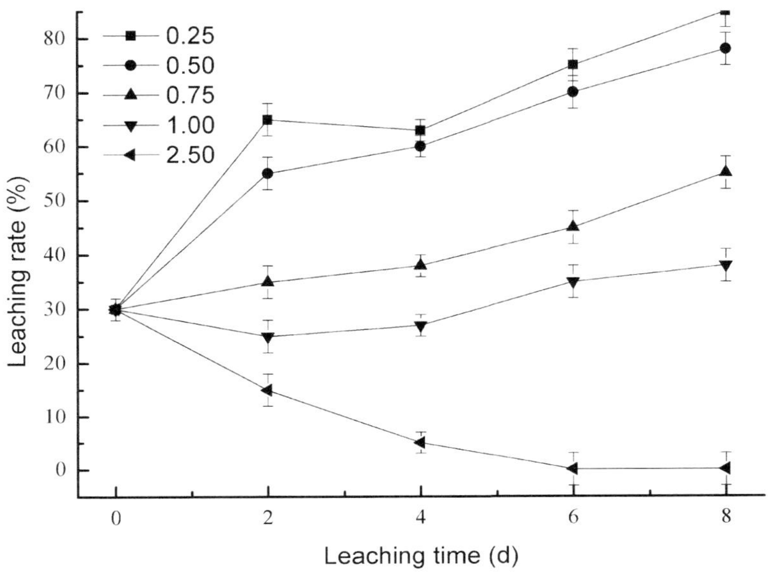

Fig. 9 Leaching rates of Zn under different initial pH.

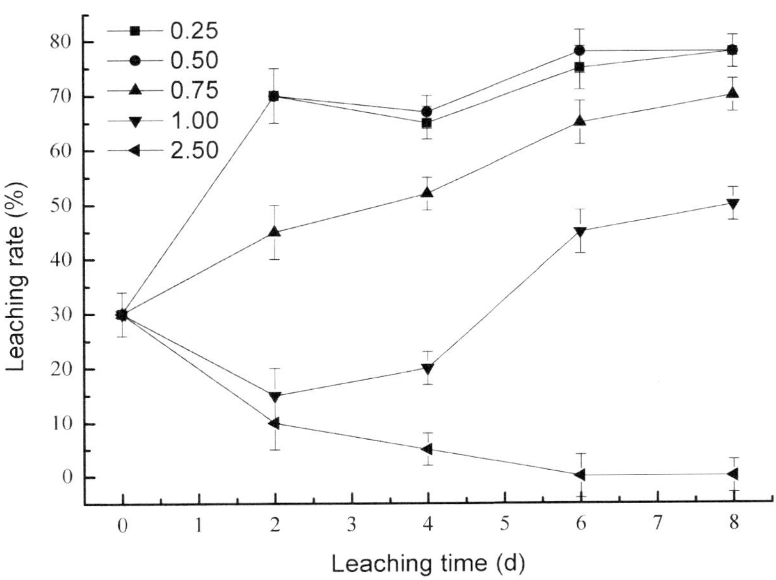

Fig. 10 Leaching rates of Cu under different initial pH.

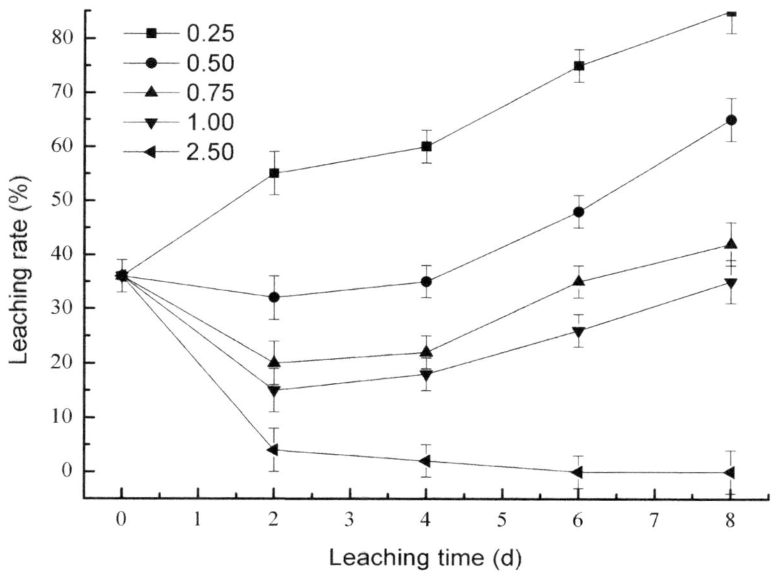

Fig. 11 Leaching rates of Ni under different initial pH.

of Cr from electroplating sludge is also not good because of its toxicity to the normal growth of *A. ferrooxidans*.

A moderate concentration of Fe^{2+} is beneficial to bioleaching heavy metals from electroplating sludge under 1/10 of the solid-liquid ratio condition. When the concentration of $FeSO_4 \cdot 7H_2O$ is higher than $60\,g\,L^{-1}$, a superfluous Fe^{2+} can inhibit the bioleaching of Ni and Zn. By doing the acidification process to the 9K culture medium in advance, this can accelerate the starting speed of the bioleaching process and speed up the bioleaching of heavy metals. The optimal bioleaching condition is $FeSO_4 \cdot 7H_2O$ $60\,g\,L^{-1}$, initial pH 0.5, no KCL and $MgSO_4$, solid-liquid ratio 1/10, after 6 days of reaction, the bioleaching rates of Cu, Ni and Zn are 81.43%, 75.32% and 78.97%, respectively.

4.4 Harmless research of stainless steel sludge by bioleaching technology

4.4.1 Material and methods

The sources of stainless steel sludge are also shown in Table 1. From this table, the concentrations of heavy metals in two kinds of stainless steel sludge are also known. Prior to the next experiment, the stainless steel sludge was naturally dried and crushed uniformly, and sieved by 60 mesh of sifter.

The microorganism used in this experiment is also isolated from the Dexing copper mine at Jiangxi Province, and is named *A. ferrooxidans* (Wang et al., 2009). Usually, as previously said, the cultivation and domestication processes of *A. ferrooxidans*, which is adopted to extract heavy metals from stainless steel sludge, are the same as the reaction processes that are adopted between *A. ferrooxidans* and electroplating sludge. Prior to the next experiment, the domestication processes are also repeated three times.

The influencing factors such as sludge property, extraction time, solid-liquid ratio, initial pH, and concentration of $FeSO_4{\cdot}7H_2O$ are almost the same as the reaction between *A. ferrooxidans* and electroplating sludge. Therefore, the operation steps are almost the same as the bioleaching reaction between *A. ferrooxidans* and electroplating sludge, and the only difference is the sample used in this experiment is stainless steel sludge. First, use ICP-OES to analyze the concentrations of heavy metals in stainless steel sludge. Then, use *A. ferrooxidans* to extract heavy metals from stainless steel sludge.

4.4.2 Results and discussion

4.4.2.1 Effect of solid-liquid ratio on the bioleaching rate

Due to the different constitutions and concentrations of heavy metals in stainless steel sludge when compared to the PCB sludge and electroplating sludge, eight days of reaction is selected as a control point in this section, and the effects of the solid-liquid ratio on the bioleaching process are investigated. The bioleaching rates of heavy metals from stainless steel sludge, which contains Cr along with the variation of the solid-liquid ratio, are shown in Fig. 12. From this figure, when the solid-liquid ratio declines, the bioleaching rates of Mn and Cr are 80% and 2%, respectively while the bioleaching rates of Ni and Sn are also increased slightly on the basis of 40%. Comparing the groups between bacteria and no bacteria, the variation of bioleaching rates remains unchanged. The bioleaching rates of heavy metals from stainless steel sludge that contains Zn along with the variation of the solid-liquid ratio are shown in Fig. 13. Seen from this figure, the bioleaching rate of Mn remains approximately 90%, and the bioleaching rate of Zn is increased from 50% to 90%. Comparing the groups between bacteria and no bacteria, the bioleaching rate of Mn remains unchanged. All the results indicate that when the solid-liquid ratio is higher than 1/50, it can inhibit the bioleaching effects of heavy metals. Using *A. ferrooxidans* to extract stainless steel sludge that contains Cr is not good while the effect of using *A. ferrooxidans* to extract stainless steel sludge that contains Zn is good.

Generally speaking, the bioleaching process of *A. ferrooxidans* also exists direct and indirect action. The direct action represents the direct oxidation of metal sulfide into metal sulfates, and dissolves the metals. The indirect action means using *A. ferrooxidans* to oxidize Fe^{2+} to Fe^{3+} in the 9K culture medium, and Fe^{3+} is reacted with the metal sulfide to generate metal sulfates. At the same time, the hydrolysis of Fe^{3+} will produce H^+, and this process can also accelerate the bioleaching process of heavy metals from stainless steel sludge. The possibly presented mechanisms of the bioleaching process are also listed as follows (Liu et al., 1998), and it is almost the same as in Section 4.2.2.1. "M" also represents any divalent metal cations.

$$MS + 2O_2 \rightarrow M^{2+} + SO_4^{\ 2-} \tag{4}$$

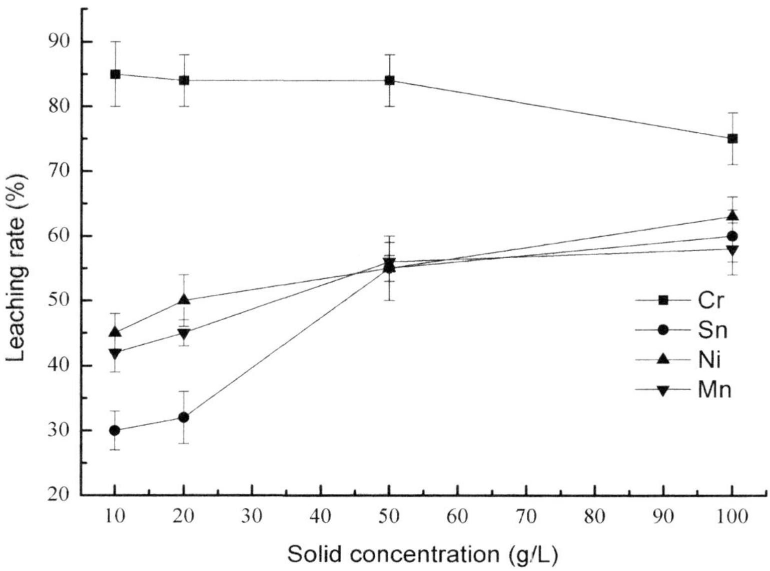

Fig. 12 Leaching rates of heavy metals under different solid-liquid ratios.

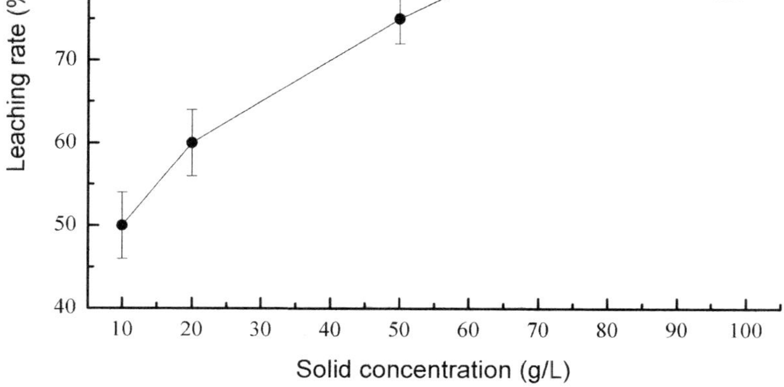

Fig. 13 Leaching rates of heavy metals under different solid-liquid ratios.

$$2Fe^{2+} + 0.5O_2 + 2H^+ \rightarrow 2Fe^{3+} + H_2O \quad (5)$$

$$2Fe^{3+} + MS + H_2O + O_2 \rightarrow M^{2+} + 2Fe^{2+} + S^0 \quad (6)$$

$$2S^0 + 3O_2 + 2H_2O \rightarrow 2H_2SO_4 \quad (7)$$

Due to the strong biotoxicity of Cr to the microorganism, the bioleaching rates of heavy metals are mainly unchanged between the vaccinated and nonvaccinated processes. For the stainless steel sludge that contains Cr, the bioleaching rates of heavy metals are mainly influenced by the chemical reaction. The bioleaching rates of heavy metals in stainless steel sludge that contains Zn are increased along with the increasing of the solid-liquid ratio. Obviously, a high concentration of heavy metals will inhibit the normal growth of *A. ferrooxidans*, and the reaction (5) can be inhibited and the oxidation rate of Fe^{2+} will be reduced, leading to the decline of heavy metal bioleaching rates.

4.4.2.2 Effect of acclimation period on the bioleaching rate

The bioleaching rates of heavy metals from stainless steel sludge that contains Zn along with the variation of the acclimation period are shown in Fig. 14. Comparing the bioleaching rates of Mn and Zn between the first and the fifth acclimation period, the bioleaching rates of Mn and Zn are increased from 83.32% and 74.31% to 90.32% and 78.47%, respectively. The increasing bioleaching rates reflect the enhancing of *A. ferrooxidans*'s metabolism. Obviously, the domestication process can select the high tolerance of *A. ferrooxidans* and enhance the bioleaching rates of heavy metals from stainless steel sludge that contains Zn through the domesticated *A. ferrooxidans*.

4.4.2.3 Effect of $FeSO_4 \cdot 7H_2O$ concentration on the bioleaching rate

Fig. 15 indicates the bioleaching rates of heavy metals along with the variation of $FeSO_4 \cdot 7H_2O$. Obviously, the bioleaching rates of heavy metals are increased with the increasing of $FeSO_4 \cdot 7H_2O$. When the concentrations of $FeSO_4 \cdot 7H_2O$ are 30 g L^{-1} and 40 g L^{-1}, the bioleaching rate of heavy metals reaches 86.35%. The results indicate that a sufficient quantity of Fe^{3+} can improve the bioleaching rates of heavy metals. Therefore, using 1/10 of the solid-liquid ratio and 30 g L^{-1} of $FeSO_4 \cdot 7H_2O$ as a control point in the bioleaching process of stainless steel sludge that contains Zn is the best of all.

4.4.2.4 Effect of initial pH of culture medium on the bioleaching rate

The bioleaching rate of Zn under a different initial pH of the 9K culture medium has shown that when the quality of H_2SO_4 is 1/80, the bioleaching rate of Zn reaches 79% after eight days of reaction. At the beginning of the bioleaching process, most Zn in the stainless steel sludge is dissolved by the chemical reaction, and the residue of Zn is dissolved by the metabolism of *A. ferrooxidans*. The results indicate that using a low initial pH can enhance the bioleaching rate of Zn, and the metabolism of *A. ferrooxidans* can reduce the usage of acid.

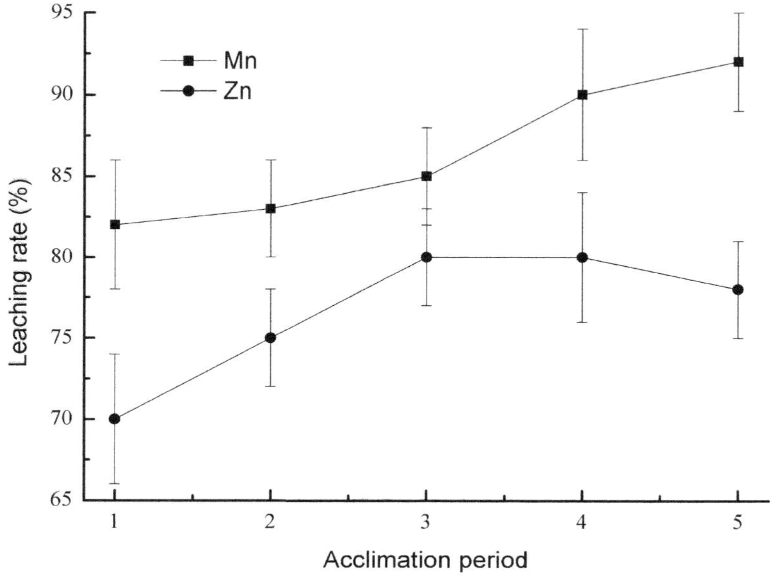

Fig. 14 Leaching rates of heavy metals under different acclimation periods.

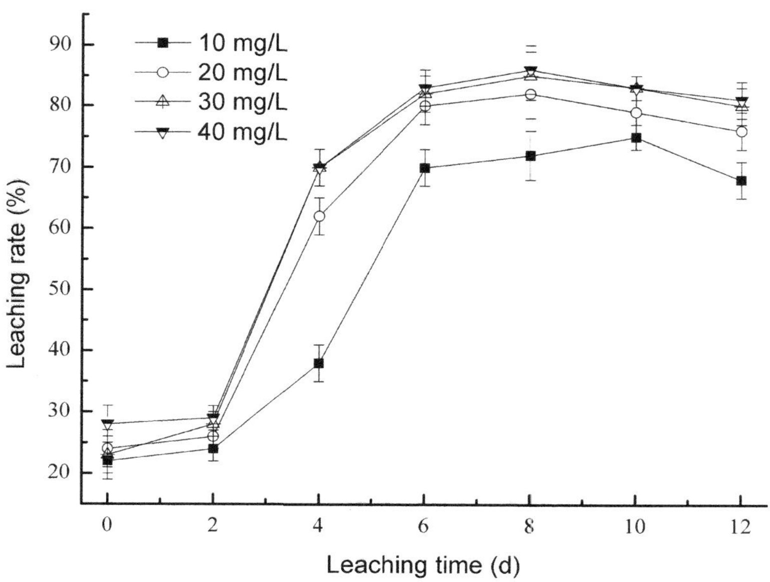

Fig. 15 Leaching rates of heavy metals under different concentration of $FeSO_4 \cdot 7H_2O$s.

Using *A. ferrooxidans* to dispose of stainless steel sludge that contains Cr is not good, and the high concentration of Cr will poison the normal growth of *A. ferrooxidans*.

When the solid-liquid ratio is higher than 1/50, there is an inhibition in the bioleaching process between *A. ferrooxidans* and stainless steel sludge that contains Zn. Through continuous domestication processes, the heavy metal tolerance of *A. ferrooxidans* can be improved. Compared with no domestication of *A. ferrooxidans*, the bioleaching rate of Cr is improved 5.16%.

A moderate concentration of Fe^{2+} is favorable to extract heavy metals from stainless steel sludge under 1/10 of solid-liquid ratio. The optimal condition is listed as follows: with $30\,g\,L^{-1}$ of $FeSO_4{\cdot}7H_2O$, 1/10 of solid-liquid ratio, and 1/40 of acidic quality, the bioleaching rate of Zn reaches 78.16% after eight days of reaction.

5 Recovery technologies

According to the descriptions above, there are several metals (Zn, Ni, Cr, and Cu) presented in PCB sludge, electroplating sludge, and stainless steel sludge, respectively. In general, after the bioleaching process, the reaction products will be filtered, and the filtrate and residue will be obtained by doing the filtration procedure. Most metals will be presented in the filtrate, and the residue contains a low concentration of metals, accordingly. Usually, many methods can be adopted to extract the metals from filtrate. Harmless technologies are also used to dispose of the residue of the bioleaching reaction. The sketch of the e-waste sludge's treatment is shown in Fig. 16. In general, before the bioleaching process, the metal and anion analysis of sludge and the domestication of *A. ferrooxidans*, are first performed. The optimization of conditions and the analysis of mechanisms will be investigated subsequently. A filtration process will be operated, and electrolysis is adopted to extract metals from the leachate. Speciation analysis of the metals inside the sample will be performed on the remaining residue. Finally, a harmless treatment process will be adopted to dispose of the residue that contains no active heavy metals.

5.1 Disposal of metals in filtrate

Here, two methods will be introduced to extract metals from the filtrate. One is precipitation by using alkali, the other is extraction by chemical reagents followed by the electrolysis procedure. More details are shown in Fig. 17. Due to the metal components in the filtrate, Zn powder will be first adopted to remove the Cu^{2+} presented in the filtrate. Previous studies have proved that different metals will be precipitated at different pH values. Through controlling the pH value by NaOH, Cr^{3+}, Zn^{2+}, and Ni^{2+} will be precipitated in turn in the filtrate. A filtration process must be adopted when precipitation emerges in the filtrate. By doing these processes, the remaining metals can be separated step by step. All precipitations obtained by using this method are sent to the metallurgical plant to do the refinement process subsequently, if the quantity of precipitation is large. Electrolysis is also used to extract metals from

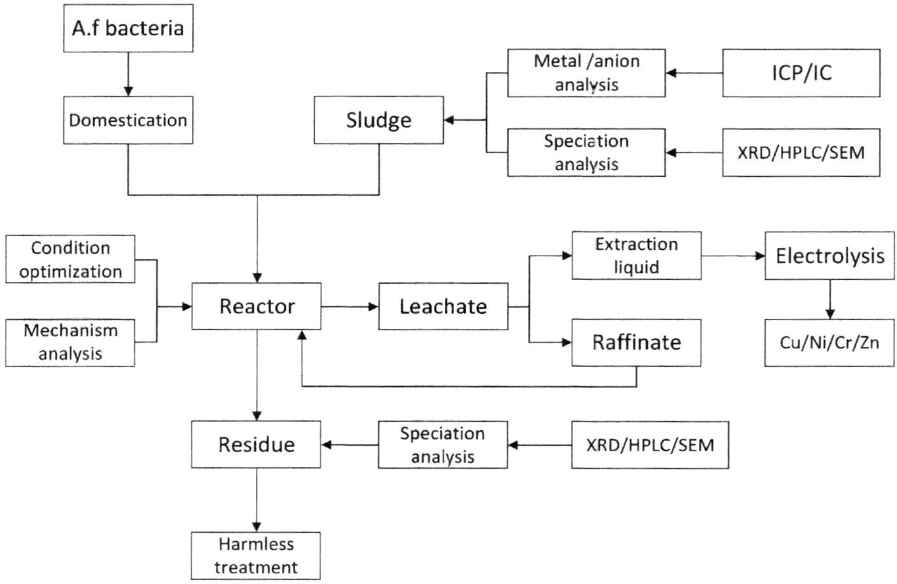

Fig. 16 Sketch of the e-waste sludge's treatment.

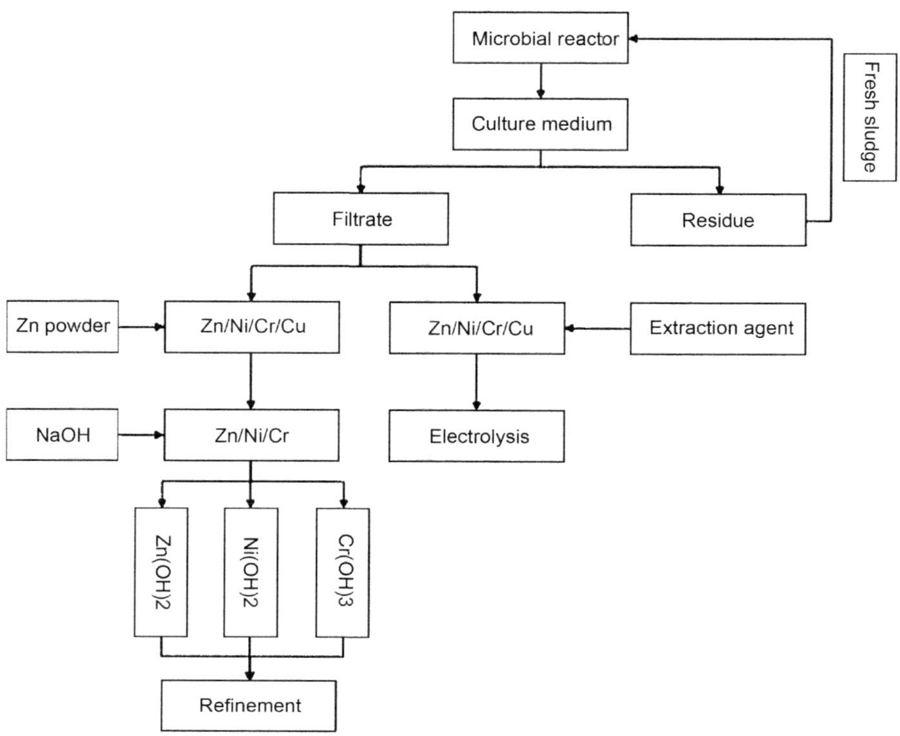

Fig. 17 Treatment sketch of the e-waste sludge's bioleaching products.

the bioleaching filtrate. More detailed processes are listed as follows. Due to the low concentration of metals obtained by the bioleaching reaction, a liquid-liquid extraction process will be performed before the electrolysis process. An extraction agent called RE609 will be adopted to condense the concentration of metals, then organic phase which contains metal ions such as Cu^{2+}, Ni^{2+} are entered the reverse extraction phase, and the remaining solution is collected to do the next cyclic utilization. A high concentration of metals is obtained after doing the reverse extraction phase, and the electrolysis process is applied for extracting metals from this solution. In general, by selecting a suitable electric current, voltage, and electrode material, a high purity of metals will be obtained through the electrolysis method.

5.2 Harmless disposal of residue

The safe disposal of bioleaching residue is very important as the final recycling period of sludge. There are many methods adopted to safely dispose of e-waste sludge that contains no metals. The residue is used as a material to manufacture adsorbent, ceramsite, cement, and biofiberboard after its stabilization and mummification processes. Through the reutilization procedure, resources and costs will be saved. When the residue is no more reused, a sanitary landfill can be adopted to dispose these residues finally. In general, a laboratory scale of e-waste sludge without metals containing will be accumulatived until its pilot scale, then harmless treatment will be adopted subsequently.

6 Conclusions

This chapter mainly investigated the research status of bioleaching heavy metals from e-waste sludge. The situation of using *A. ferrooxidans* to bioleaching heavy metals from PCB sludge, electroplating sludge, and stainless steel sludge was emphatically introduced. Besides, the technologies for disposing of filtrate and residue obtained after the bioleaching process were also introduced. The main conclusions are listed as follows.

1) Using *A. ferrooxidans* to bioleaching heavy metals from e-waste sludge is mostly performed in a laboratory scale, not in a pilot scale.
2) Using *A. ferrooxidans* to bioleaching heavy metals from PCB sludge, the optimization condition is $FeSO_4 \cdot 7H_2O$ 60 g L^{-1}, initial pH 0.5, reaction time six days, and the bioleaching rates of Cu, Ni, and Zn are 76%, 74%, and 72%, respectively. The optimum bioleaching condition for electroplating sludge is $FeSO_4 \cdot 7H_2O$ 60 g L^{-1}, initial pH 0.5, no KCL and $MgSO_4$, solid-liquid ratio 1/10, six days of reaction, and the bioleaching rates of Cu, Ni, and Zn are 81.43%, 75.32%, and 78.97%, respectively. The optimal condition for stainless steel sludge is listed as follows: 30 g L^{-1} of $FeSO_4 \cdot 7H_2O$, 1/10 of solid-liquid ratio, and 1/40 of acidic quality, and the bioleaching rate of Zn reaches 78.16% after eight days of reaction. Besides, the bioleaching rate of Cr in PCB sludge, electroplating sludge, and stainless steel sludge is not good because of its toxicity to the normal growth of *A. ferrooxidans* in the culture medium.

3) The bioleaching mechanisms of heavy metals from PCB sludge, electroplating sludge, and stainless steel sludge by *A.ferrooxidans* are almost the same. Both direct and indirect actions are presented in these bioleaching processes.
4) Electrolysis and alkaline methods are adopted to dispose of filtrate obtained after the bioleaching reaction. Reusing and sanitary landfills are adopted to dispose of the remaining residue.

Acknowledgments

The authors greatly thank the Research Center of Resource Recycling Science and Engineering, Shanghai Polytechnic University, and Gaoyuan Discipline of Shanghai-Environmental Science and Engineering (Resource Recycling Science and Engineering). They particularly thank the subsidization from the Natural Science Foundation of China (51578397) and the Program of Shanghai Technology Research Leader (17XD1420500).

References

Afroz, R., Masud, M.M., Akhtar, R., Duasa, J.B., 2013. Survey and analysis of public knowledge, awareness and willingness to pay in Kuala Lumpur, Malaysia—a case study on household WEEE management. J. Clean. Prod. 52, 185–193.

Baldé, C.P., Forti, V., Gray, V., Kuehr, R., Stegmann, P., 2017. The Global E-Waste Monitor-2017. United Nations University (UNU), International Telecommunication Union (ITU) & International Solid Waste Association (ISWA), Bonn/Geneva/Vienna.

Cui, J., Zhang, L., 2008. Metallurgical recovery of metals from electronic waste: a review. J. Hazard. Mater. 158, 228–256.

Darezereshki, E., Khodadadi Darban, A., Abdollahy, M., Jamshidi, A., 2018. Synthesis of magnetite nanoparticles from iron ore tailings using a novel reduction-precipitation method. J. Alloys Cmpd. 749, 336–343.

Das, A., Vidyadhar, A., Mehrotra, S.P., 2009. A novel flowsheet for the recovery of metal values from printed circuit boards. Resour. Conserv. Recycl. 53, 464–469.

Ding, J.D., Chang, H.M., Ding, Y., 2013. New technology for heavy metals recovery from electroplating sludge. Chin. J. Environ. Eng. 7 (5), 1969–1973.

Fei, Y.X., Bai, J.F., Deng, M.Q., 2015. Bioleaching copper from a gantry by *Acidithiobacillus ferrooxidans*. Res. Environ. Sci. 28 (8), 1267–1273.

Gu, W.H., Bai, J.F., Dai, J., 2014. Characterization of extreme acidophile bacteria bioleaching copper from flexible PCB by ICP-AES. J. Spectroscopy. 2014, 269351. https://doi.org/10.1155/2014/269351.

Gu, W.H., Bai, J.F., Dong, B., Zhuang, X.N., Zhao, J., Zhang, C.L., 2017a. Enhanced bioleaching efficiency of copper from waste printed circuit board driven by nitrogen-doped carbon nanotubes modified electrode. Chem. Eng. J. 324, 122–129.

Gu, W.H., Bai, J.F., Dong, B., Zhuang, X.N., Zhao, J., Zhang, C.L., 2017c. Catalytic effect of graphene in bioleaching copper from waste printed circuit boards by *Acidithiobacillus ferrooxidans*. Hydrometallurgy 171, 172–178.

Gu, W.H., Bai, J.F., Yao, H.Y., 2017b. Heavy metals in soil at a waste electrical and electronic equipment processing area in China. Waste Manag. Res. 35 (11), 1183–1191. https://doi.org/10.1177/0734242X17725803.

Habib, M., 2013. Recovering metallic fractions from waste electrical and electronic equipment by a novel vibration system. Waste Manag. 33, 722–729. https://doi.org/10.1016/j.wasman.2012.11.017.

Harikrushnan, B., Shreyass, G., Hemant, G., Pandimadevi, M., 2016. Recovery of metals from printed circuit boards (PCBs) using a combination of hydrometallurgical and biometallurgical processes. Inter. J. Environ. Res. 10, 511–518.

He, Y., Xu, Z., 2014. The status and development of treatment techniques of typical waste electrical and electronic equipment in China: a review. Waste Manag. Res. 32 (4), 254–269.

Ilyas, S., Lee, J.C., 2014. Biometallurgical recovery of metals from waste electrical and electronic equipment: a review. Chem. Bio. Eng. Rev. 1, 148–169.

Kaya, M., 2016. Recovery of metals and nonmetals from electronic waste by physical and chemical recycling processes. Waste Manag. 57, 64–90.

Khanna, R., Cayumil, R., Mukherjee, P., Sahajwalla, V., 2014. A novel recycling approach for transforming waste printed circuit boards into a material resource. Procedia Environ Sci 21, 42–54.

Liu, J.S., Qiu, G.Z., Wang, D.Z., Xu, J., 1998. The Fe behavior during the growth of *Acidithiobacillus ferrooxidans*. Hydrometallurgy 2, 29–31.

Nakade, D.B., 2013. Bioleaching of copper from low grade ore bornite using halophilic *Thiobacillus ferrooxidans*, N-11. Res. J. Recent Sci. 2, 162–166.

Ogunniyi, I.O., Vermaak, M.K.G., 2009. Investigation of froth flotation for beneficiation of printed circuit board comminution fines. Miner. Eng. 22, 378–385.

Sahni, A., Kumar, A., Kumar, S., 2016. Chemo-biohydrometallurgy—a hybrid technology to recover metals from obsolete mobile SIM cards. Environ. Nanotechnol. Monit. Manage. 6, 130–133.

Schipper, A., Bosecker, K., Willscher, S., Sproer, C., Schumann, P., Kroppenstedt, R.M., 2002. Nocardiopsis metallicus sp. nov., a metalleaching actinomycete isolated from an alkaline slag dump. Int. J. Syst. Evol. Microbiol. 52 (6), 2291–2295.

Tezcan, U.U., Onpeker, S.E., Ozel, E., 2017. The treatment of chromium containing wastewater using electrocoagulation and the production of ceramic pigments from the resulting sludge. J. Environ. Manag. 200, 196–203.

Tuncuk, A., Stazi, V., Akcil, A., Yazici, E.Y., Deveci, H., 2012. Aqueous metal recovery techniques from e-scrap: hydrometallurgy in recycling. Miner. Eng. 25, 28–37.

Wang, J., Bai, J., Xu, J., Liang, B., 2009. Bioleaching of metals from printed wire boards by *Acidithiobacillus ferrooxidans* and *Acidithiobacillus thiooxidans* and their mixture. J. Hazard. Mater. 172 (2–3), 1100–1105.

Wang, S., Zheng, Y., Yan, W., Chen, L., Dummi Mahadevan, G., Zhao, F., 2016. Enhanced bioleaching efficiency of metals from e-wastes driven by biochar. J. Hazard. Mater. 320, 393–400.

Xie, X.Y., Sun, P.D., Lou, J.Q., Guo, M.X., Ma, W.G., 2013. Effect of simulated heavy metal leaching solution of electroplating sludge on the bioactivity of *Acidithiobacillus ferrooxidans*. Environ. Sci. 34 (1), 209–216.

Yang, T., Xu, Z., Wen, J.K., Yang, L.M., 2009. Factors influencing bioleaching copper from waste printed circuit boards by *Acidithiobacillus ferrooxidans*. Hydrometallurgy 97, 29–32.

Zeng, X., Yang, C., Chang, J.F., Li, J., 2017. Innovating e-waste management: from macroscopic to microscopic scales. Sci. Total Environ. 575, 1–5.

Zhou, L., Xu, Z., 2012. Response to waste electrical and electronic equipment in China: legislation, recycling system, and advanced integrated process. Environ. Sci. Technol. 46, 4713–4724.

Sustainable sludge management by removing emerging contaminants from urban wastewater using carbon nanotubes

24

Binoy Sarkar[*,†], *Sanchita Mandal*[†], *Yiu Fai Tsang*[‡], *Meththika Vithanage*[§], *Jayanta Kumar Biswas*[¶], *Haakrho Yi*[||], *Xiaomin Dou*[#], *Yong Sik Ok*[**]
*Department of Animal and Plant Sciences, University of Sheffield, Sheffield, United Kingdom, †Future Industries Institute, University of South Australia, Mawson Lakes, SA, Australia, ‡Department of Science and Environmental Studies, The Education University of Hong Kong, Tai Po, Hong Kong, §Ecosphere Resilience Research Centre, University of Sri Jayewardenepura, Nugegoda, Sri Lanka, ¶International Centre for Ecological Engineering, Department of Ecological Studies, University of Kalyani, Nadia, India, ||Department of Gwangyang Research Group, RIST, Gwangyang, Republic of Korea, #College of Environmental Science and Engineering, Beijing Forestry University, Beijing, PR China, **Korea Biochar Research Centre, O-Jeong Eco-Resilience Institute & Division of Environmental Science and Ecological Engineering, Korea University, Seoul, Republic of Korea

1 Introduction

The global population is expected to rise from 7.3 billion in 2015 to 8.5 billion by 2030 (UN DESA, 2015). Accordingly, the percentage of the population living in urban areas is expected to increase significantly, from 55.3% in 2018 to 60.4% in 2030 (UN DESA, 2018). In many countries, including India and China, the production of enormous volumes of untreated wastewater and sewage sludge due to expeditious industrialization and urbanization has become a serious problem. For example, China alone produced 6.25 million tons of dry sludge from urban wastewater treatment facilities in 2013 (NBSC, 2013).

Urban wastewater and sewage sludge are often loaded with various classes of environmental contaminants, including heavy metals and metalloids, pharmaceuticals and personal care products (PPCPs), food additives, surfactants, pesticides, pathogens, and manufactured nanoparticles. Country- and region-specific regulations, such as the Urban Waste Water Treatment Directive (UWWTD) in Europe, require that urban wastewater and sludge be appropriately treated before they are used for other purposes. Regional and national regulations dictate the appropriate treatment of urban

Industrial and Municipal Sludge. https://doi.org/10.1016/B978-0-12-815907-1.00024-6

wastewater and sludge before disposal, water reclamation, and resource recovery. Sewage sludge is generated as solid or semisolid residue left over following urban wastewater treatment processes (Fijalkowski et al., 2017).

The environmental fate of contaminants in urban wastewater and sludge remains unknown. Because urban wastewater is a predominant source of emerging contaminants globally (Rosal et al., 2010; Yang et al., 2017), the behavior and toxicity of these contaminants and their transformation products have become an issue of increasing social and scientific concern. Conventional wastewater treatment processes cannot effectively remove the ultralow amount of these complex environmental contaminants. The sludge also contains a large amount of plant nutrients, but the toxic contaminant residues remaining in the sludge after conventional wastewater treatment make the product unfit for land application to grow crops. Advanced wastewater treatment technologies, such as ultrafiltration (UF), nanofiltration (NF), reverse osmosis (RO), membrane bioreactors (MBRs), advanced oxidation processes (AOPs), and combined chemical and biodegradation processes, can partially remove some of these emerging contaminants, but their high cost is an issue with their use. This chapter discusses the scope of novel materials and technologies based on carbon nanotubes (CNTs) for use in removing CECs from urban wastewater, which potentially result in sustainable sludge management practices in the water supply chain.

2 Types of contaminants in urban wastewater

A list of contaminants in urban wastewater, as well as their possible sources, is provided in Table 1. Contaminants in urban wastewater and sludge can be broadly classified into two categories: potentially toxic elements and organic pollutants. While all heavy metals and metalloids fall into the first category, organic pollutants are classed into numerous subgroups, including PPCPs, pesticides, and surfactants (Fijalkowski et al., 2017; Lamastra et al., 2018; Tran et al., 2018). The key sources of heavy metals and metalloids are industrial discharge and urban runoff, which encompasses both lithospheric and atmospheric contaminants. The organic category of contaminants mostly comes from domestic discharge and daily personal care products (PCPs), including medicines and cosmetic products (Table 1).

The categorization of contaminants in sewage sludge varies from country to country. There are only a few countries, such as Sweden, where a dedicated program undertakes systematic sampling, analysis, and banking of sewage sludge (Olofsson et al., 2012; Fijalkowski et al., 2017). Because most inorganic contaminants in the environment are nonbiodegradable, they easily accumulate through the biomagnification process once they enter the human body. Similarly, many complex organic compounds with high molecular weight (e.g., brominated diphenyl ethers, polychlorinated naphthalenes, and perfluorinated surfactants) are highly resistant to biodegradation.

The organic group of contaminants in sewage sludge can be extremely heterogeneous in terms of their physicochemical properties, such as molecular weight, water solubility, hydrophobicity, pK_a, and biodegradation potential (Harrison et al., 2006; Fijalkowski et al., 2017; Lindholm-Lehto et al., 2017). Usually, a high concentration

Table 1 Urban contaminants and their possible sources

Contaminant group	Major contaminants	Possible sources
Heavy metal and metalloids	Zn, Cu, Cd, Pb, As, Cr	Industrial and metallurgical activities, car repair, printing and paint, wood processing, hairdressing, laundry, dental surgery
Pharmaceutical compounds	Codeine, paracetamol, tramadol, venlafaxine, propranolol, fluoxetine, iopromide, carbamazepine	Treatment of common health issues (e.g., pain relief, fever, hypertension, seizure)
	Antibiotics	Treatment of bacterial and viral diseases
Endogenous estrogens	17β-estradiol, estrone	Oral contraceptives
Brominated diphenyl ethers	Multiple congeners	Flame retardants in electronic goods, furniture, and textiles
Polychlorinated naphthalenes	Multiple congeners	Incineration of waste materials
Food additives	Acesulfame	Artificial sweetener
Pesticides	2,4-D, 3,4-dichloroaniline, carbaryl, diuron, 2-methyl-4-chlorophenoxyacetic acid (MCPA), simazine	Herbicides and insecticides
Phthalates	Dimethyl phthalate, diethyl phthalate, dibutyl phthalate, benzylbutyl phthalate, diethylhexyl phthalate	Cosmetics and PCPs, plasticizer
Pathogens	Bacteria, virus, fungi, protozoa, helminths	Carcass disposal, human waste
Nanoparticles	Ag, Fe, Pt, ZnO, SiO_2, TiO_2, CeO_2, fullerene	Textiles, cosmetics, sunscreens, paints, coatings, medical uses, fuel catalysts
Perfluorinated surfactants	Perfluorinated sulfonic acid, perfluorinated octanoic acid	Aqueous firefighting foams

of these contaminants in urban wastewater reflects their elevated content in the corresponding sewage sludge collected from treatment processes (Fijalkowski et al., 2017). The physicochemical characteristics of the sludge, including pH, electrical conductivity (EC), elemental concentrations, organic contents, and microbial loads, ultimately would influence the fate, mobility, and transformation of contaminants following application of the sludge in soil (Singh and Agrawal, 2008; Alvarenga et al., 2015).

3 Urban wastewater treatment technologies

To deal with the environmental and health risks associated with urban wastewater, continuous efforts are needed to develop appropriate treatment technologies before the wastewater or sludge can be employed for secondary uses. In this regard, many developed countries of the world already have made significant progress, at least in addressing the basic stages of the problem, such as the removal of suspended solids and pathogens. On the other hand, developing and underdeveloped countries are facing a continuous challenge in both urban wastewater treatment and sustainable surface water management. Removing micropollutants and CECs from urban wastewater and stormwater using conventional treatment processes has been a major problem facing nations throughout the world (Petrović et al., 2003; Talib and Randhir, 2017; Philip et al., 2018).

Removal of CECs is such a challenge for several reasons: (1) the futility of current water-treatment facilities; (2) the high cost of upgrading or reforming municipal systems and management policies; (3) the persistent (nondegradable) properties of CECs and a general lack of belief that the CECs discharging from water treatment plants are major problems; (4) the variety of occurring CECs, their concentration, and the lack of well-developed treatment technologies; and (5) the lack of routine monitoring and awareness about the levels of CECs in the influents and effluents of water treatment plants. In addition, the lack of proper analytical methods of detecting contaminants, inadequate information regarding their toxicity, effects, and behaviors in the environment, and little knowledge of their environmental and human health risks obstruct the development and commissioning of new water treatment plants that can adequately handle the issue of CECs (Naidu and Wong, 2013).

In many cases, conventional activated sludge processes (ASPs) are inadequate to remove CECs from wastewater because they were originally designed to remove biological oxygen demand (BOD) and suspended solids. Therefore, advanced treatment processes, such as AOPs and membrane filtration, are needed. Membrane bioreactors (MBRs) have demonstrated greater efficiency of removing several pharmaceutically active compounds, such as mefenamic acid, indomethacin, diclofenac, propyphenazone, pravastatin, and gemfibrozil (Radjenović et al., 2009). González et al. (2007) reported that an MBR system outperformed an ASP system in removing nonylphenolic surfactant compounds in a municipal wastewater treatment plant. The MBR system eliminated 94% of the total nonylphenol ethoxylate (NPEO)-derived compounds, as opposed to a total elimination of 54% of these materials by the ASP system. In particular, the ASP system performed poorly at removing short-ethoxy-chain NPEOs and nonylphenoxy carboxylates (NPECs; González et al., 2007). The removal efficiencies for sulfonamide, macrolide, and trimethoprim antibiotics with the MBR system were 15%–42% better than with the ASP system, but the performance different dropped to only 20% when an ultrafiltration system was used along with the conventional activated sludge (CAS) system (Sahar et al., 2011).

Similarly, for removing microbiological contaminants from urban wastewater, disinfection systems such as chlorination, ultraviolet (UV) radiation, and reactive oxidation would be a better choice than a macrofiltration system made of pressure sand filters or disc filters. However, membrane technology could replace the advanced radiation and oxidation methods with equally superior performance (Gómez et al., 2006; Arévalo et al., 2012; Álvarez-Arroyo et al., 2015). As a result, the past decade has seen a significant upsurge in research efforts to find effective and inexpensive filters and membrane materials for wastewater treatment. The remaining sections of this chapter will discuss the role of one such material, carbon nanotubes (CNTs), in eliminating emerging contaminants from urban wastewater.

4 CNTs for CEC removal from wastewater

The unique physicochemical characteristics of CNTs make them suitable for a wide range of applications in wastewater treatment, including adsorbents, membranes, and catalysts (Sarkar et al., 2018). Fig. 1 shows the general principles and properties of CNTs that enhance the removal of CECs, including heavy metals, PPCPs, microbial

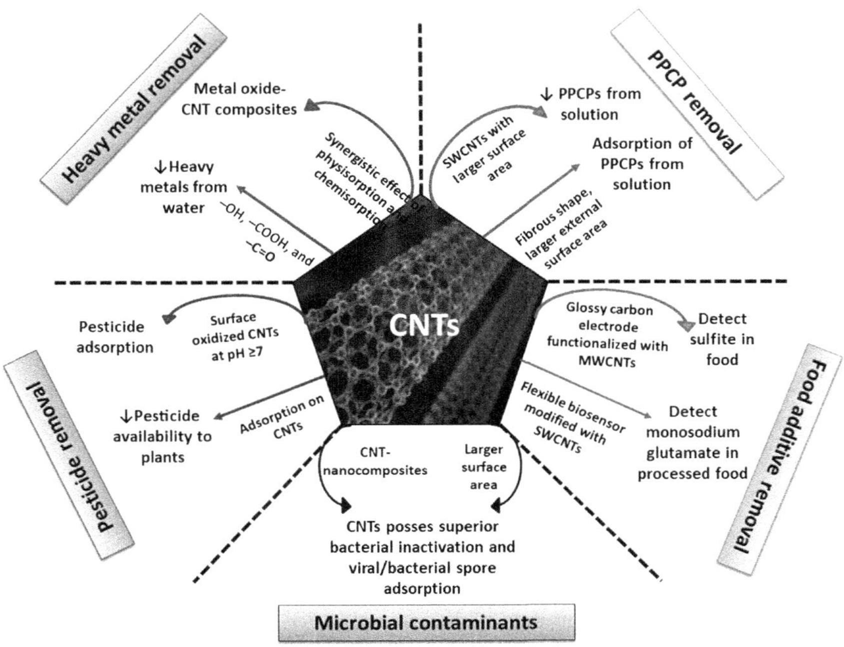

Fig. 1 Schematic representation of various properties of CNTs that are useful for removing emerging contaminants from wastewater.
Adapted from Sarkar, B., Mandal, S., Tsang, Y.F., Kumar, P., Kim, K.-H., Ok, Y.S., 2018. Designer carbon nanotubes for contaminant removal in water and wastewater: a critical review. Sci. Total Environ. 612, 561–581.

contaminants, pesticides, and food additives from aqueous solution (Sarkar et al., 2018; Yang et al., 2017).

The type of CNT, either single-walled (SWCNTs) or multiwalled (MWCNTs), can influence the affinity of materials to CECs. SWCNTs are more effective than MWCNTs in adsorbing contaminants because of their superior specific surface areas. The key functions of contaminant removal using CNTs involve adsorption, degradation, and detoxification to some extent. Furthermore, designing CNTs with surface engineering might facilitate the removal of CECs. Modification and functionalization of CNTs with magnetic compounds comprise one of the common approaches to separate nanoparticles and remove a range of CECs from water (Abdel Salam et al., 2012; Alimohammadi et al., 2017; Li et al., 2017b). Table 2 summarizes examples of CEC removal using CNT-based materials from aquatic environments.

4.1 Removal of heavy metals and metalloids

CNTs have been widely used to remove contaminants from water and wastewater due to their highly porous structure, light mass density, large surface area, and strong interactions with contaminants (Ihsanullah et al., 2016; Nyairo et al., 2018; Sarkar et al., 2018). The adsorption of heavy metals on CNTs depends on the purity, porosity, surface area, surface functional groups, and site density of CNT-based materials (Ihsanullah et al., 2016). Four possible sites for contaminant adsorption on CNTs are internal sites, interstitial channels, grooves, and outside surfaces. Among these sites, the maximum amount of heavy metal adsorption takes place on CNT's interstitial channel, outer surfaces, and grooves (Ihsanullah et al., 2016; Heroux et al., 2006; Jiang et al., 2005).

Surface modification has proved to enhance the heavy metal adsorption capacity and selectivity of CNTs. The adsorption of heavy metals and metalloids on CNTs can be improved by loading or depositing active components on CNTs and by simple oxidative modification (Cho et al., 2010; Yu et al., 2011). In addition, functionalization of CNTs with nonmagnetic or magnetic metal oxides (Addo Ntim and Mitra, 2011; Gupta et al., 2011; Daneshvar Tarigh and Shemirani, 2013; Zhao et al., 2010) and thiol and other sulfur-containing groups (Bandaru et al., 2013; Gupta et al., 2014) can improve the performance of heavy metal removal by CNT-based adsorbents.

Iron oxide particles such as goethite (α-FeO(OH)), hematite (α-Fe_2O_3), maghemite (γ-Fe_2O_3), and magnetite (Fe_3O_4) can be used to alter CNTs for the purpose of removing ultralow concentrations of heavy metals from water (Addo Ntim and Mitra, 2011). CNTs modified with oxides of magnesium (Mg), aluminum (Al), and manganese (Mn) are also efficient adsorbents of heavy metals in aqueous solution (Mubarak et al., 2014). The removal capacity of heavy metals by CNT-based adsorbents depends significantly on the type of CNT modification or functionalization because the process can alter the CNT surface area, surface charge hydrophobicity/hydrophilicity, and dispersion (Gupta et al., 2016; Ihsanullah et al., 2016). CNT-composites made of metal oxide adsorb heavy metals through a combined mechanism of physisorption and chemisorption (Addo Ntim and Mitra, 2011; Lu et al., 2017; Yang et al., 2018).

Table 2 Removal of various CECs from wastewater by CNT-based materials

Contaminants	CNT material	Removal capacity/ efficiency	References
Heavy metals and metalloids			
Hg(II)	Thiol-derivatized SWCNTs	131 mg/g	Bandaru et al. (2013)
Ni(II) and Sr(II)	MWCNT-iron oxide magnetic composite	~80% Ni(II) at pH 8; ~95% Sr(II) at pH 10.4	Chen et al. (2009a)
Sr(II) and Eu(III)	Oxidized MWCNTs	~36% Sr(II); ~96% Eu(III)	Chen et al. (2008)
Uranium	Diglycolamide-functionalized MWCNTs	133.74 mg/g	Deb et al. (2012)
As(III) and As(V)	Iron oxide-coated MWCNTs	As(III) 1723 μg/g; As(V) 189 μg/g	Addo Ntim and Mitra (2012)
Pb(II)	Alumina-coated MWCNTs	99%	Gupta et al. (2011)
PPCPs			
Norfloxacin	MWCNTs	84.7 mg/g	Yang et al. (2012)
Ciprofloxacin	SWCNTs	724 mg/g	Ncibi and Sillanpää (2015)
Ofloxacin	MWCNTs	80%	Peng et al. (2012)
Oxytetracycline	SWCNTs	554 mg/g	Ncibi and Sillanpää (2015)
Bisphenol A and 17β-estradiol	SWCNTs	7.3%–95%	Heo et al. (2012)
Triclosan	MWCNTs	157.7 mg/g	Zhou et al. (2013)
Ibuprofen and triclosan	HNO_3-refluxed SWCNTs	Ibuprofen 232 mg/g; triclosan 558 mg/g	Cho et al. (2011)
Ciprofloxacin	Acid-heat treated MWCNTs	150 mg/g	Carabineiro et al. (2011)
Pesticides			
Atrazine	Oxidized MWCNTs	17.35 mg/g	Chen et al. (2009b)

Continued

Table 2 Continued

Contaminants	**CNT material**	**Removal capacity/ efficiency**	**References**
Diuron	Oxidized MWCNTs	29.82 mg/g	Deng et al. (2012)
Isoproturon	MWCNTs	8.1 mg/g	Sotelo et al. (2012)
Dicholbenil	MWCNTs	17.5 mg/g	Chen et al. (2011a)
2-Methyl-4-chlorophenoxyacetic acid	SWCNTs	25.7 mg/g	De Martino et al. (2012)
Microbial contaminants			
Escherichia coli and *Bacillus subtilis*	SWCNTs	>90% inactivation	Ahmed et al. (2012)
E. coli K12	SWCNTs	79% inactivation	Brady-Estévez et al. (2008)
E. coli DH5α	MWCNT-Ag	96% inactivation	Su et al. (2013)
Streptococcus mutans	Modified MWCNTs	Viable cell reduced by 7.5 log	Bai et al. (2011)
Swine influenza virus (SIV); swine flu (H1N1)	SWCNTs	Virus detection limit: 180 $TCID_{50}$ mL^{-1}	Lee et al. (2011)
MS2 bacteriophage	SWCNTs	9.3 and 9.8 PFU/mL limit detection	Prieto-Simón et al. (2015)
Influenza virus	CNTs	detective limitation >3.4 PFU/mL	Ahmed et al. (2016)
MS2 bacteriophage, along with host *E. coli*	MWCNTs	5.8–7.4 log inactivation	Rahaman et al. (2012)
Food additives			
Sulfites	Glassy carbon electrodes functionalized with MWCNTs	Detection limit 4.2 μM	Sartori and Fatibello-Filho (2011)
MSG	Biosensors functionalized with SWCNTs	Detection limit 200 μM	Juntae et al. (2008)

The capacity and efficiency of CNT-enhanced adsorption capacity further depend on the heavy metal (or metalloid) properties, such as hydrolysis potential, ionic radius, and hydration energy (Hu et al., 2011). CNT-based materials have demonstrated superior heavy metal removal from wastewater, but extensive studies, especially pilot- and full-scale experiments, are needed before they can be applied in real-life situations in industry.

4.2 Removal of PPCPs

PPCPs have been discovered in surface water, groundwater, raw sewage, and treated effluents globally. Even in trace concentrations, some of these contaminants can be linked to significant ecological effects (Benotti et al., 2009; Chang and Wilton, 2009). For example, ibuprofen and triclosan, which are the commonly used PPCPs in nonsteroidal antiinflammatory drugs and commercial disinfectants, are the typical PPCPs found in aquatic environments. However, conventional drinking-water treatment plants relying on coagulation could remove only a small portion of PPCPs from aqueous solution (Westerhoff et al., 2005; Jung et al., 2015). On the other hand, CNTs showed significant improvement in the removal of these compounds from aqueous solution owing to the large surface area and O-containing functional groups of the adsorbent material (Cho et al., 2011). Operating parameters, such as pH, temperature, ionic strength, initial solution concentration, and contact time, significantly influence the adsorption rate of PPCPs onto CNTs (Jung et al., 2015). Of the various types of CNTs, SWCNTs showed a greater capacity for adsorption of ibuprofen than MWCNTs due to their larger surface area (1020 vs 283 m^2/g) and thin layer structure (Cho et al., 2011).

Usually, wastewater samples contain a mixture of pollutants, including humic acids, carbohydrates, proteins, and other biological building blocks, and the interaction of these pollutants may enhance or deteriorate the adsorption capacity of CNTs. For example, all these compounds compete with diclofenac for binding sites on the MWCNTs, while the adsorption capacity of diclofenac alone was significantly greater than the mixed wastewater sample (Sotelo et al., 2012). Contaminant properties like hydrophobicity also could affect their affinity for CNTs. For example, disinfectant triclosan showed a noticeably stronger affinity for SWCNTs and MWCNTs than ibuprofen due to the difference in hydrophobicity (i.e., $\log K_{ow} = 4.76$ and 3.97 for triclosan and ibuprofen, respectively) (Cho et al., 2011). In addition to hydrophobicity, the presence of specific functional groups (e.g., OH groups) on the surface of PPCPs could enhance their adsorption on CNTs. The OH groups in the PPCP molecules may create an attraction between the adsorbed molecules and molecules in aqueous solution, and additional hydrogen bonding may develop between the adsorbate's OH groups and the adsorbent's O-containing functional groups (Lin and Xing, 2008).

CNTs have demonstrated great potential to remove a range of PPCPs from water and wastewater due to their fibrous shape and large external surface area that are accessible to PPCP molecules (Table 2). Electrostatic attraction, partitioning effects, and π-π interactions between the aromatic molecules of PPCPs and CNT surfaces are the key mechanisms of PPCP removal by CNTs (Sarkar et al., 2018). However, the major barrier for application of CNTs in full-scale wastewater treatment plants is their high production costs. The current market price of high-quality SWCNTs and MWCNTs may reach \$300 and \$25/g, respectively (Sarkar et al., 2018), which can be prohibitive for treating thousands of gallons of wastewater in a large water treatment plant. Furthermore, it is important to understand the physicochemical properties of PPCPs and CNTs before commercialization of the materials for real-life applications in wastewater management.

4.3 Pesticide removal

The widespread use of pesticides in the agricultural industry, including urban agriculture, have significantly contaminated soil and water. Due to their heavy application and their persistence, polar nature, and water solubility, they can disperse in the environment, and their residues can cause long-term human health risks. CNTs can potentially remove a range of pesticides because of their strong adsorption affinity to a wide variety of organic compounds (Table 2). Deng et al. (2012) observed that a solution pH $\geq$7 favored the adsorption of diuron by pristine and oxidized MWCNTs. Some pesticides were successfully recovered after the CNT-based treatment, and the rate of recovery depended on the external diameter of CNTs. El-Sheikh et al. (2007) observed that the highest recovery of atrazine was accomplished by CNTs with external diameters of 40–60 nm. Furthermore, short-length CNTs (1–2 μm) had greater pesticide recovery than long-length CNTs (5–15 μm) (Pyrzynska, 2011).

Surface functionalization of CNTs can remarkably increase the adsorption capacity of pesticides (Dichiara et al., 2015; Deokar et al., 2017; Liu et al., 2018). This is mainly due to the increase in pore volume and surface area of CNTs after functionalization treatment. However, Hamdi et al. (2015) found the application of amino-functionalized CNTs could reduce chlordane and p,p′-dichlorodiphenyldichloroethylene uptake in lettuce plants by 57% and 23%, as opposed to 88% and 78% for nonfunctionalized CNT applications, respectively. Therefore, CNTs and their functionalized products can affect the availability of pesticides to plants, i.e., CNTs can potentially prevent pesticides to be taken up in plant edible parts. In another study, the adsorption capacity of pesticides (i.e., 1-pyrenebutyric acid, 2,4-dichlorophenoxyacetic acid, and diquat dibromide) on semiconducting-type SWCNTs were significantly higher than metallic-type SWCNTs (Rocha et al., 2017). However, the application of CNTs for pesticide removal is restricted to batch systems, and there is not as much information on pesticide removal by CNTs than there is on other CECs (Sarkar et al., 2018).

4.4 Microbial contaminant removal

Microbial contaminants, including bacteria, viruses, and protozoa, present a major human health issue in surface and drinking water (Smith and Rodrigues, 2015; Sharma and Bhattacharya, 2017; Stillo and Gibson, 2017). The removal of microbial contaminants by adsorption-based methods was practiced for many decades (Hijnen et al., 2010; Babi et al., 2007). Conventional adsorbents, such as activated carbon and polymers, can be used for removing microbial contaminants, but they are not always effective enough to bring the quality level to that of drinking water (Smith and Rodrigues, 2015). Infectious bacterial breakouts due to the leakage of these adsorption-based filter systems have been reported, which warrants the development of a new generation of adsorbent materials. The structural and functional properties of CNTs and their high affinitive interactions with microbial contaminants can make them suitable for this purpose (Fig. 1 and Table 1).

Owing to their larger surface areas than many other adsorbents, CNTs inactivate bacterial cells and adsorb viral/bacterial spores with remarkable efficiency (Lu and Su, 2007; Brady-Estévez et al., 2008). Brady-Estévez et al. (2008) found that the *E. coli* bacterial community was 100% retained by a poly(vinylidene fluoride) (PVDF)-based, microporous-modified membrane with a thin layer of SWCNT. The SWCNT layer increased the exclusion behavior of the modified membrane compared to normal PDVF membrane. SWCNTs have a surface area of approximately $407\,m^2/g$ and show a preferential affinity for various bacterial species, which can be beneficial in removing pathogenic over nonpathogenic species selectively (Smith and Rodrigues, 2015; Sarkar et al., 2018). CNTs also showed an immense potential for antimicrobial applications, particularly drinking water disinfection (Table 1). The various geometries of CNTs (e.g., tubes, sheets, and spheres) can affect their interactions with microorganisms in different way. For example, MWCNTs usually show less antibacterial activity than SWCNTs because the former are larger in diameter than the latter, reducing their chances to invade the nucleus and their affinity to nucleic materials. The rod-shaped, short SWCNTs have a higher affinity toward the bacterial community than MWCNTs.

Polymeric CNTs and CNT-metal oxide nanocomposites showed greater microbial disinfection efficiency in water than unmodified CNTs. The poly-*N*-vinyl carbazole-SWCNT nanocomposites could inactivate both Gram-positive and Gram-negative bacterial cells in water with an efficiency of more than 80% (Ahmed et al., 2012; Mejias Carpio et al., 2012). Nanoscale Ag particles, when deposited on CNTs using ion beams, showed excellent bactericidal activity against Gram-positive and Gram-negative species (Liu et al., 2007). However, modification of CNTs with antibacterial agents like Ag could pose the risk of developing bacterial resistance, which should be taken into consideration.

4.5 Food additive removal

To inhibit the growth of bacteria and enhance the appearance and flavor of food during preparation, processing, and storage, various chemical additives, like sulfur dioxide, sodium sulfite, and sodium and potassium metabisulfites are commonly used in food and beverage products (Gan et al., 2013; Sang et al., 2014; Li et al., 2017a). Consumption of these additives along with food on a regular basis can cause serious human health issues. The consumption of food containing sulfite additives, which are highly toxic, can cause hypersensitivity, nausea, diarrhea, gastric irritation, vomiting, food poisoning, and asthma (Sartori and Fatibello-Filho, 2011; Walker, 1985).

The tendency to use these toxic additives in foods, including dried fruits, vegetables, juice, fish, and beverages, has been steadily increasing in many countries. It is very important to develop accurate analytical procedures to identify and quantify the trace levels of food additives. Conventional analytical determination methods include conductometry, spectrophotometry, and electrophoresis, but these techniques are highly costly and time-consuming, and sometimes they also have low sensitivity and selectivity, making them unsuitable for routine food analysis (Jankovskiene et al., 2001; McLeod and Davey, 2007; Sartori and Fatibello-Filho, 2011).

The application of SWCNTs and MWCNTs as detectors of toxic food additives has been proven useful due to their suitable structural (larger surface area), chemical (functional groups), and electrical properties. Sartori and Fatibello-Filho (2011) prepared a glassy carbon electrode that was modified and functionalized with MWCNTs to determine the sulfite concentration in food samples, and the detection limit was 4.2 μM. The MWCNT-modified glass electrode successfully detected sulfites in vinegar, coconut water, shredded coconut, and pickle water (Sartori and Fatibello-Filho, 2011). Juntae et al. (2008) prepared a flexible biosensor functionalized with SWCNTs and tested its ability to detect food additives like monosodium glutamate (MSG). The CNT-functionalized sensor showed a higher sensitivity to detect MSG (200 μM) in processed food than normal sensors. Therefore, CNT-based sensors can be more useful to detect harmful food additives like MSG, which can cause a severe allergic reaction in human beings (Juntae et al., 2008). Also, sensors made of functionalized CNTs can be sensitive, accurate, and precise and exhibit good reproducibility and stability under heterogeneous working conditions. However, the production of this kind of sensor can be costly, and further detailed analyses are required to assess its large-scale application potential.

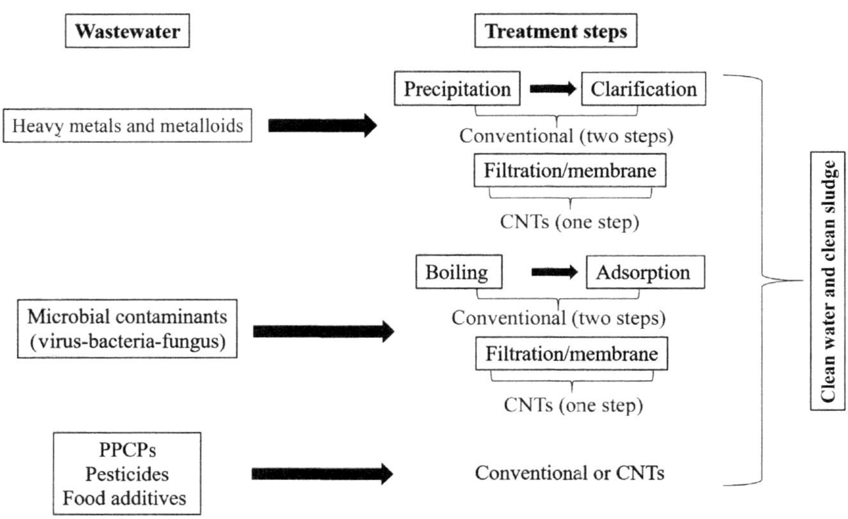

Fig. 2 Conceptual diagram showing the roles of CNT-based materials in treating wastewater and achieving clean sludge.

5 Conclusions and future perspectives

CNTs can be promising materials for removing CECs from urban wastewater, and thus they can be applied in practical water treatment plants in order to achieve sustainable sludge management. However, CNT-based membrane filtration needs to be fully validated for its performance in pilot- and large-scale wastewater treatment plants that are currently being tested. While the large-scale testing and optimization of wastewater treatment processes using CNT technologies are important, the functionalization and modification of CNT-based filtration/membrane materials are also necessary for obtaining the best performance.

Cost can be the most prohibitive factor for advocating CNTs to the operators of urban wastewater treatment plants. CNT-based wastewater treatment has not yet been well studied, but it has been proved to be a promising technology for CEC removal. In certain cases, such as when wastewater contains high loads of toxic heavy metals and mixed microbial contaminants (virus/bacterial/fungal contaminants), CNT-based treatment technologies can be economically feasible and even profitable. No existing treatment method is capable of treating such complex wastewater in one step. The treatment of heavy metal-loaded wastewater requires at least two steps (namely, precipitation followed by clarification). Thus, CNT-based methods would have tremendous potential in this scenario because they have the potential to treat such wastewater in a single step, thus overcoming the cost issue (Fig. 2). Similarly, wastewater treatment to remove mixed microbial contaminants (viral/bacterial/fungal communities) also requires at least two steps (boiling followed by adsorption), which could potentially be replaced by a single-step treatment by CNT-based methods (Fig. 2).

Finally, the unwanted migration of CNTs into sludge at the end of the treatment process also should be avoided because, like other manufactured nanoparticles, CNTs might pose risks to some environmental organisms (Vithanage et al., 2017; Sarkar et al., 2018). Thus, the removal of micropollutants and CECs from urban wastewater via CNT-based membrane filtration can make the treated effluent fit for reclamation. At the same time, it potentially makes sludge a by-product fit for sustainable uses such as growing food crops and revegetating degraded lands (Singh and Agrawal, 2008; Roig et al., 2012). Therefore, like other wastewater treatment methods, CNT-based methods require a holistic technoecological assessment prior to their large-scale deployment in water treatment plants promoting sustainable practices for sludge management.

References

Abdel Salam, M., Gabal, M.A., Obaid, A.Y., 2012. Preparation and characterization of magnetic multi-walled carbon nanotubes/ferrite nanocomposite and its application for the removal of aniline from aqueous solution. Synth. Met. 161 (23–24), 2651–2658.

Addo Ntim, S., Mitra, S., 2012. Adsorption of arsenic on multiwall carbon nanotube–zirconia nanohybrid for potential drinking water purification. J. Colloid Interface Sci. 375 (1), 154–159.

Addo Ntim, S., Mitra, S., 2011. Removal of trace arsenic to meet drinking water standards using iron oxide coated multiwall carbon nanotubes. J. Chem. Eng. Data 56, 2077–2083.

Ahmed, F., Santos, C.M., Vergara, R.A.M.V., Tria, M.C.R., Advincula, R., Rodrigues, D.F., 2012. Antimicrobial applications of electroactive PVK-SWNT nanocomposites. Environ. Sci. Technol. 46 (3), 1804–1810.

Ahmed, S.R., Kim, J., Suzuki, T., Lee, J., Park, E.Y., 2016. Enhanced catalytic activity of gold nanoparticle-carbon nanotube hybrids for influenza virus detection. Biosens. Bioelectron. 85, 503–508.

Alimohammadi, V., Sedighi, M., Jabbari, E., 2017. Experimental study on efficient removal of total iron from wastewater using magnetic-modified multi-walled carbon nanotubes. Ecol. Eng. 102, 90–97.

Alvarenga, P., Mourinha, C., Farto, M., Santos, T., Palma, P., Sengo, J., Morais, M.-C., Cunha-Queda, C., 2015. Sewage sludge, compost and other representative organic wastes as agricultural soil amendments: benefits versus limiting factors. Waste Manag. 40, 44–52.

Álvarez-Arroyo, R., Rojas-Serrano, F., Garralón, G., Plaza, F., Pérez Pérez, J.I., Gómez Nieto, M.Á., 2015. Evaluation of drinking water quality produced by ultrafiltration membranes in distribution systems. Desalin. Water Treat. 56 (13), 3447–3455.

Arévalo, J., Ruiz, L.M., Parada-Albarracín, J.A., González-Pérez, D.M., Pérez, J., Moreno, B., Gómez, M.A., 2012. Wastewater reuse after treatment by MBR. Microfiltration or ultrafiltration? Desalination 299, 22–27.

Babi, K.G., Koumenides, K.M., Nikolaou, A.D., Makri, C.A., Tzoumerkas, F.K., Lekkas, T.D., 2007. Pilot study of the removal of THMs, HAAs and DOC from drinking water by GAC adsorption. Desalination 210 (1), 215–224.

Bai, Y., Park, I.S., Lee, S.J., Bae, T.S., Watari, F., Uo, M., Lee, M.H., 2011. Aqueous dispersion of surfactant-modified multiwalled carbon nanotubes and their application as an antibacterial agent. Carbon 49 (11), 3663–3671.

Bandaru, N.M., Reta, N., Dalal, H., Ellis, A.V., Shapter, J., Voelcker, N.H., 2013. Enhanced adsorption of mercury ions on thiol derivatized single wall carbon nanotubes. J. Hazard. Mater. 261, 534–541.

Benotti, M.J., Trenholm, R.A., Vanderford, B.J., Holady, J.C., Stanford, B.D., Snyder, S.A., 2009. Pharmaceuticals and endocrine disrupting compounds in U.S. drinking water. Environ. Sci. Technol. 43 (3), 597–603.

Brady-Estévez, A.S., Kang, S., Elimelech, M., 2008. A single-walled-carbon-nanotube filter for removal of viral and bacterial pathogens. Small 4 (4), 481–484.

Carabineiro, S.A.C., Thavorn-Amornsri, T., Pereira, M.F.R., Figueiredo, J.L., 2011. Adsorption of ciprofloxacin on surface-modified carbon materials. Water Res. 45 (15), 4583–4591.

Chang, H., Wilton, S., 2009. SoC architectures for signal processing. In: 2009 IEEE Custom Integrated Circuits Conferencevols. 13–16, pp. 1–2.

Chen, C., Hu, J., Xu, D., Tan, X., Meng, Y., Wang, X., 2008. Surface complexation modeling of Sr(II) and Eu(III) adsorption onto oxidized multiwall carbon nanotubes. J. Colloid Interface Sci. 323 (1), 33–41.

Chen, C., Hu, J., Shao, D., Li, J., Wang, X., 2009a. Adsorption behavior of multiwall carbon nanotube/iron oxide magnetic composites for Ni(II) and Sr(II). J. Hazard. Mater. 164 (2–3), 923–928.

Chen, G.-C., Shan, X.-Q., Zhou, Y.-Q., Shen, X.-E., Huang, H.-L., Khan, S.U., 2009b. Adsorption kinetics, isotherms and thermodynamics of atrazine on surface oxidized multiwalled carbon nanotubes. J. Hazard. Mater. 169 (1–3), 912–918.

Chen, G.-C., Shan, X.-Q., Pei, Z.-G., Wang, H., Zheng, L.-R., Zhang, J., Xie, Y.-N. 2011a. Adsorption of diuron and dichlobenil on multiwalled carbon nanotubes as affected by lead. J. Hazard. Mater., 188(1–3), 156-163.

Cho, H.-H., Wepasnick, K., Smith, B.A., Bangash, F.K., Fairbrother, D.H., Ball, W.P., 2010. Sorption of aqueous Zn[II] and Cd[II] by multiwall carbon nanotubes: the relative roles of oxygen-containing functional groups and graphenic carbon. Langmuir 26, 967–981.

Cho, H.-H., Huang, H., Schwab, K., 2011. Effects of solution chemistry on the adsorption of ibuprofen and triclosan onto carbon nanotubes. Langmuir 27 (21), 12960–12967.

Daneshvar Tarigh, G., Shemirani, F., 2013. Magnetic multi-wall carbon nanotube nanocomposite as an adsorbent for preconcentration and determination of lead (II) and manganese (II) in various matrices. Talanta 115, 744–750.

De Martino, A., Iorio, M., Xing, B., Capasso, R., 2012. Removal of 4-chloro-2-methylphenoxyacetic acid from water by sorption on carbon nanotubes and metal oxide nanoparticles. RSC Adv. 2 (13), 5693–5700.

Deb, A.K.S., Ilaiyaraja, P., Ponraju, D., Venkatraman, B., 2012. Diglycolamide functionalized multi-walled carbon nanotubes for removal of uranium from aqueous solution by adsorption. J. Radioanal. Nucl. Chem. 291 (3), 877–883.

Deng, J., Shao, Y., Gao, N., Deng, Y., Tan, C., Zhou, S., Hu, X., 2012. Multiwalled carbon nanotubes as adsorbents for removal of herbicide diuron from aqueous solution. Chem. Eng. J. 193–194, 339–347.

Deokar, S.K., Bajad, G.S., Bhonde, P., Vijayakumar, R.P., Mandavgane, S.A., 2017. Adsorptive removal of diuron herbicide on carbon nanotubes synthesized from plastic waste. J. Polym. Environ. 25 (2), 165–175.

Dichiara, A.B., Harlander, S.F., Rogers, R.E., 2015. Fixed bed adsorption of diquat dibromide from aqueous solution using carbon nanotubes. RSC Adv. 5 (76), 61508–61512.

El-Sheikh, A.H., Insisi, A.A., Sweileh, J.A., 2007. Effect of oxidation and dimensions of multi-walled carbon nanotubes on solid phase extraction and enrichment of some pesticides from environmental waters prior to their simultaneous determination by high performance liquid chromatography. J. Chromatogr. A 1164 (1), 25–32.

Fijalkowski, K., Rorat, A., Grobelak, A., Kacprzak, M.J., 2017. The presence of contaminations in sewage sludge—the current situation. J. Environ. Manag. 203 (3), 1126–1136.

Gan, Z., Sun, H., Feng, B., Wang, R., Zhang, Y., 2013. Occurrence of seven artificial sweeteners in the aquatic environment and precipitation of Tianjin, China. Water Res. 47 (14), 4928–4937.

Gómez, M., de la Rua, A., Garralón, G., Plaza, F., Hontoria, E., Gómez, M.A., 2006. Urban wastewater disinfection by filtration technologies. Desalination 190 (1), 16–28.

González, S., Petrovic, M., Barceló, D., 2007. Removal of a broad range of surfactants from municipal wastewater—comparison between membrane bioreactor and conventional activated sludge treatment. Chemosphere 67 (2), 335–343.

Gupta, V.K., Agarwal, S., Saleh, T.A., 2011. Synthesis and characterization of alumina-coated carbon nanotubes and their application for lead removal. J. Hazard. Mater. 185 (1), 17–23.

Gupta, A., Vidyarthi, S.R., Sankararamakrishnan, N., 2014. Enhanced sorption of mercury from compact fluorescent bulbs and contaminated water streams using functionalized multiwalled carbon nanotubes. J. Hazard. Mater. 274, 132–144.

Gupta, V.K., Moradi, O., Tyagi, I., Agarwal, S., Sadegh, H., Shahryari-Ghoshekandi, R., Makhlouf, A.S.H., Goodarzi, M., Garshasbi, A., 2016. Study on the removal of heavy metal ions from industry waste by carbon nanotubes: effect of the surface modification: a review. Crit. Rev. Environ. Sci. Technol. 46, 93–118.

Hamdi, H., De La Torre-Roche, R., Hawthorne, J., White, J.C., 2015. Impact of non-functionalized and amino-functionalized multiwall carbon nanotubes on pesticide uptake by lettuce (*Lactuca sativa* L.). Nanotoxicology 9 (2), 172–180.

Harrison, E.Z., Oakes, S.R., Hysell, M., Hay, A., 2006. Organic chemicals in sewage sludges. Sci. Total Environ. 367 (2), 481–497.

Heo, J., Flora, J.R.V., Her, N., Park, Y.-G., Cho, J., Son, A., Yoon, Y., 2012. Removal of bisphenol a and 17β-estradiol in single walled carbon nanotubes–ultrafiltration (SWNTs–UF) membrane systems. Sep. Purif. Technol. 90, 39–52.

Heroux, L., Krungleviciute, V., Calbi, M.M., Migone, A.D., 2006. CF4 on carbon nanotubes: physisorption on grooves and external surfaces. J. Phys. Chem. B 110 (25), 12597–12602.

Hijnen, W.A.M., Suylen, G.M.H., Bahlman, J.A., Brouwer-Hanzens, A., Medema, G.J., 2010. GAC adsorption filters as barriers for viruses, bacteria and protozoan (oo)cysts in water treatment. Water Res. 44 (4), 1224–1234.

Hu, J., Zhao, D., Wang, X., 2011. Removal of Pb(II) and Cu(II) from aqueous solution using multiwalled carbon nanotubes/iron oxide magnetic composites. Water Sci. Technol. 63, 917–923.

Ihsanullah, A.A., Al-Amer, A.M., Laoui, T., Al-Marri, M.J., Nasser, M.S., Khraisheh, M., Atieh, M.A., 2016. Heavy metal removal from aqueous solution by advanced carbon nanotubes: critical review of adsorption applications. Sep. Purif. Technol. 157, 141–161.

Jankovskiene, G., Daunoravicius, Z., Padarauskas, A., 2001. Capillary electrophoretic determination of sulfite using the zone-passing technique of in-capillary derivatization. J. Chromatogr. A 934 (1), 67–73.

Jiang, J., Sandler, S.I., Schenk, M., Smit, B., 2005. Adsorption and separation of linear and branched alkanes on carbon nanotube bundles from configurational-bias Monte Carlo simulation. Phys. Rev. B 72 (4), 045447.

Jung, C., Son, A., Her, N., Zoh, K.-D., Cho, J., Yoon, Y., 2015. Removal of endocrine disrupting compounds, pharmaceuticals, and personal care products in water using carbon nanotubes: a review. J. Ind. Eng. Chem. 27, 1–11.

Juntae, K., Mihye, Y., Byung Yang, L., Tae Hyun, K., Joohyung, L., Young Min, J., Seunghun, H., 2008. Directed assembly of carbon nanotubes on soft substrates for use as a flexible biosensor array. Nanotechnology 19 (50), 505502.

Lamastra, L., Suciu, N.A., Trevisan, M., 2018. Sewage sludge for sustainable agriculture: contaminants' contents and potential use as fertilizer. Chem. Biol. Technol. Agric. 5 (1), 10.

Lee, D., Chander, Y., Goyal, S.M., Cui, T., 2011. Carbon nanotube electric immunoassay for the detection of swine influenza virus H1N1. Biosens. Bioelectron. 26 (8), 3482–3487.

Li, A.J., Wu, P., Law, J.C.-F., Chow, C.-H., Postigo, C., Guo, Y., Leung, K.S.-Y., 2017a. Transformation of acesulfame in chlorination: kinetics study, identification of byproducts, and toxicity assessment. Water Res. 117, 157–166.

Li, X., Lu, H., Zhang, Y., He, F., 2017b. Efficient removal of organic pollutants from aqueous media using newly synthesized polypyrrole/CNTs-$CoFe_2O_4$ magnetic nanocomposites. Chem. Eng. J. 316, 893–902.

Lin, D., Xing, B., 2008. Tannic acid adsorption and its role for stabilizing carbon nanotube suspensions. Environ. Sci. Technol. 42, 5917–5923.

Lindholm-Lehto, P.C., Ahkola, H.S.J., Knuutinen, J.S., 2017. Procedures of determining organic trace compounds in municipal sewage sludge—a review. Environ. Sci. Pollut. Res. 24 (5), 4383–4412.

Liu, T., Tang, H.Q., Cai, X.M., Zhao, J., Li, D.J., Li, R., Sun, X.L., 2007. A study on bactericidal properties of Ag coated carbon nanotubes. Nucl. Instrum. Methods Phys. Res. B 264, 282–286.

Liu, G., Li, L., Huang, X., Zheng, S., Xu, X., Liu, Z., Zhang, Y., Wang, J., Lin, H., Xu, D., 2018. Adsorption and removal of organophosphorus pesticides from environmental water and soil samples by using magnetic multi-walled carbon nanotubes @ organic framework ZIF-8. J. Mater. Sci. 53 (15), 10772–10783.

Lu, C., Su, F., 2007. Adsorption of natural organic matter by carbon nanotubes. Sep. Purif. Technol. 58 (1), 113–121.

Lu, W., Li, J., Sheng, Y., Zhang, X., You, J., Chen, L., 2017. One-pot synthesis of magnetic iron oxide nanoparticle-multiwalled carbon nanotube composites for enhanced removal of Cr(VI) from aqueous solution. J. Colloid Interface Sci. 505, 1134–1146.

McLeod, S., Davey, D.E., 2007. Rapid microstill determination of free and total sulfite in wine with conductimetric detection. Anal. Chim. Acta 600 (1), 72–77.

Mejias Carpio, I.E., Santos, C.M., Wei, X., Rodrigues, D.F., 2012. Toxicity of a polymer-graphene oxide composite against bacterial planktonic cells, biofilms, and mammalian cells. Nanoscale 4 (15), 4746–4756.

Mubarak, N.M., Sahu, J.N., Abdullah, E.C., Jayakumar, N.S., 2014. Removal of heavy metals from wastewater using carbon nanotubes. Sep. Purif. Rev. 43, 311–338.

Naidu, R., Wong, M.H., 2013. Contaminants of emerging concern. Sci. Total Environ. 463–464, 1077–1078.

NBSC, 2013. National Bureau of Statistics of China, China Statistical Yearbook. China Statistics Press, Beijing, China.

Ncibi, M.C., Sillanpää, M., 2015. Optimized removal of antibiotic drugs from aqueous solutions using single, double andmulti-walled carbon nanotubes. J. Hazard. Mater. 298, 102–110.

Nyairo, W.N., Eker, Y.R., Kowenje, C., Akin, I., Bingol, H., Tor, A., Ongeri, D.M., 2018. Efficient adsorption of lead (II) and copper (II) from aqueous phase using oxidized multiwalled carbon nanotubes/polypyrrole composite. Sep. Sci. Technol. 53, 1498–1510.

Olofsson, U., Bignert, A., Haglund, P., 2012. Time-trends of metals and organic contaminants in sewage sludge. Water Res. 46 (15), 4841–4851.

Peng, H., Pan, B., Wu, M., Liu, R., Zhang, D., Wu, D., Xing, B., 2012. Adsorption of ofloxacin on carbon nanotubes: solubility, pH and cosolvent effects. J. Hazard. Mater. 211–212, 342–348.

Petrović, M., Gonzalez, S., Barceló, D., 2003. Analysis and removal of emerging contaminants in wastewater and drinking water. TrAC Trends Anal. Chem. 22 (10), 685–696.

Philip, J.M., Aravind, U.K., Aravindakumar, C.T., 2018. Emerging contaminants in Indian environmental matrices—a review. Chemosphere 190, 307–326.

Prieto-Simón, B., Bandaru, N.M., Saint, C., Voelcker, N.H., 2015. Tailored carbon nanotube immunosensors for the detection of microbial contamination. Biosens. Bioelectron. 67, 642–648.

Pyrzynska, K., 2011. Carbon nanotubes as sorbents in the analysis of pesticides. Chemosphere 83 (11), 1407–1413.

Radjenović, J., Petrović, M., Barceló, D., 2009. Fate and distribution of pharmaceuticals in wastewater and sewage sludge of the conventional activated sludge (CAS) and advanced membrane bioreactor (MBR) treatment. Water Res. 43 (3), 831–841.

Rahaman, M.S., Vecitis, C.D., Elimelech, M., 2012. Electrochemical carbon-nanotube filter performance toward virus removal and inactivation in the presence of natural organic matter. Environ. Sci. Technol. 46 (3), 1556–1564.

Rocha, J.-D.R., Rogers, R.E., Dichiara, A.B., Capasse, R.C., 2017. Emerging investigators series: highly effective adsorption of organic aromatic molecules from aqueous environments by electronically sorted single-walled carbon nanotubes. Environ. Sci.: Water Res. Technol. 3, 203–212.

Roig, N., Sierra, J., Martí, E., Nadal, M., Schuhmacher, M., Domingo, J.L., 2012. Long-term amendment of Spanish soils with sewage sludge: effects on soil functioning. Agric. Ecosyst. Environ. 158, 41–48.

Rosal, R., Rodríguez, A., Perdigón-Melón, J.A., Petre, A., García-Calvo, E., Gómez, M.J., Agüera, A., Fernández-Alba, A.R., 2010. Occurrence of emerging pollutants in urban wastewater and their removal through biological treatment followed by ozonation. Water Res. 44 (2), 578–588.

Sahar, E., Messalem, R., Cikurel, H., Aharoni, A., Brenner, A., Godehardt, M., Jekel, M., Ernst, M., 2011. Fate of antibiotics in activated sludge followed by ultrafiltration (CAS-UF) and in a membrane bioreactor (MBR). Water Res. 45 (16), 4827–4836.

Sang, Z., Jiang, Y., Tsoi, Y.-K., Leung, K.S.-Y., 2014. Evaluating the environmental impact of artificial sweeteners: a study of their distributions, photodegradation and toxicities. Water Res. 52, 260–274.

Sarkar, B., Mandal, S., Tsang, Y.F., Kumar, P., Kim, K.-H., Ok, Y.S., 2018. Designer carbon nanotubes for contaminant removal in water and wastewater: a critical review. Sci. Total Environ. 612, 561–581.

Sartori, E.R.T., Fatibello-Filho, O., 2011. Glassy carbon electrode modified with functionalized carbon nanotubes within a poly(allylamine hydrochloride) film for the voltammetric determination of sulfite in foods. Electroanalysis 23 (11), 2526–2533.

Sharma, S., Bhattacharya, A., 2017. Drinking water contamination and treatment techniques. Appl. Water Sci. 7 (3), 1043–1067.

Singh, R.P., Agrawal, M., 2008. Potential benefits and risks of land application of sewage sludge. Waste Manage. 28 (2), 347–358.

Smith, S.C., Rodrigues, D.F., 2015. Carbon-based nanomaterials for removal of chemical and biological contaminants from water: a review of mechanisms and applications. Carbon 91, 122–143.

Sotelo, J.L., Rodríguez, A.R., Mateos, M.M., Hernández, S.D., Torrellas, S.A., Rodríguez, J.G., 2012. Adsorption of pharmaceutical compounds and an endocrine disruptor from aqueous solutions by carbon materials. J. Environ. Sci. Health B 47 (7), 640–652.

Stillo, F., Gibson, J.M., 2017. Exposure to contaminated drinking water and health disparities in North Carolina. Am. J. Public Health 107 (1), 180–185.

Su, R., Jin, Y., Liu, Y., Tong, M., Kim, H., 2013. Bactericidal activity of ag-doped multi-walled carbon nanotubes and the effects of extracellular polymeric substances and natural organic matter. Colloids Surf. B: Biointerfaces 104, 133–139.

Talib, A., Randhir, T.O., 2017. Managing emerging contaminants in watersheds: need for comprehensive, systems-based strategies. Sustain. Water Qual. Ecol. 9-10, 1–8.

Tran, N.H., Reinhard, M., Gin, K.Y.-H., 2018. Occurrence and fate of emerging contaminants in municipal wastewater treatment plants from different geographical regions-a review. Water Res. 133, 182–207.

UN DESA, 2015. United Nations, Department of Economic and Social Affairs, Population Division. World Population Prospects: The 2015 Revision, Online Edition UN DESA, New York.

UN DESA, 2018. United Nations, Department of Economic and Social Affairs, Population Division. World Urbanization Prospects: The 2018 Revision, Online Edition UN DESA, New York.

Vithanage, M., Seneviratne, M., Ahmad, M., Sarkar, B., Ok, Y.S., 2017. Contrasting effects of engineered carbon nanotubes on plants: a review. Environ. Geochem. Health 39 (6), 1421–1439.

Walker, R., 1985. Sulphiting agents in foods: some risk/benefit considerations. Food Addit. Contam. 2 (1), 5–24.

Westerhoff, P., Yoon, Y., Snyder, S., Wert, E., 2005. Fate of endocrine-disruptor, pharmaceutical, and personal care product chemicals during simulated drinking water treatment processes. Environ. Sci. Technol. 39, 6649–6663.

Yang, W., Lu, Y., Zheng, F., Xue, X., Li, N., Liu, D., 2012. Adsorption behavior and mechanisms of norfloxacin onto porous resins and carbon nanotube. Chem. Eng. J. 179, 112–118.
Yang, Y., Ok, Y.S., Kim, K.-H., Kwon, E.E., Tsang, Y.F., 2017. Occurrences and removal of pharmaceuticals and personal care products (PPCPs) in drinking water and water/sewage treatment plants: a review. Sci. Total Environ. 596-597, 303–320.
Yang, Z.-F., Li, L.-Y., Hsieh, C.-T., Juang, R.-S., 2018. Co-precipitation of magnetic Fe_3O_4 nanoparticles onto carbon nanotubes for removal of copper ions from aqueous solution. J. Taiwan Inst. Chem. Eng. 82, 56–63.
Yu, X.-Y., Luo, T., Zhang, Y.-X., Jia, Y., Zhu, B.-J., Fu, X.-C., Liu, J.-H., Huang, X.-J., 2011. Adsorption of lead (II) on O_2-plasma-oxidized multiwalled carbon nanotubes: thermodynamics, kinetics, and desorption. ACS Appl. Mater. Interfaces 3, 2585–2593.
Zhao, X., Jia, Q., Song, N., Zhou, W., Li, Y., 2010. Adsorption of Pb(II) from an aqueous solution by titanium dioxide/carbon nanotube nanocomposites: kinetics, thermodynamics, and isotherms. J. Chem. Eng. Data 55, 4428–4433.
Zhou, S., Shao, Y., Gao, N., Deng, J., Tan, C., 2013. Equilibrium, kinetic, and thermodynamic studies on the adsorption of triclosan onto multi-walled carbon nanotubes. Clean: Soil, Air, Water 41 (6), 539–547.

Further reading

Ai, L., Jiang, J., 2012. Removal of methylene blue from aqueous solution with self-assembled cylindrical graphene–carbon nanotube hybrid. Chem. Eng. J. 192, 156–163.
Chang, P.R., Zheng, P., Liu, B., Anderson, D.P., Yu, J., Ma, X., 2011. Characterization of magnetic soluble starch-functionalized carbon nanotubes and its application for the adsorption of the dyes. J. Hazard. Mater. 186 (2–3), 2144–2150.
Chen, X., Xia, X., Wang, X., Qiao, J., Chen, H., 2011b. A comparative study on sorption of perfluorooctane sulfonate (PFOS) by chars, ash and carbon nanotubes. Chemosphere 83 (10), 1313–1319.
Fang, Q., Chen, B., 2012. Adsorption of perchlorate onto raw and oxidized carbon nanotubes in aqueous solution. Carbon 50 (6), 2209–2219.
Ghaedi, M., Hassanzadeh, A., Kokhdan, S.N., 2011. Multiwalled carbon nanotubes as adsorbents for the kinetic and equilibrium study of the removal of Alizarin Red S and Morin. J. Chem. Eng. Data 56 (5), 2511–2520.
Gong, J.-L., Wang, B., Zeng, G.-M., Yang, C.-P., Niu, C.-G., Niu, Q.-Y., Zhou, W.-J., Liang, Y., 2009. Removal of cationic dyes from aqueous solution using magnetic multi-wall carbon nanotube nanocomposite as adsorbent. J. Hazard. Mater. 164 (2–3), 1517–1522.
Ji, L., Chen, W., Bi, J., Zheng, S., Xu, Z., Zhu, D., Alvarez, P.J., 2010a. Adsorption of tetracycline on single-walled and multi-walled carbon nanotubes as affected by aqueous solution chemistry. Environ. Toxicol. Chem. 29 (12), 2713–2719.
Ji, L., Shao, Y., Xu, Z., Zheng, S., Zhu, D., 2010b. Adsorption of monoaromatic compounds and pharmaceutical antibiotics on carbon nanotubes activated by KOH etching. Environ. Sci. Technol. 44 (16), 6429–6436.
Xia, X., Li, Y., Zhou, Z., Feng, C., 2010. Bioavailability of adsorbed phenanthrene by black carbon and multi-walled carbon nanotubes to Agrobacterium. Chemosphere 78 (11), 1329–1336.
Yang, G., Zhang, G., Wang, H., 2015. Current state of sludge production, management, treatment and disposal in China. Water Res. 78, 60–73.

Section D

Valorization of sludge as a resource

Gasification of sewage sludge

25

Sebastian Werle, Mariusz Dudziak*[†]
*Institute of Thermal Technology, Silesian University of Technology, Gliwice, Poland,
[†]Institute of Water and Wastewater Engineering, Silesian University of Technology, Gliwice, Poland

1 Introduction

Sewage sludge is a by-product of wastewater treatment plant operations. Due to the rapid rate of urbanization and industrialization and growth of the population, the world is facing the serious challenge of disposing of this waste. For the last 50 years, the world's population has multiplied faster than ever before, and more rapidly than it is projected to grow in the future. In 1950, the world had 2.5 billion people; and in 2005, the world had 6.5 billion people. By 2050, this number could rise to more than 9 billion (World Population Prospects, 2008). The problem of this waste material is a key issue in the majority of countries.

In the European Union (EU), the problem of sewage sludge is being tackled by general directives and indicators and national legislative requirements. As a result of the implementation of the Urban Waste Water Treatment Directive (UWWTD) 91/271/CE (Council Directive, 1991), the member-countries have improved their collecting and wastewater treatment systems. The amount of sewage sludge produced increased by more than 50%, from 6.5 million tons of dry matter (DM) in 1992 to 10.9 million tons in 2015 (http://epp.eurostat.ec.europa.eu, Accessed 27 November 2017).

Fig. 1 shows the amount of sewage sludge produced in EU member-countries in 2015 and the total amount produced in EU-28 (http://epp.eurostat.ec.europa.eu, Accessed 27 November 2017).

According to the European Union's and other countries' demographic projections, the total sludge production in the union in 2020 increased to 13 million tons DM (UN DESA, 2014).

In line with the 7th Environment Action Programme (http://ec.europa.eu/environment/action-programme/, Accessed 27 November 2017), the reduction of sewage sludge disposal in landfills by 50% from 2000 to 2050 has to be achieved. Based on these principles, the Landfill Directive (LD) 99/31/EC (Council Directive, 1999) prohibited the landfilling of both liquid and untreated waste and set restrictions. Landfill and water deposition of sewage sludge with the indicated parameters presented in Table 1 in the European Union is prohibited.

Additionally, generally the member-states are applying stricter restrictions than those determined in Sewage Sludge Directive 86/278/EEC (SSD) (Council Directive, 1986), which describes the beneficial use of sludge in soil. The majority

Industrial and Municipal Sludge. https://doi.org/10.1016/B978-0-12-815907-1.00025-8

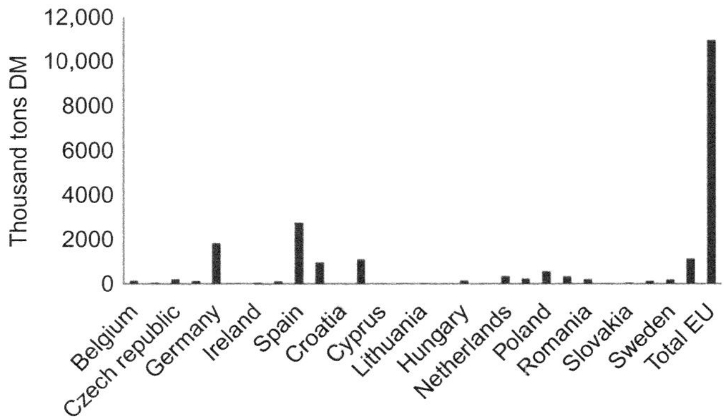

Fig. 1 The sewage sludge production in EU-28 in 2015 (http://epp.eurostat.ec.europa.eu, Accessed 27 November 2017).

Table 1 Criteria for the storage of sewage sludge in a nonhazardous waste landfill (Council Directive, 1999)

Number	Parameter	Value limit
1	Overall organic carbon (OOC) (%DM)	5.0
2	Loss at calcinations (LOC) (%DM)	8.0
3	The higher heating value (MJ/kg DM)	Maximum 6.0

of EU countries have set more stringent national requirements for heavy metal concentrations in sewage sludge then required by SSD.

Considering these facts, as well as the limitations of current disposal methods and the challenges of energy sustainability, it can be concluded that energy recovery is definitely going to be an important part of modern sewage sludge management. Consequently, there is an urgent need to explore low-cost, energy-efficient, and sustainable solutions for the treatment, management, and future utilization of sewage sludge. There are several thermal technologies for utilizing sewage sludge to obtain forms of energy through correct pretreatment (e.g., torrefaction, slow pyrolysis, and hydrothermal carbonization processes) (Wilk et al., 2016). In Fig. 2, the main processes (combustion and cocombustion, gasification, and pyrolysis) used for sewage sludge thermochemical conversion are presented.

Based on the flowchart in Fig. 2 it seems that gasification of sewage sludge has several advantages over a traditional combustion or cocombustion process and pyrolysis (Muzyka et al., 2015). Pyrolysis requires comprehensive knowledge of the physicochemical, thermal, and kinetic decomposition characteristics of feedstock (Yaman, 2004). Gasification is the partial oxidation of biodegradable material in an air-restricted environment, which yields a mix of flammable gases such as hydrogen (H_2), carbon monoxide (CO), and methane (CH_4), (known as

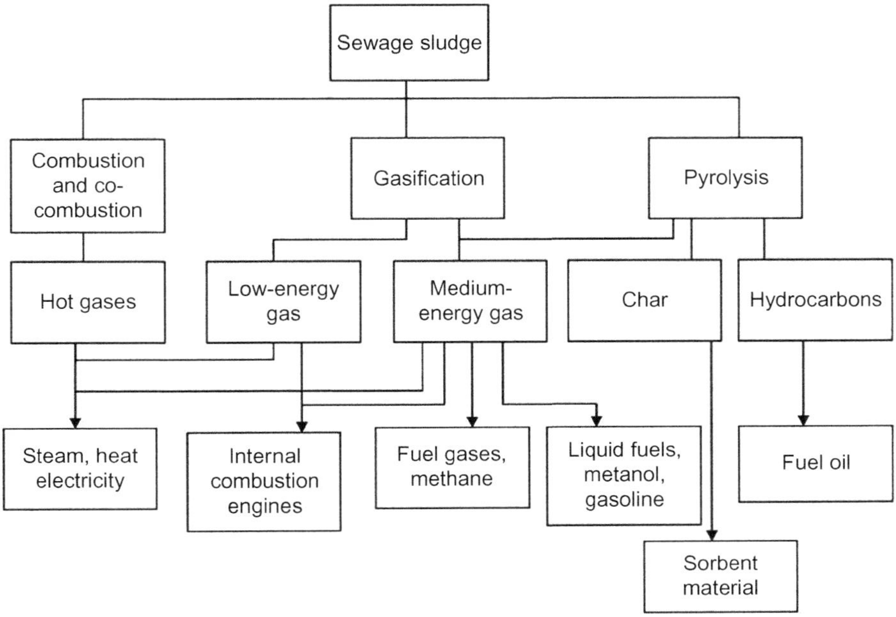

Fig. 2 Main thermochemical processes for sewage sludge conversion (Wilk et al., 2016).

producer gas or *gasification gas*) and a solid fraction of carbonaceous, ash-rich char (Skorek-Osikowska et al., 2017). Producers can be used to generate mechanical or electrical power in dedicated gas engines or can be fed into the intake manifold of diesel engines in dual-fuel operations.

Such gasification power systems are technologically mature, tolerant of diverse feedstocks, and practical on smaller scales than combustion-based steam power systems (Pinto et al., 2007). Taking into consideration that gasification is characterized by a low-level gasification agent environment, the volume of produced gas is less than the volume of flue gas from the combustion process. Consequently, sulfur in gasified material is transformed to hydrogen sulfide (H_2S), nitrogen to ammonia (NH_3), and chloride to hydrogen chloride (HCl), the formation of the dioxins sulfur dioxide (SO_2) and NO_x is prevented, and gas-cleaning installation is smaller and less expensive than classic combustion. This feature is very profitable, considering the possible use of such installations in wastewater treatment plants. A more efficient scenario for sewage sludge gasification is combining the fuel production process with phosphorus recovery. After sewage sludge conversion, the residue is a sanitized source of minerals and organic elements, including valuable fertilizing components, which makes them a potential substitute for natural phosphorus ore (Gorazda et al., 2017).

The development of gasification technology is associated with the development of gasifiers.

Gasification reactors can be divided into three main types (Gorazda et al., 2017; Werle, 2014): fixed-bed, fluidized-bed, and entrained-bed gasifiers. One of the most widely known European plants using the gasification process is a system in Güssing, Austria, built in 2001, with power in the fuel of 8 MWth (http://biomasspower.gov.in/, Accessed 27 November 2017). Similar systems can be found in Harbore, Denmark; Spiez, Switzerland; Kokemäki, Finland; and Skive, Denmark (Uchman and Werle, 2016). The installation by the Institute of Thermal Technology is another example of the sewage sludge gasification facility (Werle and Wilk, 2011).

During sewage sludge gasification, the occurrence of inorganic and organic contamination, including the waste by-products ash and tar, is also important (Werle and Dudziak, 2014). In that context, new procedures are still generated (Werle and Dudziak, 2013; Werle et al., 2016). For example, in sewage sludge and in tars from gasification, both toxic and hazardous organic substances (primarily of polycyclic aromatic hydrocarbons in sewage sludge, and phenols and their derivatives in tars) and inorganic substances (e.g., nine heavy metals) were identified. Moreover, in ash, mainly inorganic substances (heavy metals) were detected. The inorganic and organic contamination is mainly transported within the system as sewage sludge–gasification by-products. For their determination, basic instrumental methods (gas chromatography and absorption spectrometry) (Werle and Dudziak, 2014) can be used, as well as indirect methods like photoacoustic spectrometry and ecotoxicological analysis (Werle et al., 2016).

The aim of this research is a multiparametric investigation of the sewage sludge gasification in the fixed-bed gasifier. The deep characterization of sewage sludge as feedstock is presented. The influence of the gasification process parameters on the sewage sludge gasification gas quality is determined. The characterization of the gasification by-products is also included. The proposition of the reuse process of ash produced during gasification is postulated.

2 Gasification of sewage sludge in a fixed-bed reactor—research characteristics

A fixed-bed gasification facility located at the Laboratory of Fuels Combustion and Gasification, at the Institute of Thermal Technology of Silesian University of Technology, Gliwice, Poland, was used for the current study. A schematic of the facility is shown in Fig. 3.

The main part of the system was an insulated, stainless gasifier with an internal diameter of 150 mm and a total height of 300 mm. The granular sewage sludge was fed into the reactor from the fuel container located at the top. The gasification air was fed from the bottom by the fan. The temperature inside the reactor was measured by six N-type thermoelements integrated with the Agilent temperature recording system. Thermocouples were located along the vertical axis of the reactor. Additionally, the temperature of the gasification gas at the outlet of the reactor was measured. The flow rates of gasification air and of produced gas was measured by flow meters. The gasification gas Syngas was transported by the gas pipe and then was cleaned by a cyclone, scrubber, and drop separator. Volumetric fractions of the main components

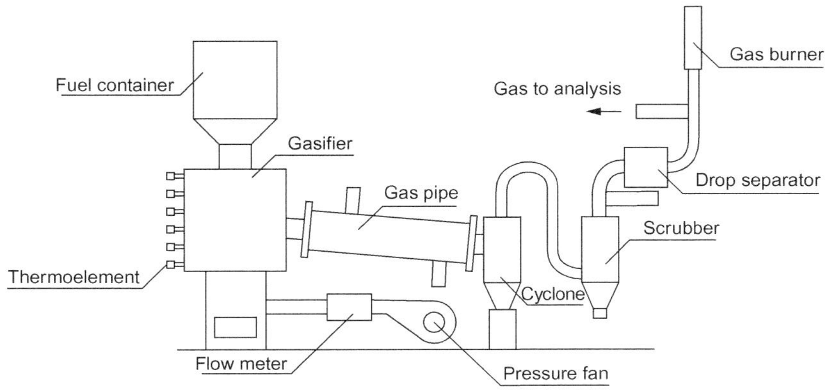

Fig. 3 Schematic diagram of the experimental system (Werle and Wilk, 2011).

Table 2 Gasification process methodology (Werle, 2014, 2015)

Sewage sludge (SS)	Gasification agent	Air ratio λ	Gasification agent temperature (K)
SS1 (from the mechanical-biological wastewater treatment plant) SS2 (from mechanical-biological-chemical wastewater treatment plant)	Atmospheric air (molar fraction of oxygen $z_{O_2}=0.21$; molar fraction of nitrogen $z_{N_2}=0.79$) and oxygen-enriched air (molar fraction of oxygen $z_{O_2}=0.25$; molar fraction of nitrogen $z_{N_2}=0.75$)	0.12, 0.14, 0.16, 0.18, 0.23, 0.27	298 (cold), 323, 373, 423, 473 (preheated)

of produced gas were measured online using the Fisher Rosemount and ABB integrated set of analyzers.

The gasification methodology is presented in Table 2.

3 Feedstock characterization

During the experiments, two types of sewage sludge feedstock were analyzed: SS1 and SS2. SS1 was taken from a mechanical-biological wastewater treatment plant, and SS2 was taken from a mechanical-biological-chemical wastewater treatment plant with phosphorus precipitation. In both analyzed cases, the biological part of the wastewater treatment plant has worked with low-load activated sludge. Such solution offers

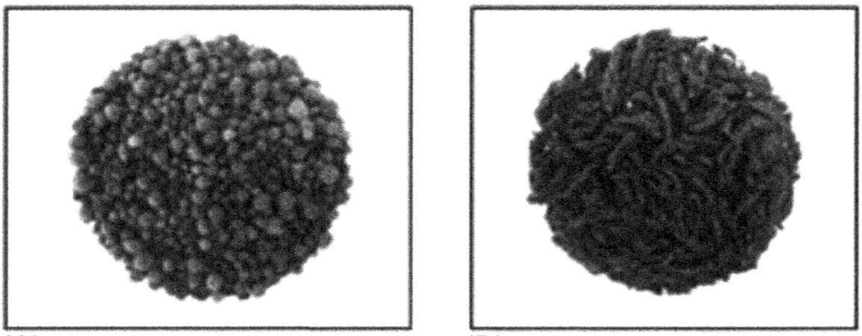

(A) (B)

Fig. 4 The (A) SS1 and (B) SS2 samples (Werle and Dudziak, 2015).

the chance for effective removal of phosphorus and nitrogen from wastewater. Additionally, in both cases, sewage sludge is stabilized by anaerobic digestion and dehydration.

After this, sewage sludge is mechanically dried. For the SS1 sample, hot air ($t_{hotair} = 260°C$) in a cylindrical dryer with a heated shelf was used. For the SS2 sample, hot air ($t_{hotair} = 150°C$) in an air belt dryer was used. As a consequence of the mechanical process of drying, the form of SS1 samples is similar to granules and the form of SS2 samples is similar to small "noodle." Fig. 4 shows the two samples.

4 Ultimate and proximate analysis and occurrences of organic and inorganic contaminants

Table 3 presents the proximate and ultimate analysis of both samples in the study. The following procedures and standards were used for the sewage sludge characterization. For the moisture PN-EN 14774-3:2010, the volatile content was determined according to the standard PN-EN 15402:2011. The ash content was obtained using PN-EN 15403:2011. The calorific value *t* was determined in accordance with the CEN/TS15400:2006 standard. The infrared (IR) spectroscopy analyzer was also used to carry out the ultimate analysis of the sewage sludge.

The organic contaminants in the sewage sludge analyzed by Werle et al. (2016) were characterized by the presence of polycyclic aromatic hydrocarbons, pesticides, and polychlorinated biphenyls (PCBs; Table 4). All of the analyzed sewage sludge also was characterized by the presence of inorganic contaminants (namely, various heavy metals). The technology of the wastewater treatment strongly influenced the profile of the contamination in sewage sludge.

In the sewage sludge from the wastewater treatment plant operating in the mechanical-biological system (SS1), nine compounds of polycyclic aromatic hydrocarbons were identified: phenanthrene, anthracene, benzo(a)fluoranthene, pyrene, chrysene, benzo(b)fluoranthene, dibenzo(a, h) anthracene, benzo (g, h, i) perylene, and indeno (1, 2,3-cd)pyrene. In sludge from the mechanical-biological-chemical wastewater

Table 3 Proximate and ultimate analysis of the samples SS1 and SS2 (Werle and Dudziak, 2015)

Parameter	SS1	SS2
Proximate analysis (%) (as received)		
Moisture	5.30	5.30
Volatile matter	44.20	36.50
Ash	49.00	51.50
Ultimate analysis (%DM)		
C	27.72	31.79
H	3.81	4.36
O	3.59	4.88
N	13.53	15.27
S	1.81	1.67
F	0.003	0.013
Cl	0.033	0.022
LHV (MJ/kgDM)	10.75	12.96

treatment plant (SS2), there were eight compounds [acenaphthene, benzo(a) fluoranthene, pyrene, benzo(a)anthracene, chrysene, benzo(b)fluoranthene, benzo(a) pyrene, and indeno(1,2 3-cd)pyrene] in this pollutant group [naphthalene, acenaphthylene, acenaphthene, fluorene, phenanthrene, anthracene, benzo(a)fluoranthene, pyrene, benzo(a)anthracene, chrysene, benzo(b)fluoranthene, benzo(k)fluoranthene, benzo(a)pyrene, dibenzo(a, h)anthracene and benzo(g, h, i)perylene, indeno(1,2,3-cd) pyrene]. However, the total concentration of polycyclic aromatic hydrocarbons was almost four times higher in SS1 than in SS2. Also, in earlier research (Werle, 2014), it was determined that polycyclic aromatic hydrocarbons are the primary form of pollutants present in sewage sludge. In addition, pesticides and PCBs were included in the identified organic pollutants in the tested sewage sludge. Content analysis shows that the total concentration of heavy metals in the sewage sludge is similar in both cases: 1841.19 mg/kg for SS1 and 1847.6 mg/kg for SS2 (Table 4).

5 Indirect methods of determining the level of sewage sludge contamination

The identification of various organic and inorganic elements in sewage sludge spurs the need to find indirect methods of determining its contamination level. Werle et al. (2016) documented that photoacoustic spectrometry (PAS) is a very effective way to determine the organic contamination level in sewage sludge. Using PAS and gas chromatography–mass spectrometry (GC-MS) analysis, a comparison of the pollution degree of SS1 and SS2 was made. Table 5 summarizes the obtained results.

These results lead to the conclusion that photoacoustic spectroscopy (PAS) analysis can be used to make a preliminary assessment of the degree of contamination in

Table 4 Concentration of organic and inorganic contaminants in sewage sludge (Wilk et al., 2016)

Group/compounds	Sewage sludge	
	SS1	**SS2**
	Concentration (μg/kg DM)	
Polycyclic aromatic hydrocarbons		
Acenaphthene	n.o.	80.84
Phenanthrene	511.12	n.o.
Anthracene	200.03	n.o.
Benzo(a)fluoranthene	44.78	126.48
Pyrene	187.22	123.86
Benzo(a)anthracene	n.o.	35.15
Chrysene	108.14	23.79
Benzo(b)fluoranthene	700.51	53.62
Benzo(a)pyrene	n.o.	46.11
Dibenzo(a,h)anthracene	101.54	n.o.
Benzo(g,h,i)perylene	209.44	n.o.
Indeno(1,2,3-cd)pyrene	370.62	131.48
Sum	2433.40	621.33
Pesticides		
Heptachlor	4.14	n.o.
Aldrin	3.13	1.28
Endrin	11.58	n.o.
Sum	18.85	1.28
PCBs		
2,20,5,50-PCB	9.75	7.90
2,20,4,5,50-PCB	33.33	n.d.
2,20,4,40,5,50-PCB	23.78	4.57
Sum	66.86	12.47
Heavy metal content, concentration (mg/kg DM)		
Zn	920.90	991.20
Cu	495.30	183.16
Pb	119.30	59.97
Ni	103.67	18.90
Cr	180.53	584.53
Cd	6.47	3.24
As	4.19	3.94
Hg	0.99	0.96
Se	9.84	1.70
Sum	1841.19	1847.6

n.o., not observed.

Table 5 Comparison of the results from GC-MS and PAS analysis of the dry mass of the analyzed sewage sludge (Wilk et al., 2016)

	GC-MS analysis	**PAS analysis**		
Sewage sludge	**Group of pollutants**	**Characteristic binding**	**Wavelength (cm^{-1})**	**Presence (+)/ absence (–)**
SS1	Aromatic hydrocarbons	C=C	1450–1610	+
	Pesticides	C=C	1450–1610	+
		C—Cl	600–800	+
	PCBs	C=C	1450–1610	+
		C—Cl	600–800	+
SS2	Aromatic hydrocarbons	C=C	1450–1610	+
	Pesticides	C=C	1450–1610	+
		C—Cl	600–800	+
	PCBs	C=C	1450–1610	+
		C—Cl	600–800	+

the samples (e.g., before conducting GC-MS analysis). It was observed that the compared techniques have very similar results in terms of pollution detection. Thus, photoacoustic spectroscopy can be successfully used in screening analysis. Another indirect method of determining the sewage sludge contamination level, including environmental safety, is ecotoxicology analysis. The comparative results of toxicity tests using various types of assays are shown in Table 6.

Based on the results of the en zymatic Microtox assay, it was determined that SS1 was characterized by low toxicity and that SS2 was toxic. Similar conclusions can be drawn from the results of the Daphtoxkit F survival test and the Phytotoxkit F growth test.

Table 6 Comparison of the toxicity analysis of the sewage sludge (Werle and Dudziak, 2015)

		Sewage sludge	
		SS1	**SS2**
Test (type/test species)	**Test time**	**Effect (%) (evaluation of toxicity[a])**	
Microtox (enzymatic with *Vibrio fischeri*)	5 min	26.1 (+)	57.0 (++)
Daphtoxkit F (survival with *Daphnia magna*)	24 h	40.0 (+)	60.0 (++)
Phytotoxkit F (growth with *Lepidium sativum*)	24 h	40.0 (+)	52.0 (++)

[a](–) nontoxic; (+) low toxic; (++) toxic; (+++) highly toxic.

The source of the SS2 toxicity was probably chromium (Cr); the Cr concentration in SS2 was three times higher than in SS1.

6 Gasification process results

6.1 *Influence of the air ratio on the gas composition, LHV of gas, and temperature distribution*

The temperature distribution in the reactor was obtained by measuring the temperatures at six characteristic points of the reactor (Fig. 5). The points, T_1, T_2, T_3, T_4, T_5, and T_6, were measured at 10, 60, 110, 160, 210, and 260mm above the grate, respectively.

Fig. 5 showed that the temperature at the T_3 point was always the highest; thus, T_3 may be located in the partial oxidation zone, which should be the hottest area in the fixed-bed gasifier. Correspondingly, the measuring points T_6 and T_5 may be located in the drying zone, T_4 in the pyrolysis zone, T_2 in the oxidation (combustion) zone, and T_1 in the ash zone. A similar tendency was observed during the experiment with the SS2 sample.

Fig. 6 shows the evolution of the CH_4, H_2, CO, and carbon dioxide (CO_2), concentrations in gasification of the SS1 and SS2 samples, with varying air ratios for each sludge.

Analyzing this graphs in this figure, it can be confirmed that throughout the range of analyzed air ratios, the volumetric fractions of the main combustible components of the gasification gases (CO and H_2) are higher in SS1 than in SS2. At lower air ratios,

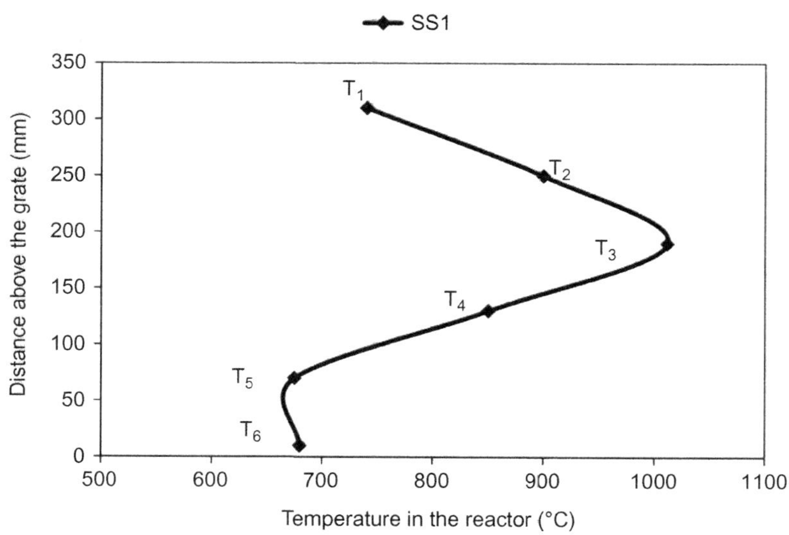

Fig. 5 Temperature distribution in the reactor for SS1 gasification (Werle, 2014).

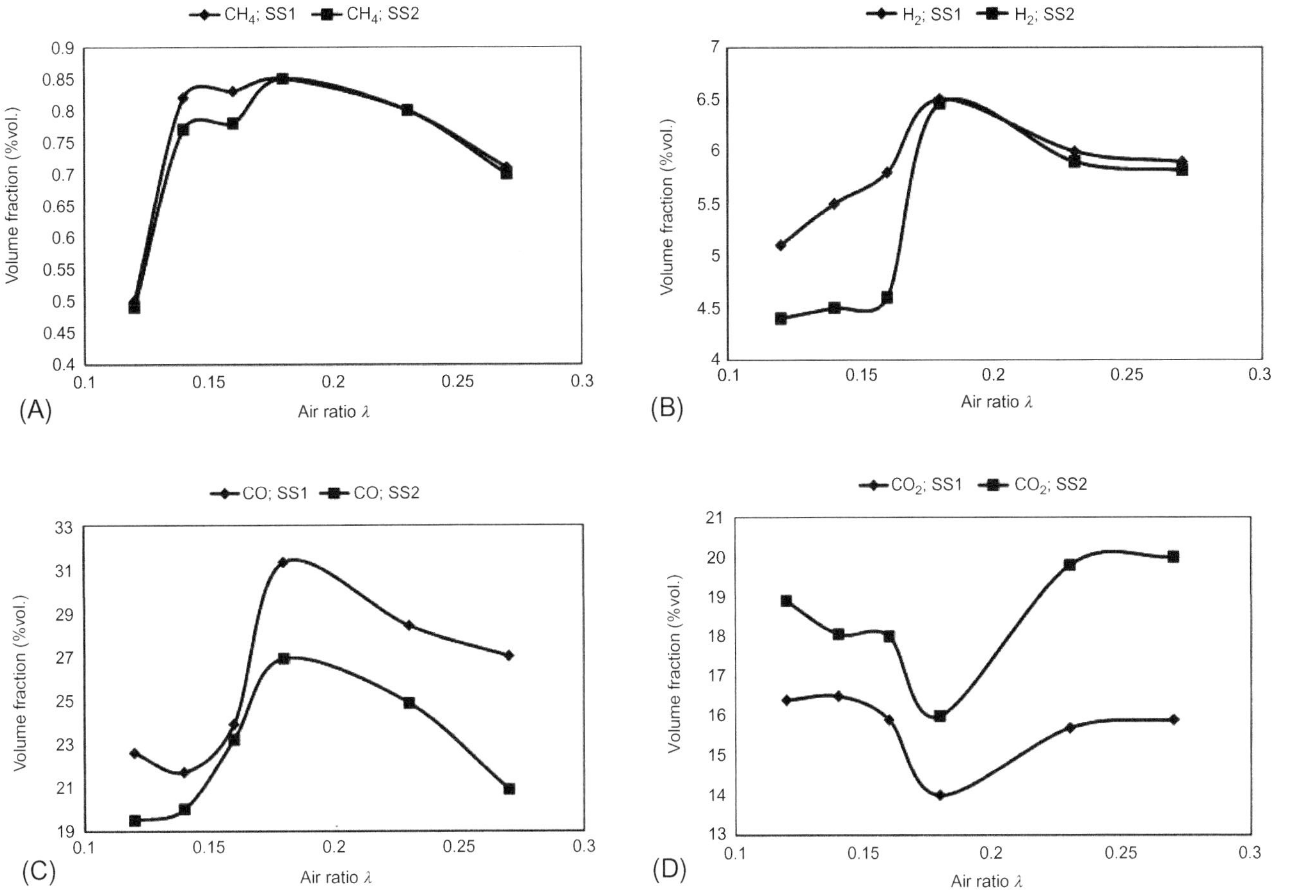

Fig. 6 The volume fraction of main components in SS1 and SS2 gasification gases as a function of air ratio; (A) methane; (B) hydrogen, (C) CO, (D) CO_2 (Werle, 2014, 2015)

the CO fraction was found to start out low, and then it rises until the optimum air ratio of 0.18 is reached, and later drops again for higher air ratios. The maximum CO average composition values of 31.3% for SS1 (and 26.9% for SS2) were obtained for gasification at $\lambda = 0.18$. CO_2 shows an inverse relation with CO, as the reactions that produce those gases are competing for the same reactant (namely, C). The concentration of CO_2 is generally expected to be minimum of the optimal air ratio range between 0.18 and 0.24.

The rapid growth of CO observed with an air ratio equal to 0.18 is caused by the dominant role of the primary water gas reaction. The reactions that can occur in the gasifier as a result of the gasification agent flow can be categorized as the reaction of gasification agents and C in the fuel and the reaction of gasification agents and CO in the gas. The reaction of gasification air and C is an endothermic reaction that generates mainly CO, whereas the reaction of gasification air and CO is an exothermic reaction that generates mainly CO_2 and H_2. When gasification air is fed with the fuel into the reactor, the endothermic reaction of air and carbon occurs first (e.g., the primary water gas reaction $CO + H_2O \rightarrow CO + H_2$), and the CO in a gaseous state produced from the fuel reacts with the residuals, causing the next reactions (e.g., water gas shift $CO + H_2O \leftrightarrow CO_2 + H_2$). Thus, the composition of H_2, CO, and CO_2 in the gasification gas changes according to the amount of the air supplied to the reactor.

Fig. 7 illustrates the dependence of the collated based on the volume fraction of the gas component's lower heating value (LHV) for obtaining the gas versus an air ratio.

The following formula was used (Kim et al., 2011):

$$LHV = 0.126 \cdot CO + 0.108 \cdot H_2 + 0.358 \cdot CH_4$$

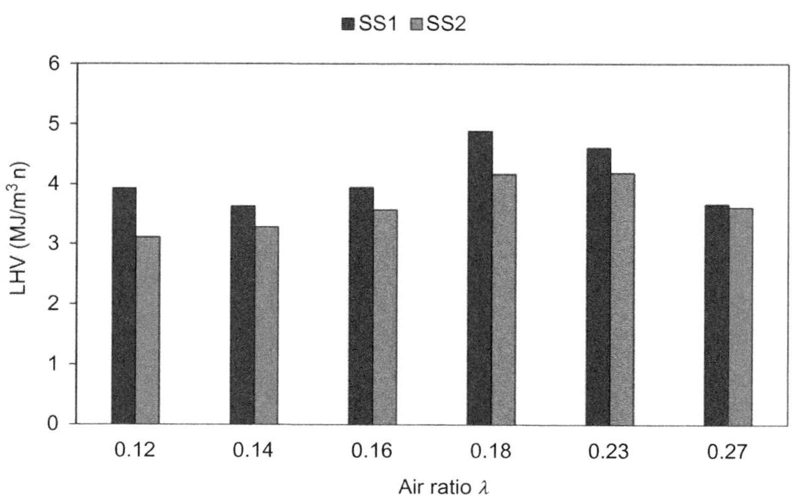

Fig. 7 The LHV of the SS1 and SS2 gasification gases as a function of the air ratio (Werle, 2014, 2015).

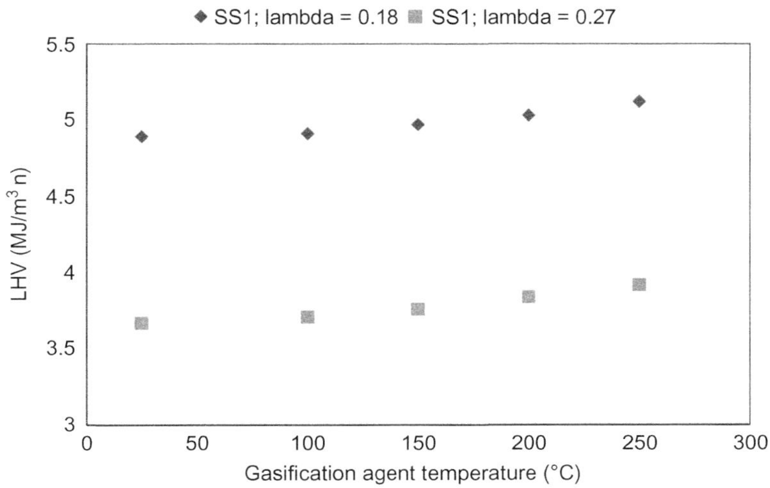

Fig. 8 The LHV of the SS1 gasification gas as a function of gasification agent temperature (Werle, 2014, 2015).

Analyzing Fig. 7, it can be concluded that taking into consideration the LHV of the gasification gas, the optimum value of the air ratio is equal to 0.18, from which the LHV takes its maximum value. This is true regardless of the sewage sludge type. Above that optimal value, the thermochemical process can shift from gasification to combustion.

6.2 Influence of the gasification temperature on the LHV of gas and temperature distribution

Fig. 8 shows the LHV of the SS1 gasification gas as a function of gasification agent temperature. Analyzing the presented data reveals that the increase in the gasification temperature causes a corresponding increase in the calorific value of the obtained gas. This is mainly because the increase in gasification gas temperature is profitable for heat-cracking reactions, which act to promote high-calorific value hydrocarbon production in the gas. Such components exert a strong influence on the LHV. Further, these reactions are exothermic so the medium reactor temperature is also increasing (Fig. 9).

6.3 Influence of the oxygen content in gasification agents on gas composition, LHV of gas, and temperature distribution

In Fig. 10, the influence of oxygen concentration of the gasification agent on the composition of the SS1 gasification gas is shown.

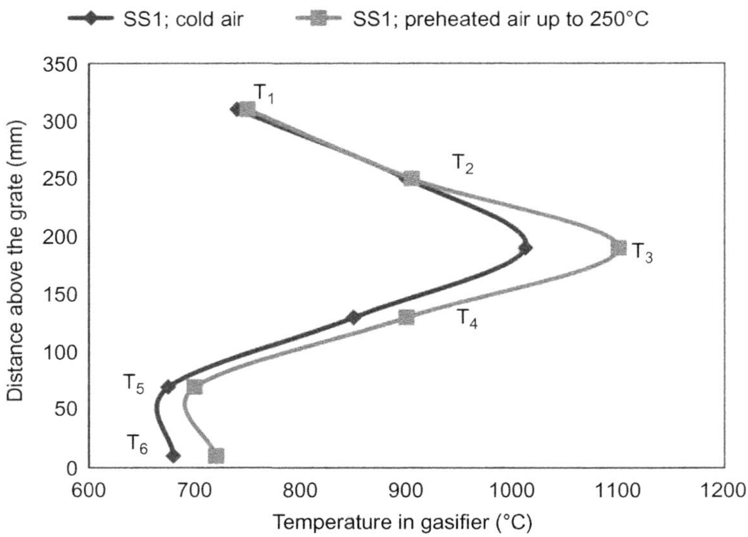

Fig. 9 The influence of the gasification agent temperature on the temperature distribution in the reactor for SS1 gasification (Werle, 2014, 2015).

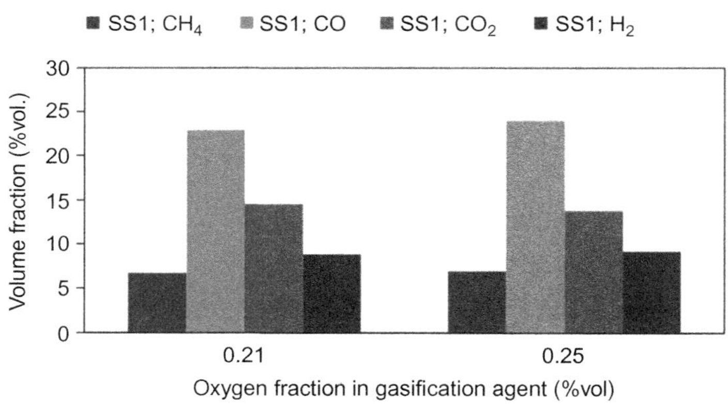

Fig. 10 The influence of oxygen concentration of the gasification agent on the SS1 gasification gas composition (Werle, 2014, 2015).

Analyzing the data presented in Fig. 10 reveals that the use of a gasifying agent with increased O_2 content ($z_{O_2}=0.25$) causes an increase in the proportion of flammable components in gasification gas. This occurs because under such conditions, the effect of dilution with nitrogen is balanced, thereby increasing the share of flammable gases in the produced gas.

Analyzing the graph shown in Fig. 11 shows that an increase in the proportion of oxygen in the gasifying agent results in an increase in the calorific value of the gas compared to when atmospheric air is used as the gasifying agent.

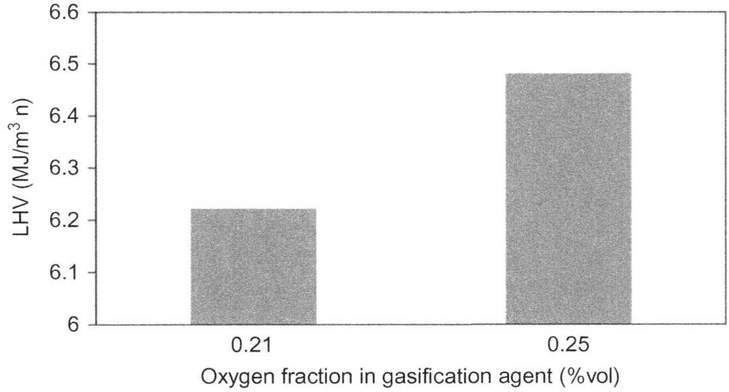

Fig. 11 The influence of the oxygen concentration of the gasification agent on the LHV of the SS1 gasification gas (Werle, 2014, 2015).

7 Solid and liquid waste by-product analysis

Werle and Dudziak (2014) concluded that both solid (ash and charcoal) and liquid (tar) gasification products do not consist of the organic contaminants that have been identified primarily in sewage sludge (as shown earlier in this chapter in Table 4). The tests of the liquid products (tar) formed during sewage sludge gasification determined TOC, which directly measures the amount of the various organic substances in the sample. The indicator was very high for the tar produced during gasification of both SS1 and SS2, being 20,950 and 22,390 mg TOC/L, respectively.

A chromatographic analysis of the tar samples showed their contamination, mainly by phenols and their derivatives (Table 7). The inorganic contaminants (heavy metals) identified in sewage sludge were also found in solid by-products from gasification (ash, charcoal). In addition, it has been documented that the concentrations of some heavy metals have increased significantly due to cumulative effects. This applied to seven of the nine heavy metals investigated (zinc, lead, nickel, copper, chromium, cadmium, and arsenic; the others were selenium and mercury), regardless of the type of sewage sludge. Due to reductive atmosphere during the gasification the migration of heavy metals from fuel to solid phase is observed. Similar conclusions were drawn by Li et al. (2012), who found an increase in the concentration of cadmium (from 0.93 to 1.67 mg/kg dry basis), chromium (from 80.82 to 247.95 mg/kg dry basis), copper (from 580.36 to 922.14 mg/kg dry basis), lead (from 78.27 to 125.09 mg/kg dry basis), and zinc (from 402.09 to 637.50 mg/kg dry basis) in sewage sludge and ash samples after gasification.

However, some differences in the accumulation of the solid waste by-products of particular heavy metals were detected. For instance, zinc concentration in an ash sample increased twofold for SS2 gasification, and even threefold for SS1. The liquid by-products (tar) produced during sewage sludge gasification demonstrated high conductivity, which proves their high contamination with various inorganic substances.

Table 7 Concentrations of phenols and their derivatives in the tar produced during the gasification of sewage sludge (Werle and Dudziak, 2014)

Compound	SS1	SS2
	Concentration (μg/L)	
2-Chlorophenol	211.84	57.20
2-Nitrophenol	89.52	n.o.
2,4-Dichlorophenol	361.56	n.o.
4-Chloro-3-methylphenol	9.27	1.98
2,4,6-Trochlorophenol	62.32	46.33
Pentachlorophenol	57.97	53.90
Sum	792.48	100.23

n.o., not observed.

The conductivity of the tar was 9800 μS/cm for SS1, and 8170 μS/cm for SS2. The hazardous inorganic contaminants occurring in the tar also included ammonia, whose concentrations were 1090 (tar from SS1) and 950 mg NH_4^+/L (tar from SS2).

Ecotoxicological analysis of solid and liquid by-products of gasification has shown that they can pose a danger to the natural environment (Table 8). By-products were obtained during the gasification process under constant conditions (temperature of the gasifying agent, 298 K, and excess air ratio $\lambda = 0.18$). The decrease in bioluminescence intensity (shown by the Microtox test) caused by the gasification by-products

Table 8 Comparison of the gasification by-products (gasification parameters: $T = 298$ K, $\lambda = 0.18$) (Werle and Dudziak, 2015)

Test (type/test species)	Test time	Gasified sewage sludge				
		SS1	SS2	SS1	SS2	SS1
		Gasification by-products				
		Ash		Tar		Charcoal
		Effect (%) (evaluation of toxicity[a])				
Microtox (enzymatic with *Vibrio fischeri*)	5 min	2.1 (–)	64.7 (++)	56.8 (++)	66.3 (++)	9.4 (–)
Daphtoxkit F (survival with *Daphnia magna*)	24 h	40.0 (+)	70.0 (++)	80.0 (+++)	100.0 (+++)	20.0 (–)
Phytotoxkit F (growth with *Lepidium sativum*)	24 h	20.0 (–)	50.0 (+)	44.0 (+)	40.0 (+)	20.0 (–)

[a](–) nontoxic; (+) low toxic; (++) toxic; (+++) highly toxic.

depended noticeably on the type of sample (ash, tar, or charcoal). The observed correlation is also greatly affected by the type of sludge that was gasified.

The samples of ash and tar produced during the gasification of SS1 turned out to be nontoxic. Comparing them to the toxicity of the sludge, their toxic effect decreased. The reverse phenomenon was observed for the SS2 sample, as well as for ash that formed during the gasification of this sludge. The tar samples were toxic regardless of the type of gasified sewage sludge. The results of the Microtox test were the opposite of the results of the Daphtoxkit F survival test and the Phytotoxkit F growth test. The tars proved to be highly toxic to the crustacean *Daphnia magna* (Daphtoxkit F test). The dicot plant *Lepidium sativum* was the least sensitive to the effect of tars (Phytotoxkit F test).

8 Management and purification of the gasification by-products

Dudziak and Werle (2016) have demonstrated that solid gasification by-products can be used as an adsorption material for the elimination of toxic organic substances from the water streams (e.g., phenol). Table 9 presents the comparison of the maximum adsorption capacity of the phenol monolayer on various adsorbents. Based on the presented data, it can be concluded that the efficiency of phenol adsorption on the ash was greater than for the other unconventional adsorbents (i.e., bagasse fly ash, neutralized red mud, and olive pomace). The adsorption of phenol was found for commercially available activated carbon and activated carbon derived from waste materials such as beet pulp or rice husk. However, only ashes that are nontoxic (Table 8) can be used as the adsorption material.

Taking into account the profile of contaminants identified in the tars from sewage sludge gasification (Table 7), they should be subjected to deepen purification processes. This also indicates the ecotoxicological analysis of the tar samples that confirms their toxicity (Table 8). Advanced oxidation methods that are effective in coking

Table 9 Comparison of the maximum monolayer adsorption capacity of phenol onto various adsorbents (Dudziak and Werle, 2016)

Adsorbents	q (mg/g)
Activated carbon fiber	110.20
Beet pulp carbon	90.61
Commercial activated carbon	49.72
Rice husk carbon	22.00
Chemically modified green macro algae	20.00
Baggase fly ash	12.00–13.00
Neutralized red mud	5.13
Olive pomace	4.00–5.00
Sewage sludge ash	42.22

wastewater treatment have been proposed (Krzywicka and Kwarciak-Kozłowska, 2014). There is some analogy between the profile of contaminants in tar from sewage sludge gasification and coke wastewater. However, the issue raised is currently in the field of new research.

9 Conclusions

- The operating conditions (amount of the gasification agent) of the sewage sludge gasification process greatly influence the gasification gas composition distribution.
- Higher values of the main components (especially C and H) in the sewage sludge plant affect increases in the LHV of gasification gases.
- Taking into consideration the LHV of the gasification gas, the optimum value of the air ratio is equal to 0.18, in which the LHV takes its maximum value.
- The yield of the main gasification gas components, CO, H_2, and CH_4, was enhanced by increasing the gasification agent temperature and the oxygen concentration of the gasification agent.
- Both standard and indirect evaluation methods can be used to assess the degree of contamination of sewage sludge and the by-products of gasification (i.e., ash, charcoal, tar). In this respect, ecotoxicological analyses should be included that allow an assessment of environmental safety.
- Originally, sewage sludge contains a variety of organic pollutants (PAHs, pesticides, PCBs) and inorganic ones (heavy metals).
- Solid waste by-products from sewage sludge (ashes) can be used as an adsorbent for the elimination of toxic organic substances from water streams such as phenol. The sludge from sludge gasification should be subjected to deep purification processes.

Acknowledgments

The paper has been partially prepared within the frame of the National Science Centre project based on Decision No. DEC-2011/03/D/ST8/04035 and within the framework of the Ministry of Science and Higher Education Iuventus Plus Programme Project No. 0593/IP2/2011/71 and within the frame of the project "Study on the solar pyrolysis process of the waste biomass," financed by the National Science Centre, Poland (registration number 2016/23/B/ST8/02101).

References

Council Directive 86/278/EEC, 1986. On the protection of the environment, and in particular of the soil, when sewage sludge is used in agriculture. Off. J. Eur. Commun. L181(6–12).

Council Directive 91/271/EEC, 1991. Concerning urban waste water treatment. Off. J. Eur. Communities. L135(40–52).

Council Directive 99/31/EC, 1999. On the landfill of waste. Off. J. Eur. Union. L182(1–19).

Dudziak, M., Werle, S., 2016. Studies on the adsorption of phenol on dried sewage sludge and solid gasification by-products. Desalin. Water Treat. 57 (3), 1067–1074.

Gorazda, K., Tarko, B., Wzorek, Z., Kominko, H., Nowak, A.K., Kulczycka, J., Henclik, A., Smol, M., 2017. Fertilizer production from ashes after sewage sludge combustion—a strategy towards sustainable development. Environ. Res. 154 (4), 171–180.

Kim, J.W., Mun, T.Y., Kim, J.O., Kim, J.S., 2011. Air gasification of mixed plastic wastes using a two-stage gasifier for the production of producer gas with low tar and high caloric value. Fuel 90 (6), 2266–2272.

Krzywicka, A., Kwarciak-Kozłowska, A., 2014. Advanced oxidation processes with coke plant wastewater treatment. Water Sci. Technol. 69 (9), 1875–1878.

Li, L., Xu, Z.R., Zhang, C., Bao, J., Dai, X., 2012. Quantitative evaluation of heavy metals in solid residues from sub- and super-critical water gasification of sewage sludge. Bioresour. Technol. 121 (9), 169–175.

Muzyka, R., Chrubasik, M., Stelmach, S., Sajdak, M., 2015. Preliminary studies on the treatment of wastewater from biomass gasification. Waste Manage. 44 (9), 135–146.

Pinto, F., Lopes, H., André, R.N., Dias, M., Gulyurtlu, I., Cabrita, I., 2007. Effect of experimental conditions on gas quality and solids produced by sewage sludge co-gasification. 1. Sewage sludge mixed with coal. Energy Fuel 21 (5), 2737–2745.

Skorek-Osikowska, A., Uchman, W., Werle, S., 2017. Modeling of energy crops gasification based on experimental data. ACEE Arch. Civil Eng. Environ. 10 (3), 135–141.

Uchman, W., Werle, S., 2016. The use of low-calorific value gases in environmental protection engineering. ACEE Arch. Civil Eng. Environ. 9 (1), 127–132.

UN DESA, 2014. United Nations Department of Economic and Social Affairs/Population Division World Population to 2030. Printed in United Nations, New York.

Werle, S., 2014. Impact of feedstock properties and operating conditions on sewage sludge gasification in a fixed bed gasifier. Waste Manage. Res. 32 (10), 954–960.

Werle, S., 2015. Gasification of a dried sewage sludge in a laboratory scale fixed bed reactor. Energies 8, 8562–8572.

Werle, S., Dudziak, M., 2013. Evaluation of toxicity of sewage sludge and gasification waste-products. Przem. Chem. 92 (7), 1350–1353.

Werle, S., Dudziak, M., 2014. Analysis of organic and inorganic contaminants in dried sewage sludge and by-products of dried sewage sludge gasification. Energies 7 (1), 462–476.

Werle, S., Dudziak, M., 2015. The assessment of sewage sludge gasification by-products toxicity by ecotoxicologal test. Waste Manage. Res. 33 (8), 696–703.

Werle, S., Wilk, R.K., 2011. Biomass gasification installation, mainly for sewage sludge. Polish Patent No. P-397225, 2.12.2011.

Werle, S., Dudziak, M., Grübel, K., 2016. Indirect methods of dried sewage sludge contamination assessments. J. Environ. Sci. Health A Tox. Hazard. Subst. Environ. Eng. 51 (9), 754–758.

Wilk, M., Magdziarz, A., Kalemba, I., Gara, P., 2016. Carbonisation of wood residue into charcoal during low temperature process. Renew. Energy 85 (1), 507–513.

World Population Prospects, 2008. The 2008 Revision. World Population Prospects, United Nations.

Yaman, S., 2004. Pyrolysis of biomass to produce fuels and chemical feedstocks. Energy Convers. Manage. 45 (5), 651–671.

From waste to resource: Sorption properties of biological and industrial sludge

26

Martina Grifoni, Francesca Pedron, Irene Rosellini, Gianniantonio Petruzzelli
Institute of Ecosystem Study, Pisa, Italy

1 Introduction

The hierarchy of waste management is based on the effects that each disposal or reuse option has on the environment and sustainability. The increase in the sludge produced from urban wastewater treatments and industrial production related to rapid urbanization and industrialization has raised environmental concerns in many countries. Sustainable management of sludge needs efficient strategies to contribute to a reduction in its environmental impact. Existing strategies need to be optimized along with the development of additional innovative technologies in order to maximize the recovery and reuse of sludge (Prasad, 2016).

Many kinds of sludge contain carbonaceous substances, oxyhydroxides of iron (Fe) and aluminum (Al), and other materials with high adsorptive properties, which can be exploited to decontaminate different environmental matrices. Using the adsorptive properties of these materials is one way of employing different residues, and thereby the waste problem becomes a catalyst for the development of new, environmentally sustainable resources.

The adsorption process (namely, the transfer of chemicals from the aqueous or gaseous phase to the solid phase) is essential for the treatment of polluted water and wastewater (Clifford et al., 1986), and it is one of the best alternatives for this treatment thanks to its convenience and flexibility. Through adsorption, it is possible to remove a wide range of residual pollutants present in contaminated media, and consequently to improve pollution management.

A lot of sludge and sludge-derived materials can be regarded as cheap, usable adsorbents that remove heavy metals and organic contaminants from water or soil. The adsorption capacity of these alternative adsorbents depends on their specific characteristics, such as surface area, functional groups, and pore size. Generally, an efficient adsorbent material should possess a high surface area, high porosity, and fast adsorption kinetics (Gupta et al., 2009). Nevertheless, the composition, structure, and chemical characteristics of adsorbent precursors are also important for ensuring that the best suitable end use of sludge is implemented. In fact, if an adsorbent has high

Industrial and Municipal Sludge. https://doi.org/10.1016/B978-0-12-815907-1.00026-X

porosity but the chemical affinity between the adsorbent and the pollutant target is low, adsorption does not occur (Smith et al., 2009).

Activated carbon is one of the most effective adsorbents used to remove pollutants from water. Because it derives from carbonaceous material, such as coal, lignite, and wood, the natural structure of the carbon gives a high degree of porosity, an extended surface area up to 1500 m^2/g, and active surfaces (Gascó et al., 2005; Otero et al., 2003; Smith et al., 2009). However, activated carbon requires expensive starting materials and high preparation costs, and thus research has focused on the production of cheaper adsorbents (inorganic and organic).

In addition to being cheap, a low-cost adsorbent should be environmentally friendly and easily available, with an easy manufacturing process (Yagub et al., 2014). The recycling method and lifetime of the adsorbent also need to be taken into account. The regeneration of adsorbents should be considered as well (Devi and Saroha, 2017), because their final disposal may cause concern due to the presence of harmful substances (Bhatnagar and Minocha, 2009; Kundu and Gupta, 2008).

The reuse of sludge and sludge-derived materials as low-cost adsorbents is a strategic approach to the problem of waste management. Sludge is widely available as a by-product of wastewater treatment plants (WWTPs) and industrial production, and thus it is economically viable. Furthermore, the reuse of sludge as an adsorbent reduces the by-product volume and cost of waste disposal and makes waste processing more sustainable (Bailey et al., 1999; De Gisi et al., 2016). As a result, processes that produce sludge-derived adsorbents from different kinds of sludge have been increasing.

Sludge also may contain potentially hazardous organic and inorganic materials, such as pathogenic bacteria and heavy metals, which affect the main recycling routes (Smith et al., 2009). Thus, the reuse of sludge as an adsorbent requires thermal, chemical, and biological stabilization techniques prior to their application, which can reduce the volume of waste by up to 50%. The preparation of sludge-derived adsorbents requires specific activation processes, by physical and chemical procedures, in order to enhance their adsorption performance (Lin et al., 2012). The preparation treatment and activation process of sludge-derived adsorbents may influence their behavior toward different adsorbates in various ways (Otero et al., 2003). Several investigations have shown the feasibility and efficacy of these activations. Usually, the activation processes are performed after the preparation steps or at the same time.

A postactivation washing of sludge with distilled water or acid such as hydrochloric acid (HCl) can generally eliminate the activation agent residue from the new free interstices formed and reduce the inorganic content in the sludge precursor (Alvarez et al., 2016; Ros et al., 2006). An acid treatment, especially preactivation washing, also enhances the surface area and the porous structure of the final adsorbents (Zou et al., 2013).

This chapter covers two issues: (1) urban and municipal sewage sludge and (2) sludge derived from industrial processes. The preparation method used and the main characteristics of sludge-derived adsorbents, such as specific surface areas and adsorption capacities, are also reported.

2 Sewage sludge as a source of low-cost adsorbent

Wastewater treatment plants can be found in all industrialized countries, and their aim is to reduce the pollution of surface and deep waters. Management of the high quantity of sewage sludge produced is subject to increasingly restrictive environmental legislation (Kacprzaka et al., 2017; Varank et al., 2014). Because sewage sludge is often enriched with heavy metals and other toxic substances (organic or inorganic), the final use of sewage sludge is determined by appropriate treatments or disposal systems (Yoshida et al., 2015), which can prevent secondary environmental pollution (Ahmaruzzaman, 2011; Smith et al., 2009). For a long time, sewage sludge was considered only as waste and thus was dealt with mainly via composting, landfilling, and incineration. Sewage sludge is also an opportunity in agricultural applications. In fact, after undergoing a stabilization process to decrease its potential pollutants, it can be used as a fertilizer or soil conditioner, thanks to its high content of organic matter, micronutrients, and macronutrients (Passuello et al., 2010; Singh and Agrawal, 2008).

One promising and effective option for the reuse of sewage sludge is converting it into a biosorbent for the low-cost removal of substances from polluted water and reduction of contaminants leaching in soil. Studies have shown the high effectiveness of sludge-based adsorbents (SBAs) for the removal of a wide range of pollutants from water, including heavy metals, organic compounds, acidic gases, phenols, dyes, and phosphates (Devi and Saroha, 2017; Hadi et al., 2015; Pan et al., 2011). These biomaterials are also characterized by high economic efficiency. Due to the wide variety of sludge biomasses available, biosorption can be carried out for a large range of pH values (3–9) and temperatures. These features make biosorption highly competitive with respect to conventional sorption materials.

Because SBAs can be used effectively for many types of pollutants, only the most important results of the experiments performed to remove the main aquatic contaminants (i.e., dyes and metals) are reported in this section, subdividing them by activation process and adsorbate. Further detailed experimental information can be found in each related source reported in the reference list.

2.1 Case studies

2.1.1 Physical activation

Generally, physical activation follows an initial pyrolysis step. Pyrolysis is one of the most common processes used to produce SBAs, based on thermal decomposition (300–700°C) of the biomass in an inert atmosphere with the production of gas (syngas) and solid material called *biochar* (Li et al., 2011). This solid residue is carbon rich and can be used as a carbon-based adsorbent (Fonts et al., 2012). The product obtained by pyrolysis may already be considered an adsorbent material because it has a relatively high specific surface area (100–200 m^2/g) (Rio et al., 2005b; Ros et al., 2006). According to Utomo et al. (2013), up to 98% of methylene blue can be adsorbed from polluted water by dried sewage sludge at 700°C for 2–5 h, obtaining similar adsorption values to those obtained with activated carbons.

The physical activation is carried out by reacting the material from pyrolysis with an oxidizing gas such as carbon dioxide (CO_2), steam, air, or N_2, at high temperatures, ranging 800–1100°C depending on the gas used. With this procedure, the micropore development and the increase in particle area can occur in several ways: by partial oxidation of the carbon atoms (Rodríguez-Reinoso et al., 1995); the release of volatile products following the thermal decomposition of inorganic compounds, organic matter, or both (Zou et al., 2013); and the loss of water (Sierra et al., 2017).

The SBAs thus obtained generally have surface areas ranging from 100–200 m^2/g, and in particular cases, the area can increase to as much as 400 m^2/g (Méndez et al., 2005). However, the characteristics, and thus the efficacy of each specific SBA, can be controlled by varying the operating parameters such as the activation time and temperature and the gas flow rate, as well as by selecting the suitable activating agent. A higher temperature generally has higher reaction rates, which create more pores, resulting in an increased surface area. Even a greater gas flow within the optimal activation value can improve the performance of the SBAs. However, an activation prolonged beyond a specific time may negatively affect the adsorbent characteristics (Devi and Saroha, 2016). Although the most common activation gas used is steam, due to its capacity to generate wider pore size distribution (Hadi et al., 2015; Wang et al., 2008), the use of CO_2 at high temperatures removes carbon atoms from the particles of the adsorbent, creating a well-developed, microporous structure. However, CO_2 is less reactive than steam and needs higher temperatures to exceed the adsorptive performance of the steam. The activation gases also influence the surface functional groups of an adsorbate, modifying their basicity or acidity (Martin et al., 2003).

Jindarom et al. (2007) compared CO_2 and N_2 activation at the same high temperature (750°C). The specific surface area of the SBA, evaluated by the Brunauer-Emmet-Teller equation (S_{BET}), was improved after pyrolysis with CO_2, reaching up to 60.7 m^2/g—about two times higher than that obtained with N_2 pyrolysis. In addition, better dye-adsorption capacities were achieved with SBA activated by pyrolysis with CO_2. A maximum adsorption capacity of 588 mg/g was obtained for a basic dye (basic blue 41), which is about 5 and 23 times higher than acid (acid yellow 49) and reactive (reactive red 198) dyes, respectively. The higher basicity of CO_2 generally influences the electrostatic interactions of the adsorption mechanisms of several types of dyes.

The postwashing or prewashing of SBA also plays an important role in defining the properties of the adsorbent. Sierra et al. (2017) concluded that especially if acid washing is carried out after activation, this generally greatly increases SBA porosity. Their results showed that the S_{BET} value varied from 113 m^2/g for the sample prepared by only physical activation (with CO_2 at 800°C for 30 min and then cooled into an N_2 atmosphere), up to 339 m^2/g for the sample washed with HCl after the same physical activation. This led to an increase in the adsorption capacity. In fact, the maximum removal ability of methylene blue dye, 60.7 mg/g, was reached with the after-washed SBA, whereas the adsorption capacity of the prewashed sample, 35.0 mg/g, was only slightly higher than the maximum adsorption, 30.2 mg/g, obtained in a SBA that had only been physically activated.

Only a few studies have investigated the removal of heavy metals from water using SBAs activated by the physical method. Nevertheless, the physical activation with air seems to produce an effective adsorbent for metal removal, as well as being inexpensive and environmentally acceptable. The air activation (275°C for 4h) used by Méndez et al. (2005), after pyrolysis (450°C for 1h in N_2), produced good results for the removal of Fe from water—namely, about 65.0% and 99.7% for aerobic and anaerobic adsorbents, respectively. In addition, these authors reported a S_{BET} improvement of the tested aerobic and anaerobic sewage sludge, from 16 to 102 m^2/g and from 81 to 105 m^2/g, respectively.

Zhang et al. (2013) focused on optimizing the pyrolysis process in order to increase the metal sorption by SBA by changing the heating conditions. From their results, the largest S_{BET}, 23.7 m^2/g, was obtained by pyrolysis at 400°C for 2h, which corresponds to the highest sorption for lead (Pb) at pH 5, and for chromium (Cr) at pH 2. Similar results for Pb removal were reported by Lu et al. (2012), with Pb adsorption increasing at higher pH values, going from 16.1 mg/g at pH 2 to 30.9 mg/g at pH 5, using sewage sludge pyrolysed at 550°C for 2h under inert conditions and with a surface area of 24.7 m^2/g. Although these authors reported a strong correlation between the metal sorption capacity by SBA and solution pH, Agrafioti et al. (2014) did not find this effect. The latter studied the pyrolysed adsorbents (300°C for 60 min under an N_2 atmosphere) derived from sewage sludge, solid waste, and rice husks for the removal of Cr(III), Cr(VI), and As(V) from aqueous solutions. Despite the S_{BET} values being 155 m^2/g for rice husks, 51 m^2/g for sewage sludge, and 5 m^2/g for solid waste, the adsorbent originating from sewage sludge showed the best sorption capacity, with a removal efficiency of 99%, 89%, and 53% for Cr(III), Cr(VI), and As(V), respectively.

As shown in these studies, both the surface area and micropore fraction of an adsorbent are key factors for effective pollutant adsorption. These factors can be modified by changing both the activation agent and the operating parameters. Physical activation with oxidizing gases generally produces an adsorbent with the highest surface area and pore size. In addition, high temperatures and extended activation times increase the number and volume of pores.

2.1.2 Chemical activation

Chemical activation is based on the mixing of the precursor material with chemical reagents under various temperatures (450–700°C) in an inert atmosphere. The mixing can take place together with the pyrolysis or subsequently on the product obtained. The reagents can act as dehydrating and oxidants agents (Ahmadpour and Do, 1996). The most common chemicals used are phosphoric acid (H_3PO_4), zinc chloride ($ZnCl_2$), sulfuric acid (H_2SO_4), and potassium hydroxide (KOH). These reagents are used both to increase the chemical functional groups within the adsorbent (Dural et al., 2011; Orlandi et al., 2017) and to inhibit the formation of tar, thus increasing the surface area and adsorption capacity of SBAs (Li et al., 2011). The best specific surface area of chemically activated SBAs is generally in the range of 550 m^2/g, depending on the type of chemical agent and precursor used, although in special cases, it can exceed

1700 m^2/g, as reported by Ros et al. (2006). The influence of chemical activation on the SBA adsorption capacity depends on the process parameters, such as the concentration of the activation agent, the activation temperature, and time.

Chemically activated SBAs are frequently used to remove the dyes from the industrial effluents or water bodies. Rozada et al. (2005) studied the difference between two chemical activation agents, $ZnCl_2$ and H_2SO_4, for the removal of methylene blue and brilliant red from aqueous solution. The most efficient adsorbent was pyrolysed (650°C for 5 min) sludge activated with $ZnCl_2$, which adsorbed the brilliant red dye better than methylene blue. Also, the values of S_{BET}, 472 m^2/g, and iodine number (IN, for internal area information), 1358 mg/g, were higher than pyrolysed (650°C for 30 min) sludge activated with H_2SO_4, which reached only 216 m^2/g of S_{BET} and 536 mg/g of IN.

Lower levels of S_{BET} (298 m^2/g) and IN (326 mg/g) were obtained by Xie et al. (2008) after $ZnCl_2$ activation (followed by pyrolysis in N_2 atmosphere at 550°C for 15 min) for the adsorption of reactive brilliant red. Again, the adsorption of dye was favorable, reaching 90% efficiency in the optimal condition of pH (2–10), adsorbent amount (12–16 g/L), and adsorption time (60–140 min).

The H_2SO_4 activation of SBAs increased the removal of anionic dyes in solution, as shown by Martin et al. (2003). After pyrolysis (700°C in N_2 for 30 min) in the presence of H_2SO_4, the S_{BET} value of the SBA increased about 80-fold, changing from 3 m^2/g for raw, dried sludge to 253 m^2/g for activated sewage sludge, thus giving SBA material a mesoporous character. This led to high adsorption of the anionic-type dyes (acid brown 283, direct red 89, and direct black 168), with a removal rate that varied from about 30%–90% with the decrease in pH.

Rio et al. (2005a) also reported that the adsorption of large dyes (2–3 nm) is especially influenced by the micropore and mesopore sizes rather than the surface area. However, the SBAs prepared by these researchers through chemical activation with H_2SO_4 in order to remove acid dyes (basic violet 4 and acid red 18) from aqueous solution in the optimal conditions (700°C in N_2 for 220 min) had a wide range of S_{BET}, which varied from 58–353 m^2/g. The adsorption of basic violet 4 and acid red 18, based on surface area and micropore and mesopore volume, changed from 62–89 mg/g and from 46–53 mg/g, respectively.

Although $ZnCl_2$ and H_2SO_4 are the most common agents used in the chemical activation of sewage sludge, KOH activation has shown satisfactory results for dye removal. Wang et al. (2008) reported the removal of about 99.7% of acid brilliant scarlet GR using pyrolyzing KOH-impregnated sludge (600°C in steam for 60 min). Although the obtained specific surface area (382 m^2/g) was less than that of commercially activated carbons or other SBAs, the final physical characteristics of the SBAs produced, such as total pore volume (0.25 cm^3/g), pore diameter (5.62 nm), and iodine number (563 mg/g), made it a competitive adsorbent.

The removal of metal is often more difficult than the removal of organic compounds. The adsorption process is improved with appropriate chemicals that increase the SBA adsorptive capacity. In the case of metal adsorption, $ZnCl_2$ and H_2SO_4 are among the most studied chemical agents. Zhai et al. (2004) tested cadmium (Cd) and nitrogen (Ni) removal from an aqueous solution, treating the original sewage sludge with $ZnCl_2$ at 850°C in an N_2 atmosphere. Under optimal adsorption conditions in the pH range 5–7, the removal efficiency was very high, varying from 75%–90% for both metals. The values of the specific area and iodine number were also high (550 m^2/g and 336 mg/g, respectively), similar to the results obtained by Rozada et al. (2008), in which a $ZnCl_2$ activation followed by pyrolysis (650°C for 5 min) was carried out to remove mercury (Hg), Pb, copper (Cu), and Cr. The S_{BET} was 472 m^2/g, whereas the iodine index was 1358 mg/g. These characteristics, together with the optimal pH for each metal, led to adsorption rates of 175, 64.1, 30.7, and 15.4 mg/g of Hg, Pb, Cu, and Cr, respectively. High values of Hg adsorption (Zhang et al., 2005) from aqueous solution, about 130 mg/g in the pH range of 5–12, were also obtained using SBAs activated with $ZnCl_2$, immediately followed by SBAs activated with H_3PO_4 (about 110 mg/g) and H_2SO_4 (about 70 mg/g). The highest S_{BET}, 555 m^2/g, was achieved after $ZnCl_2$ activation of organic sewage sludge, whereas 408 m^2/g and 289 m^2/g were obtained after H_2SO_4 and H_3PO_4 activation, respectively.

A mixture of $ZnCl_2$ and H_2SO_4 in a volume ratio of 4:1 (550°C in N_2 for 40 min) has been proposed as an activating agent to develop effective adsorbents for the removal of metal ions from water (Xie et al., 2013). However, the adsorptive properties and capacities did not improve much compared to the single use of each chemical agent. In fact, the S_{BET} value was only 328 m^2/g and the adsorption rates for Cu, Pb, and Cd were 52, 65, and 5 mg/g, respectively, thus being in the normal range of results obtained by chemical activation. However, according to this study, the adsorptive performance can be enhanced by treating the SBA with a natural mineral: pyrolusite.

Additional techniques for preparing the adsorbents entail the alkali modification of sewage sludge. The alkali treatment seems to be very effective with a two-stage method (carbonization and KOH or NaOH activation), which probably promotes the formation of pores during activation (Smith et al., 2009). Al-Qodah (2006) reported an increase of 30% in Cd adsorption efficiency using NaOH without precarbonization, reaching up to 93.4 mg/g of removal. In addition, this study reported that the use of HCl activation reduced adsorption by 26% compared to untreated sewage sludge. In this case, because the physical properties of the treated and untreated sewage sludge were similar, alkaline activation seemed to influence the functional groups of the sludge more significantly.

Although in the research literature, the KOH activation method is defined as the best chemical method for developing a high specific surface area (Kang et al., 2006; Lillo-Ródenas et al., 2008; Ros et al., 2006), few studies have used this process of sewage sludge activation to remove dyes and metals. In fact, although alkaline reagents allow the development of greater porosity, these contaminants are often more influenced by the superficial functional groups. However, to obtain the maximum removal of contaminants, the optimal activation conditions for each type of reagent and precursor material need to be studied, along with the possibility of performing acid washing before the activation. Among the acidic activation agents, H_2SO_4 and H_3PO_4 require lower activation temperatures and costs than $ZnCl_2$, although they are also less effective (Smith et al., 2009; Xu et al., 2015).

2.1.3 Comparison between physical and chemical activation

Results in different studies are often due to the different sludge source materials, the activation protocols used, and the target pollutants. In general, establishing which is the best activation technique, in terms of being environmentally friendly, economical, and efficient at the same time, is rather difficult because many variables for each type of method (e.g., activation agent, activation time, and temperature) and adsorbent (e.g., starting material, chemical and physical characteristics, and water content) are involved.

Simple drying, pyrolysis techniques, and physical activation are the cheapest and most ecofriendly methods (Devi and Saroha, 2016; Li et al., 2011; Marques et al., 2011), as they do not require the use of chemical agents, which are potentially harmful for the environment. Although these methods seem to be less efficient, an increase in activation temperature or making other appropriate modifications can sometimes enhance surface areas up to values similar to those of chemically activated SBAs (Monsalvo et al., 2012). However, these modifications entail greater energy consumption, and often the carbonization step or the high pyrolysis temperatures can generate highly volatile organic compound concentrations (Chiang et al., 2012) or tar-rich polycyclic aromatic hydrocarbons (PAHs; Zhang et al., 2011), depending on the original sludge characteristics. On the other hand, pyrolysis with higher temperatures and longer dwelling times seems to effectively stabilize metals in SBAs, thus reducing the risk of harmful leaching (Lu et al., 2012; Xu et al., 2015).

Some authors thus highlight the energetic advantage of chemical activation because lower temperatures and times are used compared to physical activation (Dural et al., 2011; Hwang et al., 2008). On the other hand, chemical activation generally requires a pyrolysis step with an expensive gas such as N_2, which involves a further increase in production costs. However, although the highest surface areas are achieved, there are several environmental concerns about using activating chemical agents (Chen et al., 2002; Wang et al., 2008).

An economical evaluation of SBAs entails considering factors such as local availability and transportation cost of sewage sludge, the processing required, the use of chemicals for the activation, and the electrical energy required. Gases produced during pyrolysis can often be reused as renewable energy, thus reducing the energy costs.

Also, with regard to the adsorption performance and SBA properties, there are conflicting theories concerning the various activation techniques, including the pyrolytic carbonization of sewage sludge. However, a general assessment can be provided in terms of the control parameters that determine the performance of the adsorbent, such as the surface area, porosity, and adsorption of the target contaminant.

For example, in a comparison between dried (105–110°C under N_2), carbonized (600°C for 60 min), and physically activated (steam, 500°C for 60min, acid washing) sludge obtained under optimal conditions, the S_{BET} of activated sludge was found to be 288 m^2/g, which is approximately 4 and 48 times higher than carbonized and dried sludge, respectively. This corresponds to an enhanced adsorption by steam-activated sludge, which reached 228 mg/g in the iodine adsorptivity test, with an adsorbent yield of 77.2% (Jung and Chun, 2009).

Conversely, Otero et al. (2003) studied the differences in adsorptive capacities of SBAs by comparing simple drying (105°C), pyrolysis (625°C in N_2), and chemical activation (pyrolysis+H_2SO_4+HCl washing). The adsorption test showed that the dried sewage sludge led to the most efficient removal of methylene blue dye (up to 115 mg/g), compared to pyrolysed sewage and chemically activated sludge, which adsorbed up to 25 and 33 mg/g, respectively. However, chemical activation reached an S_{BET} of 390 m^2/g, compared to those obtained by pyrolysis (about 80 m^2/g; Calvo et al., 2001). The S_{BET} of only dried sewage sludge was not evaluated due to the highly volatile content.

In addition, Yu and Zhong (2006) reported a greater adsorptive capacity throughout the chemical process. In fact, the iodine value of chemically activated sludge (a complex of $ZnCl_2$+H_2SO_4) was about 332 mg/g, as opposed to that obtained with physical activation (water vapor) or simple carbonization, which were 13.2 and 5.50 mg/g, respectively. Of the adsorbents obtained, the sludge activated by $ZnCl_2$ exhibited the best porosity, with 332 mg/g of iodine value, whereas H_2SO_4 and KOH produced 53.6 and 1.20 mg/g, respectively. The combination of physical and chemical activations (H_2SO_4+KOH+ mixture of nitrogen atmosphere and water vapor) instead showed satisfactory values of around 306 mg/g, even if that result was lower than those obtained with chemical activation alone. However, the highest specific surface area obtained by chemical activation (only 145 m^2/g) was due to a mixture of $ZnCl_2$ and H_2SO_4.

Further research has shown that specific adsorbent properties, such as S_{BET} and micropore volume, can be improved using a combination of these methods (Khalili et al., 2000; Wang et al., 2008), or a mixture of precursors with a relatively high carbon content (Ren et al., 2012; Velghe et al., 2012; Xie et al., 2013). The best adsorptive characteristics obtainable tend to depend on the best available combination of SBA precursor material and activation agent.

Considering the whole SBA life cycle, further alternative and sustainable ways to prepare low-cost adsorbents from sewage sludge are still needed.

2.2 Sewage SBAs in soil

Besides being exploited in agriculture as soil improver and fertilizer, sewage sludge provides effective adsorption surfaces (Vaca-Paulín et al., 2006; Zhang et al., 2017). These adsorbing properties play a key role in specific situations, but their efficacy is much lower than those of biochar originating from sewage sludge (i.e., SBAs). The production of biochar leads to the creation of a powerful adsorbent that provides surfaces on which the contaminants can be permanently retained (Fig. 1).

Biochar is rich in carbon, and when applied to soil, it is an important agricultural amendment with strong adsorptive properties. Many studies have reported the positive impact of applying biochar to agricultural soil. Due to the high porosity of biochar, the water retention capacity is increased and the need for water and fertilizers reduced (Ahmad et al., 2014). The use of biochar in soil also leads to a reduction in leaching and an improvement in biological fertility.

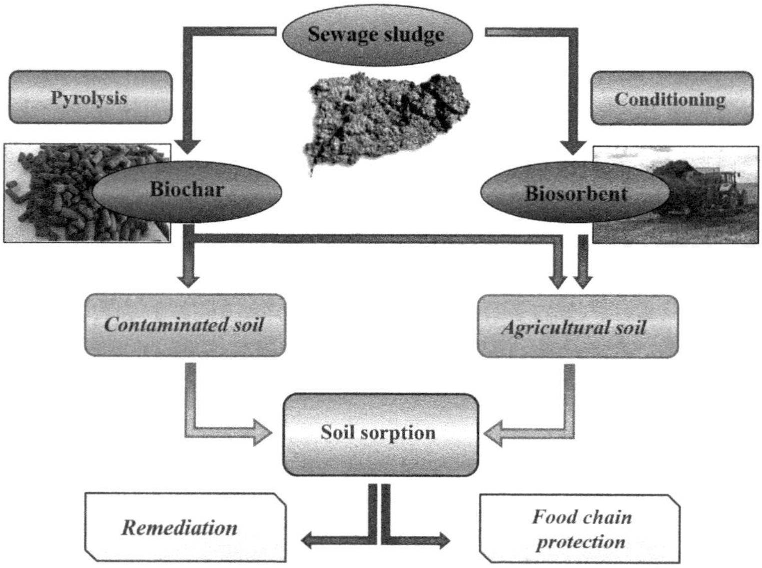

Fig. 1 Various pathways of sewage sludge use in soil, and the increase of adsorptive properties from sewage sludge to sewage SBAs.

In most cases, biochar presents negatively charged surfaces, which in contaminated soil enhance the cation exchange capacity of its phenolic and carboxylic functional groups, thus increasing the retention of contaminants. This action protects the food chain, drastically reducing the uptake of contaminants by plants (Fig. 2).

Many biochars are produced at pyrolysis temperatures that are not very high (<400°C), which creates areas that are not completely carbonized in the biochar matrix (Beesley et al., 2011). This allows biochar to maintain phenolic and carboxylic functional groups, which interact with the various contaminants in a similar way to the naturally occurring humic matter in the soil. Biochar derived from treatments at temperatures above 400°C is very efficient in removing organic contaminants due to the increase in the percentage of carbon with hydrophobic sites. However, when biochar surfaces are subjected to higher temperatures, they are less polar due to the loss of functional groups containing hydrogen (H) and oxygen (O), which may not be effective in the sorption of polar organic compounds (Ahmad et al., 2014).

3 Industrial sludge as a source of low-cost adsorbent

The use of industrial sludge (e.g., waste slurry, fertilizer, fly ash, paper mill sludge, red mud, rice husk, and steel plant waste) as a low-cost adsorbent in water remediation is a well-established technology for removing both inorganic and organic pollutants (De Gisi et al., 2016). Before applying some of these materials, however, they may

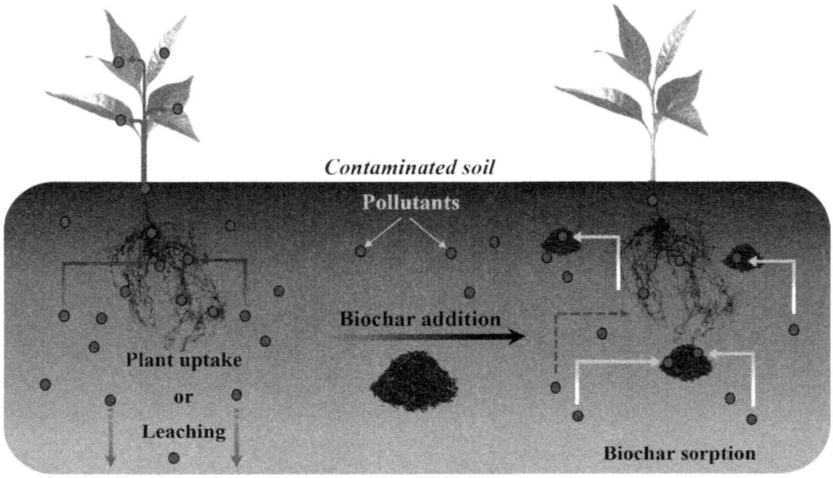

Fig. 2 Schematic illustration of the protective actions of biochar in contaminated soil. Contaminant uptake by plants and leaching is drastically reduced.

need neutralization treatments due to their toxicity, or activation methods (physical and chemical) in order to increase their adsorption efficiency (Geethakarthi and Phanikumar, 2010). However, if these methods are not overly laborious, the industrial sludge-based adsorbents (ISBAs) offer great environmental and economical benefits.

3.1 Case studies

For adsorbents derived from industrial sludge, various parameters need to be taken into account, such as pH, adsorbent dosage, sorbate concentration, and contact time, which may significantly influence the efficiency of the process. Water decontamination using these materials has been studied extensively and reported in the literature (Devi and Saroha, 2016). In this section, only case studies concerning the use of sludge from various industries that compare different types of industrial sludge for the same contaminant are discussed. Detailed information on experimental techniques and results can be found in each related source on the reference list.

Pb, Cd, and Cr are the main heavy metals derived from industrial activities that occur in high concentrations in wastewater. Pb in particular is extremely dangerous because it tends to bioaccumulate in the food chain, even in low concentrations.

Martín et al. (2005) investigated the use of blast furnace sludge for the removal of Pb, zinc (Zn), and Cd from polluted aqueous solutions. Blast furnace sludge, composed of iron oxides and coke, is a by-product of the steelmaking industry. These authors reported a specific surface area of 27.4 m^2/g for this sludge, as well as a greater affinity for Pb. In fact, Pb adsorption ranged from 64.2–79.9 mg/g, increasing with temperatures rising from 20°C to 80°C, as well as Zn and Cd adsorption, from 4.23 to 9.16 mg/g, and from 6.74 and 10.2 mg/g, respectively. In addition, they found that the blast furnace sludge adsorbed greater amounts of metals than rolling mill scale, a by-product of the same steelmaking industry.

A similar efficiency in Pb adsorption has been reported from the use of clarified sludge as a low-cost adsorbent (Naiya et al., 2009), originating from the sludge thickener of basic oxygen furnace in steel production. In this case, although the specific surface area was about 78.5 m^2/g, which is higher than that reported by Martín et al. (2005), the adsorption capacity of about 92.5 mg/g was similar. Under optimum conditions (namely, a pH value of 5, an adsorbent dosage level of 5 g/L, and 1 h of contact time), the percentage removal of Pb decreased with a rise in temperature from 30°C to 50°C.

Red mud is another widespread industrial by-product frequently used as an adsorbent for water treatment. This solid waste residue derives from the basic digestion of bauxite ore during alumina production (Bhatnagar et al., 2011). Due to its strong basic properties, red mud is often used for metal adsorption, including Pb, Cr, Cd, and Zn. Santona et al. (2006) investigated the adsorption capacities of Pb, Cd, and Zn with untreated and acid-treated red mud. The acid treatment was carried out with HCl, followed by washing with distilled water. The specific area values obtained with untreated and acid-treated red mud were 18.9 m^2/g and 25.2 m^2/g, respectively. However, the quantity of metals adsorbed by the untreated red mud was greater than that adsorbed by the treated red mud. For both adsorbents, Zn was more efficiently removed than Pb and Cd.

The application of further treatments to red mud greatly influenced the surface area. The highest S_{BET}, 28.0 m^2/g, was obtained after HCl activation plus heating at 600°C. The S_{BET} value with only HCl activation was 20.7 m^2/g, while the lowest value, 14.2 m^2/g, was obtained in red mud washed only with water. Again, the increased surface area did not reflect the adsorption capacities of the adsorbents because the maximum adsorption was reached with untreated red mud for all three investigated metals, Pb, Cu, and Cr (Apak et al., 1998).

A treatment of red mud with H_2O_2 and air seemed to produce an efficient ISBA, both in terms of adsorption (64.8 mg/g for Pb and 35.7 mg/g for Cr), and specific surface area (108 m^2/g) (Gupta et al., 2001). Red mud was also used to remove other heavy metals, such as Zn ions, from aqueous solution (Sahu et al., 2011). The treatment of red mud neutralized by the sequestration of CO_2 and calcination at 500°C obtained a maximum adsorption capacity of 14.9 mg/g, corresponding to a 96% Zn removal capacity. The authors highlighted that this activation method enhanced S_{BET}, reaching 68.2 m^2/g, which is about twice the surface area value of the untreated red mud (31.7 m^2/g). Although the specific surface area values were higher than those found in previous studies on red mud, the Zn adsorption capacity was lower than that of Pb.

Low values of Zn adsorption, in the range of 7 mg/g, were also found by Mishra et al. (2013) when using a dried steel plant waste sludge as adsorbent. Although the S_{BET} value (average 7.5 m^2/g) was much lower than that found by Martín et al. (2005), the adsorption capacities were comparable.

ISBAs are also widely used for dye removal from polluted solutions. Numerous studies on dye adsorption have been carried out with both inorganic and organic ISBAs. However, the results show that organic industrial adsorbents are more efficient than inorganic adsorbents for basic or acid dye removal (Bhatnagar and Jain, 2005; Jain et al., 2003).

Of the industrial waste examined by Jain et al. (2003), the sludge obtained from the fertilizer industry was better suited for removing basic dyes such as chrysoidine G, crystal violet, and meldola blue than those from steel plants (i.e., blast furnace slag, dust, and sludge). All this waste was activated on the basis of similar treatments. All the adsorbents showed a similar adsorption trend for each dye, but the best affinity was for meldola blue. The maximum adsorption for this dye was 170 mg/g on carbonaceous adsorbents, 67 mg/g on blast furnace sludge, 34 mg/g on blast furnace dust, and 3.7 mg/g on blast furnace slag. In addition, these values precisely followed the decreasing trend in surface area of the adsorbents (i.e., 380, 28, 13, and 4 m^2/g, respectively).

Dye can also be removed with the sludge derived from paper and cellulose manufacturing. The main advantage of this type of adsorbent is the high content of organic material and its production in large amounts by the paper industry (Jaria et al., 2017). Although this sludge may contain toxic substances and chemical additives, and thus require treatment and activation, its reuse as a low-cost adsorbent seems highly efficient.

After activation treatment, this carbonaceous adsorbent more effectively removed the cationic dye (methylene blue, 263 mg/g), than the anionic dye (reactive red, 34.3 mg/g) from aqueous solution (Li et al., 2011). In addition, the specific surface area of this adsorbent (average 135 m^2/g) obtained after carbonization at low temperatures and following physical activation with steam was about fivefold higher than that of raw sludge (25 m^2/g). Nasr et al. (2017) showed that chemical activation with potassium carbonate (K_2CO_3), followed by physical activation with steam, of a paper sludge used as a raw precursor for low-cost adsorbent preparation, achieved high levels of methylene blue dye removal from aqueous solutions. Under optimum conditions, the maximum adsorption capacity was 260 mg/g. This reflected the development of the pore structures obtained at an activation temperature of 900°C, which contributed to a specific surface area of 908 m^2/g, about 56 times higher than that of raw paper sludge. Bhatnagar et al. (2007) found that anionic dyes are removed in smaller amounts than cationic dyes by adsorbents from paper mill sludge. These results indicated that even after physical activation with air (at 500°C), the adsorption capacity of orange G dye was only 62.3 mg/g. However, the effectiveness of paper mill sludge in removing anionic dyes was slightly lower than commercial active carbons. Therefore, given its low cost, paper mill sludge still can be considered an efficient adsorbent for dye removal from water.

In conclusion, pollutant adsorption in the case of ISBAs depends on several factors related to the precursor material, as well as to the various parameters concerning the preparation method. These factors need to be studied to obtain the most effective possible water decontamination.

4 A short overview of adsorption processes

From the case studies reported thus far, adsorption on SBAs and ISBAs emerges as a potential technique for the removal of inorganic and organic pollutants from water and soil due to their porous structure, high surface area, and presence of reactive functional groups. The use of these materials as adsorbents involves many reactions, which should be defined as sorption reactions rather than adsorption reactions. Due to the retention mechanisms, it could be more appropriate to use the term *sorption* rather than *adsorption*. *Adsorption* describes the reactions that occur only at the interface between a substance (sorbate) in solution and the solid phase (sorbent). On the other hand, *absorption* is a process in which a substance can be incorporated into the solid sorbent. When the sorbent is a biological material the process can be defined as *biosorption*. However, the term *adsorption* has been used throughout in its broadest meaning to indicate any retention process, which is similar to the approach used in the articles cited here.

Adsorption reactions are regulated by forces of attraction and repulsion between the substances in solution and the solid phase. Depending on the type of interaction involved, adsorption can be physical or chemical. Physical adsorption is characterized by weak "long-range" interactions between the sorbent and the sorbate due to electrostatic or Van der Waals forces. Chemical adsorption is characterized by strong "short-range" interactions of a chemical nature, such as ion exchange or complex formation involving ionic and covalent bonds with interactions between orbitals (Dąbrowski, 2001; Yagub et al., 2014).

Adsorption processes are quantitatively describable by adsorption isotherms, which relate the concentration of a substance sorbed on the solid phase with the sorbent concentration remaining in solution. This relationship requires the adsorption/desorption reactions to be at equilibrium at a constant temperature. The isotherm is a powerful tool to evaluate and then compare the efficiency of the various sorbent materials.

There are several models that describe these isotherms, which are generally classified into four categories (S, L, H, and C) according to the adsorption mechanisms and the solute-solvent-sorbate interactions (Giles et al., 1974). The S pattern of the isotherm (*S* reflects the shape of the curve) describes a condition of unfavorable adsorption. The interactions between the solid phase and the sorbate are less strong than those of the sorbate-solution. The trend of the L isotherm indicates that the sorbent-sorbate interactions are stronger than those between the sorbate and the solution. The isotherm is characterized by a plateau that represents the maximum sorptive capacity of a sorbent, which can then be used to compare the efficiency of adsorption processes. Theoretically, the availability of sites on the sorbent surfaces decreases with the increase in adsorption and the isotherm is described by the Langmuir equation, from which the name of the isotherm is derived. The pattern of the H isotherm (*H* stands for "high affinity") describes the case in which the sorbate has a very high affinity for the sorbent. At low concentrations, the isotherm shows a very high initial slope and the sorbate is totally sorbed by the sorbent material.

The type C isotherm (constant partition) is characterized by a rectilinear trend, which indicates the constant distribution of the sorbate between the sorbent and solution. The number of sites available remains constant as adsorption proceeds; thus, new sites are formed, while those that were previously available then become saturated. This condition occurs when the sorbate-sorbent interactions are very strong and the sorbate can penetrate the structure of the sorbent.

Several equations are used to evaluate the retention/release process, as described next.

4.1 The Langmuir equation

The Langmuir equation was originally developed for the adsorption of a gas by a solid material. It is based on various assumptions:

- Sorption takes place at specific localized sites on the surface of the sorbent material.
- All sites on soil surfaces have the same affinity for the sorbing substance.
- The energy of sorption is constant and independent of the degree of surface coverage.
- There are no interaction forces between adjacent sorbed molecules.
- The sorption will proceed until the complete occupation of all the sites.

Despite the fact that these assumptions are not fully valid in many adsorption processes, the Langmuir equation is one of the most used because it adequately describes all adsorption reactions that approach saturation at increasing sorbate concentration.

The most common form of the Langmuir equation is defined by Eq. (1):

$$q = \frac{q_{\max} \times KC_e}{1 + KC_e}, \tag{1}$$

where q = the amount of metal sorbed per unit weight of sorbent (mmol/kg) and C_e = the equilibrium concentration (mmol/L). The adjustable parameters $q_{\max}$ (mmol/kg) and K (L/mmol) are linked to the maximum and to the adsorption energy; $q_{\max}$ defines the maximum adsorption capacity of the sorbent, while the constant K is related to the affinity of the compound for the sorbent solid surfaces.

The evaluation of favorability of adsorption is an essential step to optimize the sorbent features and the adsorption process. From the Langmuir equation, it is possible to derive the separation factor R_L, a dimensionless constant (Hall et al., 1966; Ho et al., 2002) defined by Eq. (2):

$$R_L = \frac{1}{1 + KC_i}, \tag{2}$$

where K is the Langmuir constant related to the enthalpy of adsorption through the van't Hoff equation, C_i is the initial sorbate concentration (mmol/L), and R_L describes the kind of isotherm that can be defined as unfavorable ($R_L > 1$), linear ($R_L = 1$), favorable ($0 < R_L < 1$), or irreversible ($R_L = 0$).

Using the R_L factor has some limitations due to the different initial concentrations that influence the values obtained. Moreover, the range of R_L $(0-1)$ is very high, and often further subgroups must be created to identify "favorable," "very favorable," or "highly favorable" adsorption options (Sun et al., 2013).

The Langmuir isotherm is widely used in its simple form with good results. However, to take into account the possible existence of various kinds of adsorption sites, a more complex Langmuir equation (Limousin et al., 2007) can be used:

$$q=\sum_{i=1}^{p} q\max_{,i}\frac{K_iC}{1+K_iC}, \tag{3}$$

where, K_i and $q_{max,i}$ correspond to each type of site.

4.2 The Freundlich equation

The Freundlich equation [Eq. (4)] is based on the assumption that the sorbent material is comprised of heterogeneous surface sites, in which the adsorption heat is reduced exponentially with the degree of saturation. Thus, there is a linear relationship between the surface energy and the sites occupied.

The general form of the Freundlich equation widely used to describe adsorption phenomena is

$$q_e=K_FC_e^{1/n}, \tag{4}$$

where q_e is the sorbate concentration on the solid sorbent, C_e is the sorbate concentration in solution, and K_F and $1/n$ are empirical parameters that are related to the degree of adsorption and to the energy of linkages between sorbent and sorbate, respectively. The equation is also conveniently used in the linear form [Eq. (5)] by taking the logarithm of both sides as

$$\log q_e=\log K_F+\frac{1}{n}\log C_e. \tag{5}$$

A value of $1/n<1$ indicates that the adsorption process proceeds via chemisorptions where $1/n>1$ is indicative of cooperative adsorption.

The favorability of an adsorption process can also be evaluated by the n factor of the Freundlich equation. According to the n values, adsorption is classified as favorable ($2<n<10$), moderately difficult ($1<n<2$), and unfavorable ($n<1$) (Hamdaoui, 2006).

The Freundlich model also can be modified to take into account the competition of several species for the same sites. Often, the equation for m competitive species [Eq. (6)] takes the following form:

$$q_i=K_{F_i}C_i\left(\sum_{j=1}^{m}a_{i,j}C_j\right)^{n_i-1}, \tag{6}$$

where $a_{i,j}$ is a dimensionless coefficient that accounts for the competition of the i species in presence of j species (Limousin et al., 2007), K_{Fi} and n_i are the coefficients of the Freundlich isotherm.

The amount of substance adsorbed differs depending on the adsorption sites on the sorbent surface, and the sites with the highest bonding energy are the first to be occupied.

4.3 The Redlich-Peterson equation

This isotherm model takes into account both the Langmuir and Freundlich equations and can be used when the adsorption mechanism does not follow ideal monolayer adsorption. The Redlich-Peterson equation [Eq. (7)] can be expressed in the following form:

$$q_e = \frac{K_R C_e}{1 + a_R C_e}\beta, \tag{7}$$

where q_e (mg/g) is the solid phase sorbate concentration at equilibrium, C_e (mg/L) is the liquid phase sorbate concentration at equilibrium, K_R (L/g) is the Redlich-Peterson isotherm constant, a_R (L^{β}/mg^{β}) is the Redlich-Peterson isotherm constant, and β is an exponent, which lies between 1 and 0.

This isotherm is similar to the Freundlich model at high sorbate concentrations, while it fits the Langmuir model at lower sorbate concentrations, and when $\beta = 1$, it becomes a Langmuir equation. It has been used in the evaluation of adsorption of basic and acid dyes on various sorbent materials (Chan et al., 2012; Parab et al., 2009).

Tables 1 and 2 list the case studies in which these equations have been used.

4.4 The main adsorption mechanisms

The main parameters that influence the adsorption process are surface area, pH, and temperature.

The larger the surface area, the greater the adsorption capacity. Most of the area available for adsorption is localized in the pores of the adsorbent that are created during the activation process. Thus, the efficiency of the process increases with the porosity of the adsorbent.

The efficiency of the adsorbent material is largely dependent on the pH of the aqueous polluted solution, which determines the interactions between the sorbate and the charges present on the surfaces of activated materials. This effect is obviously related to the various types of sorbent because the surface charges are closely related to the composition of the basic material and to the method of activation of the adsorbent.

Although the parameters described here are useful for studying the mechanisms of sorption, in the treatment of water and wastewater, normal temperature variations usually have negligible effects on the process. The main mechanisms involved in the

Table 1 Adsorption equations used in studies on SBA applications

References	Source of SBA	Activation method	Contaminant	Sorption equation
Jindarom et al. (2007)	Sewage sludge from municipal WWTP	Pyrolysis in N_2 or CO_2 atmosphere (350–750°C)	Acid yellow 49, basic blue 41, reactive red 198	Langmuir and Freundlich models
Utomo et al. (2013)	Sludge from water reclamation plant	Thermal activation (200–700°C)	Methylene blue	Langmuir and Freundlich models
Sierra et al. (2017)	Sewage sludge from urban WWTP	CO_2 activation (600–1000°C) without or with pre- or post-HCl washing	Methylene blue	Langmuir, Freundlich and Redlich-Peterson models
Rozada et al. (2005)	Sewage sludge from urban treatment plant	Pyrolysis in N_2 (650°C); pyrolysis in N_2 and H_2SO_4 or $ZnCl_2$ activation (650°C) and HCl washing	Methylene blue, sandolan brilliant red N-BG 125	Freundlich model
Otero et al. (2003)	Sewage sludge from urban WWTP	Pyrolysis in N_2 (625°C); H_2SO_4 activation and pyrolysis in N_2 (625°C) and HCl washing	Methylene blue	Langmuir and Freundlich models
Xie et al. (2008)	Sewage sludge from municipal WWTP	Pyrolysis in N_2 (550°C) and $ZnCl_2$ activation and HCl washing	Reactive brilliant red K-2BP	Langmuir and Freundlich models
Martin et al. (2003)	Surplus biological sludge from WWTP	Pyrolysis in N_2 (700°C) and H_2SO_4 activation and HCl washing	Acid brown 283, direct red 89, direct black 168	Langmuir model
Wang et al. (2008)	Sewage sludge from WWTP	KOH activation and pyrolysis with steam (600°C) and deionized water washing	Acid brilliant scarlet GR	Langmuir and Freundlich models

Table 1 Continued

References	Source of SBA	Activation method	Contaminant	Sorption equation
Zhang et al. (2013)	Sludge from municipal WWTP	Pyrolysis in N_2 (600°C)	Pb, Cr	Langmuir and Freundlich models
Lu et al. (2012)	Sludge from WWTP	Pyrolysis (550°C)	Pb	Freundlich model
Agrafioti et al. (2014)	Sewage sludge from municipal WWTP	Pyrolysis in N_2 (300°C)	As(V), Cr(III), Cr(VI)	Langmuir and Freundlich models
Zhai et al. (2004)	Sewage sludge from municipal WWTP	$ZnCl_2$ activation in N_2 (850°C)	Cd, Ni	Langmuir and Freundlich models
Rozada et al. (2008)	Sewage sludge from urban WWTP	Pyrolysis in N_2 (650°C); $ZnCl_2$ activation and pyrolysis in N_2 (650°C) and HCl washing	Hg, Pb, Cu, Cr	Langmuir and Freundlich models
Zhang et al. (2005)	Sewage sludge from disposal plant	H_2SO_4, H_3PO_4 activation (650°C) and NaOH washing; $ZnCl_2$ activation (650°C) and HCl washing	Hg	Freundlich model

Table 2 Adsorption equation used in the studies on industrial sludge application

References	Source of ISBA	Activation method	Contaminant	Sorption equation
Jain et al. (2003)	Steel production waste and carbon slurry from fertilizer plants	H_2O_2 activation (400–500°C) and HCl treatment and deionized water washing	Meldola blue, chrysoidine G, crystal violet	Langmuir model
Li et al. (2011)	Paper mill sewage sludge	Carbonization (300°C) and steam activation (850°C)	Methylene blue, reactive red 24	Langmuir and Freundlich models
Nasr et al. (2017)	Paper sludge	K_2CO_3 activation (700–900°C) and steam activation	Methylene blue	Langmuir and Freundlich models

Continued

Table 2 Continued

References	Source of ISBA	Activation method	Contaminant	Sorption equation
Zaini et al. (2014)	Palm oil mill effluent sludge	$ZnCl_2$ and KOH activation (500°C) and HCl washing	Methylene blue	Langmuir, Freundlich, Redlich-Peterson models
Bhatnagar et al. (2007)	Paper mill sludge	Thermal activation with air activation (500°C)	Orange G	Langmuir model
Santona et al. (2006)	Red mud	HCl treatment and water washing and heating (105°C)	Pb, Cd, Zn	Langmuir model
Apak et al. (1998)	Red mud	HCl treatment and water washing; HCl treatment and water washing and heating (600°C)	Pb, Cd, Cu	Langmuir and Freundlich models
Gupta et al. (2001)	Red mud	H_2O_2 treatment and air activation (500° C)	Pb, Cr	Langmuir and Freundlich models
Sahu et al. (2011)	Caustic red mud	Neutralization by CO_2 sequestration and calcination (500°C)	Zn	Langmuir and Freundlich models
Martín et al. (2005)	Rolling mill scale and blast furnace sludge	Drying (80°C)	Pb, Cd, Zn	Langmuir model
Naiya et al. (2009)	Sludge thickener of basic oxygen furnace	Drying (105°C)	Pb	Langmuir and Freundlich models
Mishra et al. (2013)	Steel plant waste sludge	Drying (105°C)	Zn	Langmuir, Freundlich, Redlich-Peterson models

biosorption processes by activated sludge materials are the ion exchange and complexation with functional groups on the sorbent surfaces with the release of H^+ ions into solution. Ion exchange is the most important mechanism in the sorption of both basic dyes and heavy metals. The ion exchange mechanism has been reported in the metal biosorption on various sorbent materials (Akar et al., 2012; Ding et al., 2012;

Fiol et al., 2006), which are determined by the electrostatic interaction between positive ions of the sorbate and the negatively charged groups on the biosorbent. In addition to ion exchange, complexation with the hydroxyl and carboxyl group at the surface of sorbents has been often reported in the sorption of Al (Qian and Chen, 2013) and Pb and Cu (Uchimiya, 2014) on various biochars (to cite one example).

The strength linkages can be evaluated on the basis of the effect of ionic strength, because in the adsorption process, inner-sphere complexes do not depend on ionic strength, while outer-spheres are greatly dependent on ionic strength (Petruzzelli and Pedron, 2017; Vithanage et al., 2015). These complexation mechanisms of adsorption recently were clarified in detail for a biochar obtained from sewage sludge (Pourret and Houben, 2018).

5 Conclusions

One of the great challenges that nations face today is managing to combine a sustainable

environment and economic growth. In a world with limited resources, a model of economic growth must consider the best use of resources. Waste management is not just an issue of growing environmental concern, but also a crucial economic question.

The increase in global consumption, with the consequent high waste production, can create significant problems for ecosystems and the climate. However, a good deal of waste material has the potential to be transformed into resources. Thus, a new technological vision is needed, which considers the transformation of waste into reusable materials, as well as the resolution of environmental issues. Therefore, the transformation of various kinds of sludge into adsorbent materials is very important for the environment and society.

The production of low-cost adsorbents derived from the sludge from urban wastewater treatments and industrial production enables materials to be obtained with high adsorptive properties, which efficiently removes organic and inorganic pollutants from various environmental matrices. Knowledge of sorption processes is the key to understanding the efficiency of decontamination procedures in relation to the characteristics of the specific sorbents and sorbates.

To best exploit the adsorptive properties of these materials, it is necessary to consider both the nature of the source sludge and the activation methods, which are often performed to enhance the adsorptive capacities.

The effects of the physical and chemical activations on adsorbent properties (i.e., surface areas) have proved to be very efficient in removing dyes and metals from polluted water. However, it is not possible to rank the adsorbents, as each has its own specific properties. Each adsorption process needs to be evaluated on the basis of the specific characteristics of the adsorbent and the contaminant. Although the possible reuse of sludge as an adsorbent still needs to be carefully evaluated, especially in environmental terms, it is an emerging low-cost and ecofriendly method for decontaminating polluted matrices.

References

Agrafioti, E., Kalderis, D., Diamadopoulos, E., 2014. Arsenic and chromium removal from water using biochars derived from rice husk, organic solid wastes and sewage sludge. J. Environ. Manag. 133, 309–314.

Ahmad, M., Rajapaksha, A.U., Lim, J.E., Zhang, M., Bolan, N., Mohan, D., Vithanage, M., Lee, S.S., Ok, Y.S., 2014. Biochar as a sorbent for contaminant management in soil and water: a review. Chemosphere 99, 19–33.

Ahmadpour, A., Do, D.D., 1996. The preparation of active carbons from coal by chemical and physical activation. Carbon N. Y. 34, 471–479.

Ahmaruzzaman, M., 2011. Industrial wastes as low-cost potential adsorbents for the treatment of wastewater laden with heavy metals. Adv. Colloid Interf. Sci. 166, 36–59.

Akar, S.T., Arslan, S., Alp, T., Arslan, D., Akar, T., 2012. Biosorption potential of the waste biomaterial obtained from *Cucumis melo* for the remal of Pb^{2+} ions from aqueous media: equilibrium, kinetic, thermodynamic and mechanism analysis. Chem. Eng. J. 185–186, 82–90.

Al-Qodah, Z., 2006. Biosorption of heavy metal ions from aqueous solutions by activated sludge. Desalination 196, 164–176.

Alvarez, J., Lopez, G., Amutio, M., Bilbao, J., Olazar, M., 2016. Preparation of adsorbents from sewage sludge pyrolytic char by carbon dioxide activation. Process. Saf. Environ. Prot. 103, 76–86.

Apak, R., Tütem, E., Hügül, M., Hizal, J., 1998. Heavy metal cation retention by unconventional sorbents (red muds and fly ashes). Water Res. 32, 430–440.

Bailey, S.E., Olin, T.J., Bricka, R.M., Adrian, D.D., 1999. A review of potentially low-cost sorbents for heavy metals. Water Res. 33, 2469–2479.

Beesley, L., Moreno-Jiménez, E., Gomez-Eyles, J.L., Harris, E., Robinson, B., Sizmur, T., 2011. A review of biochars' potential role in the remediation, revegetation and restoration of contaminated soils. Environ. Pollut. 159, 3269–3282.

Bhatnagar, A., Jain, A.K., 2005. A comparative adsorption study with different industrial wastes as adsorbents for the removal of cationic dyes from water. J. Colloid Interface Sci. 281, 49–55.

Bhatnagar, A., Minocha, A.K., 2009. Utilization of industrial waste for cadmium removal from water and immobilization in cement. Chem. Eng. J. 150, 145–151.

Bhatnagar, A., Minocha, A.K., Jeon, B.-H., Park, J.-M., Lee, G., 2007. Adsorption of orange G dye on paper mill sludge: equilibrium and kinetic modeling. Fresenius Environ. Bull. 16, 1049–1055.

Bhatnagar, A., Vilar, V.J.P., Botelho, C.M.S., Boaventura, R.A.R., 2011. A review of the use of red mud as adsorbent for the removal of toxic pollutants from water and wastewater. Environ. Technol. 32, 231–249.

Calvo, L.F., Otero, M., Morán, A., Garcı´a, A.I., 2001. Upgrading sewage sludges for adsorbent preparation by different treatments. Bioresour. Technol. 80, 143–148.

Chan, L.S., Cheung, W.H., Allen, S.J., McKay, G., 2012. Error analysis of adsorption isotherm models for acid dyes onto bamboo derived activated carbon. Chin. J. Chem. Eng. 20, 535–542.

Chen, X., Jeyaseelan, S., Graham, N., 2002. Physical and chemical properties study of the activated carbon made from sewage sludge. Waste Manag. 22, 755–760.

Chiang, H.-L., Lin, K.-H., Chiu, H.-H., 2012. Exhaust characteristics during the pyrolysis of $ZnCl_2$ immersed biosludge. J. Hazard. Mater. 229–230, 233–244.

Clifford, D., Subramonian, S., Sorg, T.J., 1986. Water treatment processes III: removing dissolved inorganic contaminants from water. Environ. Sci. Technol. 20, 1072–1080.

Dąbrowski, A., 2001. Adsorption—from theory to practice. Adv. Colloid Interf. Sci. 93, 135–224.

De Gisi, S., Lofrano, G., Grassi, M., Notarnicola, M., 2016. Characteristics and adsorption capacities of low-cost sorbents for wastewater treatment: a review. Sustain. Mater. Technol. 9, 10–40.

Devi, P., Saroha, A.K., 2016. Improvement in performance of sludge-based adsorbents by controlling key parameters by activation/modification: a critical review. Crit. Rev. Environ. Sci. Technol. 46, 1704–1743.

Devi, P., Saroha, A.K., 2017. Utilization of sludge based adsorbents for the removal of various pollutants: a review. Sci. Total Environ. 578, 16–33.

Ding, Y., Jing, D., Gong, H., Zhou, L., Yang, X., 2012. Biosorption of aquatic cadmium(II) by unmodified rice straw. Bioresour. Technol. 114, 20–25.

Dural, M.U., Cavas, L., Papageorgiou, S.K., Katsaros, F.K., 2011. Methylene blue adsorption on activated carbon prepared from *Posidonia oceanica* (L.) dead leaves: kinetics and equilibrium studies. Chem. Eng. J. 168, 77–85.

Fiol, N., Villaescusa, I., Martínez, M., Miralles, N., Poch, J., Serarols, J., 2006. Sorption of Pb(II), Ni(II), Cu(II) and Cd(II) from aqueous solution by olive stone waste. Sep. Purif. Technol. 50, 132–140.

Fonts, I., Gea, G., Azuara, M., Ábrego, J., Arauzo, J., 2012. Sewage sludge pyrolysis for liquid production: a review. Renew. Sust. Energ. Rev. 16, 2781–2805.

Gascó, G., Méndez, A., Gascó, J.M., 2005. Preparation of carbon-based adsorbents from sewage sludge pyrolysis to remove metals from water. Desalination 180, 245–251.

Geethakarthi, A., Phanikumar, B.R., 2010. Industrial sludge based adsorbents/industrial by-products in the removal of reactive dyes—a review. Int. J. Water Resour. Environ. Eng. 3, 1–9.

Giles, C.H., Smith, D., Huitson, A., 1974. A general treatment and classification of the solute adsorption isotherm I: theoretical. J. Colloid Interface Sci. 47, 755–765.

Gupta, V.K., Carrott, P.J.M., Ribeiro Carrott, M.M.L., Suhas, 2009. Low-cost adsorbents: growing approach to wastewater treatment—a review. Crit. Rev. Environ. Sci. Technol. 39, 783–842.

Gupta, V.K., Gupta, M., Sharma, S., 2001. Process development for the removal of lead and chromium from aqueous solutions using red mud—an aluminium industry waste. Water Res. 35, 1125–1134.

Hadi, P., Xu, M., Ning, C., Lin, S.K.C., McKay, G., 2015. A critical review on preparation, characterization and utilization of sludge-derived activated carbons for wastewater treatment. Chem. Eng. J. 260, 895–906.

Hall, K.R., Eagleton, L.C., Acrivos, A., Vermeulen, T., 1966. Pore- and solid-diffusion kinetics in fixed-bed adsorption under constant-pattern conditions. Ind. Eng. Chem. Fundam. 5, 212–223.

Hamdaoui, O., 2006. Batch study of liquid-phase adsorption of methylene blue using cedar sawdust and crushed brick. J. Hazard. Mater. 135, 264–273.

Ho, Y.S., Porter, J.F., McKay, G., 2002. Equilibrium isotherm studies for the sorption of divalent metal ions onto peat: copper, nickel and lead single component systems. Water Air Soil Pollut. 141, 1–33.

Hwang, H.-R., Choi, W.-J., Kim, T.-J., Kim, J.-S., Oh, K.-J., 2008. The preparation of an adsorbent from mixtures of sewage sludge and coal-tar pitch using an alkaline hydroxide activation agent. J. Anal. Appl. Pyrolysis 83, 220–226.

Jain, A.K., Gupta, V.K., Bhatnagar, A., Suhas, 2003. A comparative study of adsorbents prepared from industrial wastes for removal of dyes. Sep. Sci. Technol. 38, 463–481.

Jaria, G., Silva, C.P., Ferreira, C.I.A., Otero, M., Calisto, V., 2017. Sludge from paper mill effluent treatment as raw material to produce carbon adsorbents: an alternative waste management strategy. J. Environ. Manag. 188, 203–211.

Jindarom, C., Meeyoo, V., Kitiyanan, B., Rirksomboon, T., Rangsunvigit, P., 2007. Surface characterization and dye adsorptive capacities of char obtained from pyrolysis/gasification of sewage sludge. Chem. Eng. J. 133, 239–246.

Jung, D.H., Chun, Y.N., 2009. Characteristics on carbonization and steam activation of sewage sludge. Open Process Chem. J. 2, 12–19.

Kacprzaka, M., Neczaja, E., Fijałkowskia, K., Grobelaka, A., Grossera, A., Worwaga, M., Rorat, A., Brattebob, H., Almåsc, Å., Singhc, B.R., 2017. Sewage sludge disposal strategies for sustainable development. Environ. Res. 156, 39–46.

Kang, H.Y., Park, S.S., Rim, Y.S., 2006. Preparation of activated carbon from paper mill sludge by KOH-activation. Korean J. Chem. Eng. 23, 948–953.

Khalili, N.R., Campbell, M., Sandi, G., Golaś, J., 2000. Production of micro- and mesoporous activated carbon from paper mill sludge: I effect of zinc chloride activation. Carbon N. Y. 38, 1905–1915.

Kundu, S., Gupta, A.K., 2008. Immobilization and leaching characteristics of arsenic from cement and/or lime solidified/stabilized spent adsorbent containing arsenic. J. Hazard. Mater. 153, 434–443.

Li, W.H., Yue, Q.Y., Gao, B.Y., Ma, Z.H., Li, Y.J., Zhao, H.X., 2011. Preparation and utilization of sludge-based activated carbon for the adsorption of dyes from aqueous solutions. Chem. Eng. J. 171, 320–327.

Lillo-Ródenas, M.A., Ros, A., Fuente, E., Montes-Morán, M.A., Martin, M.J., Linares-Solano, A., 2008. Further insights into the activation process of sewage sludge-based precursors by alkaline hydroxides. Chem. Eng. J. 142, 168–174.

Limousin, G., Gaudet, J.-P., Charlet, L., Szenknect, S., Barthès, V., Krimissa, M., 2007. Sorption isotherms: a review on physical bases, modeling and measurement. Appl. Geochem. 22, 249–275.

Lin, Q.H., Cheng, H., Chen, G.Y., 2012. Preparation and characterization of carbonaceous adsorbents from sewage sludge using a pilot-scale microwave heating equipment. J. Anal. Appl. Pyrolysis 93, 113–119.

Lu, H., Zhang, W., Yang, Y., Huang, X., Wang, S., Qiu, R., 2012. Relative distribution of Pb^{2+} sorption mechanisms by sludge-derived biochar. Water Res. 46, 854–862.

Marques, R.R.N., Stüber, F., Smith, K.M., Fabregat, A., Bengoa, C., Font, J., Fortuny, A., Pullket, S., Fowler, G.D., Graham, N.J.D., 2011. Sewage sludge based catalysts for catalytic wet air oxidation of phenol: preparation, characterisation and catalytic performance. Appl. Catal. B Environ. 101, 306–316.

Martín, M.I., López, F.A., Pérez, C., López-Delgado, A., Alguacil, F.J., 2005. Adsorption of heavy metals from aqueous solutions with by-products of the steelmaking industry. J. Chem. Technol. Biotechnol. 80, 1223–1229.

Martin, M.J., Artola, A., Balaguer, M.D., Rigola, M., 2003. Activated carbons developed from surplus sewage sludge for the removal of dyes from dilute aqueous solutions. Chem. Eng. J. 94, 231–239.

Méndez, A., Gascó, G., Freitas, M.M.A., Siebielec, G., Stuczynski, T., Figueiredo, J.L., 2005. Preparation of carbon-based adsorbents from pyrolysis and air activation of sewage sludges. Chem. Eng. J. 108, 169–177.

Mishra, U., Paul, S., Bandyopadhaya, M., 2013. Removal of zinc ions from wastewater using industrial waste sludge: a novel approach. Environ. Prog. Sustain. Energy 32, 576–586.

Monsalvo, V.M., Mohedano, A.F., Rodriguez, J.J., 2012. Adsorption of 4-chlorophenol by inexpensive sewage sludge-based adsorbents. Chem. Eng. Res. Des. 90, 1807–1814.

Naiya, T.K., Bhattacharya, A.K., Das, S.K., 2009. Clarified sludge (basic oxygen furnace sludge)—an adsorbent for removal of Pb(II) from aqueous solutions—kinetics, thermodynamics and desorption studies. J. Hazard. Mater. 170, 252–262.

Nasr, J.B., Hamdi, N., Elhalouani, F., 2017. Characterization of activated carbon prepared from sludge paper for methylene blue adsorption. J. Mater. Environ. Sci. 8, 1960–1967.

Orlandi, G., Cavasotto, J., Machado, F.R.S., Colpani, G.L., Magro, J.D., Dalcanton, F., Mello, J.M.M., Fiori, M.A., 2017. An adsorbent with a high adsorption capacity obtained from the cellulose sludge of industrial residues. Chemosphere 169, 171–180.

Otero, M., Rozada, F., Calvo, L.F., Garcı´a, A.I., Morán, A., 2003. Kinetic and equilibrium modelling of the methylene blue removal from solution by adsorbent materials produced from sewage sludges. Biochem. Eng. J. 15, 59–68.

Pan, Z., Tian, J., Xu, G., Li, J., Li, G., 2011. Characteristics of adsorbents made from biological, chemical and hybrid sludges and their effect on organics removal in wastewater treatment. Water Res. 45, 819–827.

Parab, H., Sudersanan, M., Shenoy, N., Pathare, T., Vaze, B., 2009. Use of agro-industrial wastes for removal of basic dyes from aqueous solutions. CLEAN Soil, Air, Water 37, 963–969.

Passuello, A., Mari, M., Nadal, M., Schuhmacher, M., Domingo, J.L., 2010. POP accumulation in the food chain: integrated risk model for sewage sludge application in agricultural soils. Environ. Int. 36, 577–583.

Petruzzelli, G., Pedron, F., 2017. Tungstate adsorption onto Italian soils with different characteristics. Environ. Monit. Assess. 189, 379.

Pourret, O., Houben, D., 2018. Characterization of metal binding sites onto biochar using rare earth elements as a fingerprint. Heliyon 4, e00543.

Prasad, M.N.V., 2016. Recovery of resources from biowaste for pollution prevention. In: Environmental Materials and Waste. Elsevier, pp. 1–19.

Qian, L., Chen, B., 2013. Dual role of biochars as adsorbents for aluminum: the effects of oxygen-containing organic components and the scattering of silicate particles. Environ. Sci. Technol. 47, 8759–8768.

Ren, X., Liang, B., Liu, M., Xu, X., Cui, M., 2012. Effects of pyrolysis temperature, time and leaf litter and powder coal ash addition on sludge-derived adsorbents for nitrogen oxide. Bioresour. Technol. 125, 300–304.

Rio, S., Faur-Brasquet, C., Le Coq, L., Courcoux, P., Le Cloirec, P., 2005a. Experimental design methodology for the preparation of carbonaceous sorbents from sewage sludge by chemical activation—application to air and water treatments. Chemosphere 58, 423–437.

Rio, S., Faur-Brasquet, C., Le Coq, L., Le Cloirec, P., 2005b. Structure characterization and adsorption properties of pyrolyzed sewage sludge. Environ. Sci. Technol. 39, 4249–4257.

Rodríguez-Reinoso, F., Molina-Sabio, M., González, M.T., 1995. The use of steam and CO_2 as activating agents in the preparation of activated carbons. Carbon N. Y. 33, 15–23.

Ros, A., Lillo-Ródenas, M.A., Fuente, E., Montes-Morán, M.A., Martín, M.J., Linares-Solano, A., 2006. High surface area materials prepared from sewage sludge-based precursors. Chemosphere 65, 132–140.

Rozada, F., Otero, M., Morán, A., García, A.I., 2008. Adsorption of heavy metals onto sewage sludge-derived materials. Bioresour. Technol. 99, 6332–6338.

Rozada, F., Otero, M., Parra, J.B., Morán, A., García, A.I., 2005. Producing adsorbents from sewage sludge and discarded tyres: characterization and utilization for the removal of pollutants from water. Chem. Eng. J. 114, 161–169.

Sahu, R.C., Patel, R., Ray, B.C., 2011. Adsorption of Zn(II) on activated red mud: neutralized by CO_2. Desalination 266, 93–97.

Santona, L., Castaldi, P., Melis, P., 2006. Evaluation of the interaction mechanisms between red muds and heavy metals. J. Hazard. Mater. 136, 324–329.

Sierra, I., Iriarte-Velasco, U., Gamero, M., Aguayo, A.T., 2017. Upgrading of sewage sludge by demineralization and physical activation with CO_2: application for methylene blue and phenol removal. Microporous Mesoporous Mater. 250, 88–99.

Singh, R.P., Agrawal, M., 2008. Potential benefits and risks of land application of sewage sludge. Waste Manag. 28, 347–358.

Smith, K.M., Fowler, G.D., Pullket, S., Graham, N.J.D., 2009. Sewage sludge-based adsorbents: a review of their production, properties and use in water treatment applications. Water Res. 43, 2569–2594.

Sun, C.J., Sun, L.Z., Sun, X.X., 2013. Graphical evaluation of the favorability of adsorption processes by using conditional Langmuir constant. Ind. Eng. Chem. Res. 52, 14251–14260.

Uchimiya, M., 2014. Influence of pH, ionic strength, and multidentate ligand on the interaction of CdII with biochars. ACS Sustain. Chem. Eng. 2, 2019–2027.

Utomo, D.H., Ong, X.C., Lim, S.M.S., Ong, G.C.B., Li, P., 2013. Thermally processed sewage sludge for methylene blue uptake. Int. Biodeterior. Biodegrad. 85, 460–465.

Vaca-Paulín, R., Esteller-Alberich, M.V., Lugo-de la Fuente, J., Zavaleta-Mancera, H.A., 2006. Effect of sewage sludge or compost on the sorption and distribution of copper and cadmium in soil. Waste Manag. 26, 71–81.

Varank, G., Demir, A., Bilgili, M.S., Top, S., Sekman, E., Yazici, S., Erkan, H.S., 2014. Investigation of sorption characteristics of anaerobically digested dewatered municipal sewage sludge. Environ. Prot. Eng. 119–133.

Velghe, I., Carleer, R., Yperman, J., Schreurs, S., D'Haen, J., 2012. Characterisation of adsorbents prepared by pyrolysis of sludge and sludge/disposal filter cake mix. Water Res. 46, 2783–2794.

Vithanage, M., Rajapaksha, A.U., Ahmad, M., Uchimiya, M., Dou, X., Alessi, D.S., Ok, Y.S., 2015. Mechanisms of antimony adsorption onto soybean stover-derived biochar in aqueous solutions. J. Environ. Manag. 151, 443–449.

Wang, X., Zhu, N., Yin, B., 2008. Preparation of sludge-based activated carbon and its application in dye wastewater treatment. J. Hazard. Mater. 153, 22–27.

Xie, J., Yue, Q., Yu, H., Yue, W., Li, R., Zhang, S., Wang, X., 2008. Adsorption of reactive brilliant red K-2BP on activated carbon developed from sewage sludge. Front. Chem. China 3, 33–40.

Xie, R., Jiang, W., Wang, L., Peng, J., Chen, Y., 2013. Effect of pyrolusite loading on sewage sludge-based activated carbon in Cu(II), Pb(II), and Cd(II) adsorption. Environ. Prog. Sustain. Energy 32, 1066–1073.

Xu, G., Yang, X., Spinosa, L., 2015. Development of sludge-based adsorbents: preparation, characterization, utilization and its feasibility assessment. J. Environ. Manage. 151, 221–232.

Yagub, M.T., Sen, T.K., Afroze, S., Ang, H.M., 2014. Dye and its removal from aqueous solution by adsorption: a review. Adv. Colloid Interf. Sci. 209, 172–184.

Yoshida, H., Christensen, T.H., Guildal, T., Scheutz, C., 2015. A comprehensive substance flow analysis of a municipal wastewater and sludge treatment plant. Chemosphere 138, 874–882.

Yu, L., Zhong, Q., 2006. Preparation of adsorbents made from sewage sludges for adsorption of organic materials from wastewater. J. Hazard. Mater. 137, 359–366.

Zaini, M.A.A., Cher, T.Y., Zakaria, M., Kamaruddin, M.J., Setapar, S.H.M., Yunus, M.A.C., 2014. Palm oil mill effluent sludge ash as adsorbent for methylene blue dye removal. Desalin. Water Treat. 52, 3654–3662.

Zhai, Y., Wei, X., Zeng, G., Zhang, D., Chu, K., 2004. Study of adsorbent derived from sewage sludge for the removal of Cd^{2+}, Ni^{2+} in aqueous solutions. Sep. Purif. Technol. 38, 191–196.

Zhang, B., Xiong, S., Xiao, B., Yu, D., Jia, X., 2011. Mechanism of wet sewage sludge pyrolysis in a tubular furnace. Int. J. Hydrogen Energy 36, 355–363.

Zhang, F.S., Nriagu, J.O., Itoh, H., 2005. Mercury removal from water using activated carbons derived from organic sewage sludge. Water Res. 39, 389–395.

Zhang, W., Mao, S., Chen, H., Huang, L., Qiu, R., 2013. Pb(II) and Cr(VI) sorption by biochars pyrolyzed from the municipal wastewater sludge under different heating conditions. Bioresour Technol. 147, 545–552.

Zhang, X., Wang, X., Wang, D., 2017. Immobilization of heavy metals in sewage sludge during land application process in China: a review. Sustainablity 9, 2020.

Zou, J., Dai, Y., Wang, X., Ren, Z., Tian, C., Pan, K., Li, S., Abuobeidah, M., Fu, H., 2013. Structure and adsorption properties of sewage sludge-derived carbon with removal of inorganic impurities and high porosity. Bioresour. Technol. 142, 209–217.

Valorization of paper and pulp waste: Opportunities and prospects of biorefinery

27

Debkumar Chakraborty, Shikha Dahiya, Kotamraju Amulya, Venu Srivastav, S. Venkata Mohan
Bioengineering and Environmental Sciences Lab, EEFF Department, CSIR-Indian Institute of Chemical Technology (CSIR-IICT), Hyderabad, India

1 Introduction

Ever since paper was created (c.3700–3200 BC), it has become an essential part of human life. With the use of more modern papermaking techniques, it has become a cheap commodity. Meanwhile, the material demands ballooned even more, consumption exploded accordingly, and waste generation increased (Gavrilescu, 2008; Veluchamy and Kalamdhad, 2017). Before 2009, the United States and Canada were the leading participants in the paper and pulp industry (PPI). Currently, China is the largest producer of paper, followed by Japan, Germany, Sweden, Canada, and India (Bajpai, 2015). Meanwhile, for pulp production, the United States is the leader, followed by China and Canada. The highest exporters of paper are the United States, Germany, and Sweden, and for the export of pulp, Canada comes first, followed by Brazil and United States. According to the Indian Paper Manufacturers Association (IPMA), it was predicted that the consumption of paper in India would reach 14 million tons per annum by 2015. India has a total of 500–800 pulp-processing units, and it contributes to about 1.6% of the world's total paper production (Sadhasivam et al., 2010; Sharma et al., 2014). However, what is missing from these statistics is a trend that could be of greater significance: striking a balance between the declining resources and increasing demand. Also, what lurks behind these statistics is the menacing reality of waste generation and its environmental impact.

1.1 Environmental impact of paper and pulp industry

With the rise in environmental awareness and the need to reverse the trajectory of climate change, PPI is becoming increasingly aware of the necessity to make changes in its long-established papermaking process. The global PPI production per annum was approximately 307 million metric tons in 2009, and it increased to 407.6 million metric tons by 2015 (Source Statista, 2017) and is expected to reach approximately 500 million metric tons by 2020. On an average, 905.8 million m^3 of water is consumed per annum and around 695.7 million m^3 of wastewater is discharged (Elango et al., 2017).

Industrial and Municipal Sludge. https://doi.org/10.1016/B978-0-12-815907-1.00027-1

The papermaking process is complex and has far-reaching environmental impacts. It consumes huge amounts of water and natural resources (wood) and in turn discharges a variety of solid, liquid, and gaseous wastes into the environment. The processing stages, which significantly contribute to pollution, are wood embarking, chip making, pulp manufacturing, pulp bleaching, paper manufacturing, and fiber recycling (Fig. 1). Chemical processes such as calendaring (coating on paper surfaces for smoothness), krafting (for pulp recovery), acid sulfite, organosolvents, steam explosions, and ammonium-based processes also contribute to it (Robertson and Kirsten, 2000). During pulping, lignin is solubilized, releasing individual fibers and resins, alcohols, and sugars that contribute to 85% of the total effluents generated (Sumathi and Hung, 2006). Lignin fibers, energy-rich biomasses, starch, hemicellulose, cellulose, and organic acids are the major sources of organic pollutants, resulting in a chemical oxygen demand (COD) in the range of 20–125 kg per ton of pulp generated (Hagelqvist, 2013a,b). Waste-paper pulp consists of lignin (50%) and different amounts of water (60%–80%), depending on the efficiency of the dewatering procedure and other inorganic and organic components (Pečar and Goršek, 2015). Effluents containing high BODs (biological oxygen demand)/CODs deplete the oxygen for the flora and fauna in the water bodies into which they are discharged. Suspended solids (SSs), adsorbable organic halides (AOXs), pH (5–9), and color (black/brown) are the other major causes for environmental pollution. The intense coloration in the effluent is mainly due to lignin and its derivatives, such as chlorinated lignosulfonic acids, chlorinated resin acids, chlorinated phenols (trichlorophenol, trichloroguicol,

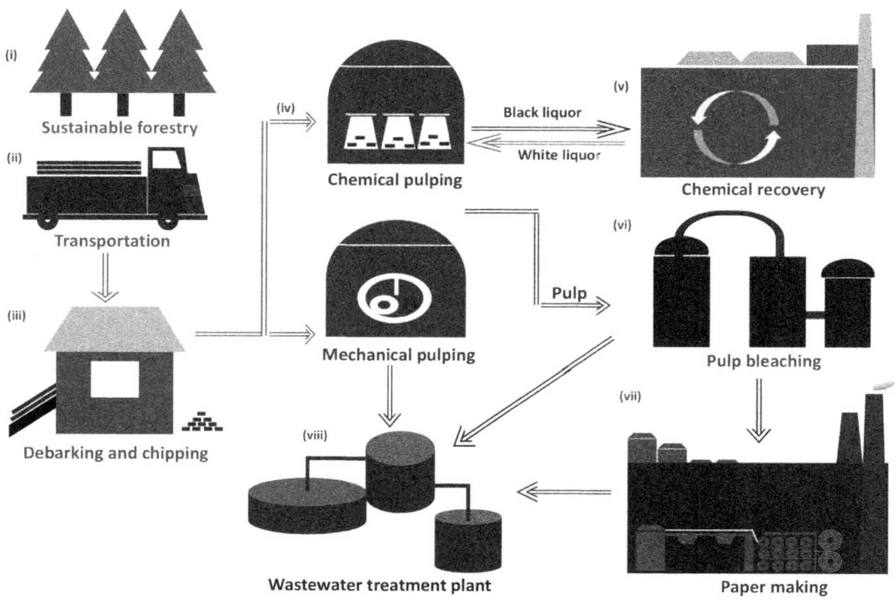

Fig. 1 Different stages of paper manufacturing.

tetrachloroguicol, dichlorophenol, dichloroguicol, and pentachlorophenol), and chlorophenolic hydrocarbons released during the pulping, bleaching, and chemical recovery stages (Badar and Farooqi, 2012; Subashini, 2015).

Before the 1970s, wastewaters from PPI were usually released into rivers without proper treatment. The high organic loads, solid content, and toxic compounds in the effluents affected the aquatic ecosystems. This promoted changes in the papermaking process. One of the most significant of these was the replacement of chlorine with chlorine dioxide as the bleaching chemical agent. However, in the majority of the PPI, wastewaters from various unit operations are mixed together and treated using conventional processes, which are not very effective in removing the recalcitrant materials.

Additionally, the current environmental standards and discharge norms of the Indian paper industry are quite stringent (pH of 5.5–9; BOD of 30 mg/L, COD of 250 mg/L, SS of 100 mg/L, and color 500 Platinum Cobalt Color Unit (PCU)), necessitating the adoption of advanced treatment technologies (Tewari et al., 2009). In recent years, the implementation of biorefinery concept has emerged rapidly, aiming at an efficient and complete utilization of waste generated from unit operations to produce fuels, power, and value-added products. Following this line of thought, this chapter discusses the valorization of various sidestreams of PPI for the production of high-value bioproducts in a biorefinery approach. Initially, various conventional treatment technologies and their limitations are outlined. Later in the chapter, the pretreatment strategies employed to reduce the complexity of waste are covered, followed by technologies that can be integrated into the mill. The chapter finally summarizes the development of a biorefinery and the main interventions that can help overcome the challenges of existing technologies.

2 Potential of PPI waste

The treatment processes employed most commonly by the PPI are primary, secondary, and tertiary (optional) treatment methods, which are broadly classified as physicochemical and biological treatment methods (Sumathi and Hung, 2006). Usually in the PPI, wastewater passes through primary treatment (e.g., sedimentation and filtration), followed by a secondary treatment (which is biological) and sometimes sequential anaerobic and aerobic treatments and tertiary treatment. Primary treatment mainly focuses on the physical removal of solids, secondary treatment on the removal of dissolved organic compounds, and tertiary treatment on effluent polishing and removal of other constituents.

The initial wastewater of PPI contains high concentrations of suspended solids and floating materials dominated by wood fibers and chips (in pulp mills, the percentage is 3%–4%, and in waste-paper mills, 15%–30%), with both inorganic and organic materials. This wastewater is put through a clarification process, wherein the sludge is captured by sedimentation in a gravity-settling tank. This sludge is often referred to as *primary sludge* or *fiber sludge*. The pulping and paper production stage mainly contributes to sludge production.

High lignin content makes pulping inefficient and hence produces 0.3–1 m^3 of primary sludge per ton of paper (Priadi et al., 2014). Limited data is available on the characteristics of primary solid-waste generation in the paper industry due to the varied composition of feed material (Veluchamy and Kalamdhad, 2017). However, primary sludge is usually dominated by lignocellulose components (cellulose, hemicellulose, and lignin), papermaking filler (kaolin and calcium carbonate), pitch, by-products of lignin, microorganisms, undigested organics, and ash (de Alda, 2008) (Table 1). Primary sludge also consists of solid particles (solid rejects and deinking sludge) from PPI having lumps of fiber and paper with relatively low moisture content and high content of cellulose fiber (Fig. 2). The most PPI participants have optimized their fiber recovery systems to increase their raw material yield and reduce operating costs, in turn reducing the generation of primary sludge.

The clarified wastewater then moves on to the activated sludge process (ASP), where aerobic microorganisms digest components of the wastewater and produce large amounts of waste-activated sludge (WAS). This process is carried out in an aeration basin with low HRT (hydraulic retention time), a sedimentation basin that removes sludge is recycled to the aeration basin, and finally, 60%–85% of COD removal can be achieved. Other types of biological treatment methods, including aerated lagoons, and sequence batch reactors, make up a fairly advanced treatment

Table 1 Typical characterization of pulp and paper mill sludge

	Pulp and paper	References
Primary Sludge		
pH	7.39 ± 0.004	Monte et al. (2009); Veluchamy and Kalamdhad (2017)
Total solid (TS) (%)	80	
VS (%)	57.94 ± 6.36	
Ash content (%)	54.06 ± 1.35	
Total COD (mg/L)	64000 ± 3000	
Lignin (%)	5.68 ± 0.23	
Hemicellulose (%)	6.53 ± 0.25	
Cellulose (%)	32.49 ± 0.37	
Secondary Sludge		
pH	6.0–7.6	Xu and Lancaster (2009)
Total solids (TS), %	1.0–2.0	
VS (% of TS)	65–97	
COD (mg/L)	11,000	
Nitrogen (% of TS)	3.3–7.7	
Phosphorus (% of TS)	0.5–2.8	
Iron (g/kg TS)	0.33–2.2	
Heating value (MJ/kg TS)	22–25	

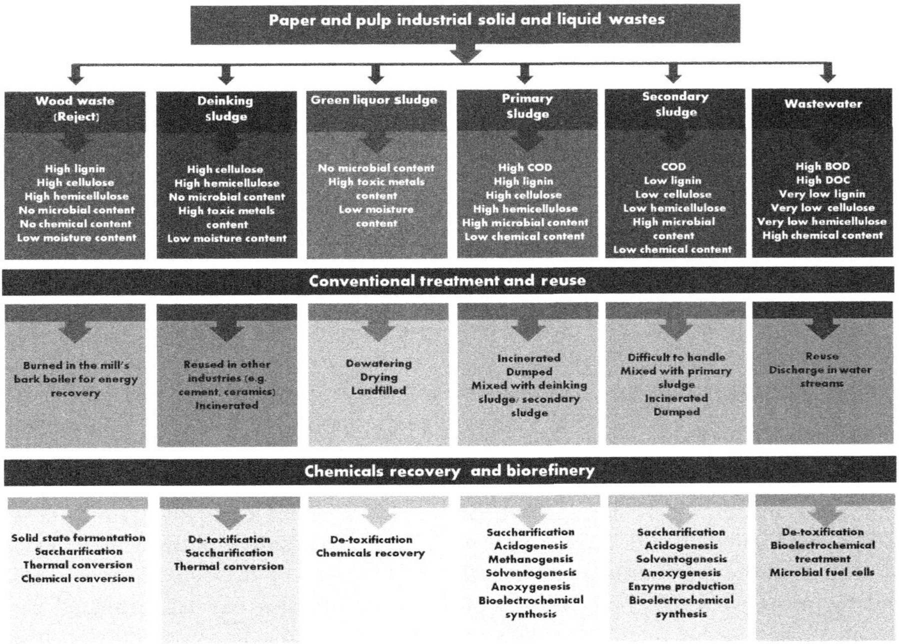

Fig. 2 Characterization and potential of paper and pulp industrial waste.

for high concentrations of lingocellulosic compounds and halogenated chemicals (Thompson et al., 2001). Aerated lagoons, although capable of reducing COD and AOX emissions, it consume a lot of space and energy for their mechanical aeration (Tsang et al., 2007).

Meanwhile, in the case of ASP, a huge quantity of activated sludge (biosludge) is often dewatered and either landfilled or incinerated to produce thermal energy. WAS consists of organic materials—predominantly excess microbial cell mass and unused chemical residuals of pulping process (Kyllönen et al., 1988; Thompson et al., 2001; Persson, 2011) (Table 1). Microorganisms (bacteria, protozoa, fungi, and rotifers) exist in the activated sludge reaction vessel and clarifier as flocs (Stoica et al., 2009; Bajpai, 2015). Additionally, after secondary treatment, the effluent and collected sludge still contain a good amount of organic material and other compounds, which often go to waste as they are discharged (Fig. 2).

Although most of the industry employs these methods for treatment, the frequency has been gradually decreasing, as they are costly. Reducing these costs, or even turning them into profits, depends on the ability of the sector to utilize valuable components in the sidestreams by reusing them internally or converting them to intermediates. The following section discusses the valorization possibilities of PPI wastewater and sludge for the generation of valuable and economically important biologically based products, with a major emphasis on the biorefinery concept.

3 Integrated strategies for adding value

The following sections discuss the industrially important chemicals and fuels that can be produced from PPI waste, with a special emphasis on various pretreatment methods and considering the fact that pretreatment is a crucial step.

3.1 Pretreatment strategies for the valorization of PPI waste

It has been determined that only 30%–50% of the COD within the PPI and WAS is biodegradable, as it contains resistant extracellular polymeric substances and microbial cells (Saha et al., 2011; Kaluža et al., 2014). Due to the recalcitrant nature of these materials, pretreatment is an important step to make the organic matter accessible for the conversion into a number of value-added products. A wide range of physical, mechanical, chemical, and biological strategies, a combination of the strategies, and alternative methods were reported for achieving this (Fig. 3). The effects of various pretreatments on lignocellulosic substances were reviewed by Nizami et al. (2009), who found that the surface area that is accessible for hydrolysis was enhanced by almost all the pretreatment technologies, and some even dissolved hemicellulose and lignin as well, enabling faster hydrolysis. Others also have investigated pretreatment as a means to enhance methane (CH_4) and bioelectricity production from pulp and paper-mill sludge (Lin et al., 2009; Yunqin et al., 2010; Saha et al., 2011; Vamsi Krishna et al., 2014).

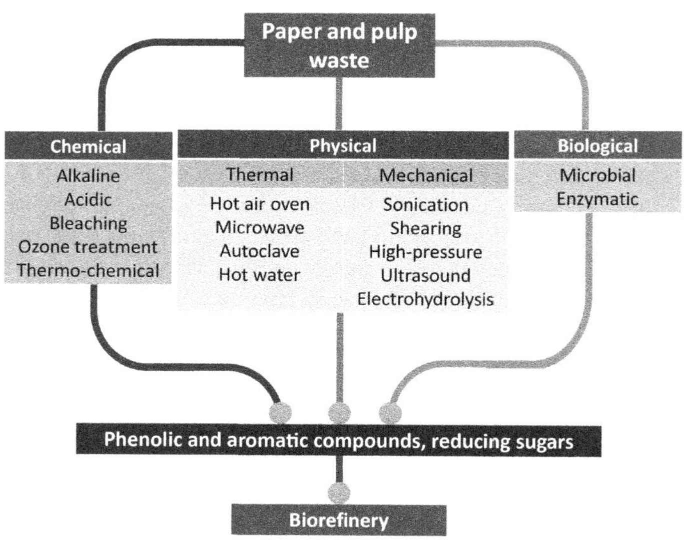

Fig. 3 Integrated pretreatment strategies for valorization of PPI waste.

3.1.1 Chemical treatments

Chemical techniques like bleaching of pulp with oxygen and extended cooking, hydrogen peroxide (H_2O_2), and ozone treatment could be options for pretreatment of paper and pulp mill (P&P) primary sludge enriched with wood chips and fibers. Alkaline or acidic are the two types of chemical pretreatment methods. Hemicellulose solubilization is promoted by acid pretreatment, which make cellulose more accessible for the acidogenic microorganisms in anaerobic digestion. On the other hand, lignocellulose material becomes swollen during alkaline pretreatment, which creates a more accessible surface area and pore size for microorganisms to act upon (Hendriks and Zeeman, 2009). PPI waste can be cooked for 2–6 h in digesters at about 10 atm pressure with caustic soda and magnesium oxide and a temperature of 140°C to result in lignin, resin, alcohol, and sugar (Jingjing, 2011). Coupled pretreatments are discussed in the following sections.

3.1.2 Physical treatments

Mechanical pretreatment methods help to physically rupture flocs within the biosludge to allow more of the COD to be accessible for biodegradation by the microorganisms. Ball milling, sonication, shearing, and high-pressure homogenization techniques are the most commonly used mechanical treatments. Using high-pressure (83-MPa) homogenization, sludge flocs can be disintegrated into relatively fine particles or soluble COD by very high shear force due to a sudden pressure drop when it released from the orifices (Elliott and Mahmood, 2012). Strong temperature and pressure gradients disintegrate sludge particles during pretreatment with high-frequency electromagnetic waves (e.g., microwave and ultrasound techniques). This ruptures the cell walls, and intercellular matter gets released in which microorganisms can act easily to make it in soluble COD (Meyer and Edwards, 2014). Cellulose crystallinity and the degrees of polymerization decrease, which further increase the accessible surface area and pore size during ultrasonic and electrohydrolysis pretreatment (Taherzadeh and Karimi, 2008; Veluchamy and Kalamdhad, 2017).

Thermal treatment is an effective alternative pretreatment method (Elliott and Mahmood, 2007). Sludge solubilization is possible at temperatures as low as 50°C, with long retention times (0–72 h) (Vlyssides and Karlis, 2004; Elliott and Mahmood, 2007), and as high as approximately 200°C. Only two studies looked at thermal pretreatment of pulp and paper mill biosludge, but results from studies of other types of sludge have been reported. It should be noted that sodium hydroxide (NaOH) can be applied with the mechanical pretreatment to enhance its effectiveness. The quantity of sludge produced can also vary depending upon the mechanical treatment; for instance, 0.1 m^3/t of sludge is produced from kraft pulping, whereas mechanical and semichemical processes yield nearly 0.6 m^3/t (Bajpai, 2015).

3.1.3 Microbial pretreatments

Biological pretreatments have important advantages, as it consumes low chemical and energy input. The screening of microorganisms capable of treating pulp and paper waste, as reported by many researchers, mainly depends on their ability to produce lingocellulosic compound–degrading enzymes such as extracellular isoenzymes, lignin peroxidase, manganese peroxidases (MnP), and laccases. Among various microorganisms used, white rot fungi like *Phanerochaete chrysosporium*, *Fomes lividus*, *Trametes versicolor*, *Coriolus versicolor*; *Pycnoporus cinnabarinus*; *Trametes villosa*, *Aspergillus fumigates*, *Pycnoporus sanguineus*, *Pleurotus ostreatus*, *Pleurotus sajor-caju*, and *Phanerochate flavido-alba* are well known to decolorize lignin/phenolic rich kraft mill effluents and reducing AOX and COD of the discharged material (Choudhury et al., 1998; Modi et al., 1998; Selvam et al., 2002).

Bacterial strains like *Streptomyces cyaneus CECT 3335 and γ-proteobacterium JB* have been reported to be laccase producers and were used for biobleaching of pulp and for removing toxic substances (Singh et al., 2015). Fungi are more capable of degrading lignocellulose due to higher C/N ratios of lignocellulosic materials (Mao et al., 2015). Apart from fungi, the use of bacteria algae was reported to treat wood-based PPI wastewater with 58% of COD, 84% of color, and 80% of AOX removal (Tarlan et al., 2002). Yeast strains (e.g., *Saccharomyces cerevisiae, Pichia pastoris,* and *Kluyveromyces marxianus*) and bacterial strains (e.g., *Geobacillus* and other thermophilic bacteria) were used to produce recombinant hydrolyzing enzymes like cellulases, which showed cellulolytic activity under extreme industrial conditions (Apruzzese et al., 2017).

3.1.4 Enzymatic pretreatments

The main component of lignocellulosic material is cellulose, which is a polymer made up of β Glucose monomers, which are in turn linked with β-1,4-glycosidic bonds, and its high tensile strength is provided by hydrogen bonds and van der Waals forces. Cellulose are attached to each other by hemicellulose and covered by lignin complexes (Fig. 4). Microbial activity during hydrolysis upon the degradation of lignocellulose material is very slow, which results in the requirement of longer solid retention time and lower degradation efficiency (Ha et al., 1998; Zheng et al., 2014). Thus, enzyme treatments might be useful before microbial action to create targeted biorefinery products and energy (Fig. 4). Sustainable, ecofriendly biobleaching (removal of hemicelluloses from cellulose), pitch removal, and deinking of paper waste by using potent biocatalysts such as xylanases, laccases, and cellulases is always beneficial.

Endoxylanases and β-xylosidases (collectively called *xylanases*) are the two key enzymes responsible for hydrolysis of xylan, the major component of hemicellulose (Chávez et al., 2006). Endoxylanases act on the homopolymeric backbone of 1,4-linked β-d-xylopyranose-producing xylooligomers (Ahmed et al., 2009), while β-xylosidases act on these xylooligomers to release xylose (Knob and Carmona, 2010). Xylans are intimately linked to cellulose and lignin and present within pulp fibers. When the covalent bonding of lignin to xylan gets degraded with the help of enzymes, lignin extraction is easy. Xylanase enzymes catalyze the hydrolysis of xylan that has reprecipitated on the fibers during alkaline pulping. Xylanase can be extracted from *Bacillus megaterium* and *Bacillus stearothermophilus* SDX (Sindhu et al., 2006; Dhiman et al., 2008; Nagar et al., 2013).

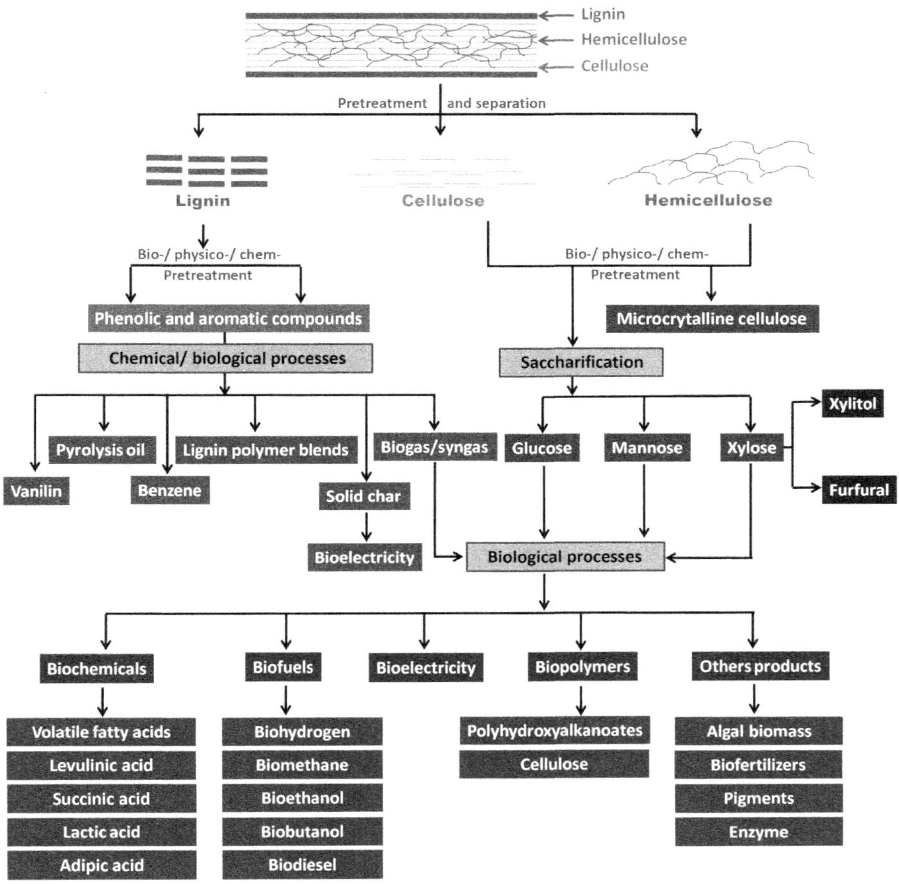

Fig. 4 Integrated strategies of lignin biorefinery from PPI waste.

Laccase is another versatile biocatalyst with a phenolic mediator, that not only plays a role in delignification but can also remove the lipophilic extractives from pitch deposition in the pulps. Therefore, laccases are capable of improving the physical, chemical, and mechanical properties of pulp by either forming reactive radicals with lignin or by functionalizing lignocellulosic fibers. They can also target the colored and toxic compounds released as effluents from PPI, which might be toxic for biorefinery processes (Virk et al., 2012). The primary sludge requires alkali-tolerant laccases for delignification purposes because primary and secondary sludges are slightly alkaline (Singh and Agrawal, 2008). In addition, cellulolytic enzymes (endoglucanases, exoglucanases, and cellobiase) are the most important industrial enzymes and have demonstrated their biotechnological potential in PPI sectors (Pandey et al., 1999; Kuhad et al., 2011). Bioconversion of lignocellulosic waste to soluble sugars relies on the synergistic action of a complete cellulase system which includes CBH (cellobiohydrolases), EG (endoglucanase), BG (beta-glucosidase), and hemicelluloses (Aro et al., 2005).

4 Valorization of PPI waste—Overview of current and future bioproducts

Focusing on the production of value-added products and bioenergy in addition to treatment will enhance the business development of biorefineries. However, it is imperative for these products to meet basic requirements in order to be produced in a biorefinery. In addition to being biodegradable, nontoxic, and durable, they should be feasible for large-scale production and able to compete with current and expected future market prices and market capacity.

Over the past few years, the producers of bio-based products have primarily targeted high-value or specialty chemical markets. However, advancements in science and technology allow the economic production of other bio-based products from waste that have equal monetary value and, more important, have a wide range of applications (Fig. 4 and Table 2). The following section discusses a number of major platform chemicals, on which the focus of research has been intensified over the past several years.

Table 2 Probable biorefinery products and their sources from PPI waste

	Components (pollutants)	Key pollutants	Products	References
Raw materials being processed	Tree debris, bark, tannin, lignin, cellulose, hemicelluloses, and organic compounds such as acidic resin	Lignin	Polymers, resins, adhesives	Park et al. (2008); Thakur et al. (2014); Wu et al. (2009)
Pulping	BOD, COD, uncooked wood, bark particles, soluble wood materials, color, resin acids (including isopimaric, sandacopimaric, levopimaric, abietic, dehydroabietic, neoabietic, and palustric acids), fatty acids, dissolved inorganics, lignin, hemicelluloses,	Cellulose, hemicellulose, organic compounds, COD, sludge	VFAs	Bengtsson et al. (2008a,b)
			Succinic acid,	Cok et al. (2014); Werpy and Petersen (2004)
			PHAs	Jiang et al. (2012)
			Biofuels (biohydrogen, biodiesel)	Venkata Mohan et al. (2016c)
			Enzymes	Cavka et al. (2013); Karn et al. (2013);
			Bioelectricity	Krishna et al. (2014); Ketep et al. (2013b)

Table 2 Continued

	Components (pollutants)	Key pollutants	Products	References
	and some chemical additives such as soluble silicates			
Bleaching	Organic halides (AOX, EOXs), polychlorinated biphenyls, polychlorinated dibenzodioxines, dioxins, furans, chlorinated lignosulfonic acids, chlorinated resin acids, residual lignin, color, COD, carbohydrates, inorganic chlorines, VOCs, and halogenated hydrocarbons	VOCs: Methanol, acetone, terpenes, phenols, COCs: Chloroform, AOX Absorbable organic halides	Found in minimal quantities and should be remediated	Kumar et al. (2014)
Pulping/ deinking	Residual lignin, bleaching agents, and some barely biodegradable organic compounds	Lignin, organic compounds	Found in minimal quantities and should be remediated	Dwivedi et al. (2004)
	Mineral additives, AOXs, resins, BOD, COD, resin acids, particulate wastes			McGuire and Suffet (1980)
	Metallic components, sand, glass, plastic, coatings, fillers, organic compounds from paints and printing inks such as 2,4,7,9-tetramethyl-5-decyne-4,7-diol,	SO_2, NOx, PO4, metals	Algal biomass, biodiesel, protein	Wang et al. (2010); Venkata Mohan et al. (2016c)

Continued

Table 2 Continued

	Components (pollutants)	Key pollutants	Products	References
	pulping additive chemicals, compounds like Si and Ca, higher amounts of organicsi, thermoplastic resinsj, TCMTB, chlorophols			
	Deinking additives and ink particles, fibers, fines, fillers, ash	Dyes, IOX	Removal of dye	Kumar et al. (2014)

4.1 Bioenergy

Energy conversion in PPI ranges from 4 to 400 Mt annually (Meyer and Edwards, 2014). From 25% of sludge produced from PPI each year, 100 TWh of electricity could be generated annually through biogas production (REF, 2012). Due to its high organic load, some researchers have described the energy potential of the primary sludge through anaerobic digestion (AD) (Bayr et al., 2012; Meyer and Edwards, 2014; Priadi et al., 2014). Thus, anaerobic treatment has become an alternative energy source to augment and supplement fossil fuel in PPI industry.

In 2004, the bioenergy production accounted for 50% of the total energy consumption in PPI in Europe. AD has many advantages over aerobic treatment, including lower sludge production, lower chemical consumption, smaller space requirements, and energy production in the form of biogas (Persson, 2011; Hagelqvist, 2013a,b). Its predominant characteristics, such as high holocellulose content, short and more fibers, low pH (near neutral), and high buffering capacity, make primary sludge the best resource for AD. However, the AD of primary sludge is virtually unexplored to date due to the lignin-rich substances in it (Meyer and Edwards, 2014).

The inherent recalcitrant characteristic of lignocellulosic content (recycled fibers) of the effluents makes the hydrolysis stage a rate-limiting step in the AD process (Meyer and Edwards, 2014). Sulfide toxicity and nitrogen deficiency are other major concerns about AD of PPI sludge (Thompson et al., 2001). Enzyme hydrolysis can be used to make the hydrolysis reaction faster and to remove the toxicity of substrates. This leads to enhancement of rate limiting by reducing the degree of crystallinity of lignocellulosic materials. Various forms of energy that can be generated from the PPI waste using suitable technology are discussed next.

4.1.1 Biogas

CH_4 is among the most explored biofuel generated through AD. Wet PPI sludge is effectively utilized for the generation of high-quality fuels like CH_4, hydrogen, and heavy oils (Zhang et al., 2010). Longer HRT and high water content in the biosludge lead to an increase in volume in the digesters, as well as an increase in capital cost. Low biogas yield from PPI bio sludge compared to other types of sludge is one of the major concerns during the AD process (Huiliñir et al., 2014). That is why CH_4 production from lignocellulosic biomass is highly dependent on the pretreatment strategies employed for its degradation. Upon degradation, lignocellulose-rich primary sludge breaks down into recalcitrant polymers to increase volatile solid (VS) reduction and biogas production. The effects of hydrothermal, enzymatic, ultrasound, and chemical pretreatments (acidic and alkaline), both alone and in combination, have been studied (discussed in Section 3.1) to increase CH_4 yield from PPI biosludge (Bayr et al., 2013).

Biological pretreatment of lignocellulosic biomasses like cellulases and cellulase-producing microorganisms enhances hydrolysis and promotes fermentable sugar production and improved biogas production (Elliott and Mahmood, 2007; Kumar et al., 2009). Many scientists have discussed codigestion as another strategy to use with various pretreatment methods to improve the biodegradation of PPI biosludge. PPI sludge can be effectively codigested with substrates with higher biodegradability, like municipal sludge, food waste, dairy farm waste, rice straw, pig slaughterhouse waste, grease trap waste, grass silage, and pulp mill biosludge (Borowski and Kubacki, 2015; Hagelqvist, 2013a,b; Mussoline et al., 2013; Trulli and Torretta, 2015; Yalcinkaya and Malina, 2015; Chakraborty et al., 2017) (Table 3). Lin et al. (2011) used codigestion of pulp mill sludge with nitrogen-source monosodium glutamate (MSG) waste liquor to maintain the C/N ratio during AD.

The addition of a small amount of food waste with PPI sludge helped to improvement in biogas production significantly, which might help with the supplementation of missing nutrients and N source in PPI (Yun et al., 2013). Natural zeolite was used as a catalyst to enhance biogas production through AD without pretreatment (Huiliñir et al., 2014). In addition to PPI sludge, kraft evaporator condensate was utilized for CH_4 production using a submerged anaerobic membrane bioreactor (SAnMBR) (Xie et al., 2010). The configuration of the reactor (namely, anaerobic filter, SAnMBR, serum bottles, etc.) was also studied for the purpose of enhancing CH_4 recovery and COD removal efficiency for effluents from PPI under mesophillic and thermophillic conditions (Kamali et al., 2016).

$$_6H_{12}O_6 + 2H_2O \rightarrow 2CH_3COOH + 2CO_2 + 4H_2 \ (\Delta G = -135.6\,\mathrm{kJ/mol})$$

$$C_6H_{12}O_6 + 4H_2O + 2NAD+ \rightarrow 2CH_3COO^- + 2HCO_3^- + 2NADH + 2H_2 + 6H^+ \ (\Delta G = -215.7\,\mathrm{kJ/mol})$$

$$C_6H_{12}O_6 \rightarrow CH_3COO^- + CH_3CH_2COO^- + CO_2 + H_2 + 2H^+ \ (\Delta G = -287.0\,\mathrm{kJ/mol})$$

$$C_6H_{12}O_6 + 2H_2O \rightarrow CH_3CH_2CH_2COO^- + 2HCO_3^- + 2H_2 + 3H^+ \ (\Delta G = -261.5\,\mathrm{kJ/mol})$$

$$C_6H_{12}O_6 + 2H_2O + 2NADH \rightarrow 2CH_3CH_2OH + 2HCO_3^- + 2NAD^+ + 2H_2 \ (\Delta G = -234.8\,\mathrm{kJ/mol})$$

Table 3 Summary of codigestion of pulp and paper mill biosludge with other substrates

Cosubstrate with pulp and paper mill biosludge	Operating conditions	Enhancement over biosludge alone/ specific CH_4 yield ($mL\,g^{-1}$ VS, COD fed)	References
Food waste	Batch	55% increase	Yun et al. (2013)
Municipal biosludge	Batch	~50% increase	Hagelqvist (2013a)
MSG waste liquid	Semicontinuous (pilot scale)	245 mL CH_4/gVS	Lin et al. (2011)
Natural zeolite	Batch	Maximum 10% increase	Huiliñir et al. (2014)
Mixture primary sludge from chemi-thermo mechanical pulp mill (70%–90% WAS)	Bench scale	90 mL/g VS, COD fed	Puhakka et al. (1992)
Mixture municipal sludge, primary sludge, and WAS from thermomechanical pulp and paper mill	Bench scale	185	Jokela et al. (1997)
Mixture Kraft pulp and paper mill PS and WAS (VS ratio 3:2)	Bench scale (thermophilic)	150–170	Bayr et al. (2012)
Pig waste and paper sludge (2:1)	Biomethane potential assay	250	Parameswaran and Rittmann (2012)
Pulp and paper sludge and food waste	Bench scale	432.3	Lin et al. (2013)
Mixture WAS from thermo-mechanical pulping/chemi-thermo mechanical pulp mill and municipal WAS	Batch	84	Hagelqvist (2013b)
Mixture of PPMS and cow dung	BMP assay	264	Veluchamy and Kalamdhad (2017)

H_2 production from PPI waste streams is reported. PPI sludge was investigated as a good carbon source for H_2 and CH_4 production (Lin et al., 2009). PPI sludge is an alkaline substrate with C/N ratio (30–35); thus, anaerobic codigestion of it with FW was found to be an effective way to obtain maximum H_2 and CH_4 (Table 3). Cellulase-form *Trichoderma reesei* was effectively used for PPI effluent pretreatment, and the hydrolysate with sugars was converted to H_2 with the help of *Enterobacter aerogenes* (Lakshmidevi and Muthukumar, 2010). Pulp and paper mill effluent (PPME), a rich cellulosic material, was pretreated with ultrasonication, which gave the highest biohydrogen yield through a photofermentation process using *Rhodobacter sphaeroides* NCIMB (Hay, 2015). Paperboard mill wastewater was used in a multiphase anaerobic reactor for H_2 and CH_4 production (Farghaly and Tawfik, 2017).

4.1.2 Bioethanol

Being a rich source of cellulosic and lignocellulosic residue, waste PPI sludge is an attractive and promising biomass feedstock for the production of fermentable sugars and bioethanol (Lin et al., 2012). Lignocellulosic biomass is pretreated with alkali, H_2O_2, or acid to get cellulose-rich biomass. Then, enzymatic treatment followed by autoclaving to enhance the recovery of sugars (e.g., glucose, xylose, arabinose) from biomass is carried out, a process known as *saccharification*. Cellulase pretreatment enhanced fermentable sugar recovery from biomass by milling and acid treatment; finally, glucose to bioethanol bioconversion was carried out using immobilized yeast cells (Pečar and Goršek, 2015). Cellulose-rich PPI primary sludge was converted to ethanol and H_2 with other by-products such as acetate, lactate, formate, and carbon dioxide (CO_2) using a strain of *Clostridium thermocellum* (Moreau et al., 2015). Solid substrate fermentation of pretreated PPI waste was carried out using the yeast *Pichia stipitis* CBS 5773 and after 48 h of incubation, ethanol was recovered (Marques et al., 2008a,b). Peng and Chen (2011) reported a good amount of ethanol yield and a good sugar conversion rate from cellulose-pretreated dry paper sludge using *Saccharomyces cerevisiae* GIM-2. Simultaneous scarification and fermentation (SSF) using cellulose-produced *Acidothermus cellulolyticus* and *S. cerevisiae* TJ14 was reported for ethanol production (Prasetyo and Park, 2013).

4.2 Biopolymers

4.2.1 Polyhydroxyalkanoates

Biodegradable plastics/polymers have an expanding range of potential applications and have been proposed as a solution for the problem of plastic waste. The biopolymers, such as polyhydroxyalkanoates (PHAs), are synthesized exclusively by bacteria. PHAs are a group of linear aliphatic polyesters composed mainly of R-(−)-3-hydroxyalkanoate units. PHAs have attracted commercial interest as a plastic material because its physical properties are remarkably similar to those of the conventional plastics (polypropylene). Two major factors that currently inhibit the widespread use of PHAs are their high cost and poor mechanical properties. Therefore, research efforts are being directed toward the utilization of industrial sidestreams as raw materials for low-cost PHA production.

PPI wastewater is one of the possible feedstocks for PHA production. The reducing sugars obtained after pretreatment of PPI waste can be subjected to acidogenesis to generate volatile fatty acids (VFAs), which can further act as substrates for PHA production using pure or mixed microbial cultures. Reports are available on the use of PPI wastewater for PHA production (Jiang et al., 2012; Jarpa et al., 2012).

Bengtsson et al. (2008a,b) reported that after acidogenic fermentation of paper mill wastewater, 74% of the soluble COD was present in the form of VFAs, with the potential to accumulate 48% PHA while providing 95% removal of soluble COD (Yan et al., 2010). They found that 15 g/L optimum concentration of activated sludge is good for PHA production and that continuous feeding of acetic acid can enhance PHA content and productivity.

It was also observed that, the composition of VFAs influences the properties of the produced polymer (Bengtsson et al., 2008a,b). In another study (Queirós et al., 2014), hardwood-spent sulfite liquor (HSSL), a complex feedstock having lignosulfonates and phenolic derivatives, sugars, and acetic acid originating from the pulp industry, was directly used as a substrate for PHA production without using initial acidogenic fermentation. It is worth noting that some early attempts are being made to utilize PHAs as a raw material in papermaking, particularly as a surface sizing/coating agent (Laycock et al., 2014).

4.3 Bioelectricity

Microbial fuel cells (MFCs) are gaining immense interest for the production of bioelectricity using PPI wastewater. MFCs are sustainable and self-reliant, as they do not require any external energy supply, and with the focus of waste remediation, they also can be considered as a bioelectrochemical treatment (BET) system (Venkata Mohan et al., 2014). The bioelectricity production from MFCs is highly dependent on the MFCs' in situ interactions among anodes and cathodes, their kinetics, diffusion, and multirate-limiting parameters (Ketep et al., 2013b). Several studies have reported on PPI wastewater treatment with simultaneous bioelectricity generation (Huang and Logan, 2008; Huang et al., 2009; Velasquez-Orta et al., 2011; Ketep et al., 2013a; Krishna et al., 2014). Maximum power densities of ca. 144 mW/m^2 and ca. 672 mW/m^2 were observed in unamended PPI wastewater and in a buffered solution, respectively (Huang and Logan, 2008). The effect of organic load accompanied by electricity production in a continuous flow MFC system was reported in a study using PPI wastewater (Huang et al., 2009). A system operated with PPI wastewater in fed batch mode was operated with bioanodes, and high current densities ($>5\,A/m^2$) were observed (Ketep et al., 2013a).

The current density of 4 A/m^2 and COD removal of up to 91% were obtained with microbial bioanodes under constant potential, without adding any nutrients or minerals (Ketep et al., 2013b). Krishna et al. reported the maximum power production using PPI wastewater in an MFC of 40.2 mW/m^2 and a current density of 180 mA/m^2 (Krishna et al., 2014). For efficient and additional resource recovery, MFCs can be further integrated with various other bioprocesses like AD but extensive studies are needed to explore this idea further.

4.4 Enzymes

Cellulosic residues, such as sugarcane bagasse, wheat bran, paper pulp, corn cob residue, and wheat straw were reported as carbon sources and inducers of cellulase production in submerged or solid-state fermentation processes (Singhania et al., 2010). The indigenous microbial load produces hydrolytic enzymes to convert cellulose and low lignin present in PPI secondary sludge into simpler compounds (Karn and Kumar, 2015). These enzymes are found to be immobilized on flocs or through hydrophobic and ionic interactions attached to microorganisms' cellular walls.

The production of enzymes like bacterial cellulase and hydrolytic enzymes from PP waste is being reported in the literature (Karn et al., 2013; Cavka et al., 2013). Belghith et al. reported 23 IU/ml filter paper activity, 21 IU/ml CMCase activity (endoglucanase) with 25 mg/ml of protein production utilizing paper pulp as an inducer substrate using a mutant strain of *Penicillium occitanis* (Belghith et al., 2001). Fiber sludge hydrolysates were also used as substrates for enzyme production through sequential fermentation with the bacterium *Gluconacetobacter xylinus* and the filamentous fungus *T. reesei* (Cavka et al., 2013). High lipase and protease activity extracted from PPI sludge using the nonionic detergent Triton X-100 at 0.1% and 1%, respectively, was reported (Karn et al., 2013).

4.5 Lignin as a feedstock

Besides the use of biodegradable fractions, lignin, a major by-product of debarking and pulping and a constituent of primary sludge, can be directly employed for the production of a spectrum of commodity and specialty chemicals. Research in this direction has become a major industrial and scientific endeavor (Pye, 2008). The destruction of lignocellulosic biomass for paper pulp production is well established and intended for the separation of lignin from cellulose and hemicellulose lysates. The major fraction of lignin from PPI is produced by kraft pulping, while some amount of lignin comes from the sulfite pulping process as lignosulfonates (Zhang et al., 2011).

The extracted lignin can be converted into a numerous aromatic/phenolic compounds, such as biofuels, adhesives, vanillin, benzene, solid char, biogas, and polymer blends (Holladay et al., 2007), and carbon fiber (Kubo and Kadla, 2005). The U.S. Department of Energy (DOE) predicts that 0.225 billion tons of lignin can be utilized for biofuel generation by 2030 (Holladay et al., 2007). Phenolic compounds derived from the depolymerization of lignin can be readily used in phenolic resin manufacturing. The conversion of lignin requires the implementation of various pretreatment strategies, like base-catalyzed depolymerization, pyrolysis, alkaline pulping, kraft pulping, and Lewis acid-catalyzed solvolysis followed by executing various chemical and biological processes like pyrolysis, hydrogenolysis, oxidation, gasification, fermentation, and enzymatic approaches (Zhang et al., 2011).

Due to the heterogeneous nature of polymeric lignin, the major challenge for conversion of lignin into standardized products is its separation, and the implementation of proper pretreatment strategies (Davis et al., 2013). The lignin produced from PPI is mainly contaminated with sulfur (mostly from sulfite pulping), therefore hindering its consumption for product generation and many other applications. Thus, it necessitates the removal of sulfur content from the lignin extract (black liquor) prior to product formation. Integrating the pulping processes with other nonsulfur utilizing processes as discussed previously and increasing paper quality by incorporating various other by-products to create a wide spectrum of products may be favorable for the establishment of an effective and ecofriendly lignin refinery.

4.5.1 Lignin-polymer blends

Lignin is the second-most-abundant natural aromatic polymer after cellulose and can be considered as an untapped biopolymer in biomass. However, due to its poor processability, it cannot be used as a polymer by itself. However, it can be incorporated into other polymeric materials as lower-cost and biodegradable fillers in order to enhance the overall properties of the polymeric materials. The hydroxyl and carboxylic groups present in lignin are compatible with most of the polymer types. Blending lignin with polymers increases the tensile strength, preventing mechanical deformation. Lignin can be blended with protein materials, starch, cellulose, PHA, polyolefins, polypropylene/ethyle, polyvinyl chloride (PVC), and natural rubber (Huang et al., 2003). The water resistance of starch can be improved by blending it with a hydrophobic polymer such as lignin (Lepifre et al., 2004) and lignin also can increase the plasticity of a starch matrix due to its lower-molecular-weight fraction (Morais et al., 2005). Çalgeris et al., 2012 observed that the addition of hazelnut lignin to starch biofilm enhanced its thermal and mechanical properties for drug release. In a similar study (Wu et al., 2009), cellulose/starch/lignin film was prepared in ionic liquid that showed thermal stability, good mechanical properties, and impermeability to gas. Where, the researchers observed that lignin can be used as a plasticizer of starch, a reinforcing agent of cellulose, and a compatibilizer of cellulose and starch in order to prepare natural biopolymer composites (Wu et al., 2009). In addition, these lignins can also be blended with PHA to reduce biodegradation of PHB films (Mousavioun et al., 2012).

The presence of lignin may have resulted in the blends being more resistant to microbial attack, and with the formation of strong hydrogen bonds with PHB, rapid degradation was prevented. Better understanding of the structure-function relationships between lignin and polymers needs to be developed to achieve further successful bioengineering developments and commercial applications.

4.6 Prospective chemicals from PPI waste

Apart from the products that have been discussed so far, PPI waste can be valorized for the production of other promising chemicals and a number of other possibilities that have been less widely studied. A large number of relevant possibilities for its practical implementation have received attention from researchers, as detailed next.

4.6.1 Volatile fatty acids (VFAs)

Volatile fatty acids (VFAs)—mainly acetic acid, propionic acid, and butyric acid—have great importance as commodity chemicals, and biological production of these acids from various waste streams is gaining a huge amount of attention. They are the major products of acidogenesis, along with cogeneration of biohydrogen, and further, they are consumed for the production of CH_4 in the complete AD process. Currently, the major focus is on the biogas production from PPI waste, which needs to be shifted towards VFA production. This can be achieved by suppressing methanogenesis.

A few studies have reported on the production of VFAs from PPI wastewater (Bengtsson et al., 2008a,b; Jiang et al., 2012), where the major focus was on the use of VFAs in the production of bioplastics, as previously discussed (Bengtsson et al., 2008b; Jiang et al., 2012). In batch reactors, 60% VFA of the soluble COD was achieved at a pH of 6 and later utilized for bioplastic production (Jiang et al., 2012). HSSL, a by-product from PPI, was effectively utilized as a substrate, and a 36% degree of acidification (VFA production) was achieved during the 71 days of operation (Queirós et al., 2017). Further, acetic acid can be extracted efficiently from prehydrolysis liquor (PHL) during the pulp production process (Yang et al., 2013; Kaur and Ni, 2015). The potential of biological VFA production from PPI waste needs to be explored, and in order to achieve that, an appropriate pretreatment strategy for the enhanced production of VFAs must be adopted.

4.6.2 Adipic acid

Adipic acid (also known as *hexadecanoic acid*) is an important, commercially available, C6 platform chemical with a market volume of 2.6 million tons per year and an annual demand growth forecast of 3%–3.5% globally (Polen et al., 2013). It is used as a precursor to the production of nylon, plasticizers, lubricants, and polyester polyols (Vardon et al., 2015). In conventional processes, adipic acid is produced via nitric acid oxidation of benzene, which is not sustainable and involves the use of fossil-based resources (Polen et al., 2013). A recent analysis of the utilization of lignin in an integrated biological and chemical catalysis suggests that it can offset the major economic and greenhouse gas effects (Davis et al., 2013). The production of adipic acid via muconic acid was proposed by Draths and Frost (Draths and Frost, 1994). In this approach, carbohydrates are converted to D-glucose, and then fermented to muconic acid, followed by hydrogenation to adipic acid. Vardon and coworkers metabolically engineered *Pseudomonas putida* KT2440 and used it to transform lignin-derived aromatics into cis,cis-muconate, which then was chemically converted to adipic acid by hydrogenation (Vardon et al., 2015). Although no literature is available on the direct use of PPI waste to produce adipic acid, the lignin fraction obtained in a PPI wastewater stream can be used as a key ingredient for making adipic acid. While it is a great challenge to realize the full value of lignin application for adipic acid production, there is also enormous potential to develop commercially feasible routes in a synergistic approach to improve the economic viability of the process.

4.6.3 Succinic acid

The DOE published a report in 2004 that listed the 15 most promising sugar-derived products (Werpy and Petersen, 2004). One of these products is a four-carbon dicarboxylic called succinic acid ($C_4H_6O_4$), which is an important platform chemical with numerous applications (Cok et al., 2014). Succinic acid can be produced via chemical routes, such as paraffin oxidation, catalytic hydrogenation, or electroreduction of maleic acid or maleic anhydride (Jong et al., 2012). Recent developments in the production of succinic acid have focused on biotechnological alternatives, particularly the use of microorganisms and renewable sources of energy as feedstock (Song and Lee, 2006; Lee et al., 2000; Cheng et al., 2012).

Succinic acid is a key intermediate of the tricarboxylic acid (TCA) cycle and can also be a fermentation end product when glucose or glycerol is used as a carbon source with the fixation of CO_2. Alexandri et al. (2016) reported the production of succinic acid by *Actinobacillus succinogenes* and *Basfia succiniciproducens* in batch cultures using crude and pretreated spent sulfite liquor produced by the sulfite pulping process. Besides *A. succinogenes*, *Escherichia coli* AFP 184 also can consume xylose. When this strain was cultivated in diluted spent sulfite liquor in dual phase fermentations with aeration, biomass was produced in the first phase and succinic acid was produced in the second phase from sugar in presence of CO_2 supply. Although PPI waste as such cannot be used for the production of succinic acid, the reducing sugars obtained during pretreatment serve as a potential substrate for succinic acid.

4.6.4 Levulinic acid

Levulinic acid (4-oxopentanoic acid), a key molecule for biofuel formation, has high demand and is projected to 42580 MT by 2025. To upsurge the supply of levulinic acid to meet the growing demands, an alternative pathway and derivatives for production could be found. The current stage of technology has the potential to process lignocellulosic PPI discharge for levulinic acid production. The conversion of waste to potential substrate molecules could be made possible using various pretreatment techniques (e.g., enzymatic and hydrothermal conversion). The chemical conversion of waste to levulinic acid is preferably carried out by the breakdown of β (1–4) glycosidic link, intrachain, and interchain H-bonds of cellulose.

The catalytic conversion of cellulose to levulinic acid at high temperatures is performed by various metal chlorides (Peng et al., 2010). This process also could use a chemoenzymatic path to levulinic acid formation, which utilizes cellulose, hemicellulose, and lignin. Heterogeneous and homogenous solvents, solvolysis (organic solvents), and microwave irradiation to convert waste to levulinic acid (Morone et al., 2015). It not only brings about new advancements and biotechnological approach for levulinic acid production, but also helps in the integration of these techniques in waste degradation to make the process cost effective and generate a commodity chemical.

4.6.5 Lactic acid

Lactic acid (LA) has a broad range of applications, such as food, beverages, cosmetics, pharmaceuticals, and the chemical industry, and as a result, its demand is increasing continuously (Jantasee et al., 2017). The worldwide consumption of LA in 2015 was

330,000 tons per annum, and according to a report by Global Industry Analysts, the market is expected to increase by 24% till 2023. Commonly, LA is being produced by chemical synthesis or by fermentation. The fermentative production of LA has received a lot of interest because it offers an alternative to environmental pollution caused by petrochemical industry (Abdel-Rahman et al., 2013).

Utilization of cellulosic materials seems promising because they are cheap, abundant and renewable. However, the process using cellulosic-based material is more complicated than that for starch-based material because pretreatment is required (Martinez et al., 2013). The enzymatic hydrolysis step is typically combined with the fermentative step to achieve simultaneous SSF. Generally, the sugars are oxidized to puruvate and adenosine triphosphate (ATP) is gained. Direct reduction of pyruvate leads to the production of LA. Many efficient production processes have been established based on a variety of lactic acid bacteria. Budhavaram and Fan (2009) have reported on the production of LA from paper sludge using thermophilic *Bacillus coagulan* strains 36D1 and P4-102B. More than 80% of LA yield and more than 87% of cellulose conversion were achieved. In addition, the presence of calcium carbonate in the sludge may be beneficial for LA fermentation, as they help in maintain the pH of the medium, reducing the addition of buffering reagents (Budhavaram and Fan, 2009). Marques et al. (2017) proposed a novel fed-batch strategy for production of LA from paper recycling mill-recycled paper sludge. Polymeric paper sludge (PPS) can be converted into monosaccharide glucose and xylose, using enzymatic hydrolysis. Then monosaccharide can be converted into LA using *Lactobacillus rhamnosus* ATCC 7469 (Marques et al., 2008a,b). Considering the numerous applications of LA, PPI waste can serve as a potential resource for its production.

4.6.6 Biodiesel

PPI effluents and other waste streams, such as activated sludge, showed potential for the sustainable production of lipids for biofuel production. By inoculating activated sludge into PP wastewater and supplementing glucose and nutrients (nitrogen and phosphorus), lipids were harvested having rich quantities of C16–C18 fatty acids. Finally, 6%–8% of biodiesel was separated from accumulated lipids (Upadhyaya et al., 2013).

PP wastewaters could be good growth media for oleaginous microorganisms for biolipid production, and 53.40% of lipid production was achieved from sonicated paper mill sludge extract (PMSE) by the oleaginous yeast *Cryptococcus vishniaccii* (MTCC 232). This result was better than lipid production from glucose synthetic media, as reported by Deeba et al. (2015). The lipids composing of oleic acid, followed by palmitic acid, linoleic acid, and stearic acid, was further transesterified into good quality biodiesel by reaction. *Nannochloropsis oculata* was cultivated in artificial seawater enriched with effluent from PPI wastewater, and an algal biomass with 42% (w/w) lipids was harvested on the 10th day of treatment (Gurumoorthy and Saravanan, 2016).

4.6.7 Biobutanol

Butanol (C_4H_9OH) can be used as a solvent in the production of antibiotics, hormones, and vitamins. It is an intermediate for the production of butyl acrylates and an ingredient in coatings and adhesives, as well as glycol ethers and butyl acetates (Niemisto et al., 2013). Paper mill sludge was investigated as a feedstock for acetone-butanol-ethanol (ABE) production through SSF by using *Clostridium acetobutylicum* strain without pretreatment and pH control (Guan et al., 2016).

4.6.8 Biofertilizer

Organic waste produced by PPI is a rich source of nutrients that might be exploited for production of biogas and biofertilizers via AD (Gavrilescu, 2008). Thus, AD can be used not only as a waste-to-energy conversion technology, but also to allow nutrients to remain in the digested material. The anaerobically digested materials contains partially degraded organic matter, residue from microflora, inorganic compounds, and nitrogen, phosphorous, and potassium in concentrations ranging from 2.3–4.2, 0.2–1.5, and 1.3–5.2 kg/t, respectively (Alburquerque et al., 2012).

4.6.9 Algal biomass

Black liquor generated from a kraft pulp mill contains 50% wood substance as a dissolved organic fraction, which can be considered as a substrate for microalgae cultivation and is beneficial in harnessing biomass by degrading waste at the same time (Tarlan et al., 2002). PPI wastewaters with the potential for 11,000 mg/l of COD and inorganic salts (NO_3 and PO_4) can be used as a suitable substrate for microalgae biomass cultivation (Thompson et al., 2001; Gentili, 2014). Biomass is a high-value product, as it is a feedstock for fuel generation (Perlack et al., 2005), enhances productivity in fisheries (Spolaore et al., 2006), and is part of a high-protein diet for cattle, among many useful functions. Microalgae species like *Chlorella vulgaris*, *Chlorella sorokaina*, *Chlorella pyrenoidosa*, *Scenedesmus* spp., and *Nannochloropsis* can be used in PPI wastewater treatment to harvest biomass using the mechanized high-rate ponds, photo bioreactors (Wang et al., 2010; Polishchuk et al., 2015). An algal biomass also has the ability to synthesize chlorophyll pigment, proteins, and lipid bodies with minimal external intervention (Harun et al., 2010). Variation in the physiochemical parameters, the algal regime of products in the cell can alter and the product of interest can be harnessed with high biomass productivity, which makes it a potential feedstock for many high-value products (Brennan and Owende, 2010).

5 Possibilities of a PPI waste biorefinery

Biorefineries are emerging industrial systems with a novel approach to attaining sustainability in the present generation. The main aim of biorefineries is to effectively utilize waste in a cascading approach and deliver multiple useful bioenergies and bioproducts. Unlike petroleum refineries, which fractionate oil into transportation fuels and produce petroleum products in separate facilities, biorefineries are designed to integrate several processes that produce multiple products in a single system, with waste as a sole feedstock. PPI waste also can be considered as a potential feedstock for developing a paper and pulp biorefinery (Figs. 2 and 5). An integrated strategy for PPI waste valorization using the biorefinery approach could be developed by interlinking multiple processes, starting with decreasing the chemical load followed by pre-treatment techniques.

Conventional PPI wastewater and sludge management methods include dewatering, landfilling, which produces leachate that contaminates groundwater and surface water, and incineration, which emits greenhouse gases. However, channelizing sludge and effluent from this process can create substrates in a number of routes, such as acidogenic fermentation, biomethanation, anoxygenesis, microbial electrocatalyzed systems (MESs), and photosynthetic routes, in order to generate a wide variety of economically important products (discussed in Section 4). Process integration could

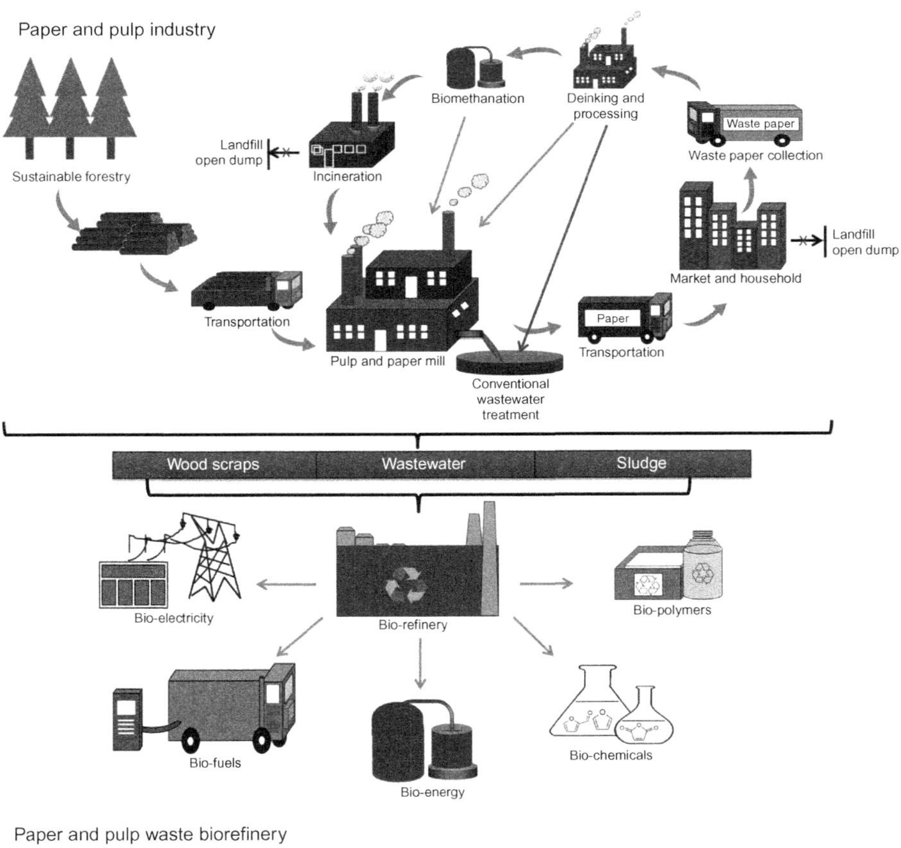

Fig. 5 Prospect of PPI waste biorefinery and value added products generation.

reduce the load of pollution by 20%–60% and could generate an approximately 50% hike in the economy (Birdie and Birdie, 2008).

Because PPI waste (mainly primary sludge and solid fiber waste) is complex (rich in lignocellulosic matter), pretreatment strategies (as discussed in Section 4) are required to obtain simple and easily accessible forms of carbon, including reducing sugars and phenolic and aromatic compounds. These simple sugars then serve as a potential substrate for the production of high-volume biofuels such as biohydrogen, biodiesel, bioethanol, or biogas. Also, high-value bioproducts obtained in comparatively low volumes, such as biochemicals, biopolymers, biofertilizers, and biomaterials, can be produced.

Promising strategies can be employed by using acidogenic fermentation as the focal point, whereas biohydrogen and VFA can be produced from reducing sugar after pretreatment. VFAs can be extracted or used as substrates for bioplastics, bioelectricity production using mixed microbial cultures, or photosynthetic machinery. Effluent from this process could be used as a substrate in the anode chamber of MES, while in the biocathode chamber, CO_2 can be utilized to electrosynthesize biobased platform chemicals. The residual VFAs and sugar obtained could be further utilized as feedstocks for microalgal photosynthesis, to generate biodiesel, polyunsaturated fatty acids, pigments, nutraceuticals, and other high-value products. The deoiled algal cake obtained from microalgal cultivation, which is rich in protein and carbohydrates, could be reused as a carbon source in these processes, thus completely recycling the carbon.

6 Conclusions and future perspectives

The PPI industry is the sixth-largest polluter in the world (Ali and Sreekrishnan, 2001), and it is estimated that global paper consumption can go up to 500 million tons/year by 2025. As paper consumption increases, waste generation is bound to increase as well, indicating that the P&P industry faces a series of obstacles ranging from availability of raw materials to the generation of large amounts of sidestreams. To overcome these challenges, attempting a multipronged intervention that includes conversion of the existing P&P industry to biorefineries could be a worthwhile step (Venkata Mohan et al., 2016a, 2018).

A review of the available literature suggests that research has largely centered around two primary areas: the development of potential new raw materials for papermaking, and the generation of biobased products from waste streams. For the first scenario, the use of new raw materials from plant fibers or from sidestreams of the agro-industrial sector seems to be technically possible. This could in turn suggest that research efforts should focus on developing simpler methods for processing alternative fibers. Conventionally, the waste originating from the P&P industry is being landfilled or incinerated, and both options bear costs for the sector. However, these costs can be minimized and even converted to profits by utilizing the valuable components in waste to create useful products (Venkata Mohan et al., 2016b).

The schematic summarized in Fig. 5 helps illustrate how wastes emerging from the P&P industry can be utilized to their full potential. Utilization of these residues for alternative energy sources and high-value products could improve the sustainability of the bioenergy chain and reduce the negative environmental impacts of inappropriate disposal.

However, the challenge of economically feasible production of bio-based products cannot be blithely dismissed. The economic performance of biorefineries can be improved by enhancing the conversion efficiency and technological advancements for the effective separation of the products. Apart from this, the degradation of lignin during carbon recycling is a rate-limiting step. It has been documented that lignin limits the accessibility of microbial enzymes to other major cell-wall polysaccharides. Because lignin is nontoxic and extremely versatile in its performance, it can be an attractive industrial option if effective and diverse treatment/utilization methods are developed. In Figs. 4 and 5, numerous applications have been highlighted, of which only some have been explored to date (even those have not been examined at scale). The complete potential of P&P biorefineries will come into play only when all energy-rich components (including lignin) are fully researched and utilized. Understanding the complexity of the lignin structure and considering that lignin polymers are equally valuable sources for fuels and chemicals are vital for the advancement of P&P-based biorefneries.

References

Abdel-Rahman, M.A., Tashiro, Y., Sonomoto, K., 2013. Recent advances in lactic acid production by microbial fermentation processes. Biotechnol. Adv. 31 (6), 877–902.

Ahmed, S., Bashir, A., Saleem, H., Saadia, M., Jamil, A., 2009. Production and purification of cellulose-degrading enzymes from a filamentous fungus *Trichoderma harzianum*. Pak. J. Bot. 41 (3), 1411–1419.

Alburquerque, J.A., De la Fuente, C., Campoy, M., Carrasco, L., Nájera, I., Baixauli, C., Caravaca, F., Roldán, A., Cegarra, J., Bernal, M.P., 2012. Agricultural use of digestate for horticultural crop production and improvement of soil properties. Eur. J. Agron. 43, 119–128.

Alexandri, M., Papapostolou, H., Vlysidis, A., Gardeli, C., Komaitis, M., Papanikolaou, S., Koutinas, A.A., 2016. Extraction of phenolic compounds and succinic acid production from spent sulphite liquor. J. Chem. Technol. Biotechnol. 91 (11), 2751–2760.

Ali, M., Sreekrishnan, T.R., 2001. Aquatic toxicity from pulp and paper mill effluents: a review. Adv. Environ. Res. 5 (2), 175–196.

Apruzzese, I., Bhardwaj, V., Degrassi, G., Bhardwaj, R.K., 2017. Biogas production from organic wastes of paper and leather industries. IOSR J. Biotechnol. Biochem. 3 (4), 8–14.

Aro, N., Pakula, T., Penttilä, M., 2005. Transcriptional regulation of plant cell wall degradation by filamentous fungi. FEMS Microbiol. Rev. 29 (4), 719–739.

Badar, S., Farooqi, I.H., 2012. Pulp and paper industry—manufacturing process, wastewater generation and treatment. In: Malik, A., Grohmann, E. (Eds.), Environmental Protection Strategies for Sustainable Development. Springer, Dordrecht, the Netherlands, pp. 397–436.

Bajpai, P., 2015. Generation of waste in pulp and paper mills. In: Management of Pulp and Paper Mill Waste. Springer, pp. 9–17.

Bayr, S., Kaparaju, P., Rintala, J., 2013. Screening pretreatment methods to enhance thermophilic anaerobic digestion of pulp and paper mill waste water treatment secondary sludge, wastes. Chem. Eng. J. 223 (1), 479–486.

Bayr, S., Rantanen, M., Kaparaju, P., Rintala, J., 2012. Mesophilic and thermophilic anaerobic co-digestion of rendering plant and slaughterhouse wastes. Bioresour. Technol. 104, 28–36.

Belghith, H., Ellouz-Chaabouni, S., Gargouri, A., 2001. Biostoning of denims by *Penicillium occitanis* (Pol6) cellulases. J. Biotechnol. 89 (2), 257–262.

Bengtsson, S., Hallquist, J., Werker, A., Welander, T., 2008a. Acidogenic fermentation of industrial wastewaters: effects of chemostat retention time and pH on volatile fatty acids production. Biochem. Eng. J. 40 (3), 492–499.

Bengtsson, S., Werker, A., Christensson, M., Welander, T., 2008b. Production of polyhydroxyalkanoates by activated sludge treating a paper mill wastewater. Bioresour. Technol. 99 (3), 509–516.

Biomass bolsters energy lead with increased electricity use, 2012. Renew. Energy Focus (REF) 13 (4), 44–46.

Birdie, G.S., Birdie, J.S., 2008. Water Supply and Sanitary Engineering. Dhanpat Rai & Sons, Delhi.

Borowski, S., Kubacki, P., 2015. Co-digestion of pig slaughterhouse waste with sewage sludge. Waste Manag. 40, 119–126.

Brennan, L., Owende, P., 2010. Biofuels from microalgae—a review of technologies for production, processing, and extractions of biofuels and co-products. Renew. Sust. Energ. Rev. 14 (2), 557–577.

Budhavaram, N.K., Fan, Z., 2009. Production of lactic acid from paper sludge using acid-tolerant, thermophilic *Bacillus coagulan* strains. Bioresour. Technol. 100 (23), 5966–5972.

Çalgeris, İ., Çakmakçı, E., Ogan, A., Kahraman, M.V., Kayaman-Apohan, N., 2012. Preparation and drug release properties of lignin–starch biodegradable films. Starch-Stärke 64 (5), 399–407.

Cavka, A., Guo, X., Tang, S.J., Winestrand, S., Jönsson, L.J., Hong, F., 2013. Production of bacterial cellulose and enzyme from waste fiber sludge. Biotechnol. Biofuel. 6 (1), 25.

Chakraborty, D., Parthiba Karthikeyan, O., Selvam, A., Wong, J.W.C., 2017. Co-digestion of food waste and chemically enhanced primary treated sludge in a continuous stirred tank reactor. Biomass Bioenergy 111, 232–240.

Chávez, R., Bull, P., Eyzaguirre, J., 2006. The xylanolytic enzyme system from the genus Penicillium. J. Biotechnol. 123 (4), 413–433.

Cheng, K.K., Zhao, X., Zeng, J., Zhang, J., 2012. Biotechnological production of succínic acid: current state and perspectives. Biofuels Bioprod. Biorefin. 6, 302–318.

Choudhury, S., Rohella, R., Manthan, M., Sahoo, N., 1998. Decolourization of kraft paper mill effluent by white-rot fungi. Indian J. Microbiol. 38 (4), 221–224.

Cok, B., Tsiropoulos, I., Roes, A.L., Patel, M.K., 2014. Succinic acid production derived from carbohydrates: an energy and greenhouse gas assessment of a platform chemical toward a bio-based economy. Biofuels Bioprod. Biorefin. 8 (1), 16–29.

Davis, R., Tao, L., Tan, E.C.D., Biddy, M.J., Beckham, G.T., Scarlata, C., Jacobson, J., Cafferty, K., Ross, J., Lukas, J., Knorr, D., 2013. Process design and economics for the conversion of lignocellulosic biomass to hydrocarbons: dilute-acid and enzymatic deconstruction of biomass to sugars and biological conversion of sugars to hydrocarbons. (No. NREL/TP-5100-60223). National Renewable Energy Laboratory (NREL), Golden, CO.

de Alda, J.A.O., 2008. Feasibility of recycling pulp and paper mill sludge in the paper and board industries. Resour. Conserv. Recycl. 52 (7), 965–972.

Deeba, F., Pruthi, V., Negi, Y.S., 2015. Converting paper mill sludge into neutral lipids by oleaginous yeast *Cryptococcus vishniaccii* for biodiesel production. Bioresour. Technol. 213, 96–102.

Dhiman, S.S., Sharma, J., Battan, B., 2008. Industrial applications and future prospects of microbial xylanases: a review. Bioresources 3 (4), 1377–1402.

Draths, K.M., Frost, J.W., 1994. Environmentally compatible synthesis of adipic acid from D-glucose. J. Am. Chem. Soc. 116 (1), 399–400.

Dwivedi, P., Gaur, V., Sharma, A., Verma, N., 2004. Comparative study of removal of volatile organic compounds by cryogenic condensation and adsorption by activated carbon fiber. Sep. Purif. Technol. 39 (1), 23–37.

Elango, B., Rajendran, P., Bornmann, L., 2017. A scientometric analysis of international collaboration and growth of literature at the macro level. Malays. J. Libr. Inf. Sci. 20(2).

Elliott, T., Mahmood, 2007. Pretreatment technologies for advancing anaerobic digestion of pulp and paper biotreatment residues. Water Res. 41 (19), 4273–4286.

Elliott, T., Mahmood, 2012. Comparison of mechanical pretreatment methods for the enhancement of anaerobic digestion of pulp and paper waste activated sludge. Water Environ. Res. 84 (6), 497–505.

Farghaly, A., Tawfik, A., 2017. Simultaneous hydrogen and methane production through multiphase anaerobic digestion of paperboard mill wastewater under different operating conditions. Appl. Biochem. Biotechnol. 181 (1), 142–156.

Gavrilescu, D., 2008. Energy from biomass in pulp and paper mills. Environ. Eng. Manag. J. 7 (5), 537–546.

Gentili, F.G., 2014. Microalgal biomass and lipid production in mixed municipal, dairy, pulp and paper wastewater together with added flue gases. Bioresour. Technol. 169, 27–32.

Guan, W., Shi, S., Tu, M., Lee, Y.Y., 2016. Acetone-butanol-ethanol production from Kraft paper mill sludge by simultaneous saccharification and fermentation. Bioresour. Technol. 200, 713–721.

Gurumoorthy, P., Saravanan, A., 2016. Biofuel production from marine microalgae using paper and pulp industry waste water. Int. J. Chem. Sci. 14 (4), 3249–3255.

Ha, M.A., Apperley, D.C., Evans, B.W., Huxham, I.M., Jardine, W.G., Viëtor, R.J., Reis, D., Vian, B., Jarvis, M.C., 1998. Fine structure in cellulose microfibrils: NMR evidence from onion and quince. Plant J. 16 (2), 183–190.

Hagelqvist, A., 2013a. Batchwise mesophilic anaerobic co-digestion of secondary sludge from pulp and paper industry and municipal sewage sludge. Waste Manag. 33 (4), 820–882.

Hagelqvist, A., 2013b. Sludge from Pulp and Paper Mills for Biogas Production; strategies to improve energy performance in wastewater treatment in sludge management. (doctoral Dissertation).

Harun, R., Singh, M., Forde, G.M., Danquah, M.K., 2010. Bioprocess engineering of microalgae to produce a variety of consumer products. Renew. Sust. Energ. Rev. 14 (3), 1037–1047.

Hay, J.X.W., 2015. Improved biohydrogen production and treatment of pulp and paper mill effluent through ultrasonication pretreatment of wastewater. Energy Convers. Manag. 106, 576–583.

Hendriks, A.T.W.M., Zeeman, G., 2009. Pretreatments to enhance the digestibility of lignocellulosic biomass. Bioresour. Technol. 100 (1), 10–18.

Holladay, J.E., White, J.F., Bozell, J.J., Johnson, D., 2007. Top Value Added Chemicals from Biomass-Volume II, Results of Screening for Potential Candidates from Biorefinery

Lignin (No. PNNL-16983). Pacific Northwest National Lab (PNNL), Richland, WA (United States). National Renewable Energy Laboratory (NREL), Golden, CO.

Huang, J., Zhang, L.N., Chen, P., 2003. Effects of lignin as a filler on properties of soy protein plastics. II. Alkaline lignin. J. Appl. Polym. Sci. 88 (14), 3291–3297.

Huang, L., Cheng, S., Rezaei, F., Logan, B., 2009. Reducing organic loads in wastewater effluents from paper recycling plants using microbial fuel cells. Environ. Technol. 30 (5), 499–504.

Huang, L., Logan, B.E., 2008. Electricity generation and treatment of paper recycling wastewater using a microbial fuel cell. Appl. Microbiol. Biotechnol. 80, 349–355.

Huiliñir, C., Quintriqueo, A., Antileo, C., Montalvo, S., 2014. Methane production from secondary paper and pulp sludge: effect of natural zeolite and modeling. Chem. Eng. J. 257, 131–137.

Jantasee, S., Marlene, K., Mungma, N., Siebenhofer, M., 2017. Potential and assessment of lactic acid production and isolation—a review. J. Chem. Technol. Biotechnol.

Jarpa, M., Pozo, G., Baeza, R., Martínez, M., Vidal, G., 2012. Polyhydroxyalkanoate biosynthesis from paper mill wastewater treated by a moving bed biofilm reactor. J. Environ. Sci. Health A 47 (13), 2052–2059.

Jiang, Y., Marang, L., Tamis, J., Van Loosdrecht, M.C.M., Dijkman, H., Kleerebezem, R., 2012. Waste to resource: converting paper mill wastewater to bioplastic. Water Res. 46, 5517–5530.

Jingjing, L., 2011. Isolation of lignin from wood. Saimaa University of Applied Sciences, Imatra Unit of Technology.

Jokela, J., Rintala, J., Oikari, A., Reinikainen, O., Mutka, K., Nyrönen, T., 1997. Aerobic composting and anaerobic digestion of pulp and paper mill sludges. Water Sci. Technol. 36 (11), 181–188.

Jong, E., Higson, A., Walsh, P., Wellisch, M., 2012. Product developments in the bio-based chemicals arena. Biofuels Bioprod. Biorefin. 6 (6), 606–624.

Kaluža, L., Šuštaršič, M., Rutar, V., Zupančič, G.D., 2014. The re-use of Waste-Activated Sludge as part of a "zero-sludge" strategy for wastewater treatments in the pulp and paper industry. Bioresour. Technol. 151, 137–143.

Kamali, M., Gameiro, T., Costa, M.E.V., Capela, I., 2016. Anaerobic digestion of pulp and paper mill wastes—an overview of the developments and improvement opportunities. Chem. Eng. J. 298, 162–182.

Karn, S.K., Kumar, A., 2015. Hydrolytic enzyme protease in sludge: recovery and its application. Biotechnol. Bioprocess Eng. 20 (4), 652–661.

Karn, S.K., Kumar, P., Pan, X., 2013. Extraction of lipase and protease and characterization of activated sludge from pulp and paper industry. Prep. Biochem. Biotechnol. 43 (2), 152–162.

Kaur, I., Ni, Y., 2015. A process to produce furfural and acetic acid from pre-hydrolysis liquor of kraft based dissolving pulp process. Sep. Purif. Technol. 146, 121–126.

Ketep, S.F., Bergel, A., Bertrand, M., Achouak, W., Fourest, E., 2013a. Sampling location of the inoculum is crucial in designing anodes for microbial fuel cells. Biochem. Eng. J. 73, 12–16.

Ketep, S.F., Fourest, E., Bergel, A., 2013b. Experimental and theoretical characterization of microbial bioanodes formed in pulp and paper mill effluent in electrochemically controlled conditions. Bioresour. Technol. 149, 117–125.

Knob, A., Carmona, E.C., 2010. Purification and characterization of two extracellular xylanases from Penicilliumsclerotiorum: a novel acidophilic xylanase. Appl. Biochem. Biotechnol. 162 (2), 429–443.

Krishna, K.V., Sarkar, O., Venkata Mohan, S., 2014. Bioelectrochemical treatment of paper and pulp wastewater in comparison with anaerobic process: integrating chemical coagulation with simultaneous power production. Bioresour. Technol. 174, 142–151.

Kubo, S., Kadla, J.F., 2005. Lignin-based carbon fibers: effect of synthetic polymer blending on fiber properties. J. Polym. Envron. 13 (2), 97–105.

Kuhad, R.C., Gupta, R., Singh, A., 2011. Microbial cellulases and their industrial applications. Enzyme Res.. 2011280696. https://doi.org/10.4061/2011/280696.

Kumar, S., Bhattacharyya, J.K., Vaidya, A.N., Chakrabarti, T., Devotta, S., Akolkar, A.B., 2009. Assessment of the status of municipal solid waste management in metro cities, state capitals, class I cities, and class II towns in India: an insight. Waste Manag. 29 (2), 883–895.

Kumar, V., Dhall, P., Naithani, S., Kumar, A., Kumar, R., 2014. Biological approach for the treatment of pulp and paper industry effluent in sequence batch reactor. J. Bioremediat. Biodegrad. 5 (3), 1–10.

Kyllönen, H.L., Lappi, M.K., Thun, R.T., Mustranta, A.H., 1988. Treatment and characterization of biological sludges from the pulp and paper industry. Water Sci. Technol. 20 (1), 183–192.

Lakshmidevi, R., Muthukumar, K., 2010. Enzymatic saccharification and fermentation of paper and pulp industry effluent for biohydrogen production. Int. J. Hydrog. Energy 35 (8), 3389–3400.

Laycock, B., Pratt, S., Halley, P., Werker, A., Lant, P., 2014. Biodegradable polymers from pulp and paper wastewater streams—a critical review. Appita J. 67 (4), 309–315.

Lee, P.C., Lee, W.G., Kwon, S., Lee, S.Y., Chang, H.N., 2000. Batch and continuous cultivation of *Anaerobiospirillum succiniciproducens* for the production of succinic acid from whey. Appl. Microbiol. Biotechnol. 54 (1), 23–27.

Lepifre, S., Froment, M., Cazaux, F., Houot, S., Lourdin, D., Coqueret, X., Lapierre, C., Baumberger, S., 2004. Lignin incorporation combined with electron-beam irradiation improves the surface water resistance of starch films. Biomacromolecules 5 (5), 1678–1686.

Lin, J., Zuo, J., Gan, L., Li, P., Liu, F., Wang, K., Chen, L., Gan, H., 2011. Effects of mixture ratio on anaerobic co-digestion with fruit and vegetable waste and food waste of China. J. Environ. Sci. 23 (8), 1403–1408.

Lin, T.H., Huang, C.F., Guo, G.L., Hwang, W.S., Huang, S.L., 2012. Pilot-scale ethanol production from rice straw hydrolysates using xylose-fermenting *Pichia stipitis*. Bioresour. Technol. 116, 314–319.

Lin, Y., Wang, D., Wu, S., Wang, C., 2009. Alkali pretreatment enhances biogas production in the anaerobic digestion of pulp and paper sludge. J. Hazard. Mater. 170 (1), 366–373.

Lin, Y., Wu, S., Wang, D., 2013. Hydrogen-methane production from pulp & paper sludge and food waste by mesophilic–thermophilic anaerobic co-digestion. Int. J. Hydrog. Energy 38 (35), 15055–15062.

Mao, C., Feng, Y., Wang, X., Ren, G., 2015. Review on research achievements of biogas from anaerobic digestion. Renew. Sust. Energ. Rev. 45, 540–555.

Marques, S., Alves, L., Roseiro, J.C., Gírio, F.M., 2008a. Conversion of recycled paper sludge to ethanol by SHF and SSF using *Pichia stipitis*. Biomass Bioenergy 32 (5), 400–406.

Marques, S., Gírio, F.M., Santos, J.A.L., Roseiro, J.C., 2017. Pulsed fed-batch strategy towards intensified process for lactic acid production using recycled paper sludge. Biomass Convers. Biorefinery 7 (2), 127–137.

Marques, S., Santos, J.A., Gírio, F.M., Roseiro, J.C., 2008b. Lactic acid production from recycled paper sludge by simultaneous saccharification and fermentation. Biochem. Eng. J. 41 (3), 210–216.

Martinez, F.A.C., Balciunas, E.M., Salgado, J.M., González, J.M.D., Converti, A., Oliveira, R.P.D.S., 2013. Lactic acid properties, applications and production: a review. Trends Food Sci. Technol. 30, 70–83.

McGuire, M.J., Suffet, I.H., 1980. Activated carbon adsorption of organics from the aqueous phase. Two Volumes. Ann Arbor Science Publishers, Ann Arbor.

Meyer, T., Edwards, E.A., 2014. Anaerobic digestion of pulp and paper mill wastewater and sludge. Water Res. 65, 321–349.

Modi, D.R., Chandra, H., Garg, S.K., 1998. Decolourization of bagasse-based paper mill effluent by the white-rot fungus Trametes versicolor. Bioresour. Technol. 66 (1), 79–81.

Monte, M.C., Fuente, E., Blanco, A., Negro, C., 2009. Waste management from pulp and paper production in the European Union. Waste Manag. 29 (1), 293–308.

Morais, L.C., Curvelo, A.A.S., Zambon, M.D., 2005. Thermoplastic starch—lignosulfonate blends. Factorial planning as a tool for elucidating new data from high performance size-exclusion chromatography and mechanical tests. Carbohydr. Polym. 62, 104–112.

Moreau, A., Montplaisir, D., Barnabé, S., 2015. Hydrogen, ethanol and cellulase production from pulp and paper primary sludge by fermentation with *Clostridium thermocellum*. Biomass Bioenergy 72, 256–262.

Morone, A., Apte, M., Pandey, R.A., 2015. Levulinic acid production from renewable waste resources: bottlenecks, potential remedies, advancements and applications. Renew. Sust. Energ. Rev. 51, 548–565.

Mousavioun, P., George, G.A., Doherty, W.O.S., 2012. Environmental degradation of lignin/poly (hydroxybutyrate) blends. Polym. Degrad. Stab. 97 (7), 1114–1122.

Mussoline, W., Esposito, G., Lens, P., Spagni, A., Giordano, A., 2013. Enhanced methane production from rice straw co-digested with anaerobic sludge from pulp and paper mill treatment process. Bioresour. Technol. 148, 135–143.

Nagar, S., Jain, R.K., Thakur, V.V., Gupta, V.K., 2013. Biobleaching application of cellulase poor and alkali stable xylanase from *Bacillus pumilus* SV-85S. Biotech 3 (4), 277–285.

Niemisto, J., Saavalainen, P., Pongrácz, E., Keiski, R.L., 2013. Biobutanol as a potential sustainable biofuel—assessment of lignocellulosic and waste-based feedstocks. J. Sustain. Dev. Energy Water Environ. Syst. 1 (2), 58–77.

Nizami, A.S., Korres, N.E., Murphy, J.D., 2009. Review of the integrated process for the production of grass biomethane. Environ. Sci. Technol. 43 (22), 8496–8508.

Pandey, A., Selvakumar, P., Soccol, C.R., Nigam, P., 1999. Solid state fermentation for the production of industrial enzymes. Curr. Sci. 149–162.

Parameswaran, P., Rittmann, B.E., 2012. Feasibility of anaerobic co-digestion of pig waste and paper sludge. Bioresour. Technol. 124, 163–168.

Park, Y., Doherty, W.O., Halley, P.J., 2008. Developing lignin-based resin coatings and composites. Ind. Crop. Prod. 27 (2), 163–167.

Pečar, D., Goršek, A., 2015. Enzymatic Treatment of Paper Sludge: Chemical Engineering Transactions. vol. 43. The Italian Association of Chemical Engineering. Available at: www.aidic.it/cet.

Peng, L., Chen, Y., 2011. Conversion of paper sludge to ethanol by separate hydrolysis and fermentation (SHF) using Saccharomyces cerevisiae. Biomass Bioenergy 35 (4), 1600–1606.

Peng, L., Lin, L., Zhang, J., Zhuang, J., Zhang, B., Gong, Y., 2010. Catalytic conversion of cellulose to levulinic acid by metal chlorides. Molecules 15 (8), 5258–5272.

Perlack, R.D., Wright, L.L., Turhollow, A.F., Graham, R.L., Stokes, B.J., Erbach, D.C., 2005. Biomass as Feedstock For a Bioenergy and Bioproducts Industry: The Technical Feasibility of a Billion-ton Annual Supply. Oak Ridge National Lab, Oak Ridge, TN.

Persson, P.O., 2011. Cleaner Production: Strategies & Technology for Environmental Production. Royal Institute of Technology—Industrial Ecology, Stockholm.

Polen, T., Spelberg, M., Bott, M., 2013. Toward biotechnological production of adipic acid and precursors from biorenewables. J. Biotechnol. 167 (2), 75–84.

Polishchuk, A., Valev, D., Tarvainen, M., Mishra, S., Kinnunen, V., Antal, T., Yang, B., Rintala, J., Tyystjärvi, E., 2015. Cultivation of Nannochloropsis for eicosapentaenoic acid production in wastewaters of pulp and paper industry. Bioresour. Technol. 193, 469–476.

Prasetyo, J., Park, E.Y., 2013. Waste paper sludge as a potential biomass for bio-ethanol production. Korean J. Chem. Eng. 30 (2), 253–261.

Priadi, C., Wulandari, D., Rahmatika, I., Moersidik, S.S., 2014. Biogas production in the anaerobic digestion of paper sludge. In: APCBEE Procediapp. 65–69.

Puhakka, J.A., Alavakeri, M., Shieh, W.K., 1992. Anaerobic treatment of kraft pulp-mill waste activated-sludge: gas production and solids reduction. Bioresour. Technol. 39 (1), 61–68.

Pye, E.K., 2008. Industrial lignin production and applications. In: Biorefineries-Industrial Processes and Products: Status Quo and Future Directions, pp. 165–200.

Queirós, D., Rossetti, S., Serafim, L.S., 2014. PHA production by mixed cultures: a way to valorize wastes from pulp industry. Bioresour. Technol. 157, 197–205.

Queirós, D., Sousa, R., Pereira, S., Serafim, L.S., 2017. Valorization of a pulp industry by-product through the production of short-chain organic acids. Fermentation 3 (2), 20.

Robertson, S., Kirsten, 2000. Water and waste-water management in the paper and pulp industry prepared for the water research commission. Water Research Commission, Johannesburg. WRC PROJECT No. 145 TT 49/90. http://www.wrc.org.za/Knowledge%20Hub%20Documents/Research%20Reports/TT-49-90).pdf.

Sadhasivam, S., Savitha, S., Swaminathan, S., 2010. Deployment of *Trichoderma harzianum* WL1 laccase in pulp bleaching and paper industry effluent treatment. J. Clean. Prod. 18, 799–806.

Saha, M., Eskicioglu, C., Marin, J., 2011. Microwave, ultrasonic and chemo-mechanical pretreatments for enhancing methane potential of pulp mill wastewater treatment sludge. Bioresour. Technol. 102 (17), 7815–7826.

Selvam, K., Swaminathan, K., Song, M.H., Chae, K.S., 2002. Biological treatment of a pulp and paper industry effluent by Fomeslividus and Trametes versicolor. World J. Microbiol. Biotechnol. 18 (6), 523–526.

Sharma, R., Chandra, S., Singh, A., Singh, K., 2014. Degradation of pulp and paper mill effluents. IIOAB J. 5 (3), 6.

Sindhu, I., Chhibber, S., Capalash, N., Sharma, P., 2006. Production of cellulase-free xylanase from Bacillus megaterium by solid state fermentation for biobleaching of pulp. Curr. Microbiol. 53 (2), 167–172.

Singh, N.K., Kazmi, A.A., Starkl, M., 2015. A review on full-scale decentralized wastewater treatment systems: techno-economical approach. Water Sci. Technol. 71 (4), 468–478.

Singh, R.P., Agrawal, M., 2008. Potential benefits and risks of land application of sewage sludge. Waste Manag. 28 (2), 347–358.

Singhania, R.R., Sukumaran, R.K., Patel, A.K., Larroche, C., Pandey, A., 2010. Advancement and comparative profiles in the production technologies using solid-state and submerged fermentation for microbial cellulases. Enzym. Microb. Technol. 46 (7), 541–549.

Song, H., Lee, S.Y., 2006. Production of succinic acid by bacterial fermentation. Enzym. Microb. Technol. 39 (3), 352–361.

Spolaore, P., Joannis-Cassan, C., Duran, E., Isambert, A., 2006. Commercial applications of microalgae. J. Biosci. Bioeng. 101 (2), 87–96.

Stoica, M., Sandberg, O., Holby, 2009. Energy use and recovery strategies within wastewater treatment and sludge handling at pulp and paper mills. Bioresour. Technol. 100 (2009), 3497–3505.

Subashini, L.M., 2015. Review on Biological Treatment processes of Pulp and Paper Industry Waste Water. Int. J. Innov. Res. Sci. Eng. Technol. 5 (4), 3721–3725.

Sumathi, S., Hung, Y.T., 2006. Treatment of pulp and paper mill wastes. In: Wang, L.K., Hung, Y.T., Lo, H.H., Yapijakis, C. (Eds.), Waste Treatment in the Process Industries. Taylor&Francis, New York. ISBN 0-8493-7233-X, pp. 453–497.

Taherzadeh, M.J., Karimi, K., 2008. Pretreatment of lignocellulosic wastes to improve ethanol and biogas production: a review. Int. J. Mol. Sci. 9 (9), 1621–1651.

Tarlan, E., Dilek, F.B., Yetis, U., 2002. Effectiveness of algae in the treatment of a wood-based pulp and paper industry wastewater. Bioresour. Technol. 84 (1), 1–5.

Tewari, P.K., Batra, V.S., Balakrishnan, M., 2009. Efficient water use in industries: cases from the Indian agro-based pulp and paper mills. J. Environ. Manag. 90, 265–273.

Thakur, V.K., Thakur, M.K., Raghavan, P., Kessler, M.R., 2014. Progress in green polymer composites from lignin for multifunctional applications: a review. ACS Sustain. Chem. Eng. 2 (5), 1072–1092.

Thompson, G., Swain, J., Kay, M., Forster, C.F., 2001. The treatment of pulp and paper mill effluent: a review. Bioresour. Technol. 77, 275–286.

Trulli, E., Torretta, V., 2015. Influence of feeding mixture composition in batch anaerobic co-digestion of stabilized municipal sludge and waste from dairy farms. Environ. Technol. 36 (12), 1519–1528.

Tsang, Y.F., Hua, F.L., Chua, H., Sin, S.N., Wang, Y.J., 2007. Optimization of biological treatment of paper mill effluent in a sequencing batch reactor. Biochem. Eng. J. 34 (3), 193–199.

Upadhyaya, A., Lama, J.P., Tawata, S., 2013. Utilization of pineapple waste: a review. J. Food Sci. Technol. Nepal 6, 10–18.

Vamsi Krishna, K.V., Sarkar, O., Mohan, S.V., 2014. Bioelectrochemical treatment of paper and pulp wastewater in comparison with anaerobic process: integrating chemical coagulation with simultaneous power production. Bioresour. Technol. 174, 142–151.

Vardon, D.R., Franden, M.A., Johnson, C.W., Karp, E.M., Guarnieri, M.T., Linger, J.G., Salm, M.J., Strathmann, T.J., Beckham, G.T., 2015. Adipic acid production from lignin. Energy Environ. Sci. 8 (2), 617–628.

Velasquez-Orta, S.B., Head, I.M., Curtis, T.P., Scott, K., 2011. Factors affecting current production in microbial fuel cells using different industrial wastewaters. Biores. Technol. 102, 5105–5112.

Veluchamy, C., Kalamdhad, A.S., 2017. Influence of pretreatment techniques on anaerobic digestion of pulp and paper mill sludge: a review. Bioresour. Technol. 245, 1206–1219.

Venkata Mohan, S., Butti, S.K., Amulya, K., Dahiya, S., Modestra, J.A., 2016a. Waste biorefinery: a new paradigm for a sustainable bioelectro economy. Trends Biotechnol. 34 (11), 852–855.

Venkata Mohan, S., Chiranjeevi, P., Dahiya, S., Kumar, N., 2018. Waste derived bioeconomy in India: a perspective. New Biotechnol. 40 (Part A), 60–69.

Venkata Mohan, S., Modestra, J.A., Amulya, K., Butti, S.K., Velvizhi, G., 2016b. A circular bioeconomy with biobased products from CO_2 sequestration. Trends Biotechnol. 34 (6), 506–519.

Venkata Mohan, S., Nikhil, G.N., Chiranjeevi, P., Reddy, C.N., Rohit, M.V., Kumar, A.N., Sarkar, O., 2016c. Waste biorefinery models towards sustainable circular bioeconomy: critical review and future perspectives. Bioresour. Technol. 215, 2–12.

Venkata Mohan, S., Velvizhi, G., Krishna, K.V., Lenin Babu, M., 2014. Microbial catalyzed electrochemical systems: a bio-factory with multi-facet applications. Bioresour. Technol. 165, 355–364.

Virk, A.P., Sharma, P., Capalash, N., 2012. Use of laccase in pulp and paper industry. Biotechnol. Prog. 28 (1), 21–32.

Vlyssides, A.G., Karlis, P.K., 2004. Thermal-alkaline solubilization of waste activated sludge as a pre-treatment stage for anaerobic digestion. Bioresour. Technol. 91 (2), 201–206.

Wang, L., Min, M., Li, Y., Chen, P., Chen, Y., Liu, Y., Wang, Y., Ruan, R., 2010. Cultivation of green algae Chlorella sp. in different wastewaters from municipal wastewater treatment plant. Appl. Biochem. Biotechnol. 162 (4), 1174–1186.

Werpy, T., Petersen, G., 2004. Top Value Added Chemicals from Biomass. USDOE, Washington, DC.

Wu, R.L., Wang, X.L., Li, F., Li, H.Z., Wang, Y.Z., 2009. Green composite films prepared from cellulose, starch and lignin in room-temperature ionic liquid. Bioresour. Technol. 100 (9), 2569–2574.

Xie, K., Lin, H.J., Mahendran, B., Bagley, D.M., Leung, K.T., Liss, S.N., Liao, B.Q., 2010. Performance and fouling characteristics of a submerged anaerobic membrane bioreactor for evaporator condensate treatment. Environ. Technol. 31 (5), 511–521.

Xu, C.C., Lancaster, J., 2009. Treatment of secondary sludge for energy recovery. In: Energy Recovery. Nova Science Publishers Inc, New York, pp. 187–211.

Yalcinkaya, S., Malina, J.F., 2015. Model development and evaluation of methane potential from anaerobic co-digestion of municipal wastewater sludge and un-dewatered grease trap waste. Waste Manag. 40, 53–62.

Yan, Q., Zhao, M., Miao, H., Ruan, W., Song, R., 2010. Coupling of the hydrogen and polyhydroxyalkanoates (PHA) production through anaerobic digestion from Taihu blue algae. Bioresour. Technol. 101 (12), 4508–4512.

Yang, G., Jahan, M.S., Ahsan, L., Zheng, L., Ni, Y., 2013. Recovery of acetic acid from pre-hydrolysis liquor of hardkraft-based dissolving pulp production process by reactive extraction with triisooctylamine. Bioresour. Technol. 138, 253–258.

Yun, Y.M., Jung, K.W., Kim, D.H., Cho, S.K., Shin, H.S., 2013. Synergistic enhancement of hydrolytic enzyme activities on anaerobic co-digestion. In: Paper Presented at World Congress on Anaerobic Digestion (Santiago de Compostela, Spain.

Yunqin, L., Dehan, W., Lishang, W., 2010. Biological pretreatment enhances biogas production in the anaerobic digestion of pulp and paper sludge. Waste Manag. Res. 28 (9), 800–810.

Zhang, L., Xu, C.C., Champagne, P., 2010. Energy recovery from secondary pulp/paper-mill sludge and sewage sludge with supercritical water treatment. Bioresour. Technol. 101 (8), 2713–2721.

Zhang, X., Tu, M., Paice, M.G., 2011. Routes to potential bioproducts from lignocellulosic biomass lignin and hemicelluloses. BioEnergy Res. 4 (4), 246–257.

Zheng, Y., Zhao, J., Xu, F., Li, Y., 2014. Pretreatment of lignocellulosic biomass for enhanced biogas production progress in energy combustion. Science 42, 35–53.

Further reading

Consulting Engineers, 2001. Water and waste-water management in the paper and pulp industry. Prepared for the Water Research Commission by WRC Project No. 145 NATSURV12. Steffen Robertson and Kirsten Inc, Johannesburg.

Diogo, Q., Rossetti, S., Serafim, L.S., 2014. PHA production by mixed cultures: a way to valorize wastes from pulp industry. Bioresour. Technol. 157, 197–205.

Kadla, J.F., Kubo, S., Venditti, R.A., Gilbert, R.D., Compere, A.L., Griffith, W., 2002. Lignin-based carbon fibers for composite fiber applications. Carbon 40 (15), 2913–2920.

Venkata Mohan, S., Rao, N.C., Sarma, P.N., 2007. Simulated acid azo dye (Acid black 210) wastewater treatment by periodic discontinuous batch mode operation under anoxic–aerobic–anoxic microenvironment conditions. Ecol. Eng. 31 (4), 242–250.

Phosphorus extraction and sludge dissolution

28

Marzena Smol
Mineral and Energy Economy Research Institute, Polish Academy of Sciences, Cracow, Poland

1 Introduction

Phosphorus (P) is an essential element for human nutrition and therefore is irreplaceable (Adam et al., 2002). It is one of the most important mineral resources, as it determines the development of agricultural production, stimulated by the food needs of the world's growing population (Le Corre et al., 2009; Arnout and Nagels, 2016). In addition, mineral phosphate minerals are a nonrenewable resource, and their stocks have gradually been depleted as a result of the current intensive level of exploitation (Cornel and Schaum, 2009; Adam et al., 2009; Krüger and Adam, 2017).

Phosphorus has great biological importance, and it is indispensable for agricultural production (Adam et al., 2002; Gorazda et al., 2003). In the fertilizer industry, phosphores, which are the main source of phosphorus in nature and which constitute the basic raw material for its production, including the production of phosphate fertilizers (superphosphate) and mineral fertilizers, are particularly significant. It should be pointed that living organisms depend on access to phosphate fertilizers. This is related to the fact that the human population is systematically growing and is expected to keep growing, but a high production of food can be achieved only by the use of fertilizers such as phosphorus (Jørgensen et al., 2018). In the European Union (EU), the population in 2016 (510,277,177) was only 1.6% higher than in 2009 (502,090,235). However, it has led to a much greater increase in the consumption of phosphorus fertilizers in EU countries in recent years; in 2016, it was 12.3% higher than in 2009. The consumption of phosphorus mineral fertilizers in the European Union in 2009–2016 is shown in Fig. 1 (Eurostat, 2018).

Europe has very limited primary phosphorus resources (Scholz and Wellmer, 2013), and so it is dependent on outside providers. Currently, approximately 88% of phosphate rock and 100% of phosphorus is imported to European countries (COM No. 490, 2017). Phosphate rock is mainly imported from Morocco, Russia, Algeria, and Syria. As for phosphorus imports, the leader is Kazakhstan (almost 80% of phosphorus imported into Europe), followed by China and Vietnam. European production of phosphorus is only carried out in Finland, but it has less than 1% of the world's phosphate resources and could cover only about 8% of the EU demand. The procurement is getting phosphorus difficult by the fact that some of the main importers are considered politically risky (e.g., EU and Russia), which can cause major problems for the European economy (Smol et al., 2016). Because about 90% of phosphorus is

Industrial and Municipal Sludge. https://doi.org/10.1016/B978-0-12-815907-1.00028-3

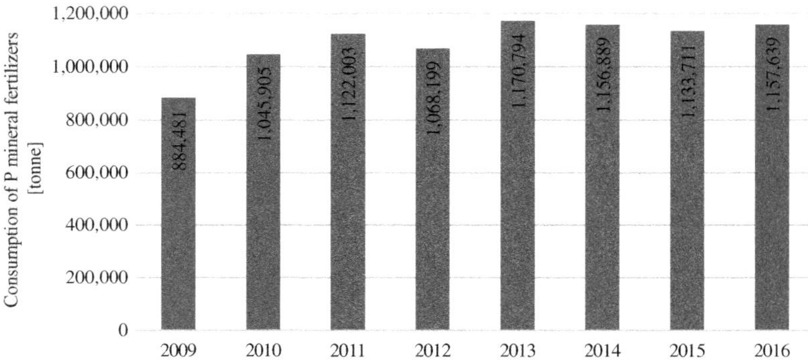

Fig. 1 Consumption of phosphorus mineral fertilizer in the European Union from 2009 to 2016.

used to produce food for humans and animals (Science for Environment Policy Report, 2017), and in order to keep the security of this material and to meet the needs of current and future generations, it is recommended to search for ways to recover phosphorus from secondary materials, which are available on the internal market (COM No. 2020, 2010).

According to the newest EU economic strategy, moving toward a more circular economy (CE) in all possible branches of industry is an essential way to achieve the agenda established by the Europe 2020 Strategy for smart, sustainable, and inclusive growth (COM No. 2020, 2010). Due to the implementation of CE assumptions (COM No. 398, 2014c; COM No. 614, 2015; Smol et al., 2017) in the wastewater and fertilizer sectors (COM No. 157, 2016), it is expected that investments designed to recover phosphorus from sewage sludge (SS) and sewage sludge ash (SSA) will be undertaken in European countries in the coming years. Moreover, due to the limited natural resources of phosphorus recorded in Europe, special attention should be paid to recovering this material from internal waste resources. The objective of this chapter is to present SS and SSA from wastewaster as possible sources of phosphorus in the European market. The structure of the chapter is as follows:

- Clarify the importance of recycling of phosphorus from waste streams.
- Give an overview of possible sources of phosphorus in the wastewater sector.
- Review phosphorus extraction from SS.
- Review phosphorus extraction from SSA.
- Present conclusions.

2 Waste as a source of phosphorus

The issue of more sustainable use of phosphorus was undertaken by the European Commission in 2013 in a report called *Consultative Communication on the Sustainable Use of Phosphorus* (COM No. 517, 2013). The objective of this document was to draw attention to the sustainability of phosphorus use and to initiate debate on the current situation and the actions that should be considered in the future. The European Union pointed out that special attention should be paid to improving the recovery

Table 1 Indicators for the evaluation of CRM—phosphorus (COM No. 490, 2017)

Raw material	Import reliance rate[a]	Substitution indexes EI/SR[b]	End-of-life recycling input rate[c]
Phosphate rock	88%	1,0/1,0	0,3%
Phosphorus	100%	0,91/0,91	17%

[a]Takes into account global supply and actual EU sourcing in the calculation of supply risk, calculated as follows: EU net imports/(EU net imports + EU domestic production).
[b]A measure of the difficulty in substituting the material, scored and weighted across all applications, calculated separately for both economic importance and supply risk parameters. Values are between 0 and 1, with 1 being the least substitutable.
[c]Measures the ratio of recycling from old scrap to EU demand for a given raw material, with the latter being equal to primary and secondary material supply inputs to the European Union.

of phosphorus from waste due to sustainable use of phosphorus goes beyond the issues around that one element. When phosphorus is wasted, the energy, water, and other valuable resources contributing to its production cycle are wasted along with it. Moreover, phosphorus causes such environmental problems as eutrophication (Tong and Chen, 2009; Sengupta and Pandit, 2011). In Europe, more and more attention has been focused on several forms of phosphorus. Since 2014, phosphate rock is listed (COM No. 297, 2014b) as one of the critical raw materials (CRMs) for the European economy. On the updated list in 2017 (COM No. 490, 2017), pure phosphorus is indicated as an important source of phosphates, second only to phosphate rock. In this document, the estimated values of the end-of-life recycling input rate for phosphate rock is 17%, and for phosphorus, it is 0.3% (Table 1).

Because phosphorus recovery is possible from the technical point of view, the end-of-life recycling input rates for Europe should be higher than at the moment. Research and development and recovery and recycling policies are needed in this area. Therefore, the inclusion of phosphorus in the EU CRM list should cause the development of EU policies promoting the sustainable management of phosphorus materials in the near future (Smol et al., 2016). It also needs to be underlined that Europe is not the only one that should take action to maintain secure stockpiles of phosphorus. There is no single solution to achieving a phosphorus-secure future: in addition to increasing use efficiency, phosphorus will need to be recovered and reused from all current waste streams for food production and consumption system throughout the world (Cordell et al., 2011; Mihelcic et al., 2011; Desmidt et al., 2015).

There are a number of secondary materials with high potential for recovering phosphorus, including municipal and industrial wastewater (Adam et al., 2002; Adam et al., 2009; Krüger et al., 2014; Egle et al., 2016), SS (Gorazda et al., 2003; Adam, 2009; Ali and Kim, 2016; Vardanyan et al., 2018), SSA (Kabbe, 2013; Krüger and Adam, 2015; Herzel et al., 2016; Fang et al., 2018), meat and bone meal (Jeng et al., 2004; Ylivainio et al., 2008), pig slurry (Kowalski et al., 2013), biomass (Tan and Lagerkvist, 2011), and industrial waste (phosphogypsum waste, to be exact) (Kulczycka et al., 2016). Domestic waste, especially SS, contains large amounts of phosphorus, which if recycled in line with a CE model, could cover about 20%–30% of the European Union's need for phosphate fertilizers (COM No. 157, 2016).

Significant discrepancies exist among the member-states in the European Union in terms of how much waste in the wastewater sector are used (either directly as SS, or in the form of SSA). It offers the potential for harmonization around best practices (COM No. 517, 2013). Currently, the related investment potential remains largely unexploited, and full-scale phosphorus recovery technologies in the wastewater sector in Europe were implemented only in Germany (6 from digestate, 2 from liquor), Belgium (5 from liquor), United Kingdom (2 from liquor), Denmark (4 from liquor), the Netherlands (11 from liquor, 2 from digestate), France (1 from ash), Italy (1 from liquor), and Spain (1 from liquor) (P-REX Project, 2018). In Poland, there is only one installation under construction for struvite production at the sewage treatment plant in Cielcza. The investment is realized by the Jarocin Waterworks Company. Although presented phosphorus recovery industrial technologies are already on-stream and used to varying degrees (Smol et al., 2015a), there is no common strategy to promote the use of such renewable sources by farmers (Smol et al., 2016). The problem stems from the fact that the price of recovered fertilizer is generally higher than the price of mineral phosphate fertilizer. The European Commission indicated that much more could be done to identify markets for recycled phosphorus and barriers to its increased use, as well as implementing the technologies that already exist (COM No. 157, 2016).

3 Phosphorus recovery in the wastewater sector

The presence of phosphorus and nitrogen (N) in discharged wastewater has become an emerging worldwide concern because it causes eutrophication in natural water (Yamashita and Yamamoto-Ikemoto, 2014). The consequence of the increasing consumption of phosphate rocks, wastewater containing large quantities of phosphorus is deemed as a potential source of phosphorus recovery. The recovery of this element from wastewater could be done by via physical, biological, and chemical approaches, including chemical precipitation, crystallization, adsorption and ion exchange processes, membrane processes, electrochemical processes, and biological processes (Peng et al., 2018). Recovery potential appears in several ways in wastewater treatment plants (WWTPs), as wastewater outflow from the treatment plant, sedimentary liquid (leachate), dehydrated SS, and SSA (Fig. 2).

In each of the subsequent processes of wastewater treatment and SS processing, a smaller volume of the substrate used for the recycling of phosphorus is obtained; however, the concentration of this element per unit volume is increasing; it reaches up to 64 grams of phosphorus per kilogram of dry matter (d.m.). The recovery potential of phosphorus from selected waste streams in WWTPs is presented in Table 2 (Podewils, 2014).

Because the major part of the phosphate from wastewater is transferred to the sludge (i.e., approximately 90%), one of the most promising phosphorus-rich residue sources are SS and SSA (Herzel et al., 2016). The recovery practices for this two types of waste are described in the following sections.

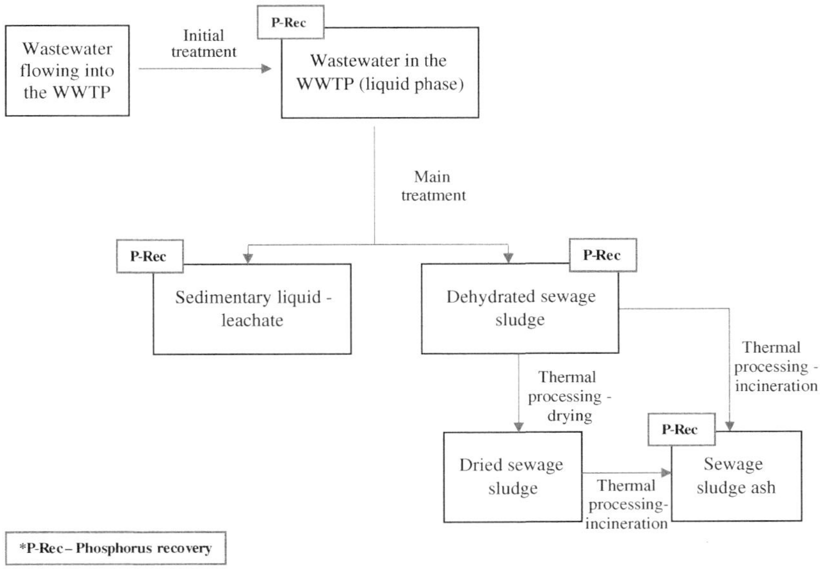

Fig. 2 Concept of phosphorus recovery in municipal WWTPs.

Table 2 Places of possible phosphorus recovery in WWTPs (Kasprzyk et al., 2017; Xu et al., 2012)

Place in WWTP	Concentration of phosphorus	Form	Recovery potential
Wastewater outflow from the treatment plant	<0.5 mg/l	Dissolved	45%–55%
Sedimentary liquid (leachate)	20–100 mg/l	Dissolved	45%–50%
Dehydrated SS	10 g/kg d.m.	Biologically/ chemically bound	50%–60%
SSA	64 g/kg d.m.	Chemically bound	About 90%–95%

3.1 Phosphorus extraction from SS

The extraction of phosphorus from sludge has high potential for meeting the growing demand of phosphorus for fertilizer and commercial products like detergent and phosphoric acid. Moreover, the side products obtained from phosphorus recovery from sludge can be used as additives for cement manufacturing (Chapagain, 2016). Phosphoric minerals may be precipitated in the form of struvites, hydroxyapatites, or calcium phosphates (Cieślik and Konieczka, 2017).

Phosphorus extraction and sludge dissolution by wastewater sludge treated by wet chemicals (acids and alkalis) was conducted by Ali and Kim (2016). These authors have indicated the following method of extraction of selected forms of phosphorus:

- Inorganic phosphorus (IP) was extracted with the use of 1 NHCl for 16h.
- The residual after IP extraction was calcined for 3h at 450°C and again extracted with 1 N hydrochloric acid (HCl) in order to identify organic phosphorus (OP).
- Nonapatite inorganic phosphorus (NAIP) associated with oxides and hydroxides of Fe, Al, Mg, or Mn (Fe/Al-P) was extracted with the use of 1 N sodium hydroxide (NaOH) for 16h. Then, part of the obtained extract was treated with 3.5 N HCl.
- The residue of this NAIP extraction was extracted for the identification of apatite phosphorus (AP) associated with Ca (Ca-P) with 1 N HCl for 16h.
- The extracted solution from the previous step was measured for total phosphorus (TP) analysis.

Phosphorus dissolution was conducted by using acid and alkali treatments of polyaluminum chloride (PAC) at five concentrations of HCl and NaOH: 0.1 N, 0.2 N, 0.5 N, 1.0 N, and 2.0 N. In the leaching experiment, 10g of dried, PAC-treated sludge was mixed with 100mL of acid or alkali, for a liquid-solid (L/S) ratio of 10, in a 300-mL flask for phosphorus extraction. After that, the solution was immediately centrifuged at 10,000g for 10min and the supernatant was separated. The residual solid was washed by the respective extractant again in order to leach any remaining phosphorus, and the TP in the supernatant was measured. The sludge residue was dried at 105°C and weighed in order to calculate sludge dissolution after acid/alkali leaching. An analysis of pH changes should be done before and after the leaching. Sequential phosphorus extraction was conducted in order to measure the acid and alkali releasable fraction of phosphorus. In a sequential extraction, dried, PAC-treated sludge residue after acid treatment was treated with alkali, and vice versa, for extracting the remaining phosphorus in the sludge residue. The following equation is used to calculate the sludge dissolution:

$$\text{Sludge dissolution}\,(\%) = \frac{S_f}{S_t} \times 100 \tag{1}$$

where S_f = sludge dry weight (g) after leaching and S_i = sludge dry weight (g) before leaching.

It is also possible to analyze the kinetics of phosphorus release and sludge dissolution with the first-order kinetic model. To determine the kinetics of phosphorus release, the following equation is used:

$$C_t = C_o\left(1 - e^{-\mathrm{k}_1 t}\right) \tag{2}$$

and for the kinetics of sludge dissolution, the following equation is used:

$$C'_t = C'_o e^{-\mathrm{k}_2 t} \tag{3}$$

where

C_t = concentration of solid (mM) at time t,
C_o = initial concentration of phosphorus (mM),
C'_t = concentration of sludge in TS solid (g/L) at time t
C'_o = initial concentration of sludge in TS solid (g/L),
k_1 = phosphorus release rate constant at room temperature
and k_2 = sludge dissolution rate constant at room temperature.

The rate constants (k_1, k_2) can be calculated by statistical data analysis and scientific graphic software—for example: SigmaPlot, V. 10.

In this study, sludge fractionation showed that NAIP was the dominant phosphorus (90.9% of TP), while AP reached only 3.7%. It was indicated that after 2 h of extraction with 1 N NaOH or 2 N HCl, 80.5% or 77.9% of total phosphorus was leached, while sludge dissolution reached 72.7% or 75.6%, respectively. The kinetic study with HCl and NaOH reported that phosphorus release and sludge dissolution follow a first-order reaction, with rate constants of 0.50 and 0.35 min^{-1} (phosphorus release) and 0.47×10^{-2} and 0.15×10^{-2} min^{-1} (sludge dissolution), respectively. Sequential extraction by NaOH/HCl leached 91.7% of the total phosphorus (Ali and Kim, 2016).

Shiba and Ntuli (2017) studied the extraction and precipitation of phosphorus from dry SS from domestic and industrial WWTPs. During the experiment, the effect of the type of acid (sulfuric acid, nitric acid, or HCl), acid concentration, solid loading, and reaction time was investigated. Approximately 82% of phosphorus was extracted from the raw SS, with 1 M of sulfuric acid at 5% solid loading for 2 h. Due to the iron phosphates obtained from SS after this reaction, a further purification step using ion exchange to remove iron (Fe) was required in order to increase the degree of phosphorus release. Magnesium oxide (MgO) and ammonium hydroxide [$(NH_4)_2O$] were used as the magnesium (Mg) and nitrogen sources, respectively, as well as pH regulators to precipitate phosphorus as a struvite. In this study, approximately 57% of struvite was precipitated. The total phosphorus content of the precipitate was 25.9%. The authors also conducted kinetic studies that showed that the leaching of phosphorus follows the Dickinson model for the first 100 min, with a reaction rate of about 2×10^{-5} s^{-1}. It was indicated that the rate-limiting step was controlled by diffusion. The obtained phosphorus solubility in 2% critic acid reached 96%, which is the amount of phosphorus available to plants if the precipitate is applied as a fertilizer. In the precipitate, Gram-positive *Bacillus subtilis* was found, which is harmless to the environment because this bacterium already exists in the soil where the precipitate can be applied as a fertilizer.

In another study (Vardanyan et al., 2018), the effects of pH, microorganisms, and sequential extraction on phosphorus dissolution from dewatered anaerobic sludge (DWAS) were studied. The authors indicated that the parameters that have high impact on phosphorus dissolution (86%) from DWAS were exposure to pH 2.5, anaerobic conditions, and sequential extraction. It was observed that the addition of chemolithotrophic, acidophilic bacteria has negatively influenced phosphorus dissolution, whereas the addition of acidophilic, heterotrophic iron reducing (HIR) bacteria increased phosphorus dissolution slightly, but it also contributed to pH maintenance at lower levels compared to no addition of HIR. The addition of acidophilic HIR bacteria at DWAS increased phosphorus dissolution to 93%, but it mainly contributed to a lower amount of H_2SO_4 (0.76 kg H_2SO_4 per kilogram of TS sludge) compared to DWAS without HIR (0.93 kg H_2SO_4 per kilogram of TS sludge). After phosphorus fractionation of the residual DWAS after sequential extraction, the organically bound phosphorus was hardly dissolved. The residual DWAS after acid treatment generated around 45% less methane (CH_4) than the initial DWAS.

Blöcher et al. (2012) studied a process of phosphorus recovery from the SS via a hybrid process of low-pressure, wet oxidation and nanofiltration. Nine SS samples were taken from seven WWTPs in North Rhine-Westphalia, Germany. In the first step of this process, the sulfuric acid was added to adjust the starting pH to about 2. In the selected experiments, ferrous sulfate heptahydrate was added as a catalyst. Then, the preconditioned SS was fed to the reactor and heated. Oxygen feeding was done depending on the oxidation of organics. Intermediate samples were taken via a cooled sampling line. In most cases, effluent samples from the first step of the experiment were collected and used in the ultrafiltration/nanofiltration investigations after cooling and storage for up to 3 weeks. During this process, phosphorus recovery of 54% was obtained. These authors also indicated that the cost of the entire process is similar to conventional SS disposal, but with the additional benefit of phosphorus recovery and reduced emissions of greenhouse gases due to avoidance of SS combustion.

Saktaywin et al. (2005) developed an advanced sewage treatment process in which excess sludge reduction by ozonation and phosphorus recovery by crystallization are incorporated to a conventional anaerobic/oxic (A/O) phosphorus removal process. Phosphorus recovery by crystallization has been developed as an alternative method for chemical precipitation in wastewater treatment. These researchers conducted sludge ozonation experiments in order to investigate the solubilization characteristics of the SS and change in microbial activity by using sludge cultured with feed made of synthetic sewage via the A/O process. Phosphorus was solubilized by the process of ozonation and by the use of organics, and acid-hydrolyzable phosphorus (AHP) comprised most of the solubilized phosphorus for phosphorus-accumulating organisms (PAOs) in the SS. At a solubilization rate of 30%, around 70% of the SS was inactivated by ozonation. These authors indicated that the process configuration that they were exploring has the potential to reduce the excess SS production, as well as to recover phosphorus in usable forms. The results show that this system is practical, in which 30% of solubilization was achieved by ozonation. In this study, 30% of solubilization was achieved at 30 mg O_3/g SS of ozone consumption.

Phosphorus-controlled crystallization as struvites is indicated to be a suitable method to recover this nutrient. Pastor et al. (2008) studied the influence of separate and mixed thickening of primary and secondary SS on struvite recovery. In this research, four pilot plants located in Carraixet Wastewater Treatment Plant (Valencia) were used (Fig. 1):

- Plant 1—Fermentation and elutriation of primary sludge
- Plant 2—Biological nitrogen and phosphorus removal
- Plant 3—Anaerobic digestion of primary and secondary waste sludge
- Plant 4—Struvite crystallization

The phosphorus precipitation in the crystallizer can be assessed by taking into account two types of efficiencies: recovery efficiency and precipitation efficiency:

$$\text{Recovery efficiency} = \frac{P_{Tin} - P_{Tef}}{P_{Tin}} \times 100 \tag{4}$$

$$\text{Precipitation efficiency} = \frac{PO_4 - P_{in} - PO_4 - P_{ef}}{PO_4 - P_{in}} \times 100 \tag{5}$$

The precipitation efficiency represents the process efficiency from a thermodynamic point of view, provided that the supersaturation can be almost completely consumed, which should be the case if there is sufficient residence time. The recovery efficiency takes into account both the precipitation and crystal growth efficiency. In this study, phosphorus precipitation in the digester was reduced from 13.7 g of phosphorus per kilogram of treated SS in the separate thickening experiment to 5.9 g in the mixed thickening experiment. This lessening of the uncontrolled precipitation leads to a reduction of the operational problems and enhances the phosphorus availability for its later crystallization. The high phosphorus precipitation and recovery efficiencies were observed in both crystallization experiments. The mixed thickening configuration showed a lower percentage of phosphorus precipitated as struvites due to the presence of high calcium (Ca) concentrations. In spite of this low percentage, the global phosphorus mass balance showed that the mixed thickening experiment produces a higher phosphorus recovery as struvite per kilogram of treated SS compared to separate thickening (i.e., 3.6 gP/kg sludge versus 2.5 gP/kg sludge).

Acelas et al. (2014) studied the gasification of dewatered SS in supercritical water (SCWG) for energy recovery combined with phosphorus recovery from the solid residue generated by this process. SCWG temperature (400°C, 500°C, 600°C) and residence time (15 min, 30 min, 60 min) were varied to test their effects on gas production and phosphorus recovery by acid leaching. The results show that the dry gas composition for this uncatalyzed gasification of the SS in SCW was mainly comprised of carbon monoxide (CO), CO_2, CH_4, H_2, and some C_2–C_3 compounds. It was shown that higher temperatures and longer residence times favored the production of H_2 and CH_4. After SCWG, more than 95% of the phosphorus was recovered from the solid residue by leaching with oxalic and sulfuric acids. The authors indicated that SCWG combined with acid leaching would be an effective method for both energy recovery and high phosphorus recovery from the SS.

3.2 Phosphorus extraction from SSA

In most European countries, the combustion of SS was commonly practiced during the last four decades (Xu and Lancaster, 2009) as a result of the banning of landfilling of SS in line with the EU Landfill Directive (99/31/EC). According to the available literature, it was shown that from 5 to 10 times more phosphorus can be recovered from the SSA than from sludge and leachates. However, those technologies are economically viable only in large WWTPs, mainly due to the large capital expenditures associated with building a facility that meets all the ecological criteria for sludge incineration (Cieślik and Konieczka, 2017). Generally, phosphorus-recovery processes aim at separating heavy metals (Kacprzak et al., 2014a,b) from the phosphorus and at converting the valuable mineral either into a plant-available form for reuse as fertilizer or into raw material for the industry. There are several methods aimed at extracting and recovering phosphorus from SSA, which include two categories of phosphorus-recovery technologies:

- Wet chemical approaches include separate phosphorus and heavy metals at temperatures of 1000–2000°C and transform phosphorus into a plant-available form or into yellow phosphorus.
- Thermal approaches include acid and alkaline dissolution techniques (Petzet et al., 2012).

Phosphorus is most often obtained from SSA through extraction using mineral or organic acids such as H_2SO_4, HCl, HNO_3, H_3PO^4, citric acid, and oxalic acid (Cieślik and Konieczka, 2017)

In the acid wet chemical phosphorus-recovery technologies, the first step is an almost complete acidic dissolution of phosphorus at pH values < 2. This process is unavoidably accompanied by the partial dissolution of metals or their compounds. The composition of raw SSA (which is Fe- or Al-rich) and the type and amount of the added acid (HCl or H_2SO_4) influences the amount of dissolved metals. The phosphorus component is predominantly present in the SSA in the form of calcium phosphates (Ca-P), aluminum phosphates (Al-P), and iron phosphates (Fe-P). The chemical demand for such a complete acidic dissolution of phosphorus can be estimated as follows:

$$Ca_9(Al)(PO_4)_7 + 21H^+ \rightarrow 9Ca^{2+} + Al^{3+} + 7H_3PO_4 \tag{6}$$

$$AlPO_4 + 3H^+ \rightarrow Al^{3+} + H_3PO_4 \tag{7}$$

$$Fe_3(PO_4)_2 + 6H^+ \rightarrow 3Fe^{2+} + 2H_3PO_4 \tag{8}$$

$$FePO_4 + 3H^+ \rightarrow Fe^{3+} + H_3PO_4 \tag{9}$$

Based on Eqs. (4)–(7), it can be seen that the theoretical chemical demand for complete acidic dissolution of each of these phosphorus compounds is 3 mol H^+/mol phosphorus. In practice, more acid is needed due to other acid-soluble components usually contained in SSA, such as CaO, $CaCO_3$, $Ca(OH)_2$, MgO, Fe_2O_3, K_2O, and Al-compounds, are also dissolved. Moreover, heavy metals are another problem; they need to be separated from the dissolved phosphorus in order to create a valuable phosphorus-recovery product. It is also recommended to separate Al and Fe because these elements impair the quality of the final product (Petzet et al., 2012).

Because Al-P cannot be directly reused as fertilizer, the precipitate may be dissolved by alkaline treatment followed by precipitation as Ca-P:

$$AlPO_4 + 4NaOH \rightarrow 4\,Na^+ + \left[Al(OH)_4\right]^- + PO_4{}^{3-} \tag{10}$$

$$3Ca^{2+} + 2PO_4{}^{3-} \rightarrow Ca3(PO_4)_2 \tag{11}$$

In the direct alkaline dissolution of phosphorus, only the amphoteric Al-P compounds dissolve [Eq. (10)], while most heavy metals remain in the SSA. In a following step, the dissolved phosphorus can be precipitated from the alkaline solution (pH > 13) as Ca-P with a very low impurity level via the addition of $CaCl_2$ [Eq. (11)]. The amount of Al-P directly leachable via alkaline treatment depends on both the Al content and the Ca content of the SSA (Petzet et al., 2012).

Wet chemical recovery of phosphorus from the SSA was investigated by Schaum et al. (2007). By acidic elution (HCl) at pH values < 1.5, phosphorus was released next to the greater percentage of the metals. Due to the formation of calcium phosphate, a release in alkaline environments is limited and mainly dependent on SSA composition. From the 15 SSA samples, the maximum release rate was 30% for phosphorus. After removing the insoluble solids, phosphorus needs to be separated from the metals. Therefore, two processes were presented: sequential precipitation and nanofiltration. Sequential precipitation of phosphorus with the use of the sequential precipitation of phosphorus (SEPHOS Process) was indicated as a promising technique due to the obtained product, an aluminum phosphate, is a valuable raw material for the phosphorus industry. After alkaline treatment of the aluminum phosphate, it is possible to precipitate phosphorus as calcium phosphate (via an advanced SEPHOS Process). Following acidic elution of the SSA, nanofiltration could be used to separate phosphorus.

The potential of the wet chemical process for phosphorus recovery from SSA by sequential elution with acidic and alkaline solutions has been investigated by Petzet et al. (2011). The developed method was SESAL-Phos sequential elution of SSA for aluminum and phosphorus recovery. The innovative aspect of this treatment is an acidic pretreatment step involving the leaching of calcium from the SSA. Thus, the percentage of alkaline soluble aluminum phosphates increased from 20% to 67%. In a later step, aluminum phosphate was dissolved under alkali conditions. Subsequently, the dissolved phosphorus was precipitated as calcium phosphate with low heavy metal content and recovered from the alkaline solution. Dissolved aluminum was recovered and may be reused as a precipitant in a WWTP.

The selective extraction of phosphorus compounds from the SSA with nitric acid was investigated by Wzorek et al. (2006). The cost of 1 mol hydrogen ions required for the SSA leaching solution under strongly acidic conditions increases in the series H_2SO_4, HCl, HNO_3, and H_3PO_4. Based on the cost estimations, nitric acid was chosen. The authors stated that if the final product of phosphorus recovery is used as a fertilizer containing phosphate and calcium nitrate, the cost of HNO_3 should be compensated for in the market value of the product.

In the first step of the investigation, 2.77 g of 65% HNO_3 was added to thoroughly mixed suspension (5.00 g Ash_{av}, in 19.8 g water). After the addition of the acid, the pH of the suspension dropped from 9.40 to 0.41, and then increased to 0.62. After 0.5 h, 1.8 cm^3 of the suspension was sampled. After another 15 min, the subsequent 0.7 g of HNO_3 was introduced, which caused the pH drop to 0.13. Then, pH continued to increase until it became stable, at the level of 0.30. After 2.7 h, when the leaching was finished, the concentration of PO_4^{3-} ions in the extract was equal to 36.6 g/dm^3. The mass ratio Fe^{3+}/PO_4^{3-} in the solution increased from 0.102 after 0.5 h to 0.176 after 23.3 h.

In the second stage of the experiment, the mass ratio of the SSA to water and nitric acid concentration in the suspension was increased twice in order to obtain higher concentrations of PO_4^{3-} in the extract. During the investigation, the concentration of phosphate and calcium ions stabilizes after about 2 h. The Fe^{3+}/PO_4^{3-} ratio systematically increases with the leaching time, from 0.0320 to 0.0500 after 0.5 and 28 h, respectively. Phosphate ion concentration in the solution after 28 h of extraction reached 85 g/dm^3. Estimating residue losses as 20% (associated with sampling about 2.3 g of slurry for analysis six times and with the transfer of cake from the filter), it was assessed that approximately 99% PO_4^{3-} initially present in the SSA was extracted.

The wet chemical method (PolFerAsh, which stands for "Polish Fertilizers from Ash") for phosphorus recovery from industrial SSAs in Poland (Cracow University of Technology) was developed by Gorazda et al. (2010, 2017). The extracts obtained from leaching the SSA with phosphoric acid are valuable products that can be used in fertilizer production. The obtained heavy metal content did not exceed levels that are characteristic of some industrial phosphoric acids. On the contrary, some elements present in the extracts can be treated as micronutrients essential for proper plant growth. The studied SSA does not change significantly in composition throughout the year, and therefore, it is possible to produce extracts with nearly constant element concentrations over a long period of time. The precipitates obtained from the neutralization process are fertilizers containing primary (P) and secondary (Ca and Mg) nutrients with the addition of micronutrients [Fe, zinc (Zn), and copper (Cu)] for field crops and grassland. The requirements for all standardized metals and Cd were fulfilled. The final products could be suspension fertilizers; precipitate; nitrogen-phosphorus fertilizer (NP); phosphorus-potassium fertilizer (PK); nitrogen, phosphorus, and potassium fertilizer (NPK).

Li et al. (2018) evaluated the potential of six different extractants (two organic acids, two inorganic acids, and two chelating agents) to recover phosphorus from the SSA. During the study, extraction of the two heavy metals in the highest concentrations (Zn and Cu) also was conducted. The extraction was performed at a liquid-to-solid ratio of 10 L/kg and 30 rpm in an end-over-end mechanical rotator. More than 70% of phosphorus was leached within 2 h of extraction by acids, and the acid types had only a small effect on phosphorus extraction. It was indicated that the fast phosphorus extraction rate can be attributed to the acid-soluble nature of the SSA and the small particle size. The authors stated that due to a similar pH range (from 0.85 to 1.70 for the four of the six extractants), the extracted phosphorus concentration changed only marginally. It was also proved by other researchers (e.g., Franz, 2008) that the extracted phosphorus reached a plateau pH level <1.8. It also should be emphasized that the heavy metals and phosphorus in the SSA are codissolved during acid extraction (Biswas et al., 2009). In a recent study (Li et al., 2018), the solubility of Zn and Cu in the SSA was much lower than that of phosphorus. This finding indicates that phosphorus is present in the SSA in an acid-soluble form, while the solubility of the Zn and Cu compounds is relatively lower. The relatively low leaching rate of Zn and Cu was also due to the fact that parts of metals were immobilized in an acid-insoluble glassy phase and present as nucleation particles in the central matrix of the SSA.

The leaching behavior of phosphorus and other metals [Fe, Ca, Al, Mg, and manganese (Mn)] was evaluated by Biswas et al. (2009). The leaching of phosphorus by using acid leach liquors was considered, based on the economic indicators provided by Wzorek et al. (2006). Therefore, sulfuric acid and HCl were chosen to extract phosphorus and other metal ions (Biswas et al., 2009). The leaching percentage of phosphorus increases with an increase in both the L/S ratio and acid concentration (0.01–0.1 mol/L). At concentrations of 0.2 mol/L and higher, the phosphorus leaching equalled almost 100% regardless of the L/S ratio. Considering the cost-effective consumption of acid and the achievement of complete phosphorus leaching, 0.05 mol/L sulfuric acid at a L/S ratio of 150 mL/g is recommended as the most suitable condition for extraction.

The authors also indicated that the solubility of phosphorus is higher than that of other metals, which indicates that phosphorus exists in the SSA as an acid-soluble form. The impact of temperature on phosphorus leaching by using sulfuric acid from the SSA was also evaluated. The leaching percentage increases very smoothly with increasing temperatures for P, Ca, Mg, Al, and Cu. This development may be related to the fact that an increase in temperature provides the necessary energy to break down the chemical bonds of the metals in the SSA (Naoum et al., 2001). In the case of Fe, the leaching percentage remains constant (about 15%) from 30°C to 50°C, but it increases to 75% at 70°C, suggesting that the iron compound in the SSA needs higher energy (in the form of either temperature or acid concentration) to be broken. The temperature of 30°C was indicated to be suitable for further extraction. The selection of 30°C avoids high solubility of other metals such as Fe and Cu (Biswas et al., 2009).

An interesting method of phosphorus recovery from the SSA by a combined two-step extraction and selective precipitation process was developed by Fang et al. (2018). In the first step, SSA was treated with ethylenediaminetetraacetatic acid (EDTA), and then with sulfuric acid. The recovery conditions were optimized [pre-extraction by 0.02 mol/L of EDTA at a liquid-to-solid ratio (L/kg) of 20:1 during 120 min, and then extraction by 0.2 mol/L sulfuric and at a liquid-to-solid ratio of 20:1 during the next 120 min]. The authors have indicated that compared with direct extraction by sulfuric acid (namely, single-step leaching), the two-step extraction method dissolved 2.34 mmol/kg SSA less phosphorus, but with 5.16 mmol/kg SSA less metalloid contamination. The potential for coprecipitation of phosphorus and the metalloid contaminates in the extracts (both the single-step extraction and the two-step extraction methods) was tested in the pH range from 2–12, adjusted by the addition of NaOH and $Ca(OH)_2$. When $Ca(OH)_2$ is used, and at the optimal pH of 4, the two-step extraction method may increase Ca-P significantly in the precipitate and notably decreased the metalloid contaminants by 50% compared to the single-step method.

Phosphorus also can be recovered by high-temperature metallurgical processes. In this approach, transformation of phosphorus into a metallurgical slag by reductive smelting at temperatures above 1,450°C in a shaft furnace is conducted or reduction of phosphorus to elemental phosphorus that is separated via the gas phase in an inductively heated shaft furnace at similar temperatures. In both cases, volatile heavy metals such as Zn, Pb, Cd, and Hg are separated from the product via the gas phase, and heavy metals with high boiling points such as Fe, Cu, nickel (Ni), and chromium (Cr) are separated and form an alloy (Herzel et al., 2016).

The thermochemical recovery of phosphorus was investigated by Herzel et al. (2016). The SSA was thermochemically treated with sodium and potassium additives under reducing conditions. Then, the phosphate-bearing mineral phases were transformed into plant-available phosphates. In such experiments, the high phosphorus-bioavailability was achieved with a molar Na/P ratio > 1.75 in the initial materials. Sodium sulfate, carbonate, and hydroxide performed comparably as additives for this calcination process. Potassium carbonate and hydroxide must to be added in a molar K/P ratio > 2.5 to achieve comparable phosphorus solubility. The authors have conducted the lab-scale and industrial-scale experiments (ash treatment with sodium sulfate). Simultaneously, the volatile transition metal arsenic (61% removal), as well as volatile heavy metals such as Cd (80%), Ag (68%), Pb (39%), and Zn (9%), were removed via off-gas treatment. The obtained product has high bioavailability and a toxic trace element mass fraction below the limit values of the German fertilizer ordinance. It can be treated and used as a phosphorous-fertilizer.

Pettersson et al. (2008) conducted leaching of phosphorus from ashes from cocombustion of SS and wood with the use of sulfuric acid. The investigated pH range was in the range of 0.5–2.5. Each experiment was started with 0.4 g of ash and 60 ml of water (i.e., a mass ratio of solid to liquid of 1:150). The addition of the amount of acid required to keep the pH at the chosen level was logged. Further, the length of this extraction time (90 min) was considered relevant for real processes. The obtained results showed that it is possible to find a pH range (0.5–1) in which one can extract a significant fraction of the phosphorus from these ashes without precipitation of secondary phosphates. The type of flocculation agent used in the WWTP where the SS is formed has a significant effect on phosphorus recovery. Ashes from combustion of the SS that was formed using aluminum sulfate as a flocculating agent released nearly all phosphorus at a pH value of 1. When iron sulfate was used as the flocculating agent, this affected the chemistry of the resulting ashes, resulting in more difficult phosphorus recovery. The yield of phosphorus from those ashes was 50%–80%.

Xu et al. (2012) investigated an effective phosphorus recovery method from SSA via struvite precipitation. They extracted more than 95% of the total phosphorus content from SSA with the use of 0.5 mol/L HCl at an L/S ratio of 50 mL/g. Although heavy metal leaching also occurred during phosphorus extraction, cation exchange resin efficiently removed the heavy metals from the phosphorus-rich solution. The orthogonal tests showed that the optimal parameters for phosphorus precipitation as struvite may be a Mg:N:P molar ratio of 1.6:1.6:1 at pH 10.0. The formation of struvite was proved by X-ray diffraction (XRD) analysis. The investigations revealed that the harvested precipitate had high struvite content (97%), high phosphorus bioavailability (94%), and low heavy metal content, which classifies it as a high-quality fertilizer.

In conclusion, the factors that affect the phosphorus recovery process are acid concentration, L/S ratio, temperature, and rate of reaction (Alamdari and Rohani, 2007). Due to the high energy associated with combustion of SS (with environmental and economic impacts on the profitability of the investment) and the necessity of leaching of heavy metals, further research should be conducted on recovering phosphorus by direct leaching of SS without incineration. Another major issue is meeting the requirements for the quality of produced fertilizers, which are more restrictive in the European Union (Smol et al., 2015b). Moreover, one of the most important limitations in the use of phosphorus-recovery methods is the fact that the recovered phosphorus is more expensive than fertilizers available on the market (Lundin et al., 2004; Shu et al., 2006; Molinos-Senante et al., 2011). However, the prospect of limited phosphorus rock deposits indicates a lessening of this difference and increased cost-effectiveness of phosphorus recycling in the coming decades.

The phosphorus-recovery techniques discussed in this chapter have been shown to be technologically feasible. However, they should be evaluated in order to meet the technological requirements, with the use of best-available techniques (BATs) and best-available techniques not entailing excessive costs (BATNEEC) (Generowicz et al., 2011a,b; Kowalski et al., 2011, 2012; Makara and Kowalski, 2018), as well as multicriteria analysis (Generowicz, 2014; Makara et al., 2018). The phosphorus recovery technologies also should be examined from the environmental (Sørensen et al., 2015; Heimersson et al., 2016) and economic (Gaterell et al., 2000; Parsons and Smith, 2008; Mihailescu et al., 2015) points of view. In those cases, life cycle assessment (LCA) (Kowalski et al., 2007a,b; Kulczycka et al., 2011; Linderholm et al., 2012; Remy et al., 2013; Svanström et al., 2017) and life cycle cost analysis (LCCA) (Kulczycka and Smol, 2016) can be used. Unfortunately, in many cases, phosphorus recovery technologies are economically unprofitable. Moreover, legislation and national policies are the major reasons why these techniques are not yet used worldwide (Desmidt et al., 2015).

It also must be underscored that phosphorus recycled in the processes discussed here should be employed further, such as for fertilizing (Adam et al., 2007; Vogel et al., 2014). This is recommended by the European Commission, according to the principles of the CE model, which promotes the recovery of raw materials from waste streams.

4 Conclusions

- According to the assumptions of the CE concept, the newest economic strategy of the European Union, alternative sources of phosphorus from waste materials in nature should be sought, as municipal and industrial wastewater, SS, SSA, meat and bone meal, pig slurry, biomass, and industrial waste.
- In the wastewater sector, the most promising phosphorus-rich residues are SS and SSA because the major part of the phosphate from wastewater is transferred to the sludge (approx. 90%).
- There is phosphorus recovery potential in several areas of WWTPs, as wastewater outflow from the treatment plant (up to 50%), sedimentary liquid (leachate) (up to 50%), dehydrated SS (up to 50%), and SS (more than 90%).
- Phosphorus extraction from SS can be conducted by wet chemical methods under acid and alkali conditions.

- Phosphorus extraction from SSA can be conducted by wet chemical methods that separate phosphorus and heavy metals at temperatures of 1000–2000°C and transform phosphorus into a plant-available form or into yellow phosphorus, and also by thermal techniques (acid and alkaline dissolution).

Acknowledgments

This work was funded by the statutory research of the MEERI PAS and the InPhos project No. 17022 (2018-2020), that is financed by EIT Raw Materials, a body of the European Union.

References

Acelas, N.Y., López, D.P., Brilman, D.W., Kersten, S.R., Kootstra, A.M.J., 2014. Supercritical water gasification of sewage sludge: gas production and phosphorus recovery. Bioresour. Technol. 174, 167–175.

Adam, C., 2009. Techniques for P-recovery from waste water, sewage sludge and sewage sludge ashes—an overview. In: Presentation at Baltic 21 Conference Phosphorus Recycling and Good Agricultural Practice, Berlin, 28–30 September 2009.

Adam, C., Gnirss, R., Lesjean, B., Buisson, H., Kraume, M., 2002. Enhanced biological phosphorus removal in membrane bioreactors. Water Sci. Technol. 46 (4-5), 281–286.

Adam, C., Kley, G., Simon, F.G., 2007. Thermal treatment of municipal sewage sludge aiming at marketable P-fertilisers. Mater. Trans. 48 (12), 3056–3061.

Adam, C., Peplinski, B., Michaelis, M., Kley, G., Simon, F.G., 2009. Thermochemical treatment of sewage sludge ashes for phosphorus recovery. Waste Manag. 29 (3), 1122–1128.

Alamdari, A., Rohani, S., 2007. Phosphate recovery from municipal wastewater through crystallization of calcium phosphate. Pac. J. Sci. Technol. 8 (1), 27–31.

Ali, T.U., Kim, D.J., 2016. Phosphorus extraction and sludge dissolution by acid and alkali treatments of polyaluminum chloride (PAC) treated wastewater sludge. Bioresour. Technol. 217, 233–238.

Arnout, S., Nagels, E., 2016. Modelling thermal phosphorus recovery from sewage sludge ash. Calphad 55, 26–31.

Biswas, B.K., Inoue, K., Harada, H., Ohto, K., Kawakita, H., 2009. Leaching of phosphorus from incinerated sewage sludge ash by means of acid extraction followed by adsorption on orange waste gel. J. Environ. Sci. 21 (12), 1753–1760.

Blöcher, C., Niewersch, C., Melin, T., 2012. Phosphorus recovery from sewage sludge with a hybrid process of low pressure wet oxidation and nanofiltration. Water Res. 46 (6), 2009–2019.

Chapagain, Y., 2016. Methods and Possibility for Recycling of Phosphorus from Sludge. Helsinki Metropolia University of Applied Sciences, Bachelor of Engineering, Helsinki.

Cieślik, B., Konieczka, P., 2017. A review of phosphorus recovery methods at various steps of wastewater treatment and sewage sludge management. The concept of "no solid waste generation" and analytical methods. J. Clean. Prod. 142, 1728–1740.

Commission of European Communities, 2010. Communication. A strategy for smart, sustainable and inclusive growth Europe 2020. COM No. 2020.

Commission of European Communities, 2013. Consultative communication on the sustainable use of phosphorus. COM No. 517.

Commission of European Communities, 2014b. Communication on the review of the list of critical raw materials for the EU and the implementation of the Raw Materials Initiative. COM No. 297.

Commission of European Communities, 2014c. Communication. Towards a circular economy: a zero waste programme for Europe. COM No. 398.

Commission of European Communities, 2017. Communication on the 2017 list of critical raw materials for the EU. COM No. 490.

Commission of European Communities. Communication No. 614, 2015. Closing the loop—An EU action plan for the circular economy. COM No. 614, 2015.

Commission of European Communities. Communication on the Circular Economy Package, 2016. Proposal for a Regulation of the European Parliament and of the Council laying down rules on the making available on the market of CE marked fertilising products and amending Regulations (EC) No 1069/2009 and (EC) No 1107/2009. COM No. 157.

Cordell, D., Rosemarin, A., Schröder, J.J., Smit, A.L., 2011. Towards global phosphorus security: a systems framework for phosphorus recovery and reuse options. Chemosphere 84 (6), 747–758.

Cornel, P., Schaum, C., 2009. Phosphorus recovery from wastewater: needs, technologies and costs. Water Sci. Technol. 59 (6), 1069–1076.

Desmidt, E., Ghyselbrecht, K., Zhang, Y., Pinoy, L., Van der Bruggen, B., Verstraete, W., Rabaey, K., Meesschaert, B., 2015. Global phosphorus scarcity and full-scale P-recovery techniques: a review. Crit. Rev. Environ. Sci. Technol. 45 (4), 336–384.

Egle, L., Rechberger, H., Krampe, J., Zessner, M., 2016. Phosphorus recovery from municipal wastewater: an integrated comparative technological, environmental and economic assessment of P recovery technologies. Sci. Total Environ. 571, 522–542.

Eurostat, 2018. https://ec.europa.eu/eurostat/data/database.

Fang, L., Li, J.S., Donatello, S., Cheeseman, C.R., Wang, Q., Poon, C.S., Tsang, D.C., 2018. Recovery of phosphorus from incinerated sewage sludge ash by combined two-step extraction and selective precipitation. Chem. Eng. J. 348, 74–83.

Franz, M., 2008. Phosphate fertilizer from sewage sludge ash (SSA). Waste Manag. 28 (10), 1809–1818.

Gaterell, M.R., Gay, R., Wilson, R., Gochin, R.J., Lester, J.N., 2000. An economic and environmental evaluation of the opportunities for substituting phosphorus recovered from wastewater treatment works in existing UK fertiliser markets. Environ. Technol. 21 (9), 1067–1084.

Generowicz, A., 2014. Multi-criteria analysis of waste management in Szczecin. Pol. J. Environ. Stud. 23 (1), 57–63.

Generowicz, A., Kowalski, Z., Kulczycka, J., 2011a. Planning of waste management systems in urban area using multi-criteria analysis. J. Environ. Prot. 2 (06), 736.

Generowicz, A., Kulczycka, J., Kowalski, Z., Banach, M., 2011b. Assessment of waste management technology using BATNEEC options, technology quality method and multi-criteria analysis. J. Environ. Manag. 92 (4), 1314–1320.

Gorazda, K., Kowalski, Z., Wzorek, Z., Jodko, M., Rzepecki, T., Kulczycka, J., Przewrocki, P., 2003. Possibilities of phosphorus recovering from municipal sewage and sewage sludge. Polish J. App. Chem. 47 (2), 51–63.

Gorazda, K., Tarko, B., Wzorek, Z., Kominko, H., Nowak, A.K., Kulczycka, J., Henlik, A., Smol, M., 2017. Fertilisers production from ashes after sewage sludge combustion—a strategy towards sustainable development. Environ. Res. 154, 171–180.

Gorazda, K.J., Wzork, Z., Kowalski, Z., 2010. R.o.P.P.a.T (Ed.), Office.

Heimersson, S., Svanström, M., Laera, G., Peters, G., 2016. Life cycle inventory practices for major nitrogen, phosphorus and carbon flows in wastewater and sludge management systems. Int. J. Life Cycle Assess. 21 (8), 1197–1212.

Herzel, H., Krüger, O., Hermann, L., Adam, C., 2016. Sewage sludge ash—a promising secondary phosphorus source for fertilizer production. Sci. Total Environ. 542, 1136–1143.

Jeng, A., Haraldsen, T.K., Vagstad, N., Grønlund, A., Tveitnes, S., 2004. Meat and bone meal as nitrogen fertilizer to cereals in Norway. Agric. Food Sci. 13 (3), 268–275.

Jørgensen, M.K., Sørensen, J.H., Quist-Jensen, C.A., Christensen, M.L., 2018. Wastewater treatment and concentration of phosphorus with the hybrid osmotic microfiltration bioreactor. J. Membr. Sci. 559, 107–116.

Kabbe, C., 2013. The limited resources of phosphorus and how to close the phosphorus cycle. In Factor X 30, 261–273.

Kacprzak, M., Grobelak, A., Grosser, A., Napora, A., 2014b. The potential of biosolid application for the Phytostabilisation of metals. Desalin. Water Treat. 52 (19-21), 3955–3964.

Kacprzak, M., Grobelak, A., Grosser, A., Prasad, M.N.V., 2014a. Efficacy of biosolids in assisted Phytostabilization of metalliferous acidic sandy soils with five grass species. Int. J. Phytoremediation 16 (6), 593–608.

Kasprzyk, M., Gajewska, M., Molendowska, S., 2017. Possibilities of phosphorus recovery from effluents, sewage sludge and ashes from sewage sludge thermal processing. Ecol. Eng. 18 (4), 65–78. (in Polish).

Kowalski, Z., Generowicz, A., Makara, A., 2012. Evaluation of municipal waste disposal technologies by BATNEEC. Przem. Chem. 91 (5), 811–815.

Kowalski, Z., Generowicz, A., Makara, A., Banach, M., 2011. The evaluation of technology for the production of phosphoric acid with BATNEEC options and multicriteria analysis. Przem. Chemi. 90 (5), 853–856.

Kowalski, Z., Kulczycka, J., Skowron, G., Sobczak, A., 2007a. Comparative evaluation of calcium feed phosphate production methods using life cycle assessment. Arch. Ochr. Środowiska 33 (1), 83–94.

Kowalski, Z., Kulczycka, J., Wzorek, Z., 2007b. Life cycle assessment of different variants of sodium chromate production in Poland. J. Clean. Prod. 15 (1), 28–37.

Kowalski, Z., Makara, A., Fijorek, K., 2013. Changes in the properties of pig manure slurry. Acta Biochim. Pol. 60 (4), 845–850.

Krüger, O., Adam, C., 2015. Recovery potential of German sewage sludge ash. Waste Manag. 45, 400–406.

Krüger, O., Adam, C., 2017. Phosphorus in recycling fertilizers-analytical challenges. Environ. Res. 155, 353–358.

Krüger, O., Grabner, A., Adam, C., 2014. Complete survey of German sewage sludge ash. Environ. Sci. Technol. 48 (20), 11811–11818.

Kulczycka, J., Kowalski, Z., Smol, M., Wirth, H., 2016. Evaluation of the recovery of Rare Earth Elements (REE) from phosphogypsum waste—case study of the WIZOW chemical plant (Poland). J. Clean. Prod. 113, 345–354.

Kulczycka, J., Kurczewski, P., Kasprzak, J., Lewandowska, A., Lewicki, R., Witczak, A., Witczak, J., 2011. The Polish Centre for Life Cycle Assessment—the centre for life cycle assessment in Poland. Int. J. Life Cycle Assess. 16 (5), 442–444.

Kulczycka, J., Smol, M., 2016. Environmentally friendly pathways for the evaluation of investment projects using life cycle assessment (LCA) and life cycle cost analysis (LCCA). Clean Techn. Environ. Policy 18 (3), 829–842.

Le Corre, K.S., Valsami-Jones, E., Hobbs, P., Parsons, S.A., 2009. Phosphorus recovery from wastewater by struvite crystallization: a review. Crit. Rev. Environ. Sci. Technol. 39 (6), 433–477.

Li, J.S., Chen, Z., Wang, Q.M., Fang, L., Xue, Q., Cheeseman, C.R., Donatello, S., Liu, L., Poon, C.S., 2018. Change in re-use value of incinerated sewage sludge ash due to chemical extraction of phosphorus. Waste Manag. 74, 404–412.

Linderholm, K., Tillman, A.M., Mattsson, J.E., 2012. Life cycle assessment of phosphorus alternatives for Swedish agriculture. Resour. Conserv. Recycl. 66, 27–39.

Lundin, M., Olofsson, M., Pettersson, G.J., Zetterlund, H., 2004. Environmental and economic assessment of sewage sludge handling options. Resour. Conserv. Recycl. 41 (4), 255–278.

Makara, A., Generowicz, A., Kowalski, Z., 2018. Assessment and comparison of technological variants of the sodium tripolyphosphate production with the use of multi-criteria analysis. Int. J. Environ. Sci. Technol. 1–14.

Makara, A., Kowalski, Z., 2018. Selection of pig manure management strategies: case study of polish farms. J. Clean. Prod. 172, 187–195.

Mihailescu, E., Ryan, W., Murphy, P.N.C., Casey, I.A., Humphreys, J., 2015. Economic impacts of nitrogen and phosphorus use efficiency on nineteen intensive grass-based dairy farms in the South of Ireland. Agric. Syst. 132, 121–132.

Mihelcic, J.R., Fry, L.M., Shaw, R., 2011. Global potential of phosphorus recovery from human urine and feces. Chemosphere 84 (6), 832–839.

Molinos-Senante, M., Hernández-Sancho, F., Sala-Garrido, R., Garrido-Baserba, M., 2011. Economic feasibility study for phosphorus recovery processes. Ambio 40 (4), 408–416.

Naoum, C., Fatta, D., Haralambous, K.J., Loizidou, M., 2001. Removal of heavy metals from sewage sludge by acid treatment. J. Environ. Sci. Health A 36 (5), 873–881.

Parsons, S.A., Smith, J.A., 2008. Phosphorus removal and recovery from municipal wastewaters. Elements 4 (2), 109–112.

Pastor, L., Marti, N., Bouzas, A., Seco, A., 2008. Sewage sludge management for phosphorus recovery as struvite in EBPR wastewater treatment plants. Bioresour. Technol. 99 (11), 4817–4824.

Peng, L., Dai, H., Wu, Y., Peng, Y., Lu, X., 2018. A comprehensive review of phosphorus recovery from wastewater by crystallization processes. Chemosphere 197, 768–781.

Pettersson, A., Åmand, L.E., Steenari, B.M., 2008. Leaching of ashes from co-combustion of sewage sludge and wood—part I: recovery of phosphorus. Biomass Bioenergy 32 (3), 224–235.

Petzet, S., Peplinski, B., Bodkhe, S.Y., Cornel, P., 2011. Recovery of phosphorus and aluminium from sewage sludge ash by a new wet chemical elution process (SESAL-Phos-recovery process). Water Sci. Technol. 64 (3), 693–699.

Petzet, S., Peplinski, B., Cornel, P., 2012. On wet chemical phosphorus recovery from sewage sludge ash by acidic or alkaline leaching and an optimized combination of both. Water Res. 46 (12), 3769–3780.

Podewils, W., 2014. Recycling of phosphorus in Germany—current status and prospects. Wodociągi—Kanalizacja 1 (119), 26–30 (in Polish).

P-REX Project, 2018. Sustainable sewage sludge management fostering phosphorus recovery and energy efficiency. Available from: www.batchgeo.com/map/0f9d56a3aa57a51379a3cb23af27d202.

Remy, C., Lesjean, B., Waschnewski, J., 2013. Identifying energy and carbon footprint optimization potentials of a sludge treatment line with life cycle assessment. Water Sci. Technol. 67 (1), 63–73.

Saktaywin, W., Tsuno, H., Nagare, H., Soyama, T., Weerapakkaroon, J., 2005. Advanced sewage treatment process with excess sludge reduction and phosphorus recovery. Water Res. 39 (5), 902–910.

Schaum, C., Cornel, P., Jardin, N., 2007. Phosphorus recovery from sewage sludge ash–a wet chemical approach. In: Proceeding of the IWA Conference on Biosolids, Moving Forward Wastewater Biosolids Sustainability: Technical, Managerial, and Public Synergy pp. 24–27.

Scholz, R.W., Wellmer, F.W., 2013. Approaching a dynamic view on the availability of mineral resources: what we may learn from the case of phosphorus? Glob. Environ. Chang. 23 (1), 11–27.

Science for Environment Policy Report, 2017. Available from: http://ec.europa.eu/Eurostat. Accessed 10 March 2018.

Sengupta, S., Pandit, A., 2011. Selective removal of phosphorus from wastewater combined with its recovery as a solid-phase fertilizer. Water Res. 45 (11), 3318–3330.

Shiba, N.C., Ntuli, F., 2017. Extraction and precipitation of phosphorus from sewage sludge. Waste Manag. 60, 191–200.

Shu, L., Schneider, P., Jegatheesan, V., Johnson, J., 2006. An economic evaluation of phosphorus recovery as struvite from digester supernatant. Bioresour. Technol. 97 (17), 2211–2216.

Smol, M., Kulczycka, J., Avdiushchenko, A., 2017. Circular economy indicators in relation to eco-innovation in European regions. Clean Techn. Environ. Policy 19 (3), 669–678.

Smol, M., Kulczycka, J., Henclik, A., Gorazda, K., Tarko, B., Wzorek, Z., 2015b. Sewage sludge ash (SSA) as phosphate fertilizer in the aspect of legal regulations. In: WASTES: Solutions, Treatments and Opportunities. CRC Press Taylor & Francis Group, Editors, Candida Vilarinho, Fernando Castro, Mario Russo, pp. 323–328.

Smol, M., Kulczycka, J., Henclik, A., Gorazda, K., Wzorek, Z., 2015a. The possible use of sewage sludge ash (SSA) in the construction industry as a way towards a circular economy. J. Clean. Prod. 95, 45–54.

Smol, M., Kulczycka, J., Kowalski, Z., 2016. Sewage sludge ash (SSA) from large and small incineration plants as a potential source of phosphorus–Polish case study. J. Environ. Manag. 184, 617–628.

Sørensen, B.L., Dall, O.L., Habib, K., 2015. Environmental and resource implications of phosphorus recovery from waste activated sludge. Waste Manag. 45, 391–399.

Svanström, M., Heimersson, S., Peters, G., Harder, R., I'Ons, D., Finnson, A., Olsson, J., 2017. Life cycle assessment of sludge management with phosphorus utilisation and improved hygienisation in Sweden. Water Sci. Technol. 75 (9), 2013–2024.

Tan, Z., Lagerkvist, A., 2011. Phosphorus recovery from the biomass ash: a review. Renew. Sust. Energ. Rev. 15 (8), 3588–3602.

Tong, J., Chen, Y., 2009. Recovery of nitrogen and phosphorus from alkaline fermentation liquid of waste activated sludge and application of the fermentation liquid to promote biological municipal wastewater treatment. Water Res. 43 (12), 2969–2976.

Vardanyan, A., Kafa, N., Konstantinidis, V., Shin, S.G., Vyrides, I., 2018. Phosphorus dissolution from dewatered anaerobic sludge: effect of pHs, microorganisms, and sequential extraction. Bioresour. Technol. 249, 464–472.

Vogel, C., Adam, C., Kappen, P., Schiller, T., Lipiec, E., McNaughton, D., 2014. Chemical state of chromium in sewage sludge ash based phosphorus-fertilisers. Chemosphere 103, 250–255.
Wzorek, Z., Jodko, M., Gorazda, K., Rzepecki, T., 2006. Extraction of phosphorus compounds from ashes from thermal processing of sewage sludge. J. Loss Prev. Process Ind. 19 (1), 39–50.
Xu, C.C., Lancaster, J., 2009. Treatment of secondary sludge for energy recovery. In: Energy Recovery. Nova Science Publishers Incorporation, New York, NY, pp. 187–211.
Xu, H., He, P., Gu, W., Wang, G., Shao, L., 2012. Recovery of phosphorus as struvite from sewage sludge ash. J. Environ. Sci. 24 (8), 1533–1538.
Yamashita, T., Yamamoto-Ikemoto, R., 2014. Nitrogen and phosphorus removal from wastewater treatment plant effluent via bacterial sulfate reduction in an anoxic bioreactor packed with wood and iron. Int. J. Environ. Res. Public Health 11 (9), 9835–9853.
Ylivainio, K., Uusitalo, R., Turtola, E., 2008. Meat bone meal and fox manure as P sources for ryegrass (Lolium multiflorum) grown on a limed soil. Nutr. Cycl. Agroecosyst. 81 (3), 267–278.

Further reading

Ciesielczuk, T., Rosik-Dulewska, C., Kusza, G., 2016. Extraction of phosphorus from sewage sludge ash and sewage sludge—problem analysis. Polish J. Sustain. Develop. 20, 21–28 (in Polish).
Commission of European Communities, 2014a. Communication on the review of the list of critical raw materials for the EU and the implementation of the Raw Materials Initiative. COM No. 297.
European Landfill Directive, 1999. European Union 1999. 99/31/EC: European Landfill Directive (99/31/EC).

Section E

Energy recovery from sludge

Volatile fatty acid (VFA) yield from sludge anaerobic fermentation through a biotechnological approach

29

Małgorzata Worwag, Anna Kwarciak-Kozłowska
Faculty of Infrastructure and Environment, Institute of Environmental Engineering, Czestochowa University of Technology, Czestochowa, Poland

1 Characterization of volatile fatty acids (VFAs)

Volatile fatty acids (VFAs) are saturated aliphatic acids that contain from two to six carbon atoms in the particle, and sometimes even eight (Chwiałkowska and Oleśkowicz-Popiel, 2016; Zielewicz, 2007).

Their structure contains an acetyl group characterized by the presence of a carbon atom that is double-bonded to oxygen. Due to the electronegativity of oxygen present in the acetyl group, carboxyl acids belong to polar compounds. Despite their high boiling point, these acids can be removed from water due to distillation with water steam. These compounds are characterized by high vapour pressure, which is utilized during the determination of their concentration in environmental samples (Zielewicz, 2007). VFAs have strong hydrophilic properties. The nomenclature of VFAs is mostly based on customary names that suggest their plant or animal origins. These acids are referred to in the literature by various names, including *low-weight carboxylic acid (LWCAs), short-chain fatty acids (SCFAs), low-molecular-weight aliphatic acids (LMWAs)*, and *volatile organic acids (VOAs)*, as well as VFAs (Banel, 2010).

Table 1 presents the names, chemical structures, and physicochemical properties of these compounds. Knowledge of the physicochemical properties of selected VFAs allows for prediction of their behavior in the environment and development of procedures for their determination (Banel and Zygmunt, 2009). Knowledge of the value of the negative log of the dissociation constant (pKa) allows for determination of the part of the compound present in the dissociated/nondissociated form in the sample. The balance of dissociation of three of the most important organic acids is presented in Fig. 1. It shows that reaction plays an important role. Analysis of the degree of dissociation of acetic acid reveals that at pH 6, there are 5% of nondissociated ethanoic acid (CH_3COOH) particles, only 0.5% at pH 7, and 0.05% at pH 8. In general, it is accepted that in the range of pH from 6.0 to 8.0, the concentrations of nondissociated forms are reduced by two orders of power. The increase in the amount of nondissociated acids causes them to react with carbonates, with a simultaneous increase

Industrial and Municipal Sludge. https://doi.org/10.1016/B978-0-12-815907-1.00029-5

Table 1 Nomenclatures, chemical compositions, and physicochemical properties of VFAs (Banel, 2010; Chwiałkowska and Oleśkowicz-Popiel, 2016)

System name (custom)	Chemical formula	M (g/mol)	Density at 25°C (g/L)	Boiling point (°C)	Solubility in water at ambient temperatures (g/L)	pKa
Methanoic acid (formic)	HCOOH	46.03	1220	100.8	Big	3.75
Ethanoic acid (acetic)	CH_3COOH	60.1	1049	117	Big	4.75
Propane acid (acid)	C_2H_5COOH	74.1	990	141	Big	4.87
Isobutane acid (isobutyric)	$(CH_3)_2CHCOOH$	88.1	960	154	210	4.85
Butanoic acid (butyric)	C_3H_7COOH	88.1	964	164	Average	4.81
Isopentanoic acid (isovaleric)	$(CH_3)_2CHCH_2COOH$	102.1	925	177	25	4.78
Pentanoic acid (valeric)	C_4H_9COOH	102.1	938	186	40	4.82
Hexane acid (caproic)	$C_5H_{11}COOH$	116.2	927	206	10	4.88
Heptanoic acid (enanthic)	$C_6H_{13}COOH$	130.2	918	223	2.6	4.89
Octane acid (caprylic)	$C_7H_{15}COOH$	144.2	910	235	0.7	4.89

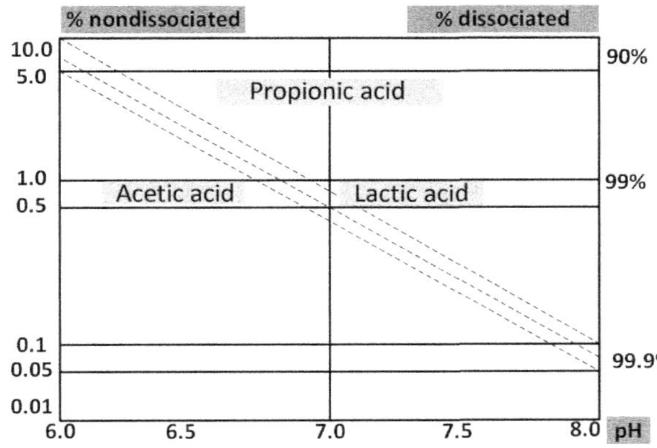

Fig. 1 Percentage of nondissociated and dissociated forms of acids for pH values ranging from 6 to 8.
Based on Sadecka, Z., 2010. Basics of Biological Wastewater Treatment, Seidel-Przywiecki Publishing.

in carbon dioxide (CO_2) content, leading to a reduction in reaction. This results in the shift in balance between dissociated and nondissociated forms of VFAs toward nondissociated forms (Banel, 2010; Sadecka, 2010).

With acidification of the reaction environment, dissociation is reversed, leading to easier transfer of nondissociated acid particles to the organic or headspace phase. In their dissociated form, VFAs are mostly soluble in water, whereas they are poorly soluble in organic solvents (nonmixable with water). Low concentrations of nondissociated forms of acetic acid and propionic acid have an inhibiting effect on the yield of the fermentation gas (Banel, 2010; Sadecka, 2010). VFAs are the compounds that are often observed in the environment. However, they rarely occur in a free state; rather, they mostly take the form of esters, salts, or amids. VFAs can be derived from both natural and anthropogenic sources. They are present in large amounts in sewage sludge, sludge generated during farming of breeding swine, and sludge from the food industry. Their sources are also organic waste and liquors generated during deposition of this waste in landfills or composting.

In the liquors generated in young landfills, VFAs account for nearly 80% of all organic components, with their concentrations reaching as much as 6000 mg/L. With the landfill aging, the amount of VFAs in liquor is substantially reduced, even to the level of 5 mg/L (Banel, 2010; Chwiałkowska and Oleśkowicz-Popiel, 2016; Banel and Zygmunt, 2009). VFAs are also generated in human and animal intestines during digestion. In the environment, VFAs also can play a negative role. The compounds from this group may lead to acidification of the environment or increased mobility of heavy metals and radionuclides. Due to their chemical composition, VFAs play a significant role in biological treatment processes, as they represent an easily available source of carbon for microorganisms (Chwiałkowska and Oleśkowicz-Popiel, 2016).

Despite the substantial interest in the problems of acids in various environmental samples, much attention has been devoted to the methodology of determination of these compounds. The choice of adequate analytic methodology used for their determination depends mainly on the accuracy and precision of determinations, matrix composition, available apparatus, predicted level of their content in the sample, and time and cost of performing analysis. The methods based on titration, often preceded by the isolation of analytes from the primary matrix by means of distillation methods, have been used to evaluate total VFA content. Determination of individual acids is conducted using separation techniques, mostly chromatographic (Banel, 2010). The use of ion chromatography often requires a complex procedure of sample cleaning in order to reduce interference and reach an adequately low determination threshold. A useful tool for determination of VFAs in both water and gas samples is gas chromatography with flame ionization detector (GC-FID). This technique allows for the identification and quantitative determination of individual fatty acids (Banel and Zygmunt, 2009). One of the main steps before analysis of the sample from a medium is isolation of the compounds to the level that allows for detection and determination. Table 2 compared the methods used for separation of VFAs from the medium studied.

Table 2 Methods of VFA separation from the medium studied (Zacharof and Lovitt, 2013)

Separation methods	Description	Benefits	Defects
Precipitation	Calcium salts are added to the purified medium, whereas the obtained solution of calcium carboxylane is solidified by vaporization and, through crystallization, separates it from the parent liquid.	• High production coefficient. • Low functional costs. • High-purity products.	Solid waste production due to the use of sulfuric acid for the release of carboxylic acids from calcium carboxylates
Distillation	Ammonia is used to neutralize acids that react in order to generate ammonium carboxylate, which is mixed with alcohol in order to produce esters, followed by division through distillation.	• High purity of products. • By-products can be used as fertilizer.	High costs of energy connected with distillation, which is used for the separation of alcohol from carboxylic acids

Table 2 Continued

Separation methods	**Description**	**Benefits**	**Defects**
Adsorption	Ion-exchange resins are used as adsorbents for exchange of carboxylate ions from the medium.	Easy operation	• High price of resins • High demand for energy due to resin regeneration • Low adsorption abilities • Separation not highly selective
Electrodialysis	With electric current in an electrodialyser, negatively charged carboxylate ions go through an anion-exchange membrane to the anode.	Carboxylate is concentrated in water solution and does not require acidification.	• Products are much contaminated, and it is necessary to continue their purification. • High demand for energy.
Extraction with solvent	Organic acids are used for extraction of carboxylic acids from the purified stream.	• Low cost • Opportunities for the production of carboxylate salts	• Initial acidification of the sample in order to obtain greater and efficient extraction. • Extracts have to be regenerated through distillation or extraction.
Membrane processes	Utilization of semipermeable barriers where VFAs are selectively separated depending on the types of membranes (pore size). Transport is possible at the instant of using the driving force (pressure difference, concentrations, temperature and electrical potential).	High selectivity	• Occurrence of concentration polarization • Reduction in the stream of permeate due to the phenomena of fouling and scaling • High functional costs

Table 3 Methods of production of carboxylic acids (Zacharof and Lovitt, 2013)

Carboxylic acid	Chemical synthesis methods	Bioprocess
Formic acid	Alkene oxidation Hydrogenation of CO_2 Methanol carbonylation	Glucose oxidation CH_4 fermentation
Acetic acid	Methanol carbonylation Acetaldehyde oxidation Ethylene oxidation	Glucose oxidation CH_4 fermentation
Propionic acid	Ethylene hydrocarboxylation Aerobic oxidation of propionaldehyde	CH_4 fermentation
Butyric acid	Chemical oxidation of butyraldehyde	Alcohol fermentation
Caproic acid	Ethylene oxidation	CH_4 fermentation
Lactic acid	Chemical synthesis	CH_4 fermentation

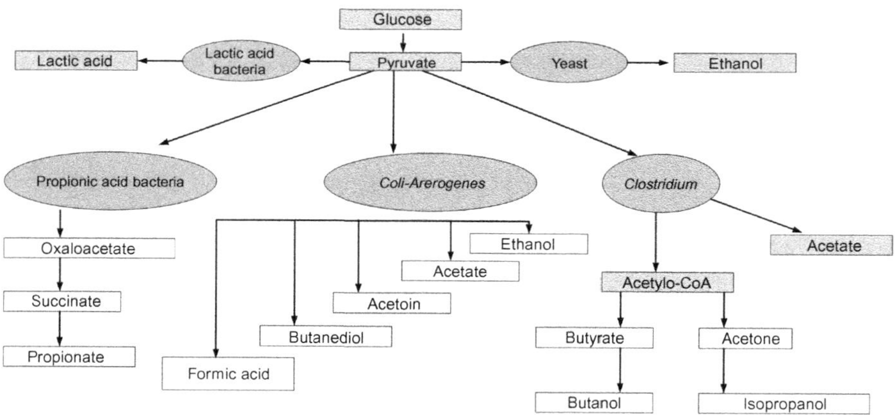

Fig. 2 Diagram of hydrocarbon fermentation processes in bacteria.
Based on Sadecka, Z., 2010. Basics of Biological Wastewater Treatment, Seidel-Przywiecki Publishing.

2 Production of VFAs

VFAs can be produced via the chemical processing of petroleum or anaerobic biological processes. Table 3 compares the most frequently used processes of VFA production.

The diagram of initiated CH_4 fermentation can be presented using the example of carbohydrate decomposition (Fig. 2). It shows that this process begins from glycolysis as it is the case in aerobic processes. The central derivative from which the pathways of aerobic and anaerobic biodegradation diverge is pyruvate. The fermentation process can be conducted depending on the orientation to a specific product. Different

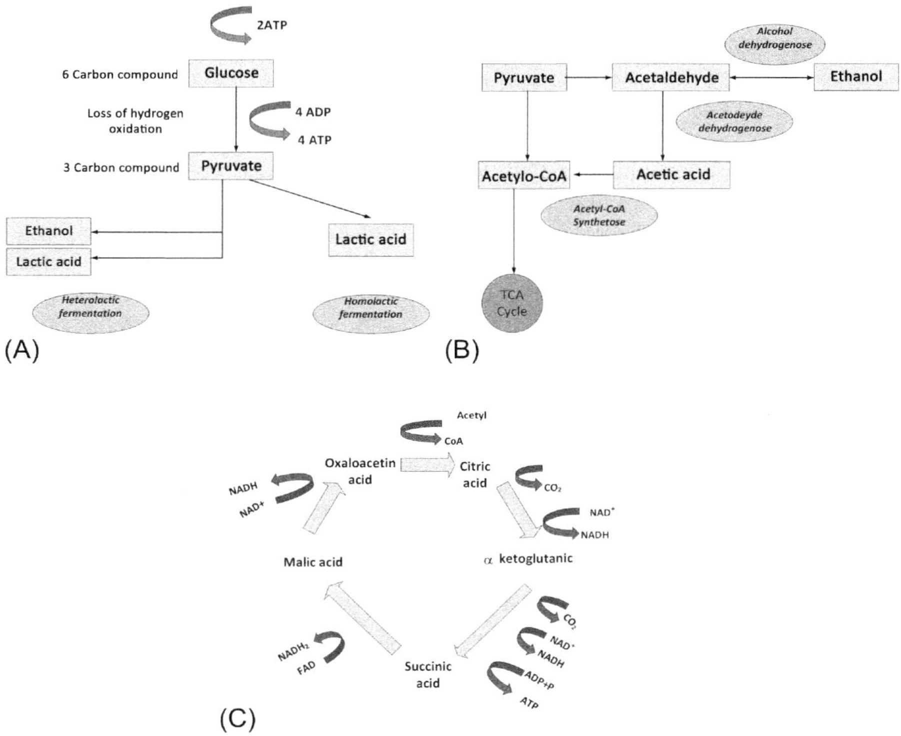

Fig. 3 Biological pathways of carbohydrate conversion into VFAs during homolactic and heterolactic fermentation (A), alcohol fermentation (B), and during the cycle of tricarboxylic acids (C).
Based on Zacharof (2013).

types of fermentation have been discovered, with their names taken from the dominant final product: alcoholic, lactic, propionic, formic, butyric, and acetic fermentation. In addition to the dominant final products, there are succinate, caproate, n-butanol, 2,3-butanodiol, acetone, isopropanol, CO_2, and hydrogen. All of the listed fermentation products can be generated separately or in different arrangements (Schlegel, 2003). Fig. 3 illustrates schematically biological pathways of carbohydrate conversion into VFAs during homolactic and heterolactic fermentation (in panel A), alcohol fermentation (B) and through the cycle of tricarboxylic acids (C).

Extensive research has been conducted in recent years on the use of mixed cultures for production of carboxylic acids. With the high flexibility of these cultures, it is possible to process organic waste, sewage sludge, and agricultural and food industry waste (Chwiałkowska and Oleśkowicz-Popiel, 2016).

In anaerobic conditions, biological decomposition of organic matter occurs in several stages. The first stage of the CH_4 fermentation process is hydrolysis; when insoluble, polymerized organic compounds are exposed to the effect of extracellular hydrolytic bacteria enzymes (e.g., amylases, lipases, or proteases), leading to their decomposition and liquidization. Polysaccharides are hydrolysed to monosaccharides, proteins to amino acids, and fats to polyols and fatty acids. The hydrolysis phase is a stage that limits the course and efficiency of fermentation (Zielewicz, 2007; Agler et al., 2011).

Another stage is the phase of acidification or acidogenesis, which is also termed *dark fermentation*. The initial substrate for acid-forming bacteria is hydrolysis products. The bacteria that participate in this process are facultative bacteria, but there are also obligate anaerobes. The most frequently used of these include *Aerobacter*, *Alcaligenes*, *Clostridium*, *Escherichia*, *Flavobacterium*, *Pseudomonas*, *Lactobacterium*, *Lactobacillus*, *Micrococcus*, and *Streptococcus*. Bacteria of the acidification phase are relatively less susceptible to temperature or pH changes. They proliferate much faster than CH_4 bacteria. With their activity, low molecular organic compounds are processed mainly (76%) into VFAs (e.g., formic acid, acetic acid, propionic acid, butyric acid, valeric acid, and caproic acid) and to alcohols (e.g., ethanol and methanol) and gaseous products (CO_2 and H_2). The remaining part is decomposed into acetates. In this phase, proteins are further decomposed into simple organic acids, thiols, and amines (Zielewicz, 2007; Rosik-Dulewska, 2007; Sadecka, 2010).

In the third stage of CH_4 fermentation, the so-called acetate phase (*acetogenesis*), ethanol and VFAs are processed into CO_2, hydrogen, and acetic acid. These transitions occur with two groups of anaerobic bacteria of the *Syntrophobacter* and *Syntrophomonas* genuses (Michalska et al., 2014). The first group of acetogenic microorganisms is responsible for hydrolysis of organic polymers and lipids to basic structure-forming compounds such as monosaccharides or amino acids. The second group is involved in fermentation and destruction of products to simple organic acids, including acetic acid, propionic acid, and butyric acid (Zielewicz, 2007)

Distribution of organic acids by bacteria is relatively specific because it requires the involvement of various bacteria species. For example, valeric acid is decomposed into acetic acid and propionic acid by *Mbacterium suboxydans*, but further decomposition of propionians into acetates, CO_2, and CH_4 is achieved with *Mbacterium propionicum*. The latter species is unable to absorb and decompose acetates; therefore, their further decomposition into CH_4 requires the presence of *Methanococcus mazei* bacteria (Fig. 4).

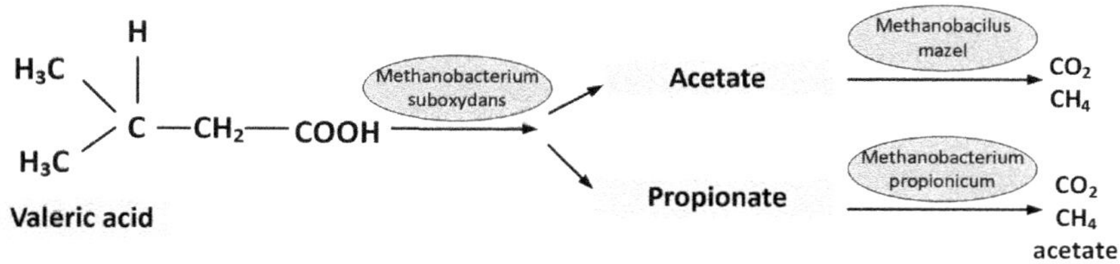

Fig. 4 Activity of bacteria to produce CH_4 from valeric acid.
Based on Sadecka, Z., 2010. Basics of Biological Wastewater Treatment, Seidel-Przywiecki Publishing.

In addition to fermentation of carbohydrates and proteins, acetic acid is generated during the anaerobic oxidation of fats. Generation of propionic acid occurs first and foremost during the degradation of carbohydrates. Another product of the fermentation of proteins and fats is butyric acid. This acid also can be generated during the fermentation of carbohydrates from pyruvate.

Oxidation of acids such as lactic acid, propionic acid, and lactic acid is possible if hydrogen produced by acetogens is received from the environment. Their oxidation is possible only at very low concentration of hydrogen and low redox potential ($<-300\,mV$). An insignificant increase in hydrogen pressure results in an immediate decline in the rate of degradation of these acids (Błaszczyk, 2009). These transitions can occur if molecular hydrogen is removed from the reaction environment. With these environmental requirements, the two groups of microorganisms (i.e., acetogenic and methanogenic bacteria) live in symbiosis (Zalewski et al., 2012). Both types of bacteria can connect to various aggregates (e.g., flocs or granules), and their symbiosis results in the decomposition of propionic acid according to Eqs. (1)–(3) (Bitton, 2005; Rosik-Dulewska, 2007; Żygadło, 1999).

- Acetogenic bacteria:

$$CH_3CH_2COO- + 3H_2O \rightarrow HCO_3^- + CH_3COO^- + H^+ + 3H_2 \quad (1)$$

- Methanogenic bacteria:

$$HCO_3^- + 4H_2 + H^+ \rightarrow CH_4 + 3H_2O \quad (2)$$

The total balancing reaction for these chemical reactions can be written as follows:

$$4CH_3CH_2COO^- + 3H_2O \rightarrow HCO_3^- + 4CH_3COO^- + 3CH_4 + H^+ \quad (3)$$

The phenomenon of this symbiosis was discovered in the 1970s and is termed *interspecies hydrogen transfer (IRT)*. Undisturbed hydrogen transfer guarantees the correct course of anaerobic distribution of organic substrates.

Isolation of pure cultures of acetic bacteria is impossible, and growth of mixed cultures is very difficult and labour-consuming. Bacteria of the acetogenic phase are characterized by a long time of generation that depends on the type of available substrate. Fig. 5 presents the times of acetic bacteria generation at mesophilic temperature depending on the type of substrate (Sadecka, 2010).

In addition to production of acetate from hydrogen and CO_2, acetogenic bacteria can utilize other simple compounds (e.g., carbon dioxide, methanol, or glucosis), generating acetate as a final product of metabolism. Production of acetic acid by acetogenic microorganisms proceeds according to the reactions given as Eqs. (4)–(7) (Błaszczyk, 2009):

$$CO_2 + 4H_2 \rightarrow CH_3COOH + 2\,H_2O \quad (4)$$

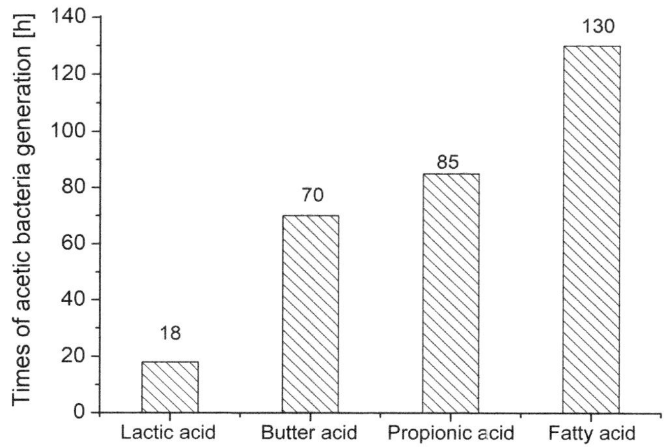

Fig. 5 Times of acetic bacteria generation at mesophilic temperatures (35–37°C) depending on the type of substrate.
Based on Sadecka, Z., 2010. Basics of Biological Wastewater Treatment, Seidel-Przywiecki Publishing.

$$\Delta G^{\circ} = -95 \text{ kJ/mol ethyl}$$

$$4\,CO_2 + 2H_2O \rightarrow CH_3COOH + 2\,CO_2 + H^+ \tag{5}$$

$$\Delta G^{\circ} = -175\,\text{kJ/mol ethyl}$$

$$4CH_3OH + 2\,CO_2 \rightarrow 3CH_3COOH + 2\,H_2O + 3H^+ \tag{6}$$

$$\Delta G^{\circ} = -7\,\text{kJ/mol ethyl}$$

$$C_6H_{12}O_6 \rightarrow 3CH_3COOH + 3H^+ \tag{7}$$

$$\Delta G^{\circ} = -104\,\text{kJ/mol ethyl}$$

Disturbance to the process is observed when instead of a specific methanogenic bacteria, hydrogen starts to be consumed by another "partner." This phenomenon is observed when sulfur bacteria from the *Desulfovibrio* genus start to be dominant in the environment. A large amount of sulfates in sewage sludge contributes to the growth of these bacteria. They oxidate hydrogen while reducing sulfates to hydrogen sulfide. Consequently, less hydrogen is supplied to CH_4 bacteria, leading to lower CH_4 production. The bacteria that reduce sulfates are active in two configurations: sometimes as acetogenic bacteria and sometimes as a competitor of acetogenic and methanogenic symbiosis (Fig. 6).

The last phase of the fermentation process (methanogenesis) is dominated by methanogenic bacteria that belong to obligate anaerobes of the *Methanobacterum*,

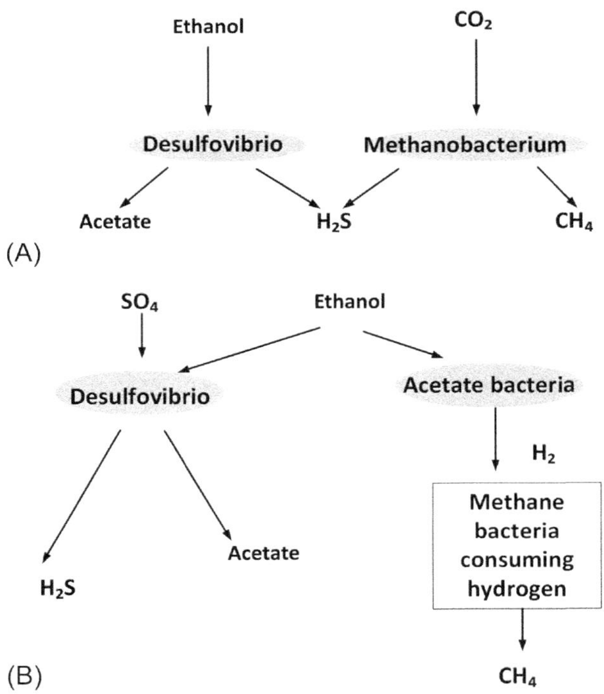

Fig. 6 Sulfur bacteria as a competitor of acetogenic and methanogenic symbiosis. The bacteria that reduce sulfates are active in two configurations: sometimes as acetogenic bacteria (B) and sometimes as a competitor of aceto- and methanogenic symbiosis (A). Based on Sadecka, Z., 2010. Basics of Biological Wastewater Treatment, Seidel-Przywiecki Publishing.

Methanococcus, and *Methanogenium* genuses. They are characterized by high susceptibility to temperature and reaction changes. The time of generation ranges from 15 to 85 h. In addition to biocoenosis composition, its duration is affected by the temperature of the process and the type of substrate, which are presented in Fig. 7 (Sadecka, 2010).

Two groups of bacteria can be distinguished, one of which generates CH_4 from formic acid, methanol, hydrogen, and CO_2. Energy yield in these reactions ranges from 106 to even 145 kJ/mol CH_4. Time of generation of these bacteria is short, whereas their tolerance to changes in reaction is high. Reduction of CO_2 with hydrogen is presented as follows:

$$CO_2 + 4H_2 \rightarrow CH_4 + 2H_2O \tag{8}$$

This reaction occurs with the participation of coenzymes and complex enzymatic systems. Synthesis of CH_4 from methanol occurs according to the following reaction:

$$CH_3OH + H_2 \rightarrow CH_4 + 2H_2O \tag{9}$$

The second group of bacteria is much poorer in type and generates CH_4 from acetic acid according to the following reaction (Sadecka, 2010):

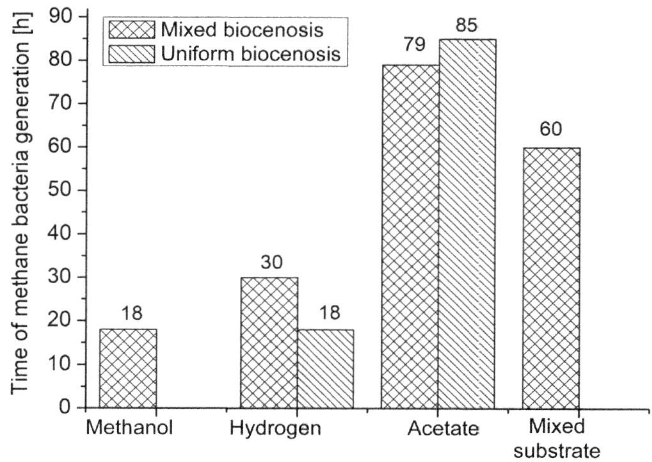

Fig. 7 Times of CH_4 bacteria generation at mesophilic temperatures (35–37°C) depending on the type of substrate.
Based on Sadecka, Z., 2010. Basics of Biological Wastewater Treatment, Seidel-Przywiecki Publishing.

$$CH_3COOH \rightarrow CH_4 + CO_2 \text{ and } \Delta G = -35.9\,\text{kJ/mol} \tag{10}$$

Bacteria that decompose acetic acid to CH_4 grow relatively slowly and are susceptible to environmental changes (Sadecka, 2010). The studies have shown that about 70% CH_4 is generated during CH_4 fermentation from acetic acid, whereas 30% is generated from hydrogen and CO_2 (Janosz-Rajczyk, 2008).

A key factor in VFA yield is inhibition of the last phase (methanogenesis). This can be achieved by adding a specific inhibitor or changing the operating conditions of the process so that they are unfavorable to methanogens.

3 Effect of various factors on VFA yield

Fermentation using mixed bacteria cultures is characterized by the fact that with adequate control of such parameters as pH, temperature, and retention time, there are opportunities for obtaining products such as VFAs, biopolymers, and long-chain fatty acids (Shen et al., 2014; Arslan et al., 2013).

3.1 pH

Another important factor in VFA yield is the value of maintained pH in the reactor. Many of the bacteria that generate these acids do not tolerate an environment that is strongly acidic environment ($pH < 3$) or strongly alkaline ($pH > 12$). The hydrolysis stage occurs most efficiently when the pH value is maintained between 5 and 7 (Arslan et al., 2013). Depending on the type of the substrate used, the most beneficial pH for generation of VFA ranges from 5.5 to 11. At the time when fermentation is used for sewage sludge, it is indicated that the reaction should be alkaline, food waste–neutral, and sewage-acidic (Shen et al., 2014).

CH_4 bacteria develop most efficiently in an environment with pH similar to neutral (Carrillo-Reyes et al., 2014). Numerous studies have demonstrated that inhibition of the methanogenesis process occurs at pH 10–11. This causes a substantial increase in VFA yield compared to the process conducted at lower pH values. It was demonstrated that the VFA yield from activated sludge at pH 10 is over 6 times higher than for pH 4. Similar findings were observed during examinations performed by Liu et al. in 2012, who demonstrated that inhibition of methanogenesis occurs at pH 3, 11, and 12, whereas the VHA yield at pH 9 is nearly 10 times higher than for pH 3 (Khiewwijit et al., 2015).

In addition to VFA generation rate, pH has an effect on their type. The most beneficial pH for generation of propionic acid during sludge fermentation ranges from 4 to 4.5, whereas for acetic acid and butyric acid, it ranges from 6 to 6.5. If the substrate is whey, an insignificant increase in pH (from 5.25 to 6) leads to an increase in the production of propionic acid and simultaneous lower production of butyric and acetic acid. Using a substrate rich in glucose, an increase in pH from 6 to 8 results in the increase in production of acetic and propionic acid, at the expense of butyric acid. This can be linked to the dominance of other groups of bacteria depending on the environment reaction (i.e., *Clostridium butyricum* (pH 6) and *Propionibacterium* (pH 8)) (Shen et al., 2014).

3.2 Temperature

A number of studies have been devoted to the effect that the temperature of the CH_4 fermentation process has on the level of production of VFAs. The examinations were performed at psychrophilic temperatures (4–20°C), mesophilic temperatures (20–50°C), thermophilic temperatures (50–60°C) and extremely thermophilic temperatures (60–80°C).

It was found that the increase in temperatures in the first two ranges affects the increase in VFA concentration, as well as the speed and efficiency of their production. Increasing the temperature from 10°C to 35°C during fermentation of activated sludge results in the increase in VFA concentration by 300%, whereas the increase in temperature from 8°C to 25°C during fermentation of initial sludge can even lead to a sixfold increase in the VFA production index. Numerous examinations have demonstrated that VFA generation at thermophilic temperature (55°C) is lower than at mesophilic temperature (37°C). The amount of VFA obtained at a temperature of 55°C is up to 40% lower than at 37°C (Shen et al., 2014).

Most of CH_4 bacteria grow optimally in the mesophilic range of temperatures. The increase in temperature of 60°C or its reduction below 20°C results in higher efficiency of production of hydrogen, hydrogen dioxide, and ethanol during CH_4 fermentation. Biogas maintained during fermentation of sewage sludge in these temperatures is characterized by lower CH_4 content (Ledakowicz and Krzystek, 2005; Shen et al., 2014). Compared to pH, the effect of temperature on VFA production is low and changes the composition of generated acids to an insignificant degree (Shen et al., 2014; Chwiałkowska and Oleśkowicz-Popiel, 2016).

3.3 Retention time

Important parameters of monitoring of anaerobic bioreactors include hydraulic retention time (HRT) and time of retention of mixed cultures of bacteria (i.e., sludge retention time (SRT)). HRT is connected with bioreactor volume, whereas SRT has an effect on the bacteria species that live in the bioreactor (Shen et al., 2014). Sufficiently long hydraulic retention times allow for longer contact of microorganisms with the substrate, resulting in greater production of VFAs. However, excessive HRT times may lead to the stagnation of carboxylic acid production. It should be also remembered that longer HRT times require bigger reactors, which is linked to higher functional costs (Shen et al., 2014). Similar to pH, control of HRT allows for the production of a specific type of carboxylic acids. The studies have shown that the increase in HRT from 20h to 95h is conducive to the production of propionic acid, but it inhibits the generation of butyric acid. Similarly, extending the HRT time during fermentation of sewage from the paper industry from 11h to 24h results in more intensive generation of propionic acid and lower generation of butyric acid. However, the changes in HRT from 1 day to 4 days do not affect VFA composition during cofermentation of sewage sludge with the waste from fruit and vegetable industry (Shen et al., 2014).

The effect of SRT on the composition of VFA generated has not been fully examined to date, while it has been found that SRT should be long enough to allow for substrate hydrolysis. Available results from a number of studies have demonstrated that shorter SRT times are beneficial to VFA generation from sewage sludge. This involves a lower rate of growth of methanogenic bacteria than acidogenic bacteria. It was found that elongation of SRT from 4 to 12days during fermentation of activated sludge yielded a greater amount of VFA. Another extension of SRT to 16days, however, caused a reduction in the amount of VFA in the reactor due to their gradual use by methanogenic bacteria (Shen et al., 2014).

Table 4 shows selected parameters and factors that disturb the anaerobic process of biodegradation of organic compounds. A frequent case is washing methanogenic bacteria from anaerobic bioreactors (Miksch, 2010). During inhibition of methanogenesis, it is possible to increase the production of carboxylic acids and continue elongation of their carbon chain while yielding products that are easier to separate from the mixture of fermentation residue (Oleskowicz-Popiel and Jankowska, 2014).

4 Substrates used for VFA production

Due to factors such as high content of organic matter, sewage sludge generated in wastewater treatment plants (WWTPs) is a valuable substrate for the generation of VFAs. The chemical oxygen demand (COD) value of this sludge ranges from 14,800–23,000mg/L. VFAs account for 3.5%–6.4% d.m. of initial sludge

Table 4 Disturbances in the process of anaerobic biodegradation of organic compounds

Parameter or disturbing factor	Biological effect	Change size		
		Gas production	VFA	pH
Increase in substrate load with solved organisms	Disturbed balance between process phases	+++	+++	- -
Increase in substrate load with organisms in the suspension	Washing the CH_4 bacteria	++	++	
Temperature fluctuations	Disturbed balance between process phases	- - -	+++	- -
Increase in pH > 8	Inhibition of methanogenesis	- - -	+++	+++
Decline in pH < 6	Inhibition of methanogenesis	- - -	+++	- - -
Toxic substances in the inlet	Inhibition of methanogenesis	- - -	+++	- -

- - - Fast decline; +++ Fast increase.
- - Moderate decline; ++ Moderate increase.
Based on Miksch, K., 2010. Wastewater Biotechnology. Warsaw, PWN.

(Malej, 2001). Greater amounts of VFAs are obtained at higher concentrations of organic dry matter. The problem occurs with the cellular structure of the initial sludge and excess sludge, which limits the stage of hydrolysis. This sludge requires initial processing that intensifies this fermentation stage (i.e., chemical, biological, and thermal processing) using microwaves, ultrasound, or combinations of these methods. The presence of VFA is not observed in fresh activated sludge because using organic acids as easily available substrates takes place quickly in biochemical transitions. The soluble part of COD in this sludge is smaller, which delays their hydrolysis. At the moment of performing cofermentation of the mixture of initial sludge and activated sludge, hydrolysis occurs most effectively. Maximal production of VFAs from initial sludge is on average 85 mg COD/$g_{d.m}$ and can even increase by 40% (to 119 mg COD/g_s) when the activated sludge is added with the ratio of 1:1. The increase in VFA production from initial sludge can be obtained by adding sludge that is rich in starch (Shen et al., 2014; Chwiałkowska and Oleśkowicz-Popiel, 2016).

There is a huge potential of VFA production from biomass and biological waste in Europe. The amount of waste and by-products of the food industry is 250 million Mt/year (Jankowska et al., 2017). Food waste and kitchen waste comprise a group that accounts for 22%–54% of municipal waste, which can be used successfully for VFA production. Their content of COD is very high, ranging from 91,900–166,180 mg/L. In the case of this substrate, the problem arises with separation of the usable part of organic waste from nonbiodegradable fraction (i.e., glass, aluminum, or plastics)

(Shen et al., 2014). Production of VFAs also can use liquid waste and wastewater from agriculture and the dairy, cellulose, and paper industries. The whey permeate and waste from paper and cellulose industry are suitable for VFA generation due to their high content of easily fermenting organic matter. Despite the high COD ($<10{,}000\,mg/L$), petrochemical waste is not a good substrate for fermentation, as it contains compounds that can have a toxic effect on microorganisms. CH_4 fermentation is also not used for municipal waste due to its low COD content, ranging from 450 to 1200 mg/L (Shen et al., 2014; Sadecka, 2010; Janosz-Rajczyk, 2008).

The lignocellulose biomass can be also used as a substrate for production of VFAs, with substantial resources on Earth. Its global production in 2009 was 916 million tons. Lignocellulose belongs to materials that are especially resistant to degradation processes. Consequently, with insufficient initial processing, its distribution in the phase of hydrolytic CH_4 fermentation is slow and incomplete (Oleskowicz-Popiel and Jankowska, 2014; Ghimire et al., 2015). Disintegration of lignocellulose biomass allows, through permeating the layer of lignin, for hydrolysis of cellulose and hemicellulose to simple sugars. The generated carbohydrates can be converted to organic acids or biohydrogen during dark fermentation (Ghimire et al., 2015). Biological initial processing can be performed by means of fungi such as *Phanerichataete chrysosporium*, *Phlebia radiata*, *Dichmitus squalens*, *Rigidosporus lognosus*, and *Jungua separabilima* (Owczuk et al., 2014). Lignin is not decomposed during hydrolysis (Mulka and Szlachta, 2013).

It is generally accepted that the best substrate for VFA production must be characterized by a high content of organic matter ($COD > 4000\,mg/L$) and the content of ammonium nitrogen should be $<5000\,mg/L$. Apart from properties of substrates used for fermentation processes, one should take into consideration their availability (Shen et al., 2014). The substrates used for production of VFAs include activated sludge, primary sludge, food waste, kitchen waste, and palm oil. Higher concentrations of VFAs can be obtained with the addition of cosubstrate (Shen et al., 2014).

5 Technological systems used for VFA production

Two technologies of bioreactors are used for production of VFAs from waste or industrial wastewater. With the first type, the growth of microorganisms occurs on porous material (e.g., activated carbon or zeolites), whereas with the second type, it occurs in tiny suspended particles (e.g., sand). For industrial or semi-industrial purposes, production of VFAs is performed by means of various types of reactors, filled with porous bed, fluidized-bed, UASB reactors, and CSTR (Fig. 8).

Sludge used in UASB reactors has a granulated form characterized by a specific composition of biocenosis. Bacteria that reduce sulfates (acidogenesis) and homoacetate bacteria that produce acetates and bind water to CO_2 develop on the external part of the granule. A dominant type of colonizing bacteria in the granule is bacteria of the *Methanosarcina* genus. They are present in both external and central parts of the granule. Homoacetate bacteria and CH_4 bacteria that consume hydrogen and CO_2 in biochemical processes develop under the external layer The use of the sludge in the

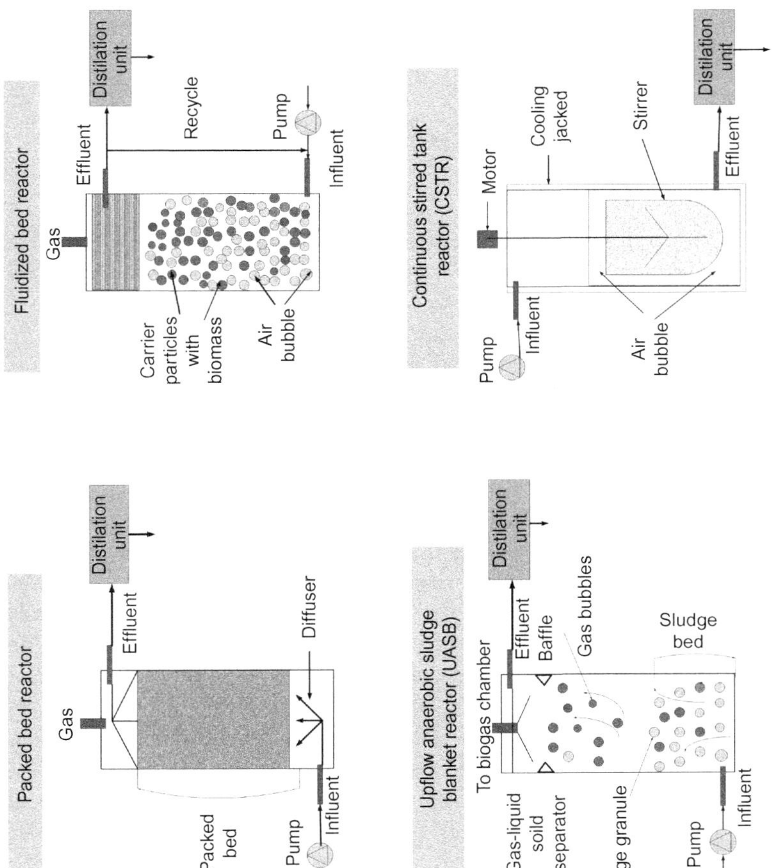

Fig. 8 Types of reactors for VFA production.
Based on Lee, W.S., Chua, A.S.M., Yeoh, H.K., Ngoh, G.C., 2014. A review of the production and applications of waste-derived volatile fatty acids. Chem. Eng. J. 235, 83–99; Bhatia, S.K., Yang, Y.H., 2017. Microbial production of volatile fatty acids: current status and future perspectives. Rev. Environ. Sci. Biotechnol. 16, 327–345.

form of granules in the anaerobic process allows for the use of higher loads of reactor with the load of contaminants, while maintaining a long retention time (Kwarciak-Kozlowska and Krzywicka, 2015). Reactors of the UASB type have opportunities for working at a high load of contaminants (from 5 to 15 kgCOD/m^3d) and sludge concentration (from 10 to 50 g d.m.o/L). CSTRs operate at a lower load of contaminants to the sludge compared to UASB reactors, not higher than 5 kgCOD/m^3d.

The abovementioned anaerobic bioreactors operate mostly in the flow mode, which results in a short retention time. To accommodate reactions that need a longer time to occur, one should use reactors operating in the periodical mode (i.e., the breaks should be made during the supply and return of their content) (Shen et al., 2014).

6 Intensification of hydrolysis in order to increase VFA yields

Generation of VFAs from initial sludge has been used virtually in many WWTPs. However, there are numerous examinations of intensification of hydrolysis by initial conditioning of wastewater, sewage sludge, or waste using physical, chemical, biological, or hybrid methods (Zielewicz, 2007; Zawieja et al., 2016).

The example of this method is ultrasonic disintegration, which significantly affects the intensification of the phase of hydrolysis of organic compounds, consequently leading to increased production of VFAs. The increased VFA production in the fermentation process results from both biological hydrolysis, representing the first stage of the process, and the phenomenon of sonochemical hydrolysis, taking place during sludge conditioning with ultrasound fields. Such compounds as nucleic acids, proteins, polysaccharides, and lipoproteids are decomposed when exposed to the effect of ultrasonic vibration. These processes are similar to the reaction of enzymatic hydrolysis, but disintegration of particles occurs irregularly during exposure to ultrasounds (Zielewicz, 2007; Zawieja and Wolny, 2014; Zawieja et al., 2015). Depending on the amount of energy supplied and sonication time, the first phase of conditioning with an ultrasound field leads to the disintegration of flocculent structures. However, this stage does not lead to the destruction of cellular walls of microorganisms. This occurs in the next phase, where free hydroxyl radicals generated during sonochemical reactions react with molecules of various organic substances, leading to their chemical decomposition and transformation of soluble forms (Zawieja and Wolny, 2014; Kwarciak-Kozlowska and Krzywicka, 2015).

In a study published in 2009, Zawieja et al. demonstrated that with sonochemical processes occurring during conditioning of sewage sludge with an ultrasonic field, extension of sonication time and increase in vibration amplitude lead to a substantial increase in VFAs. Conditioning of sewage sludge with ultrasound fields with a vibration amplitude of 36.6 μm allows for an eightfold increase in VFA over their initial value. In the case of fermentation not supported with the effect of ultrasonic fields, the increase in VFA yield was 2.5 times. The frequency of sonication in that study was 40 kHz. (Zawieja et al., 2009). A somewhat worse effect in VFA generation (i.e., only 3.2-fold increase during CH_4 fermentation of sewage sludge) was observed if the sonication frequency was 25 kHz. This study also showed that the increase in VFA for nonconditioned tests was about 2.6 times (Kuszmider and Kudela, 2015).

Another physical method that affects the increase in disintegration of organic compounds is the thermal method. It leads to the destruction of cells of microorganisms and acceleration or shortening of hydrolysis. The use of the thermal method before CH_4 fermentation leads to an increase in the speed of VFA generation and increase in its yield on consecutive days of the process. With initiation of thermal hydrolysis, a higher degree of transformation of VFAs into CH_4 is obtained during the anaerobic stabilization process (Zawieja and Wolski, 2013). After thermal conditioning at 50°C and 70°C, an increase in the VFA yield on the second day of the process compared to the initial value is observed, at 22% and 30%, respectively.

Other studies have shown that the best effects of acceleration of hydrolysis are obtained by storing sludge at temperature of 165–180°C for 10–30 min. Conducting the process at lower temperatures (60–80°C) has a beneficial effect on the composition of the obtained hydrolysate, but it requires longer reaction times (60–120 min). The hydrolysis phase also can be accelerated by alternate freezing and defrosting of the biomass. The experiments performed by Franceschini in 2010 demonstrated that the best effects are obtained for a single cycle of freezing and defrosting at freezing temperatures (below 10°C). The process is most effective if the freezing temperature is reduced at the rate of 1–3°C per min (Owczuk et al., 2014).

An alternative solution is offered by the use of microwave radiation. The effect of microwave radiation is of selective character, which means that it is effective only for substances with specific dielectric properties. Numerous studies have shown an increase in activity of enzymatic processes in systems subjected to the effect of microwave radiation compared to a study conducted conventionally (Dębowski et al., 2010). It was demonstrated that the highest increase in COD and VFAs occurs in the initial period of exposure of samples to the effect of microwaves (i.e., from 0 to 3 min) (Owczuk et al., 2014). However, microwave radiation may stimulate the formation of hardly biodegradable compounds such as melanoidins or humic acids (Shen et al., 2014).

Popular chemical methods include conditioning with bases or acids. The use of acids or bases as an initial method of processing is aimed to increase the solubility of extracellular polymeric substances and then release extracellular organic matter through the destruction of cellular walls. Another option is initial processing using strong oxidizers such as ozone (O_3) and hydrogen peroxide (H_2O_2). The use of O_3 involves higher costs than H_2O_2. However, O_3 has a higher oxidation potential (2.1 V) compared to the cheaper alternative of H_2O_2 (1.8 V). The zone dose used solubilization of solid waste is 0.16 $g/g_{s.m}$, whereas this value for H_2O_2 is about 1 $g/g_{s.m}$ (Shen et al., 2014). VFA generation from sewage sludge also can be based on peracetic acid. The use of this reactant with the amount of 0.1 cm^3 CH_3COOOH/L leads to a 15% increase in VFAs. The increase in its dose to 0.5 cm^3 CH3COOOH/L and 2.5 cm^3 CH_3COOOH/L results in a further increase in their amounts by 41% and 60%, respectively (Zawieja and Wolski, 2013).

Another method of intensification of hydrolysis is adding biological agents such as hydrolytic enzymes, pure cultures of *Cellulomonasuda*, *Cellulomonas biazotea*,

Aspergillus awamori, activated sludge, and mature compost. The dose added to the reactor is mostly a mixture of enzymes, which allows for a simultaneous hydrolysis of various organic compounds that are contained in the fermented substrate. The initial processing using enzymes and pure cultures is expensive, but a cheaper alternative can be a mature compost or activated sludge (Shen et al., 2014).

Numerous examinations that combine various methods of intensification of the hydrolysis stage have been conducted. When choosing a combination of the methods of initial support for hydrolysis, the amount of waste or sewage sludge for biodegradation and investment and functional costs also should be taken into consideration. Methods of intensification of hydrolysis contribute to the generation of VFAs. Single pretreatment methods include acid, alkaline, O3, H_2O_2, biological, microwave, ultrasound, and thermal, and combined pretreatments include slkaline + ultrasound, O3 + ultrasound, thermal + biological, and ultrasound + thermal (Shen et al., 2014).

7 The use of VFAs in industry

VFAs are mainly used in the food, pharmaceutical, petrochemical, chemical and cosmetics industries (Fig. 9). The increase in interest in the CH_4 fermentation process in recent years has led to research into the production of chemicals from a variety of biomass types. It is possible to combine CH_4 fermentation with other processes based on the so-called carboxylic platforms, where VFAs are intermediates. The main advantages of the biological acquisition of VFA are its efficiency, which does not require

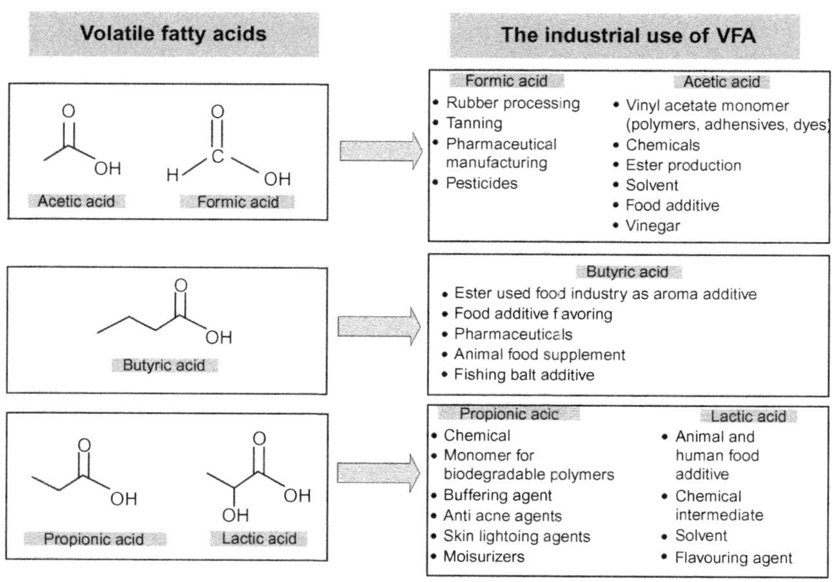

Fig. 9 The industrial use of VFAs. Based on Zacharof (2013).

sterilization, low costs, low CO_2 production, and coproduction of hydrogen (Zacharof and Lovitt, 2013).

In recent years, many studies have been carried out to maximize the production of VFA from various types of waste, along with the methods of preconditioning. Adjusting the conditions of an anaerobic reactor and proper process control allow for the production of the appropriate type of carboxylic acid. The use of VFAs in industrial processes requires the use of integrated methods of processing, and the appropriate place for this to occur is called a *biorefinery* (Oleskowicz-Popiel and Jankowska, 2014). A wide range of chemical and biochemical processes are used here, including chemical and/or enzymatic hydrolysis, dehydration, fermentation involving microorganisms, pyrolysis, thermal deoxidation, and hydrogenation (Szwach and Kulesza, 2014). The VFAs obtained in the anaerobic process as intermediates can be used for the production of biopolymers (PHA), biohydrogen, biogas, bioplastics, and biodiesel, among other products (Grzelak et al., 2015; Shen et al., 2014; Chen and Wu, 2005).

VFAs also have been used in the technology of microbial fuel cells (MFCs). The type of VFA used in MFCs has a significant impact on electricity generation efficiency. The average amount of current produced in MFC supplied with acetate is twice as high as with higher-molecular-weight carboxylic acids (Freguia et al., 2010; Markowska et al., 2013).

References

Agler, M.T., Wrenn, B.A., Zinder, S.H., Angenent, L.T., 2011. Waste to bioproduct conversion with undefined mixed cultures: the carboxylate platform. Trends Biotechnol. 29, 70–78.

Arslan, D., Steinbusch, K.J.J., Diels, L., De Wever, H., Hamelers, H.V.M., Buisman, C.J.N., 2013. Selective carboxylate production by controlling hydrogen, carbon dioxide and substrate concentrations in mixed culture fermentation. Bioresour. Technol. 136, 452–460.

Banel, A., 2010. Development and comparison of chromatographic methodologies for determination of short-chain alkan monocarboxylic acids in water samples and permanent. Doctoral Dissertation, Gdansk.

Banel, A., Zygmunt, B., 2009. Volatile fatty acids in the landfill—occurrence and determination. Ecol. Chem. Eng. 16 (S2), 192–206.

Bitton, G., 2005. Wastewater Microbiology. John Wiley & Sons, Inc.

Błaszczyk, M., 2009. Microorganisms in Environmental Protection. PWN Scientific Publisher.

Carrillo-Reyes, J., Celis, L.B., Alatriste-Mondragón, F., Razo-Flores, E., 2014. Decreasing methane production in hydrogenogenic UASB reactors fed with cheese whey. Biomass Bioenergy 63, 101–108.

Chen, G.Q., Wu, Q., 2005. The application of polyhydroxyalkanoates as tissue engineering materials. Biomaterials 26 (33), 6565–6578.

Chwiałkowska, J., Oleśkowicz-Popiel, P., 2016. The carboxylic platform. Production and use of volatile fatty acids. Chem. Ind. 95 (4), 800–808.

Dębowski, M., Zieliński, M., Krzemieniewski, M., 2010. Influence of microwave radiation on methane fermentation of post-slaughter waste in thermophilic conditions. Annu. Set Environ. Protect. 12, 381–392.

Freguia, S., The, E.H., Boon, N., Rabaey, K., 2010. Microbial fuel cells operating on mixed fatty acids. Bioresour. Technol. 101 (4), 1233–1238.

Ghimire, A., Frunzo, L., Pirozzi, Trably, E., Escudie, R., Lens, P., Esposito, G., 2015. A review on dark fermentative biohydrogen production from organic biomass: process parameters and use of by-products. Appl. Energy 144, 73–95.

Grzelak, J., Ślęzak, R., Krzystek, L., 2015. Obtaining volatile fatty acids and hydrogen from the organic fraction of municipal solid waste. Eng. Chem. Apparatus 4, 157–158.

Jankowska, E., Chwialkowska, J., Stodolny, M., Oleskowicz-Popiel, P., 2017. Volatile fatty acids production during mixed culture fermentation—the impact of substrate complexity and pH. Chem. Eng. J. 326, 901–910.

Janosz-Rajczyk, M., 2008. Research on Selected Wastewater Treatment Processes. Technical University of Czestochowa.

Khiewwijit, R., Temmink, H., Labanda, A., Rijnaarts, H., Keesman, K.J., 2015. Production of volatile fatty acids from sewage organic matter by combined bioflocculation and alkaline fermentation. Bioresour. Technol. 197, 295–301.

Kuszmider, G., Kudela, P., 2015. Impact of sonification time on VFA generation and changes of total solids during sewage sludge anaerobic digestion process. Ecol. Eng. 41, 148–152.

Kwarciak-Kozlowska, A., Krzywicka, A., 2015. Influence of ultrasonic field on the increase of biodegradability of coke wastewater. Arch. Waste Manage. Environ. Protect. 17 (3), 133–142.

Ledakowicz, S., Krzystek, L., 2005. The use of methane fermentation in the utilization of waste from the agri-food industry. Biotechnology 70, 165–183.

Malej, J., 2001. The generation of volatile fatty acids from the raw sewage stream and some problems of sewage brought by the slurry tanker. Annu. Set Environ. Protect. 3, 103–128.

Markowska, K., Grudniak, A.M., Wolska, K., 2013. Microbial fuel cells: the basics of technology, its limitations and potential applications. Microbiol. Adv. 52 (1), 29–40.

Michalska, K., Pazera, A., Bizukojć, M., 2014. Innovations for the dairy industry—plant biogas installations. Electron. Doc. 1–19.

Miksch, K., 2010. Wastewater Biotechnology. PWN, Warsaw.

Mulka, R., Szlachta, J., 2013. State of knowledge regarding modeling of anaerobic digestion processes. Agric. Eng. 145, 281–290.

Oleskowicz-Popiel, P., Jankowska, E., 2014. The use of fermentation processes in biorefinery systems. Chem. Ind. 93/3 (39), 51–354.

Owczuk, M., Matuszewska, A., Filip, A., Prachnio, P., 2014. Investigation of the efficiency of biomass disintegration for the methane fermentation process. Automotive Arch. 66 (4), 143–151.

Rosik-Dulewska, C., 2007. Basics of Waste Management. PWN Scientific Publisher.

Sadecka, Z., 2010. Basics of Biological Wastewater Treatment. Seidel-Przywiecki Publishing.

Schlegel, H.G., 2003. General Microbiology. PWN Scientific Publisher.

Shen, L.W., Seak, M., Chua, A., Koon, Y.H., Cheng, N.G., 2014. A review of the production and applications of waste-derived volatile fatty acids. Chem. Eng. J. 235, 83–99.

Szwach, I., Kulesza, R., 2014. Potential of biomass in the aspect of obtaining selected raw materials and chemical products. Chemist 68 (10), 893–900.

Zacharof, M.P., Lovitt, R.W., 2013. Complex effluent streams as a potential source of volatile fatty acids. Waste Biomass Valor 4, 557–581.

Zalewski, M., Chmielewski, A., Palige, J., Roubinek, O., Wierzchnicki, R., Usidus, J., 2012. Technological solutions for installations for biogas production from vegetable raw materials. Instal 12, 41–44.

Zawieja, I., Wolny, L., 2014. Influence of degree of disintegration of sewage sludge subjected to alkaline modification to the value of unitary biogas production. Eng. Environ. Protect. 17 (3), 503–512.

Zawieja, I., Wolny, L., Próba, M., 2016. Effectiveness of generating volatile fatty acids during mesophilic and thermophilic methane fermentation of excessive sludge. Ecol. Eng. 48, 226–232.
Zawieja, I., Wolny, L., Wolski, P., 2009. Influence of the ultrasonic hydrolysis process on the generation of volatile fatty acids in the acid fermentation of excessive sludge. Eng. Environ. Protect. 12 (3), 207–217.
Zawieja, I., Wolny, L., Wolski, P., 2015. Influence of ultrasonic pretreatment on anaerobic digestion of excess sludge from the food industry. Ann. Set Environ. Protect. 17 (1), 351–366.
Zawieja, I., Wolski, P., 2013. The influence of chemical-thermal modification of excessive sludge on the generation of volatile fatty acids in the methane fermentation process. Annu. Set Environ. Protect. 15 (3), 2054–2070.
Zielewicz, E., 2007. Ultrasonic Disintegration of Excess Sludge in Obtaining Volatile Fatty Acids. Scientific Papers Silesian University of Technology, Gliwice.
Żygadło, M., 1999. Municipal Waste Management. Kielce University of Technology Publishing.

Further reading

Bhatia, S.K., Yang, Y.H., 2017. Microbial production of volatile fatty acids: current status and future perspectives. Rev. Environ. Sci. Biotechnol. 16, 327–345.
Kwarciak-Kozlowska, A., 2007. The process of purification of leachate from municipal landfills in anaerobic membrane bioreactor supported by an ultrasonic field. Ph.D. dissertation, Czestochowa.
Lee, W.S., Chua, A.S.M., Yeoh, H.K., Ngoh, G.C., 2014. A review of the production and applications of waste-derived volatile fatty acids. Chem. Eng. J. 235, 83–99.

Biogas (methane production) and energy recovery from different sludges

30

*Anna Grosser**, *Piotr Celary*†
*Institute of Environmental Engineering, Czestochowa University of Technology, Czestochowa, Poland, †Central Laboratory for Measurements and Research, Jastrzebie-Zdroj, Poland

1 Introduction

Activated sludge process (ASP) is a common method used for the treatment of wastewater from various industrial and domestic sectors. However, this purification technology generates large amounts of sewage sludge. Depending on the stage of the treatment process at wastewater treatment plants (WWTPs), primary, secondary [waste-activated sludge (WAS)], digested sludge, and chemical-precipitated sludge are generated. This gives them different properties and significantly affects selection of their treatment and disposal methods (Buendía et al., 2009; Shi et al., 2018; Zhen et al., 2017).

Currently, the main sewage sludge-management practices include landfilling (Faubert et al., 2016; Shi et al., 2018); land applications for agricultural (Fijalkowski et al., 2018), forestry, and reclamation purposes (Grobelak, 2016; Kacprzak et al., 2014; Placek et al., 2018); energy recovery through incineration (Bień et al., 2011; Wystalska et al., 2013); anaerobic digestion (AD) (Grosser, 2018; Sadecka, 2008; Zawieja, 2017); and being recycled as construction materials (Shi et al., 2018; Zhen et al., 2017). For a very long time, landfilling was the main way to dispose of it. However, stricter waste storage regulations (such as European Union Directive 1999/31/WE, Directive 96/278/EEC, and Decision 203/31/EC) limited this route. This development, as well as the fact that issues of sewage sludge management have changed in recent years, have spurred a search for new solutions. Sewage sludge is no longer seen as waste; rather, it has become a valuable potential source of metals, nutrients, organic matter, and energy (Kacprzak et al., 2017; Tyagi and Lo, 2013).

As shown in Fig. 1, there are three main strategies of sewage sludge processing: biochemical, thermochemical, and mechanical-chemical. Due to the high cost of sewage sludge management (even up to 40% of capital cost, and about 60% of operation and maintenance cost), it seems that the choice of the treatment and disposal method should be aimed at recovering its potential value (Divyalakshmi et al., 2017; Kameswari et al., 2011). From an environmental point of view, the situation is

Industrial and Municipal Sludge. https://doi.org/10.1016/B978-0-12-815907-1.00030-1

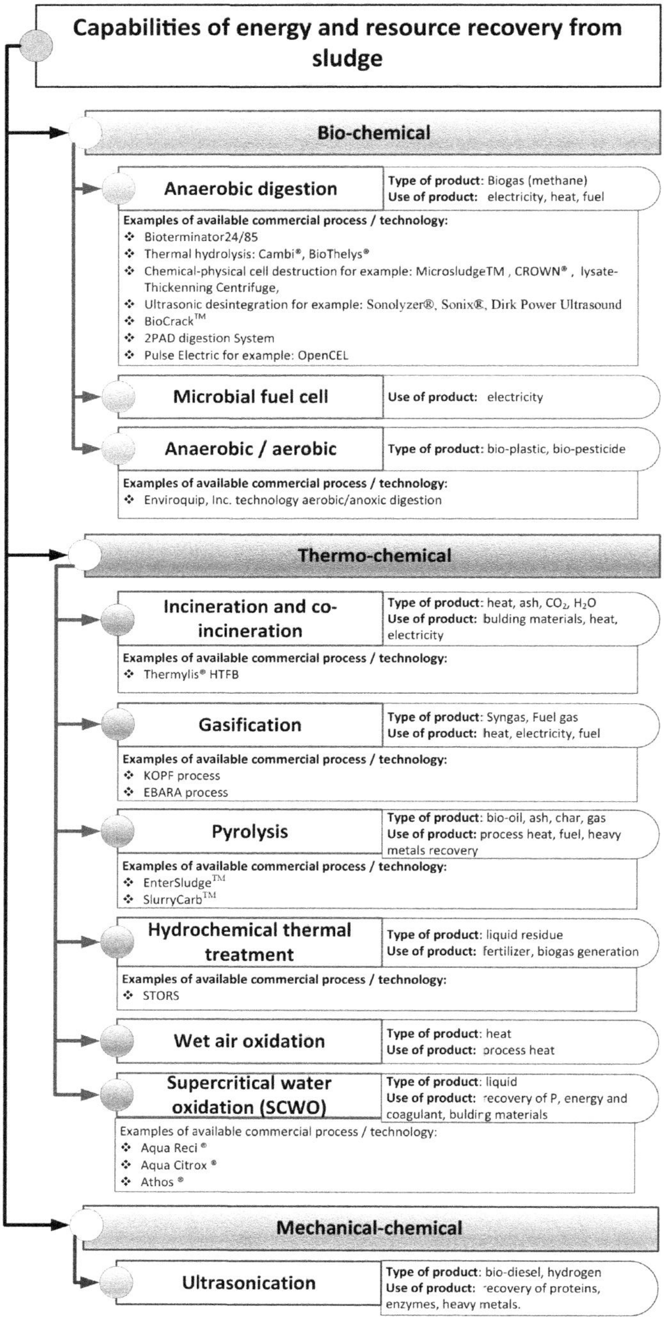

Fig. 1 Techniques for energy and resource recovery from different sludges (Tyagi and Lo, 2013).

complicated by the fact that some contaminants in the sewage sludge (pathogens, heavy metals, and organic and inorganic contaminations) make them toxic not only to biological treatment methods of sewage sludge stabilization, but also to all living organisms (Kacprzak et al., 2015; Laskowski et al., 2017; Włóka et al., 2015).

The main objective of this chapter is to investigate the issues of energy recovery from different kinds of sludge. The discussion focuses on the possibility of energy recovery at municipal WWTPs and from sludge generated in the most polluting industrial sectors such as the pulp and paper industry (PPI), tannery industry, dairy industry, and meat industry. Table 1 lists characteristics of many types of sludge. Among the available energy recovery methods, AD seems particularly interesting. On the one hand, its implementation at WWTPs allows for the reduction of sludge volume, improvement of sludge dewaterability, and destruction of most pathogens. On the other hand, the process generates two potentially useful by-products—methane (CH_4) and digested sludge. The first substance is a renewable fuel that can be used in cogeneration units for heat and electricity production, vehicle fuel after adequate purification and enchainment, a substitute for natural gas, and a resource for fuel cell production technologies (Pöschl et al., 2010; Tyagi and Lo, 2013). The second product may be applied in agriculture, forestry, and reclamation if it has been treated properly (Grobelak et al., 2017, 2018; Veluchamy and Kalamdhad, 2017a). While very promising, there are many drawbacks to AD of these sludges, most of which have been addressed in detail by presenting methods of its intensification by means of feed and substrate pretreatment, as well as the possibility of simultaneous treatment of the mentioned sludges with other organic waste in the process of anaerobic codigestion (AcD). To a lesser extent, the possibility of using other methods of energy recovery has been examined as well.

2 Energy recovery from different sludges

The following subsections evaluate the possibility of energy recovery from various sewage sludges, with the main focus being on AD and methods of its intensification.

2.1 Municipal sewage sludge (MSS)

AD is commonly used in the process of municipal waste-activated sludge (WAS) treatment. The main and most essential factors limiting the effectiveness of AD is long hydraulic retention time (HRT) (20–30 days), as well as low volatile solid (VS) removal (30%–50%) (Appels et al., 2008). As in the case of any other sludge, the bottleneck stage of the process is hydrolysis. Therefore, pretreatment of sludge prior to AD to accelerate the hydrolysis of sludge is often used (Bień et al., 2015; Zawieja et al., 2017). Such action allows for increasing municipal sewage sludge (MSS) biodegradability through the disruption of extracellular polymers and microbial cell walls and membranes enriching the fermentation broth with easily available compounds for acidogenic and methanogenic microorganisms (Carrere et al., 2016; Cesaro et al., 2014; Shah et al., 2015; Zhen et al., 2017).

Table 1 Characteristic of various types of sludge

Parameter	Unit	Kind of sludge									
		MSS[g]		DWAS[h]	TnS[i]	PPS[j]				SHS[k]	PC[l]
		PS	WAS			PS	WAS	MS	DPPS		
pH	–	5.0–8.0	6.5–8.0	6.9–7.45	7.36–8.8	5.0–11.0	6.0–7.6	3.8–8.1	7.2–9.2	5.5–8.4	7.2–7.28
TS	%	5.0–9.0	0.8–1.2	13.85–14 (8.51–25)[e]	6.98 (19–65.3)[e]	1.5–6.5 (15–57)[c]	1.0–2.0 (1–47)[c]	(19–60)[c]	(32–63)[c]	16–17	0.4–2.28
VS	%TS	60–80	59–68	6.9 (5.16–15.2)[e]	35.24 (12–19.7)[e]	51–80	65–97	62.3–83	–	11–13.8	31.67–77
Total P	%	0.8–2.8	0.5–0.7	33[f]	0.097	0.01–0.06	0.25–2.8	0.22–0.74	0.0012–0.16	(7)[f]	0.5
Total N	%	1.5–4.0	2.4–5.0	54.4[f]	0.9996	0.045–0.5	1.1–7.7	0.7–3.6	0.15–1.0	(41)[f]	6.04
TCOD	g/L	–		10.2–24.55	–	61–67				129–208	5.3
SOD	g/L	–		0.2–15.2	1.02–12	1.425–1.775				41.38	0.1
C	wt.%	15.45–52.5		–	32.04–42.97	16.46–38.76				18.91–61.9	23.25–51.2
H	wt.%	2.96–7		–	3.32	1.6–5.2				2.01–7.2	3.77–7.54
N	wt.%	2.74–5.27		–	0.03	0.7–6.69				2.97–6.5	0.52–3.77
S	wt.%	0.62–1.5		–	0.83–1.95	1.42–1.5				0.0–0.24	0.69–1.69
O	wt.%	33–78.08		–	31.90	20–22–47.85				5.3–75.87	4.05–12.98
Ash	wt.%	54.78–68.51 (10–59)[d]		–	18.77–42.32	10–49[d]	12–41[d]	20[d]	40–60[d]	19.1–29.18	35–54.95

Table 1 Continued

Parameter	Unit	Kind of sludge									
		MSS		DWAS	TnS	PPS				SHS	PC
		PS	WAS			PS	WAS	MS	DPPS		
Energy content	MJ/kg	23–29	19–23	–	8.66–10.64	5.5–20	14–25	14–19	1.5–5.7	27.7	10.36–23.60
Zn	mg/kg	Bdl–391.8		194–289.74	110–1300	170–183				870	392
Cu	mg/kg	Bdl–147.8		18.7–58.55	80–123.2	140–160				Bdl–533[a]	10.6
Cr	mg/kg	Bdl–136.2		14.1–15.99	100–474,000 (72–436.5[b])	450–1310				Bdl–222[a]	9

BDL, below detection limit; *DPPD*, de-inking primary pulp and paper sludge; *DWAS*, dairy waste-activated sludge; *MS*, mixed sludge; *MSS*, municipal sewage sludge; *PC*, petrochemical sludge; *PS*, primary sludge; *PPS*, pulp and paper sludge; *SCOD*, soluble chemical oxygen demand; *TCOD*, total chemical oxygen demand; *TS*, total solids; *VS*, volatile solids; *WAS*, waste-activated sludge.

[a] Also below the limit of detection (Cr: 4 ppb; Cu: 2 ppb).

[b] Cr^{6+}.

[c] fresh matter.

[d] % dry solids.

[e] g/L.

[f] g/kg

[g] Based on: Tyagi and Lo (2013); Shao et al. (2015); Gebreeyessus and Jenicek (2016); Xiao et al. (2015).

[h] Based on: Uma Rani et al. (2012a, 2012b, 2013, 2014); Poornima Devi et al. (2014); Gayathri et al. (2015); Sowmya Packyam et al. (2015); Jayashree et al. (2014); Kavitha et al. (2016); Eswari et al. (2016); López-Mosquera et al. (2000); Frąc et al. (2017).

[i] Based on: Bień et al. (2017); Thangamani et al. (2010); Haroun et al. (2009); Priyadarshini et al. (2015); Xu et al. (2008); Abbas et al. (2016); Dong et al. (2015).

[j] Based on: Faubert et al. (2016); Shao et al. (2015); Kamali et al. (2016); Meyer and Edwards (2014); Veluchamy and Kalamdhad (2017a); Hu et al. (2015).

[k] Based on: Sittijunda et al. (2010); Xia et al. (2012, 2015); Buendía et al. (2008, 2009); Shao et al. (2015); Oh and Yoon (2017).

[l] Based on: Haak et al. (2016); Roy et al. (2016); Conesa et al. (2014); Wang et al. (2016); Ma et al. (2013); Zhou et al. (2015); Ali et al. (2017).

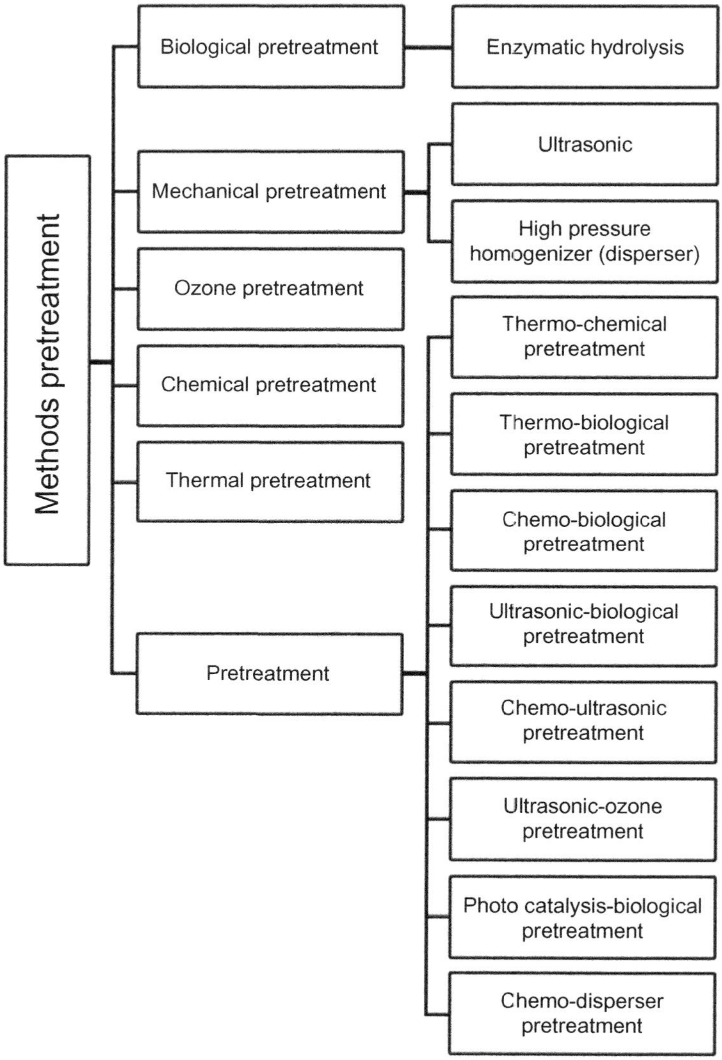

Fig. 2 Methods of sludge pretreatment (Banu et al., 2017; Veluchamy and Kalamdhad, 2017a).

Various methods are used to achieve this. Depending on the nature of the conditioning agent, pretreatment methods are generally divided into four groups: mechanical, thermal, chemical, and biological (Carrere et al., 2016; Cesaro et al., 2014; Shah et al., 2015). These detailed categories are illustrated in Fig. 2. Recently, there has been an increasing interest in hybrid methods (i.e., combinations of pretreatment methods). The impact of pretreatment methods on efficiency of AD was the subject of numerous publications, as listed in Table 2. However, proper and detailed energy

Table 2 Overview of selected pretreatment studies addressing sludge solubilization and AD performance

Type of sludge	Pretreatment method: Type, conditions	Pretreatment method: Results	AD: Conditions	AD: Results	References
Secondary wastewater sludge	Fenton treatment 60 g H_2O_2/kg TS 0.07 g Fe^{2+}/g H_2O_2	SS and VSS decrease by 21% and 25%, respectively, increment of SCOD by 8 times	35°C, 30 d, bench scale	Increase of CH_4 yield by 15%[a] Improve sludge dewaterability—CST reduction by 7.5-fold[a] Lower GHG emission by 33%[a] (0.128 Mg CO_2/Mg of TDS) Increase the net energy 3,1-fold[a]	Pilli et al. (2016)
Thickened raw WAS, as well as mixture of PS and WAS originating from different WWTPs	Full-scale HPTH plants 160°C at 6 bar.	20%–30% decrease of suspended solids and an increase of soluble COD from 0.04 to 0.4	Semicontinuous mode, HRT 1–6 d, Temperature 42°C or 37°C, 55°C, and period of operation depends on sludge range from 40 to 142 d	Increase in VFAs yield by 2–5 times[a] Increase of VFAs production rate by 4–6 times[a]	Morgan-Sagastume et al. (2011)
Mixed-sewage sludge	UD; pretreatment time 0–60 min, 150 W	Max. increment of TOC concentration and TN concentration as well as VFAs concentration by 176%, 88%, and 115%, respectively	Batch mode, mesophilic temperature (35°C); OLR = 0.9 ± 0.1 kg/(m^3d)	Increased CH_4 yield by 95%[a] Biodegradability of 81% in VS, methane production rate of 1270 L STP/(m^3 d)	Morgan-Sagastume et al. (2011)

Continued

Table 2 Continued

Type of sludge	Pretreatment method		AD		References
	Type, conditions	**Results**	**Conditions**	**Results**	
WAS	UD-specific energy equalling 3380kJ/kg TS	21% solubilization was achieved	Temperature-phased anaerobic digestion (TPAD) BMP assays	Increment of CH_4 production by 50%[a] Enhance VS removal by 13%[a]	Riau et al. (2015)
Thickened WAS	Microwave (MW) pretreatment, 1250W, 2450MHz, 100% intensity	Increase of SCOD/TCOD from 0.06 to 0.2	Semicontinuous TPAD, SRT (20, 15, 10, and 5d)	Increment of CH_4 production by 106%[a] Max. VS removal: 53.1%	Coelho et al. (2011)
Primary sludge	Pulsed-electric field; treatment intensity of 33kWh/m^3	Increase of VFAs accumulation (by max. 2.5 times) and decrease of methane production (by 2–7.5 times).	Microbial electrolysis cells	Increase of density by 2.4-fold[a]	Ki et al. (2015)

CST, capillary-suction-time; *OLR*, organic loading rate; *SCOD*, soluble chemical oxygen demand; *SS*, suspended solid; *STP*, standard temperature and pressure conditions; *TCOD*, total organic carbon; *TDS*, total dry solids; *TOC*, total organic carbon; *TN*, total nitrogen; *UD*, ultrasound pretreatment; *VSS*, volatile suspended solid

[a]Compared to the control (AD of sewage sludge without pretreatment).

assessments of proposed solutions are often nonexistent in publications. The importance of this issue is shown by Cano (2015). Based on about 70 publications, the author estimated the limiting values of energy and/or heat consumption of these pretreatment methods. Above these values, the methods are characterized by a negative energy balance (i.e., the energy demand required by the pretreatment method is much higher than the energy gained due to sludge conditioning). As Cano shows, for pretreatments requiring electric energy, the mentioned value equals about 30 kWh/m^3sludge. In turn, for heat-based pretreatments, this value is dependent on the energy integration level. For designs without heat recovery, with heat recovery from flash, with heat recovery via heat exchangers, and with recovery via combined heat and power (CHP) energy units, this value equals 205, 116, 65, and 10–15 kWh/m^3sludge, respectively.

Besides pretreatment, there are two other ways to enhance AD performance: modification of anaerobic reactors and AcD with other organic waste. The second approach offers the following advantages: improved nutrient balance, increased content of macronutrients and micronutrients, dilution of toxic compounds, (usually) enhanced VS removal and biogas/CH_4 yield, increased share of biodegradable fraction, and the possibility of digesting wastes. Until then, the latter could not be treated separately due to production of intermediates, which may destabilize the process and even inhibit it (e.g., VFAs or long-chain fatty acids, which are produced in large quantities during decomposition of food waste and fat-rich materials, respectively) (Bień et al., 2010; Grosser, 2018; Grosser and Neczaj, 2016; Mata-Alvarez et al., 2014). Predominantly, the AcD process was carried out for binary mixtures. However, the number of publications per year regarding the codigestion of multicomponent mixtures gradually increases (Grosser, 2017; Mata-Alvarez et al., 2014; Zhen et al., 2017). Most often, sludges are digested with an organic fraction of municipal waste (OFMSW) (due to complementary properties of these wastes) and fat-rich materials (due to their high CH_4 yield) (Grosser, 2017; Neczaj et al., 2013) (Table 3).

Other publications evaluated the possibility of using various kinds of organic waste, such as shredded grass (Hidaka et al., 2013; Wang et al., 2014), rice straw (Komatsu and Kudo, 2007), strawberry extrudate (Serrano et al., 2015), pig manure (Zhang et al., 2014), algae (Caporgno et al., 2015; Wang et al., 2013b), crude glycerol (Kurahashi et al., 2017; Nartker et al., 2014), food waste, cheese whey, and olive mill wastewater (Maragkaki et al., 2018). As of recently, the codigestion process has been combined with various disintegration methods. For instance, Serrano et al. (2015) demonstrated that a combination of thermal pretreatment of sewage sludge and its codigestion with strawberry extrudate may increase CH_4 yield by 24%. However, the available data is still pretty limited because pretreatment methods employed for codigestion have become more widely described in literature starting from the year 2011. Therefore, more research needs to be carried out, such as to assess how pretreatment affects process stability or how it influences microbial population diversity. Although Grosser and Neczaj (2017) reported that ultrasonic and thermochemical pretreatment improve the solubilization of organic matter, and positively affects CH_4 yield. Additionally, the authors showed that both methods are self-sufficient in terms of energy.

Table 3 Overview of MSS codigestion studies regarding biowaste, fat-rich materials

Cosubstrate	Biogas, CH_4 yield/VS removal: Without cosubstrate	Biogas, CH_4 yield/VS removal: With cosubstrate	Increment of biogas, CH_4 yield (%)	Increment of VS removal (%)	References
Biowaste					
FW	260^a/na	300^a/na	15	–	Fonoll et al. (2015)
OW	390^b/71	Max. 890^b/81	128	14	Zupančič et al. (2008)
FW	Max.237^a/38.2	Max. 400^a/70	69	83	Dai et al. (2013)
OFMSW	Max. 186^a/51	Max. 316^a/41.5	70^d	-19^d	Borowski (2015)
OFMSW	164^d/na	Max. 221^d/na	35	–	Nielfa et al. (2015)
Fat-rich materials					
GTS_A	278^a/52	463^a/67	67	29	Luostarinen et al. (2009)
GTW_B	271^a/45	344^a/58	27	29	Davidsson et al. (2008)
FOG_A	252^a/40	614^a/54	143	36	Wan et al. (2011)
FOG_B	249^a/36	369^a/52	48	44	Silvestre et al. (2011)
DAF	400^a/29	591^a/39	48	34	Girault et al. (2012)
FOG_B	294^a/52	490^a/59	67	13	Noutsopoulos et al. (2013)
FOG_B	430^b/57	640^b/59	49	3.5	Noutsopoulos et al. (2012)
GTS_B	0.83^cd/44	1.63^c/56	96	27	Razaviarani et al. (2013)
Used oil	510^b/37	630^b/41	23.5	11	Pastor et al. (2013)
FOG_B	260^a/50	300^c/46	15	−8	Silvestre et al. (2014)
GTS_B	180^a/na	752^a/na	318	–	Wang et al. (2013a)
FOG_C	159^a/25.2	473^a/45.1	197	79	Kabouris et al. (2009)
GTS_A	221^a/45	423^a/60	91	33	Grosser and Neczaj (2018)
GTS_A	300^a/50	456^a/56.4	52	13	Grosser et al. (2017)
GTS_A +OFMSW	300^a/50	547^a/64.7	82	29	
WFVO	330^b/44	165^b/36.7	–50	–16.5	Grosser et al. (2013)

Table 3 Continued

Cosubstrate	Biogas, CH_4 yield/VS removal		Increment of biogas, CH_4 yield (%)	Increment of VS removal (%)	References
	Without cosubstrate	With cosubstrate			
BFW	163^a/na	520^a/na	219	–	Martínez et al. (2016)
GTW_A	223^a/26	517^a/35	132	35	Yalcinkaya and Malina (2015a)
GTW_B	384^a/47	641^a/59	67	26	Yalcinkaya and Malina (2015b)

BFW, butcher's fat waste; *FOG_A*, fat oil grease from receiving facility; *FOG_B*, fat oil grease from the dissolved air flotation unit of a WWTP; *FOG_C*, polymer-dewatered FOG; *FW*, fruit wastes; *GTS_B*, grease trap sludge from restaurant; *GTW_A*, grease trap waste collected from receiving facility; *GTWB*, *GTW* from the transfer station of a GTW hauler; *GTS_A*, grease trap sludge from meat industry; *OFMSW*, organic fraction of municipal waste; *OW*, organic waste from domestic refuse; *WFVO*, waste fats of vegetable origin.
na, data not available.
[a]methne yield (m^3 /Mg VS_{add}).
[b]biogas yield (m^3/Mg VS_{add}).
[c]gas production rate (m^3/m^3d).
[d]compared to AD of OFMSW alone.

It is worth emphasizing that in the case of biogas, there is one more option. Typical biogas has a lower heating value ranging from 5.8 to 6.4 kWh/m^3, while for pure CH_4, this value can reach 10 kWh/m^3. The calorific value of biogas can be increased by removing impurities using biogas-upgrading technologies. The number of installations operating in the world can be proof of the potential of such methods (above 400). Water scrubbing and chemical absorption, and membrane separation are the most commonly applied commercial technologies (Angelidaki et al., 2018).

Gasification, pyrolysis, and combustion/incineration also may be used to recover energy from MSS (Mulchandani and Westerhoff, 2016; Syed-Hassan et al., 2017). The mentioned methods and their advantages and disadvantages are summarized in Table 4. The composition of gas, and gas yield during the gasification of sewage sludge are determined by the following factors: temperature, steam-to-biomass ratio, equivalence ratio, addition of catalytic materials, and gas residence time. Gas yield ranges from 0.9 to 2.9 Nm^3/kg sludge on a dry ash-free (daf) basis, while lower heating values may range between 1.3 and 6.2 MJ/m^3 (Syed-Hassan et al., 2017).

Yields of pyrolytic products depend on temperature, characteristics of raw materials, reaction time, and operating pressure (Tyagi and Lo, 2013). Sewage sludge bio-oil yields may range between 25 and 54 wt.% on a sewage sludge–fed basis (or 51–80 wt.% on a daf basis) and their high heating values are usually higher than 22 MJ kg/kg (Fonts et al., 2012). In turn, the maximum yield of char and gas is in the range of 35.6–64 wt.% on a fed basis and 3–28.9 wt.% on a fed basis, respectively (Syed-Hassan et al., 2017).

Table 4 Comparison of main conventional thermochemical methods used for energy recovery from sewage sludge (Mulchandani and Westerhoff, 2016; Syed-Hassan et al., 2017; Tyagi and Lo, 2013)

Combustion/Incineration

Exothermic process of oxidation of organic compounds to CO_2 and water in the presence of excess air

Advantages:

- Reduction of volume and mass by up 90%
- Destruction of pathogens
- Quick and total decomposition of organic matter

Disadvantages:

- Requires treatment of gases prior to their release into the environment
- High operating costs
- Process difficult to control
- Concentrated residuals
- Improper storage may result in salinity of groundwater
- Cooling of combustion residues and washing of gases generates wastewater
- Coincineration with other fuel or waste can dilute metal concentration in ash (an important parameter in the recovery of metal from sludge)

Cost:

Capital: $280 million/300000 Mg/a

O&M: $400/dry Mg

Pyrolysis

Endothermic process of thermochemical transformation of carbon-rich organic substances carried out at temperatures from about 275°C to maximum 900°C at the absence of oxygen

Advantages:

- Significant reduction of sludge mass and volume
- Versatility, i.e., the possibility of large- and small-scale use
- The possibility of storing products and their subsequent use for energy purposes; possibility of recycling secondary raw materials (mainly nonferrous metals) from pyrolytic solid residues
- High operational reliability resulting from the simplicity of the installation
- Lack of organic compounds containing chlorine in the gases
- Removal of bioactive compounds, deactivated antibiotic resistant genes
- Synthetic gases may be used as chemical feedstock or energy source

Disadvantages:

- Raw material requires preliminary preparation due to sensitivity of the process to moisture content (complicates product purification, may lead to increase production of noncondensable gases)
- Heavy metals concentrated in biochar
- Lower yield of products in comparison to fast pyrolysis of biowaste (25%–52%)

Cost:

Capital: $25 million/25 Mg/d

O&M: $50–200/dry Mg

Table 4 Continued

Gasification
Exothermic process of thermochemical transformation of organic matter conducted at high temperature (800-900°C), which is converted into syngas (mixture CO, H_2, CO, and CH_4), tar and char; their shares mainly depend on the type of sludge, its properties, temperature, sludge particle size; air, steam, air-steam, steam-O_2 and CO_2 are usually used as gasifying agents
Advantages:
• Reduction of mass up to 80%
• High thermal efficiency
• Simple to control
• Synthetic gases may be used as chemical feedstock or energy source
Disadvantages:
• Difficulty in dispensing some types of waste into reactors
• Small volume of feedstock
• Can lead to the production of hazardous residues which may require disposal
• Safety issues
Cost:
Capital: $70 million/1000 Mg/d

a, annually; *O&M*, equivalent annual operating and maintenance cost.

An interesting alternative solution is the integration of AD and pyrolysis. The purpose of the second stage is energy recovery from digested sludge (via transformation into useful energy carriers such as gas and oil). As shown by Cao and Pawłowski (2013, 2012) this pathway in comparison to pyrolysis of sludge itself allows for increased energy recovery, enhanced VS removal and decreased greenhouse gas (GHG) emissions.

2.2 Dairy waste-activated sludge (DWAS)

Dairy waste-activated sludge (DWAS) is characterized by protein-rich organic content [hence, having high chemical oxygen demand (COD) and biological oxygen demand (BOD)], low heavy metal concentration, below normative standards, and abundance of such elements as nitrogen, magnesium, calcium, and phosphorus (Frąc et al., 2017; Kwarciak-Kozłowska and Bień, 2018; Uma Rani et al., 2014). Furthermore, anaerobic decomposition of proteinaceous waste such as DWAS leads to production of an intermediate (ammonia) that may create a risk of AD inhibition (Kovács et al., 2015). Current research mainly focuses on accelerating the rate of hydrolysis and improvement of AD performance. For instance, Uma Rani et al. (2012a) carried out low-temperature, thermochemical pretreatment of DWAS. Biochemical CH_4 potential assays showed a rise in solubilization, biogas production, and suspended solid (SS) removal by up to 23%, 51%, and 22%, respectively. While trials combining the mentioned methods in lab-scale semicontinuous anaerobic reactors, led to 103%, 80.5%, 90.4%, and 117% improvement of biogas production, TS,

VS, and SS removal, respectively. In another study, Uma Rani et al. (2012b) observed an increase of digestion efficiency after chemomechanical sludge. In comparison to the control reactor, an increase of COD solubilization, SS reduction, and biogas production by 24%, 23.3%, and 76%, respectively, have been noted. In turn, Gayathri et al. (2015) investigated the effect of citric acid application followed by ultrasonic pretreatment (UD) for DWAS. In that study, combination of both actions led to a 105% improvement in biogas yield. Furthermore, in a comparison with flocculated WAS (a ultrasonically pretreated sample without the addition of citric acid), it was found that SS removal increased by 47%. In another laboratory-scale work, two-step sono-alkalization pretreatment (4172 kJ/kg TS; pH 10) led to increases in COD, solubilization, SS reduction, and biogas production by 59%, 46%, and 80%, respectively (Uma Rani et al., 2014). Uma Rani et al. (2013) also studied the application of MW irradiation at different intensities for DWAS. The optimum conditions for microwave irradiation were 70% intensity and 12 min. of exposure time, which improved SS reduction, VS reduction and biogas production significantly, by 67%, 64%, and 57% respectively, in comparison to the control reactor (untreated sludge). They also found that the method improved digestibility, which significantly reduced the amount of sludge. Kavitha et al. (2016) performed biochemical methane potential (BMP) assays and semicontinuous anaerobic sludge digestion and evaluated the potential application of chemomechanical disintegration of WAS using sodium dodecyl sulfate (SDS) and a disperser (at 5000–25,000 rpm). They noted that this hybrid pretreatment method (with optimal established conditions of 7377 kJ/kg and 0.04 g/g SS), in combination with AD at 15 days SRT (solid retention time), leads to 51% of VS removal and an increase of CH_4 yield (from 12 to 50 mL/gVS removed), and 38% COD solubization. In turn, at a specific energy of 5013 kJ/kg and SDS at an optimum dosage of 0.04 g/g SS, Poornima Devi et al. (2014) achieved 26% COD solubilization. Sowmya Packyam et al. (2015), through a combination of ultrasonic sludge deflocculation and ozonation, achieved an almost threefold higher biogas yield (an increase from 0.202 to 0.637 L/gVS). Eswari et al. (2016) reported that the application of a combination of microwave-hydrogen peroxide (H_2O_2) pretreatment in acidic conditions (18,600 kJ/kg TS, 0.3 mg/g SS) can significantly improve solubilization (up to 50%) and CH_4 yield (by about 12% and 10-fold compared to trials in nonacidic conditions and the control sample, respectively).

An interesting direction in energy conversion from DWAS are microbial fuel cells (MECs). As shown by Jayashree et al. (2014), pretreatment of DWAS using a low-temperature, thermochemical method significantly improves performance of microbial cells. In MECs fed with pretreated sludge under the same conditions, the TCOD removal, SS removal, and power density were 54%, 43%, and 43% higher than the control unit, respectively.

2.3 Tannery sludge (TnS)

TnS is mainly generated during treatment of wastewater from the leather industry. Its quantity and quality depend on the wastewater treatment technology, raw materials, and a skin-type production profile. It is estimated that on average, the processing of

1 Mg of raw hides results in the generation of 100–225 kg of total sludge (Bień et al., 2017; Kameswari et al., 2011). According to alternative data sources, the average values for WAS and PS (primary sludge) range between 15–20 kg and 50–60 kg, respectively (Priyadarshini et al., 2015). For instance, in India, the annual amount of generated sludge ranges greatly, from as little as 18,200 up to 165,200 Mg (Kanchinadham et al., 2015). This sludge poses serious toxic hazards to the environment due to its contamination with different syntans, pigments, dyes, retaining agents, and other toxins, and elevated concentration of heavy metals such as arsenic (As), cadmium (Cd), copper (Cu), cobalt (Co), chromium (Cr), iron (Fe), nickel (Ni), and zinc (Zn) (Juel et al., 2017). Trivalent Cr salts present in sludge derived from tanning agents (90% of tanneries in the world use this compound for this purpose) are a big problem. Their concentration varies greatly, ranging from 6030 up to an ore-like content of 474,000 mg/kg (Bień et al., 2017). However, significantly lower (even miniscule) levels of Cr were noted as well. For instance, Haroun et al. (2009) reported a Cr level of 500 mg/kg, While, Shukla et al. (2011) reported that the concentration of Cr was 628 mg/kg.

In turn, Kanchinadham et al. (2015) reports that due to Cr recovery of about 98% in tanneries in India, concentration of Cr in primary sludge ranges from 1.6 to 2 g/L (all of which is present in trivalent form), while in secondary sludge, digested sludge, and the fleshings, it was supposedly below the detection limit. This explains the interest in biological treatment methods in this region. The presence of tannin in this sludge, however, may negatively affect AD performance, which can manifest itself by decreased VS removal and production of biogas. The mentioned phenomenon is observed when concentration of such compounds exceeds 790 mg/L (Agustini et al., 2018).

The characteristics of TnS determine the methods of its management. High content of Cr, much higher than its permissible concentrations in compost (100 mg/kg) (Haroun et al., 2009), and concentrations inhibiting the AD process question the potential usefulness of biological methods. Strong inhibitory effects of hexavalent and trivalent Cr are observed when their concentration is higher than 200–250 (total)/3 (soluble) mg/L and 180–240 (total)/3 (soluble) mg/L, respectively (Appels et al., 2008). According to Priebe et al. (2016), the LD_{50} of the mentioned forms of Cr equals 80 and 160 mg/L, respectively, while Altaş (2009), based on trials involving CH_4-producing granular sludge, demonstrated that for Zn(II), Cr(VI), Ni(II), and Cd(II), the IC50 values equaled 7.5 mg/L, 27 mg/L, 35 mg/L, and 36 mg/L, respectively.

Despite this, few publications examined this issue. For instance, Thangamani et al. (2010) studied the feasibility of AcD of lime fleshings with primary sludge from tannery WWTPs and found that the process may be carried out successfully with VS removal and CH_4 yield comparable to ones reported in the literature for various organic waste. The noted VS removal and CH_4 yield ranged from 41%–52% and 0.419–0.635 L/g VS_{add} respectively, with the share of CH_4 in the biogas being between 71% and 77%. Kameswari et al. (2011) investigated the effects of ultrasounds and ozonation on AD of primary sludge (PS) and secondary sludge, reporting maximum solubilization of 166% and 162% due to ozonation for PS and WAS, respectively. Application of an ultrasound field led to an increase in SCOD of up to 85% and 97% for primary sludge (PS) and secondary sludge, respectively. Both methods have a positive impact on biogas yield of codigestion mixtures consist of fleshings and mixed sludge (with and without pretreatment). Compared with untreated samples, 45% and 53% increases in biogas production were noted for comixtures with ozone (O_3)-pretreated sludge and ultrasound pretreatment processes. Priyadarshini et al. (2015) also observed that UD of a codigestion mixture consisting of primary and secondary sludge with cow dung prior to AD leads to enhanced process efficiency (for pretreated samples, the released heat energy increased from 7.6 to 16.3 kJ/kg), and improved kinetic parameters. The rate constant for pretreated feedstock was 1.5 times higher than for untreated samples, while lag time was prolonged from 5.4 to 7.1 days. Kanchinadham et al. (2015) assessed the effect of different organic load rates (OLRs) on codigestion of tannery solid wastes (i.e., PW + WAS + fleshings), finding that biogas production increased gradually as the organic load increased. For the highest OLR, the biogas yield was in the range of 410–470 mL/g of VS_{add}, making it 37% higher than during the period when reactors were loaded with feedstock characterized by the lowest concentrations of organic matter.

Another route to energy recovery from TnS is its incineration. The process is often the first stage of various technologies used to recover Cr from TnS (Kavouras et al., 2015). To minimize the impact on the environment (reduction Cr oxidation and volatilization of heavy metals), before incineration of TnS, different pretreatment steps are applied. For instance, Abreu and Toffoli (2009) used Cr waste salt extraction, while Tang et al. (2008) treated TnS using 85% phosphoric acid, and in another study (Tang et al., 2013), they mixed sludge with hydroxyapatite.

Pyrolysis also has been successfully used to treat and reduce the amount of sludge and minimize volatilization of heavy metals (Xu et al., 2008; Zhou et al., 2018), while vitrification has been employed to immobilize them in a silica matrix (Celary et al., 2016; Celary and Sobik-Szołtysek, 2014; Sobiecka and Szymanski, 2014). As shown in Section 2.1, pyrolysis products are energy carriers and can be used to produce heat and electricity, while vitrification products can act as construction aggregate substitutes.

2.4 Pulp and paper sludge (PPS)

It is estimated that during the production of 1 ton of paper, 0.3–1 m^3 of primary sludge is produced (Priadi et al., 2014). Sludge production depends strictly on wastewater treatment technology. Using aerobic treatment technology results in the production of biosludge in amounts of 0.4–1.0 Mg per megagram of COD reduced. In turn, during anaerobic treatment, sludge quantities are significantly lower, on the level of 0.02 Mg per megagram of COD removed (Meyer and Edwards, 2014).

As shown in Fig. 3, two kinds of sludge are produced: primary sludge (PS) and secondary sludge (also called WAS in the case of activated sludge process, biological sludge, or biosludge), which are often mixed at highly variable ratios and depends mainly on raw materials used, production technology of pulp and paper, and type of wastewater treatment process (Bajpai, 2015; Kamali et al., 2016; Pervaiz and Sain, 2015; Stoica et al., 2009). For instance, in Canadian pulp and paper mills, the ratio of PS to WAS is 7:3 (Mahmood and Elliott, 2006); however, there are known

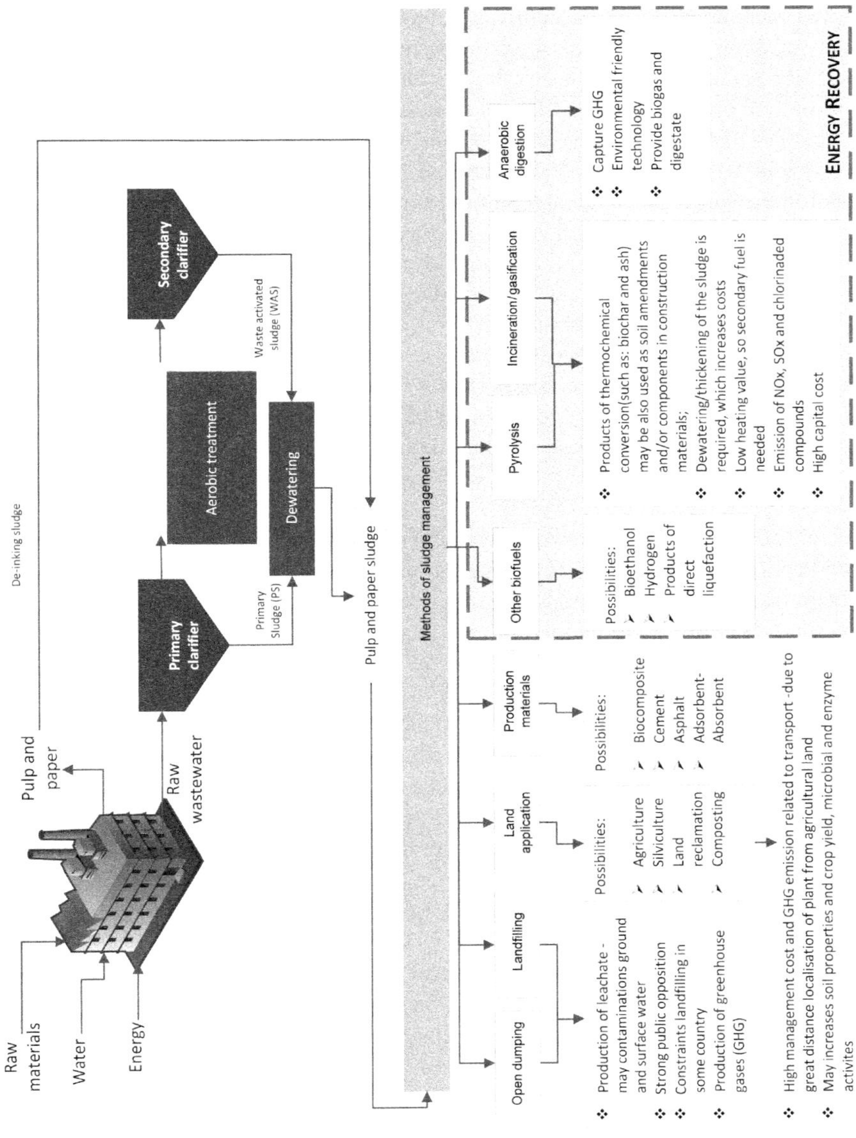

Fig. 3 Methods of sludge management used for PPS (Faubert et al., 2016; Veluchamy and Kalamdhad, 2017a).

examples of plants where only one of these sludges is produced (Pervaiz and Sain, 2015; Rashid et al., 2006; Veluchamy and Kalamdhad, 2017a). However, due to the majority of inorganic solids and fibers being removed during mechanical treatment, the volume of PS is significantly higher than WAS (Faubert et al., 2016; Kamali and Khodaparast, 2015; Veluchamy and Kalamdhad, 2017a). According to other sources, deinking sludge is also distinguished (Faubert et al., 2016). PS consists mainly of fibrous materials (lignin, cellulose, hemicellulose) and residuals from the paper production process, such as fillers (calcium carbonate and kaolin), pitch, metallic components, resin acids, and ash (Kamali et al., 2016; Ochoa de Alda, 2008; Veluchamy and Kalamdhad, 2017b). In turn, WAS consists predominantly of excess biomass produced during biological treatment wastewater from this industry, and nonbiodegradable compounds and chemical residuals from pulp and paper production processes (Kamali et al., 2016; Veluchamy and Kalamdhad, 2017a). According to Kyllönen et al. (1988), more than half of organic matter flowing into WWTPs is transformed into sludge that contains about 0.5%–2% of TS. Meanwhile, deinking sludge is produced during the deinking process. For this reason, this sludge is basically a residue characterized by high content of fibers, heavy metals, deinking agents [i.e., compounds used to remove recycled paper fiber dyes and color pigments (inks)], ink particles, and adhesive components and fillers (Gavrilescu, 2008; Monte et al., 2009). Characteristics of all type of sludges and their mixtures (PS + WAS) were shown earlier in this chapter, in Table 1.

Costs related to the disposal and management of sludge are estimated at about 60% of the total operating cost of a WWTP (Monte et al., 2009; Xu and Lancaster, 2007). In this context, it is not surprising that this sector seeks solutions that would reduce these costs. As shown in Fig. 3, current methods of sludge management are characterized by significant limitations. Furthermore, changes in Council Directive 86/278/EEC and more stringent waste storage regulations (European Union Directive 99/31/WE) forced a search for new methods of sludge management.

Of the available methods, AD seems particularly interesting. According to Meyer and Edwards (2014), the amount of COD available in the PPI for energy conversion varies from 4 to 400 Mt per year. Transforming only 25% of this COD load into biogas would allow 100 TWh of electricity to be generated annually. The potential of AD to generate energy from this type of sludge can be proved somewhat by the fact that in 2011, all biogas plants in the world generated only approximately 40% of this kind of energy (Veluchamy and Kalamdhad, 2017a).

Only a few systematic studies are available for AD of PPS alone. For instance, Bayr and Rintala (2012) evaluated the feasibility of mesophilic and thermophilic conditions for primary sludge and its mixture with biosludge in semicontinuously fed, continuously stirred tank reactors. The authors reported that AD of PS itself resulted in a CH_4 yield (Y_M) of 210 and 230 L CH_4/kg VS_{add} at 35°C and 55°C, respectively. Thus, the achieved CH_4 yields were higher than those obtained for secondary sludge (50 and 100 L CH_4/kg VS_{add}) or sludge mixtures (150 and 170 L CH_4/kg VS_{add}). As indicated by several previous studies, CH_4 yields for WAS range widely, between 40 and 200 L CH_4/kg VS_{add}, while VS removals vary from 21%–57% (Bayr et al., 2013; Park and Li, 2012; Saha et al., 2011). It is worth emphasizing that in most of the previous studies, AD of PPS without pretreatment was carried out in a batch regime. Studies in semicontinuous mode are still rare.

Long HRT (30 days–60 days) and low VS removal (30%–50%) are the main limitations of the AD process for PPS. An additional barrier in the case of primary sludge is its low biodegradability, which derives from high lignocellulosic content. Furthermore, accumulation of various toxic compounds such as adsorbable organic halides (AOX) makes it very difficult to use safely (Lin et al., 2017). The solution to these problems can be codigesting this sludge with other organic waste or using pretreatment methods (Elliott and Mahmood, 2007; Kamali et al., 2016; Meyer and Edwards, 2014; Veluchamy and Kalamdhad, 2017a). The recommended methods may enhance the digestibility of sludge, which in turn may accelerate the hydrolysis stage, increase the production rate of biogas/CH_4, positively affect process efficiency, shorten HRT, reduce digesters and production of sludge, and allow for the use of anaerobic digesters to their full capacity (Ashrafi et al., 2015; Faubert et al., 2016; Meyer and Edwards, 2014; Veluchamy and Kalamdhad, 2017a).

Numerous researchers have conducted studies to evaluate the impact of pretreatment methods on biogas/CH_4 yield, but there are not as many on the codigestion process. An overview of the methods used to enhance AD performance is shown in Table 5, while Table 6 summarizes the effects induced by various pretreatment methods on lignocellulose biowaste.

As shown by the latest studies, energy recovery from this sludge is not only performed via its thermal conversion or AD. There is a growing interest in the possibility of their bioconversion to other biofuels, such as bioethanol, hydrogen, or products of direct liquefaction. Sewage sludge also can be used as components of biocomposites, fiber additives for asphalt road pavement, and cement mortar products, as well as for synthesizing adsorbents or production of volatile fatty acids (VFAs) and short alcohols (Beauchamp et al., 2002; Li et al., 2011; Mari et al., 2009; Meyer and Edwards, 2014; Yan et al., 2011).

2.5 Slaughterhouse sludge

Slaughterhouse sludge (SHS) contains large amounts of protein, and their composition varies depending on the industrial process and production profile (Bustillo-Lecompte and Mehrvar, 2015; Oh and Yoon, 2017). The decomposition of both classes of organic compounds leads to the formation of intermediate products that can inhibit the AD process (Hamawand, 2015). In the case of protein, it is ammonia, while for fats, they are long-chain fatty acids (Chen et al., 2008; Rajagopal et al., 2013). Furthermore, fats may cause quite a few operational problems, such as sedimentation hindrance, scum formation, clogging, and flotation of biomass (Worwąg et al., 2011). There is also a potential risk of sewage contamination by prion proteins from BSE-infected cattle (Yamamoto et al., 2006). Moreover, high solid content may cause problems during loading of the anaerobic digesters (Oh and Yoon, 2017).

Despite this, as suggested by the study by (Buendía et al., 2008) WAS is characterized by good biodegradability and optimal carbon/nitrogen (C/N) and carbon/phosphorus (C/P) ratios for the anaerobic population. This is reflected in the CH_4 yield

Table 5 Overview of methods that affect AD performance

Type of sludge/ waste	Type of activates	Conditions of activates/AD	Results	References
PW:PPS	AcD	BMP, mesophilic conditions PW:PS 1:3 (v/v)	Reduction of lag phase from 23 to 14[a] In. k_h by approx. 2.5 times[a] In. k_h by approx. 11 times[b] In. Pm by 4.3 times[a]	Parameswaran and Rittmann (2012)
		BMP, mesophilic conditions PW:PS 3:1 (v/v)	Reduction of lag phase from 23 to 20[a] In. k_h by approx. 20%[a] In. k_h by approx. 6 times[b] In. Pm by 6.5 times[a]	
PW:PPS	AcD	SCm; HRT=SRT=35 d; PW:PPS 2:1 (v/v), mesophilic conditions	In. of VS removal by 28% [a] In. of MPR by 79%[a] In. of CH_4-COD conversion by 68%[a]	
BCTMP: MGWL	AcD	BS, CSTR, mesophilic conditions	VS removal: 35.21% Y_M: 200 L/g VS_{add} MDP: 0.5 m^3/(m^3 d)	Lin et al. (2011)
PPS:FW	AcD	BS, hydrogen and CH_4 coproduction, mesophilic conditions PW:FW 1:1 on VS basis	Y_M: 432.3 L/g VS_{add} Y_H: 64.48 L/g VS_{add} 87% removal efficiencies of SCOD	Lin et al. (2013)
PPS	Microbial consortium OEM1	Batch assay, 50 d, mesophilic conditions conditions of activation of microorganisms: 6 d at 28°C, 120 rpm	In. of CH_4 yield by 1.4 times[b] Fivefold increase in adsorbable organic halides degradation In. of the activity of hydrogenase by 41.6%[b]	Lin et al. (2017)

Table 5 Continued

Type of sludge/ waste	Type of activates	Conditions of activates/AD	Results	References
WAS from activated sludge plant treating wastewaters from an integrated bleached (chlorine dioxide, oxygen) kraft pulp (softwood and birch) and paper mill (producing coated magazine paper)	Various pretreatment methods	UD: 30 min, 45 kHz Al: 24 h, 22°C, pH 12 with 5 M NaOH Ac: 24 h, 22°C, pH 3 with 5 M HNO_3 E: 72 h, 50°C; commercial enzyme Accelerase HT70: 40 min, 70°C HT150: 10 min, 150°C BMP, $VS_{substrate}$ to $VS_{inoculum}$ ratio of 2, thermophilic condition	Impact of pretreatment method on YM[b]: UD 5.56% Al –20% Ac –43.5% E 5.56% UD+Al –38% HT70 3.7% HT150 24% UD+E 0.9% HT150+E 18.5% HT70+E 14.8% UD+HT150 30.5% UD+HT150+E 21.3%	Bayr et al. (2013)
WAS at the activated sludge-based pulp and paper mill effluent treatment plant	Various pretreatment methods	MW: 50°C–175°C UD: 15–60 min, 0.75 W/mL (27,551–110,204 kJ/ kg TS) Al.: pH: 9, 10, 11, 12, and 12.5 (addition of 6 N NaOH) MW+Al. UD+Al BS, 30°C, 22 d	Impact of the pretreatment methods on COD solubilization[b]: MW x 29-fold MW + Al x 70-fold UD x 52-fold UD + Al x 59-fold Al x 15-fold Impact of the pretreatment methods on CBP[b]: MW 16% MW+Al 6.3% UD 22.5% UD+Al 47%	Tyagi et al. (2014)
PPS	EH	15 V for 45 min BMP, 35 d, VSsubstrate to VSinoculum ratio of 2, at room temperature which varied from 30°C to 38°C	In. of Y_M removal of 10%[b] (increase from 274 ± 5 to 301 ± 4 mL CH_4/g VS)	Veluchamy et al. (2017)

Continued

Table 5 Continued

Type of sludge/waste	Type of activates	Conditions of activates/AD	Results	References
PPS	HOA	HAO at 80°C for 90 min BMP, 40 d, VSsubstrate to VSinoculum ratio of 2, at room temperature which varied from 30°C to 38°C	CH_4 yield and biodegradability were increased up to 303 mL of CH_4/g VS and 73%., respectively	Veluchamy and Kalamdhad (2017c)
WAS + PS (40%:60% v/v) WAS	Various pretreatment methods	MW (50°C–175°C), UD (20 kHz, 400 W) MicroSludge pretreatment BMP at mesophilic (33 ± 2°C) and thermophilic (53 ± 2°C) temperatures	MW: WAS (M) 63%[b] WAS (T) 13%[b] WAS + PS (M) 46%[b] WAS + PS (T) 38%[b] UD: WAS (M) 51%[b] WAS (T) 38%[b] WAS + PS (M) 43%[b] WAS + PS (T) 38%[b] MicroSludge: WAS (M) +34%[b] WAS (T) +16%[b]	Saha et al. (2011)

Ac, acid; *Al*, alkali; *BCTMP*, WAS from bleaching chemi-thermomechanical pulp; *BMP*, biochemical CH_4 potential assay; *BS*, batch scale; *CBP*, cumulative biogas production; *E*, enzyme; *EH*, electrohydrolysis; *HAO*, hot-air oven; *HT70*, hydrothermal 70°C; *HT150*, hydrothermal 150°C; *In.*, increment; k_h, hydrolysis constants; *M*, mesophilic; *MDP*, CH_4 daily production; *MGWL*, monosodium glutamate waste liquor; *MPR*, CH_4 production rate; *MW*, microwave; *PW*, pig waste; *Pm*, ultimate CH_4 production (mL); *SCM*, semicontinuous mode; *SCOD*, soluble chemical organic demand; *T*, thermophilic; Y_M, CH_4 yield; Y_H, hydrogen yield; *UD*, ultrasound.
[a]Compared to either PS by itself.
[b]Compared to either PPS by itself, without any pretreatment.

(500 L/kg VS). For these reasons, this kind of sludge may be a cosubstrate for codigestion with other meat industry waste. As shown by Buendía et al. (2009), the use of WAS as a cosubstrate improved the digestion performance of cow manure ruminal waste by twofold to almost fourfold. Meanwhile, in the case of pig or cow waste slurry, the addition of WAS allowed biogas to be obtained (while AD of pig or cow waste slurry alone ended in failure). Further, Xia et al. (2015) reported that the addition of beef hooves does not increase the rate of CH_4 production in any

Table 6 The effects induced by different pretreatment methods on lignocellulose biowaste (Veluchamy and Kalamdhad, 2017a)

Type of pretreatment		Increase of accessible surface area	Disruption of the crystalline nature of cellulose	Lignin solubilization	Hemicellulose solubilization	Formation of inhibitory compound[a]	Change in the structure of lignin
P	MBM/MJ	++	+	0	0	0	0
	UD	++	+	na	+	0	na
	EH	++	++	+	+	++	++
	HPH	++	na	na	na	na	na
T	HAO	++	+	++/+	+	0	++
	A	++	0	++/+	++	++	++/+
	MO	++	+	0	+	+	+
C	Ac	++	0	+	++	++	++
	Al	++	0	++/+	+	+	++
	O	++	Na	++/+	0	+	++
	OS	++	+=	0	+	0	0
	AEF	++	++	++	+	+	++
B	E	++	na	++/+	++	0	++
	Mic	++	na	++/+	++	0	++

0, Lack of effect; ++, major effect; +, minor effect; *na*, data not available.
A, autoclave; *Ac*, acid; *AEF*, ammonia fiber explosion; *Al*, alkali; *B*, biological; *C*, chemical; *E*, enzyme; *EH*, electrohydrolysis; *HAO*, hot-air oven; *HPH*, high-pressure homogenizer; *M*, mechanical; *MBM*, mechanical ball mill, *Mic*, microorganisms; *MJ*, mechanical jet; *MO*, microwave oven; *O*, ozonolysis; *OS*, organic solvent; *T*, thermal; *UD*, ultrasound.
[a]For example, furfural/hydroxymethylfurfural.

statistically significant manner. In turn, Xia et al. (2012) codigested WAS with chicken feathers. Compared to the control without chicken feathers, CH_4 yield increased by 47% (from 112 L/kg VS_{add} to 162 L/kg VS_{add}).

Similarly, in the cases of other types of sludge, modification of feedstock composition through the addition of different organic wastes (AcD process) and/or their pretreatment may be a useful tool for the enhancement of biogas production from SHS. There are few studies in the area of pretreatment. They are directed more toward stabilization of sediments than energy recovery. For instance, Masoumi et al. (2015) successfully used Fenton reactions to stabilize the excess sludge from poultry slaughterhouse WWTPs.

SHS also may be used as a source of inoculum (e.g., for hydrogen production). According to Sivaramakrishna et al. (2014), it is associated with high diversity of microorganisms and adaptation to various substrate conditions. Several studies also examined the possibility of using them for the production of hydrogen. For instance, Sittijunda et al. (2010) used poultry SHS from chicken-processing WWTP to achieve this objective. The results show that treated sludge (aerobic thermophilic digestion at 55°C at 48 h) achieved a hydrogen yield of 136.9 mL H_2/g TS and a hydrogen production rate of 2.56 mL H_2/L/h.

Thermochemical processes are also used as means of energy recovery from SHS (Shao et al., 2015). For instance, Oh and Yoon (2017) used hydrothermal carbonization (HTC) to process SHS cake. At optimal conditions (180°C), energy recovery efficiency was 95.1% (the potential energy of feedstock was 4.541 MJ/kg), while CH_4 yield and HTC-biochar production were equal to 15.2 NL/kg_{input} and 0.121 kg/kg_{input}, respectively. Thus, gross energy recovery and energy recovery for sludge were much higher than those obtained after simple drying.

2.6 Petrochemical (oil refinery) sludge

The quantity of sludge and its properties vary depending on the process used during the treatment of petrochemical wastewater (Ma et al., 2013), which in turn is linked to refinery configurations, units dealing with the processing of crude oil (Diya'Uddeen et al., 2011), and the type of oil being refined and operating parameters (Haak et al., 2016). Due to high calorific values of sludge, varying from 17,000–19,000 kJ/kg (Wang et al., 2016), thermal treatments such as incineration pyrolysis and gasification are the main ways of energy recovery in the case of petrochemical sludge (Conesa et al., 2014; Vieira et al., 2009). However, residues such as ash and dust (whose properties are similar to waste obtained from incinerated MSS) that are generated during incineration require landfilling. This makes alternative options like pyrolysis seem like a better solution.

The issues involved with the use of pyrolysis for sludge management have been the subject of numerous publications (Conesa et al., 2014; Hu et al., 2013; Vieira et al., 2009). For instance, Vieira et al. (2009), using pyrolysis around 400°C [namely, low-temperature conversion (LTC)], tested how the reactor's operating mode affects the efficiency of the process. The produced final product in batch mode consisted of 40% pyrolysis oil, 47% pyrolytic char, 8% gas, and 5% water. Meanwhile, for continuous mode, the share of the mentioned components equaled 17%, 63%, 16%, and 4%, respectively. Despite the differences in the yields of the fractions and the amounts of aromatic compounds, the obtained pyrolysis oil had similar chemical characteristics regardless of operating mode. It is worth emphasizing that the heating value of the liquid product of pyrolysis from petrochemical (oil refinery) sludge as opposed to other sewage sludge is the highest (Syed-Hassan et al., 2017) compared to commercial diesel (Karayildirim et al., 2006).

The high concentration of polycyclic aromatic hydrocarbons (PAHs) (up to 30 wt.%), persistent organic pollutants, toxic inorganic compounds, pathogens, and waste oil in this sludge are the main factors limiting the possibility of using AD processes (Ali et al., 2017; Krzywicka and Kwarciak-Kozłowska, 2014; Kwarciak-Kozłowska, 2015; Wang et al., 2016; Zhou et al., 2015). These components may inhibit the growth of microorganisms and consequently lead to process failure (Lin et al., 2014).

Only a few studies have evaluated the feasibility of AD of petrochemical sludge. However, even in these studies, the obtained results were much lower than those obtained for other kinds of sludge presented in this chapter; moreover, the best results were obtained for previously conditioned sludge. For instance, Zhou et al. (2015) found that pretreatment methods enhance the digestibility of this sludge and leads to an increase in process performance. The cumulative biogas production for samples after thermal pretreatment (1466 mL) exceeded the control sample (296 mL), as well as samples after ultrasound pretreatment (366 mL). Also, VS (volatile solid) removal of conditioned sludge was higher than for untreated WAS; the mentioned parameter for samples conditioned using thermal pretreatment (28.1%) was approx. 3.5 and 1.8 times higher than those in the control sample (8.02%) and samples conditioned using ultrasound pretreatment (15.1%). Moreover, both pretreatment methods led to enhanced degradation of PAHs and improved sludge dewaterability.

Furthermore, in a study by Wang et al. (2016) thermal treatment under 170°C was found to be the most effective in terms of sludge solubilization. For this method, the concentration of soluble chemical oxygen demand (SCOD), soluble protein, and polysaccharide were at least 1.25-fold higher than for other tested methods in the study. Furthermore, Wang et al. (2016) noted that two-phase anaerobic digestion (TPAD) boosted performance. The cumulative CH_4 production and SCOD removal efficiency were 1.6 and 2.1 times higher than single-phase AD. It was also noted that separation of the hydrolytic-acidogenic phase and methanogenesis phase accelerate oil conversion. In turn, Haak et al. (2016) recorded a doubling of biogas yield, and COD reduction when 0.05 g O_3/g TS was dosed to WAS (from 80 L/kg COD_{add} to 160 L/kg CDO_{add}, and from 14% to 28%, respectively).

As shown in several studies, this kind of sludge also can be codigested with other waste generated by the petroleum industry. However, the issue is still in its infancy. For instance, Roy et al. (2016) codigested WAS collected from biological treatment with sludge from a dissolved air flotation unit (dissolved air flotation sludge—DAF sludge). The integration of the codigestion process with ozonation enhanced the reduction of solids, and removal of organic compounds determined in DAF sludge, such as PAHs, volatile organic compounds (VOCs) such as benzene, toluene, ethylbenzene and xylene, and semivolatile organic compounds (SVOCs).

3 Conclusion

As the data presented in this chapter shows, the AD process has been successfully used for different kinds of organic wastes originating from municipal and industry sectors. This attractive, environmentally friendly technology minimizes GHG emissions, reduces the use of fossil fuels, and may play a large role in creating the production-renewable energy and energy self-sufficiency of WWTPs. Its drawbacks may be resolved through codigestion of sludge with other organic waste or pretreatment of feedstock. Furthermore, AD may be employed alone or combined with other technologies, such as pyrolysis. Pretreatment of codigestion feedstock also seems to be a very promising and developing approach.

Future research in the field of AD/AcD is still required to resolve the remaining issues. It should focus on the following areas:

- The development of adaptation strategies of microbial fauna to toxic loads, aimed at increasing the diversity and adaptive capacity of microorganisms and creating microbial populations resistant to adverse environmental conditions in terms of organic loading rate, and HRT.
- The possibilities of decomposing new anthropogenic compounds, so-called emerging contaminations, which is an increasingly important issue. Even though their concentration in sewage sludge ranges from nanograms to millograms per kilogram of dry matter, they have become a serious threat due to their potentially negative impact on the environment and various life forms (i.e., aquatic and terrestrial organisms and human health).
- Detailed analyses of the microbial community structures, especially in the case of AcD.
- Process optimization, using pretreatment methods in conjunction with codigestion.
- Verification of existing mathematical models with studies focusing on enhancing process performance.
- Implementation of laboratory scale solutions to industrial scale.
- The link between operational parameters and the dynamics of the microbial community.
- The possibility of using hybrid pretreatment methods promoting the solubilization of feedstock.
- Codigestion of sewage sludge with different waste streams.

Acknowledgments

The study was supported by two internal Czestochowa University of Technology grants, No. BS/PB-401-301/13 and BS/MN – 401-302/17.

References

Abbas, N., Jamil, N., Hussain, N., 2016. Assessment of key parameters in TnS management: a prerequisite for energy recovery. Energ. Sources Part A 38, 2656–2663.

Abreu, M.A., Toffoli, S.M., 2009. Characterization of a chromium-rich tannery waste and its potential use in ceramics. Ceram. Int. 35, 2225–2234.

Agustini, C.B., Spier, F., da Costa, M., Gutterres, M., 2018. Biogas production for anaerobic co-digestion of tannery solid wastes under presence and absence of the tanning agent. Resour. Conserv. Recycl. 130, 51–59.

Ali, A.M., Abu-Hassan, M.A., Ibrahim, R.R., Zaini, M.A.A., Abdulkarim, B.I., Hussein, A.S., Su, S.M., Mohd Halim, M.A.I., 2017. Characterization of petroleum sludge from refinery industry biological wastewater treatment unit. Preprints 1–7.

Altaş, L., 2009. Inhibitory effect of heavy metals on methane-producing anaerobic granular sludge. J. Hazard. Mater. 162, 1551–1556.

Angelidaki, I., Treu, L., Tsapekos, P., Luo, G., Campanaro, S., Wenzel, H., Kougias, P.G., 2018. Biogas upgrading and utilization: current status and perspectives. Biotechnol. Adv. 36, 452–466.

Appels, L., Baeyens, J., Degrève, J., Dewil, R., 2008. Principles and potential of the anaerobic digestion of waste-activated sludge. Prog. Energy Combust. Sci. 34, 755–781.

Ashrafi, O., Yerushalmi, L., Haghighat, F., 2015. Wastewater treatment in the pulp-and-paper industry: a review of treatment processes and the associated greenhouse gas emission. J. Environ. Manag. 158, 146–157.

Bajpai, P., 2015. Generation of waste in pulp and paper mills. In: Management of Pulp and Paper Mill Waste. Springer International Publishing, Cham, pp. 9–17.

Banu, R., Ushani, U., Kannah, R.Y., 2017. Activated sludge process and energy. In: Purohit, H., Kalia, V., Vaidya, A., Khardenavis, A. (Eds.), Optimization and Applicability of Bioprocesses. Springer, Singapore.

Bayr, S., Kaparaju, P., Rintala, J., 2013. Screening pretreatment methods to enhance thermophilic anaerobic digestion of pulp and paper mill wastewater treatment secondary sludge. Chem. Eng. J. 223, 479–486.

Bayr, S., Rintala, J., 2012. Thermophilic anaerobic digestion of pulp and paper mill primary sludge and co-digestion of primary and secondary sludge. Water Res. 46, 4713–4720.

Beauchamp, C.J., Boulanger, R., Matte, J., Saint-Laurent, G., 2002. Examination of the contaminants and performance of animals fed and bedded using de-inking paper sludge. Arch. Environ. Contam. Toxicol. 42, 523–528.

Bień, J., Celary, P., Wystalska, K., 2017. The problems in achieving sustainable development in the tannery industry in regard to sewage sludge management. J. Ecol. Eng. 18, 13–20.

Bień, J., Kamizela, T., Kowalczyk, M., Grosser, A., Zwierz, N., Zabochnicka-Swiatek, M., 2015. The effectiveness of acid fermentation of sonicated primary sludge. J. Res. Sci. Technol. 12, 1–8.

Bień, J., Neczaj, E., Worwag, M., Grosser, A., Nowak, D., Milczarek, M., Janik, M., 2011. Directions management of sludge in Poland after 2013. Eng. Protect. Environ. 14, 375–384.

Bień, J.B., Grosser, A., Neczaj, E., Worwąg, M., Celary, P., 2010. Co-digestion of sewage sludge with different organic wastes: a review. Pol. J. Environ. Stud. 2, 24.

Borowski, S., 2015. Co-digestion of the hydromechanically separated organic fraction of municipal solid waste with sewage sludge. J. Environ. Manag. 147, 87–94.

Buendía, I.M., Fernández, F.J., Villaseñor, J., Rodríguez, L., 2008. Biodegradability of meat industry wastes under anaerobic and aerobic conditions. Water Res. 42, 3767–3774.

Buendía, I.M., Fernández, F.J., Villaseñor, J., Rodríguez, L., 2009. Feasibility of anaerobic co-digestion as a treatment option of meat industry wastes. Bioresour. Technol. 100, 1903–1909.

Bustillo-Lecompte, C.F., Mehrvar, M., 2015. Slaughterhouse wastewater characteristics, treatment, and management in the meat processing industry: a review on trends and advances. J. Environ. Manag. 161, 287–302.

Cano, R., 2015. Energy feasibility study of sludge pretreatments: a review. Appl. Energy 149, 176–185.

Cao, Y., Pawłowski, A., 2012. Sewage sludge-to-energy approaches based on anaerobic digestion and pyrolysis: Brief overview and energy efficiency assessment. Renew. Sust. Energ. Rev. 16, 1657–1665.

Cao, Y., Pawłowski, A., 2013. Bioresource technology life cycle assessment of two emerging sewage sludge-to-energy systems: evaluating energy and greenhouse gas emissions implications. Bioresour. Technol. 127, 81–91.

Caporgno, M.P., Trobajo, R., Caiola, N., Ibáñez, C., Fabregat, A., Bengoa, C., 2015. Biogas production from sewage sludge and microalgae co-digestion under mesophilic and thermophilic conditions. Renew. Energy 75, 374–380.

Carrere, H., Antonopoulou, G., Affes, R., Passos, F., Battimelli, A., 2016. Bioresource Technology Review of feedstock pretreatment strategies for improved anaerobic digestion: From lab-scale research to full-scale application. Bioresour. Technol. 199, 386–397.

Celary, P., Sobik-Szołtysek, J., 2014. Vitrification as an alternative to landfilling of tannery sewage sludge. Waste Manag. 34, 2520–2527.

Celary, P., Sobik-Szołtysek, J., Tajchman, A., 2016. Heavy metal volatilization during vitrification of tannery sewage sludge with mineral waste. Eng. Protect. Environ. 19, 503–515.

Cesaro, A., Velten, S., Belgiorno, V., Kuchta, K., 2014. Enhanced anaerobic digestion by ultrasonic pretreatment of organic residues for energy production. J. Clean. Prod. 74, 119–124.

Chen, Y., Cheng, J.J., Creamer, K.S., 2008. Inhibition of anaerobic digestion process: a review. Bioresource Technol. 99, 4044–4064.

Coelho, N.M.G., Droste, R.L., Kennedy, K.J., 2011. Evaluation of continuous mesophilic, thermophilic and temperature phased anaerobic digestion of microwaved activated sludge. Water Res. 45, 2822–2834.

Conesa, J.A., Moltó, J., Ariza, J., Ariza, M., García-Barneto, A., 2014. Study of the thermal decomposition of petrochemical sludge in a pilot plant reactor. J. Anal. Appl. Pyrolysis 107, 101–106.

Dai, X., Duan, N., Dong, B., Dai, L., 2013. High-solids anaerobic co-digestion of sewage sludge and food waste in comparison with mono digestions: stability and performance. Waste Manag. 33, 308–316.

Davidsson, Å., Lövstedt, C., la Cour Jansen, J., Gruvberger, C., Aspegren, H., 2008. Co-digestion of grease trap sludge and sewage sludge. Waste Manag. 28, 986–992.

Divyalakshmi, P., Murugan, D., Sivarajan, M., Sivasamy, A., Saravanan, P., Rai, C.L., 2017. Optimization and biokinetic studies on pretreatment of sludge for enhancing biogas production. Int. J. Environ. Sci. Technol. 14, 813–822.

Diya'Uddeen, B.H., Daud, W.M.A.W., Abdul Aziz, A.R., 2011. Treatment technologies for petroleum refinery effluents: a review. Process Saf. Environ. Prot. 89, 95–105.

Dong, H., Jiang, X., Lv, G., Chi, Y., Yan, J., 2015. Co-combustion of TnS in a commercial circulating fluidized bed boiler. Waste Manag. 46, 227–233.

Elliott, A., Mahmood, T., 2007. Pretreatment technologies for advancing anaerobic digestion of pulp and paper biotreatment residues. Water Res. 41, 4273–4286.

Eswari, P., Kavitha, S., Kaliappan, S., Yeom, I., Banu, J.R., 2016. Enhancement of sludge anaerobic biodegradability by combined microwave-H_2O_2 pretreatment in acidic conditions. Environ. Sci. Pollut. Res. 13467–13479.

Faubert, P., Barnabé, S., Bouchard, S., Côté, R., Villeneuve, C., 2016. Pulp and paper mill sludge management practices: what are the challenges to assess the impacts on greenhouse gas emissions? Resour. Conserv. Recycl. 108, 107–133.

Fijalkowski, K., Rosikon, K., Grobelak, A., Hutchison, D., Kacprzak, M.J., 2018. Modification of properties of energy crops under Polish condition as an effect of sewage sludge application onto degraded soil. J. Environ. Manag. 217, 509–519.

Fonoll, X., Astals, S., Dosta, J., Mata-Alvarez, J., 2015. Anaerobic co-digestion of sewage sludge and fruit wastes: evaluation of the transitory states when the co-substrate is changed. Chem. Eng. J. 262, 1268–1274.

Fonts, I., Gea, G., Azuara, M., Ábrego, J., Arauzo, J., 2012. Sewage sludge pyrolysis for liquid production: a review. Renew. Sust. Energ. Rev. 16 (5), 2781–2805.

Frąc, M., Jezierska-Tys, S., Oszust, K., Gryta, A., Pastor, M., 2017. Assessment of microbiological and biochemical properties of dairy sewage sludge. Int. J. Environ. Sci. Technol. 14, 679–688.

Gavrilescu, M., 2008. Energy from biomass in pulp and paper mills. Environ. Eng. Manag. J. 7(5).

Gayathri, T., Kavitha, S., Adish Kumar, S., Kaliappan, S., Yeom, I.T., Rajesh Banu, J., 2015. Effect of citric acid induced deflocculation on the ultrasonic pretreatment efficiency of dairy waste activated sludge. Ultrason. Sonochem. 22, 333–340.

Gebreeyessus, G.D., Jenicek, P., 2016. Thermophilic versus mesophilic anaerobic digestion of sewage sludge: a comparative review. Bioengineering (Basel) 3 (2), 1–14.

Girault, R., Bridoux, G., Nauleau, F., Poullain, C., Buffet, J., Peu, P., Sadowski, A.G., Béline, F., 2012. Anaerobic co-digestion of waste activated sludge and greasy sludge from flotation process: batch versus CSTR experiments to investigate optimal design. Bioresour. Technol. 105, 1–8.

Grobelak, A., 2016. Organic soil amendments in the phytoremediation process. In: Ansari, A.A., Gill, S.S., Gill, R., Lanza, G.R., Newman, L. (Eds.), Phytoremediation. Springer International Publishing, Cham, pp. 21–39.

Grobelak, A., Placek, A., Grosser, A., Singh, B.R., Almås, Å.R., Napora, A., Kacprzak, M., 2017. Effects of single sewage sludge application on soil phytoremediation. J. Clean. Prod. 155, 189–197.

Grobelak, A., Rorat, A., Kokot, P., Singh, B.R., Kacprzak, M., 2018. Chapter 28—Mine waste rehabilitation case studies from Poland. for M.S.R. In: Prasad, M.N.V., de C. Favas, P.J., Maiti, S.K.B.T.-B.-G. (Eds.), Bio-Geotechnologies for Mine Site Rehabilitation. Elsevier, pp. 515–527.

Grosser, A., 2017. The influence of decreased hydraulic retention time on the performance and stability of co-digestion of sewage sludge with grease trap sludge and organic fraction of municipal waste. J. Environ. Manag. 203, 1143–1157.

Grosser, A., 2018. Determination of methane potential of mixtures composed of sewage sludge, organic fraction of municipal waste and grease trap sludge using biochemical methane potential assays. A comparison of BMP tests and semi-continuous trial results. Energy 143, 488–499.

Grosser, A., Neczaj, E., 2016. Enhancement of biogas production from sewage sludge by addition of grease trap sludge. Energy Convers. Manag. 125, 301–308.

Grosser, A., Neczaj, E., 2017. Pretreatment methods as a means of boosting methane production from sewage sludge and its mixtures with grease trap sludge. E3S Web Conf. 22, 00058.

Grosser, A., Neczaj, E., 2018. Sewage sludge and fat rich materials co-digestion—performance and energy potential. J. Clean. Prod. 198, 1076–1089.

Grosser, A., Neczaj, E., Singh, B.R., Almås, Å.R., Brattebø, H., Kacprzak, M., 2017. Anaerobic digestion of sewage sludge with grease trap sludge and municipal solid waste as co-substrates. Environ. Res. 155, 249–260.

Grosser, A., Worwag, M., Neczaj, E., Grobelak, A., 2013. Semi-continuous anaerobic co-digestion of mixed sewage sludge and waste fats of vegetable origin. Rocz. Ochr. Sr. 15.

Haak, L., Roy, R., Pagilla, K., 2016. Toxicity and biogas production potential of refinery waste sludge for anaerobic digestion. Chemosphere 144, 1170–1176.

Hamawand, I., 2015. Anaerobic digestion process and bio-energy in meat industry: a review and a potential. Renew. Sust. Energ. Rev. 44, 37–51.

Haroun, M., Idris, A., Omar, S., 2009. Analysis of heavy metals during composting of the tannery sludge using physicochemical and spectroscopic techniques. J. Hazard. Mater. 165, 111–119.

Hidaka, T., Arai, S., Okamoto, S., Uchida, T., 2013. Anaerobic co-digestion of sewage sludge with shredded grass from public green spaces. Bioresour. Technol. 130, 667–672.

Hu, G., Li, J., Zeng, G., 2013. Recent development in the treatment of oily sludge from petroleum industry: a review. J. Hazard. Mater. 261, 470–490.

Hu, S., Ma, X., Lin, Y., Yu, Z., Fang, S., 2015. Thermogravimetric analysis of the co-combustion of paper mill sludge and municipal solid waste. Energy Convers. Manag. 99, 112–118.

Jayashree, C., Janshi, G., Yeom, I.T., Kumar, S.A., Banu, J.R., 2014. Effect of low temperature thermo-chemical pretreatment of dairy waste activated sludge on the performance of microbial fuel cell. Int. J. Electrochem. Sci. 9, 5732–5742.

Juel, M.A.I., Mizan, A., Ahmed, T., 2017. Sustainable use of tannery sludge in brick manufacturing in Bangladesh. Waste Manag. 60, 259–269.

Kabouris, J.C., Tezel, U., Pavlostathis, S.G., Engelmann, M., Dulaney, J., Gillette, R.A., Todd, A.C., 2009. Methane recovery from the anaerobic codigestion of municipal sludge and FOG. Bioresour. Technol. 100, 3701–3705.

Kacprzak, M., Fijałkowski, K., Grobelak, A., Rosikoń, K., Rorat, A., 2015. *Escherichia coli* and Salmonella spp. early diagnosis and seasonal monitoring in the sewage treatment process by EMA-qPCR method. Pol. J. Microbiol. 64, 143–148.

Kacprzak, M., Grobelak, A., Grosser, A., Napora, A., 2014. The potential of biosolid application for the phytostabilisation of metals. Desalin. Water Treat. 52, 3955–3964.

Kacprzak, M., Neczaj, E., Fijałkowski, K., Grobelak, A., Grosser, A., Worwag, M., Rorat, A., Brattebo, H., Almås, Å., Singh, B.R., 2017. Sewage sludge disposal strategies for sustainable development. Environ. Res. 156, 39–46.

Kamali, M., Gameiro, T., Elisabete, M., Costa, V., Capela, I., 2016. Anaerobic digestion of pulp and paper mill wastes—an overview of the developments and improvement opportunities. Chem. Eng. J. 298, 162–182.

Kamali, M., Khodaparast, Z., 2015. Review on recent developments on pulp and paper mill wastewater treatment. Ecotoxicol. Environ. Saf. 114, 326–342.

Kameswari, K.S.B., Kalyanaraman, C., Thanasekaran, K., 2011. Effect of ozonation and ultrasonication pretreatment processes on co-digestion of tannery solid wastes. Clean Technol. Environ. Policy 13, 517–525.

Kanchinadham, S.B.K., Kalyanaraman, C., Kumarasamy, T., 2015. Optimization of organic load for co-digestion of tannery solid waste in semi-continuous mode of operation. Clean Technol. Environ. Policy 17, 693–706.

Karayildirim, T., Yanik, J., Yuksel, M., Bockhorn, H., 2006. Characterisation of products from pyrolysis of waste sludges. Fuel 85, 1498–1508.

Kavitha, S., Saji Pray, S., Yogalakshmi, K.N., Adish Kumar, S., Yeom, I.T., Rajesh Banu, J., 2016. Effect of chemo-mechanical disintegration on sludge anaerobic digestion for enhanced biogas production. Environ. Sci. Pollut. Res. 23, 2402–2414.

Kavouras, P., Pantazopoulou, E., Varitis, S., Vourlias, G., Chrissafis, K., Dimitrakopulos, G.P., Mitrakas, M., Zouboulis, A.I., Karakostas, T., Xenidis, A., 2015. Incineration of tannery sludge under oxic and anoxic conditions: study of chromium speciation. J. Hazard. Mater. 283, 672–679.

Ki, D., Parameswaran, P., Popat, S.C., Rittmann, B.E., Torres, C.I., 2015. Effects of pre-fermentation and pulsed-electric-field treatment of primary sludge in microbial electrochemical cells. Bioresour. Technol. 195, 83–88.

Komatsu, T., Kudo, K., 2007. Anaerobic codigestion of sewage sludge and rice straw. Manage. Public 16, 495–501.

Kovács, E., Wirth, R., Maróti, G., Bagi, Z., Nagy, K., Minárovits, J., Rákhely, G., Kovács, K.L., 2015. Augmented biogas production from protein-rich substrates and associated metagenomic changes. Bioresour. Technol. 178, 254–261.

Krzywicka, A., Kwarciak-Kozłowska, A., 2014. Advanced oxidation processes with coke plant wastewater treatment. Water Sci. Technol. 69, 1875–1878.

Kurahashi, K., Kimura, C., Fujimoto, Y., Tokumoto, H., 2017. Value-adding conversion and volume reduction of sewage sludge by anaerobic co-digestion with crude glycerol. Bioresour. Technol. 232, 119–125.

Kwarciak-Kozłowska, A., 2015. Fenton process supported with ultrasonic field in coke wastewater treatment Wspomaganie procesu Fentona polem ultradźwiękowym w oczyszczaniu ścieków koksowniczych. Przem. Chem. 1, 93–95.

Kwarciak-Kozłowska, A., Bień, J., 2018. Treatment of dairy wastewater in UASB-UF system. E3S Web Conf. 33, 1–8. https://doi.org/10.1051/e3sconf/20184400087.

Kyllönen, H.L., Lappi, M.K., Thun, R.T., Mustranta, A.H., 1988. Treatment and characterization of biological sludges from the pulp and paper industry. Water Sci. Technol. 20, 183–192.

Laskowski, L., Laskowska, M., Fijalkowski, K., Piech, H., Jelonkiewicz, J., Jaskulak, M., Gnatowski, A., Dulski, M., 2017. New class of antimicrobial agents: SBA-15 silica containing anchored copper ions. J. Nanomater. 2017, 1287698. https://doi.org/10.1155/2017/1287698.

Li, W., Yue, Q., Tu, P., Ma, Z., Gao, B., Li, J., Xu, X., 2011. Adsorption characteristics of dyes in columns of activated carbon prepared from paper mill sewage sludge. Chem. Eng. J. 178, 197–203.

Lin, J., Ortiz, R., Steele, T.W.J., Stuckey, D.C., 2014. Toxicants inhibiting anaerobic digestion: a review. Biotechnol. Adv. 32, 1523–1534.

Lin, Y., Liang, J., Zeng, C., Wang, D., Lin, H., 2017. Anaerobic digestion of pulp and paper mill sludge pretreated by microbial consortium OEM1 with simultaneous degradation of lignocellulose and chlorophenols. Renew. Energy 108, 108–115.

Lin, Y., Wang, D., Li, Q., Xiao, M., 2011. Mesophilic batch anaerobic co-digestion of pulp and paper sludge and monosodium glutamate waste liquor for methane production in a bench-scale digester. Bioresour. Technol. 102, 3673–3678.

Lin, Y., Wu, S., Wang, D., 2013. Hydrogen-methane production from pulp & paper sludge and food waste by mesophilic-thermophilic anaerobic co-digestion. Int. J. Hydrog. Energy 38, 15055–15062.

López-Mosquera, M.E., Moirón, C., Carral, E., 2000. Use of dairy-industry sludge as fertiliser for grasslands in northwest Spain: heavy metal levels in the soil and plants. Resour. Conserv. Recycl. 30, 95–109.

Luostarinen, S., Luste, S., Sillanpää, M., 2009. Increased biogas production at wastewater treatment plants through co-digestion of sewage sludge with grease trap sludge from a meat processing plant. Bioresour. Technol. 100, 79–85.

Ma, X., Duan, Y., Liu, M., 2013. Effects of petrochemical sludge on the slurry-ability of coke water slurry. Exp. Thermal Fluid Sci. 48, 238–244.

Mahmood, T., Elliott, A., 2006. A review of secondary sludge reduction technologies for the pulp and paper industry. Water Res. 40, 2093–2112.

Maragkaki, A.E., Vasileiadis, I., Fountoulakis, M., Kyriakou, A., Lasaridi, K., Manios, T., 2018. Improving biogas production from anaerobic co-digestion of sewage sludge with a thermal dried mixture of food waste, cheese whey and olive mill wastewater. Waste Manag. 71, 644–651.

Mari, E.L., Moran, M.S.R., Austria, C.O., 2009. Paper mill sludge as fiber additive for asphalt road pavement. Philipp. J. Sci. 138 (1), 29–36.

Martínez, E.J., Gil, M.V., Fernandez, C., Rosas, J.G., Gómez, X., 2016. Anaerobic codigestion of sludge: dddition of Butcher's fat waste as a cosubstrate for increasing biogas production. PLoS One 11, e0153139.

Masoumi, Z., Shokohi, R., Atashzaban, Z., Ghobadi, N., Rahmani, A.R., 2015. Stabilization of excess sludge from poultry slaughterhouse wastewater treatment plant by the fenton process. Avicenna J. Environ. Health Eng. 2 (1), e3239.

Mata-Alvarez, J., Dosta, J., Romero-Güiza, M.S., Fonoll, X., Peces, M., Astals, S., 2014. A critical review on anaerobic co-digestion achievements between 2010 and 2013. Renew. Sust. Energ. Rev. 36, 412–427.

Meyer, T., Edwards, E.A., 2014. ScienceDirect anaerobic digestion of pulp and paper mill wastewater and sludge. Water Res. 65, 321–349.

Monte, M.C., Fuente, E., Blanco, A., Negro, C., 2009. Waste management from pulp and paper production in the European Union. Waste Manag. 29, 293–308.

Morgan-Sagastume, F., Pratt, S., Karlsson, A., Cirne, D., Lant, P., Werker, A., 2011. Production of volatile fatty acids by fermentation of waste activated sludge pre-treated in full-scale thermal hydrolysis plants. Bioresour. Technol. 102, 3089–3097.

Mulchandani, A., Westerhoff, P., 2016. Recovery opportunities for metals and energy from sewage sludges. Bioresour. Technol. 215, 215–226.

Nartker, S., Ammerman, M., Aurandt, J., Stogsdil, M., Hayden, O., Antle, C., 2014. Increasing biogas production from sewage sludge anaerobic co-digestion process by adding crude glycerol from biodiesel industry. Waste Manag. 34, 2567–2571.

Neczaj, E., Grosser, A., Worwa¸g, M., 2013. Boosting production of methane from sewage sludge by addition of grease trap sludge. Environ. Prot. Eng. 39, 125–133.

Nielfa, A., Cano, R., Fdz-Polanco, M., 2015. Theoretical methane production generated by the co-digestion of organic fraction municipal solid waste and biological sludge. Biotechnol. Reports 5, 14–21.

Noutsopoulos, C., Mamais, D., Antoniou, K., Avramides, C., 2012. Increase of biogas production through co-digestion of lipids and sewage sludge. Global Nest J. 14, 133–140.

Noutsopoulos, C., Mamais, D., Antoniou, K., Avramides, C., Oikonomopoulos, P., Fountoulakis, I., 2013. Anaerobic co-digestion of grease sludge and sewage sludge: the effect of organic loading and grease sludge content. Bioresour. Technol. 131, 452–459.

Ochoa de Alda, J.A.G., 2008. Feasibility of recycling pulp and paper mill sludge in the paper and board industries. Resour. Conserv. Recycl. 52, 965–972.

Oh, S.Y., Yoon, Y.M., 2017. Energy recovery efficiency of poultry slaughterhouse sludge cake by hydrothermal carbonization. Energies. 10.

Parameswaran, P., Rittmann, B.E., 2012. Feasibility of anaerobic co-digestion of pig waste and paper sludge. Bioresour. Technol. 124, 163–168.

Park, S., Li, Y., 2012. Evaluation of methane production and macronutrient degradation in the anaerobic co-digestion of algae biomass residue and lipid waste. Bioresour. Technol. 111, 42–48.

Pastor, L., Ruiz, L., Pascual, A., Ruiz, B., 2013. Co-digestion of used oils and urban landfill leachates with sewage sludge and the effect on the biogas production. Appl. Energy 107, 438–445.

Pervaiz, M., Sain, M., 2015. Recycling of paper mill biosolids: a review on current practices and emerging biorefinery initiatives. Clean Soil, Air, Water 43, 919–926.

Pilli, S., More, T.T., Yan, S., Tyagi, R.D., Surampalli, R.Y., 2016. Fenton pre-treatment of secondary sludge to enhance anaerobic digestion: energy balance and greenhouse gas emissions. Chem. Eng. J. 283, 285–292.

Placek, A., Singh, B.R., Grobelak, A., Włóka, D., Almås, Å.R., Kacprzak, M., 2018. Phytosequestration of carbon in Miscanthus × giganteus and *Pinus sylvestris* L. in degraded zinc smelter and post-mining soils. Pedosphere 28, 555–560.

Poornima Devi, T., Vimala Ebenezer, A., Adish Kumar, S., Kaliappan, S., Rajesh Banu, J., 2014. Effect of deflocculation on the efficiency of disperser induced dairy waste activated sludge disintegration and treatment cost. Bioresour. Technol. 167, 151–158.

Pöschl, M., Ward, S., Owende, P., 2010. Evaluation of energy efficiency of various biogas production and utilization pathways. Appl. Energy 87, 3305–3321.

Priadi, C., Wulandari, D., Rahmatika, I., Moersidik, S.S., 2014. Biogas production in the anaerobic digestion of paper sludge. APCBEE Procedia 9, 65–69.

Priebe, G.P.S., Kipper, E., Gusmão, A.L., Marcilio, N.R., Gutterres, M., 2016. Anaerobic digestion of chrome-tanned leather waste for biogas production. J. Clean. Prod. 129, 410–416.

Priyadarshini, R., Vaishnavi, L., Murugan, D., Sivarajan, M., Sivasamy, A., Saravanan, P., Balasubramanian, N., Rai, C.L., 2015. Kinetic studies on anaerobic co-digestion of ultrasonic disintegrated feed and biomass and its effect substantiated by microcalorimetry. Int. J. Environ. Sci. Technol. 12, 3029–3038.

Rajagopal, R., Massé, D.I., Singh, G., 2013. Bioresource technology a critical review on inhibition of anaerobic digestion process by excess ammonia. Bioresour. Technol. 143, 632–641.

Rashid, M.T., Barry, D., Goss, M., 2006. Paper mill biosolids application to agricultural lands : benefits and environmental concerns with special reference to situation in Canada. Soil Environ. 25, 85–98.

Razaviarani, V., Buchanan, I.D., Malik, S., Katalambula, H., 2013. Pilot-scale anaerobic co-digestion of municipal wastewater sludge with restaurant grease trap waste. J. Environ. Manag. 123, 26–33.

Riau, V., De la Rubia, M.A., Pérez, M., 2015. Upgrading the temperature-phased anaerobic digestion of waste activated sludge by ultrasonic pretreatment. Chem. Eng. J. 259, 672–681.

Roy, R., Haak, L., Li, L., Pagilla, K., 2016. Anaerobic digestion for solids reduction and detoxification of refinery waste streams. Process Biochem. 51, 1552–1560.

Sadecka, Z., 2008. Sewage sludge as regular source of energy in WWTP. Eng. Protect. Environ. 11, 319–329.

Saha, M., Eskicioglu, C., Marin, J., 2011. Microwave, ultrasonic and chemo-mechanical pretreatments for enhancing methane potential of pulp mill wastewater treatment sludge. Bioresour. Technol. 102, 7815–7826.

Serrano, A., Siles, J.A., Gutiérrez, M.C., Martín, M.Á., 2015. Improvement of the biomethanization of sewage sludge by thermal pre-treatment and co-digestion with strawberry extrudate. J. Clean. Prod. 90, 25–33.

Shah, F.A., Mahmood, Q., Rashid, N., Pervez, A., Raja, I.A., 2015. Co-digestion, pretreatment and digester design for enhanced methanogenesis. Renew. Sustain. Energy Rev. 42, 627–642.

Shao, J., Yuan, X., Leng, L., Huang, H., Jiang, L., Wang, H., Chen, X., Zeng, G., 2015. The comparison of the migration and transformation behavior of heavy metals during pyrolysis and liquefaction of municipal sewage sludge, paper mill sludge, and slaughterhouse sludge. Bioresour. Technol. 198, 16–22.

Shi, S., Xu, G., Yu, H., Zhang, Z., 2018. Strategies of valorization of sludge from wastewater treatment. J. Chem. Technol. Biotechnol. 93, 936–944.

Shukla, O.P., Juwarkar, A.A., Singh, S.K., Khan, S., Rai, U.N., 2011. Growth responses and metal accumulation capabilities of woody plants during the phytoremediation of tannery sludge. Waste Manag. 31, 115–123.

Silvestre, G., Illa, J., Fernández, B., Bonmatí, A., 2014. Thermophilic anaerobic co-digestion of sewage sludge with grease waste: effect of long chain fatty acids in the methane yield and its dewatering properties. Appl. Energy 117, 87–94.

Silvestre, G., Rodríguez-Abalde, A., Fernández, B., Flotats, X., Bonmatí, A., 2011. Biomass adaptation over anaerobic co-digestion of sewage sludge and trapped grease waste. Bioresour. Technol. 102, 6830–6836.

Sittijunda, S., Reungsang, A., O-Thong, S., 2010. Biohydrogen production from dual digestion pretreatment of poultry slaughterhouse sludge by anaerobic self-fermentation. Int. J. Hydrog. Energy 35, 13427–13434.

Sivaramakrishna, D., Sreekanth, D., Sivaramakrishnan, M., Sathish Kumar, B., Himabindu, V., Narasu, M.L., 2014. Effect of system optimizing conditions on biohydrogen production from herbal wastewater by slaughterhouse sludge. Int. J. Hydrog. Energy 39, 7526–7533.

Sobiecka, E., Szymanski, L., 2014. Thermal plasma vitrification process as an effective technology for fly ash and chromium-rich sewage sludge utilization. J. Chem. Technol. Biotechnol. 89, 1115–1117.

Sowmya Packyam, G., Kavitha, S., Adish Kumar, S., Kaliappan, S., Yeom, I.T., Rajesh Banu, J., 2015. Effect of sonically induced deflocculation on the efficiency of ozone mediated partial sludge disintegration for improved production of biogas. Ultrason. Sonochem. 26, 241–248.

Stoica, A., Sandberg, M., Holby, O., 2009. Energy use and recovery strategies within wastewater treatment and sludge handling at pulp and paper mills. Bioresour. Technol. 100, 3497–3505.

Syed-Hassan, S.S.A., Wang, Y., Hu, S., Su, S., Xiang, J., 2017. Thermochemical processing of sewage sludge to energy and fuel: fundamentals, challenges and considerations. Renew. Sust. Energ. Rev. 80, 888–913.

Tang, P., Zhao, Y., Xia, F., 2008. Thermal behaviors and heavy metal vaporization of phosphatized tannery sludge in incineration process. J. Environ. Sci. 20, 1146–1152.

Tang, P., Zhou, Y., Xie, Z., 2013. Effects of hydroxyapatite addition on heavy metal volatility during tannery sludge incineration. Environ. Sci. Pollut. Res. 20, 4405–4413.

Thangamani, A., Rajakumar, S., Ramanujam, R.A., 2010. Anaerobic co-digestion of hazardous tannery solid waste and primary sludge: biodegradation kinetics and metabolite analysis. Clean Techn. Environ. Policy 12, 517–524.

Tyagi, V.K., Lo, S.L., 2013. Sludge: a waste or renewable source for energy and resources recovery? Renew. Sust. Energ. Rev. 25, 708–728.

Tyagi, V.K., Lo, S.L., Rajpal, A., 2014. Chemically coupled microwave and ultrasonic prehydrolysis of pulp and paper mill waste-activated sludge: effect on sludge solubilisation and anaerobic digestion. Environ. Sci. Pollut. Res. 21, 6205–6217.

Uma Rani, R., Adish Kumar, S., Kaliappan, S., Yeom, I.T., Rajesh Banu, J., 2012a. Low temperature thermo-chemical pretreatment of dairy waste activated sludge for anaerobic digestion process. Bioresour. Technol. 103, 415–424.

Uma Rani, R., Adish Kumar, S., Kaliappan, S., Yeom, I.T., Rajesh Banu, J., 2013. Impacts of microwave pretreatments on the semi-continuous anaerobic digestion of dairy waste activated sludge. Waste Manag. 33, 1119–1127.

Uma Rani, R., Adish Kumar, S., Kaliappan, S., Yeom, I.T., Rajesh Banu, J., 2014. Enhancing the anaerobic digestion potential of dairy waste activated sludge by two step sono-alkalization pretreatment. Ultrason. Sonochem. 21, 1065–1074.

Uma Rani, R., Kaliappan, S., Adish Kumar, S., Rajesh Banu, J., 2012b. Combined treatment of alkaline and disperser for improving solubilization and anaerobic biodegradability of dairy waste activated sludge. Bioresour. Technol. 126, 107–116.

Veluchamy, C., Kalamdhad, A.S., 2017a. Influence of pretreatment techniques on anaerobic digestion of pulp and paper mill sludge: a review. Bioresour. Technol. 245, 1206–1219.

Veluchamy, C., Kalamdhad, A.S., 2017b. Enhancement of hydrolysis of lignocellulose waste pulp and paper mill sludge through different heating processes on thermal pretreatment. J. Clean. Prod. 168, 219–226.

Veluchamy, C., Kalamdhad, A.S., 2017c. Enhanced methane production and its kinetics model of thermally pretreated lignocellulose waste material. Bioresour. Technol. 241, 1–9.

Veluchamy, C., Raju, V.W., Kalamdhad, A.S., 2017. Prerequisite—an electrohydrolysis pretreatment for anaerobic digestion of lignocellulose waste material. Bioresour. Technol. 235, 274–280.

Vieira, G.E.G., Romeiro, G.A., Sella, S.M., Damasceno, R.N., Pereira, R.G., 2009. Low temperature conversion (LTC)—an alternative method to treat sludge generated in an industrial wastewater treatment station—batch and continuous process comparison. Bioresour. Technol. 100, 1544–1547.

Wan, C., Zhou, Q., Fu, G., Li, Y., 2011. Semi-continuous anaerobic co-digestion of thickened waste activated sludge and fat, oil and grease. Waste Manag. 31, 1752–1758.

Wang, F., Hidaka, T., Tsumori, J., 2014. Enhancement of anaerobic digestion of shredded grass by co-digestion with sewage sludge and hyperthermophilic pretreatment. Bioresour. Technol. 169, 299–306.

Wang, L., Aziz, T.N., de los Reyes, F.L., 2013a. Determining the limits of anaerobic co-digestion of thickened waste activated sludge with grease interceptor waste. Water Res. 47, 3835–3844.

Wang, M., Sahu, A.K., Rusten, B., Park, C., 2013b. Anaerobic co-digestion of microalgae Chlorella sp. and waste activated sludge. Bioresour. Technol. 142, 585–590.

Wang, Q., Liang, Y., Zhao, P., Li, Q.X., Guo, S., Chen, C., 2016. Potential and optimization of two-phase anaerobic digestion of oil refinery waste activated sludge and microbial community study. Sci. Rep. 6, 1–10.

Włóka, D., Kacprzak, M., Grobelak, A., Grosser, A., Napora, A., 2015. The impact of pahs contamination on the physicochemical properties and microbiological activity of industrial soils. Polycycl. Aromat. Compd. 35, 372–386.

Worwąg, M., Neczaj, E., Grosser, A., 2011. Methane production from fat-rich materials. Civil Environ. Eng. Rep. 147–162.

Wystalska, K., Sobik-Szołtysek, J., Bień, J.B., 2013. Vitrification and devitrification of ash after sewage sludge combustion. Rocz. Ochr. Sr. 15, 181–191.

Xia, Y., Massé, D.I., McAllister, T.A., Beaulieu, C., Ungerfeld, E., 2012. Anaerobic digestion of chicken feather with swine manure or slaughterhouse sludge for biogas production. Waste Manag. 32, 404–409.

Xia, Y., Wang, D.K., Kong, Y., Ungerfeld, E.M., Seviour, R., Massé, D.I., 2015. Anaerobic digestibility of beef hooves with swine manure or slaughterhouse sludge. Waste Manag. 38, 443–448.

Xiao, Z., Yuan, X., Li, H., Jiang, L., Leng, L., Chen, X., Zeng, G., Li, F., Cao, L., 2015. Chemical speciation, mobility and phyto-accessibility of heavy metals in fly ash and slag from combustion of pelletized municipal sewage sludge. Sci. Total Environ. 536, 774–783.

Xu, C.C., Lancaster, J., 2007. Treatment of secondary sludge for energy recovery. Europe.

Xu, G.R., Zou, J.L., Li, G.B., 2008. Stabilization of heavy metals in ceramsite made with sewage sludge. J. Hazard. Mater. 152, 56–61.

Yalcinkaya, S., Malina, J.F., 2015a. Anaerobic co-digestion of municipal wastewater sludge and un-dewatered grease trap waste for assessing direct feed of grease trap waste in municipal digesters. Int. Biodeterior. Biodegrad. 104, 490–497.

Yalcinkaya, S., Malina, J.F., 2015b. Model development and evaluation of methane potential from anaerobic co-digestion of municipal wastewater sludge and un-dewatered grease trap waste. Waste Manag. 40, 53–62.

Yamamoto, T., Kobayashi, S., Nishiguchi, A., Nonaka, T., Tsutsui, T., 2006. Evaluation of bovine spongiform encephalopathy (BSE) infection risk of cattle via sewage sludge from wastewater treatment facilities in slaughterhouses in Japan. J. Vet. Med. Sci. 68, 137–142.

Yan, S., Sagoe-Crentsil, K., Shapiro, G., 2011. Reuse of de-inking sludge from wastepaper recycling in cement mortar products. J. Environ. Manag. 92, 2085–2090.

Zawieja, I., 2017. Energy potential of the modified excess sludge. E3S Web Conf. 22, 00200.

Zawieja, I., Lidia, W., Marta, P., 2017. Impact of the excess sludge modification with selected chemical reagents on the increase of dissolved organic substances concentration compounds transformations in activated sludge. Environ. Res. 156, 652–656.

Zhang, W., Wei, Q., Wu, S., Qi, D., Li, W., Zuo, Z., Dong, R., 2014. Batch anaerobic co-digestion of pig manure with dewatered sewage sludge under mesophilic conditions. Appl. Energy 128, 175–183.

Zhen, G., Lu, X., Kato, H., Zhao, Y., Li, Y., 2017. Overview of pretreatment strategies for enhancing sewage sludge disintegration and subsequent anaerobic digestion: current advances, full-scale application and future perspectives. Renew. Sust. Energ. Rev. 69, 559–577.

Zhou, J., Ma, H., Gao, M., Sun, W., Zhu, C., Chen, X., 2018. Changes of chromium speciation and organic matter during low-temperature pyrolysis of tannery sludge. Environ. Sci. Pollut. Res. 25, 2495–2505.

Zhou, J., Xu, W., Wong, J.W.C., Yong, X., Yan, B., Zhang, X., Jia, H., 2015. Ultrasonic and thermal pretreatments on anaerobic digestion of petrochemical sludge: Dewaterability and degradation of PAHs. PLoS One 10, 1–13.

Zupančič, G.D., Uranjek-Ževart, N., Roš, M., 2008. Full-scale anaerobic co-digestion of organic waste and municipal sludge. Biomass Bioenergy 32, 162–167.

Biogas production by thermal hydrolysis and thermophilic anaerobic digestion of waste-activated sludge

31

Ewa Neczaj, Anna Grosser
Institute of Environmental Engineering, Czestochowa University of Technology, Czestochowa, Poland

1 Introduction

One of the urgent problems associated with the treatment of municipal wastewater is the cost of sewage sludge disposal, which involved approximately 50% of operating costs of the wastewater treatment plant (WWTP). The main methods used for sludge treatment are thickening, aerobic and anaerobic stabilization, and dewatering (Bień et al., 2011; Grobelak, 2016; Kacprzak et al., 2017; Macherzyński et al., 2015). Among the available methods of sludge stabilization, biological methods are the most important, particularly methane (CH_4) fermentation and aerobic stabilization (Grobelak et al., 2013, 2018; Rorat et al., 2017; Suleiman et al., 2017). Anaerobic digestion (AD) is one of the oldest and dominant methods of sewage sludge stabilization, which allows for simultaneous energy recovery from sludge and limits its negative impact on the environment through the reduction of organic matter concentration. In medium-size and large WWTPs, AD plays a very important role because it contributes to improving their energy balance (Grosser et al., 2013; Grosser and Neczaj, 2018; Zawieja and Wolny, 2013).

The main volume of sludge directed to digesters is waste-activated sludge (WAS). In contrast to primary sludge (PS), WAS is poorly biodegradable. Only 30%–40% of organic fraction of WAS is transformed into biogas via the conventional mesophilic AD process, which requires a long sludge retention time (SRT) and results in poor CH_4 production. This is because WAS is a mixture of microorganisms and chemical compounds (organic and inorganic) agglomerated together by extracellular polymeric substances (EPSs) (Grosser and Neczaj, 2017; Wolski and Zawieja, 2014; Zawieja et al., 2017).

The fermentation process of sewage sludge is carried out mainly in mesophilic conditions in the temperature range of 30–42°C and ensures good stabilization of sludge, but it does not provide sufficient hygienization. The thermophilic digestion (50–60°C) gives better results in the reduction of pathogens, and it is conducted at shorter SRTs.

Industrial and Municipal Sludge. https://doi.org/10.1016/B978-0-12-815907-1.00031-3

However, the thermophilic AD process is more sensitive to process parameter fluctuation than the mesophilic formation and requires more energy to heat the sewage sludge to the right temperature. Proper CH_4 fermentation should ensure biochemically stabilized sewage sludge and the production of high-calorie biogas. Hydrolysis is considered as a step that limits the rate of AD. Therefore, accelerating the conversion of organic compounds in sludge into soluble form results in a significant improvement in the efficiency of subsequent fermentation phases. Increasing the degree of volatile solid (VS) degradation and biogas production can be obtained using various methods: mechanical (ultrasound, homogenization, high pressure), thermal, chemical [alkalis, ozone (O_3) oxidation], and biological. Recently, biological methods have become very popular for industrial use due to their good efficiency and lower costs compared to physical and chemical methods (Bień et al., 2015; Grosser and Neczaj, 2016; Neczaj et al., 2013).

This chapter presents a review of biogas production from activated sludge using thermal hydrolysis (TH) and thermophilic AD. The impact of various pretreatment methods on AD efficiency is also covered.

2 Biogas production from WAS

Sewage sludge is a by-product of wastewater treatment processes generated as a result of a number of physical, physicochemical, and biological processes taking place in WWTPs. The main volume of sludge generated in the wastewater treatment process consists of PS and secondary sludge after the biological treatment. PS is generated during primary treatment in primary clarifiers as a result of removed settleable solids. PS contains about 70% of organic matter with good biodegradability, and more heavy metals than other sludge generated in WWTPs. PS is a good substrate for fermentation and is subject to rapid decomposition (Khanal, 2011).

Secondary sludge is generated in secondary clarifiers, where activated sludge from biological wastewater treatment (biological reactors) is separated from treated wastewater. Some of these sediments are recycled to the bioreactor (recirculated sludge), while the excess sludge is discharged from the biological system. Excess activated sludge is called WAS or *surplus activated sludge* and is removed from the treatment system to keep the ratio of biomass to food supplied in the wastewater at the right level (Foladori et al., 2010; Khanal, 2011).

Fig. 1 shows a simplified scheme of sludge generation during the biological wastewater treatment process. Production of WAS is associated with the following steps (Foladori et al., 2010):

- The growth and death of heterotrophic biomass
- The accumulation of endogenous residues
- The accumulation of inert (neutral) solids from the influent

WAS is difficult to dewater, especially after multiphase wastewater treatment systems (e.g., A2O) or long-term anaerobic conditions. Degradation of WAS occurs slowly

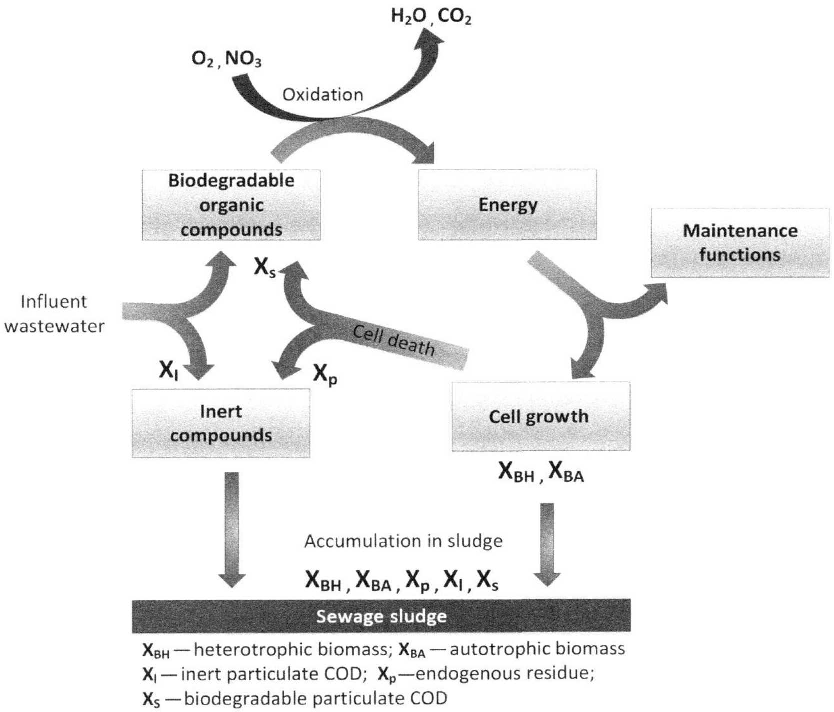

Fig. 1 A scheme of sludge generation during biological wastewater treatment (Foladori et al., 2010).

and with very limited efficiency. The extracellular polymeric substances presented in sludge flocs are mainly responsible for the poor biodegradability of WAS. EPSs are a mixture of biopolymers containing polysaccharides, proteins, nucleic acids, uronic acids, humic substances, and lipids, among others (Fig. 2) (Lin et al., 2014; Shi et al., 2017).

These substances are resistant to degradation in both aerobic and anaerobic conditions (Park et al., 2008). Studies show varying degrees of WAS biodegradation, ranging from 30%–50% (Carrère et al., 2010). The main factors determining its value are concentration of inert substances in raw wastewater, temperature of the process, length of the anoxic phase, and in particular the age of the sludge (Mottet et al., 2010). During typical AD, with a fermentation time equal to 20 days, the degree of organic matter conversion to biogas ranges from 25% for longer sludge age to 60% for shorter sludge age (Latif et al., 2017).

The theoretical production of CH_4 Y_M can be calculated based on the elemental composition of the compound being converted using the following equation, proposed by Buswell (Gallert and Winter, 2005; Khanal, 2011):

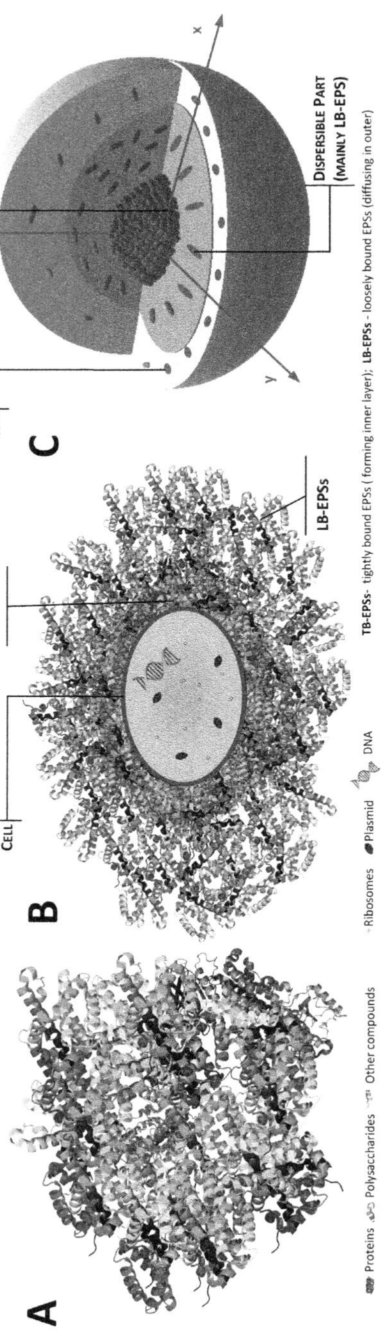

Fig. 2 EPSs structure (TB-EPSs—tightly bound EPSs (forming inner layer); LB-EPSs—loosely bound EPSs (diffusing in outer)) (Lin et al., 2014; Shi et al., 2017).

$$Y_m = C_cH_hO_oN_nS_s + \left(\frac{4c-h-2o+3n+2s}{4}\right)H_2O$$
$$\rightarrow \left(\frac{4c-h+2o+3n+2s}{8}\right)CO_2 +$$
$$+\left(\frac{4c+h-2o-3n-2s}{8}\right)CH_4 + nNH_3 + sH_2S.$$

The maximum value of the CH_4 yield calculated on the basis of the Buswell equation is shown in Table 1. It does not consider many factors affecting the production of biogas under real conditions, such as VS degradation or biomass synthesis.

3 Pretreatment strategies for enhancing WAS AD

There are two main options to enhance CH_4 fermentation: modification of the AD process (e.g., reactor design) or pretreatment of sewage sludge prior to AD to accelerate the hydrolysis of sludge (the stage identified as the rate-limiting step) (Carrere et al., 2016; Elliott and Mahmood, 2007; Zhang, 2010; Pérez-Elvira et al., 2006). Due to the slow rate of WAS hydrolysis during AD, various pretreatment methods have been suggested to improve the solid disintegration and biogas production. Some of them have been already implemented in WWTPs, and some of them are still in the laboratory phase. Mechanical, chemical, thermal, and biological processes, as well as combinations of them, could be applied for the purpose of disrupting cell walls, which causes cell lysis or disintegration and consequently leads to the release of intracellular polymers and polymers from cell walls into the liquid phase (Fig. 3) (Bien et al., 2004;

Table 1 Theoretical methane yield for selected substrates

Substrate	Reaction	Y_M (m^3/kg d.m.)	CH_4 (%)
Lipids	$C_{50}H_{90}O_6 + 24{,}5H_2O \rightarrow 34{,}75CH_4 + 15{,}25CO_2$	1.014	70
Carbohydrates	$C_6H_{10}O_5 + H_2O \rightarrow 3CH_4 + 3CO_2$	0.415	50
Proteins	$C_{16}H_{24}O_5N_4 + 14{,}5H_2O \rightarrow 8{,}25CH_4 + 3{,}75CO_2 + 4NH_4^+ + 4HCO_3^-$	0.496	50
PS	$C_{10}H_{19}O_3N + 4{,}5H_2O \rightarrow 6{,}25CH_4 + 3{,}75CO_2 + NH_3$	0.7	–
Secondary sludge	$C_5H_7O_2N + 2H_2O \rightarrow 2{,}5CH_4 + 2{,}5CO_2 + NH_3$	0.5	–

Source: Alves, M.M., Pereira, M.A., Sousa, D.Z., Cavaleiro, A.J., Picavet, M., Smidt, H., Stams, A.J.M., 2009. Waste lipids to energy: how to optimize methane production from long-chain fatty acids (LCFA). Microb. Biotechnol. 2, 538–550; Angelidaki, I., Sanders, W., 2004. Assessment of the anaerobic biodegradability of macropollutans. Rev. Environ. Sci. Biotechnol. 3, 117–129; Khanal, S., 2011. Anaerobic Biotechnology for Bioenergy Production: Principles and Applications. John Wiley & Sons (Google Books).

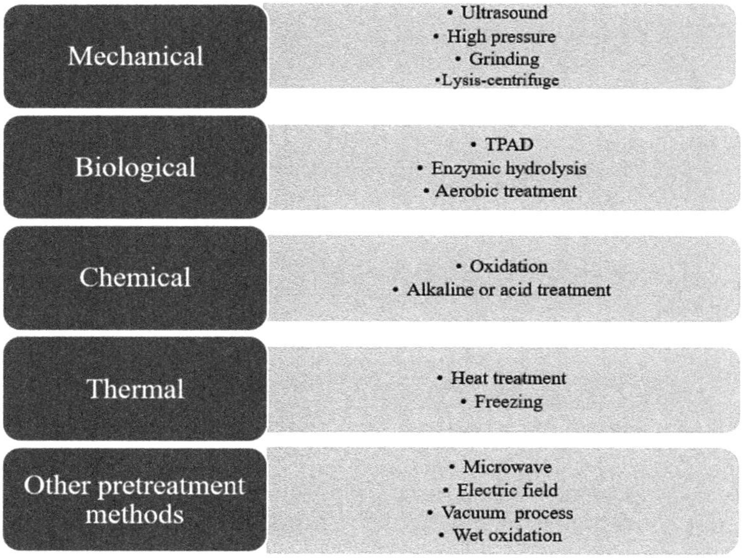

Fig. 3 Pretreatment methods in various categories; where: *AOP*, advanced oxidation processes; *TPAD*, temperature-phased anaerobic digestion (Carrere et al., 2016; Elliott and Mahmood, 2007).
Modified from Carrère, H., Dumas, C., Battimelli, A., Batstone, D.J., Delgenès, J.P., Steyer, J.P., Ferrer, I., 2010. Pretreatment methods to improve sludge anaerobic degradability: a review. J. Hazard. Mater. 183, 1–15.

Zhang, 2010; Pérez-Elvira et al., 2006). The main benefits of sludge pretreatment are improvement of the solubility of sludge solids that become better accessible to anaerobic microorganisms, increased organic solid degradation and CH_4 production rates, mass reduction and pathogens, and decreased anaerobic digester retention time.

3.1 Mechanical methods

To date, various types of mechanical disintegration methods have been developed and tested, such as high-pressure homogenizers (HPHs), grinding, lysis-centrifuge, and ultrasonication (Carrère et al., 2010; Shah et al., 2015). One of the most widely used technologies is the HPH, which consists of a high-pressure pump and a homogenizing valve. The pump compresses the sludge to a pressure of several hundred bars, which causes cavitation bubbles to form.

During the passage of sludge through the homogenizing valve, velocity increases rapidly up to 300 m/h, and the pressure suddenly drops below vapor pressure. In these conditions, the cavitation bubbles implode, which leads to cell disaggregation (Onyeche and Schäfer, 2003). Several commercial systems based on HPH have been developed, including MicroSludge, Biogest Crown Disntegration System, CELLRUPTOR, and Rapid Nonequilibrium Decompression (RND) "BUG BUSTER" technology.

Ultrasonic disintegration is a well-known method for the breakdown of microbial cells and sludge flocs. The disintegration effect is the result of ultrasonic cavitation and the formation of pulsating bubbles, inside which the temperature and pressure can reach as high as 500 K and 100 MPa, respectively, and very reactive hydroxyl radicals (H·, OH·) and hydrogen peroxide (H_2O_2) are formed. The efficiency of ultrasound disintegration depends on the sludge characteristics and ultrasound field parameters (e.g., energy input/applied, as well as frequency). The main ultrasonic technologies available for sludge disintegration are Sonolyzer, Sonix, MaXonics, and Hielscher. The main differences between these systems are based on the construction of a sonotrode (or horn), which is used to introduce energy into the medium (Pilli et al., 2011).

Microwave (MW) irradiation is recognized as an alternative method to conventional thermal pretreatment, mainly due to the production of focused direct heat and, as a consequence, lower energy losses during transmission. The mechanism of MW irradiation is associated with a thermal effect and an athermal effect caused by the polarized macromolecules that are not associated with an increase of temperature. The influence of MW irradiation on the solubilization of WAS, volatile suspended solid (VSS) destruction, and biogas production has recently been reported (Eskicioglu et al., 2009, 2007; Qiao et al., 2008). Higher digestion efficiency and biogas yield were reported for AD of MW sludge pretreated at a temperature higher than 100°C (Toreci et al., 2009).

A pulsed electric field (or electrokinetic disintegration) is also used for sludge disintegration. A high-voltage field creates charges that induce the sudden disruption of sludge flocs and cellular membranes. Vogelsang Ltd. has launched BioCrack technology based on electrokinetic disintegration that has been successfully applied at full scale in European WWTPs and pilot plants in the United States. WWTPs in which BioCrack technology is applied show about 8% higher biogas production, more than 11% VS reduction, and about 18% lower consumption of polyelectrolytes for digestate sludge dewatering. A similar technology is OpenCEL, introduced at the Northwest Water Reclamation Plant in Mesa, Arizona. The use of a pulsating electric field to sludge disintegration causes an increase in the concentration of a soluble fraction of chemical oxygen demand (COD), 60% higher biogas production, and 40% VS reduction. The implementation of OpenCEL technology has improved the energy balance of WWTPs (Neczaj, 2016).

3.2 Chemical methods

Acids, alkali, and oxidants like O_3 are the major reagents used for chemical pretreatment of sewage sludge. The pretreatment of sewage sludge with the use of alkaline compounds or acid reagents [hydrochloric acid (HCl), sulfuric acid (H_2SO_4)] leads to the hydrolysis and depolymerization of carbohydrates, fats, and proteins present in the sludge to lower-molecular-weight compounds (i.e., aliphatic acids, polymorphs, and amino acids). However, the most widely used method is alkaline treatment, which is often combined with TH. According to Foladori et al. (2010), the optimal conditions for inducing biosolid solubilization are pH >10, 50–60°C, and contact time of less than 1 h. It was reported that WAS pretreatment with sodium hydroxide (NaOH) was more efficient than other hydroxides like potassium hydroxide (KOH), magnesium hydroxide ($Mg(OH)_2$), and calcium hydroxide ($Ca(OH)_2$) (Kim et al., 2003; Shah et al., 2015). The mechanism of alkali hydrolysis is based on solvatation and saponification. Acid and alkali pretreatment technologies are simple and easy to operate and provide high CH_4 conversion efficiency.

Oxidation with O_3 indirectly or directly leads to the destruction of the cell walls of activated sludge microorganisms and the release of cytoplasm into the supernatant. The mechanism of organic compound oxidation is based on direct oxidation or generation of hydroxyl radicals. During the WAS ozonation process, the following reactions take place: floc destruction and cell lysing, solubilization, and mineralization of the soluble organic matter. It was found that the optimum dose of O_3 in order to carry out the sludge disintegration is in the range of 0.05–0.5 g O_3/g TS (Xu et al., 2010). Moreover, mineralization of biodegradable organic compounds at a higher dose of this oxidant is observed (known as the *mineralization phenomenon*). O_3 can be also applied after the AD process. The main ozonation technologies available on the market used for WAS oxidation are Aspal Sludge, and Praxair Lyso™.

3.3 Biological pretreatment

The main biological methods used in WAS pretreatment include enzymic hydrolysis (EH) and predigestion (two-stage AD), which will be discussed in Chapter 32.

The enzymes used in EH are mainly extracted from bacteria, yeast, and mold, and they catalyze the breakdown of macromolecular polymers in activated sludge flocs into shorter monomers or other easily digestible simple chemical compounds. The most commonly used enzymes in EH technology are those that break down proteins and polysaccharides. Proteases are most often used for protein degradation and pronase (a mixture of proteases isolated from *Streptomyces griseus*), while cellulases (endocellulose and exocellulose), α-amylases, and carbohydrases are responsible for carbohydrate degradation (Barjenbruch and Kopplow, 2003; Roman et al., 2006; Wawrzynczyk et al., 2007). Lipolytic enzymes are used to break down fats present in WAS (Ayol, 2005). Research is also carried out on WAS enzymatic hydrolysis using mixtures of extracted enzymes (proteases, lipases, and/or cellulases) with endogenous enzymes derived from the disintegrated cells of *Aspergillus oryzae*, *Aspergillus* sp., *S. griseus*, and various strains of anaerobic bacteria. Mayhew et al. (2003) observed a 10% improvement in biogas production for WAS pretreated with EA during a laboratory experiment.

3.4 Thermal methods

The thermal pretreatment of sludge is carried out at temperatures of 90–200°C (usually in the range of 160–190°C) for 20–30 min, and this affects the acceleration of the hydrolysis reaction and consequently improves the efficiency of the fermentation process. The carbohydrates and lipids present in the sludge have better biodegradability than proteins, which are protected against the enzymatic hydrolysis by the cell wall

(Müller, 2000). Changing the properties and structure of sludge under high temperatures is associated with the destruction of cell walls of microorganisms and the release of proteins (and other compounds present in the cytosol), which can become a substrate for hydrolytic enzymes. However, it should be noted that the use of high temperatures for the pretreatment of sludge is also associated with the formation of compounds resistant to biodegradation.

Thermal pretreatment also may be performed at low temperatures, but only a few studies have investigated the effect of freezing disintegration on AD efficiency. TH is commercial technology used to intensify the CH_4 fermentation of sewage sludge; it will be discussed in more detail in Chapter 32. Table 2 summarizes the impact of WAS pretreatment technologies on sludge properties and AD efficiency.

Each of the sewage sludge pretreatment methods has its advantages and disadvantages. The choice of method should be adapted to a specific WWTP and take into account many factors, including the efficiency of the process, the complexity of technology, and economic aspects (Fig. 4).

Table 3 presents a comparison of the degree of disintegration of cells and estimated costs of selected technologies of preliminary sludge conditioning according to Zhang (2010).

Nagler et al. (2018) provided a cost-benefit calculation of thermochemical WAS pretreatment for Zirl WWTP in Tyrol, Austria, based on net energy output and cost efficiency (Table 4). The authors have shown that it is possible to reduce electricity (22%) and disposal (27%) costs by optimization of thermochemical pretreatment. However, this pretreatment method is still not recommended due to high investment costs.

3.5 Thermal hydrolysis

TH is commonly used as a pretreatment method to increase biogas production and VS degradation in the AD process. This method is defined as the reaction occurring in aqueous solutions at a specific temperature and pressure. As the temperature in sewage sludge increases, the sequence of reactions shown in Fig. 5 takes place. One of the first mathematical models describing the sludge decomposition during the TH process was proposed by Takamatsu et al. (1970). The solid matter, evaporative and non-evaporative soluble matter, and water were taken into account in this model. The authors described the sludge decomposition reactions in terms of the solubilization of solid matter and assumed the Arrhenius type of the rate constant created the following mathematical model:

$$\frac{dA_{wt}}{dt} = -\left(0.37e^{-\frac{249}{T}} + 0.319e^{-\frac{500}{T}}\right)(1-\xi)A_{wt}$$

$$\frac{dB_{wt}}{dt} = \left(0.37e^{-\frac{249}{T}} \times (1-\xi) \times A_{wt} - \left(1800e^{-\frac{460}{T}} - B_{wt}\right) + 0.55e^{-\frac{1260}{T}} \times C_{wt}\right.$$

$$C_{wt} = (A_{wt} + B_{wt})_{init.} - A_{wt} - B_{wt}$$

$$(A_{wt} + B_{wt})_{init.} - A_{wt} - B_{wt}$$

Table 2 Impact of pretreatment technologies on properties of WAS and anaerobic digestion

	Pretreatment			Anaerobic digestion			
Method	**Conditions**	**% COD solubilization**	**Other effects**	**Conditions**	**Improvements of biogas production (%)**	**Improvements of VS removal (%)**	**Other effects**
High-temperature treatment (100–210°C)	Temperature: 100–210°C Time: 20–60 min Pressure: 3–21 bar	12–80	SVI—70% reduction CST–reduction at 130–150°C (up to 90% at 190°C) In. concentration of ammonia, soluble inert and color	Assay: B, SC Scale: L, FS Temp: M; T	25–150 (15–100)[a]	7–105	Divergent impact on CST In. color and concentration of ammonia in the effluent In. electricity production Digestate reduction
Low-temperature treatment (<100°C)	Temperature: 70–95°C Time: 15 min–7 d	1–20	SVI—40% reduction at 95°C In. CST and fecal *Streptococcus* removal, as well as VFAs and HM concentration in the soluble phase Reduced sludge	Assay: B, SC Scale: L Temp: M; T	10–984	10–40 (28–617)[b]	In. methane production rate In. methane concentration in biogas (2%–10%)

			viscosity Consumption of OM with 72h of treatment (70° C)				
Freezing/ thawing	Temperature: –90 to –10°C Time: 24h (freezing) 12–15h (thawing)	9	Consumption of OM during thawing In. settling velocity of sludge Reduction	Assay: SC Scale: L Temp: M	52	4	Reduced moisture from 75% to 62%
Ultrasound	Specific energy: 112–10,800kJ/kg TS Power: 60–480W Frequency: 20–200kHZ Time: 0.5–60min	1–40	CST—In. 15–20times In. concentrations of soluble nitrogen and VFAs Reduction particle size In. specific charge and turbidity DD 2%–47%	Assay: B, SC Scale: L, FS Temp: M; T	4–83 (14–260)[a]	6–47 (22–83)[b]	In. removal of some organic pollutants Possible net energy gain at full scale
High pressure homogenization	Pressure: 30–600bar Cycles: 1–2 Specific energy: 300–3380kJ/kg TS	3–8	DD 4%–43%	Assay: SC Scale: L Temp: M	17–115 (60)[a]	7–138	In. CST In. methane concentration in biogas Moderate control of foaming

Continued

Table 2 Continued

	Pretreatment			Anaerobic digestion			
Method	**Conditions**	**% COD solubilization**	**Other effects**	**Conditions**	**Improvements of biogas production (%)**	**Improvements of VS removal (%)**	**Other effects**
Microwave	Specific energy: 14,000–336,000 kJ/m^3 Temp.: 30–175°C Power: 800–1250 MW Frequency: 2450 MHz	7–31 (2–21)[c] (up to 214)[d]	VFAs—In. up to 3 times Possible Maillard reactions at high temperatures (>80°C)	Assay: B, SC Scale: L Temp: M; T	16–570.7 (31)[a]	23–48	Removal of *Salmonella* spp. and *E. coli*, 50% decrease of *C. perfringens*
Electric pulse electrolysis	Specific energy: 1.1–34 kWh/m^3 Field strength: 8–30 kV/cm Configuration: single and dual chamber	2–10	Reduced volume of settled sludge Possible loss of solids due to generation of colloids (42% at 19.8 kWh/m^3)	Assay: B, SC Scale: L, FS Temp: M	10–100	9	In. diversity of microbial communities of digesters Partial elimination of foaming during digestion
Alkalis	Reagents: NaOH, KOH, $Mg(OH)_2$, $Ca(OH)_2$ Doses: 20–40 meq NaOH/l 0–26.1 g NaOH/l 0.05–4.0 mol NaOH/l	Up to 50 (up to 80)[c]	Up to 33% solid removal during pretreatment In. SRF Decreased particle size	Assay: B, SC Scale: L Temp: M; T	13–120	Up to 133	Divergent impact on methane concentration in biogas In. dewaterability

	0.02–0.05 mol $Ca(OH)_2$ pH: 8–12 Time: 30 min–8 d		Possibility of reflocculation, with high doses of alkali Reduction: total coliforms, *E. coli*, *C. perfringens* DD around 60%				of sludge Possible Maillard reactions led to decreased biodegradability
Acids	Doses: 1.25–17.5 mL HCl/kg wet sludge 15–100 g PAA/kg DS pH: 1–6 Time: 1–24 h	Up to 400[d]	Reduction of *E.coli*	Assay: B, SC Scale: L Temp: M	12–32	5	High concentration of PAA led inhibition Elimination of *Salmonella* spp. and reduction of *E. coli*, Decrease polymer dose for dewatering
Ozonation	Doses: 0.05–0.2 gO_3/g COD 0.045–0.16 gO_3/g TS 10–20 mgO_3/g TSS 0.66–2.65 mgO_3/g biomass	10–29	In. CST, SRF Acidification of sludge Decreased apparent viscosity DD 18% VSS reduction 18%	Assay: B, SC Scale: L Temp: M; T	8–800 (5–80)[a]	9[e] (1.6 folds)[b]	Reduced CST In. removal: total coliforms, *E. coli*, *C. perfringens*

Continued

Table 2 Continued

	Pretreatment			Anaerobic digestion			
Method	**Conditions**	**% COD solubilization**	**Other effects**	**Conditions**	**Improvements of biogas production (%)**	**Improvements of VS removal (%)**	**Other effects**
Fenton process	Doses: 5–100 g H_2O_2/kg DS 0.4–5.0 g Fe(II)/kg DS 10–100 mg H_2O_2/l 1–5 mg Fe(II)/l	6–11.9 times[d]	DD 7%–24% Reduced particle size	Assay: B, SC Scale: L Temp: M; T	10–30	31–72	Decreased CST but not affect filterability In. net energy Reduced GHG emissions
Enzyme addition	Kind of enzyme: protease, amylase, carbohydrase, cellulase Temp: 37–50°C Time: 4 h–11 d	15.1–25[c]	Reduction CST Up to 58% VSS reduction during pretreatment	Assay: B, SC Scale: L, P Temp: M; T	Up to 12%	1–16.3	–

B, batch assay, *COD*, chemical oxygen demand; *CST*, capillary suction time; *DD*, disintegration degree; *DS*, dry sludge; *FS*, full scale; *GHG*, greenhouse gas; *HM*, heavy metals; *In.*, increased; *L*, laboratory; *M*, mesophilic; *OM*, organic matter; *P*, pilot; *PAA*, peracetic acid; *SC*, semicontinuous assay; *SCOD*, soluble chemical oxygen demand; *SRF*, specific resistance to filtration; *SVI*, sludge volumetric index; *T*, thermophilic; *TCOD*, total chemical oxygen demand; *Temp*, temperature; *TS*, total solids; *TSS*, total suspended solids.

[a]Improvements of methane production.

[b]VSS removal.

[c]SCOD/TCOD ratio.

[d]Increase in SCOD (soluble).

[e]Total suspended solid removal.

Adapted from Carrère, H., Dumas, C., Battimelli, A., Batstone, D.J., Delgenès, J.P., Steyer, J.P., Ferrer, I., 2010. Pretreatment methods to improve sludge anaerobic degradability: a review. J. Hazard. Mater. 183, 1–15; Neumann, P., Pesante, S., Venegas, M., Vidal, G., 2016. Developments in pre-treatment methods to improve anaerobic digestion of sewage sludge. Rev. Environ. Sci. Biotechnol. 15, 173–211; Zhen, G., Lu, X., Kato, H., Zhao, Y., Li, Y., 2017. Overview of pretreatment strategies for enhancing sewage sludge disintegration and subsequent anaerobic digestion: current advances, full-scale application and future perspectives. Renew. Sust. Energ. Rev. 69, 559–577.

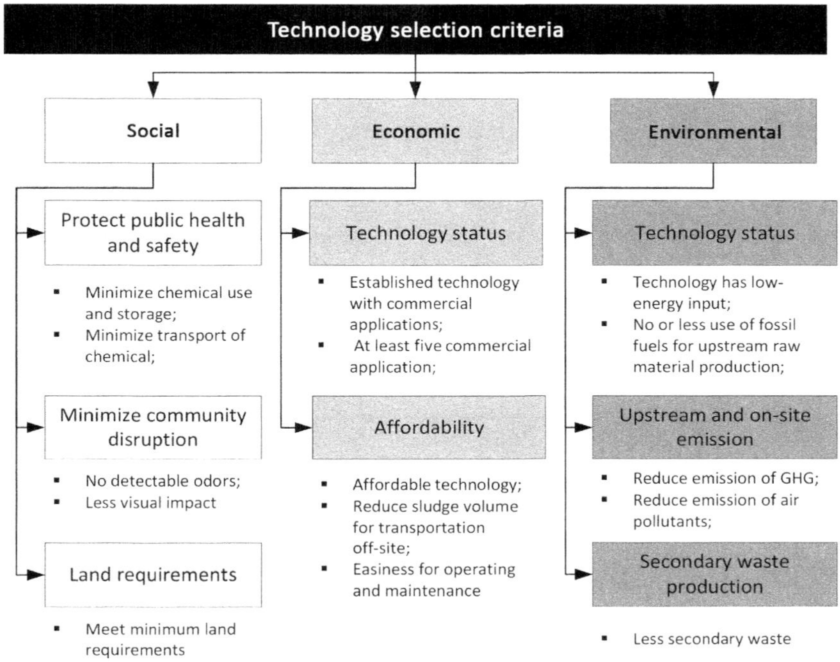

Fig. 4 Criteria for the selection of sludge pretreatment methods (Kacprzak et al., 2017; Tyagi and Lo, 2013).

Table 3 Degree of cell disintegration and costs of selected sludge pretreatment methods

Method	% Cell disintegration	Estimated cost (EURO/Mg)
Thermal	30	190
Oxidation	90	800
Thermochemical	15–60	Not available
Ball mill	90	414–2500
High pressure homogenization	85	42–146
Ultrasound	100	8330

Based on Zhang, H., 2010. Sludge treatment to increase biogas production (Trita-LWR Degree Project 10-20). Department of Land and Water Resources Engineering, Royal Institute of Technology, Stockholm, Sweden.

Table 4 Cost-benefit calculation of thermochemical WAS pretreatment for Zirl WWTP

	Unit	Native	Treated	% Change
Energy				
Methane potential	m_N^3/MgVS	174	213	22
Electricity ($\eta = 35\%$)	MWh/y	213	260	22
Heat ($\eta = 35\%$)	MWh/y	243	297	22
Degradation efficiency				
Regular degradation (%)	(w/w)	50	50	
Additional degradation (%)	(w/w)	0	11	
TR_A (%)	(w/w)	27	33	21
Affected parameters				
Electricity generation	MWh/y	213	260	22
Disposal	Mg/y	1204	878	–27
NaOH	Mg/y	0	16	
Polymer	kg/y	4128	4580	11
Calculated effective cost				
Cost/return electricity	€/y	–21,300	–26,014	22
Disposal	€/y	84,259	61,468	–27
NaOH	€/y	0	5600	Not applicable
Polymer	€/y	12,383	13,741	11
Personnel	€/y	0	4400	Not applicable
Overall costs	€/y	75,342	59,195	–21

Source: Nagler, M., Aichinger, P., Kuprian, M., Pümpel, T., Insam, H., Ebner, C., 2018. A case study for a cost-benefit-based, stepwise optimization of thermo-chemical WAS pre-treatment for anaerobic digestion. J. Mater. Cycle Waste Manage. 20, 266–273.

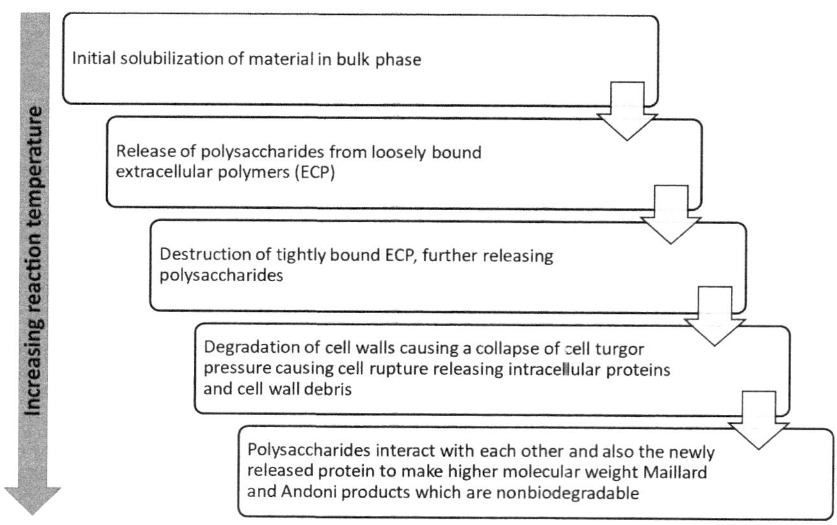

Fig. 5 Physical and chemical phenomena at sludge as temperature increases (Barber, 2016).

where A_{wt}, B_{wt}, C_{wt} is the weight of components (mg/kg total sludge), A is solid matter, B is soluble nonevaporative matter, C is soluble evaporative matter, $\xi = (A_{wt}+B_{wt})_{init.}(-0.00457 \times T - 2.323)$; T is temperature (K), and $(A_{wt}+B_{wt})_{init.}$ is initial weight of total solidants.

The main disadvantage of this mathematical model is the fact that it was based on one laboratory experiment. In the most recent studies, Imbierowicz and Chacuk (2012) proved in a kinetic model for TH of WAS that two parallel first-order reactions would take place. The first reaction is associated with the destruction and solubilization of sludge, while the second is associated with the production of a new solid phase.

Fig. 5 shows the effect of temperature increase on physical and chemical phenomena occurring in sludge.

Kim et al. (2003) showed that thermal conditioning of sludge at 121°C, at 1.5 atm, for 30 min affects the solubilization of organic matter and leads to an increase in COD concentration in the supernatant water. In addition, the authors recorded a 32% increase in biogas production during the fermentation of pretreated sludge. A significant increase in the concentration of the soluble form of organic compounds ($COD_{disolved}$) with the temperature increase was also demonstrated by studies carried out in the Zipperclave reactor (Valo et al., 2004). Heating the sludge for 1 h caused an increase in the COD concentration from 2.7% to 25.3%, 43.9%, and 59.5% for 130°C, 150°C, and 170°C, respectively. A 51% solubilization degree of organic matter and a 61% increase in biogas production were obtained at a temperature of 170°C. At the same time, no effect of higher temperature (190°C) on the organic matter solubilization was observed.

Temperature between 160°C and 180°C is recommended as the optimal one for the TH process because at the higher temperature, the sludge biodegradability drastically decreases. It is caused by increasing production of refractory compounds. The generation of Malliard and Amadori products at higher temperatures is well documented (Dwyer et al., 2008; Liu et al., 2012). They are nonbiodegradable polymers with high molecular weight. Lower temperatures in the range of 50–70°C are also used for preconditioning of sludge. However, the process is carried out with longer SRTs, ranging from about 10 h to several days (Palatsi et al., 2009). At the optimal process parameters (i.e., temperature in the range of 160–180°C and retention time of 20–40 min), the following effects of TH are observed (Barber, 2016):

- Increase of carbohydrates solubility
- Increase of protein solubility
- Negligible increase of lipids solubility
- Particle size reduction
- Decrease of apparent viscosity
- Improvement of downstream sludge anaerobic digestibility

As mentioned previously, when the TH process is carried out outside the optimal operating conditions, a significant increase of refractory compounds take place. On the other hand, further reduction of viscosity and improvement of dewaterability are observed.

It is considered that TH is more suitable for materials containing a high concentration of carbohydrates and proteins, while its effect on lipids is smaller. This confirms that it is more suitable to pretreatment of WAS than PS.

One of the first installations using the TH process on an industrial scale (Proteous process) was established in Great Britain. The process was carried out at elevated pressure at 180°C for 30 min. The efficiency of fermentation was high, but odor problems appeared, and high investment costs were also significant. Modification of the Proteous process was Zimpro process (in this process, the oxidation reactions occur at high pressure of WAS at approx. 250°C), which was fully functional in the United Kingdom, Norway, and Denmark (Kepp et al., 2000; Weisz et al., 2000). Unfortunately, the installation in the United Kingdom was closed due to odor problems, corrosion of equipment, and very high COD value in the centrate (filtrate). Developed by the company Zimpro (now Siemens Water Technologies), the Zimpro installations that have been used until today have been modified and used for low-pressure oxidation of sewage sludge at temperatures below 200°C.

Currently, the largest companies offering TH technologies include the Norwegian company Cambi and the French company Veolia Water. At Cambi, sludge is pretreated at a pressure of 4.5 bar, at a temperature of 150–180°C, and with a retention time of up to 30 min (Fig. 6). The sludge is conditioned to obtain a dry matter concentration of 17%, and after TH treatment, it is introduced to anaerobic chambers. The main disadvantages of this process include the generation of odors and the formation of compounds resistant to biological decomposition. This technology has been applied to WWTPs in many European countries and in countries around the world.

Table 5 presents a comparison of AD of sewage sludge using the technology at Cambi with classic mesophilic AD.

Veolia Water also offers TH technology operating in both a periodic and a continuous system under the names Exelys (Figs. 7 and 8) and Biothelys. In these technologies, the process temperature is 165°C, the pressure is 9 bar, and the contact time does not exceed 30 min.

Table 6 summarizes the main advantages and disadvantages of TH.

Currently, numerous studies are being conducted on technologies based on the principle of TH. One of the most promising ones is the advanced thermal hydrolysis (ATH) technology developed by Abelleira et al. (2012). This is a modification of the TH technology that involves the addition of H_2O_2 deposits to the conditioning chamber. The use of an additional oxidizer affects the high degree of sludge solubilization and improves its dewaterability. In addition, the degree of sewage sludge digestion in ATH reaches the level of 85%–92%. However, ATH technology requires further research to optimize the process parameters and increase in biogas production.

3.6 Thermophilic AD of WAS

The operating parameters have a major impact on AD efficiency. The most important of these are organic loading rate (OLR), SRT, temperature, pH, and alkalinity/LKT ratio (Donoso-Bravo et al., 2009; Myczko, 2011). The sludge concentration, the content of micro and macro elements, and the presence of inhibitors are also important. Temperature affects such elements as the enzymatic reaction rate, mass transport, and solubility of chemical compounds. The AD process can be carried out under psychrophilic (<20°C), mesophilic (30–40°C), thermophilic (45–60°C) and extremely

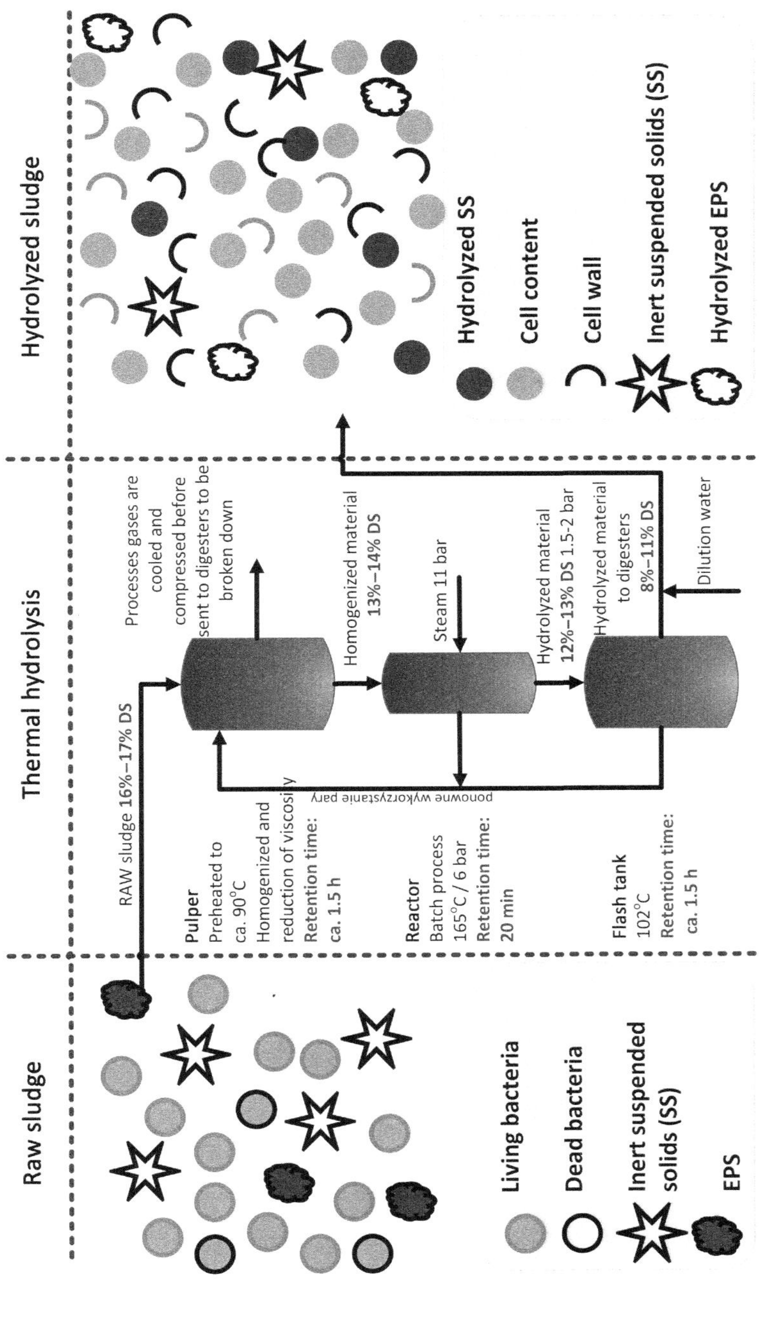

Fig. 6 Principle of technology used at Cambi (Neczaj, 2016).

Table 5 Comparison of and mesophilic AD efficiency fermentation

Parameter	Unit	Cambi reactor	Conventional digester
SRT	Days	<15	>20
Volume of reactor	–	1/3 of the volume of a conventional digester	1
VS	%	12	4–6
OLR	kg/m^3/d	5	2–3
pH	–	7.5–8.0	6.8–7.5
Temperature	°C	38–42	35–37
VFAs/alkalinity	–	0.1–0.5	0.1–0.5
Ammonia	mg/L	2500–3200	600–1000
Composition of biogas	%	65%–70% CH_4 Low concentration of H_2S	60%–65% CH_4 High concentration of H_2S
VS removal	%	70	50–55
Biogas yield	m^3/Mg VS	510	350

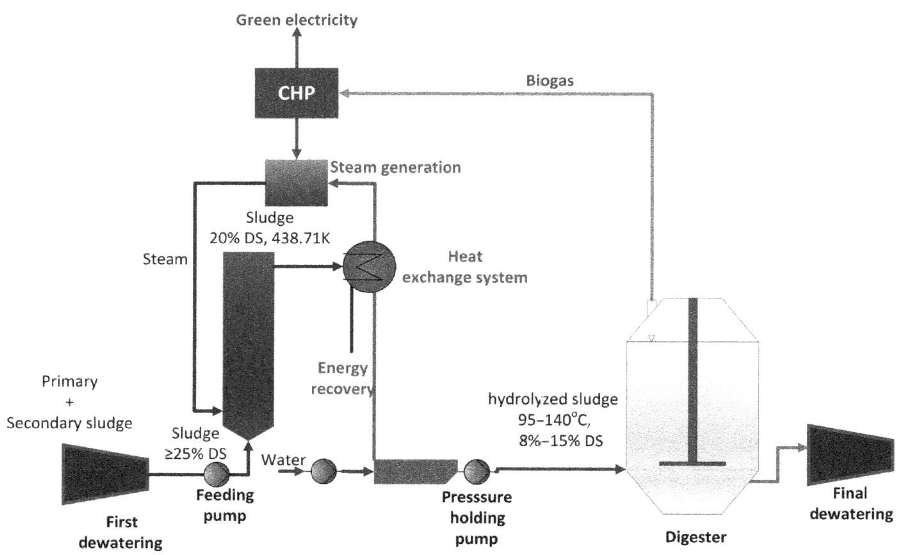

Fig. 7 Scheme used in the Exelys system (Neczaj, 2016).

thermophilic (>60°C) conditions (Gerardi, 2003). For psychrophilic bacteria, the optimal process temperature is around 25°C. Maintaining this temperature does not require heating the digester or substrates, but it reduces the efficiency of organic compound decomposition, and the production of biogas is also limited.

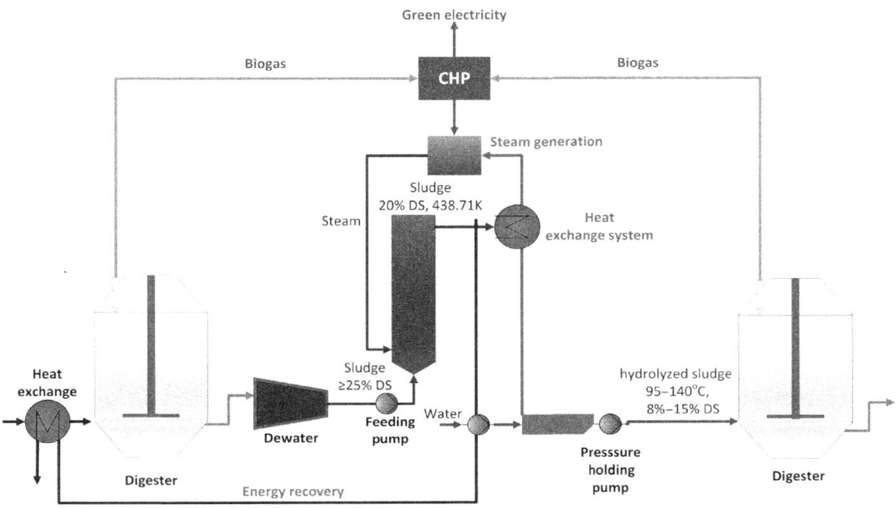

Fig. 8 Scheme of Exelys—digestion-lysis-digestion (DLD) mode (Neczaj, 2016).

Table 6 Advantages and disadvantages of thermal hydrolysis

Advantages	Disadvantages
Significant improvement of activated sludge biodegradability Improvement of primary sludge biodegradability Higher loading rates of AD	High energy demand with some configuration High ammonia concentration Possibility of refractory compounds creation
Shorter sludge retention time Increase of biogas production Reduction of sludge viscosity	More complicated equipment Sludge should be thickened to 16%–18% DS
Improvement of sludge dewaterability	High nutrient release (salt crystallization)
Sterilization of sludge Foaming reduction Reduction of downstream requirements for drying and thermal processes	

Adapted from Barber, W.P.F., 2016. Thermal hydrolysis for sewage treatment: a critical review. Water Res. 104, 53–71.

The most industrial AD installations work at temperature of 30–42°C. Mesophilic digestion results from achieving relatively high biogas production while maintaining a relatively stable process. In order to ensure a stable AD process, it is important to maintain a stable temperature level in the reactors because methanogens are very sensitive to temperature fluctuations. According to Deublein and Steinhauser (2008), daily fluctuations of temperature in the digestion chamber should not exceed 2°C. At higher temperature fluctuations, biogas production may be reduced by up to 30%.

The thermophilic AD process is carried out at temperatures in the range of 50–55° C. It should be noted that thermophilic digestion is characterized by greater sensitivity to fluctuations of operating parameters.

It is known that the biochemical processes take place faster and more efficiently at higher temperatures (Cho et al., 2010; Ge et al., 2010). It was found that the most temperature-influenced stage of AD is hydrolysis, and that process speed increases 1.5 times for each 10°C increase (Ge et al., 2010).

A higher reaction rate at higher temperatures can be expressed by the following equation:

$$r_T = r_{20\times\theta^{(T-20)}},$$

where T = temperature (°C), r_T = reaction rate at T, r_{20} = reaction rate at 20°C, and θ = temperature-activity coefficient.

Table 7 shows a comparison of the maximum growth rates of several mesophiles and thermophiles, whose values are generally higher for thermophiles than for mesophiles.

As shown at Fig. 9, the relative growth rates of thermophilic methanogens are much higher than psychrophilic and mesophilic ones.

The increase in temperature also contributes to the reduced value of parameters such as viscosity and surface tension of liquids, which results in easier mass transport at the bioreactor and an increase in the organic compound degradation (Gebreeyessus and Jenicek, 2016). However, the temperature in the digestion chamber cannot grow indefinitely; the limit is the resistance for biodegradation of organic matter. Temperatures that are too high denature a biologically active protein structure. Considering these conditions for biological processes, the temperature should not exceed 55°C. For hydrolytic bacteria involved in the production of acids, operating temperature is not a limitation because the bacteria can adapt to new conditions relatively easily. On the

Table 7 Comparison of maximum growth rates of mesophiles and thermophiles

Substrate	**Genus**	**Mesophiles (μ_{max}/day)**	**Thermophiles (μ_{max}/day)**
H_2/CO_2	*Methanobacerium*	0.26	4.80–16.6
	Methanococcus	2.16–5.52	18.2–51.1
	Methanosarcina	0.48–1.44	NA
	Methanobrevibacter	1.44–4.08	NA
Formate	*Methanobacerium*	1.44–1.92	13.4
Acetate	*Methanothrix*	0.10–0.22	0.48–0.72
	Methanosarcina	0.24–0.67	1.27–2.04
Propionate	*Syntrophobacter*	0.10–0.14	0.14–0.31
Butyrate	*Syntrophomonas*	0.36	0.48–0.77

Adapted from Van Lier, J.B., 1995. Thermophilic Anaerobic Wastewater Treatment; Temperature Aspects and Process Stability.

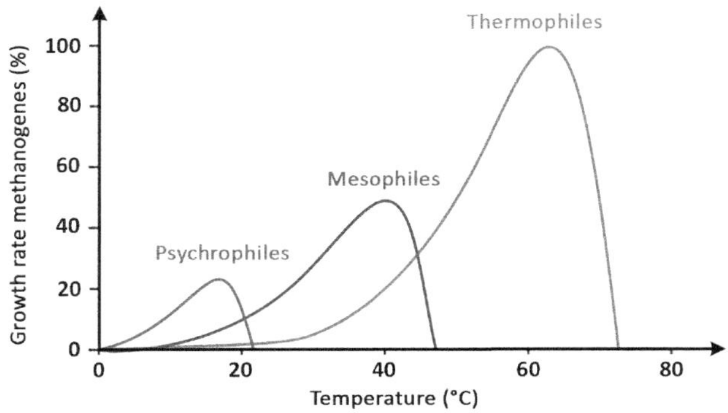

Fig. 9 Relative growth rates of psychrophilic, mesophilic, and thermophilic methanogens. Adapted from Lettinga, G., Rebac, S., Zeeman, G., 2001. Challenge of psychrophilic anaerobic wastewater treatment. Trends Biotechnol. 19, 363–370.

other hand, this is a very important parameter for methanogenic bacteria, whose enzymes are sensitive to temperature fluctuation and generally show lower resistance to high temperatures.

The AD process in the temperature range of 50–57°C also ensures high hygienization of the feed material and, according to some data, up to 25%–30% higher efficiency of gas production (Silvestre et al., 2014). Temperature also affects the degree of decomposition of organic matter and the quantity and composition of the gas produced. A slightly lower CO_2 content or higher CH_4 concentration in biogas is observed for higher fermentation temperatures (thermophilic and mesophilic). Despite the sensitivity of AD to temperature fluctuation, the tolerance of the bacterial flora increases as the organic load rate increases.

Table 8 summarizes the advantages and disadvantages of thermophilic AD compared to mesophilic AD.

Table 8 Advantages and disadvantages of thermophilic anaerobic digestion compared to the mesophilic system

Advantages	Disadvantages
Increased biogas output with a higher CH_4 and lower sulfide content	Higher energy demand
Shorter SRT	Worsened sludge water quality
Smaller volume of the digester chamber	More sensitive to operating parameters (temperature, toxic substances)
Higher pathogen destruction	
Better sludge dewatering	
Reduction of foaming	

Adapted from Ruffino, B., Fiore, S., Roati, C., Campo, G., Novarino, D., Zanetti, M., 2015. Bioresource technology scale effect of anaerobic digestion tests in fed-batch and semi-continuous mode for the technical and economic feasibility of a full scale digester. Bioresour. Technol. 182, 302–313.

The main advantages reported by Ruffino et al. (2015) comparing the efficiency of mesophilic and thermophilic AD are the speed of the biochemical process and the biogas yield. Moreover, they described better biogas quality and digestate quality.

Due to the lower energy demand and better process stability, a mesophilic AD is more widely used than thermophilic AD. On the other hand, a thermophilic AD accelerates the biochemical reaction rate, increases the growth of microorganisms, and improves biogas production at lower HRTs. It also should be noticed that after thermophilic AD, the digestate complies with the policy of the European Union regarding pathogen elimination and meets the class A biosolid product requirements of the U.S. Environmental Protection Agency (EPA).

The thermophilic system is often used in a two-stage AD, which can function in a thermophilic-mesophilic or mesophilic-thermophilic system. The thermophilic step is an acidic fermentation in which the complex organic compounds are converted to volatile fatty acids (VFAs). In the second stage, VFAs are transformed into biogases in mesophilic conditions. In the thermophilic-mesophilic fermentation of temperature-phased anaerobic digestion (TPAD), the first phase can be carried out under either thermophilic (about 55°C) or hyperthermophilic (60°C and 70°C) conditions (Ariunbaatar et al., 2014; Carrère et al., 2010; Tyagi and Lo, 2011). Compared to the classical technology, the TPAD process is characterized by greater stability, better pH control, and the possibility of work with much higher organic load rates. An increase in the specific methanogenic activity is also observed, leading to higher CH_4 production and VS reduction, as well as high pathogen removal efficiency (Ariunbaatar et al., 2014). However, this technology has higher operating and investment costs compared to conventional ones. This solution is already used on an industrial scale, such as by Aqua Solutions created by Infilco Degremont, Inc. (namely, the ODI 2 PAD™ System) (Fig. 10).

The effects of TPAD on AD efficiency are shown in Table 9. Recently, a number of studies on the intensification of the TPAD process have been carried out. They relate primarily to the use of various methods of sludge pretreatment or codigestion with other biodegradable waste (Table 10).

The main problems associated with the implementation of different methods for upgrading WWTPs to improve the AD process involve economic aspects. Gianico et al. (2015a) described an upgrading of a WWTP by the implementation of wet oxidation pretreatment for PS and thermophilic AD integrated with TH for secondary sludge. They found that biogas generation was always 18%–37% higher in thermophilic reactors where pretreated WAS was digested and VS digestion took place. Moreover, a complete inactivation of pathogens was observed. Digestate meets the requirements for agricultural use in Europe (Environment DG, EU 2000) and the United States (class A). The authors assumed that it is possible to associate the enhancement of sludge biodegradability with positive heat balances (Fig. 11).

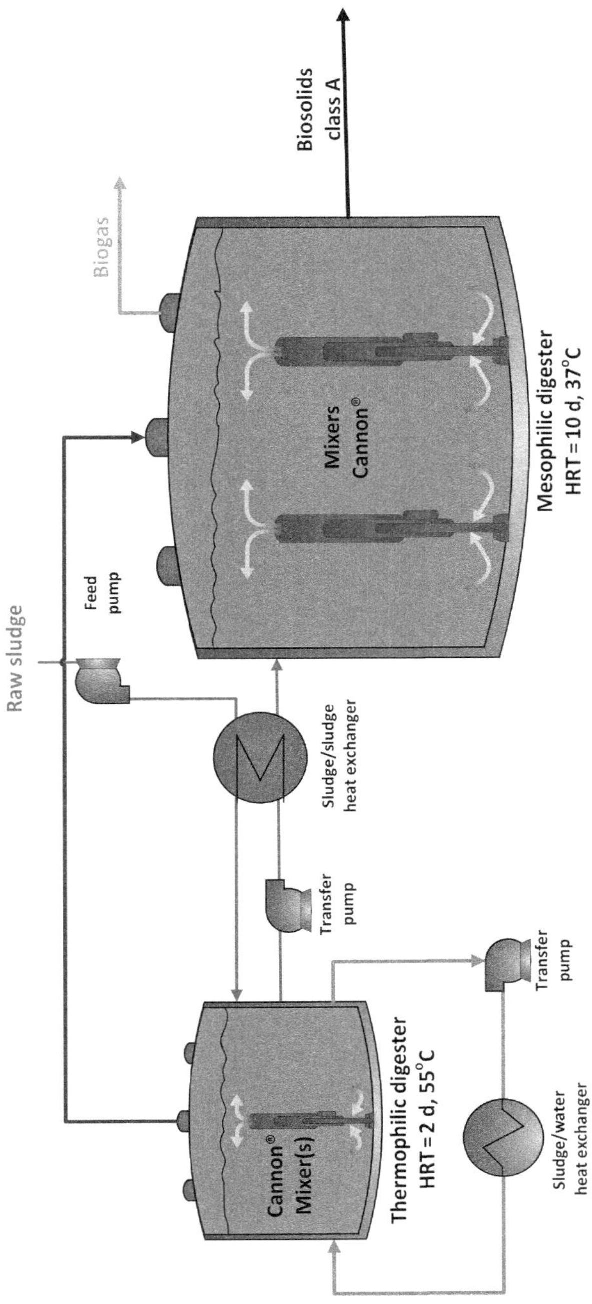

Fig. 10 A TPAD scheme (Neczaj, 2016).

Table 9 Summary of reviewed effects of TPAD

Anaerobic digestion				
Conditions	**Biogas/CH_4 production**	**VS removal (%)**	**Other effects**	**References**
PS (CSTR); $V_w=0.15\,m^3$, 55°C, HRT = 20 d; OLR = 2.2 kg TVS/(m^3 d)	0.21[a]	34 ± 3.5	Heat losses digesters—about 14% higher for TPAD systems Content of CH_4 in biogas—higher for TPAD (62% and 69% for SAD and TPAD, respectively) GPR—higher for TPAD (0.5 and 0.59 m^3/m^3 d for SAD and TPAD, respectively) Produced total energy—higher for TPAD by 15% Both digestates met the European requirements established for the use of sludge in agriculture regarding pathogens, metals, and organic contaminants	Leite et al. (2016)
PS (CSTRs); TPAD: 55°C/55°C First-stage reactor: $V_w=0.15\,m^3$, HRT = 2 d; OLR = 2.2 kg TVS/(m^3 d) Second-stage reactor: $V_w=0.23\,m^3$, HRT = 18 d; OLR = 2.2 kg TVS/(m^3 d)	0.31[a] (+47.6%)[b]	38 ± 2.7 (+11.8%)[b]		
PS; $V_w=1.3\,m^3$, 37°C, HRT = 20 d; OLR = 2.1 kg VS/(m^3 d)	0.33[c]	36	Lack of *Salmonella* in all samples; *Escherichia coli* was present at levels of some 1×10^7 CFU both in MAD and TAD while it was absent in the two-phases systems (TPS) the lowest of SRF for TPAD	Bolzonella et al. (2009)
PS; $V_w=1.3\,m^3$, 55°C, HRT = 20 d; OLR = 2.3 kg VS/(m^3 d)	0.45[c] (+36%)[b]	48 (+33%)[b]		
PS; TPAD: 70°C/55°C First-stage reactor: $V_w=0.2\,m^3$, HRT = 2 d; OLR = 15 kg VS/(m^3 d) Second-stage reactor: $V_w=1.3\,m^3$, HRT = 18 d; OLR = 2.3 kg VS/(m^3 d)	0.49[c] (+48%)[b]	55 (+53%)[b]		

CSTR, $V_w = 5l$; 35°C TPAD: 55°C/35°C First-stage reactor: $V_w = 3L$, Second-stage reactor: $V_w = 12L$, TPAD: 70°C/35°C First-stage reactor: $V_w = 3L$, Second-stage reactor: $V_w = 12L$,	0.33–0.35[c] 0.5[c] (+51.5%)[b] 0.33[c] (–6%)[b]	37.3–41 49 (+19.5%)[b] 51.8 (+39%)[b]	Compared to MAD, for TPAD higher removals particle proteins, particle COD	Qin et al. (2017)
Batch, 55°C, $V_w = 0.3L$, 32d TPAD: 70°C/55°C First-stage reactor: 9h Second-stage reactor: $V_w = 0.3L$	0.33[d] 0.52[d] (+55%)[b]	– –	VFAs concentration–increment by 43% TPAD—decreased biogas production rate by 3.6%	Climent et al. (2007)
LS, $V_w = 5L$ TPAD: 70°C/35°C First-stage reactor: $V_w = 3L$, Second-stage reactor: $V_w = 12L$, TPAD without recirculation (A) and a TPAD with recirculation (B)	No significant difference observed in CH_4 production between the single-stage system and two-stage systems, with CH_4 yield of 0.24L/g VS added	37.6 (A) 43.3(+15%)[b] (B) (+13%)[b]	Higher methane production rate in the hyperthermophilic stage was observed in TPAD with recirculation Increased share of methane in biogas; compared to MAD by 11 and 17% for TPAD without recirculation (A) and a TPAD with recirculation (B), respectively The recirculation increased the microbial community diversity in the hyperthermophilic stage.	Wu et al. (2016)
CSTR, $V_w = 50L$, HRT: 30, 20, 12d First-stage reactor: $V_w = 7.5L$, Second-stage reactor: $V_w = 42.5L$, HRT: 30, 20, 12d	0.16–0.22[e] 0.22–0.27[e] (+23%–37.5%)[b]	26.3–32.2 31.6–35.5 (+0.3%–35%)[b]	For TPAD—slightly higher share of methane in biogas Different microbial profiles in reactors; *Flavobacteriales* and *Clostridiales* was identified only in the acidogenic reactor	Maspolim et al. (2015)
TPAD: 35°C/35°C First-stage reactor: $V_w = 0.6L$, Second-stage reactor: $V_w = 4L$ TPAD: 50–70°C/35°C First-stage reactor: $V_w = 0.6L$, Second-stage reactor: $V_w = 4L$	0.07–0.11[e] 0.10–0.17[e] (+43%–89%)	37 34–48 (+11%–30%)[b]	The hydrolysis coefficient in the thermophilic/thermophilic system was significantly enhanced to 0.6 1/d at 60°C, up to 1 1/d at 65°C and 70°C compared to the mesophilic/mesophilic system. Solubilization was improved during thermophilic pretreatment.	Ge et al. (2011)

Continued

Table 9 Continued

Anaerobic digestion				
Conditions	**Biogas/CH_4 production**	**VS removal (%)**	**Other effects**	**References**
LS, $V_w = 5L$	0.34–0.38[c]	37.6–40.9	Thermophilic and hyperthermophilic stage accelerated the hydrolysis rate of the whole system, 0.053 and 0.040 gCOD/gVS/d, respectively, compared with the corresponding 0.029 and 0.024 gCOD/gVS/d, respectively, in the steady state of MAD. More surplus energy was obtained in TPAD: 55°C/35°C than that in TPAD: 70°C/35°C and MAD.	Wu et al. (2015)
TPAD: 55°C/35°C, HRT = 30 d First-stage reactor: $V_w = 3L$, Second-stage reactor: $V_w = 12L$,	0.49[c] (+44%)[b]	49.7 (+21.5%)[b]		
TPAD: 70°C/35°C First-stage reactor: $V_w = 3L$, Second-stage reactor: $V_w = 12L$, HRT = 30 d (both TPAD with recycle)	0.38 (–)[b]	52.5 (+39.6%)[b]		
LS (CSTR), 55°C, HRT = 15 d, $V_w = 3L$	40[f]	6[g]	For TPAD—higher specific propionate consumption activity and hydrogenotrophic specific methanogenic activity of digester	Skiadas et al. (2005)
TPAD: 70°C/55°C, HRT = 30 d First-stage reactor: $V_w = 0.5L$, HRT = 2 Second-stage reactor: $V_w = 12L$, HRT = 13d, VW = 3L	55[f] (+37.5%)[b]	43[g] (approx. eightfold)[b]		
CSTR, $V_w = 45L$, 35°C HRT: 12.3, 14, 16.4 and 19.7 d	0.44 ± 0.02[c]	31.5–33.8	Dual digestion improved biogas selectivity by 5% compared to MAD	Athanasoulia et al. (2012)
TPAD: 35°C/35°C First-stage reactor: $V_w = 40L$, Second-stage reactor: $V_w = 60L$, HRT: 12.3, 14, 16.4, and 19.7d	0.49 ± 0.06[c] (+9.5%–40.1%)[b]	36.2–40.7 (15%–29%)[b]	Higher volumetric gas production rate for dual digestion as well as better effluent quality	

ASBR, $V_w=2$ L, 37°C, HRT = 20 d TPAD: 60°C/37°C, First-stage reactor: $V_w=1$ L, Second-stage reactor: $V_w=2$ L, HRT = 20 d	0.33[c] 0.36[c] (+9%)[b]	47.6 60.9 (+28%)[b]	Compared to MAD, TPAD has higher methane content in biogas, as well as biogas rate	Ennouri et al. (2016)

ASBR, anaerobic sequencing batch reactors; *CSTR*, continuous stirred reactor; *LS*, laboratory scale; *MAD*, mesophilic anaerobic digestion; *PS*, pilot scale; *SAD*, single anaerobic digestion; *SRF*, specific resistance to filtration; *TAD*, thermophilic anaerobic digestion; V_w, working volume.

[a] $m^3/(kg\,TVS\,d)$.

[b] Difference in comparison to single anaerobic digestion.

[c] Biogas yield ($m^3/kg\,VS_{add}$).

[d] L.

[e] Methane yield ($L/g\,VS_{add}$).

[f] $mL/(L\,d)$.

[g] Efficiency in volatile solids removal.

Table 10 Enhancement of TPAD process efficiency by pretreatment method and/or codigestion of WAS with other organic waste

Method	Pretreatment			TPAD				References
	Conditions	COD solubilization	Other effects	Conditions	Improvement of methane production (%)	Improvement of VS removal (%)	Other effects	
Ultrasound	UP400S (Dr. Hielscher, Germany) 300 W, 24 kHz 0.4–0.5 kWh/kg dry solids Predigestion stage: Temp: 37°C Retention time: 3–5 d	9–15-fold	DD: 2%–3% Reduction of TS by 2.2%–3.5%	SM, TPAD:55°C/35°C First-stage reactor: $V_w = 7$ L, OLR: 3.9, 10 kgVS/(m^3 d) Second-stage reactor: $V_w = 7$ L, Trial 1: OLR: 1.2, 2.5 kg VS/(m^3 d)	Up to 33[a] (11)[b]	Up to 70[a] –[2]	logarithmic removal >3.5, >2.3, and >0.8, respectively, for *E. coli*, SOMCPH, and *Salmonella*. In. methane in biogas up to 8%	Gianico et al. (2015b)
HPH+NaOH	EmulsiFlex-C3, AVESTIN, Inc.) Pressure: 6000 or 12,000 psi Dose NaOH: 9–36 mg/g TS (pH 10–11) Contac time: 30 min Predigestion stage:	Chemical addition alone achieved 1.5–1.9 times higher solubilization over control (untreated) TWAS, whereas	In. CST fourfold Reduction of particle size by 35%–38% after HPH In. of percentages of soluble proteins,	SM, TPAD:35°C/55°C $V_w = 7$ L, SRT: 20, 14, 7 d Before TPAD: Acid phase (3 d, under	75–81[a] 4–33[b] 45–54[c]	14–21[a] 0.5–6.8[b] 5–9[c]	The 2PAD system decreased the specific CST value. Higher for HPH+NaOH+TPAD	Wahidunnabi and Eskicioglu (2014)

	Temp: 37°C Retention time: 3–5 d	application of HPH pretreatment attained about 3.5–5.9 times higher COD solubilization than the control TWAS (	sugars as well as humic substance by 18.5, 18.5 and 31%, respectively after HPH	thermophilic temperature)				
Ultrasound	Ultrasonic processor UP400S model (Hielscher, Germany), 400 W, 24 Hz Time: 5–30 min specific energy: 1690–10 141 kJ/kg TS	In. up to 22%	–	BMP, ISR: 0.6 TPAD:55°C/35°C Three TPAD systems; 1. Sonication-thermophilic-mesophilic; 2. Thermophilic-sonication-mesophilic; 3. Thermophilic-mesophilic	Sonication-thermophilic-mesophilic: 26[b] Thermophilic-sonication-mesophilic: 25.7[b]	Sonication-thermophilic-mesophilic: 29[b] Thermophilic-sonication-mesophilic: 29[b]	Increased soluble COD removal when ultrasound was applied	Riau et al. (2015)

Continued

Table 10 Continued

Method	Pretreatment			TPAD				References
	Conditions	COD solubilization	Other effects	Conditions	Improvement of methane production (%)	Improvement of VS removal (%)	Other effects	
MW	Panasonic NNS53W + inverter; Capacity: 0.045 m^3, Frequency:2450MHz Temp.: 96°C Retention time: 2 d	SCOD/TCOD increased from 0.06 to 0.2	In. VFAs concentration by 1.6–3.6 times In. ammonia concentration up to 71% In. rates of hydrolysis	SM, TPAD:55°C/35°C V_w = 800 mL SRT: 20, 15, 10, and 5 d	Up to 94[a] Up to 67[c]	Up to 99[a] Up to 87[c]	Reduction total coliforms and *E. coli* to levels under detection limit Reduced CST compared to mesophilic reactor	Coelho et al. (2011)
				SM, TPAD:55°C/55°C V_w = 800 mL SRT: 20, 15, 10, and 5 d	Up to 106[a] Up to 79[c]	Up to 106[a] Up to 94[c]	Reduction total coliforms and *E. coli* to levels under detection limit only for SRT equal to 20 and 15 d Reduced CST compared to mesophilic reactor	

Alkali-hydrodynamic disintegration	8.8–51.25 mmol/l NaOH (2M); pH 7–11; designed a 1.2-mm nozzle; the working pressure of the pump: 12 bar	SCOD values increased above 3500 mg/L (from <500 mg/L)	DD: 28%–35%	TPAD:35°C/55°C $V_w = 3$ L First-stage reactor: Duration: 5, 7, 9, 11, 13 and 15 d Second-stage reactor: Duration: 19, 17, 15, 13, 11, 9 d	Up to 27.8[a]	Up to 56[a]	Reduction numbers of microorganism indicators such as *Salmonella, E.coli* somatic coliphages to 0 Reduction sulfite reducing Clostridia from 9.75×10^d (MAD) to 2.46×10^2 (TPAD)	Grübel and Suschka (2015)
Ultrasound	A Fisher Scientific Model 500 Ultrasonic Dismembrator (400 W; frequency: 20 kHz)	Compared to control COD solubilization increased by 20.2%	–	TPAD:35°C/55°C $V_w = 0.25$ L First-stage reactor: Second-stage reactor: $V_w = 0.75$ L SRT:7, 14, 20 d	From –15.4 to 48.3[a] From –7.1 to 3.7[b]	From 3.6 to 16.6[a] From 3.3 to 5.7[b]	Only for TPAD systems (untreated, as well as with pretreatment), fecal coliform counts were below detection limits; while single-stage digesters (with and without pretreatment) yielded Class B biosolids. The highest extra heat was achieved for system with MW pretreatment.	Akgul et al. (2017)
MW	A Milestone Microwave Lab station (Ethos EZ, 2450 MHz, 0–1200 W; maximum temperature and pressure of 300°C and 35 bars, respectively	Compared to control COD solubilization increased by 60.6%	–	TPAD:35°C/55°C $V_w = 0.25$ L First-stage reactor: Second-stage reactor: $V_w = 0.75$ L SRT:7, 14, 20 d	From –19.9 to 40.9[a] From –11.6 to 3.7[b]	From –6.3 to 15.9[a] From –4.2 to 2.6[b]		

Continued

Table 10 Continued

	Pretreatment			TPAD				
Method	**Conditions**	**COD solubilization**	**Other effects**	**Conditions**	**Improvement of methane production (%)**	**Improvement of VS removal (%)**	**Other effects**	**References**
Biological pretreatment + codigestion process	Codigestion mixture (WAS+FW, 1:1 m/m), 35°C, Retention time: 15, 24 and 35 h	SCOD increased up to 158%	Increased hydrolysis degree and acidification degree by 167% and 290%, respectively	SM, $V=1$ L, 35°C	24.6[d]	10.1[d]	Anaerobic copretreatment reduced the abundance of filamentous bacteria of genus *Levilinea*	Zhang et al. (2017)

BMP, biochemical CH_4 potential; *CST*, capillary suction time; *DD*, disintegration degree around; *FW*, food waste; *In.*, increased; *ISR*, inoculum substrate ratio; *MAD*, mesophilic anaerobic digestion; *SCOD*, soluble COD; *SM*, semicontinuous mode; *SOMCPH*, somatic coliphages; *TCOD*, total COD; *TS*, total solids; *V*, volume of reactor; V_w, working volume of reactor.

[a]In compared to mesophilic anaerobic digestion.

[b]In compared untreated TPAD.

[c]In compared to thermophilic anaerobic digestion.

[d]Compared to codigestion mixture without pretreatment.

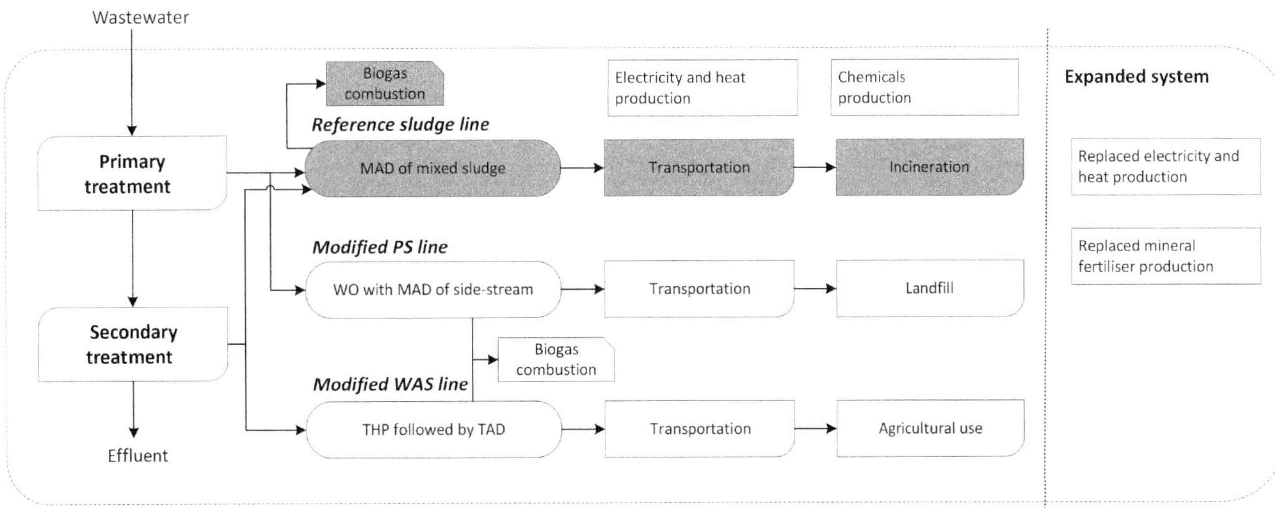

MAD — *mesophilic anaerobic digestion;* **PS** — *primary sludge;* **WAS** — *secondary sludge (waste activated sludge)*
WO — *wet oxidation;* **TAD** — *thermophilic anaerobic digestion;* **THP** — *thermal hydrolysis process*

Fig. 11 System boundaries based on the literature study and upgraded system. White boxes indicate processes in both systems; dark gray indicates reference systems; light gray indicates upgraded systems.
Adapted from Gianico, A., Bertanza, G., Braguglia, C.M., Canato, M., Laera, G., Heimersson, S., Svanström, M., Mininni, G., 2015. Upgrading a wastewater treatment plant with thermophilic digestion of thermally pre-treated secondary sludge: techno-economic and environmental assessment. J. Clean. Prod. 102, 353–361.

3.7 Conclusion

The biogas production from activated sludge using TH and thermophilic AD was investigated in this chapter. TH is a well-established technology used as a WAS pretreatment method. It leads to higher biogas production, although the energy benefit of AD is relatively natural due to its higher energy demand. The main energy benefit is connected to improving the dewaterability of the sludge and reducing the costs of downstream processes. AD following TH should not be longer than 10–12 days. Protein degradation causes an increase of ammonia, alkalinity, and pH and does not improve biogas generation at longer SRTs.

Thermophilic digestion is preferred over classical mesophilic AD, mainly due to higher biogas production at shorter SRTs and faster VS degradation. Moreover, digestate meets the EPA's class A requirements. Because thermophilic AD is very sensitive to the fluctuation of operating parameters and the quality of rejected water is poor, additional research is required to address those problems.

Acknowledgment

The study was supported by internal Czestochowa University of Technology Grant No. BS/PB-401-301/13.

References

Abelleira, J., Pérez-Elvira, S.I., Portela, J.R., Sánchez-Oneto, J., Nebot, E., 2012. Advanced thermal hydrolysis: optimization of a novel thermochemical process to aid sewage sludge treatment. Environ. Sci. Technol. 46, 6158–6166.

Akgul, D., Cella, M.A., Eskicioglu, C., 2017. Influences of low-energy input microwave and ultrasonic pretreatments on single-stage and temperature-phased anaerobic digestion (TPAD) of municipal wastewater sludge. Energy 123, 271–282.

Ariunbaatar, J., Panico, A., Esposito, G., Pirozzi, F., Lens, P.N.L., 2014. Pretreatment methods to enhance anaerobic digestion of organic solid waste. Appl. Energy 123, 143–156.

Athanasoulia, E., Melidis, P., Aivasidis, A., 2012. Optimization of biogas production from waste activated sludge through serial digestion. Renew. Energy 47, 147–151.

Ayol, A., 2005. Enzymatic treatment effects on dewaterability of anaerobically digested biosolids—I: performance evaluations. Process Biochem. 40, 2427–2434.

Barber, W.P.F., 2016. Thermal hydrolysis for sewage treatment: a critical review. Water Res. 104, 53–71.

Barjenbruch, M., Kopplow, O., 2003. Enzymatic, mechanical and thermal pre-treatment of surplus sludge. Adv. Environ. Res. 7, 715–720.

Bien, J.B., Malina, G., Bien, J.D., Wolny, L., 2004. Enhancing anaerobic fermentation of sewage sludge for increasing biogas generation. J. Environ. Sci. Health A 39, 939–949.

Bień, J., Neczaj, E., Worwag, M., Grosser, A., Nowak, D., Milczarek, M., Janik, M., 2011. Directions management of sludge in Poland after 2013. Eng. Protect. Environ. 14, 375–384.

Bień, J., Kamizela, T., Kowalczyk, M., Grosser, A., Zwierz, N., Zabochnicka-Swiatek, M., 2015. The effectiveness of acid fermentation of sonicated primary sludge. J. Resid. Sci. Technol. 12.

Bolzonella, D., Fatone, F., Di Fabio, S., Cecchi, F., 2009. Mesophilic, thermophilic and temperature phased anaerobic digestion of waste activated sludge. Chem. Eng. Trans. 17, 385–390.

Carrère, H., Dumas, C., Battimelli, A., Batstone, D.J., Delgenès, J.P., Steyer, J.P., Ferrer, I., 2010. Pretreatment methods to improve sludge anaerobic degradability: a review. J. Hazard. Mater. 183, 1–15.

Carrere, H., Antonopoulou, G., Affes, R., Passos, F., Battimelli, A., 2016. Bioresource technology review of feedstock pretreatment strategies for improved anaerobic digestion: from lab-scale research to full-scale application. Bioresour. Technol. 199, 386–397.

Cho, S.K., Kim, D.H., Shin, H.S., 2010. Combined pretreatments of various(alkaline? ultrasound, alkaline? thermal, alkaline? MW) effect on sewage sludge disintegration. In: 19th KKNN Symposium, pp. 22–45.

Climent, M., Ferrer, I., del Baeza, M., Artola, A., Vázquez, F., Font, X., 2007. Effects of thermal and mechanical pretreatments of secondary sludge on biogas production under thermophilic conditions. Chem. Eng. J. 133, 335–342.

Coelho, N.M.G., Droste, R.L., Kennedy, K.J., 2011. Evaluation of continuous mesophilic, thermophilic and temperature phased anaerobic digestion of microwaved activated sludge. Water Res. 45, 2822–2834.

Deublein, D., Steinhauser, A., 2008. Biogas from Waste and Renewable Resource: An Introduction. Wiley-VCH Verlag GmbH & Co, KGaA, Weinhem.

Donoso-Bravo, A., Retamal, C., Carballa, M., Ruiz-Filippi, G., Chamy, R., 2009. Influence of temperature on the hydrolysis, acidogenesis and methanogenesis in mesophilic anaerobic digestion: parameter identification and modeling application. Water Sci. Technol. 60, 9–17.

Dwyer, J., Starrenburg, D., Tait, S., Barr, K., Batstone, D.J., Lant, P., 2008. Decreasing activated sludge thermal hydrolysis temperature reduces product colour, without decreasing degradability. Water Res. 42, 4699–4709.

Elliott, A., Mahmood, T., 2007. Pretreatment technologies for advancing anaerobic digestion of pulp and paper biotreatment residues. Water Res. 41, 4273–4286.

Ennouri, H., Miladi, B., Zahedi, S., Alberto, L., Güelfo, F., Solera, R., Hamdi, M., Bouallagui, H., 2016. Bioresource technology effect of thermal pretreatment on the biogas production and microbial communities balance during anaerobic digestion of urban and industrial waste activated sludge. Bioresour. Technol. 214, 184–191.

Eskicioglu, C., Terzian, N., Kennedy, K.J., Droste, R.L., Hamoda, M., 2007. Athermal microwave effects for enhancing digestibility of waste activated sludge. Water Res. 41, 2457–2466.

Eskicioglu, C., Kennedy, K.J., Droste, R.L., 2009. Enhanced disinfection and methane production from sewage sludge by microwave irradiation. Desalination 248, 279–285.

Foladori, P., Andreottola, G., Ziglio, G., 2010. Sludge Reduction Technologies in Wastewater Treatment Plants. IWA Publishing, London, UK.

Gallert, C., Winter, J., 2005. Bacterial metabolism in wastewater treatment systems. In: Jördening, H.-J., Winter, J. (Eds.), Environmental Biotechnology: Concepts and Applications. Wiley-VCH Verlag GmbH & Co, KGaApp, Weinheim, pp. 1–48 (Chapter 1).

Ge, H., Jensen, P.D., Batstone, D.J., 2010. Pre-treatment mechanisms during thermophilic–mesophilic temperature phased anaerobic digestion of primary sludge. Water Res. 44, 123–130.

Ge, H., Jensen, P.D., Batstone, D.J., 2011. Temperature phased anaerobic digestion increases apparent hydrolysis rate for waste activated sludge. Water Res. 45, 1597–1606.

Gebreeyessus, G., Jenicek, P., 2016. Thermophilic versus mesophilic anaerobic digestion of sewage sludge: a comparative review. Bioengineering 3, 15.

Gerardi, M.H., 2003. The Microbiology of Anaerobic Digesters. John Wiley & Sons, New Jersey.

Gianico, A., Bertanza, G., Braguglia, C.M., Canato, M., Laera, G., Heimersson, S., Svanström, M., Mininni, G., 2015a. Upgrading a wastewater treatment plant with thermophilic digestion of thermally pre-treated secondary sludge: techno-economic and environmental assessment. J. Clean. Prod. 102, 353–361.

Gianico, A., Braguglia, C.M., Gallipoli, A., Mininni, G., 2015b. Innovative two-stage mesophilic/thermophilic anaerobic degradation of sonicated sludge: performances and energy balance. Environ. Sci. Pollut. Res. 22, 7248–7256.

Grobelak, A., 2016. Organic soil amendments in the phytoremediation process. In: Ansari, A.A., Gill, S.S., Gill, R., Lanza, G.R., Newman, L. (Eds.), Phytoremediation. Springer International Publishing, Cham, pp. 21–39.

Grobelak, A., Kacprzak, M., Grosser, A., Napora, A., 2013. Chemophytostabilisation of soil contaminated with cadmium, lead and zinc. Rocz. Ochr. Srodo. 15.

Grobelak, A., Rorat, A., Kokot, P., Singh, B.R., Kacprzak, M., 2018. In: Prasad, M.N.V., de C. Favas, P.J., Maiti, S.K.B.T.-B.-G., For M.S.R. (Eds.), Mine Waste Rehabilitation Case Studies from Poland. Elsevier, Amsterdam, Netherlands, pp. 515–527 (Chapter 28).

Grosser, A., Neczaj, E., 2016. Enhancement of biogas production from sewage sludge by addition of grease trap sludge. Energy Convers. Manage. 125, 301–308.

Grosser, A., Neczaj, E., 2017. Pretreatment methods as a means of boosting methane production from sewage sludge and its mixtures with grease trap sludge. In: E3S Web of Conferences 22, p. 00058.

Grosser, A., Neczaj, E., 2018. Sewage sludge and fat rich materials co-digestion—performance and energy potential. J. Clean. Prod. 198, 1076–1089.

Grosser, A., Worwag, M., Neczaj, E., Grobelak, A., 2013. Semi-continuous anaerobic co-digestion of mixed sewage sludge and waste fats of vegetable origin. Rocz. Ochr. Srodo. 15.

Grübel, K., Suschka, J., 2015. Hybrid alkali-hydrodynamic disintegration of waste-activated sludge before two-stage anaerobic digestion process. Environ. Sci. Pollut. Res. 22, 7258–7270.

Imbierowicz, M., Chacuk, A., 2012. Kinetic model of excess activated sludge thermohydrolysis. Water Res. 46, 5747–5755.

Kacprzak, M., Neczaj, E., Fijałkowski, K., Grobelak, A., Grosser, A., Worwag, M., Rorat, A., Brattebo, H., Almås, Å., Singh, B.R., 2017. Sewage sludge disposal strategies for sustainable development. Environ. Res. 156, 39–46.

Kepp, U., Machenbach, I., Weisz, N., Solhelm, O.E., 2000. Enhanced stabilisation of sewage sludge through thermal hydrolysis—three years of experience with full scale plant BT—disposal and utilisation of sewage sludge: treatment methods and application modalities, October 13, 1999–October 15, 1999. Water Sci. Technol. 42, 89–96.

Khanal, S., 2011. Anaerobic Biotechnology for Bioenergy Production: Principles and Applications. John Wiley & Sons (Google Books).

Kim, J., Park, C., Kim, T.-H., Lee, M., Kim, S., Kim, S.-W., Lee, J., 2003. Effects of various pretreatments for enhanced anaerobic digestion with waste activated sludge. J. Biosci. Bioeng. 95, 271–275.

Latif, M.A., Mehta, C.M., Batstone, D.J., 2017. Influence of low pH on continuous anaerobic digestion of waste activated sludge. Water Res. 113, 42–49.

Leite, W.R.M., Gottardo, M., Pavan, P., Belli Filho, P., Bolzonella, D., 2016. Performance and energy aspects of single and two phase thermophilic anaerobic digestion of waste activated sludge. Renew. Energy 86, 1324–1331.

Lin, H., Zhang, M., Wang, F., Meng, F., Liao, B.Q., Hong, H., Chen, J., Gao, W., 2014. A critical review of extracellular polymeric substances (EPSs) in membrane bioreactors:

characteristics, roles in membrane fouling and control strategies. J. Membr. Sci. 460, 110–125.

Liu, X., Wang, W., Gao, X., Zhou, Y., Shen, R., 2012. Effect of thermal pretreatment on the physical and chemical properties of municipal biomass waste. Waste Manage. 32, 249–255.

Macherzyński, B., Włodarczyk-Makuła, M., Ładyga, E., Pękała, W., 2015. The kinetics of degradation of PAHs in sewage sludge in anaerobic conditions. Engineering and protection of environment, pp. 437–448.

Maspolim, Y., Zhou, Y., Guo, C., Xiao, K., Ng, W.J., 2015. Comparison of single-stage and two-phase anaerobic sludge digestion systems—performance and microbial community dynamics. Chemosphere 140, 54–62.

Mayhew, M., Le, M.S., Brade, C.E., Harrison, D., 2003. The united utitlities "enzymic hydrolysis process"—validation of phased digestion at full scale to enhance pathogen removal. Proc. Water Environ. Fed. 2003, 1000–1013.

Mottet, A., François, E., Latrille, E., Steyer, J.P., Déléris, S., Vedrenne, F., Carrère, H., 2010. Estimating anaerobic biodegradability indicators for waste activated sludge. Chem. Eng. J. 160, 488–496.

Müller, J.A., 2000. Pretreatment processes for the recycling and reuse of sewage sludge. Water Sci. Technol. 42, 167–174.

Myczko, A., 2011. Budowa i eksploatacja biogazowni rolniczych: poradnik dla inwestorów zainteresowanych budową biogazowni rolniczych. Instytut Technologiczno-Przyrodniczy, Raszyn, Poland.

Nagler, M., Aichinger, P., Kuprian, M., Pümpel, T., Insam, H., Ebner, C., 2018. A case study for a cost-benefit-based, stepwise optimization of thermo-chemical WAS pre-treatment for anaerobic digestion. J. Mater. Cycle Waste Manage. 20, 266–273.

Neczaj, E., 2016. The Methods Used to Intensifity a Biogas Production from Sewage Sludge. Monografie—Politechnika Częstochowska, Wydawnictwo Politechniki Częstochowskiej.

Neczaj, E., Grosser, A., Worwa¸g, M., 2013. Boosting production of methane from sewage sludge by addition of grease trap sludge. Environ. Prot. Eng. 39, 125–133.

Onyeche, T.I., Schäfer, S., 2003. Sludge homogenisation as a means to reduce sludge volume and increase energy production. Elec. J. Env. Agricult. Food Chem. Title. 2.

Palatsi, J., Gimenez-Lorang, A., Ferrer, I., Flotats, X., 2009. Start-up strategies of thermophilic anaerobic digestion of sewage sludge. Water Sci. Technol. 59, 1777–1784.

Park, C., Novak, J.T., Helm, R.F., Ahn, Y.O., Esen, A., 2008. Evaluation of the extracellular proteins in full-scale activated sludges. Water Res. 42, 3879–3889.

Pérez-Elvira, S.I., Diez, P.N., Fdz-Polanco, F., 2006. Sludge minimisation technologies. Rev. Environ. Sci. Biotechnol. 5, 375–398.

Pilli, S., Bhunia, P., Yan, S., LeBlanc, R.J., Tyagi, R.D., Surampalli, R.Y., 2011. Ultrasonic pretreatment of sludge: a review. Ultrason. Sonochem. 18, 1–18.

Qiao, W., Wang, W., Xun, R., Lu, W., Yin, K., 2008. Sewage sludge hydrothermal treatment by microwave irradiation combined with alkali addition. J. Mater. Sci. 43, 2431–2436.

Qin, Y., Higashimori, A., Wu, L.J., Hojo, T., Kubota, K., Li, Y.Y., 2017. Phase separation and microbial distribution in the hyperthermophilic-mesophilic-type temperature-phased anaerobic digestion (TPAD) of waste activated sludge (WAS). Bioresour. Technol. 245, 401–410.

Riau, V., De la Rubia, M.A., Pérez, M., 2015. Upgrading the temperature-phased anaerobic digestion of waste activated sludge by ultrasonic pretreatment. Chem. Eng. J. 259, 672–681.

Roman, H.J., Burgess, J.E., Pletschke, B.I., 2006. Enzyme treatment to decrease solids and improve digestion of primary sewage sludge. Afr. J. Biotechnol. 5 (10), 963–967.

Rorat, A., Wloka, D., Grobelak, A., Grosser, A., Sosnecka, A., Milczarek, M., Jelonek, P., Vandenbulcke, F., Kacprzak, M., 2017. Vermiremediation of polycyclic aromatic hydrocarbons and heavy metals in sewage sludge composting process. J. Environ. Manage. 187, 347–353.

Ruffino, B., Fiore, S., Roati, C., Campo, G., Novarino, D., Zanetti, M., 2015. Bioresource technology scale effect of anaerobic digestion tests in fed-batch and semi-continuous mode for the technical and economic feasibility of a full scale digester. Bioresour. Technol. 182, 302–313.

Shah, F.A., Mahmood, Q., Rashid, N., Pervez, A., Raja, I.A., Shah, M.M., 2015. Co-digestion, pretreatment and digester design for enhanced methanogenesis. Renew. Sust. Energ. Rev. 42, 627–642.

Shi, Y., Huang, J., Zeng, G., Gu, Y., Chen, Y., Hu, Y., Tang, B., Zhou, J., Yang, Y., Shi, L., 2017. Exploiting extracellular polymeric substances (EPS) controlling strategies for performance enhancement of biological wastewater treatments: an overview. Chemosphere 180, 396–411.

Silvestre, G., Illa, J., Fernández, B., Bonmatí, A., 2014. Thermophilic anaerobic co-digestion of sewage sludge with grease waste: effect of long chain fatty acids in the methane yield and its dewatering properties. Appl. Energy 117, 87–94.

Skiadas, I.V., Gavala, H.N., Lu, J., Ahring, B.K., 2005. Thermal pre-treatment of primary and secondary sludge at 70°C prior to anaerobic digestion. Water Sci. Technol. 52, 161–166.

Suleiman, H., Rorat, A., Grobelak, A., Grosser, A., Milczarek, M., Płytycz, B., Kacprzak, M., Vandenbulcke, F., 2017. Determination of the performance of vermicomposting process applied to sewage sludge by monitoring of the compost quality and immune responses in three earthworm species: *Eisenia fetida*, *Eisenia andrei* and *Dendrobaena veneta*. Bioresour. Technol. 241, 103–112.

Takamatsu, T., Hashimoto, I., Ohno, H., 1970. Optimal design of a large complex system from the viewpoint of sensitivity analysis. Ind. Eng. Chem. Proc. Desig. Dev. 9, 368–379.

Toreci, I., Kennedy, K.J., Droste, R.L., 2009. Evaluation of continuous mesophilic anaerobic sludge digestion after high temperature microwave pretreatment. Water Res. 43, 1273–1284.

Tyagi, V.K., Lo, S.-L., 2011. Application of physico-chemical pretreatment methods to enhance the sludge disintegration and subsequent anaerobic digestion: an up to date review. Rev. Environ. Sci. Biotechnol. 10, 215.

Tyagi, V.K., Lo, S.L., 2013. Sludge: a waste or renewable source for energy and resources recovery? Renew. Sust. Energ. Rev. 25, 708–728.

Valo, A., Carrère, H., Delgenès, J.P., 2004. Thermal, chemical and thermo-chemical pretreatment of waste activated sludge for anaerobic digestion. J. Chem. Technol. Biotechnol. 79, 1197–1203.

Wahidunnabi, A.K., Eskicioglu, C., 2014. High pressure homogenization and two-phased anaerobic digestion for enhanced biogas conversion from municipal waste sludge. Water Res. 66, 430–446.

Wawrzynczyk, J., Recktenwald, M., Norrlöw, O., Dey, E.S., 2007. Solubilisation of sludge by combined chemical and enzymatic treatment. Afr. J. Biotechnol. 6.

Weisz, N., Kepp, U., Norli, M., Panter, K., Solheim, O.E., 2000. Sludge disintegration with thermal hydrolysis—cases from Norway, Denmark and United Kingdom. In: Proceedings of the 1st World Water Congress of the International Water Association (IWA) Paris, pp. 3–7.

Wolski, P., Zawieja, I., 2014. Hybrid conditioning before anaerobic digestion for the improvement of sewage sludge dewatering. Desalin. Water Treat. 52, 3725–3731.

Wu, L.-J., Qin, Y., Hojo, T., Li, Y.-Y., 2015. Upgrading of anaerobic digestion of waste activated sludge by temperature-phased process with recycle. Energy 87, 381–389.
Wu, L.J., Higashimori, A., Qin, Y., Hojo, T., Kubota, K., Li, Y.Y., 2016. Comparison of hyperthermophilic-mesophilic two-stage with single-stage mesophilic anaerobic digestion of waste activated sludge: process performance and microbial community analysis. Chem. Eng. J. 290, 290–301.
Xu, G., Chen, S., Shi, J., Wang, S., Zhu, G., 2010. Combination treatment of ultrasound and ozone for improving solubilization and anaerobic biodegradability of waste activated sludge. J. Hazard. Mater. 180, 340–346.
Zawieja, I., Wolny, L., 2013. Ultrasonic disintegration of sewage sludge to increase biogas generation. Chem. Biochem. Eng. Q. 27, 491–497.
Zawieja, I., Lidia, W., Marta, P., 2017. Impact of the excess sludge modification with selected chemical reagents on the increase of dissolved organic substances concentration compounds transformations in activated sludge. Environ. Res. 156, 652–656.
Zhang, H., 2010. Sludge treatment to increase biogas production (Trita-LWR Degree Project 10-20). Department of Land and Water Resources Engineering, Royal Institute of Technology, Stockholm, Sweden.
Zhang, J., Li, W., Lee, J., Loh, K.C., Dai, Y., Tong, Y.W., 2017. Enhancement of biogas production in anaerobic co-digestion of food waste and waste activated sludge by biological co-pretreatment. Energy 137, 479–486.

Further reading

Alves, M.M., Pereira, M.A., Sousa, D.Z., Cavaleiro, A.J., Picavet, M., Smidt, H., Stams, A.J.M., 2009. Waste lipids to energy: how to optimize methane production from long-chain fatty acids (LCFA). Microb. Biotechnol. 2, 538–550.
Angelidaki, I., Sanders, W., 2004. Assessment of the anaerobic biodegradability of macropollutans. Rev. Environ. Sci. Biotechnol. 3, 117–129.
Lettinga, G., Rebac, S., Zeeman, G., 2001. Challenge of psychrophilic anaerobic wastewater treatment. Trends Biotechnol. 19, 363–370.
Neumann, P., Pesante, S., Venegas, M., Vidal, G., 2016. Developments in pre-treatment methods to improve anaerobic digestion of sewage sludge. Rev. Environ. Sci. Biotechnol. 15, 173–211.
Van Lier, J.B., 1995. Thermophilic Anaerobic Wastewater Treatment: Temperature Aspects and Process Stability, Ph.D. Thesis, Wageningen Agricultural University, Wageningen, The Netherlands.
Zhen, G., Lu, X., Kato, H., Zhao, Y., Li, Y., 2017. Overview of pretreatment strategies for enhancing sewage sludge disintegration and subsequent anaerobic digestion: current advances, full- scale application and future perspectives. Renew. Sust. Energ. Rev. 69, 559–577.

Microbial fuel cell application for sludge remediation and minimization

32

Gauravarapu Navlur Nikhil
Department of Biotechnology, Dr. B.R. Ambedkar National Institute of Technology Jalandhar, Punjab, India

1 Introduction

Sewage sludge is a by-product of municipal wastewater treatment, generated from primary and secondary sedimentation. Sewage denotes both black water and gray water at the household level, where *black water* refers to wastewater generated in toilets and *gray water* to the wastewater generated in kitchen, bathroom, and laundry (Demirbas et al., 2017). Sewage sludge is the semisolid precipitate produced in wastewater treatment plants (WWTPs). A survey analysis by the nonprofit Centre for Science and Environment (CSE) reported that India generates a staggering 1.7 million tons of sewage waste every day and about 78% of the sewage generated remains untreated and is disposed of in rivers, groundwater, or lakes. The CSE analysis points out that the amount of sewage generated in 2009 was 38,255 million liters daily (MLD), while India only had the capacity to treat 11,788 MLD—a mere 30% of the total. However, the actual sewage treated was even less, at 8251 MLD, which was about 22%. It has been reported that in India only 33% of houses are connected to sewer systems, only around 38% use septic tanks. These septic tanks do not treat waste and have to be emptied periodically. The services are offered to clean septic tanks, pump out the sludge from the tanks and empty it into drains, fields—just about anywhere. Also, cities are not able to recover the high costs of supply and therefore have no money to invest in sewage treatment (Chandramouli and General, 2011).

Sewage sludge generation in India is increasing at a faster rate as more and more sewage treatment plants (STPs) are required; in addition, alternative ecotechnologies are been developed to relieve this serious issue. At the moment, sewage sludge and effluents from these STPs are frequently disposed of on agricultural land for irrigation or manure (Singh and Agrawal, 2008). In India, wastewater disposal systems are usually managed by local bodies. In a few specific cases, they are managed by departments/statutory boards set up by state governments. This service facility falls into the water supply and sanitation sector. In municipal WWTPs, the treatment and disposal of sewage sludge can comprise up to 50% of the operation costs (Appels et al., 2008).

Industrial and Municipal Sludge. https://doi.org/10.1016/B978-0-12-815907-1.00032-5

1.1 Scope and objectives

Sewage treatment is a process that removes the majority of the contaminants from wastewater or sewage and produces both a liquid effluent suitable for disposal in the natural environment and sludge. Estimates from the Ministry of New and Renewable Energy (MNRE) in India suggest that there is a potential of about 226 MW to be generated from treated and untreated sewage sludge (Vaish et al., 2016). There are several approaches for treating sludge to reduce solid content and to stabilize biomass; however, anaerobic digestion (AD) is generally preferred because of its cost-effectiveness and bioenergy production (Ge et al., 2013). Digested sludge can be further composted for agricultural use, and biogas can be converted to electricity and/or heat through combustion, thus compensating for some of the energy use in a WWTP (Rulkens, 2007). Because of a large amount of organic content, primary sludge contains about 66% of the energy content of wastewater (Ting and Lee, 2007), and about 81% of biodegradable organic energy may be converted to methane (CH_4) (McCarty et al., 2011). Despite the great energy potential of biogas production, several issues limit successful AD application; for instance, electric generators and their maintenance are costly, and biogas may need pretreatment to remove contaminants such as hydrogen sulfide (Appels et al., 2008). In addition, energy will be lost during CH_4 conversion because the common efficiency of CH_4-to-electricity conversion is about 33% (Ge et al., 2013). Efficient, noninvasive techniques are desired for restoring organic-contaminated sludge. Therefore, it is of great interest to explore alternative technologies for sludge treatment and energy recovery. This chapter comprehensively describes microbial fuel cell (MFC) technology. Considering the MFC research trends, the following objectives are pursued in this chapter: (1) to highlight MFC for sludge remediation and to provide insights into how to extend its application to a large scale; (2) to update the current status of MFC technology in India and (3) to present a vision for the future of sludge management and remediation.

2 MFCs—Construction and working mechanism

Bioelectrochemical technology, typically involving MFCs, has been widely used as an environmental remediation approach. Its applications include wastewater, sludge, sediment, and soil remediation (Butti et al., 2016). MFCs directly produce electricity from organic wastes, in which bacteria oxidize organic and inorganic compounds in the anode chamber and generate protons and electrons that travel to the cathode to reduce oxygen to water. Electron flow from the anode to the cathode generates electric current or power if a load is connected (Fig. 1).

The potential of contamination removal has numerous aspects. First, an inexhaustible anode is introduced as the electron acceptor, which solves the problem of lack of electron acceptors. Second, the biocurrent stimulates the growth and activity of functional microbes and symbiotic microflora, which in turn accelerates the biodegradation of organic pollutants. Third, the coupling effect derived from the biocurrent boosts the oriented transfer of free electrons; this speeds up the rate of the oxidation-reduction reaction. Finally, the method of synchronous power production

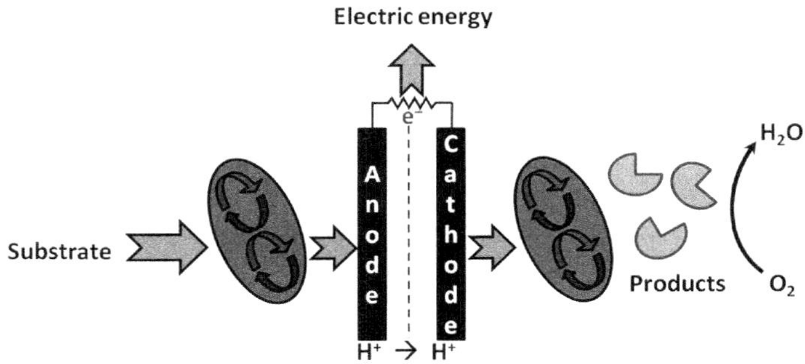

Fig. 1 Schematic diagram of microbial fuel cell during process operation.

from MFCs is the opposite of that of electrokinetic remediation, which needs power input; to some extent, this reduces the cost of environmental governance. Additionally, bioelectrochemical technology has no secondary pollution and is more specifically suited to clear low-concentration contaminants with the advantages of a high energy conversion rate, operation under normal temperature and pressure and in mild conditions, and easy to control (Wang and Ren, 2013).

Theoretically, MFCs are able to utilize any kind of biodegradable organic substance. The degradation kinetics are different and dominated by the structural characteristics of substances and their operational conditions (Pant et al., 2010). MFCs reduce the consumption of energy and the production of by-products, which are the main advantages of their application. Compared with more complicated wastewater treatment by bioelectrochemical systems, simpler sediment MFCs have been under development in the last decade. Based on a previously successful bioelectrochemical methodology in dealing with wastewater and sediment, sludge remediation has recently emerged.

MFC remediation belongs to the category of bioelectrochemical remediation, which is one of the biostimulation technologies (Lu et al., 2014; Rozendal et al., 2008). When an air-cathode MFC is used, the pollutant is biodegraded by microbial catalysis as the electron donor. Electrons are exported from the intracellular space and (1) hop to the anode directly (Fig. 2A), (2) are transferred by the mediators (e.g., flavin or thionine) to the anode (Fig. 2B), and (3) reach the anode by flowing through the *nanowires* (i.e., pili) (Fig. 2C). Subsequently, electrons are conveyed to the air cathode through the external circuit, and react with oxygen (terminal electron acceptor) to form water (Venkata Mohan et al., 2014). In MFCs for sludge remediation, the existence of anodes accelerates the capability of the electrogenic microorganism to provide more electrons to promote the metabolic reaction rates of anaerobic bacteria that degrade the contaminants.

The main advantages of using MFCs in wastewater treatment come from the savings of aeration energy and sludge disposal (Oh et al., 2010). For traditional activated sludge systems, aeration can amount to 45%–75% of plant energy costs, so the conversion of aeration tank to MFC units is very beneficial because not only does it eliminates aeration energy consumption, but studies also showed that MFCs can

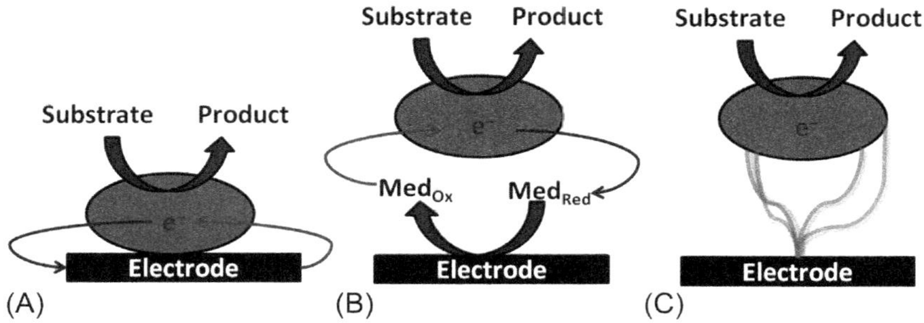

Fig. 2 Electron transfer mechanisms involved during microbial fuel cell operation where (A) direct electron transfer, (B) an electron shuttle, and (C) a solid conductive matrix (nanowires).

produce 10%–20% more energy that can be used for other processes (Huggins et al., 2013; Pant et al., 2010). The reported maximum power density from lab-scale air-cathode MFCs has reached 2.87 kW/m^3, making them promising for commercialization development (Fan et al., 2007), even though the system scale-up remains a major challenge.

Another main benefit of MFC systems is the low biomass production. The MFC is a biofilm-based system, and the cell yield of electrochemically active bacteria (0.07–0.16 gVSS/gCOD) is much less than the activated sludge (0.35–0.45 gVSS/gCOD), so it can reduce sludge production by 50%–70% (Fan et al., 2007; Huggins et al., 2013), which in turn may reduce 20%–30% of the plant operation costs. Other benefits may include nutrient removal and the production of value-added products, such as caustic solutions for disinfection, or H_2 and biogas for energy, which will be discussed more extensively in the following sections.

Diverting some organic compounds to direct electricity generation in MFCs could reduce biogas processing and conversion, resulting in some potential economic benefits, but we also need to understand the challenges of MFC application. For example, MFCs generally have much more complex structures and higher capital cost than anaerobic digesters (Pham et al., 2006). The use of high-surface-area electrodes and high-solid substrates like sludge can create problems such as reactor clogging. Unlike anaerobic digesters, which can be constructed in a single reactor with a large volume, MFCs are expected to be built in small-scale modules to form an MFC assembly; a single MFC with a very large volume will have a greater distance between the anode and the cathode electrodes, thereby increasing the internal resistance and decreasing electricity generation. With multiple small-scale MFC modules, the heating and feeding of the anolytes will be very challenging (Kondaveeti et al., 2018).

3 Applications of MFC for sludge remediation

In recent decades, the development of MFC applications has attracted increasing attention. A vast number of studies were performed in the fields of wastewater, sediment, sludge, and soil treatment. The use of MFCs is a promising approach for the direct production of electric energy or other energy carriers, such as hydrogen gas, from various organic substrates (Logan and Regan, 2006; Pant et al., 2010). Sewage sludge also has been studied in MFCs for electricity generation. A single-chamber MFC with a baffle inside its anode compartment generated low power from anaerobic sludge due to large internal resistance from the baffle (Hu, 2008). Hydrolysis is considered to be a limiting step in AD (Halalsheh et al., 2005), therefore appropriate pretreatment of solid organic waste is expected to improve the contents of soluble and small-particle organics which can be better used by microorganisms. Chandrasekhar and Venkata Mohan (2014) have reported bioelectrohydrolysis as a potential strategy for pretreatment of solid waste with simultaneous power generation. The ultrasonic and alkaline pretreatment of sludge improved its degradability and resulted in a higher power output of 12.5 W/m^3, with 61.0% and 62.9% reduction of total chemical oxygen demand (TCOD) and volatile solids (VS), respectively (Jiang et al., 2010; Jiang et al., 2009). Likewise, improved power output and total solid production were observed in an MFC after pretreatment with sterilization and alkalization (Xiao et al., 2011). When an MFC was linked to an anaerobic digester to form an integrated recirculation loop, it was found that CH_4 production was higher than the digester alone (Inglesby and Fisher, 2012) because a high concentration of ammonium/ammonia will inhibit methanogenic activity (Sung and Liu, 2003). The improved biogas production, resulting from the use of a recirculation loop, was likely due to the migration of ammonium ions from the digester to the cathode compartment of the MFC driven by electricity generation in the MFC, which was also demonstrated previously (Kim et al., 2008).

A recent study reported the performance of MFCs in treating a fermentation solution from primary sludge, in which higher power production was obtained when treating a mixture of fermentation supernatant and primary effluent because of elevated concentrations of soluble COD and volatile fatty acids (VFAs) after the fermentation process (Yang et al., 2013). In another study, long-term investigation (almost 500 days) of MFCs treating sewage sludge for energy production, organics removal, and solid reduction were reported (Ge et al., 2013). The experiment consisted of two phases: In Phase I, two tubular MFCs were operated with primary sludge and digested sludge, respectively, for more than 10 months; in Phase II, both MFCs were operated as a two-stage system to treat primary sludge for about 6 months. Biogas production in the MFCs was examined and compared energy production between MFCs and anaerobic digesters. The study concluded that the MFCs fed with the primary sludge acted mostly as modified anaerobic digesters. Alternatively, MFCs could function as a post-treatment process to polish the supernatant of digested sludge. Recently studies on MFCs for sludge remediation and energy recovery are being reported and a few are listed in Table 1.

4 Current status

As the global demand for renewable energy and organic matter increases, organic wastes, including sewage sludge, could be one of the available resources to use for this purpose. This substrate can be used as an energy resource for power and heat

Table 1 Comparison of MFC performances in systems that employed different inoculum, substrate, and designs

MFC configuration	Inoculum/ pretreatment	Substrate	Power density	References
Two-chamber	Anaerobic sludge	Pretreated activated sludge	5.6 $\mu W/cm^2$	Xiao et al. (2011)
Single-chamber	Effluent from MFC	Fermented primary sludge with 15.5 g/L soluble COD	32.0 $\mu W/cm^2$	Yang et al. (2013)
Single-chamber	Effluent from MFC	Fermented primary sludge with 15.5 g/L soluble COD	103 $\mu W/cm^2$	Yusoff et al. (2013)
Tubular MFC	Raw sludge from a primary sedimentation tank	Primary sludge and digested sludge from anaerobic digestors	1.43 and 1.80 kWh/ m^3, respectively	Ge et al. (2013)
Two-chamber	Anaerobic microbial consortium	Secondary sludge obtained from a sewage treatment plant	13.5 $\mu W/cm^2$	Passos et al. (2016)

via emerging technologies. Moreover, sewage sludge can be considered as a substrate for soil fertilization and remediation if the applied technology allows high-quality products to be obtained. Such reuses of sewage sludge are economically viable and environmentally sustainable compared to waste-handling and land-filling (Rulkens, 2007). The economic potential is found in offsetting costs related to traditional waste treatment methods and decreased energy costs by the generation of biogas and biofuels, which can partially replace the usage of traditional fuels (Kacprzak et al., 2017).

To achieve these goals, wastewater treatment operations require careful management of sewage sludge. The management of sewage sludge in an economically and environmentally acceptable manner is one of the critical issues facing modern society, due to the rapid increase in sludge production. In India, sludge management is governed by Municipal Solid Waste (MSW) Handling Rules, 2000 if the sludge is categorized as nonhazardous and the sewage sludge mostly falls into this category. However, if the sludge is characterized as hazardous for any reason, the disposal and management is governed by Hazardous Waste Handling and Management rules, as stipulated by an amended version of the regulations in 2003. All these regulations are administered by the Ministry of Environment and Forests (MoEF) (Chandrappa and Das, 2012).

A paradigm shift in waste remediation and waste management has been noticed in India since the launch of the Swachh Bharat Mission (SBM), an official campaign initiated by the government of India in October 2014, which aims to clean up the streets, roads, and infrastructure of the cities, smaller towns, and rural areas by the end of 2019. The government has a clearly defined path for achieving open defecation-free (ODF) cities and districts/villages. More critically, there is also a well-defined process, for the different phases of the mission, across the sanitation value chain called *build, use, maintain, and treat (BUMT)*.

A survey conducted by the Government of India in 2017 declared Indore as cleanest city based on various sanitation and cleanliness parameters, including waste collection, ODF status, and feedback from citizens. The survey is part of the government's initiative toward a cleaner India. Its focus on sanitation, preventing open defecation, and waste collection is significant, considering their impact on the environment and on the health of city dwellers.

The southern Indian city of Bangalore is extolled as India's "*Silicon Valley.*" It has experienced a growth of 45% of its urban population in a span of five years, crossing the 12 million population mark in 2016. The Water Supply and Sewerage Board of Bangalore has selected the French multinational corporation SUEZ to support the city's efforts to build sludge recovery plants and improve the wastewater infrastructures for two million inhabitants of Vrishabhavathi Valley. SUEZ has operations primarily in water, electricity, and natural gas supply, as well as waste management. The SUEZ group will be in charge of building a new WWTP of 150,000 m^3/day, besides, rehabilitation of an existing plant of 150,000 m^3/day, and also, building of a sewage sludge recycling and recovery plant coming from these two plants. This rapid urban rise, combined with the development of business activities, has led to an explosion in the demand for drinking water supply and sewage disposal. The common sludge treatment plant for the two treatment plants will be equipped with SUEZ's Degrémont technologies (Sedipac and Digelis), which will convert into electricity the biogas produced in order to make the plant self-sufficient. During the first three years, the group will design and build a new plant that uses activated sludge process (i.e., a bacterial treatment) to treat effluents to achieve an optimum discharge quality into the Vrishabhavathi River in accordance with the most recent standards of the Indian authorities, requiring in particular a total nitrogen content of less than 10 mg/L.

5 Future scope and perspective

The world is undergoing a new wave of urbanization, especially in developing countries, leading to multiple impacts on natural resources and the environment. A review of the research literature suggests that waste management poses various challenges and opportunities at an urban and rural scale (Reddy et al., 2018). In this context, biorefinery has emerged as a potential alternative to petroleum-based refineries, where biomass of nonedible feedstocks/biogenic wastes are used as raw materials and a range of products, such as biofuel, industrial biochemicals, and biomaterials including commercially important biopolymers, are produced (Clark and Deswarte, 2015).

Concepts such as the biorefinery create aspirations toward increasingly integrated technologies (Aresta et al., 2012). The burgeoning amount of waste (specifically sewage sludge) has tremendous potential, but to date, humankind has not exploited even a fraction of that potential. Waste as a prime target substance in the biorefinery, with a wide range of opportunities for mercantile interests rather than for utilitarian reasons (Tuck et al., 2012; Zondervan et al., 2011). A vivid, articulate review by Venkata Mohan et al. (2016) presented the possibilities of biorefinery models that could be explored for the utilization of biogenic wastes as a resource in the framework of a circular economy. The most popular microbial and photosynthetic processes are widely deployed for many technological solutions. Integrating various models involving unit operations and bioprocesses gives a much more holistic outcome through an updated "product versus energy" approach explored in the bioeconomy (Fig. 3).

An expression of interest in modern energy and material recovery technologies using waste feedstock often attempts to integrate processes and remediation. Therefore, this paradigm transition from a linear economy toward a circular economy is obligatory; the futuristic biorefinery platform should have an ambitious vision to promote a switch from the consumption of fossil reserves to the use of renewable waste as

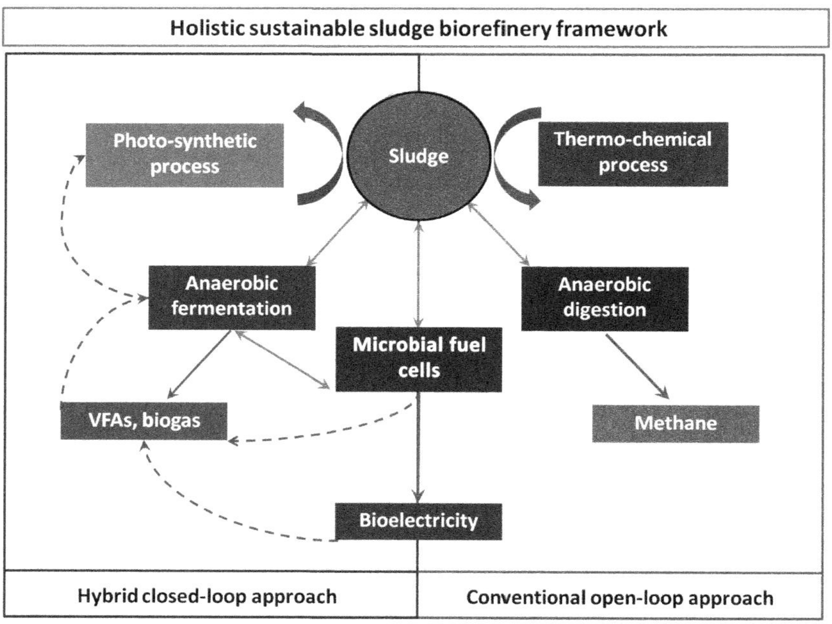

Fig. 3 Conceptualized flow sheet of sludge biorefinery that includes: Photo-synthetic process—bioelectrochemical process (microbial fuel cells) and thermo-chemical process. In the figure, the direction and *color* of *arrows* infer—bi-directional *yellow bold arrows* indicate formation and usage of excess sludge—uni-directional *grey bold arrows* indicate bioprocess yielding products—uni-directional *dashed arrows* indicate conditional channeling or by-products of that bioprocess.

resources. This will also lead to incredible employment opportunities in industry and academia, especially in the sectors of agriculture, food, chemicals, healthcare, pharmaceuticals, and logistics (Amulya et al., 2016). However, the sustainable biorefinery is viewed as a way to move fuels and other valorized coproducts onto much greener footprints, with life-cycle analysis providing the rigor needed to make the best decisions on feedstock utilization, upscaling process technology, and value-chain conformation (Liu and Shonnard, 2014)

Gottumukkala et al. (2016) reviewed opportunities and prospects of biorefinery-based valorization of pulp and paper sludge. The paper and pulp industry (PPI) generates a large amount of solid waste with high moisture content. Numerous opportunities exist for the valorization of waste paper sludge, although this paper focused on primary sludge with high cellulose content. The most mature options for paper sludge valorization are fermentation, AD, and pyrolysis. In this review, biochemical and thermal processes are considered individually and also as an integrated biorefinery. The objective of an integrated biorefinery is to reduce or avoid paper sludge disposal by landfilling, water reclamation, and value addition. The assessments of selected processes for biorefinery vary from a detailed analysis of a single process to high-level optimization and integration of the processes that allow the initial assessment and comparison of technologies. This data can be used to provide key stakeholders with a road map of technologies that can generate economic benefits and reduce carbon waste and pollution load.

Crutchik et al. (2018) recently reported on a biorefinery of cellulosic primary sludge toward targeted short-chain fatty acids (SCFAs), phosphorus, and CH_4 recovery. Cellulose from used toilet paper is a major untapped resource embedded in municipal wastewater, of which recovery and valorization to valuable products can be optimized. Cellulosic primary sludge (CPS) can be separated by upstream dynamic sieving and anaerobically digested to recover as much as 4.02 m^3/capita·year of CH_4. On the other hand, optimal acidogenic fermenting conditions of CPS allows the production of targeted SCFAs to reach as much as 2.92 kg COD/capita·year. Here, propionate content can be more than 30% and can optimize the enhanced biological phosphorus removal (EBPR) processes or the higher valuable copolymer of polyhydroxyalkanoates (PHAs). In this work, the optimized conditions were applied in the long term to a sequencing batch fermentation reactor where the highest propionate production (100–120 mg COD/g TVS_{fed}·d) was obtained at 37°C and adjusting the feeding pH at 8.

At the same time, around 88% of the phosphorus released during the acidogenic fermentation was recovered (amounting to as much as 0.15 kg of struvite per capita·year). Finally, the potential market value was preliminarily estimated for the recovered materials, which can triple over the conventional scenario of biogas recovery in existing municipal WWTPs.

Zhang et al. (2018) reported on value-added products derived from WAS in a biorefinery scenario. Substantial research has been carried out on sustainable WAS management over the last decade to endorse WAS as a feedstock for the production of bioproducts such as amino acids, proteins, short chain fatty acids, enzymes, biopesticides, bioplastics, bioflocculants, and biosurfactants. In this review, a critical assessment of the production process for a wide range of value-added products from WAS, their current limitations, and recommendations for future research to help promote more sustainable management of this underutilized and ever-growing waste stream has been discussed.

6 Conclusions

In India, management of solid wastes in general, and sewage sludge in particular, is going to be a big challenge in the near future due to rapid urbanization and economic growth. Unless technoeconomically viable alternatives are developed, sewage sludge management is going to be a formidable task in environmental management. MFCs are a promising alternative and renewable technology for the current water-waste conundrum. Based on the current literature review, MFCs have satisfactorily reduced both organics and suspended solids in sewage sludge. The total energy production from primary sludge in the two-stage MFC system was comparable to that of anaerobic digesters; however, direct electricity generation made a minor contribution, while energy from biogas still dominated the overall energy production. It will be very challenging to apply MFC technology to treating primary sludge; but MFCs may be used to polish the digested effluent from anaerobic digesters, offering potential benefits in energy savings compared with aerobic treatment. Alternatively, the proposed biorefinery approach could be a potential sludge management option, along with value addition.

Acknowledgments

The author would like to sincerely thank the editors of this book for giving this opportunity to contribute a chapter. I also give my heartfelt gratitude to Dr. S. Venkata Mohan (Book Editor) for his kind support and valuable suggestions.

References

Amulya, K., Dahiya, S., Venkata Mohan, S., 2016. Building a bio-based economy through waste remediation: innovation towards sustainable future. In: Bioremediation and Bioeconomy. Elsevier, pp. 497–521.

Appels, L., Baeyens, J., Degrève, J., Dewil, R., 2008. Principles and potential of the anaerobic digestion of waste-activated sludge. Prog. Energy Combust. Sci. 34 (6), 755–781.

Aresta, M., Dibenedetto, A., Dumeignil, F., 2012. Biorefinery: From Biomass to Chemicals and Fuels. Walter de Gruyter.

Butti, S.K., Velvizhi, G., Sulonen, M.L., Haavisto, J.M., Koroglu, E.O., Cetinkaya, A.Y., Singh, S., Arya, D., Modestra, J.A., Krishna, K.V., 2016. Microbial electrochemical technologies with the perspective of harnessing bioenergy: maneuvering towards upscaling. Renew. Sust. Energ. Rev. 53, 462–476.

Chandramouli, C., General, R., 2011. Census of India 2011. Provisional Population Totals. Government of India, New Delhi.

Chandrappa, R., Das, D.B., 2012. Solid Waste Management: Principles and Practice. Springer Science and Business Media.

Chandrasekhar, K., Venkata Mohan, S., 2014. Bio-electrohydrolysis as a pretreatment strategy to catabolize complex food waste in closed circuitry: function of electron flux to enhance acidogenic biohydrogen production. Int. J. Hydrog. Energy 39 (22), 11411–11422.

Clark, J.H., Deswarte, F., 2015. Introduction to Chemicals from Biomass. John Wiley and Sons.

Crutchik, D., Frison, N., Eusebi, A.L., Fatone, F., 2018. Biorefinery of cellulosic primary sludge towards targeted short chain fatty acids, phosphorus and methane recovery. Water Res. 136, 112–119.

Demirbas, A., Edris, G., Alalayah, W.M., 2017. Sludge production from municipal wastewater treatment in sewage treatment plant. Energ Source Part A 39 (10), 999–1006.

Fan, Y., Hu, H., Liu, H., 2007. Enhanced Coulombic efficiency and power density of air-cathode microbial fuel cells with an improved cell configuration. J. Power Sources 171 (2), 348–354.

Ge, Z., Zhang, F., Grimaud, J., Hurst, J., He, Z., 2013. Long-term investigation of microbial fuel cells treating primary sludge or digested sludge. Bioresour. Technol. 136, 509–514.

Gottumukkala, L.D., Haigh, K., Collard, F.-X., Van Rensburg, E., Görgens, J., 2016. Opportunities and prospects of biorefinery-based valorisation of pulp and paper sludge. Bioresour. Technol. 215, 37–49.

Halalsheh, M., Koppes, J., Den Elzen, J., Zeeman, G., Fayyad, M., Lettinga, G., 2005. Effect of SRT and temperature on biological conversions and the related scum-forming potential. Water Res. 39 (12), 2475–2482.

Hu, Z., 2008. Electricity generation by a baffle-chamber membraneless microbial fuel cell. J. Power Sources 179 (1), 27–33.

Huggins, T., Fallgren, P., Jin, S., Ren, Z., 2013. Energy and performance comparison of microbial fuel cell and conventional aeration treating of wastewater. J. Microb. Biochem. Technol. 6(2).

Inglesby, A.E., Fisher, A.C., 2012. Enhanced methane yields from anaerobic digestion of Arthrospira maxima biomass in an advanced flow-through reactor with an integrated recirculation loop microbial fuel cell. Energy Environ. Sci. 5 (7), 7996–8006.

Jiang, J., Zhao, Q., Wang, K., Wei, L., Zhang, G., Zhang, J., 2010. Effect of ultrasonic and alkaline pretreatment on sludge degradation and electricity generation by microbial fuel cell. Water Sci. Technol. 61 (11), 2915–2921.

Jiang, J., Zhao, Q., Zhang, J., Zhang, G., Lee, D.-J., 2009. Electricity generation from biotreatment of sewage sludge with microbial fuel cell. Bioresour. Technol. 100 (23), 5808–5812.

Kacprzak, M., Neczaj, E., Fijałkowski, K., Grobelak, A., Grosser, A., Worwag, M., Rorat, A., Brattebo, H., Almås, Å., Singh, B.R., 2017. Sewage sludge disposal strategies for sustainable development. Environ. Res. 156, 39–46.

Kim, J.R., Zuo, Y., Regan, J.M., Logan, B.E., 2008. Analysis of ammonia loss mechanisms in microbial fuel cells treating animal wastewater. Biotechnol. Bioeng. 99 (5), 1120–1127.

Kondaveeti, S.K., Seelam, J.S., Mohanakrishna, G., 2018. Anodic electron transfer mechanism in bioelectrochemical systems. In: Microbial Fuel Cell. Springer, pp. 87–100.

Liu, J., Shonnard, D.R., 2014. Life cycle carbon footprint of ethanol and potassium acetate produced from a forest product wastewater stream by a co-located biorefinery. ACS Sustain. Chem. Eng. 2 (8), 1951–1958.

Logan, B.E., Regan, J.M., 2006. Microbial fuel cells-challenges and applications. Environ. Sci. Technol. 40 (17), 5172–5180.

Lu, L., Huggins, T., Jin, S., Zuo, Y., Ren, Z.J., 2014. Microbial metabolism and community structure in response to bioelectrochemically enhanced remediation of petroleum hydrocarbon-contaminated soil. Environ. Sci. Technol. 48 (7), 4021–4029.

McCarty, P.L., Bae, J., Kim, J., 2011. Domestic Wastewater Treatment as a Net Energy Producer – Can This Be Achieved? ACS Publications.

Oh, S.T., Kim, J.R., Premier, G.C., Lee, T.H., Kim, C., Sloan, W.T., 2010. Sustainable wastewater treatment: how might microbial fuel cells contribute. Biotechnol. Adv. 28 (6), 871–881.

Pant, D., Van Bogaert, G., Diels, L., Vanbroekhoven, K., 2010. A review of the substrates used in microbial fuel cells (MFCs) for sustainable energy production. Bioresour. Technol. 101 (6), 1533–1543.

Passos, V.F., Aquino Neto, S., Andrade, A.R.D., Reginatto, V., 2016. Energy generation in a Microbial Fuel Cell using anaerobic sludge from a wastewater treatment plant. Sci. Agric. 73 (5), 424–428.

Pham, T., Rabaey, K., Aelterman, P., Clauwaert, P., De Schamphelaire, L., Boon, N., Verstraete, W., 2006. Microbial fuel cells in relation to conventional anaerobic digestion technology. Eng. Life Sci. 6 (3), 285–292.

Reddy, C.N., Sudhakar, M., Min, B., Shanmugam, P., 2018. Future perspectives on cost-effective microbial fuel cells in rural areas. In: Microbial Fuel Cell Technology for Bioelectricity. Springer, pp. 283–302.

Rozendal, R.A., Hamelers, H.V., Rabaey, K., Keller, J., Buisman, C.J., 2008. Towards practical implementation of bioelectrochemical wastewater treatment. Trends Biotechnol. 26 (8), 450–459.

Rulkens, W., 2007. Sewage sludge as a biomass resource for the production of energy: overview and assessment of the various options. Energy Fuel 22 (1), 9–15.

Singh, R., Agrawal, M., 2008. Potential benefits and risks of land application of sewage sludge. Waste Manag. 28 (2), 347–358.

Sung, S., Liu, T., 2003. Ammonia inhibition on thermophilic anaerobic digestion. Chemosphere 53 (1), 43–52.

Ting, C., Lee, D.-J., 2007. Production of hydrogen and methane from wastewater sludge using anaerobic fermentation. Int. J. Hydrog. Energy 32 (6), 677–682.

Tuck, C.O., Pérez, E., Horváth, I.T., Sheldon, R.A., Poliakoff, M., 2012. Valorization of biomass: deriving more value from waste. Science 337 (6095), 695–699.

Vaish, B., Srivastava, V., Singh, P., Singh, A., Singh, P.K., Singh, R.P., 2016. Exploring untapped energy potential of urban solid waste. Energy Ecol. Environ. 1 (5), 323–342.

Venkata Mohan, S., Nikhil, G.N., Chiranjeevi, P., Reddy, C.N., Rohit, M., Kumar, A.N., Sarkar, O., 2016. Waste biorefinery models towards sustainable circular bioeconomy: critical review and future perspectives. Bioresour. Technol. 215, 2–12.

Venkata Mohan, S., Velvizhi, G., Modestra, J.A., Srikanth, S., 2014. Microbial fuel cell: critical factors regulating bio-catalyzed electrochemical process and recent advancements. Renew. Sust. Energ. Rev. 40, 779–797.

Wang, H., Ren, Z.J., 2013. A comprehensive review of microbial electrochemical systems as a platform technology. Biotechnol. Adv. 31 (8), 1796–1807.

Xiao, B., Yang, F., Liu, J., 2011. Enhancing simultaneous electricity production and reduction of sewage sludge in two-chamber MFC by aerobic sludge digestion and sludge pretreatments. J. Hazard. Mater. 189 (1-2), 444–449.

Yang, F., Ren, L., Pu, Y., Logan, B.E., 2013. Electricity generation from fermented primary sludge using single-chamber air-cathode microbial fuel cells. Bioresour. Technol. 128, 784–787.

Yusoff, M.Z.M., Hu, A., Feng, C., Maeda, T., Shirai, Y., Hassan, M.A., Yu, C.P., 2013. Influence of pretreated activated sludge for electricity generation in microbial fuel cell application. Bioresour. Technol. 145, 90–96.

Zhang, W., Alvarez-Gaitan, J.P., Dastyar, W., Saint, C.P., Zhao, M., Short, M.D., 2018. Value-added products derived from waste activated sludge: a biorefinery perspective. Water 10 (5), 545.

Zondervan, E., Nawaz, M., de Haan, A.B., Woodley, J.M., Gani, R., 2011. Optimal design of a multi-product biorefinery system. Comput. Chem. Eng. 35 (9), 1752–1766.

Index

Note: Page numbers followed by *f* indicate figures and *t* indicate tables.

H

I

M

N

O

Q

R

W

X

Y

Z

9780128159071